The Molecular Biology of Cancer

A Bridge From Bench to Bedside

SECOND EDITION

EDITED BY

Stella Pelengaris
Pharmalogos Ltd, UK

and

Michael Khan
University of Warwick, UK

WILEY-BLACKWELL
A John Wiley & Sons, Inc., Publication

This edition first published 2013 © 2013 by John Wiley & Sons, Inc.

© 2006 by Stella Pelengaris, Michael Khan, William P. Steward, Maria Blasco, Cassian Yee, David Shima, Charles Streuli, Norbert C.J. de Wit, Nicky Rudd, Christiana Ruhrberg, Anne Thomas, Esther Waterhouse, and Martine Roussel

Wiley-Blackwell is an imprint of John Wiley & Sons, formed by the merger of Wiley's global Scientific, Technical and Medical business with Blackwell Publishing.

Registered office: John Wiley & Sons, Ltd, The Atrium, Southern Gate, Chichester, West Sussex, PO19 8SQ, UK

Editorial offices: 9600 Garsington Road, Oxford, OX4 2DQ, UK
The Atrium, Southern Gate, Chichester, West Sussex, PO19 8SQ, UK
111 River Street, Hoboken, NJ 07030-5774, USA

For details of our global editorial offices, for customer services and for information about how to apply for permission to reuse the copyright material in this book please see our website at www.wiley.com/wiley-blackwell.

The right of the author to be identified as the author of this work has been asserted in accordance with the UK Copyright, Designs and Patents Act 1988.

All rights reserved. No part of this publication may be reproduced, stored in a retrieval system, or transmitted, in any form or by any means, electronic, mechanical, photocopying, recording or otherwise, except as permitted by the UK Copyright, Designs and Patents Act 1988, without the prior permission of the publisher.

Designations used by companies to distinguish their products are often claimed as trademarks. All brand names and product names used in this book are trade names, service marks, trademarks or registered trademarks of their respective owners. The publisher is not associated with any product or vendor mentioned in this book. This publication is designed to provide accurate and authoritative information in regard to the subject matter covered. It is sold on the understanding that the publisher is not engaged in rendering professional services. If professional advice or other expert assistance is required, the services of a competent professional should be sought.

Limit of Liability/Disclaimer of Warranty: While the publisher and author have used their best efforts in preparing this book, they make no representations or warranties with the respect to the accuracy or completeness of the contents of this book and specifically disclaim any implied warranties of merchantability or fitness for a particular purpose. It is sold on the understanding that the publisher is not engaged in rendering professional services and neither the publisher nor the author shall be liable for damages arising herefrom. If professional advice or other expert assistance is required, the services of a competent professional should be sought.

Library of Congress Cataloging-in-Publication Data

The molecular biology of cancer / edited by Stella Pelengaris and Michael Khan. – Second edition.
    pages cm
  Includes bibliographical references and index.
  ISBN 978-1-118-02287-0 (hardback : alk. paper) – ISBN 978-1-118-00881-2 (pbk. : alk. paper) 1. Cancer–Molecular aspects. 2. Cancer–Genetic aspects. 3. Cancer cells. I. Pelengaris, Stella, editor of compilation. II. Khan, Michael, editor of compilation.
  RC268.4.M65 2013
  616.99'4042–dc23

                                              2012031371

9781405118149 (paperback)   9781118022870 (hardback)

A catalogue record for this book is available from the British Library.

Wiley also publishes its books in a variety of electronic formats. Some content that appears in print may not be available in electronic books.

Cover image: 5 μm tissue section from a specimen of colorectal carcinoma, imaged using TIS microscopy for colocation of more than 15 different cancer-related proteins. Courtesy of Nasir Rajpoot, Adnan Mujahid, Shan-E-Amad Raza (all from the Department of Computer Science, University of Warwick) and Michael Khan.
Cover design by Design Deluxe

Set in 8.75/11 pt Meridien by Toppan Best-set Premedia Limited
Printed and bound by CPI Group (UK) Ltd, Croydon, CR0 4YY

C9781118008812_050924

Contents

Contributors, vii

Preface to the Second Edition, ix

Reviews of the First Edition, x

Acknowledgments and Dedication, xi

About the Companion Website, xii

Introduction, 1

1 Overview of Cancer Biology, 3
Michael Khan and Stella Pelengaris

Introduction, 5
Cancer incidence and epidemiology, 8
Towards a definition of cancer, 8
Causes of cancer, 16
Cancer is a genetic disease, 21
Cancers (and Darwin's finches) evolve by mutation and natural selection, 21
Blame the parents – inherited single gene defects and susceptibility to cancer, 21
The cancer "roadmap" – What kinds of genes are epimutated in cancer?, 23
Viruses and the beginnings of cancer biology, 25
Hens and teeth or bears and woods? The hens have it – cancer is rare, 25
The barriers to cancer, 25
What is the secret of cancer developme ... "timing", 28
Location, location, location – the cancer environment: nanny or spartan state, 28
Cancer goes agricultural, 29
Cancer superhighways – blood vessels and lymphatics, 31
On your bike and turn the lights off before you go, 31
Catching cancer, 31
Hammering the hallmarks, 32
Painting a portrait of cancer, 33
The drugs don't work, 34
Mechanism of origin rather than cell of origin – towards a new functional taxonomy of cancer, 35
Is it worth it?, 36
Conclusions and future directions, 36
Bibliography, 37
Appendix 1.1 History of cancer, 40

2 The Burden of Cancer, 43
William P. Steward and Anne L. Thomas

Introduction, 43
Lung cancer, 45
Breast cancer, 49
Colorectal cancer, 53
Carcinoma of the prostate, 56
Renal carcinoma, 57
Skin cancer, 58
Carcinoma of the cervix, 60
Hematological malignancies, 60
Conclusions and future directions, 63
Outstanding questions, 63
Bibliography, 64
Questions for student review, 66

3 Nature and Nurture in Oncogenesis, 67
Michael Khan and Stella Pelengaris

Introduction, 69
Risk factors, 73
Preventing cancers, 76
Cancer genetics – in depth, 78
Cancer genomics, 87
Gene–environment interactions, 89
Mutations and treatment, 89
Chemoprevention of cancer, 90
Risk factors act in combination, 90
Environmental causes of cancer, 93
The clinical staging and histological examination of cancer, 101
Screening and biomarkers, 102
Somatic gene mutations, epigenetic alterations and multistage tumorigenesis, 105
Conclusions and future directions, 107
Outstanding questions, 107
Bibliography, 107
Questions for student review, 109

4 DNA Replication and the Cell Cycle, 111
Stella Pelengaris and Michael Khan

Introduction, 112
The cell cycle – overview, 114
Phases of the cell cycle, 120
The cell-cycle engine: cyclins and kinases, 123
Regulation by degradation, 126
Regulation by transcription, 129
MicroRNAs and the cell cycle, 131
Chromatin, 131
DNA replication and mitosis, 131
Checkpoints – putting breaks on the cell-cycle engine, 135
The DNA damage response (DDR), 136

The checkpoints, 136
Cell-cycle entry and its control by extracellular signals, 138
Changes in global gene expression during the cell cycle, 139
Cell cycle and cancer, 139
Drugging the cell cycle in cancer therapies, 141
Conclusions and future directions, 142
Outstanding questions, 143
Bibliography, 143
Questions for student review, 144

5 Growth Signaling Pathways and the New Era of Targeted Treatment of Cancer, 146
Stella Pelengaris and Michael Khan

Introduction, 147
Growth factor regulation of the cell cycle, 150
Growth homeostasis and tissue repair and regeneration, 151
Regulated and deregulated growth, 155
Cellular differentiation, 157
Tissue growth and the "angiogenic switch", 158
Cancers and nutrients, 158
Growth factor signaling pathways, 160
A detailed description of signal transduction pathways and their subversion in cancer, 160
Translational control and growth, 184
Conclusions and future directions, 185
Outstanding questions, 185
Bibliography, 186
Questions for student review, 187

6 Oncogenes, 188
Stella Pelengaris and Michael Khan

Introduction, 189
The oncogenes, 189
The discovery of oncogenes ushers in the new era of the molecular biology of cancer, 191
Overview of oncogenes, 191
Types of oncogenes, 193
Oncogene collaboration – from cell culture to animal models, 199
The c-*MYC* oncogene, 199
The RAS superfamily, 213
SRC – the oldest oncogene, 228
BCR–ABL and the Philadelphia chromosome, 232
The BCL-2 family, 235
Biologically targeted therapies in cancer and the concept of "oncogene addiction", 235
Conclusions and future directions, 235
Outstanding questions, 236
Bibliography, 236
Questions for student review, 238

7 Tumor Suppressors, 239
Martine F. Roussel

Introduction, 239
The "two-hits" hypothesis: loss of heterozygosity (LOH), 240
Haploinsufficiency in cancer, 240
Epigenetic events, 242
Definition of a tumor suppressor, 242
The retinoblastoma protein family, 242
p53/TP53, 250
INK4a/ARF, 254
The p53 and RB pathways in cancer, 257
Senescence and immortalization: Role of RB and p53, 258
Tumor suppressors and the control of cell proliferation, 258
Tumor suppressors and control of the DNA damage response and genomic stability, 260
The microRNAs and tumor suppressors, 260
Conclusions and future directions, 263
Acknowledgments, 263
Outstanding questions, 264
Bibliography, 264
Questions for student review, 265

8 Cell Death, 266
Stella Pelengaris and Michael Khan

Introduction, 267
An historical perspective, 267
Apoptosis in context, 267
Apoptosis as a barrier to cancer formation, 271
Apoptosis versus necrosis, 271
Cell death by necrosis – not just inflammatory, 272
The pathways to apoptosis, 272
The apoptosome – "wheel of death", 274
Caspases – the initiators and executioners of apoptosis, 274
The IAP family – inhibitors of apoptosis and much more, 276
The central role of MOMP and its regulators in apoptosis – the BCL-2 family, 279
Mitochondrial outer membrane permeabilization (MOMP), 281
Endoplasmic reticulum stress, 282
Stress-inducible heat shock proteins, 282
Tumor suppressor p53, 282
Oncogenic stress: MYC-induced apoptosis, 283
Autophagy – a different kind of cell death and survival, 287
Cell death in response to cancer therapy, 290
Exploiting cell death (and senescence) in cancer control, 290
Conclusions and future directions, 292
Outstanding questions, 293
Bibliography, 293
Questions for student review, 294

9 Senescence, Telomeres, and Cancer Stem Cells, 295
Maria A. Blasco and Michael Khan

Introduction, 296
Senescence, 298
Conclusions and future directions, 310
Outstanding questions, 310
Bibliography, 311
Questions for student review, 312

10 Genetic Instability, Chromosomes, and Repair, 314
Michael Khan

Introduction, 316
Telomere attrition and genomic instability, 321
Sensing DNA damage, 323
Repairing DNA damage, 325
Checkpoints, 336
Microsatellites and minisatellites, 343
Chaperones and genomic instability, 344
Cancer susceptibility syndromes involving genetic instability, 345
Genomic instability and colon cancer, 346
Conclusions and future directions, 346
Outstanding questions, 347
Bibliography, 347
Questions for student review, 349

11 There Is More to Cancer than Genetics: Regulation of Gene and Protein Expression by Epigenetic Factors, Small Regulatory RNAs, and Protein Stability, 350
Stella Pelengaris and Michael Khan

Introduction, 351
The language of epigenetics, 353
Epigenetics, 353
Methylation of DNA, 359
Acetylation of histones and other posttranslational modifications, 360
Epigenetics and cancer, 362
CIMP and MIN and the "mutator phenotype", 365
Imprinting and loss of imprinting, 366
Clinical use of epigenetics, 367
Regulation of translation, 368
Noncoding RNA and RNA interference, 369
Therapeutic and research potential of RNAi, 371
Treatments based on miRNA, 373
Regulating the proteins, 373
Therapeutic inhibition of the proteasome, 376
Receptor degradation, 377
Wrestling with protein transit – the role of SUMO and the promyelocytic leukemia (PML) body, 377
Conclusions and future directions, 380
Outstanding questions, 380
Bibliography, 381
Questions for student review, 382

12 Cell Adhesion in Cancer, 383
Charles H. Streuli

Introduction, 383
Adhesive interactions with the extracellular matrix, 384
Cell–cell interactions, 393
Critical steps in the dissemination of metastases, 395
E-cadherin downregulation in cancer leads to migration, 399
Epithelial–mesenchymal transitions, 401
Integrins, metalloproteinases, and cell invasion, 402
Survival in an inappropriate environment, 404
Conclusions, 406
Outstanding questions, 406
Bibliography, 407
Questions for student review, 409

13 Tumor Immunity and Immunotherapy, 410
Cassian Yee

Introduction, 410
Endogenous immune response, 411
Effector cells in tumor immunity, 413
Tumor antigens, 417
Antigen-specific therapy of cancer, 420
Clinical trials in vaccine therapy, 422
Cytokine therapy of cancer, 423
Tumor immune evasion, 424
Clinical trials in immunomodulatory therapy, 425
Conclusions, 425
Bibliography, 426
Questions for student review, 427

14 Tumor Angiogenesis, 429
Christiana Ruhrberg

Introduction, 429
General principles of new vessel growth, 430
Pathological neovascularization: tumor vessels, 430
Basic concepts in tumor angiogenesis: the angiogenic switch, 432
Vascular growth and differentiation factors: stimulators of the angiogenic switch, 432
Role of inhibitors in angiogenesis, 436
Clinical outcomes and future directions, 436
Acknowledgments, 437
Bibliography, 437
Questions for student review, 437

15 Cancer Chemistry: Designing New Drugs for Cancer Treatment, 438
Ana M. Pizarro and Peter J. Sadler

Introduction, 439
Historical perspective, 439
The drug discovery process and preclinical development of a drug, 442
Questions remaining, 457
Conclusions and future directions, 457
Bibliography, 458
Questions for student review, 459

16 Biologically Targeted Agents from Bench to Bedside, 461
Michael Khan, Peter Sadler, Ana M. Pizarro, and Stella Pelengaris

Introduction, 463
Targeted therapies, 465
Cancer cell heterogeneity, 466
Finding the molecular targets, 468
Tumor regression in mice by inactivating single oncogenes, 468
Targeted cancer therapies, 473
Targeting oncogenes to treat cancer?, 473

The concept of synthetic lethality and collateral vulnerability, 475
Clinical progress in biological and molecular targeted therapies, 476
Molecular targeted drugs – an inventory, 479
DNA damage responses, 490
Transcription factors, 491
Targeting epigenetic regulation of gene expression, 492
Hitting the extrinsic support network and preventing spread, 493
Gene therapy, antisense, and siRNA, 495
Resistance to targeted therapies – intrinsic resistance and emergence of secondary pathways and tumor escape, 497
Negative feedback loops and failure of targeted therapies, 500
Biomarkers to identify optimal treatments and tailored therapies, 501
Pharmacogenetics and pharmacogenomics, 505
Clinical trials in cancer, 506
Conclusions and future directions, 506
Bibliography, 507
Questions for student review, 508

17 The Diagnosis of Cancer, 509
Anne L. Thomas, Bruno Morgan, and William P. Steward

Introduction, 509
Clinical manifestations, 510
Investigations in oncological practice, 511
Non-invasive imaging techniques, 516
Future novel uses of imaging, 521
Proteomics and microarrays, 523
Circulating tumor cells, 523
Disease staging, 523
Conclusions and future directions, 524
Bibliography, 524
Questions for student review, 525

18 Treatment of Cancer: Chemotherapy and Radiotherapy, 526
Anne L. Thomas, J.P. Sage, and William P. Steward

Introduction, 526
Radiotherapy physics, 526
Radiobiology, 527
Treatment planning, 528
Recent advances, 529
Chemoradiation, 530
Conclusion, 540
Bibliography, 542
Questions for student review, 543

19 Caring for the Cancer Patient, 544
Nicky Rudd and Esther Waterhouse

Introduction, 544
Key concepts, 544
Communication with the cancer patient, 544
When is palliative care appropriate for cancer patients?, 545
Palliative care assessment, 545
Symptom control, 545
Respiratory symptoms, 547
Nausea and vomiting, 547
Bowel obstruction, 548
Constipation, 549
Fatigue, 549
Cachexia and anorexia, 549
Psychological problems, 549
The dying patient, 550
Supportive care, 550
An example of the care of a cancer patient, 551
Questions remaining, 551
Conclusions and future directions, 551
Underlying problems, 551
Comment, 551
Underlying problems, 552
Bibliography, 552
Questions for student review, 553

20 Systems Biology of Cancer, 554
Walter Schubert, Norbert C.J. de Wit, and Peter Walden

Introduction, 556
Information flow in cells, 556
Model organisms and cancer models, 557
Array-based technologies: genomics, epigenomics, and transcriptomics, 559
SNPs, the HapMap, and the identification of cancer genes, 559
Cancer mRNA expression analysis, 562
CGH arrays, CpG island microarrays, and ChIP-on-Chip, 564
Next-generation sequencing, 564
Proteomics, 566
Posttranslational modifications, 567
Protein complexes and cellular networks, 569
Clinical applications of proteomics, 570
Toponomics: investigating the protein network code of cells and tissues, 571
Processing the images from the cyclical imaging procedures, 571
Structure, code, and semantics of the toponome: a high-dimensional combinatorial problem, 573
Detecting a cell surface protein network code: lessons from a tumor cell, 575
The molecular face of cells in diseases, 576
Individualized medicine and tailored therapies, 576
Discussion and conclusion, 579
Bibliography, 579
Internet resources, 581
Questions for student review, 582
Appendix 20.1 Techniques for the generation of genetically altered mouse models of cancer, 582

Glossary, 585

Answers to Questions, 597

Index, 603

Contributors

About the Editors

Michael Khan, PhD, FRCP, is Associate Professor of Medicine at the University Hospitals of Coventry and Warwickshire and former Head of Molecular Medicine at the University of Warwick. He was elected as a fellow of the Royal College of Physicians in 2002 and as a member of the Association of Physicians in 2004. His main research interests have been in the regulation of tissue growth and plasticity during development and in adult tissue homeostasis. Currently, he is collaborating with mathematicians and others in a systems biology approach to define key functional gene and protein networks involved in regulating cell fate and to identify new biomarkers for colorectal cancer. Dr. Khan teaches cancer biology to undergraduates and runs postgraduate training courses at masters level and beyond in cancer biology and metabolism. Michael is Chief Medical Advisor to Silence Therapeutics PLC. He has co-authored four textbooks.

Stella Pelengaris, PhD, was a Senior Research Fellow in Molecular Medicine in the Department of Biological Sciences at the University of Warwick and Warwick Medical School. While working at the Imperial Cancer Research Fund, she established a series of unique model systems for studying the role of c-Myc and apoptosis in cancer initiation and reversal. From 1999 to 2008 she and Michael Khan jointly ran the Cancer Research Group at the University of Warwick, where, in collaboration with Gerard Evan, they confirmed the inherent tumor suppressor activity of c-Myc (apoptosis) as a major barrier to oncogenic activity of c-Myc. Stella is now director of Pharmalogos Ltd, in which capacity she provides advice to biotechnology and pharmaceutical companies on promising novel targets for future oncology therapy developments.

About the co-authors

Maria A. Blasco PhD is Director of the Spanish National Cancer Research Centre (CNIO) and Head of the Telomeres and Telomerase Group. She obtained her PhD from Universidad Autónoma de Madrid (Spain) in 1993. That same year, she joined Carol W. Greider's lab at Cold Spring Harbor Laboratory (New York, USA). In 1997 she returned to Spain and joined the CNIO in 2003 as Director of the Molecular Oncology Programme and Leader of the Telomeres and Telomerase Group. She was appointed CNIO Director in 2011.

Norbert C.J. de Wit PhD is a clinical chemist at Maastricht University Medical Center (The Netherlands) with a subspecialization in laboratory hematology. He undertook a PhD at Warwick University (UK) in clinical proteomics and his current research interests are in laboratory hematology and hemato-oncology.

Bruno Morgan is Professor of Cancer Imaging at the University of Leicester and University Hospitals Leicester. He studied at Oxford Medical School and subsequently has trained in both hospital medicine and radiology. He has an active research program in developing CT and MRI applications to monitor drug therapy.

Ana M. Pizarro PhD obtained her doctorate at the Universidad Autónoma de Madrid (Spain). She was awarded an Intra-European Marie Curie Fellowship to work at The University of Edinburgh, and in 2007 became a research fellow at The University of Warwick (UK).

Martine F. Roussel PhD is Professor of Molecular Oncogenesis and Co-Director of the Cancer Center Signal Transduction Program at St. Jude, USA. She is also a Professor in the Department of Molecular Sciences at UT Memphis. She is a major figure in global cancer education and research with nearly 200 publications, many in the area of tumor suppression.

Nicky Rudd is Clinical Lead, Cancer & Haematology Services and Consultant in Palliative Medicine at the Leicester Royal Infirmary and LOROS, UK. She is also chair of the Specialist Training Committee for Palliative Medicine for the UK. Her main interests are hospital palliative care teams, communication skills training and teaching.

Christiana Ruhrberg obtained her PhD in biochemistry from Imperial College London and trained as a postdoctoral fellow in neuronal biology at the National Institute for Medical Research and in vascular biology at the Imperial Cancer Research Fund, both in London. Being awarded a Career Development Fellowship from the UK Medical Research Council and then a Lectureship from University College London, she now heads a research group that investigates the mechanisms of physiological angiogenesis and neural development.

Peter J. Sadler FRS obtained his BA, MA, and DPhil at the University of Oxford. Subsequently he was an MRC Research Fellow at the University of Cambridge and National Institute for Medical Research, and Professor at Birkbeck College, University of London, Crum Brown Chair of Chemistry at the University of Edinburgh, and from 2007 Chair of Chemistry at the University of Warwick. He is a Fellow of the Royal Society of Edinburgh (FRSE) and the Royal Society of London (FRS), and a European Research Council Advanced Investigator. His research interests are centered on the design and mechanism of action of metal-based anticancer compounds.

Walter Schubert MD is Associated Professor (HD) for Medical Neurobiology and Head of the Molecular Pattern Recognition Research (MPRR) group at Otto-von-Guericke-University Magdeburg, Germany. He studied neurology/psychiatry, histology, and molecular cell biology at universities of Bonn and Heidelberg (Center for Molecular Biology), Germany and is Visiting Professor of Toponomics at the Max-Planck-CAS (CAS-MPG) Partner Institute of Computational Biology, Shanghai, China. He also founded the field of toponomics and invented the toponome imaging technologies MELC and TISTM.

William P. Steward is Professor of Medical Oncology and Head of the Department of Cancer Studies and Molecular Medicine at the University of Leicester and an oncologist at the Leicester Royal Infirmary, UK. He has a major interest in new drug development, particularly in the fields of colorectal and hepatobiliary/pancreas cancers, and is working in an extensive translational research program in chemoprevention focusing on biomarker development and identification of novel agents.

Charles H. Streuli PhD is Professor of Cell Biology at the University of Manchester, UK and Director of the Wellcome Trust Centre for Cell-Matrix Research. He is also a founding member of the Manchester Breast Centre and the Breakthrough Research Unit in Manchester. His research focuses on how cellular adhesion regulates breast epithelial cell proliferation, differentiation, and polarity.

Anne L. Thomas PhD FRCP is Reader and Consultant in Medical Oncology at the University of Leicester and an oncologist at the Leicester Royal Infirmary, UK.

Peter Walden is a biochemist with specialization in molecular and cellular immunology. He received his PhD from Tübingen University, Germany worked at MIT, Cambridge, USA and MPI for Biology, Tübingen and is Head of the Tumor Immunology Translational Research Group at Charité – Universitätsmedizin Berlin, Germany.

Esther Waterhouse is a consultant in palliative medicine at University Hospitals Leicester and LOROS Hospice in Leicester, UK. She has a particular interest in communication skills training and education.

Cassian Yee MD, PhD is an associate member of the Program in Immunology in the Clinical Research Division at Fred Hutchinson Cancer Research Center and an Associate Professor at the University of Washington School of Medicine (USA). He is a world authority on immunotherapy of melanoma and ovarian cancer.

Preface to the Second Edition

Based on our extensive experience of teaching undergraduates and postgraduates, it became clear that no single current resource covered in detail the cellular and molecular changes that give rise to cancer alongside the basic principles of biology and clinical practice, without which these cannot be readily understood. We had not intended to write a textbook at this stage in our careers, but realized that there was a real need for such a work for undergraduates, medical students, and even established researchers in the field. Very few cancer molecular biology textbooks were available that started at the beginning, using a format and language easy to digest, and included not only a comprehensive description of all aspects of cancer biology but also important chapters on diagnosis, treatment, and care of cancer patients.

Much has changed since the first edition and we have responded to the explosion in knowledge around targeted therapies and how these are developed and tested. Moreover, the emergent field of systems biology has impacted strongly on cancer biology, and may well revolutionize the way in which we view, study, and treat cancer in the near future, in particular with the inextricable association with concepts such as individualized and tailored therapies. We follow a similar structure to the first edition, but all chapters have been extensively revised, new chapters have been added, and an even stronger up-front emphasis has been placed on first presenting easy-to-digest models served up in plain English.

Students are first introduced to an overview of the cancer cell and important new concepts and those which are only just emerging (Chapter 1), and of selected human cancers (Chapter 2), following which the textbook covers in depth those key cellular processes of greatest relevance to cancer. Thus, Chapters 3–14 cover the full range of cancer-relevant biology, including highly topical and important areas such as apoptosis, telomeres, DNA damage and repair, cell adhesion, angiogenesis, immunity, epigenetics, and the proteasome, as well as traditionally important areas such cell-cycle control, growth regulation, oncogenes, and tumor suppressors. A major improvement on the first edition has been the inclusion of a detailed account of how cancer drugs are developed and brought to market. Moreover, the great strides forward in targeted treatments have allowed us to introduce Chapter 16, specifically to link the subject of each of the scientific chapters to classes of newly available treatments or to those in various stages of development. The result is that the science is put firmly into the context of treating cancer patients – the relevance becomes crystal clear.

The book then gives a description of cancer diagnosis, treatment, and care of cancer patients, which is not only essential to medical students but also important for cancer researchers and biology students who need to have a broader view of cancer and its impact. Finally, Chapter 20 concludes with a vision of how the future of cancer biology and oncology may be directed by interdisciplinary sciences, such as the exciting field of systems biology and the new technologies that underpin it.

The role of textbooks as information repositories is increasingly under threat. Yet even now that we are well into the new millennium, with students and researchers alike bathed in seemingly limitless available information on the World Wide Web, textbooks still exist. Why is this? With the near-universal availability of Internet access to students and researchers, the most current information is potentially available to any interested party almost instantaneously. No printed source can hope to provide the same immediacy of the latest breakthroughs or experimental findings, although they are free of the distractions of online gambling, 24-hour shopping, and less savory diversions that plague the Internet. However, limitless information creates new problems, namely how to evaluate, correlate, and place into context this wealth of knowledge. More than a million cancer-related publications are referenced on Medline alone, and even for the initiated it can prove daunting to attempt to construct a balanced overview of the many aspects of cell and molecular biology that impact on cancer. Because of these difficulties, one of the key aims of this book is to provide in a single source the necessary framework within which new information can subsequently be aligned and a more comprehensive, but still contextual, understanding of cancer achieved. In particular, we have taken the opportunity to highlight controversial areas and to identify areas of research promise, while establishing potential links between often diverse subdisciplines in a coordinated and accessible way. It is hoped that, having read this book, the reader will be suitably equipped to understand the significance and relevance to cancer of a new publication and be able to place the work into an overall picture of the disease. Moreover, the book should also provide established cancer researchers with valuable insights into the important questions that remain to be addressed.

The issue of references, how many and where to cite, is often difficult to judge for a textbook. One has to balance the flow of the text with the need to give pointers to the reader for further information and to highlight key studies. This textbook can be used by undergraduates in biology and medical students and can be used alongside cancer biology courses structured either for a quarter or semester system. Moreover, the book will be of value to those preparing for professional exams in medicine and oncology and for established cancer researchers seeking a single-source overview of all aspects of cancer.

Features

We have included a number of features to facilitate the use of this textbook to teach cancer biology:
- Each chapter begins with a series of bullet points which explain the key concepts and illustrate areas of controversy in plain English. This is the platform on which the more complicated and detailed processes and models will be built throughout the rest of the chapter.
- Each chapter builds on concepts learned in previous chapters and is organized in a similar fashion, starting with an introduction and ending with a "Conclusions and future directions" section, a list of key outstanding questions remaining in the field, suggestions for further reading, and questions for student review.
- All the chapters contain textboxes that provide additional and relevant information as it relates to a described concept and are fully illustrated throughout.

Reviews of the First Edition

"Pelengaris, Khan, and the contributing authors are to be applauded. The Molecular Biology of Cancer is a comprehensive and readable presentation of the many faces of cancer from molecular mechanisms to clinical therapies and diagnostics. This book will be welcomed by neophyte students, established scientists in other fields, and curious physicians." Dean Felsher, Stanford University

"The explosion of information on the molecular biology of cancer, and its widespread and immediate availability via the internet, provides major challenges for those engaged in cancer treatment and research. A single up-to-date reference textbook on this topic is needed more than ever. This book will go a long way towards meeting this need, providing a valuable resource for a range of individuals and departments." Stan Kaye, Royal Marsden Hospital, London

Acknowledgments

An enormous number of talented scientists contributed to the knowledge described in this textbook. We acknowledge the many colleagues past and present whose important work could not be referenced in the text due to space constraints. In addition, we apologize if we failed to adequately identify contributions in the reference section at the end of this text. This oversight was not intentional, but rather a reflection of the overwhelming number of contributors to this field.

We thank mentors past and present for their help and encouragement: Martin Raff and Anne Mudge for making cell biology interesting and intelligible and Gerard Evan for introducing us to the world of cancer research. We thank our friends and colleagues who took time from their hectic research and clinical commitments to contribute to this book. In addition to those mentioned in the first edition, we especially thank our dedicated research team, Sylvie, Luxian, Yi-Fang, Elena, and Liam for bearing with us while we were writing and editing this book and for their patient reading and suggestions for improving the text. A special thanks is due to David Epstein FRS, our friend and colleague, for taking on too many tasks while we were occupied with this venture as well as for reading several chapters. We also greatly appreciate the suggestions and the gentle way in which these were presented by our friend Anthony Parker. Finally, we thank freelance project manager Nik Prowse and freelance copy-editors Cheryl Adam and Harriet Stewart-Jones, who have painstakingly teased out our many abuses of English and have helped us eliminate every tortured metaphor and incomprehensible sentence. Any that remain are entirely our fault. We are also very grateful to Rosie Hayden and Kelvin Matthews at Wiley-Blackwell for the belief and support of this exciting adventure.

We also acknowledge the contributions of our outside reviewers: Stewart Martin of Nottingham University; Brian Keith of the University of Pennsylvania; S.J. Assinder of the University of Wales, Bangor; Satya Narayan of the University of Florida; Mary Jane Niles of the University of California, San Francisco; Fiona Yull of Vanderbilt University, as well as those who have chosen to remain anonymous.

Reviewing is an enormous and time-consuming activity. We greatly appreciate the time spent by our reviewers, generating insightful and helpful comments.

For the cover figure showing a multiplexed protein expression image of colorectal cancer we owe a big thank you to our friends and collaborators Adnan Mujahid, Shan-E-Ahmed Raza, Professor David Epstein FRS, and Dr Nasir Rajpoot, from the Department of Computer Science at University of Warwick, and to the much-missed former doctoral student Dr Sylvie Abouna.

Dedication

This book is dedicated to our parents, whose unceasing support and encouragement made this work possible. A much loved father who recently finally lost his brave struggle with lung cancer would be particularly proud of this legacy. We also thank our daughter Charlotte for providing the perfect balance to academic work, namely a very happy and always entertaining family life. Many friends have helped and encouraged us through this process and we particularly thank Anthony Parker, David Epstein, and Liam Jones for their tactful and valuable comments.

About the Companion Website

This book is accompanied by a companion website:

www.wiley.com/go/pelengaris/molecularbiologyofcancer

The website includes:

- References for each chapter
- Powerpoints of all figures and tables from the book for downloading

Introduction

By doubting we come to enquiry, and by enquiring we pursue the truth.

Abelard, 1079–1140

The legend went, unconfirmed and unaccredited, but still propagated.

Charlotte Bronte

Everybody said so. Far be it from me to assert that what everybody says must be true. Everybody is, often, as likely to be wrong as right.

Charles Dickens

Much has changed since the last edition. If you inadvisably type the word "cancer" into PubMed you will now be rewarded with over 2.5 million papers and reviews, a figure increasing at a rate of around 4000 per month, making a complete nonsense of any pretentions to keep up to date. Yet, some bold conclusions may be drawn from this overflowing font of knowledge.

The world of cancer remains one of relentless clonal competition and selection – a cellular "Tumor's Got Talent." Yet cancer cells, performing from their rewritten genomic libretto, are no longer the unchallenged divas of the tumor opera. The stroma, previously regarded as the backdrop against which portentous cellular events were enacted, has now finally gained recognition for staging the whole performance. Moreover, just as no rousing aria is possible without appropriate cooperation between stage and pit, so few cells will hit a high C and fulfill their malignant potential without the incendiary score orchestrated by chronic inflammation, both prefiguring and fomenting the development of most epithelial cancers.

Caught in the slipstream of inflammatory cells and mediators, junior cancer cells are propelled to their fate by the accretion of liberating epimutations. Through chronic inflammation, normal conformist cell behaviors may be cast aside. Paradoxically, those same liberating forces may also offer a chink in cancer's armour, an Achilles' heel that if correctly exploited could leave the cancer cells ripe for sacrifice on the altar of their own oncogenes.

We are also now much more aware of how a cancer cell may be fashioned not by occasional seismic molecular events but often by the infinitely subtle calibrations of cellular behavior played out over many years. These shifts occur under the auspices of inherited and sequentially acquired mutations. New high-throughput molecular techniques are enabling us to read this curriculum vitae of the cancer cell. Following the molecular clues leads us inexorably to an identification of the culpable mutations, at least where we avoid *cum hoc ergo propter hoc* fallacies by supporting conjecture with appropriate functional studies and clinical trials. We are increasingly able to differentiate propitious and "mission critical" molecular alterations from irrelevant bystanders; the practical benefit is the reduced list from which will be drawn our new treatment targets.

Successful deployment of pharmaceutical hardware requires a knowledge not only of what to target but also who to target. Thus, among the cancer cells themselves, all is not as it was. We are now seeking to identify the hardliners and the agents provocateurs – those responsible for inciting and maintaining the cancer and therefore arguably the engine for malignant behavior. This cancer elite includes the elusive so-called "cancer stem cells" (CSCs), referencing their perceived molecular and functional similarity to normal stem cells. These pernicious cells, along with the epithelial–mesenchymal transition (EMT), the molecular plan that directed their evolution, are likely to become the key targets for new cancer therapies. By implication, does this suggest that other tumor cells will henceforth be relegated to the position of subsidiary drones, meagre ciphers that will automatically reconform to the template of acceptable cell behavior once CSCs are no longer there to throw them out of kilter? We must wait and see.

Another important concept concerns the relative clinical importance of the primary tumor as compared to the infinitely more dangerous distant colonies that have been seeded from it. Patients rarely die because their primary tumors enlarge; they die because some cancer cells desire *Lebensraum*. Sallying forth from their ancestral homeland, cancer cells establish footholds within immediately adjacent territories but also in more distant and alien environments that are connected to the primary tumor by vessels and lymphatics. The road to hell is paved with endothelial cells. Initially precarious and potentially vulnerable to an effective counterattack, these estranged voyagers will eventually give rise to an increasingly malign and radicalized cadre of invasive

---

The Molecular Biology of Cancer: A Bridge From Bench to Bedside, Second Edition. Edited by Stella Pelengaris and Michael Khan.
© 2013 John Wiley & Sons, Inc. Published 2013 by John Wiley & Sons, Inc.

Introduction

and resistant cancer cells. These will expand faster, spread further, and shorten life more readily than their forebears. Particular types of cancer cell prefer to colonize specific secondary sites and the "seed versus soil" dilemma is still unresolved in most cases. It seems likely that separating the effects of liberating epimutations in the cancer cells from those of genius loci of the target tissues may help with diagnosis and development of more effective treatment strategies.

Another topical issue involves the ineffable abilities of itinerant, as well as resident, cancer cells to efface themselves and avoid detection by the immune system. The highly vigilant and efficient way in which the immune system can deal with cancer cells behind the scenes when conditions are optimal is underscored by the greatly increased risk of several cancers when the immune system is disabled by viral infection or drugs.

Continuing progress in unraveling the molecular basis of cancer has yielded new drug targets and biomarkers. The holy grail of individualized medicine would now seem to be within our grasp, but may yet slip from our fingers. Thus, the availability of several new targeted therapies complicates the choice of the most appropriate agent or combinations of agents. The increasing problem of resistance to these drugs seems poised to thwart our best efforts. Recognition that in many cases cancers may be better categorized by a shared molecular origin rather than by cells or tissues of origin may enable these new challenges to be tackled more rationally. The possibilities for new drug development are being expanded by an increasingly nuanced view of concepts such as oncogene addiction; cancer cells are held to be wholly dependent on a given mutation for their continued growth and survival. Thus, addiction to a particular oncogene product or missing tumor suppressor may in turn make the cancer cell equally dependent on other, even nonmutated, proteins ("synthetic lethality," to coin a phrase derived from yeast genetics). As a result, the repertoire of potential drug targets is greatly expanded. These co-dependent proteins are often involved in DNA damage repair or other processes required for the cancer cell to survive mutations in tumor suppressors or oncogenes and may in some cases be more readily neutralized. This is in many respects analogous to oncogene cooperation, but in this case the cooperating protein is not derived from a mutant gene. In terms of therapeutics, synthetic lethality may point to more tractable targets for new drugs than the causative mutations themselves. Tumor suppressors are really hard to put back into all cancer cells, whereas co-dependent enzymes may be easily inhibited. This concept is well illustrated by the successful use of PARP inhibitors in breast cancer cells, which rely on PARP for DNA repair as BRCA mutations render alternative processes defective. Synthetic lethal interactions will likely be very important in finding treatments for those many cancers with a mutant RAS oncogene, which has proved very difficult to target effectively with drugs directly. The reader will find that the repertoire of signaling pathways deemed to be cancer-relevant has expanded. Increasingly, we see the identification of pathways that are causal and supportive for the initiation and progression of many different cancers and also those involved in the genesis of cancer stem cells and responsible for acquired resistance to targeted drugs.

Some research areas have expanded explosively over the last 5 years and are actively supporting a new molecular taxonomy of cancers. Notably, ongoing progress in cancer genomics, epigenetics and the biology of noncoding RNAs (particularly microRNAs) is not only increasing our fundamental understanding of how tumorigenesis operates but also providing the tools for molecular classification. These areas feature extensively throughout the book. We have also included new chapters on drug development, targeted therapeutics, and clinical trials. As befits its position at the cutting edge of research, the chapter on systems biology has grown substantially.

There has been a recent vogue for describing cancer cells in terms of a transfiguration, for regarding them as somehow more perfect and unstoppable versions of ourselves and somehow indifferent to our best endeavors to thwart them. It is worth considering, however, within the compass of what kind of very nasty world could this hold true – only one in which extreme self-serving nepotism rules the roost, and all will rapidly be consigned to oblivion through the activity of a single bad parent. But let us conclude in upbeat fashion and end on a note of optimism: cancer cells are neither immortal nor perfect, they are vulnerable to a range of agents that have little effect on normal ones, they can be stopped and even killed. So let us play Delilah to the cancer cell's Samson and treat it to a number 2, close-cropped, all-over molecular hair cut.

In this edition we have used more quotations than in the previous one. The intention is to find an entertaining way of introducing basic concepts. The quotations are all clear thoughts well-expressed, which may not always be true of our own text.

> Considering how much we are all given to discuss the characters of others, and discuss them often not in the strictest spirit of charity, it is singular how little we are inclined to think that others can speak ill-naturedly of us, and how angry and hurt we are when proof reaches us that they have done so.
>
> *Anthony Trollope from Barchester Towers.*

1 Overview of Cancer Biology

Michael Khan[a] and Stella Pelengaris[b]
[a]University of Warwick and [b]Pharmalogos Ltd, UK

A new scientific truth does not triumph by convincing its opponents and making them see the light, but rather because its opponents eventually die, and a new generation grows up that is familiar with it.

Max Planck

A discovery must be, by definition, at variance with existing knowledge. During my lifetime, I made two. Both were rejected offhand by the popes of the field. Had I predicted these discoveries in my applications, and had those authorities been my judges, it is evident what their decisions would have been.

Albert Szent-Gyorgyi

Key points

- Cancer is a genetic disease characterized by the emergence of deranged versions of normal cells, born out of aberrant molecular biology.
- Cancer is the malign byproduct of an ensemble performance in which mutations in the DNA and altered gene expression are enacted against a background of conniving environmental factors such as carcinogens and chronic inflammation.
- A large number of factors are adduced to explain the genesis of cancer, including the twin pillars of incitement of primeval urges and the emancipation from normal restraining forces. Together, these produce untrammeled cell-cycle progression.
- Studies of rare familial "monogenic" cancer syndromes have had a major impact on our fundamental understanding of cell biology, but most cancers do not result from inheritance of single, potent, cancer-causing mutations.
- Instead, they are "sporadic," with cancer-causing gene mutations arising in adult somatic cells.
- Hereditary factors may, however, exert weak and subtle influences on the risk of development and subsequently the behavior of most if not all so-called "sporadic tumors," through a complex interplay between multiple, largely unknown polymorphic alleles, some of which may only be disadvantageous if the individual is exposed to particular environmental carcinogenic factors, such as tobacco smoke.
- In general, factors that cause mutations and those that increase cell replication can combine to cause cancer, which may explain the powerful role of chronic inflammation in the causality of many carcinomas.
- Cancer is a clonal disease arising by the multistep accumulation of genetic or epigenetic changes in tumor suppressor genes, oncogenes, and "caretaker" genes that favor expansion of the new clone over the old in a process akin to Darwinian evolution.
- Natural selection will favor expansion of clones with acquired characteristics advantageous to the cancer cells, often referred to as the "hallmark" features of cancer (Fig. 1.1), which have been famously distilled by Robert A. Weinberg and Douglas Hanahan as:
  (1) the capacity to proliferate irrespective of exogenous mitogens;
  (2) refractoriness to growth inhibitory signals;
  (3) resistance to apoptosis;
  (4) unrestricted proliferative potential (immortality);
  (5) capacity to recruit a vasculature (angiogenesis);
  (6) ability to invade surrounding tissue and eventually metastasize.
- Recently, the "Warburg effect," a metabolic switch towards increasing ATP production by glycolysis, along with evasion or subversion of the immune system have been championed as the seventh and eighth hallmark features, respectively.
- RB and TP53, the doyens of tumor suppressor proteins, can arrest the cell cycle or trigger apoptosis in response to assorted cellular stresses, activation of DNA damage checkpoints, or during attempted oncogenic hijacking of cell-cycle control.

(Continued)

The Molecular Biology of Cancer: A Bridge From Bench to Bedside, Second Edition. Edited by Stella Pelengaris and Michael Khan.
© 2013 John Wiley & Sons, Inc. Published 2013 by John Wiley & Sons, Inc.

- An intriguing question is exactly how a tumor cell with DNA damage retains so many varied options in the face of TP53 activation? Thus, TP53 can mitigate cell death and inspire DNA repair but, in complete contrast, if repairs fail it might drive a cell to celibacy or suicide.
- Put simply, with respect to replication, a cell with irreparably damaged DNA has to either kick the habit or kick the bucket.
- Not surprisingly, therefore, loss of tumor suppressors is a prerequisite for tumorigenesis. Although there is some overlap, broadly speaking, cancer cells without TP53 can survive an alarming rate of mutation, whereas absence of the other archetypal tumor suppressor, RB, represents a fountain of youth for the otherwise rapidly senescing cancer cell.
- When and where, in the life history of a cancer, do the genetic changes required for metastases occur? There is no satisfactory answer to date. Natural selection does not really provide an explanation as to why a clone of cancer cells with metastatic capabilities would be selected for in the primary tumor, unless the causal mutations first and foremost also provide a growth advantage. It is possible that potential metastatic behavior is serendipitously acquired early in tumorigenesis as a byproduct of mutations promoting growth of the ancestral primary tumor (supported by some gene expression profiling studies of whole tumors). Alternatively, it may be that mutations in specific metastases-suppressing genes that do not confer a growth advantage to the primary occur at a later stage, possibly once cancer cells have begun circulating.
- Recent intriguing questions have been posed regarding the ongoing evolution of cancer cells in primary and secondary tumors. Recent findings suggest that following an initial shared origin, clones with metastatic capabilities emerge in the primary. Once ensconced within a new secondary environment, the metastatic alumni follow a parallel and distinct evolutionary path that may intriguingly begin while still in transit.
- Cancers are complex and heterogeneous, comprising a series of genetically differing populations (clones) of cancer cells. In fact, the *dramatis personae* of cancer includes the cancer cells-elect, the profligate parents, and a number of libertine relatives of dubious provenance.
- Moreover, the whole ensemble is supported by a strong supporting cast of both collaborating and insurgent noncancer cells that together constitute the cancer microenvironment.
- The cell of origin for any given cancer – be it stem cells that partially differentiate or differentiated cells that partially dedifferentiate, continues to offer opportunities for spirited debate.
- Tumors are not egalitarian societies. Rather they are in most cases oligarchies run by a malign minority of so-called cancer stem cells (CSCs). Part gang master and part queen bee, CSCs lie embedded within a large cast of bit part players. CSCs were first described in hematological malignancies, where they are strongly implicated in maintaining the malignant phenotype. More recently, CSCs have been identified in solid tumors and may be responsible for invasive behaviors, treatment resistance, and recurrence. By implication, these cancer oligarchs are the target of the original cancer-causing mutations, suggesting that in the case of a tumor the fish rots from the head.
- CSCs share many properties and molecular markers with normal stem cells, but this does not constitute proof of paternity. Under the influence of relevant mutations, including those that provoke epithelial–mesenchymal transition (EMT), normal cells can have "stemness" thrust upon them.
- This departs from the more traditional view of indefatigable clonal competition; dog eats dog, the strongest prevails with the extinction of the weakest – *aut Caesar aut nihil*.
- The cancer microenvironment, including the inflammatory milieu and the tissue stroma (connective tissue, fatty tissue, blood vessels, and lymphatics), represents an *alma mater* for cancer cells, which by encouraging EMT can help to generate CSCs and support the success of tumorigenesis.
- The greater recognition of the portentous events unfolding within the purlieus of the tumor peripheries during tumorigenesis has already yielded dividends. Thus, the stroma plays society hostess to a prohibition-free orgy of concupiscent cancer cells, egged on by a small faction of attendant immunocytes and under the averted gaze of the rest.
- Remarkably, it now transpires that cancer-contributing mutations are no longer the sole preserve of cancer cells themselves. In fact, mutations in stromal cells may allow them to more effectively mentor cancer cells towards the achievement of their six or eight hallmark milestones.
- The molecular profile of a tumor constitutes a manifesto, within which its future behavior is adumbrated and from which its weaknesses might be divined. Moreover, seminal parts of this manifesto achieve remarkably widespread circulation. Therefore, for diagnosis it may be unnecessary to directly remove tumor tissue, because cells, proteins, and nucleic acids derived from it are continually being shed into more readily accessible body fluids.
- The search for clinically useful molecular biomarkers represents one of the most promising areas of cancer research. Many biomarkers are already in routine clinical practice, where they assist in disease monitoring and in treatment selection.
- However, biomarkers have, as yet, not helped us to paint more accurate portraits of tumors. Unfortunately, in most cases they fail to unambiguously identify their subjects. There is no "Habsburg lip" for cancers.
- In fact, biomarkers have had limited impact on screening the general population for most cancers.
- Given the increasing number of therapeutics in our arsenal against cancer, great efforts are being made to find biomarkers that may help select appropriate treatments for individual patients.
- Cancers may be cured by surgery, but only if the entire tumor is accessible and no cells have spread to other sites. Modern approaches to cancer drug development are increasingly moving away from traditional chemotherapeutic agents which paralyze cell division or cause DNA damage and instead are aimed at targeting specific cancer-relevant proteins, such as oncogenic tyrosine kinases.

- Oncogene addiction, the process by which cancer cells become critically reliant on the mutant signaling molecules, offers the potential of both effective and minimally toxic agents directed against such proteins. A potential realized by pioneering therapeutic successes, such as imatinib, used to such good effect to target the abnormal BCR–ABL fusion protein in chronic myeloid leukemia.
- However, use of these agents is in most cases severely limited by acquired or, on occasion, inherent resistance of cancer cells to the treatment. It is hoped that understanding the resistance mechanisms involved will allow rational development of combinations of targeted agents in the future, though further mutations may render even these ineffective over time. One could easily be forgiven for likening these efforts to cure cancers by drug therapy with the task set before Sysiphus.
- However, we may yet keep the boulder from rolling down the hill. Knowledge is power and by exploiting the potential of treatment biomarkers we may gain an edge over cancer. Thus, we can assess whether a given cancer will respond to particular drugs as exemplified by the presence of estrogen or progesterone receptors and mutant *NEU* in breast cancer, or may conversely suggest a response to be unlikely as in the presence of *KRAS* in colorectal cancer. Armed with enough of these biomarkers there is reason to suppose that the goal of individualized medicine and tailored therapy may soon be within reach.

Introduction

And yet there is something so amiable in the prejudices of a young mind, that one is sorry to see them give way to the reception of more general opinions.

Jane Austen

In this chapter we give a historical overview of cancer and go on to introduce and summarize the concepts and topics to be covered in this book. Wherever possible, we emphasize new thinking, emerging views, and novel models for studying and understanding oncogenesis. Unapologetically, this chapter aims to be stimulating and thought-provoking.

Cancer has been recognized throughout recorded history and was known to the ancient Egyptians (see Appendix 1.1 – History of cancer), but it was not until the seventeenth century that the formal study of cancer (oncology) was first documented. As with much of biology, the last 50 years has witnessed spectacular progress in describing the fundamental molecular basis of cancer following the advent of molecular biology and genetics. Frustratingly, such exponential progress in describing the biology of cancer has not yet translated into an equally impressive progress in the war against most common cancers (Fig. 1.2). We can, however, claim victory in some important skirmishes. Possibly the single greatest success has been in altering the status of cancer in many cases to that of a chronic illness. Most people now live for some time with the disease rather than rapidly succumbing to it. This is in part testimony to better treatments. At the time of going to press, the overall median survival time for the commonest 20 cancers had increased from around one year in 1971 to just under 6 years by 2007. Most of this gain, however, reflects

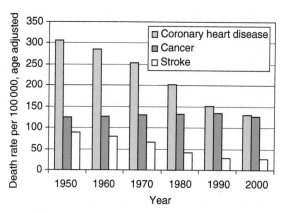

Figure 1.2 The incidence of cancer is not declining when compared to other major diseases, yet in the United States alone more than US$4.7 billion per year is spent on cancer research. Leland Hartwell and others at a meeting of the American Association of Cancer Research identified the following areas, in addition to developing new therapies, as key targets to address this major public health issue: (1) More coordinated and concerted activity between researchers. This would require establishing the necessary infrastructure for facilitating collaborative working and information exchange. (2) Testing drugs and agents in early-stage disease rather than as at present largely in end-stage cancer (we may be underestimating the potential of many drugs and therapies for this reason). (3) Real-time monitoring of treatments in early-stage cancers, though to identify earlier stages will require improved biomarkers. (4) Use of RNAi to explore combinations of targets. (5) Improved understanding of chromosomal aberration. This occurs very early in mouse tumors. (6) Exploiting genomic instability in therapeutics. Understanding more about DNA repair and repair of double-strand breaks (the latter are unusual in mouse unless telomeres are shortened). (7) Improved diagnostics from blood and body fluids – proteomics (less than 1% of proteins in blood identified, and less than 20% of these licensed for diagnostics). Data from Centers for Disease Control.

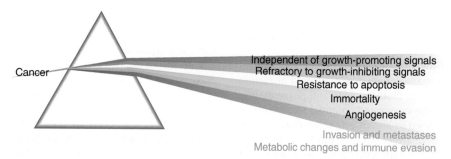

Figure 1.1 The behavior of cancers can be split into a number of common hallmark features.

the very pronounced improvements in survival from breast and colon cancers and from lymphoma. Unfortunately, over the same period, for other common cancers, notably those of lung, pancreas, and brain, improvements have been negligible and survival time is still measured in weeks.

At first glance, the biology of cancer appears straightforward. Cancer cells stop obeying the "societal" restraints imposed on individual cells within the adult organism. Instead, they multiply uncontrollably and congregate in places that should be off-limits – much like teenagers. However, in order to achieve such independence cancer cells must first be emancipated and overcome the numerous intrinsic and extrinsic barriers that seek to prevent such selfish behavior before it can threaten the survival of the entire organism.

In this book we describe the means by which normal cells are transformed into cancer cells and the key cellular processes subverted along the way. We also describe the cellular forces arraigned against the designate cancer cells and those operating on behalf of them and the weaponry available to both sides. We explain how this basic knowledge has been translated into improved diagnostics and more biologically targeted therapeutics for cancer patients. The global burden of different types of cancer is described alongside the current state of the art in diagnosis and management of cancer patients. Along the way, we make some predictions as to where new scientific and clinical breakthroughs may come from and offer our humble opinions as to why, despite some notable successes, cancer cells continue to flourish even in the face of our most sophisticated anticancer therapies. Maybe we should accept at this point that perfect theoretical proof of fact is impossible. The Münchausen Trilemma (used in philosophy to imply that it is not possible to prove any truth, even in mathematics) may provide some reassurance on this point. However, in practical terms, a good model incorporating treatments or biomarkers that work in the clinic may be a more realistic goal, even if our understanding of why the treatments work proves misguided. We are sure to continue to use the treatments until something better comes along, even if we discard the model.

Cancer poses a major threat to already overstretched healthcare services. The magnitude of the problem was summarized by Dr Gro Harlem Brundtland, former Director-General of the World Health Organization, in the Foreword to the 2003 *World Cancer Report*: "The global burden of cancer continues to increase. In the year 2000, 5.3 million men and 4.7 million women developed a malignant tumour and 6.2 million died from the disease. The number of new cases is expected to grow by 50% over the next 20 years to reach 15 million by 2020."

Cancer is responsible for more than 10% of deaths worldwide and more than 25% in some countries. Excluding the relatively frequent nonmelanoma skin cancers, lung cancer is the commonest cancer worldwide, accounting for 1.2 million new cases per year, followed closely by breast cancer and colorectal with around 1 million new cases.

The high incidence of this disease, its life-threatening nature, and often unsatisfactory management has motivated academic researchers and those from the biotechnology and pharmaceutical industries to focus on the causes and potential treatments of cancer on a scale unparalleled in almost any other disease area. Remarkably, at present there are almost 500 products in clinical trials, of which 100 are in phase III, with breast cancer and non-small-cell lung cancer receiving the most attention.

In general, cancers begin with a mutational event in a single cell and then develop in multiple stages through the acquisition of further mutations, propitious and otherwise, that are passed on to the progeny of that cell when it divides. So cancer is a clonal disease (Fig. 1.3). Aside from a few notable rare exceptions, these events arise predominantly in adult somatic cells and so are not inherited by the offspring of the affected cancer patient but only by the progeny of the affected cancer cell. In other words, transmission of the mutation ceases with the death of the patient, unless by some chance the mutant gene has been picked up by a virus, which survives and propagates. If such a virus carrying a mutated gene infects a potentially vulnerable host then the cancer-causing potential of that gene may again be unleashed upon another hapless organism. Contrary to accepted wisdom, very recent studies have suggested the astonishing possibility that cancer cells could under rare circumstances be directly inoculated from a tumor-bearing host into an unfortunate recipient. Thus, leukemias may be transmitted from mother to child and dogs may transmit cancer cells to their partners during mating.

Mutations – alterations in the coding sequence of the DNA – are not the only route to inactivation or activation of a key gene/protein. Gene expression may also be strongly influenced by a variety of **epigenetic factors** that alter chromatin structure without changing the coding DNA; these can still be passed on through successive cell generations. The term "epimutations" is often used to encompass both these major routes by which cancer cells acquire aberrant expression/activity of key genes and proteins. The average adult human has been estimated to contain as many as 10^{14} cells (i.e. 100 000 000 000 000 cells), most of which could theoretically become a cancer cell given the right sort of genetic mutations and epigenetic changes. In fact, cancer is unique in that epimutations in a single cell can give rise to a devastating disease because the resultant aberrant gene and associated antisocial behavior are transmitted to all the cellular progeny of that cell.

Because DNA replication and synthesis are essentially error prone, it is replicating cells that are most vulnerable to cancer-causing mutations. Not surprisingly, as stem cells are the main replicating cell population in the hematological system and also in epithelia, from which most cancers arise, they have long been intimated as the cell of origin for cancer. This is supported by the presence within many cancers of a side population of cells bearing stem cell characteristics known as the "cancer stem cells" (CSCs). More of this later. Although some differentiated cell types, of which adult nerve cells are a good example, are by implication unlikely to give rise to cancers because they are essentially non-proliferating in the adult, most cells either regularly replicate or can do so at a pinch. Most adult cells survive on average for 4–6 weeks and then have to be replaced. Over a hundred billion cells may die each day and are renewed either by replication of existing cells or from stem cell precursors. Given that each cell gets a substantial amount of daily DNA damage and 10^{11} or more of them will replicate each day, that is a lot of potential cancer cells!

With this in mind, a cancer might be expected to be a frequent occurrence. Yet cancer is diagnosed in only in 1 in 3 people and usually even then only after 60 or 70 years of potentially mutation-causing events. So why does a clinically apparent cancer only arise in every third individual when there are somewhere in the region of 10^{14} good potential cellular targets at risk? Moreover, we live in a world in which each of those cells is continually exposed to a myriad of avoidable and unavoidable

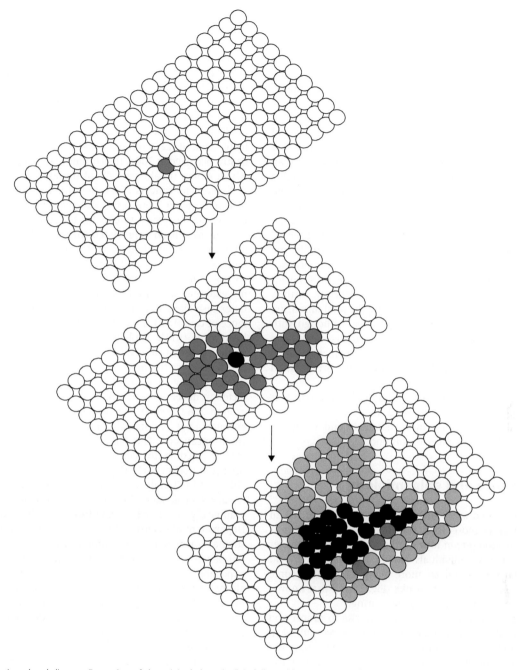

Figure 1.3 Cancer is a clonal disease. Expansion of the original clone (red) is followed by emergence of a new clone (black) which gradually replaces the original. Subsequently, a further clone (orange) emerges and expands.

DNA-damaging agents. Let us state this in the boldest terms possible. **At a cellular level, cancer is very, very rare. This surprising observation can only be accounted for by the existence of some extraordinarily effective barriers to cancer cell development. These barriers are clearly not infallible, but they must be tremendously powerful.** It is also worth reflecting on what the purpose of many of these conserved anticancer mechanisms may have originally been, given their presence in short-lived and even oligo-cellular "organisms" that are at no risk of cancer. One possibility is that processes such as senescence and apoptosis are a byproduct of archaic processes involved in balancing nutrient supply, growth and repair, and energy that fortuitously also limit cancer in longer lived multicellular organisms.

Cancers may well originate from a single bad cell, but are self-evidently not clinically detectable at that stage either by direct observation or conventional investigations such as X-ray,. This requires the presence of a small nodule, at which point replication has increased the number of cancer cells to around one billion (10^9). In other words, by the time a cancer is discovered, the original cancer cell has proceeded through some 30 or more cell divisions, and acquired a host of further epimutations. This situation is compounded by the near universal loss of normal DNA damage surveillance and repair processes. Not surprisingly, this has complicated studies attempting to unravel the initial causes of cancer in humans. We are detectives investigating a crime that occurred some 30 generations in the past. Imagine today trying to identify the particular something, potentially quite innocuous, that happened to one of my antecedents at the time of the battle of Hastings, something which may in some unexplained way have been propagated through the ages and finally made me write this book.

Safely squirreled away within a stromal nursery, those normal cells that have been successfully emancipated will give rise to a new dynasty of proto-cancer cells. This new found freedom may result from chronic inflammation but once rendered immutable by epimutations a door to a malignant future has been forced open. The resultant unfettered cell can indulge in previously proscribed behaviors, such as unrestrained growth, and be afforded unprecedented opportunities for travel and preferment. Gradually, successive generations, honed by exposure to the hostile forces arraigned against them, will witness the emergence of increasingly malign elites that begin to dominate and supplant their forbears. If we can stretch the societal analogy further, then we may claim that normal tissues exemplify totalitarianism, whereas cancerous ones are essentially pluralist. Although with time, one or more clones may become first among equals and even have imperialist aspirations, it is now recognized that in many cases tumors continue to harbor substantial representation of earlier clonal dynasties.

A schematic of how we believe cancers arise and progress is provided in Fig. 1.4. This, can be used as an overview to be referred to while reading the more detailed (and complex) description of the basis of cancer in this book.

Cancer incidence and epidemiology

In the United Kingdom and North America, the lifetime risk of developing cancer is more than one in three, and cancer is responsible for around one in four deaths. Yet, the fear of cancer experienced by many individuals should be balanced by an appreciation that one is still far more likely to die or become disabled due to a heart attack or stroke (Fig. 1.5), if that knowledge may be in any way regarded as reassuring.

Given that almost every cell type can give rise to cancer and that more than 200 different types of cancer are recognized, it is notable that four – breast, lung, large bowel (colorectal), and prostate – account for over half of all new cases. It should also be noted that although nonmelanoma skin cancer (NMSC) is very common, with 100 000 new cases recorded each year in the United Kingdom, this data is likely incomplete and the disease usually curable, so the NMSC statistics are now routinely omitted from the overall totals. In 2006 in the United Kingdom, 293 601 people were diagnosed with cancer, excluding NMSC.

Different cancers affect people at different ages, but not surprisingly the overall risk of developing a cancer rises sharply with increasing age, with 65% of cancers in the United Kingdom occurring after the age of 65 years and 35% above age 75 (Fig. 1.6). In children, leukemia is the most common cancer (around 30% of all pediatric cancers); in young men aged 20–39 it is testicular cancer.

The incidence of cancer has changed over the last 20 years; there has been a decline in lung cancer in the United Kingdom and North America in men (but an increase in women), mainly as a result of changes in smoking habits, and an increase in breast and prostate cancer. Yet despite this, an estimated 160 000 people died from lung cancer in the United States alone in 2009. In 1981 there were 78 cases of breast cancer per 100 000 women in Great Britain, and 38 cases of prostate cancer per 100 000 men. By 2009 rates were 124 and 106, respectively (http://info.cancerresearchuk.org/cancerstats/incidence).

The International Agency for Research on Cancer (IARC) has released figures on global cancer incidence for 2008 and made predictions for the next decades (http://globocan.iarc.fr). Globally, around 12.7 million new cases and 7.6 million cancer deaths occurred in 2008, the commonest being lung (1.6 million), which makes up almost 13% of the total, breast (1.38 million), and colorectal cancers (1.23 million). The most common causes of cancer death were lung, stomach, and liver, indicating the relative success of treatments for breast and colon cancers.

It has long been appreciated that there is a geographical variation in cancer incidence and deaths. Importantly, 56% of cases and 63% of cancers and deaths were in the developing world. Of the estimated 371 000 new cases of cervical cancer in 1990, for example, around 77% were in developing countries. This latter case likely reflects socioeconomic pressures and the prevalence of causal factors such as certain strains of the human papilloma virus (HPV). Globally, the most common cancer affecting women is breast cancer, followed by cervical cancer. However, in North America the most common cancer in women after breast cancer is lung or colorectal. Around 226 870 women are predicted to develop breast cancer in the United States in 2012 and around 226 160 men and women will develop lung cancer, and around 143 460 colorectal cancer (www.cancer.gov/cancertopics/commoncancers). The data for this period should soon be available. Predictions for global cancer make sobering reading: it is predicted that by 2030 there will over 21 million new cases and above 13 million deaths each year from cancer.

Race and gender also influence rates of cancer and this is graphically illustrated by data from 1999 from the United States Department of Health and Human Services. Some of the findings, such as lower rates of melanomas in men and women of Afro-Caribbean origin, attributed to inherent protection from UV exposure, are predictable. Others, however, are less so. Thus, although prostate cancer is the most frequent cancer in males, rates are 1.5 times higher in Afro-Caribbean men than in white men. Similarly, the leading cancer in women, regardless of race, is breast cancer, followed by lung/bronchus and colon/rectal in white women and colon/rectal and lung/bronchus in Afro-Caribbean women. Breast cancer rates are about 20% higher in white women. Multiple myeloma and cancer of the stomach are among the top 15 cancers for Afro-Caribbean women but not white. Recent data have become available for the United States from 2005, which shows that the rate of all cancers combined for black, white, Hispanic, Asian/Pacific islander, and Native American Indian are 591, 526, 406, 314, and 280 thousand per annum respectively (http://apps.nccd.cdc.gov/uscs/).

Towards a definition of cancer

A definition of poetry can only determine what poetry should be and not what poetry actually was and is; otherwise the most concise formula would be: Poetry is that which at some time and some place was thus named.

Karl Wilhelm Friedrich Schlegel

Overview of Cancer Biology

Intrinsic factors	**The would-be cancer cell**	**Extrinsic factors**

Inherited susceptibility:
High-penetrance genes: rare. Low-penetrance: likely; polymorphisms at multiple alleles (100s or 1000s) may all confer a degree of sensitivity or resistance to cancer (however slight the effect).

Initiation:
Spontaneous mutation in an oncogene, tumor suppressor gene or caretaker gene. (Could be 'Knudson's second hit in rare familial cancers.) DNA repair genes and p53 pathway will try to protect if intact.

Promotion:
Selective growth advantage leads to start of clonal expansion. Anti-apoptotic lesion probably required before a "mitogenic" lesion, in order to block "default" cell death. Properties acquired: minimal platform, deregulated cell proliferation, and avoidance of apoptosis. Genetically homogeneous clone. May be "pre-malignant"

Progression:
Further mutations confer additional growth advantage to successive clones. Genetic instability and aneuploidy. Properties acquired: deregulated proliferation, avoidance of apoptosis (and senescence), loss of differentiation, loss of cell adhesion, invasiveness, and angiogenesis. Invasion of lymphatics and vasculature. Clones genetically heterogeneous.

Metastatic spread:
Mutations in "metastasis" suppressor genes (possibly some already acquired earlier); eventually cancer cells entering lymphatics and vessels are able to colonize distant organs or tissue.

Carcinogens (mutagens) may increase risk of DNA damage and mutation.

Carcinogens (mutagens) may increase risk of mutation. Important cancer-causing mutations may also occur in stromal cells (i.e., not necessarily in the cancer cell). Cross-talk with microenvironment also critical.

Carcinogens (mitogens) may support promotion. Stroma may actively support tumor growth by providing survival and growth factors. Immune surveillance may try to eliminate cancer cells.

Carcinogens may support progression. Important cancer-causing mutations may occur in stromal cells also. Stroma may actively support tumor growth by providing survival and growth factors; angiogenesis; MMPs facilitate invasion and may provoke DNA damage.

Inflammatory cells may help "convey" cancer cells. "Seed" and "soil" may determine where a given cancer cell can establish colonies. Gross factors such as sites of lymph drainage will also dictate sites of metastases.

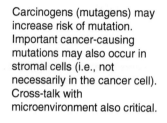
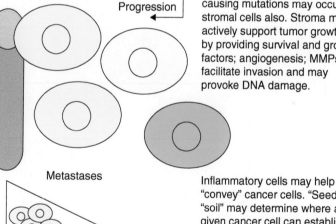

Figure 1.4 A highly stylized potential "life history" of a cancer cell. Cancer cells are shown in yellow (different shades denote subclones); stromal cells are green, vessels are orange.

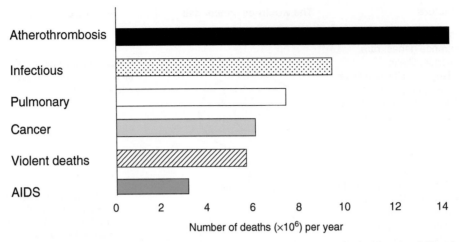

Figure 1.5 Leading causes of death worldwide, 2004. Redrawn from Murray CJL, Lopez AD (1997) Mortality by cause for eight regions of the world: Global. Burden of Disease Study. *Lancet*, **349**: 1269–1276.

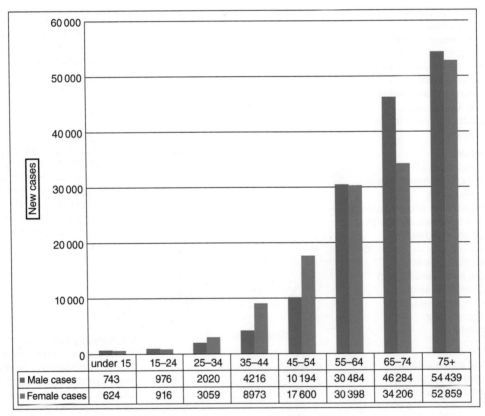

Figure 1.6 Cancer development by age for 2007 in the United Kingdom. Modified from International Agency for Research on Cancer (IARC) and Cancer Research UK data.

The terms "tumor" or "neoplasm" are used interchangeably to describe a diverse group of conditions associated with uncontrolled cell replication. Tissue mass is normally tightly controlled to serve the needs of the organism. This control is achieved by the balancing of various and often opposing cellular processes (Fig. 1.7). Disturbing the balance of these processes results in diseases; if cell losses exceed renewal this results in degeneration/involution, whereas the converse results in tissue expansion, hyperplasia, or neoplasia. If the expansion in cell numbers is confined locally then it is described as "benign," but if this unscheduled cell replication is accompanied by invasion of surrounding tissues or spread to distant sites ("metastasis"), then it is unambiguously described as malignant.

These terms are relatively straightforward as they are descriptive and based on gross observations. It should be remembered, however, that the pathological definitions of benign and malig-

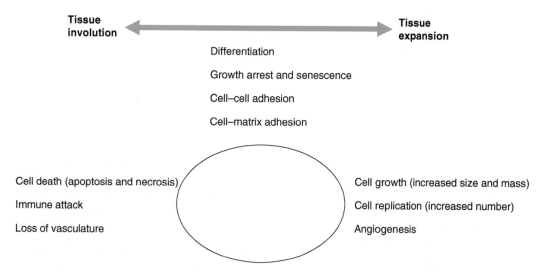

Figure 1.7 Processes contributing to regulation of tissue mass. Cell mass is determined by the balance of various cellular processes including, at the two extremes, growth/replication and cell death.

nant do not always translate into similarly benign or malign outcomes for the patient. Thus, a benign brain tumor causing severe neurological disturbance may be inoperable or require potentially life-threatening surgery, whereas a malignant prostatic cancer or microscopic metastases may have had no clinical impact and be discovered accidentally at post mortem. Adenomas are benign tumors originating in glandular or secretory tissues (such as lactotroph adenomas of the pituitary, which secrete prolactin, or parathyroid adenomas, which secrete parathyroid hormone – PTH). Such adenomas can result in substantial morbidity as a result of deregulated secretion of hormones and may also progress to become malignant, when they are termed "adenocarcinomas."

Classification of cancer

Classification of cancer is complicated by the variety of human cancers, with hundreds of different tumor types arising from almost every tissue and in every organ. This is further complicated by the ability of a cancer cell to invade surrounding tissues and metastasize to distant organs. Cancer biologists and oncologists have agreed on a classification based on the tissue of origin, regardless of organ location, focusing on the similarities in cellular structure and function among these tumors. Tumors are generally classified as either liquid or solid. The former includes leukemias and lymphomas comprising neoplastic cells whose precursors are usually motile. Solid tumors comprise either epithelial or mesenchymal cells that are usually immobile. Pathologically, cancers are classified as:

- **carcinoma**, originating from epithelial cells in skin or in tissues that line or cover internal organs and typically represent over 80% of human cancers;
- **sarcoma**, originating in bone, cartilage, fat, muscle, blood vessels, or other connective or supportive tissue;
- **leukemia**, a cancer originating in blood-forming tissues, such as the bone marrow, causing large numbers of abnormal blood cells to be produced and enter the bloodstream; and
- **lymphoma**, originating in the cells of the immune system.

It is worth emphasizing that the purposes behind disease classification are to help make the most accurate predictions about prognosis and response to particular therapies in the clinic and in the laboratory to ensure that as far as is possible like is studied alongside like. As discussed later, this objective may increasingly be better served by grouping cancers on the basis of their shared molecular pathoetiology rather than by tissue of origin.

"Carcinoma *in situ*" refers to lesions regarded as cancer that remain localized to the tissue of origin, often constrained by intact basement membrane. Such tumors often respond well to treatment, with good prognosis for the patient. In contrast, "invasive carcinomas," by disrupting basement membranes and growing into surrounding tissues, are more difficult to treat successfully. In addition, since invasion is usually a prerequisite for metastasis, the ultimate cause of most cancer-related deaths, even when the local lesion is treated, the prognosis is often poor.

Importantly, disease classification is not written in stone, as technical advances are made and larger numbers of individuals with a given disease are studied, it is often possible to recognize previously unappreciated "subclasses" of disease that can readily be detected and further improve accuracy of prognosis and prediction of treatment responses. Most recently, advances in postgenome era technologies such as oligonucleotide arrays and proteomics (Chapter 20) are allowing a subclassification of cancers in terms of molecular profile termed "tumor fingerprinting." At the same time genomics (essentially reading the DNA) is being increasingly applied to look for cancer susceptibility genes in patients and for mutations in tumors. It is hoped that in the future such powerful tools will ultimately allow more accurate determination of prognosis and even "tailored" therapy, whereby each patient can be uniquely classified and treated on the basis of such tests. These aspirations are often referred to as "individualized medicine", reflecting the ideal of being able to treat each individual in a uniquely appropriate way, based on variation in one or more of the following parameters: gene alleles, gene expression/protein expression and mutations in tumor cells, proteins in the blood.

It is surprisingly difficult to define cancer in practice

Cancer is a difficult term to define accurately. Put simply, cancer is synonymous with malignancy, and refers to a group of conditions that have manifested malignant behavior, namely

> **Box 1.1 Cancer screening**
>
> In 1968, Wilson and Jungner of the WHO set down ten principles that should govern a national screening program:
>
> 1. The condition is an important health problem.
> 2. Its natural history is well understood.
> 3. It is recognizable at an early stage.
> 4. Treatment is better at an early stage.
> 5. A suitable test exists.
> 6. An acceptable test exists.
> 7. Adequate facilities exist to cope with abnormalities detected.
> 8. Screening is done at repeated intervals when the onset is insidious.
> 9. The chance of harm is less than the chance of benefit.
> 10. The cost is balanced against benefit.
>
> The aim of screening is to identify at-risk individuals for whom effective interventions or treatments are available, and should also be limited to situations where that treatment is more effective if administered early and before the condition to be treated becomes readily apparent. If the above criteria are satisfied, then in general, the ideal screening test for any given condition should be highly sensitive (few false negatives – patients deemed normal who actually have the condition) and highly specific (few false positives – normal patients deemed to have the condition). In many cases, increasing sensitivity may result in decreasing specificity, and often health policy decisions have to be made that take account of the prevalence and severity of the condition to be screened for, economic factors relating to the cost of screening and the subsequent proposed interventions, and also both efficacy and the safety of the available interventions (risk/benefit ratio).
>
> Broadly, two types of screening are applied: (1) **population screening**, where mechanisms are put in place to ensure that all appropriate individuals are screened at given times/intervals – largely the responsibility of public health organizations; (2) **opportunistic screening**, where healthcare workers undertake screening when individuals present to them for whatever reasons – this is largely the responsibility of healthcare professionals. The latter approach is cheaper, but will provide less cover of the population.
>
> In the vast majority of cancers there is little doubt as to the potential severity of the condition, and in some cases where the treatment offered is fairly innocuous (e.g. lasering of cervical lesions) one can afford to treat a number of so-called false positives. However, if the treatment involves bowel resection or mastectomy, for example, this calls for much greater accuracy in prediction and a smaller number of false positives can be accepted. In clinical practice this is often reflected in how early in the evolution of a potential cancer such treatments are offered and therefore also on the extent to which the given cancer can be prevented. In general, earlier is better, but this requires much greater ability to predict the behavior of a given tumor or lesion.

unscheduled and uncontrolled cell growth leading to invasion and/or metastases. There is no ambiguity in this case as the definition is "retrospective" and based on the readily observable behavior of the "cancer." Such a narrow definition is of limited practical value in the laboratory, however, and particularly in the clinic, as it precludes true preventative or even early treatment. This seemingly abstract issue is placed in context when it is remembered that for those cancers where rates of death have actually been reduced over the last few decades, this has resulted primarily from improvements leading to earlier diagnosis and earlier administration of treatment.

It is clear that certain features at a microscopic level can accurately be employed to identify a tumor as cancer before it manifests overtly malignant behavior clinically (metastasizes to lymph nodes or other organs or has on imaging or surgery been shown to have invaded local structures). In other words, a cancer is a cancer before it necessarily declares itself by behaving as one. In most cases, this requires the demonstration of evidence of penetration of a basement membrane or invasion into surrounding tissue (which means you need to look at a piece of tissue that includes the tumor – histological examination) and/or the presence of "cancer cells," namely cells exhibiting defined changes, which from experience are the same or similar to those seen in circumstances which are incontrovertibly cancer (which means you need to have acquired some cancer cells from body fluids, sputum, or via a smear- cytological examination). Clearly, the latter is often quicker and less invasive in clinical practice.

In a clinical setting, where the primary purpose is to identify a tumor or lesion that requires surgical excision or other treatment, it may be sufficient to know that a particular lesion (based on gross appearance or histological examination) poses a risk of proceeding to an invasive cancer. A lesion may already be regarded as a cancer, on the basis of abnormal growth or appearance and the near inevitably of progression to invasion (carcinoma *in situ*), or its potential may not be yet realized/manifested but risk of progression is high ("precancerous" or "premalignant"). This forms the basis for identifying "high-risk" lesions such as breast carcinoma *in situ*, Barrett's esophagus (a precursor of esophageal cancer), colonic polyps (a precursor of colon cancer), and others. Cytological examination may identify premalignant cells and is employed where such cells can readily be obtained, including cervical screening for the early detection and prevention of cervical cancer (Box 1.1).

For a research scientist, these distinctions are also of critical importance. The ability to define the point at which a premalignant benign lesion ends and a malignant cancer begins is a prerequisite to understanding the initiation and key early events in cancer formation. In the laboratory the progressive behavior of transformed cells or tumor progression can be investigated in animal models, as long as the necessary investigative tools are available, but this opportunity is self-evidently usually lacking in the study of cancer in humans. The cancer researcher can validate predictions made about the future behavior of a given lesion by prospectively tracking the eventual emergence of invasive metastatic cancer but, as will become clear later, the actual stage of evolution at which cancer cells emerge and acquire ability to become invasive and metastasize is still contentious and quite difficult to detect.

Cancers may not always be clinically apparent

Difficulties of definition notwithstanding, the clinical situation is further complicated by the increasing awareness that microscopic colonies of cancer cells (*in situ* tumors) can be detected in different tissues (thyroid, breast, prostate for example) at autopsy in most older individuals. In fact, such clinically irrelevant *in situ* cancers may be a 100- to 1000-fold more common than clinically apparent cancers arising in those same tissues during life. For example, most older individuals have *in situ* thyroid carcinomas at autopsy, whereas only around 0.1% of similarly aged individuals are found to have thyroid cancer during life. Although biologically intriguing and testifying to the potential effectiveness of innate anticancer defenses (such as antiangiogenic factors), such findings may increasingly be problematic in the clinic.

Until recently, we have generally not detected the vast majority of such *in situ* tumors during life, largely because we do not routinely biopsy tissues in apparently healthy individuals. However, one area in which detection of such *in situ* tumors may pose difficult and as yet unresolved clinical dilemmas, is increasing use of diagnostic prostatic biopsy in older men, and discovery of so-called "incidentalomas" during routine imaging procedures such as CT and MRI scanning. The now ubiquitous presence of privately run "walk-in" imaging centers offering the dubious benefits of whole-body scans will undoubtedly compound this problem. For example, what do you do about the incidental lump that is not self-evidently cancerous – particularly as benign irrelevant lesions will be considerably more numerous? The patient will likely be anxious and may well push to undergo potentially dangerous invasive diagnostic steps in order to be as certain as possible that they do not have cancer. Guidelines have had to be developed and will continue to be needed to assist clinicians in deciding which individuals with such findings actually require any form of treatment or just reassurance.

However, this is far from straightforward, as there are many cases where the actual ability to predict the risk of future invasive cancer based on the appearances of a given lesion are not yet sufficiently mature. A good example is the readily visible dysplastic white lesion in the mouth that may in some cases – but by no means all – herald the development of an oral squamous cell carcinoma. Ironically, at least in some cases, where the lesion may be less likely to come before the eagle eyes of dentists and GPs or it is not technically possible to detect the early lesion let alone examine it, this may be for the best until our ability to more accurately predict the future behavior of these early lesions improves and/or we greatly increase our current arsenal of sufficiently well-tolerated and nonharmful therapies to exploit the potential benefits of early diagnosis.

This interesting debating point notwithstanding, it is abundantly clear that in order to prevent or cure cancer effectively it is essential to diagnose disease as early as possible, and nothing should distract us from our efforts to progress in this goal. Failure to do so will inevitably mean that potentially life-saving early treatment for some individuals destined to develop clinically important cancer will be delayed. To resolve this conundrum is theoretically simple – we just need to distinguish early lesions that will never progress to disease from those that will progress to cancer. Although, routinely screening apparently healthy individuals for certain cancers has been well-validated and has become accepted best practice for cancers of breast (mammography), cervix (Pap smear), and colorectum (fecal occult blood) in many countries, for most cancers we urgently need better tests and tools. Fortunately, the research community has responded to this challenge and much progress is being made in finding new tests and "biomarkers" for various cancers that might give important information about prognosis and treatment response (see below and Box 1.2).

Box 1.2 Cancer biomarkers

Leland Hartwell, in his keynote address at the 2004 meeting of the American Association for Cancer Research (AACR), suggested that earlier diagnosis and improved monitoring of cancer progression by noninvasive means could dramatically improve the outcome for many patients. Early detection represents one of the most promising approaches to reducing the growing cancer burden and has been revolutionized with the advent of postgenome era technologies that can identify cellular changes at the level of the genome or proteome and new developments in data analyses and modeling. Gene expression profiling of various human tumor tissues has led to the identification of expression patterns related to disease outcome and drug resistance, as well as to the discovery of new therapeutic targets and insights into disease pathogenesis. However, techniques requiring removal of cancer tissues can only be employed once a tumor has been detected and are unsuitable for earlier diagnosis and for general screening. A noninvasive test would have numerous advantages. Therefore considerable efforts are now directed at finding biomarkers in blood tests. These are obtained relatively noninvasively and rapidly, and could be employed in screening. Biomarkers could also be useful in posttreatment follow-up for disease recurrence. Most current tumor biomarkers are lacking in sensitivity and specificity, and more effective ones are required.

Therefore, considerable efforts are now directed at finding "biomarkers" in blood or urine tests that can be obtained relatively noninvasively and rapidly, and could much more readily be employed in screening large numbers of individuals. Their role could also be extended into surgical surveillance for potentially operable disease and postoperative follow-up for disease recurrence.

Broadly, three overlapping technologies can be employed to look for cancer biomarkers:

1. Analyses of proteins by: (a) immunoassay of single known proteins predicted to be of interest; (b) proteomics, including 2D gel-based separation or liquid chromatography followed by mass spectrometry to identify potentially thousands of different proteins; (c) proteomic pattern analysis or "fingerprinting," which relies on the pattern of proteins observed and does not rely on the identification of individual traceable biomarkers.

2. Analyses of free RNA, including miRNA, in the circulation some of which derives from the cancer.

3. Isolation and study of circulating tumor cells, which can in turn be profiled for gene expression by microarrays.

As mentioned earlier, in order to improve our predictive/diagnostic abilities, traditional examination of patients in the clinic, application of imaging techniques, and cytology/histology of the tumor are increasingly being supported by newer techniques, such as molecular profiling. Traditionally, genetic analysis looks for single susceptibility genes that confer a high risk of cancer formation, but in future this may include more complex genomic testing (of multiple polymorphic alleles – see below), or direct analyses of gene/protein expression in the tumor by various techniques including gene chip microarrays and proteomics. Considerable enthusiasm has been generated by the possibility of using relatively noninvasive tests to identify cancer biomarkers in blood samples or other body fluids from patients with cancer or at risk of cancer. Thus, proteins, mRNA, or miRNA derived from the tumor or from the body's response to it might be analysed in body fluids. In many cases it has also proved straightforward to isolate and examine cancer cells (or their DNA) from blood or topically. If such information can be correlated with the presence or absence of cancer in the healthy population, or with clinical outcome or treatment response in known cancer patients, then these will be useful biomarkers.

The best-known currently used serum biomarker is prostate-specific antigen (PSA), elevated levels (or progressively rising levels) of which are associated with significant risk of prostate cancer. However, this falls short of the ideal in several respects, in particular the number of false positives (the test wrongly suggests the possibility of prostate cancer) and false negatives (a cancer fails to be diagnosed). This means that even clinical trials disagree on the benefits of general screening with PSA. In fact, there is a more fundamental flaw in the notion of simply detecting presence of prostate cancer by screening: it gives no insights into prognosis. This quandary is easiest to appreciate if we assume an ideal performance for the test and thus have in some way eliminated false positives (without compromising sensitivity). So now we use the test and it unambiguously tells us which patient has prostate cancer. What it does not tell us is what to do with the patient. Why not? Because recent trials have suggested that PSA screening results in overtreatment because the prognosis of occult prostate cancer is so variable and often does not affect mortality or morbidity (see Chapter 3). This does not mean that PSA screening is without value, it makes a major contribution to the investigation of patients with symptoms of prostatic enlargement (difficulty in micturition) and in the follow-up of prostate cancer patients following treatment. However, how suitable PSA is for screening the general population is controversial.

Tumor-derived biomarkers are already in routine use in the clinic. Thus, the presence of estrogen and progesterone receptors or of a *HER2* mutation in breast cancer defines patients who will likely benefit from hormone-based therapies or trastuzumab, respectively. Commercial biomarker assays which measure expression of multiple genes and mutant versions from tumor samples by reverse transcription polymerase chain reaction (RT-PCR), such as the Oncotype DX test for breast cancer (measures *HER2*, *ER*, and *PR* status as well as 13 other cancer-relevant genes including Ki-67 and survivin) are now available. The presence of mutant *KRAS* in a colorectal tumor identifies a subgroup who will respond poorly to drugs targeting epidermal growth factor receptor (EGFR).

Genotype may also be helpful. Thus, a recent large study has confirmed that breast reduction surgery in women who are carrying germline mutations in *BRCA1* and *BRCA2* can markedly prevent breast cancer in these individuals. In a recent study, there were no diagnosed cancers in 247 women with risk-reducing mastectomy compared with 98 women of 1372 diagnosed with breast cancer who did not have risk-reducing mastectomy. Moreover, women undergoing risk-reducing salpingo-oophorectomy had improved survival.

If we find a cancer what do we do with it?

Not only do we often not know who to treat, we are often unsure what treatments to use, particularly before the development of an obviously life-threatening cancer. This situation has not been helped by the fact that the majority of therapeutic trials have focused largely on the end stages of cancer, by definition, the point at which these therapies are least likely to successfully cure the disease. Why? Because regulatory approval requires a lengthy series of clinical trials (see Chapters 15 and 16) and these are most readily conducted in patients with advanced cancer for whom no further treatments are available. Use earlier in the disease process is often left until after marketing and then interpretation may be confused by the need to use the new drug alongside existing best practice.

Improved ability to predict treatment response is fundamental to avoiding the morbidity and mortality associated with cancer while also restricting potentially harmful or even life-threatening treatments to those individuals most likely to benefit. Most treatments are justifiable when a life-shortening cancer is prevented, but would be very undesirable if employed in an individual never destined to develop cancer and who will eventually die of some unrelated other cause and whose life would have been affected less by the cancer than by the treatment. In practice, what is needed are clinical measures or new biomarkers that correlate with prognosis and that ideally also assist in selecting the best treatment or combination of treatments (from among watchful waiting, surgery, radiation, and drugs).

One thing is clear: early treatment offers the best chance of a successful outcome. This problem is addressed by various screening programs aimed at identifying premalignant or early stage cancers (see Box 1.1). Importantly, in these cases suitable treatment strategies have been defined.

As discussed in the previous section, it is hoped that detailed molecular analyses of tumor samples or body fluids will not only improve our understanding of the "roadmap" to cancer for any given cancer, which might in turn guide us to the application of specific drugs to target particular genes/proteins, but may also improve our ability to predict therapeutic responses. Such detailed analysis of individual tumors starts to realize the potential of post-genome era science and may finally deliver the ultimate goals of "individualized medicine" and "tailored therapy" – where treatment is fitted specifically to an individual.

The best treatment is prevention

Prevention requires a combination of activities involving different organizations, including public health strategies aimed at the whole population and exemplified by activities targeting adverse lifestyles, including smoking and poor diet. More targeted advice and possibly interventions may be needed for individuals at the highest predicted risk of disease. A new discipline of chemoprevention has been established with the sole purpose of designing the perfect weapon for a pre-emptive strike against future cancer cells. However, with the rare exceptions of individuals with known familial cancer syndromes, this has proved far more difficult a strategy for cancer prevention than it was for preventing coronary heart disease (CHD); there are no statin equivalents for cancer prevention.

At one extreme, no complex tests are needed to spot obese patients and smokers, and accurately predicting which of these will get early cancers as a result may be unnecessary, because encouraging all to change behavior appears a reasonable approach, particularly as in these cases such a lifestyle treatment is not likely to have any "off target toxicity" (prevent cancer but cause something as bad or worse). In other words, assuming that everybody is at risk of smoking and obesity-related disease may be good enough. Recent guidelines have placed vaccination of young women to prevent cancer-causing infection with HPV in this category. Being female is considered a sufficient risk of being infected with HPV and developing cervical cancer in the future and no subclassification is deemed desirable. Indeed, further selection could actually compromise the efficacy of vaccination as it might reduce the chance of developing herd immunity, although one could use sexual activity as an additional screening test and restrict vaccination to the noncelibate. In the case of invasive treatments, clearly it is preferable to narrow down the risk much more before offering preventative surgery or toxic drugs unless these are targeted correctly. Thus, before offering mastectomy to a woman to prevent breast cancer we need to know a lot more than just her gender. In this case the presence of very high-risk mutations in *BRCA1* and *BRCA2* identify a small subset of women who will benefit from surgery. However, there really is not much in between these two extremes. In other words, for most of us there are no simple tests that can be used to predict our risk of future cancer.

Robust tools have been developed allowing reasonably accurate estimation of future risk of heart attacks or strokes based on using simple information such as age, sex, blood pressure, and level of circulating fats (readily determined in the clinic) in order to calculate a risk score. For cancer the hope is that improved genetic testing (Box 1.3), measurement of new disease biomarkers, and improved clinical investigational tools will match these successes in time. Screening is discussed in more detail below and in Box 1.1.

What is next best?

The early detection of cancer or precancer syndromes is self-evidently the next best to prevention, based on the assumption that small numbers of well-localized cells of a potentially less advanced state of malignancy will prove easier to treat or cure. This forms the basis of screening for cervical, breast, and colon cancers (Box 1.1). Improved early detection also involves the speedy selection of patients with appropriate symptoms or signs for early application of diagnostic tests (including X-rays, blood tests, biopsy, etc.). The nature of such tests is continually evolving, with great hope placed on the identification of cancer biomarkers and resultant possibility of molecular diagnostics gradually supplanting or complementing more traditional morphological assessments. Biomarkers may derive from a variety of sources, including serum proteins or nucleic acids, circulating cancer cells singly, or as part of complex molecular signatures. Not all such new diagnostic tests will necessarily result from ever more advanced molecular and cellular biology. Some of the ideas of the original pioneers of cancer biology still have potential and are being evaluated (see Jean Astruc, in Appendix 1.1), even highly creative or eccentric ideas such as training "sniffer dogs" to identify bladder cancer from the smell of a person's urine (though with any dog I've met the trick appears to be to stop them publicly "screening" everybody within reach!).

Currently available treatment options

The number of treatment options has expanded dramatically in recent years with the emergence of specific therapies targeting individual cancer-relevant molecules or signaling pathways. Thus, knowledge that a cancer is possessed by a particular malign oncogenic mutation can be exploited by the administration of a suitable therapeutic exorcism. However, choice of appropriate treatment regimens for any given patient remains challenging. In general, the first decision to be made is whether the cancer may be cured by surgical resection and radiation or drugs, or both. A more detailed discussion of cancer therapies is presented in Chapter 16, but a few interesting aspects will be highlighted here.

Achieving lasting remission in patients suffering from nonlocalized malignancies remains elusive. We are rarely, if ever, able to kill all the cancer cells in the primary tumor and metastatic lesions. Such failures may be the result of poor access of effective treatments to all tumor locations, varying susceptibility to conventional DNA-damaging anticancer agents, or the rapid evolution of resistance. A particular problem is posed by cancers where cells spread early via the circulation to establish micrometastases in the bone marrow or elsewhere. While increasing drug dosage can overcome some of these barriers it also increases toxicity to normal cells; to paraphrase Paracelsus, "The dose makes the poison."

Traditional cytotoxic treatments aim to kill all cancer cells, whereas some newer approaches may be directed at disabling cancer cells (inducing growth arrest, differentiation, etc.), without necessarily killing them. Therapeutic resistance is a major issue in cancer treatments and may arise by cancer cells acquiring new routes of signaling that bypass the drug-targeted protein or even by developing ways of blocking the drugs access to the cell. Cancer stem cells are also a potential explanation for treatment resistance and recurrence, as such cells may be inherently more resistant to agents or may occupy environments such as hypoxic niches which protect them.

Despite some notable successes, concerns remain about potential adverse effects of traditional radio- and chemotherapy on normal tissues, and intriguingly also on the surviving cancer cells themselves. Cancer progression is an evolutionary process driven by acquisition of epimutations, which provide a selective growth advantage to particular cell populations. Therapies that induce irreparable damage to cell DNA may have undesirable consequences on cancer cells – they may fail to undergo apoptosis and

---

Box 1.3 Genetic testing

The identification of disease-related genes has led to an increase in the number of available genetic tests that detect disease or an individual's risk of disease. Gene tests are available for many disorders, including Tay–Sachs disease and cystic fibrosis, in cancer testing for the *BRCA1* genes and breast cancer, *MEN1* and *RET* in endocrine tumors, and as more disease genes are discovered, more gene tests can be expected.

go on to survive the onslaught. In fact, one mechanism of resistance in cancer cells may be the increased mutation rate and selection pressure provided by such drugs. The net effect of unsuccessful cancer therapies could be to speed the progress of the disease, as more mutated cells expand without the competition of their less aggressive predecessors and their offspring. We might actually help to select for more aggressive clones or enrichment of more malign cells such as CSCs. Increasingly, therefore, new combinations of drugs are employed to reduce the likelihood of cancer cells surviving to become resistant to all these agents.

An interesting parallel may be drawn here between evolution of species and evolution of a cancer. Evolution is driven not only by mutations and natural selection, but also by catastrophic extinctions which, by removing less-hardy competitors, clear the path for the survivors to fill the vacuum. It may be that subtotal cancer cell killing with chemotherapy, radiotherapy, or even surgery is the cancer equivalent of a meteor impact. Given, that only some 1 in 10 000 of the estimated 50 billion or more species that have evolved on earth still exist, and that if we are anything to go by the survivors include some of the hardiest and nastiest, then perhaps extinctions of some of the less able to survive may be undesirable if you do not cull the lot. Moreover, the situation in cancer therapy is likely a lot worse, as cancer cells are repeatedly selected for their ability to not be killed by cancer therapy, whereas species have not necessarily been selected largely on their ability to survive repeated meteor impacts or volcanic activity but probably somewhat more randomly. The risks inherent to increasing "selection pressure" have been ably demonstrated in the case of emergence of antibiotic resistance in bacteria.

So what might be an effective alternative to treatments involving chemo-, radiotherapy or surgery? Theoretically, arresting replication in cancer cells might be a good alternative or addition to traditional treatments that offer anything other than complete extinction of cancer cells, as this would prevent expansion of an aggressive surviving clone and might instead foster "stagnation" of the cancer cell population. Assuming that no treatment will ever immediately kill all cancer cells – what proportion effectively constitutes total extinction? The 90% extinction of species believed to have occurred in the Permian era was followed by a substantially slower recovery (based on fossil records and therefore really only applies accurately to "big organisms") than after those in different eras which resulted in 60–70% extinction – but they still eventually recovered. Arguably, we might wish to know what proportion of cancer cells need to be killed in an individual for no symptomatic recurrence of the tumor to take place during that individual's lifespan!

Causes of cancer

Much has been learned about the causes of cancer, including the role of genetic predisposition, gene–environment interactions, and infectious agents. Intriguingly, recent research points to the considerable overlap between the behavior of cancer cells and that of cells during normal physiological wound healing and during embryogenesis. Similarities include replication, less differentiated state, invasion/migration, with the major differences reflecting the lack of control and the unscheduled nature of replication that characterize cancer. One intriguing question, addressed later, is how the organism is able to distinguish between normal growth and tissue repair (normal cell cycles) on the one hand and neoplastic growth (cancer cell cycles) on the other.

The clonal evolution theory

Most cancers derive from an individual somatic cell in the adult organism, with the initiation and progression of tumorigenesis dependent on the accumulation of genetic or epigenetic changes that determine the emerging cancer phenotype. Initiation is believed to be through DNA damage, which renders the cell capable of forming a cancer; initiated precancer cells then multiply during a promotion phase. One way of looking at this is to assume that the first mutation in some way liberalizes the would-be cancer cell, thereby generating an underclass of uniquely susceptible cells among which some may subsequently become increasingly radicalized. Cancer cells are created by the assimilation of epimutations that promulgate increasingly individualistic and sociopathic behaviors at odds with the best interests of the organism and, moreover, this process may proceed through recognizable stages.

A "multistage" model of carcinogenesis, based largely on epidemiological observations, was first articulated by Peter Armitage and Richard Doll in the 1950s. The rapid expansion of knowledge about the molecular genetic basis of disease then allowed Nowell, in 1976, to suggest that cancers arise by a process of multistep clonal evolution. He proposed that most neoplasms arise from a single cell of origin, and tumor progression results from acquired genetic variability within the original clone, allowing sequential selection of more aggressive sublines. He also stated, rather prophetically, that acquired genetic instability may result in apparently similar individual advanced tumors being very heterogeneous both at a molecular and behavioral level and might require individual specific therapy. He also predicted that therapy could be thwarted by the emergence of genetically variant resistant sublines.

Becoming a cancer cell – multistage carcinogenesis

A wealth of data has supported the view that cancers are multistage diseases progressing via protracted accumulation of multiple genetic and/or epigenetic changes (lesions) that compromise control of cell proliferation, survival, differentiation, migration, and social interactions with neighboring cells and stroma.

Hanahan and Weinberg have recently updated their seminal review of 2000, in which they originally construed the axiomatic requirements of cancer cells as:

(1) the capacity for self-sufficient proliferation, independent of exogenous growth signals;
(2) refractoriness to growth inhibitory signals;
(3) resistance to apoptosis;
(4) unrestricted proliferative potential (immortality);
(5) capacity to recruit a vasculature (angiogenesis);
(6) ability to invade surrounding tissue and eventually metastasize.

Not surprisingly, other cancer-critical processes have been vying for the dubious accolade of becoming the seventh hallmark feature of cancer ever since. The Warburg effect, a shift in energy production from oxidative phosphorylation to glycolysis, is currently leading the polls, but is being hard pressed by avoidance of immune surveillance, tissue remodeling, and a variety of forms of stress or stress phenotype. In most cancers the presence of chronic inflammation alongside subversion of expected interactions with immune and stromal cells also appear to be common features.

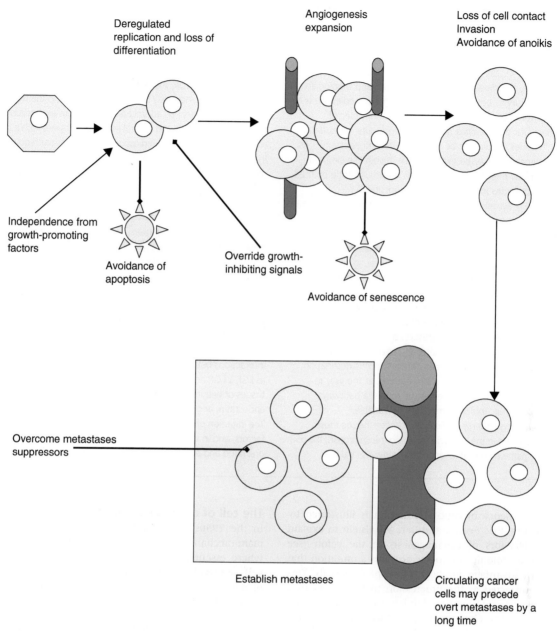

Figure 1.8 Processes contributing to cancer formation. The "hallmark" features of cancer are shown, appearing in a potential sequence. It should be noted that this does not imply that this is the actual sequence in which such features are acquired in any particular cancer.

By implication, tumor progression proceeds by the acquisition of lesions that provide the tumor cell with these attributes and which thereby shape the complex phenotype of the tumor cell (Fig. 1.8). Mostly, these lesions are acquired in somatic cells, but in the inherited cancer syndromes (see Chapter 3), one of the lesions is inherited and is present in all somatic cells – the would-be cancer cell has a headstart in life. It is important to note that seemingly phenotypically similar cancers may arise through differing combinations of lesions: there are likely many different routes to cancer, even in the same cell type (Box 1.4). Many key cancer-relevant signaling pathways may be activated or inactivated by mutations at various different points that could result in largely identical cell behaviors. Many of the "hallmark" features of cancer cells may be the consequence of reactivating embryonic developmental programs by different routes and will be discussed later with respect to CSCs and EMT.

Genetic alterations conferring the hallmark features generally involve gain-of-function mutations, amplification, or overexpression of cancer-driving genes (oncogenes) or loss-of-function mutations, deletion, or epigenetic silencing of cancer-restraining genes (tumor suppressors) or DNA-repair genes (caretakers). Although the genes involved show considerable overlap between individual patients and types of cancers (mutations in some 14 or more genes, including *RAS*, *RB*, *p53*, *PI3K*, are frequent offenders) they are found alongside a wide variety of other much more "individualized" low-frequency alterations involving several hundred distinct genes that give each tumor its often unique blueprint.

Box 1.4 Two steps to seven? The roadmap for cancer

"Pluralitas non est ponenda sine necessitate."

William of Ockham, the most influential philosopher and theologian of the fourteenth century, is best known for applying the medieval rule of parsimony to formulate one of the best-known principles of science, Ockham's razor: *Pluralitas non est ponenda sine necessitate*, translated as "entities should not be multiplied beyond necessity." As a principle in science this may be expressed as "favour the simplest model which explains the observations." Even earlier, Aristotle pointed out that "nature operates in the shortest way possible."

It has been widely assumed that since (i) human solid tumors when examined carry a plethora of genetic and epigenetic alterations and (ii) it is genetically difficult to transform cells under tissue culture conditions, cancer formation can only occur under the influence of multiple (possibly 7 or more) genetic lesions. However, in some cases the situation may be much simpler. Namely, that the key requirements for tumorigenesis are deregulated cell proliferation and suppression of cell death, and that mutations enabling these may constitute the "minimal platform" for the development of a cancer, at least where one of those lesions is deregulated expression of c-MYC. It is clear that there are far fewer "pathways" implicated in cancer than genes. Therefore, some cancer cells may indeed "arrive" at this destination via a protracted route involving multiple mutations, as the way in which a given cell activates or suppresses the requisite pathways needed to complete this "journey" may be very variable.

Some of the pathways strongly implicated in cancers include those regulating G_1/S transition in the cell cycle, including the Rb protein, the p53 tumor suppressor pathway and other apoptosis pathways, and the angiogenesis/HIF1 pathway. In fact, there are now numerous examples of only two genetic lesions fulfilling these requirements and promoting neoplastic progression, suggesting that at least in some cases the genetic basis of a given cancer may be remarkably simple. In this model, the genetic complexity of an advanced tumor is more a reflection of evolutionary pressures and natural selection of clones with a growth advantage, rather than an indication of the mutations required to initiate that tumor. The "mission critical" mutations are concealed within the plethora of mutations, many of which are likely irrelevant to tumorigenesis.

This minimal platform model may be reconciled with studies of cell transformation *in vitro* – it may be much harder to establish transformation and immortality in a cultured cell than to produce a cancer cell within the organism. The intact organism comprises a network of usually highly effective anticancer barriers, but once these become breached they may instead support the developing tumor. This is not pure conjecture, it is clear that the organism provides the developing tumor with a blood supply as long as it is instructed to do so; in some cases this may require an "angiogenic switch" (an acquired mutation which allows the tumor to "request" to stromal cells for angiogenesis), but might also be an inevitable accompaniment of tissue growth, no matter how inappropriate. In fact, much is now known about the interactions between proangiogenic factors produced by the tumor (such as FGF, VEGF, and PDGF) and antiangiogenic factors produced in the tissues or within the circulation (such as thrombospondin, tumstatin, endostatin, angiostatin, and interferons alpha and beta, respectively). The initiation of angiogenesis is likely dictated by the balance of these factors, and in turn by the genes expressed by a given cancer cell on the one hand and by the tumor microenvironment on the other.

The multistage theory of cancer formation is illustrated by models proposed by Eric Fearon and Bert Vogelstein to explain the observed behaviors of carcinogenesis in the colon (see Fig. 3.3). A normal colonic enterocyte acquires a mutation that confers a growth advantage and begins to expand clonally. This stage may be protracted as the progression to full malignancy may require not just one mutation, but between 8 and 12 independent mutations. The chances of a single mutation occurring among the billions of gut cells over a 70-year or more lifespan is substantial. However, the chance of two mutations occurring in one cell is much less (the square of the original probability) and for all 8 or more mutations to occur in one cell in the lifetime of an individual is vanishingly small. However, if one also assumes that each mutation results in clonal expansion, then these odds begin to narrow rapidly (a second mutation is clearly going to be more likely in a few million proliferating cells than in one).

An alternative explanation for the infrequency of cancer development is that interlocking combinations of mutations might be required from the outset; in other words, more than one mutation is needed for the initial expansion of a clone of cells. Recent work by several laboratories has supported this notion by finding that in certain cases the mutational route to cancer is rather short (in molecular terms), with as few as two interlocking mutations required for initiation and progression of cancers in animal models, and – at least where one of these lesions involves particularly "dangerous" oncogenes such as c-MYC – also in humans (Box 1.4).

The cell of origin in cancer

In the 1950s, the histologist Charles Leblond described three main mechanisms by which adult organs are maintained: **static**, where essentially no replication occurs (e.g. nervous system); **self-renewal**, where stem cells compensate for rapid losses of differentiated cells (e.g. gut and skin epithelia, blood); and **simple duplication**, where tissues are maintained by proliferation of their own differentiated cells (tissues with slower turnover, such as pancreas, liver, kidney, blood vessels). Interestingly, this early view has been largely discarded in recent decades in favor of the notion that essentially all adult tissues are maintained primarily from a local minority subpopulation of progenitor cells, which retain a strong proliferative capacity, as well as the ability to differentiate into the required mature cell types after dividing – the so-called stem cells (see Box 5.2 – Stem cells). Only recently, with seminal studies employing direct lineage tracking using "pulse-chase" techniques (see Chapter 20), have experimental data actually provided unambiguous support for Leblond's original idea at least with respect to simple duplication being important in pancreas.

It is a widely held view that cancers originate primarily in stem cells. In fact, the stem cell origin of cancer originates from mid-nineteenth century microscopic observations, which showed the similarity between embryonic tissue and cancer, leading to the suggestion that tumors arose from embryo-like cells. The later demonstration in the late nineteenth century of so-called "embryonic remnants" in adult tissues that could become activated in cancer gave rise to the "embryonal rest" theory of cancer – now understood as the origin of cancer from adult stem cells.

Given their longevity and unique abilities to self-renew and proliferate, it is not surprising that cancers might originate in stem cells. Importantly, the evidence for this is strongest for cancers of the blood and epithelial cells; tissues usually maintained by stem cell replication. The "cancer stem cell" model has recently been supported by a study with another tissue where progenitor cells are the major or only source of cell renewal in the adult, the brain. It was noted that only a subpopulation of brain cancer cells expressing a marker indicating their progenitor cell status were able to generate tumors when implanted into mice. Such xenograft studies have shown similar results for other solid tumors. However, these studies are not without critics. One major confounder is that what we think of as a cancer stem cell population might simply be those cells which make the right kinds of unnatural relationships with cells in the alien environment and avoid immune interactions. When these influences are accounted for by homotypic grafting in very immunocompromised hosts, as many as 20% of cancer cells can give rise to new tumors in the host. It is likely that different cancers will follow different pathways – some will be driven by a very small number of stem cell–like cells, whereas in others a substantial clone of cancer cells will generate new cancer cells as the tumor grows.

A major factor often cited in support of the stem cell origin theory of cancer is the observed similarity between many cancer cells and various embryonic or adult stem cells. However, it is frequently observed that overexpression of many different oncogenes, such as c-*MYC* or *RAS*, may result in a rapid loss of differentiation and re-entry into the cell cycle for various previously differentiated cell types (see Chapter 6). Moreover, various signaling molecules can confer "stemness" on previously differentiated cells by activating EMT programs. In other words, the initiating mutation could equally well occur in a postmitotic differentiated cell as long as such mutations confer or capitalize on the potential of that cell to re-enter the cell cycle. In this scenario, the phenotypic similarities between cancer cells and primitive precursors or stem cells arises not necessarily because this reflects the nature of the cell of origin but rather one of the associated consequences of the initiating oncogenic lesion, whatever the original state of differentiation of the cell involved (Fig. 1.9).

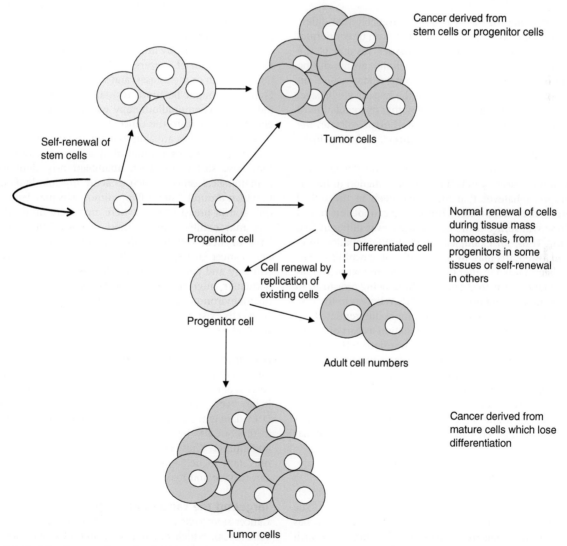

Figure 1.9 Cell of origin of cancer. Cancers probably originate most frequently in progenitor or stem cells, but may also arise from more differentiated cells that lose differentiation as part of the oncogenic process.

If "dedifferentiation" is an inevitable accompaniment of cancer-causing mutations, then the preferential role of stem cells in the initiation of cancer may instead reflect the higher intrinsic rate of replication or their longevity in adult organisms. This is more plausible as it is extremely likely that mutations would occur more frequently during cell division because of the vulnerability during DNA replication. However, this is by no means the only way in which mutations occur (see Chapter 3), and it is not only stem cells that replicate in the adult organism. The observation that "promotion" of an epidermal cancer may be accomplished months or even years after the initial exposure to carcinogen ("initiation") is often taken to imply that the original carcinogenic event occurs in a long-lived epithelial stem cell population. While this is highly likely in skin, where mature cells are continually removed by shedding at the surface, it is equally plausible in other tissues that the original mutation conferred longevity (particularly likely given the repeated observation in mouse models that an antiapoptotic lesion may be among the earliest required mutational events in cancer formation), or that cell turnover of differentiated cells in a given tissue is usually slow (thus, unless the mutation conferred an immediate growth advantage, it would only be passed on to a small number of progeny).

It must be remembered that it is now unarguable that differentiated cells can and do replicate in the adult even under normal physiological circumstances, and in some tissues this may be the sole or major source of new cells. The cellular events during development of liver cancer suggest that cancers may arise from cells at various stages of differentiation in the hepatocyte lineage. Much experimental data support the view that dysregulation of specific genetic pathways, rather than cell of origin, dictates the emergence and phenotype of various cancers, including high-grade glioma and others.

Whatever the actual outcome of these scientific debates, it is clear that treating cancer by inducing its differentiation (differentiation therapy), whatever that may have been in the cell of origin, offers considerable promise. However, it cannot be assumed that this alone will suffice if the cell of origin was differentiated to begin with. Thus, for example, inducing differentiation in c-*MYC*-induced osteosarcomas by transiently inactivating c-*MYC* has recently been shown to alter the epigenetic context surrounding c-*MYC* signaling so as to change this from being procancer to instead becoming proapoptotic (anticancer). Whereas in the case of a c-*MYC*-induced tumor arising from a more differentiated cell type, which in consequence loses differentiation as part of c-*MYC* activation, inducing "redifferentiation" by transient c-*MYC* inactivation does not change the context, and reactivation of c-*MYC* results in further tumor progression. Once again, a general rule holds true, namely, that most things related to cancer are a matter of timing and are also determined by numerous factors including the cell or origin, the mutations accumulated and the cancer environment – together referred to as the molecular "roadmap" of that cancer.

The cancer stem cell and niche

The cell of origin notwithstanding, considerable interest is developing in the existence of CSCs within some if not all tumors. The concept of CSCs was first proposed over 100 years ago, but has only recently hit the mainstream, with the identification of such cells in a variety of human cancers. In fact, there is even interest in the possibility that the particular resistance of CSCs to chemoradiotherapy may in part explain the failure to cure most metastatic cancers.

Although the clonal nature of cancers is well-established, there are some unanswered questions. Our incomplete understanding is illustrated by studies which suggest that hundreds of thousands of cancer cells may have to be transplanted in order to establish a new tumor from an existing one. Clearly this could simply reflect the chance nature of cell replication and survival, but it is also open to an alternate interpretation, namely that only a small number of cancer cells in the original tumor are capable of initiating a new tumor. When examined at a molecular level, these different possibilities would suggest that in the "chance" model, most if not all cancer cells contain the necessary epigenetic changes needed for tumorigenesis and some get lucky or make the right connections in the new location, whereas the alternative model presupposes that cells in the cancer are very heterogeneous, with only a select minority group of "Über-cancer cells" capable of recapitulating tumorigenesis – the cancer stem cells. These tumor stem cells are a rare population of cells that can reconstitute a new tumor comprising all the cell types present in the original cancer. It is tempting to blame such cells for the formation of metastases and of new tumors following inoculation of cancer cells in a different host organism (xenografts). The CSC hypothesis states that a minority of transformed stem cells, or progenitors with acquired self-renewal properties, are the source of new tumor cells. By implication, such cells are also responsible for the behavior of cancers, such as rate of growth or proliferation, invasion or metastases, and sensitivity to various treatments. Stem cells might be more resistant, for example, to apoptosis induced by chemoradiotherapy when compared to more differentiated cells within the cancer.

Tumor stem cells are akin to adult and embryonic stem cells in that they undergo self-renewal by asymmetric cell division, but they have so far only been unambiguously identified in some hematological cancers, such as acute myeloid leukemia (AML), in which around 1 per million tumor cells may be a tumor stem cell, and in breast cancer, where anywhere up to 2% of tumor cells may exhibit some of these characteristics. The molecular basis of stem cell behavior may prove useful in developing new cancer drugs, and with this in mind the Wnt-signaling pathway and polycomb genes, discussed in later chapters are of particular interest. As with other stem cells, the immediate microenvironment comprising stromal cells (niche) within which such cells exist is just as interesting as the nature of the stem cells themselves.

A chemotherapy-resistant niche

The tumor microenvironment is also a critical determinate of the success of chemotherapy. In a mouse model of Burkitt lymphoma it has been shown that survival of cancer cells in the face of DNA-damaging agents is influenced not just by cell intrinsic factors but also by local secretion of paracrine factors, such IL-6 and Timp-1. These create what the authors describe as a "chemoresistant niche," within which a small number of cancer cells can survive and may be able to repopulate a recurrent cancer.

Targeting the cancer-initiating cells

Cancer stem cells are sometimes referred to as tumor-initiating cells (TICs), which neatly avoids any presuppositions about the nature or origin of the cell. The controversies alluded to earlier notwithstanding, there are numerous points of interest in the model. Such cells have been proposed in large number of human cancers, though not incontrovertibly by any means, including hematological malignancies and tumors of the breast, prostate, brain, pancreas, head and neck, and colon. Their presence in the tumor may worsen prognosis, may partly account for resistance to conventional chemoradiotherapy, and may provide a specific target within the cancer for new drugs. The latter depends on the identification of unique markers on the cell surface which may allow such TICs to be isolated and studied. However, despite early promise, various markers such as CD133, CD44, and CD166 have not unambiguously defined malignant from normal stem cells in various cancers and moreover do not entirely define the nasty subset of cancer cells within a given tumor. There is much hope that new techniques for concurrent determination of multiple surface markers might address these limitations.

In order to eradicate TICs in cancers we will need to unravel the molecular mechanisms regulating processes such as self-renewal, differentiation, and escape from therapy. Pathways involved in self-renewal and cell fate have been described and include those important in normal stem cells, such as Wnt, Notch, and Hedgehog, but also, tumor suppressor genes such as *PTEN* and *TP53*. Once these pathways are deregulated in TICs they can drive uncontrolled self-renewal, resulting in treatment-resistant cancers, because some rare TICs will survive even if the bulk of the tumor is annihilated. The CSC model implies that curing cancer requires new cancer therapeutics that target and eradicate these CSCs. Reactivation of embryonic/developmental signaling pathways such as Notch, WNT–β-catenin, BMI-1, sonic hedgehog, and EGFR, when combined with drug-resistant mechanisms such as efficient DNA-repair processes, checkpoint regulation and ABC transporter–mediated drug efflux, shown in a variety of TICs may represent new targets for treatment of resistant cancers. The local microenvironment of CSCs, or niche, may also be a target as such location-specific cues may not be critical for other cells.

With this in mind, recent studies have identified the *PTEN* tumor suppressor as a key regulator of TICs in leukemia, brain, and gut, and suggest that drugs such as rapamycin, which targets the PI3K–AKT–mTOR pathway normally suppressed by *PTEN*, might at least in transgenic animals deplete TICs without damaging normal stem cells.

The latent niche

It has been suggested that CSCs may form cell–cell interactions similar to those that have been described for normal stem cells and stem cell niches. Recent studies in the nematode worm have suggested that under some conditions, differentiated cells that do not normally contact stem cells nor act as a niche can promote ectopic self-renewal, proliferation, or survival of competent cells, with which they form aberrant contacts. The authors have described this as a "latent niche." One of the important implications of this mechanism for tumor initiation is that it does not necessarily require genetic changes in the tumor-initiating cell itself. It will be interesting to see if such a mechanism occurs in human cancers.

Cancer is a genetic disease

> Scientists have found the gene for shyness. They would have found it years ago, but it was hiding behind a couple of other genes.
>
> *Jonathan Katz*

With the availability of the reference genome for humans and mouse, the last decade has witnessed an explosion of new knowledge in human genetics. Our understanding of the genetic basis of disease has grown dramatically, with nearly 5000 diseases identified as heritable. Moreover, it is now known that genes contribute to common conditions such as heart disease, diabetes, and many types of cancer.

Currently, more than 1% of all human genes are "cancer genes," of which approximately 90% exhibit somatic mutations in cancer, 20% bear germline mutations that predispose to cancer, and 10% show both somatic and germline mutations. A recently published "census" of cancer genes (see the Sanger Institute website – www.sanger.ac.uk/genetics/CGP/Census/) is dominated by genes that are activated by somatic chromosomal translocations in leukemias, lymphomas, and mesenchymal tumors. Interestingly, the protein kinase domain was the most frequently represented domain encoded by cancer genes, providing support for the development of therapies targeting this domain in cancer, followed by domains involved in DNA binding and transcriptional regulation.

Cancers (and Darwin's finches) evolve by mutation and natural selection

Broadly, cancers arise due to genetic (or epigenetic – see Chapter 11) alterations in three types of genes: oncogenes (see Chapter 6), tumor suppressor genes (see Chapter 7), and caretaker genes, such as DNA-repair genes (see Chapter 10). Combinations of epimutations in these classes produce tumors. Genetic (but most probably not epigenetic) alterations may occur in the germline, resulting in inherited cancer predisposition, or more commonly either occur in somatic cells, giving rise to sporadic tumors. The first somatic epimutation in an oncogene or tumor suppressor gene that enables clonal expansion may be regarded as the initiating insult. Unfortunately, in the vast majority of human cancers this key early step is not known. Tumors progress through the acquisition of further somatic epimutations, which allow further rounds of clonal expansion. Broadly, therefore, tumor cells evolve, with those cells with a growth advantage selected for at each mutational event. Individuals with an inherited abnormality in any of these genes are cancer-prone presumably because they are one step ahead of those without such germline abnormalities.

Blame the parents – inherited single gene defects and susceptibility to cancer

> Children begin by loving their parents; after a time they judge them; rarely, if ever, do they forgive them.
>
> *Oscar Wilde*

Most cancers are not the result of hereditary high-penetrance mutations. In those cancers where inherited mutations are an important contributor they often involve inactivation or silencing of a "caretaker" or tumor suppressor gene. Inherited forms of cancer represent perhaps about 5–10% of all cancers and include two rare inherited cancers, studies of which have resulted in disproportionately spectacular insights into cell and cancer biology in general: a childhood eye cancer known as retinoblastoma (caused by loss of the *RB* tumor suppressor) and the Li–Fraumeni syndrome (caused by loss of the *p53* tumor suppressor), in which children and young adults of the family develop an assortment of cancers, including sarcomas, brain tumors, acute leukemia, and breast cancer.

More recently, gene mutations associated with common cancers, including colon cancer and breast cancer have been identified. The familial adenomatous polyposis coli gene (*APC*) has been identified as a cause of inherited precancerous polyps, and a contributor to colon cancers. Another inherited form of colorectal cancer, Lynch syndrome, is caused by loss of mismatch repair genes. Possibly the most clinically important hereditary cause of cancer involves mutations in the *BRCA1* or *2* genes and predisposes affected women to both breast and ovarian cancers. It is estimated that as many as 1 in 300 women may carry inherited mutations of breast cancer susceptibility genes. People who inherit cancer genes are more likely to develop cancer at a young age because the predisposing gene damage is present throughout their lives. Recently, a further ovarian cancer susceptibility gene, *RAD51*, has been identified, which is also involved in the DNA-damage response.

Loss of heterozygosity and comparative genome hybridization

Deletion of genetic material is a very common event in human cancer. Indeed, it is the most frequently observed genetic abnormality in solid tumors. There are several mechanisms through which a somatic cell, with an inherited mutated gene allele, can lose the normal gene copy and become vulnerable to cancer (Fig. 1.10, also see Chapter 10). These mechanisms may result in what has been described as loss of heterozygosity (LOH). LOH can occur by deletion of the normal allele, deletion of part of or the entire chromosome (referred to as aneuploidy), possibly followed by duplication of the chromosome containing the mutated allele, or by mitotic recombination (crossing over) with genetic recombination in mitosis (it is a normal part of meiosis). Thus, a particular chromosomal region might be found in 0, 1, 2, or many copies, whereas the similar region in normal cells always have two copies. These extreme genetic aberrations in cancer cells (loss or gain of chromosomal regions) may be readily detectable during cytological examination and such abnormalities can form the basis of diagnostic and prognostic decisions.

Haploinsufficiency

Alfred Knudson's two-hit model of tumor suppressor genes, first proposed in 1971, supposes that two mutations are required to cause a tumor, one occurring in each of the two alleles of the gene (see Chapter 7). Recently, however, tumor suppressors that do not conform to this standard definition have been described, including genes requiring inactivation of only one allele (also referred to as "haploinsufficient"), and genes inactivated by epigenetic silencing (see Chapters 7 and 11).

Blame everyone – complex polygenic mechanisms and inherited susceptibility to cancer

What remains to be uncovered is how low-penetrance genetic variants (polymorphisms) contribute to the risk of developing so-called "sporadic cancers." Polymorphism refers to a gene that exists in more than one version (allele), where the rare allele can be found in more than 2% of the population. The term broadly encompasses any of the many types of variations in DNA sequence found within a given population. Specific subtypes of polymorphisms include mutations, point mutations, and single-nucleotide polymorphisms (SNPs) (see Chapter 10). Although this is an oversimplification, polymorphisms may be regarded as having less dramatic or overt functional effects than mutations.

Although we are a long way from describing variations in these multiple potential gene alleles, we know that polymorphisms contribute to response to carcinogens, variations in drug responses,

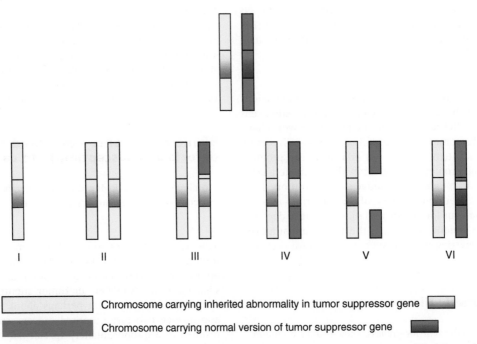

Figure 1.10 Loss of heterozygosity through various genetic events for a tumor suppressor gene. I, Nondisjunction; II, nondisjunction and reduplication; III, mitotic recombination; IV, gene conversion; V, gene deletion; VI, point mutation.

and undoubtedly to many other aspects of cancer. Recently, much interest has been sparked by the identification of polymorphisms, which may contribute to the risk of lung cancer by influencing the susceptibility to carcinogens in tobacco smoke.

An apology to Jean-Baptiste Lamarck: cancer is an epigenetic disease (but you were wrong about giraffes)

Epigenetic information is not contained within the DNA sequence itself, but is transmitted from one cell to all its descendants. Such a control is referred to as "epigenetic," as the DNA sequence is not altered. This is a major potential flaw inherent in attempts to understand diseases by sequencing genomes, as these epigenetic factors will be missed. Some altered gene expressions may be driven by environmental factors such as nutrient levels or hypoxia, and others by means of changes in, for example, methylation of genes. Such changes, as mentioned, do not alter the DNA sequence and yet can be passed on to subsequent generations of cells. But before we all rush out and put parents on the rack to produce the next generation of basketball players, there is as yet little real evidence that such epigenetic factors can be inherited through the germline, as it is generally accepted that most epigenetic information is wiped clean in the germ cells. The closest we get to Lamarck's view of giraffe necks is that the fetus may be conditioned by the intrauterine environment, likely by epigenetic effects. Although this is an example of early environmental conditioning of the individual, there is as yet no evidence that this can affect subsequent generations and thus be truly hereditary.

The importance of epigenetic factors in cancer was first articulated by Feinberg and Vogelstein, who noted generalized hypomethylation of DNA in tumor samples (see Chapter 11). Although the focus of attention is now more on the selective hypermethylation of certain genes such as tumor suppressors, these studies were of crucial importance. Many key genes may be silenced by epigenetic changes during successive cell differentiation stages during development, and two epigenetic events in particular have been associated with transcriptional silencing in cancer cells: methylation of CpG islands in gene promoter regions and changes in chromatin conformation involving histone acetylation. Genes known to be epigenetically silenced in cancers include more than half of all known tumor suppressors, with much data in particular available for *p53* and *PTEN*, and the *MLH1* mismatch repair gene, silencing of which can cause genetic instability thus linking epigenetic and genetic factors. Studies in the Min mouse (*APC*-defective mutation) revealed that reducing DNA methylation with an inhibitor of a key enzyme, DNA methyltransferase (DNMT), reduced intestinal polyp formation directly, establishing the key role of epigenetic factors and tumorigenesis.

Loss of imprinting (LOI) – the silencing of active imprinted genes or the activation of silent imprinted genes – is frequently observed in human cancers and is responsible for overexpression of the gene encoding insulin-like growth factor (IGF)-2 in the pathogenesis of Wilms tumor, in Beckwith–Weidemann syndrome, and in some epithelial cancers, including colon cancer.

"So it isn't really junk after all." Noncoding DNA

Having long been regarded as largely junk, it now turns out that the large amount of DNA that does not actually encode instructions for making a specific protein actually contains important regions involved in regulating gene expression, DNA structure, and cell fate. Some of the noncoding DNA has long been known to contain key regulatory elements for the gene, such as gene promoters, which control gene expression. Perhaps surprisingly, most of the eukaryotic genome is actually transcribed, resulting in a confusing jumble of RNA transcripts that include tens of thousands of microRNAs (miRNAs), long noncoding RNAs, and others with little or no protein-coding capacity. Small RNAs (see Chapter 11) can silence various genes, in part by forming dsRNAs which target mRNAs for destruction.

Most long noncoding RNAs remain uncharacterized but many are likely to represent more than just transcriptional "noise." Some are already known to be differentially expressed amongst differing cell types and conditions and to be localized to specific subcellular compartments. The potential role played in cellular function is far from clear, but might include processing to yield small RNAs; in some cases noncoding RNA transcription itself may affect the expression of adjacent genes and in other cases noncoding RNAs may function in a similar way to proteins and directly influence activity or localization of proteins.

The last few years have seen a huge increase in the amount of information available about the critical role played by miRNAs in posttranslational regulation of gene expression. miRNAs are short, single-stranded RNAs, typically in the size range 19–25 nucleotides. Essentially all cell biological processes are influenced in some way by miRNA, because most if not all signaling pathways are in some way regulated by miRNAs as well as other factors. In cancer, those many miRNAs which can act as oncogenes or tumor suppressors are collectively referred to as "oncomirs." Distinct clusters have distinct functions, and to give you some idea of how complex these regulatory mechanisms are take a look at Fig. 6.8 in Chapter 6, showing the relationship between one transcription factor, c-Myc, and miRNAs. Oncomirs can influence essentially all cellular processes altered during tumorigenesis, and many specific miRNAs with central roles have been identified. These include the mir-17–92, a polycistronic miRNA cluster that contains multiple miRNA components, also known as oncomir-1, which is amplified in several human B-cell lymphomas and can promote proliferation and survival, inhibit differentiation, and increase angiogenesis. Overexpression of miRNAs LIN28 and LIN28B is found in many human cancers and is associated with repression of Let-7 family miRNAs. In turn, loss of Let-7 releases inhibition on targets such as *HMGA2*, *KRAS*, and c-*MYC*, which drive tumorigenesis. Other miRNAs are functionally important targets of *p53* while others regulate the activity and function of *p53*. Other tumor suppressors are also regulated by miRNA, including *PTEN* (see Chapter 7).

Mutations can result in activation of oncomirs. Chronic lymphocytic leukemia (CLL) is typified by chromosomal deletions on 13q, 17p, and 11q, sites at or near the miR-15a/miR-16-1 cluster, p53, and miR-34b/miR-34c clusters, respectively. A miRNA/TP53 feedback loop is involved in CLL pathogenesis and outcome.

Biomarkers in diagnosis and subclassification of cancers and miRNAs can now be detected and measured in serum. The RNA interference mechanism is being used in new therapies with siRNAs, miRNA analogs and antagonists of miRNAs (antagomirs). See Chapter 16.

The cancer "roadmap" – What kinds of genes are epimutated in cancer?

Broadly, three classes of genes are involved in cancer:
- **Oncogenes** – These are usually variants of normal genes that are involved in promoting behaviors such as replication that are

essential drivers of cancer. Unlike their normal cellular counterparts, the proto-oncogenes (a term which rather underplays the important role played in normal cell growth/expansion and rather erroneously conveys the impression that their role is to wait around until they go bad and cause cancer), oncogenes are either abnormally activated or overexpressed versions that can drive aberrant growth in the absence of normal regulatory controls. Not surprisingly, most oncogenes are related to growth factors or more usually their receptors, downstream signaling molecules activated by them, or ultimately the nuclear targets of such signaling pathways and the drivers of the cell-cycle machinery.

- **Tumor suppressors** – Conversely, these normally act to restrain the oncogene signaling described briefly above either by acting as restraints of growth factor signaling or in general ways as guardians of cell stress, DNA damage, or abnormal oncogene-driven growth, to which they respond by promoting apoptosis or senescence or blocking cell-cycle progression. The tumor suppressors must be inactivated in order for cancers to develop. As genes such as *p53* appear in evolution before cancer was likely to have posed any problems to the organism, it is believed that the original role, at least of this tumor suppressor, was something else. In fact, recent studies suggest that *p53* may play a normal physiological function in meiotic recombination.
- **Caretaker genes** – These are involved in sensing and repairing DNA damage. They include the important mismatch repair genes, which may be damaged as a relatively early event in some cancers such as those of the colon, thereby accelerating the development of mutations that activate oncogenes or inactivate tumor suppressors.

It has been estimated that up to seven rate-limiting genetic or epigenetic events are needed for the development of common human epithelial cancers, and these may be ordered in multiple different combinations depending on which particular tissue or cell-specific "anticancer" barriers need to be circumvented and because there may be a number of different effective "routes" available for getting around any given obstruction. And as those of us with satellite navigation systems are only too aware, many of these are far from direct. It should also be borne in mind that the actual number of "mission critical" epimutations needed to initiate cancer may differ depending on which cancer we are considering and which genes are deregulated. Thus, several studies suggest that fewer mutations may be needed if one of the lesions results in persistent or sustained deregulation of activity of the oncoprotein c-MYC (see Box 1.4). Figure 1.11 shows how MYC can cooperate with RAS in tumorigenesis.

Importantly, many key molecular contributors to cancer progression may not themselves be deregulated at the gene level. Thus, downstream signaling proteins may become upregulated because of alterations upstream in growth factor signaling genes, altered catabolism, genes inactivated by epigenetic factors, protein expression altered by enzyme activity, degradation, chaperones,

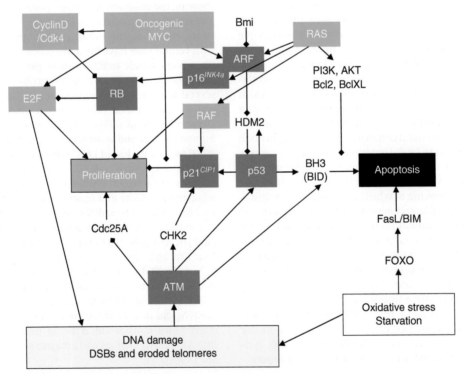

Figure 1.11 Linkage between signaling regulating replication, DNA damage, apoptosis, and growth arrest. Several links exist between mitogenic signaling and that regulating growth arrest and apoptosis. Moreover, DNA damage response pathways may be involved in linking oncogenic cell cycles with growth arrest and apoptosis. Activation of RAS and c-MYC (MYC) via growth factor signaling results in potential engagement of both replication and growth but also of apoptosis and possibly growth arrest. If either MYC or RAS levels are excessive (as might occur during oncogenesis) or other proapoptotic signals are received, then the balance may be tipped away from replication. Oncogenic RAS can promote senescence through either p16^{INK4a} or ARF, which activate the RB or p53 pathways, respectively. Intriguingly, MYC may activate apoptosis through activation of ARF, possibly at least in part via DNA damage responses. Although it remains unclear as to how the cell can distinguish between a normal cell cycle and an aberrant "cancer cell cycle," one possibility is that the latter may be more likely to result in DNA damage. Apoptosis may be blocked by RAS activation of the PI3K and AKT pathways. DSB, double-strand break.

etc. Again, it should be noted that events contributing to cancer are not restricted to the cancer cells. Thus, for example, expression of key cancer-contributing proteins in the cancer cell, such as NF-κB in hepatocytes, may be upregulated through changes in expression of TNF-α in neighboring stromal inflammatory cells.

Viruses and the beginnings of cancer biology

The identification of the genetic mechanisms of transformation owes much to the study of transforming viruses, in which the transforming effect could be attributed to specific oncogenes. DNA viruses express proteins analogous to key proliferation factors that substitute for or replace the function of the cellular factors. In contrast, the oncogenes of RNA retroviruses are derived from the hijacking of critical cellular regulatory genes with the addition of gain-of-function mutations (see Chapters 3 and 6). In fact, many normal cellular genes involved in growth were first identified as viral oncogenes, with the normal cellular counterparts or proto-oncogenes discovered subsequently.

Knowledge gained about DNA tumor viruses and the molecular biology of viral transformation has played a major role in furthering understanding of oncogene and tumor suppressor function and in the development of cancer biology in general, although the actual contribution of viruses to the formation of most human cancers is by comparison rather modest. The studies of SV40 large T antigen and HPV E6/E7 proteins, together with studies of the familial cancers, have proved critical in understanding the importance of the *RB* and *p53* tumor suppressor genes. This is one example of how several fields of study often converge in science to help illuminate a key process (see Chapter 3).

Michael Bishop and Harold Varmus won the Nobel Prize for Medicine in 1989 for their work in showing that the chicken Rous sarcoma virus (RSV) carried an oncogene called v-*src*, a version of a normal chicken gene called c-*src* but without introns, which the virus had hijacked from a chicken host some time during its evolution. This study transcended the identification of a cause of a chicken cancer when it was subsequently shown that many other retroviruses contained oncogenes that had important normal cellular counterparts involved in growth signaling, many of which were discovered in this way.

The role of infection in human cancers has become much better understood in the last decade or so. We now have active vaccination programs in many countries to prevent infection of women with cervical cancer-causing HPV, for example. We have also identified several other less common cancers in which viruses may be important and have found that infection with the bacteria *Helicobacter pylori* causes gastric inflammation and ulcers and contributes to gastric cancer.

Hens and teeth or bears and woods? The hens have it – cancer is rare

Adversity has the effect of eliciting talents, which in prosperous circumstances would have lain dormant.

Horace

Given the evolutionary nature of cancer, it is perhaps surprising that three lifetimes are required to generate an effective cancer cell. The mutation rate has been estimated at 1 in 2×10^7 per gene cell division. Given, that there are around 10^{14} target cells in the average adult human, with a myriad of potential target genes involved in regulation of cell expansion, and that the chances of further mutations are greatly increased by clonal expansion of those cells carrying the initial lesion, highly effective innate barriers to cancer must exist. Some of these barriers are now well described and include the coupling of oncogenic proliferative signals to those which induce apoptosis, senescence, or differentiation and the tumor suppressor pathways involving *p53* and *RB*. Large, long-lived animals like humans have a large potential somatic mutational load. It has been estimated that point mutations resulting in activation of *RAS* occur in thousands of cells daily in the average human. As the vast majority of these do not result in neoplasia, it is assumed that the usual outcome of such mutations is apoptosis, differentiation, or growth arrest. It should also be remembered that epithelia, such as gut, have the unique advantage of being able to shed potential cancer cells from the surface into the outside world. Once estranged from their usual nurturing environment, they undergo a form of apoptosis (anoikis) before ending life in the bath, lavatory, or waste disposal. In fact, the perceived ubiquity of cancer in humans is simply a product of the truly mind-boggling numbers of cells in our bodies and the fact that they have to divide so many times during our three score and ten years. If, however, these mechanisms are disabled then cancer may become inevitable.

The barriers to cancer

Your silence gives consent.

Plato

The tumor suppressors

Two key pathways, those involving the tumor suppressors p53 and RB, are among the most critical barriers to cancer development. Not surprisingly, the p53 and RB pathways are frequently inactivated in human tumors and may be disrupted at different points. Thus, genetically, the RB pathway (cyclin D, CDK4, p16^{INK4A}, RB), a critical determinate of the G_1/S transition in the replication cell cycle, acts as one "critical target" in cancer cells, but the mechanism of disruption varies according to tissue. Thus, for example, cyclin D is overexpressed by amplification in breast cancer and by translocation in parathyroid cancer; CDK4 is mutated or overexpressed in melanoma; p16^{INK4A} is inactivated by deletion or silencing in melanoma and pancreatic cancer; RB expression is lost by mutation or deletion in retinoblastoma and soft tissue sarcomas. Such patterns may not be random. Specific associations of events are seen within individual tumors, and these presumably reflect the evolution of the tumors along particular pathways.

The p53 tumor suppressor protein is a major component of the natural defenses against cancer. The p53 protein acts by arresting the cell cycle and promoting apoptosis (programmed cell death) in response to DNA damage, hypoxia, or unscheduled activation of oncogenes such as c-*MYC*. The *p53* gene is altered in more than half of all human cancers and, because of its role in mediating growth arrest or apoptosis in response to DNA damage, referred to as genotoxic stress, has been termed the "guardian of the genome." However, given the equally important (and from recent studies, controversially the more important) tumor suppressive role of p53 activation in response to inappropriate oncogene activation, referred to as "oncogenic stress," this term is somewhat underrepresentative. It might be more accurate to view p53 as the universal overseer of cell stress – a kind of intracellular barman. Thus, recent findings that p53 may also shut down key metabolic processes that allow aerobic glycolysis (the Warburg effect) and the pentose shunt support this view. Mediators and regulators of p53 activities are also targeted in cancer, and inactivation of $p21^{CIP1}$ or ARF or activation of MDM2 (an inhibitor of p53) are all observed in cancers.

Over the last decade, numerous links between the p53 and RB tumor suppressor pathways have been identified, including regulation of the G_1/S transition and its checkpoints. This has highlighted the crucial role of the E2F transcription factor family in these pathways. Virtually all human tumors deregulate either the RB or p53 pathway or both. Many other tumor suppressors are known and are discussed in Chapter 7.

One area that has greatly excited the research community and, incidentally, the pharmaceutical and biotechnology sectors in recent years, has been the unveiling of the crucial role played by noncoding DNA and miRNAs in regulation of gene expression, and in particular how this gets derailed in cancer. Thus we now know of miRNAs which contribute to oncogene activity, to tumor suppressor pathways, and even to regulation of CSCs and EMT. It is extremely likely that miRNAs will be found to contribute to the regulation of essentially everything over the next few years by providing another level at which the activity of genes is controlled.

Avoiding suicidal urges

When God desires to destroy a thing, he entrusts its destruction to the thing itself. Every bad institution of this world ends by suicide.

Victor Hugo

In 1972, John Kerr, Andrew Wyllie, and Alistair Currie published a description of an unusual form of cell death distinctly different from necrosis, which they termed "apoptosis." This is now one of the most published areas of biology (see Chapter 8). Robert Horvitz who, along with Sir John Sulston, was awarded the Nobel Prize for his work on apoptosis has rather succinctly summarized the three stages of apoptosis as follows: "First, killing the cell, then getting rid of the body and then destroying the evidence."

Perhaps the single most critical barrier against cancer is the "selfless" suicide (apoptosis) of a potential cancer cell, which, either because it has been unable to repair damaged DNA or because it is being inappropriately pushed into the cell cycle, disassembles and repackages itself as an energy-giving snack for its neighbors, rather than pose a threat to the whole organism. Apoptosis offers several distinct advantages to the organism, not least of which is a relative absence of inflammation (which might well result if the body had been required to "murder" the potential cancer cell – necrosis). Such an absence of collateral damage during apoptotic death is largely because of the ability of neighboring cells and phagocytes to swiftly recognize and cannibalize the apoptotic cell (usually before it has actually "died"). Moreover, when operating correctly, this also prevents the release of viruses or harmful cellular contents into the environment, instead seamlessly passing them from the apoptotic cell to another cell where they can be neutralized. Arguably, the ability to undergo apoptosis is one of the major hallmarks of moving from a unicellular to being part of a multicellular organism, where "social responsibility" among constituent cells becomes paramount for the survival of the whole organism.

Cells are continually receiving and integrating a variety of both positive and negative growth signals. One intriguing result of much research over the last 20 years has been the appreciation that cells seem only too willing to commit suicide. In fact, cells require continuous signals from neighboring cells in order to survive. Loss of these normal "survival" signals or an increase in negative growth signals will tip the balance and a cell will undergo apoptosis. Two major pathways of apoptosis are known: one is **intrinsic** and is integrated by a variety of signals operating at the mitochondria and the other **extrinsic**, triggered by activation of cell surface receptors such as FAS or TNF receptor. Both pathways eventually activate cascades of caspases, expressed as inactive zymogens, which when activated in cells destined to undergo apoptosis execute the necessary steps for apoptosis. However, the initiating caspases (apical caspases) differ – the intrinsic pathway commences with activation of caspase-9, while the extrinsic starts with caspase-8.

In cancer, the intrinsic pathway of apoptosis may be triggered by "sensors" that determine the presence of irreparably damaged DNA or inappropriate attempts to engage the cell-cycle machinery, which in turn may be modulated by external signals, which either prevent or provoke apoptosis. In general, these mechanisms are largely integrated at the mitochondria. Although, the body rarely "murders" would-be cancer cells it can certainly drive these cells to suicide. The extrinsic pathway is utilized by the immune system to engage the apoptotic machinery via surface "death receptors." These death receptors, which include those for TNF and FASL, respond to some secreted inflammatory cytokines and to some populations of T cells. The pathways activated by these receptors include those able to trigger caspase cascades independently of the mitochondria.

Apoptosis can also be executed by caspase-independent death effectors, such as apoptosis-inducing factor (AIF), endonuclease G, and a serine protease (Omi/HtrA2), released from mitochondria during permeabilization of the outer membrane. It is worth noting that many of these proteins have important or even essential roles in cellular processes unrelated to cell death. AIF and Omi/Htra2 are involved in redox metabolism and/or mitochondrial biogenesis; caspase activation is essential in some cells for terminal differentiation, lipid metabolism, inflammatory responses, and proliferation. This has important ramifications, as it implies that certain key parts of the apoptotic response could not be ablated therapeutically without impeding normal cellular functions, unless drugs can be designed to target only the lethal (and not vital) role of these proteins.

Necrosis is the form of death once thought to be the major if not only cause of death of cells. Apoptosis is a friendly form of cell elimination as collateral damage is slight and free of inflammatory consequences, largely because the corpses are removed fast and intact. Moreover, gorging on apoptotic corpses leaves macrophages sated and quiescent and may even sooth a previously activated and inflammatory cell. Sadly, necrosis does not share these soothing qualities and involves the release of proinflammatory molecules, which can be extremely damaging particularly if the necrotic cell is loaded with destructive agents, such as macrophages and neutrophils. Further inflammatory cells are recruited and healing may be delayed, potentially contributing to chronic inflammation if either the cause of the necrosis is not removed or phagocytosis is impeded. With this in mind, it is worth noting that in some types of chronic inflammation apoptosis may end in necrosis if the phagocytosis of the apoptotic corpses is delayed – a situation that arises in the presence of high levels of oxidized LDL cholesterol (ox-LDL), due to competition for scavenger receptor-mediated uptake in macrophages.

Chronic inflammation is a contributor to many epithelial cancers and underpins the cancer association between ulcerative colitis and colon cancer and the origin of some gastric cancers, esophageal cancers and probably most non-small-cell lung cancers (NSCLCs). Why? Because inflammatory proteins such as IL-1 may promote proliferation and angiogenesis; inflammatory cells can facilitate spread by producing matrix-degrading enzymes and through the formation of a cancer-supporting stroma. This begs the question as to whether a build-up of ox-LDL, characteristic of adverse lifestyle, obesity, and diabetes may also contribute to cancer and whether statin drugs might be protective.

Other forms of death

Apoptosis and necrosis are not the only forms of cell death described; others include anoikis, endoplasmic reticulum stress, and autophagy. **Autophagy** is essentially self-cannibalization, in which cells collect some of their own organelles and cytoplasm and then proceed to digest them within lysosomes, subsequently using the breakdown products to generate energy and construct new proteins. We have all have seen movies in which the protagonists survive bitter cold by burning the furniture – well this is the cellular equivalent. The cell survives adversity and also gets to replace old and damaged organelles, such as ribosomes and mitochondria. Although autophagy may help the organism survive adverse conditions and may restrict degenerative diseases, it can also be exploited by cancer cells, which may use autophagy to survive in preangiogenic conditions until nutrient delivery can be secured. Autophagy can be stimulated by most forms of cellular stress, including nutrient or growth factor deprivation, hypoxia, DNA damage, and damaged organelles, and is integrated with other cellular stress responses by multifunctional stress-signaling molecules such as p53 and mTOR. Thus, autophagy may be triggered by downregulation of key metabolic sensor signals such as mTOR and can be regulated by p53 through a new family of proteins known as damage-regulated autophagy modulators (DRAM). Beclin, a member of the BH3-only family, triggers autophagy and provides some interconnection with apoptosis. Autophagy appears to be another potential barrier to tumorigenesis that must be overcome. However, autophagy may also be a contributory factor to tumor cell dormancy, which, if released, could give rise to recurrence after therapy.

Anoikis is a form of homicidal homesickness that specifically refers to a variant of apoptosis noted in cells that have become estranged from their ancestral homelands.

Avoiding senescence

In 1961, Leonard Hayflick and Paul Moorhead found that many human cells, such as fibroblasts, had a limited capacity to replicate themselves in culture. In fact, they observed that cells can undergo between 40 and 60 cell divisions, but then can divide no more, a process described as senescence, or they die. This number is often referred to as the "Hayflick limit." Cellular senescence is associated with aging and longevity and has also been termed "replicative senescence." The Hayflick limit for dividing cells may in part be determined by the length of telomeres, which are noncoding regions at the tips of chromosomes (see Chapter 9). Cell division requires the duplication of chromosomes, but each time a chromosome reproduces itself, it loses a part of the telomere (telomere attrition). Once a cell's telomeres reach a critically short length, the cell can no longer replicate its chromosomes and thus will stop dividing. Such cells are termed "senescent." Cells taken from older humans divide fewer times before this occurs, as the "chromosome clock" has been ticking throughout adult life (*vide infra* – stem cells appear less bound by these restrictions).

A key feature of cancer cells is that they have found the means to avoid death and senescence, a form of cellular immortalization. In testimony to their remarkable longevity, cancer cell lines are routinely distributed, cultured, and studied in laboratories across the globe. In most cases these are cells derived from a human or animal cancer that continue to divide under appropriate cell culture conditions with scant regard for the Hayflick limit; because they essentially never stop dividing, such cell lines constitute a limitless supply of cancer cells for laboratory use.

In a spectacular illustration of the resilience and fecundity of cancer cells, the famous HeLa cell line has been dividing ceaselessly since the progenitors were first harvested from a cervical tumor biopsy of a single patient, Henrietta Lacks, in 1951. This was the first human cell line and, in large part because of the generosity of George Otto Gey, who made these cells available to any interested researchers, it has quietly revolutionized cell biology. Interestingly, much as the original cells would have done during the life of the patient, HeLa cells growing in culture plates in different laboratories continue to evolve and several variant strains are now known. What they share is the ability to keep dividing as long as they are appropriately nourished and kept free of infection. Although, clearly, cancer cells do die through hypoxia, extensive DNA damage/chromosomal instability, etc. Cellular senescence may have evolved as one mechanism to avoid cancer, which clearly increases in frequency with aging. Several studies have shown that the induction of cellular senescence can inhibit particular cancers.

Importantly, the majority of cancer cells seem able to avoid telomere attrition (shortening). Thus, expression of the telomere-stabilizing enzyme telomerase is induced in tumors and effectively allows cancer cells to rewind their odometer and enjoy unrestrained replication. However, this situation is not as straightforward as it might at first appear. First, inactivating telomerase in some models of viral oncogene-induced cancers does not impede tumorigenesis or growth potential, suggesting that alternative methods for telomere maintenance are also important. Moreover, in other cancer models, where p53 is inactivated, telomere shortening, instead of promoting apoptosis or senescence, may instead lead to a more genetically unstable cancer as chromosome rearrangements are favored.

Oncogene-induced senescence

As if this were not already complex enough, senescence can also be triggered by activation of various signaling pathways (see Chapter 9). Long appreciated as a major restraint to replicative potential *in vitro*, several recent studies have now confirmed that oncogene-induced senescence (OIS) is also a key inherent restraint to tumorigenesis (along with apoptosis) *in vivo*. Although the exact signaling pathways most critical for OIS may vary for different cell types and cancers, there are common features and overlap with activation of DNA-damage responses such as those seen with telomere attrition and variously engagement of either the ARF–p53–p21^{CIP} and/or p16^{INK4a}–Rb pathways. What remains unclear is for how long such senescent cells persist before being culled and whether this state is truly and always irreversible.

One intriguing question in biology is why damaged cells under some circumstances undergo growth arrest or senescence rather than apoptosis – they forsake Eros rather than embrace Thanatos.

Oncogenes as tumor suppressors
Studies over the last two decades have revealed another crucial antineoplastic mechanism, namely that many signaling networks promoting cellular replication also possess intrinsic growth-suppressing activities. Under normal growth conditions, such as tissue maintenance and repair, signaling networks are activated in a coordinated fashion by appropriate extracellular signals, which can block the growth-suppressing pathways and the cell replicates and survives. However, inappropriate activation of a potentially powerful replicative signal such as c-*MYC*, for instance by mutation, occurs without activation of those other key collaborative pathways; so instead of unscheduled replication the mutated cell dies by apoptosis, thereby eliminating the risk of further mutations and cancer. This "intrinsic tumor suppressor" activity is manifested by several mitogenic proteins; the resultant apoptosis or growth represents a critical "failsafe" mechanism in the avoidance of cancer. By implication, therefore, the inherent growth-suppressing activities of oncogenes such as c-*MYC* must first be suppressed if cancers are to develop or progress – an example of oncogene cooperation discussed in detail in later chapters.

What is the secret of cancer developme... "timing"

The exact role of any given protein may be largely a matter of timing with respect to the stage of a cancer's evolution and likely also the developmental stage of the cell under consideration. Thus, even individual proteins within the cancer cell can exert widely differing effects on phenotype. Mitogenic proteins like c-MYC may prevent the initiation of cancer through their inherent apoptotic activity, but once the cancer cell has acquired the ability to avoid apoptosis, or the environment provides sufficient survival signals, it may instead confer a wide range of cancer-promoting behaviors.

A recent study has shown that brief inactivation of c-MYC was sufficient for the sustained regression of c-MYC-induced invasive osteogenic sarcomas in transgenic mice; subsequent reactivation of c-MYC led to extensive apoptosis rather than restoration of the neoplastic phenotype. Possible explanations for this outcome include changes in epigenetic context that may have occurred within the cell type, that is, between the immature cell in which c-MYC was originally activated and the differentiated cell resulting from subsequent (brief) inactivation of c-MYC. In this tumor model, although c-MYC expression is initiated in immature osteoblasts during embryogenesis, subsequent inactivation of c-MYC in osteogenic sarcoma cells induces differentiation into mature osteocytes. Therefore, reactivation of c-MYC now takes place in a different cellular context and induces apoptosis rather than neoplastic progression.

TGF-β was initially identified in culture media from transformed cells as part of a factor that could produce a transformed phenotype in a nontransformed cell line. The observations that TGF-β1 inhibited the growth of epithelial cells, and that inactivating mutations within the TGF-β1-signaling pathway occurred in many cancers, supported the view of TGF-β1 signaling as a tumor suppressor pathway for early stages of cancer. However, many human carcinomas overexpress TGF-β1 and it is associated with a poor prognosis and metastasis. Similar results pertain to tumor cell lines and animal models. Together, this suggests that TGF-β1 switches from tumor suppressor to oncogene as the context changes, probably due to genetic or epigenetic alterations in tumor cells or stromal cues. Thus, the role of TGF-β1 in cancer is stage-specific.

Location, location, location – the cancer environment: nanny or spartan state

Numerous studies now point to the crucial interplay between the cancer cell and its local and systemic microenvironment. It is often assumed that the body is largely a hostile environment for an incipient cancer, with hostilities beginning upon recognition of the errant cells with the express aim being to kill, contain, or starve them into submission. In this Nietzschean power struggle, immune and inflammatory responses are mobilized to eliminate the cancer cells, stromal cells form an impenetrable barrier to contain the spread of cancer cells, and both blood supply and nutrients are withheld from the growing tumor. By implication, cancers will need to overcome these hostile forces in order to progress. As in ancient Sparta, newborn cancer cells are left exposed to die – and it is worth noting that this was an experience that made any survivors strong and nasty.

However, it now appears that for many cancer cells the new infrastructure of a growing tumor may actually represent a *locus amoenus* – a safe haven and nursery in which they may be cosseted and eventually fledged.

Cancer cells as "cuckoos"
It is entirely possible that the rareness of cancer (at a cellular level) reflects the success of these extrinsic hostile forces as well as of intrinsic tumor suppressors in eliminating the inchoate (rudimentary and not fully formed) tumor cells. However, recent studies have increasingly challenged this heroic view in favor of a more nuanced one that acknowledges the sometimes ambiguous relationship between cancer and noncancer cells. Thus, at least once a tumor has become established, cancer cells find ready allies to their cause and environmental interactions that actively support their expansion and spread and that might even offset suicidal urges (see Chapter 12). In fact, the developing tumor may well be – or at least become with time – a nanny state in which newborn cancer cells want for nothing and are fed, sanitized, and cosseted, perhaps because, like unfledged cuckoos, they are not recognized as different.

Cancers, chronic inflammation, and tissue remodeling
In some circumstances, such an ideal microenvironment may precede the cancer rather than evolve alongside it. Thus, chronic inflammation has long been known to increase risk of many cancers, possibly by increased mitogenesis (and thereby mutagenesis) or through paracrine effects from inflammatory cells. In

fact, cells enlisted to serve in wound healing or inflammatory engagement are allowed *interregnum* privileges denied to their "peacetime" counterparts, including a license to migrate and proliferate. Perhaps not surprisingly, such liberated cells may be peculiarly susceptible to becoming cancer cells. In fact, once corrupted by epimutations they may fail to relinquish the extraordinary freedom they enjoyed, even when calm has been restored – a big headstart to cancer.

However, even in the absence of preceding inflammation, malignant transformation takes place within the context of a dynamically evolving "microenvironment" and is accompanied by fibroblast proliferation and transdifferentiation, extracellular matrix deposition and remodeling, increased matrix metalloproteinase expression and activity, infiltration of immune cells (see Chapter 13), and angiogenesis (see Chapter 14). It is readily appreciated how such a milieu may actively support tumor cell invasion, survival, and growth and this is particularly important in epithelial carcinogenesis (see Chapter 12).

Liaisons dangereuses encourage tumorigenesis

Recent studies in epithelial tumors extend the pernicious repertoire of matrix activities during tumorigenesis beyond that of a supporting role. Thus, matrix cells and others may conspire together to initiate and encourage designate cancer cells to participate in promiscuous behaviors conducive to cancer, including proliferation, EMT and invasion, and may even permanently damage the DNA. Moreover, these permissive changes may extend even to normal epithelial cells. Thus, matrix can trigger production of matrix metalloproteinases (MMPs) and reactive oxygen species (ROS) in epithelial cells and the stiffer, more fibrotic stroma present in tumors when compared to normal connective tissue can provoke activation of Rho family members.

Interactions between cancer cells and other cells in their environment are thus key determinants of tumor progression.

Policemen or agent provocateur – immunocytes in cancer

Interactions between tumor-infiltrating leukocytes and tumor cells are also of key importance given that immune cells might either interfere with tumor progression or actively promote tumor growth. Certainly, context is likely to be a critical factor, when it is considered that many cytokines and inflammatory products may not only act as anticancer barriers but could also support cancer behaviors such as growth and invasion. The roles played by stage of cancer evolution and the ability of cancer cells to resist the negative and yet benefit from the positive aspects of immune responses are now being unravelled.

Despite the existence of tumor-specific immune cells, most tumors appear to have acquired a means to avoid immune attack. In recent years a considerable interest has developed in "immune privilege" (see Chapter 13). Foreign antigens that enter immunologically privileged sites, of which the eye, brain, and testis are examples, can survive for an extended period of time, whereas the same antigens would normally be swiftly eliminated elsewhere. It has been proposed that the tumor microenvironment may become a site of immune privilege, possibly through factors produced by the tumor, which might impair immune surveillance. Immune privilege could provide a "safe haven" for cancer cells. Recent studies in ovarian cancer have suggested that one means of immune privilege is recruitment of regulatory T cells by the tumor. These regulatory T cells can block the activity of those T cells that are reactive to tumor antigens, thereby interfering with tumor-specific T-cell immunity and enabling progression of ovarian cancers *in vivo*. Other possibilities include production by the cancer cells of cytotoxic or inhibitory factors for tumor-reactive T cells, such as galectin-1, TGF-β, or Fas ligand.

Neutrophils may play a role in facilitating the metastatic capabilities of circulating cancer cells, for example those that become trapped in small blood vessels within the lung. Thus, neutrophils may play lifeguard and actually help anchor these cancer cells within the capillary endothelium. Interestingly, release of IL-8 by cancer cells may attract the attention and assistance of neutrophils, thereby representing a potential target for drug therapy in cancer.

Location also affects tumorigenesis in other ways. Thus recent studies have started to unravel differences between sites in the way in which key tumor suppressor pathways are activated and regulated. Thus, oncogenic Ras strongly activates the Ink4a/Arf locus, in some cases promoting cell-cycle arrest or senescence. Lung tumors form independently of p19^{Arf}, whereas p19^{Arf} must be disabled for formation of sarcomas. These differences in behavior between tissues may in part reflect the action of Polycomb-group complexes, which repress Ink4a/Arf in lung tumors.

Cancer goes agricultural

The field effect

D.P. Slaughter and colleagues first introduced the notion of a "field effect" following studies on oral squamous carcinoma in 1953. They identified the presence of histologically abnormal tissue surrounding the carcinoma. This field effect was proposed to underlie development of multiple primary tumors, in the absence of familial predisposition, in the same tissue and possibly also recurrence locally following treatment. According to the multistep carcinogenesis model of Fearon and Vogelstein, propitious genetic alterations accumulate in a more or less stepwise fashion by natural selection, so that clones emerge sequentially, each with growth advantages over the preceding one and thus evolve eventually into cancer. One implication of this model is that precancerous cells in proximity to the cancer will represent earlier "less successful" clones and will have some, but minus one or more, of the genetic alterations present in the adjacent cancer.

This model is supported by studies in a variety of human cancers, including lung, gut, cervix, and prostate, which show genetic alterations in the vicinity of the cancer. More recently, epigenetic alterations in methylation have been shown to contribute to this field effect in premalignant conditions such as Barrett's esophagus and in colonic mucosa affected by ulcerative colitis, and also in prostate cancer and NSCLC. In one recent study of colorectal cancer, a field effect comprising MGMT (O6-methylguanine DNA-methyltransferase) promoter methylation was shown in normal-looking mucosa 1 cm from the tumor margin and not 10 cm distant.

As we have discussed already, paracrine interactions between epithelial cancer cells and adjacent stroma are important and may in some cases actually boost the tumorigenic potential of the cancer cell. Furthermore, tumor-associated stroma is notably heterogeneous in terms of fibroblast behavior, gene expression and may itself demonstrate increased motility and invasive potential. Recent studies using laser-capture microdissection to examine the stromal and epithelial compartments of primary breast cancers have shown that the stroma bears mutations and loss of heterozygosity of the tumor suppressor gene *TP53* different to those present in the epithelium. In fact, surprisingly, in more than a quarter of breast cancers the stroma had *TP53* mutation even when none could be demonstrated in the cancer cells. In fact, as there was no overlap in the loss-of-heterozygosity profile between the cancer and the stroma, different pathways of clonal expansion must have been involved.

The intriguing fact that tumor-associated stromal fibroblasts may themselves have oncogenic mutations raises many interesting possibilities. Thus, a common epithelial progenitor cell may have given rise to both the tumor and the associated stromal cells. Such EMT has been shown in generating tumor-associated myofibroblasts, which therefore share a common genetic lineage and carry the same mutations. So what about when the mutations are different and lineage must differ? One possibility is that the cancer microenvironment is mutagenic due to ROS from immunocytes and possibly any carcinogens that contributed to the development of the tumor in the first place. The field effect may also explain these findings, with disease causing epimutations present in both the tumor and surrounding "field."

The seed and the soil: metastatic spread

As tumors progress, cells within them develop the ability, or the inclination, to invade into surrounding normal tissues and through tissue boundaries to form new growths at sites distinct from the primary tumor. The seeding and growth of cancer cells in distant organs is termed "metastasis" and is the ultimate cause of death in around 90% of cancer patients. Metastasis was first described in 1839 by the French gynaecologist Joseph Recamier, and soon thereafter, physicians found that certain cancers were most likely to spread to certain organs. Breast and prostate cancer, for example, move to lymph nodes, bones, lung, and then the liver. Skin cancer tends to spread to the lungs, colon cancer targets the liver, and lung cancer typically moves to the adrenal glands and the brain.

In 1889, Stephen Paget proposed that cancer cells shed from an initial tumor were dispersed randomly throughout the body by the circulatory system. He called these circulating cancer cells "seeds" and proposed that only some seeds fall onto "fertile soil" – organs where they can grow. About 30 years later, a researcher named James Ewing proposed an alternative nonrandom model by which circulating cancer cells become trapped in the first small blood vessels, or capillaries, they encounter and then grow in the surrounding organ.

While much is now known about molecular alterations that contribute to tumorigenesis, the genetic and epigenetic alterations that result in metastatic spread of the disease are less well understood. Although as with initiation and progression of other cancer behaviors it now seems that inherited as well as acquired factors may contribute to the likelihood or not of developing metastases, analogous to the hallmark features of cancer, there are hallmark features specifically related to metastasis. These include the abilities to:
- escape from the primary tumor,
- intravasate into local blood vessels or lymphatics,
- survive within the blood or lymphatic fluid,
- extravasate into a distant tissue,
- proliferate within the new environment (metastatic colonization), and
- evolve in parallel to cells within the primary tumor.

Metastasis conferring mutations, and at least some of the resultant behaviors, are believed to be atavistic. As a result, it is often assumed that the ability to invade or metastasize is a chance byproduct of mutations that were originally selected for because they gave cancer cells in the original tumor locus a growth advantage. Over the last decade the view that occasional cancer cells might elope from the primary tumor and settle down to start a family in some distant site has been challenged. Rather, it appears that many solid cancers may experience the exit of large numbers of cells from their homeland in a mass "Volker Wanderung" to seek pastures new even if few will succeed to establish new colonies.

In fact, millions of tumor cells can be shed into the vasculature daily, so why are so few secondary tumors formed? The general explanation for this has relied on the assumption that a number of additional genetic events had to occur in order for a small subclone of cells to arise with the capabilities to enter, navigate, and exit the vasculature and thence to colonize a distant site. However, some recent studies suggest that genes required for metastatic spread may already be expressed in primary tumors and before any metastatic spread, suggesting that metastatic ability might be preprogrammed in tumors by the initiating oncogenic mutations. One problem with such data is that even though multiple genes were aberrantly expressed in such primary tumors, they may not all have been so in any individual cell (gene expression profiles were generated from mushed up whole tumors and epigenetic factors were not addressed).

In the past decade much has been learned about how cancers metastasize. Key findings have included the observation that cancer cells are subject to growth regulation at the secondary site and moreover the molecular characterization of proteins that can suppress the metastatic phenotype. These proteins are encoded by metastasis suppressor genes (MSGs), defined as genes that suppress *in vivo* metastasis without inhibiting primary tumor growth when transfected into metastatic cell lines and injected into experimental animals. To date, over 20 such MSGs have been identified and may represent novel disease biomarkers as well as therapeutic targets. Among the best described of these are *NM23*, *PEBP1*, *RECK*, *KAI1*, *RHOGD12*, *KISS1*, and *CTGF*.

Key processes required for metastatic spread include migration and invasion of tumor cells, requiring cancer cells to detach from the primary tumor and then travel to secondary sites via the lymphovascular systems. Cancer cells are able to secrete MMPs and alter expression of cell adhesion molecules (see Chapter 12) that facilitate invasion by degrading extracellular matrix and disrupt cell–matrix and cell–cell interactions. Once in the maelstrom of the circulation, cancer cells must survive being buffeted by blood flow shear forces and the full broadside of immune assault. Finally, once entrapped within capillary networks they must find the means to extravasate into the ambient tissue and establish a foothold. Various proteins have been implicated in these processes, including cell adhesion molecules, proteolytic enzymes, and members of the RHO family, including RHO, RAC, and CDC42, that are involved in cytoskeletal organization.

Recent exciting data suggest that invasive and metastatic potential is related to reactivation of general embryonic pathways involved in morphogenesis and might include mutations that deactivate E-cadherin and other cell adhesion molecules, those that activate transcription factors and signaling molecules such as NF-κB and TWIST, which might promote EMT. EMT, originally described *in vitro* as dedifferentiation of epithelial cells to fibroblastoid, migratory, and more malignant cells, with an accompanying altered mesenchymal gene expression program, correlates well with late-stage tumor progression. Typical phenotypic features of EMT include loss of E-cadherin and acquisition of vimentin immunoreactivity. EMT also occurs during embryonic development and is regulated by a complex network of signaling pathways, including the RAF–MEK–MAPK pathway, PI3K–AKT pathway, NF-κB, and TGF-β. In various animal models systems, metastatic potential strictly correlates with the ability of epithelial tumor cells to undergo EMT. Importantly, it is now likely that EMT may also promote the development of CSCs and may provide a further link between inflammation and cancer.

Other recent studies have now added to the complexity of metastasis biology. As discussed earlier, metastatic tumors can secrete factors into the circulation that prepare a distant site for colonization. More recently, it has also been shown that some nonmetastatic human tumor cells can secrete factors, such as prosaposin, that conversely, in part by inducing thrombospondin-1 expression in fibroblasts, renders the microenvironment in distant tissues resistant to colonization.

A question that is currently of tremendous interest is at what time cancer cells acquire the capabilities to undergo metastatic spread. This is addressed in the next section. As in so many other areas of cancer biology, miRNAs have also been shown to have a profound influence on metastasis. Specific networks of miRNAs have been described which affect tumor metastasis, EMT, and invasion through posttranslational alterations in gene expression and epigenetic changes.

Another underexplored area of research is how cancer cells first gain entry into the systemic circulation by directly intravasating into venous capillaries or indirectly via lymphatics.

Treatments based on the identification of MSGs are available; clinical trials of drugs targeting NM23 as an antimetastatic therapy are in progress, although the challenges inherent in trying to restore missing function are substantial (much easier to try and inhibit an overactive protein than replace a missing one).

Cancer superhighways – blood vessels and lymphatics

The metastatic spread of tumor cells is most often the lethal aspect of cancer and frequently occurs via the lymphatic system. Many tumor types, including breast and prostate cancers and melanoma, first metastasize via lymphatic vessels to regional lymph nodes. The presence of lymph node metastases is associated with poor prognosis, but that the lymphatic system might actively participate in cancer metastasis has only been unravelled recently. In fact, tumor-induced lymphangiogenesis may precede lymph node metastases and might therefore be a novel target for prevention. Lymphangiogenic growth factors, such as VEGF-C and VEGF-D, act on cognate receptors such as VEGFR-3 on the surface of lymphatic endothelial cells to promote lymphangiogenesis and metastases to lymph nodes. Interestingly, recent studies suggest that lymphangiogenic growth factors from the primary tumor can induce lymphangiogenesis in nearby lymph nodes before the arrival of metastasizing tumor cells.

On your bike and turn the lights off before you go

One area of considerable general interest is the role played by light–dark and sleep–wake cycles (diurnal and circadian rhythms) in various aspects of cellular biology. At a whole-animal level, it has long been known that many hormonal processes, arousal/alertness, and mood are strongly influenced by sleep–wake patterns and that under usual circumstances in humans these are inextricably linked to day–night cycles. However, when this goes awry, as in shift work or in those who frequently cross time zones, these sleep–wake and light–dark cycles become disconnected and ill-health may result. At the benign end this may cause transient jet-lag, but recent studies have suggested that in some cases there may be more serious consequences, including an increased risk of cancer.

Normally, diurnal and circadian rhythms and cell proliferation are coupled in humans. Various animal studies have shown that exposing rats and mice to light at night can accelerate cell cycle and this is associated with increased IGF-1R/PDK1 signaling and accelerated tumorigenesis. Perhaps it is time to discard the night light?

Catching cancer

Recent studies have confirmed some long-suspected and intriguing notions about cancer cells, namely, that they might be spread between individuals (i.e. you might be able to "catch" cancer like a cold). It is crucial to note the difference between being infected with a cancer-causing virus from another individual, not at all controversial and well exemplified by HPV infection and cervical cancer, and being infected by another person's cancer cells directly. Thus, it now seems that cancer cells do not necessarily perish along with their host but might carry on through generations by spreading to further individuals. In canine transmissible venereal tumor (CTVT), tumor cells are implanted from one animal into a new host, where a new tumor grows – effectively analogous to a transplanted "graft." This raises interesting questions as to how cancer grafts avoid rejection; in CTVT, tumor cells downregulate expression of major histocompatibility complex (MHC) molecules involved in immune recognition, though in many cases an immune response against the tumor eventually does occur and eradicates the cancer. A similar infectious cancer has been described in Tasmanian devils.

In both these cases the infectious nature of the cancer has been revealed by genotyping the cancer cells from numerous different animals from different geographical areas (at least with CTVT-carrying dogs) and showing that these are more genetically similar to one another than they are to the host cells and less genetically variable than even very inbred dogs are to one another. Although such infectious cancers are yet to be demonstrated in humans it is worth noting that certain types of cancer transmission are known. For example,

- During pregnancy, transplacental transmission of leukemia, lymphoma and melanoma to the fetus has been demonstrated.
- Organ transplants carrying occult cancer cells have been shown and might be facilitated by immunosuppression aimed at limiting rejection, although this route may result in a detectable cancer in under 0.05% of graft recipients – usually melanoma.

Hammering the hallmarks

The hallmark features referred to previously not only distinguish normal from cancerous cells but thereby also represent attractive drug targets for treating or even curing cancers. In the modern era we now have a range of targeted anticancer drugs that specifically antagonize important molecular targets, such as growth factor signaling (BCR–ABL, EGFR, HER2 to name a few). In fact, an entire new vocabulary has been established to describe the application of these treatments and the changes in the cancer cell that accompany them. We will describe a select few here.

Cancer – Achilles' heel and Paris' arrow

The last decade has witnessed the beginnings of what is predicted to become a sea-change in cancer chemotherapeutics and arguably the single biggest paradigm shift since metaphor and hyperbole were first successfully mangled and combined in the cancer literature. What has driven so many of us to wax lyrical about a new dawn, about "Achilles'; heels," "oncogene addiction," and "personalized medicine"?

In a nutshell – we are excited by the identification in cancers of key signaling proteins essential for the maintenance of the cancer and the availability of drugs and ever more drugs that can relatively selectively inactivate those proteins. This is the realization in cancer therapy of the "magic bullet" model first proposed by Paul Ehrlich in the nineteenth century. Because this may arguably represent one of the biggest changes in thinking about drug design since the use of multidrug regimes first became mainstream in the 1950s and 1960s, we will devote the next few sections to this subject. To be fair, there have been examples of targeted therapies based on molecular grounds in the past, but they have never come so thick and fast and so specifically fuelled by detailed knowledge of the cancer-causing mutations and hypothesis-driven drug development. Thus, use of hormone manipulation such as anti-estrogens in breast cancer and antiandrogens in prostate cancer, somatostatin treatment for neuroendocrine tumors, and the use of HCG as a marker for treatment monitoring choriocarcinoma, all paved the way for today's targeted therapies and diagnostics.

Much current cancer research is directed towards finding and studying those specific molecular targets that are essential to the continued growth and survival of the cancer because these are obvious points of vulnerability that can be exploited in drug development. One, perhaps unexpected finding in cancer models that has excited great interest is that cancer cells often become highly dependent on some mutated growth signaling pathways. In other words, the cancer cells are said to manifest "oncogene addiction." Thus, constitutively active signaling through EGFR, for instance, suppresses other growth signaling pathways (oncogene amnesia) by various feedback mechanisms, leaving the cancer cell critically reliant on this one particular growth factor pathway (oncogene addiction). Importantly, normal cells either do not have these aberrant pathways or if they do they have other options and are relatively unaffected by their removal or inhibition. This explains why a targeted agent can have such initially powerful effects and comparatively little toxicity.

Incidentally, as oncologists are all obviously well versed in the classics, the weak spots of a cancer are often referred to as its "Achilles'; heel," and presumably by extension targeted therapies aiming to skewer that particular vulnerability should then be referred to as "Paris'; arrow" – well we'll see if the name Styx!

At present, the two main classes of new therapies which exploit such molecular knowledge are the humanized monoclonal antibodies and the tyrosine kinase inhibitors (TKIs). Although it took some time to convert hypothesis into reality, the successful treatment of the hitherto resistant chronic myeloid leukemia associated with the *BCR–ABL* oncogene with the TKI imatinib confirmed that cancers could respond to the specific antagonism of a single aberrant protein. Moreover, this acted as a proof-of-concept for the translation of progress in cancer molecular biology into new treatments and biomarkers. However, lest in mourning the plumage we forget the dying bird, it may prove salutary to remind ourselves that for most cancer sufferers, systemic treatments will still largely consist of DNA-damaging chemotherapy and – despite the often substantial associated side effects – usually to good effect.

Getting the GIST of oncogene addiction

Gastrointestinal stromal tumor (GIST) is a rare cancer but one that highlights many of the issues surrounding targeted therapies in cancers in general. Along with chronic myeloid leukemia, it was one of the original cancers treated with the then novel TKI imatinib mesylate (Gleevec). For many years, GIST was notorious for its lack of response to conventional chemoradiotherapy, yet much was known of the causative mutations, with most GIST-bearing mutations in *KIT* or, occasionally, *PDGFRA* or *BRAF* genes. Appreciating the pivotal role of KIT and the availability of imatinib, an inhibitor of KIT kinase, clinical trials soon followed and achieved a quite remarkable response in about 80% of patients with metastatic GIST. Along with parallel studies targeting the BCR–ABL kinase in chronic myeloid leukemia, these were the first examples of targeted therapy determined by genotype and have been followed by herceptin for breast cancers with HER2 mutations and others discussed later.

Cooking with ERBBs

An exemplar of the identification of key signaling pathways essential for cancer growth and survival is that regulated by members of the wider EGF receptor family. The ERBB family of proto-oncogenes comprise four closely related receptor tyrosine kinases, which include the epidermal growth factor receptor (EGFR), ERBB2 (also known as HER2), and ERBB3. These are powerful mediators of growth and survival signals in normal cells and in many human cancer cells. A further member, ERBB4, may actually be an inhibitor of growth. They become activated by ligand binding, which leads to dimerization of these receptors in homo- or heterodimers. EGFR is itself overexpressed in many non-small-cell lung cancers (NSCLCs) and this knowledge has been exploited in the increasing use of TKIs, such as erlotinib and gefitinib, in their treatment. In fact, finding the presence of EGFR mutations in some NSCLCs may identify those patients in whom TKI may be more effective than platinum-based chemotherapy.

ERBB2 is unusual in being constitutively in the active formation ready to bind to other ERBBs that have bound a ligand, and is aberrantly overexpressed in the evolution of many breast as well as subsets of gastric and ovarian cancers and may become aberrantly activated in some NSCLCs alongside EGFR. In fact, when present at high levels it can form spontaneously active homodimers and heterodimers resulting in ligand-independent replication- and survival-promoting signals which together potently drive growth of the tumor. ERBB2 is amongst the best characterized and studied specific targets in cancer drug development, and much will be learned about how such knowledge has been exploited in developing new drugs for treating breast and other cancers. Thus ERBB2 signaling can be targeted by antibodies that prevent ligand binding (and may trigger immunity), TKIs, inhibitors of downstream signaling pathways, and cytotoxic antibodies.

Recently, it has been suggested that ERBB3 may be required for activation of the PI3K–AKT survival pathway and, moreover, may become overexpressed in some cancers or during therapy against other ERBBs, for example by amplification of MET, thus bypassing the need for other ERBBs. It is now the target of new drug development. Thus, knowledge of the presence of oncogenes such as NEU/HER, EGFR mutations/copy number, estrogen receptors, and others are already being used to guide treatments for individual cancer patients. Below we discuss some of the pioneering studies in targeted drugs.

Painting a portrait of cancer

Recent landmark studies have indicated that molecular analyses and gene expression profiling can identify key disease- and treatment-relevant molecules and even more complex relatively unique tissue "molecular signatures" that can be employed to improve our ability to predict disease prognosis and response to therapy. In fact, increasingly a combination of imaging techniques and molecular assays are being used to paint a portrait of cancer that brings out its true nature and reveals its particular obsessions and vulnerabilities.

Savile Row tailoring of cancer therapies
One aspect of recent progress in the area of biomarkers is the increasing realization that defining expression for single molecules may not be enough to accurately determine the optimal treatments and schedule for many patients, and great progress is being made in the simultaneous analysis of expression of multiple genes/proteins or in looking at genetic variation.

Earlier studies have employed a variety of high-throughput tools such as gene arrays, proteomics, and others to analyze changes and differences in expression of hundreds or thousands of genes/proteins between normal and cancer cells.

Thus, a gene expression signature was identified by global gene expression analyses in breast cancer that conferred a high risk of early development of metastases. Importantly, this "signature" was able to identify those individuals likely to progress among those otherwise generally regarded as low risk. This "poor prognosis signature," was shown to include genes regulating cell cycle, invasion, metastasis, and angiogenesis. Such studies provide support for the current "holy grail" of postgenome era medicine – namely disease fingerprinting and individualized medicine. There are still a number of barriers to overcome, and this is exemplified by the creation of whole new fields of scientific endeavor, badged under the heading of "systems biology" (see Chapter 20), which seek to develop the new techniques needed to derive more accurate and detailed information alongside new analytical techniques needed to handle large volumes of data.

One current hurdle to overcome relates to the loss of anatomical and topological information that accompanies many of today's high-throughput techniques. The differences between the output from destructive techniques that "mash up" a tissue in order to describe the molecular contents and that from microscopy techniques enabling co-localization of protein expression in the context of the intact topography and anatomy are akin to the differences between hearing a painting described on the radio and seeing it on television. Thus, many studies generally examine gene expression in "lumps" of tumor, which contain multiple cancer cells, stromal cells, and others and end up with a list of contents. This does not lessen the clinical utility of such whole-tumor studies, but imagine the "power" of a similar study looking at gene or, indeed, protein coexpression in individual cells on the canvas – with clear visibility not just of different cancer cells, but of stromal cells and vascular endothelial cells, and without disruption of the tissue anatomy.

Some preliminary steps have been made towards this ultimate goal but in general they have still ended up destroying the portrait to see what it is made of. Today, researchers are doing it one very small piece at a time. Thus, small bits of tumor can be isolated from tissue sections by means of laser capture microdissection. One study on breast cancer using this technique allowed identification of expression signatures that were remarkably similar across seemingly different clinical stages of cancer progression. This has fuelled notions that gene expression alterations conferring the potential for invasive growth might already be present in early preinvasive stages. In contrast to tumor stages, different tumor grades were found to be associated with distinct gene expression signatures, particularly between preinvasive and invasive.

Despite this progress, in most cases we are still unable to fully explain cancer behavior by such studies, and prognostic and treatment decisions are still often empirical. Even basic questions regarding cancer cell behavior and interactions with the microenvironment are unanswered. In particular, it is still far from clear just how clonal metastatic tumors actually are, and how individual metastases in the same patient are "related" to one other. Furthermore, the location in which evolution of the various mutations detected in metastases has occurred is still debated – in the primary or after spread. This likely varies from one cancer to another. Thus, in a study of over 200 human hepatocellular carcinomas and 7 metastatic liver lesions a MET-regulated gene expression signature (MET is associated with invasive behavior) was found in a subset of primary tumors and in all liver metastases, suggesting that the metastatic cells in this case originated from a clone within the primary and at least this metastasis-supporting mutation occurred before the cells left the primary tumor. The MET signature also correlated with increased vascular invasion and decreased mean survival time of hepatocellular carcinoma patients. Such poor prognosis signatures have also recently been reported for colon cancer, endometrial cancer, and NSCLC.

It must be remembered, however, that even in such studies "groups" of cells rather than single cells have been profiled; it is by no means certain that all the genes expressed apply to any individual cell. Tumors are usually genetically heterogeneous, and therefore tumor profiling, unless supplemented by single-cell analyses, may lead to erroneous conclusions, particularly if the assumption is made that all abnormalities detected apply to all individual tumor cells. Clonal expansion does not equate to all cancer cells being identical, simply that all cells will in some way carry the initiating genetic lesions alongside those additional mutations acquired during "cancer evolution." Moreover, with the increasing acceptance of the stem cell theory of cancer (discussed earlier), which implies that a small side population of cancer-initiating cells carry the replicative and invasive potential, this is increasingly pertinent.

With this in mind, recent studies suggesting that single cancer cells from primary tumors may indeed carry the "poor prognosis" signature for metastases are very exciting, but will need confirming. Two recent papers using DNA sequencing to look at pancreatic cancers provide further food for thought. Both looked at clonal relationships between primary tumors and metastases in a number of different patients. In these studies, initiating mutations or rearrangements were identified in the primary tumors, including some that might drive amplification of cancer genes, such as telomere dysfunction and checkpoint disturbances. One study demonstrated that genomic instability frequently persisted after spread, driving parallel and even convergent evolution within cancer cells in different metastases. This also suggested that metastasis-initiating cells were genetically heterogeneous, supporting the contention that seeding metastasis requires mutations different to those supporting growth in the primary tumors. Furthermore, they also found that phylogenetic trees across metastases showed branches specific for a given secondary location. In the second paper it was also shown that clonal populations that seed the distant metastases were represented within the primary carcinoma, and had evolved from the original parental, nonmetastatic clone. Much of the genetic heterogeneity of metastases is simply a mirror for that already present in the primary carcinoma.

The two papers largely differ in terms of the degree of heterogeneity resulting within the primary or after spread. Mathematical analysis suggested that a decade or more might be required between the occurrence of the initiating mutation and the birth of the parental, nonmetastatic founder cell and a further 5 years for the acquisition of metastatic ability.

The pitfalls of tumor profiling

Increasingly, it is apparent that a good understanding of major genetic and epigenetic factors will still provide only a partial picture of disease. In practical terms, cancer patients with ostensibly identical clinical stages of disease (and probably even those with apparently similar genetic factors) may have markedly different treatment responses and overall outcome. In the same way that genomics offers the possibility of a more complete understanding of disease by describing multiple polymorphisms, so advances in molecular biology raise the possibility of going a step further. Cell behavior and disease pathogenesis ultimately arise through the differential expression of multiple genes and in turn by their protein products, in the diseased cells and also in other cells, neighboring and more distant, within the affected organism. At best, genomic sequences will have only a partial relationship to gene/protein expression, particularly as they will largely overlook epigenetic factors and moreover large-scale identification of polymorphisms may be far more difficult to comprehend than a molecular profile from a given cell/tissue.

A cancer protein expression profile

> You have made your way from worm to man, and much within you is still worm.
>
> *Friedrich Nietzsche*

Ultimately, it is proteins that determine phenotype. Not all genes are expressed in any given cell, and even of those genes expressed (for which mRNA is formed), alternative regulatory events may still take place after transcription that determine protein levels. Although considerable correlation exists between gene and protein expression, there are far more proteins than genes (Fig. 1.12). Thus, alternative splicing of RNA, posttranslational modifications, and enzyme activities can all contribute to the generation of a multitude of different proteins. Importantly, not all of these different proteins can therefore be directly inferred from examination of either the genome or even the transcriptome of a given cell at any given time. This has been the major impetus behind efforts to describe the cell proteome using mass spectrometry and other techniques (see Chapter 20).

There are now several contenders in the race to provide a tool that can enable the examination of molecular phenotype at high resolution and for multiple proteins simultaneously in their normal cellular or anatomical context. These include microscopy-based techniques such as the toponome imaging system (described in Chapter 20), in which thin-tissue sections are examined the co-localization of 30–100 proteins at a cellular and subcellular level, and variations of mass spectrometry imaging, which have lower resolution but do not necessarily require specific reagents to identify each protein (discovery techniques). Such techniques may bring us close to being able to finally look at the genuine portrait of cancer or at least a high-quality broadcast version! In Chapter 20 (systems biology) we discuss some of the exciting new techniques being used to look at single cancer cells within tumors and how systems biology will contribute to one day making individualized medicine and tailored therapy a reality.

The drugs don't work

Unfortunately, an addictive personality characterizes cancer cells and if they cannot get their normal EGFR fix then they either get it from somewhere else (a different drug-resistant mutation in EGFR occurs) or they get their growth hit from activating mutations in some other pathway, such as MET! In other words, oncogene addiction and acquired resistance to targeted treatment appear inextricably linked.

Lest we forget, however, it is worth noting that resistance to chemotherapy is not a new finding. The early chemotherapy pioneers, using generally cytotoxic drugs (at the opposite end of the targeted spectrum to imatinib), struggling to cure acute lymphoblastic leukemia in young children, soon realized that combinations of four drugs were needed to achieve remissions and that these needed to be repeated to achieve cures. Why? Because leukemia cells stopped responding to individual drugs or even small numbers of drugs, presumably by acquiring resistance. Even small numbers of surviving cells would then sooner or later repopulate. Two concepts were introduced – first, drugs kill proportions of cells and therefore many rounds of treatments may be needed to eliminate essentially all cancer cells, and second, cancer cells develop resistance. Resistance may relate to a variety of factors, including cancer cells finding sanctuary in areas where the drugs do not reach or work, acquiring mutations that enable them to avoid, exclude, or destroy the drugs, getting tougher and failing to die or even finding ways to avoid the activity of cancer-unfriendly immune cells. The concepts of specific pathway activation conferring resistance is not therefore really a new one, it is simply a byproduct of the specific nature of the new drugs,

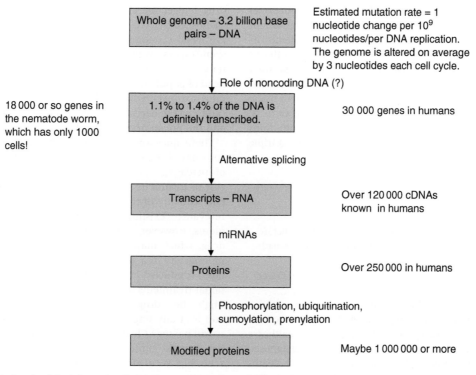

Figure 1.12 The complexity of cellular information flow in cancer. Although the issue of information flow seems hopelessly complex, there is much reason for hope. First, the availability of the reference genome for humans and many experimental models, alongside new technologies for analyzing the expression of multiple genes and proteins and appropriate techniques for analyzing and distributing experimental data will, it is hoped, result in major progress in "discovery science." Second, as many key genes/proteins have homologs in more primitive and experimentally amenable organisms we should have a much greater scope for functional studies.

which allow a simpler and more obvious route of escape for evolving cancer cells.

The addictive personality of cancer – synthetic lethality and non-oncogene addiction

To stretch the analogy further, an addictive personality may also result in cancer cells being addicted to more than one protein, many of which may not be oncogenes or even mutated at all. A good example of this is that breast cancer cells with defective homologous recombination (a form of DNA damage repair) are very sensitive to inhibitors of the enzyme poly(ADP-ribose) polymerase (PARP), whereas normal cells are not. This specific vulnerability to a drug inhibiting a specific target is referred to as "synthetic lethality" (a term shamelessly purloined from yeast genetics). In its original usage, synthetic lethality referred to the ability of a combination of mutations in two or more genes to kill a cell when a mutation in any one alone did not. In yeast cells a scientist would start with a cell carrying a nonlethal mutation and then test additional mutations one by one to find killing combinations. In some cases, such synthetic lethal interactions would identify how a cell may protect itself from the effects of the original mutation. This same technique has now been adopted to find new drug targets in cancer cells by using rapid gene knockdown screens using siRNA libraries. Synthetic lethality has been used in cancer to describe the killing of a cancer cell by a drug targeting the oncogene to which the cancer cell is addicted, but more usually refers to the targeting of a second protein to which the cancer cell is rendered dependent by another recognized mutation. Thus, the synthetic lethal interaction between defective BRCA and PARP has been exploited in breast cancer by use of PARP inhibitors; only breast cancer cells with defective BRCA are killed by these drugs.

On a related note, the ability of traditional chemotherapies to kill cancer cells more readily than normal cells has been referred to as genotype-dependent lethality; the totality of the cancer cell genotype/phenotype makes the cell vulnerable to DNA damage or cell-cycle paralysis.

Mechanism of origin rather than cell of origin – towards a new functional taxonomy of cancer

As we have discussed, cancers are traditionally classified on the basis of tissue of origin and this can be further refined to include cell of origin. However, as you will appreciate by now we are increasingly able to describe cancers according to the molecular alterations responsible for their development and required for their continued survival, and it will not have escaped your attention that these may sometimes be shared by cancers in different tissues. In fact, we are moving inexorably towards a new taxonomy of cancer in which diseases may be grouped not by tissue of origin but rather by common underlying disease mechanisms. The obvious exemplar would be breast and ovarian cancer, particularly those related to genomic, and therefore inherited, mutations in BRCA genes or, more recently, *RAD51D*. In these cases the mutations may illuminate the means by which tumorigenesis has proceeded, namely through an apical defect in the DNA damage response (once a cell has lost the remaining functional allele). Moreover, this insight may also point to specific treatment target – a form of personalized medicine. Finally, the identification of the causative inherited mutation will enable the offer of genetic testing to relatives of affected cases, which may be used to predict family members at future risk of both types of cancer.

Arguably, this is a far more useful clinical definition than tissue of origin. In keeping with this new way of classifying cancers, ovarian and breast cancer related to inherited BRCA mutations will share more common features than will, for example, a triple negative and a *HER2*-related breast cancer.

Is it worth it?

> Now I saw, though too late, the folly of beginning a work before we count the cost, and before we judge rightly of our own strength to go through with it.
>
> *Daniel Defoe, Robinson Crusoe*

No discussion of diagnosis and treatment can take place without consideration of the overarching importance of economic considerations. There is no doubt about the challenge facing healthcare systems; around 12 million new cases of cancer were diagnosed in 2008 and cancers accounted for nearly 15% of all deaths globally. How are we meeting this challenge? First, by spending money on research; large pharmaceutical companies alone spend around US$100 billion per year, which is incidentally roughly the same as the annual cost to healthcare providers across Europe for treating cancer.

How do we quantify the cost to cancer patients? Measures have been devised which include both mortality and disability suffered by survivors. One composite used by WHO is the DALY (disability adjusted life years lost), which effectively equates to the loss of a healthy year of life. Another similar measure is that of quality-adjusted life years (QALY). These are particularly important in the United Kingdom, which, unlike the United States, widely uses health economics to ration available treatments in order to keep spending within often narrow budgetary constraints (save money). Thus DALYs can be balanced against treatment costs in order to decide which therapies will be provided by the state. Obviously, there is a risk of establishing a two-tier system, as the well-heeled can simply pay privately for the drugs not thought sufficiently good value for money by the state. It is not hard to imagine that the patient may put a rather higher value on their life and health than the state!

Thus, available treatments for cancer patients in many countries are not dictated simply by the speed with which academics and pharmaceutical companies can get new drugs delivered to cancer units, but much more by the willingness of healthcare commissioners and providers to pay for them (and let us not pass the buck completely to politicians, but also our own willingness – or that of our insurers – to pay for them directly or through increased taxation). Inevitably, economics raises the big question – how much is a life worth? And, lest we naively assume that everybody, even in the United Kingdom, gets the same level of healthcare, related questions such as how much is somebody else's life worth, how much is my life or that of my family worth? There are already large differences in views on this across different countries. For instance, cancer drugs account for around 10–20% of the direct costs of cancer care (about 5% of the total drug budget for all diseases). At present the United Kingdom spends effectively less on cancer drugs than any other large European economy, and this trend is increasing as uptake and spend on new drugs continues to be less than for France, Germany, Italy, or Spain. In fact, the United Kingdom spends 50% less per head of population than France and even Spain, in which cancer rates are lower, which reflects the relatively limited role of health economics in decision-making in these countries on the one hand and the preeminent role of this in the United Kingdom.

These questions are particularly pertinent to cancer, where costs have been spiraling out of control under the twin influences of increasing incidence and survival of cancer patients and high cost of treatments. Moreover, many feel that the marginal benefits of many of these costly treatments should encourage us to re-evaluate existing practice and closely scrutinize any new treatments. However, how easy is it to assess the value of a cancer drug, which may have shown a mean 6-month improvement in longevity in a group of cancer patients? Remember, trials are often conducted in high-risk groups who have failed on conventional treatments and often have late- or even endstage disease. Might these drugs not do better in the real world if used earlier and isn't any improvement in life expectancy or quality of life worth having? Clearly, as health resources are limited, somebody has to make difficult decisions or put another way implement rationing. What factors do you include in these decisions? Simple metrics – cost of treatments versus life years gained, societal benefits from returning somebody to work? Should the affluent avoid these compromises by simply paying for the drugs etc. themselves? Is this equitable? Do you treat those whose failure to comply with preventative advice on obesity, smoking, etc. has contributed to their eventual illness differently to those who become ill despite a healthy lifestyle? This happens already: active smokers and alcoholics are very unlikely, respectively, to get a coronary artery bypass graft or a liver transplant should they need it. How do you compare the value of renal dialysis in adults, stroke care for the elderly, and chemotherapy for children? Who is involved in these decisions?

Broadly, government and their representative organizations with varying degrees of political autonomy (such as NICE in the United Kingdom), professional societies, such as the American Society of Clinical Oncology or Cancer Research UK, licensing authorities such as the Food and Drug Administration, insurance companies, and, most of all, cancer specialists must fight for their corner as do representatives of all other medical and surgical specialities with varying degrees of success. Practitioners should not have to make individual rationing decisions day to day in their practices, as this will compromise the doctor–patient relationship, but should instead lobby and discuss to influence policy overall. But all too often this is unavoidable.

Conclusions and future directions

Early diagnosis is essential for most effective treatment and it is likely that advances in this area will produce the most extensive and immediate benefits for cancer patients. At least as important is reversing the more self-destructive lifestyle choices such as smoking and obesity, which account for a substantial number of cancers. In some cases where lifestyle change is undesirable or unlikely we may be able to prevent some cancers by vaccination or drug treatments.

Greater biological understanding of tumorigenesis is also important. Cancers arise by the stepwise accumulation of mutations and epigenetic factors that alter gene expression to confer the so-called "hallmark features" of cancer. The presence of inherited cancer-causing mutations will give a would-be cancer cell a headstart, but somatic mutations and epigenetic alterations are still needed for cancer development (Fig. 1.13). Variation in multiple genes when coupled with poor lifestyle choices (your own or those of others) increase risk of developing some cancers. It is likely, given the increasing susceptibility of progressing cancer cells to mutations, that not all such mutations are actually cancer-relevant. It is anticipated that improved knowledge about these various processes regulating aberrant gene expression and gene–environment interactions will lead to new preventive strategies and treatments aimed at specifically targeting the expression of genes/proteins "mission critical" for the initiation and progression of cancer.

The identification of key proteins to which the cancer cell has become addicted is already being translated into new therapies, as is the way in which resistance to these evolves during treatment. Increasingly, focus will likely shift towards an assault on a limited subset of specific cancer-promoting signaling pathways involved in survival, self-renewal/replication, and spreading and directing these at the ring-leaders within the tumor. In fact, it is hoped that a cancer could be arrested or even eliminated by assassinating a subpopulation of particularly malign cancer stem cells and/or nontumor cell collaborators within the stroma.

Of course, a note of caution is always recommended.

"I am afraid," replied Elinor, "that the pleasantness of an employment does not always evince its propriety."

Jane Austen

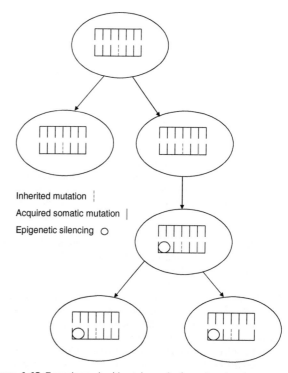

Figure 1.13 Tumorigenesis ultimately results from disordered gene expression. Tumor cells arise through aberrant expression of genes and the proteins they encode. This may result from mutations in the coding or noncoding regulatory regions of genes, which can be either inherited or acquired in somatic cells or even by major rearrangements of the chromosomes; epigenetic factors such as altered patterns of methylation and acetylation, which control the "accessibility" of genes for transcription. These events may in turn affect the stability and processing of RNA or proteins.

Bibliography

General

Alberts, B., Johnson, A., Lewis, J., Raff, M., Roberts, K., and Walter, P. (2002). *Molecular Biology of the Cell*, 4th edn. New York: Garland Publishing.

Fojo, T., and Grady, C. (2009). How much is life worth: cetuximab, non-small cell lung cancer, and the $440 billion question. *Journal of the National Cancer Institute*, Jun 29.

Kimball, J.W. *Kimball's Biology Pages*. users.rcn.com/jkimball.ma.ultranet/BiologyPages/

Kroemer, G. and Pouyssegur, J. (2008). Tumor cell metabolism: cancer's Achilles'; heel, *Cancer Cell*, 13: 472–82.

Kufe, D.W., Pollock, R.E., Weichselbaum, R.R., Bast, R.C., Jr., Gansler, T.S., Holland, J.F., and Frei, E. (ed.) (2003). *Cancer Medicine*, 6th edn. Hamilton (Canada): BC Decker Inc.

Luo, Ji, Solimini, N.L., and Elledge, S.J. (2009). Principles of cancer therapy: oncogene and non-oncogene addiction. *Cell*, 136(4): 823–37.

Genes and cancer

Balmain, A., Gray, J., and Ponder, B. (2003). The genetics and genomics of cancer. *Nature Genetics*, 33(Suppl): 238–44.

Esquela-Kerscher, A. and Slack, F.J. (2006). Oncomirs – microRNAs with a role in cancer. *Nature Reviews Cancer*, 6(4): 259–69.

Futreal, P.A., Coin, L., Marshall, M., et al. (2004). A census of human cancer genes. *Nature Reviews Cancer*, 4: 177–83.

Hunter, K.W. and Crawford, N.P. (2006). Germ line polymorphism in metastatic progression. *Cancer Research*, 66(3): 1251–4.

Knudson, A.G. (2002). Cancer genetics. *American Journal of Medical Genetics*, 111: 96–102.

Lord, C.J., Martin, S.A., and Ashworth, A. (2009). RNA interference screening demystified. *Journal of Clinical Pathology*, 62(3): 195–200.

Ma, X.J., Salunga, R., Tuggle, J.T., et al. (2003). Gene expression profiles of human breast cancer progression. *Proceedings of the National Academy of Sciences of the U.S.A.*, 100(10): 5974–9.

Pietsch, E.C., Humbey, O., and Murphy, M.E. (2006). Polymorphisms in the p53 pathway. *Oncogene*, 25(11): 1602–11.

van 't Veer, L.J., Dai, H., van de Vijver, M.J., et al. (2002). Gene expression profiling predicts clinical outcome of breast cancer. *Nature*, 415(6871): 530–6.

Wilusz, J.E., Sunwoo, H., and Spector, D.L. (2009). Long noncoding RNAs: functional surprises from the RNA world. *Genes & Development*, 23(13): 1494–504.

Multistage carcinogenesis

Fearon, E.R. and Vogelstein, B. (1990). A genetic model for colorectal tumorigenesis. *Cell*, 61: 759–67.

Hanahan, D. and Weinberg, R.A. (2000). The hallmarks of cancer. *Cell*, 100: 57–70.

Nowell, P.C. (1976). The clonal evolution of tumour cell populations. *Science*, 194(4260): 23–8.

Rangarajan, A., Hong, S.J., Gifford, A., and Weinberg, R.A. (2004). Species- and cell type-specific requirements for cellular transformation. *Cancer Cell*, 6(2): 171–83.

Vogelstein, B. and Kinzler, K.W. (2004). Cancer genes and the pathways they control. *Nature Medicine*, 10(8): 789–99.

Tumor suppressors

Bourdon, J.C., Fernandes, K., Murray-Zmijewski, F., et al. (2005). p53 isoforms can regulate p53 transcriptional activity. *Genes & Development*, Aug 30.

Burrows, A.E., Smogorzewska, A., and Elledge, S.J. (2010). Polybromo-associated BRG1-associated factor components BRD7 and BAF180 are critical regulators of p53 required for induction of replicative senescence. *Proceedings of the National Academy of Sciences of the U.S.A.*, **107**(32): 14280–5.

Christophorou, M.A., Martin-Zanca, D., Soucek, L., et al. (2005). Temporal dissection of p53 function in vitro and in vivo. *Nature Genetics*, **37**(7): 718–26.

Dai, C. and Gu, W. (2010). p53 post-translational modification: deregulated in tumorigenesis. *Trends in Molecular Medicine*, **16**(11): 528–36.

Feng, Z. and Levine, A.J. (2010). The regulation of energy metabolism and the IGF-1/mTOR pathways by the p53 protein. *Trends in Cell Biology*, **20**(7): 427–34.

Foulkes, W.D., Smith, I.E., and Reis-Filho, J.S. (2010). Triple-negative breast cancer. *New England Journal of Medicine*, **363**(20): 1938–48.

Kuilman, T., Michaloglou, C., Mooi, W.J., and Peeper, D.S. (2010). The essence of senescence. *Genes & Development*, **24**(22): 2463–79.

Lowe, S.W. and Sherr, C.J. (2003). Tumor suppression by Ink4a-Arf: progress and puzzles. *Current Opinion in Genetic Development*, **13**(1): 77–83.

Lu, W.J., Chapo, J., Roig, I., and Abrams, J.M. (2010). Meiotic recombination provokes functional activation of the p53 regulatory network. *Science*, **328**(5983): 1278–81.

Robanus-Maandag, E., Giovannini, M., van der Valk, M., et al. (2004). Synergy of Nf2 and p53 mutations in development of malignant tumours of neural crest origin. *Oncogene*, **23**(39): 6541–7.

Vousden, K.H. and Prives, C. (2005). P53 and prognosis: new insights and further complexity. *Cell* **120**(1): 7–10.

Whyte, P., Buchkovich, K.J., Horowitz, J.M., et al. (1988). Association between an oncogene and an anti-oncogene: the adenovirus E1A proteins bind to the retinoblastoma gene product. *Nature*, **334**(6178): 124–9.

Young, N.P. and Jacks, T. (2010). Tissue-specific p19Arf regulation dictates the response to oncogenic K-ras. *Proceedings of the National Academy of Sciences of the U.S.A.*, **107**(22): 10184–9.

miRNA

Inui, M., Martello, G., and Piccolo, S. (2010). MicroRNA control of signal transduction. *Nature Reviews Molecular Cell Biology*, **11**(4): 252–63.

Newman, M.A. and Hammond, S.M. (2010). Emerging paradigms of regulated microRNA processing. *Genes & Development*, **24**(11): 1086–92.

Siomi, H. and Siomi, M.C. (2010). Posttranscriptional regulation of microRNA biogenesis in animals. *Molecular Cell*, **38**(3): 323–32.

Telomeres, instability, and chromosomal abnormalities

DePinho, R.A. and Polyak, K. (2004). Cancer chromosomes in crisis. *Nature Genetics*, **36**(9): 932–4.

Feldser, D.M., Hackett, J.A., and Greider, C.W. (2003). Telomere dysfunction and the initiation of genome instability. *Nature Reviews Cancer*, **3**: 623–7.

Jiricny, J. (2006). The multifaceted mismatch-repair system. *Nature Reviews Molecular Cell Biology*, **7**(5): 335–46.

Rafnar, T., Sulem, P., Stacey, S.N., et al. (2009). Sequence variants at the TERT-CLPTM1L locus associate with many cancer types. *Nature Genetics*, **41**(2): 221–7.

Oncogene-induced senescence

Braig, M., Lee, S., Loddenkemper, C., et al. (2005). Oncogene-induced senescence as an initial barrier in lymphoma development. *Nature*, **436**(7051): 660–5.

Braig, M. and Schmitt, C.A. (2006). Oncogene-induced senescence: putting the brakes on tumor development. *Cancer Research*, **66**(6): 2881–4.

Chen, Z., Trotman, L.C., Shaffer, D., et al. (2005). Crucial role of p53-dependent cellular senescence in suppression of PTEN-deficient tumorigenesis. *Nature*, **436**(7051): 725–30.

Gorgoulis, V.G. and Halazonetis, T.D. (2010). Oncogene-induced senescence: the bright and dark side of the response. *Current Opinion in Cell Biology*, Aug 29.

Michaloglou, C., Vredeveld, L.C., Soengas, M.S., et al. (2005). BRAFE600-associated senescence-like cell cycle arrest of human naevi. *Nature* **436**(7051): 720–4.

Wajapeyee, N., Serra, R.W., Zhu, X., Mahalingam, M., and Green, M.R. (2008). Oncogenic BRAF induces senescence and apoptosis through pathways mediated by the secreted protein IGFBP7. *Cell*, **132**(3): 363–74.

Apoptosis and cell death

Fleming, A., Noda, T., Yoshimori, T., and Rubinsztein, D.C. (2011). Chemical modulators of autophagy as biological probes and potential therapeutics. *Nature Chemical Biology*, **7**(1): 9–17.

Goss, P.E. and Chambers, A.F. (2010). Does tumour dormancy offer a therapeutic target? *Nature Reviews Cancer*, **10**(12): 871–7.

Lowe, S.W., Cepero, E., and Evan, G. (2004). Intrinsic tumour suppression. *Nature*, **432**(7015): 307–15.

Pelengaris, S., Khan, M., and Evan, G.I. (2002). Suppression of MYC-induced apoptosis in beta cells exposes multiple oncogenic properties of Myc and triggers carcinogenic progression. *Cell*, **109**(3): 321–34.

Rabinowitz, J.D. and White, E. (2010). Autophagy and metabolism. *Science*, **330**(6009): 1344–8.

Ryan, K.M. (2011). p53 and autophagy in cancer: guardian of the genome meets guardian of the proteome. *European Journal of Cancer*, **47**(1): 44–50.

Yu, L., McPhee, C.K., Zheng, L., et al. (2010). Termination of autophagy and reformation of lysosomes regulated by mTOR. *Nature*, **465**(7300): 942–6.

Microenvironment, inflammation, immunity, and cancer

Bissell, M.J. and Labarge, M.A. (2005). Context, tissue plasticity, and cancer: are tumor stem cells also regulated by the microenvironment? *Cancer Cell*. **7**(1): 17–23.

Curiel, T.J., Coukos, G., Zou, L., et al. (2004). Specific recruitment of regulatory T cells in ovarian carcinoma fosters immune privilege and predicts reduced survival. *Nature Medicine*, **10**(9): 942–9.

Hill, R., Song, Y., Cardiff, R.D., and Van Dyke, T. (2005). Selective evolution of stromal mesenchyme with p53 loss in response to epithelial tumorigenesis. *Cell*, **123**(6): 1001–11.

Karin, M., Lawrence, T., and Nizet, V. (2006). Innate immunity gone awry: linking microbial infections to chronic inflammation and cancer. *Cell*, **124**(4): 823–35.

Kopfstein, L. and Christofori, G. (2006). Metastasis: cell-autonomous mechanisms versus contributions by the tumor microenvironment. *Cellular and Molecular Life Sciences*, **63**(4): 449–68.

McGovern, M., Voutev, R., Maciejowski, J., Corsi, A.K., and Hubbard, E.J. (2009). A "latent niche" mechanism for tumor initiation. *Proceedings of the National Academy of Sciences of the U.S.A.*, **106**(28): 11617–22.

Milicic, A., Harrison, L.A., Goodlad, R.A., et al. (2008). Ectopic expression of P-cadherin correlates with promoter hypomethylation early in colorectal carcinogenesis and enhanced intestinal crypt fission in vivo. *Cancer Research*, **68**(19): 7760–8.

Orimo, A., Gupta, P.B., Sgroi, D.C., et al. (2005). Stromal fibroblasts present in invasive human breast carcinomas promote tumor growth and angiogenesis through elevated SDF-1/CXCL12 secretion. *Cell*, **121**(3): 335–48.

Paszek, M.J., Zahir, N., Johnson, K.R., et al. (2005). Tensional homeostasis and the malignant phenotype. *Cancer Cell*, **8**(3): 241–54.

Radisky, D.C., Kenny, P.A., and Bissell, M.J. (2007). Fibrosis and cancer: do myofibroblasts come also from epithelial cells via EMT? *Journal of Cell Biochemistry*, **101**: 830–9.

Slaughter, D.P., Southwick, H.W., and Smejkal, W. (1953). Field cancerization in oral stratified squamous epithelium; clinical implications of multicentric origin. *Cancer*, **6**: 963–8.

Sneddon, J.B., Zhen, H.H., Montgomery, K., et al. (2006). Bone morphogenetic protein antagonist gremlin 1 is widely expressed by cancer-associated stromal cells and can promote tumor cell proliferation. *Proceedings of the National Academy of Sciences of the U.S.A.*, **103**: 14842–7.

Thiery, J.P. and Sleeman, J.P. (2006). Complex networks orchestrate epithelial-mesenchymal transitions. *Nature Reviews Molecular Cell Biology*, **7**(2): 131–42.

Whiteside, T.L. (2006). The role of immune cells in the tumor microenvironment. *Cancer Treatment and Research*, **130**: 103–24.

Wu, J., Dauchy, R.T., Tirrell, P.C., et al. (2011). Light at night activates IGF-1R/PDK1 signaling and accelerates tumor growth in human breast cancer xenografts. *Cancer Research*, **71**(7): 2622–31.

Zhu, Y., Ghosh, P., Charnay, P., Burns, D.K., and Parada, L.F. (2002). Neurofibromas in NF1: Schwann cell origin and role of tumour environment. *Science*, **296**(5569): 920–2.

Stem cells, EMT, and cancer-initiating cells

Hayashi, K. and Surani, M.A. (2009). Resetting the epigenome beyond pluripotency in the germline. *Cell Stem Cell*, **4**(6): 493–8.

Goldstein, A.S., Huang, J., Guo, C., Garraway, I.P., and Witte, O.N. (2010). Identification of a cell of origin for human prostate cancer. *Science*, **329**(5991): 568–71.

Moore, K.A. and Lemischka, I.R. (2006). Stem cells and their niches. *Science*, **311**(5769): 1880–5.

Pantel, K., Alix-Panabières, C., and Riethdorf, S. (2009). Cancer micrometastases. *Nature Reviews Clinical Oncology*, **6**(6): 339–51.

Roesch, A., Fukunaga-Kalabis, M., Schmidt, E.C., et al. (2010). A temporarily distinct subpopulation of slow-cycling melanoma cells is required for continuous tumor growth. *Cell*, **141**(4): 583–94.

Rosen, J.M. and Jordan, C.T. (2009). The increasing complexity of the cancer stem cell paradigm. *Science*, **324**(5935): 1670–3.

Singh, A. and Settleman, J. (2010). EMT, cancer stem cells and drug resistance: an emerging axis of evil in the war on cancer. *Oncogene*, **29**(34): 4741–51.

Wang, Y., Ngo, V.N., Marani, M., et al. (2010). Critical role for transcriptional repressor Snai12 in transformation by oncogenic RAS in colorectal carcinoma cells. *Oncogene*, **29**(33): 4658–70.

Wodarz, A. and Gonzalez, C. (2006). Connecting cancer to the asymmetric division of stem cells. *Cell*, **124**(6): 1121–3.

Yang, M.H., Hsu, D.S., Wang, H.W., et al. (2010). Bmi1 is essential in Twist1-induced epithelial-mesenchymal transition. *Nature Cell Biology*, **12**(10): 982–92.

Yeung, T.M., Gandhi, S.C., Wilding, J.L., Muschel, R., and Bodmer, W.F. (2010). Cancer stem cells from colorectal cancer-derived cell lines. *Proceedings of the National Academy of Sciences of the U.S.A.*, **107**(8): 3722–7.

Yilmaz, O.H., Valdez, R., Theisen, B.K., et al. (2006). Pten dependence distinguishes haematopoietic stem cells from leukaemia-initiating cells. *Nature*, **441**(7092): 475–8.

Metastases

Campbell, P.J., Yachida, S., Mudie, L.J., et al. (2010). The patterns and dynamics of genomic instability in metastatic pancreatic cancer. *Nature*, **467**(7319): 1109–13.

Condeelis, J., Singer, R.H., and Segall, J.E. (2005). The great escape: when cancer cells hijack the genes for chemotaxis and motility. *Annual Review of Cell and Developmental Biology*, **21**: 695–718.

Crnic, I. and Christofori, G. (2004). Novel technologies and recent advances in metastasis research. *International Journal of Developmental Biology*, **48**(5–6): 573–81.

Gupta, P.B., Kuperwasser, C., Brunet, J.P., et al. (2005). The melanocyte differentiation program predisposes to metastasis after neoplastic transformation. *Nature Genetics*, **37**(10): 1047–54.

Kang, S.Y., Halvorsen, O.J., Gravdal, K., et al. (2009). Prosaposin inhibits tumor metastasis via paracrine and endocrine stimulation of stromal p53 and Tsp-1. *Proceedings of the National Academy of Sciences of the U.S.A.*, **106**(29): 12115–20.

Kang, Y., He, W., Tulley, S., et al. (2005). Breast cancer bone metastasis mediated by the Smad tumor suppressor pathway. *Proceedings of the National Academy of Sciences of the U.S.A.*, **102**(39): 13909–14.

Kaposi-Novak, P., Lee, J.S., Gomez-Quiroz, L., Coulouarn, C., Factor, V.M., and Thorgeirsson, S.S. (2006). Met-regulated expression signature defines a subset of human hepatocellular carcinomas with poor prognosis and aggressive phenotype. *Journal of Clinical Investigation*, **116**(6): 1582–95.

Ma, L., Young, J., Prabhala, H., et al. (2010). miR-9, a MYC/MYCN-activated microRNA, regulates E-cadherin and cancer metastasis. *Nature Cell Biology*, **12**(3): 247–56.

Martello, G., Rosato, A., Ferrari, F., et al. (2010). A microRNA targeting dicer for metastasis control. *Cell*, **141**(7): 1195–207.

Steeg, P.S. (2003). Metastasis suppressors alter the signal transduction of cancer cells. *Nature Reviews Cancer*, **3**(1): 55–63.

Yachida, S., Jones, S., Bozic, I., et al. (2010). Distant metastasis occurs late during the genetic evolution of pancreatic cancer. *Nature*, **467**(7319): 1114–17.

Yang, J., Mani, S.A., Donaher, J.L., et al. (2004). Twist, a master regulator of morphogenesis, plays an essential role in tumor metastasis. *Cell*, **117**(7): 927–39.

Cancer profiling and "tailored" medicine

Araujo, R.P., Liotta, L.A., and Petricoin, E.F. (2007). Proteins, drug targets and the mechanisms they control: the simple truth about complex networks. *Nature Reviews Drug Discovery*, **6**(11): 871–80.

Ben-Porath, I., Thomson, M.W., Carey, V.J., et al. (2008). An embryonic stem cell-like gene expression signature in poorly differentiated aggressive human tumors. *Nature Genetics*, **40**(5): 499–507.

Blackford, A., Serrano, O.K., Wolfgang, C.L., et al. (2009). SMAD4 gene mutations are associated with poor prognosis in pancreatic cancer. *Clinical Cancer Research*, **15**(14): 4674–9.

Carr, K.M., Rosenblatt, K., Petricoin, E.F., and Liotta, L.A. (2004). Genomic and proteomic approaches for studying human cancer: prospects for true patient-tailored therapy. *Human Genomics*, **1**(2): 134–40.

Domchek, S.M., Friebel, T.M., Singer, C.F., et al. (2010). Association of risk-reducing surgery in BRCA1 or BRCA2 mutation carriers with cancer risk and mortality. *Journal of the American Medical Association*, **304**(9): 967–75.

Garman, K.S., Acharya, C.R., Edelman, E., et al. (2008). A genomic approach to colon cancer risk stratification yields biologic insights into therapeutic opportunities. *Proceedings of the National Academy of Sciences of the U.S.A.*, **105**(49): 19432–7.

Gilbert, L.A. and Hemann, M.T. (2010). DNA damage-mediated induction of a chemoresistant niche. *Cell*, **143**(3): 355–66.

Iorns, E., Turner, N.C., Elliott, R., et al. (2008). Identification of CDK10 as an important determinant of resistance to endocrine therapy for breast cancer. *Cancer Cell*, **13**(2): 91–104.

Luchini, A., Fredolini, C., Espina, B.H., et al. (2010). Nanoparticle technology: addressing the fundamental roadblocks to protein biomarker discovery. *Current Molecular Medicine*, **10**(2): 133–41.

Mathews, L.A., Crea, F., and Farrar, W.L. (2009). Epigenetic gene regulation in stem cells and correlation to cancer. *Differentiation*, **78**(1): 1–17.

Ning, C. and Karantza, V. (2011). Autophagy as a therapeutic target in cancer. *Cancer Biology and Therapy*; **11**(2): 157–68.

Yu, S.L., Chen, H.Y., Chang, G.C., et al. (2008). MicroRNA signature predicts survival and relapse in lung cancer. *Cancer Cell*. **13**(1): 48–57.

Zhang, H., Li, Y., and Lai, M. (2010). The microRNA network and tumor metastasis. *Oncogene* **29**(7): 937–48.

Screening and prevention

Andriole, G.L., Crawford, E.D., Grubb, R.L. III, et al. (2009). Mortality results from a randomized prostate-cancer screening trial. *New England Journal of Medicine*, **360**: 1310–19.

Miller, A.B. (1992). *Cervical cancer screening programmes*. Geneva: WHO.

Schröder, F.H., Hugosson, J., Roobol, M.J., et al. (2009). Screening and prostate-cancer mortality in a randomized European study. *New England Journal of Medicine*, **360**: 1320–8.

Stewart, B.W. and Coates, A.S. (2005). Cancer prevention: a global perspective. *Journal of Clinical Oncology*, **23**(2): 392–403.

Sturgeon, C.M., Lai, L.C., and Duffy, M.J. (2009). Serum tumour markers: how to order and interpret them. *British Medical Journal*, **339**: b3527–b3527.

Appendix 1.1 History of cancer

(see also: http://press2.nci.nih.gov/sciencebehind/cioc)

The difficulty in identifying traces of cancer in ancient remains and fossils inevitably makes a chronological survey of cancer difficult, and in particular largely precludes a reliable estimate of the prevalence of cancer until relatively recent times. Cancer has clearly existed for a very long time and skeletal metastases have been identified in archaeological specimens and a rectal cancer was found recently in an Egyptian mummy. At least one convincing report of a metastatic cancer has been reported in a dinosaur fossil, suggesting that cancer may have existed as long as complex organisms, but such findings are rare. There are few, if any, convincing fossil remains suggestive of cancer in Neanderthals or early humans.

The key question is whether this scarcity of cancer-containing specimens is a result of the technical challenges in diagnoses and therefore the vagaries of paleopathology or, on the other hand, represents confirmation of the central importance on cancer pathogenesis of a modern lifestyle replete with environmental carcinogens, aversion to physical activity, and, ironically, an extended lifespan. The answer is not clear.

The widespread mummification of bodies in Ancient Egypt alongside the availability of written records offers greater opportunities to consider cancer in antiquity. Early Egyptian papyri from around 1600 BC, such as the "Edwin Smith" and "George Ebers" papyri, include descriptions of benign and malignant tumors and treatments based on castor oil and various animal parts, including pigs'; ears. Not that the Edwin Smith papyrus was particularly encouraging, as illustrated by this extract from case 45:

> If you examine a man having tumours on his breast . . . if you put your hand upon these tumours and you find them very cool, there being no fever at all therein . . . they have no granulation, they form no fluid, they do not generate secretions of fluid, and they are bulging to thy hand. *There is no treatment.* If you find tumours in any member of a man, you shall treat him according to these directions.

Nevertheless, many of the early written descriptions of cancer originate from the Classical Greek and Roman physicians Hippocrates and Galen, who laid the foundations for modern medicine by emphasizing that diseases were natural physical processes. In fact, we owe our names for cancer to Hippocrates, who first applied the terms *karkinos* and *karkinoma* (Ancient Greek for "crab") to various diseases, including cancers of the breast, uterus, stomach, and skin. Cancer is the Latin equivalent. Interestingly, although Galen performed some early surgical interventions for cancer, he maintained that cancer was generally best left untreated, a view that appears to still find favor with some health economists. However, Galen also believed that diseases resulted from imbalances in the four bodily "humors" (blood, phlegm, yellow bile, and black bile), which were also responsible for differing temperaments such as melancholy!

Humoral theory, first raised by Hippocrates around 2500 years ago, and extended by Galen, remained the central tenet of essentially all Western medicine until the 1800s. Given the prevalence of this view for around 2000 years, it bears a brief diversion to

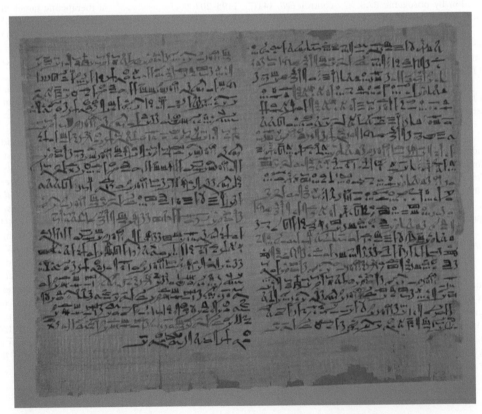

Section of the Edwin Smith papyrus. From the National Library of Medicine http://archive.nlm.nih.gov/proj/ttp/smith_home.html.

discuss it. The human body was believed to comprise a mix of the four humors: black bile (or melancholy), yellow or red bile, blood, and phlegm. The balance of these varied from individual to individual and as long as they were in the correct balance for you, you remained healthy (the first example of individualized medicine perhaps). The humors were directly linked to temperaments: melancholic, sanguine, choleric, and phlegmatic – thus also encompassing the links between mind and body.

Unfortunately, relatively little progress was recorded during the so-called Middle Ages (from the fall of the Roman Empire until the Renaissance). Although clearly in the Arab world, Moorish Spain, Constantinople, and in the West in monastic communities, much classical learning was preserved and recorded for the future benefit of Renaissance scholars. This generally negative view of human progress in the Middle Ages as being largely the copying and preservation of classical texts for the future benefit of Renaissance scholars is rather overstated, as illustrated by an intriguing quotation from Theodoric, Bishop of Cervia (1267) – "The older a cancer is, the worse it is. And the more it is involved with muscles, veins and nutrifying arteries, the worse it is, and the more difficult to treat. For in such places incisions, cauteries and sharp medications are to be feared."

Much important scholarship was also taking place in the Arab world, not least of which was laying the foundations for modern mathematics. With respects to cancer, the insightful writings of two prominent Arab scholars have been recorded. Thus, to quote Avicenna (981–1037):

The difference between cancerous swelling and induration. The latter is a slumbering silent mass, which . . . is painless, and stationary. . . . A cancerous swelling progressively increases in size, is destructive, and spreads roots which insinuate themselves amongst the tissue-elements;

and Albucasis (1050):

The Ancients said that when a cancer is in a site where total eradication is possible, such as a cancer of the breasts or of the thigh, and in similar parts where complete removal is possible, and especially when in the early stage and small, then surgery was to be tried. But when it is of long standing and large you should leave it alone. For I myself have never been able to cure any such, nor have I seen anyone else succeed before me.

From classical times until the late Renaissance, when Vesalius and artists such as Michelangelo and Leonardo da Vinci developed an interest in anatomy, cancer was still believed to be caused variously by Acts of God or still, in deference to Galen, by an excess of black bile. Although still believed to be incurable, a wide variety of arsenic-containing preparations were employed to treat it. Based on his observations in Austrian mines, Theophratus Bombastus von Hohenheim, better known as Paracelsus, described the "wasting disease of miners" in 1567. He proposed that the exposure to natural ores such as realgar (arsenic sulfide) and others might have been causing this condition. Paracelsus was actually among the first to consider a chemical compound as an occupational carcinogen. Paracelsus was probably the first prominent objector to Galen's humoural doctrine, and instead proposed that mineral salts when concentrated in a particular part of the body and unable to find an outlet, were the real cause of cancer.

The beginnings of recognizably modern science took place in the seventeenth century; William Harvey described the continuous circulation of the blood, finally resulting in the rejection of the humoral theory of disease, and cancer was no longer attributed to bile. A contemporary of Harvey, Gaspare Aselli identified the lymphatic system, which he suggested as a primary cause of cancer. However, on the basis of this discovery, René Descartes developed a new theory, termed the "sour lump" theory in 1652, whereby it was suggested that lymph became hard through some congealing process and formed a scirrhus. If this fermented (i.e. became acid or sour) then a cancer would develop. Surgery for cancer now began to include removal of the lymph nodes when enlarged and near the tumor site. A renowned German surgeon, Fabricius Hildanus, removed enlarged lymph nodes in breast cancer operations, but in the absence of either septic techniques or anesthetics it was an extremely hazardous procedure.

In the eighteenth century, oncology became a recognized discipline, with early experiments conducted. The French physician Claude Gendron (1663–1750) concluded after 8 years of research that cancer arises locally as a hard, growing mass, untreatable with drugs that must be removed with all its "filaments." The Dutch professor Hermann Boerhaave believed inflammation could result in a scirrhus, or tumor, capable of evolving into cancer. John Hunter, one of the earliest modern surgeons, taught that if a tumor were movable, it could be surgically removed, as could resulting cancers in proper reach. If enlarged glands were involved, he advised against surgery.

Two eighteenth-century French scientists, physician Jean Astruc and chemist Bernard Peyrilhe, conducted experiments to confirm or disprove hypotheses related to cancer. Their efforts may appear eccentric to us now, but they helped establish the discipline of experimental oncology. For example, in 1740 Astruc, a professor of medicine at Montpellier and Paris, sought to test the validity of the humoral theory by comparing the taste of boiled beef-steak with that of boiled breast tumor; he found no black bile–like taste in the tumor – he may also have had a lasting influence on French culinary practices! Peyrilhe attempted to demonstrate an infective cause for cancer by injecting human cancer tissue into a dog. The resultant infected abscess (no cancer!) resulted in a housemaid drowning the poor dog to end its misery.

Later in the same century, two English physicians – John Hill and Percival Pott – described the occurrence of cancerous alterations in the nasal mucosa and at the skin of the scrotum in a few patients, and linked it with local long-term exposure to snuff and repetitive local contamination by soot, respectively.

The nineteenth century heralded the beginnings of modern biology. Virchow focused pathology on the cell; and anesthesia and antisepsis improved surgery. Oncology progressed as Röntgen described X-rays, the Curies isolated radium, and Müller observed abnormalities of cancer cells. By the mid-nineteenth century, French and Italian researchers had found that women died from cancer much more frequently than men, and that the cancer death rate for both sexes was rising. Domenico Rigoni-Stern concluded that incidences of cancer increase with age.

Throughout the early decades of the twentieth century, researchers pursued different theories of the origin of cancer. Theodor Boveri, professor of zoology at Wurzberg, proposed that cancer was due to abnormal chromosomes. This was remarkably prescient given that it was more than 40 years before the discovery of the structure of DNA. A viral cause of cancer in chickens was documented in 1911, and both chemical and physical carcinogens were conclusively identified. Radium and X-rays were employed against cancer early in the century, and it was found that X-rays selectively damaged cancer cells, causing less harm to other tissues. As safe levels of dosage were determined, the therapy became standard. Chemical- and radiation-induced cancers were first reliably confirmed as carcinogens. While the smoking–cancer link was noted in the 1930s, causality was only proven following extensive epidemiological studies in 1950.

Molecular biology has revolutionized both medicine and cancer research; following the identification of the structure of DNA by Francis Crick and James Watson in 1953, the genetic code was soon broken, and the foundations were laid for much of what is discussed in this book.

We conclude with two quotes, illustrating how far we have progressed in cancer therapy:

Cancer is an uneven swelling, rough, unseemly, darkish, painful, and sometimes without ulceration . . . *and if operated upon, it becomes worse* . . . and spreads by erosion; forming in most parts of the body, but more especially in the female uterus and breasts. It has the veins stretched on all sides as the animal the crab (cancer) has its feet, whence it derives its name.

Paul of Aegina (625–690)

A carcinoma does not give rise to the same danger [as a carbuncle] *unless it is irritated by imprudent treatment*. This disease occurs mostly in the upper parts of the body, in the region of the face, nose, ears, lips, and in the breasts of women, but it may also arise in an ulceration, or in the spleen. . . . At times the part becomes harder or softer than natural. . . . After excision, even when a scar has formed, none the less the disease has returned, and caused death.

Aulus (Aurelius) Cornelius Celsus (25BC–AD50)

2 The Burden of Cancer

William P. Steward and Anne L. Thomas
University of Leicester, UK

All interest in disease and death is only another expression of interest in life.

Thomas Mann

Key points

- The incidence of cancer in developed countries is rising due to increased aging of the population and increasing exposure to carcinogens.
- Cancers can cause symptoms from local mass effects, leading to pain or organ dysfunction, or may cause systemic symptoms. These latter may be nonspecific, including weight loss, lethargy, muscle wasting, and debility, or predictable arising from effects of secreted proteins.
- A suspected diagnosis of cancer should be confirmed by histological assessment of biopsy specimens taken from areas of abnormal tissue and may be supported by results of imaging and serum markers.
- Lung cancer is the commonest malignancy causing death in Western nations and comprises small-cell and non-small-cell variants, which each have different biological behaviors, treatments, and outcomes.
- Breast cancer is the commonest malignancy in women and has widely differing incidences in different nations, suggesting the importance of environmental factors in its etiology.
- Prostate cancer has a 10-fold difference in incidence between Japanese and Afro-Caribbean men and carries a worse prognosis in the Afro-Caribbean population.

Introduction

One death is a tragedy, 100 000 deaths are statistics.

Albert Szent-Gyorgyi

It is estimated that worldwide there are approximately 10 million new cases of cancer per year, causing 6 million deaths. By 2012, with a worldwide population estimated at 8 billion, it is anticipated that at least 20 million new cancer patients will be diagnosed each year, causing 12 million deaths. Currently, around half of all cancers occur in developing nations, a figure that is expected to increase to 70% over the coming decade. However, this still represents a much smaller percentage of the overall population when compared to developed countries; incidence is approximately 100 per 100 000 population whereas it is 3- to 4-fold higher in Western nations. In fact if non-melanoma skin cancers are included, then around 1 in 3 of us can expect to develop some form of malignancy during our lifetime. Cancer is the number three killer in developing countries, behind infection and cardiovascular diseases, but has claimed the number two spot in developed countries after cardiovascular disease. In fact, if we consider the comparative ease with which we could forestall risk of cardiovascular diseases and with an ever aging population one would not bet against cancer reaching the gold medal position before the end of the next century.

In 2004, based on World Health Organization (WHO) reports, 7.4 million people died as a result of cancer worldwide. The commonest type of cancer was lung, which accounted for 1.3 million deaths and then stomach (803 000 deaths), colorectal (639 000 deaths), liver (610 000 deaths), and breast (519 000 deaths). Figures for men revealed that the most frequest types of cancer worldwide (in order of the number of global deaths) were lung, stomach, liver, colorectal, esophagus, and prostate, whereas for women the order of frequency was breast, lung, stomach, colorectal, and cervical cancers. Over the last 20 years there has been a global increase of malignancies ascribed to HIV/AIDS, with marked increases of lymphomas and hepatocellular carcinomas.

The incidence of cancer in Western societies has risen steadily over the last century, predominantly because of the increasing median age of the population but also because of rising exposure to carcinogens. In fact, in Europe and North America cigarette smoking and dietary factors are major or contributory factors to the development of up to two-thirds of all malignancies. Lung

---

The Molecular Biology of Cancer: A Bridge From Bench to Bedside, Second Edition. Edited by Stella Pelengaris and Michael Khan.
© 2013 John Wiley & Sons, Inc. Published 2013 by John Wiley & Sons, Inc.

Box 2.1 Key concepts of Chapter 2

Selected malignancies

- Lung
- Breast
- Colorectal
- Prostate
- Renal
- Skin
- Cervix
- Hematological
  - Lymphomas
  - Leukemias

Incidence

- Pathology
- Etiology

Presentation

- Local symptoms from primary
- Symptoms from metastases
- Systemic symptoms of malignancy
- Paraneoplastic syndromes
- Symptoms from biochemical abnormalities

Therapy

- Surgery
- Chemotherapy
  - Adjuvant
  - Neoadjuvant
  - Advanced
- Hormones
- Radiotherapy
- Biological therapies
- Modulators of cell signaling pathways

cancer is by far the commonest cause of death from malignancy in both males and females, followed by colorectal cancer in men and breast cancer in women.

This chapter will focus on the clinical features of the most important cancers and will include information on clinical manifestations, diagnosis and treatment (Box 2.1). More detailed information on diagnostic procedures is given in Chapter 17.

Any malignancy may cause symptoms, resulting from various biological effects on the host. The earliest manifestations may be caused by the local effect of the tumor, including the presence of a mass, discomfort from compression of local organs or nerves, hemorrhage from the involvement of blood vessels, or obstruction of airways, ureter, bile duct, and other key structures. Tumors may also present with nonspecific effects such as cachexia (the loss of body mass frequently seen in patients with chronic debilitating diseases), lethargy, weight loss, fever, and a variety of neuromuscular syndromes. Although more usually associated with benign hyperplasia or adenomas arising in tissues that are normally hormone-secreting (endocrine tissues), some malignancies may produce hormones acting on distant tissues via the circulation. Such hormones include ADH, PTH, erythropoeitin, ACTH, calcitonin, and even members of the IGF family of growth factors, such as IGF-II, which can all cause distant effects that may prove life-threatening. More common, however, is the near ubiquitous production of growth factors whose action is confined to the tumor, blood vessels, and surrounding stroma. Ultimately, most cancer deaths are the result of invasion and spread of the primary cancer to other more distant tissues (metastases), which may in some cases become clinically apparent before the primary tumor. Indeed, in some cases the primary may never be discovered. The diagnosis of malignancy can therefore be difficult with a wide variety of presentations, which may not necessarily indicate an obvious site of the primary tumor.

For those patients who complain of specific symptoms that raise the possibility of a particular malignancy (e.g. altered bowel habit and carcinoma of the colon), the necessary investigations and diagnostic procedures are usually self-evident. For those patients who present with nonspecific symptoms (e.g. cachexia and weight loss) the choice of investigations is difficult. The majority of such individuals will probably not have an underlying malignancy and it is essential to balance the potential morbidity and occasional mortality of investigations (together with the cost) against the likelihood of detecting a malignancy and the potential value of making an early diagnosis. The "gold standard" of diagnosis is obtaining tissue that, on histological examination, confirms the presence of neoplastic cells. In the majority of cases this is straightforward, with symptoms guiding the choice of a site for biopsies. Support for a diagnosis can be provided by tumor markers (usually proteins specifically expressed by or in response to a tumor) and abnormalities of hematological and biochemical blood results, but in a small percentage of patients no obvious site of neoplastic focus is detected and a diagnosis has to be made on the basis of probabilities.

A further occasional clinical problem arises when a metastatic site is detected (most commonly an isolated area of lymphadenopathy) and histological examination reveals malignant cells (frequently adenocarcinoma or squamous carcinoma). Investigation of common primary sites (e.g. breast, gastrointestinal tract, head and neck, lung) reveals no obvious origin and in this instance a diagnosis of "carcinoma of unknown primary" is made, which can lead to a dilemma when trying to choose appropriate therapies. Some of the reasons why cancer cells may diverge so far from their cell type of origin so as to become unrecognizable are discussed in the following chapters.

Key features, particularly the prevalence, pathogenesis, and clinical manifestations of selected cancers, are outlined below. Current approaches to diagnostic techniques and treatment are summarized. It must be stressed that approaches to diagnosis and management of cancer are changing rapidly and many new methods of diagnosis, including new imaging techniques and markers, are constantly becoming available. In addition, numerous new targeted therapies that are not chemotherapy drugs have been included in treatment regimens. The following chapters will give a fair idea of what processes and molecules we predict to become the next generation of cancer therapeutic targets, whereas here we concentrate on what is already part of clinical practice.

Lung cancer

> Smoking is hateful to the nose, harmful to the brain, and dangerous to the lungs.
>
> *King James I*

Incidence

The incidence of carcinoma of the bronchus rose steadily throughout the twentieth century in the Western world, closely mirroring the increase in cigarette smoking. Of all malignancies, lung cancer has the highest mortality rates, with the death rate approaching that of incidence. As a result in the Western world, it has long been the main cause of cancer-related death in men, and has become the main cause in women since the late 1980s. The overall annual incidence in most Western nations is approximately 100 per 100 000 males and 40 per 100 000 females. There has been a decline in the incidence among males since the late 1990s, but as smoking has become more popular in females, there has been an increasing incidence in women, such that the male:female ratio for lung cancer incidence has risen steadily from approximately 13:1 in the 1950s to 2:1. The global rise in lung cancer incidence following a marked increase in smoking in the Middle East, Asia, and Africa is of concern. It was estimated that during 2005 approximately 500 000 new cases of lung cancer were diagnosed in China.

Between 80% and 90% of all lung cancers are attributable to cigarette smoking. The first study to clearly demonstrate this association was published by Doll and Hill in 1964 and was based on the incidence of lung cancer among doctors in the United Kingdom (Box 2.2). Several subsequent studies have confirmed this link and have also shown a relationship between the number of cigarettes smoked and the risk of lung cancer. Importantly, there is also clear evidence that the cessation of smoking is associated with a subsequent fall in lung cancer risk such that approximately 12 years after discontinuation, the risk of lung cancer almost reaches that of a life-long nonsmoker. There has also been an observation that lung cancer risk rises among nonsmokers who live or work in an environment with smokers, leading to an approximately twofold rise in the risk of lung cancer ("passive smoking"). Young females who smoke appear to be at a particularly high risk of subsequent lung cancer, an extremely worrying observation considering the current vogue for cigarette smoking among young peri-pubertal females in most Western nations.

Other factors that might increase the risk of lung cancer include exposure to asbestos, radon, industrial air pollution, chromium, nickel, and inorganic arsenic compounds. Of these, asbestos exposure appears to have the greatest impact for both lung cancer and for mesothelioma (a cancer of the lung lining that is very rare in those not exposed to asbestos). The effects of carcinogens are additive, as exemplified by the risk of lung cancer in an asbestos worker who also smokes, which is a staggering 45-fold that of the normal population.

As a recent example of the potential value of identifying genetic susceptibility for cancer (see Chapter 3), there is now good supporting evidence that polymorphic alleles for genes encoding phase I/II metabolizing enzymes may influence the production and clearance of carcinogens and could be identified as biomarkers for individuals particularly vulnerable to the carcinogenic effects of inhaling tobacco smoke. Numerous studies have investigated the possible link between other genetic abnormalities and the risk of developing lung cancer (discussed in detail in Chapter 3). Thus, for instance, activating mutations in *KRAS*, upregulation of the epidermal growth factor receptor (EGFR), and inactivating mutations in tumor suppressor signaling, particularly p53, are ubiquitous in non-small-cell lung cancers (NSCLCs), but in most cases result from somatic mutations. Therefore, such knowledge may have limited value in developing biomarkers for early diagnosis but might instead increasingly in coming years be employed to determine prognosis and to guide subsequent chemotherapy. To date, although numerous genetic changes have been reported, no clear links between specific hereditary genetic abnormalities and the risk of most lung cancer have been proven.

Pathology

Lung cancers are divided into two major subgroups – small-cell lung cancer (SCLC) and non-small-cell lung cancer (NSCLC); these differ markedly in respects to their biology, therapeutic sensitivity, and outcomes. NSCLC is further separated, on the basis of pathological features, into subgroups, including squamous cell and adenocarcinoma. For reasons that remain unclear – but do not appear to include smoking behavior – the incidence of adenocarcinoma, which currently accounts for 40% of all lung cancers, is continuing to increase. Adenocarcinomas frequently originate at the lung periphery and will often invade the pleura. Squamous cell carcinoma was the most frequent subtype until the early 1990s, since when its incidence has been reducing. Squamous cell carcinomas are more likely to be centrally placed and may be more easily detected by cytology. As these tumors grow, they may obstruct major airways with resulting distal pneumonia. They appear to begin as *in situ* carcinomas with subsequent development over 3–5 years before they become clinically apparent. A third group of less differentiated large cell

---

Box 2.2 Key clinical study

Doll and Hill's study, which was reported in 1964, assessed a population of 40 637 British male medical practitioners and followed this group over many years, documenting smoking habits. They detected a striking increase in incidence of lung cancer in the smokers and correlated lung cancer risk with increasing exposure to cigarettes. The death rate per 1000 per annum from lung cancer in nonsmokers was 0.07, but rose to 0.93 in all cigarette smokers and was >25 in those smoking more than 25 cigarettes a day. During the follow-up period of the study there was a program to reduce smoking in doctors and a marked fall (25%) in lung cancer deaths was seen in the population who discontinued, with a continuing rise in deaths (26%) in the general male population who had increased smoking rates during the same period.

This was a landmark epidemiological study which had profound implications for society and defined a major carcinogen. A parallel study was reported from the United States by Hammond and Horn in 1958. Almost 190 000 men were followed for 44 months and a similar association between a history of cigarette smoking and cancer incidence, with a gradient of risk relating to the number of cigarettes, was seen. As this study had a much shorter follow-up period, an association between discontinuation and reducing risk was not described.

carcinomas comprise only some 15% of all NSCLCs and usually arise in more distal bronchi.

Small-cell lung cancer is identified by the presence of diffuse small cells, which contain fine granular nuclei. Granules are frequently seen and contain a variety of hormones, including ACTH and ADH. Markers of neural differentiation are often expressed by these tumors and overexpression of one or more of the oncogenes from the *MYC* family is frequently observed (see Chapter 6). Between 20% and 25% of all lung cancers show a mixed picture, frequently with a small and non-small-cell component. The cellular origin of lung malignancies remains controversial and this is discussed later in the book under the umbrella of the cancer stem cell debate.

Clinical features

Although lung cancer is the most common cause of paraneoplastic syndromes, most patients present with symptoms related to the primary tumor. Initially there may be a cough, which is often persistent and may be associated with a wheeze. This is frequently confused with similar symptoms induced by smoking and may simply be perceived as an exacerbation of existing chronic obstructive airways disease. Consequently, patients often delay seeking medical advice. Hemoptysis may follow, with frequent blood-streaked sputum but rarely massive hemorrhage. Dyspnea may result from obstruction of an airway or the development of a pleural effusion. This may be exacerbated by segmental collapse of one of the lungs. Pain is a frequent associated feature and may indicate mediastinal involvement or invasion of ribs or pleura.

Hoarseness of the voice is seen when tumor invasion of the recurrent laryngeal nerve occurs and dysphagia may occur with mediastinal lymphadenopathy (enlarged lymph nodes in the chest), which compresses the esophagus. Apical tumors (arising in the uppermost area of lung) may compress the superior vena cava, causing the syndrome of superior vena cava obstruction (SVCO). This is associated with a reduced return of venous blood from the head and neck, resulting in swelling of the face, arms, and neck (Fig. 2.1).

Numerous symptoms follow the development of metastases, and may relate directly to the site of the secondary tumors, including bone pain, liver capsule distension from hepatomegaly, headaches, and epileptic fits from cerebral metastases (Fig. 2.2) and discomfort at the sites of lymphadenopathy. Some patients may actually present with a mass in lymph nodes (commonly in the neck) from a metastasis.

Several paraneoplastic syndromes are associated with lung cancer, predominantly SCLC. Some of these may result from the abnormal production and release of hormones by the SCLC:

- The syndrome of inappropriate ADH secretion (SIADH) is associated with profound hyponatremia (due to water retention, secondary to the action of ADH on the kidney), causing lethargy and somnolence.
- The production of a protein related to parathyroid hormone, PTHrp, may cause hypercalcemia, leading to nausea, polyuria, and, if untreated, coma.
- The elevated and unregulated secretion of ACTH ("ectopic ACTH" – because it is not coming from its usual source in the pituitary gland) may result in dramatic increases in secretion of the steroid hormone cortisol from the adrenal gland, leading to Cushing's syndrome.

In fact, ectopic production of protein hormones by cancer cells is often far more problematic than the excess production of the same or related hormones from their usual cell type. This is primarily because mutations or altered differentiation of the cancer cells that have conferred the ability to produce these hormones are rarely, if ever, matched by the production of proteins needed

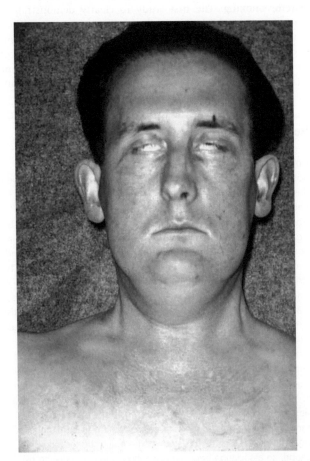

Figure 2.2 Bilateral ptosis caused by cerebral metastases from small-cell lung cancer. This patient had metastases involving the cerebral hemispheres and brainstem.

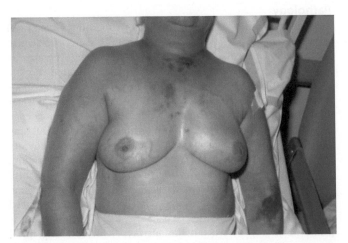

Figure 2.1 Superior vena cava occlusion leads to marked swelling of neck, arms, and dilation of vessels on the chest wall. In this patient it resulted from large apical carcinoma of the lung.

for feedback control – there are no brakes on this system. Other paraneoplastic effects can include cerebellar syndromes, muscle weakness (myasthenia), and a variety of other neurological abnormalities.

Small-cell lung cancer has a more aggressive rate of development than NSCLC and has one of the highest tumor doubling times of all solid tumors. As a result it is rarely localized at the time of diagnosis and, almost invariably, metastases can be detected. Bone marrow metastasis is frequent and can be identified in up to 90% of cases by using sensitive assays for detecting cancer cell genes such as polymerase chain reaction (PCR) or sequencing. Left untreated, SCLC results in rapidly progressing symptoms over as little as 2–3 months, whereas the rate of progression with NSCLC is much slower, with a proportion of patients having truly localized disease. This difference in the biological behavior of the disease has a significant impact on management.

Diagnostic and staging investigations

The possibility of lung cancer may first be raised when an opacity ("shadow") is seen on a chest X-ray, which may be a mass often associated with an area of pneumonia. Confirmation of a diagnosis relies on obtaining neoplastic cells and in almost all cases this is undertaken by fiberoptic bronchoscopy, which is a rapid outpatient procedure allowing direct visualization and biopsy of the majority of tumors (Fig. 2.3). Brushings may also be taken from the bronchi and can reveal the presence of cancer cells (cytological evidence) in up to 80% of cases. In some cases metastases, usually lymph nodes, may be biopsied to confirm a suspicious X-ray.

Following diagnosis, staging is undertaken to determine the extent of disease. For SCLC, surgical resection is unlikely to be performed and staging is therefore relatively simple, the aim being to obtain a rough estimate of the total tumor burden and divide patients into those who will receive more or less aggressive therapy (see Chapter 18). Staging for SCLC is divided into "limited" disease, in which tumor is confined to one side of the chest hemithorax (a third of patients), or "extensive" disease, where cancer has spread beyond this (60–70% of patients). Staging involves blood tests for routine biochemistry, liver function, and full blood count, and often isotope bone scans and imaging of the liver either by ultrasound or computed tomography (CT) scanning. If blood counts are abnormal, it is practice in many centers to undertake a bone marrow examination. If there is any clinical evidence of cerebral metastases, imaging of the brain is undertaken.

For NSCLC, staging is more important as surgical resection may be possible and it is essential to exclude metastatic disease for those who might undergo surgery. A widely accepted TNM (tumor, node, metastasis) staging system is used that divides patients into stage I (a), (b), stage II (a), (b), stage III (a), (b), or stage IV disease (Table 2.1). Survival is closely linked to stage. Staging investigations include rigid bronchoscopy in many centers for those patients considered potential candidates for resection, CT scanning of the thorax and abdomen, mediastinoscopy to directly visualize the mediastinum and, increasingly, positron emission tomography (PET) imaging. These investigations complement each other to minimize the risk of undertaking a futile major surgical procedure (thoracotomy) on a patient with metastatic disease.

Treatment

Surgery

Surgical resection is the only form of therapy that provides the possibility of cure in NSCLC. This reflects the potential for this disease to be detected at a time prior to dissemination. Unfortunately, metastases are almost invariable present in SCLC patients (albeit microscopic) and so surgery is rarely offered. An exception, in some centers, is for patients with small peripheral lesions that have been detected coincidentally.

For all patients with NSCLC, the potential for resectability is considered in the first instance. Such surgery is only undertaken in patients with localized disease (stage I/II). Approximately 30% of patients will be offered resection (many will be excluded because the cancer is deemed inoperable or the patient's general condition is poor). Surgery depends on the site of disease, and may include local segmental or wedge resections, lobectomy, or even pneumonectomy (removal of an entire lung). Success of surgery depends on the stage of cancer, with 5-year survival rates of approximately 54% in stage I, but only 24% for stage IIb. For those centers who operate on stage IIIa disease, neoadjuvant chemotherapy is now frequently offered and combined with radiotherapy. Several studies have shown this to significantly improve median survival up to 20–30% in some series. Postoperative adjuvant chemotherapy has been used increasingly since the mid-2000s after several large randomized studies demonstrated prolonged survival and reduced recurrence rates with cisplatin-based regimens in radically resected stage II and IIIa NSCLCs.

Radiotherapy

Radiotherapy is used to good effect in NSCLC to manage immediate symptoms or to delay the time until symptomatic progression rather than in an attempt to cure (palliation) (see also Chapters 18 and 19). SVCO also often responds to radiation treatment. The

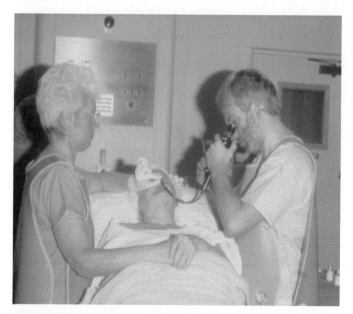

Figure 2.3 During bronchoscopy the patient can be either lying down or sitting upright. The procedure is undertaken with local anaesthetic to the throat only in most cases.

Table 2.1 TNM staging of non-small-cell lung cancer

Primary tumor (T)

TX	Primary tumor cannot be assessed, or tumor proven by the presence of malignant cells in sputum or bronchial washings but not visualized by imaging or bronchoscopy
T0	No evidence of primary tumor
Tis	Carcinoma *in situ*
T1	Tumor 3 cm or less in greatest dimension, surrounded by lung or visceral pleura, without bronchoscopic evidence of invasion more proximal than the lobar bronchus
T2	Tumor with any of the following features of size or extent: • More than 3 cm in greatest dimension • Involving main bronchus, 2 cm or more distal to the carina • Invading the visceral pleura • Associated with atelectasis or obstructive pneumonitis that extends to the hilar region but does not involve the entire lung
T3	Tumor of any size that: • directly invades the chest wall (including superior sulcus tumors), diaphragm, mediastinal pleura, or parietal pericardium • is located in the main bronchus less than 2 cm distal to the carina but without involvement of the carina • is associated with atelectasis or obstructive pneumonitis of the entire lung
T4	Tumor of any size that: • invades the mediastinum, heart, great vessels, trachea, esophagus, vertebral body, or carina • is associated with a malignant pleural effusion

Regional lymph nodes (N)

NX	Regional lymph nodes cannot be assessed
N0	No regional lymph nodes metastasis
N1	Metastasis in ipsilateral peribronchial and/or ipsilateral hilar lymph nodes, including direct extension
N2	Metastasis in ipsilateral mediastinal and/or subcarinal lymph node(s)
N3	Metastasis in contralateral mediastinal, contralateral hilar, ipsilateral or contralateral scalene or supraclavicular lymph node(s)

Distant metastases (M)

MX	Presence of distant metastasis cannot be assessed
M0	No distant metastasis
M1	Distant metastasis

Stage grouping

Occult	TX	N0	M0
Stage 0	Tis	N0	M0
Stage I	T1	N0	M0
	T2	N0	M0
Stage II	T1	N1	M0
	T2	N1	M0
Stage IIIA	T1	N2	M0
	T2	N2	M0
	T3	N0	M0
	T3	N1	M0
	T3	N2	M0
Stage IIIB	Any T	N3	M0
	T4	Any N	M0
Stage IV	Any T	Any N	M1

timing of palliative radiotherapy remains controversial, with no clear evidence that early initiation prior to the development of symptoms provides better quality of life or improved survival compared with waiting for symptoms to develop and treating at that point. Nevertheless, in most centers, early radiotherapy is initiated. Radical radiotherapy may occasionally be offered to patients with the intent of producing long-term survival. In particular, this option is favored for localized disease unsuitable for surgery, but the 5-year survival is only 10%. Continuous hyperfractionated accelerated radiotherapy (CHART) has been demonstrated to improve survival (by approximately 10%) compared with standard techniques. Improved survival has also been demonstrated with the addition of cisplatin to radiotherapy. Both approaches are now widely adopted.

Radiotherapy is frequently used in SCLC as this is one of the most radiosensitive of all tumors. Rapid reduction in tumor volume occurs in 80–90% of patients but recurrence is almost inevitable. The addition of radiotherapy to chemotherapy regimens appears to improve median and 2-year survival rates by approximately 5%. For patients with limited-stage disease, thoracic radiotherapy is widely used early during the administration of chemotherapy and prophylactic cranial radiation is also routinely offered with chemotherapy, reducing the frequency of brain metastases from approximately 40% of all patients to 8%. Local radiotherapy for the palliative treatment of painful metastases is of frequent value during the course of disease.

Chemotherapy

Chemotherapy is widely utilized in the treatment of all forms of lung cancer. A wide variety of combinations of agents are used in NSCLC as single-agent response rates are relatively low at 15–30%. The most commonly used agents are cisplatin, carboplatin, gemcitabine, paclitaxel, docetaxel, vinorelbine, topotecan, and ifosfamide (see Chapter 18). Many combinations have been compared and it appears that so long as a platinum agent is included, activity for two or more agents is similar with response rates of 40–60%. National guidelines suggest that standard of care should be with a platinum-containing doublet in patients who can tolerate platinum compounds.

Several targeted agents (see Chapter 16) have been examined in the management of NSCLC, and bevacizumab, the monoclonal antibody against vascular endothelial growth factor (VEGF), has produced significant survival benefits when combined with paclitaxel and carboplatin. It was found to cause increased rates of hemorrhage in squamous carcinomas and is therefore only approved for nonsquamous histology in patients without evidence of brain metastasis. As is the case for many solid and hematological malignancies, there is now increasing enthusiasm for the use of EGFR inhibitors (see Chapter 5). EGFR signaling is increased in 40–80% of lung cancers and two EGFR inhibitors – gefitinib and erlotinib – have been included in large randomized trials. Response rates of up to 20% have been seen with gefitinib in pretreated patients, but two large randomized trials of chemotherapy with or without gefitinib proved negative. Significantly better survival is seen in a subgroup of patients who have never smoked, are of Asian origin, have adenocarcinoma, and are female. Erlotinib has been shown to produce response rates of approximately 9% in patients who have been pretreated with chemotherapy.

Recent analysis suggests that mutations of the *EGFR* gene predict improved outcome and use of EGFR inhibitors is increasingly focused on this subgroup. There is also growing interest in many other targeted agents in NSCLC and this is in line with the increasingly prevalent philosophy of "tailored" or individualized medicine, where subgroups or even individuals are selected by molecular diagnostic techniques to identify those best suited to a given therapy.

For inoperable disease, the early introduction of chemotherapy appears to have significant, though modest, benefit in terms of median and 1-year survival and is associated with an improvement in the quality of life. Surprisingly, the early use of chemotherapy also appears to be cost-effective when compared with palliative care alone. The use of preoperative chemotherapy for those deemed surgically resectable is gaining increasing support. Large randomized trials comparing surgery alone with preoperative chemotherapy are ongoing as the results of published studies to date remain conflicting. One meta-analysis, however, has suggested an improvement of 1-year survival of approximately 5%.

Chemotherapy has a more established role in the management of SCLC. This disease is highly chemosensitive, with single-agent response rates of approximately 60% for carboplatin and etoposide and a slightly lower response rate of 30–40% for a wide variety of other agents. Limited-stage disease is usually treated aggressively with combinations of agents (including cisplatin or carboplatin for best results) and a median survival of 14–16 months is now achieved (compared with just 4 months without treatment). In fact, up to 10% may survive beyond 5 years. For those with extensive-stage disease, chemotherapy is offered for palliation and, as a result, regimens are chosen with less toxicity. Median survival rates of 8–10 months are usually achieved, with approximately 2% of patients alive at 2 years.

Breast cancer

Breast cancer is the most common malignancy in women in Western nations and the second most frequent cause of cancer death (estimated to account for 26% of cancer cases and resulting in 15% of cancer deaths). The highest frequency is in the United States with an annual incidence of 85 per 100 000. In Europe, it is commonest in the Netherlands with an incidence of 70 per 100 000/year and lowest in Spain with an incidence of 43 per 100 000/year. It is much less common in Asia, and Japan has an incidence of just 25 per 100 000. The incidence increased steadily during the twentieth century but has plateaued or reduced since then. The death rate from breast cancer decreased by approximately 24% in the United States since 1990 and similar reductions have been seen in Europe. On average, approximately one in nine women develop breast cancer in Western nations, and one in three of these will die of their disease.

Many potential risk factors for the development of breast cancer are known. A family history of carcinoma of the breast increases the risk, with one first-degree relative conferring a threefold increase and an even greater risk if the relative was premenopausal when diagnosed. Lifetime exposure to estrogen is an important risk factor, with breast cancer risk increased by early menarche or late menopause and by use of hormone replacement therapy after menopause. Women who have their first child over age 30 years experience a threefold increased risk. Previous benign breast disease, a prior exposure to radiation and high dietary fat intake all appear to increase the risk of breast cancer. A factor which has recently been shown to be an important predictor of breast cancer risk is the detection of increased mammographic breast density seen on screening.

The key role of inherited genetic factors in breast cancer was illustrated by the identification of mutations in the *BRCA* genes in patients with familial breast and ovarian cancer (see Chapters 3 and 10). Mutations of this gene are associated with a 50–85% risk of developing breast cancer during the lifetime of a woman. Other inherited mutations are being identified.

Pathology

The majority of breast cancers are adenocarcinomas, which may be either infiltrating lobular or ductal. With the increasing uptake of mammographic screening, more preinvasive tumors – ductal or lobular carcinoma *in situ* – are being detected. Rarely, primary lymphomas, sarcomas, and squamous carcinomas may be found in the breast.

Clinical features

Breast cancers vary considerably between individuals in their rate of growth and pattern of spread. Some may remain predominantly localized, infiltrating local structures, others may spread via lymphatics to draining lymph node areas, while others may disseminate widely in the bloodstream. The majority of women will present with a lump in the breast and some will have pain in this region, discharge, or bleeding from the nipples. Increasing numbers of women are being diagnosed following screening and will have no symptoms (Fig. 2.4). In some cases, local infiltration can produce a fungating tumor with erosion of the skin (Fig. 2.5). If fungation does not occur, there may be widespread skin involvement leading to thickening and a typical pattern of irregularity termed "peau d'orange" (Fig. 2.6). In this instance there is local lymphatic infiltration and obstruction leading to edema.

Presentations may occur, less commonly, from sites of metastases and include the detection of an enlarged lymph node or symptoms from distant metastases to bone, brain, or liver. These may cause pain and neurological symptoms and signs. As with all malignancies, nonspecific symptoms, including anorexia and lethargy, may predominate.

Diagnostic procedures and investigations

Physical signs associated with breast cancer include the presence of a firm irregular mass and fixation to skin or deep structures such as the chest wall. Mammography is widely used in the initial assessment of lumps thought to be possible breast cancer; the typical X-ray appearances are fine calcification with areas of irregularity. Ultimately, diagnosis depends on a histological or cytological assessment of a specimen from the mass. Fine-needle aspiration usually provides sufficient cells for cytological assessment but a core biopsy is often obtained using a percutaneous biopsy needle which will yield sufficient tissue for histological assessment. Occasionally, the diagnosis remains uncertain and an excision biopsy may be required.

In breast cancer, the importance of obtaining cells extends well beyond that of simply confirming the diagnosis of malignancy. It is also now an exemplar of the great potential of molecular phenotyping of cancer tissues for disease subclassification and also for individualizing therapeutics. The first successful application for molecular diagnosis in breast cancer was the detection of the presence of hormone receptors – both progesterone and estrogen. This is already exploited in clinical practice, because positive receptor status correlates with the likelihood of benefit from the use of hormone manipulation therapy, as exemplified by

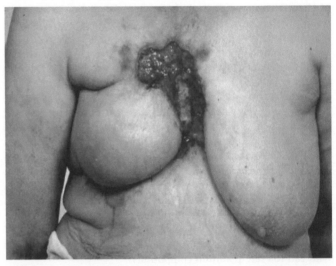

Figure 2.5 Fungating breast carcinoma. The primary tumor grew locally and eventually invaded the skin, causing local fungation.

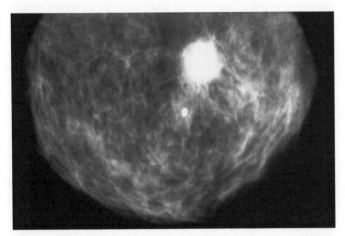

Figure 2.4 Calcification seen in mammography image of breast of patient who subsequently underwent excision of carcinoma of the breast.

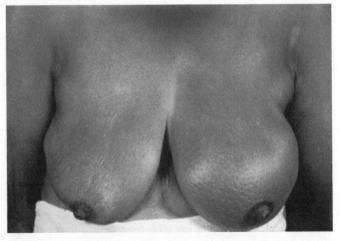

Figure 2.6 Typical features of "peau d'orange" of the breast, with thickening of the skin from edema and puckering of the skin. A large underlying tumor could be felt.

Table 2.2 TNM staging system for breast cancer

Stage	Tumor description
T1	Tumor less than 2 cm in diameter
T2	Tumor 2–5 cm in diameter
T3	Tumor more than 5 cm
T4	Tumor of any size with direct extension to chest wall or skin
N0	No palpable node involvement
N1	Mobile ipsilateral node(s)
N2	Fixed ipsilateral nodes
N3	Supraclavicular or infraclavicular nodes or edema of arm
M0	No distant metastases
M1	Distant metastases

tamoxifen. The molecular taxonomy of breast cancer now also includes detecting the presence of oncogenic HER2 receptor (found in 20–30% of invasive breast cancers). HER2 status not only improves prognostic information as it is associated with increased risk of disease recurrence and reduced survival, but importantly also clearly defines a subset of patients who will benefit most from use of therapies targeting this protein, such as trastuzumab. The importance of these tissue markers is emphasized by the description of those breast cancers that do not express either hormone receptors or HER2 mutations as triple negative breast cancers. It is likely that other molecular markers will be increasingly employed to assist choice of patients for other drugs, such as PARP inhibitors (see also discussion of synthetic lethality in Chapter 10). The use of genetic testing in familial cancer syndromes is discussed elsewhere.

Treatment decisions are based on the histological grade and size of the primary tumor, together with the presence or absence of metastases. The vast majority of patients will undergo resection of the primary tumor, and preoperative assessment of the extent of disease varies between centers. Basic investigations, including chest X-ray, liver function tests, and full blood count, would be considered standard and provide an indication of possible metastases. Most centers also perform abdominal imaging to exclude liver metastases. This may be with CT or ultrasound scanning. The presence of bone pain with elevated serum calcium or alkaline phosphatase indicate the possibility of bone metastases that can be detected with isotope bone scans. Imaging of the brain should only be performed routinely in patients who complain of symptoms suggesting possible cerebral metastases. The TNM staging system is widely used alongside the histological grade of tumor and receptor status when deciding on appropriate therapy (see Table 2.2).

Treatment

Surgery

Surgical resection of primary breast tumors aims to control local disease and prevent recurrence in regional draining lymph nodes. Our understanding of the biology of breast cancer is a major example of how clinical and fundamental research have modified a therapeutic approach. Until relatively recently, the surgical approach to breast cancer was to undertake a radical operation with removal of the breast, block dissection of the axillary nodes, and often removal of underlying muscle. Radical mastectomy is a mutilating and disfiguring operation that often led to marked long-term morbidity. However, once it was recognized that blood-borne metastases and not lymphatic and local spread were the main causes of dissemination, the role of surgery was adapted. In particular, since most patients die from metastases that may have been "seeded" prior to the opportunity for surgery, one might expect only a limited benefit from aggressively removing the primary tumor, surrounding tissues, and lymph nodes.

Today, surgery is less radical, with the widespread adoption of breast-conserving surgical approaches, such as lumpectomy (removal of the local tumor alone) when possible. The choice of surgery is affected by the size of the primary tumor and of the breast (large tumors in small breasts are often removed with better cosmetic results using a simple mastectomy than with lumpectomy) and the presence of multiple tumors within the same breast or widespread intraduct carcinoma (both providing indications for mastectomy rather than lumpectomy). Surgical reconstruction is now increasingly used for patients who have undergone mastectomy. With modern techniques, this can produce excellent cosmetic results but one concern with immediate reconstructuion is that if subsequent radiotherapy is required, increased cardiac irradiation may occur in left-sided cancers.

Radiotherapy

Conservative surgery alone is followed by local recurrence in approximately 30% of cases, but can be reduced to just 5–10% by employing postoperative radiotherapy. Recurrence is more common with tumors >5 cm in diameter or in patients with axillary node disease. The results from the United States (where radical mastectomy is still more widely used than in Europe) suggest local recurrence rates following radical surgery alone are 4–14%, indicating the equivalence with breast-conserving surgery followed by radiotherapy. Likewise, survival appears to be identical in the two groups and these results strongly support the use of breast-conserving surgery with postoperative radiotherapy whenever technically possible. Interestingly (and perhaps not surprising when considering the biology of the disease), the vast majority of randomized studies have shown no effect on survival of postoperative radiotherapy, despite a reduction in local recurrence rates. A meta-analysis (pooled statistical analysis of multiple single studies) which included 8000 patients did suggest a slight, but significant, improvement in survival for the radiotherapy arm, but these results must be interpreted with some caution, particularly as no single prospective study has demonstrated this. Meta-analyses are particularly prone to bias, because negative studies may not have been reported in the literature (and thus self-evidently are not available for the meta-analysis) and because there may be considerable variation between groups of patients in the different studies included.

For patients with large primary tumors (particularly those >5 cm or with direct extension to the chest wall or skin) and for a large proportion of patients with clear nodal involvement, surgery appears to have little role other than occasionally to debulk the tumor for symptomatic benefits. For such patients, radiotherapy has an important role, producing high local control rates (80–90%). It is usually given in combination with hormones and/or chemotherapy.

As with all malignancies, radiotherapy can be particularly useful for patients who have developed painful areas of metastasis. This is particularly the case with bone lesions, where symptomatic benefit is achieved in the majority of patients (Fig. 2.7). Between 15% and 20% of patients will develop cerebral

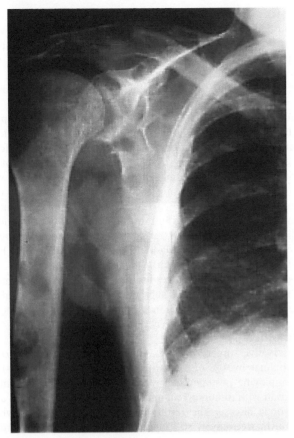

Figure 2.7 Plain X-ray showing extensive lytic metastases in the humerus of a patient with carcinoma of the right breast. This caused severe pain, requiring local radiotherapy and subsequent bisphosphonate therapy. Provided by Natalie Mahowald.

metastases, and radiotherapy can reduce tumor volume and improve symptoms in the majority of these patients. Likewise skin, lymph node, and other areas of metastasis can benefit from local radiotherapy. One final role for radiotherapy is the induction of menopause with pelvic irradiation. This appears to be as valuable as oophorectomy for the relief of symptoms from metastatic disease if patients are premenopausal.

Chemotherapy

Breast cancer is a relatively chemosensitive disease, with numerous single agents producing responses in 30–50% of patients. The anthracylines, taxanes, alkylating agents, and vinca alkaloids are most widely utilized, generally in combinations. More recently, capecitabine and gemcitabine have been used in prospective trials, commonly in combination with paclitaxel. A large proportion of patients will receive chemotherapy following a diagnosis of breast cancer, in either the adjuvant (immediate postoperative) or the metastatic setting. Several trials are examining the role of chemotherapy in the preoperative, neoadjuvant setting. This approach attempts to treat any micrometastatic disease early and potentially reduce the need for radical surgery. It now has an established role in some patients with highly aggressive locally advanced and inflammatory breast cancers. There is a proven role for chemotherapy in the adjuvant and metastatic settings.

The role of adjuvant therapy is based on the observation that even at relatively early stages of development, a proportion of patients with breast cancer already have micrometastases. In fact, when taken together with the demonstration of large numbers of circulating tumor cells (CTCs) in a patient's blood, and the demonstration of metastases-related gene expression in early primary tumors, which are present before overt metastatic disease, these observations are having a profound impact on how we think about metastases (see Chapter 12).

Even with stage I disease, metastatic breast cancer cells can be detected within the bone marrow by a variety of techniques, including gene expression analyses, and the number of such cells rises with advancing stage. Although it is likely that many of these cells would not have long-term viability, clearly in a significant proportion of patients they are responsible for local systemic relapse and death. The use of systemic therapy at an early stage is now widely adopted, with the aim of eradicating these cells whenever possible and prolonging relapse-free and overall survival. Since the introduction of adjuvant therapy there has been a steady decrease in the death rate from breast cancer, suggesting a significant impact on public health. Numerous trials have been performed in different countries and by many cooperative groups. These have all shown a reduction in both the risk of recurrence and of death. The effect of chemotherapy is more marked in younger age patients, particularly those below the age of 50. In fact, for premenopausal women, the annual odds of recurrence are decreased by approximately 35% and of death by 25–30%. The benefits of adjuvant chemotherapy appear to be largely confined to patients with node-positive disease and high-grade tumors.

Adjuvant chemotherapy is now routinely recommended for premenopausal women with stage II disease. The decision as to whether or not to offer adjuvant chemotherapy is based on the potential absolute gain for each patient, bearing in mind risk factors and comorbidity. A widely used online software program that quantifies the benefits of adjuvant chemotherapy is the imaginatively named "adjuvant" package.

Combination chemotherapy is offered to most women with metastatic disease, particularly premenopausal node-negative patients and patients who have failed hormone therapy. Although response rates tend to be high (50–60%) with improvement of symptoms and quality of life, prolongation of survival, is, at best, modest. Thus, potential benefits must be carefully weighed against toxicity. The availability of trastuzumab for tumors overexpressing the HER2 receptor (see Chapter 5) does prolong survival and is usually given in combination with chemotherapy.

A subgroup of patients (more often younger women, those with *BRCA1* mutations, and of African-American and Hispanic origin) may have what has become known as "triple negative" breast cancer. In other words, these tumors do not express the genes for estrogen, progesterone, or HER2/NEU receptors. This subtype has a more aggressive clinical course with a poor prognosis and rarely responds to hormonal therapy. Chemotherapy can effect a response but this is usually of short duration. Recently the PARP inhibitor BSI-201 has shown promise (see also Chapters 10 and 16).

A variety of other targeted drugs are now approved or under investigation. Lapatinib is a dual-kinase inhibitor that targets both the HER2 and EGFR tyrosine kinase signaling pathways (see Chapter 5). It has been approved in the treatment of advanced breast cancer. Bevacizumab, a humanized monoclonal antibody that targets VEGF, has also been approved in combination with chemotherapy for treatment of advanced disease (see Chapter 14).

Hormone therapy

It has been recognized for over 100 years that hormone manipulation can reduce tumor volume in patients with metastatic or locally advanced breast cancers. More recently, it has been found that estrogen receptor-positive cancers are most likely to respond to hormone manipulation. Knowledge of progesterone receptor status adds to the ability to predict response to endocrine therapy: 10% of those who have negative status of both receptors will respond, whereas 70% of those who have estrogen and progesterone receptor positivity will respond. Between 30% and 50% of patients who have either estrogen or progesterone positivity obtain a response to hormone manipulation. For many years, the anti-estrogen tamoxifen has been the most widely used means of altering hormone activity in tumor cells. It may also have a direct cytotoxic effect. More recently, aromatase inhibitors (e.g. anastrozole) have been widely used, particularly as they are not associated with the development of uterine malignancies, which are a rare complication of tamoxifen therapy.

Several studies and a key overview of all prospective randomized trials published in 1992 demonstrated the clear benefit for tamoxifen given in the adjuvant setting. The annual odds of death appear to be reduced by approximately 30%, with 40–50% reductions in the annual odds of recurrence. Both node-negative and node-positive patients appear to benefit, as do premenopausal and postmenopausal women. The optimum duration of treatment remains unclear but studies have shown that at least 5 years of therapy appears superior to shorter durations, and current trials are exploring the role of longer durations.

For patients with metastatic disease, hormone therapy is appropriate if receptor status is positive. For patients who are premenopausal, radiation-induced menopause or the use of luteinizing hormone-releasing hormone (LHRH) antagonists (e.g. goserelin) are often utilized, whereas tamoxifen may be the preferred first approach in postmenopausal women. Between 30% and 50% of patients will respond to first-line hormone manipulation and there is increasing evidence that the early use of aromatase inhibitors may be superior to tamoxifen, providing longer disease-free survival and reduced toxicity. Hormone therapy is preferred prior to chemotherapy for patients with receptor-positive disease even though the response rate is lower because toxicity is also less and response more durable. Chemotherapy is introduced when hormone therapy fails or first-line if the cancer may cause early complications and where a more rapid response is required.

Colorectal cancer

Malignancies arising in the large bowel are the second most common cause of cancer-related deaths in the Western world. The incidence rate varies markedly between nations, being lower in Africa and Asia than in Western Europe and North America. The annual incidence varies between 30 and 60 per 100 000 across Western nations with overall 5-year survival rates being approximately 50%. The age-adjusted incidence rates have remained relatively constant over the past 30 years, although the number of cases has risen because the world's population is increasing and living longer. Sixty percent of these tumors arise in the colon and 40% in the rectum. Tumors arise more frequently in distal sites of the large bowel, providing evidence for the suggested association of dietary intake factors and risk of colorectal cancer. As bowel contents move distally, they become more solid and their transit time is lengthened. They thus have a greater likelihood of more prolonged intimate contact with the bowel epithelium. It is therefore suggested that if potential carcinogens are present, they are more likely to have an effect in the distal bowel, explaining the greater propensity of this region to develop malignancies.

More is known about the development of colorectal cancer than any other malignancy. In fact, colon cancer has proved a paradigm for the development of "multistage" theories of carcinogenesis (see Chapter 3). Several clear stages of progression from normal epithelium through early adenomas, late adenomas, early carcinoma, and ultimately to metastatic disease have been described and many of the key genetic lesions responsible have been identified. There is a clear association between family history of colon cancer and the likelihood of developing this disease, particularly for an individual with more than one first-degree relative diagnosed prior to the age of 40 years. Several familial syndromes are known to be associated with an increased risk of developing colorectal cancer, and these include Peutz–Jeghers, Lynch and Gardner's syndromes.

Familial adenomatous polyposis coli (FAP) has been widely studied and is a clear example of an autosomal dominant inherited malignancy. Inherited loss of a single allele of the *APC* genes is inevitably followed by loss of the single remaining allele in some colon cells during adulthood, thereby removing the tumor suppressor function. Multiple polyps develop throughout the bowel, predominantly in the distal colon. There is an inevitable conversion to malignancy and a panproctocolectomy (removal of the entire colon and rectum) is recommended prophylactically when an individual is found to have inherited FAP. Sporadic polyps may also occur in individuals and will convert to malignant lesions in approximately 15% of cases. Larger lesions are associated with a greater risk and prophylactic removal at the time of colonoscopy is generally undertaken. There is increasing information becoming available on inherited risks of nonpolyposis malignancies of the colon (HNPCC) (see Chapters 3, 10, and 11).

Several chromosomal changes have been described as being associated with a risk of colorectal cancer. The familial polyposis gene is located on chromosome 21 and a loss of an allele at this site can be found in up to 40% of cases of sporadic carcinoma of the colon. Loss of an allele of the *P53* gene on chromosome 7 and of the *DCC* gene (deleted in colorectal carcinoma) on chromosome 18 have also been implicated in colorectal cancer risk. Mismatch repair genes form the genetic basis for hereditary nonpolyposis colon cancer and families with this predisposition appear to have an excess risk of many other adenocarcinomas apart from colorectal cancer. To date, at least four separate mismatch repair genes which contain mutations have been implicated in the etiology of hereditary nonpolyposis colon cancer. These include *hMSH2*, *hMLH1*, *hPMS1*, and *hPMS2* (see Chapter 10).

Several other factors have been linked with the risk of developing colorectal cancer. Dietary intake of fat and fiber appear to play a role and ulcerative colitis, particularly when the onset is at a young age, is associated with a significant risk of the development of malignancies.

As with other carcinomas, much heated debate is generated by raising the topic of "cell of origin," with the jury probably favoring a verdict of "stem cell."

Screening

Colorectal cancer screening is increasingly being instituted in many countries. Those at "normal" risk are generally offered screening after the ages of 50 or 60 (varies between nations). A fecal occult blood test is usually performed initially and endoscopy offered to those with positive results. Polyps can be removed at this procedure and potentially malignant lesions can be biopsied. Colorectal cancer screening is significantly reducing the incidence of advanced stage colorectal cancer and could have a major impact on malignancy from this disease in the next 10 years.

Diagnostic and surgical procedures

The diagnosis of colorectal cancer is usually relatively straightforward once the suspicion is raised and the patient has been referred to a specialist center. Simple rectal examination will detect approximately 75% of all tumors within the rectum (i.e. almost 30% of all large bowel tumors). With the advent of high-quality flexible endoscopes, flexible sigmoidoscopy or colonoscopy are generally the first investigation of choice. Sigmoidoscopy is rapid and will examine the distal 25–30 cm of large bowel, the site of approximately 60% of all tumors. Colonoscopy allows visualization of the entire large bowel through to the cecum, but is a more lengthy and costly investigation. However, the advantages are considerable – ease of use and speed together with the high sensitivity and ability to biopsy any abnormal lesions. Colonoscopy also allows the removal of polyps and the detection of other sites of disease. There is a small risk (approximately 0.01%) of serious complications – usually hemorrhage or perforation). Barium enemas are also widely used and allow visualization of the entire large bowel. A mass may be identified or a stricture which has typical features in a malignancy (Fig. 2.8). The disadvantage of barium investigations is the discomfort they produce and the fact that they will often miss lesions on the right side of the colon. In addition, histological confirmation is not obtained.

"Virtual" endoscopy is used in some centers and involves CT scan imaging of the colon. Results to date suggest high sensitivities for malignant lesions from this procedure.

As with all tumors, histological confirmation of malignancy is essential as abnormalities such as diverticular disease or ulcerative colitis can mimic malignancy both in their clinical presentation and, occasionally, on their appearance at investigation. The vast majority of tumors are adenocarcinomas but, rarely, other tumors can present in the large bowel and include carcinoids, sarcomas, and lymphomas.

Staging

Given that the "gold standard" is to surgically resect the primary tumor in order to cure the patient, it is important to first determine the extent of local disease and the presence of any possible metastases. A relatively simple staging system is the widely utilized Dukes' system (Table 2.3). While the Dukes' system has considerable clinical value in determining prognosis and indicating optimal therapy, it is widely recognized that it provides only a crude assessment of local involvement, particularly with Dukes' B and D disease. A more refined staging system has been developed by the American Joint Committee on Cancer (AJCC staging system) and this provides information on the degree of local involvement in the bowel wall and the number of regional lymph nodes involved.

Colorectal cancer usually metastasizes to the liver but may also disseminate to lung and para-aortic lymph nodes. Distant metastasis is more common with colon cancer than rectal malignancies,

Table 2.3 Dukes' staging system for colorectal cancer

Stage	Description of tumor extent
A	Tumor confined to bowel wall (mucosa and submucosa)
B1	Tumor penetrating the bowel wall involving muscularis propria (but not penetrating through it)
B2	Tumor penetrating through the bowel wall
C	Tumor extending into local lymph nodes
D	Distant metastases

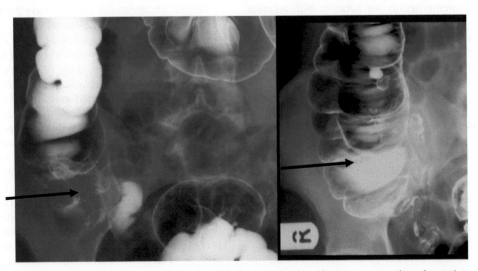

Figure 2.8 Barium enema image of patient with carcinoma of the colon. Note the area of reduced barium contrast where the carcinoma extends into the colon.

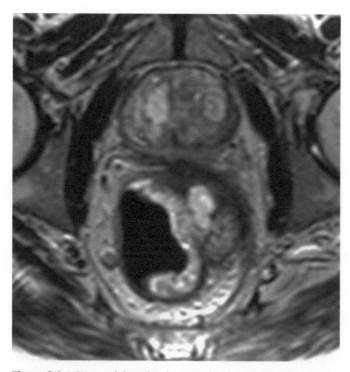

Figure 2.9 MRI scan of the pelvis showing a large rectal tumor invading into the lumen of the rectum. Anatomical detail is excellent, showing the relationship of the tumor to other organs in the pelvis.

whereas the latter will often invade locally into adjacent structures. For rectal tumors, high-quality imaging of the local extent of disease is beneficial when making a decision about surgical techniques and magnetic resonance imaging (MRI) scans of the pelvis are usually performed (Fig. 2.9). CT imaging of para-aortic nodes and ultrasound or CT of the liver is usually recommended. Imaging of the thorax is also important and CT scanning of this region is becoming a standard.

Routine biochemistry and blood count, together with measures of the tumor biomarker carcinoembryonic antigen (CEA), can provide useful information, particularly by giving an indication of possible metastases at the time of presentation or in subsequent follow-up to help screen for recurrent disease. It was widely hoped that CEA would turn out to a specific marker of colorectal malignancy, but it is elevated in many other conditions and its specificity is therefore low. It is, however, useful to obtain a preoperative measure; if it is elevated, a subsequent fall can be useful in monitoring response to treatment. PET scanning is now becoming widely used when routine CT or MRI scanning suggest the possibility of metastatic disease. It can allow the discrimination between malignant and benign abnormalities in many instances. Progress is being made in identifying other biomarkers that may assist in determining prognosis and in selecting patients for postoperative chemotherapy, including the presence of mutations in *KRAS* or *BRAF*. These are described later in the book (see Chapters 3 and 16).

Clinical manifestations

Colorectal cancer produces a variety of symptoms, the predominant nature of which depends on the site of the tumor. Right-sided tumors arise where the bowel contents are fluid and therefore tend not to produce obstruction early in the course of their growth. They are often associated with chronic hemorrhage (bleeding), which leads to symptoms of anemia but may also cause pain or a detectable abdominal mass. A change in bowel habit occurs in approximately 50% of patients and nonspecific symptoms such as anorexia and weight loss are common. In the left side of the colon, the bowel contents are more solid with a slower transit time. As a result the tumors more readily cause obstruction with subsequent colicky abdominal pain and vomiting. Altered bowel habit occurs in 60–70% of patients and blood may be noticed with the stools. Abdominal pain is common. Rectal tumors are more likely to be associated with obvious blood loss (in over 60% of patients) and altered bowel habit is almost universal. All tumors may be associated with nonspecific symptoms such as lethargy, anorexia, and weight loss and all may cause symptoms from metastases if present.

Many groups have attempted to develop scoring systems for risk factors of malignancy. Altered bowel habit, rectal bleeding, and anemia (particularly in women) are not uncommon and none are specific for a bowel malignancy. It is clearly important to balance the costs and risks of investigation against the risk of missing the diagnosis of a malignancy when a patient complains of symptoms that could indicate colorectal cancer.

Management of colorectal cancer

High-quality surgery by experienced surgeons is essential to optimize the chance of cure in colorectal cancer. Excision of the primary tumor together with the involved segment and draining lymph nodes should be undertaken and the exact extent of bowel removed depends on the site of the primary lesion. For some patients who have obstruction but in whom there is disease that is either too locally extensive for resection or who have widespread metastases, a defunctioning colostomy may be performed to control symptoms. Rectal tumors are technically more difficult to remove and require great expertise to minimize the risk of nerve damage and attempt sphincter preservation. Local recurrence is much more common with rectal cancers than for tumors in the colon and it is clearly essential to remove as wide an area of normal tissue as is possible around the primary rectal cancer to reduce this risk. Several centers now use laparoscopic surgical resection instead of open surgery for colon cancer and have demonstrated equivalent results with less morbidity and shorter inpatient stay.

Surgery is increasingly employed for patients with liver metastases. The mortality from this procedure has fallen dramatically and encouraging long-term survival rates (up to 30%) have been demonstrated in specialist centers. There is increasing experience of surgical removal of lung metastases and again long-term survival may be seen in such cases.

Radiotherapy

The role of radiotherapy in the management of large bowel tumors is relatively limited. Local recurrence is very unusual following surgery so there is no role for perioperative radiotherapy. However, radiotherapy has been shown to reduce local recurrence in rectal tumors. Survival rates have been higher in several series where radiotherapy has been added to surgery but these have not usually reached the levels of significance. Debate continues on whether to offer preoperative or postoperative radiotherapy and there is wide geographic variation in the choice of these two approaches. Chemotherapy is frequently added to radiotherapy (chemoradiotherapy) in this setting.

Chemotherapy

Unfortunately, colorectal cancer is one of the more chemo-resistant malignancies and until the late 1990s, only 5-fluorouracil (5-FU) was widely used in the treatment of this disease. Its response rate when modulated by folinic acid is modest at 20–30%. Oxaliplatin and irinotecan became available over the last decade and have been shown to significantly increase the activity of 5-FU. The oral fluoropyrimidine capecitabine has been demonstrated to show equivalence in terms of survival compared with 5-FU and is more acceptable to patients. In many centers it has replaced intravenous 5-FU as the fluoropyrimidine of choice. Chemotherapy is administered postoperatively in the adjuvant setting for patients with Duke's C disease (i.e. disease involving lymph nodes) and in many centers is often administered to patients with Duke's B disease. Randomized prospective trials have shown an approximately 30% relative risk reduction in the death rate from colorectal cancer with the use of adjuvant 5-FU-based therapy.

A large international randomized trial (X-ACT) demonstrated equivalence of oral capecitabine for intravenous 5-FU in this setting and another large international trial (MOSAIC) demonstrated superiority of oxaliplatin with 5-FU in the adjuvant setting, with highly significantly superior relapse-free survival when compared with 5-FU and folinic acid. Oxaliplatin, 5-FU, and folinic acid (or oxaliplatin with capecitabine) have become the standard of care in the adjuvant setting for patients with high-risk Dukes' B and Dukes' C disease if the patient is able to tolerate this regimen, and single-agent capecitabine is usually offered to those patients unable to tolerate the toxicity induced by this combination. Early data from the X-ACT international trial has shown at least equivalence for the substitution of oral capecitabine as adjuvant therapy and the preliminary analysis from the MOSAIC trial has suggested that the combination of oxaliplatin with 5-FU and folinic acid may produce superior relapse free and overall survival rates compared with 5-FU and folinic acid alone. More mature data will be necessary to confirm these promising early results.

The treatment of metastatic colorectal cancer has improved markedly in the last decade. The median survival for those with advanced disease was approximately 8 months with single-agent 5-FU (with folinic acid), but the use of oxaliplatin or irinotecan combined with a fluoropyrimidine used in the first- and second-line setting have improved median survival to 18–20 months. The use of chemotherapy at the time of establishment of metastases confers a survival advantage compared with waiting until the development of symptoms before commencing treatment. Oral capecitabine as a single agent is widely used for metastatic disease, particularly for patients who are unable to tolerate more aggressive regimens. This agent is also increasingly combined with oxaliplatin and irinotecan. With the advent of inhibitors of VEGF, survival was further improved and the combination of bevacizumab with oxaliplatin and a fluoropyrimidine improves the median survival to over 2 years. Cetuximab (Erbitux) is a monoclonal antibody directed against the EGF receptor and this improves the response rate when combined with chemotherapy (particularly irinotecan based). Some studies have shown improvements of survival with the addition of cetuximab to chemotherapy and it appears to have a role in the preoperative setting when added to oxaliplatin-based regimens to improve the resection rate for liver metastases. Its use is restricted to those patients with KRAS wildtype disease.

Carcinoma of the prostate

There are wide geographic variations in the incidence of prostate cancer. The United States has among the highest rates at 180 cases per 100 000 population, whereas the incidence in Asian countries is much lower (only 8 per 100 000 in Korea). In some parts of China it is a relatively rare disease. Mortality rates also vary considerably with fewer than 10% of patients dying from their disease in the United States but over one third dying in Caribbean nations. The incidence of prostate cancer increased markedly during the 1970s and the 1980s, but has declined since the early 1990s. In parallel, 5-year cancer survival rates have increased since the 1980s. It is likely that a significant contribution to these figures relates to increased diagnosis with the use of prostate-specific antigen (PSA) screening such that earlier stage disease is being detected.

Little is known about the etiology of prostate cancer, although there is some evidence that testosterone levels and androgen receptor gene activity may be related to the risk of developing this disease. There is, however, likely to be an environmental contribution in that migrant studies have demonstrated that men who move from a low-risk country to an area of high risk increase their incidence and mortality from prostate cancer within a generation. No specific gene abnormalities have consistently been related to prostate cancer risk.

PSA is a useful biomarker that is usually elevated in this disease and is increasingly used as a screening tool. There is a good correlation between extent of disease and the level of PSA found in the serum. Approximately 70% of cancer cases can be detected using a PSA cut-off level of 4 ng/ml. Recent evidence suggests that the PSA velocity, which measures changes in PSA values over time, may provide useful information and increases the specificity of a single PSA measurement. PSA testing is repeated at intervals of no less than 18 months with a comparison of changes in levels. A rise of 0.75 mg/ml per year increases the sensitivity of PSA testing alone.

Pathology

The commonest form of malignancy to arise in the prostate is adenocarcinoma although, more rarely, squamous and small-cell carcinomas may arise as may transitional cell carcinomas. The histological grade of the tumor has an impact on outcome. Patients with low-grade tumors survive significantly longer than those with high-grade disease. The Gleason system is widely used to describe the degree of differentiation and pattern of glandular histology. Grades I–V are described and correlate well with outcome. Additional information can be obtained from TNM staging. Diagnosis requires histological assessment of the tumor and increasingly this is obtained using transrectal biopsy.

Clinical features

A large proportion of patients with prostate cancer are asymptomatic until late in the course of their disease. Interestingly, autopsies performed for other reasons (e.g. following road traffic accidents) on individuals over the age of 50 years have revealed carcinoma in the prostate in up to 30% of men. This indicates the prolonged period of development of this disease in many individuals. Once the prostate has enlarged to the point of reducing bladder outflow, typical symptoms of hesitancy of micturition, nocturnal frequency, and occasionally hematuria may be experienced. Metastatic disease is often the cause of presentation and the most common site is in the bones, where metastases are often sclerotic in nature. Bone pain, most frequently in the back and pelvis, may occur.

Diagnostic and staging procedures

Serum PSA levels (and increasingly PSA velocity) are excellent early indicators of the presence of prostate cancer and should be confirmed by histological assessment of a biopsy. Isotope bone scans are frequently used to determine the extent of metastatic bony involvement (although they are not useful in low-risk patients with baseline PSA levels lower than 10 mg/ml). Abnormal areas may be assessed by plain skeletal X-rays or MRI imaging. Renal function should be assessed and should include intravenous pyelography to determine whether there is hydronephrosis (enlargement of the kidneys from outlet blockage). The extent of local disease, especially the presence of nodal involvement, can be determined by CT or MR imaging. Transrectal ultrasound may also provide useful information on the extent of disease. Increasingly, PET scanning is becoming a standard approach to determine disease extent and, in particular, the presence of lymph node metastases.

Treatment

If disease is localized to the prostate gland, many centers will take an approach aimed at cure. Radical prostatectomy or radical radiotherapy may be offered and different centers will often predominantly utilize one or other of these approaches. In general, surgical approaches are more widely utilized in North America than Europe. Radiotherapy generally involves external beam treatment using high doses. An alternative used increasingly is the use of ^{125}I seeds placed interstitially in the prostate. Cryosurgery and brachytherapy are also being used in several specialized centers. For larger tumors, some centers may offer local radical therapy after tumor volume reduction with hormone intervention.

There has been enormous debate as to the optimal management of localized disease, with several studies suggesting similar long-term survival rates from both approaches. Both approaches have significant complication rates and for surgery this involves impotence and urinary incontinence in approximately 10% of patients. Radiotherapy can result in urethral stricture and reduction in bladder capacity. There is a lower rate of impotence.

Controversy remains as to the optimal management of screen-detected early prostate cancer, with some centers recommending radical intervention and others following a "watch and wait" policy (see Chapter 1).

Hormonal therapy is of value for patients whose disease is inoperable and for those with metastatic disease. Several approaches to manipulating circulating hormone levels are available and many have been offered for several decades. Orchidectomy is often used as an alternative to pharmacological intervention. A variety of estrogens are available and are the most longstanding pharmacological agents. Gonadotropin-releasing hormone analogs are now widely used as an alternative to systemic estrogens or orchidectomy and are generally well tolerated. These interfere with gonadotropin release, leading to a fall in circulating testosterone. Goserelin is a depot form of gonadotropin-releasing hormone analog and is given on a monthly basis, leading to rapid reduction in circulating testosterone levels. Anti-androgens are now widely used and bind to the androgen receptor to serve as a competitive antagonist for androgens. Of these agents, flutamide is perhaps the most widely utilized. A feedback rise in circulating testosterone occurs with use of these agents and they are usually used in conjunction with androgen deprivation using agents such as 5-alpha reductase inhibitors.

Chemotherapy has a relatively limited role in the management of prostate cancer but relatively recently survival benefits have been demonstrated in patients with metastatic disease. Docetaxel is the only agent that prolongs survival in this setting and it is frequently used in conjunction with steroids. Bisphosphonates are increasingly used and have been shown to reduce skeletal events such as pathological fractures. Widespread bone pain may also be eased with the use of strontium-89.

Renal carcinoma

Carcinoma involving the kidney accounts for approximately 3% of adult malignancies in Western nations with a male predominance. There has been a steady increase in incidence over the past three decades with a rise of 43% since the mid-1970s.

Causes of renal carcinoma include cigarette smoking, which has been estimated to contribute to almost a third of all cases. Obesity and analgesic abuse are also risk factors, as is long-term exposure to high levels of industrial solvents. There are also some rare heredity forms of renal carcinoma including von Hippel–Lindau (VHL) syndrome.

Renal carcinoma can produce pain, hematuria, and a mass detectable in the flank. However, a large proportion of patients may not notice any symptoms until late in the course of the disease. Almost a third of patients will present with metastatic disease. Some patients may present with symptoms of anemia as a result of hematuria.

Pathology

There are a number of histological subtypes of renal carcinoma, with clear cell being the most common, making up 75% of cases. Other subtypes include papillary, chromophobe, and oncocytoma.

Diagnosis

Diagnosis is usually made from imaging with CT, MRI, or ultrasound scanning. CT is the most useful modality and is able to combine examining the extent of local disease with evaluation for potential metastases. As imaging is frequently undertaken for a variety of reasons, incidental carcinomas of the kidney are increasingly detected. The prognosis for such patients is usually more favorable than for those who present with symptoms.

Treatment

Radical resection is the most common procedure undertaken with the aim of curing this disease. Laparoscopic nephrectomy is increasingly used and reduces morbidity and inpatient stay. Once metastatic disease is present, systemic therapy with interferon-alpha may be used and the benefit appears to be enhanced by a palliative nephrectomy. Unfortunately, renal carcinoma is highly refractory to chemotherapy and this has a minimal role in the management of metastatic disease. Recently, however, several new approaches have become available and include inhibitors of angiogenesis using bevacizumab, sunitinib, and sorafenib. Temsirolimus, an inhibitor of mTOR, has also been shown to statistically prolong survival compared with interferon-alpha and has obtained regulatory authority approval in this setting.

Skin cancer

Malignancies arising in the skin are common but the vast majority are curable and therefore mortality is very low. A variety of carcinogens implicated in causing skin cancers have been described, including arsenic, chimney soot, sunlight, and ultraviolet and ionizing radiation. Patients who have received kidney transplants have a markedly increased risk of all forms of malignancy and, in particular, of squamous carcinomas of the skin, basal cell carcinoma, and melanomas. There are a variety of inherited disorders associated with skin malignancies, including xeroderma pigmentosum, which involves a deficiency in the DNA excision repair mechanism (Chapter 10). Such individuals have an increased sensitivity to UV light, leading to the development of malignant skin tumors on exposed areas. Ultraviolet light has been shown to be linked with causing mutations in the *p53* gene and to cause defects in nucleotide excision repair genes. Basal cell carcinomas have been linked to specific gene defects, including inactivation of the patched gene. Approximately 90% of patients with squamous cell carcinoma of the skin have *p53* mutations, and *BRAF* abnormalities have been detected in 80–90% of patients with malignant melanoma. *p16* has been identified as a familial melanoma susceptibility locus (Chapters 6 and 7 discuss the concepts of oncogenes and tumor suppressors in more detail). Squamous carcinomas of the skin are more common in individuals with albinism and familial traits for malignant melanoma have been described with gene defects being defined (Chapter 3).

The major malignancies involving the skin have all increased in frequency significantly over the past two decades, presumably related to the increase in intensity of sunlight exposure in many Western nations as a result of reduction of the ozone layer and also because of increased travel to warm climates for holidays with rapid changes in sunlight exposure.

Malignant melanoma

Malignant melanoma varies in frequency between regions of the world and is generally more common in nations with high levels of sunlight exposure. It has a much higher frequency in the white population and in the United States 98.2% of cutaneous melanomas are found in the white population. In nonwhite populations there is a much higher frequency of melanomas in mucosal and acral locations. The incidence of melanoma is 10 times greater in Australasia (which has the highest incidence in the world) than Europe but the frequency is increasing at up to 7% per year in some parts of Europe. In the United States, melanoma incidence was 8.2 per 100 000 in the 1970s and more than doubled to 18.7 per 100 000 in 2003. Particular risk factors include the presence of fair skin, red hair, and pigmented nevi, and the majority of malignant melanomas occur in sites that are exposed to sunlight.

Clinical features

Most malignant melanomas arise at sites of previous nevi and are associated with enlargement of the nevus, ulceration, or bleeding in the majority of patients (Fig. 2.10). Any of these observations merits consideration of an excision biopsy. Familial melanomas have been described and alterations of pigmented lesions in first-degree relatives of patients who have had a melanoma should be treated with suspicion.

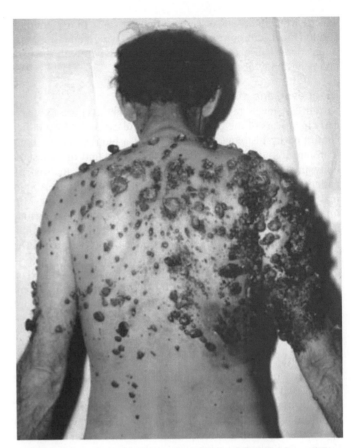

Figure 2.10 Extensive malignant melanoma in a patient who first noted pigmented lesions 12 months before seeking medical attention.

There are three well-described patterns of presentation of malignant melanoma – superficial spreading melanomas, nodular melanomas, and lentigo maligna.

Patients may present with a cutaneous lesion but a small percentage present in other sites, including the eye, mucosa of the head and neck, anal–rectal region, and female genital tract, and other sites including the bowel. Symptoms usually arise from involvement of the primary site but occasionally patients may present with symptoms of metastasis. Melanoma can spread to any organ, lymph nodes, or bones.

Treatment

The prognosis for malignant melanomas is particularly affected by the degree of invasion. The Breslow stage describes the thickness in millimeters of vertical tumor invasion into the dermis with tumors <0.75 mm rarely metastasizing and those invading more than 1.5 mm having a high risk of subsequent metastasis. The presence of lymph node or distant metastasis at the time of diagnosis confers a very poor prognosis, lymph node involvement being associated with a 5-year survival rate of <20% and distant metastasis being rarely curable. Treatment is aimed at excision of the tumor together with a margin of surrounding skin (extent of margin has been the subject of several studies but generally recommended at greater than 1 cm for thicker melanomas). This will often involve skin grafting. Removal of regional lymph nodes can provide prognostic information as survival is reduced if there is microscopic involvement. Until recently, lymphadenectomy was routinely performed in higher risk melanomas but there is increasing use of sentinel node biopsy to reduce the morbidity associated with routine block dissection of the regional nodes. There is considerable morbidity from block dissection and its routine use may not be appropriate as 5-year survival rates are poor at only approximately 10%.

Following excision of high-risk melanomas, adjuvant therapy with high-dose interferon-alpha has been shown to reduce relapse rates and increase survival. This benefit, however, is associated with significant toxicities. Other approaches aimed at improving the results of surgery alone include the use of neoadjuvant interferon and adjuvant immunotherapy.

Unfortunately, malignant melanoma is one of the most resistant tumors in terms of response to chemotherapy and is also relatively resistant to radiotherapy. Radiotherapy may, however, be beneficial for symptomatic metastases and chemotherapy continues to be used for some patients with disseminated disease, despite response rates of only 10–20% for most agents. The major reason for resistance in melanoma cells appears to be the upregulation of DNA repair enzymes and high expression of efflux pumps. They also appear to have a lower susceptibility to apoptosis as a result of activating mutations of *BRAF*. For several decades, the most widely used cytotoxic agent was dacarbazine (DTIC). Median overall survival with this agent is only approximately 8 months and it is associated with significant toxicity. Temozolamide is an oral cytotoxic agent which is metabolized into an active agent closely related to DTIC. This has a greater response rate and modest improvement in overall survival and is generally better tolerated by patients so that it is increasingly used in the treatment of metastatic melanoma.

For patients with localized disease with invasion <0.75 mm, 5-year survival rates of 70% can be achieved.

Malignant melanoma is one of the more resistant tumors to both radiotherapy and chemotherapy. Radiotherapy may be beneficial for symptomatic metastases and chemotherapy can be used for disseminated disease, although response rates are poor at only 10–20% for most agents. Dacarbazine is the most widely used drug for this disease but median survival is only 6–8 months once metastases have been demonstrated, regardless of therapies employed. When disease is apparently localized to an extremity, some centers use regional perfusion of chemotherapy agents in the hope that obtaining high concentrations will produce greater tumor cell kill but avoid the systemic toxicity that would otherwise follow. Cutaneous lesions can be dramatically reduced with this approach but survival seems to be little affected.

There has been considerable interest in the use of immunotherapy with interferons, interleukin-2, and a variety of vaccines for advanced disease but unfortunately early promise has not been followed by significant improvements in survival rates (see Chapter 13).

Basal cell carcinoma
Basal cell carcinomas, also termed "rodent ulcers," are the commonest skin malignancies, comprising approximately 75% of all skin cancers and are also the most common human malignancy. They make up almost a quarter of all cancers diagnosed in Western nations. Basal cell carcinomas may occasionally arise in inherited conditions, including the nevoid basal cell carcinoma syndrome. There has been considerable advance in our understanding of the molecular defects in this disorder and they have been shown to result from inactivation of the patched gene (*PTCH*) which is a tumor suppressor (see Chapters 5 and 7). This gene takes part in the Hedgehog signaling pathway and the PTCH protein binds in a complex with Smoothened that acts as a receptor for the secretion molecule Hedgehog. When Hedgehog binds, Smoothened is released and functions as an oncogene.

The incidence of basal cell carcinomas rises steadily with age, such that at the age of 80 there are approximately 450 per 100 000 of the male population who develop a basal cell carcinoma (BCC). These tumors are presumed to originate from stem cells in the basal layers of skin, which have the potential to produce a variety of structures including sweat glands and hair follicles. They predominantly develop in areas that are exposed to sunlight and are most frequent on the face. There is usually a characteristic clinical presentation with a firm, pink nodule that has a distinct raised edge. There may be ulceration in the center, which frequently bleeds.

Histologically, these tumors appear malignant with local infiltration and pleomorphic cells. Surprisingly, however, metastasis is very uncommon but there is local invasion that may penetrate to cartilage and occasionally to bone.

Management may involve initial incision biopsy to exclude other forms of skin lesion where doubt about diagnosis exists. When diagnosis has been made, a variety of treatment options are available. These predominantly involve surgical excision with a variety of techniques, but cryosurgery and radiotherapy are also effective. Surgical excision offers the advantage of confirming a diagnosis and margin of clearance, whereas radiotherapy has the advantage of avoiding the need for an anesthetic and often provides a better cosmetic result than surgery. Radiotherapy is usually reserved for older patients. Cure rates are high and approach 100%.

Recently, inhibitors of the Hedgehog pathway (see Chapter 5) have become available for clinical trials and these have been shown to be highly effective in producing responses in basal cell carcinomas with minimal toxicity.

Squamous cell carcinoma
Squamous cell carcinomas of the skin comprise approximately 20% of skin malignancies. They are also related to sunlight exposure but in addition they may occur at the periphery of previous areas of radiation exposure and have also been linked to exposure to arsenic. Patients taking immunosuppressive therapy following organ transplantation and those who are HIV positive also have a high incidence of squamous cell carcinoma of the skin.

These tumors have the histological appearance of keratinizing squamous carcinomas and vary in their degree of differentiation. The typical clinical presentation is with a crusted ulcer or nodular lesion, most commonly on the face, neck, and some exposed areas of the limbs. More rarely, they may arise at mucocutaneous junctions, particularly in the anus and vulva. Squamous cell carcinomas of the skin may resemble several benign conditions and it is therefore important to undertake a diagnostic biopsy initially to confirm the diagnosis. Once the diagnosis is confirmed, surgical excision, or if this is not technically feasible, radiotherapy, can be undertaken. As with basal cell carcinomas, the initial procedure is a diagnostic biopsy, which, if it confirms squamous carcinoma, should be followed by surgical excision, or, if this is not technically feasible, radiotherapy. Squamous cell carcinomas of the mucocutaneous junctions are usually treated by chemoradiotherapy initially, often followed by surgical resection. Cryosurgery and topical chemotherapy may be used but are followed by higher rates of recurrence. For squamous carcinomas of the head, neck, and limbs, cure rates exceed 90%.

Carcinoma of the cervix

In the United States and Western Europe, the incidence and mortality from cervical cancer have steadily reduced over the last 50 years. A significant contribution to this reduction is the widespread availability of cervical screening. Although deaths have reduced in these areas, carcinoma of the cervix remains the leading cause of cancer-related deaths among women in many other countries, including those in Latin America, Africa, Asia, and Eastern Europe. There is a strong relationship between infection with the human papillomavirus (HPV) and the incidence of cervical carcinoma and it is estimated that 99% of cases may be related to this virus (see Chapter 3).

Clinical features

Many patients are diagnosed from routine cervical cytology screening. Both preinvasive and early invasive disease can be detected without any associated symptoms. Abnormal vaginal bleeding is the most common early symptom and may follow coitus. A vaginal discharge may be associated with this. Pelvic pain may occur at later stage, as may pain in the flank which can result from pyelonephritis. Extensive pelvic disease may occur and lead to leg edema and sciatic nerve pain. Some patients may develop extensive pelvic involvement which can produce hematuria, incontinence, and constipation. Diagnosis is made from biopsy and features seen on CT imaging.

Treatment

Treatment is determined by the pathology findings and extent of disease. Surgery, radiotherapy, or chemoradiotherapy all have roles in management. For early-stage disease, local resection may be appropriate but more radical surgery and/or external-beam radiotherapy and brachytherapy may be necessary. For patients with loco-regionally advanced disease, the addition of cisplatin-containing chemotherapy to standard radiotherapy significantly reduces the risk of recurrence by as much as 50%. For patients with metastatic disease there is no curative option. Single-agent chemotherapy with cisplatin or combination chemotherapy may produce responses in up to 20% of patients, but these are usually of brief duration and do not have a significant effect on survival. Palliative radiotherapy may be used for localized painful lesions. The side effects of intervention in advanced disease have to be balanced against the small potential clinical benefits. It is hoped that the increasing use of vaccination against HPV in prepubertal girls will reduce the incidence of this disease and avoid the significant morbidity and mortality that can result.

Hematological malignancies

The major diseases of the hemopoietic systems are lymphomas and leukemias. The lymphomas include Hodgkin disease and non-Hodgkin lymphomas. Leukemias are acute or chronic and predominantly involve the lymphatic or myeloid series.

Information on genetic causes for the majority of hematological malignancies is becoming increasingly available. One of the earliest abnormalities to be described was the translocation between chromosomes 9 and 22, involving the *bcr* and *abl* genes, resulting in a fusion protein with deregulated growth-promoting properties (see Chapters 5 and 6).

Hodgkin disease

The incidence of Hodgkin disease follows an unusual bimodal distribution with an early peak in individuals between the ages of 10 and 20 years and a second peak after the age of 50 years. There is predominance in males. The incidence of Hodgkin disease is steadily falling in Western nations. Links between Hodgkin disease and infection with Epstein–Barr virus have been made and some clusters of disease have been seen in small geographic areas, suggesting environmental causes (see Chapter 3). There appears to be a genetic predisposition for Hodgkin disease, with an increased incidence in some Jewish populations and among first-degree relatives of those who have been diagnosed with the disease. There is also an association with certain HLA antigens. No other definite etiological factors have been identified.

Diagnosis

Biopsy of abnormal lymph nodes reveals typical changes in Hodgkin disease. The diagnosis is confirmed by the presence of the Reed–Sternberg cell, a binucleate cell with a typical appearance (Fig. 2.11). Immunophenotyping reveals B-cell characteristics for the Reed–Sternberg cell and a variety of chromosomal changes have been detected. Typical histological patterns are seen in different patients, such that the disease is divided into a variety of subgroups according to five different classifications which have been published to date. The REAL classification recognizes five subgroups: lymphocyte-rich, nodular lymphocyte-predominant, nodular sclerosis, mixed cellularity, and lymphocyte-depleted. Eighty percent of all patients have either nodular sclerosing or mixed cellularity histology.

Clinical features

Almost all patients present with a painless enlarged lymph node (Fig. 2.12), usually in the neck but may involve other sites including the inguinal region (10%) and axilla (25%). There may be a long history of slow growth of these nodes and they may fluctuate in size. Approximately a quarter of all patients will have systemic symptoms such as fever, sweats (may be drenching and predominantly during the night), and weight loss. Patients may

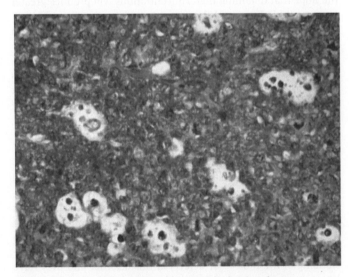

Figure 2.11 Histological section of lymph node showing changes associated with Hodgkin disease. Note large, binucleate Reed–Sternberg cells.

The Burden of Cancer

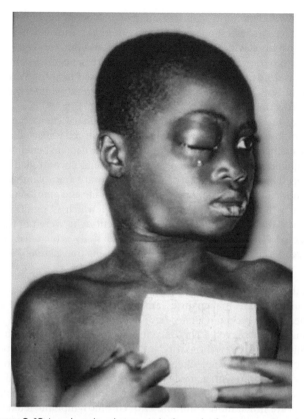

Figure 2.12 Lymph node enlargement in the neck of a young male patient. This was caused by lymphoma (in this case of Burkitt's type) and was associated with central nervous system involvement causing loss of control of movement of eyelids and excessive lacrimation.

also complain of alcohol-induced pain at the site of lymphadenopathy, pruritus, and symptoms suggestive of anemia. The enlarged nodes may compress airways, blood vessels, and bile duct leading to symptoms including cough, dyspnea, edema, and jaundice. Occasionally, skin lesions may occur and central nervous system involvement can lead to spinal cord compression.

Staging
Investigations are undertaken to determine the volume of disease and its anatomical spread. Chest X-ray, CT scan of the thorax, abdomen, and pelvis, full blood count and biochemical screen are routinely performed and provide sufficient information on staging for the majority of patients. For patients with stage IIB–IV disease, bone marrow biopsies are usually performed. PET scanning is now frequently performed when results of other conventional diagnostic procedures are inconclusive. Disease stage is described according to the Ann Arbor system, with stage I disease defining a single lymph node area of involvement, stage II being two lymph node areas on the same side of the diaphragm, stage III describing two or more lymph node areas on opposite sides of the diaphragm, and stage IV disease being dissemination to non-lymph node sites including liver, lungs, and bone marrow.

Treatment
The treatment of Hodgkin disease represents a major therapeutic advance in oncology and most patients are now cured. Both chemotherapy and radiotherapy are used; localized disease is usually managed with involved field radiotherapy whereas more widespread disease, and all patients with significant symptoms, will receive combination chemotherapy. For some patients, chemotherapy may be added to radiotherapy. Several combination chemotherapy regimens are used in different centers; one example includes adriamycin, bleomycin, vinblastine, and dacarbazine (ABVD). Radiotherapy is often given after chemotherapy to sites of previous large-volume disease. Late toxicity is important given the high cure rate and this can include infertility, cardiotoxicity, and secondary neoplasia. Second cancers may occur in up to 8% of patients after 10 years, particularly for those receiving both radiotherapy and chemotherapy or following protracted chemotherapy. A major research focus is exploring methods of reducing treatment-related second malignancies without compromising cure rates. For patients who develop relapsed Hodgkin disease, high-dose chemotherapy with autologous stem cell rescue can be curative and is increasingly becoming standard both for early and late relapse.

Non-Hodgkin lymphomas
Non-Hodgkin lymphomas (NHLs) are the most common hematological malignancies in adults and represent approximately 5% of all new cancer cases. The worldwide incidence varies considerably, with the highest rates in Western Europe and the United States and approximately fivefold lower rates in Asia. While the incidence of Hodgkin disease has been falling, that for NHL has increased steadily since the 1970s. NHLs appear to arise from different types of lymphocytes at varying stages of development, but the loss of cellular differentiation that almost inevitably accompanies tumorigenesis complicates the determination of the cell of origin – still a fundamental question to be answered in cancer biology. Their pathological classification has been the subject of several revisions over the years such that a large number of systems of classification exist. The cause of most cases of NHL is unknown.

Immunodeficiency is associated with an increase risk and in particular can occur in patients with AIDS and those who have undergone immunosuppression for organ transplantation and those with celiac disease. This is predominantly a disease of older people, with a steady rise in frequency throughout the decades following the 30th year. There is a slight male preponderance. Several etiological factors have been described, including viral infections (HTLV-I, EBV, human herpes virus type 8). Increasing information is becoming available on genetic factors relating to the etiology of NHLs. More than 80% of patients with Burkitt lymphoma have a reciprocal translocation between chromosomes t(8;14) (q24;q32). For the other 20% of patients with this disease, translocations between chromosomes 8 and 22 and 2 and 8 are demonstrated. The break on chromosome 8 occurs at the site of the c-*MYC* proto-oncogene (see Chapter 6). Over 80% of cases of follicular lymphomas have a chromosome translocation between 14q32 and 18q21. The breakpoint on chromosome 18 is the *BCL2* gene, encoding a protein that inhibits apoptosis (see Chapter 8). Increasing numbers of chromosome abnormalities are being described in association with different types of lymphoma and can be used to support diagnosis.

Pathology
The pathological classification of NHLs is complex and evolving. The most recent classifications have been developed by the World Health Organization but both the Kiel and REAL systems are widely used. These divide the NHLs into B-cell or T-cell origins

and subdivide these into histological patterns which have different prognostic significance. T-cell lymphomas comprise approximately 10% of all NHLs.

This diagnosis of NHL is made from a biopsy, which is usually derived from an enlarged lymph node. NHL may involve a wide variety of sites and tissue may be obtained from these (e.g. liver, bowel, brain). The diagnosis can be difficult and relies on obtaining enough tissue to undertake immunohistochemistry and to obtain a large enough field of cells to comment on the histological pattern. Needle biopsies should be avoided wherever possible as a result.

Clinical presentation

In the majority of patients, NHL presents as enlargement of one or more lymph node areas. The lymphadenopathy is usually painless and, as with Hodgkin disease, is most frequently found in the neck, but any lymph node area may be involved. Symptoms may thus arise from local compression of nerves, airways, or blood vessels. NHL tends to spread widely at a much earlier stage than Hodgkin disease, so several sites of disease may be present at diagnosis. Examination may reveal hepatosplenomegaly in addition to lymphadenopathy.

Staging is undertaken to determine the extent of disease and the majority of patients will have stages III or IV disease at presentation if careful investigation and examination are undertaken. Full CT scanning, biochemical assays, and blood count together with bone marrow aspirate and trephine should be undertaken. Several centers are increasingly using PET scanning to provide further information on stage.

Treatment

The management of NHL is becoming increasingly complex with differing approaches for different histological subgroups. Follicular lymphoma tends to have a chronic history with long median survival. In a majority of cases this is not curable and recurrence is inevitable after initial response to treatment. The diffuse "high-grade" lymphomas behave in a much more aggressive fashion with median survivals, if untreated, of only 4–6 months. They are, however, potentially curable. For those with follicular lymphomas, there does not appear to be survival advantage to immediate treatment for patients with low-volume disease and who have no serious symptoms. As a result, localized radiotherapy may be offered to those with stage I and II disease with the potential for 8–10 years of freedom from recurrence. For some patients a "watch" policy without any intervention may be appropriate. For those with higher volume disease or symptoms, the use of retuximab (humanized anti-CD20 monoclonal antibody) produces high levels of response and prolongs survival. The overall survival for this disease has steadily improved over the last three decades with 5-year survival improving from 60% to 87% between the 1980s and 2000s.

For patients with diffuse high-grade lymphoma, combination chemotherapy is used for all stages of disease. A variety of regimens are used in different centers but the long-established combination of cyclophosphamide, doxorubicin, vincristine, and prednisolone (CHOP) forms the most widespread backbone of therapies. Radiotherapy may be incorporated to sites of bulky disease or when localized disease is the presentation. Recent studies have demonstrated the benefit in terms of response and survival by adding retuximab to the CHOP regimen and this is now becoming a standard approach. Modern regimens can produce cure rates in excess of 60%. There is some evidence that for patients with high-risk disease the addition of high-dose chemotherapy with autologous stem cell rescue may confer a survival advantage and it is now widely used in most Western countries. A large randomized trial is currently underway to address the value of this approach.

For those patients who relapse following initial chemotherapy, high-dose chemotherapy and autologous stem cell transplantation is usually offered. Unfortunately, there is little data on benefit from randomized trials although results from several uncontrolled trials suggest that the majority of patients who are transplanted will achieve a response that improves quality of life and approximately 40% will have 3 or more years survival which is significantly greater than historical data for those patients who were not transplanted. There is one small randomized study which seems to suggest a survival advantage for patients being transplanted.

Leukemias

Leukemias are malignancies arising from the white blood cell lineage and recent evidence has demonstrated the existence of a "leukemic stem cell." This leukemic stem cell is thought to have limitless self-renewal capacity and gives rise to clonogenic leukemic progenitors. The leukemias are the most common malignancies in children with acute lymphoblastic leukemia (ALL) making up approximately 80% of all cases. In adults, the chronic leukemias are much more common, with chronic lymphatic leukemia (CLL) being a frequent coincidental finding on routine blood counts in the elderly population. Acute lymphoblastic leukemia makes up approximately 80% of all pediatric cases, whereas acute myeloid leukemia (AML) is commoner in adults.

Etiological factors that have been identified include ionizing radiation, chemotherapy for previous malignancies, the HTLV virus, smoking, and exposure to benzene. There are also a variety of inherited syndromes, including Down syndrome, that are associated with an increased risk of acute leukemia.

Chromosomal translocations have been described in several leukemias. For chronic myeloid leukemia, a translocation between chromosomes 9 and 22 (t9;22) (q34;q11) has been identified and shown to bring the *BCR* gene and the *ABL* gene together, resulting in the production of a tyrosine kinase which is central to growth regulation. This abnormality is seen in the vast majority of such cases and is used to help make the diagnosis. Knowledge about this particular mutation also proved central to the subsequent development of the tyrosine kinase class of anticancer drugs (see Chapter 16). Translocations contribute to over 50% of all leukemias.

Diagnosis and pathology

The diagnosis of leukemia is usually suspected from an abnormal peripheral blood count. This frequently shows an elevated white count and is usually accompanied by anemia. There are often leukemic blasts visible in the blood. A bone marrow aspirate and trephine will usually provide the diagnosis. The marrow is almost completely replaced by blast cells, which have typical appearances and staining patterns of lymphoid or myeloid cells. Several subtypes of myeloid and lymphoid leukemia are described, according to the histological appearance and staining of the malignant cells. Lymphoid leukemias may be of B- or T-cell origin. Although in a majority of cases diagnosis is usually made on morphology and cytochemistry, cytogenetics is frequently used to aid diagnosis and provide prognostic features.

Clinical history and management

Patients may present with a variety of features, including lethargy from anemia, hemorrhage, or bruising from thrombocytopenia, and infection from immunocompromise. Extensive involvement of bone marrow may cause bone pain. Some patients may have central nervous system involvement with associated signs and symptoms suggestive of meningeal involvement (stiff neck, headache, photophobia).

Treatment involves the use of chemotherapy for all cases of acute leukemia. This is usually divided into "an induction" phase that is intense and requires prolonged admission during periods of "pancytopenia." Subsequent "consolidation" therapy is given after induction of remission and additional therapy with central nervous system prophylaxis to prevent relapse may be offered. High-dose treatment with transplantation (allogeneic or autologous) may be offered to high-risk individuals following remission induction if they are fit enough to tolerate this.

The chronic leukemias involve older patients who may be relatively asymptomatic and have their disease detected from a routine blood count. Approximately 60% of patients with CLL will present with a variety of symptoms including fatigue, bacterial or viral infections, or they may notice lymph node enlargement. The majority of patients will be older than 70 years and may have significant co-morbidity such that immediate intervention may not be necessary or appropriate. There are no current treatment options which will cure this disease. Once symptoms occur and reduce quality of life, or bone marrow failure occurs, a variety of treatments may be offered. These include single agent chemotherapy, often with chlorambucil or the purine analogs, including fludarabine. Alemtuzumab is a humanized monoclonal antibody targeting CD52 which is highly expressed on CLL cells. It is now increasingly used to induce remission in CLL. Retuximab is also under investigation in clinical trials in combination with chemotherapy.

Chronic myeloid leukemia may also be a coincidental finding from a routine blood count but may also cause fatigue from anemia, lethargy, infections, and weight loss, together with pain over the spleen from splenomegaly. Treatment of this condition has changed dramatically in the last 10 years. Prior to this time single-agent chemotherapy was traditionally offered using hydroxyurea or busulfan which could reduce blood counts and improve symptoms but were associated with inevitable relapse and progression to an acute phase, which was rapidly fatal. Interferon-alpha was demonstrated to produce improved 5-year survival rates compared with chemotherapy but its use was limited by its toxicity. The availability of imatinib, which targets the BCR–ABL oncoprotein, dramatically changed outcome for this disease, producing remissions in approximately 90% of patients. It is well tolerated and remissions appear to be durable. Approximately 90% of patients on imatinib are alive beyond 5 years. Unfortunately, the vast majority of patients continue to have detectable levels of *BCR–ABL* transcripts and treatment with imatinib is therefore felt to be non-curative. Newer agents (such as dasatinib and nilotinib) have been approved to treat imatinib-resistant mutations. The use of imatinib has largely replaced chemotherapy. For younger patients who are fit, the only potentially curative treatment is with an allogeneic stem cell transplant following high-dose chemotherapy and total body irradiation. With transplantation, the prognosis appears to be markedly improved with 5-year relapse-free survival rates in excess of 75%.

Conclusions and future directions

The burden which is well borne becomes light.

Ovid

The rapid diagnosis and appropriate management of the common malignancies is essential to optimize the chance of cure. Unfortunately many tumors are diagnosed at a time when they have already metastasized and are likely to be incurable. It is essential that patients are well educated and able to recognize symptoms that may indicate an underlying malignancy and that primary care clinicians are also aware of patterns of presentation so that investigations and referral for a specialist opinion are organised without delay. Most malignancies present with typical symptoms and signs that should automatically trigger an appropriate series of investigations, culminating in a histological diagnosis, if an abnormal area of tissue is detected. Surgery is usually the most appropriate form of curative treatment for solid tumors, but is increasingly preceded (neoadjuvant) or followed (adjuvant) by chemotherapy and/or radiotherapy in an attempt to improve resection rates and reduce the risk of recurrence. A large number of new cytotoxic agents have been developed in recent years and often have greater activity and reduced toxicity compared with previous anticancer drugs. Of great promise is the increasing availability of targeted therapies directed at specific components of cell signaling pathways or circulating biologically active molecules. The effectiveness of therapies for malignancies should markedly improve in the next decade and the side-effects will be minimized. Unfortunately, these benefits will require huge increases in funding for healthcare providers – a challenge only just becoming apparent in many countries.

Outstanding questions

1. Will preoperative (neoadjuvant) chemotherapy improve the resectability rate and outcome in the common solid tumors?

2. How can we optimally develop and utilize signal transduction inhibitors to manage malignancies? What will be their impact on outcome?

3. Can we develop pretreatment markers that can better guide choices of optimal individualized therapy? This could allow focusing of expensive potentially toxic treatment on those who are most likely to benefit and avoid unnecessary toxicity for those who are unlikely to benefit.

4. What is the optimal method of following patients after treatment is completed? Can we develop noninvasive methods to detect recurrence?

5. Can we develop methods for effective screening of populations to detect tumors at an early stage with greater potential for cure? Do these techniques improve survival rates in the population?

Bibliography

Epidemiology of cancer

Balmain, A. (2002). Cancer as a complex genetic trait: tumor susceptibility in humans and mouse models. *Cell*, **108**: 145.

Doll, R. and Hill, A.B. (1964). Mortality in relation to smoking: ten years' observation on British Doctors. *British Medical Journal*, **1**: 1399.

Ezzati, M., Henley, S.J., Lopez, A.D., and Thun, M.J. (2005). Role of smoking in global and regional cancer epidemiology: current patterns and data needs. *International Journal of Cancer*, **116**: 963.

Ferlay, J., Bray, F., Pisani, P., and Parkin, D.M., eds (2004). *GLOBOCAN 2002: Cancer Incidence, Mortality and Prevalence Worldwide*. Lyon, France: IARC Press.

Kolonel, L.N. and Wilkens, L.R. (2006). Migrant Studies. *Cancer Epidemioly and Prevention*, 3rd edn. New York: Oxford University Press, p. 189.

Ries, L.A.G., Harkins, D., Krapcho, M., et al., eds (2006). *SEER Cancer Statistics Review, 1975–2003*. Bethesda, MD: National Cancer Institute.

Trichopoulos, D., Li, F., and Hunter, D. (1996). What causes cancer? *Scientific American*, **275**: 80.

Lung cancer

Belani, C.P., Wang, W., Johnson, D.H., et al. (2005). Phase III study of the Eastern CooperativeOncology Group (ECOG 2597): induction chemotherapy followed by either standardthoracic radiotherapy or hyperfractionated accelerated radiotherapy for patients with un-resectable stage IIIA and B non–small-cell lung cancer. *Journal of Clinical Oncology*, **23**(16): 3760.

Fritscher-Ravens, A., Bohuslavizki, K.H., Brandt, L., et al. (2003). Mediastinal lymph node involvement in potentially resectable lung cancer: comparison of CT, positron emission tomography, and endoscopic ultrasonography with and without fine-needle aspiration. *Chest*, **123**(2): 442.

Goldstraw, P. and Crowley, J.J. (2006). The international association for the study of lung cancer international staging project on lung cancer. *Journal of Thoracic Oncology*, **1**(4): 281.

Govindan, R., Page, N., Morgensztern, D., et al. (2006). Changing epidemiology of small-cell lung cancer in the United States over the last 30 years: analysis of the surveillance, epidemiologic, and end results database. *Journal of Clinical Oncology*, **24**: 4539.

Herbst, R.S., Giaccone, G., Schiller, J.H., et al. (2004). Gefitinib in combination with paclitaxel and carboplatin in advanced non–small-cell lung cancer: a phase III trial – INTACT 2. *Journal of Clinical Oncology*, **22**: 785.

Lynch, T.J., Bell, D.W., Sordella, R., et al. (2004). Activating mutations in the epidermal growth factor receptor underlying responsiveness of non–small-cell lung cancer to gefitinib. *New England Journal of Medicine*, **350**(21): 2129.

Non-small-cell Lung Cancer Collaborative Group (1995). Chemotherapy in non–small-cell lung cancer: a meta-analysis using updated data on individual patients from 52 randomized clinical trials. *British Medical Journal*, **311**(7010): 899.

Omenn, G.S. (2007). Chemoprevention of lung cancers: lessons from CARET, the beta-carotene and retinol efficacy trial, and prospects for the future. *European Journal of Cancer Prevention*, **16**(3): 184.

Sato, M., Shames, D.S., Gazdar, A.F., and Minna, J.D. (2007). A translational view of the molecular pathogenesis of lung cancer. *Journal of Thoracic Oncology*, **2**(4): 327.

Thatcher, N., Qian, W., Clark, P.I., et al. (2005). Ifosfamide, carboplatin, and etoposide with mid cycle vincristine versus standard chemotherapy in patients with small-cell lung cancer and good performance status: clinical and quality-of-life results of the British Medical Research Council multicenter randomized LU21 trial. *Journal of Clinical Oncology*, **23**: 8371.

Breast cancer

Bartelink, H., Horiot, J.-C., Poortmans, P.M., et al. (2007). Impact of a higher radiation dose on local control and survival in breast conserving therapy of early breast cancer: 10-year results of the randomized EORTC "Boost versus no Boost" trial 22881–10882. *Journal of Clinical Oncology*, **25**: 3259.

Berry, D.A., Cronin, K.A., Plevritis, S.K., et al. (2005). Effect of screening and adjuvant therapy on mortality from breast cancer. *New England Journal of Medicine*, **353**(17): 1784.

Bluman, L.G., Rimer, B.K., Berry, D.A., et al. (1999). Attitudes, knowledge and risk perceptions of women with breast and/or ovarian cancer considering testing for BRCA1 and BRCA2. *Journal of Clinical Oncology*, **17**: 1040.

Bonneterre, J., Thurlimann, B., Robertson, J.F., et al. (2000). Anastrozole versus tamoxifen as first-linetherapy for advanced breast cancer in 668 postmenopausal women: results of the Tamoxifen or Arimidex Randomized Group Efficacy and Tolerability study. *Journal of Clinical Oncology*, **18**(22): 3748.

Cuzick, J., Powles, T., Veronesi, U., et al. (2003). Overview of the main outcomes in breast-cancer prevention trials. *Lancet*, **361**(9354): 296.

Howell, A., Cuzick, J., Baum, M., et al. (2005). Results of the ATAC (Arimidex, Tamoxifen, Alone or in Combination) trial after completion of 5 years' adjuvant treatment for breast cancer. *Lancet*, **365**(9453): 60.

Mayer, E.L. and Burstein, H.J. (2007). Chemotherapy for metastatic breast cancer. *Hematology/Oncology Clinics of North America* **21**: 257.

Ravdin, P.M., Siminoff, L.A., Davis, G.J., et al. (2001). Computer program to assist in making decisions about adjuvant therapy for women with early breast cancer. *Journal of Clinical Oncology*, **19**(4): 980.

Rebbeck, T.R., Friebel, T., Lynch, H.T., et al. (2004). Bilateral prophylactic mastectomy reduces breast cancer risk in BRCA1 and BRCA2 mutation carriers: the PROSE Study Group. *Journal of Clinical Oncology*, **22**(6): 1055.

Slamon, D.J., Leyland-Jones, B., Shak, S., et al. (2001). Use of chemotherapy plus a monoclonal antibody against HER-2 for metastatic breast cancer that over expresses HER-2. *New England Journal of Medicine*, **344**(11): 783.

Walsh, T. and King, M.C. (2007). Ten genes for inherited breast cancer. *Cancer Cell*, **11**(2): 103.

Colorectal cancer

Andre, T., Boni, C., Mounedji-Boudiaf, L., et al. (2004). Oxaliplatin, fluorouracil, and leucovorin as adjuvant treatment for colon cancer. *New England Journal of Medicine*, **350**(23): 2343.

Cunningham, D., Humblet, Y., Siena, S., et al. (2004). Cetuximab monotherapy and cetuximab plus irinotecan in irinotecan-refractory metastatic colorectal cancer. *New England Journal of Medicine*, **351**(4): 337.

Doxey, B.W., Kuwada, S.K., and Burt, R.W. (2005). Inherited polyposis syndromes: molecular mechanisms, clinicopathology, and genetic testing. *Clinical Gastroenterology and Hepatology*, **3**(7): 633.

Griesenberg, D., Nurnberg, R., Bahlo, M., and Klapdor, R. (1999). CEA, TPS, CA 19–9 and CA 72–4 and the fecal occult blood test in the preoperative diagnosis and follow-up after resective surgery of colorectal cancer. *Anticancer Research*, **19**: 2443.

Haller, D.G., Catalano, P.J., Macdonald, J.S., et al. (2005). Phase III study of fluorouracil, leucovorin, and levamisole in high-risk stage II and III colon cancer: final report of Intergroup 0089. *Journal of Clinical Oncology*, **23**(34): 8671.

Hurwitz, H., Fehrenbacher, L., Novotny, W. et al. (2004). Bevacizumab plus irinotecan, fluorouracil, and leucovorin for metastatic colorectal cancer. *New England Journal of Medicine*, **350**: 2335–42.

Pickhardt, P.J., Choi, J.R., Hwang, I., et al. (2003). Computed tomographic virtual colonoscopy to screen for colorectal neoplasia in asymptomatic adults. *New England Journal of Medicine*, **349**(23): 219.

Seymour, M.T., Maughan, T.S., Ledermann, J.A., et al. (2007). Different strategies of sequential and combination chemotherapy for patients with poor prognosis advanced colorectal cancer (MRC FOCUS): a randomised controlled trial. *Lancet* **370**(9582): 143.

Tournigand, C., Andre, T., Achille, E., et al. (2004). FOLFIRI followed by FOLFOX6 or the reverse sequence in advanced colorectal cancer: a randomized GERCOR study. *Journal of Clinical Oncology*, **22**(2): 229.

Vogelstein, B., Fearon, E.R., Hamilton, S.R., et al. (1988). Genetic alterations during colorectal tumor development. *New England Journal of Medicine*, **319**: 525.

Skin tumours

Balch, C.M., Soong, S.J., Gershenwald, J.E., et al. (2001). Prognostic factors analysis of 17,600 melanoma patients: validation of the American Joint Committee on Cancer melanoma staging system. *Journal of Clinical Oncology*, **19**: 3622.

Chudnovsky, Y., Khavari, P.A., and Adams, A.E. (2005). Melanoma genetics and the development of rational therapeutics. *Journal of Clinical Investigation*, **115**: 813.

Curtin, J.A., Fridlyand, J., Kageshita, T., et al. (2005). Distinct sets of genetic alterations in melanoma. *New England Journal of Medicine*, **353**: 2135.

Demierre, M.F. (2006). Epidemiology and prevention of cutaneous melanoma. *Current Treatment Options in Oncology*, **7**: 18.

Dennis, L.K. (1999). Analysis of the melanoma epidemic, both apparent and real: data from the 1973 through 1994 Surveillance, Epidemiology, and End Results program registry. *Archives of Dermatology*, **135**: 275.

Geisse, J.K. (1995). Comparison of treatment modalities for squamous cell carcinoma. *Clinical Dermatology*, **13**(6): 621.

Kanjilal, S., Strom, S.S., Clayman, G.L., et al. (1995). p53 Mutations in nonmelanoma skin cancer of the head and neck: molecular evidence for field cancerization. *Cancer Research*, **55**: 3604.

Keilholz, U., Punt, C.J., Gore, M., et al. (2005). Dacarbazine, cisplatin, and interferon-alfa-2b with or without interleukin-2 in metastatic melanoma: a randomized phase III trial (18951) of the European Organization for Research and Treatment of Cancer Melanoma Group. *Journal of Clinical Oncology*, **23**: 6747.

Kirkwood, J.M., Manola, J., Ibrahim, J., et al. (2004). A pooled analysis of Eastern Cooperative Oncology Group and intergroup trials of adjuvant high-dose interferon for melanoma. *Clinical Cancer Research*, **10**: 1670.

Rosenberg, S.A., Yang, J.C., and Restifo, N.P. (2004). Cancer immunotherapy: moving beyond current vaccines. *Nature Medicine*, **10**: 909.

Spratt, J.S. Jr. (1999). Cancer mortality after nonmelanoma skin cancer. *Journal of the American Medical Association*, **281**(4): 325.

Von Hoff, D.D., Lorusso, P.M., Rudin, C.M., et al. (2009). Inhibition of the Hedgehog pathway in advanced basal-cell carcinoma. *New England Journal of Medicine*, **361**(12): 1164.

Prostate cancer

Beer, T.M., Ryan, C.W., Venner, P.M., et al. (2007). ASCENT Investigators. Double blinded randomized study of high-dose calcitriol plus docetaxel compared with placebo plus docetaxel in androgen-independent prostate cancer: a report from the ASCENT Investigators. *Journal of Clinical Oncology*, **25**: 669.

Gleason, D.F. and Mellinger, G.T. (1974). Prediction of prognosis for prostatic adenocarcinoma by combined histological grading and clinical staging. *Journal of Urology*, **111**: 58.

Gronberg, H. (2003). Prostate cancer epidemiology. *Lancet*, **361**: 859.

Loblaw, D.A., Virgo, K.S., Nam, R., et al. (2007). American Society of Clinical Oncology. Initial hormonal management of androgen-sensitive metastatic, recurrent, or progressive prostate cancer: 2006 update of an American Society of Clinical Oncology practice guideline. *Journal of Clinical Oncology*, **25**: 1596.

Partin, A.W., Mangold, L.A., Lamm, D.M., et al. (2001). Contemporary update of prostate cancer staging nomograms (Partin Tables) for the new millennium. *Urology*, **58**: 843.

Prostate Cancer Trialists' Collaborative Group (2000). Maximum androgen blockade in advanced prostate cancer: an overview of the randomised trials. *Lancet*, **355**: 1491.

Ross, R.K., Pike, M.C., Coetzee, G.A., et al. (1998). Androgen metabolism and prostate cancer: establishing a model of genetic susceptibility. *Cancer Research*, **58**: 4497.

Tannock, I., Osoba, D., Stockler, M., et al. (1996). Chemotherapy with mitoxantrone plus prednisone alone for symptomatic hormone-resistant prostate cancer: a Canadian randomized study with palliative end points. *Journal of Clinical Oncology*, **14**: 1756.

Valicenti, R.K., Gomella, L.G., and Perez, C.A. (2003). Radiation therapy after radical prostatectomy: a review of the issues and options. *Seminars in Radiation Oncology*, **13**: 130.

Woolf, S.H. (1995). Screening for prostate cancer with prostate-specific antigen. An examination of the evidence. *New England Journal of Medicine*, **333**: 1401.

Hematological malignancies

Akasaka, T., Lossos, I.S., and Levy, R. (2003). BCL6 gene translocation in follicular lymphoma: a harbinger of eventual transformation to diffuse aggressive lymphoma. *Blood*, **102**: 1443.

Buske, C., Hoster, E., Dreyling, M., et al. (2006). The Follicular Lymphoma International PrognosticIndex (FLIPI) separates high-risk from intermediate- or low-risk patients with advanced-stage follicular lymphoma treated front-line with rituximab and the combination of cyclophosphamide, doxorubicin, vincristine, and prednisone (R-CHOP) with respect to treatment outcome. *Blood*, **108**: 1504.

Cheson, B.D., Bennett, J.M., Kopecky, K.J., et al. (2003). Revised recommendations of the international working group for diagnosis, standardization of response criteria, treatment outcomes, and reporting standards for therapeutic trials in acute myeloid leukemia. *Journal of Clinical Oncology*, **21**: 4642.

Chiu, B.C. and Weisenburger, D.D. (2003). An update of the epidemiology of non-Hodgkin's lymphoma. *Clinical Lymphoma*, **4**: 161.

Druker, B.J., Guilhot, F., O'Brien, S.G., et al. (2006). Five-year follow-up of patients receiving imatinib for chronic myeloid leukemia. *New England Journal of Medicine*, **355**: 2408.

Duggan, D.B., Petroni, G.R., Johnson, J.L., et al. (2003). Randomized comparison of ABVD and MOPP/ABV hybrid for the treatment of advanced Hodgkin's disease: report of an intergroup trial. *Journal of Clinical Oncology*, **21**(4): 607.

Ekstrand, B.C., Lucas, J.B., Horwitz, S.M., et al. (2003). Rituximab in lymphocyte-predominant Hodgkin disease: results of a phase 2 trial. *Blood*, **101**(11): 4285.

Faderl, S., Talpaz, M., Estrov, Z., and Kantarjian, H.M. (1999). The biology of chronic myelogenous leukemia. *New England Journal of Medicine*, **341**: 164.

Feugier, P., Van Hoof, A., Sebban, C., et al. (2005). Long-term results of the R-CHOP study in the treatment of elderly patients with diffuse large B-cell lymphoma: a study by the Grouped'Etude des Lymphomes de l'Adulte. *Journal of Clinical Oncology*, **23**: 4117.

Fisher, R.I. (2002). Autologous stem-cell transplantation as a component of initial treatment for poor-risk patients with aggressive non-Hodgkin's lymphoma: resolved issues versus remaining opportunity. *Journal of Clinical Oncology*, **20**: 4411.

Fisher, R.I., Gaynor, E.R., Dahlberg, S., et al. (1993). Comparison of a standard regimen (CHOP) with three intensive chemotherapy regimens for advanced non-Hodgkin's lymphoma. *New England Journal of Medicine*, **328**: 1002.

Ghielmini, M., Schmitz, S.F., Cogliatti, S.B., et al. (2004). Prolonged treatment with rituximab in patients with follicular lymphoma significantly increases event-free survival and responseduration compared with the standard weekly 4 schedule. *Blood*, **103**: 4416.

Harris, N.L., Jaffe, E.S., Diebold, J., et al. (1999). World Health Organization classification of neoplastic diseases of the hematopoietic and lymphoid tissues: report of the clinical advisory Committee meeting – Airlie House, Virginia, November, 1997. *Journal of Clinical Oncology*, **17**: 3835.

Juweid, M.E., Stroobants, S., Hoekstra, O.S., et al. (2007). Use of positron emission tomography for response assessment of lymphoma: consensus of the Imaging Subcommittee of International Harmonization Project in Lymphoma. *Journal of Clinical Oncology*, **25**: 571.

Kuppers, R., Klein, U., Hansmann, M.L., and Rajewsky, K. (1999). Cellular origin of human B-cell lymphomas. *New England Journal of Medicine*, **341**: 1520.

Marafioti, T., Hummel, M., Foss, H.D., et al. (2000). Hodgkin and Reed-Sternberg cells represent an expansion of a single clone originating from a germinal center B-cell with functional immunoglobulin gene rearrangements but defective immunoglobulin transcription. *Blood*, **95**: 1443.

Oran, B., Giralt, S., Saliba, R., et al. (2007). Allogeneic hematopoietic stem cell transplantation for the treatment of high-risk acute myelogenous leukemia and myelodysplastic syndrome using reduced-intensity conditioning with fludarabine and melphalan. *Biology of Blood and Marrow Transplant*, **13**: 454.

Rabbitts, T.H. (1991). Translocations, master genes, and differences between the origins of acute and chronic leukemias. *Cell*, **67**: 641.

Schmitz, N. and Sureda, A. (2005). The role of allogeneic stem-cell transplantation in Hodgkin's disease. *European Journal of Haematology Supplement*, (**66**): 146.

Skinnider, B.F. and Mak, T.W. (2002). The role of cytokines in classical Hodgkin lymphoma. *Blood*, **99**(12): 4283.

Sweetenham, J.W., Carella, A.M., and Taghipour, G. (1999). High-dose therapy and autologous stem cell transplantation for adult patients with Hodgkin's disease who fail to enter remission after induction chemotherapy: Results in 175 patients reported to the EBMT. *Journal of Clinical Oncology*, **17**: 3101.

Cervical cancer

Bosch, F.X. and de Sanjose, S. (2003). Human papillomavirus and cervical cancer–burden and assessment of causality. *Journal of National Cancer Institute Monographs*, **3**.

Lanciano, R., Calkins, A., Bundy, B.N., *et al.* (2005). Randomized comparison of weekly cisplatin or protracted venous infusion of fluorouracil in combination with pelvic radiation in advanced cervix cancer: a gynecologic oncology group study. *Journal of Clinical Oncology*, **23**: 8289.

Morris, M., Eifel, P.J., Lu, I., *et al.* (1999). Pelvic radiation with concurrent chemotherapy compared with pelvic and paraaortic radiation for high-risk cervical cancer. *New England Journal of Medicine*, **340**: 1137.

Roden, R. and Wu, T.C. (2006). How will HPV vaccines affect cervical cancer? *Nature Reviews Cancer*, **6**: 753.

Renal cancer

Linehan, W.M., Walther, M.M., and Zbar, B. (2003). The genetic basis of cancer of the kidney. *Journal of Urology*, **170**(6 Pt 1): 2163.

Vogelzang, N.J., Scardino, P.T., Shipley, W.U., and Coffey, D.S. (2000). Epidemiology of renal cell carcinoma. In: Vogelzang, N.J., ed. *Comprehensive Textbook of Genitourinary Oncology*, 2nd edn. Philadelphia: Lippincott Williams & Wilkins, p. 101.

Flanigan, R.C., Salmon, S.E., Blumenstein, B.A., *et al.* (2001). Nephrectomy followed by interferon alfa-2b compared with interferon alfa-2b alone for metastatic renal-cell cancer. *New England Journal of Medicine*, **345**(23): 1655.

Motzer, R.J., Rini, B.I., Bukowski, R.M., *et al.* (2006). Sunitinib in patients with metastatic renal cell carcinoma. *Journal of American Medical Association*, **295**(21): 2516.

Motzer, R.J., Hutson, T.E., Tomczak, P., *et al.* (2007). Sunitinib versus interferon alfa in metastatic renal-cell carcinoma. *New England Journal of Medicine*, **356**(2): 115.

Ratain, M.J., Eisen, T., Stadler, W.M., *et al.* (2006). Phase II placebo-controlled randomized discontinuation trial of sorafenib in patients with metastatic renal cell carcinoma. *Journal of Clinical Oncology*, **24**(16): 2505.

Questions for student review

1) The pattern of cancer incidence
   a. Has remained relatively stable in Western nations over the last 50 years.
   b. Is related to the average age of the population in a nation.
   c. Is related to environmental factors.
   d. Has shown increases in all tumor types over the last century.

2) Lung cancer
   a. Is steadily increasing in incidence in males and females.
   b. Risk reduces after discontinuing cigarette smoking.
   c. Usually presents as a result of symptoms from metastases.
   d. Has a worse prognosis than breast cancer.

3) Breast cancer
   a. Is more common in Japanese women than in women living in United States.
   b. Is fatal in 75% of all cases.
   c. Is more common in postmenopausal women.
   d. Is managed differently in pre- and postmenopausal women.

3 Nature and Nurture in Oncogenesis

Michael Khan[a] and Stella Pelengaris[b]
[a]University of Warwick and [b]Pharmalogos Ltd, UK

Every politician, clergyman, educator, or physician, in short, anyone dealing with human individuals, is bound to make grave mistakes if he ignores these two great truths of population zoology: (1) no two individuals are alike, and (2) both environment and genetic endowment make a contribution to nearly every trait.

Ernst Mayr

This whole act's immutably decreed. Twas rehearsed by thee and me a billion years before this ocean rolled. Fool! I am the Fates lieutenant; I act under orders.

Herman Melville, Moby Dick

Key points

- Cancer is a tale of two genomes: the patient's original first edition and, derived from it but diverging over many years, the tumor's evolving work in progress.
- Although much will be discussed about the role of unavoidable genetic risk factors and acquired mutations, we must not forget the powerful influence of modifiable factors. Estimates suggest that anywhere between 35 and 45% of all cancers may be strongly influenced by ill-advised lifestyle choices such as smoking, poor intake of fruit and vegetables, excess alcohol, lack of exercise, and development of obesity.
- Without wishing to kill the suspense, much of what follows will lead the reader inexorably to the following conclusions: that cancer cells are spawned but also shaped by experience; inheritance and environment conspire together to turn an outwardly normal conformist and idealistic cell into a cancerous and solipsistic pariah.
- The central dogma of molecular biology implies that information is transmitted as follows: the DNA codes for RNA (transcription) and then the RNA codes for protein production (translation). The process is deterministic, as each stage depends completely on information encoded within the DNA and RNA. Moreover, it is also believed to be essentially unidirectional, to quote Francis Crick, "Once information has passed into protein, it cannot get out again."
- This view has required some revision since the 1950s. It has been known for some time that viruses can turn RNA into more RNA or even into DNA (hallmarks of RNA viruses and retroviruses). In fact, reverse transcription is routinely exploited in laboratory research. More recently, mammalian telomere maintenance and microRNA pathways provide further exceptions to the rule. Moreover, proteins can profoundly alter gene expression through epigenetic modifications such as methylation of DNA and acetylation of chromatin without altering the DNA code, though this falls rather short of being real reversal of information flow from protein to DNA. All of these processes have direct bearing on cancer biology.
- Any alterations to the usual flow of information can culminate in inadequate, excessive, or dysfunctional protein production and therefore in disease. Alterations to the DNA code, chromosomal rearrangements, changes in epigenetic regulation, and incorrect processing or translation of the mRNA can all disrupt normal protein production. Moreover, even subtle changes in the sequence or structure can drastically alter protein function and propensity to posttranslational modifications, both its own and that of interacting proteins.
- Cancers arise from normal cells by the accumulation of mutations or epigenetic changes (epimutations) that activate cancer-causing genes (oncogenes), inactivate cancer restraining genes (tumor suppressors), or facilitate the former by increasing DNA damage or compromising its repair (caretaker genes); mutations may be acquired by somatic cells or inherited through germline alterations in the DNA.
- Why do cancers happen in the face of all the various mechanisms designed to prevent it. Because DNA damage is unavoidable, cells affected by it innumerable and through replication cumulative – just throw in the key ingredients, nature and nurture and time.

(Continued)

The Molecular Biology of Cancer: A Bridge From Bench to Bedside, Second Edition. Edited by Stella Pelengaris and Michael Khan.
© 2013 John Wiley & Sons, Inc. Published 2013 by John Wiley & Sons, Inc.

- Currently, around 400 genes (or about 1 in 50 of our genes) are thought to be directly tumorigenic when mutated. Finding these can help unravel pathogenesis of a given cancer and could provide biomarkers to help define prognosis and even help select appropriate treatments and combinations. This knowledge is even being exploited in finding at-risk individuals before they develop cancer in order to target preventative treatments. Although there are some examples of all of these in clinical practice, there is still a long way to go before we can offer such personalized care for all.
- In some rare familial cancer syndromes, an essentially inactive copy of one of the key tumor suppressor or caretaker genes is inherited from one parent and confers a very high risk of developing cancer and often at a young age.
- In such cases, targeted screening for causative mutations in DNA, obtained by a simple blood test, is offered to other family members. This is exemplified by screening for *BRCA1* and *BRCA2* mutations in families with increased incidence of breast and ovarian cancers. However, the low incidence of mutations in these genes (less than 1 in 500 for *BRCA* genes) precludes their use in screening the general population. Genetic screening is also available for mutations in: the mismatch repair genes, *MLH1* and *MSH2*, in some hereditary colon cancers; *TP53* (Li–Fraumeni syndrome); *RB1* (retinoblastoma); *CDH1* hereditary diffuse gastric cancer; *MEN* (MEN1) and *RET* (MEN2).
- However, in most cases inherited susceptibility to cancer is determined by subtle variation in the coding and noncoding DNA of genes (polymorphisms, minor alterations in single nucleotides), each of which alone confers only a slight increase in risk of cancer. However, put a few of these together and mix in some carcinogens and you may well be several steps nearer to cancer.
- Genome-wide association studies (GWAS) that look for the presence of variation in genes by detecting single-nucleotide polymorphisms (SNPs) in large groups of patients have identified many susceptibility alleles for most common cancers. In general, the actual risk increase predicted by these is below 10% and has not translated into clinically useful screening tools to date.
- However, what does this 10% mean? Remember GWAS studies involve large numbers of patients. Does a given risk allele give everyone affected a slightly increased risk (10% closer to cancer, with the rest of the journey a chance affair driven by other inherited and somatic epimutations) or at the other extreme do 100 out of every 1000 individuals get cancer entirely because of it (i.e. this is very high risk for some people – we just don't know who), or more likely somewhere in between?
- The mitochondrial DNA, inherited entirely from the mother, is also subject to mutations, and some conferring increased risk of prostate cancer, for example, have been identified.
- In most cases, however, epimutations are somatic and are not inherited from a parent but rather are acquired by individual adult cells. Within an evolving tumor, if advantageous to the cell such mutations will be propagated to an expanding clone of progeny cells by natural selection.
- Acquired mutations are tumor-specific and are not present in germ cells. Put simply, **the patient's underlying genome and that of the tumor will differ by the acquired mutations present in the latter.** This fact is exploited in tissue-based biomarkers and in new treatments directed against those which are functionally important, one of the great success stories in modern cancer biology. Thus, pioneering successes in targeting the aberrant product of the *BCR-ABL* oncogene, in chronic myeloid leukaemia, with imatinib have been followed by other drugs directed against the constitutively active protein products of *HER2* in breast and *EGFR* in lung cancer respectively. These successes herald a new era of personalized medicine, for these and other cancers, which many authorities now consider within our reach.
- A risk factor is anything that increases the risk of developing a particular disease and may be avoidable or unavoidable, modifiable or unmodifiable. Once described and risk determined, preventive therapies may be offered to those with one or more risk factors; or with suitable bombast, "what can be foretold may be forestalled."
- However, in practice even effective and reliable preventative screening tests will not be implemented unless they are cost-effective, easily applied, and in themselves comparatively devoid of harm.
- The next best strategy is to diagnose the development of cancer at the earliest possible time so that curative treatments have the greatest chance of success. This in turn also requires the availability of reliable diagnostic tests, such as scans and biomarkers.
- Remember, biomarkers need not be functionally relevant to the disease process. In fact, the expression of any disease-irrelevant biomarker will do perfectly well so long as it recognises a cancer cell and may serve admirably as a **shibboleth** for disease classification and even in treatment selection.
- The ideal biomarker is one that can be detected in easily accessible body fluids that can be obtained with minimal invasiveness (e.g. saliva, sputum, urine, or peripheral blood) and would therefore be suitable for screening large numbers of people. Tumor cells (and other cells affected by the tumor) release proteins (and other molecules) into the circulation and these can be measured. Blood tests for prostate-specific antigen (PSA) (prostate), CEA (colon), CA-125 (ovary), and thryoglobulin (thyroid) are all in clinical use for monitoring patients after treatment and in screening for recurrence. However, their use as screening tools is limited due to poor sensitivity (they do not pick up early disease when curative surgery may be possible) and limited specificity (too many false positives). Even the best of these, PSA, is surrounded with controversy.
- Screening mammograms may be used for early diagnosis of breast cancer, a screening tool recently validated in large clinical studies.
- Geographical differences in cancer incidence, and the "migration effect" support the importance of environmental and lifestyle factors. Carcinogens are believed to act at least in part by promoting epimutations or by accelerating proliferation (and thereby propagation of cells with mutations).
- Smoking, diet, sex hormones, and increasing age influence risk of cancer. Recent studies have highlighted the risk of cancer in those with obesity and type 2 diabetes and that this risk might be reduced in part by use of metformin.

- Since groundbreaking cancer studies first showed that viruses can carry oncogenes and promote cancer by insertional mutagenesis, there have been major advances in the field of cancer-causing infections. Having fallen out of favor for some time, infection is again accepted as a significant contributor to up to 1 in 5 cancers globally – in particular the hepatitis viruses B and C, human papilloma virus (HPV), and the bacterium *Helicobacter pylori* are important causal agents in cancers of the liver, cervix/oropharynx, and stomach, respectively.
- Many countries have established vaccination programs for young women to prevent infection with cancer-causing strains of HPV to prevent cervical cancer. Other virus-related cancers are being actively studied. Various other organisms can contribute to tumor formation, with even the mysterious prion proteins implicated in pancreatic cancers.

Introduction

Heredity proposes and development disposes.

Sir Peter B. Medawar

'Tis misfortune that awakens ingenuity, or fortitude, or endurance, in hearts where these qualities had never come to life but for the circumstance which gave them a being.

William Makepeace Thackeray, The History of Henry Esmond

All cancers are caused by mutant genes. In some cases these may be inherited, but in most cases, cancer-causing mutations are acquired by non-germ cells (somatic cells) during life. Each cell experiences a unique set of environmental insults, internal malfunctions, and random setbacks. When added to a shared genetic makeup, this largely determines which cell will become a potential cancer-starter. No single cell lives through the same year as any other and to varying degrees they are exposed to a variety of DNA-damaging insults or make mistakes during the replication of their DNA. Together, the result is the emergence of cells with subtle differences in their DNA compared with the original inherited genome.

Where such mutations confer a growth advantage on a given cell, they allow it to give rise to an expanded clone and – by sequential acquisition and natural selection of further propitious mutations – to generate waves of clonal expansion culminating in the evolution of a cancer. Although the nature of these mutations varies from cancer cell to cancer cell, there are some common features. Thus, to date, just over 450 genes have been identified that may cause cancer when mutated. Of these, fewer than 20% have been shown to be inherited in the germline, and many of these appear to be involved in DNA damage responses and repair.

Inherited genetic susceptibility contributes to most if not all cancers in some way. In most cases, the cancers do not arise as a result of the powerful effects of rare mutations in single critical genes, though such cancer syndromes exist. Rather, in the majority of cancers genetic predisposition is a complex affair, operating through a range of less certain mutations and genetic variants that may increase risk of future cancer by a modest or small amount. In some cases, the risk associated with common allelic variations in peripherally relevant genes may be almost imperceptible, unless combined with multiple other gene variants or in the face of particular environmental challenges. Whereas mutations in genes acquired in somatic cells during life are the central players in cancer development and progression, inherited mutations may help to stage the process by increasing the available talent available for casting, thereby giving the whole performance a headstart on the path to cancer. This is exemplified by the earlier age of onset of common cancers in those with inherited susceptibility.

As this chapter traverses some of the most conceptually challenging terrain of cancer biology we will first try to provide the reader with an appropriate phrase book and travel guide to ease the journey. In time-honored tradition these aids will make some oversimplifications, which time spent among the chapters to come will slowly correct and embellish. By the conclusion of the chapter we hope to have given some insights into the role played in tumor initiation and evolution by genetic and environmental factors and in particular highlighted the extent to which susceptibility to cancer and responsiveness to treatment may be:
- predestined by genetic inheritance;
- influenced by environmental factors or infectious agents;
- determined by acquired somatic epimutations and natural selection.

To make sometimes difficult concepts more accessible we will on occasion take the scenic route and digress into a discussion of processes covered in depth elsewhere in the book. In some cases, such as the acquisition of genetic mutations (and epigenetic alterations) by single cells during adult life (as distinct from those inherited by all cells from the parents), this is unavoidable as it represents a major carcinogenic mechanism driven by environmental toxins and viruses. We also discuss how knowledge of these processes may actually pay dividends in the development of clinically important biomarkers and cancer screening tests for prevention, and early diagnosis of cancer and for use in a more functional subclassification of established cancers on the basis of predicted prognosis and response to particular treatments.

Cancer evolution as cellular snakes and ladders

Given the complexity of this chapter it will be helpful to bear this metaphor in mind while you read on. Cancer can be likened to a game of snakes and ladders. Viewed from the perspective of the many potential aspiring cancer cells that start as normal cells at the bottom left of life's playing board, the ladders are oncogenes and the snakes are tumor suppressors. Those with inherited susceptibility start the game further up the board and nearer to the final square, where the cancer will be realized. Stem cells and cells exposed to mitogens throw the die more often and have a greater chance of landing on a ladder, but also on a snake. Hardly any cells make it to the end of the board, because this board is overpopulated with snakes. How do any cells make it? Only because tens of millions enter the competition and because the game runs for 70–80 years inside each of us. Yet, still only one in three of us will harbor a winning cell.

The language of cancer genetics

Prophesy is a good line of business, but it is full of risks.
                                                            Mark Twain

Genes are runs of DNA that serve as blueprints for the manufacture of proteins – the functional and structural elements of the cell and of living organisms. It is known that less than 2% of genomic DNA encodes information for how a protein is put together, and another around 2% comprises regulatory regions that control how much, if any, protein will be made. This leaves a whopping 96% of DNA that, until recently, was largely, and as it turns out wrongly, dismissed as junk (more about this later).

With the exception of the sex chromosomes, you have two ostensibly identical copies of each chromosome and therefore two copies of each gene. Humans have 23 pairs of chromosomes – a pair of sex chromosomes (two X chromosomes in females and an X and a Y in males) and 22 pairs of autosomes. The number and configuration of chromosomes is referred to as the **karyotype**. Deviation from the normal karyotype is very typical of cancer cells and is referred to as **aneuploidy**. Genes usually occupy the same position on a given chromosome (**the genetic locus**) in all individuals unless something has gone badly wrong with the karyotype (note the Philadelphia chromosome in chronic myeloid leukemia and the key role of translocations as a contributory cause to cancer-relevant gene mutations). This does not mean that the two genetic loci on the two chromosome copies are necessarily completely identical.

Genes occur naturally in different versions (different nucleotide sequences), which may or may not result in the variants conveying different information, and these variants are referred to as **alleles**. Broadly, those alleles which occur at a frequency of greater than 1% in a population are termed **polymorphic** and those variants which occur at below 1% are termed **subpolymorphic**. To put this in context, most highly penetrant gene mutations (ones that almost always produce an unmistakable and clinically unpleasant outcome) occur in less than 0.1%.

In general, the genes you have in all their various allelic variants is your **genotype**; the physical manifestations of your genotype (how you look, function and what proteins you make) is your **phenotype**. Much of modern biology, and cancer biology in particular, is concerned with describing the effects of variations in genotype on resultant phenotype. Many mutations may not have a discernible effect on phenotype at all, or may indeed occasionally even improve the function of the protein product and, as a result, the overall viability of the organism in a particular environment – the "propellant of evolution." As first predicted by Charles Darwin, advantageous mutations will be retained and become increasingly common in a population, whereas disadvantageous ones will be eliminated through **natural selection**. This is because the lucky organisms with the new, better gene out-compete those with last year's model and raise more offspring, some of whom will have the better version and will themselves have more offspring and so on. Remember that natural selection is critically determined by the environmental context in which this takes place – lots of thick hair is useful in the cold but not when it is 40 degrees in the shade.

If you have one gene altered from the norm in some way you are **heterozygous** for that alteration (two different alleles); if both genes are unaltered or both similarly altered this is **homozygous**. Confusion sometimes arises when different mutations in the two alleles cause the same phenotype (e.g. neither gene makes any viable protein). In such cases the terminology will reflect the level of scrutiny that has been employed, whether this is based on phenotype observation or on sequencing of the genomic DNA. Please note that new alleles or mutations are only passed on to future generations of the organism if these are present in the germline – in sperms or oocytes or their precursors. On the other hand, mutations in somatic cells can be transmitted to cellular progeny during replication – in fact this is the predominant driver of tumorigenesis, ultimately giving rise to a clone of cells with a common ancestry – but such mutations are not passed on to the offspring of the organism.

Another important concept often referred to in the book, is that of **highly conserved** genes. Such genes appear in near-identical forms across many species and in general do not naturally occur in multiple alleles. The implication is that such genes have been around for a very long time and have long ago reached "Nirvana." The "Über"-proteins so made are the "Mary Poppins" of proteins – practically perfect in every way. Thus, any mutations will result in an inferior protein and will be eliminated over time.

On the other hand, genes that occur in several allelic variants of significant frequency in a particular population are not necessarily still a "work in progress." Recent mixing of different populations, each of which has selected the best alleles for their previous environmental context, can increase genetic variation. This is mostly a good thing because it reduces chances of nasty recessive diseases. However, matching the wrong alleles to the wrong environment can also potentially increase risk of diseases such as diabetes and some cancers if, for example, alleles best suited to low-energy diets are placed in an environment replete with energy-rich foods, either because the diet has been imported or the alleles exported!

Before we become over-enamored of our Darwinian view of cancer evolution, we should first set out the unique parameters within which this analogy holds water. In contrast to the evolutionary history of life on this planet, the whole dynastic descent of the cancer cell is played out within the space of a single human lifetime. With the dubious exception of fragments of cancer DNA pollinated by viruses, the death of the host signals the death of the cancer; the peculiar afterlife of the HeLa cancer cell line cannot in all honesty be regarded as their transcendence over Henrietta Lacks. Thus the game may follow a set of similar rules but the prize in the case of cancer is not ascendency but oblivion; there can be no clearer example of the essential difference between competing species and cancer cells, however shared the propelling force of mutations and natural selection. The latter evolve against a backdrop of cellular cooperation and selfless mutual support with other cells, because at least in part they are all closely related members of the same family – they are neither finches nor cuckoos, simply the black sheep of the family.

Cause and effect

In the strict formulation of the law of causality – if we know the present, we can calculate the future – it is not the conclusion that is wrong but the premise.
                                                            Werner Heisenberg

In most cases, the effect of any single allele appears to be very small and it is likely that discernible phenotype alteration is a product of allelic variation at many different genetic loci. Not all genotype–environment interactions necessarily change for the worse. The Japanese normally have greatly increased risks of stomach cancer compared to North Americans, but within a single generation those who emigrate to North America avoid this risk and instead experience that of the indigenous population, probably due to dietary changes. (The risks of diabetes and coronary disease show the opposite trend!) It is also worth noting that deciding to what extent these effects result from environment alone or gene–environment interactions is often challenging. Note how relatively few North Americans move to Japan and adopt that native lifestyle and diet, thereby precluding a test of increased gastric cancer risk with that direction of travel. It is often assumed that environmental factors and allelic variation may operate in the pathoetiology of the great majority of cancers. However, as we will see, in some cases genetic or environmental factors alone may be the major driving force.

At one extreme, certain genes are very important and must be in good working order for a long and healthy life (high-risk genes). Alterations in these important and often highly conserved genes almost invariably end in a very evident and very bad clinical outcome (**high-penetrance mutations**) and diseases resulting from such inherited mutations generally occur at a younger age than the sporadic counterpart. To keep things relevant, such inherited mutations may result in a complete failure of that gene to direct production of a cancer-restraining protein, or a protein product is dangerously changed or activated in a way that promotes cancer. Such changes give "would-be" cancers a headstart, explaining why these cancers often present at a younger age than the more usual sporadic version. These monogenic diseases (the alleles of a single gene determine the presence or absence of disease) are termed either **dominant** or **recessive**, depending on whether they require one or both inherited copies of the gene to be mutated. In cancer this is a bit more complex than for other inherited diseases because we have to worry not just about inheriting bad alleles in the germline, but also about somatic gene alterations (mutations) occurring in adult life in individual cells. Remember that cancer can uniquely start with genetic alterations in just one cell.

For those genes that promote cancer when activated, **oncogenes**, a mutation in a single copy may suffice, whereas for a gene which puts the brakes on cancer, a **tumor suppressor**, both gene copies must be inactivated. But, because life is complicated, there is also a half-way house in which a slightly less bad outcome may follow loss of a single gene (for example the reduced protein dose means you still have an increased risk of cancer but at an older age) – such genes are termed **haploinsufficient**.

Big changes and big consequences

Hereditary cancers resulting from major genetic defects are rare, but such families have been extensively studied and much has been learned about cancer biology as a result. Generally such hereditary cancers follow Mendelian principles and are readily diagnosed if affected family members are known. OK, if this is all true then why do the best-known familial cancers involve inheriting a single nonfunctioning copy of a tumor suppressor gene such as *RB*? There are two explanations for this example: (1) Because there are so many cells with only one working copy of the tumor suppressor and so many opportunities to acquire mutations during adult life, it is inevitable that the good copy will get lost in one cell or another (the **two-hit hypothesis**) and set the scene for tumorigenesis. (2) The tumor suppressor is haploinsufficient (you need all the protein you can make, you have none to spare, and half as much isn't enough to prevent disease).

It may be helpful here to consider these two scenarios in traditional terminology. In scenario 1, cancer is autosomal dominant at the whole organism level, as pretty much everybody with inherited loss of *RB* will get cancer, however when we start to look at the genome at a cellular level this is autosomal recessive as both copies of *RB* need to be inactivated. In scenario 2, the disease may appear autosomal dominant but there will be a dose-effect with generally worse outcome if both copies are lost. Over the last 30 years there have been numerous successes in finding high-penetrance susceptibility genes for cancer, including the mismatch repair genes in Lynch syndrome–related colorectal cancer and *BRCA* genes for breast and ovarian cancer. However, in each case these genes account for less than 90% and 80% of the hereditary risk of these cancers, respectively. At the time of going to press, a further high-penetrance susceptibility gene for ovarian cancer, *RAD51D*, was identified.

Small changes and butterfly wings

Much more common and more difficult to pin down are the effects of more subtle variation (**low- to moderate-penetrance mutations**) in one or multiple genes. As mentioned, many genes occur naturally in subtly different versions, which all still work more or less and do not cause inevitable and easily observed changes in the individual. Thus variant genes (alleles) may result in proteins that differ by as little as one amino acid, may have no discernible effect whatsoever either on protein function or on disease susceptibility or might, like the flap of a butterflies wings, conjure up a cancer in the future. Implicit in this view of risk is the subordination of simple random bad luck. In other words, we erroneously accept as fate what we lack sufficient information and understanding to have predicted. If a seemingly miniscule change can alter the course of a life, then we only need to detect those changes or conversely introduce small ones of our own design and a satisfactory outcome is guaranteed.

However, as with most of life's tribulations, responsibility rarely rests with any single factor but rather a series of events taking place in the wrong place at the wrong time may result in catastrophe; given the complex chain of happenstance it is understandable that we may often fail to identify it. Thus, even comparatively weak effects are amplified and given opportunities for fortuitous synergies with other mutations over time; this is compounded by the pervasiveness of hereditary mutations and allelic variants that affect all somatic cells from the off, and of those acquired ones propagated by clonal expansion during life. Moreover, if tiny changes can have drastic consequences over long periods of time, then we must have tools sufficiently sensitive to measure these tiny changes accurately enough, or the resources to prospectively follow the progress of patients for long enough to make the effect obvious.

This is no minor issue, as the inability to measure precisely enough limits our ability to make predictions about the future behavior of many complex systems. Weather forecasts – as an oft-cited example – offer a tolerable level of accuracy only over a week at most. In forecasting cancer, what chance do we have to determine every single relevant genetic, epigenetic, and other variation in all individuals at the outset and thence all the potential somatic variations accumulating during life? Don't get despondent though, because even if we accept that complete understanding may never be granted us, we have made, are making, and will continue to make progress in finding risk factors that we can do something about. Moving beyond politics-inspired rhetoric, progress has been made in linking allelic variants with disease risk, and some notable examples will be discussed later. All that was needed were sufficiently large numbers of individuals who could be genotyped by various means and then followed up until a detectable pathology developed in some.

There are practical ramifications to this discussion, if many cancers arise just through simple bad luck then we should stop agitating over the future and get on with life. However, if as seems likely, it is a blend of predictable and modifiable factors that may measurably increase risk and then some bad luck thrown in, then we can successfully adapt. The challenges are substantial. Even with hindsight the causes are often difficult to pin down and it is tempting to apportion blame to the most evident or prominent among these, even if this proves intractable as a prevention target, when a more nuanced understanding might have revealed a readily avoided or treated risk factor.

Predicting the future

> Physic, for the most part, is nothing else but the substitute of exercise and temperance.
>
> *Joseph Addison*

Can we state this in molecular terms? Much of the hope in personalized medicine is predicated on a belief that by determining variations in multiple genes we may be able to make fairly accurate predictions about the future experiences of disease in later life and help guide individuals towards avoiding risky behaviors and even preventative medicine. Where does random chance fit in? Try as we might to exorcise chance events from our world, it is almost certain that many of the mutations underlying disease susceptibility arise by chance events or due to unavoidable insults – so for now the difference may have no practical significance. But perhaps the single most random event in determining genotype is sexual reproduction – self-evidently you have no opportunity for selecting your parents and it is certain that they exercised little logic in selecting one another.

Let us stray briefly to consider some clinical genetics scenarios that exemplify some of these concepts. Familial hypercholesterolemia (FH) is the most common monogenic disease and is caused by a high-penetrance (high-risk) mutation that affects 1 in 500 people. It is autosomal dominant, so that on average one out of every two children born to an affected individual will also get FH and as a result will develop the very high cholesterol that typifies the disorder – so far entirely predictable. However, even if untreated, not all individuals with FH will die early through cholesterol-related complications, mainly heart attacks, though most will. Conversely, not all FH patients who are effectively treated will avoid an early heart attack, though again most will. Why not? Because variations in many many other genes, known and unknown, influence the extent of the phenotype, and others affect the risk of heart disease and then we have the influence of lifestyle thrown into the brew. Together, these can be positive and potentially offset the risk posed by a high-penetrance mutation or, on the other hand, negative and negate the benefits of cholesterol-lowering treatments.

This is important to bear in mind because one can easily become seduced by the simple view of biology derived from studying cells and animal models. Such models, although extraordinarily useful in addressing specific questions, have certain limitations. Thus, unlike humans, inbred rodent models of diseases such as FH or of cancer are very similar genetically (and inhabit a uniform, controlled, and sterile environment) and therefore if you can cause a cancer by a particular genetic manipulation(s) in one it will hold true for all, whereas this is manifestly not the case for humans, in whom genetic and lifestyle variation will powerfully modulate the effects of any high-penetrance mutations (both positively or negatively). The importance of looking at genes in functional interacting networks is likely to be even more critical when the effects on cancer risk of any single variant gene are going to be slight, particularly where there are no readily measured metabolic or physiological consequences, as is mostly the case with cancer risk. Moreover, as the variant gene may only have an effect, for example, in the presence of some external agent such as tobacco smoke, or when following a bad diet, the immensity of the task is only too apparent.

Genome-wide association studies (GWAS) are now increasingly identifying associations between common variability and cancer and have revealed, perhaps unsurprisingly, that common genetic variability does not explain all of the genetic predisposition to disease (see Box 3.1). Advances in technology as well as reduction in costs are allowing some of the apparent gap to be addressed by genome sequencing and data processing (see Chapter 20). However, huge numbers of patients may have to be studied and at great cost financially and in time in order to identify these relatively low-risk genetic susceptibilities.

Again, a clinical example here may be helpful and encouraging. Even where the disease etiology is clouded and seems to involve a plethora of complicated interacting genetic variants and lifestyle, we can take positive and useful action. In the case of type 2 diabetes, disease may only arise if at some point an at-risk individual makes a lifestyle choice ill-suited to their specific genotype (whatever that may be); thus, sloth and gluttony are particularly ill-suited to somebody with a strong family history of type 2 diabetes. Why? Because, with a 1 in 2 chance of developing diabetes the odds of surviving such a lifestyle would be stacked against you. We do not need to know anything about genotype before recommending a healthy lifestyle and avoiding obesity. It would, however, be useful to have some biomarkers that narrow down risk further, and many studies are addressing this.

Nonhereditary mutations

There is, of course, another way in which altered genes come about. Genes can also become altered in single cells during adult life – **somatic mutations** – under the influence of chance events during DNA synthesis, byproducts of oxidative metabolism, or through exposure to carcinogens. And, of course, to complicate things, this may happen because of mutations in genes that otherwise help you to avoid acquiring mutations (exemplified by genes encoding proteins needed for DNA repair). Genes may even become silenced by changes in methylation without any alterations in the DNA code (see Chapter 11). These factors all represent challenges for those attempting to find biomarkers, because:

- such changes are confined to affected cells within the tumor or proto-tumor, which may be undetected or inaccessible;
- even if cells or cell contents find their way into more accessible body fluids it may no longer be possible to quantify cancer-related proteins or mRNAs reliably and epigenetic changes will not be picked up by screening for mutations in circulating tumor cells;
- these somatic changes are essentially restricted to the individual patient and will not be inherited by any offspring, though the propensity to acquire such changes could be hereditary!

A light at the end of the tunnel

Despite the complexity of the task, there have been notable successes in cancer risk prediction and genetics. Obesity and resultant sequelae such as diabetes greatly increase risk of certain common cancers and are fairly readily screened for and preventive strategies for obesity are being developed in many countries. BRCA genes have shown their value as predictors of risk of breast and ovarian cancer, the presence of cancer-specific mutations in growth factors or estrogen receptors is now used to guide appropriate targeted treatment in breast cancer (herceptin or anti-estrogens), and, despite recent controversies, prostate-specific antigen is in widespread use as a screen for prostate cancer. A more detailed discussion of the practicalities of risk and risk screening follows below.

Risk factors

> "Winwood Reade is good upon the subject," said Holmes. "He remarks that, while the individual man is an insoluble puzzle, in the aggregate he becomes a mathematical certainty."
>
> *Sir Arthur Conan Doyle, The Sign of the Four*

A risk factor is anything that increases the likelihood (probability) of an individual developing a particular disease, such as cancer. Some risk factors are modifiable by the patient, such as tobacco and saturated fat consumption, some, such as diabetes, may be modifiable by the use of therapeutics. On the other hand, some risk factors, such as age, family history, and gender are simply not modifiable, though the risk of cancers associated with these can be reduced by targeting other risk factors that are themselves modifiable. Importantly, not everybody with risk factors gets cancer.

It is worth pausing here for a moment of quiet reflection. Risk is a difficult concept for most clinicians, and even more so for most patients. Professionals are often best at predicting what has become obvious to them, but is not necessarily so to a patient. For example, radiological studies prompted by a persistent cough may subsequently demonstrate metastatic lung cancer that will almost invariably drastically shorten life, even with most current treatments. A prognosis offered under these circumstances has considerable individual validity, and patients and carers may plan for the future with some degree of certainty. At the other extreme, however, the presence of certain risk alleles may well show a statistical association with future risk of developing a particular type of cancer, but this may be very slight and only manifest in those who pursue a particular lifestyle. In practical terms, the low risk means that individualizing any risk-modifying strategy, even one proven effective in clinical trials, is fraught and will not necessarily benefit all. Some will live a bit longer, some may live a lot longer, and some not at all – and no one will ever know which patients did what. How do you advise in such a case? It is often worth considering the concept of avoiding risk using simple analogies.

Drowning is clearly a bad thing which we should endeavor to prevent. But how? Do we offer risk-reducing strategies to everybody – or focus on those above an agreed threshold of risk? People who live near the sea are more likely to die of drowning than a Bedouin in the Sahara Desert. Wearing of a lifejacket might well be a cost-effective and accepted strategy for the former but would be wholly inappropriate for the latter. We could try and individualize the intervention more accurately by determining other risk factors, such as ability to swim and actual exposure to risk activities, such as sailing or swimming. Finally, we should have some idea of whether the intervention is devoid of harm. Making sure that everybody can swim would be a reasonable public health approach and would perhaps have additional benefits, such as fitness, and could be rolled out across whole populations. Stopping people swimming or sailing completely would be inappropriate as it would contribute to obesity, might compromise livelihoods, and would have a significant negative effect on quality of life.

As far as lifejackets are concerned, nobody would be convinced to wear one if they were never exposed to bodies of water they could ever drown in. This is not academic, as any risk-avoidance advice we give will inevitably carry with it some adverse socio-economic or health consequences. There are people now whose lives have been blighted by the knowledge that they are at risk of a particular cancer in the future. The anxiety may be worth it if an effective preventative therapy exists, but the relative risk–benefit must be assessed (see Chapter 16). It is also important to note that even where risk of a particular cancer is strongly related to the presence of a nonmodifiable risk factor, such as female gender and breast cancer, risk may be successfully modified by targeting something else, such as weight and use of exogenous estrogens, for example.

Bradford Hill and modern epidemiology

In clinical practice, how do we identify a potential causal relationship between an identified factor and a disease? In the 1960s, the British statistician Sir Austin Bradford Hill formulated a set of minimal conditions to provide evidence for causation. These criteria are as follows:
- Temporal relationship (the cause has to precede the effect)
- Strength of association (statistical tests of the relative risk associated with the factor)
- Reproducibility (the same relationship is consitent and replicated across different studies)
- Dose–response relationship (greater exposure means greater risk)
- Modifiable by experimentation (linked to above – intervention to minimize exposure reduces risk)
- Plausibility (consistent with known biological or pathological understanding)
- Coherence (within the context of overall knowledge – may be broader than plausibility)
- Specificity (a cause produces a specific effect, though this may be difficult to apply to risk factors that damage more than one tissue)

Box 3.1 Genome-wide association studies

In a genome-wide association study (GWAS), groups with disease are compared with a matched group (similar age, gender, weight, etc.) without disease. DNA is usually obtained from peripheral blood cells and the DNA, or genome, is isolated, purified, and then analyzed for preselected single-nucleotide polymorphisms (SNPs) that can demonstrate genetic variation. Those SNPs that are significantly more common among the disease group are regarded as associated with the disease and can thereby identify the region where the genetic susceptibility locus resides. The SNP may not be directly causal and all nearby genes must be considered as potential culprits until further studies have been conducted, including sequencing of the DNA in that region. GWAS have been facilitated by the Human Genome and HapMap projects and are relatively quick and cost-effective.

GWAS have confirmed the molecular complexity and heterogeneity of cancer with around 100 regions associated with various cancers and little overlap across these. Thus, in prostate cancer more than 30 susceptibility alleles have been found. In NSCLC, however, in which an environmental cause predominates, only three regions have been identified.

A catalog of published GWAS is available at www.genome.gov/gwastudies/. The Cancer Genetic Markers of Susceptibility (CGEMS) project website also contains a useful collection of data on genetic susceptibility to prostate and breast cancer (http://www.ncbi.nlm.nih.gov/gap).

- Abdominal aortic aneurysm
- Acute lymphoblastic leukemia
- Adhesion molecules
- Adiponectin levels
- Adverse response to carbamazepine
- Age-related macular degeneration
- AIDS progression
- Alcohol dependence
- Alopecia areata
- Alzheimer disease
- Amyloid A levels
- Amyotrophic lateral sclerosis
- Angiotensin-converting enzyme activity
- Ankylosing spondylitis
- Arterial stiffness
- Asparagus anosmia
- Asthma
- Atherosclerosis in HIV
- Atrial fibrillation
- Attention deficit hyperactivity disorder
- Autism
- Basal cell cancer
- Behçet's disease
- Biliary atresia
- Bilirubin
- Bipolar disorder
- Birth weight
- Bitter-taste response
- Bladder cancer
- Bleomycin sensitivity
- Blond or brown hair
- Blood pressure
- Blue or green eyes
- BMI and waist circumference
- Bone density
- Breast cancer
- C-reactive protein
- Calcium levels
- Cardiac structure and function
- Carnitine levels
- Carotenoid and tocopherol levels
- Celiac disease
- Cerebral atrophy measures
- Chronic limphocytic leukemia
- Cleft lip and palate
- Cognitive function
- Colorectal cancer
- Conduct disorder
- Corneal thickness
- Coronary disease
- Creutzfeldt–jakob disease
- Crohn's disease
- Cutaneous nevi
- Dermatitis
- Drug-induced liver injury
- Endometriosis
- Eosinophil count
- Eosinophilic esophagitis
- Erectile dysfunction and prostate cancer treatment
- Erythrocyte parameters
- Esophageal cancer
- Essential tremor
- Exfoliation glaucoma
- Eye color traits
- F-cell distribution
- Fibrinogen levels
- Folate pathway vitamins
- Follicular lymphoma
- Freckles and burning
- Fuchs coemeal distrophy
- Gallstones
- Gastric cancer
- Glioma
- Glycemic traits
- Hair color
- Hair morphology
- Handedness in dyslexia
- HDL cholesterol
- Heart failure
- Heart rate
- Height
- Homeostasis parameters
- Hepatic steatosis
- Hepatitis
- Hepatocellular carcinoma
- Hirschsprung's disease
- HIV-1 control
- Hodgkin lymphoma
- Homocysteine levels
- Hypospadias
- Idiopathic pulmonary fibrosis
- IgA leves
- IgE leves
- Inflammatory bowel disease
- Intracranial aneurysm
- Iris color
- Iron status markers
- Ischemic stroke
- Juvenile idiopathic arthritis
- Keloid
- Kidney stones
- LDL cholesterol
- Leprosy
- Leptin receptor levels
- Liver enzymes
- Longevity
- LP (a) levels
- LpPLA(2) activity and mass
- Lung cancer
- Magnesium levels
- Major moog disorders
- Malaria
- Male pattern baldness
- Matrix metalloproteinase leves
- MCP-1
- Melanoma
- Menarche and menopause
- Meningococcal disease
- Metabolic syndrome
- Migraine
- Moyamoya disease
- Multiple sclerosis
- Myeloproliferative neoplasms
- N-glaycan leves
- Narcolepsy
- Nasopharyngeal cancer
- Neuroblastoma
- Nicotine dependence
- Obesity
- Open-angle glaucoma
- Open personality
- Optic disc paramaters
- Osteoarthritis
- Osteoporosis
- Otosclerosis
- Other metabolic traits
- Ovarian cancer
- Pain
- Paget's disease
- Pancreatic cancer
- Panic disorder
- Parkinson's disease
- Periodontitis
- Peripheral arterial disease
- Phosphatidylcholine levels
- Phosphorus levels
- Photic sneeze
- Phytosterol levels
- Platelet count
- Polycystic ovary syndrome
- PR interval
- Primary biliary cirrhosis
- Primary sclerosing cholangitis
- Progranulin levels
- Prostate cancer
- Protein levels
- PSA levels
- Psoriasis
- Psoriatic arthritis
- Pulmonary function (COPD)
- QRS interval
- QT interval
- Quantitative traits
- Rrcombination rate
- Red versus nonred hair
- Refractive error
- Renal cell carcinoma
- Renal function
- Response to antidepressants
- Response to antipsychotic therapy
- Response to hepatitis C treatment
- Response to metformin
- Response to statin therapy
- Restless legs syndrome
- Retinal vascular caliber
- Rheumatoid arthritis
- Ribavirin-induced anemia
- Schizophrenia
- Serum metabolites
- Skin pigmentation
- Smoking behavior
- Speech perception
- Sphingolipid levels
- Statin-induced myopathy
- Stroke
- Systemic lupus erythematosus
- Systemic sclerosis
- T-tau levels
- Tau AB1-42 levels
- Telomere length
- TEstic germ cell tumor
- Thyroid cancer
- Tooth development
- Total cholesterol
- Triglycerides
- Tuberculosis
- Type 1 diabetes
- Type 2 diabetes
- Ulcerative colitis
- Urate
- Venous thromboembolism
- Ventricular conduction
- Vertical cup–disc ratio
- Vitamin B_{12} levels
- Vitamin D insufficiency
- Vitiligo
- Warfarin dose
- Weight
- White cell count
- YKL-40 levels

Published genome-wide associations through December 2010, 1212 published GWA at $p \leq 5 \times 10^{-8}$ for 210 traits. Taken from the National Institutes of Health GWAS website (www.genome.gov/gwastudies/). Credit: Darryl Leja and Teri Manolio.

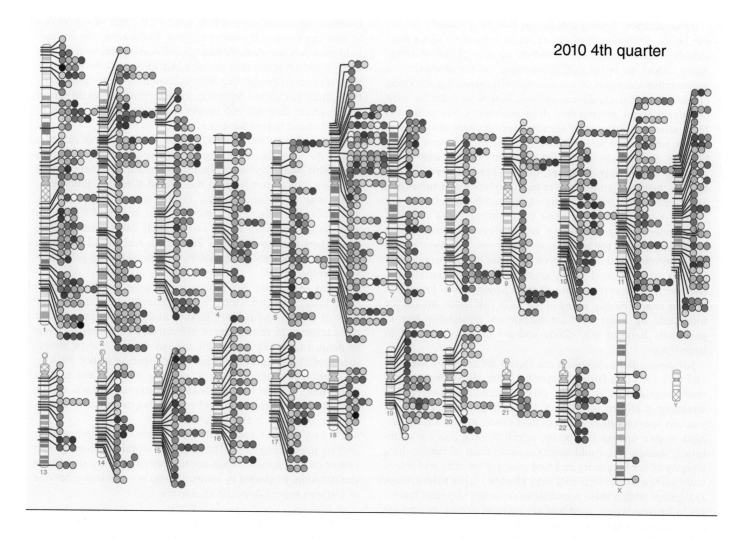

- Analogy (are there other similar examples and have other explanations been considered and tested by experimentation etc). A risk factor for cancer is just that – its presence does not invariably guarantee that an individual will go on to develop a particular cancer nor does its absence imply that the person will not. No matter how comforting the notion, a butterfly flapping its wings will not always conjure up a storm and neither will a host of frenzied eagles. Thus, being a man is a risk factor for prostate or testicular cancer but not all men will be affected. Obviously, some risk factors are much more predictive than others, but those known to predict a very high risk of developing cancer are relatively few. The best-known risk factors are not modifiable and the most important is age. The older you are, the more time you have had to acquire cancer-causing mutations and to be exposed to environmental carcinogens. Incidentally, under most circumstances gender is not regarded as a modifiable risk factor, but some aspects of gender, such as hormonal profile, are.

In terms of avoidable or modifiable risk factors, the greatest successes have resulted from the screening of whole populations or genders for signs of developing cancer or by identifying strong and avoidable or treatable environmental risk factors such as tobacco smoke, various occupational hazards (e.g. asbestos exposure), obesity, and infection with viruses (e.g. HPV). The relevant action has then proved relatively straightforward – don't smoke, follow safe working and sexual practices, eat well and exercise, accept vaccination against HPV if you are a young woman and attend for screening cytology or radiology after a certain age.

Finding the occult vulnerable before they get cancer or at least while they are still in the early curable stages of the disease is now the big challenge – that is, finding all those many people out there who are at increased risk of cancer for some as yet unknown or unsuspected reason. These groups can then be offered preventative treatments and more rigorous screening. This will require a molecular and/or genetic approach, which has to date proved much less successful. In fact, there are remarkably few reliable screening tests able to accurately identify those at risk of particular common non-inherited (sporadic) cancers, and even fewer that will accurately foretell those who will experience serious adverse effects or early death as a result of that cancer (an issue exemplified by recent controversies surrounding screening for prostate cancer in men). Actually, this becomes a matter of philosophy – determinists hold that the universe proceeds through a chain of events conforming to the laws of cause and effect.

By implication, finding risk factors that do accurately predict the future development of cancer might eventually allow individuals or particular groups of individuals at high risk of developing cancer to be identified, opening up the possibility of preventative medicine. Such an approach, involving screening of individuals for cardiovascular risk factors in order to most effectively target preventative drug treatments, has prevented heart attacks in those who already have coronary disease (secondary prevention) and in those with diabetes, adverse family histories, and even in unselected apparently healthy individuals (primary prevention). However, this has been achieved largely because high-risk groups are readily identified and prioritized for screening and subsequently cardiovascular risk factors are readily determined in the clinic by recording blood pressure, age, gender, history of smoking, family history of heart disease, and taking a blood test to measure cholesterol. Moreover, extensive epidemiological and prospective studies of huge numbers of patients have allowed us to create reasonably accurate and effective risk charts that relate all these parameters to future risk of heart attack. Finally, many of the main risk factors are readily modifiable by drugs that are themselves remarkably free of potentially harmful side effects and are of proven benefit in large trials.

For cancer, few established risk factors are currently available and they are often fairly nonspecific. Thus, all women are currently regarded as at risk of cervical and breast cancer, and screening is offered to all after a certain age (see below). All smokers are potentially at risk of lung cancer and all are encouraged to give up, but no specific screening programs are established, though one could envisage some form of regular lung imaging in smokers to try and find cancer at an early and potentially curable stage. Tests that may identify higher than average risk groups within these populations are in development but are yet to become widely used and are far from perfect. In the case of cervical screening, universal uptake of HPV vaccination may render improved screening redundant. Similarly, the increased risk of cancer in obese and diabetic patients may be managed by improving lifestyle and by application of drugs such as metformin, rather than by developing better biomarkers or actively looking for cancer. Other clinically apparent risk factors that are currently used to target at-risk groups for cancer screening, include the presence of multiple colonic polyps or ulcerative colitis, but as the screening tests then employed are designed to diagnose cancer early they are not strictly preventative. What is immediately obvious from this list is the near absence of measures of clinically silent risk factors that predict the future development of cancer (there are as yet few parallels for serum cholesterol or blood pressure readings). Finally, preventative treatments, aside from lifestyle modification, are generally similar to those used to treat established cancers and are expensive and often highly toxic.

Lifestyle could realistically be targeted without any form of screening on the basis that good diet and exercise should be good for all. Many studies have reported on the substantial increased risk of breast and colon cancer in obese people and those with type 2 diabetes. Unfortunately, in the United Kingdom, people with a healthy body weight are becoming rarer. Two in three men (67%) and more than one in two women (56%) are overweight or frankly obese, and this is increasing at an alarming rate. If current trends continue, we will all be obese by 2050! Some encouraging results suggest that some of the risk of cancer can be offset by treatment with metformin. Sadly, achieving changes in lifestyle has proved more difficult. What about less well-known lifestyle factors potentially posing a risk of cancer? Recently, the EPIC (European Prospective Investigation into Cancer and Nutrition) study has started reporting on dietary influences on cancer risk in (more than 500 000 people in Europe, including close to 100 000 in the United Kingdom).

Developing screening tests for primary prevention of cancer is clearly the ideal, but screening for recurrence or progression in those with established cancers is also important. There are around 2 million cancer survivors in the United Kingdom (2008 estimate) with breast and prostate making up the largest groups. Moreover, these figures are increasing at around 3% year on year. So this is a very sizable task in itself. Most current useful biomarkers are those such as PSA, CEA, and CA-125 that are used to monitor cancers of prostate, colon, and ovary post treatment, respectively.

In this section we will discuss recent progress in addressing these issues and also discuss what is known about the relative roles of nature and nurture in the causality of cancer. Recent rapid progress in genotyping of individuals as a potential means of predicting at risk groups for various cancers and in molecular profiling of tumors (including sequencing tumor DNA to find mutations, and looking at changes in gene and protein expression) to tailor treatments and predict outcome will also be addressed. The current status of various carcinogens and cancer-causing organisms is also discussed.

It is easy to come away with the view that our lives are spent entirely in pursuing suicidal lifestyles in a world overflowing with cancer-causing agents. This is not the case. Maybe we should, on this occasion, be guided by someone who is – arguably – not one of history's most risk-averse characters.

> The torment of precautions often exceeds the dangers to be avoided. It is sometimes better to abandon one's self to destiny.
> *Napoleon Bonaparte*

Preventing cancers

> She could not explain in so many words, but she felt that those who prepare for all the emergencies of life beforehand may equip themselves at the expense of joy.
> *E.M. Forster, Howards End*

Although there have been some notable successes in cancer prevention, including screening for breast and cervical cancer, readily determined risk factors for future development of cancer are generally not available for the common cancers. In fact, we must remember that in many cases we have no idea why a particular person has developed cancer, even if we assume the operation of the various mechanisms described in this book. However, as has become clear already, in some cases risk factors have been identified and this knowledge has helped us to develop some specific preventative strategies, ranging from public health initiatives aimed at the whole or a large part of the population through to genetic testing offered to selected individual patients with a worrying family history.

Most tests for potential "markers" of risk or of early disease are complex and invasive or require costly imaging that is not always universally applicable (cervical cytology, biopsy, X-ray, and scanning). Moreover, these largely pick up established disease, whereas the ideal screening test would identify individuals who have as yet not developed cancer at all. There are several examples of screening tests (see Chapter 18) that will be discussed later and in Chapter 17. Even where noninvasive markers that precede cancer onset have been described, the predictive strength is often not robust enough to guide use of preventative treatments, particularly where these are potentially harmful, such as surgery. Currently, mastectomy and salpingo-oophorectomy are widely used by carriers of *BRCA1* or *BRCA2* mutations to reduce risk of developing tumors of the breast and ovary, respectively, but how many would never have developed clinically meaningful cancers without such surgery is not clear.

One approach to improving the accuracy of predicting the future risk of cancer in individuals is to look at more than one risk factor concurrently (as is done for heart disease), many of which have been and continue to be described. It is also hoped that information on heritable genetic factors in the patient, as well as on how individual genes may become mutated specifically in cancer cells, may also help in prognostic and treatment decisions.

Risk factors may be environmental (lifestyle-related or unavoidable) or genetic (inherited) and are generally specific for particular cancers. Inherited factors are clearly of major importance and an historical overview of theories of heredity is given in Box 3.2. However, most cancers arise through a combination of genetic and environmental factors and less commonly by extreme exposure to radiation, viruses or cancer-causing chemicals (carcinogens) or combinations of genetic factors only (Fig. 3.1). Having a risk factor, or even several, does not guarantee that an individual will get the disease, but such guarantees rarely exist in disease prevention (where hindsight is not yet available!) and are unlikely to do so in the foreseeable future. Hence the attraction of early diagnosis, where the disease is at least present (even if outcome is less certain). What is practically required is a means of calculating risk of future cancer and cancer-related morbidity or mortality accurately enough to be able to safely apply preventative treatments. In other words, if that treatment is an entirely nontoxic and cheap drug, then treating a few individuals who would never have developed or suffered as a consequence of disease is OK, whereas if the treatment is mastectomy it is not (see Box 1.1 – Screening).

Many risk factors for cancer are avoidable. There is no need to screen individuals before advocating smoking cessation, healthy diet, and using sunscreen, for example. Such recommendations

Box 3.2 Developments of the theory of inherited traits

The Austrian monk Gregor Mendel's now famous early research into heredity in the mid-nineteenth century, using hybrid common garden peas, are among the earliest descriptions of what has in recognition been termed "Mendelian inheritance." Importantly, he also proposed a simple hypothesis for the operation of these laws, namely that observed traits are determined by discrete "factors," now called **genes**. Ironically, Mendel's work was largely ignored for 40 years, after which its "rediscovery" helped launch the new field of genetics. At the same time as Mendel was breeding peas, Charles Darwin published his work *On the Origin of Species*, describing his theory of evolution, namely that species evolved through natural selection under the pressure of environmental factors. With Darwin's theory and Mendel's laws, the foundations for genetics had been laid.

The chromosome theory of heredity (that then hypothetical entities called genes are parts of chromosomes) was originally proposed by the German biologist Theodor Boveri and an American student called Walter Sutton. In 1902 they observed that the behavior of chromosomes at meiosis was compatible with the observations and hypotheses of Mendel: genes are in pairs (and so are chromosomes); the alleles of a gene segregate equally into gametes (so do the members of a pair of homologous chromosomes); different genes act independently (so do different chromosome pairs). However, it was a further 50 years before the nature of the genetic material was discovered.

Eventually, the X-ray crystallographic work of Rosalind Franklin and Maurice Wilkins would lead to the model proposed by James Watson and Francis Crick in 1953. This model postulated that the structure of a gene lay in the sequences of the nucleotide bases (A, G, C, and T), replication of which distributed these genes to offspring and the ability of changes in the base sequence to result in mutations to the genes. Finally, a molecular explanation was beginning to emerge for the earlier work of Darwin, namely that it was the DNA upon which natural selection acted. Thus, changes in the sequence of bases in the DNA (mutations) would lead to different genetic traits that could be selected for or against by changes in the environment.

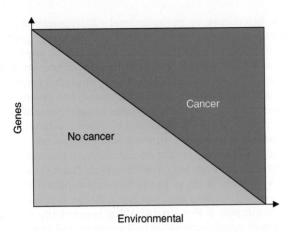

Figure 3.1 Cancer susceptibility is a result of a combination of genetic and environmental factors. The cumulative effect of both genetic and epigenetic changes and exposure to environmental factors determine the likelihood of developing cancer. In this model, only in extreme cases will cancers arise entirely due to either genetic factors alone or environmental factors alone.

are public health issues and should apply to all. It was 50 years ago that Richard Doll and A. Bradford Hill reported findings on a cohort of British doctors showing a strong link between cigarette smoking and lung cancer.

Risk factors and cancers can broadly be divided into two categories:

1. **Environmental**, including:
   a. lifestyle-related risk factors such as smoking, unhealthy diet, or unprotected exposure to strong ultraviolet (UV) radiation in sunlight;
   b. factors pertaining to the environment not under the direct control of the individual, such as high levels of radiation, cancer-causing chemicals, and certain infections;
   c. unavoidable natural processes, such as free radical generation, water-catalyzed reactions, endogenous hormones, and cosmic rays.
2. **Inherited genetic**, including
   a. single-gene (monogenic) disorders that are inherited from a parent and are direct causes of a given disease (these are generally rare and include inheriting a defective copy of the *p53* gene – Li–Fraumeni syndrome);
   b. polygenic disorders caused by variations in multiple genes, often occurring in combination with environmental influences to which an individual with a particular pattern of genes may be particularly vulnerable;
   c. acquired mutations and epigenetic alterations in key genes, risk of which may be increased by 1 and 2.

Analyzing cancer genomes can contribute to disease subclassification, improved prognosis, and even treatment selection, and many genetic variants are now employed as biomarkers in these areas. Different genomics techniques including those to determine SNPs, gene rearrangements, alterations in DNA copy number, epigenetic changes, and differential gene expression have all proved useful in this context.

Cancer genetics – in depth

You know how often the turning down this street or that, the accepting or rejecting of an invitation, may deflect the whole current of our lives into some other channel. Are we mere leaves, fluttered hither and thither by the wind, or are we rather, with every conviction that we are free agents, carried steadily along to a definite and pre-determined end?

Sir Arthur Conan Doyle

Mutations

As natural selection works solely by and for the good of each being, all corporeal and mental endowments will tend to progress towards perfection.

Charles Darwin

The term "mutation" can refer to any type of change in DNA. Exposure to genotoxic carcinogens can result in various differing forms of mutation. However, it is important to remember that mutations can also occur under the influence of unavoidable DNA damage, such as that induced by oxidative stress, background or cosmic radiation, and also by simple errors arising during DNA replication. The process by which proteins are made – translation – is based on the "reading" of mRNA that was produced via the process of transcription. Any changes to the DNA that encodes a gene will lead to an alteration of the mRNA produced. In turn, the altered mRNA may lead to the production of a protein that no longer functions properly. Even changing a single nucleotide along the DNA of a gene may lead to a completely nonfunctional protein. Mutations in one or more genes can therefore lead to disease.

The genetic changes that lead to unregulated cell growth may be acquired in two different ways – they can be inherited or they can develop in somatic cells. The phenotype of cancer cells result from mutations in key regulatory genes. The cells become progressively more abnormal as more genes become damaged, particularly when the genes that regulate DNA repair and checkpoints become damaged (see Chapters 7 and 10). Most cancers are thought to arise from a single mutant precursor cell (in other words they are "clonal"), with further clones originating by accumulation of further mutations, and those clones that gain a growth advantage will tend to take over the population (clonal expansion).

One aspect of this view of cancer is that the transition from a normal, healthy cell to a cancer cell occurs via the stepwise accumulation of mutations in multiple different oncogenes, tumor suppressor, or caretaker genes (Fig. 3.2). This model also accounts for the prevalence of cancer particularly in older individuals. Although the number, identity, and order in which mutations occur will likely vary enormously between individuals and different cancer types, attempts have been made to quantify the likely number of mutations required to generate a transformed human cell in culture. Studies from the laboratory of Robert Weinberg and others support the view of cancer formation as a multistage process, as suggested by Armitage and Doll in the 1950s, by demonstrating that at least 4–6 interlocking mutations may be needed to transform cultured human primary cells. However, whether this also equates to the requirements for formation of all cancers in the context of the intact organism remains controversial. Controversy remains because as few as two interlocking mutations may suffice to generate cancers in rodent models, a hypothesis that is difficult to test in humans, as the

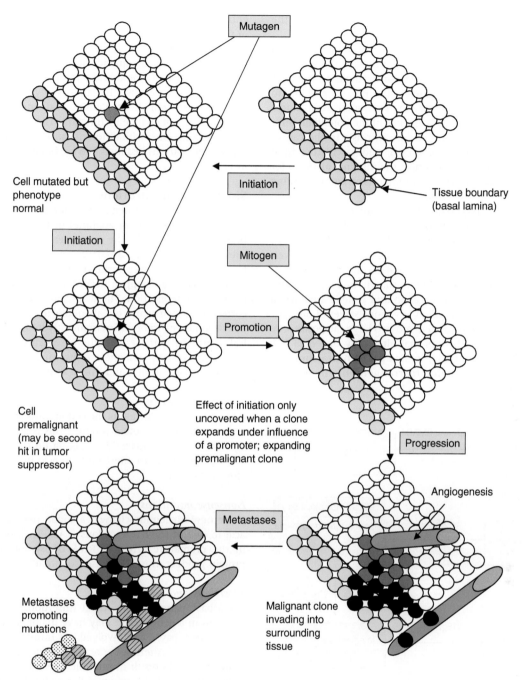

Figure 3.2 Multistage carcinogenesis – the concept of initiation and promotion. Genetic changes induced by mutagens are irreversible but may be phenotypically occult until further events such as proliferation or loss of differentiation unmask them. The mutagen or carcinogen is the "tumor initiator," but other factors (tumor promoters) affect whether mutated cells proliferate and form tumors. Promoters can contribute to cancer formation but do not alter DNA. Promoters increase the frequency of tumor formation in tissues previously exposed to a mutagen or tumor initiator. For example, skin papillomas form after exposure to carcinogens, but there may be considerable latency between the mutation in the stem cell pool and exposure to a tumor promoter that promotes proliferation, resulting in a visible lesion. This expanded clone will then itself be vulnerable to subsequent mutational events (hits) for tumor progression. Some evidence suggests that angiogenesis may accompany the growth of the tumor, or it may arise through a distinct mutational event (angiogenic switch). Finally, an invasive cancer forms with cells entering the circulation. However, it is generally believed that a further mutational event (such as inactivation of a "metastases suppressor" gene) is needed for colonies of cancer cells to establish in a distant location.

very earliest stages of cancer are generally not available to the researcher.

It is possible that intrinsic differences between human and mouse may be important and these might include the relatively longer telomeres found in mouse cells for example. However, it is also possible that the larger number of mutations required to generate a transformed cell *in vitro* than required for a cancer *in vivo*, might be explained by the absence of the potentially supportive effects of environmental factors extrinsic to the cancer cell, such as the tissue location, stroma, and vasculature from transformation assays *in vitro*. Defining the minimal platform for oncogenesis *in vivo* is likely to be an important and interesting area for study over the next decade.

The actual number required notwithstanding, mutations can occur gradually in somatic cells over a number of years, leading to the development of a "sporadic" case of cancer. Alternatively, it is possible to inherit dysfunctional genes, leading to the development of a familial form of a particular cancer. A model of cancer development in the colon is shown in Fig. 3.3. This scheme is based largely on observation and genetic analyses of tissues obtained from patient colon at various stages of disease.

Genetic alterations can be placed into two large categories. The first category comprises changes that alter only one or a few nucleotides along a DNA strand, termed **point mutations**, the second comprises various **major rearrangements** in genes or entire or parts of chromosomes (this is discussed in more detail in Chapter 10).

The structure of proteins is encoded in the nucleotide sequence of DNA. A particular sequence of nucleotides gives rise to a particular sequence of amino acids, and that in turn determines the way that that protein will function. Many changes in the nucleotide sequence will alter the amino acid sequence of the protein, and perhaps will change its function. There are several different forms of mutations, with different causes. Mutations can be classified under various headings and are briefly described below and in Fig. 3.4.

Single-base substitutions

A single base is substituted by another, termed a point mutation. If a purine (adenosine or guanine) or a pyrimidine (cytosine or thymine) is replaced by the other member of the same class, the substitution is called a **transition**, whereas if a purine is replaced by a pyrimidine or vice versa, the substitution is called a **transversion**. This can happen if DNA polymerase mismatches two bases during replication. There are "proofreading" functions that correct most such errors, but about one in a million is not detected and becomes incorporated permanently in the DNA. Some chemicals, particularly those that alter base structure, greatly increase the chance of DNA polymerase inserting an inappropriate base on the opposite strand, and thus a mutation can result.

Missense mutations

If the new nucleotide alters the codon, thus producing an altered amino acid in the protein product, this is a missense mutation. One of the three nucleotides making up the codon is replaced and this results in an altered amino acid in the protein product after translation.

Nonsense mutations

If the new nucleotide changes a codon that specified an amino acid to one of the STOP codons (TAA, TAG, or TGA), resulting in the premature arrest of RNA translation and a truncated protein product, this is a nonsense mutation.

Insertions and deletions

Extra base pairs may be added (insertions) or removed (deletions) from a gene. These may have major consequences if only one or two bases are involved, as translation of the gene is "frameshifted." Altering the reading frame one nucleotide to the right or left will result in multiple alterations in the amino acid sequence as multiple codons will now be altered and the mRNA is translated in new groups of three nucleotides. Frameshifts may also create new STOP codons and thus generate nonsense mutations. Alterations in three nucleotides or multiples of three may be less serious because they preserve the reading frame.

Silent mutations

Most amino acids are encoded by several different codons. Thus TCT, TCG, TCA, and TCC all code for the amino acid serine. Any mutation altering the base at position 3 will have no effect on the resultant protein. Such mutations are silent and detected only by gene sequencing.

Splice-site mutations

Intronic sequences are removed during the processing of pre-mRNA to mature mRNA under the influence of various proteins

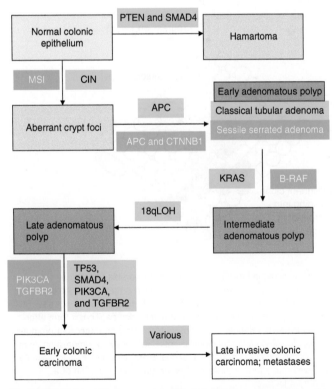

Figure 3.3 Colorectal cancer as a model of multistage carcinogenesis – the adenoma–carcinoma sequence. Sequential acquisition of mutations in various genes associated with initiation and progression of cancer are shown for the chromosome instability (CIN) model and the microsatellite instability (MSI) model. However, this does not mean that all colonic cancers arise in this sequence – activation of oncogenes and inactivation of tumor suppressor pathways are the key factors (alternative gene mutations could achieve the same net effect).

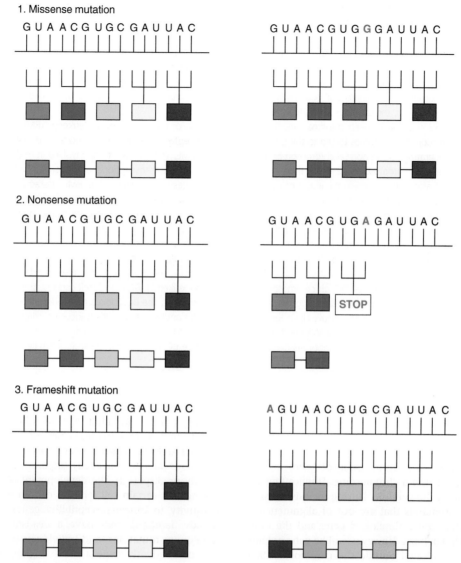

Figure 3.4 DNA point mutations. For simplicity, the small part of a mRNA transcribed from the normal or mutated genes is shown and the DNA (gene) itself is not shown. The DNA can be assumed to have a complementary sequence to the small section of mRNA shown. The mutated base is shown in red.

acting at the splice site. If a mutation alters one of these signals, then the intron is not removed and remains as part of the final RNA molecule. The translation of its sequence alters the sequence of the protein product.

More substantive DNA mutations

Chromosomal translocations

In some cases, diseases result not from changes in individual genes but in changes in the number or arrangement of chromosomes. Inherited abnormalities in chromosomes are unusual because they are generally incompatible with normal *in utero* development. One notable exception is Down syndrome, where individuals have three copies of chromosome 21 instead of the usual two. However, it is in cancers that one primarily finds altered numbers of chromosomes (see also Chapter 10). Some examples are shown in Box 3.3.

Gene amplification

Sometimes, instead of a single copy of a region of a chromosome, many copies are produced, resulting in the production of multiple copies of genes on that region of the chromosome. In extreme cases, these copies may form their own small pseudo-chromosomes called "double-minute chromosomes." This is often observed in cancer cells and can result in deregulated expression of oncogenes. Examples of this include amplification of c-*MYC* in several tumors and amplification of *NEU* in breast cancers. Gene amplification in the *MDR* gene encoding the MDR (multiple drug resistance) protein contributes to drug resistance in cancer. The MDR protein is a membrane pump capable of eliminating chemotherapeutic agents from the cancer cell, rendering them ineffective.

Inversions

DNA fragments are sometimes released from a chromosome and then reinserted in the opposite orientation. These inversions may

Box 3.3 Chromosomal rearrangements

Karyotyping (examination of chromosome numbers) of tumor cells highlights the presence of abnormal chromosomes such as chromosomal translocations, in which a part of one chromosome becomes joined to another. For example, in almost all patients with chronic myelogenous leukemia (CML), the leukemic cells show the same chromosomal translocation event between chromosomes 9 and 22. The *BCR* gene on chromosome 22 becomes joined to the c-*ABL* gene on chromosome 9, generating two abnormal chromosomes: a longer chromosome 9 and a small chromosome 22 called the "Philadelphia chromosome," after the city where the abnormality was first recorded. The resulting BCR–ABL fusion protein has the N-terminus of the BCR protein joined to the C-terminus of the ABL tyrosine protein kinase. In consequence, the ABL tyrosine protein kinase becomes inappropriately active in hematopoietic cells, driving their excessive proliferation, by activating multiple pathways normally regulated by extrinsic growth factors.

In Burkitt lymphoma, the proto-oncogene c-*MYC* on chromosome 8 is translocated to one of the three chromosomes containing the genes that encode antibody molecules: immunoglobulin heavy chain locus (chromosome 14) or one of the light-chain loci (chromosome 2 or 22). In every case, c-*MYC* now finds itself in a region of vigorous gene transcription, and it may simply be the overproduction of the c-MYC protein that turns the lymphocyte cancerous. The risk of translocations involving the heavy-chain gene locus is probably especially high because breaks in its DNA occur naturally during the synthesis of antibodies.

Fusion of the promyelocytic leukemia (PML) protein to the retinoic acid receptor-alpha (RAR-α) generates the transforming protein of acute PML. This appears to be involved in multiple functions, including apoptosis and transcriptional activation by RAR, whereas PML–RAR-α blocks these functions of PML. PML interacts with multiple corepressors (c-Ski, N-CoR, and mSin3A) and histone deacetylase 1, and this interaction is required for transcriptional repression mediated by the tumor suppressor MAD.

The *BCL2* gene is located on chromosome 18 and encodes the antiapoptotic Bcl-2 protein, which protects cells from cell death (see Chapter 8). The *BCL2* gene was discovered as the translocated locus in a B-cell leukemia (BCL). In B-cell cancers, the portion of chromosome 18 containing the *BCL2* locus has undergone a reciprocal translocation with the antibody heavy-chain locus on chromosome 14. The heavy-chain enhancer can thus regulate the *BCL2* gene, resulting in excessive BCL-2 protein in these t(14;18) cells.

---

either activate an oncogene or deactivate a tumor suppressor gene.

Duplications/deletions

Through replication errors, a gene or group of genes may be copied more than once within a chromosome. Duplications are a doubling of a section of the genome. During meiosis, crossing over between sister chromatids that are out of alignment can produce one chromatid with a duplicated gene and the other having two genes with deletions. However, unlike gene amplification, genes are not replicated outside the chromosome and only single copies are produced. Similarly, genes may become lost. Gene duplication has occurred repeatedly during the evolution of eukaryotes – genome analysis reveals many genes with similar sequences in a single organism. If two or more such paralogous genes are still similar in sequence and function, their existence provides redundancy. This may be a major reason why knocking out certain genes in yeast or mice may have little or no effect on phenotype.

Aneuploidy

Entire chromosomes may be lost or replicated during cell division if the replicated chromosomes fail to separate into the daughter cells accurately.

Inherited susceptibility to cancer

> It was the first time it had ever occurred to me that this detestable cant of false humility might have originated out of the Heep family. I had seen the harvest, but had never thought of the seed.
>
> *Charles Dickens, David Copperfield*

In inherited diseases, the disease-causing mutation is present at birth and all cells in the body have a mutation. Although most cancers are believed to arise from mutations occurring in single somatic cells in the adult, several inherited cancer syndromes have been described. In these cases, all somatic cells carry a mutation that does not cause cancer on its own, as additional somatic mutations are also needed. Importantly, inherited mutations usually predispose to the development of specific cancers, suggesting that certain cell types in certain tissues exhibit different sensitivity to cancer-susceptibility genes. Because the DNA is already damaged, cells have a headstart when accumulating subsequent mutations, and individuals with an inherited mutation consequently have a much higher risk of getting cancer compared to the general population and develop cancer at a younger age. An example of this for breast cancer is shown in Fig. 3.5.

By the sixteenth century, both physicians and families were clearly aware that certain overtly visible phenotypic features tended to cluster in families. Even though it would be some time before the scientific basis of this phenomenon would be unraveled, it was known that diseases such as cancer ran in families. Cancer syndromes, such as von Recklinghausen's neurofibromatosis, which has very evident cutaneous manifestations, were described, although both the honorific and attribution to germline inheritance of one defective copy of the NF1 tumor suppressor gene came much later (the other allele being eventually lost in some somatic cells).

Over the last few decades remarkable progress has been made, leading to the identification of numerous genes that contribute to germline inheritance of cancer susceptibility. Some examples are listed here.

- A hereditary predisposition to breast and ovarian cancer is frequently the result of germline mutations in genes involved in the DNA damage response or in repair of DNA double-strand breaks. Mutations in the *BRCA1* and *BRCA2* genes are the most important in clinical practice, but recent studies are resulting in

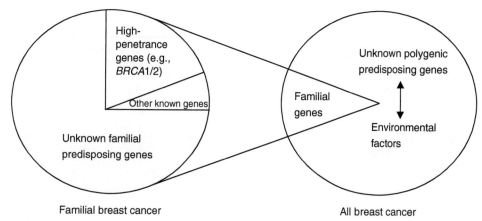

Figure 3.5 Breast cancer susceptibility. Familial breast cancer accounts for around 5–10% of all breast cancers. However, known genes, such as *BRCA1* and *BRCA2* account for only 20% of the familial risk, so that most genetic factors contributing to breast cancer are unknown. These unknown genetic variants (probably at multiple different loci) likely interact with environmental factors in the pathoetiology of around 80% or more of breast cancers. Adapted from Balmain, A., Gray, J., and Ponder, B. (2003) The genetics and genomics of cancer. *Nat Genet* 33(Suppl): 238–44.

identification of further genes, such as two RAD51 paralog-encoding genes, *RAD51C*, and for ovarian cancer, *RAD51D*. The presence of such mutations might also identify a subgroup of patients likely to respond favorably to PARP inhibitors (see Chapter 10).

- Lynch syndrome, caused by mutations in the mismatch repair genes *MSH2*, *MLH1*, and *MSH*, is associated with colorectal cancer.
- Familial adenomatous polyposis (FAP), caused by germline mutations in *APC*, is associated with colonic polyps and colorectal cancer.
- The Li–Fraumeni syndrome, caused by inherited inactivation of one copy of *p53* in the germline (occasionally *CHEK2*), increases the risk of cancers of the breast and brain as well as leukemias and sarcomas.
- Retinoblastoma is caused by inactivation of a copy of the retinoblastoma *RB* gene.

Other conditions associated with germline-inherited susceptibilities, include:

- Cowden syndrome, due to loss of *PTEN*, is associated with benign and malignant tumours in the breast, gastrointestinal tract, uterus, ovaries, and thyroid;
- Wilms tumor, associated with mutations in *WT1*;
- Werner syndrome, caused by mutations in *WRN*;
- hereditary diffuse gastric cancer, caused by mutations in *CDH1*;
- Von Hippel–Lindau syndrome, caused by mutations in *VHL*;
- variants of xeroderma pigmentosum (see Chapter 10) caused by a variety of mutations in DNA excision repair genes, such as *ERCC2–5*;
- susceptibility to types of leukemia conferred by a variety of genes encoding proteins of the Fanconi anemia complementation group;
- Peutz–Jeghers syndrome, characterized by intestinal hamartomas and increased epithelial cancers, caused by germline mutation in *LKB1* (serine/threonine kinase 11);
- familial melanoma caused by germline mutations in *CDKN2A*, both *INK4a* (p16) and *ARF* (p14), or *CDK4*.

A more comprehensive list is shown in Table 3.1.

We will now briefly describe two cancers – breast cancer and colorectal cancer – that exemplify the progress that has been made in defining susceptibility related to inherited mutations. In both cases we focus entirely on high-penetrance mutations that are already influencing routein clinical practice. A discussion of the role of lower risk alleles is reserved for later in this chapter, as the implications at this point in time are much less certain.

Colorectal cancer

Overall, around 5–10% of colorectal cancers (CRCs) are caused by inherited mutations. In general it is vital to identify affected families in order to target preventative strategies appropriately. This means that clinicians and others must look for a suggestive family history involving the earlier development of CRC and polyps (and other cancers associated with the various syndromes) and the presence of several affected first- and second-degree relatives. The commonest hereditary CRC syndromes are familial adenomatous polyposis (FAP) and Lynch syndrome (also known as hereditary nonpolyposis colorectal cancer – HNPCC). Around 1 in 100 CRCs are due to FAP, which is caused by *APC* gene mutations, resulting in the development of hundreds of polyps in early adulthood with almost inevitable progression to cancer in one or more by age 40 years. One variant of FAP, known as Gardner syndrome, is associated with multiple benign skin and connective tissue tumors. Lynch syndrome, caused by damage to the DNA mismatch repair genes *MLH1*, *MSH2*, *MSH6*, *PMS1*, or *PMS2*, is more common, causing around 5% of all CRCs. In general, Lynch syndrome is associated with far fewer polyps and arises at a slightly later age, but still earlier than sporadic CRC. In some cases, cancers affecting the endometrium, ovary, stomach, pancreas, or kidney are also found. A subgroup of families prone to CRCs and brain tumors are sometimes referred to as being affected by Turcot syndrome. As will be discussed later, Lynch syndrome tumors usually manifest a particular histopathological appearance termed "microsatellite instability" that can be recognized in diagnostic laboratories. Genetic testing can also be offered to affected families for these mutations. Two rarer syndromes associated with increased risk of CRC are Peutz–Jeghers syndrome, caused by mutations in *STK11* and associated with freckling and hamartomatous polyps, and polyposis linked to mutations in *MUTYH*.

Table 3.1 Hereditary cancer syndromes (based on OMIM data)

Syndrome	Cloned gene	Function	Chromosomal location	Tumor types
Li–Fraumeni syndrome www3.ncbi.nlm.nih.gov/entrez/dispomim.cgi?id=191170	P53 tumor suppressor	Cell cycle regulation, apoptosis	17p13	Brain tumors, sarcomas, leukemia, breast cancer
www.med.unibs.it/~marchesi/tumorsup.html – rbRetinoblastoma www3.ncbi.nlm.nih.gov/entrez/dispomim.cgi?id=180200	RB1 tumor suppressor	Cell cycle regulation	13q14	Retinoblastoma, osteogenic sarcoma
Wilms tumor www3.ncbi.nlm.nih.gov/entrez/dispomim.cgi?id=194070	WT1 tumor suppressor	Transcriptional regulation	11p13	Childhood kidney cancer
Neurofibromatosis type 1 www3.ncbi.nlm.nih.gov/entrez/dispomim.cgi?id=162200	NF1 protein=neurofibromin 1 tumor suppressor	Catalysis of RAS inactivation	17q11.2	Neurofibromas, sarcomas, gliomas
Neurofibromatosis type 2 www3.ncbi.nlm.nih.gov/entrez/dispomim.cgi?id=101000	NF2 protein=merlin or neurofibromin 2 tumor suppressor	Linkage of cell membrane to cytoskeleton	22q12.2	Acoustic neuromas, Schwann cell tumors, astrocytomas, meningiomas, ependynomas
Familial adenomatous polyposis coli www3.ncbi.nlm.nih.gov/entrez/dispomim.cgi?id=175100	APC tumor suppressor	Signaling through adhesion molecules to nucleus	5q21	Colon cancer
Tuberous sclerosis 1 www3.ncbi.nlm.nih.gov/entrez/dispomim.cgi?id=191100	TSC1 protein=hamartin tumor suppressor		9q34	Facial angiofibromas
Tuberous sclerosis 2 www3.ncbi.nlm.nih.gov/entrez/dispomim.cgi?id=191092	TSC2 protein=tuberin tumor suppressor	GTPase activation	16	Benign growths (hamartomas) in many tissues, astrocytomas, rhabdomyosarcomas
Deleted in pancreatic carcinoma 4 www3.ncbi.nlm.nih.gov/entrez/dispomim.cgi?id=600993	DPC4 also known as Smad4 tumor suppressor	Regulation of TGF-β/BMP signal transduction	18q21.1	Pancreatic carcinoma, colon cancer
Deleted in colorectal carcinoma www3.ncbi.nlm.nih.gov/entrez/dispomim.cgi?id=120470	DCC tumor suppressor	Transmembrane receptor involved in axonal guidance via netrins	18q21.3	Colorectal cancer
Familial breast cancer www3.ncbi.nlm.nih.gov/entrez/dispomim.cgi?id=113705	BRCA1 tumor suppressor	Repair of double strand breaks by association with Rad51 protein	17q21	Breast and ovarian cancer
Familial breast cancer www3.ncbi.nlm.nih.gov/entrez/dispomim.cgi?id=600185	BRCA2 tumor suppressor	DNA damage repair	13q12.3	Breast and ovarian cancer
Peutz–Jeghers syndrome www3.ncbi.nlm.nih.gov/entrez/dispomim.cgi?id=602216	STK11 tumor suppressor protein=serine-threonine kinase 11	Potential regulation of vascular endothelial growth factor (VEGF) pathway	19p13.3	Hyperpigmentation, multiple hamartomatous polyps, colorectal, breast and ovarian cancers
Hereditary nonpolyposis colorectal cancer type 1 HNPCC1 www3.ncbi.nlm.nih.gov/entrez/dispomim.cgi?id=120435	MSH2 tumor suppressor	DNA mismatch repair	2p22–p21	Colorectal cancer
Hereditary nonpolyposis colorectal cancer type 2 HNPCC2 www3.ncbi.nlm.nih.gov/entrez/dispomim.cgi?id=120436	MLH1 tumor suppressor	DNA mismatch repair	3p21.3	Colorectal cancer

Table 3.1 (Continued)

Syndrome	Cloned gene	Function	Chromosomal location	Tumor types
von Hippel–Lindau syndrome www3.ncbi.nlm.nih.gov/entrez/dispomim.cgi?id=193300	VHL tumor suppressor	Regulation of transcription elongation	3p26–p25	Renal cancers, hemangioblastomas, pheochromocytoma
Familial melanoma www3.ncbi.nlm.nih.gov/entrez/dispomim.cgi?id=600160	CDKN2A protein=cyclin-dependent kinase inhibitor 2A tumor suppressor	Inhibits cell-cycle kinases CDK4 and CDK6	9p21	Melanoma, pancreatic cancer, others
Gorlin syndrome: nevoid basal cell carcinoma syndrome (NBCCS) www3.ncbi.nlm.nih.gov/entrez/dispomim.cgi?id=109400	PTCH protein=patched tumor suppressor	Transmembrane receptor for hedgehog signaling protein	9q22.3	Basal cell skin cancer
Multiple endocrine neoplasia type 1 www3.ncbi.nlm.nih.gov/entrez/dispomim.cgi?id=131100	MEN1 tumor suppressor	Unknown	11q13	Parathyroid and pituitary adenomas, islet cell tumors, carcinoid
Multiple endocrine neoplasia type 2 www3.ncbi.nlm.nih.gov/entrez/dispomim.cgi?id=171400	RET, MEN2	Transmembrane receptor tyrosine kinase for glial-derived neurotropic factor (GDNF)	10q11.2	Medullary thyroid cancer, type 2A pheochromocytoma, mucosal hartoma
Beckwith–Wiedmann syndrome www3.ncbi.nlm.nih.gov/entrez/dispomim.cgi?id=130650	p57, KIP2	Cell cycle regulator	11p15.5	Wilms tumor, adrenocortical cancer, hepatoblastoma
Hereditary papillary renal cancer (HPRC) www3.ncbi.nlm.nih.gov/entrez/dispomim.cgi?id=164860	MET	Transmembrane receptor for hepatocyte growth factor (HGF)	7q31	Renal papillary cancer
Cowden syndrome www3.ncbi.nlm.nih.gov/entrez/dispomim.cgi?id=158350	PTEN tumor suppressor	Phosphoinositide 3-phosphatase Protein tyrosine phosphatase	10q23.3	Breast cancer, thyroid cancer, head & neck squamous carcinomas
Hereditary prostate cancer, numerous loci: HPC1(PRCA1), HPCX, MXI1, KAI1, PCAP www3.ncbi.nlm.nih.gov/entrez/dispomim.cgi?id=601518	HPC1 and PRCA1 are same designation ribonuclease L (RNaseL) maps to this locus	RNaseL involved in mRNA degradation	1q24–q25	Prostate cancer
Ataxia telangiectasia (AT) www3.ncbi.nlm.nih.gov/entrez/dispomim.cgi?id=208900	ATM	DNA repair	11q22.3	Lymphoma, cerebellar ataxia, immunodeficiency
Bloom syndrome www3.ncbi.nlm.nih.gov/entrez/dispomim.cgi?id=210900%20	BLM	DNA helicase?	15q26.1	Solid tumors, immunodeficiency
Xeroderma pigmentosum (XP) 7 complementation groups for XPA, XPC, XPD	XPA–XPG	DNA repair helicases, nucleotide excision repair	XPA=9q22.3 XPC=3p25 XPD=19q13.2–q13.3 XPE=11p12–p11 XPF=16p13.3–p13.13	Skin cancer
Fanconi's anemia	FANCA–FANCH	Components of DNA repair machinery	FANCA=16q24.3 FANCC=9q22.3 FANCD=3p25.3 FANCE=11p15	Acute myeloid leukemia (AML), pancytopenia, chromosomal instability

There are also several well-validated nonhereditary risk factors for CRC, including inflammatory bowel diseases, obesity, and a high-fat diet. Many countries now have screening programs for CRC that broadly aim to find polyps that can be removed before they become malignant. Population screening has been used for those above 50 years, whereas those with a family history or known to have a risk factor, such as ulcerative colitis, are generally screened at an earlier age. Genetic tests are also used to identify FAP and Lynch syndrome and can help select family members who may need to start colonoscopy screening and polypectomy in their teens or early twenties, or be offered colectomy (removal of the colon – offered to most people with FAP in their twenties). For population screening, various guidelines have suggested combinations of colonoscopy or imaging that might detect precancerous polyps or tests to diagnose CRC early, such as testing stool samples for blood (fecal occult blood tests) or antigens. The latter are cheaper and less invasive for the patient but will not pick up precancerous lesions.

Breast cancer

Up to 1 in 10 breast cancers may be hereditary, linked directly to mutations in *BRCA1* and *BRCA2* genes in particular. In some families the presence of such mutations may result in an 80% or greater risk of developing breast cancer and these may be bilateral and have an early onset. Rarer hereditary causes include ataxia telangiectasia (*ATM* mutations), and Li–Fraumeni and Cowden syndromes. Genetic tests are available in suspected families for mutations in *BRCA1* and *BRCA2* and also *PTEN* and *TP53*.

Risk of breast cancer can also be increased by various noninherited factors, including preexisting at-risk conditions such as hyperplasia or atypical hyperplasia in the breast, and by lifetime exposure to both endogenous sex steroids and exogenous hormones, such as the oral contraceptive and hormone replacement therapy. Not having children or having them late in life also increases risk, perhaps by indirect effects on hormones (pregnancy suppresses menstrual-associated increases in sex hormones). Excessive alcohol consumption, obesity, lack of physical activity, and previous radiation exposure also increase risk of breast cancer.

Polymorphisms and cancer

> It is curious to look back and realize upon what trivial and apparently coincidental circumstances great events frequently turn as easily and naturally as a door on its hinges.
>
> H. Rider Haggard, *Allan Quatermain*

Some of the inherited susceptibility to common cancers, which aggregate in families, may thus result from highly penetrant germline mutations in individual known genes that increase risk by at least 10-fold or more in an affected individual. But this still leaves much unaccounted for. Although in a few cases such unexplained familial risk may reflect the presence of high-penetrance mutations in as yet unidentified genes, polygenic mechanisms are increasingly being appreciated as likely contributors to inherited susceptibility to cancer (and to many other common chronic diseases). There is an intermediate scenario that is more recently being appreciated, in which rare disease-causing variants in single genes, usually known to be biologically important in tumorigenesis, can increase risk of common cancers by two- to threefold. Such variants in *ATM*, *BRIP1*, and *CHEK2* genes have been found for breast cancer and in *MUTYH* and CRC.

It is probable that most inherited susceptibility to cancer occurs through the action of polymorphic alleles. These are common variants of genes within populations, any of which may have a very slight overall effect on risk of a particular disease such as cancer (10% increase in risk or less). It is generally assumed that in most cancers that appear sporadic, several polymorphic alleles may be working in combination. This may extend as far as the notion that a particular genotype (comprising the sum of all variant alleles) may be a more cancer-susceptible one compared to another, but even this may depend on context (whether you smoke or not or become obese, for example).

In fact, most diseases may be associated with particular combinations of alleles of various genes contributing to the phenotype – known as polygenic (prefix "poly-" meaning "many"). None of the contributing alleles is disease-causing on its own, so the method of identifying and studying the genetics of these diseases is quite different from the study of mutations in single genes. In general low-penetrance polymorphic alleles are common in populations, whereas the combinations required to increase risk of cancer measurably may be very varied and individually rare, although the numbers of cancer-causing combinations overall may be numerous. Much of this is speculative and based on statistical theory as few such combinations have been identified.

Not surprisingly, therefore, and as is the case in many chronic diseases, the detection of low-penetrance susceptibility genes has proved difficult. Once one has exhausted the potential of directly looking for variations in known cancer-causing genes in patients with a specific cancer and comparing this to matched controls, using various techniques up to and including sequencing, what can one do next? In general, the next step requires the development of sufficiently large cohorts of cancer patients and controls who can then be studied for identifying unsuspected novel disease alleles. Various techniques can be employed to look for allelic variation, including genome-wide association studies (see Box 3.1). Technical advances and the availability of the reference human genome will hopefully speed progress in the coming years, with the major challenge probably being the availability of appropriately large, well-characterized patient cohorts. Some cancer-relevant polymorphic alleles are shown in Table 3.2.

A polymorphism is defined as the regular and simultaneous occurrence in a single interbreeding population of two or more alleles of a gene, where the frequency of the rarer alleles is greater than can be explained by recurrent mutation alone (typically greater than 1%). Thus, these gene variants are inherited and may be regarded as "normal" variants of genes occurring in the population. There may be as few as two or three such commonly occurring variants for individual genes in a typical population and in general none of these are strong predictors of pathology or disease. However, they may increase the risk of disease developing if both alleles are of the same polymorphic type (or occur in a particular combination), under particular environmental conditions, or if combined with several other specific polymorphisms in other genes.

The commonest polymorphisms currently studied are single nucleotide polymorphisms (SNPs). The difficulties in determining the role of these in cancer susceptibility should not be understated. Even in the case of smoking-related lung cancers, polymorphisms could influence carcinogen metabolism, DNA repair, drug responses, or even the likelihood of smoking in the first place! Separating these factors out will require vast patient cohorts that have been rigorously characterized by the use of

Table 3.2 Polymorphisms and cancer

Gene	Location	Cancer
ALDH2	12q24.2	Aldehyde dehydrogenase 2
APC	5q21–q22	Adenomatous polyposis coli. I1307K polymorphism in colorectal cancer
CCND1 (Cyclin D, PRAD1)	11q13	Cyclin D. Head and neck cancers
CDKN2A (P16, INK4A)	9p21	P16 tumor suppressor. Melanoma and mole density
KITLG		Testicular germ cell cancer
CYP17	10q24.3	Cytochrome p450 (CYP17). Breast and prostate cancers
CYP19	15q21.1	Aromatase cytochrome P450 gene (CYP19) and breast cancer risk
CYP1A1 (CYP1)	15q22-q24	Cytochrome P450, subfamily I (aromatic compound-inducible), polypeptide 1. Breast, lung
CYP1B1	2p22-p21	Cytochrome P450, subfamily I (dioxin-inducible), polypeptide 1 (glaucoma 3, primary infantile)- breast cancer
CYP2A6	19q13.2	CYP2A6; cytochrome P450, subfamily IIA (phenobarbital-inducible), polypeptide 6. Nicotine metabolism and lung cancer
CYP2E	10q24.3–qter	CYP2E; cytochrome P450, subfamily IIE (ethanol-inducible). Lung cancer
GSTM1 (GST1)	1p13.3	Glutathione S-transferase M1. Metabolism of tobacco carcinogens, lung cancer
GSTP1 (GST3)	11q13	Glutathione S-transferase pi. Breast cancer
HRAS	11p15.5	Harvey RAS – minisatellite alleles and cancer susceptibility. v-Ha-ras Harvey rat sarcoma viral oncogene homolog. Breast, ovary and lung
LTA	6p21.3	Lymphotoxin alpha (TNF superfamily, member 1). Risk of MGUS and myeloma
LM01	11p15.4	Neuroblastoma
MC1R (MSH-R)	16q24.3	Melanocortin 1 receptor (alpha melanocyte-stimulating hormone receptor). Melanoma
MTHFR	1p36.3	5,10-Methylenetetrahydrofolate reductase (NADPH). Colon cancer, acute lymphoblastic leukemia
NAT1	8p23.1–p21.3	N-Acetyltransferase 1 (arylamine N-acetyltransferase). Bladder cancer, lung cancer
SRD5A2	2p23	Steroid-5-alpha-reductase, alpha polypeptide 2. Prostate cancer, breast cancer
TNF	6p21.3	Tumor necrosis factor alpha. Myeloma

standard protocols across all centers involved in these studies. This notwithstanding, a considerable body of evidence suggests that gene polymorphisms may contribute to carcinogen metabolism (for example of tobacco smoke).

Various polymorphisms have been associated with cancer susceptibility (see Table 3.2). Genetic polymorphisms in enzymes involved either in detoxification of procarcinogens or DNA repair may affect cancer risk. Human cytochrome P450 (CYP) enzymes play a key role in the metabolism of drugs and environmental chemicals. Several CYP enzymes metabolically activate procarcinogens to genotoxic intermediates, and not surprisingly associations have been identified between SNP polymorphisms, gene duplications, and deletions and CYP enzyme activity and the risk of several forms of cancer. Strong support for an association between CYP polymorphisms and lung, head and neck, and liver cancer now exist. Polymorphism in the *TPMT* gene can result in slower than usual metabolism of the chemotherapeutic agent mercaptopurine, resulting in greatly increased toxicity to the drug. The lack of glutathione S-transferase M1 (*GSTM1*-null genotype) is associated with increased sensitivity to genotoxicity of tobacco smoke, and *GSTM1*-null smokers also show an increased frequency of chromosomal aberrations. Results of 161 meta-analyses and pooled analyses for polymorphisms in 99 genes and for 18 cancer sites published up to 2008 have been carried out recently. Around 30% of gene-variant cancer associations were found to be statistically significant, with variants in genes encoding for metabolizing enzymes among the most consistent and highly significant associations.

As will be discussed in Chapters 7 and 11, deregulation of microRNAs (miRNAs), which have powerful effects on gene expression, can contribute to cancer formation. Recent studies now suggest that SNPs in genes encoding miRNAs or those involved in their processing can also affect risk of cancer and success of various therapies.

Cancer genomics

In the struggle for survival, the fittest win out at the expense of their rivals because they succeed in adapting themselves best to their environment.

Charles Darwin

The molecular profiling of cancers (see also Chapter 18) has been set some rather lofty aims in the last decade and it is now commonplace to read or hear about how the quiet revolution that has followed the sequencing of the human genome will inevitably lead to more accurate and earlier diagnosis, better disease subclasification or staging, and prognosis, and as night follows day ultimately to individualized medicine and tailored therapy. Although progress in the whole field of 'omics technologies and systems biology has been nothing short of spectacular, the realization of these aims is not in sight.

So what does molecular profiling actually mean? In contrast to traditional biomarkers, where one or a small number of things are measured (PSA, CEA, blood count, etc.), profiling means measuring or characterizing large numbers of molecules at one go to create signatures or fingerprints. These are more complex and so might be superior to measurements of a smaller number of parameters. In cancer, molecular profiling really started with looking at gene/protein expression and for mutations in tumor samples and has progressed rapidly to studies now aiming to sequence the whole tumor and host patient genomes. In fact, in subsequent chapters we will discuss the notion of high-risk signatures of expressed genes or proteins found by tumor profiling. The hope is that these can predict those cancers most likely to spread quickly.

In this chapter we will concern ourselves with genomics – unravelling the genetic variation that underpins the development of cancer. Arguably, in this field of endeavor more than any other, we have set ourselves the lofty ambition of predicting the consequential from painstaking measurements of the seemingly inconsequential. Most common human traits and diseases have a polygenic pattern of inheritance. Put simply, variations in the sequence of DNA at many genetic loci act together to affect phenotype. Genome-wide association studies (GWAS) have identified over 600 variants associated with human traits, but typically these explain only a very small amount of variation in phenotype. Characterizing of somatic mutations in cancer genomes may help unravel the pathoetiology of the disease and contribute to improved diagnosis, subclassification, and even development of targeted treatments.

Broadly, two different approaches can be applied to examining the tumor genome: one can either look for specific, known cancer mutations in selected genes or use an unbiased approach (not burdened by any hypotheses or preconceptions about what may be found) to look at the whole or selected areas of the genome.

There have been very many publications in the last few years concerning SNPs. More recently, next generation sequencing has been used to find mutated genes in a variety of cancers. We will refer to a few of these here and the reader is referred to the reading list at the end of the chapter. The website of the Sanger Centre in Cambridge, UK is a useful resource. Using this approach 188 human lung adenocarcinomas were sequenced to find mutations in 623 known genes. Only 26 genes were mutated at significantly high frequencies, including those encoding tyrosine kinases, such as *ERBB4* and receptors for ephrin (*EPHA3*) and VEGF (*KDR*) and several tumor suppressor genes such as *NF1*, *APC*, *RB1*, *ATM*, *PTPRD*, and *LRP1B*, previously implicated in other cancer types.

The approach taken by Levi Garraway and colleagues at Dana-Farber Cancer Institute was to look for a restricted number of likely functionally important mutations (238 known mutations in 17 oncogenes) in an enormous number of different tumor samples (1000 representing 17 different types). Thirty percent of the samples carried at least one mutation and interesting combinations were identified. Andrew Futreal, Michael Stratton, and colleagues from the Cancer Genome Project and Wellcome Trust Sanger Institute in Cambridge, UK, have recently published the results of a truly Herculean labor involving resequencing cancer genomes in 210 diverse human cancers. The aim was a bold one – to find new cancer genes and to distinguish "mission critical" genes from so-called "passenger" genes (or at least as far as was possible without any functional studies). The pharmaceutical industry will be pleased to note that a greater than expected number of somatic mutations were found within the coding regions of 518 kinase genes sequenced. Moreover, as a confirmation of what has become clear to all of us involved in cancer research, there was substantial variation in the number and pattern of mutations between individual cancers. Imagine 1000 cars starting a journey with the sat nav set to a given destination but then all driving off in random directions at the start and at various times during the journey. You will all get to the same end point but it is unlikely that any two cars will follow the exact same route, even though there may be parts in common. Just as certain random directions will trigger the same response by the sat nav, certain mutations will require a specific range of other mutations in order to complete the journey.

Of the mutations identified, approximately 120 genes were deemed likely to be "mission critical," suggesting that the others were wrong turns. This supports the notion, revisited again and again in this book, that most cancer mutations (763 versus 120 in this case) are not relevant to tumorigenesis. In an interesting recent study, a single primary lung tumor was fully sequenced along with adjacent normal tissue from the same patient. By comparing the two sequences, the authors found over 50 000 single nucleotide variants with an estimated 17.7 per megabase genome-wide somatic mutation rate. Kinases once more come to the fore, and a more than expected number of amino acid-changing mutations in kinase genes were found. Again in support of the large number of "passenger" mutations, overall mutations within expressed genes were selected against.

Various other common cancers have been studied in this way and many common genetic variants or SNPs have been found. However, in most cases the functional significance is unclear. Thus, for example, recent studies have found 10 common SNPs in colorectal cancer that map to chromosomes 8q23, 8q24, 10p14, 11q23, 14q22, 15q13, 16q22, 18q21, 19q13, and 20p1. A recent published meta-analysis of three GWAS in colon cancer including 3334 affected individuals and 4628 controls has identified associations at three new CRC risk loci: 1q41, 3q26.2, and 20q13.33. What this means is not known, and some SNPs are located in poorly characterized genomic regions or gene deserts. However, functional studies follow and already it is now clear that one SNP, rs6983267, which maps to 8q24, serves as an enhancer of MYC expression by binding T-cell factor 4 (TCF4) and influencing Wnt signaling.

Several large recent GWAS have identified multiple susceptibility loci. A few will be mentioned here:
- Two loci on chromosomes 2p21 and 11q13.3 associated with renal carcinoma susceptibility have been identified. The first maps to a gene which encodes hypoxia-inducible factor 2 alpha, a transcription factor previously implicated in renal carcinoma.
- Bladder cancer susceptibility is determined by three loci on chromosomes 22q13.1, 19q12, and 2q37.1, the first of which maps to a nongenic region and the others to *CCNE1* and the

UGT1A cluster on 2q37.1, and a further four on 3q28, 4p16.3, 8q24.21, and 8q24.3,
- Ovarian cancer susceptibility loci have been found on chromosomes 9p22 and many others, such as 8q24 and 2q31, which are within or near genes encoding HOXD1, MYC, TIPARP, BNC2, and SKAP1, suggesting potential insights into tumorigenesis in the ovary.
- Pancreatic cancer susceptibility loci have been found on chromosomes 13q22.1, 1q32.1, and 5p15.33 and another large study has shown that *SMAD4* mutations in pancreas cancer may affect prognosis.
- Two major haplotypes of the gene encoding transforming growth factor beta (TGF-β) type I receptor, TGFBR1, have been found in the germline of 10–20% of patients with colorectal cancer and only 1–3% of controls.
- GWAS in testicular germ cell tumors have identified two genes – *KITLG* and *SPRY4* – involved in susceptibility.
- A recent study has employed GWAS to identify a susceptibility locus for neuroblastoma within LIM domain only 1 (LMO1) at 11p15.4 that encodes a cysteine-rich transcriptional regulator. Germline SNP risk alleles and somatic copy number gains were associated with increased LMO1 expression in neuroblastoma cell lines and primary tumors. In functional studies, shRNA-mediated depletion of LMO1 inhibited growth of neuroblastoma cells with high LMO1 expression.
- Another recent study has used a genome-wide SNP and copy number survey to answer an intriguing question in cancer biology, namely whether metastases arise from one or more clones of cancer cells in the primary tumor. The authors concluded that despite common genomic heterogeneity in primary cancers, most metastatic cancers arise from a single precursor cancer cell. Thus GWAS can be used to make predictions about tumorigenesis and these can be validated in functional studies.

To what extent might there be common features across different cancers? A study of nearly 450 different samples including breast, lung, ovarian, and prostate cancer identified 2576 somatic mutations across approximately 1800 megabases of DNA representing 1507 coding genes. Both the extent and nature of mutated genes varied substantially across tumor types and subtypes. Across the study, 77 significantly mutated genes were described, including the ubiquitous kinases, MAP2K4 (JNK signaling), G-protein coupled receptors such as GRM8, BAI3, AGTRL1 (also called APLNR), and LPHN3.

What clinical benefits might actually derive from GWAS and sequencing studies? A recent analysis published in the *New England Journal of Medicine* by Sholom Wacholder and colleagues suggests that the level of predicted breast cancer risk among most women is hardly changed at all after the addition of currently available genetic information. In this case, already a relative success story, the gains of new knowledge from 'omics techniques may be expected to be harder to show.

A further factor which may be relevant to most if not all attempts to use genetic information to understand or to predict cancer behavior, is the sheer complexity of gene variant effects. This is exemplified by a very important GWAS published in the journal *Nature* in 2010 (Lango Allen *et al.*, 2010), which has examined the effect of genetic variation on height, a known hereditable polygenic trait. By studying over 180 000 individuals, the authors found that variation at around 180 loci together account for around 10% of phenotypic variation in height. Moreover, these were not random, because genes involved in biological pathways and those that underlie skeletal growth defects were enriched. At least 19 loci had multiple independently associated variants, suggesting that allelic heterogeneity is a frequent feature of polygenic traits.

It is clear that mutations do not just affect tumorigenesis but can also have a major impact on treatment. Thus, it is well-known that mutant *KRAS* predicts resistance to anti-EGFR antibodies. Moreover, a recent study suggests that even in patients with wildtype *KRAS*, mutations in genes encoding downstream signaling molecules such as BRAF, NRAS, and PIK3CA exon 20 are also associated with a poor response rate to anti-EGFR therapy.

Gene–environment interactions

Are there, infinitely varying with each individual, inbred forces of Good and Evil in all of us, deep down below the reach of mortal encouragement and mortal repression – hidden Good and hidden Evil, both alike at the mercy of the liberating opportunity and the sufficient temptation?

Wilkie Collins

The actions of environmental factors are not independent of cancer genes. Sunlight may induce aberrant tumor suppressor genes in skin cells and cigarette smoke may make lung cells more vulnerable to carcinogenic compounds in smoke. These factors probably act directly or indirectly on the genes that are already known to be involved in cancer. Individual genetic differences also affect the susceptibility of an individual to the carcinogenic affects of environmental agents. About ten percent of the population has an alteration in a gene, causing them to produce excessive amounts of an enzyme that breaks down hydrocarbons present in smoke and various air pollutants. The excess enzyme reacts with these chemicals, turning them into carcinogens. These individuals are about 25 times more likely to develop cancer from hydrocarbons in the air than others are.

How hereditary factors may contribute to the development of sporadic cancer is still unclear. A landmark study in 2000 by Paul Leichtenstein and colleagues, involving 44 788 pairs of twins listed in the Swedish, Danish, and Finnish twin registries, provides estimates of the overall contribution of inherited genes to the development of various cancers. At least one cancer occurred in 10 803 persons among 9512 pairs of twins. Although 28 cancer types were looked at, an increased risk was found among the twins of affected persons only for stomach, colorectal, lung, breast, and prostate cancer. Statistically significant effects of heritable factors were observed for prostate cancer (42%), colorectal cancer (35%), and breast cancer (27%). These results suggested overall that inherited genetic factors make a minor contribution to susceptibility to most types of cancer, and therefore, by implication, that the environment takes the lead role in causing sporadic cancer. None the less, there were relatively large effects of heritability in some common cancers, though the genetic bases underlying these observations remain to be defined.

Mutations and treatment

In all works on Natural History, we constantly find details of the marvellous adaptation of animals to their food, their habits, and the localities in which they are found.

Alfred Wallace

Various tyrosine kinase inhibitors, including imatinib and gefitinib, can cause tumor regression in certain patients, but it has proved difficult to ascertain which patients would be responsive to the drugs. Imatinib (Gleevec) is most effective against certain leukemias and gastrointestinal stromal tumors. Importantly, the presence of specific mutations (such as the BCR–ABL fusion protein) can identify patients likely to respond. More recently, it has been shown that the presence of a specific mutation in the epidermal growth factor receptor (*EGFR*) gene in a common adult cancer, non-small-cell lung cancer (NSCLC), may identify patients most likely to respond to gefitinib. However, the situation is more complex than this would imply, because marked differences in response were observed between Japanese and American patients. In fact, Japanese women with adenocarcinoma showed the highest percentage of *EGFR* mutations and also showed the best clinical response to gefitinib. This suggests that genetic variation in different ethnic, cultural, and geographic groups might also contribute to sensitivity of NSCLC to particular drugs.

The RAS–RAF–MAPK pathway is activated in most human tumors, often through gain-of-function mutations of RAS and RAF family members. In a recent study it was found that the presence of *BRAF* mutations could predict enhanced and selective sensitivity to MEK inhibition as compared to normal cells or those bearing mutant *RAS*. Thus, combinations of inherited polygenic factors and single somatic mutations may all contribute to determine treatment responsiveness.

Chemoprevention of cancer

A man will tell you that he has worked in a mine for forty years unhurt by an accident as a reason why he should apprehend no danger, though the roof is beginning to sink.

George Eliot, Silas Marner

Finding new drugs that can prevent rather than cure cancer is an emerging area of research, but the challenges are daunting. Cancer develops over a generally very long time, under the direction of genetic changes and epigenetic alterations working together in a bewildering array of often poorly understood partnerships, including many which have no bearing on the process. In which case, do we have any chance in identifying at-risk individuals and thence those factors that can be safely modified?

Well, some notable examples of chemoprevention exist and others show promise. Chemoprevention could theoretically be deployed at various stages during the evolution of a cancer from prenatally through the development of premalignant and malignant tumors to metastatic life-shortening cancers. Obviously, the preferred option would be primary prevention – namely, to prevent the development of any malignant tumor, however small and inocuous. But blocking of progression might be adequate in many cases and may be the only "game in town" given the extreme difficulty in finding the earliest stages of cancer in humans.

So what are the available options? First, as has been demonstrated in prevention of coronary disease, the preventative agents given to otherwise completely healthy individuals must be very safe and generally free of troublesome side effects, and should not be the subject of ill-informed media scaremongering if anyone is going to actually take them. At the very least, the risk of harm must be substantially less than the predicted benefits. Moreover, they will have to target some readily identifiable high-risk precancer state. Most drugs currently approved for reducing cancer risk are aimed at high-risk groups, such as women at risk of breast cancer who can be treated with tamoxifen or raloxifene or various ointments that can be applied to skin lesions to prevent squamous cell carcinoma.

More common risk conditions are also amenable to preventative treatments, such as chronic inflammation or insulin resistance/obesity (or high-risk behaviors that could be altered by drugs – in this context nicotine replacement therapy for smokers might be considered a form of chemoprevention).

A strong body of evidence supports the idea that anti-inflammatory agents, exemplified by aspirin and COX2 inhibitors (currently off the table because of associated increased risk of cardiovascular disease) that can reduce angiogenesis and growth might prevent certain cancers such as CRC. The US Food and Drug Administration (FDA) had initially approved use of the COX2 inhibitor celecoxib for reducing risk of CRC in those with FAP. Although these drugs are currently off the market, the search for ways of exploiting the preventative benefits of COX2 inhibition for CRC and other cancers are still being explored, including by trying to develop safer agents free of adverse cardiovascular effects or even by using them to prevent oral cancers where topical use might mitigate against systemic toxicity. Aspirin is associated with dangerous gastrointestinal bleeding, which also limits use in prevention of cardiovascular disease in lower risk patients. Despite this, several trials using aspirin to prevent cancer are underway and benefits are suggested in CRC, NSCLC, and prostate cancer prevention. Recent resports from diabetic and obese patients treated with the drug metformin suggest that risk of various cancers, including NSCLC, might be reduced compared to other diabetes treatments and metformin was reported to reduce the incidence of tumors in a lung cancer-prone mouse model.

Recent research has suggested that a variety of drug combinations might be more effective than single agents. In CRC-prone mice, intermittent treatment with tumor necrosis factor-related apoptosis-inducing ligand (TRAIL) and all-*trans*-retinyl acetate could reduce tumor development. Combining the nonsteroidal anti-inflammatory drugs (NSAIDs) sulindac and difluoromethylornithine (DFMO) was also effective in CRC models and has shown promise in early clinical trials in which development of polyps and advanced adenomas was reduced by 70% and 90%, respectively. This is being further tested in phase III trials. A kinase inhibitor, genistein, derived from soya, can prevent invasion and metastases of prostate cancer cells. This is being tested in phase II clinical trials.

Risk factors act in combination

There are no big problems, there are just a lot of little problems.

Henry Ford

It is often difficult to attribute tumorigenesis to any single causative agents, but exposure to specific environmental factors is strongly predictive of the eventual development of particular cancers. Notable among these are the strong links between cigarette smoking and lung cancer and exposure to ultraviolet (UV) light and skin cancer. Even in these cases, not all individuals who smoke or experience repeated sun exposure develop cancer, but public health programs in many countries now strongly promote both smoking cessation and prevention of smoking in public to

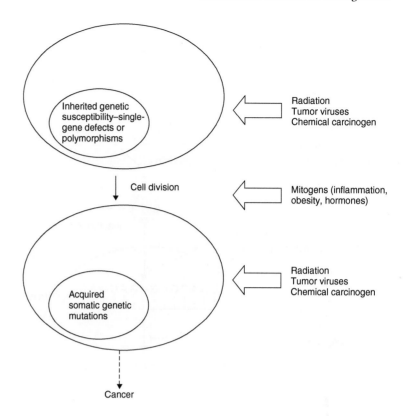

Figure 3.6 The combination of genes and environment in the genesis of a cancer cell. This diagram shows an example of how inherited and acquired genetic alterations together with environmental factors may lead to tumorigenesis and cancer.

protect nonsmokers from passive inhalation and use of sunscreens. In these cases it is accepted that infringing the rights of smokers is for the greater good and that any potential harms of sunscreen (assumed to be negligible) to those who would not have developed skin cancer anyway are outweighed by the potential benefits to those who otherwise would.

Cancers rarely develop solely because of high-risk hereditary genetic factors – inherited cancers are uncommon, but their study has contributed greatly to our understanding of the molecular basis of common forms of cancer. By implication, and as is the case for many chronic diseases, both genetic and environmental factors contribute to disease pathogenesis (Fig. 3.6). Possible ways in which mutations in various genes might directly interact with exposure to environmental factors is shown in Fig. 3.7. One aspect that is unique to cancer is the role played by acquired somatic mutations. Cancer presents a unique scenario in which mutations present within just one single cell cause disease, but only when those mutations bypass normal controls and are propagated by replication of that cell. Who would miss a single cell eliminated as a result of a lethal mutation?

Traditionally, the genetics of inherited cancer predisposition have been considered in terms of single, high-risk genes with high penetrance, such as in retinoblastoma (*RB*), Li–Fraumeni (*TP53*), familial breast cancer (*BRCA1/2*), familial colorectal cancer (*HNPCC*), and Wilms tumor (*WT1*). In fact, Knudson's hypothesis was first confirmed by the study of retinoblastoma (see Chapter 7 for a detailed account). A list of the more common single-gene inherited cancer syndromes is given in Table 3.1.

It is worth emphasizing here that the absence of an overtly inherited monogenic disorder does not imply that inherited genetic factors have not played a role in the formation or progression of that cancer; subtle variations in multiple alleles (polymorphisms) may increase disease susceptibility through numerous mechanisms, including alteration of metabolism of carcinogens, efficiency of DNA repair, free-radical handling, angiogenesis, etc. Such low-penetrance genes frequently vary by SNPs (see Table 3.2). Given the explosion of GWAS in cancer, it is not proposed to provide an exhaustive list of all these. The reader is directed to some pertinent examples in the literature review at the end of the chapter.

Such inherited factors – strong or weak predictors of risk – may in turn influence the risk of cancer after exposure to environmental carcinogens or after acquisition of various somatic mutations (Fig. 3.4). Common SNPs do contribute to the the risk of common cancers but by a relatively small amount. This should not be taken to mean that hereditary factors do not contribute to most cancers but rather that many rare, low-penetrance variants in coding and noncoding genes in the germline may also play important roles. In other words, there will be a very large number of different cancer-prone genomes and these will probably have much less in common than would once have been predicted. Thus, low-frequency polymorphisms (allele frequency <5%), subpolymorphic variants (frequency 0.1–1.0%), and even "personal" variants (frequencies of <0.1%) may all predict a future risk of cancer.

These variants will be increasingly difficult to identify and validate in population studies, even with new technologies such as less error-prone massively parallel sequencing and resequencing (see Chapter 20). Some rare variants of potential interest can also be found, because known roles in biological processes relevant to cancer directs attention specifically to a given gene.

Whatever the hereditary factors may be, there is little doubt that environmental factors – **carcinogens** – play an important role in many cancers. Moreover, the effect of carcinogens may be determined by hereditary genetic factors in a variety of ways, ranging from amplification of risk in genetically predisposed indi-

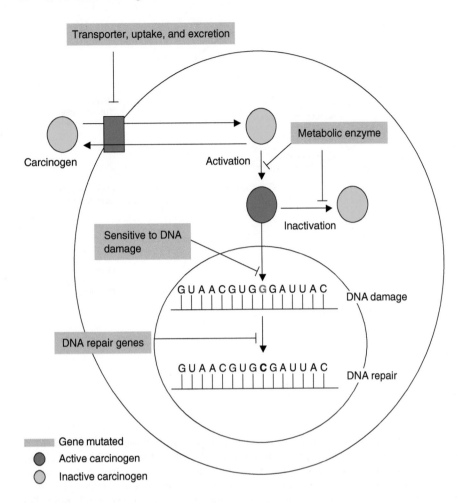

Figure 3.7 Genetic variation may influence susceptibility to carcinogens. Variation in gene alleles encoding proteins responsible for uptake, transport, and metabolism of a carcinogen may all affect the likelihood that the carcinogen will cause DNA damage. Moreover, variation in DNA repair processes will also play a role. From Balmain A, Gray J, Ponder B (2003) The genetics and genomics of cancer. *Nat Genet*, **33** Suppl: 238–244.

viduals through to a complete absence of risk in those who are genetically resistant.

Three classes of environmental agents – ionizing radiation, tumor viruses (viruses that cause cancer in animals), and chemical carcinogens (cancer-causing chemicals) – have been shown to increase the risk of cancer in both laboratory animals and people. Each of these agents may produce mutations in genes that contribute to abnormal growth. (Note, in this context we mean that carcinogens will cause somatic mutations in the cancer cells, which are not in the germline. It is easy to become confused between germline and somatic mutations.)

All cancer cells have mutations. Somatic mutations will be discussed in essentially every chapter, after all these are the dominant driving force behind essentially all cancers. In this chapter we will concentrate more on the genetic variants that are in all cells, including the germline, and are inherited. Alongside this will discuss general principles of genomics and gene–environment interactions. Now to confuse matters further, carcinogens may cause mutations in germ cells, through the same means operating in cancer cells, and these may then be passed on to subsequent offspring and generations of humans. Depending on the genetic change, increased risk of cancer may manifest in later generations.

It is also worth noting here that certain parts of the genome are frequently mutated in cancers, because they encode important tumor suppressors or oncogenes, and this applies equally to mutations that are somatic, the most common source, or inher-

ited. Thus, a 42 kb region on human chromosome 9p21 encodes for three distinct tumor suppressors – $p16^{INK4A}$, $p14^{ARF}$, and $p15^{INK4B}$ – and is altered in an estimated 30–40% of human tumors. Some regions of the genome may also be particularly vulnerable to DNA damage (see Chapter 10).

Given that only some 5% of cancers arise through inheritance of high-risk genetic variants, alongside lower risk variants, it is often concluded that environmental factors such as tobacco, alcohol, diet, infections, and occupational exposures play a major role in up to 80–90% of cancers. Carcinogens may not invariably act through causing somatic gene mutations, and at least in some cases, the "environmental" agents (such as background radiation) may not be avoidable. This notwithstanding, it has been estimated that about 75% (smokers) or 50% (nonsmokers) of all deaths due to cancer in the United States could be prevented by elimination of avoidable risk factors, making the prevention of many cancers a realistic prospect. For example, smoking cigarettes while young puts a person at 10–20 times higher risk of developing cancer later in life than persons who do not smoke. A high-fat diet and obesity contribute to the development of certain cancers.

Epidemiological studies, including studies of migrant populations, have provided the strongest support for the role of environmental factors in cancer. In particular, the fact that migrants often acquire the disease profiles of their adopted country is often taken as support for the notion that such diseases are thus more likely to be caused by environment and lifestyle rather than genetic factors. It must, however, also be appreciated that genetic factors may still play a major role. One notable example is provided by the fate of migrant South Asians in the United Kingdom and North America. In the Indian Subcontinent both diabetes and coronary heart disease used to be notably less frequent than in the United Kingdom (although this situation may be changing somewhat, particularly in urban and more affluent areas). Migrant Asian populations in the United Kingdom were then found to acquire a risk profile for these diseases thought to be more typical of the United Kingdom. However, over the last decade it has become apparent that, in fact, Asians in the United Kingdom may experience a far greater risk of diabetes and coronary heart disease than the native population – in large part due to presumed genetic factors which actually increase susceptibility to dietary and other factors to which they are exposed in the new environment. Gene–environment interactions are important!

Environmental causes of cancer

The environment is everything that isn't me.

Albert Einstein

Carcinogens are agents that contribute to the causation of cancer. Such agents may damage genes involved in cell proliferation and migration and may selectively enhance growth of tumor cells or their precursors. Free radicals, for example, may result in non-specific DNA damage; some viruses can accelerate the rate of cell division. Importantly, genetic factors and inherited predisposition may play a major role in determining how a given individual responds to these extrinsic factors. Thus, at the most obvious level, fair skin strongly predisposes to melanoma and non-melanoma skin cancers, whereas more subtle variation at individual gene loci may, for example, increase the risk of lung cancer in smokers. It is reported that as few as 3% of all cancers (WHO data) are caused by inadvertent environmental exposure to chemicals and pollutants, as exemplified by asbestos, and most of these cases are linked to occupation.

Socioeconomic differences are major contributors to patterns of disease, with poorer individuals tending to experience several specific adverse lifestyle factors, including higher rates of smoking, alcohol consumption, poor nutrition (variously malnutrition in developing countries and ironically obesity in developed countries), and exposure to certain infectious agents including *Helicobacter pylori*. Conversely, more affluent individuals are more likely to be diagnosed with cancers of the breast or prostate, the former through reduced parity and older age at time of first pregnancy, and the latter perhaps because of earlier diagnosis.

Recent strong support for the role of carcinogenic environmental factors in causing sporadic cancers has come from two large Scandinavian studies – a twin cohort from Sweden, Denmark, and Finland, and one based on the Swedish family cancer database, which includes around 10 million individuals. Although still controversial, both studies suggest that the influence of factors not shared among siblings or relatives (including inherited genetic factors and shared exposure) predominates. Such, "nonshared environmental factors" include smoking, radiation, infections, and occupational exposures, as well as sporadic mutations.

The risk of developing some cancers has declined dramatically in developed nations in this century, but unfortunately the risk of the most prominent forms has increased. This may be largely a manifestation of increasing longevity and age within populations, coupled with improved recording of disease incidence and causes of death, rather than any increases in environmental cancer-causing agents. However, the possibility of depletion in the ozone layer contributing to increased rates of UV exposure–induced cancers is a prominent example of one such explanation.

Individual environmental risk factors

The universities do not teach all things, so a doctor must seek out old wives, gipsies, sorcerers, wandering tribes, old robbers, and such outlaws and take lessons from them. . . . Knowledge is experience.

Paracelsus

Carcinogenic chemicals, such as azo dies, asbestos, benzene, formaldehyde, and diesel exhaust, are dangerous in high concentrations. Such high concentrations were once commonplace in some workplaces, but strict control of occupational carcinogens over the past 50 years has greatly reduced cancers caused by these substances.

One of the earliest recorded associations between chemical exposure and cancer was noted by the famous Renaissance intellectual and father of pharmacology Paracelsus (whose full name was the rather magnificent Philippus Aureolus Theophrastus Bombastus Von Hohenheim) (Fig. 3.8). In 1567, Paracelsus

Figure 3.8 Portrait of Paracelsus by Quentin Massys (1466–1530). "I am Theophrastus, and greater than those to whom you liken me; I am Theophrastus, and in addition I am monarcha medicorum and I can prove to you what you cannot prove . . . I need not don a coat of mail or a buckler against you, for you are not learned or experienced enough to refute even a word of mine . . . As for you, you can defend your kingdom with belly-crawling and flattery. How long do you think this will last? . . . Let me tell you this: every little hair on my neck knows more than you and all your scribes, and my shoe buckles are more learned than your Galen and Avicenna, and my beard has more experience than all your high colleges." (attributed to Paracelsus).

proposed that the "wasting disease" of Austrian miners might be caused by their exposure to natural ores such as realgar (arsenic sulfide) and was not due to being punished by mountain spirits for past misdeeds. The eighteenth century witnessed one of the earliest descriptions of a carcinogenic chemical when the link between soot and cancer of the scrotum in chimneysweeps was identified by the rather aptly named Percival Pott. Another English physician, John Hill, described the occurrence of cancer in the nose in long-term snuff users. By the end of the nineteenth century, occupational exposure of workers in the paraffin industry had been linked to certain cancers, and the development of certain skin cancers had been attributed to the use of arsenicals.

By the start of the twentieth century further examples of "occupational cancers" had been described, including bladder tumors in workers in the aniline dyestuff industry – "aniline cancer." At around the same time, tar or tarry compounds were recognized as causes of skin cancers, and this was supported by the induction of rodent skin cancer by exposure to coal tar. Over the next few decades, higher molecular weight polycyclic aromatic hydrocarbons (PAHs), such as dibenz-[a,h]-anthracene (DBA), were described and found experimentally to be cancer-causing constituents of coal tar. By the middle of the twentieth century, chemical carcinogenesis in humans was already well-established, with observational data supported by experimental studies on defined compounds in model systems. Numerous carcinogenic compounds are recognized now and will be discussed later. Importantly, many cancer chemotherapeutic agents are themselves carcinogenic. Alkylating agents are associated with a risk of several cancers; cyclophosphamide is associated with risk of bladder cancer.

Chemical carcinogens taken up by cells are usually metabolized, and the resulting metabolites are then excreted, but may on occasion be retained by the cell. Such internalized carcinogenic compounds can then directly or indirectly alter gene expression. Some carcinogens are genotoxic, forming DNA adducts or inducing various chromosomal abnormalities. For example, carcinogenic ions or compounds of nickel, arsenic, and cadmium can induce aneuploidy. Other carcinogens may act by nongenotoxic mechanisms, including promoting inflammation, suppressing immunity, forming damaging reactive oxygen species (ROS), or by activation of signaling pathways such as receptors for arylhydrocarbon receptor (AhR), estrogen receptor, PKC, and epigenetic silencing. Together, these genotoxic and nongenotoxic mechanisms may provoke the "hallmark" features of cancer.

Two terms are frequently used to classify chemical carcinogenesis:
- **Initiation** occurs as a result of exposure of a cell or cells to a carcinogen that permanently alters its genetic material but does not immediately influence phenotype. Carcinogens in this category are described as mutagens or genotoxic.
- **Promotion** occurs as a result of factors that cause tumors from cells that have already been initiated. Promoters are nongenotoxic carcinogens. The best-known promoters are the phorbol esters, which activate the PKC signaling pathway and promote mitogenesis and survival.

The classification of carcinogens in this way has limitations (Box 3.4) and is increasingly falling out of favor. However, the basic notion of cancers arising in a stepwise fashion through acquisition of successive mutations likely does apply to many human cancers, and forms a useful model for understanding how onco-

---

Box 3.4 Initiation and promotion of cancer

It can reasonably be assumed when modeling the development of cancer that all cancers are the result of the initiation, promotion, and progression phases of carcinogenesis (see Fig. 3.2). By implication there are genes that can:
- protect or predispose proto-oncogenes and tumor suppressor genes from activation or inactivation;
- bring about or suppress the growth and expansion of initiated cells;
- prevent or enhance the acquisition of genetic/epigenetic instability by the initiated cells in order for them to become malignant.

Various genetic syndromes in humans support this notion.

---

genes and tumor suppressors may cooperate in carcinogenesis. A view of multistage carcinogenesis incorporating these concepts is shown in Fig. 3.2.

Activation of carcinogens

Cursed is the man who dies, but the evil done by him survives.
Abu Bakr

Activation of carcinogens by metabolic processes is necessary for the majority of so-called procarcinogens. Although some direct-acting and activation-independent carcinogens such as alkylating agents are known, the more ubiquitous polycyclic hydrocarbons (tobacco smoke), aromatic amines, amides, azo dyes, and nitrosamines all require activation by the hepatic cytochrome P450 mixed function oxidase system in order to become carcinogenic.

Many carcinogenic compounds are mutagenic – that is they can induce mutations in DNA. Only sufficiently reactive electrophilic compounds can interact directly with DNA, but such "direct carcinogens" form a minority of known human carcinogens, including ethylene oxide, bis(chloromethyl)ether, and some aziridine or nitrogen-mustard derivatives used in anticancer chemotherapy. Most mutagenic chemicals are actually formed by metabolic activity after exposure to initially inert nucleophilic compounds, such as aromatic and heterocyclic amines, aminoazo dyes, PAHs, and N-nitrosamines – these are termed "procarcinogens."

Metabolic conversion of procarcinogens involves in particular the action of microsomal enzymes such as cytochrome P450–dependent monooxygenases (CYPs). The initial step during conversion of organic xenobiotics into hydrophilic and excretable derivatives is mainly catalyzed by CYP enzymes, of which more than 50 types are currently known. The types of CYP known to be involved in activation of carcinogens include: CYP1A1 in lung and CYP1A2, CYP2A6, and CYP2E1 in the liver. Carcinogens may also be metabolized to intermediates that can undergo transferase-catalyzed conjugation to polar molecules such as glutathione by glutathione S-transferases (GST), to glucuronic acid by glucuronosyltransferases, or to small residues such as sulfate by sulfotransferases (SULT), and acetic acid by N-acetyltransferases (NAT). These enzymes may also contribute to activation of procarcinogens into compounds that interact with DNA. Thus, N-hydroxy derivatives of procarcinogenic arylamines/amides, aminoazo dyes, or heterocyclic amines are converted by NAT or SULT enzymes into highly reactive ester intermediates that bind to DNA.

Some compounds do not bind to DNA and are not mutagenic, yet they are carcinogenic in animal models.

Carcinogenic chemicals

A number of industrial chemicals have been linked to an increased risk of cancer and in most countries use is restricted and must be identified by the inclusion of hazard warnings and guidelines for safe handling. These chemicals include:

- asbestos, used in construction and car manufacture;
- benzene, used in the petrochemical industry;
- benzidine dyes;
- arsenic, used in glass and metallurgy;
- beryllium, used in aerospace and metallurgy;
- chromium, used in paint production;
- some fertilizers and pesticides;
- many organic solvents.

Asbestos

Asbestos was extensively used in the past in construction as an insulation material for buildings. Of the three forms of asbestos, amosite (brown asbestos) and crocidolite (blue asbestos) are the most dangerous. Inhalation of microscopic asbestos fibers has been linked to both mesothelioma and NSCLC. All asbestos was banned in the United Kingdom after 1999, so it is primarily older houses that contain asbestos. This only becomes harmful when fibers are released into the air during construction or demolition work, and thus risk is largely to those conducting such work. Cancers develop up to 20–40 years following exposure to asbestos.

Dioxins

The nongenotoxic carcinogen 2,3,7,8-tetrachlorodibenzo-p-dioxin (TCDD) is produced as a byproduct during the manufacture of polychlorinated phenols. TCDD is generated during incineration of waste and has been strongly implicated in a range of human cancers by epidemiological studies. One famous case took place in 1976 in Seveso, Italy, when following an accident at a chemical plant, several thousand kilograms of chemicals were released into the air – including 20 kg of dioxins. Over 30 000 people were exposed to these chemicals, and decades later cancer rates among those exposed were significantly increased compared with control populations.

TCDD is also carcinogenic in animal models, promoting cancer in skin, lung, and liver, among others. TCDD is an agonist of the arylhydrocarbon receptor (AhR), a cytosolic protein involved in induction of microsomal arylhydrocarbon hydroxylase (AHH) activity. The AhR belongs to the basic helix–loop–helix/PAS family of transcription factors that upon ligand binding translocates to the nucleus and forms heterodimers with AhR nuclear translocator, to form an active transcriptional activator. Binding to so-called xenobiotic responsive elements (XREs) regulates transcription of a variety of genes encoding, among others, protein enzymes involved in xenobiotic metabolism, such as CYP1A1, CYP1B1, CYP1A2, and GST. Recent studies suggest that TCDD binding to AhR also influences expression of factors that regulate cell growth, proliferation, differentiation, and apoptosis. Studies using gene microarray analysis of gene expression in human hepatoma cells exposed to TCDD demonstrated that a few hundred genes were differentially expressed, including *CYP1A1* and *KRAS*. AhR–ligand complexes also affect pathways involving SRC.

PAHs

Higher molecular weight polycyclic aromatic hydrocarbons, such as dibenz-[a,h]-anthracene (DBA), require metabolic activation to become carcinogens. Present in coal tar, PAHs can be formed by incomplete combustion of most organic materials, including tobacco leaf. Following activation of PAHs, diol-epoxide metabolites have been implicated in mediating the DNA-binding activity, and they are also highly carcinogenic. Disruption of genes that encode enzymes and factors that are involved in this activation route, such as CYP1B1, microsomal epoxide hydrolase (mEH), or AhR, renders mice resistant to PAHs-induced cancers. PAHs can induce both somatic mutations in crucial genes (tumor "initiation") and subsequent growth of transformed cells (tumor "promotion") in mice. Such agents are described as "complete carcinogens." Among genes known to be mutated by PAHs, *HRAS* is of particular interest.

Arylamines/amides

Aromatic amines, such as 4-aminobiphenyl (ABP) or the arylamide AAF, induce bladder tumors in dogs, and occasionally in rodents. The primary amines are known human bladder carcinogens. In rodent tumor models, these compounds primarily induce tumors in the liver, lung or mammary gland, because of a high rate of N-hydroxylation in rodent hepatocytes, whereas humans and dogs primarily produce N-glucuronides in the liver, which are then transported to the kidney. These amines are then released in the acid urine and converted into genotoxic derivatives in bladder epithelial cells.

Tobacco smoke

Tobacco smoke causes 30% of all cancer deaths in developed countries, particularly lung, but also upper respiratory tract, esophagus, bladder, and pancreas and probably others. Despite this, over a billion people continue to smoke, making tobacco smoke the single most lethal carcinogen. Carcinogens in tobacco smoke link nicotine addiction and lung cancer. People smoke because of addiction to the nicotine in tobacco smoke, but nicotine is not itself a prominent carcinogen. The cigarette poses numerous problems as a nicotine delivery device because carcinogens are inhaled alongside nicotine in every "satisfying" puff. The cumulative exposure to carcinogens in a lifetime of smoking is substantial, even though the actual carcinogen content of each cigarette is tiny.

The risk may be determined by the number of "pack years" (number of cigarettes smoked per year and duration of habit) and debatably by tar content. However, this latter perception may falsely reassure smokers of so-called "light" cigarettes that they may as a result experience only equally "light" heart attacks or cancers. However, the risk of some chronic respiratory conditions (chronic airflow limitation, emphysema) may be reduced by lowered tar content.

PAHs, including 3-methylcholanthrene, benz[a]anthracene and benzo[a]pyrene are believed to be key carcinogens in tobacco smoke. Tobacco carcinogens in most cases are processed by enzymes to form reactive carcinogens (metabolic activation) that bind to DNA, forming covalent binding products called DNA adducts. Importantly, tobacco smoke creates a "field of injury" throughout the airway, a fact which has recently been exploited by Spira and colleagues to identify the transcriptional effects associated with smoking (and to what extent these are reversible with smoking cessation) and more recently to identify a genomic signature from airway epithelial cells that can distinguish smokers with and without lung cancer with around 90% accuracy. Individuals vary in the way they metabolically activate tobacco carcinogens and in the ways that they process DNA adducts. These and other factors may determine cancer susceptibility.

A tremendous amount of publicity has surrounded the ongoing debate about passive inhalation of secondhand smoke and cancer. The balance of evidence would seem to strongly support avoiding such "passive smoking" as a means of reducing the risk of cancer, although this risk is considerably lower than that experienced by active smokers. Importantly, here we must make a distinction between individuals who have chosen to smoke (no matter how much we may sympathize with them as victims of tobacco advertising and peer pressures), and those who have chosen not to smoke and yet may be put at risk by the actions of other individuals. It is not just individuals who should be encouraged to act responsibly. Tobacco advertising, tax revenues, and air pollution may be deemed more a matter for governments and yet all play a role in health.

Air pollution

Air pollution has been linked to a small but measurable increase in risk of lung cancer and may contribute to around 3% of cases (smoking contributes to more than 90%), mainly in road workers or those exposed to excessive diesel exhaust fumes. Nitrogen dioxide in car exhaust fumes is known to cause DNA damage, and particulates present in polluted air may cause airway irritation and increased delivery and retention of bound carcinogens present on the surface. Legislation in Europe has resulted in a reduction in carcinogens in car exhaust fumes due to use of catalytic converters and changes in car design.

Radon gas

Radon is a naturally occurring radioactive gas which, if inhaled, can increase risk of lung cancer. Radon occurs at low levels outdoors but can accumulate in residential buildings and in the workplace indoors. Areas in the United Kingdom, for example, with large numbers of potentially affected properties can be identified on maps available at www.ukradon.org/, and can then be individually tested with readily available detectors specifically for this purpose. Overall, around 1% of all cancer deaths in the United Kingdom, mainly from lung cancer, may involve radon gas exposure and this mainly involves smokers.

Ionizing radiation

Ionizing radiation is the high-energy emissions either released by radioactive materials or by electromagnetic means, such as X-rays. These are also extensively used in diagnostic radiology, in industry, and in military applications. X-rays and nuclear radiation can damage DNA, and the risk of DNA damage from ionizing radiation depends on lifetime exposure. Marie Curie, who discovered radium, paving the way for radiotherapy and diagnostic X-rays, died of cancer herself as a result of radiation exposure in her research. Nuclear radiation is clearly linked to risk of various cancers and in those cases where radiation exposure has been concentrated to specific tissues (in diagnostic or more usually therapeutic situations or inadvertently following nuclear disasters) then cancer risk is highest in those same tissues. Fortunately, relatively few individuals receive such exposure, with workers in power plants, laboratories, hospitals, etc. closely monitored for the cumulative exposure to ionizing radiation.

Much controversy has surrounded the role of low-frequency electric and magnetic fields and radiofrequencies, particularly those associated with overhead power lines, mobile telephone masts, and mobile phones themselves. No empirical evidence supports the view that mobile phones may increase the risk of cancers. Although individual studies have suggested a possible increase in risk of some brain tumors, other studies have failed to show this.

Ultraviolet radiation is primarily part of the sun's radiation and is also generated artificially in sunlamps and tanning facilities. The most harmful of this type of radiation are high-frequency, DNA-damaging ultraviolet B rays. These are the rays that cause 90% of all skin cancers.

Dietary and lifestyle factors

Adam was but human – this explains it all. He did not want the apple for the apple's sake, he wanted it only because it was forbidden. The mistake was in not forbidding the serpent; then he would have eaten the serpent.

Mark Twain

Diet may play a role in up to 20% of fatal cancers. A diet high in saturated fat may correlate with risk of cancers of the colon and rectum, and has recently been associated with postmenopausal invasive breast cancer. The role of fiber-deficient diet remains contentious; the European Prospective Investigation into Cancer and Nutrition observational study showed that higher dietary fiber from foods was associated with an estimated 25% reduction in risk for large bowel cancer, but recent pooled analyses of trials have failed to show a significant relationship between fiber intake and colorectal cancer risk. Obesity *per se*, however, is a major risk factor for both diabetes and some cancers, including breast and colon. Diets lacking in antioxidant vitamins correlate with risk of heart disease and cancer, yet are difficult to disassociate from the commonly associated problems of low socioeconomic class in general.

Carcinogenic foods remain contentious. Salted, pickled, and smoked foods, such as pickles or smoked fish, and meats treated with nitrites have been implicated. Hot drinks have also been shown to be associated with higher esophageal cancer risk. Aflatoxin is contained within grain and peanuts that have become moldy with an *Aspergillus* species endemic in Africa, and is associated with hepatocellular carcinoma. Pickled fish and vegetables have been associated with risk of nasopharyngeal and esophageal cancers in China.

Alcohol is estimated to contribute to about 3% of deaths from cancer. People who drink alcohol heavily have a higher risk of mouth, throat, esophagus, stomach, and liver cancer. Interestingly, in contrast to more substantive intake (more than 3 drinks per day), modest alcohol intake may confer some protection from oral cancer.

A very large prospective study has looked at risk factors for gastroesophageal cancers. The EPIC cohort consists of around 520 000 participants aged 35–70 years with diet and lifestyle information collected at the time of recruitment. Following a mean follow-up of 6.5 years, there were 268 cases with adenocarcinoma of the stomach and 56 of the esophagus. Results suggest a possible adverse effect of red meat consumption and a protective effect of fruit and vegetables.

Reducing dietary fat intake, with only a modest influence on body weight, may improve relapse-free survival in patients with treated breast cancer. People who ate the most fiber had 25–40% lower risk of bowel cancer compared to people who ate the least. Also those consuming the highest amounts of fruit and vegetables had lower risks of oral, esophageal, gastric, and lung cancers, but had no effect on prostate, ovarian, or kidney cancers. It is worth noting that the "eat five portions of fruit and vegetables a day" mantra was first used in 1991 and has been largely supported by trial evidence since that time. The EPIC study found that eating excessive amounts of red or processed meat could increase the risk of stomach and pancreatic cancer. In fact, a diet comprising above 100 g of meat a day increased risk of stomach cancer threefold. Conversely, eating above 80 g of fish a day reduced risk of colorectal cancer by a third.

According to the WHO, obesity is now second only to tobacco as a potentially avoidable cause of cancer. Around 7–15% of breast cancer and 14% of bowel cancers in developed countries are caused by obesity. Cancers of the uterus, esophagus, pancreas, kidney, and gallbladder are all increased by obesity and possibly also other cancers such as leukemia, liver, lymphoma, ovary, and thyroid. It appears that it is the presence of abdominal fat and the metabolic changes associated with this that are the greatest risk in obesity (as they appear to be for heart disease). Thus, insulin resistance, hyperinsulinemia, alterations in active levels of IGFs, as well as changes in circulating adipokines may all contribute to risk of cancer. Indirect effects such as gastric acid reflux damaging the esophagus may also contribute.

However, the difficulty in demonstrating that modifying diet, by targeted intervention, affects subsequent cancer risk is emphasized by the Women's Health Initiative Dietary Modification Trial, a randomized controlled trial of 48 835 postmenopausal women aged 50–79 years, which has failed to show significant effects of changing diet on breast or colorectal cancer risk. Perhaps surprisingly, despite a near 11% reduction in energy production from fat at year 1 (maintained in 8.1% of the women at year 6) and significant increases in vegetable, fruit, and grain servings, there was no evidence of reduced risk of invasive colorectal or breast cancer, although there was a nonsignificant trend to benefit. On the other hand, in the Nurses' Health Study II, an observational study of 90 659 premenopausal women aged 26–46 years, red meat intake was associated with breast cancer risk over a 12-year follow-up; 1021 cases of invasive breast carcinoma were documented with a greater red meat intake (more than three servings per week) related to elevated risk of breast cancers that were estrogen and progesterone receptor positive but not to those that were estrogen and progesterone receptor negative.

Although the role of red meat intake in promoting colorectal cancer is contentious, a protective role for fish intake has been suggested. The association of diabetes with lack of exercise is also important. Three large studies in Italy and the United States estimated that physical inactivity could be responsible for up to 15% of bowel cancers and 11% of breast cancers. Thus, Cancer Research UK states that keeping active could help to prevent about 9000 cases of cancer in the United Kingdom every year.

An important study of around 40 000 men demonstrated that walking or cycling an hour a day could reduce cancer risk by 16%. Even half an hour a day could reduce risk of dying from cancer by a third. The risk of developing a range of solid tumors is increased in type 2 diabetes, and may be influenced by glucose-lowering therapies. Those on insulin or insulin secretagogues were more likely to develop solid cancers than those on metformin, and combination with metformin abolished most of this excess risk. Metformin use was associated with lower risk of cancer of the colon or pancreas, but did not affect the risk of breast or prostate cancer.

Medical treatments

Do not free a camel of the burden of his hump; you may be freeing him from being a camel.

Gilbert K. Chesterton

Medical treatments may increase risk of cancers. Radiation and chemotherapy, for example, may result in secondary cancers and, intriguingly, even in increased aggressiveness of those cancers not fully cured by such treatments. Some immunosuppressive drugs used for chronic inflammatory diseases or after transplantation can cause lymphoma.

Women exposed to excess estrogen are placed at increased risk for some gynecological cancers (e.g. breast, uterus), because estrogens can stimulate cell proliferation in these tissues. The level of exposure to estrogen is determined by several factors, including age at menarche (first period), pregnancy and age at pregnancy, age at menopause, weight, physical activity, and diet. Thus, a woman with an early age at menarche and late age at menopause would have a greater exposure to estrogen. Obviously, taking postmenopausal HRT and premenopausal oral contraceptives will increase overall exposure, whereas conditions associated with early menopause, or low estrogen levels will reduce exposure. Excessive sex hormone exposure may also occur in those undergoing gender reassignment.

Carcinogenesis and oxidative stress

The process of aging is believed to be in part due to reactive oxygen species (ROS) produced as byproducts of normal metabolism. These ROS, such as superoxide and hydrogen peroxide, are the same mutagens produced by radiation, and can cause damage to DNA, proteins, and lipids. In rodent models, the DNA in each cell accumulates around 100 000 oxidative lesions per day, which are usually repaired by DNA-repair enzymes. However, this system is not foolproof: a young rat has about one million oxidative lesions in the DNA of each cell, and this increases to about two million in an old rat. Luckily, human cells receive about ten times less damage than a rat cell, compatible with the higher incidence of cancer and shorter lifespan in a rat.

The degenerative diseases of aging, such as cancer and coronary heart disease (CHD), all share an oxidative origin. It has often been suggested that intake of dietary antioxidants, such as vitamins C and E, might reduce the incidence of degenerative conditions, but to date clinical trials using antioxidant supplementation have been disappointing, possibly in part due to the relatively short duration of such trials. Epidemiological studies show that the incidences of most types of cancer, as well as CHD, are much higher in populations where intake of antioxidant-containing vegetables and fruits is low. Smokers may be particularly at risk as their antioxidant pools are depleted (cigarette smoke is high in oxidants).

The three main potentially avoidable causes of cancer are smoking, poor diet (excess fat and calories; inadequate intake of fruits, vegetables, and fiber), and chronic infections leading to chronic inflammation (hepatitis B and C viruses, *Helicobacter pylori* infection, etc.). All of these may in some way result in increased ROS and oxidative stress.

Testing for carcinogens

For obvious reasons, such tests are not conducted in humans but in animal models, primarily the rat. Animal cancer tests are generally conducted at the maximum tolerated dose (MTD) of carcinogens, which creates a certain amount of difficulties when interpreting results for human cancer. Formation of cancer in rats at MTD does not necessarily imply that a low dose of the chemicals tested will cause cancer in humans. In fact, this issue is highlighted by the fact that in animal cancer tests around 50% of all naturally occurring chemicals are carcinogens for rats. In an illuminating example from Bruce Ames, it was noted that of the over 1000 chemicals in a cup of coffee, only 26 have been tested in animal cancer tests and more than half are rodent carcinogens. One explanation for such results may be that high-dose animal cancer tests are largely measuring increased rates of cell division in some cell types induced in response to nonspecific toxic cell killing by the high doses used.

However, low doses of rat carcinogens (those we are usually exposed to) may not have these same effects in humans because innate defense mechanisms, such as DNA repair and oxidant defenses such as glutathione transferases, may cope at these levels of exposure and may be induced to cope with increasing exposure.

It is important to put the results of animal tests in context. Again, to quote Bruce Ames, "The effort to eliminate synthetic pesticides because of unsubstantiated fears about residues in food will make fruits and vegetables more expensive, decrease consumption, and thus increase cancer rates. The levels of synthetic pesticide residues are trivial in comparison to natural chemicals, and thus their potential for cancer causation is extremely low."

The Ames test

A widely used test for determining if a chemical is a mutagen is named after its developer, Bruce Ames. The test was devised in 1975 by Ames and his colleagues at the University of California at Berkeley. The Ames test is based on the assumption that substances that are mutagenic for a strain of bacterium (*Salmonella typhimurium*) might also be carcinogenic in humans. Some substances that cause cancer in laboratory animals (dioxin, for example) do not give a positive Ames test (and vice versa). However, the test is cheap and easily performed and has proved extremely useful in rapidly screening substances in our environment for possible carcinogenicity.

The bacterium used in the Ames test carries a gene mutation making it unable to synthesize the essential amino acid histidine from the culture medium. However, this mutation can be reversed, a back mutation, with the gene regaining its function. These revertants are able to grow on a medium lacking histidine. If cultures of these bacteria are exposed to mutagenic chemicals then increased numbers of bacteria might regain the ability to grow without histidine, forming visible colonies (above those spontaneously reverting in control dishes). Many chemicals are not mutagenic (or carcinogenic) in themselves, but are converted into mutagens (and carcinogens) as they are metabolized by the body. This is the reason the Ames test includes a mixture of liver enzymes.

The Ames test yields a number, specifically, the number of growing bacterial colonies, which is a measure of the mutagenic activity (potency) of a treatment chemical. This value is often expressed as the number of revertants per microgram of a pure chemical (mutagen) or per gram of food containing that mutagen.

Microorganisms and cancer

God in His wisdom made the fly
  And then forgot to tell us why.

Ogden Nash

In 1911, Peyton Rous discovered that sarcomas in chickens could be transmitted between animals using a "cell-free" filtrate. The active agent was subsequently identified as a virus now known as the Rous sarcoma virus; and a single gene, *src*, was identified as responsible for these cancers. In 1981, Michael Bishop and Harold Varmus discovered that normal cells contain a gene homologous to the viral *src* gene, termed c-*src*. These were key steps in defining the genetic basis of cancer (see Chapter 6).

Tumorigenesis can be driven by the actions of specific tumor viruses. Viral causation of cancer is well-documented in the lab, and is important in some (but probably not most) human cancers. Tumor viruses comprise two distinct types – viruses with DNA genomes (e.g. papilloma and adenoviruses) and those with RNA genomes (retroviruses) (Table 3.3).

DNA viruses

Cellular transformation by DNA tumor viruses, in most cases, has been shown to be the result of protein–protein interaction. Proteins encoded by the DNA tumor viruses, the tumor antigens or T antigens, can interact with cellular proteins. This interaction effectively sequesters the cellular proteins away from their normal functional locations within the cell. The predominant types of proteins that are sequestered by viral T antigens have been shown to be of the tumor suppressor type (see Chapter 7). It is the loss of their normal suppressor functions that results in cellular transformation.

Hepadnaviruses

Hepatocellular carcinoma (HCC) is the fifth commonest cancer and in bronze medal position among lethal cancers globally. Most cases of HCC cases are from Asia and Africa, with over half

Table 3.3 Examples of viral oncogenes

Oncogene	Nature of proto-oncogene	Virus-induced tumor
Src	Tyrosine kinase	Chicken sarcoma
erb-B	EGF-receptor/tyrosine kinase	Chicken fibrosarcoma
Abl	Tyrosine kinase	Mouse leukemia
Myc	Transcription factor	Chicken myelocytoma
Fos	Transcription factor	Mouse osteosarcoma
Jun	AP-1 transcription factor	Chicken fibrosarcoma
bcl-2	Antiapoptotic factor	Lymphoma
H-ras	GTP protein	Rat sarcoma
sis	Platelet-derived growth factor	Monkey sarcoma

coming from China. The vast majority of hepatocellular carcinomas are attributable to chronic hepatitis B (HBV) and hepatitis C (HCV) infections. In areas of Asia and Africa with high rates of HBV infection there are also high rates of hepatocellular carcinoma, and age of onset is often younger. Around 400 million people suffer from chronic HBV infection, which is the main cause of HCC worldwide. HBV, as a DNA virus, can integrate into host DNA and directly transform hepatocytes, whereas HCV is an RNA virus and cannot.

No HBV genes have been clearly shown to be transforming, with cancer risk believed to arise primarily as a result of chronic inflammation and triggering of hepatocyte regeneration pathways. Both HBV and HCV proteins may disturb key signaling pathways, and potentially important HBV proteins have been identified. The HBV X (HBx) protein can regulate endoplasmic reticulum (ER) stress, which itself is a factor in activation of COX2 via ATF4 and also inflammation and truncated Pre-S2/S. Recent suggestions that HBV infection can also increase risk of non-Hodgkin lymphoma is interesting and supports the view that a more general alteration in immunity might be driven by HBV infection.

In animal models, such as the woodchuck, virtually 100% of the infected animals with chronic hepatitis develop liver cancer.

Papillomavirus

Cervical carcinoma bears the classic epidemiological hallmarks of an infectious sexually transmitted disease; it is rare in the sexually inactive and conversely is commoner in women with multiple sexual partners. Cervical cancer is now known to be caused by infection with oncogenic types of human papillomavirus, mainly HPV16 and 18. In fact, HPV DNA is found in almost all invasive cervical cancers globally. HPV is the commonest cause of infection-related cancers worldwide, in particular cervical and oropharyngeal/oral cancers. Vaccination against HPV prevents development of cervical dysplasia and cancer in women who have not already been infected by the vaccine-specific HPV types.

Since 2006, when vaccination was first approved in the United States, many nations have started offering HPV vaccines to young women. In 2008, the European Parliament adopted a resolution on combatting cancer in the enlarged EU that recommended the "cost-effective integration of appropriate HPV testing for cervical cancer screening and HPV vaccination to protect young women from cervical cancer." Yet, the Vaccine European New Integrated Collaboration Effort (VENICE) and the European Cervical Cancer Association reports show that over half of EU countries still do not have vaccination programs. Two vaccines composed of HPV L1 proteins in virus-like particles (VLPs) are available, one containing VLPs of HPV types 6, 11, 16, and 18, the other HPV16 and 18.

There is no evidence that vaccination of girls will eliminate all relevant HPV infections or necessarily substantially reduce oropharyngeal/oral cancer, which largely affects men. Moreover, the benefits of vaccination will take some time to be realized, maybe another 20–30 years.

Extending vaccination more widely, and to boys, could potentially be a highly cost-effective strategy to create "herd immunity" and eliminate or decrease considerably the incidence of oropharyngeal/oral cancers, which is reaching "epidemic proportions" in Europe (in 2011 it was estimated that 100 000 Europeans would get oropharyngeal/oral cancer and 50 000 would die of it).

The HPV16 and 18 strains are most frequently identified in cervical carcinoma and also display the greatest capacity to transform cells *in vitro*. HPV oncoproteins contribute to cancer initiation and progression and continued expression is necessary for tumor maintenance. The HPV proteins E7 and E6 can subvert key cell cycle checkpoints by binding and inactivating the p53 and RB tumor suppressors. In fact, binding of cell tumor suppressor genes by viral proteins is a common property of many DNA tumor viruses. HPV16 E7 has other effects, including reducing E2F6-containing polycomb repressive complexes and an ability to activate demethylases, KDM6A and B, and thereby remove repressive trimethyl marks on lysine-27 of histone 3, needed for polycomb repressive complexes to bind and to induce expression of the cervical cancer biomarker $p16^{INK4a}$.

Herpesviruses

Epstein–Barr virus (EBV) was the first human tumor virus identified in 1964. EBV, the etiological agent of infectious mononucleosis, has been associated with the genesis of Burkitt lymphoma and nasopharyngeal carcinoma (viral DNA, and sometimes virus, is present in these cancer cells). Furthermore, EBV has been detected in the Reed–Sternberg cells in a high percentage of Hodgkin lymphomas. EBV transformation is multigenic and at least five different viral genes appear to have some involvement. Recent studies have suggested that, like other DNA viruses, one of EBV's gene products can bind with tumor suppressor proteins. EBV nuclear antigen 3C (EBNA3C) is an essential transcription factor for growth in EBV-transformed lymphoblast cell lines and regulates multiple genes, including *RAC1*, *LYN*, and *TNF*.

One of the first viruses shown to be involved in cancer was an avian herpesvirus known as Marek's disease virus. In chickens this virus causes a fatal T-cell lymphoma in virtually all infected birds. Herpesviruses in humans, including the oncogenic Kaposi's sarcoma herpesvirus (KSHV), encode cyclin D homologs. These viral cyclins differ from their cellular counterparts in that they elicit holoenzyme activity independent of activating phosphorylation by cyclin-dependent kinase (CDK) and resist inhibition by CDKIs.

In patients with HIV infection, KSHV can cause this specific tumor type. KSHV is also associated with two B-cell tumors: multicentric Castleman disease (MCD) and primary effusion lymphoma (PEL). The KSHV Orf63 protein is homologous to human NLRP1 and can block NLRP1-dependent innate immune responses, including caspase-1 activation and interleukin processing that may assist lifelong persistence of herpesviruses.

Cytomegalovirus (CMV) infection is very common in humans and the virus infects a variety of cells, including intestinal epithelial cells. The CMV protein US28, a constitutively active chemokine receptor, activates β-catenin and Wnt signaling by inhibiting glycogen synthase kinase 3 beta (GSK-3β) and promotes intestinal dysplasia and cancer in transgenic mice *in vivo*. This suggests the possibility that CMV infection could also contribute to human intestinal cancer.

Adenoviruses

These viruses have never been associated with cancer in humans, but they transform cultured cells and cause cancer in animals. At least two gene products (E1A and E1B) have been implicated in transformation studies. The two genes alone are sufficient to transform cells in culture. Studies have shown that the E1A protein binds Rb and that the E1B protein binds p53.

Poxviruses
Poxviruses cause cancerous myxomas in rabbits.

Polyomaviruses
Polyomaviruses are oncogenic in laboratory animals and can transform human cells in culture. The human JC virus has been detected in numerous brain tumors of various pathologies, but no clear correlation has been established. Polyomavirus and SV40 are linked to a variety of animal tumors. The latter is known to inactivate both the p53 and Rb tumor suppressors. So not surprisingly it can strongly predispose to cancer formation; in fact, many early animal models of cancer were founded on the platform of first overexpressing the SV40 large T antigen responsible for this activity.

Merkel cell carcinoma (MCC) is a rare but aggressive human skin cancer that has long been suspected of infectious origin as it has a predilection for the immunosuppressed and elderly. Recent studies have identified a new form of polyomavirus (MCV) that may contribute to this rare cancer. A fusion transcript between an MCV-derived T antigen and a human receptor tyrosine phosphatase was found and the clonal pattern of viral DNA integration within the tumor genome strongly suggested that MCV infection preceded clonal expansion of tumor cells.

Retroviruses

RNA tumor viruses are common in chickens, mice, and cats but rare in humans. The only currently known human retroviruses are the human T-cell leukemia viruses (HTLVs) and the related retrovirus, human immunodeficiency virus (HIV).

Tumorigenic retroviruses (oncoviruses) are grouped into three categories based on their mechanism of oncogenicity: transducing retroviruses, *cis*-activating retroviruses, and *trans*-activating retroviruses.

Two features of the retrovirus life cycle permit the acquisition and activation of oncogenes. Namely, integration into the cell chromosome provides the opportunity to hijack cellular genes and allows genetic permanence once an oncogene has been transduced and, furthermore, most retroviruses do not kill the cells they infect so that genetic alterations are transmitted to daughter cells. When a retrovirus infects a cell, its RNA genome is converted into DNA by the viral-encoded RNA-dependent DNA polymerase ("reverse transcriptase"). The DNA can then integrate stably into the genome of the host cell and be copied as the host genome is duplicated during normal cell division. Contained within the sequences at the ends of the retroviral genome are powerful transcriptional promoter sequences, termed "long terminal repeats (LTRs)." The LTRs promote the transcription of the viral DNA, leading to the production of new virus particles. The process of integration may result in rearrangement of the viral genome and the consequent incorporation of a portion of the host genome into the viral genome, termed "transduction."

HTLV-I causes adult T-cell leukemia (ATL), a disease with a uniformly poor prognosis. Activating mutations in Notch signaling are known in at least a third of patients with ATL and in many with T-cell acute lymphoblastic lymphoma (T-ALL). Such mutations include frameshift and single-substitution mutations in the PEST domain which prevent CDC4/Fbw7-mediated degradation of the active cleaved intracellular domain of Notch1 (see Chapter 5). HIV-1 Rev-binding protein (Hrb), part of the endocytosis machinery, promotes Notch-induced T-ALL development in mice. In this case Notch activates the normal mammalian gene as an essential part of tumorigenesis; whether Hrb deriving from HIV infection can in any way recapitulate this is an interesting conjecture.

Transducing retroviruses
Occasionally, transduction results in the virus acquiring a host gene normally involved in cellular growth control. The transduced gene is always altered either by point mutation or deletion of protein sequences. In many instances the gene is fused to a viral gene and during transduction all introns are lost from the cellular gene. These alterations serve to activate the transduced gene and the end result of this process is unrestricted cellular proliferation. The transduced genes are termed oncogenes (see Chapter 6). The expression of the transduced gene is driven by the virus promoter/enhancer region (LTR) and its expression is no longer under cell control. Transducing retroviruses cause tumors at high efficiency (100% of animals) and with short latency periods (days). In addition, they readily transform cells in culture with almost 100% of the cells being transformed. In very rare instances the virus transduces the cellular oncogene without the concomitant deletion of viral coding sequences. In these unusual cases the virus is replication competent.

Numerous oncogenes have been discovered in the genomes of transforming retroviruses. Typically, viral oncogenes are mutated, have lost regulatory sequences or are amplified, and are capable of causing cancer by themselves. Aside from the key role that retroviruses carrying oncogenes have played in the discovery of cancer-causing genes, they probably have no role in the great majority of human cancers (see Chapter 6).

cis-*Activating retroviruses*
The second mechanism for retroviral transformation of cells is through the transcription-promoting actions of the LTRs. Retroviral genomes integrate randomly into the host genome, which can occasionally result in the viral LTRs being in proximity to a gene that encodes a growth-regulating protein. Thus, if the growth-regulatory gene is now overexpressed it can result in cellular transformation. This is termed "retroviral integration–induced transformation" and is exemplified by the induction of certain cancers by HIV infection.

These viruses activate a cellular proto-oncogene by integrating adjacent to it and increasing or altering its expression (promoter or enhancer insertion model). Tumors formed by these viruses take longer to occur and not all animals form tumors. These viruses are replication competent.

trans-*Activating retroviruses*
These viruses may upregulate cell oncogenes through the action of a viral transactivator protein. The latency period is long (years) and the efficiency of tumor induction is very low (1%). They are replication competent and do not transform cells in culture. HTLV-I was first isolated by Gallo and coworkers in 1980 and is linked to adult T-cell leukemia in Japan. All cases of ATL show evidence of HTLV-I infection, but the lifetime risk of developing ATL is only 1% for infected individuals. The viral *tax* gene can

activate transcription of important T-cell growth factors (e.g. IL-2), suggesting that an autocrine mechanism of transformation may be operative, but verification of this theory awaits further studies.

Bacteria and others

Helicobacter pylori is a causative agent in peptic ulceration and in gastric cancer. Gastric adenocarcinoma is the second leading cause of cancer-related death in the world with over 700 000 deaths related to it each year. Survival rates are poor and generally 5-year survival is below 15%. *H. pylori* colonization causes persistent inflammation and gastritis, which is now considered the preeminent risk factor for gastric cancer globally. Not all patients with *H. pylori* develop cancer, and risk is modified by strain of the bacterium and by host factors such as gastric acid output and others. *H. pylori* may cause loss of two tumor suppressors in gastric mucosa, namely p53 and RUNX3. It can also cause epigenetic alterations, among which methylation of the E-cadherin promoter and activation-induced cytidine deaminase may be important.

The *H. pylori* gene *vacA* encodes a secreted protein (VacA) that can induce vacuolation and apoptosis in epithelial cells and is associated with gastric cancer. VacA can bind several important epithelial cell–surface proteins, including epidermal growth factor receptor (EGFR) and CD18. VacA can also interfere with T-cell–mediated immunity, allowing allowing *H. pylori* to evade the adaptive immune response needed for long-term colonization. A strain-specific determinant important in tumorigenesis is the cag pathogenicity island, and cag+ strains significantly augment the risk for distal gastric cancer compared with cag− strains. CagA can promote epithelial cell proliferation and loss of polarity. It affects several signaling pathways that can promote inflammation and proliferation in gastric epithelium, including RAP1A–BRAF–ERK, SHP2- and GRB2-regulated growth factor signals, EGF, E-cadherin/β-catenin, MET, the phospholipase Cγ, and MARK2.

Cholangiocarcinoma (CCA), a cancer originating from the neoplastic transformation of the biliary epithelium, is strongly linked to chronic inflammation induced by parasites (liver fluke) infestation prevalent in developing countries. It may also be related to HBV and HCV infection or noninfectious causes such as hepatolithiasis and primary sclerosing cholangitis.

Transient infection of eukaryotic cells with group B2 *Escherichia coli* blocks mitosis, causes DNA double-strand breaks and activates DNA damage responses. What role this might have clinically is unclear.

Prion proteins (PrP^{Sc}) are infectious agents comprising only a misfolded protein that can induce misfolding in the normal cellular protein counterparts (PrP^C), converting them into the prion form. The infectious agent PrP^{Sc} is thereby replicated and arguably information is transferred directly between proteins (another challenge to the central dogma). Epimutations can alter the structure of PrP^{Sc}, giving some potential for natural selection, and familial forms, where disease-causing mutations in the gene encoding PrP^C result in production of a version of PrP^{Sc}, are known. In all cases described to date, accumulation of prion proteins causes degenerative neurological conditions and for long it was believed that this would be one of a very select group of cellular processes not directly involved in cancer. However, some reports have suggested that prion proteins might be biomarkers for pancreas or colorectal cancers, and in the latter may even have a functional role. To date, a functional role for prions in cancer remains highly speculative.

The clinical staging and histological examination of cancer

But our beginnings never know our ends.

T.S. Elliott

Together, radiological imaging and microscopically observed histology constitute a pathological prospectus, a cancer "curriculum vitae" if you will, in which the full complement of current misbehaviors is catalogued. This "cancer charge sheet" allows some divination of future behaviors based on the tumor's existing criminal record. Thus, convictions for lymphovascular invasion, poor differentiation, and deep invasion into surrounding tissues all herald a strong likelihood of a bad outcome and progression to more serious offenses, such as metastatic spread in the future. However, predictions based solely on such anatomical or morphological criteria may often prove erroneous, as exemplified by CRC. Tumors classed as Dukes' B have an around 50% chance of producing either a poor or good future outcome. This may be considered – somewhat unkindly – as little better than the reading of tumor-derived tea leaves.

Even the improved TNM classification still struggles to make accurate clinically useful predictions. Thus, around 30% of patients with Stage II (equivalent to Dukes' B) will experience a poor outcome, but others have a comparatively good outcome. The hunt is on, therefore, to find molecular markers that can complement or enhance the predictive value of current staging techniques so that we can unambiguously decide who needs more aggressive treatments and meticulous follow-up and, just as importantly, who does not. It is anticipated that the molecular phenotype of cancer cells will reveal the cancer manifesto, in which the tumor's intentions are more clearly outlined and from which future misdemeanors can be predicted with greater accuracy.

So how does current traditional morphological staging work? Most tumors are classified on the basis of their tissue of origin, a matter that is usually anatomically self-evident, unless the primary tumor has remained occult and the disease has only come to light through the manifestation of distant metastases. In most cases, histological examination of a tumor can separate primary tumors from secondary deposits that have originated elsewhere, and may even give important clues as to the location of the primary tumor when only a metastasis has been examined. Histology can also allow a functional subclassification of a tumor that may assist clinical decision-making, which is important when the same tissue can give rise to tumors of very different prognosis and treatment responsiveness. Therefore, molecular markers are seldom employed for diagnosis. However, in some ways this is a circular argument. It is now increasingly clear that a molecular taxonomy of cancers may have greater bearing on outcome and on treatment selection than this more traditional "geographical" description. Moreover, there are numerous examples where even after histological examination diagnosis remains uncertain and could be assisted by new molecular markers.

Cancers are traditionally graded by a combination of clinical and histological parameters. Each tumor type has its own clinical and/or histological grading system (Chapter 2 provides some specific examples of this). Here we will concentrate on the classification of cancers by histology and molecular techniques. Aside from confirming tissue or even cell of origin for the cancer, histology is also important in assessing morphological/anatomical features that may have clinical ramifications. A variety of formal classifications have been developed, including the Dukes' system for CRC, but the so-called TNM classification is the most widely used staging system at the present time. Staging criteria follow universally agreed standards in most cases and there are TNM Committees that define staging criteria for most anatomical sites. As discussed in Chapter 2, the basic principles involve the determination of T (tumor), N (node), and M (metastases) separately and then grouped, allowing most cancers to be separated into one of four stages (I–IV), sometimes with subdivisions. The aim is to provide a rule of thumb to predict the natural history of the cancer and guide treatment.

Clinical staging includes physical examination, results of imaging (X-ray, CT, MRI, PET) and more or less invasive procedures (endoscopy, biopsy, and surgery). Agents that enhance imaging techniques are already in clinical use. These range from simple contrast-enhancing agents used for anatomical localization to highly selective ones targeted at biomarkers. The majority of imaging agents are radioisotopes, detected by nuclear medicine imaging, including single-photon emission computed tomography (SPECT), or by positron emission tomography (PET). These can detect the presence of receptors – exemplified by somatostatin receptor status in neuroendocrine malignancies like carcinoid tumors – that are useful in detecting the tumor and also in selecting treatments (e.g. octreotide) that target them. Techniques that allow imaging *in vivo* without removal of tissues are distinct from those used when looking for such molecules in cancer tissues being examined by the pathologist. Other more general imaging agents include the use of radiolabeled glucose and others to generate functional information which can be imaged at high resolution and even combined with CT scanning (PET-CT). Some tumors, such as cancers of the thyroid, can even be targeted by specific radiolabeled ligands in therapy.

Pathological staging by the pathologist complements clinical staging. Broadly, wherever possible cancer spread to lymph nodes and other tissues is confirmed by biopsies or surgical removal of tissues for histological examination. However, with one important caveat; namely, that there is little good served by undertaking such potentially discomfiting investigations if clinical management will not be altered no matter the result. Wherever possible, TNM staging is based on histological confirmation of spread not just on radiological investigations, as the latter are less sensitive and less specific. Moreover, histopathology can reveal other important features of a tumor that can predict prognosis. Thus, it is helpful to assess degree of differentiation: low-grade, well-differentiated tumors have a better prognosis than high-grade poorly differentiated ones, which grow and spread more rapidly. Moreover, increasing depth of penetration of tumor cells into underlying tissues or structures and invasion into the lymphovascular system all predict increased risk of metastatic disease and incremental worsening of prognosis.

Tumor grade is included in formal TNM staging in prostate and brain tumors (see Chapter 2) where it assists clinical decision-making, altering prognosis or treatment options. In current clinical practice, molecular markers are used to support histopathology in certain well-known circumstances – exemplified by ER/PR and HER2 for breast cancer. This area may in future be supported by imaging-based biomarkers (see Chapter 20) and multiplexed molecular imaging tools and computer-aided diagnostics and pattern recognition. For most solid tumors, anatomical staging aims to distinguish a cancer that will remain localized from that which might metastasize, because this predicts prognosis and will influence treatment. With this in mind, additional information from staging and histological examination and more recently molecular biomarkers can support these endeavors.

Some markers have now become a routine part of cancer subclassification, even if we are still a long way from a complete new molecular taxonomy of cancer. Serum α-fetoprotein (AFP), human chorionic gonadotropin-β (β-HCG), and lactate dehydrogenase (LDH) are all in use for testicular cancer. Histological staging is important for prostate cancer as well as those of thyroid, brain, and for sarcomas. In breast cancer, histology and molecular diagnostics are of clear value in treatment selection (see Chapter 16). These issues are under constant surveillance and although, at present, the TNM system regards nodes with cancer cell clumps less than 0.2 mm in diameter as node-negative, irrespective of histological or molecular investigations, this will almost certainly change, albeit slowly, as clinical follow-up data accumulates. In keeping with this, the presence of so-called "micrometastases" in CRC has been shown to signify a worse prognosis. Biomarker expression may complement TNM staging, particularly when treatment decisions are being made. Thus, HER2/NEU-positive breast cancers will receive trastuzumab and CD20-positive lymphomas are treated with rituximab, and so on (as covered in Chapter 16).

Screening and biomarkers

The definition of the individual was: a multitude of one million divided by one million.

Arthur Koestler

This and the following section will complement that in Chapter 16, which addresses biomarkers in treatment selection. Improved approaches for the detection of common cancers are urgently needed to help reduce the morbidity and mortality caused by these diseases. Screening can involve use of imaging techniques to look for tumors in internal organs, direct examination of cells obtained from potential sites of cancer, such as cervix, or finding molecules that indicate the presence and or behavior of a cancer (biomarker). The molecular approach is particularly attractive as biomakers may be measured more cheaply and easily and might also provide clues as to the likely behavior of a tumor and its response to given treatments. A comprehensive list of current biomarkers that are in use clinically to some extent is given in Table 16.5.

The recent acceptance that "personalized medicine" is an acheivable aim has further driven initiatives to find biomarkers that can help clinical decision-making. These biomarkers should, in a perfect world, be secreted by the cancer or other cells influenced by it and thereby be readily detected in blood or other accessible body fluids. Included in this category are inherited mutations conferring susceptibility to cancer, as these can be detected by analyzing DNA obtained from white blood cells. Tissue-based biomarkers that require access to tumor tissue may also be very helpful in prognosis and selection of drugs, though they will not necessarily be the way forward for early diagnosis. One could envisage, however, exploiting tissue-based biomarkers to find cancer cells in stool, urine, or blood that could under some circumstances assist diagnosis. In some cases, screening may seek to identify cancer or precancer cells, as exemplified by cervical cytology, or even radiological apparent tumors, as in screening mammography (see below).

Established biomarkers for cancer include a number of proteins that can be measured in blood, such as PSA (prostate), CEA (CRC), CA-125 (ovary), and thyroglobulin (thyroid), among others. All have been valuable in screening patients after definitive treatments for recurrence, and to a degree have been used to assist diagnosis of a suspected tumor in a symptomatic patient. However, they lack sufficient sensitivity or specificity to be of use in screening the general population. Newer biomarkers, developed as knowledge of cancer biology has advanced, have recently entered clinical practice and, as will be discussed below, have transformed treatments for some cancers. However, the hunt is still on for further biomarkers, particularly for early diagnosis and also to progress subclassification of cancer patients on the basis of predicted disease outcome and treatment benefit.

The challenges are immense. First, clinical validation requires not only a highly promising molecular marker and an easily applicable and standardizable analytical test, but then needs to cut the mustard in a large clinical trial, where it must prove itself both specific and sensitive. The reader will need to appreciate that even well-established biomarkers such as PSA and CEA would struggle to meet this challenge.

The on-going debate surrounding prostate screening is worth following. Recent studies have questioned the value of PSA-based screening for prostate cancer and have highlighted the very real problem of overdiagnosis, though benefits in monitoring for recurrence are unquestioned. This is compounded by uncertainties over how best to restrict local and systemic treatment to those patients who truly need it.

Screening mammography

Mammograms are X-ray images taken of the breast and can be used in screening to detect clinically silent breast cancers in women. The US National Cancer Institute recommends that all women above age 40 years should have a screening mammogram every 2 years. In the United Kingdom, this is recommended every 3 years after age 50, but is extending to women aged between 46 and 73. Mammography involves taking two X-ray images of each breast and will inevitably result in a significant and cumulative exposure to radiation. However, clinical studies suggest that this procedure can reduce the mortality from breast cancer among women aged 40–74 years, because the condition is diagnosed earlier than if one waited for clinical signs. In fact, for the United Kingdom, conservative estimates suggest that the number of deaths from breast cancer can be reduced by about 15% in women who are regularly screened – about 500 deaths prevented each year. Those with a family history may be screened earlier and more intensively (*vide infra*). Treatment can therefore be started earlier and maybe before cancer cells have spread.

Teaching women to examine their own breasts for lumps is also an important screening tool, but it should be noted that screening mammograms can detect microcalcifications associated with breast cancers that could not be detected by clinical examination.

We will now discuss some of the negative features of screening, which illustrate drawbacks with screening tests in general. First, a mammogram may fail to detect a cancer that is there (**false-negative**). This occurs in up to 1 in 5 screening mammograms and is most likely in women who have high breast density (a greater proportion of glandular and connective tissue than fat). Conversely, screening may suggest the presence of a cancer that is not there (**false-positive**), and for this reason an abnormal mammogram is always followed up by additional investigations, including ultrasound and/or biopsy. Clearly, the process can cause considerable stress in women who turn out not to have cancer. According to figures produced by Cancer Research UK, 5% of women are called back for further tests on the basis of a suspicious screening mammogram, but only 1 in 8 of these will turn out to have cancer. For every 1000 women screened only 7 will have cancer. These figures are contested and arguments about the real benefits of screening and the costs are never fully resolved.

Screening mammograms can also detect tumors that might never have shortened life (**overdiagnosis**), leading to unnecessary and potentially harmful **overtreatment**. This raises a general problem with many screening tests. Namely, that all will result in overtreatment, unless we also have tests that can individually separate those in whom treatment will prolong life and improve quality of life. Note that the latter will become a self-fulfilling prophecy; once a patient has been informed that they have cancer, few will be brave enough to accept not being given any treatment under those circumstances.

Screening must also be devoid of harm, and in this case the risk of repeated exposure to X-rays causing a new breast cancer must be weighed against the risk of delayed diagnosis of a preexisting one. Some women require more intensive screening, including those:

- with family history of breast cancer in a first-degree relative, particularly if it occurred before age 50 years;
- known to have *BRCA1/2* mutations;
- with early menarche or late menopause or on prolonged oral contraceptives or HRT (greater duration of lifetime oestrogen exposure)
- who have never been pregnant or had their first full-term pregnancy after age 30;
- with previous chest wall irradiation for lymphoma or other reason.

Obesity, type 2 diabetes, lack of physical activity also increase risk but are not generally factored in for more intensive screening.

Cervical screening

In contrast to mammography, which aims to detect established cancers (as well as some precancerous conditions such as ductal carcinoma *in situ* (DCIS)), cervical screening aims to prevent cancer by detecting abnormalities which, if left untreated, could progress to cancer. Using liquid-based cytology (LBC), a sample of cells is removed by means of a special brush that is used to sweep around the cervix. These cells are then sent to a cytologist for examination. Early detection and treatment can prevent 75% of cancers, but like all screening tests both false negatives and false positives arise.

In the United Kingdom, the HPV vaccination program started in September 2008 with all 12- to 13-year-old and 17- to 18-year-old girls being offered the vaccine. All girls born on or after September 1, 1990 have also been offered vaccination (www.nhs.uk/Conditions/HPV-vaccination).

However, the benefits of this program will not be manifest for many years, so at present there has been no revision of the cervical screening protocol, which in the United Kingdom will continue to offer all women between the ages of 25 and 64 a free cervical screening test every 3–5 years. The effectiveness of the program can be judged by coverage; if 80% of women can be screened over 5 years then death rates from cervical cancer could be reduced by as much as 95%.

Biomarkers and screening for prevention and early diagnosis of cancer

> It is impossible to trap modern physics into predicting anything with perfect determinism because it deals with probabilities from the outset.
>
> *Arthur Eddington*

There are very considerable challenges to operating and monitoring screening programs that are as resource-intensive as imaging or cytology. Moreover, for many cancers such tests may not be practicable or even possible. Regional variations are also factors in deciding which tests are cost-effective; gastric cancer is considered a sufficient priority in Japan for endoscopic screening programs, several countries are considering the best screening tests for use in CRC. These initiatives notwithstanding, there is a great deal of interest in developing screening tests based on analyzing proteins or other molecules in readily accessible body fluids, particularly peripheral blood. We will discuss some of these in current use here – those aimed squarely at treatment selection are discussed in Chapter 16.

Prostate cancer is the most common cancer in humans, yet controversy continues to surround screening and treatment. Recent evidence from two large trials has heightened the debate and suggests that current protocols may result in a high rate of overdiagnosis and overtreatment of prostate cancer, in part because of the low mortality rate relative to the incidence rate for the disease. It is worth noting here that the problem is not simply with the biomarker, but also with uncertainty over the natural course of the disease. Thus, up to a third of men may harbor microscopic prostate cancer when looked at histologically, and most will never experience any adverse effects. The European Randomized Study of Screening for Prostate Cancer randomized over 150 000 men aged 50–74 years to PSA screening or none. Overall, PSA-based screening reduced the rate of death from prostate cancer by 20% but was associated with a high risk of overdiagnosis. In real terms, 1410 men would need to be screened and 48 additional cases of prostate cancer treated in order to prevent one death from prostate cancer over 9 years, although these figures might improve over longer time periods. In contrast, the Prostate, Lung, Colorectal and Ovarian (PLCO) Cancer Screening Trial published apparently negative results.

Even though the conflicting results might be explained by shorter follow-up, by prescreening of patients, and by failure to biopsy some patients with PSA >4 ng/ml, it still suggests that the predictive value of positive biopsies triggered by positive PSA are low and many detected cancers are of low risk.

A bona fide screening tool able to detect reliably the minority of prostate cancers that will actually progress to life-threatening disease is still awaited. A recent meta-analysis of several screening trials in the *British Medical Journal* also concluded that there is little value in PSA-based screening with or without digital examination. There are a number of important issues that still need to be addressed. The usefulness of PSA screening is often undermined by the adherence to the **value** of the PSA score, when evidence suggests that the **doubling time**, only calculable by several successive blood tests (at least three, and preferably more than five) may be more sensitive and specific. A small doubling time indicates an aggressive cancer. A high PSA value and large doubling time can indicate "watchful waiting." PSA is very helpful in monitoring of patients after treatment for recurrence as this is associated with a recovery of PSA levels that had previously declined with effective treatment.

New tests are currently under investigation and include looking for the presence of prostate cancer gene 3 (*PCA3*) in the urine or for the abnormal *TMPRSS2-ERG* gene in prostate cells found in urine after a rectal exam.

Ovarian cancer, which carries a poor prognosis, largely due to the paucity of recognizable symptoms and comparatively advanced stage at diagnosis, is potentially curable in a large percentage of patients if only it could be detected earlier or prevented entirely. Currently, of the available biomarkers, screening for genomic mutations in *BRCA1* and *BRCA2* is useful only in the small number of cases where there is a strong family history of breast or ovarian cancers. The others all require examination of the tumor genome or measurement of tumor-related proteins – self-evidently not applicable for cancer prevention. Screening of the whole population, or even of half of it, for *BRCA* mutations is difficult because there are hundreds of different mutations and the consequences or effects on phenotype of many of these are not well characterized. Thus, it is relatively simple to find a *BRCA* mutation in a patient/family with high preponderance of early-onset breast cancers and then look for the same mutation in family members; the contribution to risk of future breast cancer here is implicit. Whereas randomly screening the normal population for *BRCA* mutations will inevitably lead to problems in assessing the effect on risk for those with no family history. Thus, we need not only to identify cancer-relevant genes but also which are the cancer-relevant mutations or polymorphisms within those genes.

One protein biomarker that has been used in screening more generally is CA-125 (produced by the mucin 16 (*MUC16*) gene), but this also has a number of problems that preclude its use in screening the general population. CA-125 is raised by a number of nonmalignant conditions and may be falsely positive in around 30% of cases or more in younger women, though these decrease post menopause. The biggest drawback is that CA-125 is often not raised in early-stage cancers and is not raised in 21% of women with ovarian cancer. It is, therefore, not sufficiently specific or sensitive. Rather disappointingly, recent studies examining panels of other potential biomarkers have failed to improve on CA-125 in this respect. Despite these limitations, CA-125 may be used as an adjunct to imaging in women who have symptoms consistent with ovarian cancer. The potential risks of women with falsely positive CA-125 undergoing unnecessary diagnostic surgery still needs to be resolved. At present, normal imaging in women without family history may best be followed by further imaging after a period of time rather than laparoscopy, but this is contentious.

Biomarkers and treatment selection

Biomarkers used for treatment selection that are already established in clinical practice include the presence of mutations in *HER2* and the presence of hormone receptors (ER and PR) in breast cancer, *KRAS* and *BRAF* in CRC, and a variety of others in hematological cancers. Many more are in use in some centers and are being tested in trials. The important area of **pharmacogenetics**, relating to the effect of genetic variants on the metabolism, action, and toxicity of specific drugs, is discussed in depth in Chapter 16 so will not be further addressed here. The role of the tumor genome in therapeutic selection is often referred to as **pharmacogenomics**, and is also covered in Chapters 16 and 18.

Screening for genomic mutations

In general, these tests are restricted to use in patients with a family history of rare cancer-susceptibility syndromes, such as retinoblastoma (*RB*), Li–Fraumeni syndrome (*TP53* or rarely *CHEK2*), multiple endocrine neoplasia (MEN) type 1 (*MENIN*) or type 2 (*RET*). Genetic tests are also useful in identifying at-risk family members with increased incidence of common cancers such as breast and ovary (*BRCA1* and *BRCA2*), colorectal such as Lynch syndrome CRC (*MLH1* and *MSH2*) and familial adenomatous polyposis (*APC*). In all these examples, the presence of an inherited gene mutation may lead either to more intensive screening (imaging, biomarkers, stool analysis, colonoscopy, etc.) or even to preventative surgery.

This was not intended to be an exhaustive list, as there are several other examples, many of which will be discussed elsewhere in this book in the context of the effects of these on biological processes.

Somatic gene mutations, epigenetic alterations and multistage tumorigenesis

> We do not know, in most cases, how far social failure and success are due to heredity, and how far to environment. But environment is the easier of the two to improve.
>
> *John Haldane*

Cancer cells are spawned by the stepwise accumulation of genetic alterations in somatic cells under the influence of many of the processes discussed so far and, of course, due to simple bad luck. This process is accelerated if a permissive foundation has been established by hereditary factors that give the incipient cancer cell a headstart or quicken the slide into malignancy by increasing the rate at which mutations accunulate. As we will discuss later in the book, those primordial genetic or epigenetic changes that occur early during malignant transformation may have particular importance as therapeutic targets as they will include oncogenes, to which cancer cells are addicted (see Chapter 5). Subsequent epimutations may well have provided another turn of the oncogenic screw, but may be less central to the actual endurance of the cancer cell. Nevertheless, these may still prove suitable therapeutic targets where they confer malignant behaviors such as invasion or metastasis.

We have made enormous progress in defining key cancer-critical signaling modules, in particular those involving growth factor receptor tyrosine kinases and downstream mediators such as RAS and RAF. However, aside from a few notable exceptions, pinpointing of cancer-causing mutations in specific genes has proved far more challenging. Thus, increasingly detailed analysis of the genomes of many cancers has revealed that:

- for a given tumor type there are many infrequently mutated genes and only a few frequently mutated genes;
- many mutations and genomic rearrangements are acquired early in the evolution of the cancer;
- clones from which distant metastases are generated arise in the primary carcinoma and are descendants of the original parental, nonmetastatic clone;
- there is ongoing, parallel and even convergent evolution among clones of cancer cells in the primary tumor and in the metastatic lesions driven by genomic instability;
- new mutations are brought forth by the hour of need in the face of cancer therapies.

The ramifications of this are that the curriculum vitae of the cancer remains visible in established cancers and may be read by cancer researchers in order to understand at least the later stages of the process by which a normal cell has become a cancer cell.

Thus, we keep returning to the same theme – cancer is very complex at all levels and shows substantial genetic heterogeneity. Cataloguing of somatic mutations in individual cancers can help us unravel the bewildering geography of each cancer's unique journey, even if the entire sequence may still prove elusive. So how do we translate the record of all the various insults to the cancer's genome into an understanding of tumor initiation and progression? First, we can painstakingly make cancers in cultured cells and animal models by combining various mutations in pathways or genes known to be mutated in humans. Even though the individual somatic mutational events are infrequently repeated across individuals and differing cancers, a smaller number of key defined biological pathways drive the development and progression of many human cancers. Thus, EGFR, MAPK, and PI3K pathways are often turned on, and p53 and RB pathways turned off.

We can also learn much by the study of those cancers in which much of the life history is played out in front of us. This scenario is exemplified by colorectal cancer (CRC), in which tumors at various stages of clinical evolution can be readily obtained and molecular phenotype and genotype determined, enabling us to unravel the time sequence in which somatic mutations are acquired. In other cases where this is not possible and only mature advanced cancers are available, various mathematical and statistical techniques akin to **phylogenetics** have been applied to genomic data to try to identify the temporal sequence in which somatic mutations in many common cancers are accumulated.

The concept of **multistage tumorigenesis** was raised in Chapter 1 and will be returned to frequently during successive chapters. Here we will discuss CRC in some detail as the exemplar of this model of carcinogenesis. It will also serve as a useful aide-memoire for much of what has been discussed in this chapter so far. CRC is the second most common cancer in England and Wales and third in the United States, with more than 30 000 new cases diagnosed each year, more than 25% of which are Dukes' C (advanced) at presentation.

CRCs are believed to originate by a stepwise accumulation of epimutations in normal glandular epithelial cells that then progress through the clinically recognizable adenoma–carcinoma sequence first described by Fearon and Vogelstein (see Fig. 3.3). This model is the visible embodiment of the molecular evolution of cancer cells: clinical stages mirror successive esteps in Darwinian mutation–selection cycles, the whole process culminating in the formation of a dominant clone as, in an example of extreme cellular nepotism, the progeny of a successful parent outcompete and supplant their more distant relatives.

In the last few years we have witnessed not only the validation of the multistage model for CRC development but also the renaissance of an older debate surrounding the exact cellular origin of cancer cells. In fact, the study of CRC has underpinned the rapid acceptance of the cancer stem cell (CSC) model of tumorigenesis, first linked to hematological malignancies, as a serious and possibly even general mechanism underlying the development of solid tumors (discussed in Chapters 5 and 9). Briefly, as discussed elsewhere, CSCs are held to be a subpopulation of cancer-initiating cells (CICs) within the tumor, originating either from normal stem cells or other cells that become stem cell–like during their evolution, within which much of the malign behaviors of the cancer reside. Thus, CSCs may be responsible for the more pioneering activities of the cancer, such as invasion and metastases, but may also, like queen ants, populate the tumor with lesser cancer cells and intriguingly also other types of cell such as vascular endothelial cells and paneth cells, thereby producing both the cancer and a supportive stromal environment. CSCs may also, in part, be responsible for treatment resistance and recurrence.

This notwithstanding, in the majority of cases CRC-causing epimutations arise in somatic cells (stem cells or otherwise) of the adult. Hereditary factors also play a key role, with some cells starting their journey towards cancer with facilitating germline mutations that inactivate *APC* (familial adenomatous polyposis) or mismatch repair genes (Lynch syndrome), giving rise to CRC at a younger than usual age. In fact, a propensity to mutation is thought to underpin the evolution of most CRC; broadly, the original model postulated two overlapping causative mechanisms, referred to as chromosomal instability (CIN) and microsatellite instability (MSI), which are described in Chapter 10, either of which may explain the presence of genetic or epigenetic instability frequently observed in CRC. CIN is found in 85% of CRCs and is manifested clinically as aneuploidy. MSI, related to epimutations in DNA mismatch repair (MMR) genes, is seen in around 15% of CRC tumors and is associated with unstable gene loci (30% or more) rather than chromosomal abnormalities.

Consistent with this view, Lynch syndrome, associated with early development of colorectal and uterine cancers, is caused by inherited loss of function of MMR genes and MSI. Tumors develop when the remaining functional allele is lost through somatic epimutation. Sporadic but not inherited MSI CRCs also often have *BRAF* V600E mutations, a feature that has been used in selection of chemotherapeutic agents.

Mutations in the *APC* gene are found in over two thirds of sporadic CRCs, where they are often linked to CIN and, by definition, in all those related to FAP. Mutations in other components of the WNT signaling pathway, including activation of the β-catenin gene (*CTNNB1*) are also sometimes found. The importance of gene silencing in MSI and CIN is attested to by their association with the CpG island methylator phenotype (CIMP) and global DNA hypomethylation respectively (see Chapter 11). In order to develop a unified standard for diagnosis, at least in a research or trial setting, a panel of five mononucleotide markers (BAT-25, BAT-26, NR-21, NR-24, and MONO-27) have been proposed, along with mutations in *BRAF* to define MSI. There is a clinical point to this as MSI has a better prognosis than CIN.

Having established that accretion of CRC-causing mutations is contingent upon genomic and epigenomic instability, we must next identify the genes involved and how these unleash the hallmark features of cancer. Among the best-known signaling cascades deregulated in CRC are those regulated by p53, APC–WNT–β-catenin, TGF-β, EGFR–MAPK and PI3K. *KRAS* mutations occur after *APC* mutations in the adenoma–carcinoma progression sequence, but are still a relatively early event in tumorigenesis. The *BRAF* gene is mutated in 15% of CRCs in which mutations in *KRAS* are absent; knowledge of the signaling pathway (see Chapters 5 and 6) suggests why these would be unlikely to coexist, as no growth advantage would follow the second mutation over the first.

Most CRCs have deregulated TGF-β signaling, largely through mutations in genes encoding receptors or SMAD signaling proteins. Loss of the long arm of chromosome 18 (18qLOH), in which the genes for SMAD2, 4 and 7 as well as DCC are located, occurs in two thirds of CRCs. Mutations in the tumor suppressor gene *TP53* are found in about 50% of CRCs and may be important in the adenoma–carcinoma transition. Mutations in PI3K pathway genes are observed in around 40% of CRCs, the most common being in the p110α catalytic subunit *PIK3CA*, and are also implicated in driving the transition from adenoma to carcinoma. Mutations are also observed in *PTEN*.

As has been discussed, knowledge of molecular pathways underlying tumorigenesis is already informing diagnosis and treatment for many cancers. Here we will mention two of these employed for CRC. The presence in tumor samples of *KRAS* mutations identifies a subgroup unresponsive to EGFR antibody therapy. As will be discussed in Chapters 5 and 6, KRAS is a downstream effector of EGFR that activates BRAF and the MAPK pathway, promoting cell growth and survival. Therefore, inhibiting EGFR with antibodies will not be effective if this apical signaling step is essentially bypassed through direct activation of KRAS or BRAF. Testing of CRCs for mutations in *BRAF* can help differentiate sporadic CRC from Lynch syndrome. CRCs with *BRAF* mutations will also respond poorly to EGFR inhibitors and may have prognostic importance.

Well these are the mutations, but in which cells does this sequence take place and what is the role of the mysterious cancer stem cells in CRC? The cell of origin for CRC remains controversial but the jury will probably turn in a guilty verdict for stem cells. The trial evidence and lawyers' concluding remarks will be presented in Chapter 9, so no further discussion of stems cells and cancer stem cells will be made here.

Conclusions and future directions

In recent decades, environmental factors have been accepted as predominant causes of human cancer, largely on the strength of epidemiological studies showing that cancer incidence varied widely depending on geographical factors. Moreover, immigrants tended to acquire the cancer risk of their new adopted location. Inherited factors have been largely explored in rare familial cancer syndromes, and genes identified from the study of such families, notably *RB* and *p53*, have made major contributions to the understanding of cancer and to biology in general. Yet, such high-penetrance genes contribute little to inherited susceptibility to the common cancers, though they are frequently the subject of cancer-causing somatic mutations. Advances in molecular biology over the last two decades have resulted in increasing experimental support for the role of low-penetrance genetic variants (polymorphisms) to the inherited susceptibility to common cancers, even if as yet the role in clinically improving risk prediction is unclear. Notably, the demonstration of genetic linkages in breast cancer families strongly supports the role of inherited genetic factors in a common cancer.

Whether we will rapidly move closer to the holy grail of personalized medicine and tailored therapeutics using a comprehensive database of mutations, epigenetic changes, and gene/protein expression gleened from the tumor and from the patient is a matter of conjecture. This author is optimistic about our future abilities to process and exploit such vast volumes of data (discussed in Chapter 20).

Together with the study of inherited cancer syndromes, the study of viruses in tumorigenesis has also had a disproportionately large impact on cancer biology. Thus, although relatively few human cancers are caused by viruses, with the notable exception of HPV and cervical and oral cancer, the study of virus-induced cancers in animal models and of those few human examples have contributed substantially to the discovery of several important oncogenes and also tumor suppressors. Major health benefits are expected from preventative vaccination programs of young women for HPV infection.

Major concerns still exist about how much weight can be placed on animal models of carcinogen exposure–induced cancer. In general, the levels in these studies greatly exceed those to which most humans will ever be exposed. Moreover, such studies almost invariably address single carcinogens, whereas humans will be exposed often to complex mixtures. Even tobacco smoke is believed to contain upwards of 50 different carcinogenic compounds, and the effects or potential synergistic effects of such mixtures are poorly understood. Improved understanding of the mechanisms underlying the genotoxic effects of carcinogens and how such lesions are recognized and repaired is often incomplete. This information would increase our ability to determine the minimal levels of carcinogen exposure needed to cause cancer, and thereby to set safe thresholds for these factors in the environment.

Interindividual variability in susceptibility to carcinogens and other cancer-contributing factors is also an important issue. With this in mind, we must continue to explore the interactions between susceptibility and resistance genes targeted by carcinogens or influencing the consequences of exposure to carcinogens. The ability to incorporate large numbers of genes and lifestyle factors into the accurate calculation of risk will result from our increasing ability to understand the clinical implications of human DNA sequence variability. At present, this goal is still some way ahead for cancer, although one should take heart from the spectacular progress made in the area of coronary heart disease. Advances in cancer biology, bioinformatics, systems biology, and mathematics may finally allow us to exploit the results of the Human Genome Project in order to improve the accuracy of cancer risk evaluation and benefit the patient by the implementation of tailored chemopreventative, screening, and lifestyle strategies.

> What nature does in the course of long periods we do every day when we suddenly change the environment in which some species of living plant is situated.
>
> *Jean-Baptiste Lamarck*

Outstanding questions

1. To define which individual genes/proteins involved in tumorigenesis are "mission critical" for cancer development and moreover which are most critical for maintaining the cancer – as these will be the optimum targets for new drug development.

2. To understand how combinations of various low-penetrance genetic variants (polymorphisms) interact to protect from or predispose to different types of cancer, determine prognosis once cancer develops, and dictate the likely success of particular therapies.

3. To improve the predictive value of traditional toxicity studies of carcinogens, involving relatively short-term exposure to high doses of single compounds, to more accurately define the risk of exposure to lower levels of carcinogens and combinations of carcinogens over more protracted periods in humans.

Bibliography

Crick, F. (1970). Central dogma of molecular biology. *Nature* **227**(5258): 561–3.

Parkin, D.M. (2011). The fraction of cancer attributable to lifestyle and environmental factors in the United Kingdom in 2010. *British Journal of Cancer Supplement*.

Multistage oncogenesis and gene–environment interactions

Attolini, C.S., Cheng, Y.K., Beroukhim, R., et al. (2010). A mathematical framework to determine the temporal sequence of somatic genetic events in cancer. *Proceedings of the National Academy of Sciences of the U.S.A.*, **107**(41): 17604–9.

Armitage, P. and Doll, R. (1954). The age distribution of cancer and a multi-stage theory of carcinogenesis. *British Journal of Cancer* **8**: 1–12.

Goel, A. and Boland, C.R. (2010). Recent insights into the pathogenesis of colorectal cancer. *Current Opinion in Gastroenterology*, **26**(1): 47–52.

Kan, Z., Jaiswal, B.S., Stinson, J., *et al.* (2010). Diverse somatic mutation patterns and pathway alterations in human cancers. *Nature*, **466**(7308): 869–73.

Lichtenstein, P., Holm, N.V., Verkasalo, P.K., *et al.* (2000). Environmental and heritable factors in the causation of cancer – analyses of cohorts of twins from Sweden, Denmark, and Finland. *New England Journal of Medicine*, **343**(2): 78–85.

Peto, J. (2001). Cancer epidemiology in the last century and the next decade. *Nature*, **411**(6835): 390–5.

Vineis, P. (2004). Individual susceptibility to carcinogens. *Oncogene*, **23**, 6477–6483.

Carcinogens and the Ames test

Ames, B.N., Durston, W.E., Yamasaki, E., and Lee, F.D. (1973). Carcinogens are mutagens: a simple test system combining liver homogenates for activation and bacteria for detection. *Proceedings of the National Academy of Sciences of the U.S.A.*, **70**: 2281–5.

Luch, A. (2005). Nature and nurture – lessons from chemical carcinogenesis. *Nature Reviews Cancer*, **5**(2): 113–25.

McCann, J., Choi, E., Yamasaki, E. and Ames, B.N. (1975). Detection of carcinogens as mutagens in the salmonella/microsome test: assay of 300 chemicals. *Proceedings of the National Academy of Sciences of the U.S.A.*, **72**: 5135–9.

Spira, A., Beane, J.E., Shah, V., *et al.* (2007). Airway epithelial gene expression in the diagnostic evaluation of smokers with suspect lung cancer. *Nature Medicine*, **13**(3): 361–6.

Genetics and genomics (see Chapter 7 references for tumor suppressors and inherited cancers)

Armitage, P. and Doll, R. (2004). The age distribution of cancer and a multi-stage theory of carcinogenesis. *International Journal of Epidemiology*, **33**: 1174–9.

Balmain, A., Gray, J., and Ponder, B. (2003). The genetics and genomics of cancer. *Nature Genetics*, **33**(Suppl): 238–44.

Bodmer, W. and Tomlinson, I (2010). Rare genetic variants and the risk of cancer. *Current Opinion in Genetics and Development*, **20**(3): 262–7.

Calin, G.A., Trapasso, F., Masayoshi, S., *et al.* (2005). Familial cancer associated with a polymorphism in ARLTS1. *New England Journal of Medicine*, **352**(16): 1667–76.

Campbell, P.J. Yachida, S., Mudie, L.J. *et al.* (2010). The patterns and dynamics of genomic instability in metastatic pancreatic cancer. *Nature*, **467**(7319): 1109–13.

Ding, L., Getz, G., Wheeler, D.A., *et al.* (2008). Somatic mutations affect key pathways in lung adenocarcinoma. *Nature*, **455**(7216): 1069–75.

Ding, Z. (2011). SMAD4-dependent barrier constrains prostate cancer growth and metastatic progression. *Nature*, **470**(7333): 269–73.

Goode, E.L. (2010). A genome-wide association study identifies susceptibility loci for ovarian cancer at 2q31 and 8q24. *Nature Genetics*, **42**(10): 874–9.

Greenman, C. *et al.* and colleagues from the Cancer Genome Project, Wellcome Trust Sanger Institute, Cambridge, UK (2007). Patterns of somatic mutation in human cancer genomes *Nature*, **446**(7132): 153–8.

Hall, J.M., Lee, M.K., Newman, B., *et al.* (1990). Linkage of early-onset familial breast cancer to chromosome 17q21. *Science*, **250**(4988): 1684–9.

Hindorff, L.A., Sethupathy, P., Junkins, H.A., *et al.* (2009). Potential etiologic and functional implications of genome-wide association loci for human diseases and traits. *Proceedings of the National Academy of Sciences of the U.S.A.*, **106**(23): 9362–7.

Houlston, R.S. and Peto, J. (2004). The search for low-penetrance cancer susceptibility alleles. *Oncogene*, **23**(38): 6471–6.

Hudson, T.J., Anderson, W., Artez, A., *et al.* (2010). International network of cancer genome projects. *Nature*, **464**(7291): 993–8.

Ji, H., Ramsey, M.R., Hayes, D.N., *et al.* (2007). LKB1 modulates lung cancer differentiation and metastasis. *Nature*, **448**(7155): 807–10.

Jiao, Y., Shi, C., Edil, B.H., *et al.* (2011). DAXX/ATRX, MEN1, and mTOR pathway genes are frequently altered in pancreatic neuroendocrine tumors. *Science* **331**(6021): 1199–203.

Kan, Z., Jaiswal, B.S., Stinson, J., *et al.* (2010). Diverse somatic mutation patterns and pathway alterations in human cancers. *Nature*, **466**(7308): 869–73.

Lango Allen, H., Estrada, K., Lettre, G., *et al.* (2010). Hundreds of variants clustered in genomic loci and biological pathways affect human height. *Nature*, **467**(7317): 832–8.

Lee, W., Jiang, Z., Liu, J., *et al.* (2010). The mutation spectrum revealed by paired genome sequences from a lung cancer patient. *Nature*, **465**(7297): 473–7.

Liu, W., Laitinen, S., Khan, S., *et al.* (2009). Copy number analysis indicates monoclonal origin of lethal metastatic prostate cancer. *Nature Medicine*, **15**(5): 559–65.

Meyerson, M., Gabriel, S., and Getz, G. (2010). Advances in understanding cancer genomes through second-generation sequencing. *Nature Reviews Genetics*, **11**(10): 685–96.

Moolgavkar, S.H. and Knudson, A.G. (1981). Mutation and cancer: A model for human carcinogenesis. *Journal of the National Cancer Institute*, **66**: 1037–52.

Park, J.Y., Park, S.H., Choi, J.E. *et al.* (2002). Polymorphisms of the DNA repair gene xeroderma pigmentosum group A and risk of primary lung cancer. *Cancer Epidemiology, Biomarkers and Prevention*, **11**: 993–7.

Parsons, D.W., Li, M., Zhang, X., *et al.* (2011). The genetic landscape of the childhood cancer medulloblastoma. *Science* **331**(6016): 435–9.

Petersen, G.M. (2010). A genome-wide association study identifies pancreatic cancer susceptibility loci on chromosomes 13q22.1, 1q32.1 and 5p15.33. *Nature Genetics*, **42**(3): 224–8.

Puente, X.S., Pinyol, M., Quesada, V., *et al.* (2011). Whole-genome sequencing identifies recurrent mutations in chronic lymphocytic leukemia. *Nature*, **475**(7354): 101–5.

Purdue, M.P., Johansson, M., Zelenika, D. *et al.* (2011). Genome-wide association study of renal cell carcinoma identifies two susceptibility loci on 2p21 and 11q13.3. *Nature Genetics*, **43**(1): 60–5.

Ryan, B.M., Robles, A.I., and Harris, C.C. (2010). Genetic variation in microRNA networks: the implications for cancer research. *Nature Reviews Cancer*, **10**(6): 389–402.

Singleton, A.B., Hardy, J., Traynor, B.J., and Houlden, H. (2010). Towards a complete resolution of the genetic architecture of disease. *Trends in Genetics*, **26**(10): 438–42.

Tsuge, M., Hamamoto, R., Silva, F.P., *et al.* (2005). A variable number of tandem repeats polymorphism in an E2F-1 binding element in the 5′ flanking region of SMYD3 is a risk factor for human cancers. *Nature Genetics*, **37**(10): 1104–7.

Turnbull, C., Ahmed, S., Morrison, J., *et al.* (2010). Genome-wide association study identifies five new breast cancer susceptibility loci. *Nature Genetics*, **42**(6): 504–7.

Wacholder, S. (2010). Performance of common genetic variants in breast-cancer risk models. *New England Journal of Medicine*, **362**(11): 986–93.

Wang, K., Diskin, S.J., Zhang, H., *et al.* (2011). Integrative genomics identifies LMO1 as a neuroblastoma oncogene. *Nature*, **469**(7329): 216–20.

Yachida, S., Jones, S., Bozic, I., *et al.* (2010). Distant metastasis occurs late during the genetic evolution of pancreatic cancer. *Nature*, **467**(7319): 1114–7.

Screening

Andriole, G.L., Crawford, E.D., Grubb, R.L., 3rd, *et al.* (2009). Mortality results from a randomized prostate-cancer screening trial. *New England Journal of Medicine* **360**: 1310–19.

Domchek, S.M., Friebel, T.M., Singer, C., *et al.* (2010). Association of risk-reducing surgery in BRCA1 or BRCA2 mutation carriers with cancer risk and mortality. *Journal of the American Medical Association*, **304**(9): 967–75.

Hu, Z., Chen, J., Tian, T., *et al.* (2008). Genetic variants of miRNA sequences and non-small cell lung cancer survival. *Journal of Clinical Investigation*, **118**(7): 2600–8.

Maheswaran, S., Sequist, L.V., Nagrath, S., *et al.* (2008). Detection of mutations in EGFR in circulating lung-cancer cells. *New England Journal of Medicine*, **359**(4): 366–77.

Paik, S., Shak, S., Tang, G., *et al.* (2004). A multigene assay to predict recurrence of tamoxifen-treated, node-negative breast cancer. *New England Journal of Medicine*, **351**: 2817–26.

Pantel, K. and Alix-Panabières, C. (2010). Circulating tumour cells in cancer patients: challenges and perspectives. *Trends in Molecular Medicine*, **16**(9): 398–406.

Schröder, F.H., Hugosson, J., Roobol, M.J., et al. (2009). Screening and prostate-cancer mortality in a randomized European study. *New England Journal of Medicine*, **360**(13): 1320–8.

Wang, Q., Chaerkady, R., Wu, J., et al. (2011). Mutant proteins as cancer-specific biomarkers. *Proceedings of the National Academy of Sciences of the U.S.A.*, **108**(6): 2444–9.

Lifestyle and cancer

Ames, B.N. (2006). Low micronutrient intake may accelerate the degenerative diseases of aging through allocation of scarce micronutrients by triage *Proceedings of the National Academy of Sciences of the U.S.A.*, **103**(47): 17589–94.

Bingham, S.A., Day, N.E., Luben, R., et al. (2003). Dietary fibre in food and protection against colorectal cancer in the European Prospective Investigation into Cancer and Nutrition (EPIC): an observational study. *Lancet*, **361**(9368): 1496–501.

Chlebowski, R.T., Blackburn, G.L., Thomson, C.A., et al. (2006). Dietary fat reduction and breast cancer outcome: interim efficacy results from the Women's Intervention Nutrition Study. *Journal of the National Cancer Institute*, **98**(24): 1767–76.

Cho, E., Chen, W.Y., Hunter, D.J., et al. (2006). Red meat intake and risk of breast cancer among premenopausal women. *Archives of Internal Medicine*, **166**(20): 2253–9.

Danaei, G., Vander Hoorn, S., Lopez, A.D., Murray, C.J., Ezzati, M. and the Comparative Risk Assessment collaborating group (Cancers) (2005). Causes of cancer in the world: comparative risk assessment of nine behavioural and environmental risk factors. *Lancet* **366** (9499): 1784–93.

Orsini, N., Mantzoros, C.S., and Wolk, A. (2008). Association of physical activity with cancer incidence, mortality, and survival: a population-based study of men. *British Journal of Cancer*, **98**(11): 1864–9.

Popkin, B.M. (2007). Understanding global nutrition dynamics as a step towards controlling cancer incidence. *Nature Reviews Cancer*, **7**(1): 61–7.

Renehan, A.G., Tyson, M., Egger, M., Heller, R.F., and Zwahlen, M. (2008). Body-mass index and incidence of cancer: a systematic review and meta-analysis of prospective observational studies. *Lancet*, **371**: 569–78.

WHO/FAO (2003). *Joint WHO/FAO Expert Consultation on Diet, Nutrition and the Prevention of Chronic Diseases*. WHO Technical Report Series. Geneva: WHO, pp. 95–104.

Viruses, infection, and cancer (see also Chapters 6 and 7)

Ang, K.K., Harris, J., Wheeler, R., et al. (2010). Human papillomavirus and survival of patients with oropharyngeal cancer. *New England Journal of Medicine*, **363**(1): 24–35.

Baltimore, D. (1970). RNA-dependent DNA polymerase in virions of RNA tumour viruses. *Nature*, **226**, 1209–11.

Bishop, J.M. (1995). Cancer: the rise of the genetic paradigm. *Genes and Development*, **9**: 1309–15.

Bongers, G., Maussang, D., Muniz, L.R., et al. (2010). The cytomegalovirus-encoded chemokine receptor US28 promotes intestinal neoplasia in transgenic mice. *Journal of Clinical Investigation*, **120**(11): 3969–78.

Feng, H., Shuda, M., Chang, Y., and Moore, P.S. (2008). Clonal integration of a polyomavirus in human Merkel cell carcinoma. *Science*, **319**(5866): 1096–100.

Frazer, I.H., Leggatt, G.R., and Mattarollo, S.R. (2011). Prevention and treatment of papillomavirus-related cancers through immunization. *Annual Reviews of Immunology*, **29**: 111–38.

Giannakis, M., Chen, S.L., Karam, S.M., Engstrand, L., and Gordon, J.I. (2008). *Helicobacter pylori* evolution during progression from chronic atrophic gastritis to gastric cancer and its impact on gastric stem cells. *Proceedings of the National Academy of Sciences of the U.S.A.*, **105**(11): 4358–63.

Gregory, S.M., Davis, B.K., West, J.A., et al. (2011). Discovery of a viral NLR homolog that inhibits the inflammasome. *Science*, **331**(6015): 330–4.

Hellner, K. and Münger, K. Human papillomaviruses as therapeutic targets in human cancer. *Journal of Clinical Oncology*, **29**(13): 1785–94.

Ibeanu, O.A. (2011). Molecular pathogenesis of cervical cancer. *Cancer Biology Therapy*, **11**(3).

Khwaja, S.S., Liu, H., Tong, C., et al. (2010). HIV-1 Rev-binding protein accelerates cellular uptake of iron to drive Notch-induced T cell leukemogenesis in mice. *Journal of Clinical Investigation*, **120**(7): 2537–48.

Lane, D.P. and Crawford, L.V. (1979). T antigen is bound to a host protein in SV40-transformed cells. *Nature*, **278**, 261–263.

McLaughlin-Drubin, M.E., Crum, C.P., and Münger, K. (2011). Human papillomavirus E7 oncoprotein induces KDM6A and KDM6B histone demethylase expression and causes epigenetic reprogramming. *Proceedings of the National Academy of Sciences of the U.S.A.*, **108**(5): 2130–5.

Mehanna, H., Paleri, V., West, C.M., and Nutting, C. (2010). Head and neck cancer – Part 1: Epidemiology, presentation, and prevention. *British Medical Journal*, **341**: c4684.

Moore, P.S. and Chang, Y. (2010). Why do viruses cause cancer? Highlights of the first century of human tumor virology. *Nature Reviews Cancer*, **10**(12): 878–89.

Nougayrède, J.P., Homburg, S., Taieb, F., et al. (2006). *Escherichia coli* induces DNA double-strand breaks in eukaryotic cells. *Science*, **313**(5788): 848–51.

Polk, D.B. and Peek, R.M. Jr. (2010). Helicobacter pylori: gastric cancer and beyond. *Nature Reviews Cancer*, **10**(6): 403–14.

Tsai, W.L. and Chung, R.T. (2010). Viral hepatocarcinogenesis. *Oncogene*, **29**(16): 2309–24.

Varmus, H.E. (1990). Retroviruses and oncogenes. I. *Bioscience Reports*, **10**: 413–30.

Whyte, P., Buchkovich, K.J., Horowitz, J.M., et al. (1988). Association between an oncogene and an anti-oncogene: the adenovirus E1A proteins bind to the retinoblastoma gene product. *Nature*, **334**: 124–29.

Yamashita, K., Sakuramoto, S., and Watanabe, M. (2011). Genomic and epigenetic profiles of gastric cancer: potential diagnostic and therapeutic applications. *Surgery Today* **41**(1): 24–38.

Zhao, B., Mar, J.C., Maruo, S., et al. (2011). Epstein-Barr virus nuclear antigen 3C regulated genes in lymphoblastoid cell lines. *Proceedings of the National Academy of Sciences of the U.S.A.*, **108**(1): 337–42.

Questions for student review

1) The following processes contribute to the majority of cancers:
   a. Inherited mutations in a single major cancer-causing gene.
   b. Polymorphic alleles.
   c. Viral infection.
   d. Inactivation of a tumor suppressor gene.
   e. Exposure to mutagenic factors.

2) The following statements are true of breast cancer:
   a. Most breast cancers arise through inherited abnormalities in the *BRCA1* gene.
   b. Most breast cancers are not familial.
   c. Breast cancers only affect females.
   d. Most breast cancers at diagnosis are genetically unstable.
   e. Most familial breast cancers arise through mutations in the *ATM* gene.

3) Cancer incidence is influenced by:
   a. Geographical factors.
   b. Genetic predisposition.
   c. Diet.
   d. Gender.
   e. Age.

4) The following may significantly influence the cancer risk associated with cigarette smoking:
   a. Polymorphisms in genes involved in metabolising components of tobacco smoke.
   b. Numbers of cigarettes smoked.
   c. Age.
   d. Gender.
   e. Time following smoking cessation.

5) Viruses may contribute to tumorigenesis in the following ways:
   a. By introducing oncogenes.
   b. By introducing functional tumor suppressor genes.
   c. By inactivating tumor suppressor genes.
   d. By adversely affecting aspects of immune function.
   e. By promoting apoptosis in the nascent cancer cell.

4 DNA Replication and the Cell Cycle

Stella Pelengaris[a] and Michael Khan[b]
[a]Pharmalogos Ltd and [b]University of Warwick, UK

It has not escaped our notice that the specific pairing we have postulated immediately suggests a copying mechanism for the genetic material.

Francis Crick and James Watson

An error does not become truth by reason of multiplied propagation, nor does truth become error because nobody sees it.

Mohandas Gandhi

The *ouroboros* is an ancient symbol representing, among other things, the cyclical nature of the universe. The *ouroboros* eats its own tail to sustain its life, in an eternal cycle of renewal.

Key points

- The cell cycle is an assiduously monitored and policed process, during which cells replicate themselves in order to produce two genetically identical daughter cells.
- The cell-cycle machinery and the quality-control processes that are imposed on it are highly conserved across species, working together in a complex sequence to ensure the faithful duplication and precise partitioning of the DNA between the progeny.
- The cell cycle progresses through four sequential and distinct phases: a gap period G_1 (growth phase); S (DNA synthesis phase); a gap period G_2 (growth phase); M (mitosis – nuclear and cell division). These phases vary greatly in length, depending on the cell type and upon the signals received. Moreover, cells can enter a "resting" nonproliferative state, which can be very prolonged or even permanent- termed G_0 that equates with terminal differentiation.
- In order to complete mitosis, alongside the DNA various organelles must be replicated or at least evenly partitioned, including the mitochondria that undergo fission and duplication.
- Each phase of the cell cycle is propelled forwards by sequential activation of cyclin-dependent kinases (CDKs) in association with regulatory subunits – the cyclins. The cell cycle may also be arrested by a number of powerful inhibitory forces that militate against unscheduled proliferation.
- Cell-cycle entry and progression can be stopped at multiple checkpoints, in particular by the functionally implicit cyclin-dependent kinase inhibitors (CKIs), including $p16^{INK4A}$, $p21^{CIP1}$, and $p27^{KIP1}$.
- Unscheduled cell division is the cardinal feature of cancer. It is therefore no surprise that most cancer-causing mutations are concentrated around cell-cycle control steps. The proponents of cancer that impel the cell cycle are encoded by oncogenes, whereas the restraints and quality-control steps that should have prevented it are littered with the defective products of tumor suppressor and caretaker genes.

(Continued)

The Molecular Biology of Cancer: A Bridge From Bench to Bedside, Second Edition. Edited by Stella Pelengaris and Michael Khan.
© 2013 John Wiley & Sons, Inc. Published 2013 by John Wiley & Sons, Inc.

- Orderly cell-cycle progress is ensured by keeping the various processes that affect protein activity and abundance on a tight leash. Conversely, any laxity in the control of expression (at the gene and transcript levels), degradation, or of the activation/inactivation of key cell-cycle regulators can result in inappropriate levels of cell proliferation.
- The means of removal of surplus proteins, in particular the complex operation of the ubiquitin–proteasome system, has become one of the most exciting areas of modern cell biology research. Once thought to be an insatiable and indiscriminate protein consumer, the ubiquitin–proteasome pathway has been shown to prefer dining á la carte. Specific proteins are ordered from the menu at specific times and in a highly specific way. Once coated in the appropriate ubiquitin condiments, the relevant proteins are avidly consumed by the proteasome and the degradation products recycled.
- The cyclins are degraded in a cyclical fashion (as implied by their name). This is largely directed by the stringently regulated activities of ubiquitin ligase complexes – the two best described being SCF (Skp1/cullin/F-box protein) and APC (anaphase-promoting complex). These operate largely during G_1/S transition and M phase, respectively, and determine which cyclins will be on the menu and when.
- The G_1/S transition is the key point for cell-cycle entry. Importantly, this represents something of a point of no return and mitogenic stimuli are no longer required once the cell cycle has passed this point – like a cart pushed over the hump of a hill.
- Unsurprisingly, all cancers find a way to navigate past the many obstacles standing between them and the freedom of an untrammelled passage into S phase.
- The key obstacle to G_1/S transition that must be overcome is the RB tumor suppressor. When RB is in the nonphosphorylated state, it can sequester and render inactive a number of key transcription factors required for expression of S-phase genes, particularly members of the E2F family (needed for DNA synthesis and cell-cycle progression).
- Under normal circumstances, progression through G_1 and then G_1/S transition is under the control of mitogenic stimuli and their downstream signaling pathways. These pathways sequentially activate complexes of CDK4 (and CDK6) in association with cyclin D, followed by those of CDK2 in association with cyclin E. Together, these complexes hyperphosphorylate RB, thereby freeing E2F to activate S-phase genes. Conversely, various negative signals prevent G_1/S transition by preventing RB phosphorylation. In particular, the p16^{INK4A} CKI, which blocks activity of CDK4 (and CDK6) and p27^{KIP1}, which interferes with CDK2, can both prevent RB phosphorylation and S-phase entry.
- Essentially all cancers circumvent the RB pathway in some way.
- All key stages in cell-cycle progression are subject to intense scrutiny and all-powerful "checkpoints" are activated if any of several steps in the cell cycle, such as DNA replication, duplication, and partitioning of chromosomes, have not been completed satisfactorily. If a problem is detected then the cell cycle is arrested until it is corrected, or failing that, immitigable draconian measures may be required; the cell may opt to die (apoptosis) or irrevocably exit from cell cycling (senescence).
- These checkpoints are policed by, among others, important tumor suppressors, including RB and p53 (product of the *TP53* gene in humans) aided and abetted by various CKIs, and at least two major DNA damage-limitation systems – the ATM–CHK2 and the ATR–CHK1 signaling pathways.
- A very important checkpoint, involving p53, ensures that S phase is blocked after DNA damage. This key control step prevents replication of mutated DNA until the damage has been repaired or the cell is eliminated.
- The p53 protein is also activated in response to "oncogenic stress" (cancer cell cycles driven by deregulated expression of oncogenes such as c-MYC) in this case via the ARF (p14^{ARF} in humans and p19^{Arf} in mouse) protein.
- Failure of such checkpoints can prevent the cell-cycle arrest or cell death (apoptosis) that should otherwise accompany deregulated oncogene activity or DNA damage. Together, aberrant proliferation and propagation of mutated DNA to daughter cells provides the platform for cancer.
- Mutations in genes that promote the cell cycle (oncogenes) or checkpoints that arrest the cycle or promote apoptosis (tumor suppressors) are ubiquitous and central to cancer development.

Introduction

The principal result of my investigation is that a uniform developmental principle controls the individual elementary units of all organisms, analogous to the finding that crystals are formed by the same laws in spite of the diversity of their forms.

Theodore Schwann

Essentially, in order to complete embryonic development and to get bigger, we have to make more cells. Enlarging existing cells may suffice for temporary adaptations to demand, such as to meet a requirement for more hormone production, but is almost invariably also an antecedent to cell division. With the exception of the rather exotically named "asymmetrical cell division" (performed by stem cells) and "meiosis" (reduction division that yields germ cells with only one of the original pair of chromosomes), cell division generates two identical copies of the parent cell. The processes through which cells fashion exact replicas of themselves and faithfully duplicate their DNA are so central to life and to the development and existence of complex organisms that they are subject to more intensive scrutiny and control than any other cellular process. Cell division must be precisely executed and rigorously patrolled in order that new cells can be

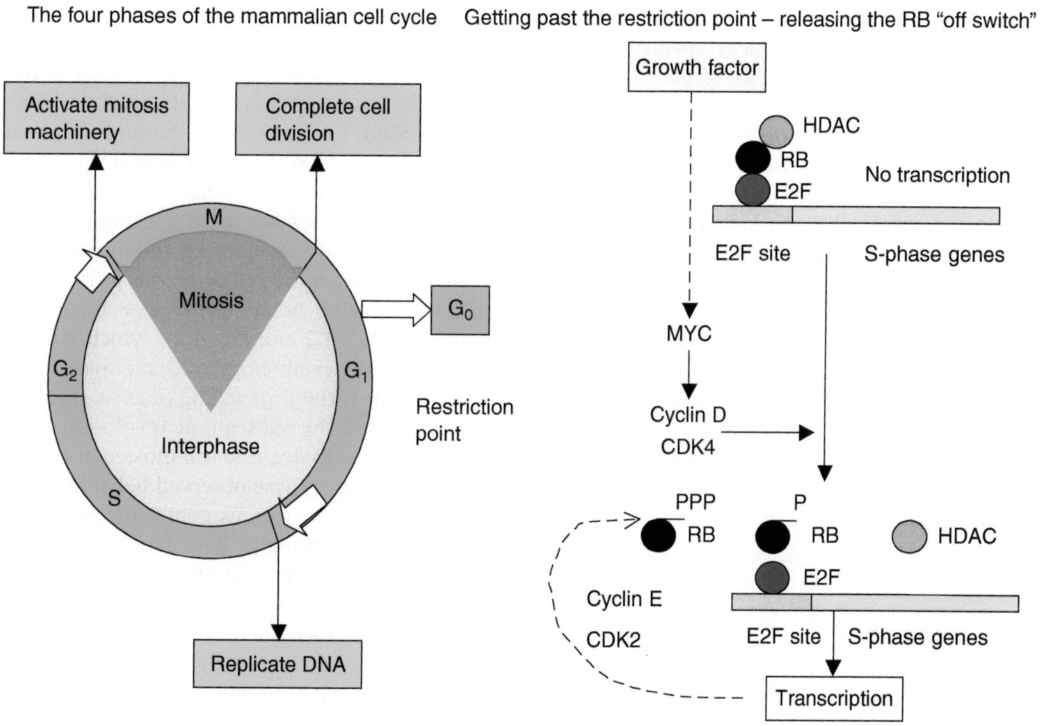

Figure 4.1 The phases of the mammalian cell cycle. The cell cycle consists of the following phases: G_1 – gap phase for growth and preparation of the chromosomes for replication; S – synthesis of DNA and centrosomes (may last 6–8 hours); G_2 – gap phase for growth and preparation for mitosis (may last 3–5 hours); M – mitosis (nuclear division) and cytokinesis (cell division), may last 1 hour or more. The M phase comprises five observable stages: prophase (chromosomes condense and become visible), metaphase (chromosomes align at the "equator" and attach to mitotic spindle), anaphase (chromatids segregate to opposite poles), telophase (chromosomes become invisible and new nuclear membranes), and cytokinesis (daughter cells form by division of cytoplasm). After going through mitosis, a normal cell that re-enters G_1 can either start cycling again or exit the cell cycle and enter a resting state. It may stay in this quiescent state, termed "G_0," for days, weeks, or years, until the balance of growth-stimulatory and inhibitory signals from outside the cell indicates that it is time to divide again; only then will the cell become active again, re-entering the cell cycle in G_1. The point at which cells commit to replicate their DNA and enter the cell division cycle is controlled by the RB protein, the product of the tumor suppressor gene *RB*. RB acts as an "off switch" for entry into S phase by binding the transcription factor E2F required for expression of S-phase genes. After appropriate growth factor stimulation, the inhibitory effects of RB are removed by phosphorylation, which allows E2F to activate S-phase genes, during early G_1, required for cell-cycle progression. One such S-phase gene is cyclin E, which together with CDK2 then acts to further phosphorylate RB and amplify the process in part by fully activating transcription factors normally held inactive by RB.

generated to contribute appropriately to growth, repair, and adaptation. At the same time the establishment of rogue cells with damaged DNA must be avoided. How is this achieved?

- The decision to initiate the cell cycle is exquisitely sensitive to extrinsic and intrinsic factors, including the balance of positive and negative growth stimuli being received by the cell and the capacity of the cell to interpret these signals and to act upon them.
- Once the cell cycle is initiated, overlapping processes operate to prevent genomic instability, which might otherwise result in cancer. These include:
  – orderly and coordinated progress through the cell cycle,
  – high-fidelity DNA replication,
  – correct chromosome segregation, and
  – largely error-free repair of DNA damage.

The development and existence of a multicellular organism depends entirely on successfully completing an almost inconceivable number of cell divisions throughout life. Embryonic development and continued growth after birth are the result of wave upon wave of new cells originating from the sequential division of the previous generation. Even during adult life, new cells are continually needed to maintain tissue homeostasis – to replace damaged or dying cells or to adapt tissue mass to varying demands. An adult human being consists of around 100 trillion (10^{14}) cells, all originating from a single cell, the **embryonic stem cell**. Furthermore, in an adult human, millions of new cells are made every day to replace dying ones and to adapt tissue mass to new demands. In fact, it has been estimated that around 10^{16} or more cell divisions will take place during our lifetime.

Cells replicate themselves by duplicating their contents and then dividing in two – a process known as the **mitotic cell cycle** (Fig. 4.1). It is imperative that two genetically identical daughter cells are produced during each cell cycle and this depends on faithful copying of the DNA and the correct segregation of the replicated chromosomes into two separate cells. Apart from DNA replication, all the cytoplasmic organelles are duplicated during each cell cycle, and most cells will double their mass. Not surprisingly, there are multiple quality-control steps in place to ensure that mistakes are not made during cell replication, which might otherwise lead to daughter cells receiving abnormal DNA, or either too many or too few chromosomes (see Chapter 10). Cells containing such damaged DNA may pose a neoplastic risk to the host, for example if a mutation allows cells to proliferate excessively (see Chapters 6 and 7) or to avoid cell death (apoptosis) (see Chapter 8).

The degree of precision with which DNA synthesis and mitosis are carried out is remarkable, but occasional errors are unavoidable, potentially resulting in mutations in the genome or aneuploidy (lack or excess of chromosomes). Such mutations will in general be harmless if confined to a single cell – after all how could just one out of 10^{14} cells jeopardize the whole organism? However, if a cell with a mutation replicates and so do the progeny, then we have a whole different kettle of fish, because the mutation, and resultant aberrant behavior, are propagated, thereby laying the foundations for the development of cancer. For this reason, cells have developed very complex and usually effective means of arresting replication and initiating DNA repair when mutations are detected. It should be emphasized, however, that without a certain minimal capacity for changes in genetic information we would not be here. Mutations passed through the germline are as essential to evolution of life as somatic mutations are to evolution of a cancer; this is the raw material upon which natural selection will act.

The cell cycle – overview

Life is like riding a bicycle – in order to keep your balance, you must keep moving.

Albert Einstein

Many students find the cell cycle conceptually difficult and for this reason numerous helpful analogies have been employed to illustrate some of the cardinal points of cell-cycle regulation (Box 4.1). Confusion is often generated by the use of different terminology when describing events during cell replication. This becomes less confusing if one remembers that historically different tools have been available to experimenters studying cell replication, and current terminologies reflect this legacy. During the late nineteenth century, using microscopy, Walther Flemming observed structures he called "threads" (now called **chromosomes**) and how these threads change during cell replication, a process he called "mitosis." In fact, the first use of the term "chromosome" (*chromosom* in the original German) is attributed to Wilhelm Waldeyer, who suggested the term as an alternative to "chromatic elements" that had been coined by Boveri to describe these threads, "in which there occurs one of the most important acts in karyokinesis, the longitudinal splitting."

For many years the cell cycle was primarily studied using light microscopy and other techniques to observe chromosomal behavior (Box 4.2 and Fig. 4.2), which has left a lasting legacy in the specific terminology still employed in cytogenetics today. More recently, the unravelling of DNA structure and replication (Box 4.3) and the subsequent revolution in molecular biology has enabled us to describe the molecular machinery that underlies and regulates these observed behaviors and identify the key role of changes in various genes/proteins in regulating the cell cycle. Not surprisingly, key events observed microscopically during mitosis are now known to equate with specific molecular events during cell-cycle transitions, each regulated by complex signaling networks. The cell cycle is normally very strictly regulated, with cell division only taking place under appropriate circumstances, such as to replace dying cells or allow tissue growth. Self-evidently, most cancer-causing gene or epigenetic alterations act in some way to deregulate normal cell-cycle control and result in inappropriate cellular replication. In fact, given the central role in cancer of processes that drive or stop the cell cycle on the one hand and those that monitor the accuracy of the process on the other, it is not surprising that essentially all mutations in cancer impact on the cell cycle either by driving it relentlessly onwards, by removing the restraints that might otherwise arrest it, or by promoting the accumulation and propagation of damaged DNA. For this reason, the cell cycle and its control feature in most chapters in this book.

Box 4.1 An analogy for the cell cycle – "It will all come out in the wash!"

It is often helpful when thinking about important but complex processes such as the cell cycle to have a more easily appreciated analogy based on common experience to hand, which proves reasonably representative of the process being considered. Such helpful analogies have been widely used; variously, the cell cycle has been likened to an automobile or an automatic washing machine cycle (or a relay race). In all these cases the common theme is that the various operations can be switched on or off and then follow a set pattern that directs the various processes occurring in a predetermined sequence, hopefully to a successful outcome (driving the car out of the garage then along the road; clothes washing then drying). Thus, the processes – whether driving or washing – may be taken to represent the active processes of the cell cycle, namely DNA replication followed by chromosome segregation.

Moreover, because all these processes involve a complicated machinery that is prone to intrinsic errors as well as adverse environmental influences (low engine oil, improperly closed doors, locked brakes, pets, and small children – DNA copying errors, radiation, improper chromosome separation) occurring at various stages, there will be disastrous consequences (engine damage, a crash, flooding – replication of damaged DNA, aneuploidy, cancer) unless a means of stopping the process exists. Therefore, all these processes have in built "failsafe" mechanisms that aim to stop the process until the process is resolved (lights flash, engine won't start – cell-cycle checkpoints). Equally, as driving a car inappropriately or with dangerous faults may damage other neighboring cars and drivers, various external controls also operate (codes of behavior, various traffic signals, traffic jams, and ultimately a policing system – body plan, homeostasis, local negative and positive growth signals, contact inhibition by neighboring cells, circulating regulators of growth), all designed to protect society (the organism) from selfish drivers or dangerous cars.

It should also be remembered that this society usually has a very strong honor code, with very poor drivers (cells) encouraged to commit suicide (apoptosis) rather than risk damaging others. This is crucial because the offspring will inevitably inherit the same poor driving skills as the parent, thus amplifying the problem until society is overrun with bad drivers. At this point it would seem in "bad taste" to extend the driving analogy further downstream from suicide – as this society also embraces cannibalism.

Box 4.2 Chromosomes

In eukaryotes, chromosomes consist of a single molecule of DNA together with associated positively charged proteins, the histones, tightly bound to negatively charged phosphate groups in the DNA. The molecule of DNA in a single human chromosome ranges in size from 50 to 250×10^6 nucleotide pairs. Stretched end-to-end, the DNA in a single human diploid cell would extend over 2 m. In the intact chromosome, however, this molecule is packed into a much more compact structure, so that during mitosis a typical chromosome is condensed into a structure about 5 μm long (a 10 000-fold reduction in length).

Five main types of histone are known, all of which are rich in lysine and arginine residues and associate with DNA. Before mitosis, each chromosome is duplicated (during S phase). Then as mitosis begins, the duplicated chromosomes, termed **dyads**, condense into short (~5 μm) structures. The duplicates are held together at the centromere, which in humans comprises over 3 million base pairs of mostly tandem arrays of repeated short sequences of DNA. While they are still attached, it is common to call the duplicated chromosomes **sister chromatids**. The **kinetochore** is a complex of proteins (11 in budding yeast) that forms at the centromere and helps to separate the sister chromatids as mitosis proceeds into anaphase. The shorter of the two arms extending from the centromere is called the **p arm**; the longer is the **q arm**.

Dyads occur in homologous pairs, one member of each pair having been acquired from one of the two parents of the individual whose cells are being examined. All species have a characteristic number of homologous pairs of chromosomes in their cells called the diploid number. The complete set of chromosomes in the cells of an organism is its **karyotype**. In humans, females have 23 pairs of homologous chromosomes (22 pairs of autosomes and one pair of X chromosomes) and males have the same 22 pairs of autosomes, plus one X and one Y chromosome. Karyotype analyses can reveal abnormal copy numbers of chromosomes (**ploidy**). Karyotype analysis can also reveal translocations between chromosomes. A number of these cause cancer, for example the Philadelphia chromosome in chronic myelogenous leukemia (CML).

Box 4.3 DNA replication in the cell cycle

Briefly, the coordinate activity of several enzymes, working together in replication complexes, is required to unravel the DNA strands and for new copies of each strand to be synthesized. These replication complexes initiate DNA replication in association with regulatory proteins required for recognition of the site of origin and which ensure that DNA is replicated only once during S phase (licensing).

Prior to cell division, the DNA is duplicated during S phase. First a portion of the double helix is unwound by a helicase enzyme, following which a DNA polymerase (delta) binds to one DNA strand and begins moving along it. This strand then acts as a template for the assembly of a leading strand of nucleotides and the regeneration of a double helix. Because DNA synthesis only occurs in 5′ to 3′, a molecule of a second type of DNA polymerase (epsilon) binds to the other template strand as the double helix opens. This molecule synthesizes discontinuous polynucleotide segments termed Okazaki fragments, which are then joined together by the enzyme DNA ligase to form the lagging strand (Fig. 4.9). DNA replication is semi-conservative, as each strand of the DNA is retained and serves as a template for the newly synthesized complementary strands.

The average human chromosome contains 150×10^6 nucleotide pairs copied at a rate of almost 50 base pairs per second. To enable this process to be completed within a short timespan (6–10 hours in mammalian cells), there are multiple sites of origin on each chromosome. Replication begins at some replication origins earlier in S phase than at others, but the process is completed for all by the end of S phase. Given these multiple origins, several control steps are required to ensure that all original DNA is replicated and that freshly synthesized DNA is not replicated again. The integrity of the genome is maintained by licensing; DNA is only replicated if each origin of replication is bound by an origin recognition complex of proteins (ORC). Also required are various licensing factors, such as CDC6 and CDT1, which accumulate during G_1 and coat DNA with MCM proteins, without which DNA is not replicated. The protein geminin is also an important negative regulator of replication that can prevent the assembly of MCM proteins on freshly synthesized DNA; when mitosis has been completed, geminin is degraded in order that DNA of the two progeny will be able to replicate their DNA at the next S phase by reacting to licensing factors.

Endoreplication

Endoreplication is the replication of DNA during S phase without subsequently completing mitosis and or cytokinesis. Endoreplication is a feature of certain types of cells in both animals and plants and takes distinct forms, including: replication of DNA with completion of mitosis without cytokinesis; repeated DNA replication but no new nuclei form in telophase resulting in either polyploidy (multiple chromosome copies – cells are polyploid if they contain more than two haploid sets of chromosomes) or polyteny (multiple chromosomes remain aligned as giant chromosomes); mitosis without cytokinesis, producing multiple nuclei. Polyploidy is observed normally in liver cells, placental trophoblasts and some hematopoietic lineages.

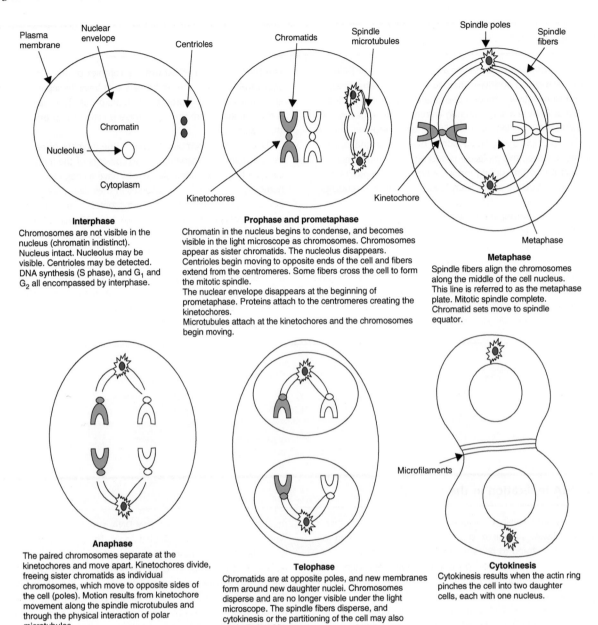

Figure 4.2 Mitosis. During the majority of the cell cycle (interphase), chromosomes are not visible under light microscopy. However, this is not the case during the nuclear division phase of mitosis, which is traditionally categorized, on the basis of early microscopy-based studies, into distinct observable stages that constitute the "chromosome cycle" as summarized here. Research over the last two decades has shown that the chromosome cycle is inextricably linked with the "cell cycle" (also referred to as the cyclin-dependent kinase or CDK cycle) shown in Figure 4.1. During mitosis, the centrosomes duplicate, separate, and generate a microtubular spindle between them, thus providing a bipolar framework for the rest of mitosis. The chromosomes, comprising two sister chromatids, condense and become attached to microtubules arising from opposing spindle poles. Sister chromatids then separate and move apart to form two nuclei, which are ultimately separated by cytokinesis. Briefly, mitosis is divided into various phases based on early microscopy studies that correlate with distinct observable events. During **prophase** the two centrosomes, each with a pair of centrioles, move to opposite poles of the cell, and the mitotic spindle forms and the chromosomes shorten and compact. During **prometaphase** the nuclear envelope disappears and the kinetochore forms at each centromere. Spindle fibers attach to the kinetochores as well as to the arms of the chromosomes, for each dyad, one of the kinetochores is attached to one pole, the second (or sister) chromatid to the opposite pole. At **metaphase**, all dyads are arranged midway between the poles called the "metaphase plate." The chromosomes are at their most compact at this time. At **anaphase**, sister kinetochores separate (cohesions are broken down by separase, normally kept inactive by the chaperone securin; anaphase is initiated when the APC targets securin for proteasomal degradation). Each kinetochore moves to its respective pole along with its attached chromatid (chromosome). During **telophase** a nuclear envelope reforms around each group of chromosomes, which re-adopt a less compacted form. Finally, during **cytokinesis**, the cell divides into two, through the action of actin filaments that constrict at the "waist" of the cell.

Box 4.4 Key studies in the cell-cycle field – cell-cycle regulation "from yeast to beast"

The appreciation that essentially all cells arise by the binary division of pre-existing cells was eloquently stated by the pathologist Rudolf Virchow: "Omnis cellula e cellula." Cell division is now thought of as a process that has been endlessly perpetuated ever since the first cell came into existence – the beginning of life on this planet.

Originating from a single cell, the fertilized ovum, an adult human consists of approximately 10^{14} cells and needs to produce about three million new cells every second simply to maintain normal tissue homeostasis.

The cell cycle includes key cellular processes, including the growth of the cell, replication of the genome, and the distribution of the chromosomes during mitosis. Not surprisingly, cell-cycle progression is subject to extensive "checkpoint" monitoring and regulation by various factors that ensure that cell replication is appropriate. Even single-cell organisms such as yeast must be able to link replication to the availability of nutrients in the environment. In multicellular organisms these controls are even more important, since cell division must be adapted to meet the needs of the whole organism. Importantly, it is clear that the regulatory processes underlying cell division are extremely similar in all animals and plants. Particularly in the two decades it has been recognized that certain "pacemakers" of the cell cycle have been conserved through several hundred million years of evolution.

It was for the characterization of these cell-cycle pacemakers that the 2001 Nobel Prize in Physiology or Medicine was awarded to Leland Hartwell, Paul Nurse, and Tim Hunt. They discovered protein kinases that are periodically switched on and off during the cell cycle, and control practically all important stages, particularly the replication of DNA and mitosis. Since all of them depend on special regulatory subunits, the cyclins, they are known as "cyclin-dependent kinases" (or CDKs).

These landmark studies on cell-cycle regulation started in the late 1960s. Hartwell used budding yeast to identify mutants that blocked specific stages of cell-cycle progression. Nurse, working in fission yeast in the 1970s, went on to isolate mutants that could also speed up the cell cycle, thus focusing his attention on the original CDK kinase, cdc2. In the 1980s, Hunt identified proteins in sea urchin extracts, the levels of which varied through the cell cycle hence "cyclins." Cyclins were first identified by work on sea urchin eggs. Noting that the concentration of cyclins went up rapidly through most of the cell cycle, then suddenly dropped to zero halfway through the M phase, it was suggested that cyclins might act as a sort of switch, turning on mitosis whenever their concentration reaches a certain level. Similar cyclins were later found in budding yeast. It soon became clear that cyclins bound to, and activated, a type of protein kinase called the cyclin-dependent kinases (CDKs). This activation was required for cells to move from one stage of the cell cycle to the next.

A large part of contemporary cell-cycle research has been conducted on mutant strains of the fission yeast (*Schizosaccharomyces pombe*) and the budding yeast (*Saccharomyces cerevisiae*) that have genetic lesions in some phase of the cell cycle. The cell division cycle (cdc) mutant strains have been a major contributor to our current understanding of how the cell cycle is regulated. The yeast cell cycle has two points at which it commits to progressing into the next phase in the cycle: start near the end of G_1 at which the cell commits to DNA synthesis (S phase); and at the beginning of M phase when the cell commits to chromosome condensation and mitosis. Seminal studies have shown that key protein complexes are formed at each of the two committal points, comprising a cyclin, and a protein kinase called p34. The existence of such a complex was described biochemically when a factor called maturation promoting factor (MPF) was isolated that could initiate mitosis in certain mutant yeast strains whose cell cycle was arrested at this stage. It was the coupling of this type of biochemical research with genetics that defined and elucidated many of the steps in the cycle.

The molecular mechanisms regulating the cell cycle are highly conserved across species, a fact which has enabled seminal discoveries using various model organisms, such as yeasts, sea urchins, and frogs to be readily applied to understanding cell-cycle control in humans, and are exemplified by the recent awards of the Nobel Prize for Physiology or Medicine to Leland Hartwell, Tim Hunt, and Paul Nurse (Box 4.4).

The cell cycle is divided into four distinct and microscopically recognizable phases (Fig. 4.1): the replication of chromosomal DNA during the synthesis phase (S phase), the partitioning of replicated chromosomes during mitosis (M phase), and two gaps, one before and one after S phase, that are referred to as G_1 and G_2, respectively. Broadly, M phase in the cell cycle includes the various microscopically observed stages of nuclear division and cytokinesis (mitosis) and is itself divided into phases termed prophase, prometaphase, metaphase, anaphase, and telophase (Fig. 4.2). Interphase is the term that encompasses stages G_1, S, and G_2 of the cell cycle.

The duration of the cell cycle can vary remarkably between cell types. For example, cells in early embryos can proceed through continuous cycles with each cell cycle completed in a mere half hour. This is in contrast with cells of the adult, where a fairly rapidly dividing mammalian cell would have a cycle time of 12–24 hours, whereas the cell cycle of a human liver cell can last longer than a year! The much longer duration of cell-cycle transit observed in adult tissues, in contrast to early embryonic cells, is due to the gap phases, G_1 and G_2, which allow for growth and importantly the repair of DNA damage and replication errors (see below). It is tempting to speculate that the large amount of apoptosis that normally accompanies ontogeny might in part reflect the consequences of sacrificing some accuracy in the need for rapid cell expansion.

Normally, cell-cycle events are subject to rigorous quality control steps. First, commitment to DNA synthesis is dictated by exposure of the cell to a variety of growth-promoting and inhibitory factors; G_1/S transition in normal cells only occurs if the balance favors growth promotion (in fact a large part of Chapter 5 concerns how this step is regulated by growth factors and signaling molecules). Second, at different steps in the cell cycle the cell determines if an earlier event has occurred correctly before proceeding to a further step. These are referred to as **checkpoints** (Fig. 4.3) and will be discussed again later and in Chapter 10. Failure of checkpoints can result in the propagation of cancer-causing mutations and even in extreme phenotypes, such as

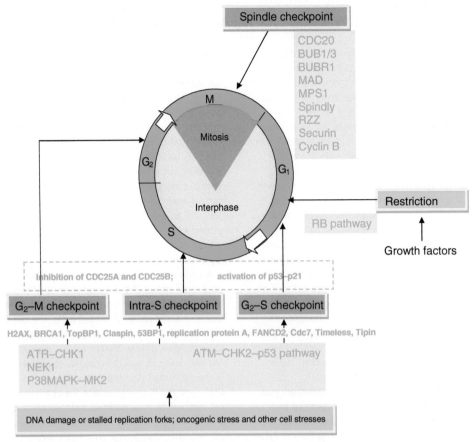

Figure 4.3 Major checkpoints activated by DNA damage and mitotic defects in the cell cycle. The first three checkpoints – G_1/S, intra-S, and G_2/M – largely respond to different types of DNA damage and cell stress more generally. They are a critical part of the DDR, which involves damage sensors, various effectors and interacts with key tumor suppressor pathways, such as p53–p21. These are activated by overlapping groups of kinases that ultimately inactivate CDC25, and activate p53–p21. Active checkpoints arrest the cell cycle to allow repair. If the genotoxic insult exceeds repair capacity, cell death signaling is activated. CHK2 and CHK1 are the main effectors in the DDR downstream of ATM/ATR. ATM/ATR activation allows recruitment and activation of a plethora of mediators, including H2AX, 53BP1, Claspin, BRAC1, TopBP1, MDC1, SMC1, FANCD2, Timeless, and Tipin. These remain located to the site of damage while CHK1 and 2 are released to activate a downstream signaling cascade that will arrest the cell cycle at different phases by inactivation of CDC25 and/or activation of p53 effectors. The spindle checkpoint ensures that each daughter cell gets the right number and configuration of chromosomes and involves a separate group of sensors and effectors. ATM (ataxia telangiectasia mutated) detects among others DSB. The MRN complex (Mre11, Rad50, and Nbs1) is involved in initiating ATM activation. In turn, ATM interacts with γH2AX and Tip60 to activate CHK2. ATR (ataxia telangiectasia related) detects stalled replication forks and oncogenic stress. The 9-1-1 complex (Rad9-Rad1-Hus1) is involved in initiating ATR activation. In turn, ATR interacts with TopBP1 and ATRIP to activate CHK1. NEK1 (never-in-mitosis A-related protein kinase 1) can activate CHK1 and 2 independently of ATM/ATR. CDC25A is a key CDK activator. G_2/M checkpoint prevents cells from entering mitosis by activating systems such as that regulated by CHK1 and NEK11 that degrades CDC25A. PLK1 (polo-like kinase 1) downregulation by p53 contributes to the G_2/M checkpoint. The spindle checkpoint inhibits Cdc20, an activator of APC/C, thus delaying anaphase onset until all chromosomes are attached correctly to spindle microtubules. RZZ (Rod/ZW10/Zwilch) complex. MAD (mitotic arrest deficient) detects attachment of microtubules to kinetochores and prevents aneuploidy. If this step fails then anaphase entry is prevented (other important proteins acting at the spindle checkpoint include MPS1, BUB1, and BUB3). CHFR is a ubiquitin ligase which can target polo-like kinase for proteasome degradation, thus arresting cells at G_2/M.

multiple rounds of DNA replication without cell division; in endoreplication, cells with replicated genomes bypass mitosis and instead replicate their DNA again, resulting in polyploidy. Cell-cycle regulators are also important in avoiding centrosome over-duplication (sometimes denoted as the "centrosomal cycle" to distinguish it from the nuclear cycle). Nuclear cell-cycle events, although critically important, are not the only pivotal processes during cell division. In the cytoplasm, various cell organelles must also be duplicated, including the mitochondria.

The G_1/S transition is the point at which cells commit to replicate their DNA and enter the cell division cycle. This is controlled by the RB protein, the product of the tumor suppressor gene *RB*. RB family proteins (including pRB/p105, pRB2/p130, and p107) are crucial cell-cycle regulators that are almost invariably lost or inactivated in human cancers, and this has been noted in lung, breast, and bladder cancers. RB (see Chapter 7) was first identified as the key missing factor in hereditary retinoblastoma, a malignant childhood tumor originating from the developing retina. RB acts as a red light for cell-cycle progression through G_1 into S phase, by binding E2F family transcription factors that are required for expression of S-phase genes (Fig. 4.1). However, after appropriate growth factor stimulation the inhibitory effects of RB are neutralized by sequential phosphorylation steps that switch the lights through amber to green. Once RB is hyperphosphorylated, E2F is given the green light to activate S-phase genes and the cell cycle moves forward (Fig. 4.4). In fact, beyond this

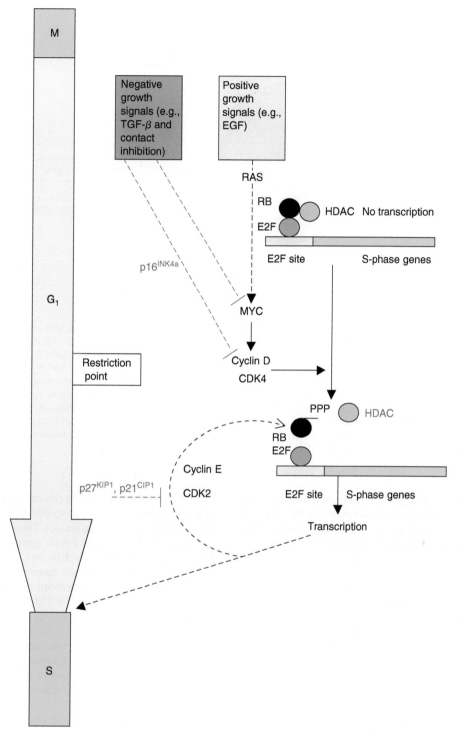

Figure 4.4 Balancing of growth-regulating signals received in G_1 determines entry into S phase. Growth factor mitogens primarily act on the cell during the G_1 phase of the replication cycle.

point the cell cycle can proceed under its own volition even if growth factor stimuli are no longer present. RB or associated signaling pathways are frequently inactivated in cancers and contribute to inappropriate and growth factor-independent cellular replication (see Chapter 7).

The engine which powers the cell-cycle machinery consists of various proteins acting together and in a predetermined sequence. In fact, it was the observation that the levels of the cyclins changed markedly through different phases that gave them their name and "put the cycle in cell division" (Fig. 4.5). Conversely, various negative inputs into cell-cycle progression, including inhibitory growth factors and triggering of various checkpoints, operate in part via the activity of a key group of proteins – the **cyclin-dependent kinase inhibitors** (CKIs).

There are at least four key stages at which the cell cycle is subjected to particularly rigorous scrutiny – the **cell-cycle**

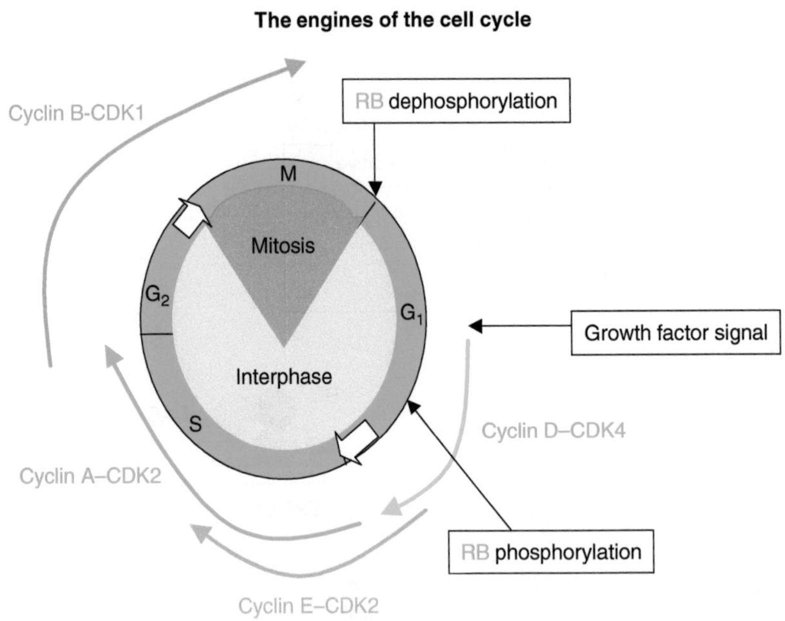

The engines of the cell cycle

Figure 4.5 The cyclins and CDKs during the cell cycle – running the cell cycle "relay." The cell cycle resembles a relay race with a cyclin–CDK "runner" responsible for each of the four legs of the race. There are four classes of cyclins, each defined by the stage of the cell cycle at which they bind CDKs and function: (1) G_1/S cyclins bind CDKs at the end of G_1 and commit the cell to DNA replication. (2) S cyclins bind CDKs during S phase and are required for the initiation of DNA replication. (3) M cyclins promote the events of mitosis. (4) G_1 cyclins help promote passage through start or the restriction point in late G_1. In G_1 the runners of the first leg (cyclin D–CDK4/6) are in the blocks waiting for the start gun (in G_0 they are stretching or milling around watching the long jump, some may have retired). The race starts in G_1, when the appropriate signal is received and the first baton is handed over by cyclin D–CDK4/6 to the S phase cyclins after the restriction point. Once the race is on, each of the four runners drops out in sequence and another takes over for each of the four legs. Like in a race, the runners of the next leg sometimes start running before they get the baton and the previous runners take a while to stop running after the baton is handed over. If something minor goes wrong (a runner trips over) the race stalls but it can restart once the problem is resolved. However, if something irretrievable happens (a baton is dropped), then the race is over.

checkpoints. The first very important checkpoint involves both the RB and p53 proteins and determines whether the cell cycle will leave G_1 and start to replicate DNA (the **G_1/S checkpoint**). The second two, the **intra-S-phase and G_2/M checkpoints**, respond to DNA damage occurring or recognized during DNA replication and will prevent the replication of mutated DNA until the damage has been repaired, or if irreparable the cell either exits permanently from replication or is eliminated through its own death (Figs 4.6 and 4.7). The importance of these two extreme outcomes of checkpoint activation is emphasized by the devotion of Chapters 9 and 8 to them. Lastly, the **spindle checkpoint** arrests mitotic progression and prevents anaphase if the spindle is not assembled or chromosomal orientation is disordered (Fig. 4.8). Failure of these checkpoints promotes genetic instability and can allow cells with damaged DNA or incorrectly partitioned chromosomes to divide and thereby propagate cells with potentially cancer-causing mutations.

This chapter will describe the phases of the cell cycle, the key regulatory proteins that ensure cell replication proceeds without mistakes, and some examples of how this complex process becomes derailed in cancer cells.

Phases of the cell cycle

> To do the same thing over and over again is not only boredom: it is to be controlled by rather than to control what you do.
>
> *Heraclitus*

The passage of a cell through the cell cycle is regulated by multiple cytoplasmic proteins, described below, which are remarkably well conserved among species. Thus, seminal work involving the study of budding and fission yeasts, sea urchins, and frogs has substantially contributed to our understanding of the cell cycle in mammals. In fact, the apparent complexity of terminology used to describe genes/proteins involved in cell-cycle regulation reflects the organisms in which they were first described or studied (Table 4.1).

It has been appreciated since the 1920s that cell-cycle progression is linked to cell growth, with early theories speculating that cells divided if they reached a particular cytoplasmic size that in some way rendered them unstable (see Box 5.1).

The cell cycle consists of the following recognizable phases (see Fig. 4.1):
- G_1 is the gap phase for growth and preparation of the chromosomes for replication;
- S phase is for synthesis of DNA (and centrosomes);
- G_2 is the gap phase for growth and preparation for mitosis;
- M is for mitosis (nuclear division) and cytokinesis (cell division).

The M phase (division phase) is relatively short, lasting about 1 hour for a cell-cycle time of 24 hours, compared to the rest of the cycle (interphase), during which time much cell growth takes place. Although lengths of each phase vary from cell to cell, the greatest variation is in G_1: cells that have not committed to DNA synthesis can pause and enter a resting state termed G_0, where they can remain for days, weeks, or years before resuming pro-

DNA Replication and the Cell Cycle

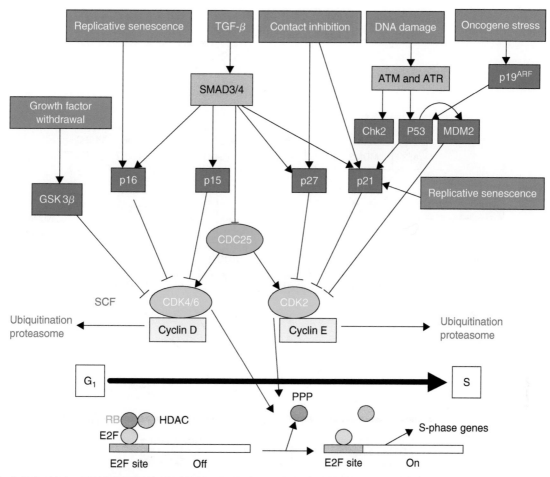

Figure 4.6 The G_1/S checkpoint. The G_1/S cell cycle checkpoint controls the passage of eukaryotic cells from the first "gap" phase (G_1) into the DNA synthesis phase (S). Two cell cycle kinases, CDK4/6–cyclin D and CDK2–cyclin E, and the transcription complex that includes RB and E2F are key regulators of this checkpoint. In G_1, the RB–HDAC repressor complex binds to the E2F transcription factors, inhibiting transcription of S-phase genes. Phosphorylation of RB by CDK4/6, and CDK2 dissociates the RB–repressor complex, permitting transcription of S-phase genes (encoding proteins that amplify the G_1- to S-phase switch and that are required for DNA replication). Different stimuli exert checkpoint control, including (1) TGF-β, (2) DNA damage, (3) contact inhibition, (4) replicative senescence, (5) oncogenic stress, and (6) growth factor withdrawal. The first five (1–5) act by inducing INK4 or KIP/CIP families of cyclin-dependent kinase inhibitors (CKIs). In addition, TGF-β inhibits the transcription of CDC25. Growth factor withdrawal activates GSK3β, which phosphorylates cyclin D, leading to its rapid ubiquitination and proteosomal degradation. Ubiquitination, nuclear export, and degradation are mechanisms often employed to rapidly reduce the concentration of cell cycle control proteins.

liferation. This state is often used to refer to cells that are growth-arrested or terminally differentiated, but does not imply that these cells cannot re-enter the cell cycle at some later stage.

Compared to DNA replication (S phase) and mitosis (M phase), which follow canonical steps that vary little from cell to cell, the regulation of entry and progression through G_1 largely depends on cell type and context. For instance, different signals are received by a stem cell of the intestinal lining compared to a lymphocyte stimulated by an antigen, or an angioblast responding to vascular injury, all of which ultimately proceed through the G_1 phase.

Through G or not through G that is the question?

The transition from the G_1 phase into the S phase is the defining moment in the cell cycle because it is the point at which a cell commits to cell division. Under normal conditions, G_1 is an important gap phase, during which the cell can ponder on the various stimulatory and inhibitory signals that it is receiving and integrate these into a decision either to embark on cell division or not (see Fig. 4.3). Given that this is also a point at which the cell must consider whether it can survive the effects of the "whips and scorns of time" on its DNA, the analogy with Hamlet's soliloquy is rather apt. Although these signals can take many forms, the growth factors – and the clue is in the name – are critically important (discussed in Chapter 5), as are those resulting directly from cell–cell and cell–matrix contacts (discussed in Chapter 12).

The net balance of positive and negative signals received ultimately dictate whether or not the cell proceeds into S phase or instead remains in G_1. Depending on the cells concerned and the context, such external signals may also be critical in regulating the differentiation state or even the continued survival of the cell. Importantly, growth signals are interpreted by the cell in a highly context-dependent fashion. Thus, during G_1 many and diverse inputs relating to metabolism, stress, and environment are integrated and interpreted in the light of potentially multiple and

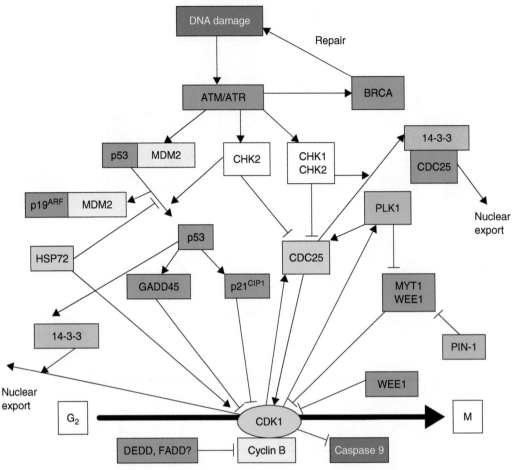

Figure 4.7 The G_2/M checkpoint. In general, the green-shaded boxes represent "Go" signals and the red/orange-shaded boxes "stop" signals. The G_2/M DNA damage checkpoint prevents the cell from entering mitosis (M phase) if the genome is damaged. The CDK1 (Cdc2)–cyclin B kinase is a key regulator of this transition. During G_2, CDK1 is maintained in an inactive state by the kinases WEE1 and MYT1. As M phase approaches, the phosphatase CDC25 is activated by the polo-like kinase PLK1, and possibly also through MAPK signaling. CDC25 then activates CDK1 (Cdc2), establishing a feedback amplification loop that efficiently drives the cell into mitosis. WEE1 is itself inhibited by CDK1 and the prolyl isomerase PIN-1 for M phase to progress. DNA damage inhibits CDK1 kinase activity at the G_2/M checkpoint in large part by activation of WEE1 and CHK1 kinases and inhibition of CDC25A and C phosphatases. DNA damage activates the ATM/ATR kinases, initiating two parallel cascades that inactivate CDK1–cyclin B. The first cascade rapidly inhibits progression into mitosis: the CHK kinases phosphorylate and inactivate CDC25, which can no longer activate CDK1. Phosphorylation of p53 dissociates it from MDM2, activating its DNA-binding activity. Acetylation by p300/PCAF further activates its transcriptional activity. The genes that are activated by p53 include 14-3-3s, which bind phosphorylated CDK1–cyclin B promoting nuclear export; *GADD45*, which apparently binds to and dissociates the CDK1–cyclin B kinase; and $p21^{Cip1}$, an inhibitor of a subset of the cyclin-dependent kinases including CDK1. HSP72 is overexpressed in some cancers and can promote activation of CDK1 and also prevent p53 activation by stabilizing HDM2 (MDM2 in mouse). In addition to preventing apoptosis, BCL-X_L colocalizes and binds to CDK1 after DNA damage and may stabilize the G_2/M DNA damage checkpoint and senescence in surviving cells after DNA damage by inhibition of CDK1 activity. Interestingly, CDK1 may also be inhibited by proteins containing the death-effector domain (DED), such as DEDD and possibly FADD, which can bind directly to cyclin B. A recent study suggests that apoptosis, often linked to the cell cycle, may be suppressed by cycling cells through inhibitory phosphorylations of caspase 9 by CDK1. Whether this is true in normal cell cycles or only in DNA damage-associated or oncogenic cycles (see Chapter 10) is unclear. One ramification might be that cycling cells could trigger MOMP and are thus potentially apoptosis-sensitized, but that apoptosis is inhibited downstream as long as mitosis is activated. The study presented shows that inducing G_2/M arrest also inhibited caspase 9, via CDK1–cyclin B; so maybe we can only assume that this mechanism may be a means of channeling cells with DNA damage towards senescence and survival to allow repair rather than apoptosis.

even antagonistic specific growth regulators. Moreover, the response will also ultimately depend on the condition, age, and nature of the receiving cell at that given time. Under the influence of the same cocktail of external growth-promoting cues, immature cells and stem cells may be more likely to enter cell cycle than terminally differentiated ones. Once a decision has been reached, the activity of various cell cycle regulatory proteins is altered (discussed below) and the cell cycle proceeds.

The source of extrinsic signals is also very diverse and includes local cells (paracrine) or from the circulation (endocrine) or even from the cell itself (autocrine). Although various intrinsic factors may influence the cell cycle, cell differentiation and replicative age as well as stresses arising from DNA damage, altered metabolism and excessive oncogenic stimuli are most relevant to understanding tumorigenesis. Given the task of interpreting a flood of signals in order for the correct outcome – cell-cycle progression, arrest, differentiation, or death – it is perhaps not surprising that mistakes in this process can lead to cancer. In fact, it is hard to imagine any cancer developing at all without deregulation of some aspects of cell-cycle control.

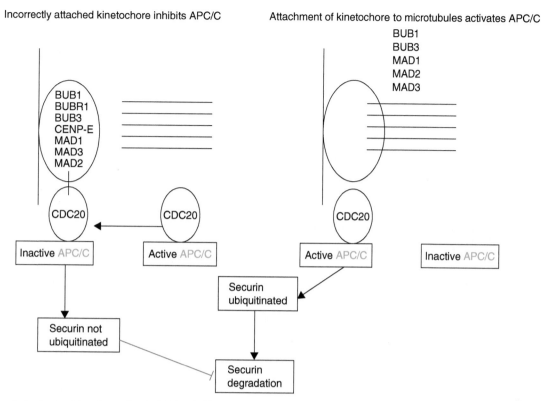

Figure 4.8 The spindle checkpoint pathway. On the left-hand side incorrectly attached kinetochores recruit BUB and MAD proteins that form an inhibitory complex catalyzing inactivation of the anaphase-promoting complex (APC/C) through MAD2 binding to CDC20. On the right, correct attachment and "tension" favors release of the checkpoint proteins from the kinetochore, releasing inhibition of APC/C and allowing anaphase. This allows APC/C to ubiquitinate securin, which is then targeted for degradation by the proteasome. The APC/C also regulates degradation of B cyclins for mitotic exit. In mammalian cells securin at sites other than the centromere is removed by the action of polo-like kinase during prophase.

The cell-cycle engine: cyclins and kinases

I saw the angel in the marble and carved until I set him free.
Michelangelo

Once the cell has received appropriate signals for entry into the G_1 phase of the cell cycle, various regulatory proteins called **cyclin-dependent kinases** (CDKs) must be activated in order for the cell to proceed and enter S phase. CDKs are serine/threonine protein kinases that become catalytically active once bound to their regulatory subunits, the **cyclins**. Active CDKs are universal regulators of the cell cycle that alter the biological functions of regulatory proteins through phosphorylation, and have been described as a "cell-cycle engine" (see below).

For full activity, CDKs require cyclin binding and T-loop phosphorylation, mediated by activating kinases (CAK), of which CDK7 is the best described. Different members of the CDK family are activated by specific appropriate cyclins during certain phases of the cell cycle (Table 4.2). While levels of CDKs, albeit often inactive, remain fairly stable throughout the cell cycle, levels of cyclins rise and fall (with the exception of cyclin D which rises early in G_1 and remains constant thereafter), thus determining at what stage of the cycle the CDK partner is activated.

Cyclins

Levels of cyclins rise and fall during the progression of the cell cycle, and each cyclin is categorized with respect to the stage at which it is elevated. In general levels of cyclins are determined by rate of synthesis on the one hand and by degradation through the ubiquin–proteasome pathway on the other (described later and in Chapter 11). Table 4.2 gives a fairly comprehensive list of currently known metazoan cyclins and CDKs. Key players in the cell cycle include:

- G_1 phase: cyclins D1, D2 and D3;
- S phase: cyclins E1, E2 and A;
- mitotic: cyclins B and A.

Cyclin-dependent kinases (CDKs)

The progression of a cell through the cell cycle is promoted by CDKs, whose periodic activation is driven by cyclins and negatively regulated by **CDK inhibitors** (CKIs). CDKs are categorized numerically. CDK protein levels remain relatively stable during the cycle, but activity is dependent on each CDK forming an active holoenzyme complex by binding to its relevant cyclin. Cyclin–CDK complexes in turn promote phosphorylation of target substrates needed for cell-cycle progression. Substrate specificity is likely determined by both subcellular localization and structural factors, with the key players comprising:

- G_1 phase: cyclin D–CDK4/6 complex (key target primarily RB);
- S phase: cyclin E or A–CDK2 complex (key target RB and others such as Cdc6);
- M phase: cyclin B or A–CDK1 (key target not well defined but include WARTS and PAK1).

Not surprisingly, many of the cyclins and CDKs are involved in cancer (Table 4.3).

Table 4.1 Nomenclature of proteins involved in cell cycle in budding yeast, fission yeast, and mammals

Factor	Function	*Saccharomyces cerevisiae*	*Schizosaccharomyces pombe*	Mammalian
CDK	Cyclin-dependent kinase	Cdc28	Cdc2	Multiple CDKs: CDK1–6
G_1 cyclin	Regulatory subunit of CDK for cell-cycle entry	Cln1,2 and 3		CDK4/6–cyclin D
S-phase cyclin	Regulatory subunit of CDK for S-phase entry	Clb5, 6	Cig2	CDK2–cyclin E
Late S-phase cyclin	Regulatory subunit of CDK for S-phase progression	Clb3, 4		CDK2–cyclin A
M-phase cyclin	Regulatory subunit of CDK for mitosis	Clb1, 2	Cdc13	CDK1(Cdc2)–cyclin B
APC (anaphase-promoting complex)	Multicomponent E3 ubiquitin ligase required for degradation of substrates in mitosis and G_1	Many proteins	Many proteins	Many proteins
APC specificity factors	Target the APC towards different substrates	Cdc20 Hct1	Slp1 Srw1	CDC20, fizzy HCT1, FZR
Securin	An APC target, inhibits sister chromatid separation	Pds1	Cut2	Securin
Separase	The securin target, a protease that degrades cohesin	Esp1	Cut1	Separase
Cohesin	A complex of proteins that holds sister chromatids together	Scc1	Rad21 Psc3 Psm1 Psm3	Rad21 SCC1 SCC3 SMC1 SMC3
SCF	Multicomponent ubiquitin ligase required for degradation of phosphorylated substrates in G_1	Skp1 Cdc53 Cdc4	Skp1 Pop1, 2	S is SKP1 C is cullin F is F box protein
CKIs	CDK inhibitors	Sic1	Rum1	$p16^{Ink4a}$, $p15^{Ink4b}$, $p18^{Ink4c}$, ARF, $p21^{Cip1}$, $p27^{Kip1}$, and $p57^{Kip2}$
ATM/ATR	Master kinase regulators of checkpoint pathways	Mec1 Te11	Rad3 Te11	ATR ATM
Checkpoint sensor	Complex of proteins consisting of a clamp loader and a clamp that binds DNA and monitors damage	Rad24 Mec3 Rad17 Ddc1	Rad17 Hus1 Rad1 Rad9	RAD17 RAD9, RAD1, HUS1 (9-1-1 complex) (also in mammals the MRN complex MRE11, RAD50, and NBS1, Chapter 10).
Effector kinases	Downstream of sensor kinase, respond to different challenges	Chk1 (damage) Rad53 (hydroxy urea)	Cchk1 Cds1	CHK1 CHK2 BRCA1
Targets	Downstream of kinases regulators of the cell cycle	Cdc5 Pds1	Wee1 Cdc25	CDC25 p53 HDM2
Transcription factors	Regulated transcription; the ones here are active for synthesis of S-phase genes	Swi6 Swi4 Mbp1	Cdc10 Res1	E2F
preRC	Prereplication complex, which marks a replication origin as ready to fire	Orc1–6 Cdc6 Mcm2–7	Orp1–6 Cdc18 Mcm2–7	ORC1–6 CDC6 MCM2–7
Cdc7	Origin-activating kinase, which may play other roles in maintaining genome integrity. Requires a subunit (DBF4) which does not look like, but acts like, a cyclin	Cdc7 Dbf4	Hsk1 Dfp1	CDC7 DBF4/ASK1

After S.L. Forsburg, University of Southern California.

Table 4.2 A summary of metazoan cyclins and Cdks

Cyclin	Cdk	Role
Cell cycle		
A1,2	CDC2, CDK2	DNA replication and mitotic entry
B1,2	CDC2	Mitosis
B3		Unlike B1 and B2, enters the nucleus during interphase
D1,2,3	CDK4, CDK6	Required for G_1/S transition
E1,2	CDK2, CDK3	DNA replication and centrosome duplication
F	?	Binds to cyclin B
Transcription		
C	CDK8	Phosphorylates C-terminal of RNA polymerase II large subunit. Target of vitamin D
H	CDK7	May phosphorylate and activate other CDK/cyclin complexes, modifies transcription machinery
T1,2a,2b	CDK9	Transcription elongation factor (Tef-2b). Binding of cyclin T1 to HIV Tat is required for efficient binding to the Tar element of HIV RNA
Other/unknown		
p35	CDK5	Found in postmitotic neurons. p35 proteolysis associated with neurodegenerative diseases
?	CDK10	Unknown
?	PCTAIRE	Unknown
?	PFTAIRE	Unknown
?	PITSLRE	Unknown, possible role in inducing apoptosis
G1,2	?	Induced by proapoptotic stimuli. Reported association with CDK5
I		Unknown, related to cyclin G
L1,2		Unknown. Homology to transcriptional cyclins. Induced by dopaminergic stimulation
M	?	Homology to transcriptional cyclins
O	?	Described as uracil-DNA glycosylase. Homology to cyclins
P	?	Distant relative of A and B type cyclins. Function unknown. Restricted to vertebrates?

Based on Murray, A.W. and Marks, D. (2001). Can sequencing shed light on cell cycling? *Nature*, **409**(6822): 844–6.

Table 4.3 Cyclins, CDKs and CDKIs in cancer

Cyclin/CDK/CDKI	Cancer
Cyclin A	Liver, breast
Cyclin B	Colon, breast
Cyclin C	Melanoma, leukemia
Cyclin D1	Lymphoma, breast
Cyclin D2	Colon, testicular, leukemia
Cyclin D3	Lymphoma, leukemia
Cyclin E	Colon, breast, prostate, ovary, lung, stomach, pancreas, leukemia
Cyclin E2	Breast, lung, cervix
Cyclin K	Kaposi's sarcoma
CDK2	Colon
CDK4	Melanoma, colon, breast
CDK6	Glioma
p16^{INK4a}	Melanoma, leukemia, glioma, lung, breast, oesophagus, pancreas
p15^{INK4b}	Melanoma, leukemia, lung
p18^{INK4c}	B cell Lymphoma, testicular cancer
p19^{INK4d}	Testicular tumors
p21^{WAF1}	Brain, colon, leukemia, melanoma
p27^{KIP1}	Breast, colon, melanoma
P57^{KIP2}	Lung, breast, mesothelioma

As mentioned earlier, CDK activity during the cell cycle is either high or low; the change in state underlies temporal control of DNA replication and links it to mitosis. Initiation of DNA replication in S phase requires the formation of a prereplication complex (pre-RC) at chromosomal sites known as "replication origins," and the activation of DNA-unwinding and polymerase functions (Fig. 4.9). The formation of a pre-RC occurs when CDK activity is low, whereas recruitment of DNA helicases (to unwind DNA) and polymerases occur when CDK activity is high. The transition from low to high CDK activity is essential for correct DNA replication. Moreover, the high CDK activity prevents the formation of further pre-RC until the completion of mitosis, when CDK activity is reduced. These events ensure that DNA will be replicated once and only once per cell cycle.

Inhibitors of the cell-cycle engine

> If he overcame his shyness, caution applied the foot-brake. If he succeeded in forgetting caution, shyness shut off the gas.
>
> *PG Wodehouse*

Untrammelled cell division is not consistent with metazoan societies and it is essential that the means of cell production are effectively tamed, subjugated, and organized, to serve the greater good. This vital task is assigned to a series of negative regulators of the cell-cycle engine, operating under the combined auspices of inhibitory factors from without and a variety of quality-control steps from within the cell. Cyclin–CDK complexes are held in check by two classes of CKI – the **INK4 family** and the **CIP/KIP family** – described below.

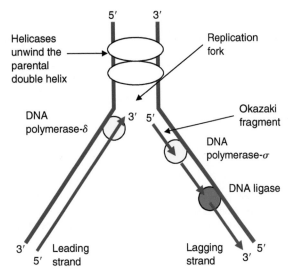

Figure 4.9 DNA synthesis during S phase. DNA is synthesized from 5′ to 3′ by the action of DNA polymerases.

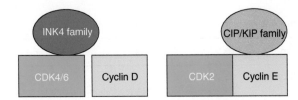

Figure 4.10 The two families of CKI proteins. The INK4 family of CKI (p15^{INK4b}, p16^{INK4a}, p18^{INK4c}, and p19^{Arf}) inhibit cyclin–CDK action by binding to the CDK, whereas CIP/KIP family members (p21^{CIP1}, p27^{KIP1}, and p57^{KIP2}) bind to the cyclin component of the cyclin–CDK complex.

These proteins – many of which are known to be tumor suppressors in humans – are able to inhibit cell-cycle entry and progression and prevent replication of abnormal DNA by allowing cells to stall at appropriate points during the cell cycle so that DNA damage can be repaired. These so-called "checkpoints" are discussed below, but briefly are activated by various (maybe most) forms of "cell stress," including two particularly cancer-relevant and partially overlapping triggers – "genotoxic stress" (DNA damage) and "oncogenic stress" (inappropriate and excessive activation of oncogenes). Although discussed in depth in Chapter 7, it should be noted here that CKIs are key participants in both RB and p53 tumor suppressor pathways and as such are frequently inactivated in human cancers.

The INK4 (inhibitor of cyclin-dependent kinase 4) family of CKIs

p15^{INK4b}, p16^{INK4a}, p18^{INK4c}, and p19^{INK4d} (not to be confused with p19^{Arf}, the mouse homolog of human p14^{ARF}) is a highly conserved and critical group of cell-cycle regulatory tumor suppressors that all inhibit the activity of cyclin D–CDK4/6 complexes and prevent RB phosphorylation. However, at least one member is also involved in activating another very important tumor suppressor, p53. ARF (p19^{Arf} in mouse and p14^{ARF} in humans) is produced from an alternative reading frame of the same gene locus, *CDKN2A*, as p16^{INK4a}, hence the name ARF. ARF can sequester MDM2 (a protein which otherwise binds and targets p53 for degradation) thus stabilizing the p53 protein in response to "oncogenic stress" (see below). This results in G_1/S or G_2/M-cycle arrest, at least in part by activating the p21^{CIP1} CKI discussed below, but ARF–p53 may also promote cell death, whereas p16^{INK4a} directly induces G_1 arrest by inhibiting the cyclin D-dependent kinases CDK4 and CDK6, thereby preventing phosphorylation of RB. The role of RB–p16^{INK4a} signaling in regulating senescence is discussed in detail in Chapters 7 and 9.

The CIP/KIP family of CKIs

p21^{CIP1}, p27^{KIP1}, and p57^{KIP2} are negative regulators of cyclin E– and cyclin A–CDK2 and of cyclin B–CDK1 and are discussed in Chapter 7. Of this group, p27^{KIP1} is a key inhibitor of the G_1/S transition and acts downstream of various inhibitory growth factors such as TGF-β, whereas, p21^{CIP1} is one of the major transcriptional targets of p53. High levels of p21^{CIP1} and p27^{KIP1} arrest the cell cycle by inhibiting cyclin E–CDK2 activity, thereby preventing RB hyperphosphorylation, whereas lower levels may instead stabilize cyclin D–CDK complexes. As cyclin D–CDK4/6 complexes form in mid-G_1, p27^{KIP1} and p21^{CIP1} redistribute from cyclin E–CDK2 to cyclin D–CDK4/6, thereby allowing active cyclin E–CDK2 complexes to complete phosphorylation of RB. Phosphorylation of p27^{KIP1} by cyclin E–CDK2 then targets the protein for degradation – a prerequisite for S-phase entry (see Figs 4.10 and 4.11 and also Box 7.4).

Recent studies show that p21^{CIP1} may also bind DNA and repress expression of c-MYC and CDC25A, thus linking DNA damage responses with inactivation of oncogenic signaling pathways. Among the most interesting questions in cancer biology has been to unravel exactly how the balance of reversible cell-cycle arrest, required to allow recovery from stress and DNA repair, on the one hand and senescence/apoptosis, to eliminate the threat posed by a rogue cell, on the other is achieved. This will be discussed later in more detail but either scenario appears to involve the same key drivers, namely the big two tumor suppressor pathways, RB–p16 and p53–p21, with outcome determined by DNA damage response and repair pathways, checkpoint kinases, and various pro- and antiapoptotic proteins (see Chapter 10).

Regulation by degradation

Destruction, hence, like creation, is one of Nature's mandates.

Marquis de Sade.

The effective and timely degradation of proteins is an essential part of the cell cycle and most signaling pathways and the processes by which this is achieved are discussed in Chapter 11. Here we will confine our attention to how protein levels and longevity are regulated post transcriptionally in the cell cycle.

The level of many cell-cycle regulators is determined by the activity of a family of enzymes responsible for "marking" specific proteins, through ATP-dependent polyubiquitylation, for subsequent degradation in the 26S proteasome. Among the first such ubiquitylating enzymes described were two key cell-cycle regulators, the ubiquitin ligases called **SCF** (after its three main protein subunits – SKP1/CUL1/F-box protein) and **APC/C** (the anaphase-promoting complex/cyclosome) (thus continuing a long tradition of engaging student interest by spattering scientific writing with memorable abbreviations!).

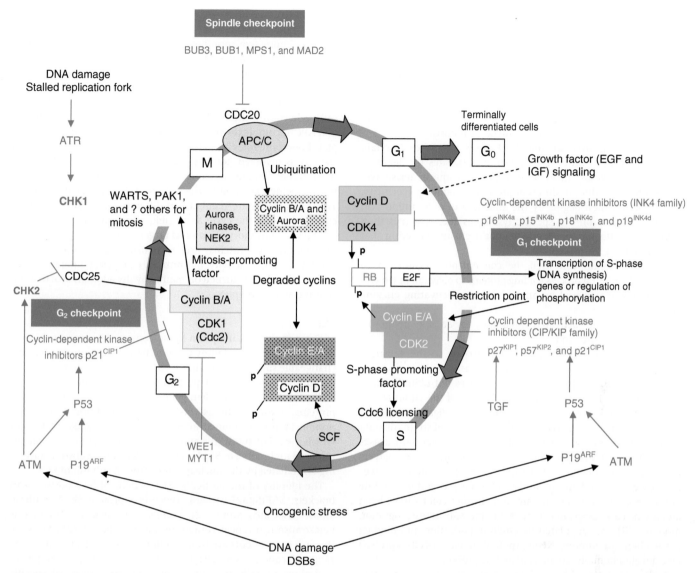

Figure 4.11 The cell cycle – showing interactions between cyclin–CDKs and CKI. The cell cycle consists of four phases: G_1, S, G_2, and M. Entry and progression through each stage is regulated by positive and negative factors. The key step in cell cycle entry is the G_1/S transition (passage through the restriction point), because after this the cell cycle will proceed even if growth factor stimuli are no longer present unless errors in DNA replication, chromosome segregation, or other factors trigger various checkpoints that arrest the cycle. Various checkpoints are shown in red text and red stop arrows. TGF-β can inhibit G_1/S transition via p27^{KIP1}, whereas stimulatory growth factors such as IGF1 activate signaling pathways (such as RAS–MAPK–ERK) that culminate in expression of c-*MYC*. c-*MYC* in turn enhances expression of cyclin D and CDK4 and promotes degradation of CKI, such as p27^{KIP1}. Together with cyclin E–CDK2, the net effect is phosphorylation of RB and release of E2F for transcription of key genes needed for S phase. Key CKI proteins active at the G_1/S transition are the INK4 and CIP/KIP families, which inhibit cyclin D–CDK4/6 and cyclin E–CDK2 holoenzymes, respectively. A recent study in mouse suggests that in some cases CDK1 (Cdc2) may replace CDK2 as a cyclin D partner. During S phase progression the SCF ubiquitin ligase complex targets D and E type cyclins for proteasome degradation. Cyclin E–CDK2 may also help the licensing of prereplicative complexes, for DNA synthesis by stabilizing Cdc6. The cyclin B–CDK1 complex is inactive through G_2, due to phosphorylation by WEE1 and MYT1; at the end of G_2 CDK1 is dephosphorylated and activated by CDC25. The G_2 checkpoint is activated by DNA damage such as DNA double-strand breaks (DSBs), which trigger various kinases and activate p53 and p21^{CIP1}. If DNA damage is successfully repaired the cell cycle may be re-entered, if not then the cell undergoes apoptosis or growth arrest. The APC/C is a multisubunit ubiquitin ligase that, analogous to SCF earlier in the cycle, targets B-type cyclins for proteasome degradation during passage through anaphase and telophase, and also limits accumulation of B-type cyclins during G_1. Inactivation is required for timely S-phase entry. The APC/C also targets cohesins for degradation, allowing separation of sister chromatids in a CDK-dependent manner at the metaphase-to-anaphase transition. The APC/C is itself regulated by the spindle or kinetochore checkpoint, which has an important role in maintaining genomic stability by preventing sister chromatid separation until all chromosomes are correctly aligned on the mitotic spindle. The spindle checkpoint regulates the APC/C by inactivating CDC20, an important coactivator of the APC/C.

SCF and APC/C are important in controlling the cell cycle because they are important in controlling the levels of several important cell-cycle regulators, including cyclins. They are ubiquitin ligases that promote the ubiquitylation of cell-cycle regulators, thereby directing them to the proteasome for destruction (see Chapter 11). Protein degradation is a critical step for many key events in the cell cycle, with SCF and APC/C acting at distinct stages – SCF mainly regulates the G_1/S transition, whereas the APC/C mainly targets proteins involved in mitosis.

Degradation is first initiated by kinases that phosphorylate protein residues in order to establish docking sites (or degrons) for ubiquitin ligases. In general, a given ubiquitin ligase will target similar residues in different proteins, producing specificity. Ubiquitylation then begins by the attaching of a 76-amino-acid ubiquitin moiety onto a specific lysine residue, which is then followed by attachment of polyubiquitin chains rendering that protein marked for proteolysis.

Polyubiquitylation can itself, be conveniently regarded as a sequential three-step process leading proteins to oblivion, in which:
1. ubiquitin is activated by an E1 ubiquitin-activating enzyme followed by,
2. transfer of the activated ubiquitin to an E2 ubiquitin-conjugating enzyme, and
3. transfer of ubiquitin from the E2 enzyme to the substrate through the activity of an E3 ubiquitin ligase.

Although the E2-conjugating enzyme carries out ubiquitylation of the target, it is actually various members of the cullin family of scaffold proteins, within the E3 multisubunit ligase complexes, that are responsible for recognizing the specific substrate (directly or through an adaptor). Thus, cullin–RING E3 ubiquitin ligases (CRLs) are key regulators of protein stability and are key players in cell-cycle regulation. All CRLs are multisubunit complexes composed of a catalytic site and a substrate-recognition module around a cullin scaffold. To date, five distinct cullins have been identified, each associated with a unique substrate-recognition module. CRLs are regulated by cullin dimerization and binding of the ubiquitin homolog Nedd8 (neural-precursor-cell-expressed and developmentally downregulated 8) protein.

Cullin activity is also regulated by the activity of the highly conserved COP9 (constitutive photomorphogenesis 9) signalosome (CSN), a protein complex involved in embryonic development, cell-cycle regulation, and the DNA damage response (DDR). The CSN multisubunit protease is important in positive regulation of the cell cycle and also in the DDR, in part by limiting the neddylation of cullins 1 and 4. CSN is often overexpressed in cancers where it deneddylates cullins and can counteract ubiquitin-mediated degradation through a CSN-associated deubiquitinating enzyme. Deubiquitinating enzymes are proteases that counteract the actions of E3 ligases on protein ubiquitination, and are themselves often deregulated in human cancers. The COP9 subunit 6 (*CSN6*) gene is amplified in human breast cancer and CSN6 is increased in breast and thyroid cancers. In mouse models CSN6 stabilizes MDM2, increasing degradation of p53, thereby disabling tumor suppressor activity and preventing apoptosis during tumorigenesis and DDR.

A large number of E3 ubiquitin ligases, including SCF, APC/C, Cbl, Smurf1, Smurf2, HDM2, BCA2, and XRNF185, have been found, which play roles in a variety of cancer processes, including the cell cycle, but also cell adhesion and migration. One way of classifying the E3 ligases is based on their homology to two archetypal ligase families – HECT (homologous to E6-AP C-terminus) and RING (really interesting new gene) family. In all, more than 1000 unique E3 ligases are known, and given the equally impressive number of substrate recognition factors (SRFs), you will forgive me if I do not describe them all here.

CRLs are the most numerous family of E3 ligases and include cullins 1–3, 4A, 4B, 5, and 7. SCF (or CRL1) ubiquitin ligase is the best studied to date. Given the pivotal role of SCF and APC/C in cell-cycle control these will be discussed in detail first. Both SCF and APC/C complexes are critical regulators of cellular growth but play important roles in many other processes in which protein stability is involved. Rapidly increasing information about how these complexes are able to confine their activity to specific proteins is being exploited in the design of new targeted therapies. These complexes are regulated in different ways. SCF activity is constant during the cycle and unbiquitinylation is regulated by changes in phosphorylation of its target proteins: only specifically phosphorylated proteins are recognized, ubiquitinated, and destroyed. APC/C activity, by contrast, changes at different stages of the cell cycle. It is turned on mainly by the addition of activating subunits to the complex.

SCF

SCF is a multi-unit ubiquitin ligase comprising several subunits: one or another F-box family protein, which functions as an interchangeable substrate recognition component for the protein to be ubiquitinated, an adaptor protein SKP1 and a catalytic core of cullin 1 (CUL1). Mammalian genomes encode 70 or more F-box proteins, mostly with as-yet-unknown functions. The F-box protein family are named after cyclin F (also called Fbx01), in which the F-box motif (the approximately 40-amino-acid domain required for binding to Skp1) was first identified. Substrate targeting is mediated by the nuclear F-box protein subunit of SCF.

The identity of the particular SCF is defined by a postscript in brackets. SCF(Skp2) contains the F-box protein Skp2 (S-phase kinase-associated protein 2), whereas SCF(beta-TrCP) contains beta-transducin repeat-containing protein (beta-TrCP). Some of the SCF components are tumor suppressors inactivated in a wide range of cancers (see Chapter 11). The specific phenotypes of mutations in SCF are due to deregulated activity of specific protein substrate-recognition components (such as the F-box proteins) within the SCF complex.

SCF(Fbxw7/hCdc4), containing the the F-box protein Fbxw7/hCdc4, was first recognized by its ability to bind cyclin E, but is now known to also recognize the oncogene products c-MYC, c-JUN, and Notch as substrates for degradation. FBXW7/hCDC4 negatively regulates several key oncoproteins, and the gene is inactivated by mutation in various human tumors and *Fbxw7* is a haploinsufficient tumor suppressor in mice. In G_1 and S phase, SCF(Skp2) is responsible for destruction of CKI proteins, such as p21^{CIP1} and p27, that control S-phase initiation. FBX4 is involved in degradation of cyclin D by SCF(Fbx4). Fbx4 is itself regulated by the Ras–Akt–GSK-3β pathway (see Chapter 5), which regulates phosphorylation and dimerization of the SCF(Fbx4) E3 ligase. Dimerization-defective mutations in Fbx4 are found in human cancers, which have increased cyclin D levels as a result. Another ubiquitin ligase often overexpressed in cancer is MDM2, which targets p53 for degradation.

FBW7 is a tumor suppressor inactivated in several human cancers, including colon and breast cancer and T-cell acute lymphoblastic leukemia (T-ALL). It can assist in destruction of several cancer-causing proteins, including JUN, MYC, cyclin E, and

Notch 1. SCF(FBW7) also regulates apoptosis by targeting MCL1, a pro-survival BCL2 family member. Loss of FBW7 can therefore deregulate oncogene expression while suppressing oncogene-related apoptosis.

APC/C

First identified 15 years ago by Avram Heshko and Marc Kirschner, APC/C plays a central role in cyclin degradation and cell-cycle progression. It is an inordinately complex E3 ubiquitin ligase, comprising at least 13 subunits in yeast, and regulates mitosis probably in large part by degradation of cyclin B and securin. The APC/C and E2 ubiquitin-conjugating enzyme UbcH10 regulates progress through mitosis by earmarking key mitotic regulators for destruction by the 26S proteasome. In M phase, APC/C is responsible for initiation of anaphase and exit from mitosis. Once activated by, among others, CDC20, CDH1/HCT1, and greatwall orthologs, APC/C can freely initiate the destruction of M-cyclins (cyclin A and B and associated CDK1) and other regulators of mitosis, such as securin and the spindle assembly checkpoint. Interestingly, the APC/C also promotes degradation of geminin, a key protein involved in preventing re-replication of newly synthesized DNA in S phase, and may also promote synthesis of G_1 cyclins for subsequent cycles.

Ubiquitination of cyclin B and securin depends on a destruction box (D box) sequence in these proteins, which is recognized by APC/C bound to CDC20 or CDH1. Cyclin B1–CDK1 maintains mitosis by allowing sufficient time for the chromosomes to be captured by spindle microtubules and for incorrect chromosome–spindle attachments to be fixed. Securin, an inhibitor of separase, secures cohesion between sister chromatids, preventing anaphase onset. The APC/C coactivator Cdc20 is targeted by the spindle assembly checkpoint to restrict APC/C activity until metaphase. The spindle checkpoint is triggered by improperly attached chromosomes. In turn, spindle checkpoint proteins that include Mad1 and Mad2, which inhibits Cdc20, prevent activation of the APC/C, thereby delaying mitotic exit and anaphase until all chromosomes reach bipolar spindle attachments. The APC/C is activated by the human ortholog of the greatwall protein, microtubule-associated serine/threonine kinase-like (MASTL). MASTL inhibits protein phosphatase 2A (PP2A), which otherwise dephosphorylates cyclin B–CDC2 substrates. MASTL is critical for balancing the activity of cyclin B–CDC2 and PP2A to ensure correct mitotic entry and exit.

Exactly how the early APC/C substrate cyclin A avoids the spindle checkpoint and is still earmarked for degradation whereas cyclin B is not is it at least in part due to the direct binding of Cdc20 by the N-terminus of cyclin, which simply outcompetes the spindle assembly checkpoint proteins. Interestingly, the Von Hippel–Lindau (VHL) tumor suppressor and the stress protein JNK are also substrates of the APC/C. Microtubule inhibitors are well-known anticancer drugs which induce mitotic arrest by activating the spindle assembly checkpoint and inhibiting activity of APC/C. The tumor suppressor PTEN, discussed in Chapters 5 and 7, restrains PI3K–AKT pathway activation by dephosphorylating PI3. However, PTEN also has a nuclear role whereby it promotes formation of an APC/C–CDH1 tumor suppressor complex, which regulates key downstream targets such as PLK1 and aurora kinases.

BCR ligase

Also known as CRL4 E3 ligase, BCR ligase is structurally related to SCF and comprises scaffold proteins (Cul4A or Cul4B), Ddb1 (damage-specific DNA binding protein-1 analogous to Skp1), substrate recruiting factors DCAFs (Ddb1 and Cul4 associated factors), and the RING finger protein (Rbx1/2) that recruits the E2 ubiquitin-conjugating enzyme (UBC). DCAFs serve the same function as the F-box proteins in the SCF. The BCR ligase regulates stability of cyclin E. Given that overexpression of cyclin E is a common cause of genetic instability and aneuploidy in cancers, it is not surprising that deletion of BCR components might produce deregulated proliferation. BCR may also help prevent genomic instability by directing degradation of Cdt1, a replication initiation protein that recruits minichromosome maintenance proteins (MCMs) to replication origins. Failure to degrade Cdt1 can result in DNA re-replication and triggering of the G_2/M checkpoint. MCMs are described later.

UbcH10 E2 ligase

UbcH10 E2 ligase is overexpressed in many human cancers, and when overexpressed in transgenic mice leads to accelerated degradation of cyclin B by the APC/C. This is associated with serious abnormalities such as centriole reduplication, lagging chromosomes, and aneuploidy, and a range of spontaneous tumors. UbcH5c, a member of the UbcH5 family of E2 ligases, earmarks IκB, TP53, and cyclin D1 for proteasomal degradation. Interestingly, UbcH5c expression is inhibited by the metastasis regulator protein SLUG through chromatin remodeling. Other enzyme complexes have been identified.

With inhibitors of the proteasome pathway already in clinical use, it is predicted that future therapeutic developments will target specific ubiquitin ligases and thus more specifically interfere with this pathway, for instance by preventing degradation of tumor suppressors but increasing that of oncogenes.

Regulation by transcription

"Festina lente" (make haste slowly).

Augustus

Again, maybe a brief reminder will be helpful. Transcription, step one in gene expression, is, as may have been anticipated, controlled by transcription factors, ably assisted by an army of binding partners, cofactors, RNA polymerases, and chromatin-modifying proteins. However, it now seems clear that the already swollen ranks of the transcriptional "regulars" must accommodate a series of new recruits before the mRNA is translated into protein, including microRNAs and alternative splicing (see Chapter 11).

Several transcription factors are central operators in the cell cycle and many will be discussed later in the book. The oncoprotein c-MYC features in Chapter 6, as it has important actions that extend well beyond driving of the G_1/S transition. *TP53*, one of the cardinal tumor suppressor genes, is covered in Chapter 7. Here we will focus on a few key transcription factors that operate predominantly in the cell cycle.

Arguably, the highly conserved E2F family are among the most important transcriptional regulators of cell-cycle progression, because they are required for DNA synthesis in S phase. Moreover, they are now known to also play key roles in apoptosis and differentiation. As already discussed, E2F transcription factors are held in check by the RB tumor suppressor, an interaction which defines the G_1/S transition. Typically, E2F proteins regulate gene expression by forming a heterodimeric complex in association with a dimerization partner (DP) needed for promoter binding.

Six classical E2F proteins have been described in mammals (E2F1 to E2F6), which all share a DNA-binding domain (DBD) and a dimerization domain for DP interaction. These proteins have different actions, with E2F1 to E2F3 being transcriptional activators, whereas E2F4 to E2F6 are suppressors in mammals.

Playing for the other team we have the "pocket proteins," pRB, p107, and p130, which oppose E2F activity by shielding of the transactivation domain and recruiting chromatin-modifying proteins such as histone deacetylases, SWI/SNF, and CtBP–polycomb complexes, methyltransferases, and other corepressors to gene promoters. Under some circumstances the activator E2Fs may act as suppressors, which may in part be dictated by interactions with RB, but may also be determined by epigenetic context.

Hyperphosphorylation of RB by CDKs allows RB to dissociate from E2F, thus liberating the protein for transcription of E2F target genes. E2F-regulated genes include many important for DNA synthesis and cell-cycle progression, including cyclins, CDKs, prereplication complex factors, such as MCMs and Orc6, and DNA synthesis genes, including DNA polymerase, topoisomerases, dihydrofolate reductase (DHFR), and thymidylate synthase (TS). E2F activators are expressed at high level at the G_1/S transition, whereas repressors are expressed at pretty constant levels throughout the cell cycle. E2F1 may also be regulated by other factors. Two further genes with some similarities to those encoding classical E2Fs have been found in mammalian genomes, *E2f7* and *E2f8*, whose protein products, referred to as atypical E2Fs, are repressors (like E2F4–E2F6) involved in preventing mitotic progression in endoreduplicating cells. Recent studies support an interesting interaction between E2F7 and E2F8 and E2F1. The former may reduce levels of E2F1 expression, in turn lower levels of E2F1 may activate DNA repair genes in favor of those needed for S phase. In other words, E2F7 and E2F8 may act in oncogenic stress or DDR checkpoints. In fact, several recent studies have revealed a key role for E2F target genes beyond the G_1/S transition and DNA replication. Thus, E2Fs may help regulate DNA repair as far into the cycle as G_2 and even during mitosis. Different E2Fs have different roles and interact with specific transcription factors and DPs in G_1/S and G_2/M.

As we will discuss in more detail in Chapter 6, mitogenic proteins may also be powerful inducers of apoptosis – oncogenic stress – and this is also true of E2F1. Although the other activator E2Fs may also induce apoptosis if deregulated, this may ultimately be mediated through E2F1. E2F1 induces transcription of multiple proapoptotic genes, including *ARF, Atm, Nbs1, Chk2, p73, APAF1, Casp3, Casp7, Bnip3, Bok, Bim, PUMA, Noxa,* and *Hrk/DP5* and can downregulate the antiapoptotic gene *Mcl1*. Some of these may link E2F1 and the DDR and p53 pathway.

E2F1 contributes to the mitotic checkpoint by affecting transcription of a number of key regulators of the APC/C. Thus, a variety of genes encoding APC/C inhibitors, such as *MAD2, Emi1,* and *BubR1,* are E2F targets. MAD2 has a vital role in chromosomal stability and can arrest cell-cycle progression in metaphase, because by inhibiting APC/C activation, target cyclins and securin are no longer degraded. Thus, in cancer a combination of E2F-mediated transcription and inappropriate protein stability collaborate in promoting genomic instability when RB is inactivated.

As will be discussed for c-MYC later, it appears that activation of the PI3K–AKT survival pathway is required to buffer against the proapoptotic effects of E2F1, though level of activation is likely also important. Another factor may be related to DP interactions. Thus, proapoptotic function of E2F1 is also increased by binding of Jab1/Csn5, which does not promote proliferation. A newly described E2F-associated phospho-protein (EAPP) appears to be a key DP for activating E2Fs. EAPP can activate $p21^{CIP1}$ following DNA damage and retard apoptosis. RB is a key player in the DDR, in part by helping regulate expression of E2F targets needed for DNA repair, but also by initiating and maintaining growth arrest and even senescence.

RB can induce cell-cycle arrest after DNA damage by interacting with the zinc finger-containing transcriptional repressor ZBRK1, which inhibits *GADD45A*. The *ZBRK1* promoter contains a binding site for E2F1 which forms an inhibitory complex with chromatin-remodeling proteins CtIP and CtBP to prevent expression. RB can also regulate a number of key differentiation-regulating transcription factors in either a positive or negative way, depending on developmental context. In fact, pRB may determine cellular fate in mesenchymal tissue development by influencing choice between bone and brown adipose tissue *in vivo*.

Various genes regulated by the p53 tumor suppressor may influence cell cycle. $p21^{CIP1}$ has already been discussed, but there are numerous others (see also Chapters 7 and 10). In response to various forms of cellular stress, including DNA damage, and aberrant mitogenic signals, p53 becomes transcriptionally active to promote either apoptosis (see Chapter 8) or growth arrest (see Chapter 9), which together represent the main tumor suppressor actions of p53. Induction of G_1/S cell-cycle arrest is largely attributed to upregulation of $p21^{CIP1}$ and possibly also to the transcriptional repression of c-MYC. In fact, it is becoming increasingly apparent that DNA damage-induced cell-cycle arrest depends on downregulation of MYC. In a recent study from the laboratory of Martin Eilers, MYC downregulation during a DDR was achieved by a p53-independent signaling pathway whereby MAPKAPK5 phosphorylated FoxO3a, allowing translocation of this transcription factor to the nucleus, where it induced expression of miR-34b and miR-34c, which inhibit c-*Myc* expression. As MAPKAPK5 expression is directly activated by MYC, this forms a novel negative feedback loop; one which may be inactivated in human colon carcinomas, where MAPKAPK5 expression is reduced.

Induction of the G_2/M DNA damage checkpoint involves a complex web of interacting proteins involved in growth arrest, DNA damage sensing and repair (see Chapter 10), all of which may influence the cell cycle (cyclin B, cdc2, cdc25C) or specifically mitosis (topoisomerase II, B99/Gtse-1, and MAP4).

As mentioned above, the INK4/ARF locus encodes three tumor suppressors, $p15^{INK4b}$, $p16^{INK4a}$, and ARF, which regulate both RB and p53 pathways. In fact, this is such a key growth inhibiting "hub" that it may be no great surprise that INK4/ARF is also among the most frequently inactivated gene loci in human cancers. Studies from the Centro Nacional de Investigaciones Oncológicas (CNIO) in Madrid have identified a DNA-replication origin at the INK4/ARF locus that is regulated by a multiprotein complex containing Cdc6 as well as Orc2 and MCMs (minichromosome maintenance). Cdc6 is overexpressed in human cancers, where, in addition to effects on cell cycle, it also results in transcriptional repression of the INK4/ARF locus by promoting recruitment of histone deacetylases. Cdc6 can also cooperate with oncogenic RAS in tumorigenesis. ARF also has additional p53-

independent functions, including the ability to inhibit gene expression by a number of other transcription factors, including c-MYC, by promoting the sumoylation of several ARF-interacting proteins.

Forkhead transcription factors contribute to cell-cycle regulation and may in turn be regulated by cell-cycle proteins. FoxM1 is an important regulator of gene expression in the G_2 phase of the cell cycle and in this situation is activated by cyclin A–Cdk2. FoxM1 transcriptional activity is also required for recovery from DNA damage–induced G_2 arrest and checkpoint recovery. Cdk activity regulates FoxM1 activity and expression of promitotic targets such as cyclin A, cyclin B, and Plk1. FOXK2, a transcriptional repressor, is progressively phosphorylated by CDK1–cyclin B and also by CDK2–cyclin A during the cell cycle.

Transcription factors of the SP/KLF family are often underexpressed in those tumors that also have mutations in *KRAS*. To date, 24 members of the SP/KLF family have been identified and have diverse functions. Of these, SP5, SP8, KLF2, KLF3, KLF4, KLF11, KLF13, KLF14, KLF15, and KLF16 all inhibit the cell cycle and can inhibit the activity of oncogenic KRAS. These actions, best studied for KLF11, result from inhibitory interaction with the cyclin A promoter and result in cells arresting in S phase.

The CCAAT-binding transcription factor NF-Y also regulates expression of cell-cycle genes, such as those encoding cyclin A, cyclin B1, cyclin B2, cdc25A, cdc25C, and cdk1. NF-Y directly binds the E2F1 promoter and may also promote apoptosis mediated by p53–E2F1 in this way.

MicroRNAs and the cell cycle

Posttranscriptional regulation of proteins is also important and will be discussed in depth in Chapter 11. MicroRNAs (miRNAs) are molecules of single-stranded RNA around 20–23 nucleotides in length. They act by reducing the stability of mRNAs in essentially all major cellular processes. Targeting resides in the specific base-pairing interactions between the 5′ end ("seed" region) of the miRNA and sites within coding and untranslated regions (UTRs) of mRNAs, particularly at the 3′ UTR. The last decade has witnessed a remarkable increase in knowledge regarding the role of noncoding RNAs; given that miRNAs in particular appear to play key role in essentially all areas of cell biology and tumorigenesis, it will come as no surprise that they will feature prominently throughout the book.

The miRNAs are important components of the p53 transcriptional network. miR-34a, commonly deleted in human cancers, is a transcriptional target of p53 and, like more traditional p53-regulated gene products, such as Noxa, promotes apoptosis. In fact, genes that are regulated by miR-34a extend beyond apoptosis and include those involved in cell-cycle progression, DNA repair, and angiogenesis. Inactivation of miR-34a strongly attenuates p53-mediated apoptosis in cells exposed to genotoxic stress, whereas overexpression of miR-34a mildly increases apoptosis. Altered miR-34a expression is seen in some human cancers, and could conceivably contribute to tumorigenesis by impeding p53-dependent apoptosis. A large number of miRNAs can impact on the cell cycle, and the complexity of this is exemplified by Fig. 6.8 that shows just the subset currently known to regulate MYC functions.

miR-302 can promote cell proliferation by silencing of p21 (*CDKN1A*) in mice. In humans, however, miR-302 simultaneously inhibits both the cyclin E–CDK2 and cyclin D–CDK4/6 pathways to inhibit G_1/S cell-cycle transition, but does not appear to silence $p21^{CIP1}$.

Chromatin

This area is covered in depth in Chapter 11, so will only be touched upon briefly here. Essentially, chromatin compaction influences everything to do with DNA; chromatin is highly condensed during the onset of mitosis and much less so during S phase. Histone modification and transcriptional repression by deacetylation may direct this condensation. The Gcn5-containing histone acetyltransferase complex ATAC (Ada two A containing) localizes to the mitotic spindle and controls cell-cycle progression through direct acetylation of cyclin A–Cdk2, which is then targeted for degradation.

DNA replication and mitosis

Wisely, and slow. They stumble that run fast.

William Shakespeare

G_1/S transition

Once a cell progresses past a certain point in late G_1 termed the **restriction point** (analogous to start in yeast), it becomes irreversibly committed to entering S phase and replicating DNA. Once the cell has passed the restriction point, cell-cycle progression no longer depends on stimulation by external growth factors (see Chapter 5). Cyclins, CDKs, and RB are all important regulators of the restriction point. The transit of cells from G_1 to S phase is regulated by both proto-oncogenes and tumor suppressors and will therefore also feature strongly in Chapters 6 and 7.

Prior to the restriction point, accumulation of $p16^{INK4a}$ and $p27^{KIP1}$ can bind and inhibit cyclinD–CDK and cyclin E–CDK activity, arresting cells in G_1 – the **G_1/S checkpoint**. The G_1/S transition is promoted by mitogenic signals through several mechanisms. Sequentially, activation of key growth regulators such as the transcription factor c-MYC first induces expression of G_1-cyclins (cyclin D) and at least one of its partners CDK4 (remember c-MYC also drives downregulation of $p27^{KIP1}$ by increasing expression of cullin 1 and CKS1B, which degrade the protein and through cyclin D1, which sequesters $p27^{KIP1}$ protein and may also inhibit transcription of *CDKN1B*, the gene encoding $p27^{KIP1}$, although the latter may be much less prominent). Cyclin D–CDK4/6 constitutes an active kinase complex and phosphorylates distinct substrates, such as the tumor suppressor retinoblastoma (RB) pocket proteins (see Fig. 4.5). In the unphosphorylated state, RB binds to E2F transcription factors preventing the expression or regulation of several genes required for DNA replication, such as that encoding DNA polymerase.

The repression of target genes due to sequestration of E2F transcription factors is now thought to be removed in stages, with the low-level phosphorylation of RB by cyclin D–CDK4 (or cyclin D–CDK6) allowing limited transcription of a subset of E2F targets, including cyclin E. In a positive-feedback loop, subsequent further phosphorylation of RB (**RB hyperphosphorylation**) is completed by cyclin E–CDK2 complexes, an essential step in enabling full activation of E2F target genes needed to promote entry into S phase. This step is under the inhibitory control of the CKI $p27^{KIP1}$. RB hyperphosphorylation is a key event in cell replication as it allows full transcriptional activation of E2F target genes, such as those for S-phase DNA synthesis. In normal cells, RB phosphorylation is mediated by cyclin D1–CDK4 in response to growth factor stimulation. However, mutations inactivating RB or the RB pathway are ubiquitous in cancer cells, resulting in inappropriate cell cycling. Thus, the CDKN2A gene encoding the p16^{INK4a} CKI is inactivated in many human cancers, thus removing an inhibitor of cyclin D–CDK4. Similarly, genes encoding CDK4 or cyclin D1 are amplified in some glioblastomas and breast cancers, all of which may inappropriately promote replication.

Leaving G_0
In contrast to the G_1/S, less is known about regulation of the G_0/G_1 transition though both involve phosphorylation of RB. In fact, RB inactivation is sufficient for cells growth arrested in G_0 to re-enter the cell cycle. A recent study suggests that RB phosphorylation can be mediated by cyclin C–CDK3 complexes during the G_0/G_1 transition, in similar fashion to the action of cyclin D–CDK4 for S-phase entry.

Licensing and replicating DNA in S phase
For obvious reasons, DNA must be replicated only once per cell cycle in order to avoid genome instability. This delicate operation, which may take more than 8 hours to complete, is strictly policed and enforced by a variety of controls impinging on the assembly of the prereplication complex (pre-RC), a process known as **replication licensing**. As these pre-RC will serve as origins for DNA replication it is not surprising that these should be a central site for the action of a series of regulatory proteins that can inhibit or promote initiation or "firing" of DNA synthesis. If these processes fail then the cell karyotype may become fatally altered.

Placed during mitosis or early G_1, the pre-RCs mark specific chromatin regions for loading of helicase, some, but not necessarily all, of which will in turn become activated for DNA replication in S phase. The permission to begin DNA replication is granted by the sequential assembly of the various components of the pre-RC. In yeast, the earliest step appears to be the loading of two proteins, Cdt1 and Cdc6, synthesized during G_1 onto the origin recognition complex (ORC) that leads to recruitment and binding of the all-important heterohexameric minichromosome maintenance protein complex (MCM2–7) to the pre-RC, after which we may regard the origins for DNA replication as licensed. Three further proteins are involved in human cells, MCM10, ORC2, and AND-1, with the former playing an important role in replication initiation by recruiting DNA polymerase alpha. The MCM2–7 license is loaded onto chromatin in early G_1 phase and will act as a helicase during DNA replication following activation in S phase.

A potentially calamitous further round of replication is avoided by the sequential removal of licensing proteins. Thus, Cdc6 is degraded or exported from the nucleus after DNA helix unwinding begins, followed by Cdt1 after recruitment of DNA polymerase. Once replication has taken place the MCM2–7 license is revoked. It is worth noting that MCM2–7 can remain bound if replication is temporarily arrested by a DDR. Various other protein components of the licensing factor have been identified, and it is perhaps misleading to regard this as a single license; maybe a series of overlapping permits would be more apt.

A number of proteins contribute to the overall regulation of this key process. As discussed, Cdt1 helps recruit MCM2–7 and Cdc6 to the pre-RC. Conversely, another protein, geminin, inhibits licensing and stops re-initiation of DNA replication by blocking reloading of MCM2–7, at least in part by inhibiting Cdt1.

Once licensed, DNA replication itself is driven by a series of kinases and other proteins. Activation at the G_1/S transition is driven by S-phase cyclin-dependent kinases (CDKs) and the Cdc7/Dbf4 kinase (DDK) that recruits Cdc45 to activated origins. It is not fully understood why some origins fire and why those that do fire do so in a particular sequence at different times after onset of S phase. Replication appears to start close to areas of the genome that recruit transcription factors adjacent to RNA polymerase II-binding sites.

Three types of kinase are responsible for controlling the initiation of DNA replication: CDKs, DDK, and the DNA damage checkpoint kinases. Interaction between CDK–cyclins and licensing factors is essential for S phase; cyclin E–CDK2 can phosphorylate and stabilize the licensing factor Cdc6 by preventing APC/C ubiquitination. Intriguingly, this enables Cdc6 to accumulate before the licensing inhibitors geminin and cyclin A (also APC/C substrates), thus establishing a "window of time" prior to S phase when pre-RCs can assemble. CDC6 can interact with ORC1, and a key role in human cell cycle regulation is evidenced by the severe phenotype of Meier–Gorlin syndrome type 5, caused by inherited defects in CDC6. This syndrome is associated with intra-uterine defects, including aplasia/hypoplasia of the patellae, and severe growth retardation with short stature, and poor weight gain. Under the influence of E2F transcriptional activity, rising levels of S-phase proteins (also referred to as S-phase promoting factor – SPF), including cyclins E and A bound to CDK2, promote duplication of DNA and centrosomes. In fact, on CDK2 activation, the initiation factors Sld2 and Sld3 are phosphorylated and can interact with Dpb11. As a result DNA helicases and polymerases are recruited to unwind the double helix and to replicate DNA.

DDK phosphorylates subunits of the MCM2–7 helicase. As DNA replication proceeds, cyclin E is destroyed, and the level of mitotic cyclins rise in G_2. Importantly, CDK2 levels remain elevated from the G_1/S transition through to M phase, when chromosomes have segregated. Together with another protein, geminin, elevated CDK2 levels play absolutely critical roles in preventing the reloading of MCM2–7 and formation of new pre-RC during the S, G_2, and M phases. During G_1 phase, both CDK and DDK are downregulated, which allows origin licensing and prevents premature replication initiation. CDKs are therefore key factors that restrict DNA replication origin firing to once per cell cycle by preventing the assembly of pre-RC outside of G_1 phase. Thus, DNA gets replicated once only in each cell cycle. If this process is impeded then one can get relicensing of already fired replication origins and re-replication of parts of the genome that have already been replicated – resulting in increased gene copy number. Resultant DNA damage or replication fork stalling will activate the DDR and checkpoint kinases will inhibit origin firing (intra-S-phase checkpoint) (see below and Chapter 10).

Faults can be detected by monitoring of various products or damage that will trigger the activation of checkpoints. Thus, for example, the presence of Okazaki fragments is monitored on the lagging strand during DNA replication and cycling is arrested until these have disappeared. However, failure to activate these responses will result in propagation of cells with altered gene copy numbers – a frequent cause of increased oncogene activity in cancer. In order to complete lagging-strand DNA synthesis, around 50 million Okazaki fragments have to be processed for each cell cycle. RNA/DNA primers are removed and Okazaki fragments joined into an intact lagging strand by the sequential activity of polymerase δ, structure-specific nucleases such as FEN1, and DNA ligase I. Mutations that prevent complete removal of the RNA primers can generate unligated nicks and DNA double-strand breaks (DSBs).

Translocations disrupting the MLL gene cause aggressive leukemias. Degradation of MLL by SCF (Skp2) E3 ligase is blocked by ATR activation in the intra-S-phase checkpoint. Stabilized MLL stalls the cell cycle by methylating histone H3 lysine-4 at late replication origins, preventing loading of CDC45. MLL fusions in leukemia act as dominant negatives to block ATR-mediated stabilization of wild-type MLL in the DDR and impede the intra-S-phase checkpoint.

An origin-activation checkpoint

Problems in initiation of DNA replication may trigger arrest of the cell cycle in G_1, a potential origin activation checkpoint. This process is regulated by different pathways that converge on the transcription factor FoxO3a. FoxO3a activates the ARF–HDM2–p53 pathway, upregulates $p15^{INK4B}$ and activates the Wnt–β-catenin signaling antagonist Dkk3, thereby downregulating Myc and cyclin D1. Together, the Rb–E2F pathway is inhibited and the G_1/S transcriptional program is aborted. Inhibition of CDC7 can provoke impaired initiation of DNA replication and activate this checkpoint.

Mitosis: the spindle and chromosome segregation

Once they have been successfully replicated, genomes must be correctly partitioned during cell division, because failures here will lead to chromosomal instability (one or other cell will have too few or too many chromosomes or parts thereof). Accurate partitioning depends on the successful attachment of chromosomes to the opposing poles of mitotic spindles, a process described as **bi-orientation**. The mitotic spindle is a highly dynamic structure, comprising microtubules originating from the centrosomes, which provides the structural framework for chromosome segregation in cell division.

If all goes to plan, then during mitosis, cells will undergo an orderly sequence of centrosome maturation, chromosome condensation, nuclear envelope breakdown, centrosome separation, bipolar spindle assembly, chromosome segregation, and cytokinesis. Mitosis is initiated and maintained under the influence of activation of the key mitosis-regulatory kinase CDK1 (cyclin B–CDK1). Conversely, mitosis is inhibited by the ATR–CHK1 kinase pathway and by protein phosphatases, such as PP2A, which prevents activation and also downstream functions of CDK1. PP2A indirectly prevents activation of CDK1 by dephosphorylation and inactivation of the kinases Wee1 and Myt1 and the phosphatase CDC25; CDK1 functions are antagonized by dephosphorylation of its mitotic substrates. PP2A is itself inhibited by greatwall orthologs and their targets, among which α-endosulfine and cyclic adenosine monophosphate-regulated phosphoprotein 19 (Arpp19) appear most prominent. Transcriptional repression through chromatin remodeling also contributes to inhibition of mitosis and indeed most everything else (see Chapter 11).

The so-called M-phase-promoting factor (comprising mitotic cyclin–CDK complexes: cyclin B or A–CDK1) initiates assembly of the mitotic spindle, breakdown of the nuclear envelope and condensation of chromosomes. These events take the cell to metaphase, at which point the M-phase-promoting factor and proteins such as CDC20 and CDH1/HCT1 activates the APC/C discussed earlier. Degradation of the protein securin then leads to separation and movement of sister chromatids to the poles (anaphase), completing mitosis.

To labor the point, accuracy of mitosis is essential for cell survival and is thus subject to tight control and is monitored by important checkpoints that can arrest mitosis if errors are detected. Errors in mitosis can lead to genomic instability, a major factor in progression and even initiation of cancers (see Chapter 10). As mentioned, microtubules of the mitotic spindle are the key drivers of chromosome segregation and during metaphase contribute to the "search and capture" of chromosomes for bipolar alignment on the spindle. The taxanes and vinca alkaloids, which target tubulin, perturb mitotic spindle microtubule dynamics and thus interfere with cell cycling and have proved successful in the treatment of a number of human cancers (see Chapter 18). Accurate chromosome segregation depends on proper assembly and function of the kinetochore and the mitotic spindle. Mitotic kinases are important in this process and also play a role in many cancers (*vide infra* and Chapter 10).

In early mitosis, the two sister chromatids of the chromosome are held together along their entire length by multisubunit complexes of cohesin proteins. Whereas during anaphase cohesion between sister chromatids is rapidly disrupted by the protease separin; sister kinetochores then separate and with an attached chromatid, move along the microtubules, powered by dynein minus-end motors, while the microtubules shorten. The overlapping spindle fibers move past each other (pushing the poles farther apart) powered by kinesin plus-end motors. Sister chromatids thereby end up at opposite poles of the spindle. Until anaphase, separin is part of an inactive complex with the protein securin. Anaphase starts once the APC/C covalently ubiquitinates securin, targeting it for proteasome degradation. Separin is then free to cleave its cohesin target and sister chromatids can separate. In mammalian cells, securin is targeted for APC/C ubiquitination during prophase through phosphorylation by polo-like kinase (PLK1) – see below. However, cohesin complexes remain at the centromere until anaphase.

Cancer cells often have more than the normal number (one or two depending on the stage of the cell cycle) of centrosomes as well as chromosomes. Mutations in *p53* (also referred to as the *TP53* gene) predispose the cell to excess replication of the centrosomes. Each centrosome contains a pair of centrioles, made up of a cylindrical array of nine triplets of microtubules. Each centriole is duplicated during G_1/S transition. Centrioles appear to be needed to organize the centrosome in which they are embedded.

Anaphase must not begin until all chromosomes are correctly attached to the spindle or mitosis may produce cells with altered chromosome content (ploidy). Monitoring of this critical stage falls to the spindle checkpoint (see below). We will now concentrate on two important aspects of mitosis regulation – the role of centrosomes and the mitotic kinases.

The centrosomes and the mitotic kinases

Centrosomes in mammalian cells each contain a pair of centrioles and are responsible for organizing of microtubules and directing the construction of the bipolar mitotic spindle in the cell cycle. Centrosome duplication is tightly coupled to DNA replication in part because both are triggered by cyclin E–Cdk2 activated in late G_1. It is crucial that centrosomes are only duplicated once, because abnormal numbers of centrosomes can be catastrophic and result in aberrant chromosome segregation and genomic instability that can drive tumorigenesis.

The centrosome is attached to the outside of the nucleus and duplicates just prior to mitosis. The two centrosomes then move apart to opposite sides of the nucleus (Fig. 4.2). As mitosis proceeds, clusters of microtubules, called **spindle fibers**, grow out from each centrosome with plus ends growing toward the metaphase plate, and play a key role in the assembly of the chromosomes at metaphase. Spindle fibers growing from opposing centrosomes attach to one of the two **kinetochores** (protein complex assemblies on the centromere or "waist" of condensed metaphase chromosomes in higher eukaryotes) of each sister chromatid pair (dyad), and some bind to the chromosome arms. The two kinetochores thus capture microtubules emanating from opposite spindle poles to produce the so-called **metaphase plate** in which all the chromosomes are bilaterally attached and aligned at the equator of the spindle.

Fibers also extend from the two centrosomes in a region of overlap. Microtubules attached to opposite sides of the dyad shrink or grow until they are of equal length. Microtubule motors attached to the kinetochores move them toward the minus end of shrinking microtubules (**dynein**) and toward the plus end of lengthening microtubules (**kinesin**). The chromosome arms use a different kinesin to move to the metaphase plate.

Centrosomes are now known to be more generally important in cell-cycle control. In fact, G_1/S and G_2/M transitions require a functional centrosome replete with a variety of interacting and colocalizing proteins, including cyclin–Cdk complexes.

Cyclins A and E both localize to centrosomes, which may be a common feature of proteins with modular centrosomal localization signal (CLS) motifs, providing intimate connections between the nuclear and centrosomal cycles. In fact, the interaction of DNA replication factors with the centrosome is needed to prevent centrosome overduplication. The location of the CLS motif in different cyclins may influence regulation by other cell-cycle proteins. Thus, the cyclin A CLS interacts with p27^{KIP1}, allowing dissociation of cyclin A but not cyclin E from centrosomes. Cyclin E-Cdk2 regulates both DNA replication and centrosome duplication during the G_1/S transition. Moreover, centrosome disruption can block S phase, as can preventing interaction of cyclin E–Cdk2 with the centrosome. CLS also appears important for the loading of essential DNA replication factors such as Cdc45 and proliferating cell nuclear antigen onto chromatin. The DNA replication factors MCM5 and Orc1 (a subunit of the origin recognition complex (ORC) that is a key component of the DNA replication licensing machinery) can prevent centrosome reduplication. Both cyclin A and cyclin E can bind and localize MCM5 and Orc1, and may function sequentially in order to prevent centrosome and centriole reduplication throughout interphase. Nucleophosmin (NPM/B23) is phosphorylated by CDK2 and helps to trigger centrosome duplication by activating Rho-associated kinase (ROCK-II).

The formation and function of the mitotic spindle during M phase of the cell cycle is regulated by protein phosphorylation, involving multiple phosphatases and mitotic kinases. The mitotic kinases are a group of highly conserved serine/threonine kinases that include CDK1, PLKs, aurora/Ip11-related kinases, NIMA-related kinases, and the WARTS/LATS1-related kinases. The combined activities of three of these kinases appear to be of central importance in steering the cell cycle: CDK1–cyclin B, PLK1, and aurora kinases. The former are discussed extensively elsewhere, so here we will concentrate on the other two.

The aurora mitotic kinases

Three classes of aurora kinases have been identified in humans: aurora kinase (AURK)-A, -B, and -C. Aurora-A (AURKA) and AURKB are expressed in most normal cell types and are involved in cell-cycle progression from G_2 through to cytokinesis, though with different location and timing of activation during the cell cycle. AURKA is a serine/threonine-specific protein kinase important for spindle assembly. Chromosome biorientation at prometaphase is driven by the chromosomal passenger complex (CPC), composed of a pivotal catalytic kinase, AURKB, and various regulatory elements, such as inner centromere protein (INCENP), survivin, and borealin. Correct localization of the CPC at the inner centromere (the center of paired kinetochores) is essential for phosphorylation of kinetochore substrates; this is largely due to Cdk1-dependent phosphorylation of borealin (survivin in yeast), which allows interaction of the complex with centromere adaptor proteins, such as shugoshin. This is by no means the only factor in directing the CPC to the correct site; histone kinase regulation with phosphorylation of histones H3-threonine 3 (H3-pT3) and 2A-serine 121 (H2A-S121) by haspin and Bub1, respectively, are also important. In concert with the previously described PP2A, shugoshin 1 prevents cohesin from dissociation during prophase. Various other important AURKA and AURKB substrates may help protect centromeres, including:

- the outer kinetochore proteins – KNL1/Mis12 complex/Ndc80 (KMN) complex (key controller of kinetochore–microtubule attachments),
- shugoshin 2, which interacts with PP2A and MCAK,
- the kinetochore motor CENP-E, MYBBP1A,
- the telomeric protein TRF1, and
- the spindle regulator TPX2.

Aurora kinases and protein phosphatases have opposing actions during mitosis. Phosphorylation of CENP-E by AURKA is crucial for towing of initially polar chromosomes toward the cell center, an effect antagonized by phosphatases. Another important regulatory complex operating during spindle assembly in mitosis is the microtubule depolymerase mitotic centromere-associated kinesin (MCAK). MCAK is activated by AURKB and by PLK1-mediated phosphorylation.

AURKA plays a central role from late S phase through M phase, with peak activity in prometaphase, regulating among others, centrosome maturation and mitotic entry and various roles in bipolar spindle assembly and chromosome alignment through to mitotic exit. AURKA is activated by phosphorylation, involving TPX2 and also an autophosphorylation amplification loop with ajuba proteins, and inactivated by phosphatases. MST1 (mammalian sterile 20-like kinase 1) can limit activity of aurora B and is needed for accurate kinetochore–microtubule attachment. MST1 directly phosphorylates aurora B and inhibits kinase activity.

AURKB is a passenger protein involved in chromosomal segregation and cytokinesis, with expression peaking at G_2/M. As discussed earlier, AURKB acts together with various proteins including INCEP, survivin, and borealin.

The polo-like mitotic kinases

Polo-like kinases (PLKs) are Ser/Thr kinases involved in several aspects of cell-cycle regulation, including directing mitotic entry. To date, four mammalian PLKs are described – PLK1, PLK2, PLK3 (FNK), and PLK4 – all of which have distinct roles in the cell cycle. PLK1 operates largely during G_2, PLK2 during G_1, PLK3 in the DDR, and PLK4 during M-phase progression. PLK1 helps to coordinate centriole duplication with the cell replication cycle and promotes maturation of the procentriole. Centriole duplication is mediated via another polo-like kinase, PLK4, which interacts with the centrosome protein Cep152. PLK1 also regulates several other processes required for G_2/M transition, including nuclear envelope breakdown.

PLK1 is restricted to dividing cells, with an essential role during M phase, by phosphorylating cyclin B1 and Cdc25C. As already discussed, PLK1 is also important in centrosome maturation, microtubule dynamics, and the spindle checkpoint, where it has several key cancer-relevant targets including the APC/C inhibitor, early mitotic inhibitor 1 (EMI1), ninein-like protein (Nlp), and translationally controlled tumor protein (Tctp).

Oncogenic *PLK1* is overexpressed in many human tumors, including breast, gastrointestinal tract, non-small-cell lung, and ovary and correlates with adverse prognosis. PLK1 likely contributes to tumor progression by bypassing key checkpoints, leading to chromosome instability and aneuploidy. PLK1, by activating CDK1, is also an essential part of a "restart" mechanism following repair of damaged DNA at the G_2/M transition.

Although all PLKs share two conserved elements, the N-terminal Ser/Thr kinase domain and a highly homologous C-terminal region termed the "polo-box motif," their functions differ markedly. PLK1 contributes to anaphase entry by phosphorylating target proteins such as securin that are then recognized by the APC/C. PLK1 is normally inhibited by various checkpoints in the cell cycle, whereas PLK2 and PLK3 are activated. Deregulation of PLK1 activity contributes to genetic instability, which in turn leads to oncogenic transformation. In contrast, PLK2 and PLK3 are involved in checkpoint-mediated cell cycle arrest to ensure genetic stability, thereby inhibiting the accumulation of genetic defects. Several interacting partners of PLK1 have been identified that are tumor suppressor gene products. During mitosis in yeast, Plk1 promotes anaphase progression by phosphorylating Cdc6, which can then in turn inhibit Cdk1 and release separase; interfering with this results in binucleated cells and incompletely separated nuclei.

Checkpoints – putting breaks on the cell-cycle engine

The first draft of anything is shit.

Ernest Hemingway

We will now concentrate on the all-important quality-control steps that monitor and ensure fidelity of genome replication and partitioning at each successive stage of the cell cycle and the ultimate sanctions at their disposal should these be imperfect and irredeemable.

To ensure the accurate transmission of fully replicated and undamaged DNA, the cell cycle is under constant surveillance. The cell has several options for arresting the cell cycle if something goes wrong – the so-called checkpoints (see Fig. 4.3). For instance, if a cell sustains DNA damage following exposure to radiation or chemical agents, or replication errors occur during the cell cycle, then the duplication of that cell might pose a potential cancer risk to the organism. Far better for the organism to temporarily halt the cell cycle and, if possible, effect a repair of the DNA before continuing and if not then to ruthlessly require the cell to eliminate itself (apoptosis) or to permanently exit from the cell cycle (senescence). There are various checkpoints throughout the cell cycle, emphasizing the critical role of quality control in cell cycle regulation. Thus, DNA damage checkpoints operate to prevent S-phase entry (G_1/S checkpoint), delay S-phase progression (intra-S-phase checkpoint), and prevent mitotic entry after DNA replication (a G_2/M checkpoint) (see Figs 4.3, 4.6, and 4.7). The purpose of these checkpoints is to allow phase-specific DNA repair.

The repair processes are equally numerous, if not readily memorable, and are described in more detail in Chapter 10. The main ones invoked during DDR include base excision repair (BER), nucleotide excision repair (NER), mismatch repair (MMR), and, specifically for the hard-to-fix DSBs, nonhomologous end joining (NHEJ) in S phase and homologous recombination (HR) in G_2 phase.

These checkpoints serve to detect DNA damage and stall G_1 or G_2 until the damage is repaired, or else trigger the elimination of the cell by apoptosis. In addition, mitotic (or spindle) checkpoints arrest the cell in metaphase if chromosomes are not properly aligned on the spindle prior to cell division (cytokinesis). As mentioned above, various CKIs are instrumental in allowing the cell to stall at a particular checkpoint, and these are highlighted in several figures in this chapter, including Figs 4.3, 4.6, 4.7, 4.8, 4.10, and 4.11.

A number of cell cycle regulators are defective in cancer, including key proteins such as P53, pRB (and related proteins, p107 and pRB2/p130), their pathways and various CKIs, such as $p15^{INK4b}$, $p16^{INK4a}$, $p18^{INK4c}$, $p19^{INK4d}$, $p21^{CIP1}$, and $p27^{KIP1}$. Importantly, all of these can arrest the cell cycle in response to DNA damage, an undesirable outcome for a cancer cell. Virtually all human tumors deregulate either the pRB or p53 pathways and in most cases both.

Recently, a further important cancer-restraining checkpoint has been identified, whereby deregulated expression of an oncogene (with or potentially even without accompanying DNA damage) triggers a stress response in the cell that is referred to as oncogenic stress. Several alternative checkpoint pathways may be involved downstream of oncogenic stress, including the stabilization/activation of the p53 tumor suppressor via either ARF or alternatively via a DNA damage–ATM/ATR- dependent pathway, or directly through activation of proapoptotic protein such as BIM (see also Chapters 8 and 10). What remains unclear is how the aberrant oncogene activity is sensed and distinguished from the normal cell-cycle-related actions of its non-deregulated proto-oncoprotein counterpart. For instance, why does normal c-MYC activation in a normal cell cycle not trigger ARF, whereas deregulated c-MYC does? One possibility is that the abnormal duration or persistence of a deregulated oncogene has different effects compared with pulsatile or oscillatory changes in expression or activity that might follow normal growth-factor-driven expression of the proto-oncoprotein, and that this activates ARF and/or causes DNA damage.

The DNA damage response (DDR)

> The difference between the right word and the nearly right word is the same as the difference between lightning and the lightning bug.
>
> *Mark Twain*

As already mentioned, we now know of three key checkpoints which a dividing cell must negotiate before being allowed to enter mitosis – the G_1/S, intra-S, and G_2/M cell-cycle checkpoints. The DDR involves activation of a kinase-based signaling network that rapidly inhibits CDC25 family proteins to arrest the cell cycle by inactivating cyclin–CDK complexes and recruits the DNA-repair machinery. There is also a slower and more measured transcriptional response that involves downstream targets such as p53, and $p21^{CIP1}$ that also inhibit CDKs.

These checkpoints are activated by various forms of damage-detecting sensors that respond to differing types of DNA damage and possibly alterations in chromatin structure. Thus, minor DNA damage and stalled replication forks are detected by the 9-1-1 complex (Rad9–Rad1–Hus1), whereas more serious DNA damage such as DSBs are detected by the MRN complex (Mre11, Rad50, and Nbs1). The source of such damage is very varied and includes many of the genotoxic mutagens described in Chapter 3. It is also a good time to note that DNA damage is caused by many of the traditional anticancer agents and the interaction between these agents and DDR is now being explored as a potential further treatment opportunity. Checkpoints allow the cell time to repair DNA damage before continuing the cell cycle, and much is now known about how this arrest is regulated; an area of great current interest is exactly how signaling pathways direct the restart of the cell cycle when effective repair has been completed. On the other hand, if the repair capacity is insufficient then the cell can succumb to the "slings and arrows of outrageous fortune" and undergo apoptosis.

The DNA damage sensors activate two critically important upstream PI3K-like kinases – ATM and ATR – and their respective downstream effector kinases, CHK2 and CHK1, which control all three DDR checkpoints. Recent studies have revealed a further checkpoint effector that works alongside CHK1, p38MAPK/MAPKAP-K2 (MK2). CHK1 is activated by a variety of insults, including single-strand DNA (ssDNA) breaks, replication stresses, stalled replication forks, and DNA-damaging agents, through both ATM and ATR. CHK2 on the other hand is specifically activated by DSBs via ATM. The repair of DNA damage is covered in detail in Chapter 10 and is intimately connected to what we are discussing here.

The checkpoints

> To insure the adoration of a theorem for any length of time, faith is not enough, a police force is needed as well.
>
> *Albert Camus*

Two checkpoint pathways involved in DNA damage have been described. In the first, DSBs activate the ATM (ataxia telangiectasia mutated) pathway, which can arrest the cell cycle in G_1 in a p53-dependent manner (discussed in Chapter 10 and Fig. 4.5). The second, involves the ATR (ATM- and Rad3-related) and CHK1 proteins, discussed in the next section that responds to "less extreme" DNA damage by inducing G_2 arrest, independently of p53.

The checkpoint signaling pathways can be categorized conceptually as sensors, transducers, mediators, and effectors. The **sensors** (MRN and 9-1-1 are the best known) detect DNA damage and recruit the **transducers** ATM and ATR to further activate further transducers CHK2 and CHK1 and also MAPKAP kinase 2 (MK2), respectively. ATM/ATR activation allows recruitment and activation of a plethora of mediators, including H2AX, 53BP1, Claspin, BRAC1, TopBP1, MDC1, SMC1, FANCD2, Timeless, and Tipin. These remain located to the site of damage while CHK1 and 2 are released to activate a downstream signaling cascade that will arrest the cell cycle at different phases by inactivation of CDC25 and/or activation of p53 effectors.

The **mediators** are becoming of great interest because they may represent druggable targets. Claspin directly binds to CHK1, allowing recruitment of BRCA1, TopBP1 directly activates ATR/ATRIP and promotes ATR-mediated CHK1 phosphorylation. Timeless and Tipin complex to RPA, which allows claspin to associate with and phosphorylate CHK1. Two major signaling pathways are activated by checkpoints: that leading from ATM/ATR to inhibition of CDC25s and a slower p53-dependent pathway, which may be preferentially activated by less reparable DNA damage. Numerous other **effectors** have been identified also and include a variety of kinases and repair proteins. CHK1 can activate DNA repair kinases such as DNA-PK, which partners Ku70-K80 in DSB repair. CHK1 phosphorylates FANCE and Rad51, which are essential steps in driving FA (Fanconi anemia)/BRCA-directed or alternatively homologous recombination repair (HRR), respectively.

G_1/S checkpoint

We have already considered the role of RB at the G_1/S checkpoint, but we must also reflect on the key contribution made by TP53 (Fig. 4.6), also discussed in Chapter 7. A key player in stress responses and in the DDR, TP53 indirectly senses DNA damage (see Chapter 10) and can either arrest the cell cycle in G_1 by an ATM-dependent expression of $p21^{CIP1}$, until the damage is repaired, or if repair is not possible can trigger apoptosis, via induction of various proapoptotic effectors (PUMA, BAX, NOXA) (see Chapters 7 and 8). A more rapid process is also in operation at this phase and involves inactivation of CDK2–cyclin E/A via Cdc25A-mediated dephosphorylation, which also involves CHK1.

Importantly, *p53* is the most frequently mutated tumor suppressor gene in human cancer. Cells lacking p53 can survive levels of DNA damage and mitogenic stimulation that would otherwise kill the cell and can enter and progress through the cell cycle, thus propagating DNA damage. The p53 protein is also activated via a complex system that responds to "oncogenic stress." Discussed in more detail in later chapters, this process broadly involves a series of processes activated in a cell in response to deregulated expression of oncogenes such as c-*Myc* or *Ras*, but intriguingly not to any significant extent during the activity of these proteins in normal growth. This also raises the important concept that cells can distinguish between "normal" cell cycles and "cancer" cell cycles. For example, in response to oncogenic c-MYC, the cell activates the $p19^{Arf}$ tumor suppressor, which in turn can both indirectly, via MDM2 and p53, and probably directly promote apoptosis or cell-cycle arrest.

Interestingly, recent work suggests that even early in the evolution of a neoplasm, oncogenic stress and aberrant cell cycles may often also result in DNA damage and activation of the ATR–CHK1 DDR pathway (see Chapter 10). The explanation for this is not clear, but at least in the case of c-MYC might arise through oxidative stress. Another key tumor suppressor pathway involves the RB pathway. The protein p16^{INK4a} inhibits CDK4 activity in G$_1$ and thereby induces cell-cycle arrest and may also contribute to replicative senescence in response to oncogenic *Ras* or other oncogenes. Once again, as seen for *p53*, the *INK4* locus is frequently altered by mutations that inactivate one or both of its protein products, p16^{INK4a} and p14^{ARF} (p19^{Arf} in the mouse) (see Chapter 7). Interestingly, the reason that oncogenic c-MYC is more likely to engender apoptosis, whereas RAS may promote senescence may be due to MYC inhibiting the CKI p21^{CIP1} that mediates growth arrest downstream of ARF–p53.

Intra-S-phase checkpoint

DNA replication is initiated at multiple origins throughout S phase and orderly progress requires that individual origins fire on cue and that replication forks progress unimpeded, while ensuring that genomic sequences are replicated once and only once per cell cycle. Replication forks can stall and collapse at sites of DNA damage and this must be recognized and remedial action undertaken to forestall genetic instability resulting from uncopied chromosomal regions. It is the responsibility of the intra-S-phase checkpoint to support replication forks and, if needed, to reduce the rate of DNA replication by delaying late origin firing and replication fork progress to allow DNA repair and avoid further damage. The checkpoint is activated by DNA damage, stalled replication forks, and DSBs in S phase. There are obvious parallels and overlap with the other DNA damage–regulated checkpoints.

This checkpoint is mainly controlled by the ATM and ATR kinases and CHK1, which phosphorylate and target CDC25A for degradation. CDC25A may also be downregulated transcriptionally by this checkpoint under the influence of ATF3. The ATR–CHK1 pathway is activated by single-stranded DNA, which binds a specific recognition protein (RPA). DSBs can also trigger this checkpoint following recognition of the lesion by the MRN complex (specifically by targeting of the NBS1 component to DNA damage by the protein MDC1) and then activation of the ATM–CHK2 pathway. The checkpoint can also be activated by an alternative route involving Nbs1 and Smc1 or FANCD2. Fork collapse can also be prevented by translesion DNA synthesis (TLS) as this allows replication forks to progress through some types of DNA damage. This is also regulated by CHK1.

G$_2$/M checkpoint

The G$_2$/M DNA damage checkpoint prevents the cell from entering mitosis (M phase) if the genome is damaged (Fig. 4.7). The cyclin B–CDK1(Cdc2) complex regulates entry into mitosis. During G$_2$, the kinases CHK1, WEE1, and MYT1 act directly or indirectly to phosphorylate CDK1 (Cdc2), thereby inhibiting activity. Cdc25 proteins activate cyclin/cyclin-dependent protein kinase complexes by removing inhibitory phosphates from conserved threonine and tyrosine groups. As M phase approaches, the CDC25 phosphatases are activated by the polo-like kinase PLK1. CDC25 then activates CDK1, in part by reversing inhibitory phosphorylation, thus establishing a feedback amplification loop that efficiently drives the cell into mitosis. In other words, at G$_2$ the activity of WEE1 family kinases exceeds that of CDC25 phosphatase. However, at M phase entry the situation is reversed, such that the activity of CDC25 exceeds that of WEE1.

As mentioned previously, DNA damage activates two parallel cascades (the ATM–CHK2 and ATR–CHK1). There is redundancy and overlap between these pathways, but broadly, among other actions discussed in Chapter 10, they negatively regulate cyclin B–CDK1 thus arresting the cycle; CHK kinases phosphorylate and inactivate CDC25, preventing both progression of S phase and entry into M phase. Phosphorylated Cdc25A is targeted by APC/C and SCF for proteasomal degradation and Cdc25C is excluded from the nucleus by binding to 14-3-3 proteins (Rad24 and Rad25). Again, if DNA damage is extensive then a sustained G$_2$/M-phase checkpoint is achieved by increased transcription of Cdk1 inhibitors such as p21^{CIP1}, Gadd45, and 14-3-3σ, via p53 or BRCA1. Chk1 also phosphorylates and stabilizes Wee1.

The ATM pathway is largely responsible for detecting major damage such as DSBs and arrests cells in G$_1$ through activation of the p53–p21^{CIP1} pathway. The ATR–CHK1 pathway is primarily involved in responding to less pronounced DNA damage, possibly triggered by stalling of the DNA polymerase during DNA replication. This process is complex, and for ATR to activate CHK1 likely requires several additional proteins, including those involved in detection of single-stranded DNA, the RAD9–RAD1–HUS1 complex (9-1-1) and others discussed in detail in Chapter 10. M-phase entry is also regulated by a p53-dependent pathway, triggered by a DDR or via stabilization of p53 by p19^{Arf}. Various p53 transcriptional targets are required for this inhibition of M phase including:

- 14-3-3σ that binds phosphorylated CDK1–cyclin B, promoting nuclear export,
- GADD45, which apparently binds to and dissociates the CDK1–cyclin B kinase, and
- p21^{CIP1}, an inhibitor of a subset of the cyclin-dependent kinases, including CDK1.

Intriguingly, the ATR–CHK1 pathway may be important in maintaining the survival of a cell with potentially repairable DNA damage. It is likely that this will in some way balance signals involving p53 and/or p19^{Arf}, which in addition to promoting cell-cycle arrest also promote apoptosis. CHK1 can also help restart stalled replication forks and may interact with DNA repair pathways. Faults can be detected by monitoring of various products or damage that will trigger the activation of checkpoints. Thus, for example, during S phase, the presence of the Okazaki fragments is monitored on the lagging strand during DNA replication and cycling arrested until these have disappeared.

Much is known, therefore, about how DNA damage triggers multiple checkpoint pathways to arrest cell-cycle progression. However, comparatively little is known about the mechanisms that allow resumption of the cell cycle once checkpoint signaling is silenced. In a recent study from Medema's group in Amsterdam, PLK1 was shown to be essential for mitotic entry following recovery from DNA damage. At least in part this effect was mediated by targeting of WEE1 for degradation at the onset of mitosis, thus favoring CDC25-mediated activation of CDK1.

The spindle checkpoint, centromeres, and kinetochores

Temptation is an irresistible force at work on a movable body.

H.L. Mencken

The metaphase-to-anaphase transition is regulated by multiple proteins that together comprise the spindle (mitotic) checkpoint, essential for ensuring the fidelity of chromosome segregation.

Activated by failure of kinetochore occupancy by microtubules or through lack of tension, this checkpoint delays anaphase by inhibiting the APC/C, and also the mitotic exit network required for exit from mitosis (Fig. 4.8). Conversely, failure of the spindle checkpoint is a major contributor to cancer as it can result in genomic instability and aneuploidy (see Chapter 10).

The centromere is a chromosomal locus required for the correct delivery of one copy of each chromosome to each daughter at cell division. During mitosis (and meiosis) the spindle attaches to chromosomes at the centromeres, which are critical for ensuring that the daughter cells inherit the correct number of correctly segregated chromosomes, and not the alternative – aneuploidy. Mammalian centromeres are substantially more complex structures than those in yeast, and centromeric repeats are among the most rapidly evolving DNA sequences across organisms. Centromeres contain a key protein, a variant of conventional histone H3 known as CENP-A. CENP-A forms a heterotetramer together with another protein, H4, which can interact with another human protein, HJURP, that localizes CENP-A to the centromeres.

Accurate segregation of chromosomes is achieved by the assembly of a multiprotein complex, the **kinetochore**, at each centromere. The kinetochore serves as the attachment site for spindle microtubules and the site at which motors generate forces to power chromosome movement. The kinetochore integrates mechanical force and chemical energy from dynamic microtubule tips into directed chromosome motion. Segregation proceeds uneventfully as long as bioriented kinetochore–microtubule attachments come under tension and are then stabilized by inactivation of AURKB.

Overlapping processes ensure accurate chromosome segregation: the mitotic checkpoint prevents anaphase onset until all sister kinetochores are attached to opposite poles and incorrect attachments that fail to generate tension at sister kinetochores are recognized. AURKB, the catalytic subunit of the chromosomal passenger complex (CPC), is a central regulator in both cases. Unattached kinetochores are the signal generators for the mitotic checkpoint, which inhibits the APC and arrests mitosis until all kinetochores have correctly attached to spindle microtubules, thereby representing the major cell-cycle control mechanism protecting against loss or gain of a chromosome. The spindle checkpoint prevents the onset of anaphase before all chromosomes are attached to spindle microtubules. The mitotic checkpoint must be disabled at the transition from metaphase to anaphase in order for sister chromatids to split without engaging the checkpoint. This is achieved through dephosphorylation of INCENP (inner centromere protein – another CPC component), thereby allowing the CPC to move away from centromeres. Segregation defects can cause cancer and Down syndrome, the most frequent inherited birth defect.

At the other end to the centromere, the mitotic spindle attaches to the centrosomes, which are key regulators of cell polarity, regulation of cell cycle and chromosomal stability. Centrosome abnormalities are frequently found in cancers and contribute to chromosomal instability (including aneuploidy, tetraploidy, and/or micronuclei) in daughter cells through the assembly of multipolar or monopolar spindles during mitosis. Many cancer-causing mutations may result in hyperamplification of centrosomes.

Key proteins essential for metaphase arrest in the presence of a disrupted mitotic spindle were initially identified in the budding yeast *Saccharomyces cerevisiae*. The *mad1–3* (mitotic arrest defective) and the *bub1–3* (budding uninhibited in benzimidazole) gene products are essential for this process. Subsequently, MAD1, MAD2, BUBR1 (MAD3), BUB1, and BUB3 homologs have been identified in vertebrates. A further protein, the kinesin CENP-E, is also an essential component of kinetochore-acting regulators of this checkpoint.

The checkpoint is generally accepted as operating by the generation of a diffusible stop signal in response to unattached kinetochores. This stop signal then prevents the onset of anaphase. Unattached kinetochores recruit a complex of BUB and MAD proteins (Fig. 4.8). The proteins BUB1, MAD1, and a portion of MAD2 comprise a catalytic platform that recruits, activates, and releases the stop signal, which may itself be part of the MAD2 protein. The MAD2 component can block entry into anaphase by inhibiting the APC/C (MAD2 binds to the CDC20 subunit required for APC/C activation). Conversely, the release of MAD1 and MAD2, but not BUB1, from kinetochores upon attachment separates the elements of the catalytic platform, thereby avoiding generation of the anaphase inhibitor despite continued rapid cycling of MAD2 at spindle poles.

Mutations in *MAD* do not prevent mitosis but damage this checkpoint, resulting in daughter cells with aberrant chromosome numbers (aneuploidy) and tumorigenesis arising from chromosome missegregation (see Fig. 4.8). Loss of one copy of *Mad2* results in aneuploidy, but intriguingly, near complete elimination of MAD2 protein in tumor cells results in p53-independent cell death. However, mutations in the checkpoint genes *BUB1*, *BUBR1*, and *MAD2* are rarely found in cancers. But alternative mechanisms may be able to deregulate the checkpoint. Thus, amplification of AURKA, overexpressed in some epithelial cancers, might override the checkpoint; AURKA is able to induce defects in chromosome–spindle attachment if overexpressed and can also prevent MAD2 from effectively preventing APC/C activation.

Approximately one in twenty patients infected with the human T-cell leukemia virus-I (HTLV-I) develop adult T-cell leukemia. HTLV-I encodes a protein, TAX, that can sequester MAD proteins, thereby damaging the spindle checkpoint; leukemic cells in these patients show many chromosome abnormalities, including aneuploidy.

Cell-cycle entry and its control by extracellular signals

what alone is not useful helps when accumulated.

Ovid

So far, we have discussed the engine that drives the cell cycle (cyclin–CDKs), the inhibitory proteins (CKIs) that play a key role in inhibiting activation of CDKs, and also a large number of key regulators of DDR and of cell-cycle checkpoints that prevent the propagation of cells with damaged or abnormal DNA. Now we turn to some of the external signals a cell might receive that will subsequently determine whether it will grow and divide, or indeed arrest, differentiate, or die. These are discussed in detail in Chapters 5 and 12, so only a broad outline of principles will be discussed here.

Growth factors stimulate cell division

As mammalian cells are not easily accessible to detailed analyses in the intact animal, most studies on mammalian proliferation derive from cells grown in culture. When cells are grown in culture, they are bathed in medium usually containing fetal calf serum, which contains a cocktail of growth factors such as PDGF, EGF, IGF, essential for cell growth and division. Without serum, cells stop dividing and become arrested, usually in G_0. Thus, in the intact animal, the availability of growth factors, either from the circulation or from neighboring cells, is likely to play an important role in regulating cell proliferation. Once a growth factor has bound to its appropriate cell surface receptor, signaling pathways are transduced within the cell that will ultimately drive the cell cycle. Some of these pathways are described in the next chapter and include induction or activation of proto-oncogene products, such as c-MYC, FOS, JUN, and RAS. In contrast, in the absence of appropriate growth factors, the cell is unable to progress past the G_1 checkpoint, arresting in G_0 until it is awakened by further signals (see Chapter 5).

In contrast to growth factor mitogens, which are positive regulators of the cell cycle, negative regulators such as TGF-β are also important in cancer. Importantly, the exact role of TGF-β may depend on the timing during tumorigenesis, as variously TGF-β has been shown to inhibit cancer, but less intuitively can also support tumor growth (though this may largely be a product of the interactions with stromal cells rather than the cancer cells themselves). TGF-β acts through surface receptors, which in turn activate SMAD2 or SMAD3 proteins by serine phosphorylation. Activated SMAD2 and 3 can then bind to SMAD4, with the resultant complexes entering the nucleus and activating transcription of genes including $p15^{INK4b}$, a G_1 CKI that can mediate G_1 arrest. What can we conclude? That cell growth is determined by the overall cumulative balance of all those positive and negative signals received by a cell at any given point.

Cell adhesion to neighboring cells and the extracellular matrix regulates cell division

Cells cultured *in vitro* require adherence to a substratum in order to proliferate. When normal fibroblasts or epithelial cells are grown in suspension unattached to any solid surface they almost never divide. This contrasts with many cancer cells that no longer require these focal contacts for cell division. Focal contacts are sites of adhesion for intracellular actin filaments and extracellular matrix (ECM) molecules. Binding of ECM molecules, such as fibronectin or laminin, to transmembrane integrin proteins can generate signals inside the cell to control cell proliferation (see Chapter 12).

Nutrients and oxygen supply

Particularly in single cell organisms such as yeasts, the major environmental regulators of cell division are the presence or absence of nutrients, which ensure that cells replicate under suitable environmental conditions. Although such signals are less relevant to cell-cycle regulation in metazoans, where growth factors are more prominent one must remember one important thing. Namely, that replication for a single-cell organism in a pond or other environment over which it has no control must be determined by environmental factors; the emphasis is on the word "organism" rather than "cell." On a metazoan scale, this scenario is closer to reproduction (replication of the whole organism), where recent advances in understanding of hormones that reflect energy homeostasis such as leptin (and thereby nutrient availability in the environment) have a profound impact on fertility (i.e. ability to reproduce), thus linking environmental nutrients to replication of the whole organism as it is in yeasts, the difference being the nature of the regulatory signal. Although, the cell-cycle machinery may be similar, there must be major differences in the controls required over replication of somatic cells within a metazoan (in a regulated environment at least in part divorced from immediate changes in nutrient availability in the environment) and replication of an entire organism, no matter how few cells it may comprise, which is much more directly dependent on environmental support.

Changes in global gene expression during the cell cycle

Most attention has been directed to regulation of cell-cycle protein activity by phosphorylation and protein levels by degradation. However, complex transcriptional networks also help to maintain the cyclic behavior of dividing cells and progress has been made in describing these in budding and fission yeast. Maybe surprisingly, little is understood about the biological significance – with most attention directed at regulation of expression of a handful of genes by c-MYC and E2F family members during G_1/S transition and S phase. Recently, global gene expression analyses of proliferating cells, using gene microarrays and other high-throughput techniques (see Chapter 18), have identified hundreds of periodically expressed genes occurring in three major transcriptional waves approximating to three main cell-cycle transitions (initiation of DNA replication, entry into mitosis, and exit from mitosis). In fact, nearly 1000 genes demonstrate significant oscillations in expression and as many as 2000 genes undergo slight oscillations during the cell cycle. The functional role of most of these is unknown and may, in fact, be irrelevant, driven as a byproduct of chromatin condensation. This notwithstanding, many of these oscillating genes are concentrated in early/mid G_2 phase and near the G_2/M transition and some can be grouped by function into those involved in ribosome biogenesis and protein synthesis and those involved in mitosis, mitotic exit, and cell separation. It is predicted that much important progress will result from work in this area in mammalian cells in the coming years. In fact, such studies involving the elucidation of genes regulated by c-MYC have already provided intriguing results (see Chapter 6).

Cell cycle and cancer

There is only one ultimate and effectual preventative for the maladies to which flesh is heir, and that is death.

Harvey Cushing

In cancer, alterations in the genetic control of cell division result in unrestrained proliferation. In general, genetic or epigenetic alterations occur in two classes of genes: proto-oncogenes and tumor suppressor genes, discussed in detail in Chapters 6 and 7, respectively. In normal cells, the proteins encoded by proto-oncogenes act in signaling pathways that stimulate cell proliferation. Mutated versions of proto-oncogenes or oncogenes promote tumorigenesis. Inactivation of tumor suppressor genes such as *RB* and *p53* results in loss of regulatory pathways that can inhibit cell cycle. Aberrant expression of oncogenes and tumor suppressors need not necessarily arise by mutations or epigenetic changes in the genes encoding these proteins, but can occur secondary to other cancer-causing epimutations. Thus, p27 protein is frequently depleted in cancer cells, but rarely is this due to deletion or inactivation of the *CDKN1B* gene. Similarly, c-Myc is often overexpressed, but with exceptions such as some lymphomas, this is not usually to gene mutations. In cancer, mutations have been observed in genes encoding CDK, cyclins, CDK-activating enzymes, CKI, CDK substrates, and checkpoint proteins (Table 4.3).

G_1 is the phase of the cell cycle wherein the cell is responsive to growth factor-dependent signals. As already discussed in general, unless something goes awry, once the cell has passed the restriction point and makes the transition from G_1 into S phase, the cell cycle will proceed under its own steam. Not surprisingly, control of the G_1/S transition is often disrupted in cancer cells, which can thus potentially replicate autonomously and independently of mitogenic stimuli. As mentioned earlier, the key event in the G_1/S transition is the phosphorylation of RB and the release of E2F for the transcription of S-phase genes. In human cancer, the RB pathway is frequently nonfunctional, though this is rarely due to a mutation of the *RB* gene itself, with the notable exception of some childhood cancers. In fact, loss of RB function may usually result from inappropriate activation of CDK–cyclins or loss of CKIs (discussed below and in Chapter 7). It should also be appreciated that cancer cells must also have overcome the inherent tumor suppressor activity of oncogenes such as MYC and RAS and/or have evaded the downstream oncogenic stress pathways.

The CDKs and cancer

Occasionally, though infrequently, the CDKs themselves may be mutated in cancers. Amplification of CDK4 may occur in melanoma, sarcoma, and glioma. Mutations in *CDK4* and *CDK6*, rendering these unresponsive to CDKIs, can also occur. CDK1 and CDK2 may be overexpressed in some colonic tumors. Some activating proteins of CDKs are implicated in cancers. As mentioned already, CDK activation is in part determined by dephosphorylation by the CDC25 phosphatase family, important regulators of the G_1/S transition. Other family members are important during S phase and entry into mitosis. Deregulation of CDC25 can result in inappropriate activation of CDK–cyclins and has been shown in some cancers. CDC25 is overexpressed in 32% of primary breast cancers. Interestingly, CDC25 is a MYC target gene (see Chapter 6).

CDKs are by implication universally active in human cancer and their inhibition can promote cell-cycle arrest and apoptosis. Inhibition of CDK4/6 predictably results in G_1 arrest and tumor regression, inhibiting of CDK2/1 induces E2F transcription factor-dependent cell death. Drugs able to selectively target particular CDKs are in trials and under development.

Cyclins and cancer

Cyclin D1 binds to CDK4 and CDK6 in early G_1, and is a key mediator of growth factor activation of cell cycling. Since cyclin D1 was first linked to tumors of the parathyroid gland in 1991, deregulated expression of cyclin D1 has been reported in many other human cancers. A cyclin D1 gene translocation is now known to occur in B-cell tumors, including a type of lymphoma. In mantle cell lymphoma a characteristic t(11;14) translocation juxtaposes the cyclin D1 gene to the immunoglobulin heavy-chain gene, leading to cyclin D1 overexpression. Amplification or overexpression of cyclin D1 is found in cancers of the breast, bladder, gastrointestinal tract, and lung. A common polymorphism in the human cyclin D1 gene is alternatively spliced, producing two protein products termed cyclin D1a and D1b. Although both are often upregulated in human cancers, cyclin D1b appears to be more tumorigenic. One explanation for this may be the recent demonstration that these proteins are functionally different, with cyclin D1a overexpression specifically able to induce G2/M arrest and promote features of a DDR.

Rhabdoid tumors are rare cancers of childhood with a uniformly poor outlook and no effective treatments. They carry two defective copies of the *INI1* tumor suppressor gene. INI1 inhibits *CCND1* and activates p16^{Ink4a} and p21^{CIP1}. In mouse models, the pan-cdk inhibitor flavopiridol resulted in complete regression of some tumors, although many were resistant, perhaps due to overexpression of cyclin D1.

The S-phase CDK2 interacting cyclins are key regulators of DNA replication and cellular proliferation. The human genome encodes two E-type cyclins (E1 and E2) and two A-type cyclins (A1 and A2). Dysregulation of the CDK2-bound cyclins plays an important role in the pathogenesis of cancer.

The E-type cyclins (cyclin E1 and cyclin E2) are expressed during the late G_1 phase of the cell cycle until the end of S phase. Cyclin E is limiting for passage through the restriction point, the point of no return for cells progressing from G_0 or G_1 into S phase. Expression of cyclin E is not only regulated by E2F transcription factors but also by proteasome-mediated degradation. Cyclin E binds and activates the kinase CDK2 and by hyperphosphorylating RB may further enhance the E2F-dependent transcription of S-phase genes. High levels of cyclin E are associated with the initiation or progression of breast cancer, leukemia, and others, and is associated with aggressive disease and poor prognosis. In addition to potential deregulation of G_1/S transition, overexpressed or deregulated cyclin E may also provoke chromosome instability and loss of heterozygosity at tumor suppressor loci (see Chapter 7). In part, therefore, the tumorigenic activity of cyclin E may arise from the aberrant assembly of prereplication complexes and incomplete DNA synthesis and also from resultant mitotic defects – increased risk of nondisjunction, impaired chromosome segregation, and ploidy.

Cyclin A2 is associated with cellular proliferation and can be used for molecular diagnostics as a proliferation marker. In addition, cyclin A2 expression is associated with a poor prognosis in several types of cancer, including some lung cancers. Cyclin A1 is a tissue-specific cyclin that is highly expressed in acute myeloid leukemia and in testicular cancer.

Mitotic kinases and cancer

AURKA is overexpressed in various human cancers and is linked to aneuploidy and chromosomal instability. In fact, the chromosome locus of *AURKA*, 20q13.2, is frequently amplified in colorectal cancer and basal breast cancers. Deregulation of AURKA can provoke mitotic defects and result in both tetraploidy and aneuploidy and in mouse models overexpression results in defects in the G_2 checkpoint and in disordered spindle assembly. Recent studies have revealed that the kinase effects of AURKA are more widespread than previously thought. Thus, aurora A can direct the alternative splicing of a variety of apoptosis regulators such as Bcl-xL and Mc11, suggesting a link between mitotic arrest and apoptosis. This effect appears to be at least in part mediated by turnover of the splicing regulator ASF/SF2, which regulates these mRNAs and also regulates apoptosis. AURKA may also contribute to cancer as a result of inhibition of p53 tumor suppressor signaling and activation of AKT. AURKA also impedes the repair of DNA DSBs, in part through preventing recruitment of RAD51, and inhibiting homologous recombination. Impairment of RAD51 function requires inhibition of CHK1 by polo-like kinase 1 (PLK1).

PLK1 is a proto-oncogene frequently overexpressed in cancer cells. In some cases this may be secondary to inactivating mutations in RB. Predicted effects of oncogenic PLK1, such as aneuploidy, result from defective mitosis and centrosome amplification and from deregulated cell-cycle progression. As discussed previously for AURKA, PLK1 can also deregulate key tumor-signaling pathways and can inhibit transactivation and proapoptotic functions of p53.

Although somewhat oversimplifying matters, PLK1 and PLK2/PLK3 have opposing actions with respect to regulation of p53 signaling, with PLK1 being inhibitory and suppressed in a normal DDR, which has led to a quest for PLK1 selective inhibitors. PLK1 inhibition has been shown to reverse tumorigenesis in mouse models, with surprisingly little effect on normal cells at doses used.

CKIs, checkpoints, and cancer

CKIs largely suppress cell cycle and growth by activation of RB and prevent G_1/S transition. Approximately 90% of human cancers have abnormalities in some component of the RB pathway (including RB itself, p16^{INK4a}, or cyclin D–CDK4/6). Cancer-promoting mutations of the E2F family of transcription factors have not yet been described.

Loss of cell-cycle checkpoints is extremely common in human cancer. The *p53* gene is the most frequently mutated gene in human cancers (see Chapter 7). However, mutations in other genes in the p53 pathway may also result in avoidance of normal regulation. The *INK4a-ARF* locus is often the target of inactivating mutations in human tumors and can affect either or both protein products – p16^{INK4a} and p14^{ARF}. In a recent study of a mouse model of pancreatic ductal adenocarcinoma, mutant KRAS (Kras(G12D)) initiated formation of premalignant pancreatic ductal lesions, and loss of either INK4A/ARF or p53 enabled their malignant progression. Further specific studies showed that cancers developed in *KRAS* mutant mice with homozygous deletion of either p53 or p16^{Ink4a} (with intact p53/ARF).

Loss of expression of another key regulator of the cell cycle, p27^{KIP1}, has been reported for a number of human tumor types (lung, breast, bladder) and has been correlated with poor prognosis and tumor aggressiveness. It has been shown in colorectal carcinomas that increased proteasome-dependent proteolysis, rather than gene deletion, is responsible for p27^{KIP1} downregulation. This may result from activating mutations in SCF components required for degradation of p27^{KIP1} or by inactivating mutations required for degradation of proteins, such as c-MYC, that can inhibit p27^{KIP1}.

In normal cells, DNA damage if detected results in the induction of p53 and one of its downstream targets p21^{CIP1}, which results in G_1 and/or G_2 growth arrest. Cancer cells that are deficient in p53 or p21^{CIP1} will fail to arrest and continue through the cell cycle into mitosis. p21^{CIP1}, originally described as a universal inhibitor of CDKs, is now known to have other key roles. In fact, p21^{CIP1} may also help maintain cell survival following DNA damage and subsequent p53 induction, in order for the cell to effect repairs. Thus, the increase in p21^{CIP1} seen in some cancers may impart these cells with a survival advantage. Another way in which cancer cells may avoid p53 regulation is by overexpression of MDM2, the negative regulator of p53, and this has been shown in leukemia, breast carcinoma, and glioma.

Checkpoint proteins involved in DDR are discussed in more detail in Chapter 10. Ataxia telangiectasia (AT) is a rare human disease caused by mutations in the AT-mutated gene (*ATM*). It is characterized by sensitivity to radiation and predisposition to cancer, a phenotype also shared with many other diseases. At least in part, because of the numerous proteins involved in the ATM-signaling pathway; activation of the ATM kinase is a critical factor in triggering of all three cell-cycle checkpoints following DSBs. AT-like diseases are due to inactivating mutations in genes encoding other key proteins in the DNA DSB repair, such as those in the MRN protein complex (Mre11, Rad50, and Nbs1), the DNA-dependent protein kinase complex consisting of the heterodimer Ku70/Ku80 and its catalytic subunit DNA-PKcs, H2AX, or p53.

H2AX functions primarily as a downstream mediator of ATM, whereas NBS1 operates as both an activator of ATM and as a coactivator of ATM downstream processes (but may not be essential for the former). NBS1, H2AX, and p53 play synergistic roles in ATM-dependent DDR and tumor suppression. The number of genes involved in DDR implicated in cancer is increasing at a steady pace. In addition to the ATM pathway described above, the ATR–CHK1 pathway is also important in activating checkpoints after genotoxic stress. CHK1 deficiency results in stabilization of the CDC25A protein and overriding of G_1/S and G_2/M arrest, which can lead to genomic instability and accumulation of DSBs. In a recent study, RAD9 (part of the 9-1-1 complex), involved in activation of ATR, was shown to be a potential oncoprotein in breast cancer. Increased levels may arise by gene amplification or differential methylation of Sp1/3-binding sites within the *RAD9* gene, and correlate with tumor size and local recurrence.

Abrogation of checkpoints through epigenetic factors is well described (see Chapters 7 and 11) and may be a key means by which tumor suppressors are inactivated in cancer cells. Loss of heterozygosity may occur through epigenetic, as well as genetic, loss of the wildtype allele of *RB*, *p53*, *MLH1*, and others. In fact, a wider role for chromatin remodeling (see Chapter 11) in checkpoint failure is underlined by several recent observations. The hSNF5 subunit of human SWI/SNF ATP-dependent chromatin remodeling complexes is a tumor suppressor which, when inactivated in malignant rhabdoid tumors, leads to chromosomal instability. hSNF5 normally operates through the RB growth arrest pathway, possibly to maintain silencing of E2F responsive genes, thus blocking cell-cycle progression.

Drugging the cell cycle in cancer therapies

Given that aberrant proliferation in cancer cells is synonymous with genetic or epigenetic alterations in key cell cycle regulators, such as the CDKs, the hunt is well and truly on for pivotal cell-cycle regulators as targets for new agents to treat cancer (see

Chapter 16). Agents that can inhibit some cell-cycle kinases, such as the cyclin-dependent kinases (CDKs) or the aurora and polo mitotic kinases, are already in several preclinical and clinical trials and have already been discussed. More generally, agents that affect mitosis, such as the microtubule-poisoning taxanes and those that target mitogenic pathways such as the tyrosine kinase inhibitors, will be considered in chapters on DNA damage and growth factors, respectively. Ongoing research is revealing new potential drug targets, including phosphatases and other mitotic regulators such as microtubule motor proteins (kinesins).

CDKs

Drugs inhibiting the CDKs can arrest the cell cycle or even induce apoptosis. However, the efficacy of specific drugs is impeded by redundancy of the various CDK proteins. As expected, inhibiting CDK4/6 leads to arrest in G_1, and reversal of tumors. Again, inhibiting CDK2/1 mostly interferes with S and G_2 and may also induce apoptosis. Transcriptional CDKs, such as CDK7/9, activate RNA polymerase II, required for transcriptional initiation and elongation; inhibiting CDK9 results in apoptosis at least in part by activating p53.

Recently, a highly specific inhibitor of CDK4/6 activity (PD-0332991) has been developed. Breast cancer lines appear to respond well to this agent at first but become resistant in part because chronic RB inactivation is followed by emergence of CDK4/6 independence.

Checkpoints

ATM inhibitors (e.g. KU-55933 and KU-60019; Kudos) are at early preclinical stages of development. CHK1 is an attractive target as it is involved in all the cell cycle checkpoints. Inhibiting CHK1 in the face of DNA damage can precipitate "mitotic catastrophe" and cell death in p53-deficient tumor cells.

Suppressing DNA damage and replication checkpoint responses by inhibiting CHK1 can enhance the lethal effects on cancer cells of multiple genotoxic agents, suggesting use of CHK1 inhibition as an adjunct to traditional chemotherapeutic agents. Numerous CHK1 and WEE1 inhibitors are in preclinical development. Some have proved effective in xenograft models and against cancer cell lines and some early-phase clinical trials have just been established. Context may again be a critical factor in choice of drugs. Defective HRR in *BRCA1*- and *BRCA2*-deficient tumors confers sensitivity to cisplatin and inhibitors of poly(ADP-ribose) polymerase (PARP), a DNA damage repair protein. A CHK2 kinase inhibitor, CCT241533. has also been studied preclinically. Although no useful potentiation of genotoxic agents was identified, the cytotoxicity of PARP inhibitors against cancer cell lines was increased.

Mitotic kinases

Drugs targeting mitosis (antimitotics), including the vinca alkaloids and taxanes (see Chapter 18), have long been central to the treatment of a variety of human cancers. However, as with many traditional cancer treatments, these agents hit normal cells as well as cancer cells and may also interfere with nondividing cells. In order to try to target cancer cells more effectively, much attention has turned to specific kinases regulating mitosis, and in particular PLK1 and AURKA and AURKB.

PLK1 is a promising drug target in cancer and several small-molecule inhibitors are in drug development and some are in clinical trials. More are likely to follow, but ensuring specificity of action and avoiding "off-target toxicity," will be critically important. It is worth noting the presence of closely related family members that have opposing roles in cancer; PLK2 and PLK3 both act as tumor suppressors in the p53 pathway.

Numerous inhibitors of PLK1 are in clinical trials. Phase I or II studies for four PLK1 inhibitors – BI2536, GSK461364, ON-01910, and HMN-214 – have shown promise in a variety of cancers, particularly refractory non-Hodgkin lymphoma and acute myeloid leukemia, with bone marrow suppression being the major dose-limiting toxicity. Interestingly, PLK1 inhibitors appear effective against cells derived from *RAS* mutant cancers, such as colon and NSCLC, an example of a synthetic lethal interaction (see Chapter 16). Recent studies also suggest that PLK1 inhibitors may be effective in tumors with inactivated *TP53*, setting these apart from more traditional chemotherapies which require wildtype p53 to be effective.

Recently, potent and selective small-molecule inhibitors of aurora kinases have been described that block cell-cycle progression and induce apoptosis in a diverse range of human tumor cell types. Inhibition of AURKA causes mitotic arrest, whereas inhibition of AURKB overrides DDR and spindle checkpoints and drives cells through aberrant mitosis. Drugs targeting AURKA and B, with varying degrees of specificity, are in phase I and II clinical studies and have shown early promise in a variety of solid tumors and some lymphomas and leukemias. As has been noted with PLK1 inhibitors, usage is limited by bone marrow suppression. Aurora kinase B inhibitors induce apoptosis secondary to polyploidy.

MLN8237, a novel aurora A kinase inhibitor, is under investigation in multiple phase I and II studies. A phase I trial of ENMD-2076, an oral aurora kinase and VEGFR inhibitor, has just been completed and included 67 patients with ovarian cancer, colorectal cancer, and refractory solid tumors. The drug showed promising antitumor activity, particularly in ovarian cancer.

It is likely that aurora kinase inhibitors will have particularly strong effects under certain circumstances. In a study from the laboratory of Mike Bishop, it has been shown that the aurora kinase inhibitor VX-680 selectively kills cells and mouse tumors that overexpress Myc – a synthetic lethal interaction (see Chapters 10 and 16) due to disabling of the CPC by aurora B inhibition. This led to DNA replication in the absence of cell division and thereby apoptosis and autophagy independent of p53.

Conclusions and future directions

During the following chapters we will frequently return to discussion of regulation of the cell cycle as a central process in cancer development and treatment. Many cell-cycle proteins are now known to be directly involved in cancer and will be discussed in Chapters 6 and 7 in particular. Although most is known about proteins regulating the G_1/S transition and various cell-cycle checkpoints, essentially all cell-cycle proteins will likely play a role in cancer and some may have either therapeutic or diagnostic potential.

Until recently, most developments in cancer prevention and therapy have resulted from trial and error. But this is changing. Through basic research, scientists are acquiring a detailed understanding of how cancer cells originate and grow. Armed with this new knowledge, scientists can now design therapies that are directly targeted at key molecules critical to the behavior of cancer cells. Given, that abnormal replication is one of the "hallmark" features of cancer, recent approaches are increasingly being directed towards key cell-cycle regulators. Thus, much work is now ongoing to examine the role of genes/proteins that regulate the G_1/S transition of the cell cycle and either promote or inhibit RB phosphorylation, respectively, and thereby control S-phase entry.

Many human cancers express increased amounts of cyclin D1, including cancers of the breast, prostate, esophagus, stomach, and colon. Recent studies have also begun to focus on other cell-cycle proteins. Cyclin E plays a critical role for G_1/S transition and high levels of cyclin E are found in many types of cancer. Overexpression of cyclin E may arise by gene amplification and transcriptional mechanisms, but may arise largely through impaired degradation by the ubiquitin–proteasome pathway. In addition, proteolytically cleaved forms of cyclin E that show oncogenic functions have been described. Overexpression of cyclin E is now linked to both aberrant proliferation and also to a more malignant phenotype associated with chromosomal instability.

Increasingly, advances in the basic understanding of cell-cycle regulation will be translated into novel therapeutic approaches, and will likely include drugs that reconstitute aspects of regulation lost in cancers.

Several new drugs targeting cell-cycle proteins are currently being studied, including nonselective inhibitors of multiple cyclin–CDK holenzymes or those more specifically targeting cyclin D–Cdk4/6 activity.

Outstanding questions

1. To translate increasing knowledge about cell-cycle regulation into improved new therapeutics, earlier cancer diagnosis, and improved accuracy of prognosis.

2. To define at what point and in what respects signaling pathways regulating cell division and those regulating apoptosis diverge.

3. To increase knowledge about the mechanisms that regulate and monitor DNA synthesis and the accurate distribution of chromosomes during mitosis.

4. To exploit the human genome reference database and new high-throughput techniques for examining gene expression to identify and functionally model new potential cell-cycle regulatory genes/proteins.

5. To develop specific therapeutics to target cell-cycle regulatory proteins.

Bibliography

We apologize to all our colleagues past and present whose important work could not be references here due to space constraints. Please refer to website for more detailed reference list.

Ceccaldi, R., Briot, D., Larghero, J., *et al.* (2011). Spontaneous abrogation of the G2 DNA damage checkpoint has clinical benefits but promotes leukemogenesis in Fanconi anemia patients. *Journal of Clinical Investigation*, **121**(1): 184–94.

Cho, H.S., Suzuki, T., Dohmae, N., *et al.* (2011). Demethylation of RB Regulator MYPT1 by histone demethylase LSD1 promotes cell cycle progression in cancer cells. *Cancer Research*, **71**(3): 655–60.

D'Angiolella, V., Donato, V., Vijayakumar, S., *et al.* (2010). SCF(Cyclin F) controls centrosome homeostasis and mitotic fidelity through CP110 degradation. *Nature*, **466**(7302): 138–42.

da Fonseca, P.C., Kong, E.H., Zhang, Z., *et al.* (2010). Structures of APC/C(Cdh1) with substrates identify Cdh1 and Apc10 as the D-box co-receptor. *Nature*, **470**(7333): 274–8.

Di Fiore, B. and Pines, J. (2010). How cyclin A destruction escapes the spindle assembly checkpoint. *Journal of Cell Biology*, **190**(4): 501–9.

Gavet, O. and Pines, J. (2010). Activation of cyclin B1-Cdk1 synchronizes events in the nucleus and the cytoplasm at mitosis. *Journal of Cell Biology*, **189**(2): 247–59.

Gharbi-Ayachi, A., Labbé, J.C., Burgess, A., *et al.* (2010). The substrate of Greatwall kinase, Arpp19, controls mitosis by inhibiting protein phosphatase 2A. *Science*, **330**(6011): 1673–7.

Gonzalez, S., Klatt, P., Delgado, S., *et al.* (2006). Oncogenic activity of Cdc6 through repression of the INK4/ARF locus. *Nature*, **440**(7084): 702–6.

Helmink, B.A., Tubbs, A.T., Dorsett, Y., *et al.* (2011). H2AX prevents CtIP-mediated DNA end resection and aberrant repair in G1-phase lymphocytes. *Nature*, **469**(7329): 245–9.

Huang, H.C., Shi, J., Orth, J.D., and Mitchison, T.J. (2009). Evidence that mitotic exit is a better cancer therapeutic target than spindle assembly. *Cancer Cell*, **16**(4): 347–58.

Hunt, T. and Sassone-Corsi, P. (2007). Riding tandem: circadian clocks and the cell cycle. *Cell*, **129**(3): 461–4.

Knoblich, J.A. (2010). Asymmetric cell division: recent developments and their implications for tumour biology. *Nature Reviews in Molecular Cell Biology*, **11**(12): 849–60.

Larochelle, S., Merrick, K.A., Terret, M.E. *et al.* Requirements for Cdk7 in the assembly of Cdk1/cyclin B and activation of Cdk2 revealed by chemical genetics in human cells. *Molecular Cell*, **25**(6): 839–50.

Li, Z., Jiao, X., Wang, C., *et al.* (2010). Alternative cyclin D1 splice forms differentially regulate the DNA damage response. *Cancer Research*, **70**(21): 8802–11.

Lu, Y. and Cross, F.R. (2010). Periodic cyclin-Cdk activity entrains an autonomous Cdc14 release oscillator. *Cell*, **141**(2): 268–79.

Ma, C.X., Janetka, J.W., and Piwnica-Worms, H. (2010). Death by releasing the breaks: CHK1 inhibitors as cancer therapeutics. *Trends in Molecular Medicine*, 2011; **17**(2): 88–96.

Manchado, E., Guillamot, M., de Cárcer, G., *et al.* (2010). Targeting mitotic exit leads to tumor regression in vivo: Modulation by Cdk1, Mastl, and the PP2A/B55α,δ phosphatase. *Cancer Cell*, **18**(6): 641–54.

Marine, J.C. (2011). MDM2 and MDMX in cancer and development. *Current Topics in Developmental Biology*, **94**: 45–75.

Metzger, E., Imhof, A., Patel, D., *et al.* (2010). Phosphorylation of histone H3T6 by PKCbeta(I) controls demethylation at histone H3K4. *Nature*, **464**(7289): 792–6.

Mochida, S., Maslen, S.L., Skehel, M., and Hunt, T. (2010). Greatwall phosphorylates an inhibitor of protein phosphatase 2A that is essential for mitosis. *Science*, **330**(6011): 1670–3.

Orpinell, M., Fournier, M., Riss, A., *et al.* (2010). The ATAC acetyl transferase complex controls mitotic progression by targeting non-histone substrates. *EMBO Journal*, **29**(14): 2381–94.

Reinhardt, H.C. and Yaffe, M.B. (2009). Kinases that control the cell cycle in response to DNA damage: Chk1, Chk2, and MK2. *Current Opinion in Cell Biology*, **21**(2): 245–55.

Sage, J. and Straight, A.F. (2010). RB's original CIN? *Genes and Development*, **24**(13): 1329–33. Comments on: *Genes and Development*, 2010; **24**(13): 1377–88; 1351–63; 1364–76.

Tsukahara, T., Tanno, Y., and Watanabe, Y. (2010). Phosphorylation of the CPC by Cdk1 promotes chromosome bi-orientation. *Nature*, **467**(7316): 719–23.

Valerio-Santiago, M. and Monje-Casas, F. (2011). Tem1 localization to the spindle pole bodies is essential for mitotic exit and impairs spindle checkpoint function. *Journal of Cell Biology*, **192**(4): 599–614.

Wang, F., Dai, J., Daum, J.R., et al. (2010). Histone H3 Thr-3 phosphorylation by Haspin positions Aurora B at centromeres in mitosis. *Science*, **330**(6001): 231–5.

Wang, R., He, G., Nelman-Gonzalez, M., et al. (2007). Regulation of Cdc25C by ERK-MAP kinases during the G2/M transition. *Cell*, **128**(6): 1119–32.

Wiebusch, L. and Hagemeier, C. (2010). p53- and p21-dependent premature APC/C-Cdh1 activation in G2 is part of the long-term response to genotoxic stress. *Oncogene*, **29**(24): 3477–89.

Yamagishi, Y., Honda, T., Tanno, Y., and Watanabe, Y. (2010). Two histone marks establish the inner centromere and chromosome biorientation. *Science*, **330**(6001): 239–43.

Zegerman, P. and Diffley, J.F. (2010). Checkpoint-dependent inhibition of DNA replication initiation by Sld3 and Dbf4 phosphorylation. *Nature*, **467**(7314): 474–8.

Zha, S., Guo, C., Boboila, C., et al. (2011). ATM damage response and XLF repair factor are functionally redundant in joining DNA breaks. *Nature*, **469**(7329): 250–4.

Cell cycle – general

Nurse, P. (2000). A long twentieth century of the cell cycle and beyond. *Cell*, **100**: 71–8.

Kittler, R., Putz, G., Pelletier, L., et al. (2004). An endoribonuclease-prepared siRNA screen in human cells identifies genes essential for cell division. *Nature*, **432**(7020): 1036–40.

Vogelstein, B. and Kinzler, K.W. (2004). Cancer genes and the pathways they control. *Nature Medicine*, **10**, 789–99.

Cell cycle entry at G_1

Aleem, E., Kiyokawa, H., and Kaldis, P. (2005). Cdc2-cyclin E complexes regulate the G1/S phase transition. *Nature Cell Biology*, **7**(8): 831–6.

Buchkovich, K., Duffy, L.A., and Harlow, E. (1989). The retinoblastoma protein is phosphorylated during specific phases of the cell cycle. *Cell*, **58**: 1097.

Massague, J. (2004). G1 cell-cycle control and cancer. *Nature*, **432**(7015): 298–306.

Sherr, C.J. and Roberts, J.M. (2004). Living with or without cyclins and cyclin-dependent kinases. *Genes and Development*, **18**(22): 2699–711.

Cyclins, CDK, CKI, and checkpoints

Bardeesy, N., Aguirre, A.J., Chu, G.C., et al. (2006). Both p16(Ink4a) and the p19(Arf)-p53 pathway constrain progression of pancreatic adenocarcinoma in the mouse. *Proceedings of the National Academy of Sciences of the U.S.A.*, **103**(15): 5947–52.

Castro, A., Bernis, C., Vigneron, S., Labbe, J.C., and Lorca, T. (2005). The anaphase-promoting complex: a key factor in the regulation of cell cycle. *Oncogene*. **24**(3): 314–25.

Drayton, S., Rowe, J., Jones, R., et al. (2003). Tumor suppressor p16INK4a determines sensitivity of human cells to transformation by cooperating cellular oncogenes. *Cancer Cell*, **4**(4): 301–10.

Evans, T., Rosenthal, E.T., Youngblom, J., Distel, D., and Hunt, T. (1983). Cyclin: a protein specified by maternal mRNA in sea urchin eggs that is destroyed at each cleavage division. *Cell*, **33**: 389.

Gil, J. and Peters, G. (2006). Regulation of the INK4b-ARF-INK4a tumour suppressor locus: all for one or one for all. *Nature Reviews in Molecular Cell Biology*, **7**(9): 667–77

Gonzalez, S., Klatt, P., Delgado, S., et al. (2006). Oncogenic activity of Cdc6 through repression of the INK4/ARF locus. *Nature*, **440**(7084): 702–6.

Kastan, M.B. and Bartek, J. (2004). Cell-cycle checkpoints and cancer. *Nature*, **432**(7015): 316–23.

Kossatz, U., Breuhahn, K., Wolf, B., et al. (2010). The cyclin E regulator cullin 3 prevents mouse hepatic progenitor cells from becoming tumor-initiating cells. *Journal of Clinical Investigation*, **120**(11): 3820–33.

Kress, T.R., Cannell, I.G., Brenkman, A.B., et al. (2011). The MK5/PRAK kinase and Myc form a negative feedback loop that is disrupted during colorectal tumorigenesis. *Molecular Cell*, **41**(4): 445–57.

Liu, H., Takeda, S., Kumar, R., et al. (2010). Phosphorylation of MLL by ATR is required for execution of mammalian S-phase checkpoint. *Nature*, **467**(7313): 343–6.

Loog, M. and Morgan, D.O. (2005). Cyclin specificity in the phosphorylation of cyclin-dependent kinase substrates. *Nature*, **434**(7029): 104–8.

Lygerou, Z. and Nurse, P. (2000). Cell cycle. License withheld Geminin blocks DNA replication. *Science*, **290**: 2271–3.

Mailand, N. and Diffley, J.F. (2005). CDKs promote DNA replication origin licensing in human cells by protecting Cdc6 from APC/C-dependent proteolysis. *Cell*, **122**(6): 915–26.

Moroy, T. and Geisen, C. (2004). Cyclin E. *International Journal of Biochemistry and Cell Biology*, **36**(8): 1424–39.

Paulovich, A.G., Toczyski, D.P., and Hartwell, L.H. (1997). When checkpoints fail. *Cell*, **88**(3): 315–21.

Shapiro, G.I. (2006). Cyclin-dependent kinase pathways as targets for cancer treatment. *Journal of Clinical Oncology*, **24**(11): 1770–83.

Skaar, J.R. and Pagano, M. (2009). Control of cell growth by the SCF and APC/C ubiquitin ligases. *Current Opinion in Cell Biology*, **21**(6): 816–24.

Tudzarova, S., Trotter, M.W., Wollenschlaeger, A., et al. (2010). Molecular architecture of the DNA replication origin activation checkpoint. *EMBO Journal*, **29**(19): 3381–94.

Zegerman, P. and Diffley, J.F. (2010). Checkpoint-dependent inhibition of DNA replication initiation by Sld3 and Dbf4 phosphorylation. *Nature*, **467**(7314): 474–8.

Zhao, R.J., Yeung, S.C., Chen, J., et al. (2011). Subunit 6 of the COP9 signalosome promotes tumorigenesis in mice through stabilization of MDM2 and is upregulated in human cancers. *Journal of Clinical Investigation*, **121**(3): 851–65.

The ubiquitin proteasome pathway in cell cycle regulation

Crusio, K.M., King, B., Reavie, L.B., and Aifantis, I. (2010). The ubiquitous nature of cancer: the role of the SCF(Fbw7) complex in development and transformation. *Oncogene*, **29**(35): 4865–73.

Mitosis

Cleveland, D.W., Mao, Y., and Sullivan, K.F. (2003). Centromeres and kinetochores: from epigenetics to mitotic checkpoint signaling. *Cell*, **112**(4): 407–21.

Hoyt, M.A. and Geiser, J.R. (1996). Genetic analysis of the mitotic spindle. *Annual Review of Genetics*, **30**: 7–33.

Mitchison, T.J. and Salmon, E.D. (2001). Mitosis: a history of division. *Nature Cell Biology*, **3**: E17–E21.

Potapova, T.A., Daum, J.R., Pittman, B.D., et al. (2006). The reversibility of mitotic exit in vertebrate cells. *Nature*, **440**(7086): 954–8.

Rieder, C.L. and Maiato, H. (2004). Stuck in division or passing through: what happens when cells cannot satisfy the spindle assembly checkpoint. *Developmental Cell*, **7**(5): 637–51.

Shah, J.V., Botvinick, E., Bonday, Z., Furnari, F., Berns, M., and Cleveland, D.W. (2004). Dynamics of centromere and kinetochore proteins; implications for checkpoint signaling and silencing. *Current Biology*, **14**(11): 942–52.

Xie, S., Xie, B., Lee, M.Y., and Dai, W. (2005). Regulation of cell cycle checkpoints by polo-like kinases. *Oncogene*, **24**(2): 277–86.

Yim, H. and Erikson, R.L. (2010). Cell division cycle 6, a mitotic substrate of polo-like kinase 1, regulates chromosomal segregation mediated by cyclin-dependent kinase 1 and separase. *Proceedings of the National Academy of Sciences of the U.S.A.*, **107**(46): 19742–7.

Questions for student review

1) The cell cycle is:
   a. Driven by the action of CKIs.
   b. Subject to various checkpoints.
   c. Driven by cyclin–CDK complexes.
   d. Linked to cell growth.
   e. Not invariably deregulated in cancer.

2) In the cell cycle, the G_2/M checkpoint:
   a. Prevents DNA synthesis.
   b. Prevents mitosis.
   c. Is triggered by DNA damage.

d. Involves the activity of the ATM and ATR protein kinases.
   e. Primarily involves the RB tumor suppressor pathway.

3) The spindle checkpoint:
   a. Responds to aberrant chromosomal segregation.
   b. Blocks cell cycle by activating the APC/C.
   c. May be triggered by abnormal tension on kinetochores during mitosis.
   d. If defective can result in chromosomal instability.
   e. Is rarely involved in human cancers.

4) The following have little influence on cell division.
   a. Interactions with extracellular matrix.
   b. Cell size.
   c. Telomeres.
   d. Secreted growth factors.
   e. The p53 signaling pathway.

5) In the cell division cycle, G_1/S transition is promoted by:
   a. Hypophosphorylation of the RB protein.
   b. E2F transcriptional activation.
   c. Mitogenic growth factors.
   d. Cyclin D–CDK4 complexes.
   e. The product of the *p16* gene.

5 Growth Signaling Pathways and the New Era of Targeted Treatment of Cancer

Stella Pelengaris[a] and Michael Khan[b]
[a]Pharmalogos Ltd and [b]University of Warwick, UK

What is the most rigorous law of our being? Growth. No smallest atom of our moral, mental, or physical structure can stand still a year. It grows – it must grow; nothing can prevent it.
Mark Twain

Life is occupied in both perpetuating itself and in surpassing itself. If all it does is maintain itself, then living is only not dying.
Simone de Beauvoir

Key points

- All cells in a multicellular organism must coordinate their behavior and subordinate their individual requirements for the overall good of the organism. This is only possible because cells have a variety of ways to communicate with one another and an elaborate machinery that allows them to interpret and integrate often diverse messages into responses (e.g. grow–don't grow; live–die).

- The size of an organism and constituent tissues is to a large extent dictated by cell number and is subject to a variety of often draconian measures, which ensure tissue homeostasis by counterbalancing cell division and survival with the opposing processes of cell death and senescence.

- The size of the adult organism is largely genetically determined, but for many organs, size is far from immutable. In fact, numerous adult tissues exhibit a remarkable degree of plasticity and can adapt to changes in the environment, such as nutrient availability, or changing demands of the organism.

- The actual means by which "appropriate organ size" is monitored by the organism and, if deemed necessary, cell number and size adjusted are poorly understood, but require the coordination not only of cellular growth and survival but often also cell adhesion, pattern formation, and polarity. The whole can usefully be thought of as an "organ-size checkpoint," analogous to the cell-cycle checkpoints discussed in Chapter 4. The interestingly named "Hippo pathway" is emerging as a strong contendor for the role of organ size controller.

- Secreted factors that promote or inhibit growth allow adaptive changes in tissue mass and can originate from local cells (paracrine) or more distant ones that release hormones into the circulation (endocrine).

- Growth factors influence cellular replication by binding to specific receptors on, or occasionally within, the cell to form active complexes that in turn unleash a cascade of downstream signaling events. In general, sequential activation of kinase enzymes by addition of phosphate groups eventually carries the growth message to the nucleus, where it triggers expression of genes needed for cell cycle. Growth factors are critically important in coercing cells to leave G_1 and enter S phase of the cell cycle – the so-called G_1/S transition.

- As the signal passes onwards, the previous participants are inactivated by phosphatases that remove the added phosphate groups, readying the system for the next stimulus. Signaling cascades are rarely if ever linear, despite the often misleading appearance given by diagrams in textbooks (including this one), and fall subject to a bewildering array of inhibitory inputs and convergence points with other pathways as well as negative feedback and positive feedforward loops.

- Not surprisingly, such convoluted pathways offer numerous opportunities for epimutations to aberrantly stimulate growth or remove normal growth-inhibiting restraints. Fortunately, however, we have in-built protective mechanisms that will prevent unalloyed growth resulting from any single mutation; the quotidian apoptosis and senescence will effectively cancel out the aberrant growth-promoting effect – a molecular embodiment of Newton's third law.

- By implication, normal cell replication requires that mitogenic signals be accompanied by those that prevent apoptosis or senescence, even if these derive from the same growth factor.

The Molecular Biology of Cancer: A Bridge From Bench to Bedside, Second Edition. Edited by Stella Pelengaris and Michael Khan.
© 2013 John Wiley & Sons, Inc. Published 2013 by John Wiley & Sons, Inc.

- Tumors are highly complex structures concocted of often multiple distinct subpopulations of cancer cells, endothelial cells, and stroma and imbued with the paraphernalia of chronic inflammation.
- It is often said that cancers are forged in the fire of chronic inflammation, steeped in growth-promoting factors and cytokines, but thereafter are burnished by mutations that generate growth signals from within. Mutations can activate growth factor signaling at various different levels including:
  - enhanced production of growth factors, either directly by the cancer cells themselves or indirectly by stimulating their production by other cells in the microenvironment,
  - increased expression of growth factor receptors (particularly the receptor tyrosine kinases),
  - expression of constitutively active receptors that trigger downstream signaling even in the absence of a ligand,
  - activation of downstream signaling pathways directly or indirectly by loss of a normal negative feedback loop or inhibitory protein,
  - deregulated expression or activity of transcription factors that promote cell cycle.
- Of the growth-promoting signaling pathways, the overlapping RAS–RAF–ERK and PI3K–AKT pathways are among the most pivotal as they transduce mitogenic and growth or survival signals, respectively. They also illustrate the schemata described above. Thus, most human cancers have activating mutations at one or more of the following levels:
  - growth factor receptor tyrosine kinases (RTK), such as HER2 and epidermal growth factor receptor (EGFR), are constitutively activated or upregulated,
  - RAS may become activated by loss of inherent or facilitated inactivating GTPase activity,
  - BRAF mutations can activate signaling downstream of RAS,
  - loss of tumor suppressors, such as NF-1 and PTEN, can contribute to aberrant activation of the pathway,
  - transcription factors, such as Myc, may be deregulated.
- It is rare for redundant mutations to coexist, so in general either RAS or RAF are mutated not both. Moreover, mutations in either will remove the requirement for EGFR activation and render such cancer cells unresponsive to EGFR inhibitors.
- Growth factors are not usually oncogenes or produced in excess as a direct result of mutations, although some tumor cells produce and secrete growth factors. In many cases growth factors are synthesized by nurturing stromal cells or as part of inflammatory processes that often underlie the development of a carcinoma.
- In order to activate signaling pathways, growth factors usually require integrin-mediated cell adhesion. This ensures activation only occurs when cells are in their correct environment.
- Two highly cancer-relevant processes, epithelial–mesenchymal transition (EMT) and formation of cancer stem cells, as well as the signaling pathways that regulate them, are being explored as new drug targets.

Introduction

So divinely is the world organized that every one of us, in our place and time, is in balance with everything else.

Johann Wolfgang von Goethe

The size of an organism or organ is determined by its total cell mass – in other words, the total number of cells and their size. Four overlapping processes regulate cell mass and the control points in these processes are the preeminent targets for cancer-causing mutations:
- cell growth,
- cell division,
- cell differentiation, and
- cell death.

These processes usually combine together in a well-orchestrated and exquisitely balanced series of performances throughout ontogeny and adult life, ensuring that everything is in the right place at the right time and of the appropriate size. The responsibility for interpreting and conducting the repertoire rests with the specific and shared regulatory factors and signaling pathways that will be discussed in this chapter. The duration and timing of the performances is of central importance – an overenthusiastic conductor leading an unending series of encores will result in overgrowth and cancer.

The signals regulating cell size and cell number, originating from distant or neighboring cells, activate appropriate responses by "docking" with specific cell surface receptors or enter the cell to bind proteins inside the cell. Not all such extracellular signals are soluble secreted proteins but also include proteins bound either to the surface of other cells or to components of the extracellular matrix (ECM). Moreover, these can operate in various different ways and combinations to regulate cell division and number as shown in Table 5.1.

The factors regulating cell mass are conveniently divided into three classes based on their predominant cellular action, although individual factors may regulate more than one process.
- **mitogens** promote cell division largely by allowing G_1/S transition in the cell cycle;
- **growth factors** promote an increase in cell mass (cell growth) by enhancing protein synthesis;
- **survival factors** promote cell survival by suppressing apoptosis.

Students should be aware that the term "growth factor" (GF) is frequently employed to describe a protein that has any of these properties – as evidenced by the names by which almost all of them are known.

Platelet-derived growth factor (PDGF) was among the first proteins found to have a mitogenic action. Originally described as a serum mitogenic factor, PDGF was subsequently identified as the product of platelets released during blood clotting. Now more

Table 5.1 Some regulatory processes affecting cell proliferation

Growth factor dependence	Cell proliferation requires tissue-type specific mitogenic and survival factors, the absence of which may also trigger apoptosis. Cells are also exposed to a variety of growth-inhibitory factors such as TGF-β
Anchorage dependence	Cell proliferation requires transmembrane proteins called integrins to interact with components of the extracellular matrix (ECM) – again such contacts may be required to prevent apoptosis
Contact inhibition	Contact with like cell types inhibits cell movement and proliferation
Senescence	Observations of vertebrate somatic cells in culture have identified a limitation on the number of times a cell can divide (around 50–70 divisions for human cells) before the cells cease further division – the Hayflick limit. This is difficult to examine in the intact organism, although loss of telomeres, which is a major factor in determining the Hayflick limit, is also clearly a mediator of replicative arrest *in vivo*. What is clear is that senescence can be coupled to signals that regulate replication and, together with apoptosis (cell death), may act as a counterbalance, particularly to inappropriate or excessive activation of mitogenic signals within the cell
Apoptosis	Many mitogenic signals intracellularly are also coupled to pathways, which can induce apoptosis
Nutrient and oxygen supply	Expansion of cell numbers may result in a tissue outgrowing its blood supply and could result in deprivation of oxygen (hypoxia) or nutrients or accumulation of damaging waste products, all of which might in turn limit further tissue growth. Moreover, the endothelial cells lining the inside of the vessels may secrete survival factors also contributing to regulation of tissue growth. Thus, cells may undergo apoptosis or necrosis unless tissue growth is accompanied by new vessel development or expansion (angiogenesis)

All of these processes are also inherent barriers to neoplasia and must be overcome in order for tumors to form and progress.

than 50 different protein mitogens have been identified with different degrees of selectivity for certain cell types. In addition, there are factors, exemplified by members of the transforming growth factor-β (TGF-β) family, which also inhibit cell replication (cytostatic). Many factors have properties additional to those on cell division and can promote cell growth, survival, or motility. Table 5.2 gives a description of some key growth factors.

Once a given signaling process has been triggered, the message is propagated and disseminated within the cell in order to ultimately elicit some kind of response, which in the context of cancer might be to divide or become motile. In most cases, proteins involved in signaling cascades activate their targets transiently by joining together and then apart. Interactions require collisions and often result in altered phosphorylation, with the whole facilitated by having the interacting participants arranged in close proximity within the cell – a kind of molecular Morris dance.

In this chapter we will commence by identifying the links between growth factors and the cell cycle before discussing general concepts of growth and differentiation. We will then outline the normal regulation of cell proliferation by growth factors, the receptors through which they act, the downstream signaling pathways activated by these receptors, and finally the activation of nuclear transcription factors, which regulate genes involved in cell cycling (see Fig. 5.13 for a simple schematic of growth factor signaling). As all of these steps involve proteins whose deregulated activity has been shown to contribute to cancer (in particular the oncogenes), we will proceed to discuss how such processes become derailed during cancer formation.

Over the last few years, there has been an explosive increase in knowledge of the key roles played by microRNAs (miRNAs) in posttranscriptional control of gene expression (see Chapters 7 and 11), which is impacting essentially all areas of cell biology. Signaling pathways, which are often exquisitely sensitive to dose effects, appear to be particularly sensitive to miRNA-mediated regulation and this has been recently shown for epidermal growth factor (EGF), TGF-β, and WNT. Finally we will conclude by discussing how knowledge of growth factor signaling has been exploited to develop a new generation of targeted anticancer agents, including the tyrosine kinase inhibitors (TKIs). Signaling pathways feature in every chapter, but to avoid duplication a detailed description of a few key proteins and their upstream and downstream partners feature in other chapters. Thus, RAS, MYC, and SRC are described in Chapter 6 (oncogenes), survival pathways in Chapter 8 (cell death), tumor suppressors and DNA damage in Chapters 7 and 10, and cell adhesion and motility in Chapter 12. Drug treatments based on growth signaling are covered in Chapter 16.

It is worth noting here that signaling pathways are often described and illustrated as linear chains or isolated cascades, with activation of an upstream protein triggering activation of one or more proteins down the chain and so on. This is a simplification to aid understanding. Most steps have critical regulatory inputs, including not just negative and positive feedback loops but also extensive crosstalk between pathways. After all, it is highly improbable that cells will be exposed to growth factors individually. It is much more likley that cells are continually having to simultaneously integrate any number of different extracellular stimuli. In fact, crosstalk and integration is a way by which complexity of responses can result through the presence of key nodes, as exemplified by the G_1/S transition, where the cell must reach a "decision" based on integrating all the information being delivered to it. As will be seen, the result of crosstalk can be to aggregate a number of weak, below-threshold signals into a response or for one mutually antagonistic signal to shut off another if sufficiently strong. Moreover, by focusing on concentration and amplitude of a signal, the importance of dynamics, such as rates of change, oscillations, and frequency, to signaling may be lost and foster the often erroneous impression that cell fate is solely determined by proteins fixed in an active or inactive conformation.

Table 5.2 A short list of some known growth factors

Growth factor	Principal source	Main physiological role	Role in cancer
PDGF	Platelets, endothelial cells	Proliferation of connective tissue cells, glia, and smooth muscle cells	Two different protein chains form three distinct dimer forms; AA, AB and BB. The product of the c-Sis proto-oncogene has been shown to be homologous to the PDGF A chain
EGF	In large quantities in salivary glands. Produced by a variety of other cell types	Proliferation of mesenchymal, glial and epithelial cells	The EGF receptor is a member of a family of four receptors (EGFR (HER1 or ERBB1), ERBB2 (HER2/neu), ERBB3 (HER3), and ERBB4 (HER4)). EGFR promotes deregulated growth in many types of cancer, including breast and lung, and activating mutations are known. Two new drugs that target the receptor or its downstream kinase activity are now in use in cancer – cetuximab and gefitinib, respectively. The Neu gene was identified as an EGF receptor-related gene in an ethylnitrosourea-induced neuroblastoma and is implicated in breast cancer and others. This is also targeted by a monoclonal anticancer agent called trastuzumab
TGF-α	Transformed cells	Normal wound healing	Often produced by cancers. May activate similar receptors and signaling pathways as EGF
FGF	Multiple cell types, protein is associated with the ECM	Promotes proliferation of many cells; inhibits some stem cells; induces mesoderm to form in early embryos	At least 19 family members, four distinct receptors. Kaposi's sarcoma cells (prevalent in patients with AIDS) secrete a homolog of FGF encoded by the K-FGF proto-oncogene. The Flg gene (Fms-like gene) encodes a form of the FGF receptor
M-CSF	Monocytes, granulocytes, endothelial cells, and fibroblasts	Proliferation and differentiation of hematopoietic cells	The c-Fms gene encodes the M-CSF receptor (CD115), first identified as a retroviral oncogene
NGF	Schwann cells, fibroblasts, sites of sensory and sympathetic innervation	Promotes neurite outgrowth, neural cell survival	NGF was identified as secreted by adrenal medullary tumors. The Trk genes encode the NGF receptor-like proteins. The first Trk gene was found in a pancreatic cancer. It is now known to comprise several related genes: trkA, trkB, and trkC. Numerous other "neurotrophic" factors are now known, including BDNF, CNTF, GDNF
HGF	Many cell types including stromal cells.	Potent mitogen for mature parenchymal hepatocyte cells. Involved in cell–matrix interactions and cellular migration in a wide range of tissues and cell types	HGF and its receptor c-Met are involved in cancer invasion and metastasis. Paracrine activation of c-Met by stromal-derived HGF facilitate invasion and metastasis. Autocrine or mutational activation of c-Met is associated with the progression of malignant tumors
Erythropoietin	Kidney	Promotes proliferation and differentiation of erythrocytes	Used therapeutically in some cases of anemia
TGF-β	Inflammatory cells, stroma	Anti-inflammatory, promotes wound healing, inhibits proliferation of many adult cell types	May either be anti- or procancer depending on timing
IGF-1	Liver, endothelial cells	Promotes survival of many cell types and proliferation of some	Related to IGF-2 and proinsulin, also called Somatomedin C
IGF-2	Variety of cells	Promotes proliferation of many cell types primarily of fetal origin	Related to IGF-1 and proinsulin. Secreted by some cancers as an autocrine/paracrine survival factor

So we should appreciate that dynamics of oscillations, as opposed to simply amplitude (as has been demonstrated for the NF-κB pathway) may generate different cell responses, even if for now we file away that knowledge for future reference. After all, with an estimated 150 000 bona fide binary protein interactions alone in the human protein universe we have our work cut-out just to determine the members and sequence of entire functional protein networks.

Another issue often clouded in diagrams of signaling pathways relates to how proteins that occupy adjacent positions in a signaling cascade ever manage to get together within the comparatively vast and fluid volume of the cell. It is little use if an activated protein never manages to bump into a target protein in order to propagate the signal. Imagine throwing grains of rice around inside a gothic cathedral. How often will they collide with one another? Clearly, steps must be taken to maximize the chances of such interactions. Well, one way would be to flick the grains along the floor or upon a single tile space, thus increasing the probabilities of a "collision" by constraining molecules to an effectively two-dimensional playing field. An alternative would be to line them along the pillars from floor to ceiling, which could also ensure that the direction of travel is anatomically apt. The cellular counterparts of this would be localizing prenylated proteins to the inner aspect of the cell membrane and the lining up of signaling molecules along cellular scaffolds, such as cytoskeletal elements from surface to nucleus.

These issues notwithstanding, for most purposes the simplified model of signaling is sufficient, as it allows us to understand the context of cancer-causing mutations within functional categories and also to unravel and predict how new and existing drugs alter cell behavior. In the future we may also learn how to target cancer signaling by interrupting localization (the rationale behind farnesyl transferase inhibitors to impede RAS signaling) and even to alter dynamics.

Growth factor regulation of the cell cycle

> People are like earth. They can either nourish you and help you grow as a person or they can stunt your growth and make you wilt and die.
>
> *Plato*

Cells in early embryos replicate very rapidly (a mere half hour) and can start DNA replication as soon as mitosis is completed. This is due to the absence of gap phases G_1 and G_2 (normally present in cycling cells of the adult; see Chapter 4), and likely serves the needs for rapid expansion of cell numbers following fertilization. However, later in development and particularly in the adult, this replication potential is restrained and subject to tighter controls. As discussed in the previous chapter, the adult cell cycle contains two gap periods: G_1 phase between M (nuclear division) and S (DNA synthesis) and a G_2 phase between S and M. Not only do these gaps allow for a period of monitoring and correction of any DNA damage or errors that may have arisen during DNA replication, but critically G_1 also provides the opportunity for many diverse signals to be integrated into a decision as to whether the cell will actually replicate, pause, or indeed remain alive. To a degree, a similar scenario applies to terminally differentiated cells in G_0, seeking to re-enter the cell cycle (Fig. 5.1). Importantly, it is increasingly apparent that any given cell is probably exposed simultaneously to a variety of stimulatory and inhibitory signals, arising locally and via the circulation (Fig. 5.2), all of which have to be integrated by that cell before it decides whether to replicate. This integration is performed by complex interacting networks of intracellular signals generated in response to these extrinsic regulating factors. It is the net balance of all these responses that determines whether a cell enters the cell cycle (Fig. 5.3).

Growth factor signaling systems are now believed to operate as complex **networks** rather than the traditional view of signaling as linear pathways. Complex signaling networks enable integration of multiple inhibitory and stimulatory inputs into binary choices by the cell, including: replicate or arrest; survive or die. Moreover, there are numerous sites of overlap between signals regulating progression through the cell cycle and those that can permanently prevent a cell from replicating by promoting either senescence or even cell death.

Although there is considerable overlap and crosstalk, broadly two of the most important signaling pathways activated by various growth factors are:
- the mitogenic RAS–RAF–ERK (MAPK) cascade and a key transcription factor, c-MYC (Fig. 5.4) and
- the PI3K–AKT survival pathway.

Together, these pathways may act synergistically to allow cell-cycle entry and progression while also avoiding various failsafe mechanisms seeking to trigger cell death and growth arrest. Not surprisingly, RAS, c-MYC (see Chapter 6) and various activators of the PI3K–AKT survival pathway are frequently involved in cancer and will often feature in subsequent chapters.

Mammalian cells must pass through a restriction point, or "R" point (equivalent of the start point in yeast cells), before committing themselves to entering the S phase and duplicating their entire DNA. Importantly, once the cell cycle has progressed beyond this point (G_1/S transition) it will proceed even if the mitogenic stimuli are removed. The net result of many positive and negative growth-regulating signals normally determines whether a cell makes the G_1/S transition and proceeds to DNA replication, and, unsurprisingly, such controls are usually lost in cancer cells. Diverse growth-regulating factors operate largely at the level of activation of cyclin-dependent kinases (CDKs), which drive the "cell cycle engine." Prominent among these regulatory factors are the polypeptide growth factors, which either promote or inhibit the G_1/S transition (Fig. 5.5).

Factors that promote the cell cycle (mitogens) generate signals resulting in increased activity of cyclin D–CDK holenzyme complexes during the G_1 phase, which, as described in Chapter 4 (Fig. 4.2), promote phosphorylation and inactivation of the tumor suppressor retinoblastoma (RB) protein a prerequisite for releasing the brakes on the cell cycle and allowing expression of S-phase genes. Conversely, inhibitory "cytostatic" signals, in particular the transforming growth factor-β (TGF-β), among other actions, activate inhibitors of CDK4, such as $p15^{INK4b}$, thus preventing RB phosphorylation. Cytostatic signals may also activate two other CDK inhibitors (CKIs) (see Chapter 4), $p21^{Cip1}$ and $p57^{Kip2}$.

It is also important to reiterate here that apart from controlling cell growth and cell cycle, some growth factors also activate signaling pathways that affect other cellular processes including cell survival, migration, and differentiation – cellular processes that if disrupted inappropriately are important in the behavior of the cancer cell (see Table 5.1).

Growth Signaling Pathways and Targeted Treatment

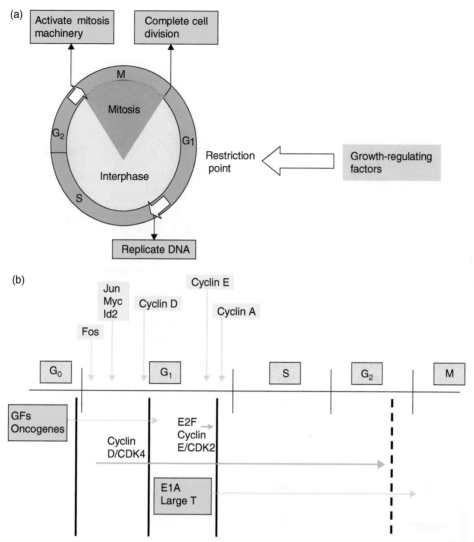

Figure 5.1 Growth factors act particularly at the G$_1$ phase of the cell cycle. (a) The point at which cells commit to replicate their DNA and enter the cell division cycle is controlled by the RB protein, the product of the tumor suppressor gene *RB*. RB acts as an "off switch" for entry into S phase by binding the transcription factor E2F required for expression of S-phase genes. After appropriate growth factor stimulation, the inhibitory effects of RB are removed by phosphorylation, which allows E2F to activate S-phase genes during early G$_1$ required for cell cycle progression. (b) A special case for differentiated cells re-entering cell cycle. Blocks to replication in "terminally differentiated" quiescent cells are somewhat different to those generally described for the regulation of cell cycle in other cells, probably because less is known, but clearly show some overlap. Three blocks to cell cycle progression in "quiescent cells" are shown in thick black lines, which can be overcome by different mitogenic stimuli. The mid-G$_1$ block is overcome by cyclin D–CDK4 activity induced by growth factor-regulated genes such as c-*MYC*. Viral proteins such as E1A and T antigen are not susceptible to these blocks and are capable of pushing previously quiescent cells through one or more cell cycles, but then result in apoptosis, which restrains oncogenic potential. E2F or cyclin E–CDK2 expression alone cannot overcome the late G$_1$ block, while cyclin D1 and CDK4 alone cannot take the cycle beyond G$_2$.

Growth homeostasis and tissue repair and regeneration

In every animal which has not passed the limit of its development, a more frequent and continuous use of any organ gradually strengthens, develops and enlarges that organ, and gives it a power proportional to the length of time it has been so used; while the permanent disuse of any organ imperceptibly weakens and deteriorates it, and progressively diminishes its functional capacity, until it finally disappears.

Jean-Baptiste Lamarck

Multicellular organisms, like human societies, rely on the cooperative behavior of their constituents. This is particularly important with respect to ensuring that cell numbers, functions, and localization are appropriately maintained. In general, such control over cell behaviors becomes increasingly stringent as the organism completes its development – the majority of cells sacrificing a large part of their potential to replicate and becoming more specialized in their functions. Thus, cells stop cycling and differentiate to give rise to populations of cells that exhibit specialized functions, such as skin, nervous system, etc. Importantly, within the adult organism, cells repress their innate potential to grow and divide beyond their normal boundaries and contrary

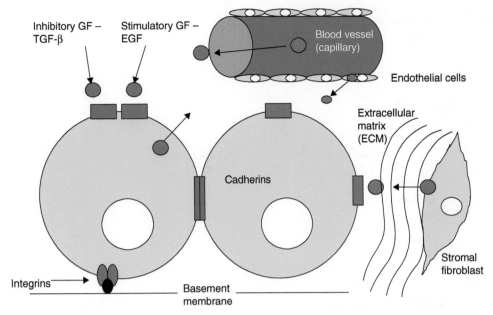

Factors regulating cellular replication and survival include secreted GFs as well as cell-cell and cell-matrix interactions. GFs may be produced locally by the cancer cell or cells of a different type and act by an autocrine (on the same cell) or paracrine (on a neighbouring cell) route. GFs can be delivered via the circulation (endocrine) or be secreted from the vascular endothelial cells or by stromal cells. Stromal cells can also produce MMPs that can act to liberate GFs from matrix or can directly affect production of ROS and thus DNA damage in the cancer cell. Recruited immune cells, including activated T lymphocytes and macrophages may also produce growth-regulating cytokines.

Figure 5.2 Growth-regulatory factors derive from local and circulating sources. Factors regulating cellular replication and survival include secreted growth factors as well as cell–cell and cell–matrix interactions. Growth factors may be produced locally by the cancer cell or cells of a different type and act by an autocrine (on the same cell) or paracrine (on a neighboring cell) route. Growth factors can be delivered via the circulation (endocrine) or be secreted from the vascular endothelial cells or by stromal cells. Stromal cells can also produce MMPs that can act to liberate growth factors from matrix or can directly affect production of ROS and thus DNA damage in the cancer cell. Recruited immune cells, including activated T lymphocytes and macrophages may also produce growth-regulating cytokines.

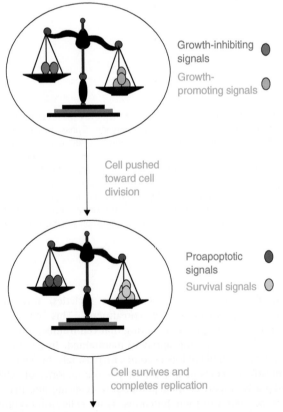

Figure 5.3 Cell division depends on the balance of different growth-regulating signals. Ultimately, the decision whether a cell lives or dies, replicates, or does not is dependent on the balance of positive and negative signals received.

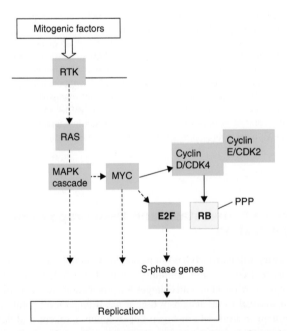

Figure 5.4 Main signaling pathways for mitogens. Mitogenic factors largely, but not exclusively, act via RTK to activate in particular the MAPK/ERK cascade.

Growth Signaling Pathways and Targeted Treatment

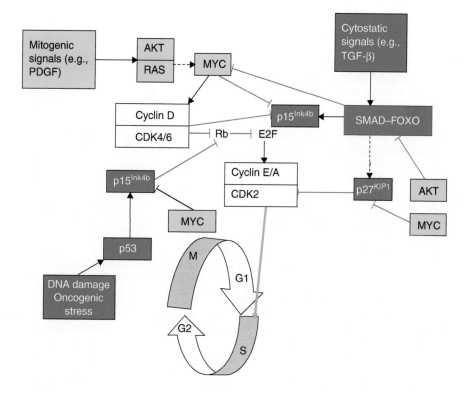

Figure 5.5 Positive and negative growth regulators determine G_1/S transition and cell cycling. SMAD and FOXO together mediate inhibitory action of TGF-β on cell replication by activating CDK inhibitors (CKI). In contrast, growth factors and c-MYC act to promote cell replication by upregulating cyclin D and inhibiting activation of CKIs such as p15^{INK4b} and p27^{KIP1}.

to patterns dictated by the overall developmental plan of the organism. This can be a particularly tricky business given that the rules normally restraining these behaviors can be relaxed when required for wound healing, regeneration of damaged tissue, or to allow adaptive growth of a particular tissue or cell population to changing demands (e.g. growth and modulation of breast tissue during lactation; growth of uterine tissue during the menstrual cycle) (Fig. 5.6).

Regenerative and adaptive capacity differs greatly across organs and organisms. Under such circumstances, many cells may be given a degree of freedom to expand numbers and to migrate into what is for them an unusual location, where they would not normally be allowed. The major challenge, given the obvious similarities in behavior to those of cancer cells, is to ensure that enhanced cell replication and motility are temporary. Therefore, when the tissue in question has returned to its normal state (e.g. the wound has healed, or lactation has ended), the normally strict controls are reapplied successfully. A simple, clear example of such a case is given in Chapter 12 (see Box 12.4).

Regenerative processes that involve unleashing unrestrained cellular migration and replication carry with them an inherent risk of cancer that is counterbalanced by several restraining processes. Important among these are limiting the extent and duration of regenerative capacity (note the cancer risk of chronic

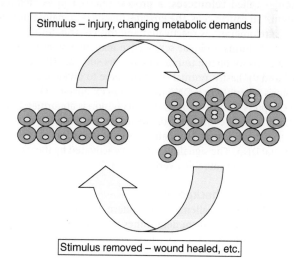

Figure 5.6 Restraints on cell replication and motility may be temporarily released. During injury to tissues cells may be allowed to replicate and migrate, but this is transient.

inflammation) by initiating and terminating signals, and tight control over stem cells, dedifferentiation, transdifferentiation, and proliferation and patterning.

Thus, in multicellular organisms, cellular cooperation is normally ensured by numerous extrinsic and intrinsic factors, which together regulate cell growth, proliferation, motility, and survival in order to best serve the requirements of the whole organism (see Fig. 5.2). It is not surprising then that disruption of the tight controls normally imposed on these processes in the adult can have drastic effects on cells and tissues of the body, leading to diseases such as cancer. In fact, one of the common defining features of cancer is that of uncontrolled and unscheduled cellular proliferation; cancer cells divide relentlessly and are able to resist the usual mechanisms regulating replication in normal cells. Normal cells require external growth factors (mitogens) to divide – at least in order to make the G_1/S transition. When synthesis of these growth factors is inhibited or access restricted by some other means, then cells stop dividing. Normal cells also respond to growth-inhibiting signals, such as TGF-β.

One of the "hallmark" features of cancer is autonomous cell division, such that cancer cells are no longer dependent on such positive or negative growth factors. Cell behavior is not just regulated by secreted factors but also by molecules on the surface of neighboring cells, such as integrins and adhesion molecules (see Chapter 12). Normal cells show contact inhibition *in vitro*; that is, they respond to contact with other cells by ceasing cell division. This characteristic is lost in cancer cells, as demonstrated in cell culture where cancer cells continue to divide and "pile up" to form transformed colonies *in vitro* (see Chapter 6 and Fig. 6.3).

Normal cells also age and die, and are then replaced in a controlled and orderly manner by new cells when possible. It is widely accepted that there are limits on the number of times that normal cells can divide (at most 50 times in cell culture) before they exit permanently from cell division either by becoming senescent or by dying. This intrinsic cell division "counter" is operated at least in part by the progressive shortening of structures at the ends of chromosomes called **telomeres**; a process referred to as "telomere attrition." In cancer cells, the enzyme involved in maintaining telomeres, telomerase, is reactivated and this contributes to continued cell divisions. Cellular senescence may also be promoted by mechanisms involving tumor suppressors such as RB (see Chapters 7 and 9). Lastly, normal cells will cease to divide and undergo apoptosis when there is DNA damage too extensive to be repaired (see Chapters 8 and 10), or when cell division is abnormal. In contrast, cancer cells continue to divide under these conditions (e.g. in cells that have lost p53), enabling the propagation of cells with abnormal DNA. The signaling pathways activated by these various factors show considerable overlap (Fig. 5.7).

An organ-size checkpoint

One of the great challenges for modern biology was to unravel exactly how an organism determines that a given tissue is of the correct size (or for that matter how an individual cell determines how big it is). A checkpoint, analogous to those described in Chapter 4, has been proposed; this "organ-size checkpoint" in some way determines the size of an organ and, if required, triggers corrective processes that will adjust cell size and number. Broadly, organism or tissue size is regulated through balancing of positive and negative growth regulatory signals within cells.

Much remains to be learned about this fundamental process but studies in fruit-flies and nematode worms have revealed

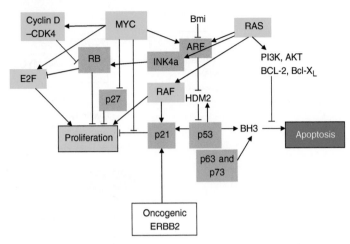

Figure 5.7 Cooperation between oncogenic signaling pathways that regulate cell replication, and apoptosis. Activation of RAS and MYC results in potential engagement of both replication and growth but also of apoptosis and possibly growth arrest. MYC triggers G_1/S transition and cell cycle entry by regulating expression of cyclin D and CDK4 (upregulated) and p21^{CIP1} and p27^{KIP1} (downregulated). Together with cyclin E–CDK complexes (not shown), the net effect is hyperphosphorylation of RB and release of E2F for transcription of S-phase genes. If either MYC or RAS levels are excessive (as might occur during oncogenesis) or other proapoptotic signals are received, then the balance may be tipped away from replication. RAS can promote senescence through either p16^{INK4a} or ARF, which activate RB or p53 pathways respectively, while MYC can inhibit growth arrest/senescence by inhibiting p21^{CIP1} and inducing TERT (telomerse reverse transcriptase). RAS may also activate p21^{CIP1} via RAF activation. Although MYC may activate the apoptotic pathway (e.g. via ARF), RAS is able to suppress apoptosis by activating the PI3K pathway and, subsequently, AKT. It can readily be appreciated how oncogenic MYC and RAS may conspire in oncogenesis. The combination of RAS and MYC acting together provides a potential means of avoiding apoptotic and growth arrest mechanisms, activated by either acting individually. Moreover, it can also be appreciated how inactivating mutations in RB or p53 may collaborate in tumorigenesis by enabling the cancer cell to avoid either growth arrest or apoptosis or both.

some basic principles. First, nutrient signals are important players, and, possibly via hormones such as insulin and insulin-like growth factors (discussed later), these activate intracellular signaling pathways that regulate both cell size and replication/survival. One protein activated by such signaling, mTOR (mammalian target of rapamycin), is of particular interest and is discussed later. Very important regulatory processes also involve cell death control. The two most important in this context are (1) activity of proteins such as c-MYC which can activate both cell replication/growth on the one hand and also apoptosis on the other and (2) a particular form of apoptosis, termed anoikis, which is triggered in cells that have inadvertently left their usual tissue location, have breached a tissue boundary, and have wandered into an inappropriate location (see Chapter 8). Finally, recent studies suggest that the organization of a tissue plays an important role in controlling cell growth. Thus, cellular polarity (many cells, particularly in epithelia, usually face in a particular direction – "up or down") and interactions between neighboring cells, basement membranes, and extracellular matrix all influence cellular behavior, including differentiation, replication, and survival (see Chapter 12).

Specific signaling pathways other than those traditionally linked to replication and survival have been identified as important in determining size. These include the aforementioned mTOR, but also the regulators of stem cells and "stemness" in cancer cells (Wnt, Notch, and Hedgehog pathways) and the conserved Hippo tumor suppressor pathway (all discussed in more depth later in the chapter).

Regulated and deregulated growth

There are some trees, Watson, which grow to a certain height and then suddenly develop some unsightly eccentricity. You will see it often in humans. I have a theory that the individual represents in his development the whole procession of his ancestors, and that such a sudden turn to good or evil stands for some strong influence which came into the line of his pedigree. The person becomes, as it were, the epitome of the history of his own family.

Sir Arthur Conan Doyle, The Adventure of the Empty House

In the living intact organism changes in cell number reflect the balance of cell proliferation and cell death, so that an increase in tissue mass can result from either suppression of cell death (apoptosis) or increase in proliferation. However, processes regulating cell volume or size are increasingly appreciated as independent contributors to cell or tissue mass (see Box 5.1).

Self-evidently, in order for oncogenic (cancer-causing) mutations (see Chapter 6) to give rise to cancer, the mutated cells must divide or else the mutation will not be propagated. It is widely accepted that cancers are most likely to originate from the most replicatively active cells, such as stem cells or progenitor cells (see Box 5.2). In part this reflects the increased probability of acquiring mutations when DNA is being copied during the cell cycle and also the fact that this will then be propagated. However, many other properties of stem cells resemble those found in cancer cells, such as longevity and self-renewal.

In general, the turnover of cells in most adult organs is slow, with small numbers of cells lost being replaced by replication of surviving cells of the same type or by neogenesis from stem cell precursors (Fig. 5.8) or even from cells of another type (transdifferentiation). Despite rapid progress in this area, it may be surprising to learn that there is still considerable debate in many situations as to the relative contributions of these various processes of cell renewal to normal tissue homeostasis and, moreover, to situations where more substantive cell renewal is required, such as after injury or during regeneration. During development, in general cell replication exceeds cell death and the organism grows, whereas in adult life the processes are balanced and tissue size is maintained. When mutations occur in nondividing cells, such as neurons, they rarely induce cancer, which accounts for the rarity of such tumors in adults. However, some tissues with usually low rates of turnover exhibit considerable plasticity during adult life; for example, many endocrine glands can undergo adaptive increases in mass in response to increased demands for their hormone output.

Importantly, such adaptive growth is not inevitably a response to a pathological process, but is a normal part of physiology, for example during pregnancy, where demands for various hormones such as insulin increase. Particularly in this latter scenario, adaptive growth of insulin-secreting pancreatic β cells is usually followed by β-cell apoptosis and involution of the previously expanded islet tissue post partum, when the demand for insulin is reduced. Intriguingly, recent studies employing a variety of tools for *in vivo* lineage-tracking suggest that both replication of pre-existing β cells and differentiation from progenitor cells or cells of another lineage contribute to β-cell mass regulation with preference determined by the nature of the damaging insult and relative availability of suitable precursors. Adaptive growth may also occur in response to nutrient deprivation or disorders in metabolism. Thus, the parathyroid glands may undergo hyperplasia (increase beyond normal cell mass) in response to reducing serum calcium (secondary hyperparathyroidism).

Importantly, under some circumstances, such as that following a prolonged or frequently repeated period of adaptive growth, hyperplasia of a tissue may become autonomous and continue even after the stimulus has been removed. A well-known example of this occurs in the parathyroid glands – a condition known as **tertiary hyperparathyroidism**. In other circum-

---

Box 5.1 Regulation of cell size – additional information

Tissue mass may increase by increases in cell volume as well as increases in cell number.

Although clearly linked, it has become increasingly apparent that the processes of cell replication and cell growth (increase in cell volume) may be independently regulated, although growth is under normal circumstances a usual precursor to replication. Increased cell proliferation requires a general increase in protein synthesis and a specific increase in the synthesis of replication-promoting proteins. Transient increase in the general protein synthesis rate, as well as preferential translation of specific mRNAs coding for growth-promoting proteins (e.g. cyclin D1), occur during normal mitogenic responses. Several signal transduction pathways involved in growth such as RAS–MAPK, PI3K, and mTOR-dependent pathways activate the translational machinery. Cell growth (an increase in cell mass and size through macromolecular biosynthesis) and cell-cycle progression are generally tightly coupled, allowing cells to proliferate continuously while maintaining their size.

The target of rapamycin (TOR) is an evolutionarily conserved kinase that integrates signals from nutrients (amino acids and energy) and growth factors (in higher eukaryotes) to regulate cell growth and cell cycle progression coordinately. In mammals, mTOR is best known to regulate translation through the ribosomal protein S6 kinases (S6Ks) and the eukaryotic translation initiation factor 4E-binding proteins.

In cancer, oncogenic activation of RAS or MYC may lead to continuous upregulation of key elements of translational machinery. On the other hand, tumor suppressor genes (p53, pRB) downregulate ribosomal and tRNA synthesis. The RAS and AKT signal transduction pathways play a critical role in regulating mRNA translation through activation of the initiation factor eIF4E, which binds the 5′ cap of mRNAs. Not surprisingly, therefore, eIF4E is overexpressed in many human cancers and controls the translation of various malignancy-associated mRNAs, which are involved in polyamine synthesis, cell-cycle progression, cell survival, angiogenesis, and invasion.

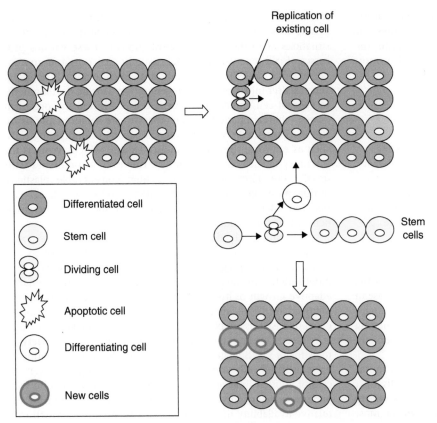

Figure 5.8 Tissue mass homeostasis. Tissue mass in the adult is maintained by balancing cell losses with cell renewal. Cells lost through apoptosis (necrosis or other cell death) or by shedding are normally replaced by replication of existing differentiated cells of the same, or occasionally even of another, cell type or by replication and differentiation of adult stem cells. Self-evidently if cell losses in any tissue exceed the ability for renewal then the tissue will shrink (good examples being loss of insulin-producing beta cells in human type 1 and type 2 diabetes), conversely if cell gains exceed losses then the tissue will expand (as exemplified in the initial response of beta cells to conditions demanding more insulin production). Many tissues are capable of switching between these conditions, thus enabling adaptation to demands – tissue plasticity. Some tissues are regarded as static, with effectively no turnover in the adult. Any losses in such tissues will result in progressive involution or atrophy of that tissue with time.

Box 5.2 Stem cells – additional information

Stem cells generate the differentiated cell types within many organs throughout the lifespan of an organism. Stem cells have been defined as clonogenic cells that undergo both self-renewal and differentiation to more committed progenitors and functionally specialized mature cells. Stem cells characteristically self-renew and differentiate into a variety of cell types. It is believed that of the two progeny of stem cell division, one normally undergoes differentiation to the required mature cell and the other to replenish the stem cell compartment. This is also described as **asymmetrical cell division**. Some stem cells, described as **totipotent cells**, have tremendous capacity to self-renew and may differentiate into all cell types. Embryonic stem cells have pluripotent capacity, able to form tissues of all three germ layers, but cannot produce a complete organism.

Embryonic stem cell research has opened up many exciting possibilities, including cell or tissue replacement therapy for diseases such as diabetes, reproductive cloning, and an improved understanding of many aspects of developmental biology. However, these remarkable opportunities are matched by difficult ethical issues, primarily relating to the harvesting of stem cells from human embryos. Adult stem cells are therefore a very attractive option as their acquisition and application is not associated with the same degree of ethical objections. However, such adult stem cells may have a restricted capacity to differentiate – although this view is currently being challenged.

The existence of multipotent somatic stem cells in bone marrow has been reported. Under appropriate experimental conditions, some bone marrow stem cells may differentiate (or transdifferentiate) into a variety of non-hematopoietic cells of ectodermal, mesodermal, and endodermal origins. However, this area is controversial, with detractors suggesting that promising *in vivo* data using adult stem cells to repair various tissues may not be transdifferentiation, but instead fusion of adult stem cells with tissue cells and subsequent adoption of their phenotype.

There are important links between stem cells and cancer cells; many cancers may originate in stem cells, a not surprising fact given their greater replicative potential at the outset. Mammalian aging occurs in part because of a decline in the restorative capacity of tissue stem cells. Replicative senescence is under the control of the p53 and RB tumor suppressor pathways. Stem cells may be protected from malignant transformation by tumor suppressor mechanisms including $p16^{INK4a}$–Rb, ARF–p53, and the telomere, which may also limit the lifespan of the stem cell and contribute to aging. Moreover, cellular senescence, by limiting the proliferation of damaged cells, may act as a natural barrier to cancer progression. One mechanism for senescence may be the recruitment of heterochromatin proteins and the Rb tumor suppressor to E2F-responsive promoters and repression of E2F target genes. Interestingly, recent studies suggest that acute loss of Rb in senescent cells can lead to a reversal of cellular senescence. Given the importance of stem cell replication throughout life, processes may exist to limit senescence. Bmi1 is required for the maintenance of adult stem cells in some tissues partly because it represses genes that induce cellular senescence and cell death. The area of stem cells and cancer is a major area, which is covered at length in other chapters, particularly Chapter 9.

stances, hyperplasia may indeed progress to formation of a benign (adenoma) or malignant (adenocarcinoma) neoplasia. Presumably in these cases, the growth-promoting environment may also provide a fertile one for the accumulation of oncogenic mutations and for expansion of mutated cells. In this scenario, such oncogenic mutations (e.g. activated c-*Myc* or *Ras*; loss of *p53* or *Rb*) would drive excessive cell replication independently of the normal regulatory signals (see Chapters 6 and 7).

Cancers occur in many tissues populated predominantly by nondividing differentiated cells. Stem cells are the major or only usual source of new cells in such tissues, which include epidermis and gut epithelium where cell losses are extensive (frequent mechanical damage, normal regular sloughing of cells from the surface), or when the normally minimal cell losses in a given tissue are increased after injury or by inflammation. It is generally believed that stem cells may be the major origin of cancers in these cases, though many recent studies have shown that terminally differentiated cells can readily become less differentiated and re-enter the cell cycle under certain circumstances and may even be induced to acquire stem cell characteristics (stemness) by undergoing EMT. Together this supports the notion that not all cancers necessarily arise in the stem cell compartment (discussed in more detail under cancer stem cells). In general, differentiation of a cell greatly reduces its replication potential. Under some circumstances, such as during inflammatory conditions, even differentiated epidermal keratinocytes may replicate, though this is associated with some loss of or delay in acquiring of the normal fully differentiated phenotype. The association between differentiation and replication is clearly seen in human tumors, with the most poorly differentiated cells (for example in the metastases) exhibiting the shortest cell division times.

Cellular differentiation

> The elementary parts of all tissues are formed of cells in an analogous, though very diversified manner, so that it may be asserted, that there is one universal principle of development for the elementary parts of organisms, however different, and that this principle is the formation of cells.
>
> *Theodore Schwann*

The cells comprising the various tissues of a multicellular organism often exhibit dramatic differences in both structure and function. In fact, if you consider the differences between a mammalian nerve cell and a lymphocyte it becomes difficult to imagine that the two cells have identical genomes. In fact, for many years differentiation was held as evidence against the central dogma of genetics and indirectly contributed to the beginnings of what we know as **epigenetics** (see Chapter 11) – namely, information that can be passed on to the progeny during cell division not encoded in the DNA. Originally, it had been suggested that differentiation might actually involve the irreversible loss of some genes (DNA). This is now known not to be the case, rather differentiation involves the progressive alteration of gene expression. Various factors may produce the unique patterns of gene expression characteristic of any given cell type, including silencing of some genes by methylation and chromatin modification (epigenetics) and selective activation of others.

Importantly, it is increasingly apparent that many cell types do not irreversibly alter gene expression during differentiation, rather they may actually retain the ability to adopt a more primitive developmental state if required and stimulated by the appropriate signals or potentially under the influence of cancer-causing mutations or epigenetic alterations. In fact, it may be a normal feature of cells that retain the potential to replicate in the adult (and this does not just apply to stem cells), to "de-differentiate" somewhat in order to be able to replicate. This is well illustrated in cancer cells, which are highly replicative and almost invariably show some loss of the normal differentiated features of the cell of origin. Thus, in several tumor models activation of c-Myc may cause loss of differentiation alongside cell-cycle entry.

It has long been appreciated, and is being increasingly explained at a molecular level, that embryonic development, tissue injury and repair, chronic inflammation, and cancer all have many common features that distinguish them from normal tissues. It is not intended to give an exhaustive list here but rather to indicate some general principles. Common features include a greater replicative and motility potential of cells, increased presence of cytokines and growth factors, more cells undergoing EMT, and showing stemness and reactivation of many signaling pathways usually confined to ontogeny or adult stem cells. Two examples that show reactivation of signaling otherwise confined to development are described here:

- Re-expression of IGF-2 due to somatic epimutations has often been shown in rodent and human cancers. It is functionally important in some cancers and in extreme cases can be sufficiently raised to increase IGF-2 protein in the circulation and mimic the effects of insulin causing low blood glucose levels. Whether re-expression of IGF-2 might actually contribute to initiation of cancer has been addressed in some studies. It has been shown that loss of imprinting (LOI) of *Igf2* (see also Chapter 11) in a mouse model doubles the risk of developing intestinal tumors and this is associated with a less differentiated phenotype of cells, even in the normal colonic mucosa. Similar observations have also been made in humans, with LOI of *IGF2*, suggesting that altered maturation of non-neoplastic tissue may be one mechanism by which epigenetic changes affect future cancer risk.
- The Hedgehog protein (Hh), which controls cell proliferation and differentiation in normal limb formation and bone differentiation, is also implicated in the self-renewal of stem cells in adult tissues. Persistent Hh pathway activation has been shown to contribute to various cancers, including some forms of skin cancer.

Differentiation therapy

The rationale behind differentiation therapy is the assumption that cancer cells have been fixed in an immature or less differentiated state and that this contributes to accelerated replication and avoidance of apoptosis. Differentiation therapy aims to drive cancer cells back into a more differentiated state and resume the process of maturation. It is therefore not in itself cytotoxic but instead restrains growth and may sensitize the cells to potential apoptotic triggers or chemotherapy. The first successful differentiation agent was all-*trans*-retinoic acid (ATRA) used to treat acute promyelocytic leukemia (APL). APL is the result of a translocation (an exchange of chromosome material) between chromosomes 15 and 17, which results in formation of a chimeric protein comprising the promyelocytic leukemia protein (PML) and the retinoic acid receptor alpha (RARa), which regulates myeloid differentiation. PML–RAR causes an arrest of maturation in myeloid cells at the promyelocytic stage, contributing to expansion of promyelocytes. Treatment with ATRA causes the promyeloctes to differentiate and may result in remission rates of up to 70%.

The concept of terminal differentiation

Traditionally, cells that have irreversibly lost the ability to proliferate during the process of developing specialized functions (differentiation) have been described as **terminally differentiated**. However, this term can be misleading because it implies that a given adult cell type cannot proliferate. In fact, many cell types previously believed terminally differentiated have subsequently been shown to be capable of proliferation; in the previous chapter the concept was introduced that some cells may remain quiescent (in G_0) for extremely long periods of time waiting for an appropriate stimulus to re-enter the cell cycle. So in general it may be preferable to think about differentiated cells that are not replicating as quiescent unless one is absolutely certain. In some cases, of which skin keratinocytes are a notable example, cells lose their nuclei during terminal differentiation and are as a consequence unambiguously irreversibly growth arrested. However, such an obvious example aside, one must be cautious in adopting the definition of terminal differentiation, as has been done in the past, to all those cell types that are perceived to be unable to proliferate to any useful extent and therefore would not usually contribute to tissue growth, maintenance, or repair.

One illuminating example would be that of the insulin-secreting β cells of the pancreatic islets. In this case it was long accepted that the β-cell mass was maintained and underwent adaptive increases in growth through the replication of a progenitor cell or cells, but, recent studies using "lineage tracking" have clearly shown that β cells can themselves re-enter the cell cycle and are, surprisingly, the predominant usual source of new β cells in the adult. Crucial to resolving the seeming paradox, namely, that differentiation and replication are generally incompatible, has been the repeated demonstration that differentiated cells can shake off some of their specialized functions (and change gene and protein expression to more closely resemble that of an earlier developmental stage) – termed cell "plasticity" – in order to replicate. Figure 5.1 shows a schematic representation of how cell-cycle entry and progression may be regulated in a well-studied, usually quiescent cell type, the myotube.

In fact, in most cancer cells, at least some of the "markers" of the presumed mature cell type from which the cancer cell may have originated are absent or reduced (compared to the normal cell counterpart). This also reflects directly on an important question in cancer biology – what is the cell of origin for various cancers? This is addressed in other chapters, but is worth noting here. Because cancer cells are in general less differentiated, does this imply that they have originated in stem cells that may have "differentiated" a bit, or conversely in mature cells that may have "de-differentiated" a bit?

These intriguing questions aside, what is clear is that increasing or restoring differentiation (specialized behaviors) in cancer cells may be one means of reducing or even removing their replicative or invasive potential. Thus, understanding the mechanisms that cause permanent loss of proliferative capacity or at least prolonged quiescence in differentiated cells is of major interest and may lead to the design of new therapeutic strategies aimed at targeting differentiation of cancer cells. Differentiation is not the only means by which cells may lose their ability to proliferate, however, and replicative senescence, which may be induced by activation of tumor suppressor pathways is discussed in more detail in Chapter 9.

Tissue growth and the "angiogenic switch"

During development and under some circumstances in the adult (e.g. wounding), tissue adaptation and regeneration are accompanied by the growth of new blood vessels, a process known as **angiogenesis** (see Chapter 14). The resultant supply of blood to the growing tissue – aside from delivering nutrients and oxygen and removing waste products and carbon dioxide – establishes an access route for circulating regulatory factors (endocrine route – Fig. 5.2) and immune cells (see Chapter 13). Importantly, it now seems increasingly likely that the vascular endothelial cells also provide growth factor signals that have a direct local influence on cell survival and growth of the tissue.

The notion that tissue expansion per se may be associated with the concurrent provision of an appropriate vasculature is not a novel one. For angiogenesis to take place, an appropriate increase in angiogenesis-promoting factors and or a decrease in angiogenesis-suppressing factors must occur. However, for many years based on landmark studies by Judah Folkman, Doug Hannahan, and others the prevailing view has been that during tissue expansion in cancer an "angiogenic switch" was required and probably conferred by a specific and distinct mutagenic event that enabled a would-be cancer to become angiogenic and thereby for cancer to progress. By implication, such a switch would activate angiogenesis-promoting factors or deactivate suppressors, thus enabling a tumor to become angiogenic and expand. However, although this may be the case in some cancers, recent results challenge the idea that this is a general requirement. Thus, tissue growth under the influence of some oncogenes, notably c-*Myc* and *Ras*, may automatically be associated with an angiogenic response without a further "oncogene driven switch." In other words, the activity of these proteins includes activation of angiogenesis. In some cases this may relate to the development of a deficiency in oxygen supply (hypoxia) as a growing tissue begins to outstrip its existing blood supply or the direct secretion of angiogenic factors such as VEGF. A more detailed discussion of angiogenesis is given in Chapter 14.

Cancers and nutrients

> The cause of nutrition and growth resides not in the organism as a whole but in the separate elementary parts – the cells.
>
> *Theodore Schwann*

As discussed above, cancer-causing epimutations ultimately enable cells to overcome environmental signals that normally regulate their behavior and to replicate and grow uncontrollably even to the detriment of the organism. Many cancer-causing epimutations involve deregulation of ancient regulatory pathways that are present in primitive organisms such as yeast, where they may largely respond to changes in environmental nutrient availability and coordinate decisions about growth and reproduction accordingly. In general, abundant nutrients activate mTOR, whereas nutrient-poor conditions favor activation of inhibitors of mTOR, such as the tumor suppressors tuberous sclerosis complex 1 (TSC1), TSC2, and liver kinase B1 (LKB1). Interestingly, inherited loss of any of these or of upstream inactivators of the RAS–RAF–ERK (MAPK) or PI3K–AKT pathways that otherwise activate mTOR, such as NF-1 or PTEN, produce clinical cancer syndromes with similar features (see Chapter 3).

Tumor cell metabolism – the "metabolic switch"

Tumor cells require a fuel supply that is able to support the increased replication and growth characteristic of a progressing cancer. Broadly, there are two main ways by which cells produce energy (as adenosine triphosphate – ATP): **oxidative phosphorylation** in the mitochondria and **glycolysis** in the cytoplasm. Although both processes are present in essentially all cells, oxidative phosphorylation predominates, with glycolysis increasing at times of oxygen deprivation, such as in exercising muscles.

In 1930, the German biochemist and Nobel Prizewinner Otto Warburg suggested that cancer might be caused by deranged energy processing in the cell. This ran counter to then (and largely still now) prevailing views that such changes are rather symptomatic of cancers rather than causes.

Warburg first described the reliance of tumors on glucose metabolism for energy nearly a century ago. In fact, the ability of tumors to switch from an aerobic metabolism of glucose within the tricarboxylic acid cycle to an anaerobic metabolism of glucose is essential for the progress and maintenance of many cancers, and Warburg showed that such cancers relied on glycolysis even in the presence of oxygen – the **Warburg effect**. Thus cancer cells use an altered metabolic program to that employed by normal differentiated adult cells; they take up more glucose and process it largely by aerobic glycolysis, producing large quantities of secreted lactate, ATP, water, and carbon dioxide. The Warburg effect thus provids substrates and energy for tumor growth because cancer cells take up increased amounts of glucose and channel it largely through aerobic glycolysis rather than oxidative phosphorylation. The Warburg effect is triggered by various oncogenes, such as c-Myc and can be arrested by tumor suppressors.

This metabolic switch is regulated by multiple signaling pathways (Fig. 5.9). Thus, tumor cells frequently upregulate proteins involved in glucose uptake, such as the glucose transporter GLUT-1 and various enzymes involved in anaerobic glucose metabolism, including hexokinase, required for initial phosphorylation of glucose to glucose-6-phosphate (G6P). G6P can then be further

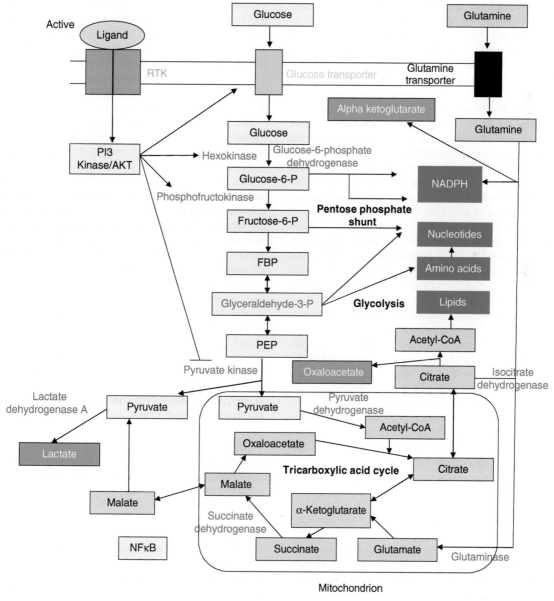

Figure 5.9 The Warburg effect and metabolic switch.

catabolized to produce energy or can be stored as glycogen. Tumor cells also upregulate enzymes involved in glycolysis. The M2 isoform of pyruvate kinase (PKM2) promotes the metabolism of glucose by aerobic glycolysis and contributes to anabolic metabolism. Phosphoenolpyruvate (PEP) is the usual substrate for pyruvate kinase and can act as a phosphate donor by phosphorylating phosphoglycerate mutase (PGAM1), a key glycolytic enzyme. Not surprisingly, then phosphorylated PGAM1 correlates with PKM2 expression in tumors. Interestingly, however, this alternative glycolytic pathway may actually be driven by reduced pyruvate kinase enzyme activity in PKM2-expressing tumor cells.

As intimated earlier, as tumors expand they often outstrip their oxygen supply (vascular supply) and become hypoxic. As a result, proteins such as HIF-1 are activated and may induce gene expression changes, including those regulating glycolysis. It remains to be shown whether glycolysis happens completely independent of hypoxia in cancer cells.

Warburg's hypothesis is currently experiencing a renaissance and recent studies are suggesting that a shift in energy production from oxidative phosphorylation to glycolysis may be the seventh hallmark feature of cancer. By implication, if the Warburg effect is essential to cancer cells then novel targeted therapies might be designed to exploit this. Although Warburg attributed this process to defective mitochondrial function, it now seems more likely that instead it is a product of mutations in various signaling pathways, such as AKT and MYC, which can in various ways direct increased glucose uptake, which together might render the cancer cell independent of regulatory signals, such as insulin. Glycolysis is less efficient (produces two molecules of ATP for every glucose molecule, whereas complete oxidation produces 38), but increasing the rate of glycolysis on top of oxidative phosphorylation may easily provide extra energy supplies needed to drive proliferation and growth of cancer cells. Moreover, what may seem a rather poor and inefficient way to generate ATP (energy), is actually a potential route by which the cancer cell can accumulate key intermediates, such as nucleotides, amino acids, and lipids, needed to support growth and cell division. Enhanced glycolysis is also exploited in a variety of diagnostic procedures in clinical practice and underlies the ability of imaging techniques such as positron emission tomography (PET) to identify cancers, because these more avidly take up the radiolabeled glucose analog (see Chapter 2).

Several novel targeted treatments are under investigation based on the perceived importance of the Warburg effect for cancers – irrespective of whether this is secondary to hypoxia, and include glucose analogs, such as 2-deoxy-D-glucose, which can downregulate glycolysis, AKT inhibitors such as the asymmetric bisimidazoacridones, inhibitors of the proton pump v-ATPase, which acts to deacidify the cytosol in cancer cells, other metabolic targeting agents such as glufosfamide, and those which can block key glycolytic enzymes.

Growth factor signaling pathways

Ships that pass in the night, and speak each other in passing, only a signal shown, and a distant voice in the darkness; So on the ocean of life, we pass and speak one another, only a look and a voice, then darkness again and a silence.

Henry Wadsworth Longfellow

Arguably, this is the single most important area in modern cancer biology and is the main focus of most initiatives aiming to develop new drugs, improved diagnostics, and deliver the holy grail of personalized medicine.

Growth factor signaling pathways affect essentially all biological processes and are particularly important in multicellular organisms, although many are highly conserved. Growth factor signaling is invariably disturbed in cancers, either by epimutations in key receptors or downstream signaling molecules which confer aberrant activation and sometimes deactivation of the pathways, thus unleashing some aspect of cancer behavior. The central role of signaling in cancer is emphasized by the fact that this and the following two chapters are largely concerned with nothing else.

The balance between cell death and survival or replication in the organism is determined in large part by growth factors that dictate whether a given cell will live or die. Epidermal growth factor (EGF) is a prototypic extracellular growth factor that regulates cell replication and survival through binding to cell surface receptor tyrosine kinases and activation of a network of signal transduction pathways, including the PI3K–AKT, RAS–MAPK–ERK, and JAK–STAT pathways. These pathways in turn activate or inhibit various transcription factors that regulate expression of genes/proteins involved in growth regulation. In cancer, such signaling pathways are among the most frequently deregulated, and in the case of EGF and family members, they are one of the most important targets for treatments.

The transmission and transduction of extracellular growth signals into the expression of genes involved in cell replication is fundamental to our understanding of cancer biology. Not surprisingly, most of the major cancer-causing mutations (but by no-means all) relate in some way to these signaling pathways, with mutations that either activate signaling molecules which promote growth (oncogenes – see Chapter 6) or disable molecules required for arresting cell replication or inducing apoptosis (tumor suppressors – see Chapter 7). Moreover, there is no shortage of signaling molecules to be deregulated – 539 kinases alone are encoded in the human genome.

Broadly speaking, proteins encoded by cancer-relevant genes can be ordered or classified according to the level they occupy within a growth factor signaling pathway and also by the nature of the key functional domains of the oncoprotein. One such scheme is outlined in Table 5.3 and will be followed in Chapter 6, where oncogenes are discussed in more detail.

A detailed description of signal transduction pathways and their subversion in cancer

The footsteps of Nature are to be trac'd, not only in her ordinary course, but when she seems to be put to her shifts, to make many doublings and turnings, and to use some kind of art in endeavouring to avoid our discovery.

Robert Hooke

Growth factors

Growth factors, as with oncogenes, are generally named after experimentally observed activities associated with their original identification – fibroblast growth factor (FGF), platelet-derived growth factor (PDGF), transforming growth factor-TGF-α and TGF-β, insulin-like growth factor (IGF), and so on.

Table 5.3 Classification of cancer-relevant genes

Growth factors	Including platelet-derived growth factor (PDGF), epidermal growth factor (EGF), and vascular endothelial growth factor (VEGF)
Receptors	Can be either extra- or intracellular and are activated once bound by a suitable ligand usually originating from another source outside the cell
Receptor tyrosine kinases	The cell membrane receptors for various growth factors, such as PDGF and EGF, and oncogenes such as HER2/NEU are receptor tyrosine kinases (RTKs). In general, ligand binding leads to dimerization of receptors and activation on intrinsic tyrosine kinase activity that regulates downstream signaling. These tyrosine kinases have proved to be spectacularly amenable to therapeutic inhibition and are the target of a vast number of new anticancer agents – the tyrosine kinase inhibitors (TKIs)
G-protein coupled receptors	Including the angiotensin receptor
Nuclear receptors	Not all growth factors/extracellular ligands act via receptors on the cell surface, but some can enter the cell and bind to nuclear receptors. These are ligand-dependent transcription factors that regulate cell growth and differentiation in many target tissues and include receptors for steroid hormones, such as the estrogen receptor, and other ligands, such as the peroxisome proliferator-activated receptor (PPAR) family and farnesoid X receptor (FXR)
Signaling transducers and pathways	Primarily involving enzyme cascades that transduce, amplify, and allow crosstalk and integration of inputs from diverse sources. Most such enzymes activate or deactivate their target proteins, which may themselves be enzymes, by either adding or removing phosphate groups
Membrane-associated G-proteins	Including three different homologs of the c-RAS gene, each of which was identified in a different type of tumor cell. RAS, which is one of the central oncogenes in human cancer, is discussed in depth in Chapter 6. For this reason, discussion of RAS and related family members will be delegated to the major subsection in Chapter 6 devoted entirely to this. However, signaling upstream and downstream of RAS is covered in this chapter
Membrane-associated nonreceptor tyrosine kinases	Including the archetypal oncogene SRC and ABL, discussed in depth in Chapter 6
Serine/threonine kinases and the mitogen-activated protein kinase (MAPK) cascade	Sequential activation of a family of proline-directed serine/threonine kinases in MAPK cascades is an evolutionary-conserved signal transduction system currently known to comprise four distinct modules named after the main MAPK active in each. These are the extracellular signal-regulated kinase 1/2 (ERK1/2), c-Jun N-terminal kinase (JNK), p38, and ERK5 cascades, respectively. These modules follow a common pattern of organization, with a core comprising three levels of kinases referred to as MAPK, MAPKK, and MAPKKK (or MAP3K), and an upstream MAP4K and a downstream MAPKAPK. Signals are transmitted through the cascade by sequential phosphorylation and activation each component in sequence. This pathway includes important oncogenes and treatment targets, including RAF, involved in the signaling pathways of most RTK, and responsible for threonine phosphorylation of the MAP kinase kinase (MEK) following receptor activation. To date, more than 70 genes and several splice variants are known that encode members of this cascade. As these are important mediators of growth and survival signals they are frequently involved in cancer and are the target for a variety of new anticancer treatments
Lipid kinases	Including PI3K involved in activating AKT
Phosphatases	Less well studied than the kinases and act to remove phosphate groups from proteins, thus opposing the effects of kinases. The phosphatase tumor suppressor PTEN antagonizes the effects of PI3K and thus prevents activation of AKT
Nuclear DNA-binding/ transcription factors	Including FOS and c-MYC, discussed in Chapter 6
Specific regulators of survival	Including the pro- and antiapoptotic members of the BCL-2 family of proteins (discussed in Chapters 6 and 8). It is worth noting that pathways discussed above can all impact on survival-regulating proteins as exemplified by PI3K–AKT–survivin and MEK–ERK–BIM pathways targeted by TKI in NSCLC

Growth factors are polypeptides that bind to cell surface receptors, with the major action of promoting cellular proliferation, but may also have major effects on differentiation, migration and survival (when they are sometimes referred to as "survival factors"). Many growth factors are very versatile and act on multiple cell types, while others are specific to a particular cell-type. As can be seen from Table 5.1, which contains a partial list of growth factors implicated in cancer, these factors are produced by various cell types, including stromal cells, such as fibroblasts and adipocytes, and even more distant tissues that can secrete growth factors, which act via the circulation (Fig. 5.10 gives a summary of the various sources of growth factors). Although many tumor cells become a source of growth factor production, tumor cells can also influence stromal cells to secrete growth

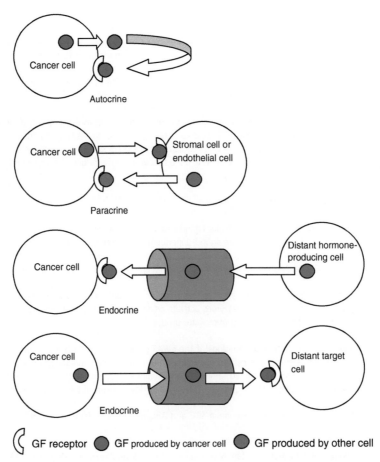

Figure 5.10 Growth factors act via various routes. Secreted growth factors may act locally on the same cell (autocrine), on neighboring cells (paracrine), or at a distance (endocrine).

factors that, in turn, can promote tumor cell behavior (see Fig. 12.7, Chapter 12). Many growth factors are synthesized by stromal cells and are normally sequestered by proteins within the extracellular matrix (ECM). In this way, tight control over their activity can be maintained within normal tissues. However, enzymes known as matrix metalloproteinases (MMPs) are often secreted by tumor cells, and can release the sequestered growth factors (e.g. IGFs, TGF-α, EGF, TGF-β) (see Fig. 5.20). Such growth factor activity can promote tumor behavior, as described in Chapter 12. It is also worth remembering that growth factors usually require integrin-mediated cell adhesion in order to activate signaling pathways, ensuring that activation only occurs when cells are in their correct environment. This homeostatic function becomes disrupted in cancer.

Analogous to this tissue-based regulation, circulating growth factors can be sequestered by various binding proteins in plasma which also regulate activity by various means. This has been particularly well studied for the IGFs and IGF-binding proteins (IGFBPs), many of which have been described. Insulin and the related IGFs may be unique among growth factors implicated in cancer as they are hormones, and therefore circulating levels are clearly physiologically and clinically important. This is in contrast to EGFs, which largely operate at an autocrine/paracrine level localized to the specific tissue. Ramifications of this are very significant. For example, in conditions such as type 2 diabetes, the persistently increased levels of circulating insulin may exert an adverse effect on cancer risk by providing growth stimuli via insulin and IGF-1 receptors on the tumor. Conversely, drugs that reduce insulin levels in type 2 diabetes, such as metformin, may protect against some cancers. Raised circulating IGFs contribute to the excess cancer risk in acromegaly. This does not mean that local production of IGF is not important, rather that both local and whole-body production must be considered.

Cytokines are a distinct category of growth factors that are secreted by leukocytes, but also by other cells, and are involved in directing a wide range of processes, including immune responses and activation of phagocytic cells. The interleukins are cytokines secreted by multiple cell types and act as growth factors for cells of hematopoietic origin, as well as other functions. The list of identified interleukins grows continuously with at least 33 recognized at the time of going to print. A more detailed discussion of cytokines in the immune system is provided in Chapter 13.

Normal cells growing in culture will not divide unless they are stimulated by one or more growth factors present in the culture medium. Extracellular polypeptides and proteins act via cell membrane receptors whose signals are transduced to the nucleus via intermediary signals, including protein–protein interactions, phosphorylation, and cytoskeletal changes (Fig. 5.11). Examples of such signaling transduction pathways are described later. Lipid-soluble regulators, such as steroid hormones and retinoic acid, readily traverse cell membranes and bind to intracellular

Growth Signaling Pathways and Targeted Treatment

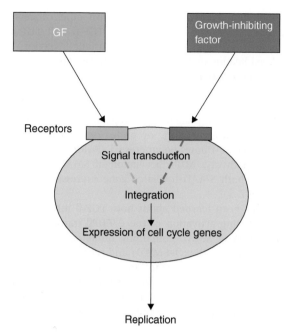

Figure 5.11 Signaling through membrane-bound growth factor receptors. Positive and negative signals are integrated at various levels before a decision on whether to respond or not is completed.

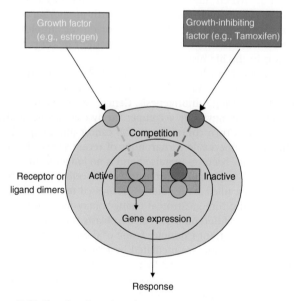

Figure 5.12 Signaling through nuclear receptors. Some hormones may act as ligands for nuclear receptors which may then act as transcription factors.

receptors to form transcription factor complexes (Fig. 5.12). The source of these extracellular signals can be endocrine (hormones which act on distant targets reached via the circulation), paracrine (produced by nearby cells of a different type and acting locally), or autocrine (produced by the same cells on which they act) (see Fig. 5.10). In cancer each of these is important, and such signals may be acting directly on the cancer cell or on other cells in its environment, such as blood vessel endothelium (see Chapter 14) or the extracellular matrix, which can potentially be induced to support aspects of neoplastic growth.

Growth factors and cancer

Although most growth factors are not oncogenes, the excess production of (or increased sensitivity to) existing growth factors has a mjor role in supporting cancer growth (see Table 5.2). Such examples include insulin-like growth factors IGF-1 and IGF-2. IGF-1 is produced largely by the liver under the influence of growth hormone from the pituitary gland. IGF-1 levels are increased in the disease acromegaly, characterized by excess production of growth hormone by pituitary tumors and generalized expansion of soft tissues. Growth hormone increases secretion of IGF-1 into the circulation, primarily from the liver, which in turn promotes expansion of connective tissue cells among others, and may increase the likelihood of some cancers.

IGFs are important activators of the IRS–PI3K–AKT pathway, which mediates cell proliferation, survival, differentiation, and transformation growth and survival-promoting signals in many cell types (see below). Insulin may also act via an circulating endocrine route in cancers, but unlike the IGFs does not have a paracrine/autocrine role except possibly in β cell–derived tumors such as insulinoma, which arise from those cells which actually produce insulin.

IGFs activate signaling by binding and releasing tyrosine kinase activity of IGF1R or insulin receptor isoform A (IR-A). The IGF1R discussed below is well known to regulate tumor growth and survival, and re-expression of IGF-2 and overexpression of IR-A have been found in many cancers and might explain emergence of resistance to IGF1R-targeting therapy. IGF-2 may be increased in many cancers, in particular colorectal cancer, by loss of imprinting (see Chapter 11). Tumor cells expressing excess insulin receptor-A, IGF-1 receptor, or hybrid receptors may transduce increased mitogenic signals in the presence of excess insulin, IGF-1, or IGF-2. Small-cell lung cancers can produce growth factors such as bombesin; some tumors produce IGF-2; and circulating estrogens can support the growth of estrogen receptor-positive breast cancer cells. Autocrine transforming interactions have been identified in a number of human malignancies (Fig. 5.2). At least one PDGF chain and one of its receptors have been detected in a high fraction of sarcomas and in glial-derived neoplasms. Tumor cells cultured from some of these cancers demonstrate a functional autocrine loop, in which persistent PDGF receptor activation is shown.

TGF-β is often detected in carcinomas that express high levels of EGF receptor. However, the role of acidic or basic FGF (aFGF or bFGF) in tumors is still unclear, as neither molecule possesses a secretory signal peptide sequence. However, recent studies have implicated bFGF in melanoma; human melanoma cell lines express bFGF, which is not observed with normal melanocytes. Moreover, bFGF antagonists can inhibit growth of melanoma cells. Amplification of fibroblast growth factor receptor 1 (FGFR1) and overexpression has been found in around 10% of breast cancers and is associated with poor prognosis. FGFR1 overexpression show enhanced ligand-dependent signaling, with increased activation of MAPK and PI3K–AKT pathways in response to FGF2.

Although several growth factors have been shown to induce transformation by an autocrine mode, it is also worth considering the possible role that growth factors might have in predisposing to cancer. It can be hypothesized that overexpression of growth factors by a paracrine mode might increase the proliferation of a polyclonal target cell population (Fig. 5.2). By such a model, increased growth factor production could act in a manner analogous to that of a tumor promoter. For example, continuous stimulation by growth factors in paracrine as well as autocrine modes during chronic tissue damage and repair associated with cirrhosis and inflammatory bowel disease may predispose to tumors. Finally, paracrine-acting angiogenic growth factors such as the VEGFs (vascular endothelial growth factors) are produced by some tumor cells. Such growth factors cause the paracrine stimulation of endothelial cells, inducing new blood vessel growth (angiogenesis) and lymphangiogenesis, which contribute to tumor progression.

Growth factors may also act via endocrine routes (Fig. 5.2). With respect to human cancers this particularly includes estrogens produced from the ovary acting on breast cancers and IGF-1 from the liver acting on diverse tumor types. It is worth reiterating here that these growth factors may exert equally important effects on the tumor's environment as on the tumor directly (for detailed discussion, see Chapter 12).

Binding proteins, which are not receptors, exist in the circulation and may bind and inactivate a ligand or serve as a pool of available ligand, because binding is reversible. Protein binding acts as an important reservoir in the circulation for IGFs and also for hormones such as thyroxine. Various IGF-binding proteins (IGFBPs) are known and generally act as negative regulators of IGF signaling by sequestering the growth factor away from cell surface receptor tyrosine kinases. The so-called IGF-2 receptor may play an analogous role in negatively regulating action of IGF-2. Increased levels of IGFBP3 are negatively linked to cancers in some studies.

Growth factor oncogenes

The v-*sis* gene was the first viral oncogene found to encode a growth factor and was subsequently identified as having homology to a known cellular gene c-*sis*. Sis is the oncogene incorporated into the genome of the simian sarcoma virus, and it encodes the PDGF B chain. This virus is able to form tumors in its host as a result of overproduction of PDGF-B under the control of viral regulatory elements rather than its own promoter.

As mentioned earlier, viruses can also integrate in the vicinity of a proto-oncogene and activate transcription of the host gene, leading to formation of the tumor. The *Int2* gene is a common site of integration of the mouse mammary tumor virus (MMTV), and encodes an FGF-related growth factor. MMTV has also been shown to integrate in the vicinity of various members of the FGF family, such as FGFs 3, 4, and 8, and consequently can activate transcription of the particular FGF in the mammary gland, as the host gene is now under the control of the viral promoter.

Potentially, viral proteins may also activate growth factor receptors. For example, the human papilloma virus (HPV) protein E5 can result in aggregation of PDGF receptors, mimicking dimerization and resulting in sustained receptor autophosphorylation and activation.

TGF-β – "playing both sides" in cancer

The cytokine TGF-β regulates cellular growth, differentiation, and migration during ontogeny and adult tissue homeostasis. The role of TGF-β in growth regulation and cancer is sometimes seen as confusing. TGF-β is a ubiquitous cytokine that is known to inhibit proliferation of numerous cell types, particularly those of epithelial origin. Moreover, mutations inactivating various components of the TGF-β signaling pathway are found in many cancers, particularly of gut origin, supporting the role of TGF-β as a tumor suppressor. In complete contrast, other cancers manifest enhanced growth in response to TGF-β and TGF-β may contribute to promoting EMT and metastatic behavior. TGF-β, and related ligands such as activin and BMPs, binds to receptors that in turn phosphorylate and activate SMAD2 and 3, members of the SMAD receptor family, which then recruit and oligomerize with SMAD4 and enter the nucleus to regulate gene expression. Deactivation of TGF-β signaling is regulated by SMAD dephosphorylation, nuclear export, or proteasomal degradation. The protein poly(ADP-ribose) polymerase-1 (PARP-1) is a key inhibitor of this pathway; ADP-ribosylation of SMAD3 and SMAD4 antagonizes both SMAD-mediated gene expression and TGF-β-induced EMT.

Much has been learned about how TGF-β inhibits cell cycle transit, including activation of the C/EBPβ transcription factor, and formation of SMAD-containing inhibitory complexes that can induce $p15^{INK4b}$ (FoxO–SMAD) and repress c-MYC (E2F4/5–SMAD). Importantly, such cytostatic effects are often disabled in cancer cells. In fact, a recent study shows that these responses are missing in up to 50% of patients with metastatic breast cancer, at least in part due to the presence of the inhibitory LIP isoform of C/EBPβ.

However, paradoxically, TGF-β has also been implicated as a contributor to tumor progression. Recent studies have shed light on the potential "switch" of TGF-β from growth inhibitor to promoter during oncogenesis for some epithelial cancers. Thus, the TGF-β "switch" may be thrown by impaired CKI activity (e.g. loss of $p27^{KIP1}$), which unmasks the growth-promoting actions of this cytokine. This is one of the best-described examples of how important the timing (as well as the nature) of cancer-causing events is in the life history of a tumor.

As discussed in Chapter 12, epithelial tissues require interactions with the stroma both in normal tissue homeostasis and particularly during tumorigenesis. Cancer-related changes in the epithelium are inevitably accompanied by changes in the tissue stroma that in turn contribute to the cancer phenotype. In one of the most remarkable turnarounds of recent years, the role of the stroma has been re-appraised and is no longer seen as a "no-man's land," implacably hostile to cancer cells but rather as a helpful annex filled with votaries harnessed to serve the needs of the growing tumor. Stromal changes may foster invasion, of which a "transdifferentiation" of fibroblasts into myofibroblasts may be an important example. Some of these cancer-promoting effects of the stroma are mediated by release of factors, such as TGF-β, by the tumor that act in a paracrine fashion to contribute to angiogenesis, "immune privilege" (avoidance of immunosurveillance), and recruitment of myofibroblasts. TGF-β may also have an autocrine effect on the cancer cells themselves, particularly if CKIs are inactivated (TGF-β switch) or the cancer cell has activated β-catenin or RAS mutations. A model for TGF-β signaling is shown in Fig. 5.5.

Recent studies have shown that in glioblastoma TGF-β inhibitors currently under clinical development can target the cancer stem cell (CSC) population (see later in this chapter) and by decreasing levels of inhibitors of DNA-binding protein (Id)-1 and -3 transcription factors can limit the ability of CSCs to initiate glioblastomas in model systems.

Receptors

Results from the Human Genome Project suggests that there may be around 1000 protein tyrosine kinases in the genome. Deregulated activation of tyrosine kinases are involved in most cancers and, in fact, over 70% of the known oncogenes and proto-oncogenes involved in cancer encode tyrosine kinase.

There are two main classes of tyrosine kinase: receptor tyrosine kinase (RTK) and cellular, or nonreceptor, tyrosine kinase. Of the known tyrosine kinases, around two thirds are RTKs. Not surprisingly, receptor and nonreceptor tyrosine kinases have emerged as clinically useful drug target molecules for treating an ever-increasing number of different types of cancer.

Receptor tyrosine kinases share an extracellular ligand-binding domain, a transmembrane domain and an intracellular catalytic domain. Different RTKs have different extracellular domains that confer ligand specificity, including one or more recognized structural motifs, such as cysteine-rich regions, fibronectin III-like domains, immunoglobulin-like domains, EGF-like domains, cadherin-like domains, kringle-like domains, Factor VIII-like domains, glycine-rich regions, leucine-rich regions, acidic regions, and discoidin-like domains.

Activation of the tyrosine kinase is achieved by ligand binding to the extracellular domain, which induces dimerization of the receptors. Activated receptors autophosphorylate tyrosine residues outside the catalytic domain, stabilizing the receptor in the active conformation and recruiting various adaptor proteins required for signaling.

Unlike RTKs, nonreceptor tyrosine kinase are located in the cytoplasm, nucleus, or anchored to the inner leaflet of the plasma membrane. They are divided into various families each comprising several members: SRC, JAK, ABL, FAK, FPS, CSK, SYK, and BTK. Aside from carrying homologous kinase domains (SRC homology 1 or SH1 domains), and some protein–protein interaction domains (SH2 and SH3 domains), these families are structurally unrelated. Biological activities are also variable; many, exemplified by SRC, are involved in cell growth (see Chapter 6). In contrast, FPS is involved in differentiation, ABL in growth inhibition (see Chapter 6), and FAK with cell adhesion. Some members of the cytokine receptor pathway interact with JAKs, which phosphorylate the transcription factors STATs (see below).

Growth factor receptor tyrosine kinases

Growth factor ligands bind to membrane receptors on the external surface of the cell membrane and subsequently generate a signal on the inside, which is then ultimately relayed to the nucleus via a signal transduction pathway involving multiple proteins and a series of phosphorylation events.

The receptors for many growth factors either contain intrinsic tyrosine kinase within their cytoplasmic domain (RTK) or, like the cytokine receptors, can recruit tyrosine kinase to the active receptor. Although growth factor receptors can be constitutively activated by autocrine loops as discussed above, various other mechanisms are known. Oncogenic activation of growth factor receptors invariably results in forced dimerization of the receptor. The outcome is activation of signal transduction pathways without the requirement for growth factor stimulation. This uncontrolled activation may result in cell proliferation, survival, or indeed other cell behaviors, such as increased motility.

Here we will describe a model RTK signal transduction pathway, the basics of which are illustrated in Fig. 5.13. For many growth factor receptors, autophosphorylation is mediated by dimerization following ligand binding (mutual phosphorylation of receptors) (Fig. 5.14). Autophosphorylation generates key phosphotyrosine sites, which enable binding of various signaling

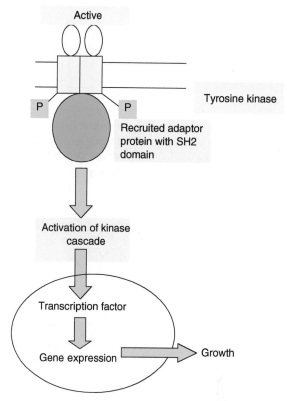

Figure 5.13 Simple schematic of receptor tyrosine kinase (RTK) signal transduction. Dimerization of RTK is triggered by ligand binding and activates the receptor. Activation involves tyrosine autophosphorylation and recruitment of various adaptor proteins which then activate downstream signaling.

adaptor proteins, many of which contain the SRC homology domain 2 (SH2 domain) (Fig. 5.15). SRC is a nonreceptor tyrosine kinase that was the first tyrosine kinase to be extensively studied, allowing it to achieve precedence in nomenclature ever more. Multiple signaling pathways are regulated by RTK (see Fig. 5.16 and below). The SH2 domain is a sequence originally described in several tyrosine kinases such as c-SRC and ABL and is distinct from the kinase catalytic domain (SH1), but is required for interaction with targets. In addition to the primary phosphorylation sites in receptors, there are secondary sites that bind other SH2 domain proteins. Several SH2-containing proteins have been shown to interact with growth factor receptors, such as PLCγ (phospholipase C), SRC, and GRB (growth factor receptor bound). GRB2 consists of a single SH2 domain flanked by two SH3 domains. SH3 domains act as binding sites for proline-rich domains in effector molecules acting downstream in the signaling pathway. Many of these aspects are illustrated in the RAS–RAF–ERK (MAPK) and ABL signaling pathways shown in Fig. 5.17.

GRB2 also recruits SOS (from the *sos* gene son of sevenless – a *Drosophila* gene responding to the sevenless RTK required to specify the seventh receptor cell of the fly eye). This localizes SOS to the membrane, where it acts as a guanine nucleotide exchange factor (GEF) for activation of RAS (exchange of GDP for GTP). Conversely, GTPase-activating proteins (GAPs) negatively regulate RAS by stimulating GTPase activity (GTP is converted to GDP). Key signaling components such as RAS and c-SRC are anchored to the membrane by isoprenyl and myristoyl chains, respectively (see Chapter 6). Recruitment of the signaling adaptor

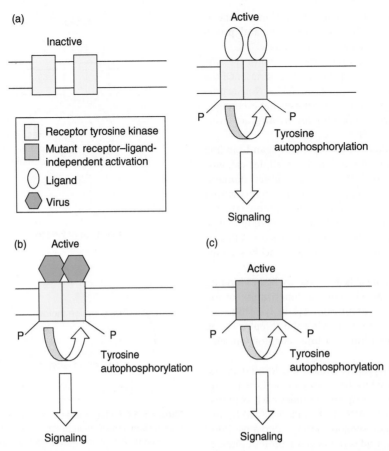

Figure 5.14 Dimerization of ligand-bound RTK activates intrinsic tyrosine kinase activity. RTKs are activated by ligand binding which induce conformational changes and receptor dimerization that activate intrinsic tyrosine kinase activity (a). RTK can also become activated by viruses (b) or by mutations (c).

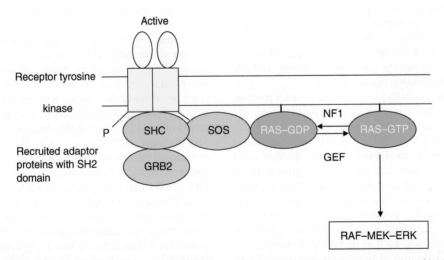

Figure 5.15 Activation of RAS by RTK. Binding of growth factor ligand to the receptor results in dimerization and activation of intrinsic tyrosine kinase activity and tyrosine autophosphorylation of the receptor. The resultant activated receptor recruits the GRB2/SOS complex to the receptor via SH2 domains. These adaptor proteins interact with the inactive GDP bound RAS to catalyze exchange for GTP, a process facilitated by the guanine nucleotide exchange factor (GEF), thus activating RAS and downstream signaling. RAS is subsequently inactivated by reversion to the GDP-bound form by intrinsic GTPase activity, which is in turn facilitated by the NF-1 protein.

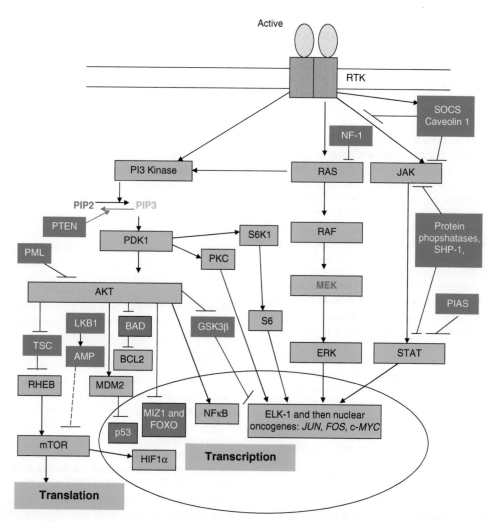

Figure 5.16 Major growth-regulating signaling pathways activated by RTK. Among others, RTK can activate three major signaling pathways, RAS-RAF-MAPK (ERK), JAK–STAT, and PI3K–AKT, all of which eventually signal to the nucleus and regulate gene expression. The details of the RAS–RAF–ERK and JAK–STAT pathways are shown elsewhere. Some key tumor suppressors regulated by or regulating this pathway are shown in red. It is worth noting that patients with inherited loss of *PTEN* (Cowden syndrome), *TSC1 or TSC2* (tuberous sclerosis), or *LKB1* (Peutz–Jeghers syndrome) all develop similar types of tumors – termed hamartomas.

proteins close to the membrane increases their local concentration, allowing the signal to pass effectively. RAS in turn activates RAF, a serine/threonine protein kinase, which is the starting point for the ERK (MAPK) pathway (Fig. 5.17) controlling cell proliferation, and is discussed in more detail in Chapter 6. Activating BRAF(V600E) kinase mutations have been found in up to 7% of human malignancies and over half of all melanomas.

Receptor tyrosine kinases and cancer

> Of all men's miseries the bitterest is this: to know so much and to have control over nothing.
>
> *Herodotus*

RTKs regulate growth and survival by activating diverse signaling cascades mediated by another important group of enzymes, the serine/threonine kinases (see below), such as the RAS–MAPK, PI3K–AKT, and mTOR pathways. The RSK (ribosomal S6 kinase), AKT, and p70 S6 family of protein kinases activated by these three pathways, respectively, in turn phosphorylate various target substrates. These same pathways are variously activated by different growth factors, including those best known in human cancer, such as EGFR, MET, IGF1R, and PDGFRA.

As is the case for other proto-oncogenes, viral oncogenes encoding mutant growth factor receptors have been identified, the best known of which is v-*erbB*, the viral oncogene of avian erythroblastosis virus. v-*erbB* is the oncogenic counterpart of the epidermal growth factor receptor (EGFR). The c-*Fms* gene encoding the colony-stimulating factor-1 (CSF-1) receptor was also first identified as a retroviral oncogene.

RTKs can also be activated as a result of mutations. The *Neu* gene was originally isolated from a chemically induced (ethylnitrosourea) neuroblastoma in the rat, and it encodes the HER-2 receptor, encoded by an EGFR-related gene (see Box 5.3). Conversion of proto-oncogenic *Neu* to oncogenic *Neu* requires only a single amino acid change in its transmembrane domain – this subsequently leads to the spontaneous dimerization (and thus activation) of the HER-2 receptor. In humans, oncogenic *NEU* is now known to be a very important contributor to the pathogenesis

> **Box 5.3 ERBB family of receptor tyrosine kinases – additional information**
>
> Growth factors and their transmembrane RTKs regulate cellular processes, particularly proliferation and migration, during embryogenesis, during tissue mass regulation in the adult and in cancer. The ERBB/HER family of receptors play essential roles in the development of epithelial cell lineages and serve as a useful model of RTK in growth regulation and deregulation in cancer.
>
> Four major forms of ERBB RTK are known and together with a variety of ligands comprise a complex signaling network in mammals. This network is present, though in a more simplified form, in primitive organisms such as *C. elegans*, but has likely evolved through gene duplication and diversification to the more complex system found in higher organisms. This greater level of organization may allow for more highly tuned responses and integration of growth-regulatory inputs. For instance, ERBB2/HER2 can only signal effectively in the presence of active ERBB3. Moreover, ERBB2/ERBB3 heterodimers result in amplified responses to growth factors, largely through their ability to engage both the RAS–MAPK cascade and the PI3K–AKT pathway, thus stimulating cell proliferation and blocking apoptosis.
>
> Heterodimers may also be more oncogenic because they can avoid receptor endocytosis, a major means of inactivating RTK. Rac1 and effectors mediate mitogenic and motility signaling by ERBB and are themselves regulated by small GTPase-exchange factors (Rac-GEFs) such as PIP3-Gβγ-dependent Rac-GEF (P-Rex1), required for Rac1 activation and ERBB tumorigenesis in breast cancer cells. Interestingly, the GPCR CXCR4 mediates activation of P-Rex1/Rac1 by ERBB ligands.
>
> ERBB proteins and various EGF-like ligands are major contributors to several human cancers. Three major mechanisms have been identified whereby ERBB is involved in cancer:
>
> 1. Through autocrine loops, whereby coexpression of a receptor and the respective ligand (e.g. ERBB1/EGFR and the transforming growth factor alpha) are found in the same cells.
> 2. Genetic aberrations, which affect primarily ERBB1/EGFR, and result in loss (deletion) of the regulatory domains. In fact, such mutant forms of ERBB1 are found in both brain and lung tumors.
> 3. Overexpression of ERBB2/HER2. Once overexpressed, excessive ERBB2/HER2 may favor the formation of receptor heterodimers – that is to say, form dimers with other growth factor receptors. Dimerization normally only happens after growth factor stimulation (ligand binding). Thus, in cancer ERBB2 can become a preactivated receptor, serving to amplify a variety of different growth signals, without the need for binding of its own ligand.
>
> Overexpression of ERBB2 identifies aggressive subsets of mammary and other tumors, testifying to the potentially wide-ranging effects of inappropriate activation of growth-signaling pathways.
>
> Recent studies have suggested the intriguing notion that EGFR may regulate gene expression independently of signaling pathways discussed. EGF induces the translocation of EGF receptor (EGFR) from the cell surface to the nucleus and activates transcription at gene promoters with AT-rich sequence (ATRS). As EGFR does not have a DNA-binding domain this action is mediated by interactions with RNA helicase A (RHA), which actually binds the ATRS.
>
> Several novel treatments involve the targeting of ERBB signaling, in particular the humanized monoclonal antibodies to ERBB2 and to ERBB1 (herceptin). These antibodies accelerate receptor internalization, thereby reducing availability of receptors for heterodimerization. Herceptin has recently shown in several important clinical trials that it can effectively improve outcome in both advanced and likely also in less advanced breast cancer. Alternatives include the use of low-molecular-weight antagonists of the tyrosine kinase domain shared by all four ERBB proteins (e.g. gleevec). These analogs of ATP bind to the nucleotide-binding site of the tyrosine kinase and block all signaling events that depend on tyrosine phosphorylation. A more novel approach entails promoting degradation of ERBB in the proteasome.

of many breast cancers and NEU is the target of the drug herceptin, used successfully in the treatment of appropriately selected cancers. Deletions in the EGFR external domain can result in ligand-independent activation of RTK activity. Several other receptors, including MET (the hepatocyte growth factor(HGF)/scatter factor (SF) receptor), PDGF, RET, c-KIT, ALK, and the neurotropin receptors (NGF receptor–like proteins) TRKA (encoded by *NTRK1*) and TRKC, have been shown to be oncogenically activated in human malignancies by gene rearrangements that lead to fusion products containing the activated tyrosine kinase domain.

RTKs that have not been structurally altered by mutation also play a role in many cancers. Thus, normal RTKs may be upregulated or amplified in human cancers, examples of which include EGFR, ERBB2, and c-MET. *ERBB2* was initially identified as an amplified gene in a primary human breast carcinoma and a salivary gland tumor and was shown to transform NIH/3T3 fibroblasts when overexpressed *in vitro*. The normal *ERBB2* gene is now known to be often amplified or overexpressed in human breast carcinomas and in ovarian carcinomas, and high *ERBB2* levels correlate with poor survival. Overexpression of normal EGFR has been found in squamous cell carcinomas and glioblastomas, where it is often activated by autocrine stimulation by TGF-α or amphiregulin, both of which are endogenous ligands for the receptor. Binding of the ligand to EGFR triggers a conformational change that allows homo- or hetero-dimerization with other members of the ERBB/HER receptor family. Dimerization promotes autophosphorylation and releases the RTK activity. Amplification and activating mutations of *EGFR* are characteristic of glioblastomas. Recently it has also been shown that activation of EGFR signaling in this tumor can also arise by deletion of the inhibitory *NFkBIA* gene and is associated with a poor outcome.

Various growth factors and their receptors operating in angiogenesis are discussed in the relevant chapter and will not be mentioned below. However, they play critical roles in tumorigenesis and as targets for therapies.

ERBB/HER2

The ERBB/HER protein tyrosine kinases, which include the EGFR, are implicated in many cancers (Box 5.3), and as such will feature extensively throughout this book.

MET

MET has recently received a lot of attention as a cancer-causing protein and as a therapeutic target, with several MET inhibitors in clinical trials. Not only is deregulated MET involved in pathoetiology of many cancers, secondary activating mutations in MET signaling are an increasingly recognized cause of acquired resistance to EGFR inhibitors.

It is interesting, therefore, to note that the same issue also dogs treatment of tumors with MET inhibitors, as these also become resistant either by secondary mutations in MET, such as gene amplification, which bypass the drugs used or by amplification and overexpression of KRAS, which enables cancer cells to lose dependence on MET.

The putative tumor suppressor Sprouty inhibits receptor tyrosine kinase signaling by interfering with Ras–ERK pathways. However, at least in colon cancer, Sprouty-2 may act as an oncogene by upregulation of c-Met protein and thereby downstream signaling.

IGF-1 receptor and insulin receptor

Insulin-like growth factors (IGFs) and the IGF-1 receptor (IGF1R) have pivotal roles in regulation of cellular proliferation and survival during embryonic development, after injury and during regulation of normal tissue mass. They are also implicated in hyperproliferative disorders such as acromegaly and in cancer. There are now various inhibitors of IGF1R in different stages of clinical development, which show great promise in treating some cancers. Those cancer cells dependent on AKT activation for maintaining downstream signaling may be particularly sensitive to IGF1R inhibition. Proapoptotic effects of treatments have been described, including hypophosphorylation of BAD (which reduces Bcl-X_L levels – see Chapter 8) and activation of downstream BAX. Interestingly, cells overexpressing Bcl-2 are resistant to IGF1R inhibitors.

It is important to note that insulin and the insulin receptor, which show some functional overlap with IGF–IGF1R, can also contribute to tumorigenesis, which raises some important issues about the links between obesity and insulin resistance and cancer risk on the one hand and resistance to IGF1R treatments on the other. In a mouse model of β-cell tumorigenesis previously shown to depend on IGF-2 and IGF1R expression, an inhibitory monoclonal antibody to IGF1R did not significantly impact tumor growth. However, levels of insulin receptor were found to be increased during progression from hyperplasia to islet tumors. By knocking out insulin receptor, tumors became sensitive to anti-IGF1R therapy. Similarly, high insulin receptor-to-IGF1R ratios marked resistance to IGF1R inhibition in human breast cancer cells.

These results have been confirmed in studies of Ewing's sarcomas, which may become resistant to IGF1R targeting by antibodies or TKI because they can form insulin receptor-A (IR-A) homodimers and secrete IGF-2. Resistant cells switch from IGF-1 and IGF1R to insulin/IGF-2 and IR-A dependency to sustain activation of AKT and ERK1/2. This also represents a switch towards insulin pathways sustaining proliferation and malignancy, rather than metabolism.

Other IGF1R targets are also important for the avoidance of apoptosis in cancer cells. Thus, survivin, an oncofetal antiapoptotic protein, is a known transcriptional target of both IGF-1 and EGF in cancer cells.

Anaplastic lymphoma kinase

The anaplastic lymphoma kinase (ALK) is an RTK belonging to the insulin receptor superfamily. ALK is mutated in a variety of cancers, both rare and common. ALK mutations are present in anaplastic large cell lymphomas, ALK locus rearrangements on chromosome 2p23 are found in more than half of inflammatory myofibroblastic tumors (IMTs) and oncogenic fusion genes consisting of the echinoderm microtubule-associated protein-like 4 (EML4) and ALK are found in around 5% of non-small-cell lung cancers (NSCLC). Activating mutations have also been found as a cause of hereditary neuroblastoma, and somatic mutations have been found in sporadic neuroblastomas. As is the case with EGFR signaling, the aberrant activation of ALK signaling leads to "oncogene addiction" and sensitivity to ALK inhibitors such as crizotinib. Early-phase clinical trials of crizotinib – an orally available small-molecule inhibitor of the ALK tyrosine kinase – have been conducted in patients with ALK mutations and show tumor shrinkage or stable disease in most patients with this form of NSCLC, and also in some with IMT.

Although a ligand, jelly belly, has been identified in fruitflies, the ligand for ALK in manmmals is still debated, thereby consigning ALK to the stasus of orphan receptor. Several candidates have, however, been proposed, including variously the cytokines pleiotrophin, the strongest contender, and midkine or zinc.

RET

Glial cell line-derived neurotrophic factor (GDNF), a ligand for the RET tyrosine kinase, promotes the survival and differentiation of a variety of neurons. GDNF–RET signaling is crucial for the normal development of the kidney and various populations of neurons in the peripheral nervous system, including sympathetic, parasympathetic, and enteric neurons. RET can activate several signaling pathways, including the RAS–RAF–ERK (MAPK), PI3K–AKT (see below), p38 MAPK, and c-Jun N-terminal kinase (JNK) pathways. These signaling pathways are activated via binding of adaptor proteins to intracellular tyrosine residues of autophosphorylated RET, as for other RTKs.

The RET/papillary thyroid carcinoma 1 (PTC1) oncogene is frequently activated in human papillary thyroid cancers. It results from fusion of the intracellular kinase domain of RET to the first 101 amino acids of CCDC6. Truncation of CCDC6 prevents repression of CREB1 transcriptional activity. In turn, phosphorylated active CREB1 transactivates target genes, such as AREG and cyclin A, which contribute to tumorigenesis.

The MEN syndromes

Multiple endocrine neoplasia (MEN) types 1 and 2 syndromes are rare hereditary cancer syndromes manifesting a variety of endocrine and non-endocrine tumors. MEN1 results from inherited inactivating mutations in a putative tumor suppressor gene encoding the protein menin, and is discussed below. MEN2 arises through various activating mutations in the RET RTK. Constitutive activation of RET by somatic rearrangement with other partner genes or germline mutations results in human papillary thyroid carcinomas or MEN (type 2A and 2B), respectively, whereas the dysfunction of RET by germline missense and/or nonsense mutations causes Hirschsprung's disease, a disorder of nervous development in the gut resulting in motility problems. The disease phenotype is affected by the nature of the mutant RET protein. For example, the MEN2B-activating mutation also alters the substrate specificity of the RET tyrosine kinase, and thus induces a different set of gene targets to RET carrying the MEN2A-activating mutation, which conversely primarily results in deregulated activity of the usual repertoire of RET targets. The different downstream activities of the mutant RET account for the phenotypic differences between MEN2A and MEN2B and thus produce a different portfolio of endocrine tumors.

MEN1 is associated with hyperplasia or tumors of several endocrine glands, including the parathyroids, the pancreas, and the pituitary. The most frequent problem in MEN1 is hyperparathyroidism. Hyperplasia of the parathyroid glands and excess production of parathyroid hormone result in elevated blood calcium levels, kidney stones, constipation, weakened bones, and depression. Almost all MEN1 patients show parathyroid symptoms by age 40. Various gut and pancreatic endocrine tumors are also common in MEN1 (including gastrinomas that can result in excess secretion of gastrin and gastrointestinal ulcers and insulinomas that cause inappropriate secretion of insulin resulting in lowered blood glucose levels). The anterior pituitary and the adrenal glands can also be involved.

MEN2A and MEN2B have some features in common, likely due to overlapping downstream activities of the two different RET mutant proteins. Two conditions associated with both MEN2A and MEN2B are medullary thyroid cancer (MTC), in which there is usually excess secretion of the hormone calcitonin, and a tumor of the adrenal gland medulla known as pheochromocytoma, in which various adrenal catecholamine hormones are overproduced, resulting in high blood pressure. However, the two forms of MEN2 can be distinguished by additional features, suggesting the activation of unique downstream targets specific to the different mutations in RET. MEN2A is associated with parathyroid gland hyperplasia or tumors of the parathyroid gland in around 20% of individuals. MEN2B is associated with more aggressive tumors and several morphological abnormalities, often detectable by 5 years of age, including a characteristic facial appearance with swollen lips; tumors of the mucous membranes of the eye, mouth, tongue, and nasal cavity; enlarged colon; and skeletal abnormalities.

Currently, DNA testing can facilitate early diagnosis of MEN syndromes by identification of germline mutation in asymptomatic mutant gene carriers.

Children who are identified as carriers of the RET gene can be offered total thyroidectomy on a preventative (prophylactic) basis to prevent the development of MTC. Prevention is the key goal, as although MTC can be treated by surgical removal of the thyroid, metastatic spread of the cancer may already have occurred and neither chemotherapy or radiotherapy are effective in controlling spread. Pheochromocytoma in both types of MEN2 can be cured by surgery.

Others – FGFR, PDGFR, KIT, and FLT3

Fibroblast growth factor receptors (FGFRs) have central roles during ontogeny and on cell differentiation, proliferation, and in angiogenesis (discussed in more detail in Chapter 14). Amplification of FGFR1 occurs in approximately 10% of breast cancers and is associated with poor prognosis. FGFR1 amplification also often accompanies squamous cell lung cancer. Unlike, adenocarcinoma, in which addiction to mutant EGFR and ALK are well known, squamous cell carcinoma was thought not to exhibit this phenomenon. FGFR1 dependence has been shown in FGFR1-amplified lung cancer cell lines, and use of a small molecule FGFR1 inhibitor led to significant tumor shrinkage in models *in vivo*. Mutational activation of FGFR2 is a frequent event in endometrioid endometrial cancer and occasionally also in ovarian cancer. FGFR3 is located on chromosome 4p16.3 and mutations located in exon 7 at A248C and S249C and in exon 10 at G372C and T375C have been found in bladder cancers. In a phase I study of patients with advanced/metastatic cancer, brivanib, a dual FGFR and VEGFR TKI, showed both antiangiogenic and antitumor activity.

The platelet-derived growth factors (PDGF) are a pleiotropic family of peptide growth factors that signal through cell surface RTKs (PDGFr) and stimulate various cellular functions, including growth, proliferation, and differentiation. To date, PDGF expression has been demonstrated in a number of different solid tumors, from glioblastomas to prostate carcinomas. In these various tumor types, the biologic role of PDGF signaling can vary from autocrine stimulation of cancer cell growth to subtler paracrine interactions involving adjacent stroma and vasculature.

The *KIT* gene encodes the cellular homolog of the feline sarcoma viral oncogene v-*kit*. KIT is a RTK for stem cell factor (SCF), also known as mast cell growth factor (MGF). KIT and PDGF receptors (PDGFRs) are oncogenic in a variety of hematologic and solid tumors. These RTKs, as well as the nonreceptor tyrosine kinase ABL and BCR–ABL, are inhibited by imatinib (see below). Gastrointestinal stromal tumor (GIST) is the most common human sarcoma and is characterized by activating mutations in the KIT or PDGFRA receptor tyrosine kinases. The interstitial cells of Cajal are believed to be the cell of origin for GIST and express high levels of KIT. Why only these cells are affected in patients with familial germline activating KIT mutations is not clear, but may be due in part to presence of the ETS family member ETV1, which increases susceptibility to oncogenic KIT, by cooperating in activation of an oncogenic ETS transcriptional program, whereas in other cancers, such as prostate, ETS is activated by translocation or amplification.

Although rarely found in common melanoma skin cancers, activating *KIT* mutations and increased copy number have been found in melanomas from acral lentiginous, mucosal, and chronic sun-damaged sites. Mutations in the *KIT* gene are also found in most cases of the mast cell-derived tumor mastocytosis. Seminomas often manifest activating mutations in exon 17 of the *KIT* gene. Rearrangements in the *PDGFRA* gene, including fusion between kinase insert domain receptor (KDR) (VEGFRII) and the *PDGFRA* gene and a more common *PDGFRA* (Δ8,9) intragenic deletion have been found in PDGFRA-amplified glioblastoma multiforme. Both rearrangements showed constitutive tyrosine kinase activity reversed by PDGFR antagonists.

FMS-like tyrosine kinase 3 receptor (FLT3) and the FTL3 ligand are important in hematopoiesis and development of dendritic cells. FLT3 mutations are found in some leukemias and internal tandem duplications are frequently observed in acute myeloid leukemia (AML). Although some promise has been shown in newly diagnosed AML patients, FLT3 inhibitors such as lestaurtinib have not proved successful in clinical trials of patients relapsing after conventional therapy.

The TKI imatinib mesylate (Gleevec) blocks activity of the BCR–ABL oncoprotein (see Chapter 6) and the cell surface receptor tyrosine kinase c-KIT and PDGFRA, and is used in the treatment of chronic myeloid leukemia (CML) and of the commonest human sarcoma GIST. Given the numerous mechanisms by which increased growth factors may arise during tumorigenesis, the PDGF pathway may represent a therapeutic target in tumors even when it is not in itself the target of an activating mutation.

G-protein coupled receptors and their ligands

In some cell types, such as Schwann cells of the peripheral nervous system, activation of G-protein coupled receptors

(GPCRs), which couple to activation of adenylate cyclase and elevate intracellular cAMP, may stimulate replication in synergy with traditional growth or survival factors. Conversely, in these types of cells activation of GPCRs, such as somatostatin receptors that negatively regulate cAMP production, will reduce growth. In some cell types, such as those involved in growth hormone production in the pituitary gland, elevation of cAMP via GPCRs (e.g. GHrh) is prosecretory. Human pituitary adenomas express multiple somatostatin receptor subtypes, the pattern of which may in part determine the response to medical treatments with the somatostatin agonists octreotide and lanreotide, which preferentially activate the type 2 receptor. Human growth hormone-secreting (somatotroph) pituitary adenomas express both type 2 and type 5 receptors that are involved in the regulation of growth hormone secretion. These adenomas are most responsive to octreotide treatment. Prolactinomas (secreting the pituitary hormone prolactin) rarely express the type 2 receptor, explaining the lack of efficacy of octreotide in lowering the elevated prolactin in patients. However, these tumors frequently express dopamine receptors and may be sensitive to treatment with dopamine agonists that can suppress growth and abnormal hormone secretion.

Amidated and non-amidated gastrins (gastrin precursors) may contribute to growth of gastrointestinal and pancreatic cancers. Progastrin and gastrin are antiapoptotic and amidated gastrins may promote migration of epithelial cells. In support of this, targeting gastrins therapeutically via a vaccine has had some limited success in treating gastrointestinal and pancreatic cancers.

It has long been appreciated that the GPCRs can activate the ERK (MAPK) signaling pathway and thereby contribute to cell replication. However, the mechanisms underlying this crosstalk have remained elusive. Work from the group of Axel Ullrich and others have identified crosstalk between GPCRs and EGFR signaling systems in a variety of normal and transformed cell types. GPCR agonists can promote activation of the RAS–RAF–ERK pathway and cell proliferation via the EGFR in fibroblasts and some cancer cells. The metalloprotease-disintegrin tumor necrosis factor-alpha-converting enzyme (TACE/ADAM17) may be important in determining this interaction. In part, activation of stromal cell EGFR signaling by TACE may be mediated by proteolysis-induced release of the ligand amphiregulin.

Regulation occurs at multiple levels in these pathways, including at the level of the G-proteins themselves. G-protein alpha-subunits (Gα) appear to be capable of interacting with different effectors, leading to engagement of distinct signaling pathways. Expression of activated Gα subunits have been shown to cause transformation of cells, and they can activate MAPK signaling. Moreover, Gαo and Gαi subunits can signal through both SRC and STAT3 pathways to contribute to cell transformation.

Effective new cancer therapies targeting key signaling molecules in the RTK signal transduction pathway are now available, whereas GPCR-mediated disorders are still largely treated with receptor-specific agonists or antagonists. Progress in understanding GPCR signaling may, however, in the future yield new therapies based on targeting GPCR activity at different molecular levels. Thus, not only could we make the targeting of specific receptor subtypes more selective, we may in the future also target various regulatory factors in G-protein signaling pathways. These include, receptor-associated proteins such as receptor activity-modifying proteins (RAMPs) and receptor activated solely by synthetic ligands (RASSLs); and also proteins which modify G-protein activity such as activators of G-protein signaling (AGS) and regulators of G-protein signaling (RGS).

Over the last few years it has become apparent that estrogen may activate signaling independently of transcriptional activation by classic nuclear estrogen receptors (ERs). A novel transmembrane receptor, G-protein coupled receptor 30 (GPR30), can mediate rapid nongenomic signaling by estrogen and can contribute to tumorigenesis by activating downstream targets such as the MAPK pathway even when nuclear ERs are blocked.

Another GPCR ligand implicated in tumorigenesis is lysophosphatidic acid (LPA). LPA is produced by the enzyme autotaxin (ATX/LysoPLD) aberrantly expressed in multiple cancer cells, and ATX and LPA receptors contribute to initiation and progression of breast cancer in transgenic mice.

Nuclear receptors

The nuclear receptor family comprises a group of ligand-activated transcription factors, with just under 50 members in humans, in most part activated by binding of small lipophilic ligands including steroid and thyroid hormones, fatty acids, and retinoids. Nuclear receptors regulate target gene transcription by binding to regulatory regions as either homo- or heterodimers. These receptors regulate multiple genes involved particularly in cell growth and differentiation and in energy metabolism. A large number of currently used drugs target nuclear receptors, including activators of PPARα and γ used in dyslipidemia and diabetes, respectively, estrogen receptor agonists used as hormone replacement and contraception, and antagonists used in treating breast cancer and various negative regulators of testosterone used to treat prostate cancer. Some, such as thyroxine, are used to correct deficiencies in endogenous hormones. Corticosteroids are widely used in inflammatory diseases, some cancers, and as immunomodulators.

Nuclear receptors are often subdivided into four functional classes, types 1–4. On ligand binding, type 1 receptors are displaced from cytoplasmic complexes with heat shock proteins, enabling formation of active homodimers and translocation from the cytoplasm into the nucleus. Once in the nucleus, type 1 receptors bind to hormone response elements to regulate gene expression. This group includes a number of very important targets in cancer therapy, such as the androgen receptor, estrogen receptors, progesterone receptor, and the glucocorticoid receptor. Type 2 receptors are located in the nucleus, even in the absence of ligand, in association with a corepressor protein. This corepressor is displaced by ligand binding, allowing formation of an active regulatory heterodimer containing retinoid X receptor (RXR). This group of nuclear receptors includes RXR, the retinoic acid receptor and thyroid hormone receptor. Less is known about type 3 and type 4 receptors, where ligands often remain unknown.

Nuclear receptors are strongly influenced by various cofactors involved in chromatin remodeling, histone modification (acetylation–deacetylation), and regulation of the transcription machinery, as well as by a variety of phosphatases, including the protein tyrosine phosphatases (PTPs), which may represent new druggable targets in the future. Many nuclear receptor ligands exist as bioactive food components and, as such, nuclear receptors and signaling can be expected to be influenced by dietary factors as well as by therapeutics.

Estrogens control growth and differentiation in target tissues via two main receptors, ERα and ERβ. ERα is mainly expressed in expressed in breast, uterus and cervix, whereas ERβ expression is largely expressed in ovary, prostate, and testis. In the presence of estrogen, the ER is released from bound heat shock proteins, allowing it to form homodimers and become transcriptionally active. As most breast cancers require estrogen signaling for growth there have been substantial successes in treating women with selective estrogen receptor modulators (SERMs), of which tamoxifen is the archetype. Tamoxifen is an unusual drug in having agonist and antagonist activity, depending on the context: it is an antagonist in breast and agonist in bone and vasculature. This balance is actually advantageous as it reduces risk of osteoporosis and premature vascular disease arising from deficient estrogen activities in these tissues. New SERMs, such as raloxifene, are now available.

Following binding of the ligand 1α,25-dihydroxycholecalciferol D3, the vitamin D receptor (VDR) controls genes involved in bone turnover and calcium homeostasis. Deficient signaling results in rickets in children, and osteomalacia and increased fractures in adults. However, in vitro vitamin D also has effects on a variety of cancer-relevant cellular processes such as replication and survival.

Unopposed estrogen action has long beeen known to greatly increase risk of endometrial cancer, which is largely negated by combined use of estrogen and progesterone. However, progesterone appears not to protect women against an increased risk of breast cancer resulting from estrogen-containing treatments. Endometrial cancer risk is also increased by tamoxifen treatment and this should be considered during treatment of breast cancer. Endometrial cancer treatments may include progestagens, and levels of progesterone receptors have important prognostic value in this cancer.

PPARs are very important in dyslipidemia and diabetes and PPARα and γ are the targets of fibric acid derivatives and glitazones, respectively. The endogenous ligands for PPARs are fatty acids and fatty acid-derived compounds. Interestingly, PPARγ has been shown to reduce proliferation in various cancer cell lines and mouse models in vivo and in colorectal cancer particularly.

High-fat diets result in increased excretion of bile acids through the bowel. Lithocholic acid (LCA) in particular is poorly recirculated and enters the large bowel. LCA may contribute to colon cancer development in animal models and can cause DNA damage in cell cultures. It is possible that this is one explanation for the increased risk of colorectal cancer in obese patients and those with high fat intake. Conversely, vitamin D intake which is negatively associated with colorectal cancer risk, may reduce LCA levels.

Androgens, such as testosterone, are analogous in cancer of the prostate to estrogens in breast cancer, and can promote hypertrophy (enlargement of the prostate) and also promote growth. In fact, the major nonsurgical approach to treatment of prostate cancer involves pharmacological and/or surgical anatagonism or ablation of androgen production. Testosterone production can be impeded by use of drugs that interfere with synthesis, that downregulate the pituitary hormones needed to drive testosterone production in the testes (chemical castration), or by surgical removal of the testes.

Signaling pathways

Nonreceptor tyrosine kinases

Nonreceptor tyrosine kinases, including SRC, ABL, and JAK, are major components of signaling networks implicated in diverse processes, including cell growth, replication, survival, differentiation, migration, and genome maintenance. JAK is discussed below under cytokine signaling, and SRC and ABL are discussed in more detail in Chapter 6.

The mitogen-associated protein kinase (MAPK) pathway

The RAF–MEK–ERK pathway was the first identified mitogen-associated protein kinase (MAPK) cascade and is also one of the best described of all the cancer-causing signaling pathways (Fig. 5.17). Because this will be covered in depth in the section on RAS in Chapter 6, only an overview will be given here.

RAS GTPases control a substantial signaling network involved in multiple key cell processes, including many involved in cancer such as proliferation, survival, differentiation, migration, and senescence. RAS–MAPK signaling is normally tightly regulated but also frequently undergoes somatic mutations in human cancers. Hereditary mutations in various RAS regulatory genes have been described in a number of developmental disorders which share facial dysmorphism, cardiac defects, and susceptibility to certain malignancies.

RAF is the first member of a three-kinase modular sequence, with the generic designation MAPKKK, MAPKK, and MAPK, standing for mitogen-activated protein kinases. At least six distinct kinase pathways follow this general pattern in mammalian cells, including the JUN kinase/stress-activated kinase (SAPK).

The mitogenic pathway involves RAF, which in turn activates the MAPK/ERK kinases (MEK1 and 2), which then activate ERK1 and 2 (MAPK).

Sequential kinase cascades may function to amplify the input signal. However, the linking of some of these signaling molecules to scaffolding proteins such as MP1 (MEK partner) and JIP1 (JUN N-terminal kinase interacting protein) could prevent such amplification and might also act to prevent undesired crosstalk. In fact, scaffolding proteins may be a general mechanism whereby the potentially undesirable spread of signaling across multiple pathways can be prevented, as only the appropriate targets of signaling components are brought together. The end result of the MAPK sequence is relocation of active ERK to the nucleus, and phosphorylation of regulated transcription factors of the ETS family. Ultimately, this leads to enhanced expression of other key transcription factors such as FOS and c-MYC, which is key to cell-cycle control (see Chapter 6). Apoptosis signal-regulating kinase 1 (ASK1) is an upstream MAPK involved in apoptosis, inflammation, and carcinogenesis.

The proto-oncoprotein c-JUN belongs to the AP-1 group of transcription factors and is a key regulator of proliferation and tumorigenesis. AP-1 stimulation is mediated by N-terminal phosphorylation of the c-JUN transactivation domain by the JNKs. This allows c-JUN to dissociate from an Mbd3-histone deacetylation (NuRD) repressor complex, and releases inhibition on transcription of target genes, such as the gut stem cell marker lgr5.

Phosphatidylinositol-3 kinase signaling pathway

On activation by receptors, the ubiquitously expressed class IA isoforms (p110alpha and p110beta) of phosphatidylinositol-3 kinase (PI3K) generate lipid second messengers, which trigger multiple signaling pathways. The PI3K signaling pathway is as important as the RAS–MAP kinase pathway in cell survival and proliferation, and hence its potential role in cancer is of great

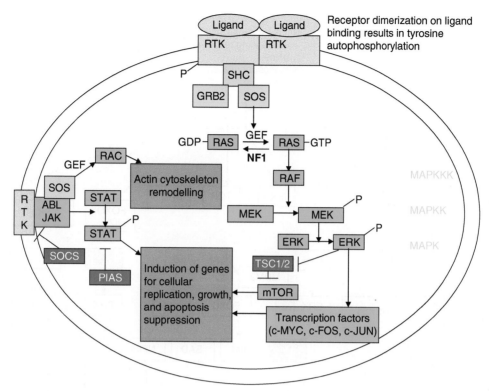

Figure 5.17 RAS–RAF–ERK (MAPK), a model growth factor receptor-activated signaling pathway. Receptor dimerization on ligand binding induces tyrosine autophosphorylation, binding of GRB2 to the receptor via the SH2 domain and translocation of SOS to the cell membrane. SOS in turn promotes RAS activation by enhancing GDP–GTP exchange. A third adaptor protein, SHC (also carrying an SH2 domain), may be phosphorylated by growth factor binding and can recruit the GRB2–SOS complex. Activated RAS by either of these two overlapping pathways in turn phosphorylates the serine/threonine kinase RAF. RAF phosphorylates and activates MAP kinase/ERK kinase (MEK). MEK then activates ERK, which can then translocate to the nucleus and activate transcription factors such as c-MYC, c-FOS, and c-JUN. This is a prototypic kinase cascade generically referred to as MAPKKK-MAPKK-MAPK. ERKs can also phosphorylate and inactivate the TSC1 and TSC2 proteins, thus activating the mTOR (raptor) complex. Growth factors (particularly cytokines) can also activate the Janus cytoplasmic protein tyrosine kinases (JAKs). These kinases lack SH2 or SH3 domains and phosphorylate the receptors themselves after ligand binding. JAK phosphorylated receptors then bind STAT transcription factors via SH2 domains, which become activated oligomers. STAT dimers can then translocate to the nucleus and activate transcription. The ABL cytoplasmic tyrosine kinase is also important in mediating growth factor signals. In particular ABL may link to RAS family members such as RAC involved in cell motility and also plays a role in regulation of the cell cycle through interactions with RB.

interest (Fig. 5.18 shows an integrated view of these signaling pathways).

PI3Ks constitute a lipid kinase family characterized by their ability to phosphorylate phosphatidylinositol and related inositol ring 3'-OH groups in inositol phospholipids, to generate the second messenger phosphatidylinositol 3,4,5-trisphosphate (PIP3). The p110beta possesses kinase-independent functions in regulating cell proliferation, whereas the kinase activity of p110beta is needed for oncogenic transformation.

PIP3, in turn, regulates signaling pathways involved in cell replication, growth, and survival, among others, in part by regulating activation of downstream kinases such as the AKT (v-akt murine thymoma viral oncogene homolog) family (also known as protein kinase B or PKB) of serine/threonine kinases and the mammalian target of rapamycin (mTOR). PIP3 activates AKT by promoting translocation to the inner membrane, where it is phosphorylated and activated by PDK1 and PDK2 (Fig. 5.19). In turn, activated AKT modulates the phosphorylation and activity of numerous substrates involved in the regulation of cell survival, cell-cycle progression, and cellular growth (Fig. 5.20), including IKK/NF-κB, mTOR (and thence S6K and HIF1α), Forkhead (FOXO), MIZ1, BAD, GSK-3β, and MDM2 (Fig. 5.20). Growth factors such as insulin and IGFs can activate the PI3K–AKT pathway via the IRS-2 adaptor protein recruited to the activated RTK. This is an important pathway in normal regulation of β-cell growth regulation and defects in this pathway may favor loss of β cells in diabetes.

Deregulation of the PI3K pathway is a feature of many common cancers, either by loss of the tumor suppressor protein PTEN (phosphatase and tensin homolog deleted on chromosome 10) discussed further in Chapter 7, or by constitutive activation of PI3K isoforms or downstream targets, such as AKT. AKT and mTOR activation is regulated by two tumor suppressors – PTEN (a lipid phosphatase involved in breaking down PIP3 to PIP2), which inhibits upstream of AKT, and the TSC1/TSC2 heterodimer, which acts downstream of AKT and upstream of mTOR. The PI3K–AKT pathway is also negatively regulated by a newly identified protein serine/threonine phosphatase called PHLPP (pleckstrin homology domain and leucine-rich repeat protein phosphatase), containing a pleckstrin homology domain.

Mutations in the PI3K pathway may account for as many as 30% of all human cancers. Direct evidence for a role of PI3K deregulation in human cancer came from discovery of amplification of the *PIK3CA* genes encoding the p110alpha catalytic

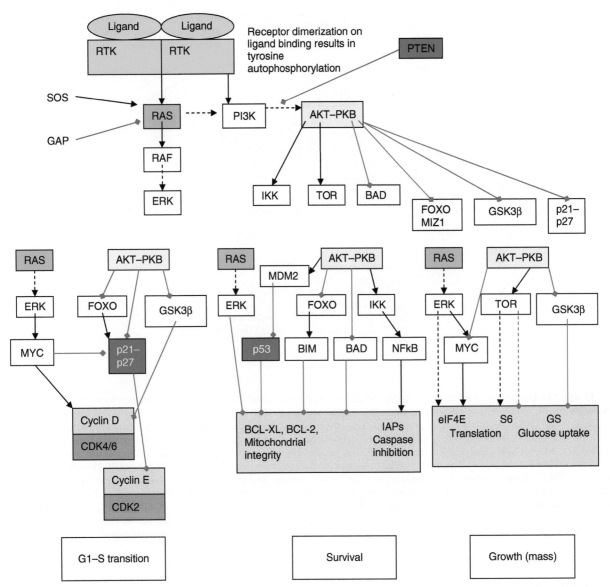

Figure 5.18 Integrated networks involving RAS and PI3K act downstream of RTKs to regulate replication, survival and growth. RAS and AKT coordinately regulate growth, survival, and proliferation. Key sites for integrated signaling are occupied by the kinases ERK, TOR, and GSK-3β, the CDK inhibitors p21^{CIP1} and p27^{KIP1} and the transcription factors c-MYC, FOXO, and NF-κB. Oncogenic RAS may also activate ARF and p16^{INK4a} (not shown), suggesting that in some way oncogenic cell cycles trigger additional responses which may seek to minimize the effect of such mutations. After Massague, J. (2004) G1 cell-cycle control and cancer. *Nature* 432: 298–306.

subunit of PI3K and the *AKT2* gene in ovarian, breast, and pancreatic cancers. Mutations in the gene encoding the regulatory subunit p85alpha, which contains the SH2 domains allowing interactions with growth factor RTK and substrate adaptor proteins, have now been shown in some colon and ovarian tumors. This notwithstanding, loss of PTEN is the most common mechanism of activation of the PI3K pathway in human cancers (see Chapter 7); *PTEN* is second only to *p53* as the commonest tumor suppressor inactivated in human cancers.

The PI3K pathway mediates multiple cancer-relevant cellular behaviors, including cell survival and proliferation and also cytoskeletal deformability and motility – key elements in tumor invasion (see Chapter 12).

The *PTEN* and *p53* tumor suppressors are among the most commonly inactivated or mutated genes in human cancer. Importantly, acute *PTEN* inactivation induces growth arrest through the p53-dependent cellular senescence pathway both *in vitro* and *in vivo*, suggesting that p53 is an essential failsafe protein preventing tumorigenesis in PTEN-deficient tumors. Hereditary mutations of *PTEN*, in Cowden disease, are associated with increased tumor susceptibility and defects in angiogenesis. The implication of the PI3K pathway in angiogenesis is relatively novel and recent studies have explained the phenotype observed in Cowden disease, as the PI3K pathway is important for angiogenesis both in normal development and in tumorigenesis.

As technological advances have increasingly allowed us to look in detail at single cells, rather than measuring averages across cell populations, in various contexts it has become increasingly apparent that there is considerable cell to cell variability in situations where this was not previously noted. This variability may

Growth Signaling Pathways and Targeted Treatment

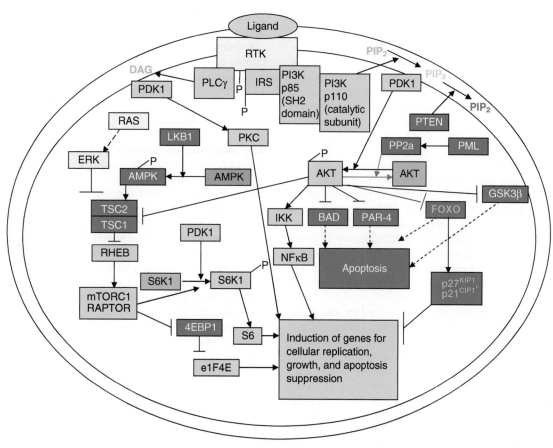

Figure 5.19 Lipid signaling from receptor tyrosine kinases (such as IGF1R, HER2/Neu, VEGF-R, PDGF-R): the PI3K and DAG–PKC pathways. RTK become activated by tyrosine autophosphorylation. Activated RTKs associate with the p85 SH2 domain-containing subunit, which recruits and activates the catalytic domain of PI3K. Activated PI3K in turn phosphorylates membrane inositol phospholipids in the 3' position, which can then act as docking sites for proteins with pleckstrin homology (PH) domains, such as phospholipase cγ (PLCγ), PDK1, and AKT. PLCγ activates protein kinase C (PKC), which is important for transducing mitogenic signals, via diacylglycerol (DAG). The serine/threonine kinase AKT is activated by phosphorylation by PDK1/PIP$_3$. AKT regulates phosphorylation of several downstream effectors, such as NF-kappa B (NF-κB), mammalian target of rapamycin (mTOR), FOXO1, BAD, glycogen synthase kinase (GSK-3β), and MDM-2, which in turn influence cell growth, proliferation, protection from proapoptotic stimuli, and stimulation of neo-angiogenesis. AKT activates mTOR by phosphorylating the tuberous sclerosis 2 protein (TSC2); the resultant complex of TSC1/TSC2 is inactivated (loss of GTPase activity for RHEB), thus increasing active RHEB–GTP levels, which activate mTOR leading to cell growth and replication. GSK-3β is a constitutively active proline-directed serine/threonine kinase that may induce apoptosis through multiple pathways. AKT can inactivate GSK-3β via phosphorylation, thus reducing apoptosis. The promyelocytic leukemia protein (PML) tumor suppressor inactivates pAKT inside the nucleus, in part by recruiting the AKT phosphatase PP2a as well as pAKT into PML nuclear bodies. From Massagué J (2004) G1 cell-cycle control and cancer. *Nature*, **432**: 298–306.

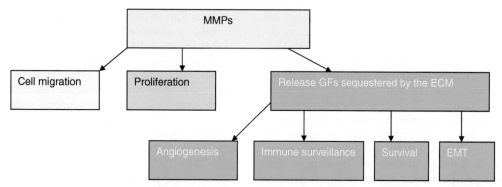

Figure 5.20 Growth factor activation by matrix metalloproteinases. Some growth factors are inactive while bound to extracellular matrix, but can be released by activity of matrix-degrading enzymes such as the matrix metalloproteinases.

in many cases be a simple product of random stochastic events (chance) but could also under certain circumstances be a regulated process required for cell fate determination, optimal growth, and tissue homeostasis (see also super competition in Chapter 6) and may potentially help restrict activation of replication-promoting pathways, such as AKT, to a subset of cells to avoid senescence (i.e. some cells replicate others keep their telomeres longer).

The AKT substrates

AKT regulates several pivotal cellular processes, including survival, replication, growth, and metabolism, by phosphorylation of a variety of known substrates. Because AKT and its upstream regulators are involved in key cancer behaviors and are deregulated in a wide range of cancers the AKT pathway is being explored as a potential prognostic marker and as a target for novel anticancer therapies. Activation of the PI3K–AKT pathway renders cancer cells resistant to apoptotic signals and promotes tumor growth.

Glycogen synthase kinase 3 (GSK-3) is inhibited by AKT phosphorylation; GSK is known to phosphorylate and inhibit multiple key cell-cycle regulators, including c-MYC, cyclin D1, and cyclin E, and transcription factors involved in cell differentiation, including c-JUN, GLI, Notch, Snail, and β-catenin. GSK-3 may help form a docking site for the ubiquitin ligases such as FBW7 (see Chapters 4 and 11). AKT could promote the activity of these factors. GSK-3 may also contribute to tumorigenesis and is required for maintenance of a poor prognosis human leukemia, defined by mutations of the *MLL* proto-oncogene.

However, exactly how AKT prevents apoptosis is not fully understood. Inhibition of GSK-3β is one important contributor, as is phosphorylation and inactivation of BAD, but recent studies suggest that AKT may prevent apoptosis in some cases by phosphorylation of another proapoptotic protein, PAR-4. In a study by Goswami and colleagues (2005), suppressing AKT activation by PTEN or other means caused apoptosis in cancer cells. Importantly, apoptosis resulting from inhibition of AKT was blocked by inhibition of PAR-4 expression, but not by inhibition of other apoptosis agonists that are AKT substrates. AKT can also prevent apoptosis by activating the IκB kinase–NF-κB pathway.

The FOXO family of forkhead transcription factors is also a substrate of AKT; phosphorylation of FOXO proteins enables binding of the 14-3-3 protein and resultant sequestration in the cytoplasm. Inactivation of FOXO by AKT contributes to resistance to stress-induced apoptosis.

It is increasingly likely that cellular proliferation requires the coordinated activation of signaling pathways: mitogenic, involving RAS-MAPK, as well as pro-survival, involving PI3K-AKT, and various downstream targets of these signaling cascades. This is illustrated by a recent publication from the Rosen laboratory (She *et al.*, 2005). In this study, both pathways combine to prevent apoptosis by phosphorylation of the proapoptotic protein BAD. EGFR–MEK–MAPK may phosphorylate BAD on serine 112, whereas PI3K–AKT phosphorylates BAD on serine 136 phosphorylation; either is sufficient to sequester BAD to 14-3-3, blocking proapoptotic activity. Interestingly, apoptosis is only triggered if BAD is dephosphorylated on both serines in response to inhibition of both pathways.

AKT and mTOR (see below) can also activate the transcription factor HIF-1α (hypoxia-inducible factor 1α), which promotes several metabolic effects often observed in tumors, including upregulation of glucose uptake and glycolytic enzymes.

Activating somatic mutations in AKT are often observed in human cancers such as breast, colorectal, and ovarian cancers, resulting in a glutamic acid to lysine substitution at amino acid 17 (E17K) in the lipid-binding pocket of AKT1. This mutation aberrantly locates AKT1 to the plasma membrane and activates downstream signaling.

The mammalian target of rapamycin (mTOR)

The LKB1–AMP-activated protein kinase (AMPK)–TSC–mTOR complex (mTORC1) cassette is a key signaling pathway that integrates metabolic and nutrient status with cell growth. The TOR pathway operates as a key growth-control node in all eukaryotes and in mammals mTOR can simultaneously sense stress, energy, and nutrient inputs such as glucose, amino acids, and hypoxia while integrating stimulatory signals from growth factors via the PI3K and RAS pathways. This pathway is deregulated by hyperactivation of mTORC1 in many different cancers and hereditary hamartoma syndromes, including Peutz–Jeghers syndrome and tuberous sclerosis.

As the name would suggest, mTOR is a well-known target of a known chemotherapeutic agent (rapamycin!), and is involved in a variety of cancer-relevant processes such as cell size, survival, and proliferation. mTOR is normally involved in regulating many aspects of cell growth, including membrane traffic, protein degradation, protein kinase C signaling, ribosome biogenesis, and transcription, in part by activating both the 40S ribosomal protein S6 kinases (S6K1 and S6K2) and the eukaryotic initiation factor 4E (EIF4E)-binding protein 1. Survival signals generated by PI3K and phospholipase D target mTOR, which in turn contributes to suppression of apoptotic pathways in cancer cells. Conversely, inhibitors of mTOR prevent G_1/S transition by nonspecific inhibition of cell growth and also by preventing CDK activation.

There are now known to be at least two distinct mTOR complexes, defined by the presence of two key substrate determining subunits:

- raptor (the mTORC1 complex), which is inhibited by rapamycin, and
- rictor (the mTORC2 complex) which is not.

Most of what is known about mTOR in mammals refers to the mTORC1 complex and its substrates, the growth-promoting S6 kinases, and the translation activators such as the 4E-binding proteins. Recent data suggest that the rictor complex my play a role in activating AKT in order to allow activation of the raptor complex by AKT. The RAS–RAF–ERK pathway can also activate mTOR, at least in part by interfering with negative regulators of the complex (discussed below).

Tuberous sclerosis (TSC) is an autosomal dominant human genetic syndrome caused by germline mutations in either *TSC1* or *TSC2* genes, and is characterized by the development of specific tumors called hamartomas. The *Tsc1* and *Tsc2* genes, encoding the proteins hamartin and tuberin respectively, have been shown to contribute to regulation of cell and organ size in several organisms, including the fruitfly. The *TSC* genes are in the PI3K–AKT–mTOR–S6K pathway. Activated AKT phosphorylates TSC2 in the TSC1/TSC2 protein complex, inactivating it; while TSC1/TSC2 has GAP activity for the RHEB GTPase (a member of the RAS family), and activated RHEB GTP activates mTOR. Thus, in cells lacking TSC1 or TSC2 there are increased levels of RHEB GTP, which leads to activation of mTOR, leading to cell size increase and growth; signaling through ERK may have a similar effect.

PTEN mutations are associated with constitutive activation of the PI3K–AKT–mTOR pathway; such tumors are relatively resistant to apoptosis and may be particularly sensitive to mTOR inhibitors (see Chapter 7).

LKB1 is a tumor suppressor that is inactivated by mutations in a familial human cancer, Peutz–Jeghers syndrome. LKB1 is the activating kinase for the AMP-activated kinase (AMPK), which is now being intensively studied because of its role in glucose homeostasis and diabetes. Briefly, during hypoxia or when glucose levels are low, AMPK is activated via LKB1. Interestingly, LKB1 inactivates mTOR (the raptor complex), in part through AMPK which can phosphorylate tuberin. In fact, the effects of LKB1 on mTOR signaling may explain the tumor suppressor activity, particularly given the phenotypic similarity between Peutz–Jeghers syndrome, Cowden syndrome (PTEN deficiency), and tuberous sclerosis discussed above – they all share formation of hamartomas.

The serine/threonine Pim kinases, overexpressed in many cancers, can promote mTORC1 activity by stimulating phosphorylation and activation of the inhibitory AMPK.

Specific mTORC1 inhibitors have been developed for clinical use and preclinical and clinical studies performed with rapamycin and derivatives have been promising.

Cytokine signaling

Cytokines and growth factors activate the MAP kinase pathways, resulting in the stimulation of ERK1/2, JNK, and p38 kinases, which in turn activate transcription factors like AP-1 and ATF-2. Other proinflammatory agents such as TNF-α and IL-1 can activate the transcription factor NF-κB that in turn regulates the expression of immediate early genes involved in immune, acute phase, and inflammatory responses. Besides the transcription factors NF-κB and AP-1, which are immediate-early transcriptional activators, components of the JAK–STAT pathway play an important role in the transcriptional activation of many inflammatory genes.

JAK–STAT pathways

Interleukins (IL)-6, -4, and -8 are raised in many patients with cancer of colon, prostate, and breast, and chronic inflammation contributes to formation of these cancers. Cytokines such as IL-6 can regulate proliferation and apoptosis via action on members of a family of key transcription factor such as STAT3, which is a known oncogene. Other cytokines such as IL-4 and -8 may act through inhibitors such as SOCS (suppressors of cytokine signaling).

The JAK–STAT signaling pathway comprises three families of proteins: Janus kinases (JAKs), signal transducers and activators of transcription (STATs), and their endogenous inhibitors – the SOCS family. JAK–STAT pathways are used by a wide variety of cytokines and growth factors, whereas inappropriate function of this pathway is implicated in many human cancers. Defective JAK–STAT–SOCS pathways may impair tumor responses to immunotherapy. The multiplicity of JAKs (four members) and STATs (seven members) enable formation of a large number of different STAT homo- and heterodimers and transcriptional complexes, which may mediate specific patterns of gene expression. In particular, processes controlled by this pathway include cell growth, differentiation, senescence, and apoptosis. STAT family proteins are latent cytoplasmic transcription factors that convey signals from cytokine and growth factor receptors to the nucleus (see Figs 5.16 and 5.17). They were originally discovered and characterized through the study of interferon-induced responses. Binding of cytokines to cell surface receptors results in two categories of signaling: either activation of cytoplasmic tyrosine kinases (particularly JAK or SRC kinase families) or activation of receptor-intrinsic tyrosine kinase activity (such as discussed for PDGF and EGF above). Tyrosine phosphorylation activates STAT monomers, which dimerize through interaction of SH2 domains. The resultant STAT dimers translocate to the nucleus and bind to STAT-specific sites known as gamma-activated sites (GAS) of target genes to induce transcription.

There are seven known STAT proteins, each encoded in a distinct gene but subject to alternative RNA splicing or posttranslational proteolytic processing, which may result in additional STATs found in malignant cells. Tumor cells acquire the ability to proliferate uncontrollably, resist apoptosis, sustain angiogenesis and evade immune surveillance. STAT proteins, especially STAT3 and STAT5, regulate all of these processes and are persistently activated in a number of human cancers. Epigenetic silencing of the *SOCS1* gene, an inhibitor of the JAK–STAT pathway, has been described in liver cancer.

Constitutive activation of the JAK–STAT pathway is known to occur in HTLV-I-transformed T cells, Sezary's syndrome, and v-*abl* or v-*src* transformation. Mutations of *STAT* genes have not been described in human cancers, rather oncogenic potential of STAT3 and STAT5 is acquired by activation through upstream pathways. Constitutive activation of STAT proteins are frequently observed in cancers, and active forms have been detected in cancers of the head and neck, lung, kidney, prostate, breast, ovaries, and blood. Aberrant signaling through a number of upstream pathways can result in constitutively activated STATs in tumor cells. Cellular transformation with v-src or BCR–ABL activates STAT3 and STAT5 and constitutively active STAT3 will transform immortalized fibroblasts. STAT5 can control transcription of various factors implicated in cancer, including BCL-X_L, PIM-1, and cyclin D1, suggesting a role both in resistance to apoptosis and in cell replication.

Deregulated STAT3 is oncogenic and promotes uncontrolled growth and survival through aberrant activation of, among others, cyclin D1, c-Myc, Bcl-x_L, and survivin.

In contrast to STAT3 and STAT5, STAT1 negatively regulates cell proliferation and angiogenesis and thereby may be a tumor suppressor. In fact, STAT1 and its downstream targets are reduced in various human cancers and STAT1 knockout mice are tumor-prone.

Again, recent studies have begun to explore the role of miRNAs in this pathway. During even transient inflammation, IL-6 and STAT3 can activate miR-21 and miR-181b-1, which in turn can inactivate the PTEN and CYLD tumor suppressors, leading to increased NF-κB activity and transformation of normal cells. Thus, STAT3 miR-21, miR-181b-1 can provide a potential link between inflammation and cancer.

Numerous inhibitors/regulators of JAK–STAT pathways that can dampen down cytokine signaling have now been identified, including SOCS, which bind to JAK or cytokine receptor SH2 domains; phosphotyrosine phosphatases (SHPs, CD45, PTP1B/TC-PTP); and the protein inhibitors of activated STATs (PIAS). At least some SOCS proteins may act as tumor suppressors. These inhibitors act at different points in JAK–STAT pathways: tyrosine phosphatases dephosphorylate activated JAK, STAT, or cytokine RTKs; PIAS prevent activated STAT binding to DNA, in part through a SUMO E3-ligase activity (see Chapter 11); and SOCS, which, unlike the first two groups, are not constitutively present but are induced by cytokines, act in a negative feedback loop to inhibit JAK and cytokine receptors or to prevent STAT recruitment.

JAK tyrosine kinases are activated by interleukins and other growth factors, and promote survival and proliferation of cells in multiple tissues and are constitutively active in many hematopoietic malignancies and certain carcinomas. JAK can activate both STAT and PI3K signaling. JAK activity is essential for lymphoma invasion and metastasis, independent of its role in survival and proliferation, and independent of STAT and PI3K signaling. Fusion proteins arising by translocation in hematological malignancies have been described that involve JAK; thus the JH1 domain of JAK2 has been fused to other proteins to generate TEL–JAK2 and BCR–JAK2, both of which deregulate JAK2 action. Silencing by methylation of two JAK–STAT inhibitors – SOCS-1 and SHP-1 – has been described in some cancer cells, supporting their role as tumor suppressors. JAK2 V617F and other JAK–STAT-activating mutations have been described in BCR–ABL1-negative myeloproliferative neoplasms. Together this has fueled development of small-molecule ATP mimetics that inhibit wildtype and mutant JAK. Disappointingly, these have failed to show disease-modifying activity although some have improved symptoms by decreasing splenomegaly and constitutional symptoms in myelofibrosis patients. Both JAK2 (TG101348) and JAK1/2 (INCB018424, CYT387) inhibitors have been examined. It is possible that targeting other proteins in the pathway involved in tumorigenesis could also be effective and HSP90, for which JAK2 is a chaperone client, has recently been suggested as one such target.

IL-6–JAK signaling drives persistent STAT3 activation in cancer by transactivating the gene for sphingosine-1-phosphate receptor 1 (S1PR1), a G-protein coupled receptor for the lysophospholipid sphingosine-1-phosphate (S1P).

The nuclear factor-κB signaling pathway

It has long been appreciated that chronic inflammation predisposes patients to a number of cancers, as first articulated by Rudolf Virchow 200 years ago. It is now well known that persistent airway irritation in tobacco smokers, ulcerative colitis, chronic hepatitis B and C virus infection, and *Helicobacter pylori* infection predispose to lung, colon, liver, and stomach cancers, respectively. NF-κB is one of the best-known transcription factors and is centrally involved in many cellular processes, including inflammation, proliferation, and apoptosis and not surprisingly also in many additional cancer-relevant behaviors, such as angiogenesis, metastasis, and response to therapy. NF-κB transcription factors consist of five subunits: RelA (p65), c-Rel, RelB, p50/NF-κB1, and p52/NF-κB2. Interestingly c-Rel is the cellular homolog of the v-*rel* oncogene. Most NF-κB dimers are retained inactive in the cytoplasm by binding to specific inhibitors – the IκBs. When appropriate signals are received, the IκB kinase is activated (comprising IKK-α and IKK-β catalytic and regulatory NEMO subunits). This IKK complex phosphorylates the bound IκBs, targeting them for ubiquitin–proteasome degradation. The free NF-κB dimers then translocate to the nucleus where they regulate multiple genes.

There are several ways by which NF-κB signaling becomes activated in tumors, probably most common being a response to inflammation driven by viruses, bacteria, and cytokines. But various growth factor pathways can cross and activate NF-κB or enhance degradation of inhibitory regulators, such as IκBα. NF-κB is constitutively active in many cancer cells and can be added to our ever-expanding list of genes/proteins to which cancer cells may become addicted.

Stem cells, cancer stem cells, and EMT and related signaling

Before embarking on a discussion of the next group of signaling pathways, we will briefly digress in order to introduce a few key concepts relating to cancer stem cells (CSCs) and the often associated epithelial–mesenchymal transition (EMT). Although, discussed in more detail in Chapter 9, some basic facts will be mentioned here.

The similarities between embryonic signaling processes and those often reinvoked in tissue injury/repair with those acquired by cancers as they evolve from normal tissues have not escaped notice. There are many possible explanations for this, encompassing the association between chronic inflammation or repeated tissue injury and causation of cancer on the one hand, and the acquisition of EMT-promoting mutations and expansion or appearance of cells akin to normal stem cells during the evolution of the cancer on the other. It is increasingly accepted that normal stem cells may prefigure their malignant counterparts; the remarkable similarities between them at a molecular and behavioral level renders such association inevitable. The delineation of a side population of cancer cells as CSCs, in which these similarities are accentuated, draws further attention to their putative stem cell origin. Many cancers are thought to derive from a subpopulation of CSCs that may re-express embryonic genes; a network of Wnt, Notch, Hedgehog (Hh), and Bmp signaling (discussed below) determines the balancing of stem cell renewal on the one hand with cellular differentiation on the other, and this appears true of stem cells and CSCs. Stem cell renewal pathways such as Hh and Wnt are upregulated in chronic tissue injury, such as liver cirrhosis, with resultant expansion of stem cells, some of which may eventually become tumor-initiating cells for primary liver cancer.

EMT is central to ontogeny and increasingly also to tumorigenesis. EMT can endow a cell with "stemness" (EMT maketh the CSC) and can drive metastasis. Epithelial cells are typically arranged as sheets of cells orientated in space with defined apical-basal polarity. Such cells are tightly adherent through a variety of structures, such as adherens junctions, tight junctions, and desmosomes, and remain separated from adjacent tissues by a basal lamina, all of which help the epithelium to fulfill surface barrier and selective transport functions. In stark contrast, mesenchymal cells are embedded within the extracellular matrix and form the connective tissue or stroma of a tissue. They are typically arranged in a seemingly disorganized way and are neither polarized nor tightly adherent to other cells and are capable of migration. The conversion of epithelial cells to mesenchymal cells (EMT) necessitates substantial changes, including loss of cell adhesions and polarity and adoption of a migratory and invasive phenotype, and is associated with expression of stem-cell markers. EMT is observed at the invasive front of many cancers, including colon and breast. Inducers of EMT such as SNAIL1, SNAIL2, FOXC, and ZEB1 are associated with worse prognosis in cancer patients.

Another property of CSCs that may contribute to tumorigenesis is differentiation. In fact, CSCs may give rise to more than just cancer cells; in cancers, tumor endothelial cells and angiogenesis and Paneth cells and a supportive stem cell niche may be the progeny of CSCs. This lineage plasticity may account for the heterogeneity of tumor cells and nontumor cells in the cancer microenvironment. In glioblastoma, CSCs have been known to interact with the vascular niche and promote angiogenesis through the release of vascular endothelial growth factor (VEGF) and stromal-derived factor 1. Two recent publications in the journal *Nature* have now extended the connections between CSCs and angiogenesis. It has been shown that endothelial cells in glioblastomas carry the same somatic mutations as are found in the cancer cells, supporting a common origin, and moreover that some CSCs also express vascular markers, suggesting that these may be CSCs differentiating into vascular endothelial cells.

Growth signaling via cell adhesion – the Wnt–β-catenin pathway

Cell–cell adhesion determines cell polarity and is critical for tissue homeostasis. Cell–cell adhesiveness is decreased in cancer, enabling deregulated proliferation, migration, invasion, and metastases, but in general only if cells can survive loss of cell–cell contacts (see Chapter 12). E-Cadherin, and interacting catenins, which connect the cadherins to the cytoskeleton, are located at adherens junctions of epithelial cells and establish firm cell–cell adhesion. Silencing of the E-cadherin gene by DNA hypermethylation around the promoter region occurs frequently, even in precancerous conditions. In diffuse infiltrating cancers, mutations are found in the genes for E-cadherin and α- and β-catenins. At the invading front of cancers, the E-cadherin cell adhesion system is inactivated by tyrosine phosphorylation of β-catenin. β-Catenin also connects the E-cadherin system to the Wingless/Wnt signaling pathway.

The Wnt–β-catenin pathway is involved in specification and localization of new cell types during tissue differentiation, primarily by regulating interactions between neighboring cells (Fig. 5.21). Wnt ligands bind to their target frizzled membrane receptor and interfere with the multiprotein destruction complex, resulting in downstream activation of gene transcription by β-catenin. The multiprotein destruction complex requires the presence of axin and the adenomatosis polyposis coli (APC) protein, which act as scaffolds, to facilitate phosphorylation of β-catenin by the enzyme GSK-3β. Phosphorylated β-catenin is degraded in proteasomes, whereas unphosphorylated β-catenin accumulates and associates with nuclear transcription factors such as TCF/LEF, leading to expression of growth-regulating target genes such as c-*Myc*, c-*Jun*, *Fra*, and *cyclin D1* (Fig. 5.21). Aberrant WNT signal

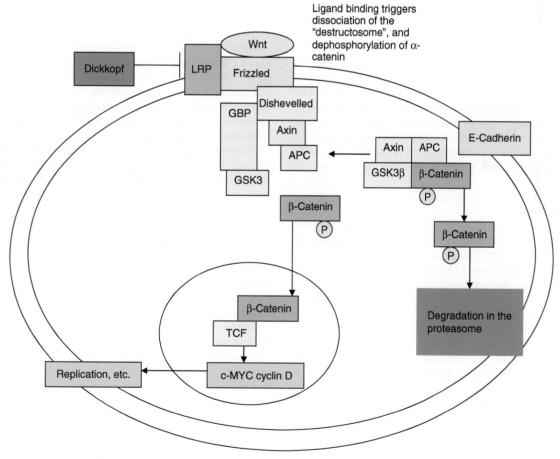

Figure 5.21 The Wnt signaling pathway. The Wnt ligand is a secreted glycoprotein that binds Frizzled receptors, resulting in displacement of GSK-3β from the APC/Axin/GSK-3β complex (the "destructosome"). In the absence of Wnt, β-catenin is phosphorylated and targeted for proteasome degradation by the APC/Axin/GSK-3β complex. In the presence of Wnt binding (on state), Dishevelled is activated and recruits GSK-3β away from the degradation complex, which stabilizes β-catenin levels and allows Rac1-dependent nuclear translocation, where β-catenin together with LEF/TCF activates transcription of growth-regulating genes such as c-*MYC* and cyclin D. In the off state, β-catenin can also be bound to E cadherin.

transduction is common in some cancers, and activation of β-catenin, the major effector of this pathway, is an important contributor to both colorectal cancer and melanoma. In fact, mutations resulting in active β-catenin are found in at least 80% of colon cancers and include activating mutations in the β-catenin gene, increased protein stability by inactivation of degradative proteins or loss of the APC tumor suppressor. Activation of β-catenin/TCF in turn results in deregulated expression of key target genes, including c-*MYC* and *cyclin D*. Recently, a potentially important positive feedback loop has been found involving phospholipase D (PLD) and its enymatic product phosphatidic acid, frequently upregulated in cancers. PLD is a target of β-catenin and also enhances β-catenin signaling. Links between WNT signaling and cell-cycle control are clearly important and it has now been shown that the mitotic CDK14/cyclin Y complex can activate WNT signaling via phosphorylation of the LRP6 coreceptor.

Recent studies have suggested that the *WTX* tumor suppressor, mutated in Wilms tumors, encodes a protein that promotes β-catenin ubiquitination and degradation and thereby acts as a negative regulator of WNT signaling.

Recent appreciation of similarities between developmental morphogenetic processes and cancer has had a major impact on cancer biology. Wnt signaling regulates morphogenetic processes such as gastrulation, which requires cells to undergo EMT (but these are usually transient), enabling the cells to dissociate and migrate. Cell dissociation and migration are also essential for tumor cell invasion and metastases, and recent data suggest that EMT is a major determinate of the latter. CSCs are believed by many to be a self-renewing side-population of cancer cells, at least in part responsible for driving some of the malignant behaviors of cancers, such as metastases, treatment resistance, and tumor recurrence. Given that WNT, and also Notch and Hedgehog (see below), pathways are important in stem cell signaling, a great deal of research effort has been devoted to potential therapies targeting these pathways. EMT is one way by which CSCs are generated, though some may also originate from normal stem cells.

Notch signaling

The history of Notch signaling begins almost exactly 100 years ago, when Thomas Hunt Morgan first identified a mutant strain of *Drosophila* fruitflies with "notched" wings. We now know that Notch is a receptor protein that can interact with a ligand protein, Delta, presented on the surface of a neighboring cell. This juxtracrine interaction releases an intracellular part of the Notch protein (icNotch) that is a transcription factor. In mammals, multiple core components of the Notch signaling pathway are now recognized, including ligands (Delta-like 1, -3, -4 and Jagged-1 and -2) and a family of four receptors (Notch 1, -2, -3, and -4), which all share an extracellular ligand-binding domain comprising a series of EGF-like repeats and a cytoplasmic domain that includes a transcriptional activation domain and a PEST (Pro Glu Ser Thr) sequence regulating protein stability.

Ligand binding results in proteolytic cleavage of the receptor; the ligand-binding domain is removed by the metalloproteinase enzyme TACE and the cytoplasmic domain (icNotch) released by an enzyme complex comprising γ-secretase and presenillin (Fig. 5.22). Gene expression is then promoted by icNotch in part by binding to the CSL (CBF1, suppressor of hairless, Lag-1) transcriptional repressor and recruiting of other coactivator proteins,

such as those of the Mastermind family. Notch-regulated genes include various regulators of differentiation, such as members of the hairy/enhancer of split (HES) family, and various genes involved in cell cycle and apoptosis. The Notch transcriptional activation complex can bind DNA either as a monomer or as a dimer; these appear to have distinct roles on gene expression and function. Dimerization is needed to activate genes encoding c-Myc and pre-T-cell antigen receptor α (Ptcra) but not those for Hey1 and CD25. Dimeric Notch transcriptional complexes are required for T-cell maturation and initiation of leukemia but not for differentiation from precursors, raising the possibility of selective theraputics targeting Notch dimerization.

As for most signaling pathways, Notch may also be activated through loss of the inhibitors of this pathway – NUMB or Deltex. A number of stages in Notch signaling are subject to ubiquitinylation and degradation by the proteasome pathway (see Chapter 11). In fact, ligand-bound Notch receptor signaling may in part be terminated by this means. Interestingly, recent studies have now shown that ubiquitinylation may also be a means of maintaining inactivity of the unbound (inactive) Notch receptor. The ubiquitin ligase Nedd4 can mimic the nicked wing phenotype characteristic of Notch loss in *Drosophila* and inhibition of Nedd4 results in ligand-independent activation of the Notch pathway. A further putative ubiquitin ligase, Deltex, can bind to and activate the Notch receptor, suggesting that Nedd4 and Deltex compete in regulating Notch activity. This also raises the possibility that Notch signaling could be manipulated therapeutically in cancer by targeting of the ubiquitin–proteasome pathway.

Notch signaling diversity is a product of the various tissue-specific intermediaries required for signaling. In mice, knockout of the Notch effector RBP-Jk increases all intestinal secretory lineages and effects dependent on Math1. Thus, in intestine it appears that all Notch effects operate through the tissue-restricted factor Math1, which acts upstream of Hes1 to promote secretory differentiation and cell-cycle exit.

Cell–cell communication via Notch is a key factor in determining cell fate, particularly well described in lineage emergence during pancreas development. Notch can either inhibit or delay differentiation of neighboring cells and is a powerful means of ensuring that we end up with the right balance of different cell types in organs such as pancreas, where cells of many different differentiated types exist in close proximity in the adult. Notch was largely considered an embryonic developmental pathway, which is usually inactivated in adult cells, but as has also been discussed for IGF-2, it can be reactivated in several human cancers. It is also important that Notch signaling can be reactivated after tissue injury and it appears to be one of the key signaling pathways involved in regulating stem cell and CSC maintenance.

Many cancers have been shown to have deregulated Notch signaling. Active Notch signaling has been found in some pre-T-cell acute lymphoblastic leukemias due to a chromosomal translocation involving the *NOTCH* receptor gene and the T-cell receptor β (TCR-β). The effect of this is to allow NOTCH to be activated by TCR-β, which pushes cells into the T-cell lineage at the expense of the B-cell lineage, but prevents these T cells from fully differentiating. Large numbers of immature T cells provide an excellent platform from which to develop leukemia. Activated Notch also causes mouse mammary tumor virus (MMTV)-induced breast cancer in mice and HTLV-induced leukemia in humans. The role of Notch activation in virus-induced cancers is

Growth Signaling Pathways and Targeted Treatment

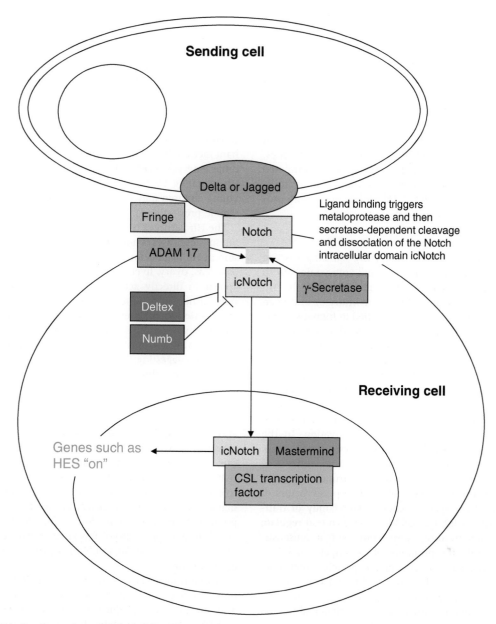

Figure 5.22 The Notch signaling pathway. Notch signaling is an evolutionarily conserved pathway that regulates cell fate determination during development and in stem cells. Notch receptor consists of an extracellular part containing multiple EGF-like repeats and an intracellular part that mediates Notch signal transduction. Binding of the ligands Delta or Jagged with Notch results in cleavage by metalloprotease and secretase activities, releasing intracellular icNotch, which binds to CSL and recruits coactivators such as Mastermind in the nucleus to regulate genes such as HES1 (hairy enhancer of split 1). Notch can be inhibited by action of Deltex1 and Numb in the cytoplasm and Notch activation is regulated extracellularly by Fringe proteins.

discussed in more detail in Chapter 3. NOTCH signaling in human B-cell lymphomas, including classical Hodgkin lymphoma (cHL) cells, can also be activated by coactivators of the Mastermind-like (MAML) family, which activate genes such as HES7 and HEY1. Notch receptor signaling pathways contribute to breast cancer development and progression and Notch4 signaling was particularly active in breast CSCs (as defined by enrichment for anoikis-resistant cells carrying stem cell markers). Interestingly, Notch1 signaling activity was lower in the stem cell-enriched cell populations, even though both appeared functionally important in regulating stemness, with Notch4 inhibition being particularly effective at inhibition of tumor initiation in model systems *in vivo*. Inactivation of the pathway inhibitor NUMB may be responsible for NOTCH1 activation in many human breast cancers. Interestingly, an important crosstalk between the EGFR–RAS–MAPK system and NOTCH might be a contributor to loss of differentiation of pancreatic cells in pancreas cancer.

Paradoxically, loss of Notch might also contribute to tumorigenesis. In some tissues Notch, rather than blocking differentiation, may do the opposite; in rodent models, inactivating Notch in skin results in keratinocyte overgrowth and carcinoma, in part by blocking differentiation.

Recent studies have shown that the Aes (Grg5) gene is an endogenous metastasis suppressor. Aes inhibits Notch signaling by converting active Rbpj transcription complexes into repression complexes. Notch signaling can be triggered in cancer cells by ligands on adjoining blood vessels to promote transendothelial migration. Although, this might suggest Notch1 as a potential cancer drug target, caution is needed. Pan-Notch receptor inhibition is fraught with complications and the potentially more promising use of selective inhibitors of Notch1 or its ligand Delta-like 4, have revealed potentially dangerous long-term complucations, such as development of vascular tumors and lethal hemorrhage.

Loss of miR-8/200 miRNAs frequently accompanies advanced and aggressive tumors and correlates with stem-like properties. Notch-induced overgrowth and tumor metastasis are inhibited by miR-8 via action on levels of the ligand Serrate, whereas miR-200c and miR-141 directly inhibited JAGGED1, preventing proliferation of human metastatic prostate cancer cells.

Hedgehog

The evolutionary conserved Hedgehog (Hh) signaling cascade controls cell proliferation, differentiation, and embryonic pattern formation, and is suppressed in the adult. However, the Hh pathway can become reactivated after tissue injury or in cancer. Three secreted Hh ligands have been identified in humans: Sonic hedgehog (Shh), Indian hedgehog (Ihh), and Desert hedgehog (Dhh); these are known to signal in an autocrine and paracrine manner. In response to Hh ligand, Smoothened (Smo), a seven-pass transmembrane protein, with similarities to the G-protein coupled receptor Frizzled, signals to an intracellular multiprotein complex containing the kinesin-related protein Costa12 (Cos2), the protein kinase Fused (Fu), and cubitus interruptus (Ci), which in turn regulate transcription factor activation. In the absence of Hh the receptor Patched (Ptc) indirectly inhibits Smo. Binding of Hh to Ptc blocks Ptc-mediated Smo inhibition, whereas in the absence of Hh, Ci is cleaved to a truncated repressor protein, Ci75. Three Smo-regulated transcription factors of the glioma-associated oncogene homolog (Gli) family of transcription factors – Gli1, Gli2, and Gli3 – are known that regulate target gene expression by direct association with a consensus sequence located within the promoter of the target genes. The Gli proteins are Krüppel-like zinc-finger proteins that play central roles in proliferation of a variety of tumor cell types. A negative regulator of Gli signaling is the chromatin remodeling protein SNF5, inactivated in human malignant rhabdoid tumors (MRT).

During development, Shh signaling is important in specification of liver rather than pancreas fate; FGF signaling promotes production of Shh and BMP and the BMP inhibitor noggin enhances pancreatic gene expression and suppresses hepatic gene expression. Forkhead box A proteins (FoxA) are transcription factors found in tissues that derived from endoderm and are important for specification of hepatic, pancreatic, and pulmonary fate. Transcriptional Hh-regulated Gli proteins are transactivators of FoxA genes. Hh activity declines as progenitors mature into hepatocytes but is reactivated in the progenitor and stromal cell compartments following liver injury and regeneration in the adult. Gli also links into p53 signaling via NANOG; NANOG can inhibit p53 in orthotopic xenografts (one useful type of cancer model, in which transplanted human tumor cells are transplanted into immunocompromised mice to study tumorigenesis *in vivo*) of glioblastoma multiforme.

Loss of function mutations in the Hh receptor PTCH were first identified in Gorlin syndrome, which is associated with several developmental defects and increased incidence of benign and malignant tumors, such as basal cell carcinomas (BCCs) and medulloblastomas. Most sporadic tumor equivalents also have enhanced and ligand-independent Hh signaling due to mutations in the signaling components PTCH, SMO, or, less frequently, suppressor of fused (SUFU), a Hh pathway inhibitor. SMO inhibitors have potent antitumor activity in patients with BCC. GDC-0449, a small-molecule inhibitor of SMO, has been shown to have antitumor activity in locally advanced or metastatic basal-cell carcinoma and a rapid (although transient) regression of a malignant medulloblastoma in a single patient. Somewhat surprisingly, it has emerged recently that the famous poison arsenic can block Hh signaling by targeting of Gli family proteins. (Those with young children may find association of Glee and arsenic an irresistible proposition!)

The Hh pathway is activated in a variety of other solid tumors, but in these cases is not driven by mutations but rather by expression of Hh ligands found in various carcinomas, including colon, pancreas, ovary, liver, and lung. Interestingly, Hh ligands do not appear to directly activate signaling in some tumor epithelial cells, rather they indirectly promote the development of a more cancer-friendly microenvironment by paracrine activation of signaling in stromal cells and by influencing expansion and maintenance of CSCs. It is worth noting that there is also a body of evidence suggesting Hh signaling in some tumor types, though to what extent this could involve CSCs is unknown.

Nodal

Another key pathway involved in embryonic development and also malignant transformation is that regulated by the embryonic morphogen Nodal. The Nodal pathway has been strongly implicated in the regulation of "laterality" (specification of the left right axis, as exemplified by having a liver on the right and a spleen on the left) during ontogeny. In fish, amphibians, and mammals, a cilia-driven leftward flow of extracellular fluid may produce asymmetric activation of the Nodal signaling cascade, which in *Xenopus* embryos has been shown to be due to downregulating a key Nodal inhibitor, Coco, on the left side of the growth plate.

Cripto-1, together with its coreceptor Nodal, regulate differentiation state in mouse and human embryonic stem cells and may also stimulate cell proliferation, migration, CSC maintenance, and EMT. Nodal is inhibited by the protein Lefty, a member of the TGF-β family. Downstream signaling of the pathway activates gene expression via members of the SMAD family. Similarities with other developmental pathways already discussed are self-evident and the fact that Cripto-1 is involved in initiation and progression of several types of human tumors will come as no surprise. Crosstalk between the Notch and Nodal pathways has been demonstrated in development and now also in tumorigenesis. Nodal reactivation is a key factor in driving metastatic melanoma and can be regulated via Notch4 through an RBPJ-dependent Nodal enhancer element. Cripto-1/Nodal signaling may represent another target for eliminating CSCs.

Other CSC signaling

More than 70% of breast cancers have high levels of Kruppel-like factor 4 (KLF4), which has been proposed as either an oncogene or tumor suppressor in different tissues. KLF4 is enriched in CSCs and has been shown to regulate EMT and stemness, both in a positive and negative way in different studies, suggesting that, as for TGF-β, effects may be contextual.

Hippo signaling and size control

The conserved Hippo, also known as the Salvador/Warts/Hippo, signaling pathway has recently come to prominence as a key regulatory network involved in determination of organ size and growth control via activities on key processes, such as replication, apoptosis, and stem cell homeostasis. The name "hippo" references the abnormally large fruitflies, hippopotamus-like, that result from inactivating mutations in the *hippo* gene. Hippo signaling inhibits cell proliferation and promotes apoptosis and operates as a tumor suppressor pathway in humans (Fig. 5.23). The Hippo signaling pathway is highly conserved, with most of the components first described in *Drosophila* (Fig. 5.24). The core regulatory kinase components Hippo, Salvador, Warts, and Mats in *Drosophila* (Fig. 5.24) and Mst1/2, WW45, Lats1/2, and Mob1 and the angiomotin protein in mammals (Fig. 5.23) phosphorylate and inactivate downstream transcriptional targets such as the Yes-associated protein (YAP – known as Yorkie in *Drosophila*), a transcriptional coactivator that promotes proliferation and inhibits apoptosis. Another target is transcriptional coactivator with PDZ-binding motif (TAZ).

Loss of normal Hippo pathway activity is associated with a wide range of human cancers. Hippo (MST1/2 in mammals) phosphorylates and activates Warts (LATS1/2 in humans), which is supported by the interaction of two other core proteins, Salvador (SAV1 in humans) that acts as a scaffold and Mob as tumor suppressor (MOBKL1A/B in humans) that boosts the kinase activity of Warts. The net effect is that activated Warts phosphorylates and inactivates the transcriptional coactivator Yorkie (YAP and TAZ in mammals). When activated, YAP and TAZ bind to several transcription factors involved in promoting growth and increasing size, such as p73, RUNX2, and TEADs.

As one would expect for a key growth regulatory pathway, the Hippo kinase cascade is itself regulated by a variety of inputs, including the planar cell polarity cadherins Fat and Dachsous, and a key complex comprising a trio of proteins – Expanded (FDM6 in mammals), Kibra (KIBRA in mammals), and Merlin (also known as NF-2). This complex may help to activate Hippo signaling by locating the core kinase cacade to the plasma membrane. Other regulatory proteins include the Ras-associated family protein RASSF, and, at least in *Drosophila*, the apicobasal polarity proteins lethal giant larvae (Lgl), atypical protein kinase, and a putative cell surface receptor for Hippo signaling called Crumbs (Crb).

YAP is inhibited by two main phosphorylation-dependent routes. First, by Ser381 phosphorylation, which in turn initiates CK1δ-dependent phosphorylation of Ser384, and Ser387, targeting YAP for SCFβTRCP-mediated ubiquitination and proteasomal degradation, and second, through 14-3-3 binding and cytoplasmic retention.

Interestingly, several recent reports from *Drosophila*-based studies suggest that Hippo may control fate and expansion of stem cells in the intestine and that Yorkie (Yap) may stimulate growth through transcription of Myc during formation of the imaginal disk. Myc, in turn, can repress Yorkie, allowing for the

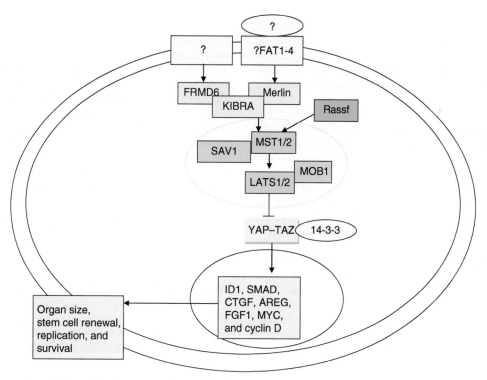

Figure 5.23 The Hippo pathway in mammals. The Hippo signaling systems acts as a growth-restricting system by phosphorylating and allowing sequestration of the stimulatory cofactor YAP in the cytoplasm, bound to 14-3-3 proteins. Most of the Hippo proteins are tumor suppressors, with the exception of YAP, which is oncogenic. Dephosphorylation of YAP allows dissociation from 14-3-3 and translocation into the nucleus, where it enhances transcription of several growth-promoting genes. Merlin is a growth-inhibiting tumor suppressor inactivated in human nervous system tumors. MST1 and 2 have reduced activity in hepatocellular carcinomas and reduced expression in colorectal cancer. LATS1 and 2 are downregulated in several human cancers. Conversely YAP and TAZ are frequently overexpressed and active in cancers. LATS1, large tumor suppressor 1; MATS, MOB as tumor suppressor; MST1, sterile-20-type kinase (ortholog of Hippo); SAV1, salvador homolog 1; YAP, YES-associated protein; FRMD6, FERM domain-containing protein 6; RASSF, RAS-associated family member.

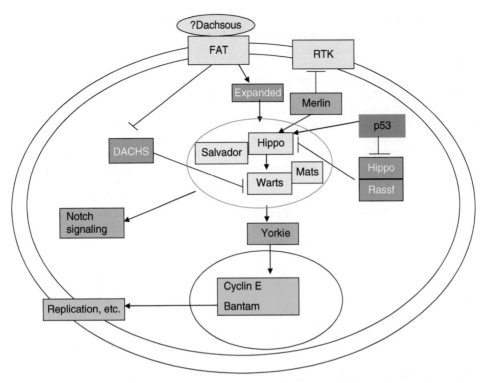

Figure 5.24 Hippo signaling system in *Drosophila*. This system is best known from studies in the fruitfly *Drosophila*. Fat activates Hippo signaling by recruiting Expanded to the membrane and by blocking Warts degradation by Dachs. p53 also promotes Hippo activity and sensitizes cells to apoptosis by inhibiting *diap1*. RASSF inhibits Hippo by preventing association with Salvador. Growth inhibition through Hippo signaling also involves downregulation of the miRNA *bantam*, and reduced levels of RTKs via Merlin and Expanded.

coordinated activity of these proteins to control organ size. In mammals, Yap activity must be tightly constrained in order to allow repair of injured intestine without promoting tumorigenesis and formation of polyps, which accompany deregulated Yap activity.

Both YAP and TAZ are important developmental regulators during ontogeny and regulate differentiation of stem cells and maintainance of stem cell pluripotency by activating expression of TGF-β and BMP target genes following binding to Smad1 and Smad2/3, respectively. TAZ inhibits the Hippo–LATS tumor suppressor pathway and is an oncogene in breast cancer and NSCLC. A recent study has provided important information linking the neurofibromatosis type 2 (NF-2) tumor suppressor gene product Merlin and regulation of stem cells. Liver-specific deletion of NF-2 resulted in an expansion of progenitor cells throughout the liver and universally resulted in both cholangiocellular and hepatocellular carcinoma in all surviving mice. Interestingly, in this case Merlin did not have much effect on YAP in liver progenitors, which instead were driven to proliferate by aberrant EGFR activity, confirming that Merlin can control the abundance and signaling of membrane receptors, such as EGFR. YAP is an oncogene and is overexpressed in numerous cancers including breast and colorectal cancer, where it may promote growth and prevent contact inhibition.

Nuclear proto-oncogenes and transcription factors

Ultimately, signals from growth factor receptor–signaling pathways control growth by regulating gene expression in the nucleus. Thus, ERK and STAT can enter the nucleus where they in turn activate the proto-oncogenes c-MYC, c-FOS, and c-JUN and others involved in cell-cycle regulation, protein synthesis, cell survival, etc. described in Chapter 6.

The AP-1 transcription factor c-JUN is essential for cellular proliferation in many cell types and is a target of the JNK signaling pathway. JUN combines with FOS (Finkel osteosarcoma), which is a product of c-MYC-induced transcription, to give a heterodimer, which activates genes with tumor response element enhancers. Long-lived or overexpressed JUN homodimers can themselves activate transcription of *FOS*. JUN–FOS heterodimers regulate genes involved in cell growth and may result in loss of differentiation. Growth factor signaling activates c-JUN–AP-1 to promote expression of target genes such as *cdc2* and *cyclinD1* via MEK/ERK-dependent stabilization of a coactivator called RACO-1 (RING domain AP-1 coactivator 1). RACO-1 can also amplify Wnt-induced intestinal tumors in transgenic mice.

Translational control and growth

Translational control is a critical factor in cell growth and proliferation and the mRNA 5' cap-binding protein eIF4E is a key regulator of this process. Increased cell proliferation requires a general increase in protein synthesis and a specific increase in the synthesis of replication-promoting proteins. The RAS–MAPK and PI3K–AKT–mTOR pathways are particularly well known to activate the function and expression of various components of the translational machinery. Protein synthesis in eukaryotic cells following binding of mRNA to small ribosomal subunits is catalyzed by a family of eukaryotic translation initiation factors (eIF). One such family, the eIF4, comprises at least four (but with completion of genome sequencing projects many more are likely) members: eIF4E, eIF4G, eIF4A, and eIF4B. Polypeptide chain initiation involves the assembly of the 43S initiation complex catalyzed by polypeptide chain initiation factor eIF2 and the binding of eIF4E to eIF4G during the recruitment of mRNA to the ribosome. eIF2 activity is controlled by phosphorylation of the alpha subunit of this factor by various kinases (including GCN2, and eIF2alpha kinase 4), whereas eIF4E is regulated by phosphorylation of a small family of binding proteins (the 4E-BPs).

Importantly, these phosphorylation steps are regulated by numerous oncogenes and tumor suppressors in cancer. The situation is complex, different eIF4 proteins can promote or inhibit translation of specific mRNAs, and may be active in different tissues and developmental stages. As discussed above, mTOR regulates ribosome biogenesis, protein synthesis, and cell growth in large part by controlling the translation machinery through activation of S6K and eIF4E (by inactivating the binding protein 4E-BP1). Deregulation of gene expression at the level of mRNA translation contributes to cell transformation and the malignant phenotype and aberrant expression and activity of eIF4E are found in many cancers and can promote tumorigenesis in model systems. Interestingly, eIF4E cooperates with c-MYC in B-cell lymphomagenesis; in this case c-MYC could overcome growth arrest triggered by deregulated eIF4E and eIF4E could prevent c-MYC-dependent apoptosis, identifying another protein in the PI3K–AKT pathway able to suppress apoptosis.

eIF4E overexpression has been demonstrated in various different human tumors, including breast, head and neck, colon, prostate, bladder, cervix, and lung. Although it regulates the recruitment of mRNA to ribosomes, and thereby globally regulates cap-dependent protein synthesis, eIF4E also selectively enables translation of a select number of directly cancer-relevant mRNAs, such as those encoding cyclin D1, c-MYC, MMP-9, and VEGF. However, upregulating eIF4E is not the only means by which cancer cells increase general protein synthesis. In fact, in many even aggressive cancers eIF4E may be barely detectable. Eukaryotic cells can respond to diverse stresses by activating remedial and damage-limitation responses. eIF2 and various regulatory kinases (eIF2 kinases) play a key role in the regulation of protein synthesis in response to such stresses. Phosphorylation of eIF2 reduces general protein synthesis, but increases translation of a defined number of specific mRNAs that encode transcription factors.

A greater appreciation of regulatory processes for protein translation, particularly how specific mRNAs are selectively targeted for enhanced or reduced translation will certainly be a very important area in cancer research and drug design in the next few years. Activity of eIF4E is regulated by phosphorylation, which appears necessary for translational upregulation of several cancer-relevant proteins, suggesting that interference with phosphorylation of eIF4E could be exploited in design of new anti-cancer agents.

Conclusions and future directions

> Every component of the organism is as much of an organism as every other part.
>
> *Barbara McClintock*

Growth factor signaling pathways are self-evidently deregulated in all cancers. Most but not all known oncogenes activate some

---

Box 5.4 Classic experiments

In 1956, Rita Levi-Montalcini and Viktor Hamburger performed the seminal studies that culminated in the first identification of a growth factor. NGF was the first growth factor identified, for its action on the morphological differentiation of neural crest–derived nerve cells. Later, its effect on neuronal cells of the peripheral and central nervous systems and on several non-neuronal cells was also determined.

In these studies a particular mouse tumor, known as a sarcoma, was implanted into developing chick embryos. These tumors became heavily innervated, with sympathetic and some sensory neurons, and it was reasoned that a factor was being produced by the tumor that promoted nerve growth. Extracts from these tumors were next found to promote neuron outgrowth in explant cultures. Further studies demonstrated that a similar activity could be observed in some snake venom and finally in submandibular salivary glands, allowing the factor to be isolated and purified. The factor was later named nerve growth factor (NGF). Rita Levi-Montalcini was awarded the Nobel Prize for this work in 1986.

Aloe, L. (2004). Rita Levi-Montalcini: the discovery of nerve growth factor and modern neurobiology. *Trends in Cell Biology*, **14**(7): 395–9.

---

parts of these signaling pathways and several tumor suppressors have negative effects on them. Growth factor signaling is not simply a regulator of cell replication but plays a key role in numerous other important processes in cancer cell and in their neighbors. Thus, growth factor signaling may prevent apoptosis and increase motility and invasion by durect signaling and also may be in part via induction of changes in cell phenotype such as EMT and stemness. Aberrant activation of growth factor signaling pathways can arise through a variety of mechanisms (mutations, epigenetic alterations, upregulation) and act at different levels, including the growth factor itself, the RTK, signaling pathways, or downstream transcription factors. Given the central role of growth factor signaling in cancer, it is not surprising that a new generation of small-molecule drugs and monoclonal antibodies that focus on elements of these pathways have emerged as the "new frontier" in cancer therapy.

Outstanding questions

1. To understand how the size or proportion of various tissues or organs (cell size and number) is determined during development and maintained during adult life.

2. To further understand how growth factor signaling is regulated, from receptor numbers and desensitization down to the nuclear transcription factors that ultimately control gene expression.

3. To progress beyond the current view of signaling pathways as largely linear cascades, towards a more integrated view and in particular to begin to unravel the complex interacting functional protein networks within which growth factor signaling pathways participate.

4. To develop further small molecule drugs specifically targeting various growth factor signaling pathways involved in cancer.

5. To further understand the means by which TGF-β may be both a growth inhibitor and also a promoter of cancer behaviors.

6. To further define and employ biomarkers, as exemplified by *EGFR* mutation analysis, *EGFR* copy number, and EGFR expression, in order to identify the specific individuals most likely to benefit from targeted drugs.

Bibliography

Growth factors

Foulstone, E., Prince, S., Zaccheo, O., et al. (2005). Insulin-like growth factor ligands, receptors, and binding proteins in cancer. *Journal of Pathology*, **205**(2): 145–53.

Hermansson, M., Nistér, M., Betsholtz, C., Heldin, C.-H., Westermark, B., and Funa, K. (1988). Endothelial cell hyperplasia in human glioblastoma: coexpression of mRNA for platelet-derived growth factor (PDGF) B chain and PDGF receptor suggests autocrine growth stimulation. *Proceedings of the National Academy of Sciences of the U.S.A.*, **85**: 7748–52.

Growth factors and cancer

Avraham, R. and Yarden, Y. (2011). Feedback regulation of EGFR signalling: decision making by early and delayed loops. *Nature Reviews Molecular Cell Biology*, **12**(2): 104–17.

Benitah, S.A., Valeron, P.F., van Aelst, L., Marshall, C.J., and Lacal, J.C. (2004). Rho GTPases in human cancer: an unresolved link to upstream and downstream transcriptional regulation. *Biochimica Biophysica Acta*, **1705**(2): 121–32.

Ebos, J.M., Lee, C.R., Cruz-Munoz, W., et al. (2009). Accelerated metastasis after short-term treatment with a potent inhibitor of tumor angiogenesis. *Cancer Cell*, **15**(3): 232–9.

Inoki, K., Corradetti, M.N., and Guan, K.L. (2005). Dysregulation of the TSC-mTOR pathway in human disease. *Nature Genetics*, **37**(1): 19–24.

Roberts, A.B. and Wakefield, L.M. (2003). The two faces of transforming growth factor beta in carcinogenesis. *Proceedings of the National Academy of Sciences of the U.S.A.*, **100**(15): 8621–3.

Sporn, M.B. and Roberts, A.B. (1985). Autocrine growth factors and cancer. *Nature*, **313**: 745–7.

Thomas, D.A. and Massague, J. (2005). TGF-beta directly targets cytotoxic T cell functions during tumor evasion of immune surveillance. *Cancer Cell*, **8**(5): 369–80.

mTOR, nutrients, and cancer

Carling, D. (2006). LKB1: a sweet side to Peutz-Jeghers syndrome? *Trends in Molecular Medicine*, **12**(4): 144–7.

Levine, A.J. and Puzio-Kuter, A.M. (2010). The control of the metabolic switch in cancers by oncogenes and tumor suppressor genes. *Science*, **330**(6009): 1340–4.

Peterson, T.R., Laplante, M., Thoreen, C.C., et al. (2009). DEPTOR is an mTOR inhibitor frequently overexpressed in multiple myeloma cells and required for their survival. *Cell*, **137**(5): 873–86.

Wullschleger, S., Loewith, R., and Hall, M.N. (2006). TOR signaling in growth and metabolism. *Cell*, **124**(3): 471–84.

Growth factors and the cell cycle

Nasmyth, K. (1996). Viewpoint: Putting the cell cycle in order. *Science*, **274**:1643–5.

Growth regulation, size control, and differentiation

Bjorklund, M., Taipale, M., Varjosalo, M., Saharinen, J., Lahdenpera, J., and Taipale, J. (2006). Identification of pathways regulating cell size and cell-cycle progression by RNAi. *Nature*, **439**(7079): 1009–13.

Halder, G. and Johnson, R.L. (2011). Hippo signaling: growth control and beyond. *Development*, **138**(1): 9–22.

Owens, D.M. and Watt, F.M. (2003). Contribution of stem cells and differentiated cells to epidermal tumours. *Nature Reviews Cancer*, **3**: 444–51.

Poss, K.D. (2010). Advances in understanding tissue regenerative capacity and mechanisms in animals. *Nature Reviews in Genetics*, **11**(10): 710–22.

Reya, T., Morrison, S.J., Clarke, M.F., and Weissman, I.L. (2001). Stem cells, cancer and cancer stem cells. *Nature*, **414**: 105–11.

Sakatani, T., Kaneda, A., Iacobuzio-Donahue, C.A., et al. (2005). Loss of imprinting of Igf2 alters intestinal maturation and tumorigenesis in mice. *Science*, **307**(5717): 1976–8.

Scaglia, L., Smith, F.E., and Bonner-Weir, S. (1995). Apoptosis contributes to the involution of beta cell mass in the post partum rat pancreas. *Endocrinology*, **136**(12): 5461–8.

Teitelman, G., Alpert, S., and Hanahan, D. (1988). Proliferation, senescence, and neoplastic progression of beta cells in hyperplasic pancreatic islets. *Cell*, **52**(1): 97–105.

Zhao, B., Li, L., Lu, Q., et al. (2011). Angiomotin is a novel Hippo pathway component that inhibits YAP oncoprotein. *Genes and Development*, **25**(1): 51–63.

Translational control

Furic, L., Rong, L., Larsson, O., et al. (2010). eIF4E phosphorylation promotes tumorigenesis and is associated with prostate cancer progression. *Proceedings of the National Academy of Sciences of the U.S.A.*, **107**(32): 14134–9.

Signaling

Chen, Z., Trotman, L.C., Shaffer, D., et al. (2005). Crucial role of p53-dependent cellular senescence in suppression of Pten-deficient tumorigenesis. *Nature*, **436**(7051): 725–30.

Fang, J.Y. and Richardson, B.C. (2005). The MAPK signalling pathways and colorectal cancer. *Lancet Oncology*, **6**(5): 322–7.

Goswami, A., Burikhanov, R., de Thonel, A., et al. (2005). Binding and phosphorylation of par-4 by akt is essential for cancer cell survival. *Molecular Cell*, **20**(1): 33–44.

Guertin, D.A. and Sabatini, D.M. (2005). An expanding role for mTOR in cancer. *Trends in Molecular Medicine*, **11**(8): 353–61.

Shaw, R.J. and Cantley, L.C. (2006). Ras, PI(3)K and mTOR signalling controls tumour cell growth. *Nature*, **441**(7092): 424–30.

Shaw, R.J., Lamia, K.A., Vasquez, D., et al. (2005). The kinase LKB1 mediates glucose homeostasis in liver and therapeutic effects of metformin. *Science*, **310**(5754): 1642–6.

She, Q.B., Solit, D.B., Ye, Q., O'Reilly, K.E., Lobo, J., and Rosen, N. (2005). The BAD protein integrates survival signaling by EGFR/MAPK and PI3K/Akt kinase pathways in PTEN-deficient tumor cells. *Cancer Cell*, **8**(4): 287–97.

Teitell, M.A. (2005). The TCL1 family of oncoproteins: co-activators of transformation. *Nature Reviews Cancer*, **5**(8): 640–8.

Vanhaesebroeck, B., Ali, K., Bilancio, A., Geering, B., and Foukas, L.C. (2005). Signalling by PI3K isoforms: insights from gene-targeted mice. *Trends in Biochemical Science*, **30**(4): 194–204.

Oncogene addiction

Dornan, D. and Settleman, J. (2010). Cancer: miRNA addiction – depending on life's little things. *Current Biology*, **20**(18): R812–13.

Sharma, S.V. and Settleman, J. (2010). Exploiting the balance between life and death: targeted cancer therapy and "oncogenic shock." *Biochemistry and Pharmacology*, **80**(5): 666–73.

Tyrosine kinases and fusion proteins

Penserga, E.T. and Skorski, T. (2007). Fusion tyrosine kinases: a result and cause of genomic instability. *Oncogene*, **26**(1): 11–20.

JAK/STAT pathway

Pardanani, A. and Tefferi, A. (2011). Targeting myeloproliferative neoplasms with JAK inhibitors. *Current Opinion in Hematology*, **18**(2): 105–1.

Marubayashi, S., Koppikar, P., Taldone, T., et al. (2010). HSP90 is a therapeutic target in JAK2-dependent myeloproliferative neoplasms in mice and humans. *Journal of Clinical Investigation*, **120**(10): 3578–93.

WNT signaling

Major, M.B., Camp, N.D., Berndt, J.D., et al. (2007). Wilms tumor suppressor WTX negatively regulates WNT/beta-catenin signaling. *Science*, **316**(5827): 1043–6.

Nelson, W.J. and Nusse, R. (2004). Convergence of Wnt, beta-catenin, and cadherin pathways. *Science*, **303**(5663): 1483–7.

van Amerongen, R. and Berns, A. (2005). Re-evaluating the role of Frat in Wnt-signal transduction. *Cell Cycle*, **4**(8).

Hippo signaling

Benhamouche, S., Curto, M., Saotome, I., et al. (2010). Nf2/Merlin controls progenitor homeostasis and tumorigenesis in the liver. *Genes and Development*, **24**(16): 1718–30.

Pan, D. (2010). The hippo signaling pathway in development and cancer. *Developmental Cell*, **19**(4): 491–505.

Hedgehog signaling

Ayers, K.L. and Thérond, P.P. (2010). Evaluating Smoothened as a G-protein-coupled receptor for Hedgehog signalling. *Trends in Cell Biology*, **20**(5): 287–98.

Barakat, M.T., Humke, E.W., and Scott, M.P. (2010). Learning from Jekyll to control Hyde: Hedgehog signaling in development and cancer. *Trends in Molecular Medicine*, **16**(8): 337–48.

Ogden, S.K., Fei, D.L., Schilling, N.S., Ahmed, Y.F., Hwa, J., and Robbins, D.J. (2008). G protein Galphai functions immediately downstream of Smoothened in Hedgehog signalling. *Nature*, **456**(7224): 967–70.

Rangwala, F., Omenetti, A., and Diehl, A.M. (2011). Cancer stem cells: repair gone awry? *Journal of Oncology*, **2011**: 465343.

Von Hoff, D.D., LoRusso, P.M., Rudin, C.M., et al. (2009). Inhibition of the hedgehog pathway in advanced basal-cell carcinoma. *New England Journal of Medicine*, **361**(12): 1164–72.

Yauch, R.L. (2008). A paracrine requirement for hedgehog signalling in cancer. *Nature*, **455**(7211): 406–10.

Notch signaling

Harrison, H., Farnie, G., Howell, S.J., et al. (2010). Regulation of breast cancer stem cell activity by signaling through the Notch4 receptor. *Cancer Research*, **70**(2): 709–18.

Koch, U. and Radtke, F. (2010). Notch signaling in solid tumors. *Current Topics in Developmental Biology*, **92**: 411–55.

Köchert, K., Ullrich, K., Kreher, S., et al. (2010). High-level expression of Mastermind-like 2 contributes to aberrant activation of the NOTCH signaling pathway in human lymphomas. *Oncogene*, **30**(15): 1831–40.

Liu, H., Chi, A.W., Arnett, K.L., et al. (2010). Notch dimerization is required for leukemogenesis and T-cell development. *Genes and Development*, **24**(21): 2395–407.

Liu, Z., Turkoz, A., Jackson, E.N., et al. (2011). Notch1 loss of heterozygosity causes vascular tumors and lethal hemorrhage in mice. *Journal of Clinical Investigation*, **121**(2): 800–8.

Mazur, P.K., Einwächter, H., Lee, M., et al. (2010). Notch2 is required for progression of pancreatic intraepithelial neoplasia and development of pancreatic ductal adenocarcinoma. *Proceedings of the National Academy of Sciences of the U.S.A.*, **107**(30): 13438–43.

Sonoshita, M., Aoki, M., Fuwa, H., et al. (2011). Suppression of colon cancer metastasis by Aes through inhibition of notch signaling. *Cancer Cell*, **19**(1): 125–37.

Nodal signaling

Bianco, C., Rangel, M.C., Castro, N.P., et al. (2010). Role of Cripto-1 in stem cell maintenance and malignant progression. *American Journal of Pathology*, **177**(2): 532–40.

Strizzi, L., Hardy, K.M., Seftor, E.A., et al. (2009). Development and cancer: at the crossroads of Nodal and Notch signaling. *Cancer Research*, **69**(18): 7131–4.

Questions for student review

1) Cell numbers in adult metazoans are:
   a. Tightly controlled by regulatory processes.
   b. Determined solely by rates of cell division.
   c. Normally fixed in most adult tissues.
   d. Regulated by circulating as well as local factors.
   e. Maintained only by balancing losses through replication of stem cells.

2) Mitogens:
   a. Act primarily by promoting G_1/S transition in the cell cycle.
   b. Only act via cell surface receptors.
   c. Often act via the Ras/MAP kinase cascade.
   d. Do not influence differentiated cells.
   e. Can be produced by the same cells on which they act.

3) Signaling pathways activated by growth factor mitogens:
   a. Involve a series of steps regulated by phosphorylation of various proteins.
   b. May be initiated by autophosphorylation of growth factor receptors on tyrosine residues.
   c. Do not include serine/threonine kinases.
   d. Include the PI3K signaling pathway.
   e. Are rarely mutated in cancers.

4) The PI3K signaling pathway
   a. Is far less frequently involved in cancer than the Ras/MAPK pathway.
   b. May be important in regulating cell death.
   c. Is not involved in regulating cell replication.
   d. May be activated by inactivation of *PTEN*.
   e. Can result in activation of AKT and mTOR.

5) The Ras/MAPK pathway:
   a. Eventually results in activation of nuclear factors such as c-Myc.
   b. In cancer, is only known to be activated by mutations in *Ras*.
   c. Is not involved in cell death regulation.
   d. Is often activated via receptor tyrosine kinases.
   e. Is not involved in replicative senescence.

6 Oncogenes

Stella Pelengaris[a] and Michael Khan[b]
[a]Pharmalogos Ltd and [b]University of Warwick, UK

> We have seen that cells are formed and grow in accordance with essentially the same laws: hence that these processes must everywhere result from the operation of the same forces.
>
> *Theodore Schwann*

> Undermine their pompous authority, reject their moral standards, make anarchy and disorder your trademarks. Cause as much chaos and disruption as possible but don't let them take you alive.
>
> *Sid Vicious, late of the Sex Pistols*

Key points

- Oncogenes (cancer-causing genes) are essentially anarchic versions of normal cellular genes. Once their antisocial behavior is unleashed, as a result of mutations or other means, they either produce inappropriate amounts of protein or an altered protein that is fixed in the "on" position.
- The normal gene versions of oncogenes are referred to as proto-oncogenes and their products are proto-oncoproteins, giving the misleading impression that they are defined entirely by their potential to cause cancer. Care is needed lest we wrongly relegate the often pivotal roles played by proto-oncoproteins in normal cellular homeostasis to a footnote; this terminology is historical and references the route by which they were first recognized.
- The modern era of cancer biology was ushered in by the discovery of oncogenes and is heavily indebted to the insightful studies of sarcomas in chickens that turned out to be the result of an infectious agent (named the Rous sarcoma virus, after Peyton Rous, who identified it). The viral oncogene v-*src* was subsequently pinpointed as the key culprit and the story was nicely rounded off when Mike Bishop and Harold Varmus found the normal cellular counterparts of the viral oncogene, c-SRC.
- With the recognition that viruses were not required to deliver activated oncogenes to would-be cancer cells and that these could be generated in-house, it was not long before the processes by which this might occur were described. We now know that oncogenes become activated by genetic mutations, chromosomal translocations, or gene amplification. In some cases, such as HPV infection, viruses are still central players, but as we have seen in Chapter 3, mostly the DNA damage occurs *in situ* within the nucleus of the target cell.
- The proto-oncogenes encode proteins that direct such pivotal cellular functions as cell replication, growth, survival, loss of differentiation, and motility. Thus, by definition, the oncogenic variants will drive these processes independently of normal regulatory signaling pathways.
- In general, an **activating** mutation in a single copy of an oncogene is sufficient to evoke a tumorigenic process, whereas, in contrast, both gene copies of a tumor suppressor must usually be **inactivated** in order for the restraints on that process to be removed.
- It is also important to note that oncogenes, as is ironically often the case for c-*SRC* and also c-*MYC*, may become upregulated (activated) and help drive tumorigenesis without mutations in either the coding or regulatory regions of their own genes. Thus, levels of expressed proteins can be inappropriately increased by epigenetic factors such as loss of imprinting (see Chapter 11), increased stability of the oncoprotein (due to alterations in processing pathways), or by deregulated upstream signaling, which will not be revealed by looking for changes in the coding sequence (mutations) in that oncogene.
- In fact, the relative paucity of mutations in c-*MYC* and c-*SRC* resulted in gross underestimation of what we now know are the central roles played by these oncogenes in the evolution and maintenance of arguably most human cancers.
- Following the structure used in the previous chapter, we will consider oncogenes under five familiar functional headings: growth factors, receptors, signal transducers, transcription factors, and those that primarily act on prevention of the mitochondrial (intrinsic) pathway of cell death – apoptosis (e.g. BCL-2 and BCL-X_L).

The Molecular Biology of Cancer: A Bridge From Bench to Bedside, Second Edition. Edited by Stella Pelengaris and Michael Khan.
© 2013 John Wiley & Sons, Inc. Published 2013 by John Wiley & Sons, Inc.

- Individual oncogenes do not cause cancer on their own but rather act in concert with other oncogenes or loss of tumor suppressors – termed "oncogene collaboration" or "oncogene cooperation." Propitious epimutations that provide a growth advantage by complementing those acquired previously will be selected for and the affected cell will give rise to a new clone that will usurp its predecessors.
- Oncogene cooperation has been intensively studied in cell lines and using various models of cancer *in vivo*. More recently, the generation of regulatable expression systems has allowed the role of various oncogenes in tumor maintenance to be explored and "proof of hypothesis" studies for oncogene-targeted therapies to be conducted before attempting the complex task of designing and testing new small molecule drugs (see Chapters 15 and 16).
- Recent work has indicated the importance of differing levels of oncogene expression or activation on phenotype and thereby also on tumor evolution. Paradoxically, high levels of expression and activation of oncogenes, such as c-MYC and RAS, can actually stall tumor progression by engaging their own "tumor suppressor" properties to promote cell death or senescence, respectively, whereas low levels can drive oncogenesis without engaging tumor suppression.
- Taken together, this creates the novel counterintuitive possibility of treating cancers by actually "overdriving" oncogenic pathways as an alternative to antagonizing them. This approach might prove more tractable than trying to restore a malfunctioning tumor suppressor pathway that has by definition failed in its duty to restrain aberrant RAS or MYC activity. However, there is a critical caveat – namely, that the "intensified" oncogenic signal is still capable of engaging a functioning death/senescence pathway.

Introduction

Turning and turning in the widening gyre. The falcon cannot hear the falconer; Things fall apart; the centre cannot hold; Mere anarchy is loosed upon the world, The blood-dimmed tide is loosed, and everywhere the ceremony of innocence is drowned.

William Butler Yeats

Having digested the previous chapter, you will now be aware that many of the genes impinging on signaling pathways are designated as oncogenes or tumor suppressors, depending on whether they promote or inhibit growth. This chapter and the following one will discuss these central regulators in more detail and illustrate key concepts through a very detailed discussion of specific examples, singled out because of their particular importance or because these genes/proteins have wide-ranging functions beyond those of traditional views of signaling molecules.

As has been discussed in Chapters 4 and 5, somatic cells only enter the cell cycle on demand and then under the strict control of mitogenic signals delivered in the appropriate way. These signals, usually in the form of secreted growth factors, bind to cell surface receptors and trigger the cascades of intracellular signals that enable cells to progress through the G_1/S transition. Growth-promoting pathways are tightly regulated at multiple sites along the cascade, creating complex signaling networks that ensure cells proliferate only when this is useful to the whole organism, including after tissue injury or to replace normal cell losses and not if the DNA is damaged.

Loss of these normal restraints can result in anarchy, as cells and their offspring increasingly dance to their own tune with scant regard for the needs of the organism. Mutations in key regulatory genes can disrupt signaling networks, resulting in unscheduled cell division, growth and avoidance of cell death, in defiance of normal controls – setting the scene for cancer. At the present time over 450 genes are known to contribute to the development of cancer in humans. Although the specific genes affected and the order in which mutations occur differ from one cancer patient to another (the "cancer road-map"), the consequent effects of these epimutations on cancer cells impinge on a finite number of common key regulatory pathways. In fact, diversity at a molecular level does not necessarily equate with similar profligacy at the level of cell behavior, as exemplified by the limited repertoire of shared phenotypic hallmark features of cancer cells described earlier. This notwithstanding, a precise description of the molecular pathway to oncogenesis is still critically important and in notable cases has already led to the identification of prognostic biomarkers and new treatment targets that may not have been revealed in any other way.

The identification and characterization of the genes involved in cancer – the oncogenes, tumor suppressors, and most recently caretaker genes, has been one of the great triumphs of molecular biology. This concept has already been discussed in general terms in Chapter 3. This chapter will concentrate on the oncogenes, Chapter 7 will discuss the tumor suppressors, and Chapter 10 the caretaker genes.

The oncogenes

Oncogenes (from Greek *onkos*, a tumor) are activated versions of normal genes (proto-oncogenes) that generally share the ability to accelerate cell division and growth, but often also variously contribute to loss of differentiation, increased cell motility, and avoidance of apoptosis and invasion. (*Note*: The nomenclature whereby a normal gene involved in important cellular functions relating to growth is defined by the role it might play in a cancer if released from normal restraints may seem at first glance a peculiar one. However, this is a historical legacy reflecting the way in which they were first discovered.)

In healthy cells, the level of expression and activity of the proto-oncogenes are tightly regulated. This is essential as these are powerful proteins with the potential to provoke excessive cell proliferation or even the inappropriate survival of a cell with DNA damage. Oncogenes are distinguished from their normal cellular counterparts because they are overexpressed or because mutations result in the formation of protein products that are abnormally active or long-lived. In this scenario an abnormality in just one of the two gene copies in a cell is sufficient to generate the aberrant phenotype and the mutation is considered "dominant." Conversely, and because most signaling pathways have both positive and negative regulators, the same net effect can be obtained by mutations which inactivate both copies of a tumor suppressor gene, thereby releasing the restraint normally imposed by the encoded tumor suppressor protein. In contrast to oncogenes, both copies of the gene must be lost or inactivated to bring about these effects, and so the mutant phenotype is considered "recessive." However, as nothing is ever simple, it must be recognized that tumor suppressors may be "haploinsufficient" (can contribute to tumorigenesis when only a single copy is inactivated and the "dose" of tumor suppressor protein reduced by around 50%). At the risk of being repetitive, identical cellular behaviors can result from inactivation of tumor suppressors as from activation of oncogenes – most of the control mechanisms in the cell involve both inhibitory (tumor suppressor) and stimulatory (proto-oncogene) components. Put simply, the behavior of a cancer cell will reflect which regulatory or signaling pathways have become derailed rather than the means by which this has been achieved.

Moreover, to complicate matters further, intriguing studies over the last 20 years have forced a major reassessment of traditional views of the role of oncogenes in cancer. Thus, despite their undisputed importance in tumorigenesis, it is now clear that oncogenes (such as c-*MYC*, *E2F*, and *RAS*) capable of driving uncontrolled cell-cycle progression also possess intrinsic tumor suppressive mechanisms that can trigger cell death (apoptosis) or cellular senescence – even if these may ultimately operate through traditional tumor suppressors such as RB, P53, and DNA damage responses. In other words, oncogenes may act as their own tumor suppressors! The ability of oncogenes to promote such contrasting cell fates in a given cell – proliferation and death – may at first seem somewhat contradictory and confusing to the reader. In fact, you may take comfort in the knowledge that the discovery of this phenomenon in the early 1990s (see Chapter 8) was no less of a surprise to the scientific community, but has since had an enormous impact on the way we now perceive the evolution as well as the future treatment of cancer. To this end, we will discuss the importance of such intrinsic tumor suppression later in this chapter (and in more detail in Chapters 8, 9, and 10), and how putative cancer cells have found ways to block tumor-suppressing activities of oncogenes, for example through oncogene collaboration. With this in mind we will also discuss members of the BCL-2 family of proteins that specifically act as oncogenes by suppressing apoptosis.

One final point that may prove among the most important discoveries of recent years is that the actual level of oncogene expression and activity has profound implications for tumor development and treatments. Briefly, while high levels of MYC and RAS promote apoptosis and senescence, respectively, low levels can actually drive oncogenesis without engaging tumor suppression. This fascinating area will be discussed later. These new insights into oncogene biology have also begun to impact on the development of new targeted therapeutics and there is thus considerable overlap between this chapter and Chapter 16.

Oncogenes include a diverse range of genes encoding proteins involved in the regulation of cell division, differentiation, death/survival, and cell motility (see Tables 6.2–6.5) and some key examples of representative oncogene protein products are shown in Fig. 6.1. Following a general overview, a select sample of key proto-oncogenes will be described in detail to illustrate the role of their protein products in normal tissue growth and how their normally tight regulation by mitogenic or survival signaling may be circumvented in oncogenesis. The previous chapters have described the basic cell-cycle machinery (see Chapter 4) and how cell-cycle entry and progression are regulated under the influence of mitogens and survival signals (see Chapter 5). Chapter 5 has also described how mutations in various signaling pathways, particularly in the tyrosine kinases, may contribute to tumorigenesis and as such overlaps with this one in many respects, not least because this chapter also describes how growth-regulating processes become derailed by the activation of oncogenes, by mutations or otherwise, in cancer. However, here we concentrate on the concept and discovery of oncogenes and will focus on two key exemplars, *MYC* and *RAS*, as well as others, which merit a more detailed discussion because of their central role in the history of cancer biology or because they involve more than just growth factor signaling pathways.

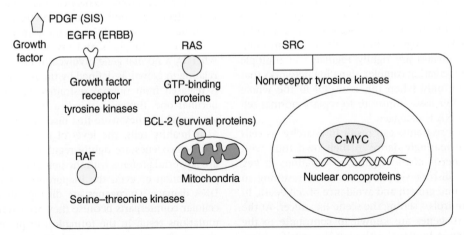

Figure 6.1 Representative oncogene products. In healthy cells, the normal nonmutated counterpart, proto-oncogenes, encode proteins that in general promote cell proliferation or cell survival but which are under tight regulation to avoid excessive proliferation, or indeed the survival of a cell that has sustained DNA damage and would normally be got rid of. Oncogenes include a diverse range of genes encoding proteins involved in the regulation of cell division, differentiation, and death or survival. The majority of oncogenes fall into five functional classes: transcription factors, growth factors, receptors, signal transducers, and survival proteins. Some key examples of representative oncogene protein products are shown here.

The discovery of oncogenes ushers in the new era of the molecular biology of cancer

I pictured myself as a virus or a cancer cell and tried to sense what it would be like.

Jonas Salk

The concept of an oncogene (cancer-causing gene) was established by the observation of viral genes that could promote cancers in animal cells. The analysis of these animal viruses was directly responsible for the subsequent identification of genes associated with human cancer. Although it still holds true that most human cancers are not the direct result of viral infection, as had once been mooted, the similarity between the molecular cancer-causing genes in viruses and endogenous cellular genes involved in growth regulation and mutated in cancer gave a tremendous headstart to cancer molecular biology. In fact, in this way the identification of virus-related cancers in other animal species has had an immense impact on the development of modern theories of oncogenesis. It is worth reminding ourselves that, although subsequently relegated to the sidelines, the contribution made by viruses to human cancers has recently seen a renaissance (see Chapter 3). The discovery of oncogenes was not only a landmark event in unraveling the molecular and genetic basis of cancer, but was a major factor in furthering knowledge concerning the regulation of normal cell proliferation. In fact, one could claim that the foundations for the modern view of cancer formation were laid by chickens!

It had been suspected for a hundred years or more that viruses could cause cancer, but the first cancer-causing elements were described in viruses infecting poultry in 1909. Early studies by Peyton Rous on neoplasia in fowl demonstrated that retroviruses could induce tumors, but the relevance of these discoveries for the understanding of human cancer was disputed. Later, it was shown that the injection of these viruses was sufficient for tumor formation. As has been true of scientific progress throughout recorded history, the development of new tools preceded the development of new understanding. In this case, the availability of DNA recombinant techniques provided the means of studying cancer-causing elements in viruses and thence their cellular homologs. Arguably, this discovery in 1976, can lay claim to being the defining moment after which the molecular biology of cancer achieved primacy. When the viral DNA was examined, regions bearing strong similarity to genes that already existed in the animal were identified, suggesting that viruses had at some point incorporated host genomic sequences into their own genomes. By incorporating genes able to promote cellular growth and mutating them, the virus could enslave its host cell, increasing proliferation, and consequently, its own replicative potential. A glance at Table 6.1 will quickly reveal just how many oncogenes were first discovered from their viral homologs.

As it was rapidly appreciated that most common cancers did not arise through viral infection, some scientists predicted that, by extension, mutant cancer-causing genes must lie within all types of tumor cells, even those that lacked evidence of tumor virus infections. The availability of techniques for gene transfection led to the identification of cellular transforming genes and even some that at the time had no known viral counterpart. As predicted, tumor cells were found to contain such genes and when these mutated genes were introduced into normal fibroblasts they were sufficient to cause cancer. This gave rise to the idea that cells contain "proto-oncogenes" – genes that are concerned with normal aspects of cell behavior. The proto-oncogenes might become mutated and thus convert to "oncogenes" – genes that have the ability to convert normal cells to a cancerous state. The key studies that led to the discovery of oncogenes are described in Box 6.1.

Overview of oncogenes

Man's yesterday may ne'er be like his morrow; Nought may endure but Mutability.

Percy Bysshe Shelley

In this section, we discuss various oncogenes that were identified initially as viral oncogenes. Some well-known viral and cellular oncogenes are listed in Table 6.1. As mentioned above, oncogenes are mutant genes that have a dominant effect on cell growth; enhanced cellular replication and avoidance of cell death are common examples, and because of their central importance these processes are discussed further in Chapters 4 and 8, respectively. Historically, oncogenes were defined using cell culture experiments as genes able to "transform" normal cells into tumorigenic ones. Such transformed cells would manifest various characteristic abnormal cell behaviors in the culture dish, or be able to form tumors if implanted into a suitably immunosuppressed mouse host. Some of the common abnormal features of transformed cells include:

- uncontrolled cell proliferation in culture medium devoid of or depleted in serum or growth factors and
- the "piling up" of cells on top of each other to form characteristic multilayered foci that provide an experimentally convenient method of identifying transformed cells from their normal counterparts, which due to a strict requirement for attachment to a solid substrate only form a single layer of cells.

However, it is worth remembering that cell behavior in a culture dish – devoid of any stroma, removed from normal contacts with neighboring cells and extracellular matrix (see Chapter 12), as well as lacking signals derived from a vascular and lymphatic supply (see Chapter 14) – may differ from that of a cell in the intact organism, and the ramifications of this will be highlighted later in the chapter. Box 6.2, later in the chapter, introduces the nomenclature used for oncogenes.

Mechanisms of converting proto-oncogenes to oncogenes

The worst enemy of life, freedom and the common decencies is total anarchy; their second worst enemy is total efficiency.

Aldous Huxley.

Much of this has been discussed in Chapter 3, but a brief reminder will be included now. A proto-oncogene can be converted to an oncogene in a number of ways. In general, such mutations either affect the coding region of the gene, resulting in the formation of an abnormal oncoprotein with enhanced stability or activity, or may affect regulatory elements, resulting in enhanced or deregulated expression of the protein (Fig. 6.2). Oncogene mutations span a spectrum of genetic alterations, from simple point mutations altering only single amino acids in the protein product, to major chromosomal rearrangements that completely alter

Box 6.1 Early studies

> Education is not the piling on of learning, information, data, facts, skills, or abilities – that's training or instruction – but is rather making visible what is hidden as a seed.
>
> *Thomas Moore*

Peyton Rous first described the virus responsible for transmissible growth of tumors in chickens and was the first to propose what was at the time a rather eccentric notion, namely that cancers could be caused by infectious agents (see also Chapter 3). This notion was strongly supported by work in the 1950s demonstrating that the Rous sarcoma virus (RSV)-induced tumor gave rise to infected cancer cells, and finally by experiments by G.S. Martin (1970) and K. Toyoshima and P.K. Vogt (1969), who showed that temperature-sensitive mutants of chicken v-*Src* failed to transform cells at nonpermissive temperatures. The v-*Src* gene was identified in the 1970s and was subsequently sequenced by Mike Bishop and co-workers (see below).

The idea that cancer cells bore mutant cellular genes was around already in the middle of the last century, but in the 1970s two separate strands of research helped to advance this idea into the forefront of cancer research. First, the work of Bruce Ames with cancer-causing chemicals (carcinogens) published in 1975 identified that cancers could be caused by carcinogens through induction of mutations, also suggesting that cells might normally carry genes that when mutated in some way would confer a growth advantage. A second strand followed studies by Harold Varmus and Michael Bishop, for which they were awarded the Nobel Prize for Medicine in 1989. In 1976, they identified the cellular homolog c-*SRC* of the v-*Src* gene, which had previously been described as the oncogene of the chicken virus referred to earlier (the Rous sarcoma virus – RSV).

Earlier genetic analysis of RSV had defined the first viral oncogene (v-*Src*) as a gene that was specifically responsible for cell transformation but was not required for virus replication, and the proposition had been put that retroviral oncogenes derive from related genes of host cells. Consistent with this proposition, normal cells of several species were subsequently found to contain retrovirus-related DNA sequences that could be detected by nucleic acid hybridization. However, it was unclear whether these sequences were related to the retroviral oncogenes or to the genes required for virus replication. It was this issue that was resolved by work from the laboratory of Varmus and Bishop. In particular, transformation-defective mutants of RSV had been isolated that had sustained deletions of approximately 1.5 kb, corresponding to most, or all, of the *Src* gene. These replication-defective mutants were employed by Dominique Stehelin, then a postdoctoral researcher in the laboratory, to prepare a cDNA probe that specifically represented *Src* sequences. The use of this defined probe in nucleic acid hybridization experiments allowed them to unambiguously show that normal avian cells contain *Src*-related DNA sequences.

It is interesting to consider the actual study in more detail. Specifically, as the viral genome is relatively compact it was possible to synthesize a radioactive cDNA probe composed of short, single-stranded DNA fragments complementary to the entire genomic RNA of RSV. This probe was then hybridized to an excess of RNA isolated from a transformation-defective deletion mutant. As would be expected, fragments of cDNA complementary to the viral replication genes, which were not deleted, hybridized to the transformation-defective RSV RNA, forming RNA–DNA duplexes. In contrast, cDNA fragments complementary to *Src* could not hybridize and so remained single-stranded. This single-stranded DNA was then isolated to provide a specific probe for *Src* oncogene sequences. A radioactively labeled *Src* cDNA was then hybridized to DNA from normal avian cells. The *Src* cDNA hybridized extensively to normal chicken, quail, and duck DNA (see figure). These landmark studies identified that normal cells contain DNA sequences that are closely related to the *Src* oncogene, supporting the hypothesis that retroviral oncogenes originated from cellular genes that became incorporated into viral genomes. The discovery of the *SRC* proto-oncogene further suggested that non-virus-induced tumors might also result from mutations in related cellular genes, leading directly to the discovery of oncogenes in human tumors. By unifying studies of tumor viruses, normal cells, and non-virus-induced tumors, the results of Varmus, Bishop, and their colleagues have had an impact on virtually all aspects of cell regulation and cancer research.

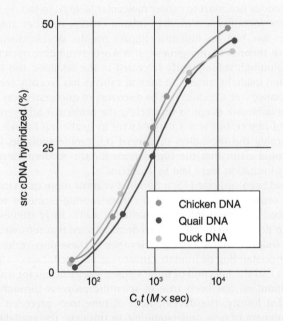

Following these studies it was becoming likely that the DNA of normal cells contained proto-oncogenes that under certain circumstances could become altered so as to promote cellular growth or transformation – the **somatic mutation theory** of cancer. This view was given a major boost by studies from the laboratory of Bob Weinberg and colleagues, who showed that transfer of DNA from cancer cells to NIH 3T3 fibroblasts could "transform" these cells. When transfected with tumor DNA, normal cells no longer exhibited contact inhibition but would pile up in a discrete area of a Petri dish, forming a clump of cells many layers thick. Such a clump, termed a "focus," contrasted to the behavior of the surrounding normal cells that stop growing once they have formed a single-cell-thick sheet termed a monolayer. The tumor DNA for these studies was prepared from mouse cells that had been exposed to methylcholanthrene (a carcinogen) and was found to induce more foci of transformed recipient cells than did control normal mouse DNA.

The similarities in behavior between the oncogenes transfected from a variety of tumor cells and those carried by retroviruses such as RSV were notable. In both cases, an agent – either a virus or a chemical carcinogen – had apparently succeeded in converting a normal cellular gene into a transforming oncogene.

Box 6.1 (Continued)

Later work on a human bladder cancer gene led to the discovery of an oncogene homologous to the *Ras* oncogene carried in the genome of Harvey sarcoma virus, a retrovirus of mixed rat–mouse origin whose v-*Ras* oncogene arose much like the v-*Src* oncogene of Rous sarcoma virus. The resulting conclusion again was that the same normal cellular gene could be activated into a potent oncogene either by a retrovirus or by a somatic mutation of the sort inflicted by mutagenic carcinogens. However, the precise nature of the mutation was only found in 1982 when several researchers found that the difference between the normal and mutant *RAS* gene was a single base substitution in the 12th codon, which caused the replacement of a glycine by a valine. Subsequently, it has been shown that around a quarter of all human tumors, derived from a variety of organs, carry point mutations in either the 12th or 61st codon of the *RAS* gene.

These discoveries led to attempts in identifying the signal transduction pathways in which RAS operated. RAS was soon found to act like the alpha subunit of heterotrimeric G-proteins: RAS switched between active GTP-bound forms to inactive GDP-bound forms. It was then found that the oncogenic form of p21^{RAS} lacked GTPase activity and thus mutant RAS remained trapped in an active state. Within months of the discovery of the point mutation that activated the *RAS* bladder carcinoma oncogene, other cellular genes such as *MYC* were also found in mutant form in human tumor DNA. We now know that cancer can be provoked by a wide variety of mutant genes.

However, tumorigenesis seemed to be much more complex than would be suggested by the development of a tumor by mutations in only a single gene; first, tumorigenesis seemed to involve a gestation period of many years, and second, histopathological analyses of tissues strongly suggested that the process of forming a tumor involved multiple steps – implying that real human tumors carried multiple mutated genes, and that a single mutated gene, on its own, was insufficient to create a malignant cell.

The earlier results in which NIH 3T3 cells could be transformed with a single oncogene seemed to contradict this view. Nevertheless, explanations for these supposedly contradictory findings were found. They derive from an immortalized cell line and thus were likely to already carry mutations that normal cells would not. This was confirmed in a landmark study by Hartmut Land in the laboratory of Bob Weinberg. Briefly, when the cloned *RAS* oncogene was transfected into normal rat embryo fibroblasts, no foci of transformants were seen. A cloned *MYC* oncogene was similarly unable to evoke transformation. However, when *RAS* and *MYC* oncogenes were cotransfected, foci of transformants arose, and could give rise to tumors in nude mice. This has been termed **oncogene cooperation**. This oncogene collaboration indicated that cellular oncogenes did not constitute a single, analogously functioning group of genes. Instead, these two oncogenes – *RAS* and *MYC* – seemed to work in distinct, complementary ways on cell phenotype.

Only later did it become apparent that the human genes that participate in mutant form in cancer pathogenesis encompass a wider spectrum, including tumor suppressor genes and those involved the maintenance of genomic integrity (see Chapters 7 and 10). In the best studied of human cancers – colon carcinoma – mutation of a *RAS* gene represents only one of multiple distinct genetic alterations that contribute to the phenotype of the malignant tumor cells.

Table 6.1 Viral and cellular oncogenes

Viral disease	Viral	Cellular	Function
Simian sarcoma	v-*sis*	PDGFB	Platelet-derived growth factor B subunit
Chicken erythroleukemia	v-*erb-b*	EGFR	Epidermal growth factor receptor
McDonough feline sarcoma	v-*fms*	CSF1R	Macrophage colony-stimulating factor receptor
Harvey rat sarcoma	v-*ras*	HRAS1	Cell signaling, activation of MAPK cascade
Abelson mouse leukemia	v-*abl*	ABL	Protein tyrosine kinase
Avian sarcoma 17	v-*jun*	JUN	Transcription factor
Avian myelocytomatosis	v-*myc*	MYC	Transcription factor
Mouse osteosarcoma	v-*fos*	FOS	Transcription factor

After Strachan, T. and Read, A. (2003). *Human Molecular Genetics*. London: Garland Science.

gene regulation (see Box 3.3). The major mechanisms involved in activating oncogenes can be classified as: **structural alterations**, from point mutations in a single gene to chromosomal translocations, or **gene amplification**, leading to overexpression of the oncogene.

Table 6.2 shows some well-known examples of oncogenes activated by amplification in different cancers. Again, it is important to emphasize that oncoproteins may become overexpressed and contribute significantly to tumorigenesis by other means not involving gene mutation, such as via activation of various upstream signaling pathways.

Types of oncogenes

> If we had a keen vision of all that is ordinary in human life, it would be like hearing the grass grow or the squirrel's heart beat, and we should die of that roar which is the other side of silence.
>
> *George Eliot*

In this section we will give a brief introduction to a selection of well-known oncogenes. These oncogenes all play an important role in various cancers and in their treatments, some have been discussed in Chapters 4 and 5, and here we have selected a small number of oncogenes, c-*MYC*, *RAS*, and *SRC*, to discuss in more detail, because they are present in a large number of human cancers and give important historical and mechanistic insights into oncogenesis in general. A fairly comprehensive list of known

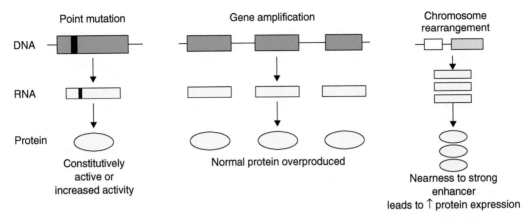

Figure 6.2 Conversion of proto-oncogenes to oncogenes. A proto-oncogene can be converted to an oncogene in a number of ways. These include point mutations in a single gene (e.g. *RAS*) that can either affect the coding region of the gene resulting in the formation of an abnormal oncoprotein with enhanced stability or activity, or may affect regulatory elements resulting in enhanced or deregulated expression. Gene amplification can lead to overexpression of the oncogene. Chromosomal translocations or rearrangements can lead to overexpression of an oncoprotein, for example, in Burkitt's lymphoma, the proto-oncogene c-*MYC* on chromosome 8 is translocated to one of the three chromosomes containing the genes that encode antibody molecules: immunoglobulin heavy chain locus (chromosome 14) or one of the light chain loci (chromosome 2 or 22). c-*MYC* now finds itself in a region of vigorous gene transcription leading to overproduction of the MYC protein. Lastly, the fusion of one protein to another might lead to its constitutive activity. For example, fusion of the promyelocytic leukemia (PML) protein to the retinoic acid receptor-alpha (RARα) generates the transforming protein of acute promyelocytic leukemias.

oncogenes is given in Table 6.3, and what immediately becomes obvious is that the vast majority of these are involved in pathways that normally serve to regulate cell division and survival.

A recent study has examined a range of different human cancers for oncogenic mutations and the following 14 were most frequently observed: *KRAS, NRAS, HRAS, BRAF, PIK3C, JAK2, EGFR, ERBB2, FGFR1, FGFR3, CDK4, RET, PDGFRA,* and *KIT* (note the absence of both c-*MYC* and c-*SRC* from this list – these are upregulated/activated in most human cancers and play key roles in tumorigenesis, but rarely have mutations in either coding or noncoding regions).

A simple classification of oncogenes

> To retain respect for sausages and laws, one must not watch them in the making.
>
> *Otto von Bismarck*

The vast majority of oncogenes are involved in pathways that normally serve to regulate cell division and survival (and other processes including cell motility and invasiveness – though it is likely that mutations influencing these behaviors would also need to confer some growth advantage in order to be selected for during cancer evolution). One way of subdividing oncogenes is, therefore, through the site in growth-regulating signal transduction pathways at which they act (we have deliberately tried to follow a similar classification to that used in Chapter 5). Thus, oncogenes may be consigned to five categories based on the functional and biochemical properties of their normal proto-oncogene/proto-oncoprotein counterparts. These are (1) growth factors, (2) receptors for growth factors and others, (3) signal transducers, (4) transcription factors, and (5) regulators of survival and others.

Growth factors

The reader is referred to Chapter 5, where this subject is covered in more depth. Oncogenic activation of growth factors invariably results from transcriptional activation of the gene, leading to overproduction of the growth factor. The best-known growth factor oncogene is *SIS*, the oncogene incorporated into the genome of the simian sarcoma virus, and it encodes the platelet-derived growth factor PDGF B chain.

Receptors for growth factors and others

Receptor tyrosine kinases (RTK)

Several mutations resulting in constitutively active receptors have been identified. For example, a single-point mutation converts the normal HER2 receptor into the NEU oncoprotein. Other RTK oncogenes have been identified and are listed in Table 6.4. The reader is again referred to Chapter 5 for a more detailed discussion.

G-Protein coupled receptors

There are few known examples of G-protein coupled receptors (see Chapter 5) acting as oncogenes, the most notable being the *MAS* gene, first identified in a mammary carcinoma. *MAS* is also believed to be involved in human epidermoid tumors and has been shown to encode an angiotensin receptor.

Signal transducers

Nonreceptor membrane-associated tyrosine kinases

Members of this group include *SRC, ABL, FGR* (see Table 6.3) which can become oncogenic following DNA rearrangements, such as gene amplification or translocation as well as genetic mutations (Tables 6.2, 6.3, and 6.4). Whichever the event, the outcome is excessive tyrosine kinase activity in affected cells, often giving rise to various human leukemias.

The v-*Src* gene is historically of great importance and was the first oncogene to be identified (see Box 6.1). Moreover, the normal cellular gene c-*SRC* (SRC) is the archetypal protein tyrosine kinase (see Fig. 6.18), and the functional domains of other

Table 6.2 List of oncogenes activated by amplification in cancer

Cellular oncogene	Location	Protein function	Type of cancer
ABL	9q34.1	Protein tyrosine kinase	Chronic myeloid leukemia
BCL1	11q13.3	G_1/S-specific cyclin D1	Breast cancer, squamous cell carcinoma of the head and neck, bladder cancer
CDK4	12q14	Cyclin-dependent kinase	Sarcomas
EGFR/ERBB1	7p12	Epidermal growth factor receptor	Glioblastoma multiforme, epidermoid carcinoma, bladder cancer, breast cancer
ERBB2(NEU)	17q12-q21	Growth factor receptor	Breast cancer, ovarian cancer, stomach cancer, renal adenocarcinoma, adenocarcinoma of salivary gland, colon carcinoma
HSTF1	11q13.3	Fibroblast growth factor	Breast cancer, esophageal carcinoma
INT1/WNT1	12q13	Probably growth factor	Retinoblastoma
INT2	11q13.3	Fibroblast growth factor	Breast cancer, esophageal carcinoma, melanoma, squamous cell carcinoma of the head and neck
MDM2	12q14.3-q15	p53-binding protein	Sarcomas
MET	7q31	Hepatocyte growth factor receptor	Amplified in cell lines from human tumors of nonhematopoietic origin, particularly gastric tumors
MYB	6q22-q23	DNA-binding protein (essential for normal hematopoiesis)	Leukemias, colon carcinoma, melanoma
MYC (c-MYC)	8q24.12-q24.13	DNA-binding protein	Small-cell lung cancer, giant cell carcinoma of lung, breast cancer, colon carcinoma, acute promyelocytic leukemia, cervical cancer, gastric adenocarcinoma, chronic granulocytic leukemia
MYCN (NMYC)	2p24.3	DNA-binding protein	Neuroblastoma, small-cell lung cancer, retinoblastoma, medulloblastoma, glioblastoma, rhabdomyosarcoma, adenocarcinoma of lung, astrocytoma
MYCL1 (LMYC) MYCLK1	1p32 7p15	DNA-binding protein	Small-cell lung cancer
RAF1 (c-RAF)	3p25	Serine/threonine protein kinase	Non-small-cell lung cancer
HRAS1	11p15.5	GTPase	Bladder cancer
KRAS2	12p12.1	GTPase	Adrenocortical tumor, giant cell carcinoma of lung
NRAS	1p13	GTPase	Breast cancer
REL	2p12-p13	DNA-binding protein	Non-Hodgkin lymphomas

subsequently identified tyrosine kinases are still referred to on the basis of homology to those in SRC (SH-domains for SRC homology). The role of oncogenic SRC in cancer has experienced something of a renaissance in recent years and is discussed in detail later.

The prototypic nonreceptor tyrosine kinase c-ABL is implicated in various cellular processes but can become oncogenic following amplification of the gene, giving rise to chronic myelogenous leukemia (CML) (see Table 6.2). In addition, c-ABL can become oncogenic as a result of translocation associated with the Philadelphia chromosome, to form the BCR–ABL fusion protein, which is discussed later under its own heading as well as in the section on RHO–GEF. Importantly, this translocation event is found in over 90% of CMLs.

Table 6.3 Some known oncogenes classified by site of action in growth-regulating pathways

Oncogene	Chromosome	Discovery	Cancer	Means of activation	Protein
Growth factors					
HST	11q13.3	Transfection	Stomach carcinoma Mammary	Constitutive synthesis	FGF family
INT2	11q13	Proviral insertion	Mammary carcinoma	Constitutive synthesis	FGF family
KS3	11q13.3	Transfection	Kaposi sarcoma	Constitutive synthesis	FGF family
v-SIS	22q12.3–13.1	Viral homolog	Glioma/fibrosarcoma	Constitutive synthesis	B-chain PDGF
Growth factor tyrosine kinases					
HER1/EGFR	7p1.1–1.3	DNA amplification	Squamous cell carcinoma	Gene amplification/ increased protein	EGF receptor
HER2/NEU	17q11.2–12	Point mutation amplification	Neuroblastoma/breast carcinoma	Gene amplification	Human EGFR-2; preferred heterodimerization partner for other EGFR family members
HER3		Amplification	Squamous carcinoma	Gene amplification	
FMS	5q33–34 (FMS)	Viral homolog	Sarcoma	Constitutive activation	CSF1 receptor
KIT	4q11–21 (KIT)	Viral homolog	Sarcoma	Constitutive activation	SCF receptor
ROS	6q22 (ROS)	Viral homolog	Sarcoma	Constitutive activation	
MET	7p31	DNA transfection	MNNG-treated human osteocarcinoma cell line	DNA rearrangement/ constitutive activation (fusion proteins)	HGF/SF receptor
ALK		chromosome	Lymphoma	fusion proteins	
RET	10q11.2	DNA transfection	Carcinomas of thyroid; MEN2A, MEN2B	DNA rearrangement/ point mutation constitutive active	GDNF receptor
TRK	1q32–41	DNA transfection	Colon/thyroid carcinomas	DNA rearrangement/ constitutive activation (fusion proteins)	NGF receptor
G-protein coupled receptor					
MAS	6q24–27	DNA transfection	Epidermoid carcinoma	Rearrangement of 5′ noncoding region	Angiotensin receptor
Signal transduction pathways					
Nonreceptor tyrosine kinases					
ABL	9q34.1	Chromosome	CML	DNA rearrangement translocation (fusion proteins)	Tyrosine kinase
FGR	1p36.1–36.2 (FGR)	Viral homolog	Sarcoma	Constitutive activation	Tyrosine kinase
BCR–ABL		chromosome	CML	fusion protein	Tyrosine kinase
FES	15q25–26 (FES)	Viral homolog	Sarcoma	Constitutive activation	Tyrosine kinase
SRC	20p12–13	Viral homolog	Colon carcinoma	Constitutive activation	Tyrosine kinase
YES	18q21–3 (YES)	Viral homolog	Sarcoma	Constitutive activation	Tyrosine kinase
Membrane-associated G-proteins					
KRAS	11p15.5	Viral homolog/ transfection	Colon, lung, pancreas carcinomas	Point mutation	GTPase
RAS	12p11.1–12.1	Viral homolog/ transfection	AML, thyroid carcinoma, melanoma	Point mutation	GTPase
NRAS	1p11–13	transfection	Carcinoma, melanoma	Point mutation	GTPase
GSP	20	DNA sequencing	Adenomas of thyroid	Point mutation	G_s alpha subunit
GIP	3	DNA sequencing	Ovary, adrenal carcinoma	Point mutation	G_i alpha subunit
GTPase exchange factor (GEF)					
DBL	Xq27	DNA transfection	Diffuse B-cell lymphoma	DNA rearrangement	GEF for Rho and Cdc42
VAV	19p13.2	DNA transfection	Hematopoietic cells	DNA rearrangement	GEF for Ras
Serine/threonine kinases: cytoplasmic					
MOS	8q11 (MOS)	Viral homolog	Sarcoma	Constitutive activation	Protein kinase (Ser/Thr)
RAF	3p25 (RAF1)	Viral homolog	Sarcoma	Constitutive activation	Protein kinase (ser/thr)
PIM1	6p21 (PIM-1)	Insertional mutagenesis	T-cell lymphoma	Constitutive activation	Protein kinase (ser/thr)
Cytoplasmic regulators					
CRK	17p13 (CRK)	Viral Homolog		Constitutive tyrosine kinase activity	SH-2/SH-3 adaptor
PRAD1			Breast		cyclin D1

Table 6.3 (Continued)

Oncogene	Chromosome	Discovery	Cancer	Means of activation	Protein
Transcription factors					
ETS2	21q24.3	Viral homolog	Erythroblastosis	Deregulated activity	Transcription factor
AM11		Chromosome	AML	Deregulated activity	Transcription factor
ERBA1	17p11–21	Viral homolog	Erythroblastosis	Deregulated activity	T3 transcription factor
ERBA2	3p22–24.1	Viral homolog	Erythroblastosis	Deregulated activity	T3 transcription factor
FOS	14q21–22	Viral homolog	Osteosarcoma	Deregulated activity	Transcription factor API
JUN	p31–32	Viral homolog	Sarcoma	Deregulated activity	Transcription factor API
c-MYC	8q24.1 (MYC)	Viral homolog	Multiple cancers, breast, stomach, lung, cervix, colon, neuroblastomas and glioblastomas. Burkitt lymphoma	Amplification, translocation, deregulation	Transcription factor
NMYC	2p24	DNA amplification	Neuroblastoma; Lung carcinoma	Deregulated activity	Transcription factor
LMYC	1p32	DNA amplification	Carcinoma of lung	Deregulated activity	Transcription factor
MYB	6q22–24	Viral homolog	Myeloblastosis	Deregulated activity	Transcription factor
REL	2p12–14	Viral homolog	Lymphatic leukemia	Deregulated activity	Mutant NF-κB
SKI	1q22–24	Viral homolog	Carcinoma	Deregulated activity	Transcription factor
TS1	11p23–q24	Viral homolog	Erythroblastosis	Deregulated activity	Transcription factor
Apoptosis regulators and others					
BCL2	18q21.3	Chromosomal translocation	B-cell lymphomas	Constitutive activity	Antiapoptotic protein
MDM2	12q14	DNA amplification	Sarcomas	Gene amplification/ increased protein	Complexes with p53

After Vogelstein, B. and Kinzler, K.W. (1998). *The Genetic Basis of Human Cancer*. New York: McGraw-Hill.
AML, acute myeloid leukemia; CML, chronic myelogenous leukemia; CSF, colony stimulating factor; EGF, epidermal growth factor; FGF, fibroblast growth factor; HGF, hepatocyte growth factor; NGF, nerve growth factor; PDGF, platelet-derived growth factor.

Table 6.4 Growth factor receptors and cancer

Growth factor receptor	Activation	Cancer
ALK	Gene rearrangement: NPM–ALK t(2;5)	Anaplastic large-cell lymphomas, NSCLC
CSF-1R/FMS	Extracellular domain mutations	AML
CSFR/KIT	Point mutations and deletions	AML, mastocytomas, gastrointestinal stromal tumors
EGFR	Autocrine activation, amplification/overexpression	Squamous cell carcinoma, glioblastoma Colorectal cancer, NSCLC
EGFR	Extracellular domain deletions	Glioblastoma
HER2/NEU	Point mutation, amplification/overexpression	Breast and ovarian carcinoma
FGFR1	Amplification	Breast cancer
FGFR1	Chromosomal translocations	Myeloid malignancies
FGFR2	Point mutations	Endometrial cancer
FGFR2	Amplification	Gastric cancer
FGFR3	Translocation	Multiple myelomas
FGFR3	Point mutations	Bladder cancer, prostate
FLK2/FLT3	Internal tandem duplication	AML, PML
C-MET	Gene rearrangement: TPR–MET t(1;7)	Stomach
C-MET	Amplification/overexpression	Thyroid, ovarian and colorectal cancers Acquired resistance in NSCLC
C-MET	Point mutations	Hereditary and sporadic papillary renal carcinoma
PDGFR	Autocrine activation	Osteosarcoma, melanoma, glioblastoma
PDGFR	TEL–PDGFRB t(5;12) translocation	CML
RET	Germline point mutations	Men2A, Men2B
TIE1, TIE2	Paracrine activation	Tumor angiogenesis and lymphangiogenesis
TRKA	Gene rearrangement: TPM–TRKA t(1;1)	Papillary thyroid carcinoma
TRKC	Gene rearrangement: TEL–TRKC t(12;15)	Congenital fibrosarcoma, AML, breast carcinoma
VEGFR1, -2, -3	Paracrine activation	Tumor angiogenesis and lymphangiogenesis

AML, acute myeloid leukemia; CML, chronic myelogenous leukemia; PML, promyelocytic leukemia; NSCLC, non-small-cell lung carcinoma.

Table 6.5 Oncogenes in signaling pathways

Signaling molecule	Activation	Cancer
H-RAS	Point mutations	Thyroid and bladder carcinoma
K-RAS	Point mutations	Pancreatic, lung and colon adenocarcinomas, non-small-cell lung carcinoma, myeloid leukemia, thyroid carcinomas
N-RAS	Point mutations	Melanoma, myeloid leukemia, thyroid carcinomas
B-RAF	Point mutations	Melanoma, colorectal carcinoma, small-cell lung cancer
SRC	Overexpression	Gastrointestinal cancers, lung, breast, ovary
BCR–ABL	Translocation	Chronic myelogenous leukemia
CYCLIN D1	Amplification/ overexpression	Breast cancer, head and neck squamous carcinoma, esophageal cancer, lymphoma
PI3K P110	Amplification/ overexpression	Ovarian and cervical cancer
STAT3		Melanoma

The RAS family

Among the first human oncogenes to be identified and frequently mutated in cancers, *RAS* genes encode members of the RAS family of membrane-associated small G-proteins. Given their importance in human cancer, we will specifically discuss RAS and related family members in detail later in the chapter. There are three different oncogenic homologs of the *c-RAS* gene (see Table 6.5), each of which was identified in a different type of tumor cell. In all cases, oncogenic *RAS* genes have undergone point mutations that eliminate the intrinsic GTPase activity of the RAS protein. The result is that RAS is permanently "switched on."

Serine/threonine kinases

The *RAF* gene was originally identified as a transforming oncogene from the rat fibrosarcoma virus, from which its name derives. Incorporation of the *RAF* gene into the viral genome resulted in truncation of the protein, with the resulting protein possessing constitutive kinase activity. Since the normal *RAF* gene product, RAF, is responsible for threonine phosphorylation of MAP kinase (MAPK) following receptor activation, oncogenic RAF leads to constitutive activation of the downstream MAPK pathway (see Fig. 6.14). The *BRAF* gene has been found to be mutated in some human cancers. Table 6.5 summarizes some of the known oncogenes in signaling pathways.

Transcription factors

A considerable number of transcription factors have been shown to possess oncogenic activity, when deregulated. Not surprisingly, given the potentially large number of genes whose expression could be deregulated as a consequence. Examples of well-known oncogenic transcription factors include *FOS, JUN, MYC, NFkB, GLI, MYB, ETS*, and others (see Tables 6.1 and 6.2).

As discussed in Chapter 3, various mechanisms of oncogene activation have been demonstrated and depend on the host and the oncogene in question. Thus, oncogenes may be activated by integration of a virus in the vicinity of the host gene or through direct incorporation of the oncogene into a transforming retroviral genome. Oncogenes may also be activated by point mutations and by chromosomal translocation (see Box 3.3). In some cases, there is evidence for mutations that lead to an increased stability of the oncoprotein (e.g. c-MYC and FOS).

The v-*fos* gene was identified as a transforming oncogene of the feline osteosarcoma virus. The product of the human homolog gene, FOS, interacts with a second proto-oncoprotein, JUN, to form a transcriptional regulatory complex. Aberrant oncogenic activation of v-*fos* results from truncations of the coding and noncoding sequences of the gene following incorporation into the viral genome. This results in the loss of "destability" sequences as well as sequences involved in transcriptional repression of *Fos*. As a consequence, a stable FOS protein is produced and, unlike the normal situation in replicating cells where expression is transient, high levels of FOS may persist throughout the cell cycle.

The ETS family (E-twenty-six) is one of the largest families of transcription factors, characterized by a conserved ETS DNA-binding domain. Under normal regulation, ETS family members are involved in a variety of cellular functions including cell cycle, differentiation, migration, apoptosis, and angiogenesis and are downstream nuclear targets of RAS–MAP kinase signaling (see Fig. 6.13). The deregulation of *ETS* genes results in the malignant transformation of cells, including invasion, metastasis, and neo-angiogenesis. Several *ETS* genes are rearranged in human leukemia (e.g. *TEL* ETS gene fused to *JAK2*) and Ewing's sarcoma (e.g. *ERG* ETS gene fused to *EWS* gene) to produce chimeric oncoproteins (fusion of elements of at least two different proteins).

The *c-MYC* gene was originally identified in the avian myelocytomatosis virus. The oncogenic functions of the transcription factor c-MYC play a key role in well-known human cancers, such as Burkitt's lymphoma as a result of chromosomal translocation (see Chapter 3, Box 3.3). In addition, this oncogene appears to be activated in the majority of human cancers at some stage during tumor development. For this reason, we have selected c-MYC for further discussion in the next section, and will highlight both its normal biological functions as well as those during oncogenic activation *in vitro* and in genetically altered mouse models. Although the mechanism for c-MYC activation remains unclear in many human cancers, there is evidence of mutations that confer increased stability of the c-MYC protein. In all cases though, the transforming effect of oncogenic c-MYC (described in the next section) is the result of its elevated or deregulated levels of expression.

Regulators of cell survival and death

Normal tissues are maintained by a regulated balance between cell proliferation and cell death (apoptosis). In fact, apoptosis is a normal process during embryonic development and tissue homeostasis in the adult – otherwise referred to as programmed cell death (Chapter 8 covers this in depth). Studies of cancer cells have shown that both uncontrolled cell proliferation and failure to undergo apoptosis can contribute to neoplasia and insensitivity to anticancer treatments. Many proto-oncogenes may affect survival in some way, usually indirectly. Of those that specifically regulate apoptosis, members of the *BCL2* family (see Chapter 8), first identified by the study of chromosomal translocations in human lymphoma (see Box 3.3) are the most studied to date.

The *BCL2* gene encodes a protein, BCL-2, localized to the inner mitochondrial membrane, endoplasmic reticulum, and nuclear membrane. This BCL-2 family is now known to include a large number of proteins with either anti- or proapoptotic activities. If expression of antiapoptotic BCL-2 family members is increased, or that of proapoptotic members lost, then cells are vulnerable to other cancer-causing mutations, in particular those which increase proliferation. Such cells will now no longer be able to activate key innate tumor suppressive mechanisms and instead of dying, as they would normally do, they are now able to survive and give rise to cancer. We highlight the importance of BCL-2 family members in tumor progression later in the chapter.

Oncogene collaboration – from cell culture to animal models

It is the long history of humankind . . . those who learned to collaborate and improvise most effectively have prevailed.

Charles Darwin

An activating mutation in a single oncogene is not sufficient to cause cancer – rather oncogenes must collaborate with one another and with inactivation of tumor suppressors in order to generate cancers. Oncogene collaboration can be demonstrated *in vitro* using cultured cells in which various oncogenes are introduced, and also *in vivo* using genetically altered mice. In different cell types, different combinations of oncogenes are required for such collaborative events to transform normal cells into cancerous ones (transformation).

If normal (wildtype) rat embryo fibroblasts cultured in serum are transfected with the c-*MYC* oncogene alone or *RAS* oncogene alone, then they behave normally. However, when both oncogenes are transfected, the cells become transformed, that is, show various features or behavior associated with cancer cells (Fig. 6.3). These include enhanced proliferation and motility, and the ability to divide without adherence to a substratum, resulting in cells piling up on top of each other to form colonies. Although much important data has been generated, it is worth emphasizing again that the behavior of cells studied *in vitro* – in this case, to identify oncogenes that cause transformation – might not always be the same as when cells are studied within their normal context in an intact organism.

For this reason, genetically altered mice are studied in which expression of particular oncogenes is targeted to certain tissues and can be examined in the context of the whole animal. The oncogene is linked to a suitable promoter and then injected into a mouse egg nucleus. The oncogenic construct becomes integrated into the mouse genome leading to the generation of a strain of genetically altered mice that carry the oncogene in all cells (see Chapter 20). Depending on the promoter selected, expression of the oncogene will ultimately be driven in a specific cell type or tissue in the animal. Historical studies show that mice carrying either a c-*MYC* or a *RAS* oncogene linked to the MMTV promoter (which ultimately drives MYC or RAS overexpression in the mammary and salivary gland) develop tumors much more frequently than normal. However, when these two strains of mice are crossbred to generate mice carrying both c-*MYC* and *RAS* oncogenes, tumors develop at a far higher rate – the result of oncogene collaboration. Nevertheless, tumors arise only after a delay and in a small proportion of cells within the tissue. This implies that a further propitious genetic alteration is needed for tumors to develop in this particular tissue. Such an event is thus "acquired" in one or a few cells through chance. We will see later in this chapter, however, that there are examples of cell types within the mouse that require only two oncogenes for tumors to arise.

Further *in vitro* work in the early 1990s showed that if rat fibroblasts transfected with c-*MYC* oncogene alone are cultured in very low amounts of serum, instead of cells ceasing growth/replication (as they would normally do), they were driven to proliferate uncontrollably. Paradoxically, cells also started to die, with the end result being overwhelming cell death by apoptosis (Fig. 6.4). However, if the oncogene *BCL2* was also overexpressed, then cells were rescued from death and allowed to proliferate excessively: *BCL2* acts as an oncogene because the encoded BCL-2 protein inhibits apoptosis. These important findings gave birth to the concept that oncogenic mutations that drive uncontrolled proliferation (as is the case with c-MYC) also possess intrinsic tumor suppressing activity. In other words, cells that acquire such oncogenic mutations have a built-in "fail-safe" mechanism, ensuring that potential cancer cells will die by apoptosis rather than survive to accumulate further genetic mutations that may otherwise lead to cancer. Such a cell population would be unable to outgrow its environment unless apoptosis was inhibited.

There is now compelling evidence *in vivo* to support this concept, as shown by the dramatic synergy between oncoproteins such as c-MYC and mechanisms that suppress apoptosis. An early example is the acceleration of lymphoma development in genetically altered mice that overexpress both c-MYC and the antiapoptotic protein BCL-2 in B lymphocytes, compared with mice expressing c-MYC alone. More recent mouse models include the overexpression of c-MYC together with an antiapoptotic protein, such as BCL-2 or BCL-X_L, or loss of p19^{ARF} or p53 tumor suppressors, some of which will be highlighted in this chapter. Later we will discuss another critical inherent tumor suppressing activity associated with oncogenic *RAS*, namely irreversible growth arrest (senescence).

A last important point to make is that such mouse models invariably have high levels of the particular oncogene(s) expressed. Recently, mouse models of cancer have been generated in which physiological (low) levels of oncogene expression more closely resemble the human disease process. These will be discussed later on.

The c-*MYC* oncogene

Choices are the hinges of destiny.

Pythagoras

The c-*MYC* proto-oncogene family ranks among the most exhaustively studied groups of genes in biology not least because it is so profoundly involved in many human and other animal cancers.

The normal cellular proto-oncogene, c-*MYC* (*MYC*), encodes the protein c-MYC (MYC) whose key biological function is to

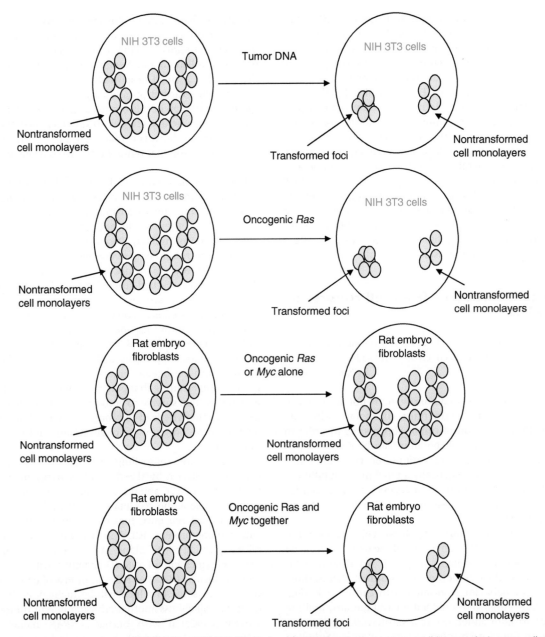

Figure 6.3 Oncogene collaboration *in vitro*. Transfection of NIH 3T3 fibroblasts with DNA from tumors or tumor cell lines results in some cells becoming transferred, as demonstrated by loss of normal contact inhibition and formation of foci (colonies) of cells piled up rather than in a monolayer. This same result could subsequently be achieved by transfecting NIH 3T3 cells with oncogenic *RAS*. In contrast, transfection of a "more normal" rat embryo fibroblast with any single oncogene (either *RAS* or *MYC* alone) did not result in transfomation. However, if both oncogenic *RAS* and *MYC* were transfected into rat embryo fibroblasts then foci did form – oncogene cooperation. By implication, NIH 3T3 fibroblasts have already undergone mutations such as those likely to occur in multistep tumorigenesis, thus enabling them to be more readily transformed in cell culture.

promote growth and cell-cycle entry. However, as you will see later, this enigmatic protein appears to be a key player in a diverse, and sometimes opposing, group of cellular processes, including proliferation, growth, cell death (apoptosis), energy metabolism, differentiation, and angiogenesis (growth of new blood vasculature). A more recent function to add to the list is the central role played by MYC in the generation and maintenance of stem cells (see Chapter 9). From this point on we will refer to c-MYC/c-*MYC* (and all species versions of these) as MYC and *MYC*, removing the prefix c-, while for the other MYC family members the prefix (e.g. n-) will be retained.

Activated MYC expression in human tumors

Our wretched species is so made that those who walk on the well-trodden path always throw stones at those who are showing a new road.

Voltaire

Before describing the mechanisms by which MYC activates or represses various biological processes in normal and in cancer cells, this section highlights what we know about MYC in human cancers and why we have had to look towards mouse tumor

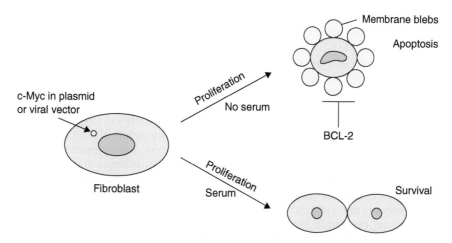

Figure 6.4 c-*MYC* promotes cell death by apoptosis (*in vitro*). In the early 1990s, several laboratories made an intriguing discovery: oncoproteins such as MYC and the adenovirus E1A – both potent inducers of cell proliferation – were shown to possess apoptotic activity. Ectopic expression of MYC in fibroblasts that were cultured in the absence (or limited supply) of survival factors (e.g. factors that are present in fetal calf serum, such as the extracellular molecule, insulin growth factor-1 (IGF-1), that mediates cell survival via its receptor) led to apoptosis, with the eventual loss of the entire cell population. Although interpreted by some as a conflict of growth signals – oncogenes activate apoptosis if the proliferative pathway is blocked in some way – the most widely held view of oncoprotein-induced apoptosis is that the induction of cell cycle entry sensitizes the cell to apoptosis. In other words, cell proliferative and apoptotic pathways are coupled. However, the apoptotic pathway is suppressed as long as appropriate survival factors deliver antiapoptotic signals. In this scenario, the predominant outcome of these contradictory processes will depend on the availability of survival factors.

Box 6.2 Nomenclature of oncogenes

The nomenclature of oncogenes is straightforward. They are described by a three-letter code relating to the cancers in which they were first identified. Thus, *Ras* oncogenes were first described in rat sarcomas. They may carry a prefix distinguishing between cellular (c-) and viral (v-) homologs, or between differing members (mutant forms) of the same oncogene family. In humans, members of the *RAS* family include *HRAS*, *KRAS*, and *NRAS*, which refer to Harvey and Kirsten murine sarcoma virus and neuroblastoma, respectively, where these variants were first described (in all cases the italicized forms refer to the genes). *Ras* genes have been incorporated into the genomes of such retroviruses and ultimately have contributed to tumor development in certain mouse tissues.

The *MYC* proto-oncogene encodes the transcription factor MYC, originally identified as the cellular homolog to the viral oncogene (v-*Myc*) of the avian myelocytomatosis retrovirus. There are now several members of the MYC family: c-*MYC*, *N-MYC* and *L-MYC*, where the prefix denotes the normal cellular (c-) gene, and the dominant tissues in which a particular homolog is expressed, neuronal (N-) and lung (L-).

---

models in order to gain further insight into MYC's oncogenic role *in vivo*.

MYC is expressed ubiquitously in cells during development and in growing tissues. Other homologs of MYC (L-MYC, N-MYC) are known (Box 6.2) and the expression of each is normally restricted to certain tissues. Likewise, in the adult, MYC is expressed in tissue compartments possessing high proliferative capacity (e.g. skin epidermis and gut), whereas it is undetectable in cells that have exited the cell cycle (quiescent). In human cancers, oncogenic *MYC* has long been known to be instrumental in the progression of Burkitt lymphoma. In this case MYC becomes oncogenic due to a translocational event (see Box 3.3), but in most cases where MYC contributes to human cancers this is actually not due to mutations in *MYC* but rather due to elevated or deregulated expression.

Activation of MYC has been detected in a wide range of human cancers, and is often associated with aggressive, poorly differentiated tumors. Such cancers include breast, colon, cervical, and small cell lung carcinomas, osteosarcomas, glioblastomas, melanoma, and myeloid leukemia. In most cases, the causes of MYC overexpression remain to be specifically described, but include gene amplification (see Table 6.2), whereby multiple copies of the *MYC* gene arise under the influence of genome instability (see Chapter 10), stabilization of *MYC* mRNA transcripts, enhanced initiation of translation due to mutation of the internal ribosomal entry site, or indirectly via activation of various upstream signaling pathways, such as WNT–β-catenin/LEF.

Levels of MYC protein are also influenced by control of degradation. Normal proliferating cells have evolved a number of mechanisms to limit the activity and accumulation of MYC. One of the most striking of these mechanisms is through the ubiquitin–proteasome pathway (see Chapter 11), which typically destroys MYC within around 20 minutes of its synthesis. Proteasome-mediated degradation of MYC is directed by at least two ubiquitin ligases – Fbw7 and Skp2. Interestingly, mutations in Fbw7 are noted in several types of tumors that manifest increased abundance of MYC. The ubiquitin-specific protease USP28 stabilizes MYC by interacting with Fbw7. USP28 is detected at high levels in colon and breast carcinomas and contributes to proliferation of these tumors.

Activation of MYC, as is the case in many human tumor cells, is defined as deregulation of the normal, highly controlled expression pattern of the *MYC* proto-oncogene. Until recently, however, distinguishing between cells that harbor normal or activated MYC has proved difficult. In normal dividing cells, MYC expression is maintained at a relatively constant intermediate level throughout the cell cycle, whereas in its oncogenic form,

MYC might be constitutively expressed at levels ranging from moderate to very high, and is nonresponsive to external signals. Alternatively, the regulated pattern of MYC expression can remain intact, but exceed normal levels of expression for the given cell type.

Historically, molecular pathologists have relied on the presence of gross chromosomal abnormalities, such as translocation or amplification, of the *MYC* locus to define activation of this oncogene in tumor cells (as is the case in Burkitt lymphoma and neuroblastoma, respectively). However, this restrictive diagnostic criterion has resulted in an underestimate of the numbers of tumors with deregulated *MYC*. Therefore, the detection of *MYC* activation in tumor cells now relies on both elevated expression of *MYC* mRNA and genetic alterations of the *MYC* locus. Advances in technology allow us to evaluate expression levels of *MYC* mRNA in tumor cells (precisely excised using laser-capture microdissection) with real-time reverse transcription-polymerase chain reaction (RT-PCR) and expression profiling using microarray technology (see Chapter 18). These approaches allow rapid, quantifiable, high-throughput screening of *MYC* activation in tumor cells. However, these techniques do not measure levels of MYC protein expression, which are largely assayed by immunohistochemical methods.

The fact that MYC activation is present in a broad range of human cancers, and is often associated with a poor prognosis, suggests that analysis of MYC deregulation in certain tumor types may be used as a diagnostic marker. Moreover, inactivating MYC, or downstream targets of MYC, may represent useful therapeutic targets. To this end, much important information has derived from regulatable mouse tumor models (see later). These models have exposed various oncogenic properties of MYC during tumor progression as well as determining whether inactivating MYC in tumors can lead to their regression. Although still in its infancy, various approaches to target MYC in human tumors are presently under investigation. If MYC inhibition were even possible, what would be the consequences for the homeostasis of normal proliferating tissues versus the fate of cancer cells? This is a crucial question as, not surprisingly, a major challenge is achieving specific delivery of the MYC inhibitor only to the nucleus of tumor cells *in vivo*. Recent studies have shed light on this important issue and there is reason to be optimistic, as will be discussed later on in this chapter.

MYC – transcription factor and "master regulator" of the genome

To rule is easy, to govern difficult.

Goethe

MYC is a transcription factor that requires dimerization with another protein, MAX, in order to become transcriptionally active (Fig. 6.5). Both MYC and MAX are basic helix–loop–helix

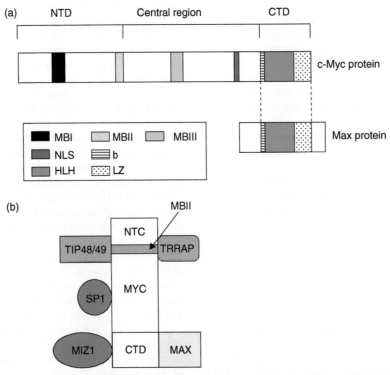

Figure 6.5 Functional domains of human MYC protein. (a) The C-terminal domain (CTD) of human MYC protein harbors the basic helix-loop-helix leucine zipper (bHLH-LZ) motif for dimerization with its partner, MAX, and subsequent DNA binding of MYC–MAX heterodimers. The N-terminal domain (NTD) harbors conserved "c-MYC boxes" I and II (MBI and MBII) essential for transactivation of c-*MYC* target genes. Recently, the MBIII situated in the central region has been found to be important for negatively regulating the apoptotic response. (b) Some major MYC-interacting proteins that may or may not bind simultaneously to MYC. These include coactivator TRRAP, part of a complex possessing histone acetylase activity (HAT), which interacts with the MBII region and mediates chromatin remodeling. TIP48 and TIP49 proteins interact with the NTD of MYC and are implicated in chromatin remodeling due to their ATP-hydrolyzing and helicase activities. Proteins involved in transcriptional regulation, such as MIZ-1, interact with the CTD of MYC, whereas SP1 interacts with the central region of MYC. NLS, nuclear localization signal.

zipper proteins (bHLHZ) and the MYC–MAX heterodimer binds the E-box sequence CACGTG of target genes to activate gene transcription – following the recruitment of multiple coactivator complexes, such as TRRAP (described in Fig. 6.5).

As discussed in Chapter 11 (see Box 11.1), DNA is not found in isolation in the cell but is associated with proteins called **histones** to form a complex substance known as **chromatin**. Thus, if chromatin is condensed, transcription factor complexes such as MYC–MAX cannot get to the DNA, and the genes will be switched off. Chromatin must be in an "open" conformation in order for gene transcription to occur and it is now known that the MYC–MAX transcription factor complex can lead to such a permissive alteration in the structure of chromatin – known as chromatin remodeling. For instance, TRRAP (transformation/transcription domain-associated protein), a MYC coactivator, is part of a complex that contains histone acetyl transferase (HAT) activity. Once recruited to MYC, TRRAP acetylates nucleosomal histone H4 at promoter sites, which alters the chromatin structure allowing accessibility of MYC–MAX complexes (see Fig. 6.7). Conversely, deacetylation of nucleosomal histones leads to switching off transcription of MYC target genes – a situation that often occurs during cellular differentiation when MAD levels are high (see Fig. 6.7b) (discussed later). Such chromatin remodeling is described as "epigenetic," since modifications to the DNA or the histones have a huge impact on gene transcription and repression without changing the nucleotide sequence of the DNA. This important and fascinating topic is described in greater depth in Chapter 11.

While the pre-eminent role of MYC is to activate gene expression, there are, in fact, many important genes that are repressed by MYC – those involved in cell-cycle arrest, cell adhesion, and cell–cell communication. This makes biological sense if MYC is to drive the cell cycle during normal cell turnover and development, and will self-evidently also increase the threat posed by oncogenic deregulation in tumorigenesis. The powerful combination of driving expression of genes that promote the cell cycle (e.g. cyclin-dependent kinases – CDKs) while repressing those that inhibit the cell cycle (e.g. CDK inhibitors – CDKIs), allows tumor cells with deregulated MYC to replicate uncontrollably; MYC can press the cell-cycle accelerator and release the brakes at the same time.

Over the last few years, technological advances, such as expression profiling (microarrays) and genomic binding studies (see Chapter 20), have led to the surprising discovery that MYC possesses a large number of gene targets (100–1000s) which, depending on the cell type, amounts to 15–20% of all genes. No wonder that MYC has been described as a "master regulator" of the genome!

Intriguingly, despite MYC binding to promoter sites of thousands of genes, only a minority are actually transcribed by MYC. In this regard, with the use of high-throughput chromatin immunoprecipitation (ChIP) to identify binding sites, a model is emerging in which multiple transcription factors are required to fire off transcription. Even with these powerful technologies, the potential complement of MYC target genes operating at any given time in any given system is bewildering, with genes involved in diverse cellular functions.

Equally important has been the recent discovery that the outcome of MYC activation (i.e. which MYC target genes are transcribed or repressed) is acutely dependent on level of expression or activity within the cell. So, for instance, at low levels, MYC–MAX binds promoter-proximal E-boxes within CpG-rich (dinucleotide sequence, cytosine-phosphate diester-guanine), H3-K4-methylated (lysine methylation of histone H3) euchromatic islands, to transcribe target genes. (The linguistic skills required to decipher this sentence can be acquired in Chapter 11, where the relevance of such DNA/chromatin modifications is detailed.) Actually, when translated, this means simply that MYC will bind to promoter regions with a more open configuration of chromatin that is transcriptionally active. However, when expressed at very high levels, MYC can bind to promoters lacking an E-box! Furthermore, a ChIP-seq analysis of relatively high abundance MYC binding showed that while 62% of the most reliable sites possess E-boxes, nearly 40% of binding loci did not.

At present we know little about these non-E-box targets, but there is circumstantial evidence that these include RNA polymerase III-transcribed genes, such as the amino acid-bearing transfer RNAs (tRNAs). MYC associates with the TFIIIB subunit of the RNA polymerase III complex and mediates recruitment of acetyl-transferase GCN5 (general control of amino acid synthesis protein 5) and the TRAPP complex described earlier. Therefore, the stimulation of polymerase III transcription is an example of MYC functioning as a coactivator, independent of its cognate DNA-binding site.

As the reader will by now be expecting, the all-pervasive microRNAs (miRNAs) are important in regulating MYC's biological functions also. The regulation of miRNAs by MYC impinges upon multiple phenotypes that contribute to tumorigenesis – cell-cycle progression, apoptosis, metabolism, angiogenesis, and metastasis (see Fig. 6.8), and is discussed later under its own heading.

To cap it all, now MYC also regulates translation

The MYC proteins were initially identified as transcription factors that activate or repress specific protein-encoding genes transcribed by RNA polymerase II. As previously mentioned, we now know that they can activate the expression of tRNAs and rRNA via RNA polymerases I and III, as well as miRNAs (described later).

To this impressive portfolio we can further add a role in translation, as it has now been demonstrated that MYC is involved in mRNA cap methylation (see below). The biological ramifications of this newly discovered MYC function could be substantial, particularly, in light of the often modest transcriptional effects of MYC on target gene expression. Thus, only a 1.5- to 2-fold increase in mRNA is typically observed, whereas, by contrast, cap-methylated mRNA may increase by 3- to 6-fold. By this means, despite the modest transcriptional response, cap-methylated mRNAs can be translated into higher levels of protein.

RNA polymerase II transcripts require modification and processing before becoming mature transcripts that can subsequently be translated into proteins. For instance, exons are spliced together: the 3' end of the transcript is cleaved, and polyadenylated RNA is exported from the nucleus to the cytoplasm, where translation factors bind to the RNA and direct recruitment of the ribosomal subunits needed for translation occurs. The methyl cap is an inverted 7-methylguanosine group joined to the first transcribed nucleotide by a triphosphate bridge, and is formed shortly after the initiation of transcription. The methyl cap – found at the 5' end of mRNA – is crucial as it protects transcripts from premature degradation by exonucleases during translation. Thus it is essential for mRNA translation and cell viability.

So, at present we know that MYC promotes cap methylation of mRNAs corresponding to most of its target genes and, surprisingly, some that are not MYC target genes! Which other transcription factors possess such nontranscriptional activities? So far, E2F1 has been shown to enhance cap methylation on its target genes, but whether this ability extends to other transcription factors remains to be seen.

MYC – cell growth and proliferation

No two men ever judged alike of the same thing, and it is impossible to find two opinions exactly similar, not only in different men but in the same men at different times.

Michel Montaigne

When a normal cell receives a signal to proliferate, such as that generated by binding of a growth factor ligand to its transmembrane receptor (see Chapter 5), signaling pathways inside the cell become activated and ultimately cascade that signal to the nucleus, leading to changes in gene expression. Among these changes will be the induction of specific genes required for cell-cycle entry, enabling the cell to leave its arrested state (G_0) and re-enter the cell cycle (described in Chapter 4). The *MYC* proto-oncogene is one of the most important of these induced genes and is one of a select few so-called "early-response genes" (induced within 15 minutes of growth factor treatment *in vitro*). MYC plays a crucial role in allowing cells to exit G_0 and progress through into S phase and proliferate. Conversely, when cells are not proliferating, *MYC* is silent. It is not difficult to picture the potential effect a gene like *MYC* might have on a cell, if it were continuously expressed or activated – that is, oncogenic *MYC*.

MYC is crucial for normal cell proliferation. Without it, cells would have a remarkably difficult time replicating themselves: normal development does not proceed in mice in which both *Myc* alleles have been inactivated, and embryos die *in utero*. In vitro, the cell-cycle effects of inactivating MYC have been investigated using a rat fibroblast cell line in which both alleles of c-*MYC* were ablated. Cells show greatly reduced rates of proliferation, mainly due to a major lengthening of G_1 phase (4- to 5-fold) and the significantly delayed phosphorylation of the retinoblastoma protein (RB) (see Chapter 4). Thus, progression from mitosis at the end of one cell cycle to the G_1 restriction point of the next cell cycle and the subsequent progression from the restriction point into S phase are both drastically delayed. Moreover, loss of MYC also leads to marked deficiency in cell growth (accumulation of mass), as demonstrated by a decreased global mRNA and protein synthesis.

Some studies in mouse knockdown models have suggested that at least under some circumstances cells may be able to replicate without normal levels of Myc, but this remains contentious as the extent of knockdown achieved and the proportion of cells affected is not clear and the role of other proteins, even other Myc family members, as substitutes for c-Myc under these circumstances is not known.

Normal healthy cells need to grow in size during the cell cycle before they divide. The first notion that MYC influenced cell growth came from the correlation between MYC and the expression of the rate-limiting translation initiation factors eIF4E and eIF2α, now known to be directly transcribed by MYC. Later studies showed that Myc has a direct role in the growth of invertebrate and mammalian cells: while diminished expression of the *Drosophila* ortholog of vertebrate *myc* (*dmyc*) resulted in smaller but developmentally normal flies, *dmyc* overexpression resulted in larger cells, with no significant change in division rate. A different outcome occurs in genetically altered mice when MYC is overexpressed in B lymphocytes: MYC induces cell growth but not cell-cycle progression in this cell type. Having said this, it is most likely that MYC induces both cell growth and proliferation in most cell types. It is indeed plausible that MYC's role in regulating cell proliferation could at least in part be mediated through its effects on cell growth.

How does MYC mediate these effects on cell growth? Although the picture is not completely clear, RNA polymerase III (pol III) which is involved in the generation of tRNA and 5S ribosomal RNA (required for protein synthesis in growing cells), is activated by MYC via binding to TFIIIB, a pol III-specific general transcription factor (mentioned earlier). In fact, recent studies show that MYC can regulate the activity of all three known nuclear RNA polymerases and thus plays a central role in mediating ribosomal biogenesis and cell growth. In addition, as explained in the previous section, a new role of MYC recently discovered is its effect on translation via mRNA cap methylation – leading to an increase in the levels of protein synthesis.

Over the last decade it has become clear that MYC can drive several key metabolic pathways required to support cell growth: glycolysis, biosynthetic pathways, mitochondrial function, and glutaminolysis. Some of this has already beeen discussed in Chapter 5 under the heading of cancer cell metabolism and the Warburg effect.

Insights into how MYC might promote cell proliferation have resulted from a number of important studies revealing MYC's ability to activate or repress target genes involved in cell-cycle progression (Fig. 6.6). As described in Chapter 4, progression of the cell cycle from G_1 phase to S phase (DNA synthesis) is controlled by the activities of the cyclin-dependent kinase (CDK) complexes: cyclin D–CDK4 and cyclin E–CDK2. MYC induces cyclin E–CDK2 activity early in the G_1 phase of the cell cycle, which is regarded as an essential event in MYC-induced G_1/S progression. Importantly, both *CCND2* (which encodes cyclin D2) and *CDK4* are direct target genes of MYC, and their expression leads to sequestration of the cyclin-dependent kinase inhibitor (CKI) p27^{KIP1} by cyclin D2–CDK4 complexes. The subsequent degradation of p27^{KIP1} involves two other MYC target genes – *CUL-1* and *CKS*. Therefore, by preventing the binding of p27^{KIP1} to cyclin E–CDK2 complexes, MYC allows inhibitor-free cyclin E–CDK2 complexes to become accessible to phosphorylation by cyclin-activating kinase (CAK) (see Fig. 6.6). As a consequence, increased CDK2 and CDK4 activities would result in RB hyperphosphorylation and subsequent release of E2F from RB (see Chapters 4 and 7).

Other important studies support the idea that MYC may also exert important influences on the cell cycle by repressing genes, such as those encoding the CKIs p15^{INK4b} and p21^{CIP1}, which are involved in cell-cycle arrest (see Chapters 4 and 7), through the MYC–MAX heterodimer interacting with transcription factors such as MIZ-1 and SP1 (Fig. 6.6). Consequently, the interaction of MYC–MAX with MIZ-1 blocks the association of MIZ-1 with its own coactivator (P300 protein), with the subsequent downregulation of p15^{INK4b}, p21^{CIP1}, and p57^{kip2}. How important then, is MYC's interaction with MIZ-1 in promoting tumor development? It has now been shown in transgenic mouse models that such interactions are critical for the induction and maintenance

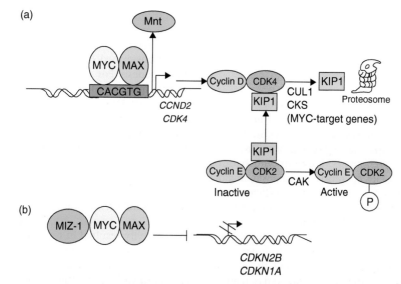

Figure 6.6 MYC induces cell cycle entry through activation and repression of target genes. (a) MYC–MAX heterodimers activate the target genes *CCND2* (cyclin D2) and cyclin-dependent kinase 4 (*CDK4*), which leads to sequestration of CDK inhibitor KIP1 (p27) in cyclin D2–CDK4 complexes. Subsequent degradation of KIP1 involves two further MYC target genes, *Cul-1* and *Cks*. In so doing, KIP1 is not available to bind to and inhibit cyclin E–CDK2 complexes, thereby allowing cyclin E–CDK2 to be phosphorylated by cyclin-activating kinase (CAK). Activation of some genes by MYC–MAX involves displacement of the putative tumor suppressor MNT from target genes. (b) MYC–MAX heterodimers repress the CDK inhibitors p15^{INK4B} and WAF1 (p21), which are involved in cell cycle arrest. By interacting with the transcription factors MIZ-1 (and/or SP1), MYC–MAX prevents transactivation of *INK4B* (*CDKN2B*) and *WAF1* (*CDKN1A*).

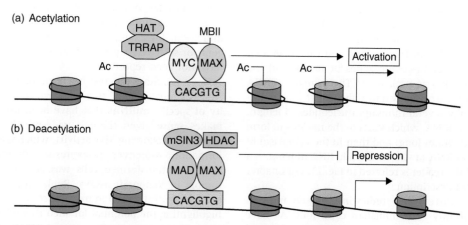

Figure 6.7 MYC–MAX and MAD–MAX heterodimers regulate gene activation through chromatin remodeling. (a) MYC–MAX heterodimers binds to an E-box sequence (CACGTG) near the promoter of a *c-MYC* target gene. The coactivator TRRAP (transformation/transcription domain-associated protein), a component of a complex that contains histone acetyltransferase (HAT) activity, is then recruited to the MYC box II (MBII) domain of MYC and acetylates (Ac) nucleosomal histone H4 at the E-box and adjacent regions. Nucleosomal acetylation alters chromatin structure, allowing accessibility of MYC–MAX transcriptional-activator complexes to target DNA, resulting in expression of the target gene. (b) Induction of MAD during terminal differentiation results in the MAD–MAX heterodimer binding to an E-box of a *c-MYC* target gene. Corepressor SIN3 and histone deacetylases (HDACs) are then recruited to MAD, resulting in local nucleosomal histone deacetylation and repression of target gene expression.

of T-cell lymphomas. Transgenic mice expressing a mutant form of *MYC* (MycV394D) that renders MYC unable to bind to MIZ1 while retaining its ability to bind to MAX and activate transcription, leads to elevated expression levels of CKIs: p15^{INK4b} and p57^{kip2}. The result is a significant delay in lymphomagenesis when compared to transgenic mice expressing wildtype Myc. Importantly, the delay in tumorigenesis is associated with growth suppression and cellular senescence induced by transforming growth factor-β (TGF-β) produced by the T-cell lymphomas.

These findings highlight the oncogenic potential of MYC–MIZ1 interactions: repressing CKI expression (thus, antagonizing growth suppression) and inhibiting cellular senescence, and therefore bypassing these tumor suppressor properties of TGF-β.

We know that MYC (with its partner MAX) activates target genes through its ability to remodel chromatin, as was mentioned earlier (Fig. 6.7). However, the mechanisms by which MYC silences gene expression are not well understood but likely include functional interference with transcriptional activators and the activity of various miRNAs. One important mechanism involves altered methylation (see Chapter 11), thus MYC binds the corepressor DNMT3a (DNA methyltransferase 3a – see Table 11.1) and associates with DNA methyltransferase activity. This means that DNMT3a can add methyl (CH_3) groups to the 5′ position of cytosine rings within CpG dinucleotides, which can directly turn off gene expression (see Fig. 11.1). Moreover, reducing DNMT3a levels results in reactivation of MYC-repressed genes, such as p21^{CIP1}, whereas the expression of MYC-activated E-box genes is unaffected. Interestingly, DNMT3a and MYC form a complex with MIZ-1, which appears to be required for methylation and repression of the p21^{CIP1} promoter.

MYC and microRNAs

To add yet another layer of complexity to MYC's biological functions, recent work has now added the transcription and repression of small noncoding RNAs, known as microRNAs (miRNAs), to the MYC target gene network. The regulation of miRNAs by MYC impinges upon multiple phenotypes that contribute to tumorigenesis – cell-cycle progression, apoptosis, metabolism, angiogenesis, and metastasis. This is described in detail in Chapter 11, but we will give a brief general summary here and then focus on the important links between MYC and miRNAs.

First identified in the context of recognition and elimination of double-stranded RNA viruses in plants, it was subsequently realized that the processes involved in destroying these viral nucleotides were also involved in the normal day to day regulation of gene expression in essentially all animals. Thus, miRNAs were identified, which could form double-stranded regions with endogenous mRNAs, thereby targeting these for essentially the same destructive pathway used to eliminate RNA viruses (Chapter 11 gives a detailed description). Briefly, miRNAs bind to specific complementary regions of target mRNAs through miRNA:mRNA base-pairing, most often within 3' untranslated regions (UTRs). The effects of this are remarkable, leading to accelerated destruction and translational repression of the target mRNA.

The miRNAs are encoded in the genome and can be located within the introns of protein-coding genes or in areas of the genome previously known as junk DNA! The primary miRNA transcripts (pri-miRNAs) are usually transcribed by RNA polymerase II, and are generally several thousand nucleotides in length. The processing of pri-miRNA, which starts in the nucleus to form pre-miRNA (~70 nucleotides long) and then in the cytoplasm to produce the mature miRNA (a ~22-nucleotide RNA duplex), is a complex process and the reader is referred to Fig. 11.6 in Chapter 11 for a more detailed description.

Since their initial discovery and studies in *C. elegans* and *Drosophila* and more recently in mammalian cells, miRNAs have been shown to have profound effects on key cellular behaviors, including differentiation, proliferation, and apoptosis. Given that these are also central to cancer progression it will come as no surprise that miRNAs also play a role in human tumorigenesis. Indeed, specific miRNAs were later found to be deleted or amplified in tumor samples and miRNAs have been shown to be deregulated in specific cancer subtypes. Essentially, miRNAs are subject to the full panoply of altered expression, as are the protein-encoding genes, from mutations to epigenetic silencing.

Of relevance to this section, miRNAs first arrived at center stage in cancer when it was recognized that MYC directly activates transcription of the miR-17-92 cluster, subsequently shown to be a bona fide oncogene. The miR-17-92 cluster produces at least six mature miRNAs (miR-17, miR-18a, miR-19a, miR-19b-1, miR-20a, and miR-92-1) that are frequently amplified and/or overexpressed in B-cell lymphomas and several solid tumors, including breast, colon, lung, pancreas, prostate, and stomach (discussed in Chapter 11). This is now known to be the tip of the iceberg as regards MYC and miRNAs. It is now clear that MYC regulates an extensive network of miRNAs and is itself in turn regulated by others (Fig. 6.8). Numerous miRNAs are repressed by MYC, including the Let-7 family, known to exhibit antiproliferative, proapoptotic, and/or antitumorigenic activities. Importantly, rescuing expression of several of these miRNAs in MYC-transformed B lymphoma cell lines dramatically inhibited tumorigenesis. For an overview of the role of miRNAs in regulating and mediating the actions of MYC, see Fig. 6.8.

How does MYC repress miRNAs? Given the fact that reversing these effects can inhibit tumor progression, elucidating the mechanisms could have great therapeutic impact on cancer treatment. Although much has still to be done, initial studies indicate that MYC uses both transcriptional and nontranscriptional mechanisms to repress miRNA expression. MYC is also able to block the maturation of specific miRNAs without affecting the transcription of the pri-mRNAs, as was seen for Let-7. In these studies, expression of Let-7 primary transcripts remained unchanged while MYC activity resulted in repression of mature Let-7 miRNAs. To this end, direct MYC targets, Lin28 and Lin28b, are both required for MYC-mediated repression of Let-7, leading to cellular proliferation, invasion, and metastasis.

Given the ability of miRNAs to potently influence cancer phenotypes, there is much excitement in the potential for targeting various miRNAs to treat cancer. As was discussed earlier, Myc drives both expression and repression of miRNAs linked to various oncogenic processes: cell cycle, apoptosis, metabolism, angiogenesis, and metastasis. The miR-17-92 cluster, which is transcribed by Myc, and possesses pleiotropic oncogenic functions, would understandably be a good target for therapeutic inhibition, for example, by systemic delivery of antisense reagents or siRNAs. As yet, such studies using *in vivo* tumor models have not been reported.

However, more progress has been made in the development of "miRNA replacement" therapies, which aim to restore the activity of specific miRNAs that have been lost in cancer cells. This makes sense given there is widespread repression of miRNA expression following Myc activity, which is associated with transformation. Moreover, re-expression of individual miRNAs in such Myc-transformed cells was shown to inhibit growth of tumors in xenograft assays. Replacement of miRNA has now been successfully used to treat cancer in multiple mouse models, highlighting the potential for such therapies in human cancers.

Regulating MYC activity

Life is the sum of all your choices.

Albert Camus

Cells have evolved a number of mechanisms to limit the activity and accumulation of MYC, so as to avoid potentially dangerous cell behaviors. MYC activity is normally tightly controlled by conflicting external signals, including growth factors, mitogens, and β-catenin, which promote its activity, and factors such as TGF-β, which inhibit such activity (also mentioned in Chapters 4 and 5). Much of this regulation is achieved by upstream signaling pathways, which influence MYC expression and activity. However, MYC, like another key transcription factor oncogene c-JUN, is a highly unstable protein and rapidly removed by ubiquitin-dependent degradation (this process is described in Fig. 11.7 of Chapter 11). The turnover of MYC is at least in part dependent on phosphorylation of two highly conserved residues in MB1 (MYC box I – see Fig. 6.5) that are mutated in v-*myc* (the viral oncogenic Myc) and in various cancers; phosphorylation at

Oncogenes

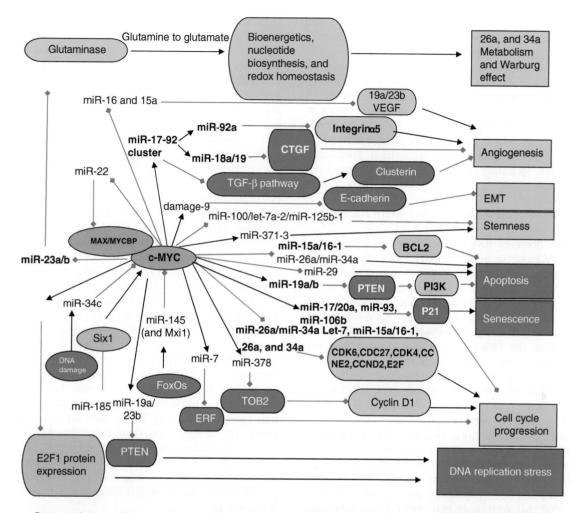

Gene products and tumor promoting cell behaviour are highlighted in green (for "go"), whereas those associated with tumour suppression are highlighted in red (for "stop"). Inhibitory arrows are in red with diamond ends, and stimulatory ones in black with normal arrow heads.

Figure 6.8 Role of miRNAs in regulating and mediating the actions of MYC. The first demonstration that miRNAs could have an important role in oncogenesis came with the discovery that MYC directly activates transcription of the miR-17-92 cluster to produce at least six mature miRNAs (miR-17, miR-18a, miR-19a, miR-19b-1, miR-20a, and miR-92-1), now known to be frequently amplified and/or overexpressed in many human cancers. As seen in the figure, MYC regulates an extensive network of miRNAs that affects many aspects of cell behavior, such as cell-cycle progression, apoptosis, metabolism, angiogenesis, and metastases. Only a few selected miRNAs are discussed here. **Cell cycle:** Both induction and repression of specific miRNAs by MYC facilitate transit through the cell cycle via complex mechanisms. For example, induction of miR-17 and related family members (miR-20a, miR-93, and miR-106b) promote cell-cycle progression by repressing levels of the cyclin-dependent kinase inhibitor P21 (CDKN1A), an essential protein involved in DNA damage/p53-mediated cell cycle arrest. Numerous miRNAs are repressed by MYC, including the let-7 family, known to exhibit antiproliferative, proapoptotic and/or anti-tumorigenic activities. For example, let-7 family members (miR-15a/16–1, miR-26a, and miR-34a) have a potent inhibitory effect on the cell cycle: miR-15a/16-1 inhibits expression of cell cycle regulators such as CDK6 and CDC27; miR-34a regulates CDK4, CDK6, CCNE2, and E2Fs, whereas miR-26a represses CCND2 and CCNE2. Thus, by downregulating expression of these miRNAs, MYC greatly facilitates cell-cycle progression. **Cell survival/apoptosis:** MYC's upregulation of the miR-17-92 cluster inhibits apoptosis. In particular, miR-19 mediates cell survival by targeting the tumor suppressor *PTEN*, an antagonist of the PI3K–AKT survival pathway (see Chapter 7). Consequently, downregulation of PTEN leads to induction of PI3K signaling and prevention of apoptosis. In addition, transcripts of proapoptotic proteins such as BIM are targeted by miRNAs within the miR-17-92 cluster including miR-19a/b, thus inhibiting apoptosis. MYC also suppresses miRNAs to promote cell survival (e.g. the miR-15a/16-1 cluster that leads to downregulation of the antiapoptotic protein BCL-2). **Angiogenesis:** The miR-17-92 cluster plays a key role in angiogenesis. For example, miR-18a and miR-19 family members directly target negative regulators of angiogenesis such as CTGF (connective tissue growth factor) and TSP1 (thrombospondin-1), respectively. The reduced expression of these proteins ultimately allows angiogenesis to proceed. In contrast, the miR-92a member inhibits angiogenesis by targeting and thus downregulating the angiogenesis-promoting factor integrin α5. Future studies will shed light on whether specific members of the miR-17-92 cluster are present at different levels in specific cell types and/or cellular contexts and thus will dictate the phenotypic outcome. **Metabolism:** Glucose and glutamine metabolism are essential for cell survival and proliferation. MYC contributes to this activity through miRNA regulation: activation of the miR-17-92 cluster stimulates signaling through the PI3K–AKT pathway via downregulating expression of PTEN (described above). The PI3K–AKT pathway strongly promotes glucose metabolism and fatty acid synthesis, for example, by increasing glucose transporter surface expression and enhancing glycolytic enzyme activity. MYC promotes glutamine metabolism through several mechanisms, including direct transcriptional activation of glutamine transporters, as well as by repressing transcription of miR-23a/b, which target mitochondrial glutaminase (GLS). This leads to an increase in expression of GLS, which is a key enzyme that converts glutamine to glutamate – a substrate for the production of ATP.

Ser62 stabilizes MYC, whereas phosphorylation at Thr58 and dephosphorylation at Ser62 are needed prior to ubiquitination of MYC and degradation in the proteasome.

The RAS pathway has been identified as a key determinate of MYC stability. Phosphorylation of MYC (and c-JUN) by the enzyme GSK-3β creates a high-affinity binding site for the E3 ligase FBW7, which targets MYC for polyubiquitination and proteasomal degradation. Conversely, the RAS–PI3K–AKT pathway (described later under the RAS section) is a major inhibitor of GSK-3β activity and can reduce MYC degradation. Specifically, AKT inhibits GSK-3β during early G_1, preventing phosphorylation of Thr58 and stabilizing MYC. Furthermore, MYC can be directly phosphorylated on Ser62 via the RAF-MAPK cascade (also described later under the RAS section), which also stabilizes MYC.

Paradoxically, however, RAS-regulated enzymes – the protein phosphatase PPA2 and the isomerase PIN1 – may also destabilize MYC by removing the phosphate at Ser62. As mentioned, the final step in MYC degradation is ubiquitination, which targets the protein to the proteasome for degradation (see Fig. 11.7). The human tumor suppressor FBW7/hCDC4 acts in a complex with SCF ubiquitin ligases to catalyze the ubiquitination of MYC. One particular isoform, the FBW7 gamma, may regulate nucleolar MYC accumulation and ubiquitination by binding the MB1 domain. However, studies have indicated how complex the regulation of MYC may prove to be, as not all MYC is necessarily subjected to the same rates of turnover. Thus, a pool of MYC that is metabolically stable has been identified.

As described earlier, the ability of MYC to activate transcription depends on the recruitment of several cofactor complexes, including histone acetyltransferases. In fact, acetylation of nucleosomal histones has long been recognized as a major regulator of gene transcription (see Chapter 11), but only recently has it been discovered that acetylation may also regulate subcellular localization and protein turnover. For example, it has been demonstrated that degradation of MYC may also be regulated by acetylation by the acetyltransferases mGCN5/PCAF and TIP60, which enhance protein stability. As these same enzymes, alongside others, seem important in mediating transcriptional activation by MYC, this raises the intriguing possibility that at least in part this may involve stabilizing of protein levels.

MYC and cell differentiation – role of the MXD protein family

For a proliferating cell to become terminally differentiated, it is usually required to exit the cell cycle and express various genes whose protein products are associated with establishing the differentiated phenotype. For this reason, the general rule is that activation of MYC and subsequent cell-cycle entry is incompatible with terminal differentiation. In other words, MYC blocks cell differentiation. Thus, in most cell types MYC would need to be downregulated in order for differentiation to occur. If this is the case, then one can begin to see that unregulated levels of MYC in a given cell might not only lead to excessive proliferation, but also keep the cell in an undifferentiated state. Of key importance to this process are the MXD family of proteins (formerly known as the MAD family) and MNT, which as you will see below, behave as transcriptional repressors by dimerizing with MYC's partner, MAX. The MXD family includes MXD1 (MAD1), MXD3 (MAD3), MXD4 (MAD4), and MXI1 (MAX interactor-1; MAD2). The expression of various members of this family (described below) usually coincides with downregulation of MYC expression, and cells begin to exit the cell cycle and acquire a terminally differentiated phenotype – although there are some exceptions. In tissues that are regionally compartmentalized with respect to cell proliferation and differentiation, such as the crypts of the gastrointestinal system and stratified squamous epithelium of the skin, MYC protein is readily detected in immature proliferating cells, whereas MXD proteins are restricted to postmitotic differentiating cells.

Intriguingly, some studies indicate that onset of differentiation does not always involve cell-cycle arrest. MYC may play a role in advancing cells along pathways of epidermal and hematopoietic differentiation. For example, activation of MYC in stem cells that reside in the basal keratinocytes of the epidermis causes them to undergo terminal differentiation following an initial proliferative burst, to form interfollicular epidermis and sebaceous glands at the expense of hair follicles. It was shown in this mouse model that MYC activation led to the downregulation of genes associated with cell adhesion and cell motility. As a result, such reduced adhesive interactions with the local microenvironment might allow epidermal stem cells to exit their compartment and subsequently receive differentiation signals from their environment. In a more recent study, low, intermediate, or high levels of MYC activity were induced in basal keratinocytes, and showed that MYC drives proliferation at all levels, but only high levels promote keratinocyte differentiation. Based on these observations, it has been proposed that promotion of differentiation may be a fail-safe mechanism acting against neoplastic conversion of epidermal stem cells. Whether an increase in cell growth and metabolism induced by MYC is important for lineage commitment awaits further investigation. With this in mind, recent studies have suggested that MYC activity may play an important part in maintaining stem cell pluripotency and self-renewal.

A last and important note to add here is the more recent work showing that in certain tissues MYC can facilitate terminal differentiation through cleavage of endogenous MYC protein. During differentiation of certain cell types, such as skin keratinocytes, muscle cells, and hematopoietic cells, calcium-activated cysteine proteases known as calpains are upregulated and activated. In a study using muscle cells, MYC protein was proteolytically cleaved by calpains at its C-terminus to generate a large N-terminal segment (termed "MYC-nick") while the cleaved C-terminus (containing DNA-binding domain, bHLH-LZ, essential for transactivation) was rapidly degraded. This event takes place in the cytoplasm and, interestingly, the transcriptionally inactive MYC-nick has been found to actually play a role in the differentiation process, promoting changes in the cytoskeleton/cell morphology by inducing α-tubulin acetylation.

How do the MXD and MNT proteins antagonize MYC function? These proteins heterodimerize with MAX and subsequently repress transcription by recruiting a chromatin-modifying corepressor complex to E-box sites on the same target genes as MYC–MAX, such that MYC–MAX complexes can no longer activate target genes. MXD–MAX recruits HDACs (histone deacetylases) to its specific target DNA to result in local histone deacetylation within nucleosomes, thereby decreasing the accessibility of DNA to transactivation factors (see Fig. 6.7). Given that MYC–MAX and MXD–MAX as well as MNT–MAX dimers bind a common set of target genes, it suggests that the balance between these protein dimers is likely to control major aspects of cell behavior through modification of chromatin structure governing activation or repression of particular genes.

Having said that, results from the laboratory of Robert Eisenmann have suggested that the situation may be more complex than previously appreciated; MYC and MXD, although possessing identical *in vitro* DNA-binding specificities, do not have an identical set of target genes *in vivo*. In particular, apoptosis is one biological outcome in which the transcriptional effects of MYC are not directly antagonized by those of MXD.

The importance of tight control over MYC activity by members of the MXD family is emphasized by the phenotypes of genetically altered mice in which the *Mad1(Mxd1)* gene or the *Mxi1(Mxd2)* gene has been "knocked out." Although quite distinct, the phenotypes displayed by both knockouts include an increase in cellular proliferative capacity. Absence of MXI1 leads to hyperplasia in multiple tissues and tumor growth, whereas absence of MXD1 results in altered differentiation of granulocytes and compensating decrease in cell survival. It is possible that the highly tissue-specific defect observed in *Mxd1*-knockout mice is a result of functional redundancy among the various members of the MXD family. In other words, it is the expression of other members of the family in those tissues that normally would not express them, that compensates for the loss of *Mxd1*. However, the results from *Mxi1*-knockout mice support the contention that MXI1 has the potential to act as a tumor suppressor in humans. Targeted deletion of the S-phase-specific MYC antagonist MAD3 (MXD3) has been shown to sensitize neuronal and lymphoid cells to radiation-induced apoptosis. Effects of disrupting other members of the MXD family await future investigation. Related to the effects of MYC on differentiation, studies have begun to unravel the key role played by MYC in regulation of stem cells. Together with other key genes such as *nanog*, *oct4*, *klf4*, and *sox2*, *MYC* is crucial to induction of pluripotent stem cells and also probably to cancer stem cells (see Chapter 9).

MYC and apoptosis – intrinsic tumor-suppressing activity (oncogenic stress)

Life and death are balanced on the edge of a razor.

Homer

Putative cancer cells must avoid death (apoptosis) in order for tumors to arise; the net expansion of a clone of transformed cells is achieved by both an increase in proliferative index and by a decreased apoptotic rate. In the early 1990s, several laboratories made an intriguing discovery: oncoproteins such as MYC and the adenovirus E1A, both potent inducers of cell proliferation, also possessed apoptotic activity (see section on "Oncogene collaboration"). Ectopic expression of MYC in fibroblasts, cultured in the absence (or limited supply) of survival factors such as IGF-1, resulted in apoptosis with the eventual loss of the entire cell population (see Fig. 6.4). At the time, these findings were interpreted by some of the research community as the result of "a conflict of growth signals," namely that oncogenes activate apoptosis if the proliferative pathway is blocked in some way, akin to an obstructed river breaking its banks. The alternate and ultimately prevailing view was that oncogene-driven cell-cycle entry, somehow "sensitized" the cell to apoptosis. In other words, cell proliferative and apoptotic pathways are coupled. These important findings gave birth to the concept of "oncogenic stress," whereby oncogenic mutations that drive uncontrolled prolifera-

tion (as is the case with MYC) are somehow recognized as aberrant by the cell and result in activation of intrinsic tumor suppressing activity. The central importance of this, and the related oncogene-induced senescence, in cancer biology is unarguable and a detailed discussion of the underlying mechanisms is given in Chapter 8 (Cell death), Chapter 9 (Senescence), and Chapter 10 (DNA damage responses).

By implication, therefore, a cell population with a deregulated oncogene, such as MYC, would be unable to outgrow its environment unless apoptosis and senescence were inhibited, for example, by excess IGF-1 signaling, overexpression of the oncogene *BCL2*, activation of RAS (see later), or loss of tumor suppressors, such as p53/Rb (see Chapter 8). As highlighted in the section on "oncogene collaboration," early *in vivo* experiments supported this notion, but, with the development of more sophisticated mouse models, the case may be regarded as proven. This is graphically illustrated by the dramatic synergy between oncoproteins such as MYC and mechanisms that suppress apoptosis (see "Mouse models of tumor progression," below). The importance of intrinsic tumor suppressive mechanisms accompanying oncogenic activation cannot be overstated. This may be a critical "fail-safe" mechanism serving to protect the organism from an immediate threat of cancer if an oncogene were to become deregulated. Having said that, intriguing recent discoveries highlight the importance of oncoprotein levels within cells, such that MYC can promote different cell behavior depending on whether levels are low or high. Due to its significance in cancer development, we will discuss the issue of oncogene levels later under its own heading "Levels of MYC matter."

Finally, although the mechanisms by which MYC induces or "sensitizes" cells to apoptosis are not completely understood, more and more knowledge is being gained. We will discuss such discoveries in Chapter 8 (Cell death) and, because DNA damage may be a central sensor for oncogenic stress, also in Chapter 10.

Next we will discuss some of the important insights into MYC biology that have resulted from the study of mouse models of cancer.

Mouse models of Myc-induced tumorigenesis

Every sin is the result of a collaboration.

Seneca

Deregulated MYC expression is often associated with aggressive, poorly differentiated tumors. However, given that most human tumors are quite advanced by the time they are seen in the clinic or surgery – often possessing many genetic alterations – it is difficult to ascertain at which stage of tumor progression MYC became activated. This is an important point if we wish to understand what part MYC has to play in the initiation and evolution of a tumor, or indeed whether it would serve as a therapeutic target at later stages of progression. Until recently, it was assumed from the majority of *in vitro* and *in vivo* data, that the predominant role of deregulated MYC in tumor initiation and progression *in vivo* is through uncontrolled cell proliferation concomitant with loss of terminal differentiation (in most cases). Although this may be an important part of the picture, it is now known that there are other attributes afforded to MYC (e.g. angiogenesis – the formation of new blood vasculature in the growing tissue mass) that had remained occult in the past due to the limitations of the study systems previously available: cell culture (*in vitro*) and conventional genetically altered mouse tumor models *in vivo* (discussed below).

Studies employing such conventional mouse models, in which the oncogene is continuously expressed in a given cell type by means of a tissue-specific promoter, have supported the view that deregulated MYC is important for the formation of certain cancers, albeit with a long latency (see section on "Oncogene collaboration"). The prolonged period of time it takes for tumors to develop implies that other mutations have occurred along the way – in genes that collaborate with MYC to transform these cells. Examples of such collaboration with MYC have been demonstrated in mouse models, where antiapoptotic proteins such as BCL-2 or BCL-X_L are overexpressed, or *ARF* or *p53* tumor suppressor genes are lost. In these genetically altered mice, dramatically accelerated tumor development is likely to be due to the inhibition of MYC-induced apoptosis as described in Chapter 8, but as will be seen below, the generation of more sophisticated mouse models has allowed us to conclusively answer this.

The importance of conventional mouse models in cancer biology is undisputed, but it is worth bearing in mind some of the limitations; the oncogene in question is continuously expressed and such expression often starts in tissues of the developing embryo. Since the majority of human cancers arise within the adult (sporadic tumors), continuous expression during embryonic development may not recapitulate the situation of the adult – the signals derived from the tissue environment are likely to differ from the adult counterpart. Moreover, how can we determine precisely what early effects the oncogene has on that tissue? By the time the mouse is born, the oncogene is likely to have been expressed for a considerable period, during which time additional changes (e.g. genetic mutations or epigenetic changes – see also Chapter 11) may have occurred.

In response to these concerns, researchers began to develop regulatable (or conditional) mouse models in which oncogene expression or activation can be switched on or off at will within the target tissue. Although the widespread expression of given oncogenes (in all cells of the target tissue rather than a single cell) does not precisely recapitulate the process likely pertaining to development of sporadic tumors, the ability to activate oncogenes at any time desired by the experimenter in the adult provided a significant advance. Apart from being able to determine the effects of a particular oncogene on cell behavior shortly after it is expressed as well as later on during tumor development in several mouse models, the oncogene activity could be switched off in the tumor. The exciting advantage of being able to turn off the initiating oncogene in tumors that have subsequently developed allowed us to determine whether tumors could regress or not, with unprecedented opportunities to conduct functional "proof of hypothesis" studies on new drug targets (this is discussed in detail in Chapter 16).

Two major approaches have been widely used to generate conditional transgenic mouse models: the tetracycline ("tet") system and the modified estrogen receptor ER^{TAM} (see Chapter 20, Appendix 20.1). The "tet" system requires the drug doxycycline to regulate expression of the gene of interest, while the ER^{TAM} system relies on the administration of tamoxifen to regulate activity of the expressed protein.

One such model employs a switchable form of the MYC protein, called MycER^{TAM} (Fig. 6.9a), which has been used to investigate the effect of "switching on" MYC activation in distinct tissues of the adult such as skin epidermis and pancreatic islet β cells. The differences in cell behavior observed following MYC activation in skin epidermis and pancreatic islets was striking, and serves to highlight the importance of validating data derived from *in vitro* work with studies in the context of an intact organism. It also serves to remind us that what is learned in one cell type does not necessarily apply to all other cell types.

Briefly, activation of MYC promoted rapid cell-cycle entry of pancreatic β cells, which quickly proceeded to undergo apoptosis, leading to severe β-cell depletion and diabetes (Fig. 6.10). In striking contrast, although MYC also promoted proliferation of keratinocytes in suprabasal epidermis, apoptosis was absent which allowed the development of epidermal hyperplasia and papillomatous lesions – precursor of squamous cell carcinoma (see Fig. 6.9b).

In addition, there appeared widespread induction of angiogenesis (growth of new blood vessels) in these early skin lesion, which led to the notion that MYC might act as an "angiogenic switch" (see Chapter 14) during the development of tumors. We now know that MYC can indeed promote new blood vessel growth in multiple tissues, examples of which are touched upon below.

What then of the intrinsic tumor-suppressing activity (i.e. apoptosis) of MYC discussed earlier? The fact that MYC-activated keratinocytes in suprabasal epidermis are already migrating outwards to the skin surface, rather than toward the underlying dermis make them less likely to pose a cancer risk to the host given that they will ultimately be sloughed off from the skin surface. Although it is not clear at present, it is likely that MYC-induced apoptosis in intact skin is suppressed by the presence of excess survival signals, such as those arising from contacts with neighboring cells or secreted factors (see Chapter 12). This indicates a crucial role for tissue context and the surrounding microenvironment in determining cell fate.

The acute sensitivity to induction of apoptosis by MYC in β cells suggests that this cell type is only modestly buffered against cell death by survival signals or intrinsic antiapoptotic mechanisms, such as BCL-2/BCL-X_L expression (see Chapter 8) *in vivo*. In fact, when MYC-induced apoptosis was blocked (by coexpressing BCL-X_L), MYC triggered rapid and uniform carcinogenic progression, generating many of the "hallmark" features of cancer, including immediate and sustained β-cell proliferation, resulting in islet expansion, loss of differentiation, angiogenesis, loss of cell–cell contacts resulting from loss of E-cadherin (see Chapter 12), and local invasion of β cells into surrounding exocrine pancreas (Fig. 6.10). The dramatic and immediate oncogenic progression observed in islets *in vivo* strongly supported the unorthodox notion that complex neoplastic phenomena involving the tumor cell, and its interactions with normal surrounding tissues, may both be induced and maintained by a simple combination of two genetic lesions – "the minimal platform." Although more mutations are likely to be necessary in the equivalent human counterparts, these results suggest that the complexity of tumor phenotype need not always be the result of an equivalent complexity of genetic alteration. This may in fact depend on the tissue type, and the combination of oncogenes.

Recent data now gives us a clear idea of how MYC drives angiogenesis in these islet beta-cell tumors (which may extend to other tissue types): MYC activation rapidly induces expression of the proinflammatory cytokine interleukin 1β (IL-1β), which triggers release of sequestered VEGF from the islet extracellular

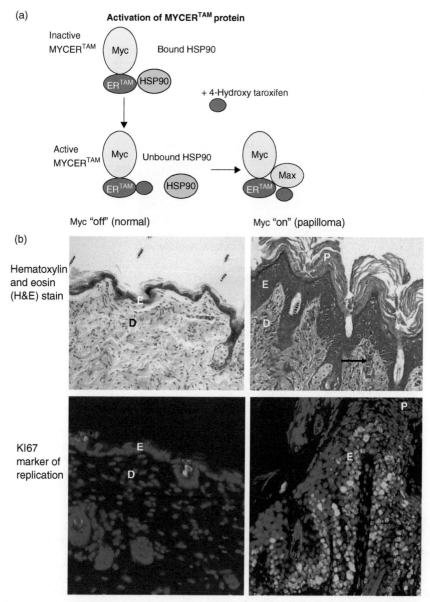

Figure 6.9 (a) A system has been developed allowing the ectopic activation of MYC in various tissues *in vivo*. A transgene encoding a chimeric protein, MycER^{TAM} (MYC fused to the ligand-binding domain of a modified estrogen receptor ER^{TAM} that responds only to the synthetic drug 4-hydroxytamoxifen – 4-OHT) is placed under a tissue-specific promoter directing expression to a predetermined tissue (involucrin (Inv) for suprabasal keratinocytes or insulin (pIns) for pancreatic β-cells). Activation of MYC is achieved by administration of 4-OHT, which binds to the ER^{TAM}, thus displacing heat shock proteins that otherwise hold the protein in an inactive state. (b) Normal adult wildtype skin (or skin in transgenic mice prior to MYC activation) is shown in the left-hand panel. In general the epidermis (E) is only two cells thick (more in humans), comprising a basal layer (containing stem cell precursors) and a suprabasal layer containing differentiating skin keratinocytes. Underlying the epidermis is the dermis (D). The upper panels are hematoxylin and eosin (H&E) stained to show structural features. The lower panels show cell nuclei in blue (DAPI), the nuclei of replicating cells are shown in green (FITC), as they are stained with an antibody to a cell cycle marker – Ki67. The right-hand panels show the effects of activating MYC for two weeks. The epidermis is greatly expanded and now contains large numbers of replicating cells. Not only is the epidermis expanded but there are now large keratotic spires (keratinized dead or dying cells, which remain nucleated – normally cells lose their nuclei at the surface and are then shed) forming papillomas (P). These papillomas become vascularized by angiogenesis (black arrow). Although not shown here, deactivating MYC results in complete reversal of this aberrant phenotype with restoration of normal-appearing skin.

matrix. In turn, VEGF homes to islet blood vessels where it induces endothelial cell proliferation (see Chapter 14) and islet angiogenesis. Further work has also shown that MYC activation induces expression of various chemokines that are involved in the recruitment of mast cells, macrophages, and neutrophils to the tumor. Interestingly, mast cells play an essential part in the growth and maintenance of islet tumor vasculature.

Regulatable "tet" mouse models of lymphomagenesis have also shown the oncogenic potential of c-Myc in T cells. Furthermore, recent studies have specifically highlighted the importance of MYC's interaction with MIZ-1 in promoting lymphoma development (described earlier) – by repressing CKI expression (thus, antagonizing growth suppression) and inhibiting cellular senescence induced by TGF-β.

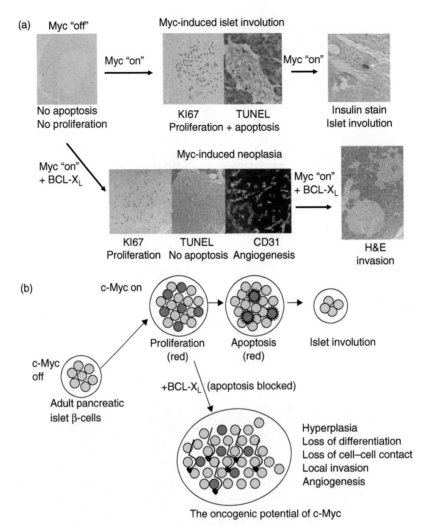

Figure 6.10 The oncogenic potential of MYC is exposed in pancreatic β cells when apoptosis is blocked. (a) Activation of MYC in β cells of the pancreatic islets results in cell cycle entry (Ki67-positive cells stained brown), but then also results in β-cell apoptosis (TUNEL positive cells stained brown). In fact, the net effect is almost complete ablation of the β-cell population within 10 days, resulting in diabetes. In this tissue, unlike in skin, the net effect of deregulated MYC expression is apoptosis, and avoidance of neoplasia. If, however, MYC apoptosis of β cells is prevented by concurrent overexpression of the antiapoptotic protein BCL-X$_L$, then the devastating oncogenic properties of unapposed MYC action are unmasked, culminating in relentless replication, avoidance of apoptosis, loss of differentiation, loss of cell–cell contacts, angiogenesis, and invasion – the "hallmark features" of cancer. (b) This intrinsic "tumor suppressor" function of MYC is illustrated schematically.

Levels of MYC matter – what is the effect on tumor development?

Can you do Division? Divide a loaf by a knife – what's the answer to that?

Lewis Carroll

We know that MYC is deregulated and/or elevated in most human cancers, and in the previous section the importance of MYC in generating tumors was described in some mouse models. However, the argument as to whether levels of MYC protein have to be high in order for tumors to arise or progress has been debated for years. In other words, is it deregulation or overexpression that is required for MYC oncogenic activity?

From cell culture and mouse model experiments over the years, it has been shown that high levels of MYC alone engage tumor suppressor activity within cells, ultimately leading to cell death – a mechanism to suppress untoward proliferation (described earlier). This raises an important problem: since MYC is required to integrate the proliferative programs of all normal cells, how is MYC-induced apoptosis confined only to tumor, and not normal, cells? Is it a matter of MYC levels or deregulation of MYC activity within the cell?

Fortunately, we now have some answers thanks to elegant work using a variant MYCER^{TAM} mouse model, in which MYC-ER^{TAM} is driven by the very weak but constitutively active *Rosa26* promoter. Expression of MYC is deregulated but expressed at low physiological levels. Activation of MYCER^{TAM} (in the presence of 4-OHT) in tissues of such animals, drives proliferation without accompanying apoptosis. But the question remained as to whether only high levels of MYC activity could drive tumorigenesis. The surprising outcome is that even at such low levels, MYC is still possessed of potent tumorigenic activity. The evidence arose from studies using a conditional mouse model expressing the *RAS* oncogene in lung epithelium. When MYC was activated alone, early bronchioalveolar lesions of the lung had developed after 6 weeks. However, low deregulated levels of MYC cooperated with RAS to profoundly accelerate lung tumor progression such that by 6 weeks adenocarcinomas were poorly differentiated, some having already become invasive.

These studies have important implications for tumor development and demonstrate that distinct threshold levels of MYC govern its effect on tumor evolution. High levels of MYC promote apoptosis in order to stall tumor progression. Such tumor suppressor mechanisms need to be blocked for MYC-activated cells to survive and proliferate uncontrollably. We now know that that low-level MYC deregulation can also drive oncogenesis without engaging tumor suppression. This phenomenon is starkly demonstrated by the differences between indirect activation of MYC in intestinal epithelium through activation of the WNT–β-catenin pathway, which engages only MYC's proliferative programs versus direct transgenic activation, and concomitant overexpression of MYC, which triggers activation of p53-mediated tumor suppression. These observations offer an explanation for why MYC is so frequently activated indirectly rather than through direct mutation.

Importantly, we will see in the next section that levels of oncogenic RAS also have a profound effect on cell behavior and tumor development.

The RAS superfamily

No man is an island entire of itself, every man is a piece of the continent.

John Donne

With over 150 known members in humans, the RAS superfamily is a diverse group of monomeric small (20–25 kDa) GTPases, participating in many normal cellular processes, including proliferation, survival, actin cytoskeletal organization, cell polarity and movement, and vesicle and nuclear transport. The RAS superfamily of GTPases is divided into five main families on the basis of sequence identity and function: RAS, RHO, RAB, ARF, and RAN (Box 6.3). There is now overwhelming evidence that aberrant activity of numerous members of the RAS superfamily contributes to cancer growth, invasion and metastasis. Because of their unquestioned importance in cancer, this section will concentrate on two particular families, RAS and RHO, and their downstream effectors.

All RAS superfamily proteins possess a conserved ~20 kDa core G domain (corresponding to RAS residues 4–166) involved in GTP binding and hydrolysis to GDP. Briefly, RAS GTPases are activated by growth factor signals, among others, that induce the active GTP-bound state. Once GTP-bound, RAS can interact with multiple proteins (effectors) to activate a complex network of downstream signaling pathways that mediate distinct cellular processes. Conversely, in their GDP-bound state, RAS GTPases remain inactive. Regulation of the GTP–GDP cycling process (Fig. 6.11) will be described in more detail later, as not only is it crucial for normal RAS GTPase function, but also its oncogenic activity.

RAS beginnings

The RAS GTPases are the founding members of the RAS superfamily and were among the first proteins identified that possessed the ability to regulate cell proliferation. They were discovered as proteins encoded by retroviral oncogenes that had been hijacked from the host genome by the Kirsten (K-) and Harvey (H-) rat sarcoma (ras) viruses (Table 6.3 and Box 6.2). The K-Ras and H-Ras retroviruses were able to produce cancers in mice, and were aptly named after the discovering scientists.

In 1981, the cellular homologs of the viral Harvey- and Kirsten-transforming (to a tumorigenic state) *Ras* sequences were identified in the rat genome and subsequently in the mouse and human genomes. Thus, like the *Src* oncogene of Rous sarcoma virus (see Box 6.1), *RAS* proto-oncogenes residing in the genomes of normal cells can be activated into potent oncogenes by retroviruses which acquire these sequences and convert them into active oncogenes.

It was not long before mutations were identified in *RAS* genes of human cancer cell lines, such as bladder, colon, and lungs. DNA sequence analyses invariably revealed point mutations in the oncogenic alleles, when compared to wildtype, which affected the reading frames of the *RAS* genes. The resulting amino acid replacements most commonly involved residue 12 (glycine to valine; Val12). However, at this time it was also discovered that mutated *RAS* was in fact unable to transform freshly isolated rodent embryo cells into tumor cells *in vitro*. At first glance, this result would seem to challenge the whole emerging concept of oncogenes (the astute reader will of course already be shouting "What about oncogene collaboration!?"). These concerns were laid to rest when reports published in 1983 described the ability of H-RAS Val12 to transform primary cells that had previously been immortalized (proliferate indefinitely) by either carcinogens or by introducing another oncogene such as *MYC* or *E1A* (see Box 6.1). In other words, mutant RAS proteins can only transform cells that have already undergone some changes – usually those that enable uncontrolled proliferation.

These findings underscored the concept of "multistep tumorigenesis" described in Chapter 3, in which cells have to acquire several oncogenic mutations in order to convert a normal cell into a cancerous one.

Importantly, the concern that *RAS*-activating point mutations could have arisen as a consequence of culturing cancer cells *in vitro* could be dismissed. By 1984, mutant *K-RAS* alleles were

---

Box 6.3 RAS superfamily

The RAS superfamily has more than 150 members. They are small (20–25 kDa) GTPases and can be subdivided into five main families on the basis of sequence identity and function: RAS, RHO, RAB, ARF, and RAN.

The RAS family (36 members) is involved in signal transduction, the regulation of gene expression, cell proliferation, survival, and differentiation. The first and most famous members discovered were K-RAS, H-RAS, and subsequently N-RAS. RHO GTPases (20 members) are involved in signal transduction, the regulation of actin organization, cell shape and polarity, movement, and cell–cell interactions. RAB GTPases (61 members) are involved in vesicular trafficking, regulating endocytosis, and secretory pathways. ARF GTPases (27 members) are involved in the same processes as RAB GTPases as well as microtubule dynamics. Finally, the RAN GTPase (only 1 member) is involved in nuclear–cytoplasmic transport and mitotic spindle organization.

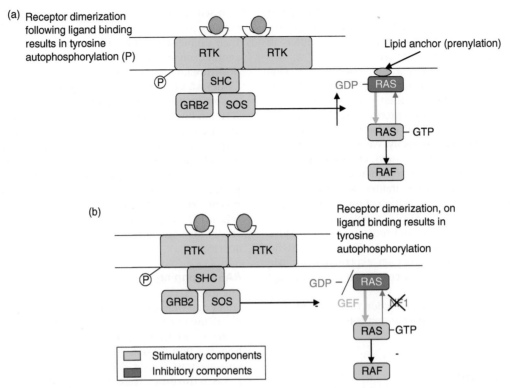

Figure 6.11 RAS activation depends on regulation of GTP/GDP exchange. (a) In a healthy cell, RAS proteins regulate diverse cellular processes by cycling between biologically active GTP- and inactive GDP-bound conformations. In the active state, RAS proteins are bound to GTP and are thus able to engage downstream effectors that activate signaling pathways controlling several aspects of cell behavior mentioned above. When bound to GDP, RAS proteins are inactive and so fail to interact with these effectors. The conversion of bound GTP to GDP, and vice versa, is catalyzed within the cell: the nucleotide exchange by guanine nucleotide exchange factors (GEFs), such as SOS, and the nucleotide hydrolysis by GTPase-activating proteins (GAPs), such as NF-1. It is the balance between these proteins that will ultimately determine the activation state of RAS and its downstream pathways. (b) Activation of RAS in tumors: mutations in RAS can prevent the intrinsic GTPase activity and impede exchange of GTP for GDP, thus favoring the active conformation. A similar outcome can result from loss of the NF-1 tumor suppressor protein, which normally amplifies the intrinsic GTPase activity of RAS. Although, growth factor stimulation will still play some role in phenotype these alterations can result in constitutive RAS activation.

detected in human lung carcinoma samples and, moreover, there appeared associations between particular *RAS* oncogenes and tumor type (e.g. *K-RAS* mutations in pancreatic and colonic carcinomas, *HRAS* mutations in bladder carcinomas, and *NRAS* in lymphoid cancer and melanomas) (Table 6.5).

In addition, carcinogen-induced cancers in mice were shown to have acquired mutations in *Ras* genes (e.g. *HRas* mutations were found in mammary and skin tumors, whereas *N-Ras* mutations were found in thymomas).

Why were mutations in *Ras* genes usually localized to a small number of sites in the reading frames of these genes? The answer would be a key step in understanding the potent transforming ability of this oncogene, and are indeed discussed shortly.

It is now known that human tumors very frequently express activated HRAS proteins (Table 6.5). In fact, an activating mutation in one of the *RAS* genes (particularly the aforementioned single-base substitution in the 12th codon that causes the replacement of a glycine with a valine) is found in around 30% of all tumors. In other cases, *RAS* gene amplification is seen (Table 6.2). Importantly, many tumors that lack *RAS* mutations still activate the same signaling pathways, or at least some parts of the networks regulated by RAS, through other means, for example EGFR overexpression, activating mutations in *BRAF*, amplification of *AKT* or PI3K, some of which are discussed later. Not surprisingly, therefore, considerable efforts have been devoted to try and develop therapies targeting the RAS proteins or the signaling pathways that they control. Numerous therapeutic agents are in clinical trials at present and more are under development, some of which are discussed later.

RAS and GTPase activity

As mentioned earlier, the prototypical RAS proteins, H-RAS, K-RAS, and N-RAS (N-RAS was the third member to be identified from neuroblastoma and leukemia cell lines) were originally identified in tumors as the hyperactive products of mutant *RAS* genes, which promote cancer by disrupting the normal controls on cell proliferation and differentiation. These proteins (in their nonmutated form) were subsequently found to contribute to cell-cycle regulation in normal, nontransformed cells. H-RAS, N-RAS, and K-RAS share 85% identity, with divergence confined largely to the C-terminus. Broadly, all three proteins have similar activities on cell-cycle progression and replication, though subtle differences may be observed in cellular distribution and aspects of downstream signal transduction pathways activated. Out of the three, however, K-RAS is the only one that is essential for embryonic development, as deduced from knockout studies. This is most likely due to K-RAS being expressed more ubiquitously compared to H-RAS and N-RAS.

In a healthy cell, typical members of the RAS superfamily function as regulators of diverse cellular processes by cycling between biologically active GTP- and inactive GDP-bound conformations. In the active state, RAS proteins are bound to GTP and are thus able to engage downstream effectors, such as the serine/threonine kinase RAF, to activate signaling pathways controlling several aspects of cell behavior mentioned above. However, when bound to GDP, RAS proteins are inactive and so fail to interact with these effectors (Fig. 6.11). Thus, in normal cells, the activity of RAS proteins is controlled by the ratio of bound GTP to GDP. The hydrolysis of bound GTP leads to deactivation of RAS, as it does also for the Gα subunit of receptor-coupled G-proteins (see Chapter 5), however, the average lifetime of RAS–GTP is far longer than that of Gα–GTP; the latter is rapidly inactivated by intrinsic GTPase activity, whereas for RAS other proteins are required to speed the otherwise slow rate of cycling between the GTP- and GDP-bound forms. The conversion of RAS-bound GTP to GDP, and vice versa, are catalyzed by interactions with other proteins within the cell, which are therefore key regulators of RAS signaling: activators that catalyze nucleotide exchange – guanine nucleotide exchange factors (GEFs) – and inactivators that facilitate nucleotide hydrolysis – GTPase-activating proteins (GAPs). It is the balance between these proteins that will ultimately determine the activation state of RAS and its downstream pathways. Perhaps not surprisingly, there is increasing interest in GEFs and GAPs as therapeutic targets for future cancer treatments.

In tumor cells, oncogenic *RAS* mutations result in loss or diminution in GTPase activity, with the result that RAS proteins remain constitutively in the active GTP-bound form (Fig. 6.11). In many human tumors, mutations may occur in genes encoding proteins involved in regulating RAS activity rather than in RAS itself. One such example is loss of the tumor suppressor gene *NF1*, which encodes a GTPase-activating protein called neurofibromin, resulting in upregulated RAS signaling. Mutations in *NF1* cause neurofibromatosis type 1, a condition immortalized, though maybe incorrectly, by the story of the elephant man. Neurofibromatosis is characterized by the development on peripheral nerves of neurofibromas, which are complex, primarily benign, tumors composed of Schwann cells and other cell types. Defective NF-1 also contributes to childhood CML.

Activation of RAS by receptor tyrosine kinases

The importance of RAS is underscored by its position downstream of most known RTKs. In contrast to G-protein coupled receptors, in which G-proteins are directly activated by ligand binding to a receptor, RAS activation requires the recruitment of adaptor proteins which link RAS to the activated growth factor-bound RTK (described in Chapter 5 and Fig. 5.15). SOS and GRB2 are amongst the best studied adaptor proteins; the SH2 domain of GRB2 binds to activated (tyrosine phosphorylated) RTKs. The SH3 domains in GRB2 then bind to SOS (a GEF), bringing this protein into proximity with RAS. As mentioned earlier, GEFs (guanine nucleotide exchange factors) catalyze the exchange of GDP for GTP, thereby activating RAS. It is believed that this proximity-based process is facilitated by lipid modification of RAS, which helps localize RAS to the cell membrane (see next section).

GTP-bound RAS subsequently enables high-affinity interactions with a number of downstream target proteins that are called "effectors." Such interactions initiate several signal-transduction cascades that regulate various cellular functions, such as proliferation (via the RAF–MEK–ERK signaling pathway), survival (via PI3K–AKT), and motility (via RHO).

Posttranslational modification of RAS

The normal function of RAS proteins requires them to be posttranslationally modified by the covalent attachment of a lipid isoprenyl group to their hypervariable (HV) C-termini by farnesyltransferase (FTase) or geranylgeranyltransferase type I (GGTase-I) (Fig. 6.12). Of the RAS superfamily, RAS and RHO family GTPases have C-terminal HV sequences that commonly terminate with a CAAX motif. The CAAX motif is a C-terminal tetrapeptide sequence comprising a cysteine, followed by two aliphatic amino acids and a terminal X residue that dictates specificity for FTase- or GGTase-I-catalyzed addition of a farnesyl or geranylgeranyl isoprenoid lipid to the cysteine residue. Interestingly, such lipid groups derive from the same intracellular sterol biosynthesis pathway that cells use to produce cholesterol. Such isoprenylation of RAS proteins appears to be an essential prerequisite for RAS function and for the transforming activity of oncogenic mutants, as it enables direction of RAS proteins to the appropriate subcellular compartment, principally the inner face of the plasma membrane, and is necessary for their dynamic sorting and trafficking. Aberrant localization of RAS impedes activity, most likely by reducing the chance of interaction with its usual targets.

Studies have shown that farnesylation of RAS is the first obligatory step in a series of posttranslational modifications leading to membrane association, which, in turn, determines the switch from an inactive to an active RAS–GTP-bound form. The enzyme responsible for this farnesylation of RAS is FTase. Inhibiting this reaction might be able to reverse the transformed phenotype, and provided the rationale for the testing of apical inhibitors of sterol biosynthesis (exemplified by the HMG-CoA reductase-inhibiting statin drugs) and the development of farnesyltransferase inhibitors (FTI) as anticancer drugs. The initial promise for the use of FTIs in cancer treatment has unfortunately not been realized and

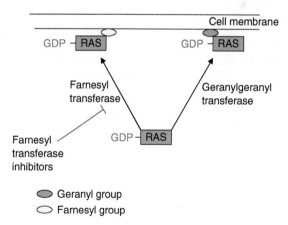

Figure 6.12 Lipid modification of RAS. RAS activation (exchange of bound GDP for GTP) requires localization of RAS–GDP to the inner cell membrane, presumably because in this location it is more readily available for interaction with growth factor receptor-associated proteins. Usually this is achieved by addition of enzymatic prenyl groups, primarily farnesyls, but if farnesyl transferase is inhibited at least for Ki-RAS then these can be replaced effectively by geranylgeranyl groups, which appear to serve the same or similar purpose.

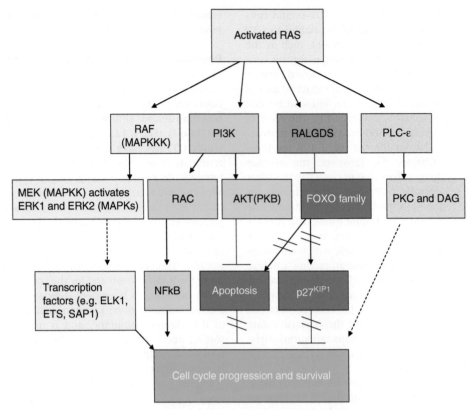

Figure 6.13 Signaling pathways regulated by RAS. Active GTP-bound RAS will interact with several families of effector proteins, with the most important shown. RAF protein kinases initiate the MAPK cascade, which leads to ERK activation. ERK has numerous substrates, including ETS family transcription factors such as ELK1 that regulate cell-cycle progression. Phosphoinositide-3 kinases (PI3Ks) generate lipid messengers such as phosphatidylinositol-3,4,5-trisphosphate (PIP3), which activate the kinase AKT/PKB, involved in survival. PI3K also stimulates RAC, a RHO family protein involved in regulating the actin cytoskeleton. RALGDS proteins are guanine nucleotide exchange factors (GEFs) for RAL, a RAS-related protein. Downregulation of RALGDS targets (e.g. Forkhead transcription factors; FOXO family) leads to p27^{KIP1} downregulation. Phospholipase C (PLC) catalyzes the hydrolysis of phosphatidylinositol-4,5-bisphosphate to diacylglycerol (DAG) and inositol trisphosphate, resulting in protein kinase C (PKC) activation and calcium mobilization from intracellular stores. Strikethroughs indicate blocked pathways.

will be discussed later in the section on RAS-directed anticancer treatments.

The view of RAS signaling being confined to the inner surface of the plasma membrane is increasingly being challenged by observations indicating that RAS proteins interact dynamically not just with specific microdomains of the plasma membrane but also with other internal cell membranes and intracellular organelles. Importantly, the location of RAS may be a crucial determinate of downstream signaling. For example, plasma membrane-tethered K-RAS can induce cellular transformation, whereas mitochondrial K-RAS induces apoptosis. Secondly, although activated H-RAS is associated with both the Golgi apparatus and the endoplasmic reticulum (ER), only the ER-associated form can activate the downstream signaling pathway RAF1–ERK, and consequently induce transformation. These observations strongly suggest that subcellular localization of RAS proteins is important in determining subsequent activation of their downstream effectors.

One important point to make here though, is that the overabundant expression of mutant RAS in such experimental systems compared to that of their endogenous counterparts might account for the seeming promiscuity of RAS subcellular localization in these studies. For instance, it has been reported that growth factor-induced activation of RAS occurs at the plasma membrane and not at the ER or Golgi.

Signalling downstream of RAS

As has been mentioned, RAS is activated by a large variety of extracellular stimuli, largely through activation of RTKs, for example following binding of mitogens EGF or PDGF (see Chapter 5). Once in the active GTP-bound form, RAS can activate a number of effector proteins, each representing distinct signaling pathways (Fig. 6.13). These effector proteins are members of the RAF family (RAF1, A-RAF, and B-RAF), phosphatidylinositol-3 kinase (PI3K), RAS-like guanine nucleotide dissociation stimulator (RALGDS), and RALGDS-like gene (RGL1 and RGL2). More recently, other RAS effectors have been identified with diverse cellular functions. These include phospholipase Cε (PLCε), which activates protein kinase C (PKC) and mobilizes calcium from intracellular stores, and T-cell lymphoma invasion and metastasis 1 (TIAM1), shown in some mouse tumor models to be required for RAS transformation. Other less well-characterized RAS effectors include AF-6, RIN1, and RASSF.

The archetypal signaling pathway downstream from RAS is the canonical serine/threonine kinase RAF–MEK–ERK pathway (also known as the mitogen-activated protein kinase (MAPK) pathway) that regulates cell-cycle progression (described in Chapter 5, Fig. 5.17).

Following close on its heels is the important PI3K–AKT pathway (see Chapter 5, Figs 5.18 and 5.19) that regulates cell survival and also stimulates RAC, a RHO family protein that is involved in regulating the actin cytoskeleton (described later).

The third well-studied RAS effector family includes the RAL proteins, involved in vesicle transport and cell-cycle progression which are also described later

We will next discuss some of the RAS signaling pathways in the context of normal cell behavior and in tumor cells in which RAS signaling is constitutively active (oncogenic).

RAS cell growth and proliferation

The ability of RAS family proteins to be rapidly switched "on" and "off" (a property incidentally shared by all G-proteins) makes them ideally suited for their involvement in a wide range of key cellular behaviors. The prototypical RAS GTPases activate several signaling pathways that control the cell cycle (see Fig. 6.13). In studies, blocking RAS activity prevents growth factor-induced G_1/S transition, whereas conversely, quiescent cells can be induced to replicate by introducing RAS protein. The RAS proteins help relay signals from RTK (see Chapter 5) to the nucleus by regulating a number of distinct but overlapping signaling pathways. By this means RAS can influence multiple cell processes, including proliferation and growth but also others such as survival, differentiation, senescence, or even cell death (apoptosis), which will be discussed later. Given the complexity of RAS signaling in regulation of cell proliferation, we will first introduce the key players involved in this cell process and conclude with the most intensively studied RAF–MAPK and PI3K–AKT pathways.

A major role of RAS is to transduce growth factor signals required for G_1/S transition in the cell cycle. RAS activates numerous signaling pathways downstream of activated RTK growth factor signals, including the RAF–MAPK and PI3K–AKT pathways (see Fig. 6.13 and also Chapter 5, Figs 5.16–5.19). Together, these signals enhance cyclin D1 expression and help inactivate the retinoblastoma (RB) protein. Such actions are mediated both by transcriptional effects, including activation of transcription factors such as MYC (cyclin D2 is a direct MYC target gene), and by altering protein stability. As described in Chapter 4, several cell-cycle regulatory proteins are degraded by the ubiquitin–proteasome pathway (see Chapter 11). In the case of cyclin D1, phosphorylation by a key enzyme, glycogen synthase kinase-3β (GSK-3β), allows ubiquitination by SCF (Skp1/cullin/F-box protein) (see Chapter 4) targeting cyclin D1 for degradation. Conversely, RAS promotes cyclin D1 stability by inhibiting GSK-3β activity. RAS activates the PI3K–AKT pathway, which inhibits GSK-3β.

RAS pathways may also promote cell division by reducing levels of the CKI p27^{KIP1} (Chapters 4 and 7); RAS inhibits synthesis of p27^{KIP1} via downregulation of forkhead family transcription factors (FOXO family), and increases p27^{KIP1} degradation. This places p27^{KIP1} at a crucial integration point of positive and negative growth factor signaling for G_1/S transition. Put simply, a cell can decide if it will commit to replication depending on the balance of positive and negative growth signals (see Chapter 5). Deregulated or oncogenic RAS is capable of both enhancing the positive signal and inhibiting some of the negative signal, thus strongly favoring cell replication.

RAS may also contribute to cell replication by enhancing cellular growth (size) by upregulating the translational machinery and protein synthesis. One of the master regulators of protein synthesis and translation control is the mammalian target of rapamycin (mTOR) protein, which controls the translational apparatus through protein phosphorylation. The mTOR–S6K pathway is regulated by signals that are transmitted by PI3K in response to mitogen stimulation and nutrient supply (see Chapter 5). Briefly, PI3K functions through AKT–PKB-mediated phosphorylation and inhibition of a suppressor complex composed of tuberous sclerosis complex 1 (TSC1) (also known as hamartin) and TSC2 (also known as tuberin) that are negative regulators of the mTOR pathway. Inactivating TSC gene mutations results in a predisposition to at least two cancer-related diseases: tuberous sclerosis and lymphangioleiomyomatosis.

It seems increasingly likely that other members of the RAS and RHO families of GTPases also influence cell-cycle progression and growth. In fact, irrespective of the effects these proteins may have on cell motility etc., they share the property of increasing cellular replication and likely do so by activating similar or overlapping pathways.

RAS signaling through the RAF–MAP kinase pathway

The serine-threonine kinase RAF–MEK–ERK pathway is a key regulator of cell-cycle progression (Fig. 6.14 and Chapter 5, Figs 5.15–5.19). This is one of a series of kinase cascades with the generic description of MAPKKK–MAPKK–MAPK. The protein serine/threonine kinase RAF1 was the first bona fide RAS effector to be identified in 1993. RAF, by the above designation a mitogen-activated protein kinase kinase kinase – MAPKKK, is bound by GTP–RAS, facilitating relocation to the plasma membrane, required for activation. Three closely related RAF proteins (C-RAF, B-RAF, and A-RAF) all appear to share this means of activation. RAS binding also induces the transition of RAF from a closed to an open conformation, which enables RAF to bind and then phosphorylate and activate MEK1 and MEK2 (mitogen-activated protein kinase kinases – MAPKK) – dual-specificity kinases that are capable of phosphorylating and activating ERK1 and ERK2 (the mitogen-activated protein kinases – MAPK).

Key substrates for ERK1/2 (extracellular signal-regulated kinases 1 and 2) include a variety of cytosolic but also nuclear proteins, reflecting the fact that ERKs can be transported into the nucleus following activation. ERK phosphorylates ETS family transcription factors such as ELK1, part of the serum response factor (SRF) that regulates the expression of FOS. In addition, ERK phosphorylates c-JUN. This is a critically important and apical step in growth factor signaling as it culminates in activation of the all-powerful AP-1 transcription factor, which is made up of FOS–JUN heterodimers. AP-1 is a key convergence point of growth factor signaling in the nucleus, and regulates expression of multiple essential genes downstream of growth factor activity, including MYC. Together, these transcription factors influence expression of several key cell-cycle regulatory proteins, including increases in cyclin D and inhibition of INK4a CKIs required for the cell to transit through G_1/S in the cell cycle. The ERK group of MAPKs is thus essential for normal mitogen-regulated cell proliferation as they are responsible for the expression of key growth-regulating transcription factors – ETS, FOS–JUN, and thence MYC.

In addition to the well-described role of the RAF–MAPK pathway in transduction of the growth-promoting or mitogenic activity of numerous growth factors, it has recently also been shown to mediate antiapoptotic survival signals, distinct from those through another RAS effector – the PI3K–AKT pathway described later. In addition, this pathway may also couple cell replication/growth with induction of angiogenesis, though this may require activity of MYC. Some studies also suggest that ERK may promote transcriptional upregulation of angiogenic factors,

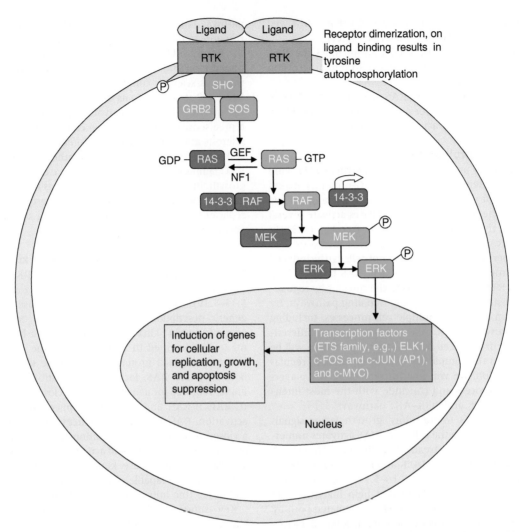

Figure 6.14 The RAS–RAF signaling pathway. Receptor dimerization on ligand binding induces tyrosine autophosphorylation of the RTK, binding of GRB2 to the receptor via the SH2 domain and translocation of the guanine nucleotide exchange factor SOS to the cell membrane. SOS in turn promotes RAS activation by enhancing GDP–GTP exchange. A third adaptor protein, SHC (also carrying an SH2 domain), may be phosphorylated by growth factor binding and can recruit the GRB2–SOS complex. Activated RAS by either of these two overlapping pathways in turn phosphorylates the serine/threonine kinase RAF, in part by a conformational effect and replacement of 14-3-3 protein. RAF phosphorylates and activates MAP kinase/ERK kinase (MEK). MEK then activates ERK, which can then translocate to the nucleus and activate transcription factors such as ETS family, c-FOS and c-JUN (which together form the AP1 transcription factor), MYC, and others.

and promote increased invasiveness as a result of ERK-mediated expression of matrix metalloproteinases (see Chapter 12) and RAC-mediated effects on the cytoskeleton.

So it would appear that aberrant signaling through the RAS–MAPK pathway has the potential to promote several of the "hallmark" features characteristic of malignant transformation. However, the actual cellular responses operating *in vivo* are likely to depend upon various factors, such as cell type, cell location, and contact with other cell types. It is crucially important not to forget that a single oncogene by itself is not sufficient for cellular transformation (as discussed earlier in "Oncogene collaboration") as additional oncogenic mutations are necessary – more needed in humans compared to mice. Later, we will indicate possible explanations as to why oncogenic *RAS* alone is unable to transform cells – due to its "intrinsic tumor suppressive" activities (analogous to those described for MYC). Importantly, collaboration between oncogenic *RAS* and *MYC* may overcome the tumor suppressor activity of each, thereby promoting tumorigenesis (see Fig. 6.15 and also Chapter 5, Fig. 5.7). Finally, aberrant activation of the RAS–MAPK pathway occurs in many human cancers through mutations that activate RAS (as discussed earlier), or other proteins (RAF, PI3K, PTEN, AKT), which serve to regulate its downstream activity. For this reason, inhibitors of RAS, RAF, MEK, and some downstream targets have been developed and many are currently in clinical trials.

RAS signaling through PI3K–AKT pathway

In addition to the RAF–MAPK pathway described above, RAS has also been found to activate several other effector pathways, the best characterized of which are shown in Fig. 6.13. Although only briefly described here, later sections will relate some of these pathways to cell behavior in more detail.

RAS activation of the PI3K–AKT pathway is of key importance in normal cell growth and in cancer (Chapter 5, Figs 5.16, 5.18, and 5.19) Tumor cells have found ways to avoid cell death, and this hallmark feature of cancer is often linked to hyperactive signaling through PI3K–AKT. RAS can interact directly with the catalytic subunit of type I PI3K, leading to activation of the lipid kinase (Fig. 5.16). In turn, activated PI3K phosphorylates PIP2 (phosphatidylinositol-4,5-bisphosphate or PtdIns(4,5)P_2) to produce PIP3 (phosphatidylinositol-3,4,5-trisphosphate or PtdIns(3,4,5)P_3). PIP3 is a key second messenger that binds to a large number of proteins through the pleckstrin homology and other domains. In this way, PI3K controls the activity of a large number of downstream enzymes, including 3-phosphoinositide-dependent protein kinase 1 (PDK1) and AKT (otherwise known as PKB – protein kinase B). PDK1 is important for the activation of a large number of protein kinases, and AKT is an important mediator of survival signals generated by RAS that prevent cells from dying by apoptosis (see Chapter 8). Moreover, AKT is a key regulator of protein stability of multiple important cell cycle regulatory proteins, because AKT can inhibit the activity of GSK-3β, preventing ubiquitination of proteins (such as MYC, c-JUN, and cyclins D and E) by E3 ligases, required for proteasomal degradation.

PI3K also activates RAC, a RHO family protein not only involved in regulating the actin cytoskeleton important in cell motility (see Chapter 12), but also in activating transcription factors such as nuclear factor-κB (NF-κB) involved in cell survival and inflammation. Regulatory subunits of NF-κB are also inhibited by GSK-3β so stability of this transcription factor will likely also be enhanced by PI3K–AKT signaling. RAS signalling in cell survival and death is highlighted in Fig. 6.16.

RAS, PLCγ, and PKC

In addition to activation of RAS, RTK signaling can also activate the enzyme phospholipase Cγ (PLCγ; Fig. 6.13) which also plays a key role in cellular growth and proliferation. Upon growth factor binding, RTKs promote phosphorylation of PLCγ at three known tyrosine residues: Tyr771, Tyr783, and Tyr1254, and its enzymatic activity is upregulated. In turn, activated PLCγ can then trigger downstream signaling (see Chapter 5) by hydrolysis of PIP2 to diacylglycerol (DAG) and inositol-1,4,5-trisphosphate (IP3). Both DAG and IP3 are important second messengers in growth signaling as they may activate protein kinase C (PKC) and calcium-dependent signaling pathways. Intriguingly, another member of the PLC family, PLCε, has now been shown to be a RAS effector, linking RAS with activation of PKC (see Chapter 5 and Fig. 5.19).

The PKC family of serine/threonine kinases regulate multiple cell functions, including growth, proliferation, differentiation, cytoskeletal organization, motility, and apoptosis. PKC signaling activates the ERK1 and 2 MAP kinases, triggering cell replication. PKC has also been implicated in mediating survival signals from various growth factors such as IGFs, and inhibition of PKC can trigger apoptosis in many cells. PKCε may activate the AKT and mTOR signaling pathways also, showing the considerable overlap between signaling downstream of PLC and RAS. Aberrant PKC, for instance by drugs such as phorbol esters, has long been known to promote cancer in skin and other tissues, at least in part by driving aberrant cell cycling. Moreover, mutant PKC proteins have been found in several different human cancers. RAS and RAS-like proteins (RAL GTPases):

The RAS-like (RAL) GTPases – RAL-A and RAL-B – were discovered over 20 years ago and for a long time their functional relevance remained unknown. Their importance came to light after recognition that a class of guanyl nucleotide exchange factors (GEFs; described earlier for RAS activation – the conversion of RAS-bound GTP to GDP) with specificity for RAL proteins were, in fact, direct effectors of oncogenic RAS. These RAS effectors include three exchange factors for the RAS-related RAL proteins: RAL guanine nucleotide dissociation stimulator (RALGDS), RALGDS-like gene (RGL1), and RGL2, through which RAS activates RAL-A and RAL-B. In turn, activation of downstream RAL effectors occurs, such as ZONAB (ZO-1-associated nucleic acid-binding protein) and RLIP (also known as RAL-A binding protein 1), respectively.

[It is important to note here that there are a variety of upstream activators of RAL proteins, which are not signaled via RAS.]

Activation of RAL proteins in normal cells engages multiple effector proteins that direct various biological processes within the cell, such as trafficking of secretory vesicles to the plasma membrane, regulation of gene expression, and protein translation, whereas chronic sustained activation of RAL-A and RAL-B as noted in human tumor cells promotes cell proliferation and survival, characteristics of oncoproteins. Thus, the depletion of RAL-A in a large panel of human pancreatic cancer cell lines (by shRNA technology) significantly diminished the potential of these cells to form tumors following subcutaneous inoculation of nude mice. Importantly, it has also been shown that loss of RAL-A activity prevents anchorage-independent proliferation of cancer cells *in vitro*. Thus, chronic activation of RAL-A appears to provide cancer cells with a hallmark feature of tumorigenesis, allowing cells to proliferate following detachment from their normal tissue environment – a prerequisite for development of metastatic lesions.

Although the depletion of RAL-B in pancreatic cancer cells had no impact on preventing generation of primary tumors in mice, it prevented the formation of lymph node metastases. RAL-B appears to be essential for the survival of a variety of tumor cell lines, probably through downstream activation of its effector protein TANK-binding kinase 1 (TBK1).

In a genetically modified mouse model, in which the gene for *RALGDS* is deleted (and consequently activation of RAL is inhibited), offspring are viable and development proceeds normally. However, using a model of topical carcinogen-induced skin papillomas, the absence of RALGDS resulted in delayed onset and decreased incidence of papillomas. Histological examination showed that apoptosis rates were higher in lesions from *RALGDS*-null mice while proliferation rates were not affected, suggesting that chronic activation of RALGDS and thus RAL promotes tumor cell survival.

The RALGDS pathway – together with AKT – inhibits downstream transcription factors including the forkhead transcription factors JNK and AFX. Indeed one of the main functions of AFX is to keep cells in G_1 by increasing the levels of the CKI p27^{KIP1} involved in repressing S-phase entry. However, it is also involved in the induction of the proapoptotic proteins BIM and FasL (see Chapter 8). Thus, by inhibiting both growth arrest and cell death, the RALGDS pathway contributes to the induction of cell proliferation and together with AKT blocks apoptosis.

RAS: cellular senescence versus transformation

Old age isn't so bad when you consider the alternative.

Maurice Chevalier

As discussed earlier for the *MYC* oncogene, cell death by apoptosis provides an important intrinsic mechanism to suppress tumor growth – at least in cells expressing high levels of MYC. However, it is not the only way that oncogenes can stop potentially harmful cells replicating. Some oncogenes, exemplified by RAS, can trigger cellular senescence – a state characterized by permanent cell-cycle arrest and specific changes in morphology and gene expression that distinguish the process from quiescence (see Chapter 9). Cells take on a flattened morphology and the nucleus becomes enlarged with a prominent nucleolus. Consistent with its role in tumor suppression, cellular senescence is regulated by a number of tumor suppressor genes, the most crucial of these encoding the p53 or $p21^{CIP1}$, pRB or $p16^{INK4a}$, $P19^{ARF}$, BMI, and PML proteins. However, unlike oncogene-induced apoptosis, the relevance of oncogene-induced senescence as a tumor suppressor mechanism was less clear because it had not been observed definitively *in vivo*. Moreover, even *in vitro*, expression of oncogenic *Ras* in primary cells does not always trigger senescence – especially when expressed from its endogenous locus, an important point that we will come back to later. Despite these concerns, studies in certain mouse models provided evidence that senescence might play a part in suppressing tumor growth. For example, chemically induced skin cancers in mice show that the initiating oncogenic mutations occur in the endogenous *HRas* gene of keratinocytes. However, these benign hyperplastic lesions only progress to malignant tumors when secondary mutations, in the *p53*, $p16^{Ink4a}$, $p19^{Arf}$, or $p21^{Cip1}$ genes (see Chapter 7) occur – precisely those genes that mediate RAS-induced growth arrest in cultured keratinocytes.

How does RAS trigger cellular senescence? *In vitro* studies using human and rodent cells have shown that oncogenic RAS signals via the MAPK pathway to induce $p16^{Ink4a}$ and/or ARF (depending on cell type), which ultimately activate RB and p53, respectively (Fig. 6.15). For example, mouse embryo fibroblasts depend primarily on ARF–p53, whereas human fibroblasts also rely on $p16^{INK4a}$–RB functions. In turn, p53 and RB promote senescence by regulating a number of effectors, including $p21^{CIP1}$, PML, and various chromatin-modifying factors.

PML is a tumor suppressor first identified in a mouse model for acute promyelocytic leukemia (see Chapter 11). It regulates responses of p53 to oncogenic signals from RAS. Expression of RAS causes p53 to accumulate and PML expression to increase; PML overexpression acetylates p53 at lysine 382, and this makes p53 biologically active. The outcome is senescence. RAS stimulation causes p53 and the acetyltransferase CBP (CREB-binding protein) to form a trimeric p53–PML–CBP complex within the nuclear bodies, a site where PML occurs even in normal cells. Knockout experiments have now shown that *PML*-null fibroblasts lose RAS-induced p53 acetylation, p53–CBP complex stability, and senescence. These data establish a strong link between PML and p53 and moreover emphasize the central role of PML in mediating the effects of RAS in this context.

Recent studies now strongly support the notion that diverse potentially cancer-promoting genes can trigger senescence *in vivo*, including oncogenic *RAS* and *BRAF*. The mechanism as may have been predicted by the earlier studies involves induction of $p16^{INK4a}$ via the ERK pathway and activation of AKT or inactivation of PTEN, which activate the ARF–p53–$p21^{CIP1}$ pathway.

If, as now seems likely, replicative senescence operates *in vivo*, then avoidance or even escape from oncogene-induced senescence would be crucial for the transformation of cells and tumor development. Perhaps this explains the collaboration between RAS and other oncogenic lesions, such as MYC, observed in transformation assays *in vitro* (see "Oncogene collaboration"). In this scenario, MYC may block RAS-induced senescence by inhibiting $p21^{CIP1}$. On the other hand, MYC-induced apoptosis may be suppressed by RAS acting via the survival protein AKT (see Fig. 6.15). This simplified view may also explain why antiapoptotic proteins, such as BCL-2 and BCL-X_L, collaborate more effectively with MYC than with RAS, since they serve to block MYC-induced apoptosis, but probably play relatively little part in activating senescence pathways (at least in the presence of activated MYC). Other examples of collaboration include mouse embryo fibroblasts and skin keratinocytes, in which loss of *p19^Arf* or *p53* prevents RAS from inducing cell-cycle arrest and promotes transformation. In human cells, at least *in vitro*, the situation is often more complex, requiring additional mutations to overcome RAS-induced senescence, for example, loss of $p16^{INK4a}$ (see Chapter 7). Indeed, it is the various combinations of oncogenes and loss of tumor suppressor genes that dictate whether cancer cells are more responsive or resistant to cancer therapy (see Figs 8.17 and 8.18 in Chapter 8). Thus both senescence and apoptosis represent inherent tumor suppressor mechanisms invoked by oncogene deregulation that may restrain tumor progression by inhibiting propagation of the mutated cell.

Recent data have now highlighted the importance of gene dosage, that is, the levels of RAS expressed within a cell may dictate how the cell behaves – senesce or proliferate/transform. *In vitro* experiments in which high levels of RAS were expressed in mouse fibroblasts led to senescence concomitant with activation of RAS effector pathways and stabilization of p53. In contrast, physiological expression of mutant K-*RAS* from its endogenous promoter resulted in uncontrolled cell replication (immortalization) and partial transformation of mouse fibroblasts. Moreover, physiological expression of mutant K-*RAS in vivo* led to tumorigenesis in mice with no obvious signs of senescence.

Further work in support of this concept came from the use of transgenic mice in which a comparison of low and high expression of mutant H-*RAS* in mammary glands was performed. Similarly, high levels of H-*RAS* ultimately induced senescence in mammary epithelial cells following a brief period of cell proliferation, accompanied by increased levels of the senescence markers p53, $p19^{ARF}$, p16, and PML. Importantly, in the absence of p16/p19 expression, high levels of RAS did not induce senescence, indicating that p16 is the key player in determining whether cells become senescent or transformed. In contrast, low levels of mutant H-*RAS* expression promoted continuous cell proliferation and hyperplasia, eventually leading to tumors. As would be expected, senescence markers were not detectable.

So it appears that oncogene-induced senescence may be the result of high levels of the oncogene. Although we know that RAS overexpression and gene amplification do feature in many human cancers, it is still unclear as to whether senescence occurs in these. But the acquired mutations of *p53*, $p16^{INK4a}$, and $p14/p19^{ARF}$ in such tumors may allow cells to avoid RAS-induced senescence, thus promoting proliferation and transformation.

RAS and endoplasmic reticulum stress

Dysfunction of the endoplasmic reticulum (ER) is increasingly recognized as a key mediator of cell stress (see Chapter 8) and is important in a number of major human diseases, including diabetes and cancer. Recent studies show that in a situation analogous to apoptosis induced by oncogenic stress, some oncogenes, notably RAS when deregulated, may also trigger ER stress and

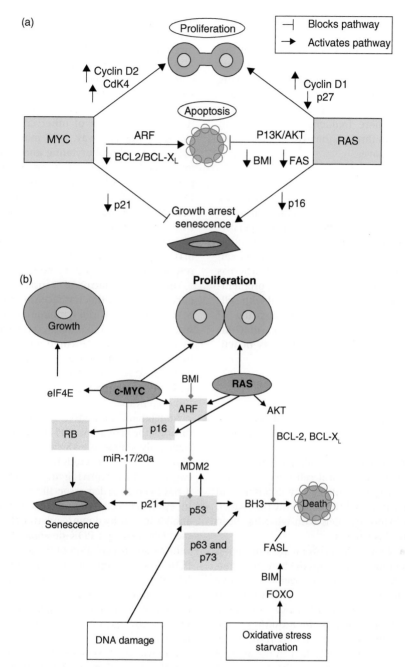

Figure 6.15 Cooperation between RAS and MYC. (a) Activation of mitogenic proteins such as MYC and RAS triggers pathways that not only lead to cell cycling but also those that promote cell death (apoptosis) and growth arrest/senescence, respectively. Deregulated activation of MYC alone may preferentially lead to apoptosis rather than cell division, while activated RAS alone may lead to cellular senescence – these outcomes serve to protect the organism from cancer-inherent tumor suppressor activity. However, when RAS and MYC are both activated, these "in-built" tumor suppressor activities are lost: RAS suppresses MYC-induced apoptosis and MYC suppresses RAS-induced growth arrest. It is such cooperation between oncogenes that promote tumor development. (b) Activation of RAS and MYC results in potential engagement of both replication and growth but also of apoptosis and possibly senescence. If either MYC or RAS levels are excessive (as might occur during oncogenesis) or other proapoptotic signals are received then the balance may be tipped away from replication. RAS can promote senescence through either p16^{INK4a} or ARF, which activate RB or p53 pathways, respectively, while MYC can inhibit growth arrest/senescence by inhibiting p21^{CIP} and inducing TERT (telomerse reverse transcriptase). Although MYC may activate the apoptotic pathway (e.g. via ARF), RAS is able to suppress apoptosis by activating the PI3K pathway and, subsequently, AKT. It can readily be appreciated how oncogenic MYC and RAS may conspire in oncogenesis. The combination of RAS and MYC acting together provides a potential means of avoiding apoptotic and senescence mechanisms activated by either acting individually. Moreover, it can also be appreciated how inactivating mutations in RB or p53 (or their pathways involving p19^{Arf}, p16^{INK4a}, p21^{CIP1}, etc.) may contribute to tumorigenesis by enabling the cancer cell to avoid either senescence or apoptosis or both.

the so-called unfolded protein response (UPR), as discussed in Chapter 8. In a cell culture model of melanoma, oncogenic forms of HRAS triggered massive vacuolization and expansion of the ER as well as the expected triggering of cell-cycle arrest. Importantly, at least in this case, p53, p16^{INK4a}, and senescence markers appeared not to explain mutant HRAS(G12V)-driven senescence. Rather, it was driven by ER stress, although this mechanism is less relevant for NRAS, the more frequent mutant RAS in melanoma, which likely operates through the p16^{INK4a}-dependent mechanisms already outlined above.

RAS and differentiation

The RAS pathway transduces divergent signals determining normal cell fate and, importantly, may have differing effects on differentiation depending on both cell type and on the persistence or extent of RAS activity. For example, sustained activation of the RAS–MAPK pathway in some neuronal precursors can induce cell-cycle arrest and may promote differentiation, whereas in primary erythroid cells differentiation is blocked and proliferation is enhanced. Differentiation is generally associated with cell-cycle arrest, but does not necessarily require silencing of MAPK–ERK. In fact, suppression of cell proliferation after differentiation can be achieved by restricting nuclear entry of activated MAPK.

RAS – survival or death

'Tis not the meat, but 'tis the appetite makes eating a delight.
John Suckling

In this section we will revisit the notion of seemingly paradoxical outcomes arising from apparently identical cell signaling pathways and will see how this may be explained by differences in the extent or duration of signaling and the context within which this occurs. There will be obvious parallels with what has been discussed for MYC earlier in this chapter.

Although the mechanisms by which oncogenic RAS promotes uncontrolled cellular proliferation are perhaps the best characterized and understood, it is now clear that oncogenic RAS (like MYC) can also deregulate processes that control cell death, otherwise known as apoptosis (see Chapter 8). Large, long-lived animals like humans have a high propensity for acquiring mutations. It has been estimated that point mutations resulting in activation of *RAS* occur in thousands of cells daily in the average human. As the vast majority of these mutations do not result in tumor growth, it is assumed that the usual outcome of such mutations is apoptosis, differentiation, or growth arrest.

The powerful stimulatory effect exerted by oncogenic RAS on proliferation has long been considered equally critical to RAS transformation. However, as was discussed for MYC earlier, no oncogene is an island and they must find collaborating partners in order to realize their oncogenic potential. Thus, tumor development and maintenance also depend on the ability of a cell to avoid apoptosis and this property must be provided by a cooperating lesion. As will be familiar by now, it is increasingly apparent that RAS, as for MYC, paradoxically induces both pro- and antiapoptotic signaling. The ultimate outcome of these contradictory signals is strongly dependent on the cell type and context. For example, in normal cells, a high level of activated RAS has been thought to be more likely to induce a protective proapoptotic response to prevent oncogenesis in response to hyperproliferative signals. However, in cells that are already transformed (e.g. with MYC and RAS), the activity of oncogenic RAS is likely to mediate survival rather than death.

How does RAS mediate cell survival? The key pathways involved in RAS-mediated cell survival are outlined in Fig. 6.16. The best known of these is the RAS–PI3K–AKT pathway, already discussed as an important facilitator of cell replication and growth. However, a number of other critical substrates are phosphorylated by this pathway. AKT may primarily suppress apoptotic signaling by phosphorylating and inactivating BAD, a proapoptotic member of the BCL-2 family (see Chapter 8). Once phosphorylated by AKT, BAD preferentially binds to 14-3-3 proteins to form an inactive complex; BAD is no longer available to sequester and inactivate the antiapoptotic proteins BCL-2 and BCL-X$_L$. In addition, AKT survival signaling is potentiated by its effects on cellular energy homeostasis and its modulation of the mTOR pathway described earlier, which controls the cell's response to nutrients (metabolism).

The initiation factor of translation (eIF-4E) is a downstream effector of mTOR that has recently been shown to cooperate with MYC in forming B-cell lymphomas in mouse models. In a recent study it was found that MYC could override senescence activated by eIF-4E, and eIF-4E could antagonize MYC-dependent apoptosis *in vivo*; this obviously mirrors the interactions shown between MYC and RAS and suggests that activation of eIF-4E may be an important mediator of PI3K and AKT during tumorigenesis.

Attachment to matrix normally provides a survival signal through the activation of PI3K, which is lost in invading or metastatic cells; activated RAS can overcome this loss by restoring the PI3K signal. In recognition of the specific features of apoptosis engendered by depriving an epithelial cell of matrix interaction it has been awarded a specific title – **anoikis** (see Chapter 12). In epithelial cell types oncogenic RAS restores resistance to anoikis in part by downregulating the proapoptotic BCL-2 family member BAK in a PI3K-dependent manner and in part by preventing downregulation of the antiapoptotic protein BCL-X$_L$ in a PI3K- and RAF-independent manner. In general, PI3K–AKT and NF-κB signaling are considered to protect against apoptosis. It is then perhaps not surprising that many human tumors have mutations that amplify signaling through the IGF pathway, which converge on sustained activation of the PI3K survival pathway.

Confusingly, RAF–ERK signaling can be either anti- or proapoptotic, depending on the circumstances. Thus, the combined outcomes of RAS signaling determine whether cells live or die. RAS specifically utilizes a PI3K–AKT pathway to promote cell survival in the presence of apoptotic signals such as those induced by MYC (Fig. 6.15). Under some circumstances, oncogenic RAS may be able to overcome its own proapoptotic signaling by inducing a concurrent antiapoptotic signal via RAC or NF-κB (see Fig. 6.16). This is obviously a dangerous and undesirable outcome and to what extent this might happen in cancer is not clear. Defects or interference in the ability of RAS to activate the antiapoptotic signals can allow its proapoptotic activity to take over.

In many contexts, excessive signaling due to oncogenic RAS provokes a response from p53 designed to cause cell-cycle arrest or apoptosis and thereby remove the threat of unbalanced oncogenic stimuli (described previously in RAS and cellular senescence). To complicate matters further, oncogenic RAS can suppress p53 by inducing its degradation via MDM2 in an RAF-

Oncogenes

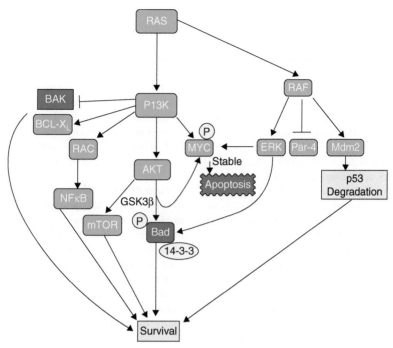

Figure 6.16 Oncogenic RAS pathways to survival or death. Tumor development and maintenance depends on the ability of a cell to avoid apoptosis. Although RAS is known to promote cell survival by preventing apoptosis, in some circumstances it has been shown that RAS can also mediate proapoptotic signals. The ultimate outcome of these contradictory signals depends greatly on the cell type and context. RAS can promote cell survival through a number of signaling pathways. A key pathway that predominantly mediates survival is the PI3K signaling cascade, which activates the serine/threonine (S/T) kinase AKT. In turn, AKT phosphorylates a number of substrates including BAD – a proapoptotic member of the BCL-2 family. AKT phosphorylation of BAD causes it to bind preferentially to 14-3-3 proteins in an inactive complex, thereby preventing it from sequestering and inactivating the antiapoptotic proteins BCL-2 and BCL-X_L. However, AKT may also phosphorylate the Thr58 residue on MYC protein, leading to its increased stability. As discussed in the main text, MYC can promote apoptosis. RAS signaling through PI3K can also mediate survival by downregulating the proapoptotic BCL-2 family member BAK, and by preventing downregulation of the antiapoptotic protein BCL-X_L. In addition, PI3K can promote survival through the transcription factor NF-κB. RAS signaling through RAF/ERK can be either anti- or proapoptotic, depending on the circumstances. This pathway can lead to stabilization of MYC through phosphorylation of its Ser62 residue (proapoptotic) or through phosphorylation of BAD as described above, which promotes survival. Finally, although in many cases oncogenic RAS can provoke a response from p53 designed to cause cell-cycle arrest or apoptosis, to complicate matters it can also suppress p53 by inducing its degradation via MDM2 in a RAF-dependent manner. RAF may also contribute to the ability of oncogenic RAS to provide a pro-survival function by downregulating transcription of PAR-4, a proapoptotic transcriptional repressor. RAF and PI3K signaling may also converge downstream of oncogenic RAS to prevent apoptosis.

dependent manner, and thus contribute to the resistance of RAS-transformed cells to p53-mediated apoptosis (Fig. 6.16). RAF may also contribute to the ability of oncogenic RAS to provide a pro-survival function by downregulating transcription of Par-4, a proapoptotic transcriptional repressor. Both RAS and RAF can alter the expression and activity of BCL-2 family members. For example, loss of K-RAS protein expression following targeting by an anti-K-RAS ribozyme in capan-1 pancreatic cancer cells results in reduced levels of BCL-2 and increased apoptosis. RAF and PI3K signaling may also converge downstream of oncogenic RAS to prevent apoptosis (Fig. 6.16). One obvious convergence point is the phosphorylation of Bad. Another point of convergence for RAF and PI3K signaling downstream of oncogenic RAS is through effects on MYC protein degradation. Phosphorylation at Ser62 stabilizes MYC, whereas subsequent phosphorylation at Thr58 is required for its degradation. Phosphorylation on Ser62 may be mediated via ERK, whereas Ser62 is dephosphorylated by protein phosphatase 2A (PP2A), which is itself regulated by the PIN1 prolyl isomerase – all potentially RAS-regulated signaling molecules. By these means, RAS-regulated signaling cascades can limit the level of MYC accumulation and duration of action. Conversely, if these regulatory mechanisms are interfered with then this may contribute to oncogenesis.

In light of the diversity of downstream effector targets known to facilitate RAS function, it is perhaps not surprising that RAS regulation of cell survival is complex, involving the balance and interplay of multiple signaling networks. While our understanding of these events is still far from complete, and is complicated by cell type and signaling context differences, several important mechanisms have begun to emerge. Moreover, an appreciation of the interplay of RAS with various signaling molecules that could either block or amplify the apoptosis signal may enable more rational design of new drugs to treat RAS-related cancers.

It is now apparent that some of the growth-inhibitory properties of RAS are mediated via the RASSF family of RAS effector/tumor suppressors. To date, five members of this family have been identified (NORE1, RASSF1, RASSF2, RASSF3, and RASSF4). This family are involved in cell-cycle arrest and in apoptosis in response to RAS. Specific effectors may regulate antiapoptotic (RAF, PI3K, and TIAM1) and apoptotic (NORE1 and RASSF1) actions of oncogenic RAS. RASSF1A is one of the most frequently inactivated genes described in human cancers and, as observed for other tumor suppressors, inactivation is often through epigenetic gene silencing through methylation of the promoter and CpG island.

RAS and cancer

RAS mutations are the most frequent cause of aberrant signaling through RAS pathways in many human cancer cells. RAS mutations are the most common oncogenic event identified in human tumors, being found in 30% of human cancers. The most frequently mutated member is *KRAS*, which is present in over 90% of ductal pancreatic cancers, 40–50% colorectal cancers, and 30% of non-small-cell lung cancers (see Tables 6.1, 6.2, and 6.4). These mutations all compromise the GTPase activity of RAS, preventing GAPs from promoting hydrolysis of GTP, thereby leading to accumulation of the active GTP-bound form of RAS (see Fig. 6.11). Almost all RAS activation in tumors is accounted for by point mutations in codons 12 (most common), 13, and 61, which affect the reading frames of each gene and thus result in amino acid replacements. However, RAS can also be activated in tumors by loss of GAPs, as exemplified by the loss of neurofibromin, encoded by the *NF1* gene, mentioned earlier. Loss of one allele is inherited by individuals with type 1 neurofibromatosis, with the second "hit" presumably generated by somatic mutations in cells destined to give rise to the characteristic benign, and occasionally malignant, tumors in tissues of neural-crest origin. Loss of both copies of *NF1* results in activation of RAS.

RAS signaling pathways are also commonly activated in tumors in which growth factor RTKs have been overexpressed (see Chapter 5). Two notable examples are overexpression of EGFR and NEU, observed in numerous human cancers including breast, ovary, and stomach. EGFR family tyrosine kinases are also commonly activated by the autocrine production of EGF-like factors, such as TGF-α, in tumors. Also, elevated signaling through the IGF pathway occurs in many tumor types (see "RAS – survival or death").

Mutations in *BRAF* have now been found in 7% of human cancers, most notably in melanomas (64%) and, to a lesser extent, in thyroid (37%), ovarian (13%), and colon cancer (13%). The related *ARAF* and *CRAF* appear not to be mutated in cancer. The vast majority of *BRAF* mutations represent a single nucleotide change of T–A at nucleotide 1796, resulting in a valine to glutamic acid change at residue 599 (V599E) within the activation segment of B-RAF, thereby constitutively activating the MAPK pathway. *BRAF* and *RAS* mutations are rarely both present in the same cancers but the cancer types with *BRAF* mutations are similar to those with *RAS* mutations. This has been taken as evidence that the inappropriate regulation of the downstream ERKs (the p42/p44 MAPK) is a major contributing factor in the development of these cancers (see Fig. 6.14). In fibroblasts and melanocytes *in vitro*, mutant B-RAF stimulates constitutive ERK signaling, leading to cell proliferation, survival, and transformed morphology and allows these cells to grow as tumors in nude mice. Conversely, if mutant B-RAF is depleted in melanoma cells, ERK activity is now blocked, leading to inhibition of proliferation while inducing cell death by apoptosis (described in Chapter 8).

Since the identification of *BRAF* mutations in human cancers, novel mouse models have been developed, which possess the same mutations and provide experimental evidence that mutant B-RAF is able to initiate tumor growth. For example, in one mouse model, activation of mutant B-RAF in lung promoted adenoma formation, although these rarely progressed to adenocarcinomas. Rather, adenomas appeared to demonstrate a senescent-like growth arrest, which was overcome following the loss of P53 or P16/P19. Such cooperation enabled mutant B-RAF to induce adenocarcinomas. It will be interesting to see the effects of expressing mutant B-RAF on transformation in other cell types relevant to human cancers, such as melanoma, thyroid, and colon cancer.

The PI3K pathway is frequently activated in tumors as a result of deletion of the tumor suppressor gene *PTEN* (phosphatase and tensin homolog), which encodes a lipid phosphatase that removes the phosphate from the 3′ position of PIP3 and PIP2, so antagonizing the effects of PI3K activity. Loss of PTEN thus results in activation of the PI3K–AKT pathway. *PTEN* is deleted in almost half of human tumors, making it the second most significant tumor suppressor gene after *p53* (see Chapter 7).

Mouse models of Ras-induced tumorigenesis

In recent years, the development of genetically modified mouse models have been instrumental in helping us understand how mutations in the RAS signaling pathway lead to cellular transformation and tumorigenesis *in vivo*. Given the enormous repertoire of RAS signaling and the effects of context, it is not surprising that we have had to rely on mouse models to unravel some of the key functional consequences of targeting specific signaling molecules. Since *K-RAS* mutations are common in many different human tumor types, mouse models in which physiological levels of oncogenic K-RAS are expressed in different tissues have helped to recapitulate the human disease process. Conditional K-RAS (*K-RAS^{G12D}*) mutant mice have been generated by crossing with several tissue-specific Cre transgenic mouse strains, such as pancreas. Expression of oncogenic *RAS* (*K-RAS^{G12D}*) in duct cells of the pancreas led to development of a preinvasive neoplasm of pancreatic ductal adenocarcinoma (PDA) – termed "pancreatic intraepithelial neoplasia" (PanIN) – a small proportion of which spontaneously progressed to metastatic PDA. This disease process was accelerated in the presence of a variety of cooperating genetic lesions and different contexts. Thus, metastatic spread was encouraged if mice were also deficient in *p16/p19* or *p53* tumor suppressors but also if TGF-β signaling was perturbed, TGF-α was activated, or in the presence of the inflammatory disorder pancreatitis.

In a different mouse model that recapitulates the stages of lung cancer development, expression of physiological levels of oncogenic RAS (K-RAS^{G12D}) led to atypical adenomatous hyperplasia, epithelial hyperplasia, and adenomas that spontaneously progressed to pulmonary adenocarcinoma through chance acquisition of cooperating lesions. Later studies showed the cooperation of K-RAS^{G12D} with loss of either *p53* or *PTEN*, which accelerated the disease process. Interestingly, activation of the MAPK pathway was not detected in the early lesions prior to loss of p53 but only in later stage lesions. It is possible that in these earlier lesions, oncogenic RAS signaling through the MAPK pathway remains low due to negative feedback loops, but p53 loss (leading to further mutations) in some way promotes MAPK signaling and thus tumor progression.

These mouse models demonstrate that physiological expression of oncogenic K-RAS in pancreas and lung is instrumental in initiation and progression of these cancers. In contrast, oncogenic K-RAS (K-RAS^{G12V}) alone appeared not to be sufficient to initiate the disease process in colon, except when accompanied by loss of the APC (adenomatous polyposis coli) tumor suppressor. This cooperating event allowed cellular transformation and tumor progression in the colon. However, other models in which oncogenic K-RAS^{G12D} was expressed in the colon differ in terms of phenotype, exhibiting the onset of hyperproliferation induced by

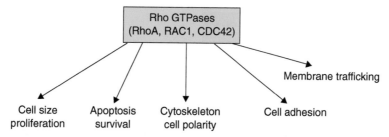

Figure 6.17 RHO GTPases mediate a diverse range of cellular effects. The high incidence of overexpression of some members of the RHO family of GTPases in human tumors suggests that these proteins are involved in cancer onset, and that they are potential candidates for a therapeutic intervention. RHO GTPases, the most widely studied of which are RHOA/B, RAC1/2, and CDC42, mediate a diverse range of cellular effects such as proliferation, motility, and adhesiveness, although the precise mechanisms by which RHO GTPases participate in carcinogenesis are still not fully understood. However, it is becoming more evident that the specific role of RHO overexpression in tumor initiation, progression, and metastasis may be linked to the activation of specific signaling pathways that result in transcriptional regulation (see Figure 6.18).

the MAPK pathway. In these models, progression of colon cancer is accelerated following the loss of APC.

RAS and angiogenesis

The family of RAS oncogenes promotes the initiation of tumor growth by stimulating tumor cell proliferation, but, like MYC, also ensures tumor progression by stimulating tumor-associated angiogenesis – growth of new vasculature (see Chapter 14). Oncogenic RAS proteins stimulate a number of effector pathways that culminate in the transcriptional activation of genes that control angiogenesis. Tumor angiogenesis is postulated to be regulated by the balance between pro- and antiangiogenic factors. A critical step in establishing the angiogenic capability of human cells is the repression of the antiangiogenic factor thrombospondin-1 (TSP-1). This repression may be essential for tumor formation. A novel mechanism has recently been described, whereby the cooperative activity of the oncogenes RAS and MYC may lead directly to angiogenesis and tumor formation. In this case, RAS may lead to TSP-1 repression via the sequential activation of PI3K, RHO, and ROCK, leading to activation of MYC through phosphorylation; phosphorylation of MYC via this mechanism enables it to repress TSP-1 expression.

Other RAS family members

RHEB and its closest relative RHEBL1 (RHEB-like-1, 52% identity) form a divergent branch of the RAS family. A replication-promoting role for RHEB was originally postulated following the observations that RHEB synergizes with RAF1 to transform NIH 3T3 fibroblasts and because RHEB is upregulated in RAS-transformed cells. Ectopic expression of RHEB is sufficient for phosphorylation of both mTOR and S6K1 (Fig. 5.19). Analysis of cell-cycle components that are influenced by TSC1/2 or mTOR indicates that potential RHEB targets might include $p27^{KIP1}$ and cyclin E.

RHO family of GTPases

> Many a trip continues long after movement in time and space have ceased.
>
> *John Steinbeck*

It is now clear that RHO GTPases have a major role in many different aspects of tumorigenesis. Studies suggest that the RHO family plays a role in cancer onset and in invasion and they are thus potential candidates for therapeutic intervention. RHO GTPases mediate a diverse range of cellular effects, such as proliferation, motility, and adhesiveness via cell–cell and cell–matrix interactions (Fig. 6.17). Some members of the RHO family of GTPases are highly expressed in human tumors. However, unlike the RAS family, which is mutated in a large proportion of human cancers, mutations in RHO GTPases are rare. Instead, hyperactivation of GTPases can occur through overexpression, or through deregulated expression/activity of their regulatory proteins, such as, loss of GAP-mediated inactivation of RHO GTPases, or overexpression of RHO GEFs. Several examples of deregulated GEFs and GAPs in human cancers will be described later.

Although the importance of RHO GTPases in cancer is undisputed, the precise mechanisms by which RHO GTPases participate in carcinogenesis are still not fully understood. In other words, the specific downstream effectors that mediate the cancer cell phenotype – proliferation, survival, invasion, and metastasis – are incompletely understood. One thing appears to be clear in that the effects of RHO overexpression on initiation, progression, and metastasis are highly context dependent and are influenced by the means through which such overexpression takes place in any given human tumor. There are at least 20 RHO family proteins in humans, the most widely studied of which are the following: RHOA/B, RAC1/2, and CDC42. Like all members of the small GTPases superfamily, in healthy cells these proteins are ubiquitously expressed and cycle between an inactive GDP-bound and an active GTP-bound state (Fig. 6.11). However, in many cancer cells RHO GTPases become hyperactivated by mechanisms mentioned above, and have been shown to be crucial for RAS-induced transformation.

These proteins exert their effects on cell behavior indirectly by activating a multitude of effector proteins. In fact, over 70 effectors have been described to date, which include transcription factors such as STAT3/5 and NF-κB (see below). The best-characterized RAC1 and CDC42 effectors are the p21-activated kinases (PAKs), whereas Rho-associated coiled-coil domain kinases (ROCK-I and II) represent the best-characterized RHOA effectors. RHO GTPases are best known as master regulators of the actin cytoskeleton, associated with motility and other processes that involve actin organization, such as cell shape and polarity. RAC1 and CDC42 remodel the actin cytoskeleton at the leading edge of the cell, resulting in lamellipodial or filopodial protrusions, respectively, whereas RhoA, B, and C are largely responsible for orchestrating focal adhesion assembly and

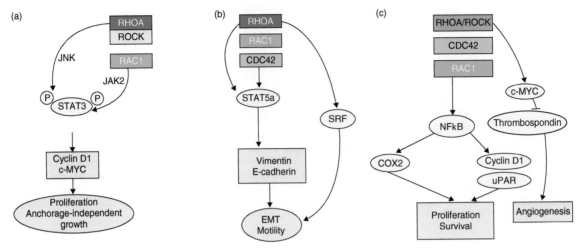

Figure 6.18 RHO promotes tumor cell properties through activation of specific transcription factors. (a) RHO GTPases (RHOA and RAC1 and CDC42) can activate STAT3 (signal transducer and activator of transcription 3) transcription factor by indirectly inducing its phosphorylation (via JNK2). It has recently been shown that RHOA (bound to the effector protein ROCK) signals through STAT3 for transformation, by inducing proliferation and anchorage-independent cell growth – hallmarks of transformed cells. Active STAT3 may promote cell proliferation by inducing the expression of MYC or cyclin D1. Although RAC1 and CDC42 also activate STAT3, it is not yet known whether these GTPases are able to induce transformation via STAT3. (b) RHOA, RAC1, and CDC42 proteins also activate STAT5a. RHOA activates STAT5a by indirect phosphorylation, causing cells to become more motile and inducing epithelial–mesenchymal transition (EMT), features associated with tumor cell behavior. STAT5a may mediate such cellular effects by upregulating the expression of vimentin which is associated with EMT, while downregulating E-cadherin expression with loss of adherens junctions (these mediate cell–cell adhesion). RHOA can also induce EMT by activating another transcription factor, SRF. (c) RHOA, RAC1, and CDC42 proteins induce activity of the transcription factor, NF-κB. NF-κB can promote cell proliferation, cell survival, invasion, and motility of cancer cells, presumably through expression of its target genes. One target gene is the urokinase plasminogen receptor gene (uPAR), whose expression results in enhanced invasiveness due to degradation of the extracellular matrix. Other target genes include the cell cycle protein cyclin D1 and the proinflammatory gene cyclooxygenase 2 (COX2), which is associated with cancer. Finally, the RHOA–ROCK pathway can activate MYC, leading to repression of the antiangiogenic factor thrombospondin-1 (TSP-1), which in turn allows proper angiogenesis to take place within the tumor.

actomyosin-mediated cell contraction at the rear of the cell, thus allowing cell movement across these adhesive contacts and subsequent detachment by the trailing end of the cell (see Chapter 12).

RHO GTPases also contribute to cell-cycle progression: inactivation of RHO by the *Clostridium botulinum* C3 ADP-ribosyltransferase or microinjection with dominant-negative forms of RAC1 or CDC42 block mitogen-stimulated G_1/S transition in Swiss 3T3 fibroblasts. Conversely, microinjection of active RHOA, RAC1, or CDC42 into quiescent cells induces G_1/S transition. RHO activation allows mitogen-stimulated cells to progress through the cell cycle. However, RHO is inactive in the absence of adhesion (see "SRC – the oldest oncogene") or under conditions of cell confluence, and $p21^{Cip1}$ (cell cycle inhibitor) expression remains high. Thus, RHO may function as an important "monitor" of the cellular environment and as an anchorage-dependent cell-cycle checkpoint.

RHO GTPases regulate the activity of many transcription factors

RHO-regulated transcription factors and their specific effects on cell behavior have recently been identified.

Due to the complexity of the signaling pathways mediated by RHO proteins, we will only describe a few of these pathways linked to the transformation of cells (albeit simplified) here.

STAT transcription factors

RHO, RAC1, and CDC42 can activate the transcription factor STAT3 (signal transducer and activator of transcription 3) by indirectly inducing its phosphorylation. In contrast to normal STAT3 signaling, which is transient, hyperactive STAT3 is present in a large number of cancers and is associated with malignant transformation: tumor cell growth and survival, tumor angiogenesis and metastasis, and tumor immune evasion. For example, RHOA (bound to the effector protein ROCK) signals through STAT3 to induce proliferation and anchorage-independent cell growth (Fig. 6.18). Anchorage-independent growth and the ability to avoid detachment-induced apoptosis (anoikis – see Chapter 12) are hallmarks of transformed epithelial cells. STAT3:STAT3 dimers translocate to the nucleus where they bind to target sequences in specific promoters leading to upregulation of genes involved in tumor promoting behavior, such as *BCLxL*, *MCL1*, *Survivin*, *AKT*, *VEGF*, *HGF*, *MYC*, *CyclinD*, and *HIF1*, while the *p53* tumor suppressor is downregulated by STAT3 activity.

RHOA, RAC1, and CDC42 activate another member of the STAT family of transcription factors – STAT5a (Fig. 6.18). Activation of STAT5a by RHOA by indirect phosphorylation causes cells to become more motile as well as inducing epithelial-to-mesenchymal transition (EMT), features associated with tumor cell metastatic behavior (see Chapter 12). STAT5a may promote such cellular effects by upregulating the expression of vimentin which is associated with EMT, while downregulating E-cadherin expression with loss of adherens junctions (these mediate cell–cell adhesion). In addition to STAT5a, RHOA can induce EMT by activating another transcription factor, SRF.

NF-κB and MYC

RHOA, RAC1, and CDC42 proteins induce the transcriptional activity of the transcription factor NF-κB, a protein important in inflammatory responses and tumorigenesis (see Chapter 5). NF-κB can promote cell proliferation, cell survival, invasion, and motility of cancer cells. As it is a transcription factor, it is likely to promote such cellular effects through expression of downstream genes. For example, one target gene is the urokinase plasminogen receptor gene (*uPAR*), whose expression results in enhanced invasiveness due to degradation of the extracellular matrix. Other target genes of NF-κB include the cell cycle protein cyclin D1 (see Chapter 4) and the proinflammatory gene cyclooxygenase 2 (*COX2*). Importantly, NF-κB signaling through COX-2 protein links inflammation and cancer (Fig. 6.18).

An important finding was the link between RAS (signaling via RHO GTPases) and the MYC transcription factor in inducing growth of tumor vasculature (angiogenesis), at least in the context of human epithelial tumors. Here, RAS signaling through the RHOA–ROCK pathway to activate MYC leads to repression of the antiangiogenic factor thrombospondin-1 (TSP-1), which in turn allows proper angiogenesis to take place within the tumor.

RHO GTPases and cancer

It is now clear that RHO GTPases play a major role in many different aspects of tumorigenesis. However, the link between RHO proteins and human cancer remained hidden for a long time because, unlike for the RAS family, no activating mutations were found within the coding sequence of these proteins. However, overexpression of RHO proteins has been detected in many human tumors. For example, in colon and breast tumor, although mRNA levels are not elevated compared to normal tissue, RHO protein levels are much higher. In contrast, there is an increase in RHO mRNA in testicular germ tumors. Importantly, aside from deregulated expression of RHO GTPases in cancer, hyperactivation of RHO GTPases can occur through deregulated expression of their regulatory proteins, GEFs (increase) and GAPs (decrease), described previously in GDP–GTP cycling of members of the RAS family.

The fact that most RHO GTPases promote tumorigenesis implies that hyperactivation of their GEFs would likewise be oncogenic and this is sometimes the case. There is now much evidence to show hyperactivation of RHO GEFs in human cancer that in turn leads to hyperactivation of RHO signaling pathways. The following section describes some of the most relevant RHO GEFs linked with human cancer. For example, in head and neck squamous cell carcinoma (HNSCC), the RHO GEF VAV2 is hyperactivated via EGFR, and knockdown of VAV2 inhibited RAC1 activation and EGFR-mediated cell invasion *in vitro*.

RHO GEFs associated with cancer

As mentioned earlier, RHO GTPases are often hyperactivated through overexpression of their regulatory proteins, GEFs. Here we will highlight the importance of particular RHO GEFs linked with cancer progression. As you will see, the hyperactivation of RHO GTPase signaling is often associated with cell migration, invasion, and metastases.

VAV RHO GEFs are associated with several types of cancer. For example, *VAV1* is overexpressed in pancreatic carcinomas via promoter demethylation, and is associated with decreased survival. VAV1 overexpression leads to activation of RAC and downstream signaling through PAK (p21-activated kinase)–NF-κB (nuclear factor- κB) and cyclin D1 upregulation. When VAV1 was depleted using RNA interference (RNAi), cancer cells were unable to grow in the absence of adhesion *in vitro*, and also in mouse tumor xenografts. A related RHO GEF, VAV2, is hyperactivated in HNSCC via EGFR and similarly activates RAC1, while VAV3 is overexpressed in glioblastomas. The effect of inhibiting VAV2 or VAV3 in cancer cells is the reduced ability to migrate and invade *in vitro*. A RAC-specific GEF, PREX-1 (phosphatidylinositol-3,4,5-triphosphate-dependent RAC exchange factor 1) is highly expressed at the gene and protein level in metastatic prostate cancer cell lines and metastatic prostate tumor tissue. Suppression of PREX-1 expression in such cancer cells inhibited RAC activity and consequently reduced cell migration and invasion *in vitro*. Conversely, overexpression of PREX-1 in prostate cancer cell xenografts in mice promoted metastasis to lymph nodes.

Another RHO GEF, and also an oncogene, is ECT2 (epithelial cell transforming sequence 2) which activates RHOA primarily but also RAC and CDC42. This oncogene is overexpressed in various human tumor cell lines and tissues, including lung and esophageal squamous cell carcinomas, and is associated with poor prognosis. Similarly, ECT2 mRNA and protein overexpression is found in glioblastoma samples as well as in samples of non-small-cell lung carcinoma (NSCLC), and RNAi-mediated suppression of ECT2 mRNA reduced or blocked migration/invasion of human tumor cells using *in vitro* assays.

There have been several reports of mutation and overexpression of TIAM1 (T-cell lymphoma invasive and metastasis 1) in various human cancers. This RAC-specific GEF is also a downstream effector of RAS, and its importance is linked with tumor initiation rather than progression and metastasis. These findings have arisen from various mouse models of cancer, such as squamous cell skin carcinoma, APC-induced colon cancer, and ERBB2 (neu)-induced mammary cancer. In general, in the absence of TIAM1, tumor formation was impaired with fewer and smaller tumors present, but those that did form were more invasive or metastatic. These observations suggest that TIAM1 may function as a metastases suppressor, so inhibiting its activity in some settings may not be beneficial!

A last but important RHO GEF includes the breakpoint cluster region (BCR)–ABL1 fusion protein that is encoded by the translocation associated with the Philadelphia chromosome described later in this chapter (see Box 6.4). This translocation event is found in over 90% of CMLs. BCR possesses a RHO GEF and a RHO GAP domain. Following the translocation event, the resulting BCR–ABL1 fusion protein retains the RHO GEF (but not the RHO GAP domain) which has become fused to truncated ABL1. Consequently, ABL1, a protein tyrosine kinase, is now constitutively active and thus able to mediate oncogenesis. Aside from the transforming potential of ABL1, RHO GEF activity also contributes to anchorage-independent growth by activating RHOA.

RHO GAPs in cancer

Surprisingly perhaps, in contrast to the many RHO GEFs with deregulated activity in cancer cells, there is very little evidence for the role of RHO GAPs in cancer. GAPs have an opposite function to GEFs, generally behaving as tumor suppressors by their ability to deactivate GTPases, such that loss of GAP function is likely to confer oncogenic properties on a cell. For example, we previously mentioned in the RAS section, that loss of NF-1, which encodes the RAS GAP neurofibromin, is found in patients with neurofibromatosis type 1. In fact, recent sequence studies have found frequent somatic mutations in *NF1* in glioblastomas as well.

One RHO GAP that stands out as having a central role as a tumor suppressor in several different cancer types is DLC1 (deleted in liver cancer 1). *DLC1* gene is deleted or transcriptionally silenced by promoter methylation. Importantly, a comprehensive analysis of the genomic loss of *DLC1* showed that heterozygous loss in tumors occurs at a similar rate to that of tumor suppressor *p53* mutation or loss.

The substrates of DLC1 are RHOA, RHOB, RHOC, and, to a lesser degree, CDC42, but not RAC. Various studies using DLC1-deficient human cancer cell lines *in vitro* have shown that re-expression of DLC1 suppressed proliferation, anchorage-independent growth, and invasion through Matrigel. Conversely, knockdown of endogenous DLC1 in an *ex vivo* mouse model of Myc-induced liver tumorigenesis accelerated the onset of tumor development and resulted in more aggressive tumors resembling human hepatocellular carcinoma. To conclude, the high occurrence of reduced DLC1 expression in many human tumors together with the functional evidence that DLC1-3 are tumor suppressors, make inactivation of this RHO GAP family the most common mechanism of altering RHO GTPase activity in human cancer.

Therapeutics – targeting RAS signaling in tumors

Since *RAS* mutations are the most frequently mutated oncogenes in human cancers, intensive efforts have been put into trying to directly inhibit RAS activity, at least for tumors harboring mutant RAS. However, to date, no successful "anti-RAS" strategies have reached the clinic, chiefly because of the difficulty in developing a suitable effective small molecule inhibitor. For this reason, the focus is now aimed at indirectly targeting RAS function: the inhibition of components that regulate RAS membrane association, and the inhibition of downstream effector signaling.

Another promising area is to identify synthetic lethal interactions (whereby a *RAS* mutation renders a cancer cell critically dependent on some other protein which could be more readily targeted with little if any effect on a normal cell). Many laboratories are using techniques such as siRNA screens to identify synthetic lethal targets in RAS-transformed cells. Thus, *KRAS* mutant cells are sensitive to the loss of the protein Survivin (see Chapter 8), APC/C proteins such as PLK1, STK33, and TBK1, all of which might be druggable targets in KRAS mutant cancer cells. This area is discussed in more detail in Chapter 16.

SRC – the oldest oncogene

> I have lived eighty years of life and know nothing for it, but to be resigned and tell myself that flies are born to be eaten by spiders and man to be devoured by sorrow.
>
> *Voltaire*

The SRC nonreceptor tyrosine kinase is the original archetype for oncogenes and among the most studied. SRC is overexpressed and activated in a large number of human cancers (though few mutations have been found), although its actual role in initiation or progression of any given cancer is not fully understood. Importantly, despite a period of relative inactivity in the field, interest has resurfaced with the availability of several new compounds targeting SRC and an appreciation of the role played by SRC in acquired resistance to other targeted therapies. Several SRC inhibitors are in clinical trials and will be discussed later.

So, what is SRC's role in cancer? In recent years, *in vitro* observations have led to the hypothesis that aside from increasing cell proliferation, SRC plays a key role in immunity and in regulating cell adhesion, motility, and invasion in cancer.

SRC was the first oncogene discovered (see Box 6.1), and was originally identified as the transforming agent (v-*src*) of the Rous sarcoma virus (RSV), a retrovirus that infects chickens and other animals. RSV is an acutely transforming virus that inserts its own genes into the cellular DNA, rapidly promoting the development of cancer. In fact, once infected, chickens develop large tumors within two weeks. With further developments in molecular biology and genetics, Bishop and Varmus discovered that v-*Src* had a cellular counterpart, the proto-oncogene *c-SRC* (*SRC*). The SRC protein is a tyrosine kinase that activates downstream signaling through the addition of phosphate groups to tyrosine residues on target proteins.

There are at least nine different known *SRC* family genes, including *SRC*, *Blk*, *Fgr*, *Fyn*, *Hcy*, *Lck*, *Lyn*, *Yes*, and *Yrk*, which through different mRNA processing can encode at least 14 different proteins, collectively referred to as SRC family kinases (SFK).

Regulation and activation of c-SRC

Low levels of c-SRC (SRC) are detected in most cells but are increased or activated in certain human cancer types, including bladder, neuroblastoma, small-cell lung, colon, and breast. The fact that both high levels of SRC protein and SRC kinase activity have been observed within cancer cells indicates the potential importance of both protein levels and protein activity in various phases of tumor development.

Human SRC protein contain four SRC homology (SH domains) comprising the SH1 kinase domain, which contains the autophosphorylation site (Tyr419); the SH2 domain, which interacts with the negative-regulatory Tyr530 and binds to the platelet-derived growth factor receptor (PDGFR); the SH3 domain, which enhances interactions with the kinase domain; and the SH4 domain, which contains the myristoylation site required for membrane localization (Fig. 6.19a). Interactions between the C-terminus and the SH2 domain, and between the kinase domain and the SH3 domain, cause the SRC molecule to assume a closed configuration that prevents substrate interaction. When the C-terminal tyrosine is phosphorylated, SRC is inactive and the protein resides at a perinuclear site. However, following binding of PDGF ligand to its receptor or after integrin engagement (see Chapter 12), SRC becomes active and translocates to the cell membrane (Fig. 6.19b). This activity is the result of dephosphorylation of Tyr530, which in turn leads to a change in conformation.

Inactivation of SRC by phosphorylation is now known to be performed by CSK (C-terminal SRC tyrosine kinase) and there is now evidence that reduced expression of CSK might have a role in SRC activation in human cancer. While CSK overexpression suppressed metastasis in animal models of colon cancer, CSK levels are decreased in patients with hepatocellular carcinoma compared with matched cirrhotic controls. Conversely, protein tyrosine phosphatases (PTPs) have been shown to dephosphorylate the terminal tyrosine residue of SRC. The most direct evidence for a role in SRC activation in cancer among these phosphatases is PTP1B, which is present at higher levels in breast cancer cell lines. So, like RAS, translocation and activation of SRC are inextricably linked, and the actin cytoskeleton is required for both catalytic activation and peripheral membrane targeting of this protein.

Activated SRC is translocated to the cell periphery, often to sites of cell adhesion, where myristoylation of its SH4 domain

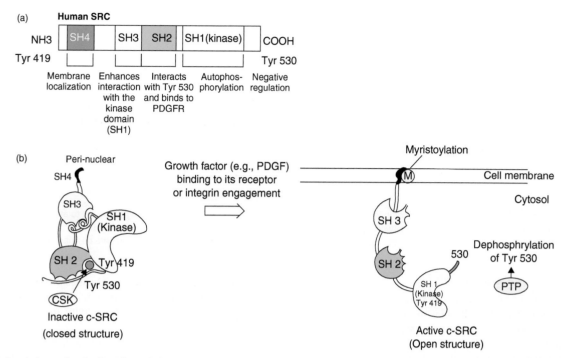

Figure 6.19 Regulation and activation of c-SRC. (a) The functional domains of human c-SRC: Human c-SRC protein contains four SRC homology (SH domains) comprising the SH1 kinase domain, which contains the autophosphorylation site (Tyr419); the SH2 domain, which interacts with the negative-regulatory Tyr530 and binds to the platelet-derived growth factor receptor (PDGFR); the SH3 domain, which enhances interactions with the kinase domain; and the SH4 domain, which contains the myristoylation site required for membrane localization. (b) Activation of c-SRC: When the C-terminal tyrosine Tyr530 is phosphorylated, SRC is inactive and assumes a "closed" conformation as a result of interactions between the C-terminus and the SH2 domain, and between the kinase domain and the SH3 domain. This closed configuration prevents substrate interaction, and the inactive c-SRC protein resides at a perinuclear site. Inactivation of c-SRC by phosphorylation is performed by CSK (c-SRC tyrosine kinase). Following binding of PDGF ligand to its receptor or after integrin engagement, SRC becomes active and translocates to the cell membrane. This activity is the result of dephosphorylation of Tyr530, which leads to an "open" conformation of the protein, thus allowing substrate interaction. Protein tyrosine phosphatases (PTP) have been shown to dephosphorylate the Tyr530 of c-SRC.

mediates attachment to the inner surface of the plasma membrane. This tethered location allows for interactions with membrane-bound RTKs and integrins associated with adhesion functions (see Chapter 12). The catalytic activity of SRC then initiates signal transduction pathways involved in cell growth, adhesion, and motility. Deregulation in cancer cells may therefore enhance tumor growth and/or stimulate migratory or invasive potential in cells that would normally be relatively nonmotile. The direct binding of focal adhesion kinase (FAK), described in Chapter 12, to the SH2 and SH3 domains of SRC also results in the open, active configuration of SRC.

Importantly, SRC can also be activated through receptor-mediated signaling: ligand-activated RTKs, such as EGFR, PDGFR, ERBB2 (or HER2/NEU), and FGFR. In a wide range of tumors in which SRC is overexpressed, RTKs are also overexpressed. In fact, there are various lines of evidence indicating that SRC cooperates with RTKs to promote tumorigenesis, probably by disrupting the intramolecular interactions that hold SRC in a closed configuration. For example, when EGFR and SRC are cotransfected into fibroblasts, their synergistic action results in increased proliferation, invasiveness, and tumorigenesis. In mouse models, overexpression of the RTK ERBB2 (HER2/NEU) also leads to SRC activation.

Other mechanisms of SRC regulation include ubiquitylation, with subsequent degradation by the proteosome. Recent evidence indicates that this regulatory pathway is disrupted in some cancer cells, thus allowing SRC activation. Finally, although extremely rare, there are two reports of SRC activation in colon cancer and endometrial cancer resulting from naturally occurring point mutations that truncate SRC just C-terminal to the regulatory Tyr530.

The SRC phenotype

As previously mentioned, much of the experimental data on the oncogenic effects of SRC comes from infecting or transfecting fibroblasts *in vitro* with the highly activated v-*src* oncogene. Apart from increasing proliferation rates of normal cells, v-*src* causes fibroblasts to round up, dissagregate and begin to float in the culture medium. These effects result from loss of cell–cell contact (through loss of E-cadherin – see later), and decreased substrate adhesion through loss of integrin-based cytoskeletal attachments (described in Chapter 12), which normally ensure that they bind to the substratum in an ordered monolayer. In addition to these cellular changes, v-*src*-transformed cells become more motile and more invasive. After several weeks in culture, foci can be seen in the culture dish, where cells pile up on top of each other (as described in "Oncogene collaboration"). All these changes correlate with the various hallmarks of a cancer cell that enable a normal cell to become a cancer cell – increased proliferation, dissagregation from the primary tumor, invasion into the surrounding tissue, and, ultimately, metastasis to a distant site. Apart from the cellular effects observed in the culture dish, if v-*src*-transfected cells are injected into mice, the cells grow rapidly to form visible tumors within days. Moreover, these tumor cells are capable of local invasion and metastasis to distant sites.

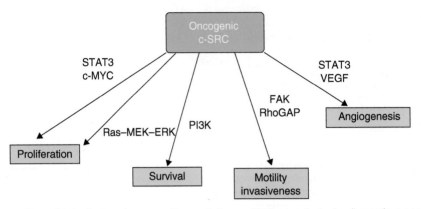

Figure 6.20 Oncogenic SRC can affect cell behavior in various ways. Oncogenic forms of SRC can activate signaling pathways to alter cell structure, in particular the actin cytoskeleton and the adhesion networks that control cell motility and invasion. In addition, SRC can also transmit signals that regulate cell proliferation and survival, as well as angiogenesis – the growth of new blood vessels. Details of these signaling pathways are described in Fig. 6.21.

How activated SRC does it

Oncogenic forms of SRC can alter cell structure, in particular the actin cytoskeleton and the adhesion networks that control cell motility and invasion, and also transmit signals that regulate cell proliferation and survival (Fig. 6.20).

Adhesion, invasion, and motility, although representing independent cellular functions, are related events that require several well-orchestrated molecular interactions. In order for cells to move, whether it be normal cells or cancer cells, alterations in both cell–cell and cell–matrix adhesion is required. Invading cells, as is the case for cancer cells, require alterations in both adhesion and motility.

Two principal subcellular structures regulate adhesion, invasion, and motility in normal healthy cells – **focal adhesions** and **adherens junction** – both of which are regulated by SRC. These structures are described in Chapter 12, and so will not be covered in any significant detail here. Briefly, focal adhesions provide the structural and mechanical properties that are necessary for cell–matrix attachment, while adherens junctions enable neighboring cells to adhere to each other: cell–cell attachment (Fig. 6.21). Focal adhesions form at the sites where integrins link the actin cytoskeleton to extracellular matrix (ECM) proteins. In addition to their role in adhesion, they also participate in cell-signaling processes that regulate proliferation and gene transcription. Focal adhesions are composed of over 50 different proteins, such as talin, vinculin, α-actinin, SRC, FAK, CAS, and paxillin, which assemble into supramolecular structures (see Chapter 12). These cytoskeletal proteins are recruited to focal adhesions to mediate cellular migration. They are associated with cytoskeletal stress fibers, composed of actin and myosin, which control the shape and, ultimately, the motility of the cell.

The assembly and disassembly of focal adhesions is associated with both significant cytoskeletal alterations and integrin signaling, which allow morphological changes and alterations in motility and invasiveness. So, for example, focal adhesions assemble in order to allow cells to adhere to the ECM, whereas disassembly of these structures promotes detachment from the ECM, a property that is acquired by cancer cells.

Activated SRC is now known to play a key role in focal adhesion disassembly, and it does this in various ways as shown in Fig. 6.21. One way is through its binding and activation of FAK: activated FAK phosphorylates substrates such as $p190^{RhoGAP}$, which in turn inhibits RHOA, leading to focal adhesion disruption. SRC can also destabilize focal adhesions through its tyrosine phosphorylation of R-RAS protein: R-RAS and v-src form a complex that suppresses integrin activity and reduces cell–matrix adhesion.

Apart from the ability of activated SRC to promote the release of cells from the ECM, it also acts to release cells from each other. As described in Chapter 12, in order for healthy cells to adhere to each other, they must form stable adherens junctions mediated by homotypic interactions between E-cadherin molecules on neighboring cells (Fig. 6.21). Cell motility and invasiveness depend on the loss of cell–cell adhesion mediated by E-cadherin. Activated SRC disrupts adherens junctions by inducing tyrosine phosphorylation and ubiquitylation of the E-cadherin complex. As a result, cadherin molecules are internalized. Phosphorylation of FAK by SRC is also required for the disruption of E-cadherin cell–cell adhesion.

In addition to SRC's role in promoting invasion through loss of E-cadherin, SRC might affect invasion by regulating matrix metalloproteinases (MMPs) and tissue inhibitors of MMPs (TIMPs) – described in Chapter 12. MMPs are a group of enzymes that can break down ECM proteins, and are involved in physiological processes such as wound healing, as well as in promoting tumor cell invasion and angiogenesis. TIMPs are secreted proteins that play a crucial role in regulating the activity of MMPs. In cancer, the regulated expression of both families of proteins can become disrupted, such that expression of MMPs is increased and/or expression of TIMPs is decreased. With respect to oncogenic SRC, activation of FAK stimulates the c-Jun N-terminal kinase (JNK) pathway, ultimately leading to the expression of MMP2 and MMP9 (Fig. 6.21).

Lastly, SRC proteins seem to promote angiogenesis – growth of new blood vasculature (see Chapter 14). For example, as was shown for MYC, v-src induces vascular endothelial growth factor (VEGF) expression. SRC achieves this through activating the transcription factor STAT3 (Fig. 6.21). In hypoxia-induced models of angiogenesis, SRC activation induces VEGF expression. Treatment with 4-amino-5-(4-chlorophenyl)-7-(t-butyl)pyrazolo[3,-d]pyrimidine (PP2), a potent and selective inhibitor of SRC, inhibits angiogenesis *in vivo* and blocks endothelial cell differentiation *in vitro*. SRC activation is also associated with increased expression of another proangiogenic cytokine, interleukin 8 (IL-8), thereby promoting activation of the VEGF receptor 2 and vascular permeability.

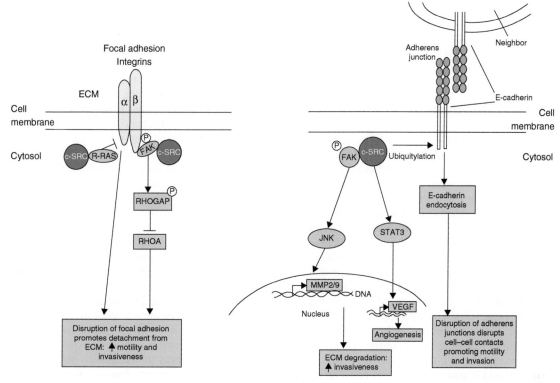

Figure 6.21 Signaling pathways activated by oncogenic SRC to promote tumor cell behavior. The two principal subcellular structures that regulate adhesion, invasion, and motility in normal healthy cells – focal adhesions and adherens junction – are regulated by c-SRC. Focal adhesions provide the structural and mechanical properties that are necessary for cell–matrix attachment and form at the sites where integrins link the actin cytoskeleton to extracellular matrix (ECM) proteins. Adherens junctions enable neighboring cells to adhere to each other: cell–cell attachment. Activated SRC plays a key role in disassembly of focal adhesions which promotes detachment from the ECM, a property that is acquired by cancer cells. It does this through its binding and activation of FAK (focal adhesion kinase): activated FAK phosphorylates p190^{RhoGAP} which in turn inhibits RHOA, leading to focal adhesion disruption. SRC can also destabilize focal adhesions through its tyrosine phosphorylation of R-RAS protein. Activated SRC can also promote the release of cells from each other by disrupting adherens junctions by inducing tyrosine phosphorylation and ubiquitylation of the E-cadherin complex. As a result, cadherin molecules are internalized. Phosphorylation of FAK by c-SRC is also required for the disruption of E-cadherin cell–cell adhesion. SRC might also promote invasion by regulating matrix metalloproteinases (MMPs), enzymes that can break down ECM proteins and thus promote tumor cell invasion and angiogenesis. SRC activation of FAK stimulates the c-Jun N-terminal kinase (JNK) pathway, ultimately leading to the expression of MMP2 and MMP9. Lastly, SRC proteins can promote angiogenesis through the expression of vascular endothelial growth factor (VEGF), by activating the transcription factor STAT3.

c-SRC and human cancer

Much redundancy is built into this complex signaling system, and hundreds of protein tyrosine kinases are now known, several of which are nearly identical to SRC. In fact, specifically blocking the action of SRC protein in normal laboratory animals has relatively little effect as other proteins fill in for the lost function, including ABL.

Of the numerous SFKs, SRC is the most clearly implicated in cancer, particularly advanced gastrointestinal tract cancers such as colorectal cancer. Increased SRC activity has also been demonstrated in several other gastrointestinal malignancies, including liver, pancreas, and esophagus, and also in breast, bladder, endometrium, and ovary. Most notably, gastrointestinal tract cancers show a progressive increase in SRC activity with advancing tumor stage, such that metastatic lesions often have the highest levels of SRC activity. Indeed, SRC is expressed at high levels in colonic polyps and adenomas early in the course of colon cancer development, but there are also large increases in SRC-specific activity in later stages of cancer progression. This would indicate that SRC plays a role in later stages of tumorigenesis.

Paradoxically though, it has been noted that the most aggressive tumors, which are poorly differentiated, often show lower levels of both SRC protein and SRC activity, when compared to more differentiated tumors. One potential explanation is that in these aggressive poorly differentiated tumors there are high levels of receptor tyrosine kinases such as EGFR, which can synergistically activate SRC (see Chapter 5). Therefore, low levels of SRC might be compensated for by high levels of a RTK.

The actual role SRC plays in tumor progression is unclear. Much experimental data on SRC and its effects on cell behavior have derived from fibroblasts grown in culture. The fact that SRC promoted fibroblast cell division led to the notion that this oncogene contributes to tumor progression by stimulating proliferation of precancerous cells. Although this may be true for certain cell types, as is the case for fibroblasts, it is possible that SRC exerts different effects such as motility and invasion on other cell types. For example, overexpression of SRC in human colon cancer cells does not affect cell growth, but can instead enhance motility. Similarly, the cooperation of SRC and EGFR regulates the invasiveness of colon cancer cells, but does not seem to influence proliferation.

c-SRC as a target of drug therapy

SRC is now the target of new drug development for cancer, in particular because it is commonly overexpressed in a large number of human cancers and because its mechanisms of activation are now better understood (this is discussed in more detail in Chapter 16).

BCR–ABL and the Philadelphia chromosome

The quest that led to discovery of the *BCR–ABL* oncogene began in 1960 and is particularly significant as it also provided the first evidence of a genetic link to cancer. The triggering event was the detection of a very small chromosome, known as the Philadelphia (Ph) chromosome, when the blood leukocytes of patients with CML were observed under a microscope. The story of the Philadelphia chromosome should be seen as a model for how a clinical observation can be translated successfully into clinical practice through rigorous basic research and is described in Box 6.4.

The prototypic nonreceptor tyrosine kinase c-ABL (the cellular homolog of the Abelson murine leukemia virus that causes leukemias in mice) has been discussed in Chapter 5. ABL is involved in a wide range of key cellular processes, including reorganization of the actin cytoskeleton, cell replication, differentiation, and apoptosis following DNA damage. There is also evidence of an important role for c-ABL in T-cell signaling. There are obvious structural and functional similarities between c-ABL and c-SRC (described previously). Along with other nonreceptor tyrosine kinases they share conserved SH3, SH2, and kinase domains. However, whereas SRC is kept in an inactive state by phosphorylation of its SH1 kinase domain (leading to a closed conformation of the protein – see Fig. 6.19), ABL is kept inactive by myristoylation of its N-terminal cap locking to the C-lobe of its kinase domain. In their activated states, SRC and ABL are able to affect cell behavior by activating signaling pathways. This is achieved through binding and activation (via tyrosine phosphorylation) of various cellular proteins as previously described for oncogenic SRC (Fig. 6.21).

In contrast to the proto-oncogene *ABL*, its oncogenic counterpart, BCR–ABL fusion protein, is the result of a translocation event created by juxtapositioning the ABL1 gene on chromosome 9 to a part of the BCR (breakpoint cluster region) gene on chromosome 22 (see Box 6.4). This event causes CML and some acute lymphoblastic leukemias (ALL) in humans. Myeloid leukemias are frequently associated with translocations and mutations of tyrosine kinase genes (Tables 6.1 and 6.3). The transforming potential of the products of these oncogenes, including BCR–ABL, TEL–PDGFR, FLT3, and c-KIT, is the result of elevated tyrosine kinase activity no longer tightly regulated within the cell. Not surprisingly then, the kinase domain makes an ideal target for therapeutic intervention and inhibitors have been successfully used to treat patients with myeloid leukemias (described previously in SRC section and also in Chapter 16, Table 16.2).

CML is a stem cell disorder of hematopoietic lineage. Clinically, there are three distinct phases: chronic, accelerated, and blast, which is inevitably fatal. Most patients present in the chronic phase, a stage that is typically indolent in nature, with increase in the number of myeloid progenitor cells found in the blood. If left untreated, the disease progresses to an accelerated phase followed by the fatal blast crisis, during which hematopoietic differentiation is blocked and blast cells accumulate in the bone marrow and peripheral blood.

The constitutive tyrosine kinase activity of BCR–ABL is responsible for this disease process, a result of uncontrolled activation of many cell signaling pathways linked to cell proliferation, cell survival, disrupted differentiation, and DNA damage among others (discussed in the next section).

Importantly, in a regulatable transgenic mouse model of CML, it was shown that deactivating BCR–ABL caused almost complete reversal of the phenotype (Table 16.1), indicating that targeted therapy aimed at blocking oncogene function, could lead to apoptosis or differentiation of tumor cells. As discussed in Chapter 16, imatinib mesylate (Gleevec, STI571, or CP57148B) is a direct inhibitor of several tyrosine kinases, including ABL (ABL1), ARG (ABL2), KIT, and PDGFR, and has had a major impact on the treatment of CML as well as other blood neoplasias and some solid tumors.

Following the translocation event, the BCR portion of the fusion protein retains the RHO GEF (but not the RHO GAP) domain, which has become fused to truncated ABL (see RHO GEF section of RAS superfamily). In addition, the N-terminal myristoylation cap of ABL (which normally keeps ABL inactive when locked to the C-terminal kinase domain) is lost, being replaced by BCR. Consequently, ABL is now constitutively active and thus able to mediate oncogenesis. Aside from the transforming potential of ABL, RHO GEF activity also contributes to anchorage-independent growth by activating RHOA.

With the advances in technology, we now know that three predominant products are formed, depending on which breakpoint located within the BCR gene is fused with exon a2 of ABL. The corresponding BCR–ABL fusion protein will encode either a p190, p210, or p230 molecular weight protein, each of which is oncogenic and is associated with a specific type of leukemia. This suggests that their respective transforming capacities are distinct. However, in the majority of patients with CML and in approximately a third of those with ALL, the fusion protein responsible is the p210 BCR–ABL product. Importantly, transgenic mouse models, each expressing one of these fusion proteins, closely resemble the distinct disease processes seen in patients: mice expressing p190 had the shortest latency period and developed B-cell origin leukemia exclusively. In contrast, p210 transgenic mice typically developed leukemia of B-, T-lymphoid, or myeloid origin, while those expressing p230 showed the longest latency period and developed a less aggressive tumor.

BCR–ABL activated signaling pathways

The BCR–ABL protein is cytoplasmic which, unlike the predominantly nuclear ABL protein, allows it access to many substrates, ultimately leading to the activation of various signal transduction pathways, which include the well-known PI3K–AKT (survival) and RAS–RAF–MEK–ERK (cell replication) pathways. As depicted in Fig. 6.22, BCR–ABL causes the recruitment of adaptor molecules, such as GRB2 and P85, which in turn lead to activation of RAS and PI3K pathways, respectively.

We know from many experimental studies, including the use of PI3K inhibitors, that BCR–ABL activation of PI3K is essential for myeloid and lymphoid cell transformation. Moreover, these inhibitors induce cell death selectively in BCR–ABL leukemia cells compared with normal hematopoietic cells. Activation of the PI3K–AKT signaling pathway is well known for protecting cells against death (apoptosis) and so is often referred to as a **survival pathway**. The importance of this pathway in regulating cell death is discussed in more detail elsewhere, in Chapter 5 (Figs

Box 6.4 Chromosome rearrangements in cancer – the Philadelphia chromosome

An important experimental approach in the detection of cancer cells, especially those derived from the hematopoietic system, has relied on the identification and characterization of clonal and recurrent cytogenetic abnormalities often resulting from chromosomal rearrangements. The presence of chromosome regions anomalously stained, may represent DNA rearrangements such as gene amplification (Table 6.2), translocations, and inversions (Tables 6.1 and 6.3).

The application of chromosome-banding techniques in the early 1970s enabled the precise cytogenetic characterization of many chromosomal translocations in human leukemia, lymphoma, and solid tumors. The subsequent development of molecular cloning techniques then enabled the identification of proto-oncogenes at or near chromosomal breakpoints in various neoplasms. Some of these proto-oncogenes, such as *MYC* and *ABL*, had been previously identified as retroviral oncogenes. Since then many new oncogenes have been identified by cloning of chromosomal breakpoints in cancers.

after the city in which the discovery was made in 1960 by Peter Nowell and his graduate student David Hungerford. The importance of this discovery cannot be overstated; although a number of previous studies had shown chromosomal abnormalities in human cancer, the Philadelphia chromosome was the first documentation of a bona fide genetic link to cancer and was to start a new era of cancer research. Nowell hypothesized that this genetic alteration might somehow provide a growth advantage to the abnormal cells.

This indeed would turn out to be the case but crucial work needed to be done over the coming years by various research groups. Important questions needed to be answered: first, was this chromosomal abnormality the cause or the consequence of CML? If causal, then the identification of the affected gene(s) might lead to new therapeutic interventions. Second, was this genetic change linked to other malignancies or only to CML – an important but still relatively uncommon human cancer?

In 1973, with the use of improved cytogenetic staining techniques, Janet D. Rowley identified the mechanism underlying formation of the Philadelphia chromosome – the long arms of two chromosomes, 9 and 22, swap places, known as a reciprocal translocation. This translocation creates an elongated chromosome 9 and a truncated chromosome 22 (the Philadelphia chromosome). Later work would reveal the consequences of this chromosomal rearrangement, Namely, a fusion gene (*BCR–ABL*) is created by juxtapositioning the *ABL1* (cellular homolog of the Abelson leukemia virus) gene on chromosome 9 (region q34) to a part of the *BCR* (breakpoint cluster region) gene on chromosome 22 (region q11). Hence, the translocation is designated as t(9;22)(q34;q11).

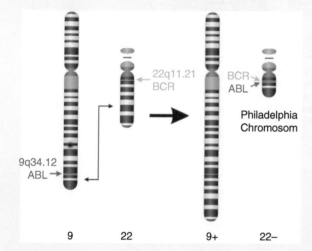

Drawn by A. Obeidat.

Recurring karyotypic abnormalities are observed in large numbers of hematologic and solid tumors and include chromosomal rearrangements, gain or loss of whole chromosomes, or chromosome segments.

The first such abnormality identified in a human cancer was a small chromosome found in leukocytes of patients with chronic myelogenous leukemia (CML) but not in normal leukocytes. This unusually small chromosome was called the Philadelphia chromosome

The potential functional significance of the fusion protein BCR–ABL became apparent with the discovery of tyrosine kinases, a class of enzymes that were shown to be critical in the control of cellular growth and differentiation (see Chapter 5). Moreover, dysregulated activity of these enzymes could lead to abnormal cell growth as seen in cancer cells. It was not long before the link to CML became clear – sequence homology and enzymatic assays showed that the ABL component itself was a protein tyrosine kinase and that the fused BCR–ABL protein had lost its appropriate regulation, that is, it was constitutively active. The 1990s saw a series of important studies using animal models which demonstrated that BCR–ABL is an oncogene and is the cause, not the consequence, of CML. The search was now on to identify inhibitors of ABL's tyrosine kinase activity that could be used in the clinic as a targeted cancer therapy. This is discussed in detail in Chapter 16.

5.16, 5.18, and 5.19) as well as in Chapter 6, hence is only briefly touched upon here.

PI3 kinase, a heterodimeric protein, consists of an 85-kDa regulatory subunit (p85) containing one SH3 domain and two SH2 domains, and a 110-kDa catalytic subunit (p110). p85 requires activation via its SH2 domain with phosphorylated tyrosine, and although unclear at present, most evidence indicates this occurs via an autophosphorylation site located on the BCR region of BCR–ABL. The 110-kDa subunit converts the second messenger PIP2 to PIP3 at the plasma membrane, which can then activate AKT (see Fig. 5.19, Chapter 5). A major role of AKT is to suppress cell death by phosphorylating proteins such as Forkhead family (FOXO), BAD, and GSK-3β, which is discussed more

fully in Chapter 5 (Figs 5.16, 5.18, and 5.19) and Chapter 6. Briefly, phosphorylation of FOXO proteins mediates binding of the 14-3-3 chaperone protein, which occurs in the nucleus and allows the efficient export of the complex to the cytoplasm. It is the cytoplasmic sequestration of this complex that effectively inhibits the transcriptional activation of several proapoptotic molecules, including Bim and Trail. Importantly, several investigators have shown that BCR–ABL inhibitors causes an increase in Bim levels and that siRNA directed at Bim induces resistance to BCR–ABL inhibitors. Similarly, AKT phosphorylation of BAD, a proapoptotic member of the Bcl-2 family, leads to it being sequestered in the cytosol by 14-3-3. In this way, since BAD is no longer free to bind antiapoptotic BCL-2 family members, this enables such

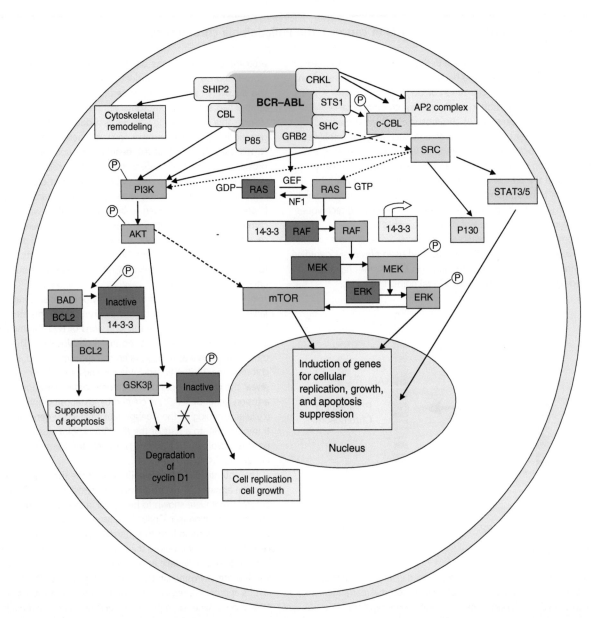

Figure 6.22 BCR–ABL signaling in chronic myelogenous leukemia (CML). The constitutive tyrosine kinase activity of BCR–ABL results in uncontrolled activation of many cell signaling pathways linked to cell proliferation, cell survival, disrupted differentiation, and DNA damage among others. A core network of seven related proteins is responsible for activation of downstream signaling in CML. BCR–ABL causes the recruitment of adaptor molecules, such as GRB2 and P85, which in turn lead to activation of the RAS and PI3K pathways, respectively (see main text). PI3K activation of downstream target AKT suppresses cell death by phosphorylating proteins, such as Forkhead family (FOXO), BAD, and GSK-3β. Phosphorylated FOXO proteins bind 14-3-3 chaperone protein in the nucleus, allowing efficient export of the complex to the cytoplasm, thus effectively inhibiting transcriptional activation of proapoptotic molecules, including Bim and Trail. Similarly, AKT phosphorylation of the proapoptotic protein BAD sequesters it in the cytosol by 14-3-3, preventing BAD from binding antiapoptotic BCL-2 family members. Phosphorylated GSK-3β renders it inactive and thereby prevents the degradation of downstream targets of GSK-3β, such as cyclin D1 and β-catenin, which are important in cell growth and replication. The second major signaling pathway activated by BCR–ABL is RAS–RAF–ERK (MAPK). The recruitment of adaptor molecule GRB2 by BCR ultimately leads to RAS activation by enhancing GDP–GTP exchange (see main text). This signal transduction pathway culminates in the activation and subsequent translocation of ERK to the nucleus, where it activates transcription factors, such as c-JUN, c-MYC, and c-FOS, which are important in the cell cycle. BCR–ABL also activates STAT proteins, predominately STAT5, leading to increased expression of antiapoptotic proteins MCL-1 and BCL-X_L, and increased expression of cyclin D1, which is important in cell-cycle progression. BCR–ABL phosphorylation of Crkl and subsequent phosphorylation of c-Cbl may contribute to activation of PI3K.

members to sequester BAX and thus prevent this proapoptotic member inducing apoptosis (described further in Chapter 8).

Lastly, GSK-3β is phosphorylated by AKT, which renders it inactive. Two downstream targets of GSK-3β include cyclin D1 and β-catenin, which are important in cell growth and replication. Phosphorylation of these proteins by GSK-3β targets both of these proteins for proteasome-mediated degradation, and so AKT-mediated inactivation of GSK-3β would promote cell growth and replication.

The second major signaling pathway activated by BCR–ABL is RAS–RAF–ERK (MAPK), which is crucial for cell replication. This pathway has already been discussed previously in the RAS section

(Fig. 6.14), and is also described in Chapter 5 (Figs 5.16, 5.17, and 5.18). As previously mentioned for the P85 subunit of PI3K, the same autophosphorylation site on BCR is critical for GRB2 binding via its SH2 domain. This leads to recruitment of adaptor protein SOS, which in turn promotes RAS activation by enhancing GDP–GTP exchange. This signal transduction pathway culminates in the activation and subsequent translocation of ERK to the nucleus, where it activates transcription factors, such as c-JUN, c-MYC, and c-FOS, important in the cell cycle.

BCR–ABL also activates signal transducer and activator of transcription (STAT) proteins, predominately STAT5 and to a lesser extent STAT3 and STAT1 (Fig. 6.22). These proteins are normally regulated by cytokine receptors and are critical for driving transcription necessary for growth, survival, and differentiation of hematopoietic cells. However, BCR–ABL expression results in constitutive activation of STAT5 and essentially bypasses such regulation. The outcome is increased expression of the antiapoptotic Bcl-2 family members MCL-1 and BCL-X_L (see Chapter 8) as well as increased expression of cyclin D1, a critical molecule that facilitates G_1/S phase cell-cycle progression (see Chapter 4).

Finally, there is evidence linking BCR–ABL with DNA damage. It has been surmised that the increase in expression of BCR–ABL that is seen during the fatal blast phase of CML may account for the presence of secondary genetic changes. It is thus feasible that such DNA changes promote transition to blast phase, presumably by allowing BCR–ABL-independent growth and so resistance to BCR–ABL inhibitors. However, a clearer insight into the mechanism of transition to blast phase awaits further investigation.

Studies have shown BCR–ABL kinase activity increases the levels of endogenous reactive oxygen species (ROS) and increases the amount of ROS-induced DNA damage. Furthermore, following DNA damage, BCR–ABL was shown to translocate to the nucleus, physically interact with ataxia telangiectasia and Rad3-related protein (ATR), and inhibit ATR-dependent phosphorylation of CHK1 (see Chapter 10).

The BCL-2 family

Do not fear death so much but rather the inadequate life.
Bertolt Brecht

The BCL-2 family of proteins comprises both proapoptotic and antiapoptotic members, the balance of which determines whether or not a cell commits suicide – apoptosis. These proteins have emerged as fundamental regulators of mitochondrial outer membrane permeabilization (MOMP), which is necessary for cytochrome *c* release – a crucial event that takes place during apoptosis. Given the importance of this family in cancer – through blocking apoptosis – we have devoted a significant portion of Chapter 8 to describing these proteins.

BCL-2 (B-cell lymphoma 2) in human follicular lymphoma involves a chromosome translocation event that moves the *BCL2* gene from chromosome 18 to 14 (t14;18) linking the *BCL2* gene to an immunoglobulin locus (see Table 6.3). A link between cell death by apoptosis (see Chapter 8) and cancer emerged when BCL-2 was found to inhibit cell death rather than promote proliferation. This important work and others showed that cell survival is controlled separately from cell proliferation and that inhibition of apoptosis is a central step in tumor development (see Fig. 6.4) – a concept we touched upon earlier in "Oncogene collaboration." This unexpected discovery gave birth to the idea, now widely accepted, that impaired apoptosis is a crucial step in tumorigenesis, and is strongly supported by several transgenic mouse models of cancer as previously described. For example, mimicking human follicular lymphoma (t14;18), *BCL2* transgenes placed under the control of a B-cell promoter in transgenic mice give rise to B-lymphoid tumors. However, the stochastic onset of tumors in these mice implied that additional genetic changes were needed for lymphoma development. Further experiments in which both MYC and BCL-2 were coexpressed in B cells of transgenic mice showed a dramatically accelerated onset of lymphomas in these mice. These results revealed a strong cooperation between MYC and BCL-2, whose elevated and constitutive expression enforced both cell proliferation and cell survival. This strong cooperation, or synergy, has been shown more recently to promote tumor formation in other tissues *in vivo*, such as breast. It is likely that all BCL-2 prosurvival family members are oncogenes. BCL-X_L, for example, has been implicated in mouse myeloid and T-cell leukemias and insulinomas as discussed in Chapter 16 (Table 16.1).

Apoptotic defects are also important in established cancer: in both solid (e.g. prostate) and non-solid (e.g. leukemia) tumors, BCL-2 overexpression is associated with poor prognosis and often (although not always) resistant to cytotoxic therapy. Conversely, proapoptotic members of the BCL-2 family, exemplified by BAX and BAK, are mutated in some human gastric and colorectal cancers as well as in leukemias; these mutations may lead to loss of proapoptotic tumor suppressor activity, thus inhibiting death of these neoplastic cells. In support of this, experimental data show that loss of BAX can enhance tumor formation and that losses of both BAX and BAK appear to synergize, leading to further enhancement of tumor formation when compared to loss of either one alone.

Biologically targeted therapies in cancer and the concept of "oncogene addiction"

The concept of oncogene addiction and targeted therapeutics is of such central importance in the development of cancer treatments in the current era that this will be discussed in depth in Chapter 16. Some of the basic concepts have already been covered in Chapter 1 and 5, so no further discussion will be included here and the reader is directed to Chapter 16.

Conclusions and future directions

So how do we summarize this wealth of information? Some basic principles have emerged and we can begin by delineating the most important ones. No oncogene, as far as we know, can induce cancer without cooperating with other oncogenes and/or inactivation of tumor suppressors. At one level such cooperation is essential in order to negate the inherent tumor suppressor activity, either apoptosis or senescence, associated with oncogenic deregulation. At further levels, additional epimutations may be necessary to unleash the full repertoire of cancer behaviors. Much has been learned about the signaling mechanisms through which oncogenes alter normal cell behavior, in particular by perverting control of cell cycle and survival. Although it is tempting to arrange many oncogenic mutations along a linear pathway leading inexorably from growth factor receptor, through RAS–RAF–ERK to FOS–JUN and MYC, we must resist this. Rather it should by now be clear that various oncogenes and indeed other cancer-relevant proteins operate within complex webs of often linked opposing processes, with outcome determined by a balance of intrinsic and extrinsic inputs and restraining forces.

Different branches of diverging signaling cascades may become activated depending both on the hierarchical position of the oncogenic lesion and by other factors, including the level of deregulated oncogene activity, tissue context, and the nature of other epimutations. How can we unravel this complexity in order to make some practical use of the information? Ultimately, the functional consequences of targeting any particular molecule must be tested in a relevant context by recourse to experimental manipulation in cultured cells, animal models, and then hypotheses verified in humans, possibly ultimately in a clinical trial.

> Now my own suspicion is that the Universe is not only queerer than we suppose, but queerer than we can suppose.
>
> *J.B.S. Haldane*

Outstanding questions

1. To further understand the innate tumor suppressor actions of some key oncogenes such as c-Myc, Ras, and others – how the different outcomes (growth/replication versus apoptosis/senescence) are determined and how they might be manipulated by small molecule drugs or even siRNA.

2. To continue to develop and study new oncogene targeted therapeutics, synthetic lethal interactions, and non-oncogene addiction, as well as the optimal conditions surrounding their application.

3. To apply a "systems biology" approach to describing the functional gene/protein networks acting downstream of key oncogene signaling during tumor evolution and during regression *in vivo*. Moreover, to attempt to develop an integrated understanding of the interactions between signaling pathways often concurrently deregulated in cancers.

4. It is abundantly clear that a dynamic interaction occurs between the cancer cell and its local and systemic microenvironment. The "tumor microenvironment," includes the stroma and comprises numerous different cell types, is rich in growth factors and enzymes, and includes parts of the blood and lymphatic systems. Probably the greatest challenge faced is to understand how interactions with these various factors determine the outcome of oncogene activation *in vivo*.

Bibliography

General

Bild, A.H., Yao, G., Chang, J.T., *et al.* (2006). Oncogenic pathway signatures in human cancers as a guide to targeted therapies. *Nature*, **439**(7074): 353–7.

Hanahan, D. and Weinberg, R.A. (2011). Hallmarks of cancer: the next generation. *Cell*, **144**: 646–74.

Martin, G.S. (1970). Rous sarcoma virus: a function required for the maintenance of the transformed state. *Nature*, **227**: 1021–3.

Moore, P.S. and Chang, Y. (2010). Why do viruses cause cancer? Highlights of the first century of human tumour virology. *Nature Reviews Cancer*, **10**: 878–89.

Paul, J. (1986). Roles of oncogenes in carcinogenesis. In O.H. Iverson (ed.) *Theories of Carcinogenesis*. New York: Hemisphere Publishing, pp. 145–160.

Roberts, R.B., Arteaga, C.L., and Threadgill, D.W. (2004). Modeling the cancer patient with genetically engineered mice: prediction of toxicity from molecule-targeted therapies. *Cancer Cell* **5**: 115–20.

Shih, C. and Weinberg, R.A. (1982). Isolation of a transforming sequence from a human bladder carcinoma cell line. *Cell*, **29**: 161–9.

Stehelin, D., Varmus, H.E., Bishop, J.M., and Vogt, P.K. (1976). DNA related to the transforming gene(s) of avian sarcoma viruses is present in normal avian DNA. *Nature*, **260**: 170–3.

Toyoshima, K. and Vogt, P.K. (1969). Temperature sensitive mutants of an avian sarcoma virus. *Virology*, **39**: 930–1.

Oncogene cooperation

Finch, A., Prescott, J., Shchors, K., *et al.* (2006). Bcl-xL gain of function and p19 ARF loss of function cooperate oncogenically with Myc in vivo by distinct mechanisms. *Cancer Cell*, **10**(2): 113–20.

Hahn, W.C. and Weinberg, R.A. (2002). Modelling the molecular circuitry of cancer. *Nature Reviews Cancer*, **2**(5): 331–41.

Land, H., Chen, A.C., Morgenstern, J.P., Parada, L.F., and Weinberg, R.A. (1986). Behavior of myc and ras oncogenes in transformation of rat embryo fibroblasts. *Molecular and Cellular Biology*, **6**(6): 1917–25.

Land, H., Parada, L.F., and Weinberg, R.A. (1983). Tumorigenic conversion of primary embryo fibroblasts requires at least two cooperating oncogenes. *Nature*, **304**(5927): 596–602.

Strasser, A., Harris, A.W., Bath, M.L., and Cory, S. (1990). Novel primitive lymphoid tumours induced in transgenic mice by cooperation between myc and bcl-2. *Nature*, **348**(6299): 331–3.

c-Myc

Adams, J.M., Harris, A.W., Pinkert, C.A., *et al.* (1985). The c-myc oncogene driven by immunoglobulin enhancers induces lymphoid malignancy in transgenic mice. *Nature*, **318**(6046): 533–8.

Amati, B. (2004). Myc degradation: dancing with ubiquitin ligases. *Proceedings of the National Academy of Sciences of the U.S.A.*, **101**(24): 8843–4.

Ayer, D.E., Lawrence, Q.A., and Eisenman, R.N. (1995). Mad-Max transcriptional repression is mediated by ternary complex formation with mammalian homologs of yeast repressor Sin3. *Cell* **80**(5): 767–76.

Berta, M.A., Baker, C.M., Cottle, D.L., and Watt, F.M. (2009). Dose and context dependent effects of Myc on epidermal stem cell proliferation and differentiation. *EMBO Molecular Medicine*, **2**: 16–25.

Brenner, C., Deplus, R., Didelot, C., *et al.* (2005). Myc represses transcription through recruitment of DNA methyltransferase corepressor. *EMBO Journal*, **24**(2): 336–46.

Bui, T.B. and Mendell, J.T. (2010). Myc: maestro of microRNAs. *Genes and Cancer*, **1**: 568–75.

Chang, T.C., Yu, D., Lee, Y.S., *et al.* (2008). Widespread microRNA repression by Myc contributes to tumorigenesis. *Nature Genetics*, **40**: 43–50.

Coller, H.A., Grandori, C., Tamayo, P., *et al.* (2000). Expression analysis with oligonucleotide microarrays reveals that MYC regulates genes involved in growth, cell cycle, signaling, and adhesion. *Proceedings of the National Academy of Sciences of the U.S.A.*, **97**(7): 3260–5.

Conacci-Sorrell, M., Ngouenet, C., and Eisenman, R.N. (2010). Myc-nick: a cytoplasmic cleavage product of Myc that promotes alpha-tubulin acetylation and cell differentiation. *Cell*, **142**: 480–93.

Dews, M., Homayouni, A., Yu, D., *et al.* (2006). Augmentation of tumor angiogenesis by a Myc-activated microRNA cluster. *Nature Genetics*, **38**(9): 1060–5.

Eilers, M. and Eisenman, R.N. (2008). Myc's broad reach. *Genes and Development*, **22**: 2755–66.

Evan, G.I., Wyllie, A.H., Gilbert, C.S., et al. (1992). Induction of apoptosis in fibroblasts by c-myc protein. *Cell*, **69**(1): 119–28.

Flores, I., Evan, G., and Blasco, M.A. (2006). Genetic analysis of myc and telomerase interactions in vivo. *Molecular and Cellular Biology*, **26**(16): 6130–8.

Gomez-Roman, N., Grandori, C., Eisenman, R.N., and White, R.J. (2003). Direct activation of RNA polymerase III transcription by c-Myc. *Nature*, **421**(6920): 290–4.

Guccione, E., Martinato, F., Finocchiaro, G., et al. (2006). Myc-binding-site recognition in the human genome is determined by chromatin context. *Nature Cell Biology*, **8**(7): 764–70.

Herbst, A., Hemann, M.T., Tworkowski, K.A., Salghetti, S.E., Lowe, S.W., and Tansey, W.P. (2005). A conserved element in Myc that negatively regulates its proapoptotic activity. *EMBO Report*, **6**(2): 177–83.

Herold, S., Wanzel, M., Beuger, V., et al. (2002). Negative regulation of the mammalian UV response by Myc through association with Miz-1. *Molecular Cell*, **10**(3): 509–21.

Knoepfler, P.S., Zhang, X.Y., Cheng, P.F., Gafken, P.R., McMahon, S.B., and Eisenman, R.N. (2006). Myc influences global chromatin structure. *EMBO Journal*, **25**(12): 2723–34.

Lawlor, E.R., Soucek, L., Brown-Swigart, L., Shchors, K., Bialucha, C.U., and Evan, G.I. (2006). Reversible kinetic analysis of Myc targets in vivo provides novel insights into Myc-mediated tumorigenesis. *Cancer Research*, **66**(9): 4591–601.

Li, F., Xiang, Y., Potter, J., Dinavahi, R., Dang, C.V., and Lee, L.A. (2006). Conditional deletion of c-myc does not impair liver regeneration. *Cancer Research*, **66**(11): 5608–12.

Muncan, V., Sansom, O.J., Tertoolen, L., et al. (2006). Rapid loss of intestinal crypts upon conditional deletion of the Wnt/Tcf-4 target gene c-Myc. *Molecular and Cellular Biology*, **26**: 8418–26.

Nilsson, J.A., Maclean, K.H., Keller, U.B., Pendeville, H., Baudino, T.A., and Cleveland, J.L. (2004). Mnt loss triggers Myc transcription targets, proliferation, apoptosis, and transformation. *Molecular and Cellular Biology*, **24**(4): 1560–9.

Oskarsson, T., Essers, M.A., Dubois, N., et al. (2006). Skin epidermis lacking the c-Myc gene is resistant to Ras-driven tumorigenesis but can reacquire sensitivity upon additional loss of the p21Cip1 gene. *Genes and Development*, **20**(15): 2024–9.

Patel, J.H., Loboda, A.P., Showe, M.K., Showe, L.C., and McMahon, S.B. (2004). Analysis of genomic targets reveals complex functions of c-Myc. *Nature Reviews Cancer*, **4**(7): 562–8.

Pelengaris, S., Khan, M., and Evan, G. (2002). c-Myc: more than just a matter of life and death. *Nature Reviews Cancer*, **2**: 764–76.

Zindy, F., Eischen, C.M., Randle, D.H., et al. (1998). Myc signaling via the ARF tumor suppressor regulates p53-dependent apoptosis and immortalization. *Genes and Development*, **12**(15): 2424–33.

RAS

Boguski, S. and McCormick, F. (1993). Proteins regulating Ras and its relatives. *Nature*, **366**: 643–54.

Downward, J. (2003). Targeting RAS signalling pathways in cancer therapy. *Nature Reviews Cancer*, **3**: 11–22.

Fisher, G.H., Wellen, S.L., Klimstra, D., et al. (2001). Induction and apoptotic regression of lung adenocarcinomas by regulation of a K-Ras transgene in the presence and absence of tumour suppressor genes. *Genes and Development*, **15**: 3249–62.

Frame, S. and Balmain, A. (2000). Integration of positive and negative growth signals during ras pathway activation in vivo. *Current Opinion in Genetics and Development* **10**: 108–13.

Goldstein, J.L. and Brown, M.S. (1990). Regulation of the mevalonate pathway. *Nature*, **343**: 425–30.

Hancock, J.F. (2003). RAS proteins: different signals from different locations. *Nature Reviews: Molecular Cell Biology*, **4**: 373–84.

Johnson, L., Mercer, K., Greenbaum, D., et al. (2001). Somatic activation of the K-ras oncogene causes early onset lung cancer in mice. *Nature*, **410**: 1111–16.

Malumbres, M. and Barbacid, M. (2003). RAS oncogenes: the first 30 years. *Nature Reviews Cancer*, **3**: 7–13.

Pylayeva-Gupta, Y., Grabocka, E., and Bar-Sagi, D. (2011). RAS oncogenes: weaving a tumorigenic web. *Nature Reviews Cancer*, **11**: 761–74.

Rodriguez-Viciana, P., Warne, P.H., Khwaja,A., et al. (1997). Role of phosphoinositide 3-OH kinase in cell transformation and control of the actin cytoskeleton by Ras. *Cell*, **89**: 457–67.

Serrano, M., Lin, A.W., McCurrach, M.E., Beach, D., and Lowe, S.W. (1997). Oncogenic ras provokes premature senescence associated with accumulation of p53 and p16ink4a. *Cell*, **88**: 593–602.

Sweet-Cordero, A., Mukherjee A, Subramanian A, et al. (2005). An oncogenic Kras expression signature identified by cross-species gene expression analysis. *Nature Genetics*, **37**(1): 48–55.

Vakiani, E. and Solit, D.B. (2011). KRAS and BRAF: drug targets and predictive biomarkers. *Journal of Pathology*, **223**: 219–29.

Yeh, E., Cunningham, M., Arnold, H., et al. (2004). A signalling pathway controlling c-Myc degradation that impacts oncogenic transformation of human cells. *Nature Cell Biology*, **6**(4): 308–18.

Rac and Rho

Coleman, M.L., Marshall, C.J., and Olson, M.F. (2004). RAS and RHO GTPases in G1-phase cell-cycle regulation. *Nature Reviews: Molecular Cell Biology*, **5**(5): 355–66.

Radisky, D.C., Levy, D.D., Littlepage, L.E., et al. (2005). Rac1b and reactive oxygen species mediate MMP-3-induced EMT and genomic instability. *Nature*, **436**(7047): 123–7.

Sahai, E. and Marshall, C.J. (2002). RHO-GTPases and cancer. *Nature Reviews Cancer*, **2**: 133–42.

Vega, F.M., Fruhwirth, G., Ng, T., and Ridley, A.J. (2011). RhoA and RhoC have distinct roles in migration and invasion by acting through different targets. *Journal of Cell Biology*, **193**: 655–65.

BCR–ABL

Bartram, C.R., de Klein, A., Hagemeijer, A., et al. (1983). Translocation of c-abl oncogene correlates with the presence of a Philadelphia chromosome in chronic myelocytic leukaemia. *Nature*, **306**(5940): 277–80.

Melo, J.V. and Barnes, D.J. (2007). Chronic myeloid leukaemia as a model of disease evolution in human cancer. *Nature Reviews Cancer*, **7**(6): 441–53.

Nowell, P.C. and Hungerford, D.A. (1960). A minute chromosome in human chronic granulocytic leukemia. *Science*, **132**: 1497.

SRC

Aleshin, A. and Finn, R.S. (2010). SRC: A century of science brought to the clinic. *Neoplasia*, **12**: 599–607

Yeatman, T.J. (2004). A renaissance for SRC. *Nature Reviews Cancer*, **4**(6): 470–80.

Bcl-2 family

Azmi, A.S., Wang, Z., Philip, P.A., Mohammad, R.M., and Sarkar, F.H. (2011). Emerging Bcl-2 inhibitors for the treatment of cancer. *Expert Opinion on Emerging Drugs*, **16**: 59–70.

Cory, S. and Adams, J.M. (2002). The Bcl2 family: regulators of the cellular life-or-death switch. *Nature Reviews Cancer*, **2**(9): 647–56.

Egle, A., Harris, A.W., Bouillet, P., and Cory, S. (2004). Bim is a suppressor of Myc-induced mouse B cell leukemia. *Proceedings of the National Academy of Sciences of the U.S.A.*, **101**(16): 6164–9.

Llambi, F., Moldoveanu, T., Tait, S.W., et al. (2011). A unified model of mammalian BCL-2 protein family interactions at the mitochondria. *Molecular Cell*, **44**(4): 517–31.

Cancer models and tumor reversal

Blakely, C.M., Sintasath, L., D'Cruz, C.M., et al. (2005). Developmental stage determines the effects of c-Myc in the mammary epithelium. *Development*, **132**(5): 1147–60.

Boxer, R.B., Jang, J.W., Sintasath, L., and Chodosh, L.A. (2004). Lack of sustained regression of c-MYC-induced mammary adenocarcinomas following brief or prolonged MYC inactivation. *Cancer Cell*, **6**(6): 577–86.

Chin, L., Tam, A., Pomerantz, J., et al. (1999). Essential role for oncogenic Ras in tumour maintenance. *Nature*, **400**(6743): 468–72.

D'Cruz, C.M., Gunther, E.J., Boxer, R.B., et al. (2001). c-MYC induces mammary tumourigenesis by means of a preferred pathway involving spontaneous Kras2 mutations. *Nature Medicine*, **7**: 235–9.

Felsher, D.W. (2010). MYC inactivation elicits oncogene addiction through both tumor cell–intrinsic and host-dependent mechanisms. *Genes Cancer*, **1**: 597–604.

Fisher, G.H., Wellen, S.L., Klimstra, D., et al. (2001). Induction and apoptotic regression of lung adenocarcinomas by regulation of a K-Ras transgene in the presence and absence of tumour suppressor genes. *Genes and Development*, **15**: 3249–62.

Jain, M., Arvanitis, C., Chu, K., et al. (2002). Sustained loss of a neoplastic phenotype by brief inactivation of MYC. *Science*, **297**(5578): 102–4.

Junttila, M.R., Karnezis, A., Garcia, D., et al. (2010). Selective activation of p53-mediated tumour suppression in high-grade tumours. *Nature*, **468**: 567–71.

Moody, S.E., Sarkisian, C.J., Hahn, K.T., et al. (2002). Conditional activation of Neu in the mammary epithelium of transgenic mice results in reversible pulmonary metastasis. *Cancer Cell*, **2**: 451–61.

Murphy, D.J., Juntilla, M.R., Pouyet, L., et al. (2008). Distinct thresholds govern Myc's Biological output in vivo. *Cancer Cell*, **14**: 447–57.

Pao, W., Klimstra, D.S., Fisher, G.H., and Varmus, H.E. (2003). Use of avian retroviral vectors to introduce transcriptional regulators into mammalian cells for analyses of tumor maintenance. *Proceedings of the National Academy of Sciences of the U.S.A.*, **100**: 8764–9.

Peeper, D. and Berns, A. (2006). Cross-species oncogenomics in cancer gene identification. *Cell*, **125**(7): 1230–3.

Pelengaris, S., Khan, M., and Evan, G.I. (2002). Suppression of Myc-induced apoptosis in beta cells exposes multiple oncogenic properties of Myc and triggers carcinogenic progression. *Cell*, **109**(3): 321–34.

Pelengaris, S., Littlewood, T., Khan, M., Elia, G., and Evan, G. (1999). Reversible activation of c-Myc in skin: induction of a complex neoplastic phenotype by a single oncogenic lesion. *Molecular Cell*, **3**(5): 565–77.

Shachaf, C.M., Kopelman, A.M., Arvanitis, C., et al. (2004). c-Myc inactivation uncovers pluripotent differentiation and tumour dormancy in hepatocellular cancer. *Nature*, **431**: 1112–17.

Soucek, L., Whitfield, J., Martins, C.P., et al. (2008). Modelling Myc inhibition as a cancer therapy. *Nature*, **455**: 679–83.

Zender, L., Spector, M.S., Xue, W., et al. (2006). Identification and validation of oncogenes in liver cancer using an integrative oncogenomic approach. *Cell*, **125**(7): 1253–67.

Questions for student review

1) Oncogenes:
   a. Are activated versions of proto-oncogenes.
   b. Mutated versions of genes that usually resist cancer formation.
   c. Usually contribute to cancer formation only if both alleles are mutated.
   d. Are often involved in deregulating cell division.
   e. Are always mutated versions of cellular proto-oncogenes.

2) c-Myc:
   a. Is usually mutated in human cancers.
   b. Is frequently overexpressed in human cancers.
   c. Is primarily a transcriptional activator when dimerized with Mad.
   d. May promote apoptosis as well as cell cycle transit.
   e. May cooperate with Ras in cell transformation.

3) Oncogenic RAS:
   a. May result from increased RAS GTPase activity.
   b. May result in increased signaling through the MAPK cascade.
   c. Was first described by studies of the Rous sarcoma virus.
   d. May suppress cell death by activation of AKT.
   e. May activate replicative senescence mechanisms.

4) Regarding RAS:
   a. *HRAS* is the most frequently mutated in human cancers.
   b. RAS activity in tumors is usually due to mutations in the *RAS* gene.
   c. RAS may be activated by inactivation of NF-1.
   d. RAS signaling may be activated by NEU.
   e. RAS activates RAF by serine/threonine phosphorylation.

5) SRC
   a. Was the first oncogene to be described.
   b. Is a receptor tyrosine kinase.
   c. Is only involved in avian cancers.
   d. Is not known to be mutated in human cancers.
   e. May be activated by receptor tyrosine kinases.

7 Tumor Suppressors

Martine F. Roussel
Cancer Center Signal Transduction Program, St. Jude and UT Memphis, USA

How much must I overcome before I triumph?

Pierre Corneille

Key points

- Tumor suppressors are guardians against DNA damage induced, for example, by ultraviolet (UV) or irradiation (X-rays) or an excess of proliferative signals.
- Tumor suppressors that monitor DNA damage can help to repair the damage before cells divide. They can induce growth arrest at two DNA damage checkpoints: the first gap phase (G_1) and the second gap phase (G_2) of the cell cycle. This enables cells to check the integrity of the chromosomes before proceeding into DNA replication (S phase) or division (M phase). In the absence of these checkpoints, as a result of the loss of tumor suppressor function, the damage is not repaired and cells accumulate mutations that ultimately lead to cancer.
- Genetic (gene mutation or deletion) or epigenetic (promoter methylation, mutations in the promoter of genes) alterations in tumor suppressors such as RB, p53, or the genes that regulate them, are part of the life history of all cancer cells.
- The retinoblastoma protein, RB, and the transcription factor p53 control growth arrest, apoptosis, and differentiation, and their pathways collaborate in tumor prevention.
- RB binds to transcription factors to regulate cell proliferation. Its phosphorylation state affects its ability to bind these factors, E2Fs being the major ones. Phosphorylation of RB releases transcription factors that transcribe genes encoding proteins required for the initiation of DNA synthesis.
- Oncogenic and hyperproliferative stresses cause the activation of p53, which in turn restrains uncontrolled cell growth by blocking the cell cycle and inducing programmed cell death (apoptosis).
- RB and p53 are regulated by $p16^{INK4a}$ and $p14^{ARF}$ ($p19^{Arf}$ in mice).

Introduction

All that makes existence valuable to any one depends on the enforcement of restraints upon the actions of other people.

John Stuart Mill

Tumor suppressors are the cell's guardians against DNA damage induced, for example, by ultraviolet (UV) exposure from sunlight, gamma irradiation (X-rays), chemotherapeutic drugs, or an excess of inappropriate proliferative signals. They prevent incipient cells from becoming malignant by arresting their proliferation or inducing them to commit suicide (apoptotic cell death). Tumor suppressors also monitor critical cellular checkpoints that govern the mitotic cycle, DNA repair, transcription, apoptosis, and differentiation. Some tumor suppressors prevent the inappropriate activation of signaling pathways involved in cell growth. The functional inactivation of tumor suppressors by mutation, deletion, or gene silencing creates an imbalance between proliferation, cell death, and differentiation programs that facilitates tumorigenesis. Some rare individuals are born with mutations in tumor suppressor genes that predispose them to develop cancer. Tumor suppressors, to date, represent about 0.1% (30) of the total number of genes (around 30 000) that make up the entire mammalian genome. This number is likely to increase as novel technologies are used to study cancer cells. It is now clear that cells have evolved tumor suppressors as part of multiple elaborate defense systems against unscheduled cellular proliferation.

As their name indicates, tumor suppressor proteins regulate cell proliferation by eliminating cells that are damaged (by cellular stresses such as UV, X-rays, chemotherapeutic drugs) or that abnormally proliferate in response to hyperproliferative

---

The Molecular Biology of Cancer: A Bridge From Bench to Bedside, Second Edition. Edited by Stella Pelengaris and Michael Khan.
© 2013 John Wiley & Sons, Inc. Published 2013 by John Wiley & Sons, Inc.

signals. They are expressed by all mammalian cells and thus play a critical role in cancer prevention. Alfred G. Knudson and others, who proposed the "two-hits" hypothesis, introduced the concept of tumor suppression. Knudson theorized that the development of retinoblastoma, a rare cancer that occurs in the eyes of children, could be explained in familial cases by the acquisition from one parent (the first hit) of a mutated allele of a gene later called the retinoblastoma (*RB*) gene, and the occurrence of another mutation on the second allele (second hit) during adolescence. Nonfamilial or sporadic retinoblastoma, however, would occur as a result of mutations on both alleles in somatic cells, and in this case, the disease is not transmitted to the next generation. In each case, tumor development is recessive, since both copies of the gene must be inactivated for cancer to occur. This concept was later confirmed by studies on the genomic DNA from patients with retinoblastoma and other types of tumors. Since the discovery of RB, the first tumor suppressor identified, many tumor suppressor proteins have been characterized and shown to conform to the "two-hits" model proposed by Knudson.

The second tumor suppressor identified was the transcription factor p53. The realization that p53 was a tumor suppressor came from studies showing that enforced expression of wildtype p53, but not of mutant forms, induces a block in cell proliferation and transformation. This proliferation block is achieved, in part, by the activation of the growth arrest gene $p21^{Cip1/Waf1}$, a cyclin-dependent kinase inhibitory protein, whose expression affects RB function. p53 was later found to be activated by DNA damage. The first link between p53 and the induction of apoptotic cell death, depending on the strength of the signals that activates it, was recognized in the early 1990s.

The RB and p53 tumor suppressor pathways are intertwined and mutations in genes in either pathway are likely events and even perhaps mandatory alterations in the life history of most tumor cells.

While these first two tumor suppressors continue to be the primary research interest of many cancer biologists, several other proteins have been found that share tumor-suppressing properties. Not surprisingly, many of these proteins participate in the regulation of cell proliferation and can be mapped to pathways that induce cell death (apoptosis) or growth arrest. In this chapter, we will discuss the function of several tumor suppressors that participate in the RB and p53 pathways, focusing on RB and p53, some of their upstream or downstream regulators, as well as tumor suppressors that regulate the flow of signals through critical growth stimulatory pathways. These tumor suppressors have been identified in many signaling pathways involved in multiple processes, including cell proliferation in response to receptor signals, DNA repair, transcriptional regulation, and cytoskeletal structure (see Table 7.1 and discussed in Chapters 4, 5, and 10).

In addition to the common loss of both tumor suppressor alleles, tumor suppression can manifest itself in a haploinsufficient manner, meaning that loss of function of only one allele, or expression of only half of the amount of protein, is sufficient to sensitize cells to tumor formation. In these cases, the second allele remains intact and there is no loss of heterozygosity (also called LOH). Examples of haploinsufficient tumor suppressors include two cyclin-dependent kinase (CDK) inhibitory (CKI) proteins, $p27^{Kip1}$ (CDKN1B in human) (Kip1, cdk inhibitor protein 1) and $p18^{Ink4c}$ (CDKN2C in human) (Ink4c, inhibitor of CDK4), reduced expression of which impairs the tumor suppressor activity of RB and p53, as well as PTEN, MSH2 (mismatch repair protein 2), and NF-1 (neurofibromatosis 1) (see Table 7.1, and discussed in Chapter 4).

Tumor suppressors have been conserved throughout evolution, and several have been found in species other than mammals, including frogs, fish, and flies. The focus of this chapter, however, will be the role of the major tumor suppressors in the RB and p53 pathways in preventing human cancer, as it relates to their structure and function in mammalian cells. All the concepts and principles attributed to RB and p53 also apply to other tumor suppressors. Indeed about 30 tumor suppressors are known, however, the vast majority of human cancers carry p53 and RB mutations or mutations in genes that regulate them.

The "two-hits" hypothesis: loss of heterozygosity (LOH)

To lose one parent may be regarded as a misfortune; to lose both looks like carelessness.

Oscar Wilde

In 1971, Knudson advanced his "two-hits" hypothesis to explain how inherited (familial) cases of a rare cancer, called retinoblastoma, develop in the eyes of children (Fig. 7.1). Since then, this hypothesis has been confirmed in many cancers that occur as a result of the loss of a tumor suppressor. Knudson proposed that a child who inherits retinoblastoma acquires a mutated allele from one parent (the first hit) in a gene later called the retinoblastoma gene, *RB*. Heterozygosity for the *RB* allele is not sufficient to cause tumor formation and the child is initially asymptomatic. However, during early childhood, a second mutation or deletion on the other *RB* allele (second hit) occurs, inducing the loss of function of both alleles, loss of heterozygosity (LOH), in the retinal cells and the development of a tumor in this tissue. Nonfamilial or "sporadic" retinoblastoma also occurs as a result of mutations on both alleles (two hits) in retinal cells, but in this case, the disease is not transmitted to the next generation. Such mutations are termed "somatic mutations." Regardless of the way the mutations occur, tumor development is recessive, since both copies of the gene must be inactivated for cancer to occur.

Mutation of *RB* genetically transmitted in the germline induces bilateral and multifocal retinal tumors, whereas sporadic retinoblastoma usually occurs in one eye only as a focal tumor. This concept was validated by studies on the genomic DNA from patients with retinoblastoma as well as DNA from patients with other cancers in which RB or other tumor suppressors were lost.

Many human tumors contain mutations in the *RB* gene, most of which occur in somatic cells, meaning that the mutation is not transmitted from one generation to the next.

Haploinsufficiency in cancer

Haploinsufficiency is defined by the appearance of a phenotype in cells or an organism when only one of the two gene copies (also called alleles) is inactivated. For some tumor suppressor genes, the loss of a single allele is sufficient to induce susceptibility to tumor formation. In other words, these are haploinsufficient tumor suppressor genes. In this scenario, inactivation of one allele can be achieved by genetic (i.e. mutation or deletion) or

Table 7.1 Tumor suppressors

Gene	Nomenclature	Function	Chromosome location (human)	Tumor types
RB1	Retinoblastoma	Cell cycle regulator	13q14.1–q14.2	Retinoblastoma, sarcomas
p53/TP53		Cell cycle and a apoptosis regulator. Haploinsufficient	17p13.1	Lymphomas, sarcomas, brain and breast cancers
CDKN2A/Ink4a	Cyclin-dependent kinase inhibitory protein p16^{Ink4a}	Cell cycle regulator	9p21	Melanoma, many cancers
CDKN2A/Arf	Alternative reading frame	Cell cycle regulator	9p21	Sarcomas, lymphomas, many cancers
Tob1		Transcriptional corepressor		Liver cancers
APC	Adenomatous polyposis (familial)	Signaling	5q21–q22	Colon cancer
BRCA1	Breast cancer (familial)	DNA repair	17q21	Breast and ovarian cancers
BRCA2	Breast cancer (familial)	DNA repair	13q12.3	Breast and ovarian cancers
CDKN2C/Ink4c	Cyclin-dependent kinase inhibitory protein p18^{Ink4c}	Cell cycle regulator. Haploinsufficient	1p21	Testicular cancers
CDKN1B/Kip1	Cyclin-dependent kinase inhibitory protein p27^{Kip1}	Cell cycle regulator. Haploinsufficient	17p	Breast, prostate, many cancers
MSH2	Nonpolyposis colon cancer (hereditary)	DNA mismatch repair. Haploinsufficient	2p22–p21	Colon cancer
MLH1	Nonpolyposis colon cancer (hereditary)	DNA mismatch repair	3p21.3	Colon cancer
VHL	Von Hippel–Lindau syndrome	Transcription elongation regulator	3p26–p25	Renal cancers, hemangioblastoma, pheochromocytoma
PTCH	Medulloblastoma	Development and differentiation regulator	9	Gorlin syndrome, cerebellar tumor=medulloblastoma, basal cell skin cancer
PTEN	Phosphatase and tensin homolog deleted on chromosome 10	Signaling	10q23.3	Cowden disease, breast, brain and prostate cancers, hyperkeratinosis
MEN	Multiple endocrine neoplasia type 1	Unknown	11q13	Parathyroid and pituitary adenomas islet cell tumors
WT1	Wilms tumors	Transcription regulator	11p13	Kidney cancer (children)
NF1	Neurofibromatosis type1/ neurofibromin 1	Ras inactivation. Haploinsufficient	17q11.2	Neurofibromas, sarcomas, gliomas
NF2	Neurofibromatosis type 2/ merlin/ neurofibromin 2	Cytoskeleton	22q12.2	Schwann cell tumors, astrocytomas, meningiomas, ependymomas
TSC1	Tuberous sclerosis 1		9q34	Facial angiofibromas
DPC4	Deleted in pancreatic carcinoma 4	TGF-β/BMP signaling	18q21.1	Pancreatic carcinoma, colon cancer
ATM	Ataxia telangiectasia mutated	Cell cycle checkpoint regulation	11q23.1	Cerebellar ataxia, cancer predisposition
DMP1	Cyclin D-binding Myb-like protein 1	Positive regulator of Arf expression. Haploinsufficient	7q21	Acute myeloid leukemia (AML) and myelodysplastic syndrome (MDS)

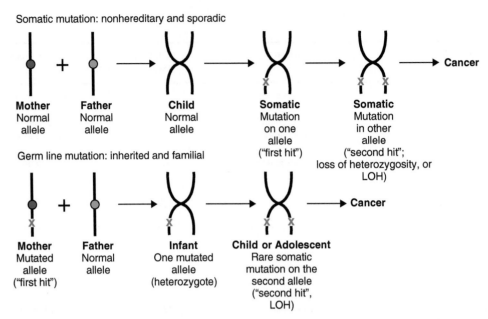

Figure 7.1 The "two-hits" hypothesis. **Somatic mutation:** In nonhereditary and sporadic mutations, genes inherited by both parents are normal, or wildtype. During early childhood or later, a mutation occurs in the allele of a gene, which represents the "first hit." This is followed by the mutation or loss of the second allele of the same gene, the second hit. Loss of function of the second allele is called loss of heterozygosity or LOH. **Germline mutation:** In hereditary or familial cancers, one parent (in this example the mother) transmits a mutated gene. The child is born as heterozygote for that gene. During childhood – adolescence in retinoblastoma (*RB*) – a second mutation occurs in the second allele of the same gene, inducing LOH and cancer.

epigenetic (i.e. transcriptional silencing by methylation, repressor complexes or mutations in the promoter region) events.

Epigenetic events

Epigenetic regulation of genes can be achieved by methylation, repressor complexes, or mutations of the promoter regions or introns, thereby reducing or inducing a loss of messenger RNA transcription, and subsequent reduction or loss of protein expression. Methylation occurs on the cytosine nucleotide of CpG pairs, usually located in promoter regions. These pairs often cluster in gene regulatory regions referred to as CpG islands, (see Chapter 11). Methylation of CpG islands is associated with the formation of nuclease-resistant and compact chromatin structures, resulting in transcriptional silencing. Mutations in the promoter region can affect the efficiency with which transcription factors bind to specific DNA-binding domains. Transcriptional repression can be achieved by repressor protein complexes that bind to the promoter regions of affected genes.

Although the end result of genetic and epigenetic events is the same, that is, a given protein is not expressed, one event (epigenetic) involves the promoter or noncoding regions (introns) while the other (genetic) affects the coding regions of the gene. Epigenetic events are linked to cancer development and are equally important in the process of carcinogenesis. For example, expression of the tumor suppressor p16^{INK4a} or p14^{ARF} is often silenced by methylation of their promoters or by active transcriptional repression by repressor protein complexes that bind to their promoter rather than deletion or mutation of the genes (Box 7.1).

Thus, DNA methylation and gene silencing are linked to growth and differentiation regulatory pathways that are disrupted in nearly all cancer cells.

Definition of a tumor suppressor

Law; an ordinance of reason for the common good, made by him who has care of the community.

Thomas Aquinas

A tumor suppressor is a protein usually expressed in all cells at very low levels that slows or inhibits cell proliferation and prevents damaged or abnormal cells from becoming malignant if they have been exposed to DNA-damaging agents, such as ultraviolet (UV) or gamma-irradiation (X-rays), or to hyperproliferative signals induced by oncogenes (see Chapter 6). Enforced expression of tumor suppressors in normal cells invariably induces growth arrest or apoptosis, depending on the level of expression and the type of cell. Conversely, loss of tumor suppressor function sensitizes cells to tumor formation by creating a proliferative advantage that in turn usually leads to the accumulation of genetic alterations, and ultimately aggressive cancers. Of the ~30 000 genes that make up the entire mammalian genome, tumor suppressors represent approximately 0.1%, although this number may increase as new technologies for studying tumor DNA become available and more tumor suppressors are identified (Table 7.1).

The retinoblastoma protein family

Genomic locus of *RB*: mutations in human cancers

The retinoblastoma (*RB*) locus is located on human chromosome 13q14. It comprises 27 coding segments (exons) and spans a region of ~200 kilobase pairs. Retinoblastoma can occur as a result of large deletions of coding exons or of more discrete

Box 7.1 Detection of methylated promoters

Several methods exist for detecting methylated DNA, one of which relies on a method based on the amplification of DNA fragments by the polymerase chain reaction (PCR) of high-molecular-weight genomic DNA treated with sodium bisulfite. Sodium bisulfite converts cytosine (C) residues to uracyl (U) but does not affect 5′-methylcytosine. When the bisulfite-treated DNA is amplified by the polymerase, the uracil is amplified as thymidine (T) and the methylated cytosines are amplified as cytosine. Therefore, primers can be designed that allow differentiation between sequences of nonmethylated and methylated DNA by virtue of the stability of the complexes formed between the DNA and the primers. In other words, if the sequences do not match, the PCR reaction will not occur and the DNA will not be amplified. Clearly, this method requires knowledge of the sequences in the promoter region of the gene under investigation, which can be found in the genome project database or be determined by sequencing.

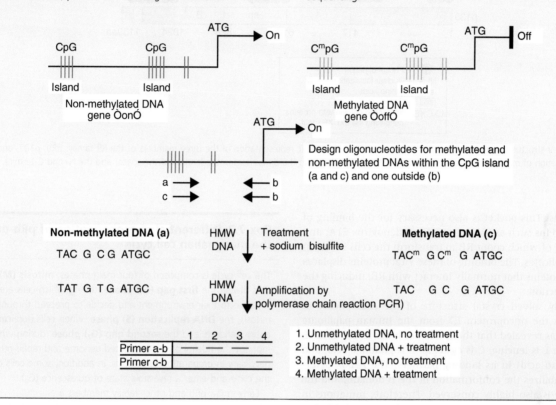

mutations (e.g. point mutation) that induce either the loss of the messenger RNA or the expression of a nonfunctional RB protein. Most inactivating *RB* mutations lead to retinoblastoma, although less disruptive mutations, such as those in the promoter region of RB that affect its transcription, induce tumors less frequently.

RB is important in the control of proliferation and development of not only retinal cells, but also other tissues. Children with inherited retinoblastoma due to germline *RB* mutations are at higher risk of developing osteosarcomas later in adolescence because they lose the remaining normal *RB* allele. Certain sporadic human tumors with *RB* mutations that have been found at high frequency include carcinomas of the bladder (33%), breast (10%), and in particular small cell lung carcinomas (85%). Thus, the retinoblastoma protein RB plays a critical role in the control of neoplasia in a variety of tissues.

The RB family of proteins

RB is part of a family of three proteins that include p107 and p130. A unique gene located on a different chromosome encodes each family member. While the *RB* locus resides on the long arm of human chromosome 13 (13q14), p107 and p130 are located on human chromosomes 20q11.2 and 16q13, respectively. Unlike RB, these family members appear not to function as tumor suppressors, despite their close structural similarities to each other and to RB. Indeed, few mutations have been reported in the gene encoding p107 (a single B-cell lymphoma,) or p130 (in isolated lung tumors).

The retinoblastoma gene was the first tumor suppressor to be identified and cloned. The gene encodes a nuclear phosphoprotein of ~110 kDa with distinct domains that are conserved between the other two family members, p107 and p130 (Fig. 7.2). The most conserved domain, termed the A–B "pocket," comprises A and B boxes linked by a spacer region. While RB has little similarity to its family members outside the pocket domain, p107 and p130 share the highly conserved spacer domain. The spacer sequences are critical for the binding of cyclin A–CDK2 and cyclin E–CDK2 complexes. The A–B pocket is a structure rich in alpha helices to which numerous proteins are known to bind, including the E2F family of transcription factors

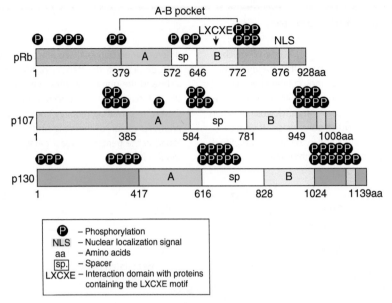

Figure 7.2 Linear structures of the RB family of "pocket proteins." Schematic representation of the three members of the RB family: pRb, p107, and p130. Note the conservation of many of the domains, including the A–B pocket, and the differences in the lengths of the spacer and the N- and C-termini.

and the cyclins. This pocket is also necessary for the binding of viral oncoproteins such as SV40 T antigen, adenovirus E1A, and E7 protein, all of which coopt RB to transform the cells in which their virus replicates. The binding of these viral proteins displaces the cellular proteins that normally interact with RB, inducing the loss of RB function.

The recently solved crystal structure of the A–B pocket in complex with the oncoprotein E7 from the human papilloma virus (HPV) has revealed that the B domain binds to an "LXCXE motif" (where L is leucine, C is cysteine, E is glutamic acid, and X is any amino acid) in its binding partners. The A box determines and stabilizes the conformation of the B domain, and the A–B interface is also highly conserved. Therefore, mutations in either domain can disrupt the structure and function of RB. The C-terminal domain of RB proteins contains a nuclear localization signal (NLS) required for localization to the nucleus, and there are several phosphorylation sites in the N-terminus, spacer, and C domain downstream of the B box that affect RB protein function (see below).

RB and its binding partners: role during the cell cycle and differentiation

The phosphorylation of RB fluctuates throughout the cell cycle (Box 7.2; see also Chapter 4). When cells are out of cycle in a state of quiescence (G_0), or are in the early part of the first gap (G_1) phase of the cell cycle, RB, p107, and p130 are underphosphorylated and in complex with other proteins, including the E2F transcription factors (E2F1–5) and their subunit partners, DP1 and DP2 (Fig. 7.3). While RB interacts principally with the "activating E2Fs," E2F1, 2, and 3a, and prevents their transcriptional activity, it interacts to some extent with E2F4 and E2F5 to repress gene transcription. In contrast, p107 and p130 form complexes with only E2F4 and E2F5, and act as repressors of transcription in G_0 and early G_1 (Fig. 7.3).

Two E2F3 proteins exist: E2F3a, the full-length protein that acts as an activating E2F, and E2F3b, a mutant form truncated at

---

Box 7.2 Differential phosphorylation of pRb during the mammalian cell cycle

The cell cycle is composed of four main phases: **mitosis (M)** when cells divide; the **first gap (G_1) phase**, during which cells are sensitive to their environment and decide to proceed through mitosis; the **DNA replication (S) phase**, when cells duplicate their entire genome; and the **second gap (G_2) phase**, during which cells monitor the fidelity of the duplicated genome and repair mistakes if necessary in preparation for mitosis. In addition, some cells can exit the cycle and enter a reversible state of quiescence (G_0).

Early in G_1, pRb and other family members are underphosphorylated and in complex with E2F transcription factors and their partner proteins, DP, repress transcription. Upon phosphorylation by cyclin D–Cdk4/6 and cyclin E–Cdk2 complexes, pRb repression of transcription is relieved by the release of E2F–DP complexes that induce the transcription of genes required for S-phase entry.

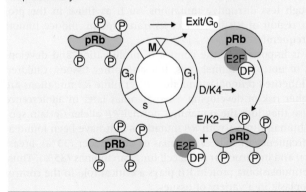

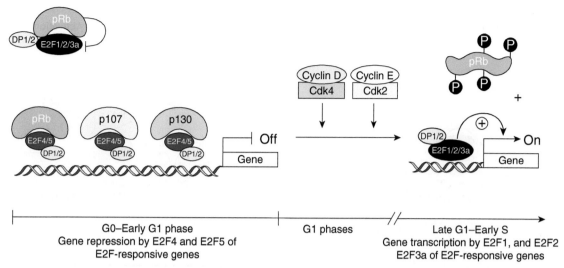

Figure 7.3 The RB family of proteins and G₁ progression. In G₀ and early G₁, pRb, p107, and p130 are underphosphorylated and in complex with E2F4, E2F5, and the DP1 and DP2 proteins repress transcription. pRb also sequesters E2F1, 2, and 3a–DP complexes without contacting DNA. In G₁, cyclin D–Cdk4/6 and cyclin E–Cdk2 holoenzymes phosphorylate pRb proteins, and relieve their repressive function, allowing E2Fs to induce the transcription of E2F-responsive genes required for S-phase entry.

Box 7.3 *ARF* expression is repressed by E2F3b

The transcription factor E2F3b is a repressor of the *ARF* gene in nonstressful conditions. It interacts with as yet unknown partners to repress transcription of ARF. In response to oncogenic stress, E2F3b is displaced by E2F1, 2, and 3a, which, by recruiting histone acetylases, induce the transcription of E2F-responsive genes, including *ARF*.

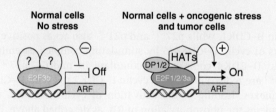

---

its N-terminus that represses transcription of proteins, including the tumor suppressor p19^{Arf} (p14^{ARF} in humans) (discussed in this chapter) by interacting with as yet unknown protein partners (Box 7.3). E2F6, another E2F family member, is part of a repressor polycomb complex that does not interact with the RB family of proteins. The most recently discovered E2F family member, E2F7, like E2F6, also acts as a transcriptional repressor and lacks an RB-binding domain.

All E2Fs form complexes with the DP1 and DP2 subunits, which confer high-affinity DNA binding on the promoter of E2F-responsive genes. As cells progress through the G₁ phase, RB becomes progressively phosphorylated on serine and threonine residues by cyclin–CDK holoenzymes that bind to RB via the LXCXE motif just described. Phosphorylation of RB is initiated by the holoenzyme cyclin D–CDK4, which forms in mid-G₁ in response to mitogenic stimulation. This stimulation induces the transcription of D-type cyclins, which bind CDK4 and CDK6. RB becomes fully phosphorylated and inactivated by cyclin E–CDK2 holoenzymes. Phosphorylation requires the interaction of the holoenzymes with RB via the LXCXE motif located on the cyclins. Mutations in this motif prevent binding of the enzymatic complexes to RB and its phosphorylation. The phosphorylation of RB alters its conformation and growth-suppressive capabilities by causing the release of bound proteins, perhaps most importantly the E2F transcription factors. Freed from RB, the E2F1, 2, and 3a proteins and their DP partners activate the transcription of genes required for the cell to commit and proceed to the DNA synthetic (S) phase of the cell cycle (Fig. 7.3). RB is dephosphorylated in mitosis by the action of the phosphatase PP1alpha 2, so that the next mitotic cycle can begin.

Overexpression of wildtype but not mutated RB blocks cells in G₁ by suppressing transcription and driving differentiation. RB, p130, and p107 interact with many proteins, but it is their binding to members of the E2F family of transcription factors that appears to be central to their role in governing DNA replication. As many as 110 proteins have been reported to interact with RB *in vitro* and *in vivo*, yet its ability to bind transcription factors, either repressing or stimulating their activity, appears to be the key to RB's ability to suppress proliferation. Unphosphorylated RB family members and the proteins with which they interact actively repress gene expression by simultaneously recruiting histone deacetylases (HDACs), other remodeling factors, and E2F–DP1 to E2F-responsive promoters, and using the E2F–DP1 complex to position the complex onto specific promoters (Fig. 7.4).

RB recruits HDACs to the B box of the pocket domain via the LXCXE motif on HDACs and in some cases by using a binding protein that bridges HDACs to RB, called RBP1 (RB-binding protein 1). By recruiting HDACs, RB not only forces cell cycle exit but also affects the expression of genes not involved in cell cycle regulation. p107 and p130 largely control the recruitment of HDACs to E2F-responsive promoters to repress their transcription, whereas RB is thought to bind to other transcription factors that regulate differentiation or senescence. In contrast, once RB is phosphorylated, gene activation proceeds by recruitment of histone acetylases to the activating E2F–DP complexes. This in

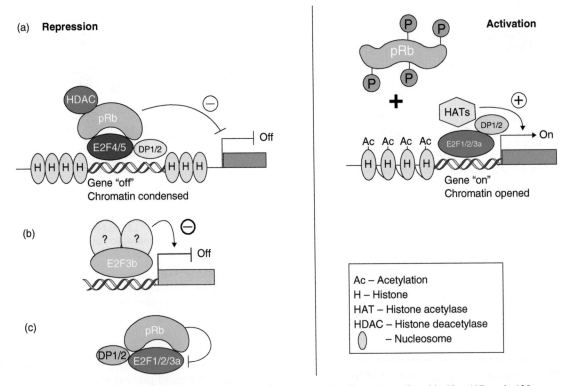

Figure 7.4 The RB and E2F family of proteins: repression and activation of gene expression. **Gene repression:** (a) pRb, p107, and p130 repress transcription (OFF) by contacting DNA via E2F4, E2F5, and DP complexes and recruiting histone deacetylase (HDAC). Histones are not acetylated and the chromatin is condensed. (b) Repression can be mediated by E2F3b bound to as yet unidentified repressor proteins. (c) pRb sequesters E2F1, 2, and 3a and their DP partners and inhibits their function without contacting DNA. **Gene activation:** Phosphorylation of pRb releases E2F1, 2, and 3a, and DP partners, which recruit histone acetylase and induce the transcription of E2F-responsive genes. Histones are acetylated; the chromatin is opened, allowing gene transcription (ON).

turn induces the acetylation of histones, unwinding of the DNA that facilitates access to the DNA of the transcriptional machinery (Fig. 7.4).

The RB signaling pathway

The RB family of proteins negatively regulates cell proliferation by repressing the transcription of the genes responsible for progression through the G_1 phase of the cell cycle and entry into S phase (Box 7.4). Relief of RB control is achieved by its phosphorylation by cyclin–CDK holoenzymes as discussed above. D-type cyclins are regulated by mitogenic signals via the Ras/Map kinase pathway that induces their transcription (Fig. 7.5). In that sense, D-type cyclins can be considered growth factor sensors. Once expressed, D-type cyclins bind to CDK4 and CDK6, a complex that is further activated by phosphorylation by a cyclin-dependent activating kinase, CAK.

These holoenzymes are themselves regulated by inhibitory proteins called cyclin-dependent kinase inhibitory proteins (CKIs) that comprise two families: the INK4 family and the CIP/KIP family. INK4 proteins bind to and specifically negatively regulate the activity of the cyclin D-dependent kinases CDK4 and CDK6. The family consists of four members – $p16^{INK4a}$, $p15^{INK4b}$, $p18^{INK4c}$, and $p19^{INK4d}$ – two of which, $p16^{INK4a}$ and $p18^{INK4c}$, act as tumor suppressors. The CIP/KIP family has three members: $p21^{CIP1/WAF1}$, a p53-responsive gene (see below), and $p27^{KIP1}$ and $p57^{KIP2}$, two tumor suppressors. All three members of this family are negative regulators of cyclin E– and cyclin A–CDK2 and of cyclin B–CDK1, while $p27^{KIP1}$ and $p21^{CIP1}$ also act as positive regulators of cyclin D–CDK4/6 by stimulating complex assembly. As cyclin D–CDK4 complexes accumulate in mid-G_1, $p27^{KIP1}$ and $p21^{CIP1}$ are redistributed from cyclin E–CDK2 to cyclin D–CDK4/6 complexes, leading to the activation of cyclin E–CDK2, which completes the phosphorylation of RB, as described above. At this time, $p27^{KIP1}$ itself is phosphorylated by active cyclin E–CDK2 and targeted for proteasomal degradation, which leads to the decrease in $p27^{KIP1}$ levels that is required for S-phase entry (Fig. 7.6).

Cell-cycle exit, maintenance of quiescence or differentiated states, and senescence are achieved, in part, by the expression of the CKIs. CKI expression is induced by antiproliferative signals that include cell–cell interaction (see Chapter 12), or antimitotic signals, such as TGF-β (transforming growth factor) or the cytokine IL-10. Mitogens also upregulate INK4 proteins in mid-G_1 of the normal cell cycle (Fig. 7.6; see also Chapter 5). As INK4 proteins accumulate, they bind with high affinity to the CDK4 and CDK6 moieties, thereby displacing the D-type cyclins from the complex, leading to their rapid degradation. Similarly, $p27^{KIP1}$ and $p21^{CIP1}$ reassort into the active cyclin E–CDK2 complexes and inhibit their activity. This prevents RB phosphorylation by both the cyclin D-containing and cyclin E–CDK holoenzymes and induces cell-cycle exit.

In addition to its role in regulating the cell mitotic cycle, the retinoblastoma protein participates in the DNA damage response (see Chapter 10), apoptosis (see Chapter 8), differentiation, and senescence (see Chapter 9).

Tumor Suppressors

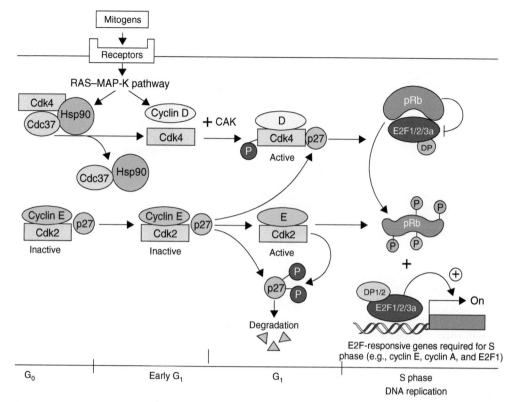

Figure 7.5 The Rb signaling pathway. In G_0, Cdk4 is bound to the chaperone protein Hsp70, and Cdc37 while cyclin E–Cdk2 is bound to $p27^{Kip1}$. Both kinases are inactive. In G_1, upon mitogen activation, receptors activate the RAS–MAPK pathway that regulates cyclin D transcription and the release of Cdk4 from Hsp70 and Cdc37. Cdk4 kinase activity is induced by binding to D-type cyclins, its phosphorylation by the cyclin-dependent kinase (CAK) and $p27^{Kip1}$ that serves as an assembly factor. The active cyclin D–Cdk4/6 holoenzyme initiates the phosphorylation of Rb. Binding of $p27^{Kip1}$ to increased levels of cyclin D–Cdk4/6 complexes relieve the inhibition of cyclin E–Cdk2 complexes that complete the phosphorylation of Rb and phosphorylate $p27^{Kip1}$, which is then degraded by the proteasome machinery. Phosphorylation of Rb releases E2F1, 2, and 3a that transactivate the expression of E2F-responsive genes, including cyclin E, cyclin A, and E2F1 and genes required for the initiation of DNA replication (S phase).

Box 7.4 The Rb pathway

Schematic linear representation of the Rb pathway. Mitogens induced the Ras/MAP kinase pathway induce the transcription of D-type cyclins that form complexes with Cdk4/6. Cyclin D–Cdk4/6 phosphorylates Rb and releases E2F transcription factors, which in complex with DP proteins induce entry of the cells into S phase. Cyclin D–Cdk4/6 complexes are negatively regulated by Ink4 (inhibitors of Cdk4) and positively by Cip/Kip (Cdk inhibitory proteins) proteins.

$$\text{Mitogens} \rightarrow \text{Ras} \rightarrow \text{MAPK} \rightarrow \text{cyclin D/cdk4} \rightarrow \text{Rb} \rightarrow \text{E2F/DP1} \rightarrow \text{S phase entry}$$

with Ink4 inhibiting cyclin D/cdk4 and Cip/Kip activating it.

RB and human cancers

There is now compelling evidence that several components of the RB pathway are mutated in human cancers, leading to the suggestion that the disabling of this pathway may be an inevitable event in the formation of most or all tumor cells. Components of this pathway act either as tumor suppressors that have been inactivated (e.g. INK4a, INK4c, KIP1, or RB), or as oncogenes that have been amplified (e.g. CDK4 or cyclin D1) (Table 7.2). While in cells containing a wildtype RB gene, the pRB protein represses the activating E2Fs via a sequestration mechanism in G_0 and early G_1 phase, loss of a functional RB protein unleashes the activating E2Fs that now transcribe genes at the inappropriate time, inducing either hyperproliferation or apoptosis (Box 7.5).

As discussed above, mutations in RB occur in several types of cancers, not just retinoblastoma. Similarly, while loss of $p16^{INK4a}$ functions by point mutations were initially discovered in familial melanoma, $p16^{INK4a}$ mutations, deletions, or epigenetic events have been found in many other cancers. Loss of the $p18^{INK4c}$

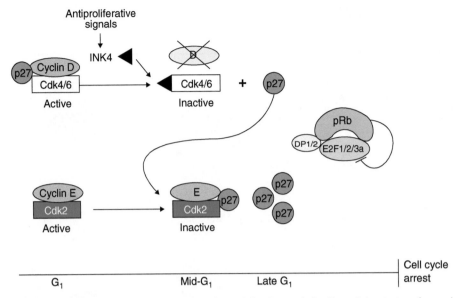

Figure 7.6 The cyclin-dependent kinase inhibitory (CKI) proteins from the Ink4 and Cip/Kip protein families collaborate to enforce cell cycle arrest. In mid-G_1 and in response to antiproliferative signals, Ink4 proteins are induced. Ink4 proteins bind to Cdk4/6 and free cyclin D that is rapidly degraded. p27^{KIP1} is reassorted to cyclin E–Cdk2 complexes to inhibit their kinase activity. Rb phosphorylation is inhibited, cells can no longer enter S phase and cells are arrested in G_1.

Table 7.2 The RB pathway in human cancers

Cancer types	INK4a loss	Cyclin D and CDK4 overexpression	RB loss
Small cell lung	15%	5% cyclin D1	80%
Non-small-cell lung	58%		20–30%
Pancreatic	80%		
Breast	31%	50% cyclin D1	
Glioblastoma	60%	40% CDK4	
T-ALL	75%		
Mantle cell lymphomas		90% cyclin D1	

protein, presumably by epigenetic events affecting gene expression rather than the loss of the gene itself, is found in testicular cancers and correlates with the most aggressive forms of the disease. More recently, loss of p27^{KIP1} expression has been shown to correlate with a poor prognosis in not only colon and breast cancer, but in many other tumors, including prostate, bladder, lung, liver, ovary, stomach, and other organs.

RB function can also be abrogated by viral proteins while leaving the gene intact. The molecular mechanisms by which these viral proteins inactivate RB are still under intense investigation, although it is clear that RB disruption is required for both viral replication and transformation of mammalian cells. Indeed, small DNA tumor viruses, including human papilloma viruses (HPV) have evolved several mechanisms to inhibit the function of RB (discussed in Chapter 13). Women infected with HPV16 and HPV18 develop cervical cancers several years after the infection occurs. HPV16 encodes a protein (E7) that disrupts RB function by mimicking the binding of cyclin D1 to RB, thereby releasing E2Fs and possibly other proteins inhibited by pocket proteins. Indeed, RB/E2F complexes are absent in HPV-positive tumors and in cells expressing E7 alone.

E7 binds RB via its LXCXE motif. Very much like cyclin D1, mutations in the LXCXE domain prevent E7 from binding to RB, and block its ability to bypass RB-dependent cell-cycle arrest. By releasing E2Fs, the virus, which infects nondividing cells, forces induction of S phase, and essentially highjacks its host's cell factors for its own replication. E7 also binds to p21^{Cip1}, causing the release of proliferating cell nuclear antigen (PCNA), which also induces transcription of genes required for S-phase entry.

RB family and mouse models

Mouse models in which RB and its family members are deleted have provided important information regarding the functions of the RB family in development and in adult tissues. Mice lacking two copies of Rb ($Rb^{-/-}$) in the germline die during embryogenesis (E) between days 12.5 and 14.5 because of defects in red blood cell and skeletal differentiation, associated with apoptosis in the central nervous system (CNS) and liver (Table 7.3). More sophisticated genetic experiments were recently reported in which Rb was deleted conditionally or by using tetraploid aggregation. These experiments reveal an unexpected role for Rb in the placenta (i.e. outside the embryonic lineages) in the control of embryonic viability and development. They also demonstrate that the loss of Rb in the environment, rather than in the embryo proper, can affect CNS and red blood cell development.

Mice heterozygotes for Rb function ($Rb^{+/-}$) survive embryogenesis but are highly prone to the spontaneous development of neuroendocrine tumors of the pituitary and thyroid. Examination of these tumors invariably reveals the loss of the other Rb allele, which is analogous to the pattern of RB loss in human cancers. Of interest, no retinoblastoma develops in these mice, despite the similarity in etiology of both the mouse and human disease progression. In contrast, transgenic expression in the mouse retina of oncogenic proteins from DNA tumor viruses, such as SV40T antigen or E6 and E7 of the human papilloma viruses, do induce retinoblastomas. In these mouse models, the functions of both Rb and $p53$ are simultaneously inhibited, sug-

Tumor Suppressors

Box 7.5 Rb and cancer

In normal cells expressing wildtype pRb, underphosphorylated Rb sequesters E2F1, 2, and 3a and their DP partners. Mitogenic signals induce pRb phosphorylation and release of E2Fs, which recruit histone actelylase (HAT) to induce the transcription of E2F-responsive genes and S-phase entry.

In cells in which pRb is lost or mutated, E2Fs constitutively activate transcription, leading to inappropriate entry in S phase and proliferation or apoptosis.

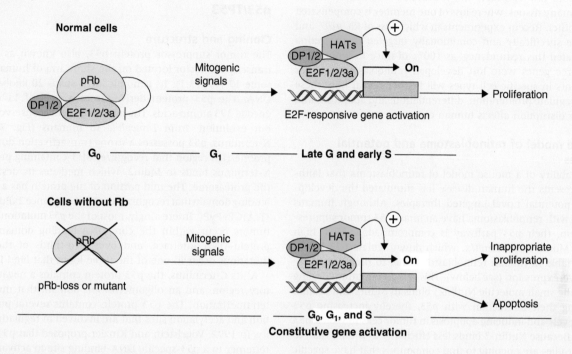

Table 7.3 Mouse models of the *Rb* gene family

Genotype	Phenotype
$Rb^{-/-}$	Embryonic lethality (E12.5–14.5); CNS and liver apoptosis; erythropoiesis and skeletal defects
Rb tetraploid	Viable
$Rb^{-/-}$ (chimera)	Perinatal death; skeletal defects
$Rb^{+/-}$	Neuroendocrine tumors by 12 months; pituitary adenocarcinomas (100%) and thyroid tumors with LOH
$p107^{-/-}$	Tumor-free; viable
$p130^{-/-}$	Tumor-free; strain-specific survival
$p107^{-/-}$, $p130^{-/-}$	Perinatal death and limb defects
$Rb^{+/-}$, $p107^{-/-}$	Retinal hyperplasia; pituitary and thyroid tumors with LOH for Rb
$Rb^{-/-}$ (chimera), $p107^{-/-}$	Retinoblastomas; pituitary; and thyroid tumors
$Rb^{+/-}$, $p130^{-/-}$	No retinoblastomas or retinal defects; pituitary and thyroid tumors

gesting that both tumor suppressors are required to suppress the proliferation of retinal cells. Indeed, the loss of *Rb* induces apoptosis, whereas the loss of *p53* prevents it, thus demonstrating cooperation between these two tumor suppressors in retinal cell proliferation.

Mice that lack *p107* or *p130* do not develop tumors, confirming that, despite their structural similarities to Rb, p107 and p130 are not tumor suppressors. The mouse genetic background or strain of *p107* and *p130* in which deletions have been programmed influences the phenotype associated with the loss of these genes. For example, loss of *p107* induces growth retardation and myeloid proliferation in Balb/c mice but not in C57BL/6 mice. Similarly, *p130*$^{-/-}$ Balb/c animals die *in utero*, whereas *p130*$^{-/-}$ C57BL/6 mice do not. The effects of specific backgrounds on phenotypes are an active focus of research. It refers to the identification of genes encoding modifiers of gene function that exist in some but not other strains.

Deletion of both *p107* and *p130* induces perinatal lethality associated with limb defects, which has led to the suggestion that there is functional overlap between these two proteins in normal limb development. When mice heterozygotes for Rb ($Rb^{+/-}$) are bred onto a *p107*-null background (i.e. the genotype of the mice is: $Rb^{+/-}$, $p107^{-/-}$), they develop retinal hyperplasia. Chimeric mice, in which the embryonic stem cells are $Rb^{-/-}$, $p107^{-/-}$ but the recipient mouse is $Rb^{+/+}$, $p107^{+/+}$, develop retinoblastomas and endocrine tumors with high incidence, suggesting that the development of retinoblastoma in these chimeras may be a function of the retinal cell environment in which the loss of both *Rb* and *p107* occurs.

Mouse models have thus revealed the complexity of functions associated with the RB family of proteins, and highlighted the functional redundancy that exists during mouse development and in adult tissues. Given that RB is expressed by all cells, it is surprising that such a restricted number of tissues are affected by its loss, until one considers the functional redundancy of this family in many tissues, where loss of one member is compensated for by another. Recent experiments in which loss of *Rb*, *p107*, and *p130* were specifically and conditionally targeted to the retina demonstrated this redundancy, as 100% of mice in whom two of the three genes were lost developed retinoblastoma. Other experiments in specific cell types will further reveal how these proteins regulate proliferation, differentiation, and apoptosis, and how their disruption affects human cancer.

A mouse model of retinoblastoma and potential therapies

The availability of a mouse model of retinoblastoma that faithfully represents the human disease has stimulated the development of potential novel targeted therapies. Although humans and mice with retinoblastoma have an intact p53 tumor suppressor protein, their p53 pathway is compromised. Indeed, high levels of Mdm2 and Mdm4/x, which downregulate p53 levels, are detectable in human retinoblastoma due to their amplification or overexpression (see below). This was an important finding because the small molecule Nutlin-3 binds to Mdm2 and Mdmx, preventing their interaction with p53, thereby increasing p53 protein levels and inducing apoptosis in tumor cells in which p53 is intact. Because Nutlin-3 binds less efficiently to Mdmx than to Mdm2, studies are ongoing to find compounds that have specific and high affinity for Mdmx. In the future, retinoblastoma could be treated with small-molecule inhibitors of Mdm2 or Mdmx, in combination with reduced levels of chemotherapy, to improve outcomes and reduce the secondary effects of high-dose chemotherapy.

p53/TP53

Cloning and structure

The tumor suppressor protein p53, also known as TP53, is a transcription factor located on the short arm of human chromosome 17 (17p13.1). Its genomic locus spans 20 kilobase pairs of DNA. The p53 protein derives from a total of 11 exons that encode 393 amino acids. The protein is highly conserved throughout evolution, from *Drosophila* to humans (Fig. 7.7). At its N-terminus, p53 possesses a strong transactivation domain and a proline-rich region that recognizes SH3-containing proteins. The N-terminus binds to Mdm2, which mediates its destruction by the proteasome. The mid portion of the protein has a core DNA-binding domain that recognizes the DNA sequence 2xPuPuPu(A/T)(T/A)GPyPyPy. Interestingly, most of the *p53* mutations in human tumors occur within the core DNA-binding domain near the protein–DNA interface, and over two-thirds of the missense mutations occur in one of the three loops that bind DNA.

At its C-terminus, the p53 protein contains a negative regulatory region, and an oligomerization domain that mediates p53 tetramerization. The p53 protein contains several phosphorylation and acetylation sites that are involved in regulating its activity. In 1992, Vogelstein and Kinzler proposed that p53 binds as a tetramer to a p53-specific DNA-binding site to activate the tran-

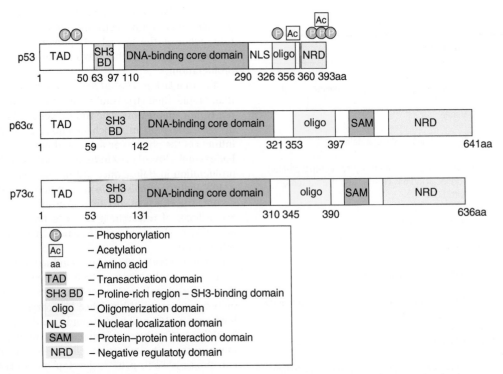

Figure 7.7 Schematic of the linear structure of the p53 family of proteins: p53, p63, and p73. Note that all three proteins share common domains but contain negative regulatory domains (NRD) of different lengths.

scription of p53-responsive genes. The 3D structure of the tetramerization domain subsequently confirmed that the protein does indeed form a tetramer through its oligomerization domain and supported the prediction that mutations in one *p53* allele would, in a *trans*-dominant manner, abolish the ability of the tetrameric p53 to bind to its recognition sequence. Indeed, mutation of a single allele of *p53* has been shown to be sufficient to cripple p53 function.

Activation and function of p53

When overexpressed, p53 inhibits cell proliferation and transformation. Therefore, low levels of p53 are maintained in cells to allow normal proliferation. p53 is a sensor of multiple cellular stresses that include genotoxic stress (e.g. ultraviolet, gamma-irradiation, carcinogens, and cytotoxic chemotherapeutic drugs) or oncogenic stresses caused by hyperproliferative signals induced by constitutive expression of oncogenes due to spontaneous mutations (Fig. 7.8).

In response to cellular stress, the p53 protein is stabilized and activated to induce a set of target genes involved in cell-cycle arrest, DNA repair, or apoptosis, depending on the strength of the signal, and cell context. Phosphorylation of p53 causes its activation through three mechanisms: stabilizing the protein by disrupting its interaction with Mdm2, regulating its transactivation activity, and promoting its nuclear localization.

p53 restrains uncontrolled cell growth by blocking the cell cycle, and inducing programmed cell death (apoptosis) (Fig. 7.8 and discussed in Chapter 8). It acts as a tetramer that recognizes p53-responsive elements in the promoters of p53-responsive genes, many of which have been catalogued by gene chip microarray analysis. Interestingly, p53 positively regulates the transcription of at least two of its negative regulators, Mdm2 and cyclin G. Mdm2 has a ring domain that possesses E3 ligase activity. Direct association of p53 with the N-terminal domain of Mdm2 results in the ubiquitination of p53 and its subsequent degradation by the ubiquitin-dependent proteasome pathway. Upon DNA damage, p53 is phosphorylated on serine residues in its transactivation domain, in particular on serine 15 by the Chk2 kinase. Serine 15 phosphorylation, in turn, induces a conformational change that prevents Mdm2 binding to p53 and results in the relief of the inhibitory effect of Mdm2 on p53. However, mice in which a S15A mutation was inserted in the p53 genomic locus were not tumor prone. This result in the mouse suggests that this Ser15 phosphorylation alone is not sufficient to release p53 from regulation by Mdm2. A negative feedback loop exists whereby p53 upregulates Mdm2 while Mdm2 downregulates p53. Of note, the *MDM2* gene is amplified in a significant proportion of the most common human sarcomas, where overexpression of the protein interferes with p53 activity. Thus, overexpression of Mdm2 equates to loss of p53 and increased propensity to tumor formation. p53 is also regulated by Mdmx (Mdm4), although the mechanism by which this occurs is less clear. Mdmx was found to be a key regulator of p53 in mice because its loss is lethal to the embryo, but co-deletion of p53 rescues the phenotype.

A similar negative feedback loop exists between cyclin G and p53. Cyclin G is thought to downregulate the activity of p53 by recruiting a PP2A phosphatase to activate Mdm2 by dephosphorylation, such that active Mdm2 may, in turn, stimulate p53 degradation. Loss of cyclin G in mice renders them susceptible to liver cancer.

Activation of p53 is regulated by upstream mediators of the DNA damage response, which include a number of kinases (ATM, ATR, DNA-PK, Chk1, and Chk2) (see also Chapter 10). ATM and ATR are members of the phosphatidylinositol-3 kinase (PI3K) family. Like p53, these kinases are activated by DNA damage and are recruited to DNA lesion where they phosphorylate proteins

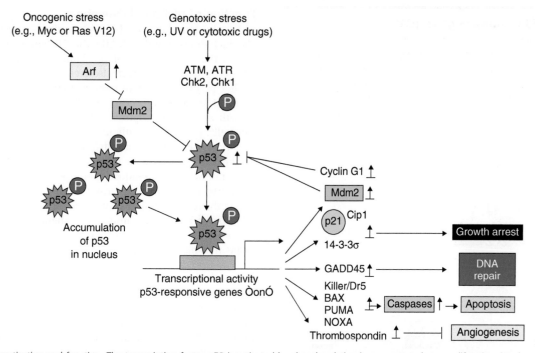

Figure 7.8 p53 activation and function. The transcription factor p53 is activated by phosphorylation in response to hyperproliferative signals and genotoxic stress. This induces an accumulation of p53 in the nucleus and the transcription of multiple p53-target genes that induce growth arrest, apoptosis, DNA repair, or a block in angiogenesis.

including p53 (Fig. 7.8). The ATM (ataxia telangiectasia mutated) protein is responsible for the disorder ataxia telangiectasia (Table 7.1, and Chapter 4). This is a rare, autosomal recessive inherited disease, in which patients are extremely sensitive, and show abnormal responses, to ionizing radiation because of the loss of the DNA damage checkpoint induced by p53 and, as a result, suffer increased chromosomal breakage and telomere-end fusion, leading to an increased rate of cancers. Normal ATM is recruited to X-ray-induced double-strand breaks and activated to phosphorylate p53 on Ser15, thereby preventing Mdm2 binding. In addition, ATM phosphorylates Mdm2 on Ser395, decreasing the ability of Mdm2 to shuttle p53 from the nucleus to the cytoplasm, thereby allowing the stabilization and nuclear accumulation of p53 protein. In individuals with ataxia telangiectasia, ATM can no longer perform these functions, which leads to the lack of functional p53.

Many of the p53-responsive genes that are induced in response to stress signals can arrest cell proliferation at the G_1/S and G_2/M transitions, allowing cells to repair any defects due to DNA damage before they proceed to DNA replication or mitosis. Thus, p53 regulates these two critical cellular checkpoints. One of the major transcriptional targets of p53 is the $p21^{CIP1}$ gene (Box 7.6). $p21^{CIP1}$ (also called WAF-1) is a cyclin-dependent kinase inhibitory protein whose overexpression induces G_1 arrest by preventing the cyclin E–CDK2 holoenzyme from phosphorylating RB (see RB section).

Other proteins with antiproliferative capabilities that are regulated by p53 include several members of the Btg family of proteins. This family of proteins is thought to suppress cell growth as part of transcriptional corepressor complexes that include histone protein deacetylase, HDAC, and protein arginine methyltransferase, which methylates histones. One member of the Btg family, named Tob1, was recently found to have tumor suppressor activity in mice and in humans (Table 7.1). Tob1 loss in the mouse predisposes the animal to spontaneous tumor development between 6 and 22 months of age. *Tob1*-null mice develop mostly hepatocellular adenoma and lung carcinoma, the onset of which is accelerated in a *p53*-null background, confirming that p53 and Tob1 cooperate in murine tumor development. An assessment of 18 tumor samples from human lung cancers revealed that in 13 of the samples, the *TOB1* gene was not mutated or deleted, but the expression of TOB1 mRNA was decreased, suggesting that TOB1 is a tumor suppressor in humans. The p53 protein also triggers cell-cycle arrest at the G_2/M boundary by inducing the transcription of genes that encode proteins that inhibit cyclin B–CDK1 activity and thus mitosis (e.g. 14-3-3 sigma, reprimo, and b99).

Oncogenic and hyperproliferative stresses can cause the activation of p53, which, in turn, induces apoptotic cell death by activating many proapoptotic genes (Fig. 7.8). Many of these p53-dependent proapoptotic genes have been identified by differential gene expression approaches such as the serial analysis of gene expression (SAGE) and gene chip microarray analysis. The proteins encoded by these proapoptotic genes ultimately activate caspases, including caspase 9, which induces cell death (see Chapter 8). Recently, a direct transcriptional target of p53 called Bbc3 (or PUMA) was found to play a key role in p53-dependent apoptotic cell death. PUMA localizes to mitochondria, where its overexpression has been shown to effectively kill cells. Remarkably, mice that lack *Puma* in the germline (created by deletion of the gene by homologous recombination) are deficient in all of the phenotypes associated with p53-dependent apoptosis. These phenotypes include the inability of the *Puma*-null cells to die in response to genotoxic stresses even in the presence of a functional, nonmutated p53 protein, which suggests that PUMA is the major downstream effector of p53-dependent apoptosis.

Function of p53 family members

p53 is the founding member of a family of three proteins that also includes p63 and p73. Like the RB family described earlier, p53 is the sole member of the family with tumor suppressor activity. p63 and p73 were isolated in the last 10 years, and studies performed to date indicate that p73 and p63 are rarely mutated in human cancers. Similar to p53, p73 can induce growth arrest and apoptosis when overexpressed in some *p53*-null cells. p73 was recently identified as a chemosensitizer for a variety of chemotherapeutic drugs, and an important contributor to the cell's response to cytotoxic agents. Both p63 and p73 cooperate with p53 to induce apoptosis, suggesting that although these two proteins are not true tumor suppressors, they have a role in the regulation of DNA damage-induced cell death. Because p73 and p63 are rarely mutated or silenced in human tumors, the possibility remains that they may be inactivated by mechanisms that have yet to be elucidated.

p53 in human cancer

Sir David Lane dubbed the p53 protein "the guardian of the genome" because its loss leads to genomic instability and increased mutagenesis. Unlike normal cells, tumor cells often have high levels of p53 that is almost always mutated. Of note, p53 was first identified in tumor cells where it was expressed at high levels but was unable to arrest cell proliferation. It was only later that scientists realized that the gene that everyone had isolated was

---

Box 7.6 p53-responsive gene, p21^{Cip1}

p53 is activated by phosphorylation in response to upstream signals. Active p53 binds DNA and activates the transcription of p21^{Cip1} (referred to here also as p21). p21^{Cip1} inhibits the cyclin E–Cdk2 holoenzyme by interacting with the complex, thereby inducing growth arrest.

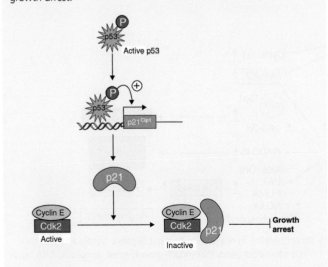

in fact mutated *p53*, at which point p53 was recognized as a bona fide tumor suppressor. Since its discovery in 1979, and its definition as a tumor suppressor in 1989, more than 1700 p53 mutations have been identified in ~150 different tumor types. Indeed, p53 inactivation is found in more than half of all human cancers.

It is believed that the most important function of p53 in protecting us from cancers is its ability to induce cell death. Loss of p53 function abolishes growth arrest or apoptosis, which prevents cells from properly responding to stress or damage, leading to genomic instability and the accumulation of additional genetic abnormalities. Inactivation of p53 can occur via numerous mechanisms: (1) missense mutations acting in a dominant-negative manner to reduce the function of the p53 tetramers; (2) deletion of one or both *p53* alleles; (3) binding of p53 to viral proteins, such as E6 from HPV or SV40 T antigen; and (4) degradation by Mdm2, the negative regulator of p53.

The updated *p53* mutation database contains 3200 variants for ~18 000 tumors listed, an impressive number that highlights the importance of this gene in cancer (www-p53.iarc.fr and www.ebi.ac.uk). Most of these *p53* mutations have been mapped to the core DNA-binding domain, although several mutations outside the core are also relevant to specific types of cancers (Fig. 7.9).

p53 mutations can be inherited via the germline, as is the case with Li–Fraumeni syndrome, where families are predisposed to diverse types of cancer, including breast cancers, bone, soft tissue, brain, adrenal, and colorectal carcinoma, or less frequently melanoma. Interestingly, in families affected by this syndrome, the *p53* mutations are transmitted from one generation to the next, and the cancers generally occur at an increasingly earlier age.

A germline mutation in the oligomerization domain of p53, where the arginine (R) residue at position 337 is mutated to histidine (H), was found initially in young Brazilian patients, although a few cases have been reported in other countries. These patients develop adrenocortical carcinoma early in life despite no family history or increased predisposition for cancer. In individuals with this *p53* mutation, the mutant protein forms tetramers that are destabilized in basic conditions. It is believed that elevated pH is a characteristic of adrenal cells, although a causal relationship between the pH of the adrenal cells and the adrenocortical carcinomas developed by individuals with this *p53* mutation has not yet been confirmed.

Nonhereditary or somatic *p53* mutations or *p53* loss have been described in many different types of cancers, including tissue-specific cancers, such as liver (hepatocellular carcinomas), bone (osteogenic sarcomas), muscle (rhabdomyosarcoma), lung, colon, bladder, cervix and anus, pancreas, esophagus, brain, and skin (squamous cell carcinoma) (see NBCI OMIM website www.ncbi.nlm.nih.gov/gene/7157).

Although p53 function is most often lost as a result of mutations or deletions, p53 can be inactivated while retaining a normal sequence. This is achieved by its rapid degradation by the negative regulator MDM2, which is often overexpressed in human tumors, or by viral proteins expressed by viruses that are etiologically linked to specific cancers. For example, functional loss of p53 in cervical cancers is rarely associated with *p53* mutations, but is often linked to HPV, a high-risk human papilloma virus that encodes the proteins E6 and E7 that neutralize the function of p53 and RB, respectively. E6 associates with p53 and with E6AP (a ubiquitin E3 protein ligase (E3A)), which catalyzes ubiquitination and degradation of p53. Other tumors lose p53 function via the overexpression of high levels of MDM2 protein that, as described earlier, binds to p53, catalyzing its degradation by the proteasome-dependent degradation machinery.

p53 family and mouse models

Each p53 family member has been deleted by homologous recombination in the mouse, providing insights into its role in

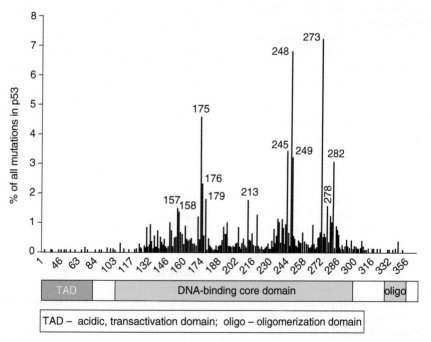

Figure 7.9 *p53* mutations in human cancers. Mutations in the *p53* gene have been found throughout the entire coding sequence but with increased frequency in the DNA-binding domain. Mutations affect *p53* transcriptional activity. From the IARC database www-p53.iarc.fr and www.ebi.ac.uk.

Table 7.4 Mouse models of the p53 gene family

Genotype	Phenotype
p53$^{-/-}$	Highly tumor prone; mostly lymphomas and sarcomas with other tumors with less penetrance
p53$^{+/-}$	Tumor prone with LOH
p53 R172H	Higher penetrance of carcinomas and metastases
"Super p53"	No tumors
p63$^{-/-}$	No tumors. Limb defects, craniofacial and epithelial development defects; no skin
p73$^{-/-}$	No spontaneous tumors; hydrocephaly and chronic infections

tumorigenesis and development (Table 7.4). Mice lacking one or two copies of p53 are highly tumor prone and develop a range of cancers, including lymphomas, sarcomas, as well as lung and brain tumors. This spectrum of tumors resembles the one seen in patients with Li–Fraumeni syndrome, characterized by the inheritance of a *p53* mutant allele and loss of heterozygocity, or the loss of the second allele in the tumor cells. These *p53*-null mice have proven useful for the testing of potential new therapies, since human tumors often develop resistance to therapy as a result of the loss of p53 function. They can also be used to screen potential carcinogenic compounds. Because they are "sensitized" to cancer, the test is more sensitive and fewer mice are needed. As expected, tumors from these animals are resistant to treatment with gamma-irradiation or adriamycin (a chemotherapeutic drug), since in the absence of functional p53, the cells can no longer trigger their stress-induced and p53-dependent apoptotic program. Since their generation, *p53*-null mice have been used extensively to define collaboration between the p53 apoptotic pathway and other tumor suppressors or oncogenes. In combination with other mice that have also been genetically engineered to lose other tumor suppressors or to express oncogenes, the *p53*-null mice develop a wide range of tumors at an accelerated rate.

More sophisticated models of cancers have now been generated in which *p53* missense mutations found in human cancers have been reprogrammed in the mouse genomic locus to mimic more closely what happens in human tumors. In addition, p53, as well as other tumor suppressors, are being deleted in a temporal and tissue-specific manner by using the Cre-Lox system. For example, a single amino acid substitution in the mouse p53 allele at residue arginine 172 to histidine (p53^{R172H}), a mutation equivalent to the same substitution at residue 175 in human tumors, in combination with loss of a guanine nucleotide at the splice junction that reduces the level of the mutated protein to wildtype levels, induces an increase in the number of carcinomas and in metastases not found in the *p53*$^{+/-}$ mice (Table 7.4). This suggests that these point mutations induce a gain rather than a loss of p53 function since the tumor spectrum is worse when the protein is mutated compared to when it is lost. These more refined mouse models may better reflect events in human cancers.

Finally, the group of Manuel Serrano demonstrated the concept that functional p53 protects from cancer. This group found that the addition of one copy of p53 to mice with the normal gene protected the animals from developing tumors. These investigators inserted in the mouse genome a large fragment of genomic DNA containing the *p53* gene and its upstream and downstream sequences required for proper regulation of gene expression. The concept is that in these "super-p53" animals, the probability that both genes become mutated is infinitely low, and therefore the mice are completely protected from the development of cancer induced by chemotherapy or other stresses.

INK4a/ARF

The *INK4a/ARF* locus

The *INK4a/ARF* locus, identified on the short arm of human chromosome 9 (9p21), is frequently mutated or deleted in human tumors. ARF negatively regulates p53 by sequestering HDM2 (Mdm2 in the mouse), thus preventing HDM2 from degrading p53. Cloning of this locus both in mice and in humans, revealed that the same exon encodes two proteins by using an alternative reading frame for each protein (Fig. 7.10). p16^{INK4a} derives from a transcript that comprises three exons – exon 1α, exon 2, and exon 3. p19^{Arf} in the mouse (p14^{ARF} in humans) derives from a transcript that begins ~13 kilobase pairs upstream of exon 1α, at the alternative exon 1β, and splices into the same acceptor site on exon 2 used by p16^{INK4a} but in an alternative reading frame (hence its name, ARF). This reading frame contains a stop codon 105 amino acids downstream in exon 2 that terminates the protein.

Another INK4 member, INK4b, is also genetically linked and located 5' to exon 1β. This unusual organization is conserved in humans. A puzzling question remains as to the origin of ARF. Indeed, to date, ARF has been found in all mammals, opossum, and birds (chicken), but it has not been identified in amphibians (*Xenopus*) or fish (*Fugu*), where only *INK4a* and *INK4b* homologs exist at the locus. Furthermore, ARF has not been found in lower organisms, such as the worm *C. elegans*. In birds, *Arf*, rather than *Ink4a*, is present at the locus, suggesting that in evolutionary terms *Ink4a* may have arisen more recently than *Arf*, from the duplication of an ancestral *Ink4a,b* gene.

Because only the first 37 amino acids of ARF are functionally relevant to its growth-suppressive function, it is possible that ARF exists in lower organisms, but that its sequence is not sufficiently conserved to allow its identification from the *Drosophila* or *C. elegans* genome databases. p19^{Arf} was recently found to inhibit ribosomal processing. This suggests that the function of the primordial *ARF* gene may be to negatively regulate growth rather than to induce growth arrest. This novel function of ARF is intriguing as it places *ARF* in opposition to the oncogene *MYC*, which drives the expression of ribosomal proteins (discussed in Chapter 6).

Ink4a and Arf functions

The p16^{INK4a} and p19^{Arf} proteins suppress tumor development by regulating the two best-known tumor suppressors in cancer, pRB and p53 (Fig. 7.11 and elsewhere in this chapter). Expression of p16^{INK4a} induces G$_1$ arrest by binding to and inhibiting the cyclin D-dependent kinases CDK4 and CDK6, thus preventing the phosphorylation of RB and attenuating its function (see section

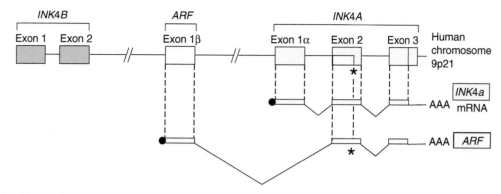

Figure 7.10 The *INK4a/ARF* genomic locus. Schematic representation of genomic organization of the genes that are encoded by the locus. p16^INK4a is encoded by three exons: exon 1α, exon 2, and exon 3. Exon 2 also encodes another protein, but in an alternative reading frame p19^Arf (p14^ARF in humans). Arf transcription is initiated from exon 1β. Another protein, p15^INK4b, is encoded by two exons that are genetically linked to the *INK4a/ARF* locus.

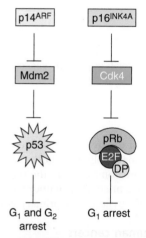

Figure 7.11 The *INK4a/ARF* genomic locus encodes two tumor suppressors. The same exon 2 encodes two tumor suppressor proteins, p14^ARF (p19^Arf in the mouse) and p16^INK4a. Whereas the tumor suppressor p14^ARF activates p53 by binding to and sequestering its negative regulator Mdm2, the tumor suppressor p16^INK4a binds to and inhibits Cyclin D–CDK4 activity and Rb phosphorylation. Expression of p14^ARF induces G_1 and G2 arrest while enforced expression of p16^INK4a induces G_1 arrest. Thus remarkably, the *INK4a/ARF* locus encodes to tumor suppressors that regulate two of the most important tumor suppressors in human cancers.

on RB). The p19^Arf protein does not affect Rb but does interact with Mdm2. Arf sequesters Mdm2, the negative regulator of p53, thereby preventing p53 degradation, which leads to an increase in p53 protein levels in the nucleus and to transcriptional activity (Fig. 7.12). This in turn induces cell cycle arrest or apoptosis (discussed in the p53 section in this chapter).

ARF function

ARF is a nucleolar protein that limits cell-cycle progression in response to an excess of proliferative signals induced by oncogenes, such as *MYC*, *E2F1*, *RasV12*, and *v-ABL* (Fig. 7.13). The molecular mechanisms by which oncogenic signals induce Arf expression are not completely understood. To date, for example, there is no evidence that Myc transactivates Arf by directly binding to its promoter region, although Myc rapidly induces p19^Arf protein levels. E2F1 was reported to bind directly to the Arf promoter and to directly regulate its mRNA expression.

Recent genetic experiments with the Eμ-Myc mouse model for Burkitt lymphomas have demonstrated that E2F1 is not required for Myc-induced Arf-dependent apoptosis but rather for Myc-induced proliferation by regulating the levels of p27^Kip1. In contrast, the transcription of p19^Arf is directly upregulated by DMP1, a transcription factor that directly binds to the Arf promoter. DMP1 is a tumor suppressor in acute myeloid leukemia and myelodysplastic syndrome in humans (Table 7.1). Mice that lack Dmp1 develop tumors spontaneously due to their low cellular levels of p19^Arf protein. Once induced, p19^Arf activates p53 by sequestering Mdm2 and causing its relocalization to the nucleolus or the nucleoplasm, depending on the cell context. Free from Mdm2, p53 is stabilized and transcriptionally active (Fig. 7.12).

As with Ink4a, expression of Arf is induced in response to stress signals that include oncogenic stress (as described above) and "culture shock," which occurs when mouse embryo fibroblasts are dissociated and established in culture. Of interest, Arf mRNA and protein are undetectable in embryos, but can be detected by polymerase chain reaction (PCR) amplification in adult tissues, suggesting that Arf expression is actively repressed in most normally proliferating cells. The active repression of Arf is thought to be mediated by several proteins, which include Tbx2 and Tbx3 (both transcriptional repressors), Twist and Dermo (two proteins that were identified in a screen for proteins that bypass Myc-induced apoptosis), Jun D (a transcription factor), Bmi1, a member of the polycomb transcriptional repressors, and E2F3b (see RB section in this chapter, Box 7.4) (Fig. 7.13). Genetic studies have demonstrated that the defects induced by the loss of Bmi1 in mice can be rescued in an *Ink4a/Arf*-null animal, which suggests that Arf and Ink4a are downstream targets of Bmi1-dependent repression, and that Arf repression is important for mouse development.

Enforced Arf expression induces acute G_1/S and G_2/M cell-cycle arrest in a p53-dependent manner. However, Arf can also induce growth arrest in the absence of p53 or Mdm2, albeit with slower kinetics (Fig. 7.14). Mice lacking Arf, Mdm2, and p53 develop more tumors per animal and more aggressive tumors than mice lacking any one of these genes alone. These genetic experiments underscored the nonlinearity of this pathway and suggest that each member of the pathway interacts with other proteins to protect cells against cancer. When the effect of enforced Arf expression in mouse embryo fibroblast cells lacking p53 and Mdm2 was compared to that in cells expressing wildtype levels of p53 and Mdm2, a family of genes induced by Arf in the

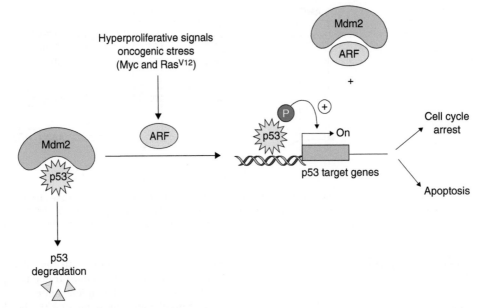

Figure 7.12 ARF activates p53 by sequestration of Mdm2. Mdm2 is a negative regulator of p53 that induces its degradation. In response to hyperproliferative signals due to oncogene activation, including overexpression of Myc or expression of a constitutively active RasV12 protein, ARF is induced, it binds to Mdm2 that results in the accumulation and activation of p53, leading to cell cycle arrest or apoptosis.

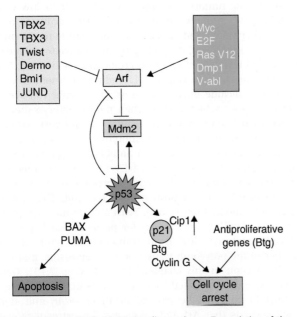

Figure 7.13 The Arf/Mdm2/p53 signaling pathway. Transcription of the tumor suppressor p19^{Arf} is induced by Myc, E2F (in green box). Arf expression is actively repressed by repressor protein complexes, including Bmi1 (in red box). Arf activates p53 by sequestration and inhibition of Mdm2 to induce cell cycle arrest or apoptosis. p19^{Arf} also induces growth arrest in a p53-independent manner by inducing many anti-proliferative genes (including the Btg proteins) which in concert induce cell cycle arrest.

absence of p53 was identified. This family is called the Btg family, and one of its members, Tob1, is a bona fide tumor suppressor, since its loss in mice has been shown to induce spontaneous tumors (see section on p53 target genes in this chapter). Thus, Arf has p53- and Mdm2-independent functions, some of which are beginning to emerge. Mice lacking Arf are blind due the hyperproliferation of the hyalovasculature that irrigates the lens during the first 10 days following birth. These mice revealed a p53-independent function for Arf in the regression of the vasculature in the eye. This phenotype in mice mimics a human condition called persistent hyperplastic primary vitreous (PHPV).

INK4a/ARF in human cancers

The *INKa/ARF* locus is often targeted by mutation, deletion, or epigenetic gene silencing in human cancers with frequencies approaching those seen for *p53*. While mutations that do not affect *ARF* have been reported in the *INK4a* gene, specifically in familial melanoma, no point mutations have been reported for *ARF* in human cancers. Instead, *ARF* is either deleted with *INK4a* and *INK4b* with the loss of the entire locus, or its expression is silenced by methylation of or repressor complexes on its promoter. Interestingly, cases in which *ARF* expression is silenced while *INK4a* remains intact, and vice versa, have been reported in certain cancers. In several types of human cancers, repressors of *ARF* expression are found overexpressed. For example TWIST is overexpressed in rhabdomyosarcoma, a type of muscle tumor and TBX2 is overexpressed in breast cancers. In these scenarios, the *ARF* locus remains intact, but p14^{ARF} expression is suppressed to a level analogous to that achieved by gene silencing or deletion.

Mouse models of *Ink4a* and *Arf* loss

The tumor suppressor properties of p16^{Ink4a} and p19^{Arf} have been assessed in knockout mice, in which each gene was deleted independently or together (Table 7.5). Mice lacking Ink4a and Arf (like *Arf*-null animals) are highly susceptible to tumors and die of cancer by 14 months of age. Mice lacking *Ink4a* alone, by deletion of exon1α (see Fig. 7.10), however, develop tumors spontaneously in only 25% of cases, and develop lung and skin cancers after exposure to carcinogens. In contrast, mice carrying an *Ink4a* mutant that can no longer bind its catalytic partner, *cdk4*, do not

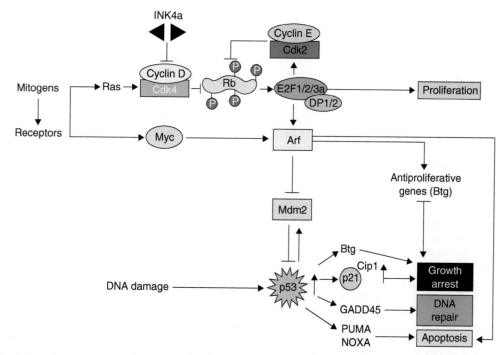

Figure 7.14 Arf links the retinoblastoma and p53 signaling pathways. Arf expression is induced in response to hyperproliferative signals. Ras induces cyclin D transcription. Cyclin D forms active complexes with Cdk4/6, phosphorylates Rb, and releases E2Fs. E2Fs and Myc induce Arf expression. Arf activates p53 by sequestering Mdm2. p53 in turn activates the transcription of p53-responsive genes responsible for growth arrest, apoptosis (cell death), or DNA repair.

develop tumors spontaneously in the first year of life. In this regard, the latter knockout mouse does not mimic tumor development in humans, because in humans, mutations in *INK4A* that abolish p16^{INK4A} binding to CDK4 and CDK6 are a characteristic of inherited familial melanoma. Melanoma is a skin cancer caused by prolonged and repeated exposure to UV and thus mutation of Ink4a was not sufficient in itself to cause skin cancers in mice. Indeed, melanomas occur in rodents lacking Ink4a function only when tumors are induced experimentally in Ink4a-null mice by either UV exposure, breeding with RasV12 transgenic mice, or when Ink4-null mutant mice retain a single copy of the *Arf* gene (Table 7.5).

The p53 and RB pathways in cancer

As described elsewhere in this chapter, *p53* and *RB*, as well as their upstream regulators and downstream targets, are often targeted in cancers, which suggest that genetic anomalies in these pathways may be a part of the life history of most if not all cancer cells. In mice and in many human tumors, *p53* mutations or *Ink4a/Arf* deletions are often mutually exclusive, suggesting that when the function of one tumor suppressor is disrupted, there is no selective pressure to inhibit the other. The p53 and Rb pathways were functionally linked by the discovery of the tumor suppressor Arf (Fig. 7.14). Following this discovery, it became clear that enforced Arf expression induces acute growth arrest in both the G_1 and G_2 phases of the cell cycle, but only in cells containing functional p53. In the presence of p53, enforced Arf expression induces growth arrest within 24 hours. Later experiments showed that Arf could induce growth arrest in the absence of p53 and Mdm2, but with slower kinetics, taking 3 days instead of one. Unlike p53, Arf is not induced by DNA damage, although

Table 7.5 Mouse models of *Ink4a* and *Arf* genes

Genotype	Phenotype
Ink4a$^{-/-}$	Spontaneous tumors in first year of life with 25% penetrance. Lung and skin tumors after carcinogen treatment. Low incidence of spontaneous melanoma
Ink4a$^{mut/-}$ (knockin of an Ink4a mutant)	No spontaneous tumors in the first year of life
Arf$^{-/-}$	Spontaneous tumors by 38 weeks (lymphomas and carcinomas) accelerated onset with carcinogens (X-rays) treatment
Ink4a$^{-/-}$, Arf$^{-/-}$	Spontaneous tumors by 34 weeks (lymphoma and sarcomas) accelerated onset with carcinogens treatment
Ink4a-null mutant, Arf$^{+/-}$	Melanomas and other tumors

it can act as a modifier of ATM function, since loss of Arf has been shown to rescue *ATM*-null induced senescence in mouse embryo fibroblasts, but not CNS apoptosis sensitivity.

Deletion of both p53 and Rb in just the cerebellum of mice was achieved by breeding mice whose *Rb* and *p53* genes were flanked by Lox sites, which are recognized by Cre recombinase, with mice expressing a transgene in which Cre recombinase was expressed from a cerebellar-specific promoter (GFAP-Cre mouse). These mice develop medulloblastoma, one of the most frequently seen

Senescence and immortalization: Role of RB and p53

Senescence represents an arrest of proliferation from which cells rarely escape (see Chapter 9). It is, in itself, a potent antitumor mechanism. Senescent cells are usually large and flat, upregulate a senescence-associated activity, β-galactosidase, and although metabolically active, do not divide, even in the presence of serum and mitotic signals. These features distinguish senescence from quiescence, from which cells can re-enter the mitotic cycle upon mitogenic stimulation. Although "replicative" senescence is triggered by telomere attrition, this state can be induced by activated oncogenes, DNA damage, oxidative stress, and suboptimal culture conditions. Cells that reach senescence upregulate many cell-cycle inhibitory proteins, including $p16^{Ink4a}$ and $p19^{Arf}$ ($p14^{ARF}$ in humans), which, as we have discussed in this chapter, regulate Rb and p53, respectively. $p16^{Ink4a}$ and $p14^{ARF}$ accumulate in senescent cells and can induce senescence when overexpressed. Conversely, loss of $p19^{Arf}$, p53, or the three RB family members (Rb, p107, and p130) induces the immortalization of mouse embryo fibroblasts.

Interestingly, loss of *Rb* in the germline does not induce immortalization because of compensating high levels of p107 that substitute for Rb, thus leading to cell senescence. Recently, it was demonstrated that acute loss of *Rb* in cells from conditional *Rb*-knockout mice relieves them from the senescence imposed by unphosphorylated Rb, stimulates S-phase entry, and eventually leads to immortalization, presumably by acquisition of mutations in the p53 pathway. These results suggest that both Rb and p53 play critical roles in the regulation of the senescence process. Rb, whose activity is induced by high levels of $p16^{Ink4a}$, is thought to be critical to promote senescence or permanent growth arrest by altering the chromatin state of growth regulatory genes. Much experimental data now implicate the Rb and p53 pathways in cellular senescence, confirming their critical roles in preventing tumor development.

Tumor suppressors and the control of cell proliferation

> The end of law is not to abolish or restrain, but to preserve and enlarge freedom. For in all the states of created beings capable of law, where there is no law, there is no freedom.
>
> *John Locke*

The phosphatase PTEN

The next most frequently mutated gene in many types of cancers is the tumor suppressor protein PTEN (phosphatase and tensin homolog), a protein phosphatase that negatively regulates the activity of the PI3K pathway (Fig. 7.15, discussed in Chapter 5). PTEN, like other tumor suppressors involved in ligand-dependent signaling, negatively regulates the strength of signals that activate transcription factors involved in cell proliferation. Many growth factors, by binding to their cognate receptors, induce signals that affect the PI3K pathway, critical for the activation of the serine/threonine kinase AKT. AKT, in turn, phosphorylates and modulates the activity of a number of proteins important in cell cycle

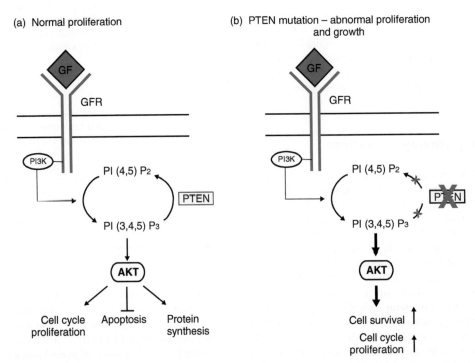

Figure 7.15 The PI3K–AKT survival pathway. By binding to their respective receptors (GFR), growth factors activate the phosphatidylinositol-3 kinase (PI3K) which converts phosphatidylinositol 4, 5-bisphosphate (PIP_2) to phosphatidylinositol 3, 4, 5-trisphosphate (PIP_3). PIP_3 activates AKT, which blocks apoptosis and induces protein synthesis and cell proliferation. PTEN is a phosphatase that antagonizes PI3K activity. In many human tumors, PTEN mutations lead to increased cell survival and cell proliferation.

control and cell survival. Consistent with being a tumor suppressor, enforced expression of PTEN in tumor cell lines induces growth arrest and/or apoptosis and in some cases prevents motility. Loss of PTEN function is common in several cancer types. Somatic inactivating mutations in PTEN are frequently found in glioblastoma, endometrial carcinoma, and prostate adenocarcinoma, but less frequently in other tumors, such as melanoma, renal cell carcinoma, and head and neck squamous cell carcinoma. In addition, PTEN mutations are found in sporadic cancers of the breast, thyroid, lung, stomach, and blood.

Inherited *PTEN* germline mutations characterize three familial cancer syndromes: Cowden syndrome, Lhermitte–Duclos disease, and Bannayan–Zonana syndrome. In mice, deletion of *PTEN* in the germline induces early embryonic lethality, but conditional deletion of *PTEN*, for example in the cerebellum, leads to a phenotype that mimics Lhermitte–Duclos disease. Loss of *PTEN* is often associated with loss of the *INK4a* and *ARF* genes, suggesting that these key tumor suppressors functionally collaborate in cancer prevention.

PATCHED and WNT signaling pathways

Patched (PTC), the receptor for sonic hedgehog (Shh), and Wnt, the ligand for frizzled were discovered in *Drosophila* and found to regulate patterning during development (Figs 7.16 and 7.17; discussed also in Chapters 5 and 12). Several genes that regulate both pathways are disrupted in cancers. PTC is an unusual receptor because in the absence of its ligand, Shh, it suppresses the activity of another transmembrane protein, smoothened, which itself represses the function and the translocation of a transcription factor Gli1 (Fig. 7.16a). Gli1 is the major transcription factor responsible for the expression of genes required for Shh-dependent proliferation, including Myc, and cyclins D1 and E. GLI itself is negatively regulated by a protein, SuFu, (suppressor of fused) that is also a tumor suppressor in cerebellar granule neural precursors (Fig. 7.16b). Cells expressing the constitutively activated PTC protein or carrying mutations in SuFu no longer require Shh to induce GLI-mediated transcription and, as a consequence, receive unregulated signals for growth, which leads to cancer development (Fig. 7.16c).

Germline mutation of *PTC1* is responsible for a familial cancer, Gorlin syndrome, which predisposes patients to a childhood cerebellar tumor (medulloblastoma) and a high incidence of basal cell carcinoma, a skin cancer (Fig. 7.16c). PTC mutations are also found in rhabdomyosarcoma, a muscle tumor. By the same token, mice lacking one copy of *Ptc* also develop medulloblastoma, but all tumors lose the wild-type allele, demonstrating that Ptc is, like Rb, a bona fide tumor suppressor. Interestingly, SuFu mutations are also associated with medulloblastoma, and like *PTC* mutations cause the constitutive activation of Shh signaling.

The WNT family is composed of conserved secreted proteins that play a crucial role in patterning during development by regulating cell–cell contacts. WNT binds to its receptor frizzled, activates the disheveled protein, which regulates the transcriptional activator β-catenin, through a complex containing the adenomatous polyposis coli (APC) tumor suppressor protein, the kinase GSK-3β and axin (Fig. 7.17a,b). This complex prevents β-catenin from entering the nucleus. Mutations in APC, AXIN, and the β-catenin gene are responsible for familial adenomatous

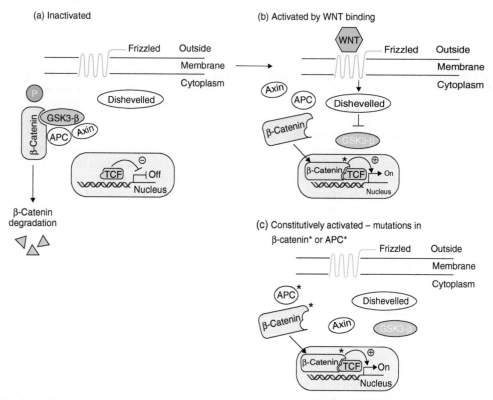

Figure 7.16 The WNT pathway. (a) In the absence of its ligand, the seven-transmembrane receptor Frizzled is inactive. The kinase GSK-3β phosphorylates β-catenin in complex with APC. β-Catenin is rapidly degraded. (b) When activated by binding of its ligand Wnt, activated Frizzled activates Dishevelled that inhibits GSK-3β. This results in the dissociation of the APC complex, accumulation of β-catenin and its translocation to the nucleus, where, in complex with the transcription factor TCF, it induces transcription of TCF-responsive genes. (c) Mutations on APC or β-catenin induces the constitutive translocation of β-catenin in the nucleus and transcription.

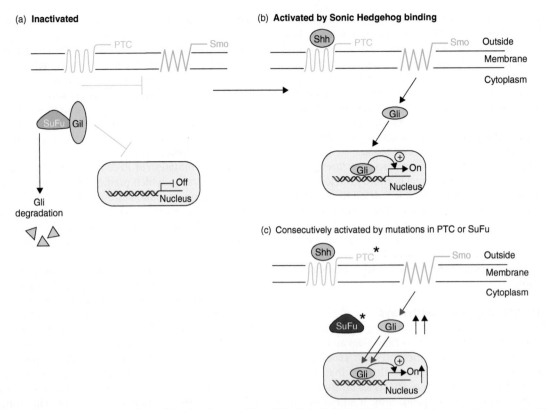

Figure 7.17 The Sonic Hedgehog Patched pathway. (a) In the absence of ligand (Sonic hedgehog), Patched, a seven transmembrane protein, inhibits Smoothened activity and the transcription factor Gli is inhibited by SuFu. (b) Upon Shh binding, Patched is activated, which relieves its suppression on Smoothened. Activated Smoothened induces the translocation of Gli to the nucleus and transcription of Gli-responsive genes. (c) Mutations in Patched or in SuFu, a negative regulator of Gli, activates the pathway constitutively in the absence of Shh.

polyposis (FAP) and colorectal cancers, as well as sporadic medulloblastoma (Fig. 7.17c).

The GTPase NF-1

NF-1 is a GTPase that negatively regulates Ras signaling. It is found mutated in patients with neurofibromatosis, hence its name, but also in sarcomas and gliomas. Ras is a critical mediator of proliferative signals in response to receptor activation, which explains why its unrestricted action induces hyperproliferative signals (see Chapter 6 on oncogenic Ras). Ras regulates cell proliferation and cell survival and links the cell cycle and the p53 pathway. Note that PTEN negatively regulates the survival pathway activated by Ras (Fig. 7.18).

Tumor suppressors and control of the DNA damage response and genomic stability

Besides controlling proliferation and repairing DNA damage, cells must also maintain the integrity of their genome to avoid the accumulation of undesirable mutations, gene duplication, or chromosomal rearrangements (see also Chapter 10). In addition to the roles of p53 and the ataxia telangiectasia kinase ATM in these processes, other tumor suppressor proteins also respond to DNA damage. They include the BRCA1 and BRCA2 proteins, which are recruited to DNA during homologous recombination and are found mutated in familial breast and ovarian cancers,

and Chk2, a protein kinase activated by ATM that controls the G_1 checkpoint in the cells of patients with Li–Fraumeni syndrome, in which germline *p53* is mutated (also discussed in Chapter 10).

MSH2 and *MLH1*, involved in DNA mismatch repair, are mutated in hereditary nonpolyposis colorectal cancer, also called Lynch syndrome. Mutations in these two genes also induce other types of tumors, including endometrial, gastric, ovarian, and bladder cancers. The proteins NSB1 and Fanconi anemia (FA) are involved in DNA repair and the control of DNA replication (S phase). Both proteins are inactivated in T-cell lymphoma and Nijmegen breakage syndrome, NSB1 is also inactivated in lymphoreticular malignancies, whereas FA is inactivated in acute myelogenous leukemia.

The microRNAs and tumor suppressors

Biology of microRNAs

Over the last 10 years, as has been previously noted, microRNAs (miRNAs) have emerged as novel regulators of many aspects of cellular physiology; as such, they are now implicated in many diseases, including cancers. Although discussed in more detail in Chapter 11, a brief general overview is given here before focusing on the role played by miRNAs in tumor suppression.

The miRNAs are small, noncoding RNAs, ~22 nucleotides in length, that negatively regulate protein expression by interacting with the 3' untranslated regions of mRNA targets to inhibit translation or induce mRNA degradation. They are encoded as long pri-miRNAs that are processed in the nucleus to pre-miRNAs by a complex that includes Drosha and DCGR8 (see Fig. 7.19 and Chapter 11). These pre-miRNAs are then processed by the enzyme Dicer in the cytoplasm to mature miRNAs that form complexes with Argonaute proteins, called RISCs (RNA-induced silencing complexes). The RISC complex recognizes complementary sequences on the mRNA targets within their 3' untranslated regions, and less often in the 5' ends and coding sequences.

The 5' portion of a miRNA, comprising nucleotides 2–8, is called the "seed" region and mediates mRNA recognition; the other nucleotides can vary among miRNAs that share the same seed sequence, allowing recognition of multiple targets. In fact, miRNAs each regulate hundreds of mRNA targets to modulate protein levels. They are encoded in human and mouse genomes, and the human genome encodes about 700 miRNAs, which are estimated to regulate at least 30% of all mRNA transcripts.

In response to stress or DNA damage, tumor suppressors, such as TP53, regulate miRNAs that in turn will alter the repertoire of downstream targets. In tumors, some miRNAs can be overexpressed and act as oncogenes, whereas others are underexpressed, suggesting that their expression might suppress tumor development. As such, miRNAs may be suitable targets for therapy. Because one miRNA can regulate multiple mRNA targets, several algorithms have been developed to identify such targets. However, these algorithms do not always accurately predict true *in vivo* targets. This issue has been resolved with the development of CLIP-Seq (crosslinking immunoprecipitation sequencing), a technique that involves the immunoprecipitation

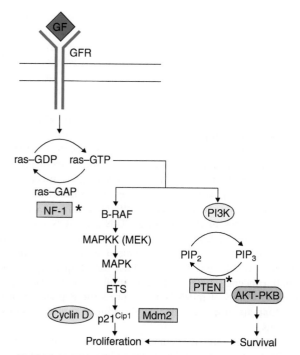

Figure 7.18 The RAS signaling pathway: NF-1 and PTEN. Ras is activated by growth factor receptor (GFR) signaling. Activated Ras-GTP is inactivated to Ras-GDP by a GAP protein such as NF-1, a tumor suppressor protein mutated in neurofibromatosis. Activated Ras-GTP induces two pathways; the RAF–MAPK and the PI3K pathways. The RAF–MAPK pathway induces proliferation by activating the transcription of D-type cyclins and Mdm2. The PI3K pathway activates AKT to enforce cell survival. The phosphatase PTEN opposes the activity of PI3K.

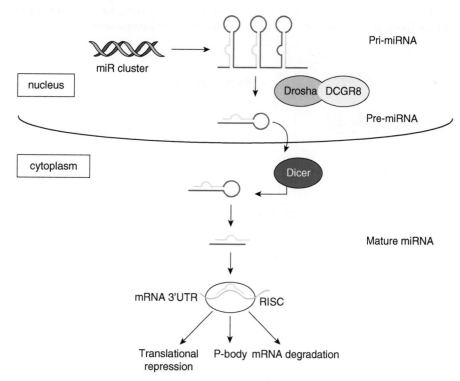

Figure 7.19 Simplified schematic showing biogenesis of miRNAs. Shown in more detail in Chapter 11, this shows a simplified model for the nuclear expression and further cytoplasmic processing of miRNA transcripts to yield the bioactive matured miRNAs.

of the RISC–miRNA–mRNA target complex after the crosslinking of cells or tissues, followed by PCR amplification of the protected RNA sequence and deep sequencing by using next generation sequencing. Briefly, a tagged Argonaute protein is transfected into cells with the miRNA of interest and the cells or tissue are then crosslinked. The mRNAs that are protected by the miRNAs–RISC complex are then amplified by using PCR and sequenced by use of high-throughput sequencing. Regions of RNAs bound to the miRNAs–RISC complexes are thus enriched for targets of the specific miRNA of interest, identifying physiologically relevant mRNA targets.

The contribution of miRNAs to critical physiological functions is currently being evaluated in mice by use of conditional deletion of specific miRNAs or repression of miRNA expression by enforced or regulated expression of specific anti-miRNAs or "antigomiRs." Based on the findings of these studies, miRNAs are already being used therapeutically in clinical trials. This rapidly moving field will no doubt yield an increasing number of miRNAs that can be used as diagnostic markers and targets for therapeutic intervention.

TP53 and miRNAs

A flurry of papers in 2007 revealed that the transcription factor and tumor suppressor p53 (TP53 in humans, Trp53 in mice) not only regulates gene expression, but also modulates miRNA levels to induce G_1 arrest, apoptosis, or senescence. MiRNAs of the miR-34 family (miR-34a, miR-34b, and miR-34c in vertebrates, with a single ortholog in invertebrates) were the first miRNAs identified as direct transcriptional targets of p53 in response to DNA damage. The miR-34 miRNAs downregulate the expression of proteins involved in cell-cycle progression, including E2Fs, cyclin E, and Cdks, during apoptosis and senescence (Fig. 7.20).

Not surprisingly, given their potential tumor suppressive functions, miR-34 family members are often downregulated in human cancers, and therefore their enforced expression may prove to be a novel therapeutic approach for tumors in which p53 tumor suppressor activity may be compromised. p53 can indirectly affect tumor proliferation via the loss of blood vessel development (angiogenesis). For example, miR-107 inhibits tumor angiogenesis by downregulating the vascular endothelial factor VEGF, via direct inhibition of HIF-1α (Fig. 7.20).

Once a link between p53 and miRNAs was established, p53 was shown to exert its tumor suppressive functions by regulating miRNAs via multiple mechanisms. One such mechanism involves p53-mediated regulation of the processing of miRNAs that have tumor suppressor activity, such as miR-16–1, miR-143, and miR-145, in response to DNA damage. This process is mediated by binding to Drosha. Interestingly, the interaction domain between Drosha and p53 is often compromised by mutations in TP53, including R175H and R273H, which are frequently found in human cancers. These mutations inhibit the levels and thus the effectiveness of miRNAs to suppress the expression of proteins that are important for TP53-induced tumor suppressive functions.

In addition to directly regulating the transcriptional expression of miRNAs, p53 itself can be regulated by miRNAs, leading to the downregulation of p53 protein levels and thus the loss of its tumor suppressive function. For example, miR-504 directly binds to two sites in the 3' untranslated region of the human TP53 mRNA, thereby influencing the ability of the TP53 protein to promote tumorigenesis. Indeed, enforced expression of miR-504 in cells induces graft tumor growth in mice.

Thus, miRNAs and p53 are intimately linked, regulating each other and increasing the complexity of gene regulation in cancers and in tissue homeostasis.

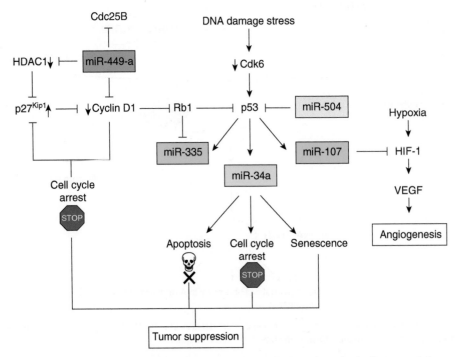

Figure 7.20 Interaction between miRNAs and tumor suppressor pathways. The miRNAs play key roles as both effectors and also regulators of the activity of traditional tumor suppressors, such as RB and p53. In this way they impact on essentially all major growth and cancer-restraining processes.

RB and miRNAs

Recent studies in mice and humans have identified an miRNA network that controls activity along the RB pathway. Two miRNAs – miR-15a and miR-16 – the loss of which was first detected in 68% of chronic lymphocytic leukemias, are now implicated in many cancers in which they are deleted or negatively regulated. These two miRNAs also regulate Rb-dependent cell-cycle progression (Fig. 7.20). In fact, miR-15a and miR-16 suppress the expression of genes that drive G_0/G_1 progression via Rb phosphorylation, such as cyclin D1 and cyclin E, in non-small-cell lung cancers. As expected, enforced expression of miR-15a/miR-16 induces cell-cycle arrest in G_0/G_1, but only in cells with functional Rb protein, indicating that they suppress tumorigenesis by, in part, inhibiting Rb phosphorylation and, in turn, forcing cell-cycle arrest.

Another miRNA, miR-449, similarly regulates Rb phosphorylation by downregulating cyclin D1 and CDK6, as well as histone deacetylase 1 (HDAC-1), which, in turn, induces the upregulation of the cyclin-dependent kinase inhibitory protein $p27^{KIP1}$ to enforce cell-cycle arrest and senescence in prostate cancer cells. Yet another miRNA, miR-335, directly controls Rb protein expression by binding with high affinity to the 3' untranslated region of Rb's mRNA.

Like mRNAs, miRNAs are regulated both transcriptionally and epigenetically. An example of a miRNA involved in RB control being regulated by epigenetic silencing is miR-124a. In acute lymphoblastic lymphoma (ALL), miR-124a expression is suppressed by CpG methylation and histone modifications, an event that confers a poor prognosis. The miRNA miR-124a acts as a tumor suppressor by downregulating CDK6 expression and, thus, its downregulation induces CDK6 levels, RB phosphorylation, and proliferation.

MiRNAs and other tumor suppressors

Besides controlling the RB and p53 tumor suppressors, miRNAs also regulate PTEN, histone deacetylase (HDAC), and the cyclin-dependent kinase inhibitory proteins $p27^{KIP1}$, $p21^{CIP1}$, and $p16^{INK4a}$, demonstrating an additional level of protein regulation.

As different human tumors are evaluated for miRNA expression and as miRNAs are tested *in vivo*, it is likely that even more miRNAs that regulate other tumor suppressors will be found. The complexity of miRNA networks is also revealed by the interactions between a single transcription factor and oncogene, c-MYC, discussed in the previous chapter. Figure 7.21 illustrates this by focusing on just one miRNA cluster regulated by MYC and showing how this can impact on a wide range of signaling processes involved in apoptosis and cell-cycle entry.

Conclusions and future directions

> The chess-board is the world, the pieces are the phenomena of the universe, the rules of the game are what we call the laws of Nature. The player on the other side is hidden from us.
>
> *Thomas Henry Huxley*

Since their discovery in the late 1970s and early 1980s, tumor suppressors have been recognized as critical contributors to cancer prevention. They participate in signaling pathways that regulate cell proliferation, DNA repair, programmed cell death (apoptosis), and the cell's architecture (cytoskeleton). They often collaborate to repress tumor formation and are found genetically

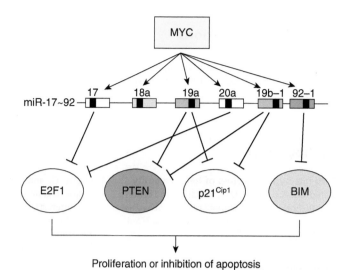

Figure 7.21 The interaction between c-MYC and one important miRNA cluster, miR17, illustrates the complexity of miRNA networks in cell signaling. Various miRNAs influenced by c-MYC contribute to cell-cycle entry/progression and to suppression of apoptosis and thereby to MYC oncogenic activity.

altered in most cancers. Although many tumor suppressors have been characterized in the last 10 years, they are significantly outnumbered by the proto-oncogenes, which positively regulate cell growth. However, more tumor suppressors will likely emerge now that the complete genome sequence of humans, and other organisms, is available.

Genetic experiments in the mouse have and will continue to provide confirmation of tumor suppressor functions and insight in to many types of human cancers. Indeed, mouse models for cancers are now used in preclinical trials to evaluate these inhibitory molecules as potential therapeutic agents. The advent of RNA interference, a new technology that allows specific, isolated gene suppression *in vivo*, will no doubt lead to the identification of new tumor suppressors that may serve as novel targets for therapeutic interventions in cancers and improve our current options for the prevention and treatment of these diseases.

Acknowledgments

I would like to apologize to all my colleagues whose critical scientific contributions could not be cited due to space restriction. My thank you to the members of my laboratory who have provided me with encouragement and provided keen insights and help during the writing of this chapter. I specifically would like to thank Drs. David Lagaesparda and Gerard Zambetti who offered many helpful comments and key criticisms. My thanks also go to Dr. Susan Watson who provided invaluable scientific editing. I would be remiss if I did not thank my collaborator and husband Dr. Charles J. Sherr and my son Jonathan R. Sherr, who encouraged and supported me throughout.

This work was possible thanks to funding from NCI grants CA71907 and CA96832, a Cancer Center support grant CA21765, the Children's Brain Tumor Foundation, the American Brain Tumor Association, the James McDonnell Foundation and the American Lebanese-Syrian Associated Charities (ALSAC) of St. Jude Children's Research Hospital.

Outstanding questions

1. To know how many tumor suppressors exist, which ones regulate with specific cell type and how loss of one influences the loss of another in cancer progression.

2. The exact order with which genetic and epigenetic events influence cell proliferation and cell death by affecting tumor suppressors. Knowing exactly what happens to a cell from the time it receives hyperproliferative signals to the time it loses a tumor suppressor would help identified the best targets for therapeutic intervention.

Bibliography

Two-hits hypothesis/loss of heterozygosity

Knudson, A.G., Jr. (1971). Mutation and cancer: statistical study of retinoblastoma. *Proceedings of the National Academy of Sciences of the U.S.A.*, **68**: 820–23.

Epigenetic events

Herman, J.G., Graff, J.R., Myohanen, S., Nelkin, B.D., and Baylin, S.B. (1996). Methylation-specific PCR: A novel PCR assay for methylation status of CpG islands. *Proceedings of the National Academy of Sciences of the U.S.A.*, **93**: 9821–6.

Jones, P.A. (2003). Epigenetics in carcinogenesis and cancer prevention. *Annals of New York Academy of Sciences*, **983**: 213–19.

The retinoblastoma protein

Classon, M. and Harlow, E. (2002). The retinoblastoma tumour suppressor in development and cancer. *Nature Reviews Cancer*, **2**: 910–17.

Dunn, J.M., Phillips, R.A., Becker, A.J., and Gallie, B.L. (1988). Identification of germline and somatic mutations affecting the retinoblastoma gene. *Science*, **241**: 1797–800.

Dyer, M.A. and Bremner, R. (2005). The search for the retinoblastoma cell of origin. *Nature Reviews Cancer*, **5**: 91–101.

Dyson, N. (1998). The regulation of E2F by pRB-family proteins. *Genes and Development*, **12**: 2245–62.

Friend, S.H., Bernards, R., Rogelj, S., et al. (1986). A human DNA segment with properties of the gene that predisposes to retinoblastoma and osteosarcoma. *Nature (London)*, **323**: 643–6.

Harbour, J.W. and Dean, D.C. (2000). The Rb/E2F pathway: expanding roles and emerging paradigms. *Genes and Development*, **14**: 2393–409.

Helt, A.M. and Galloway, D.A. (2003). Mechanisms by which DNA tumour virus oncoproteins target the Rb family of pocket proteins. *Carcinogenesis*, **24**: 159–69.

Jacks, T., Fazeli, A., Schmitt, E.M., Bronson, R.T., Goodell, M.A., and Weinberg, R.A. (1992). Effects of an Rb mutation in the mouse. *Nature*, **359**: 295–300.

Lee, J.O., Russo, A.A., and Pavletich, N.P. (1998). Structure of the retinoblastoma tumour-suppressor pocket domain bound to a peptide from HPV E7. *Nature*, **391**: 859–65.

Lipinski, M.M. and Jacks, T. (1999). The retinoblastoma gene family in differentiation and development. *Oncogene*, **18**: 7873–82.

Mulligan, G. and Jacks, T. (1998). The retinoblastoma gene family: cousins with overlapping interests. *Trends in Genetics*, **14**: 223–9.

Nevins, J.R. (1998). Toward an understanding of the functional complexity of the E2F and retinoblastoma families. *Cell Growth and Differentiation*, **9**: 585–93.

Saenz Robles, M.T., Symonds, H., Chen, J., and Van Dyke, T. (1994). Induction versus progression of brain tumour development: differential functions for the pRB- and p53-targeting domains simian virus 40 T antigen. *Molecular and Cellular Biology*, **14**: 2686–98.

Sherr, C.J. (1994). The ins and outs of RB: Coupling gene expression to the cell cycle clock. *Trends in Cell Biology*, **4**: 15–18.

Sherr, C.J. and McCormick, F. (2002). The RB and p53 pathways in cancer. *Cancer Cell*, **2**: 103–12.

Sherr, C.J. and Roberts, J.M. (1995). Inhibitors of mammalian G_1 cyclin-dependent kinases. *Genes and Development*, **9**: 1149–63.

Slingerland, J. and Pagano, M. (2002). Regulation of the cdk inhibitor p27 and its deregulation in cancer. *Journal of Cell Physiology*, **183**: 10–17.

Trimarchi, J.M. and Lees, J.A. (2002). Sibling rivalry in the E2F family. *Nature Reviews Molecular Cell Biology*, **3**: 11–20.

Weinberg, R.A. (1995). The retinoblastoma protein and cell cycle control. *Cell*, **81**: 323–30.

Wu, L., de Bruin, A., Saavedra, H.I., et al. (2003). Extra-embryonic function of Rb is essential for embryonic development and viability. *Nature*, **421**: 942–7.

Yamasaki, L. (2003). Role of the RB tumor suppressor in cancer. *Cancer Treatment Research*, **115**: 209–39.

The p53 tumor suppressor

Benard, J., Douc-Rasy, S., and Ahomadegbe, J.C. (2003). TP53 family members and human cancers. *Human Mutation*, **21**: 182–91.

DiGiammarino, E.L., Lee, A.S., Cadwell, C., et al. (2002). A novel mechanism of tumorigenesis involving pH-dependent destabilization of a mutant p53 tetramer. *Nature Structural Biology*, **9**: 12–16.

Donehower, L.A. (1996). The p53-deficient mouse: a model for basic and applied cancer studies. *Seminars in Cancer Biology*, **7**: 269–78.

El-Deiry, W.S., Harper, J.W., O'Connor, P.M., et al. (1994). WAF1/Cip1 is induced in p53-mediated G_1 arrest and apoptosis. *Cancer Research*, **54**: 1169–74.

Evans, S.C. and Lozano, G. (1997). The Li-Fraumeni syndrome: an inherited susceptibility to cancer. *Molecular Medicine Today*, **3**: 390–5.

Flores, E.R., Tsai, K.Y., Crowley, D., et al. (2002). p63 and p73 are required for p53-dependent apoptosis in response to DNA damage. *Nature*, **416**: 560–4.

Garcia-Cao, I., Garcia-Cao, M., Martin-Caballero, J., et al. (2002). "Super p53" mice exhibit enhanced DNA damage response, are tumor resistant and age normally. *EMBO Journal*, **21**: 6225–35.

Kastan, M.B. and Lim, D.-S. (2000). The many substrates and functions of ATM. *Nature Reviews Molecular Cell Biology*, **1**: 179–86.

Lane, D.P. (1992). p53, guardian of the genome. *Nature*, **358**: 15–16.

Levine, A.J., Momand, J., and Finlay, C.A. (1991). The p53 tumor suppressor gene. *Nature*, **351**: 453–6.

Li, F.P. and Fraumeni, J.F., Jr. (1969). Rhabdomyosarcoma in children: epidemiologic study and identification of a familial cancer syndrome. *Journal of National Cancer Institute*, **43**: 1365–73.

Marino, S., Vooijs, M., van der Gulden, H., Jonkers, J., and Berns, A. (2000). Induction of medulloblastomas in p53-null mutant mice by somatic inactivation of Rb in the external granular layer cells of the cerebellum. *Genes and Development*, **14**: 994–1004.

Matsuda, S., Rouault, J.P., Magaud, J.P., and Berthet, C. (2001). In search of a function for the TIS21/PC3/BTG1/TOB family. *FEBS Letters*, **497**: 67–72.

Michael, D. and Oren, M. (2003). The p53-Mdm2 module and the ubiquitin system. *Seminars in Cancer Biology*, **13**: 49–58.

Parant, J., Chavez-Reyes, A., Little, N.A., et al. (2001). Rescue of embryonic lethality in Mdm4-null mice by loss of Trp53 suggests a nonoverlapping pathway with MDM2 to regulate p53. *Nature Genetics*, **29**: 92–95.

Parant, J.M. and Lozano, G. (2003). Disrupting TP53 in mouse models of human cancers. *Human Mutation*, **21**: 321–6.

Ribeiro, R.C., Sandrini, F., Figueiredo, B., et al. (2001). An inherited p53 mutation that contributes in a tissue-specific manner to pediatric adrenal cortical carcinoma. *Proceedings of the National Academy of Sciences of the U.S.A.*, **98**: 9330–5.

Sherr, C.J. and Weber, J.D. (2000). The ARF/p53 pathway. *Current Opinion in Genetics and Development*, **10**: 94–9.

Soussi, T. (2003). P53 mutations and resistance to chemotherapy: A stab in the back for p73. *Cancer Cell*, **3**: 303–5.

Urist, M. and Prives, C. (2002). p53 leans on its siblings. *Cancer Cell*, **1**: 311–3.

Vogelstein, B. and Kinzler, K.W. (1994). Tumour-suppressor genes. X-rays strike p53 again. *Nature*, **370**: 174–5.

INK4/ARF

Aslanian, A., Iaquinta, P.J., Verona, R., and Lees, J.A. (2004). Repression of the Arf tumor suppressor by E2F3 is required for normal cell cycle kinetics. *Genes and Development*, **15**: 1413–22.

Hanahan, D. and Weinberg, R.A. (2000). The hallmarks of cancer. *Cell*, **100**: 57–70.

Lowe, S.W. and Sherr, C.J. (2003). Tumour suppression by Ink4a-Arf: progress and puzzles. *Current Opinion in Genetics and Development*, **13**: 77–83.

Maestro, R., Dei Tos, A.P., Hamamori, Y., et al. (1999). twist is a potential oncogene that inhibits apoptosis. *Genes and Development*, **13**: 2207–17.

Martin, A.C., Thornton, J.D., Liu, J., et al. (2004). Pathogenesis of persistent hyperplastic primary vitreous in mice lacking the arf tumor suppressor gene. *Investigative Ophthalmology and Visual Science*, **45**: 3387–96.

Quelle, D.E., Zindy, F., Ashmun, R.A., and Sherr, C.J. (1995). Alternative reading frames of the INK4a tumor suppressor gene encode two unrelated proteins capable of inducing cell cycle arrest. *Cell*, **83**: 993–1000.

Ruas, M. and Peters, G. (1998). The p16INK4a/CDKN2A tumor suppressor and its relatives. *BBA Reviews in Cancer*, **1378**: F115–F177.

Sherr, C.J. (1996). Cancer cell cycles. *Science*, **274**: 1672–7.

Sherr, C.J. (1998). Tumour surveillance via the ARF-p53 pathway. *Genes and Development*, **12**: 2984–91.

Sherr, C.J. (2001). The INK4a/ARF network in tumor suppression. *Nature Reviews Molecular Cell Biology*, **2**: 731–7.

PTEN

Baker, S.J. and McKinnon, P.J. (2004). Tumour suppressor function in the nervous system. *Nature Reveiws Cancer*, **4**: 184–96.

Steelman, L.S., Bertrand, F.E., and McCubrey, J.A. (2004). The complexity of PTEN: mutation, marker and potential target for therapeutic intervention. *Expert Opinion on Therapeutic Targets*, **8**: 537–50.

Senescence

Campisi, J. (2001). Cellular senescence as a tumour-suppressor mechanism. *Trends in Cell Biology*, **11**: S27–S31.

Lundberg, A.S., Hahn, W.C., Gupta, P., and Weinberg, R.A. (2000). Genes involved in senescence and immortalization. *Current Opinion in Cell Biology*, **12**: 705–9.

Shay, J.W. and Wright, W.E. (2002). Telomerase: a target for cancer therapeutics. *Cancer Cell*, **2**: 257–65.

Sherr, C.J. and DePinho, R.A. (2001). Cellular senescence: mitotic clock or culture shock. *Cell*, **102**: 407–10.

Questions for student review

1) A tumor suppressor is defined by:
   a. Loss of function by mutations.
   b. Loss of function by promoter methylation.
   c. Loss of function by deletion of the gene.
   d. Loss of function by genetic alterations.
   e. Loss of function by epigenetic alterations.

2) The role of tumor suppressors is to:
   a. Increase cell division.
   b. Restrain cell proliferation.

3) Tumor suppressors are induced by:
   a. Gamma-irradiation.
   b. DNA damage.
   c. Chemotherapeutic drugs.
   d. Ultraviolet UV damage.
   e. Excess of inappropriate proliferative signals.

4) Retinoblastoma in humans require the loss of function of:
   a. Only one allele.
   b. Both alleles.

5) Epigenetic gene regulation is defined by:
   a. Silencing of gene transcription by promoter methylation.
   b. Gene mutation.
   c. Gene deletion.

8 Cell Death

Stella Pelengaris[a] and Michael Khan[b]
[a]Pharmalogos Ltd and [b]University of Warwick, UK

Death is a very dull, dreary affair, and my advice to you is to have nothing whatever to do with it.

William Somerset Maugham

I don't want to achieve immortality through my work . . . I want to achieve it through not dying.

Woody Allen

Key points

- Programmed cell death is a normal physiological process in multicellular organisms and contributes to the efficient clearance of cells that have become damaged, aged, or simply surplus to requirements. It is a prominent feature of normal development but also contributes to tissue mass homeostasis throughout the life of the animal.
- Disrupting the normally harmonious equilibrium between cell proliferation and cell death can have a profound effect on the host. Excessive cell death can result in degenerative diseases, whereas cell death is invariably suppressed during the development of tumors.
- Apoptosis is a specialized form of programmed cell death that serves as a natural barrier to cancer development. Cancer cells have found ways to suppress apoptosis and thus avoid being killed.
- The machinery of apoptosis has been extensively described and comprises both an "extrinsic" and "intrinsic" pathway involving distinct upstream regulators and downstream effectors: the extrinsic pathway involves activation of death receptors (e.g. FAS), while the intrinsic pathway involves the release of cytochrome *c* from the mitochondria.
- In response to any of these apoptotic signals the final stages that lead to dismantling of the cell are executed by a subfamily of proteases known as caspases. Caspases are cysteine proteases that cleave hundreds of cellular proteins and ultimately lead to a series of morphological changes characteristic of apoptotic cell death.
- Cells in the final throws of apoptosis display "eat me" signals, like phosphatidylserine, that are recognized by phagocytes such as macrophages that then dispose of the corpse, without provoking an inflammatory response.
- A large and increasing number of proteins and protein families regulate apoptosis, including various tumors suppressor pathways and DNA damage responses on the proapoptotic side and growth factor signals on the antiapoptotic side.
- Inhibitors of apoptosis proteins (IAPs) are a family of proteins that are able to inhibit apoptosis by directly binding and inhibiting specific caspases. One of the mechanisms by which tumor cells are believed to acquire resistance to apoptosis is by overexpression of IAPs.
- The BCL-2 family of proteins comprises both proapoptotic and antiapoptotic members, the balance of which determines whether or not a cell commits suicide – apoptosis. These proteins have emerged as fundamental regulators of mitochondrial outer membrane permeabilization (MOMP), which is necessary for cytochrome *c* release. Overexpression of antiapoptotic BCL-2 family proteins is common to many human tumors, and confers resistance to apoptosis induced by standard anticancer therapies.
- The tumor suppressor p53, can prevent cells from becoming malignant by inducing growth arrest or apoptosis. There are several proapoptotic transcriptional targets of p53, such as BAX, NOXA, and PUMA that promote cytochrome *c* release from the mitochondria.
- Oncogenic proteins, such as c-MYC, that possess mitogenic action also induce apoptosis (or growth arrest) unless a survival signal is also received. This mechanism may operate as a "failsafe" to prevent cancer formation if an oncogene becomes deregulated.
- Most recently, another form of programmed cell death, termed "autophagy" (eating of self), has been postulated as a barrier to cancer development. Paradoxically, however, autophagy can also enable cancer cells to survive a variety of otherwise lethal stresses; the cancer

---

The Molecular Biology of Cancer: A Bridge From Bench to Bedside, Second Edition. Edited by Stella Pelengaris and Michael Khan.
© 2013 John Wiley & Sons, Inc. Published 2013 by John Wiley & Sons, Inc.

cell lives to fight another day, by temporarily degrading cell organelles in lysosomes and reutilizing certain components until the stress is removed.
- Anticancer therapy can induce cellular senescence, differentiation, and/or cell death by apoptosis or by nonapoptotic mechanisms such as necrosis, autophagy, and mitotic catastrophe. However, since defects in apoptosis cause resistance to such therapy, restoring or activating apoptosis in tumors is an active area of cancer research.

Introduction

While I thought that I was learning how to live, I have been learning how to die.

Leonardo Da Vinci

All cells live under the shadow of death and when the inevitable happens, at least under normal physiological conditions, this usually occurs by apoptosis. A staggering 50 billion or more cells die each day in the human adult. Just to balance the books each one of us replaces around 70 kg of cells every year (enough to make a full-sized adult human clone – all be it minus some neurons).

Apoptosis, a characteristic form of programmed cell death often termed "cell suicide," is a fundamental process that is essential for development, maintenance of tissue homeostasis, and the elimination of damaged cells. However, too much or too little apoptosis may lead to diseases such as neurodegenerative disorders, diabetes, or cancer. It is now widely accepted that putative cancer cells must avoid apoptosis in order for tumors to arise (see Chapters 6 and 7) and this is regarded as one of the "hallmark" features of cancer. This knowledge has fueled global research efforts into elucidating signal transduction pathways that mediate apoptosis as well as the mechanisms that have enabled cancer cells to avoid it (for example, through the loss of *p53* tumor suppressor or overexpression of BCL-2 protein). For obvious reasons, a key objective in cancer research is to develop candidate drug molecules that can rekindle the suicidal urges of cancer cells and restore their sensitivity to apoptosis.

An historical perspective

There is only one ultimate and effectual preventive for the maladies to which flesh is heir, and that is death.

Harvey Cushing

Although studies on cell death are thought to have been performed centuries ago by Aristotle, and later by Galen, who described the regression of larval and fetal structures during development (and probably first used the term "necrosis") it was not until after the formulation of the "cell theory" by Jacob Schleiden and Theodore Schwann in 1838 that nineteenth-century pathologists, in particular those of the "German school," started to take interest in the process of cell death as a physiological phenomenon. Initially, it was appreciated that cells can die (in fact this was implicit from the time it was first realized that cells existed and were alive), but such death was assumed to be a passive response with cells as the victims of circumstances, such as poisons, and trauma, including death of the organism, largely beyond their control. This view began to change in the latter half of the nineteenth century, although it would be over a century later that cell death was first recognized by modern biology as a normal feature of multicellular organisms, and moreover, that this might be a process involving the active participation of the cell. In 1842, Carl Vogt suggested that cell death could be an important part of normal development based on observations in amphibian metamorphosis. This was followed by Rudolf Virchow, widely regarded as the father of modern pathology, who in 1858 described what he called "degeneration, necrosis and mortification."

It was anatomist Walther Flemming who in 1885 proposed that cells might actually die spontaneously. During studies of regressing ovarian follicles, he observed nuclei that appeared to be breaking apart – a process he called "chromatolysis." Intriguingly, Flemming's sketches, appearing nearly a century before the concept of apoptosis was introduced and named, are probably the earliest clear example of cells undergoing apoptosis. In 1914, Ludvig Gräper suggested that mitosis would need to be balanced by processes such as "Flemming's chromatolysis" that could keep proliferation in check. Every Yin needs its Yang. By the 1950s, embryologists such as Glucksmann had clearly described physiological cell death and Christian de Duve first proposed the concept of "cell suicide." The term "programmed cell death" was introduced in 1964, to reflect the view that cell death during development was not an accidental occurrence but rather was part of a locally and temporally orchestrated plan.

Probably the key event in recent times was when in 1971, John F. Kerr, Andrew Wyllie, and Sir Alistair Currie introduced the term "apoptosis" (from the Greek for falling of leaves) to describe this phenomenon of cell suicide, and thereafter founded the field of modern cell death research – although it was not until almost 20 years later that the idea that cells carried within them an intrinsic "suicide" program became generally accepted (see Box 8.1). A molecular explanation of physiological cell suicide was provided in the 1990s by H. Robert Horvitz (Nobel Prize winner – 2002) and colleagues who identified an intrinsic signaling pathway controlling the cell death of a group of specific neuronal cells during development in the worm *Caenorhabditis elegans* (Fig. 8.1).

Apoptosis in context

Apoptosis is an orchestrated cell death process in which the entire cell is dismantled within the orderly context of membrane-enclosed vesicles, thereby preventing any untoward spillage of intracellular components from the dying cell that might otherwise provoke an immune response. In fact, the risk of inadvertent leakage of any noxious contents is avoided by the rapid clearance of the apoptotic corpses and their membrane-bound fragments through phagocytosis by macrophages and neighboring cells (Fig. 8.2). In contrast, cell death by necrosis incites an inflammatory response, because the cell and its organelles swell and rupture, showering cellular contents into the surrounding tissue. Necrosis

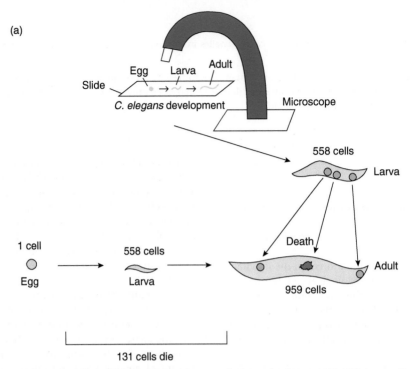

Figure 8.1 Key experiment(s): Studies from H. Robert Horvitz, Nobel Prize winner (2002) and colleagues. The rediscovery of apoptosis as a generally important mechanism using *Caenorhabditis elegans*, the tiny, transparent nematode worm.

Caenorhabditis elegans is a 1-mm-long nematode (a roundworm) that lives in temperate soils. Research into the molecular and developmental biology of *C. elegans* was first pioneered by Sydney Brenner in Cambridge, UK in the 1960s. *C. elegans* is widely used as a model organism and has proved enormously useful in studying cellular differentiation and cell death. It was also the first metazoan to have its genome sequenced and surprised many researchers as it was found to contain more than 19 000 genes (more than 50% of the number of genes subsequently found in the human genome).

The relative simplicity of the organism has facilitated developmental studies and in the 1970s and 1980s John Sulston and Robert Horvitz, working with Brenner, traced the entire *C. elegans* embryonic cell lineage. We now know where every cell comes from and where they end up. In fact, the developmental fate of all 959 somatic cells has been mapped out. One aspect of the cell lineage particularly caught the attention of the Cambridge group: in addition to the 959 cells generated during worm development and found in the adult, another 131 cells are generated but are not present in the adult. These cells are absent because they undergo programmed cell death (see Nobel lecture by Horvitz).

a Cell lineage studies in the developing worm (John Sulston and Robert Horvitz). Newly hatched *C. elegans* larvae were placed on a glass microscope slide dabbed with a sample of the bacterium *Escherichia coli* (nematode food). Then, using Nomarski differential interference contrast optics, individual cells within the living animal could be closely observed and followed as they migrated, divided, and died. In this way, the fate of every single cell from the larval stage to the adult worm could be determined and a cell lineage map was generated.

Subsequently, the pattern of cell divisions between the single-celled fertilized egg and the newly hatched larva were tracked. This was far more difficult than tracking cell fates in larvae, in part because the process of embryonic morphogenesis involves a major cellular rearrangement of a ball of cells to generate a worm-shaped larva. Imagine watching a bowlful of hundreds of grapes, trying to keep your eye on each grape as it and many others move! Nevertheless, all 558 nuclei were followed.

Together, these studies defined the first, and to date only, completely known cell lineage of an animal.

Box 8.1 Apoptosis – the birth of a new concept in biology

Developmental biologists have long been familiar with cell death in carving a vertebrate's digits and in insect metamorphosis. But today's cell-death community credits a paper by University of Edinburgh researcher Andrew Wyllie and his colleagues as the seminal work in the field (Kerr *et al.*, 1972). They coined the term "apoptosis," writing that it plays "a complementary but opposite role to mitosis in the regulation of animal cell populations."

The paper created little excitement initially – it remained dormant for over 15 years. Then it was gradually rediscovered and gained recognition as a generally important mechanism. What catapulted apoptosis into "hot topic" status was its meticulous demonstration in *Caenorhabditis elegans*, the tiny, transparent nematode worm, whose cell-death program removes precisely 131 of 1090 cells to form the adult (Sulston and Horvitz, 1977). This allowed Horvitz and colleagues to look at the process in the worm; the fact that you have a certain number of cells and can trace their developmental fates under a microscope, and ultimately isolate genes made the field flourish (see Fig. 8.1). The three critical genes regulating this cell fate included *ced3* and *ced4*, which are required for the execution of the death program such that inactivation of either gene by mutation prevents all 131 deaths. The third gene, *ced9*, acts to suppress *ced3/ced4*-dependent apoptosis. Inactivation of *ced9* leads to massive ectopic cell death.

Meanwhile, little was known about cell death in other animals. When David Hockenbery, Stanley Korsmeyer, and colleagues discovered that the proto-oncogene *BCL2* (the mammalian homolog of the worm's *ced9* gene) blocks programmed cell death, this and other work on BCL-2 refocused attention on apoptosis, contributing to the second wave of interest in the early 1990s.

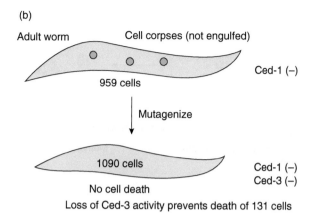

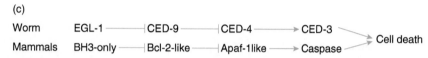

Figure 8.1 (*Continued*) **b Identifying cell death genes**. The next step was to identify the genes responsible for causing or preventing programmed cell death during worm development. Researchers mutagenized a *ced1* (cell death abnormal) mutant worm in which the process of engulfment, or phagocytosis, that normally removes dying cells from the body of the animal, is defective. In *ced1* mutants, although programmed cell death still occurs, dead or dying cells are not engulfed and so cell corpses persist and can be easily visualized in living individuals using Nomarski optics. By mutagenizing *ced1* animals, it was hoped that mutants possessing abnormalities in the pattern of programmed cell deaths (as seen with Nomarski optics) would be generated. In this way, mutants could be identified in which the process of programmed cell death had not been initiated or in which the pattern of programmed cell deaths was altered.

Indeed, a mutant was found in which no cell corpses could be seen. The gene defined by this mutant was called *ced3*. It was shown that if *ced3* activity is reduced or eliminated by mutation, essentially all 131 cells that normally die instead survive–hence the absence of cell corpses. Ced-3 protein, it was later discovered, was in fact a caspase.

A second mutant was then discovered (using different mutant animals) that prevented cell death. These worms proved to be defective in a new gene with properties essentially identical to those of *ced3*. This killer gene was named *ced4*. Some years later, in 1997, a protein similar to Ced-4 was identified, called Apaf-1 (apoptotic protease-activating factor), with a domain with significant similarity to Ced-4. APAF-1 is a proapoptotic human protein similar in both sequence and function to the *C. elegans* programmed cell death killer protein Ced-4.

In a similar manner, the *ced9* gene was discovered, and was later shown to be homologous to the human proto-oncogene, *BCL2*, whose protein product serves to protect cells against cell death.

c Genetic pathway for programmed cell death. From these studies, the core molecular genetic pathway for programmed cell death in *C. elegans* has been identified which shows great similarity to the mammalian pathway: Egl-1 (egg-laying abnormal) is similar to mammalian "BH3-only" proapoptotic proteins; Ced-9 is similar to mammalian antiapoptotic Bcl-2 protein; Ced-4 is similar to Apaf-1 (important for the activation of procaspase-9 on the apoptosome); Ced-3 is similar to caspases.

is often seen as a more passive cell response to profound physical, chemical, or genotoxic insults.

The elimination of cells by apoptosis is an essential feature of normal physiology and is observed throughout animal development and tissue homeostasis in the adult. During development, apoptosis occurs during the sculpting of somatic structures – for example, the loss of interdigital webs during the formation of the digits, and the hollowing out of solid structures to create lumina such as the gut. In the developing nervous system, only half of the neurons formed receive sufficient survival signals from their target cells; the remaining neurons die by apoptosis. Although this appears an inefficient and wasteful process, in fact this developmental mechanism is believed to be vital for ensuring the correct innervation of target cells by appropriate neurons. Similarly, during development of the immune system, up to 95% of all cells die by apoptosis because they make unproductive or autoreactive antigen receptors. Apoptosis in development is often referred to as "programmed cell death." An example of when the regulation of apoptosis plays a key role in the adult is following childbirth – in preparation for lactation there is growth of breast tissue (proliferation > apoptosis). However, following lactation, dramatic regression of tissue occurs by apoptosis.

Apoptosis continues throughout the life of an animal in order to maintain tissue homeostasis, that is, a balance between cell proliferation and cell death. In fact, in an adult human, billions of cells die every hour! Maintaining tissue homeostasis is particularly important in those tissues that have a high cell turnover, such as epithelial (e.g. skin epidermis and gut) and hematopoietic tissues. It is therefore not surprising that perturbing the balance between cell proliferation and cell death can have a profound effect on the host. For example, excessive apoptosis is associated with degenerative diseases such as Alzheimer disease, spinal muscular atrophy, Huntington disease, and Parkinson disease. Conversely, suppression of apoptosis is essential during the development of tumors – although highlighted at various points throughout the chapter. This deleterious effect is discussed in

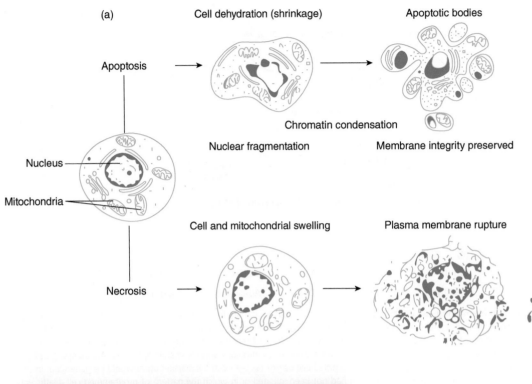
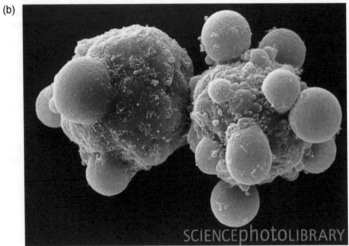

Figure 8.2 Cell death by apoptosis versus necrosis.
  a Schematic representation of cell death by apoptosis and necrosis. Once the apoptotic program has been initiated within a cell, several morphological characteristics can be seen using high-power microscopy: the cell loses contact with its neighbours; chromatin condenses at the edge of the nucleus; DNA fragmentation, cell shrinkage, and dilatation of the endoplasmic reticulum occur. "Blebs" appear on the cell surface, leading to budding of cell membrane and packaging of cellular components into vesicles – known as "apoptotic bodies." These apoptotic bodies are phagocytosed by macrophages and neighboring cells, thus avoiding an inflammatory response. In contrast, death by necrosis proceeds with loss of function of mitochondria and endoplasmic reticulum resulting in a dramatic breakdown of energy supply. The outcome is cellular, nuclear, and organellar swelling. The ultimate rupture of the plasma membrane allows the release of lysosomal enzymes, which attack neighboring cells and surrounding tissue, thus triggering an inflammatory response. Based on image from Google Images.
  b Scanning electron micrograph of liver cells dying by apoptosis. Note the obvious "blebs" on the cell surface. Image reproduced with permission from Science Photo Library.

Chapter 6 in relation to oncogenes, the so-called "cancer-causing" genes.

Here we will describe the process of apoptosis, the signal transduction pathways controlling this form of cell death, and highlight ways in which normal apoptosis regulation becomes derailed in cancer. We will also discuss other forms of programmed cell death, including the more recently described "autophagy" that appears to play a role both as a barrier to cancer and paradoxically as a means by which cancer cells could survive stress. We will conclude by discussing the potential of targeting cell death pathways as cancer therapies.

Apoptosis as a barrier to cancer formation

> As is the generation of leaves, so is that of humanity.
> The wind scatters the leaves on the ground, but the live timber
> Burgeons with leaves again in the season of spring returning.
> So one generation of men will grow while another dies.
>
> *Homer*

As described in earlier chapters, cancers arise as a result of accumulation of multiple genetic and epigenetic lesions that allow cells to proliferate uncontrollably irrespective of exogenous mitogens, resist apoptosis, recruit a blood vasculature, invade surrounding tissues, and eventually metastasize – *vide infra* the hallmark features of cancer.

In normal adult tissues, apoptosis ensures orderly homeostasis by eliminating cells that are damaged (e.g. DNA damage, hypoxia, nutrient limitation) or displaced (cells that have moved out of their normal environment – "anoikis," see Chapter 12). However, the acceptance of apoptosis as one of the central tumor suppressive mechanisms has gradually effloresced over the last decade. It is now unquestioned that evolving cancer cells must avoid cell death for tumors to arise. We now know that cells acquiring a growth-promoting mutation, exemplified by overexpression of the oncogenes c-*MYC*, *E1A*, or *E2F*, become "sensitized" to apoptosis (see Chapter 6). The importance of this cannot be overestimated, as such cells can be eliminated from the body and thus avoid becoming cancerous cells. The notion that a growth-deregulating oncogenic mutation *in vivo* could also possess an "in-built" tumor suppressor function to hinder expansion of potentially malignant cells is one of the most important concepts in cancer biology (and is featured in several chapters; see Chapters 6, 7, and 10). In particular, through the development of regulatable mouse models (see "Mouse models of tumor progression" in Chapter 6), the importance of apoptosis serving as a barrier to cancer formation is all too clear – when apoptosis is blocked, tumor progression is accelerated. As the best-studied example of an oncogene with intrinsic tumor suppressor activity, the various signaling pathways that c-MYC may activate to promote apoptosis are described later.

Proto-oncogenes that specifically regulate apoptosis and which are often deregulated in human cancers include members of the *BCL2* family, the *BCL2* oncogene being the first identified by the study of chromosomal translocations in human lymphoma (see Chapter 3, Box 3.3). The BCL-2 family is now known to include a large number of proteins with either anti- or proapoptotic activities which are described later. If expression of antiapoptotic BCL-2 family members is increased, or that of proapoptotic members lost, then cells are able to survive and consequently become vulnerable to other cancer-causing mutations, in particular those which increase proliferation (e.g. *MYC*). Instead of dying, as they would normally do, these cells are now able to survive and give rise to cancer.

It is clear that highly conserved pathways regulating cellular growth/replication and apoptosis are coupled at numerous levels, even under normal physiological conditions, and that this plays a key role in the regulation of appropriate cell numbers and organ size during development and in maintaining these in the adult. In addition to the role of c-MYC, E2F, and others in this process, studies in the fruitfly (*Drosophila*) have revealed another important signaling pathway involving the large tumor suppressor LATS/WARTS protein kinase, which alongside two other proteins, Hippo and Salvador, is also a regulator of both cell-cycle exit and apoptosis (see Chapter 5; Figs 5.23 and 5.24). At least two mammalian homologs of LATS/WARTS are known, LATS1 and LATS2, and disruption of LATS1 or in proteins, such as the Hippo homolog MST2 or the MATS1 (Mob as tumor suppressor), required to activate LATS1 prevent apoptosis and increase tumor formation in mice. ABL can also exert dual roles, but in this case may be determined by cellular location. In the cytoplasm, ABL induces replication (the BCR–ABL protein in chromic myelogenous leukemia is confined to the cytoplasm), whereas in the nucleus ABL promotes apoptosis (see Chapter 6). Moreover, some of these have now been shown to be disrupted in human cancers. It is worth noting here that an analogous role may be played by senescence (see Chapters 6 and 9) as a barrier to RAS-induced tumorigenesis.

As this chapter focuses mainly on the molecular pathways specifically involved in apoptosis, we urge the reader to look at Chapters 4–7 for a broader overview of the regulation of cell cycle, growth, and the contributions of oncogenes and tumor suppressors, cell death (or its avoidance) and cancer.

Lastly, apart from elucidating mechanisms that have enabled cancer cells to escape apoptosis, it also provides an avenue for future development of candidate drug molecules to target cancer (see last section on "Exploiting cell death (and senescence) in cancer control").

Apoptosis versus necrosis

> It is possible to state as a general principle that the mesodermic phagocytes, which originally (as in the sponges of our days) acted as digestive cells, retained their role to absorb the dead or weakened parts of the organism as much as different foreign intruders.
>
> *Metchnikoff*

Apoptosis typically affects single cells that are aged, dysfunctional, or damaged by external stimuli. Unlike necrosis, it is an active, energy-requiring process leading to a characteristic series of morphological changes that accompany the degradation of the cell. Key differences between the two forms of cell death are described below.

During apoptosis, gross morphological changes – as shown schematically in Fig. 8.2(a) – can been seen under a microscope. Early features of apoptosis occur within minutes of the apoptotic trigger, and during this phase mitochondria, lysosomes, and cellular membranes remain fully intact. These features include: chromatin condensation, DNA fragmentation (multiples of 180bp), cell shrinkage, and dilatation of the endoplasmic reticulum.

Later features of apoptosis are completed within hours, depending on cell type and tissue: budding of cell membrane ("blebs") leading to the packaging of cellular components into vesicles – known as "apoptotic bodies," which are phagocytosed by macrophages and neighboring cells, thus avoiding an inflammatory response. Figure 8.2(b) shows a scanning electron micrograph of apoptotic cells with obvious "blebs" on the cell surface.

In contrast, necrosis affects groups of cells or whole tissue after extended damage induced by external stimuli such as trauma, ischemia, and high-dose irradiation. This energy-independent form of cell death occurs over 12–24 hours, with loss of function of mitochondria and endoplasmic reticulum, resulting in a dramatic breakdown of energy supply leading to cellular, nuclear, and organellar swelling, and ultimately rupture of the plasma membrane (Fig. 8.2a). This final event allows the release of lysosomal enzymes which attack neighboring cells and surrounding tissue, thus triggering an inflammatory response. In contrast to apoptosis, there is nonspecific degradation of DNA.

Cell death by necrosis – not just inflammatory

Although necrosis has been acknowledged for many years as a form of cell death in which cells become bloated and explode, until recently it has not featured as being of particular relevance to tumorigenesis.

As mentioned above, necrosis completely differs from apoptosis, in which the dying cell contracts into a tiny corpse that is soon "eaten" by neighboring cells. Once exploded, necrotic cells release their contents into the local tissue microenvironment and induce a proinflammatory response, including recruiting of inflammatory cells such as neutrophils and macrophages. These inflammatory cells serve an important function in damaged tissue by dealing with potential pathogens, and then ultimately phagocytes (e.g. macrophages) remove the necrotic debris, terminating the inflammation and allowing tissue repair. However, in the context of tumor growth, there is increasing evidence that immune inflammatory cells can be actively tumor promoting (touched upon in the section on "The IAP family" and in Chapter 13) being capable of inducing angiogenesis, cancer cell proliferation and invasiveness.

In addition, necrotic cells can release regulatory factors like interleukin 1α (IL-1α), which can directly stimulate neighboring cells to proliferate. In this scenario, tumors in which necrotic cell death is supposedly beneficial in counterbalancing cancer-associated hyperproliferation, may ultimately do more damage by recruiting inflammatory cells that promote growth of the surviving tumor cells. Thus, tumors can appear as wounds that do not heal – chronic cell death leads to inflammation that does not resolve, and instead enhances tumor growth.

The pathways to apoptosis

In general, there are two pathways that trigger apoptosis – the extrinsic and intrinsic (mitochondrial) apoptotic pathways. The **extrinsic pathway**, otherwise known as the "death receptor pathway" is activated by the engagement of death receptors on the cell surface. In contrast, the **intrinsic pathway** involves the release of cytochrome c (and other proteins) from the mitochondria. Whichever pathway is taken (Fig. 8.3), both lead to the activation of various caspases – enzymes responsible for the demise of the cell. These crucial enzymes are discussed later in their own section. It is important to emphasize that in vertebrates the majority of apoptosis proceeds through the mitochondrial pathway.

Death receptor pathway (extrinsic)

This pathway is initiated by the binding of ligands to cell surface receptors and as such are the antithesis to growth factors and their receptors. However, the actual role played by this mechanism in normal tissue mass homeostasis is unclear, and most evidence suggests that this system is primarily involved in immune cell killing of viral infected host cells and in other pathological states. Extracellular hormones or agonists that belong to the tumor necrosis factor (TNF) superfamily, including TNF-α, FAS/CD95 ligand, and APO2 ligand/TRAIL, bind to their specific corresponding receptors (TNF/NGF receptor family), such as TNFR1, FAS/CD95, and APO2, respectively. The resultant ligand–receptor binding activates downstream signaling cascades. For example, as shown in Fig. 8.3, binding of FAS ligand (FASL) to the FAS receptor (FAS) induces clustering of FAS, which in turn recruits the adaptor protein FADD (FAS-associated death domain) to form a complex called the death-inducing signaling complex (DISC). DISC then recruits and activates the initiator procaspase-8, probably by bringing the procaspases close enough in proximity ("close proximity" model) so that they can cleave each other. Subsequently, these activated initiator caspases trigger a caspase cascade, activating downstream executioner caspases, such as caspase-3 and caspase-7, that ultimately kill the cell.

When large amounts of caspase-8 are formed at the DISC, then apoptosis proceeds via direct cleavage of procaspase-3 independently of the mitochondria (see mitochondrial pathway). However, caspase-8 can also activate the proapoptotic BCL-2 family protein BID, which promotes release of cytochrome c from the mitochondria – the "intrinsic" pathway described below (Fig. 8.3). This "crosstalk" between the death receptor and mitochondrial pathways can occur in cells when DISC formation and active caspase-8 are insufficient to activate procaspase-3 independently of the mitochondria. Cells in which the extrinsic pathway induces apoptosis only when MOMP is intact are referred to as **type II cells**, whereas cells in which this is not required and the extrinsic pathway alone is sufficient to induce cell death are referred to as **type I**.

Apoptotic cell death mediated via death receptors is critical for normal immune system function, for example, mutations in *FAS* and *FASL* ligand in humans can lead to a complicated immune disorder known as the autoimmune lymphoproliferative syndrome (ALPS), a similar phenotype to that seen in mice with *Fas* and *FasL* mutations.

Importantly, in some cell types, apoptotic response to anticancer therapy implicates the death receptor FAS (see last section). Various anticancer drugs can activate the death receptor pathway by enhancing the expression of FAS and FASL. Interaction of FASL with FAS at the cell surface defines an autocrine/paracrine pathway similar to that observed in activation-induced cell death in T lymphocytes. However, it is possible that FASL is not crucial for drug-induced apoptosis since apoptosis is not suppressed by antagonist antibodies or molecules that prevent FASL binding to FAS. It is likely that certain anticancer drugs exert their apoptotic effects by inducing clustering of FAS receptor at the cell surface of tumor cells in the absence of FASL.

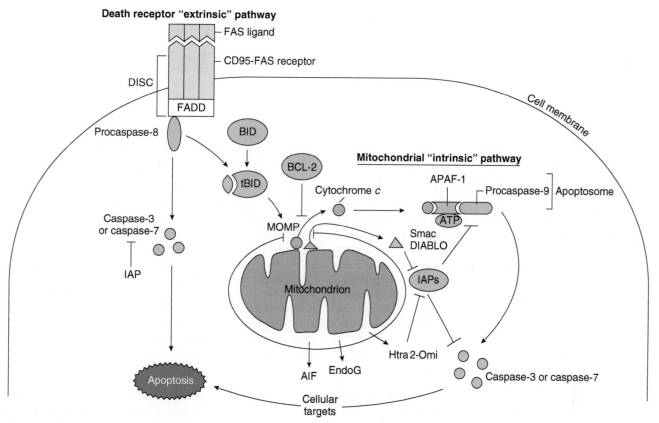

Figure 8.3 The two pathways of apoptosis: death receptor (extrinsic) and mitochondrial (intrinsic) pathways. Here, the death receptor pathway is triggered following binding of FAS ligand (FASL) to the FAS receptor (FAS). FASL binding induces clustering of FAS, which in turn recruits the adaptor protein FADD (FAS-associated death domain) to form a complex called the death-inducing signaling complex (DISC). DISC then recruits and activates the initiator procaspase-8. Subsequently, activated procaspase-8 triggers a caspase cascade, activating downstream executioner caspases, such as caspase-3 and caspase-7, that ultimately kill the cell. However, caspase-8 can also activate the proapoptotic protein BID (by cleavage to become truncated tBID), which promotes release of cytochrome c from the mitochondria – the "intrinsic" pathway. The mitochondrial pathway can be triggered by a variety of cellular stresses (e.g. DNA damage, hypoxia, depleted survival factors or deregulated oncogenes). Once released into the cytosol, cytochrome c associates with APAF-1 to create the apoptosome, a complex that activates procaspase-9. In the presence of cytochrome c and the nucleotide dATP/ATP, procaspase-9 is autocatalytically activated and can now go on to activate downstream executioner caspases, such as caspase-3 and caspase-7. Other proteins that are released from the mitochondria include Smac/DIABLO (and probably Htra2/Omi), which bind to and inhibit inhibitors of apoptosis proteins, thus preventing caspase-9 and caspase-3 inhibition.

The mitochondrial pathway (intrinsic)

It is now known that the mitochondria play a central role in most apoptotic cell death in vertebrate cells. The key and defining event in the mitochondrial pathway is MOMP (mitochondrial outer membrane permeabilization), as a result of which, various proteins normally confined to the mitochondrion are released and trigger the subsequent dismantling of the cell (Fig. 8.3). Cytochrome c was the first of these mitochondrial proteins to be detected in the cytoplasm of apoptotic cells. At first, this observation was rather puzzling, not least because cytochrome c is an essential protein in energy production within cells and is usually located inside the mitochondria (in the space between the inner and outer mitochondrial membranes). Subsequently, however, the role of cytochrome c in activating apoptosis was confirmed by two findings: the first was the identification of its downstream binding partner, APAF-1 (apoptotic protease-activating factor 1) present in the cytosol (described next under "The apoptosome"), and the second was the demonstration that the antiapoptotic protein BCL-2 (described later) inhibits cell death by preventing release of cytochrome c from the mitochondria.

Pressaging the success of Downton Abbey on both sides of the Atlantic, Doug Green, in a review article in the journal Science, posited a very apt analogy to describe the mitochodrial pathway – "we have an upstairs/downstairs situation where at first pass most of the aristocratic decisions are made before MOMP (upstairs) and the workmanlike consequences occur thereafter (downstairs)." Implicit in this is the notion that executive decisions and balancing of factors for and against apoptosis are made before MOMP. Once MOMP has been "signed off," the order for apoptosis is largely carried out regardless. This does not mean that there is no debate about matters after MOMP – the various interactions between other mitochondrial proteins and IAPs (see below) can modulate the decision to commit apoptosis. A key question is: what happens to promote MOMP and thence the release of cytochrome c and other proteins from mitochondria?

Without a doubt, the BCL-2 family of proteins are key regulators of MOMP and will be discussed in more detail later under their own heading, but it is worth continuing the upstairs/downstairs analogy at this point. As we will see, BAX and BAK

Figure 8.4 Functional domains of the APAF-1 protein. APAF-1 contains an N-terminal caspase recruitment domain (CARD) by which it interacts with procaspase-9, a nucleotide binding domain (NBD), and a long C-terminal extension containing 13 repeats of the WD40 motif. The WD40 motif is a conserved protein domain approximately 40 residues long that has a characteristic tryptophan–aspartate motif, and is thought to negatively regulate APAF-1.

are the "butlers" of this piece; they operate above and below stairs and most importantly we know by literary convention that they are culpable. We also know something of the motive; the normally affable BAX and BAK are driven to "commit MOMP" by imbalances between pro- and antiapoptotic BCL-2 family members upstairs. Stress of various kinds can trigger activity of proapoptotic (BH3-only) proteins in part by removing the calming embrace of antiapoptotic BCL-2 and BCL-X_L. The nature of the upstairs activating signals that lead to changes in expression and/or activation of various BCL-2 family members will vary depending on cell type and what stresses the cell is under, for example, hypoxia, depleted survival factors, DNA damage, or if the cell has acquired an oncogenic mutation (e.g. in c-*MYC* leading to deregulated expression).

Once MOMP occurs, cytochrome *c* can associate with APAF-1 in the cytoplasm to create the **apoptosome**, a complex that activates procaspase-9. Since formation of the apoptosome plays a crucial role in mediating apoptosis, it deserves a more thorough explanation (see below). In the presence of cytochrome *c* and the nucleotide dATP/ATP, procaspase-9 forms oligomers and becomes autocatalytically activated; active caspase-9 can now go on to activate downstream executioner caspases, such as caspase-3 and caspase-7 (Fig. 8.3), by cleavage at specific sites.

The regulation of apoptosis downstairs is complex and involves numerous activation steps and also protein stability. Several different inhibitors of caspases, the inhibitor of apoptosis proteins (IAPs), have been described that inhibit apoptosis by targeting caspases for proteasome degradation. Conversely, various proteins released from the mitochondrial intermembrane space, alongside cytochrome *c* (Fig. 8.3), allow MOMP to circumvent the action of the apoptosis-inhibiting IAPs. Thus, MOMP releases two inhibitors of IAPs, DIABLO/Smac (second mitochondria-derived activator of caspase) and Omi/HtrA2. In fact, other important proteins are released by MOMP, including two which can contribute to apoptosis even in the absence of caspase activation – AIF (apoptosis-induction factor), involved in chromatin condensation and large-scale DNA degradation, and endonuclease G (EndoG), which might aid the caspase-activated DNase (CAD) in nucleosomal DNA fragmentation. In some circumstances MOMP may even result in release of a small proportion of procaspase molecules.

The apoptosome – "wheel of death"

The three-dimensional structure of the apoptosome has now been solved using cryoelectron microscopy technology. The structure has given insight into how the apoptosome assembles, how it might activate procaspase-9 and why activation of this procaspase is distinct from the conventional caspase activation mechanism (that is, proteolytic cleavage of effector caspase), which is discussed in more detail in the below.

Once cytochrome *c* is released from mitochondria into the cytosol, it associates with APAF-1 to create the apoptosome, a complex that activates procaspase-9. In mammalian cells, APAF-1 is cytosolic and, importantly, its activity is restrained by its long C-terminal extension containing 13 repeats of the WD40 motif (Fig. 8.4). The WD40 motif is a conserved protein domain approximately 40 residues long that has a characteristic tryptophan–aspartate motif. In an individual APAF-1 molecule, two groups of WD40 repeats in the C-terminal region are thought to keep the protein inactive until cytochrome *c* engages the repeats (Fig. 8.5). After association with cytochrome *c*, APAF-1 switches from a rigid conformation ("closed") to a more flexible one ("open") such that the nucleotide dATP/ATP binding activity is greatly facilitated. This binding triggers formation of the active seven-span symmetrical wheel-like structure ("wheel of death") – the apoptosome – via interaction among the N-terminal caspase recruitment domains (CARD) of the individual APAF-1 molecules (Fig. 8.5). The apoptosome subsequently recruits procaspase-9 into its central hub through CARD–CARD domain interaction between procaspase-9 and APAF-1 molecules, which brings about a conformational change of procaspase-9. This enzyme is now in its active form and can go on to activate downstream executioner caspases, such as caspase-3 and caspase-7, which will eventually lead to cell death.

Caspases – the initiators and executioners of apoptosis

Once apoptotic signals are received by the cell, the final stages that lead to dismantling of the cell are executed by a subfamily of proteases known as caspases. Caspases are highly selective cysteine proteases that have a preference for cleaving proteins after aspartate residues. This specificity ensures that apoptosis is primarily a set of limited proteolytic cleavages, and not a generalized degradative process. Caspase-dependent cleavage is of a defined subset of cellular proteins comprising around 1 in 20 of all proteins (1000 or so) that together culminate in the various morphological changes characteristic of apoptotic cell death.

It is important to note that caspases play roles in other processes, including those involving cellular remodeling such as spermatid individualization, macrophage and skeletal muscle differentiation, cornification of skin, and erythropoiesis. Understanding how this is confined and apoptosis avoided under these circumstances is an important area of research. In some cases this involves specific caspases restricted to certain tissues (such as caspase-14 in skin) that may also be activated by routes separate from apical caspases. Interactions with the ubiquitin–proteasome system (see Chapter 11) are also believed to be important regulators of caspases.

Caspases can be grouped into two categories: **initiator caspases** and **effector** (or executioner) **caspases**. As mentioned

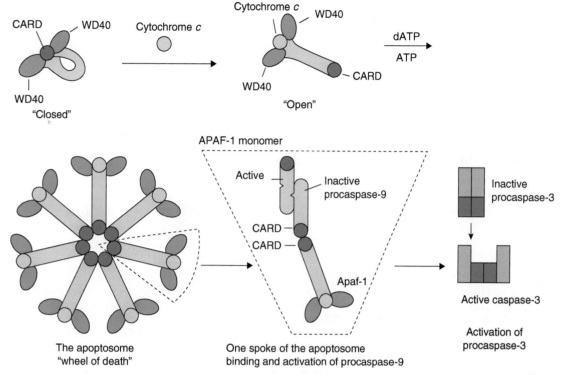

Figure 8.5 Formation of the apoptosome. In an individual APAF-1 molecule, two groups of WD40 repeats in the C-terminal region are thought to keep the protein inactive until cytochrome c engages the repeats. Association with cytochrome c causes APAF-1 to convert from a "closed" conformation to a more "open" one, thus allowing the nucleotide dATP/ATP binding activity to be greatly facilitated. This binding triggers formation of the active seven-span symmetrical "wheel of death" – the apoptosome – via interaction among the N-terminal caspase recruitment domains (CARD) of the individual APAF-1 molecules. The apoptosome subsequently recruits procaspase-9 into its central hub through CARD–CARD domain interaction between procaspase-9 and APAF-1 molecules. An inactive procaspase-9 monomer on one "spoke" of the apoptosome is thought to recruit another monomer to create a dimer with a single active site. This active caspase activates downstream executioner caspases, such as caspase-3 and caspase-7.

earlier (see Fig. 8.3), initiator caspases (e.g. caspases-2, -8, and -9) are activated as they bind to their appropriate adaptor molecules (APAF-1, FADD; see below), after which they cleave and thereby activate downstream ("effector") procaspases and various proteins. The downstream "effector" (or executioner) caspases (e.g. caspases-3, -6, and -7) go on to degrade many cellular substrates (Box 8.2) that result in the characteristic features of apoptotic cell death.

Caspases are synthesized and stored as inactive precursors, **procaspases**, in order to protect the cell from untimely proteolytic events. Each procaspase consists of a prodomain at its N-terminus, a large subunit, and a small subunit (Fig. 8.6). The length of the prodomain varies among caspases (Table 8.1) – long prodomains are characteristic of initiator caspases and contain a death effector domain (DED in procaspases-8/10) or caspase recruitment domain (CARD in procaspase-9). These domains mediate the homophilic interactions (i.e. CARD–CARD and DED–DED) between procaspases and their adaptor proteins (e.g. APAF-1 with procaspase-9; FADD with procaspase-8). An example of this can be seen in Fig. 8.5 during formation of the apoptosome.

It is now recognized that activation of initiator caspases occurs following oligomerization of adaptor proteins (APAF-1, FADD) which in turn leads to recruitment and autoactivation of procaspases (procaspase-8 and -9, respectively). This mode of activation differs from that required for executioner procaspases, which typically require proteolytic cleavage. Executioner procaspases

---

Box 8.2 Cellular substrates of effector caspases (in particular caspase-3)

Effector caspases: (i) cleave structural components of the cytoskeleton and the nuclear membrane such as actin, cytokeratins, and lamins; (ii) cause phosphatidylserine to be exposed on the outside of the cellular membrane, thus promoting phagocytosis by macrophages and neighboring cells; (iii) counteract the apoptosis-inhibiting effect of BCL-2 and BCL-X_L proteins; (iv) inhibit genes that regulate the repair of DNA lesions during cell cycle such as *MDM2* and *RB*; (v) inactivate enzymes responsible for stability, integrity, and repair of DNA, such as poly-(ADP-ribose)polymerase (PARP), DNA-PK, and DNA replication factor 140; and (vi) cleave ICAD (inhibitor of the caspase-activated DNase) and inactivate CAD (caspase-activated DNase), which causes the characteristic oligonucleosomal fragmentation of the chromatin.

(procaspases-3, -6, and -7) have short prodomains which seem to inhibit caspase activation. These caspases exist constitutively as homodimers – both before and after activation cleavage. Executioner procaspases are activated by initiator caspases; for example, procaspase-3 and -7 can be activated by caspases-6, -8, -9, and -10. During activation, the executioner procaspase is cleaved at specific Asp residues to yield a short inhibitory prodomain, a large

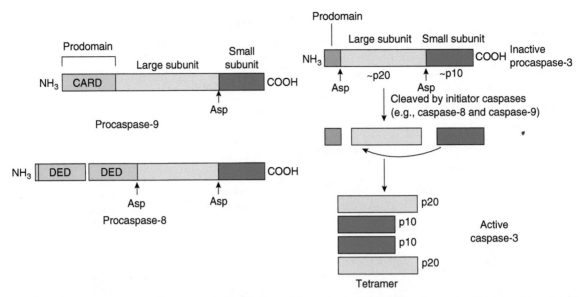

Figure 8.6 Initiator and executioner caspases. Caspases are highly selective cysteine proteases that have a preference for cleaving proteins after aspartate residues (Asp). Initiator procaspases such as procaspase-8 and procaspase-9 possess long prodomains which contain a protein interaction domain, such as the death-effector domain (DED) and caspase-recruitment domain (CARD), respectively. These domains mediate the homophilic interactions (i.e. CARD–CARD and DED–DED) between procaspases and their adaptor proteins (e.g. APAF-1 with procaspase-9; FADD with procaspase-8) and are indispensible to the activation of initiator caspases. Executioner procaspases, such as procaspase-3, have short prodomains which seem to inhibit caspase activation. Procaspase-3 is activated by initiator caspases: it is cleaved at specific Asp residues to yield a short inhibitory prodomain, a large and a small subunit, termed homodimer p20 and p10, respectively. Subsequently, p20 and p10 homodimers interact in a heterodimer – the association of two heterodimers forms the proteolytic tetramer with the two adjacent p10 subunits surrounded by two large subunits.

Table 8.1 Procaspases subgrouped on the basis of domain composition

Long prodomain (DED, CARD) – apoptotic initiator type	Caspases-2,8,9,10
Short prodomain – apoptotic effector (executioner) type	Caspases-3,6,7
Long prodomain (CARD) – cytokine activator type	Caspases-1,4,5

The human genome encodes 11 caspases, and these can be divided into subgroups depending on inherent substrate specificity, domain composition, or presumed role *in vivo*.

and a small subunit, termed homodimer p20 and p10, respectively (Fig. 8.6). Subsequently, p20 and p10 homodimers (refers to subunit size, in kDa) interact in a heterodimer; the association of two heterodimers forms the proteolytic tetramer with the two adjacent p10 subunits surrounded by two large subunits. The catalytic activity of an executioner caspase is increased by several orders after such cleavage.

Unlike executioner procaspases, our understanding of how initiator procaspases are activated is far from complete, but appears to require an appropriate scaffold in most cases. With regard to the activation of procaspase-9 by the apoptosome (described previously), two models have been proposed: the "induced proximity" model and the "proximity-induced dimerization" model. The first model states that the initiator procaspases autoprocess themselves when they are brought into close proximity. However, the mechanism underlying such caspase activation is not known. The second model states that procaspase-8 and procaspase-9 are activated on dimerization, which is facilitated by the apoptosome and DISC complexes, respectively. Procaspase-9 can be activated without processing, apparently because of the flexibility allowed by the unusually long linker between its large and small subunits. An inactive procaspase-9 monomer on one spoke of the apoptosome is presumed to recruit another monomer (homodimerization) in an antiparallel fashion to create the asymmetric dimer having a single active site (see Fig. 8.5). This is a unique feature as only one of the two constituent polypeptides becomes active.

Importantly, most of the active caspase-9 remains complexed to APAF-1 and the released caspase-9 has minimal activity, thus indicating that APAF-1 is not merely a scaffold but within the apoptosome acts as a holoenzyme. In fact, after associating with the apoptosome, there is a dramatic increase in the catalytic activity of caspase-9 (up to 2000-fold) of both processed (cleaved) and unprocessed caspase-9. Thus, for caspase-9 at least, activation has little to do with cleavage; rather, it occurs through apoptosome-mediated enhancement of the catalytic activity of caspase-9.

Most chemoradiotherapy triggers apoptosis that will ultimately induce caspase activity; for example, DNA-damaging drugs activate MOMP via p53 (see later). There have been very recent efforts to develop drugs that can directly activate apical or effector caspases in order to trigger apoptosis in cancer cells directly. Another approach is to target therapeutically any number of regulator processes affecting caspases (see Chapter 16).

The IAP family – inhibitors of apoptosis and much more

Caspases kill cells by irreversibly cleaving a host of cellular components. It is important then for healthy cells to prevent aberrant activation of these enzymes and thus unwanted suicide. To achieve this, cells have evolved a system of checks and balances, at the heart of which is the IAP (inhibitors of apoptosis proteins) family of proteins. IAPs are structurally related proteins that were initially identified in the genome of a baculovirus strain on the basis of their ability to suppress apoptosis in infected host cells. The first IAP, called OpIAP, was identified by L.K. Miller and colleagues in 1993. Cellular IAPS were subsequently found in mammals and fruitflies, but not in nematodes. To date, eight human IAPs have been identified: cIAP1, cIAP2, NAIP (neuronal apoptosis-inhibitory protein), survivin, XIAP (X-linked IAP), Bruce/Apollon, ILP-2 (IAP-like protein-2), and livin/ML-IAP (melanoma IAP). These proteins are able to inhibit apoptosis induced by a variety of stimuli, by directly binding to and inhibiting specific caspases (see Figs 8.3 and 8.10). Interestingly, although caspase-9 binds to several IAPs, it is primarily inhibited by XIAP.

IAPs – cell survival by inhibiting caspases

As their name implies, IAPs were originally identified by their ability to suppress apoptosis. However, as is often the case for newly identified proteins, it is now emerging that IAPs are involved in many important cellular processes that are often deregulated in cancer, such as proliferation, inflammation, immunity, cell migration, and metastases. Moreover, the fact that alterations in IAPs are found in many human cancers and are associated with resistance to current cancer therapies, disease progression, and poor prognosis has sparked great efforts into developing small therapeutic compounds termed "Smac mimetics" which are discussed later.

The hallmark of IAPs is the baculoviral IAP repeat (BIR) domain, a zinc-binding domain of about 70 amino acids that mediates interaction with other proteins. It is these BIR domains that are essential for the antiapoptotic potential of most IAPs, and there are between one and three BIR domains (BIR1, BIR2, and BIR3), depending on the particular IAP. XIAP, cIAP1, and cIAP2 contain three BIR domains in their N-terminus, each domain having a distinct function. For example, in XIAP, the third BIR domain (BIR3) potently inhibits the activity of processed caspase-9, whereas the linker region between BIR1 and BIR2 specifically targets caspase-3 and caspase-7. On the other hand, Survivin contains a single BIR domain and does not inhibit caspase activity *in vitro* (Fig. 8.7).

BIR domains can be classed as type I or type II depending on the presence or absence of a deep peptide-binding groove. BIR1 domains are type I, which means they do not possess this groove, and cannot bind caspases and IAP antagonists. However, they can associate with a different range of proteins from type II BIRs, such as tumor necrosis factor receptor-associated factor 1 (TRAF1) and TRAF2. In contrast, BIR2 and BIR3 domains are classed as type II as they possess this groove and are thus able to bind caspases and IAP antagonists, such as Smac (discussed later). Importantly, for these proteins to bind to the groove they must possess an IAP-binding motif (IBM) at their N-terminus with an exposed alanine or serine. This motif is hidden in non-active caspases (procaspases) but becomes exposed when caspases are activated following cleavage, described earlier. Thus, the BIR2 domains of XIAP binds to the IBMs of caspase-3 and caspase-7, while the BIR3 domain binds to the IBM of caspase-9, preventing its dimerization – a prerequisite for this initiator caspase's activity.

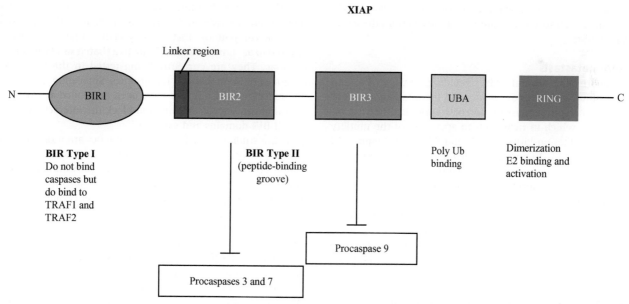

Figure 8.7 Domains of inhibitors of apoptosis proteins (IAP) family proteins. XIAP, an IAP family member, contains three baculoviral IAP repeat (BIR) domains. The third BIR domain (BIR3) potently inhibits the activity of processed caspase-9, whereas the linker region between BIR1 and BIR2 specifically targets caspase-3 and caspase-7. In contrast, Survivin contains a single BIR domain and does not inhibit caspase activity *in vitro*. BIR2 and BIR3 domains are classed as type II as they possess a deep peptide-binding groove, allowing them to bind caspases and IAP antagonists. While the linker region of XIAP's BIR2 domain binds to and inhibits caspase-3 and caspase-7, the strategy through which its BIR3 domain inhibits caspase-9 is fundamentally different: by binding to the homodimerization surface of caspase-9, BIR3 prevents caspase-9 dimerization, and consequently this initiator caspase's activity. BIR1, a type I domain, does not possess the peptide-binding groove, and thus cannot bind caspases and IAP antagonists. However, it can associate with a different range of proteins to that of type II BIRs (e.g. TRAF1 and TRAF2). UBA, ubiquitin (Ub)-binding domain, binds polyubiquitin (polyUb); RING, the C-terminal RING domain is required for Ub ligase activity and is a dimerization interface and docking site for Ub-conjugating enzymes (E2s).

Importantly, studies *in vitro* show that of all the mammalian IAPs, XIAP is the only one that functions as a direct caspase inhibitor (inhibiting caspase-3, caspase-7, and caspase-9). Other IAPs, such as cIAP1 and cIAP2, are not very efficient at directly inhibiting caspases *in vitro*. Overexpression of XIAP inhibits caspase activation and apoptosis induced by both intrinsic and extrinsic apoptosis pathways (described earlier), whereas in the absence of XIAP, cells are sensitized to apoptosis.

IAPs – cell survival by activating NF-κB and immune responses

Although IAPs were initially thought to function primarily by regulating caspases, we now know that they also influence many other cellular processes. One of these important functions that link IAPs (e.g. cIAP1, cIAP2, and XIAP) with cell survival and tumorigenesis is the activation of nuclear factor-κB (NF-κB) transcription factors, which in turn drive the expression of genes important for inflammation, immunity, cell survival, and migration (see Chapter 5). There is much evidence out there showing that constitutive activation of NF-κB and chronic inflammation play a major part in tumor expansion and are seen in most tumor types, such as leukemia, lymphomas, and solid tumors. Tumor cells secrete factors that attract inflammatory cells, which in turn support tumor development by releasing growth, trophic, and chemotactic factors. For example, mast cells produce a large amount of VEGFA (vascular endothelial growth factor A), which induces growth of new blood vessels, fibroblast growth factor 2 (FGF-2), matrix metalloproteinase 9 (MMP9), and tumor necrosis factor α (TNF-α). Importantly, constitutive production of TNF-α from the tumor microenvironment is seen in many malignant tumors and there is emerging evidence indicating TNF-α as a key mediator of cancer-related inflammation. Although it is not clear at present, it seems likely that increased levels of IAPs promote cancer cell survival by protecting the cell from the lethal effects of TNF-α.

IAPs and metastasis

Recent reports link several IAPs, including XIAP, cIAP1, and cIAP2, to tumor cell invasion *in vitro* and metastasis *in vivo*. For example, XIAP in physical association with survivin drives activation of NF-κB, which in turn leads to activation of the motility kinases, focal adhesion kinase (FAK) and SRC (see Chapters 12 and 6, respectively). Importantly, XIAP-mediated cell invasion and metastasis is independent of its ability to promote cell survival through caspase inhibition. Paradoxically, another report has shown XIAP, cIAP1, and cIAP2 actually suppresses cell motility *in vitro* in response to growth factor stimulation, as a result of IAP binding to C-RAF (see Chapter 6) and targeting it for proteasomal degradation. Clearly, further work is needed to clarify such distinct behaviors, including tumor cell state and microenvironment.

Inhibition of IAPs – possible anticancer therapy

IAP antagonists are constitutively expressed in mammalian cells, but are sequestered to the mitochondria (Smac/DIABLO, and HTRA2/OMI) or the endoplasmic reticulum (GSPT1/eRF3) until a death stimulus is received. As mentioned earlier, during apoptosis sequestered IAP antagonist proteins are released and bind to BIR domains of IAPs, thus relieving IAP-mediated caspase inhibition (Figs 8.3 and 8.10). We next discuss the association of IAPs in human cancers and emerging anti-IAP therapies.

One mechanism by which tumor cells might acquire resistance to apoptosis is by overexpressing IAPs. Of all the IAPs, survivin has received most attention to date. Fetal and embryonic tissues show high expression of survivin, while it is undetectable in normal, fully differentiated tissues. This is in marked contrast to the high levels of expression observed in a wide range of malignancies. High levels of survivin have been reported in colorectal cancer, esophageal cancer, and soft tissue sarcoma and correlate with clinical features. Survivin expression has also been associated with poor prognosis in several CNS tumors, including glioma. Although the prognostic relevance is still unclear, high levels of survivin were also detected in carcinomas of the stomach, pancreas, liver, uterus, and in pheochromocytoma.

Genomic amplification of 11q22 which contains *BIRC2* (encoding cIAP1) and *BIRC3* (encoding cIAP2) is frequently observed in hepatocellular carcinoma (HCC), lung and pancreatic cancers, oral squamous cell carcinomas, medulloblastomas, and glioblastomas. In a mouse model of HCC in which Myc drives tumorigenesis, amplification of *BIRC2* and *BIRC3* also occurs frequently. Moreover, the high levels of cIAP1 expression were essential for the rapid growth of these tumors. Following on from the previously discussed importance of the immune response and cancer development, it is tempting to propose that MYC activation – which we know induces the expression of numerous chemokines that attract inflammatory cells (see Chapter 6) – leads to constitutive production of TNF-α by such cells in the tumor microenvironment. As previously mentioned, high levels of cIAP1 and cIAP2 in tumor cells function as key mediators of TNF-α-induced activation of NF-κB and protect cancer cells from the lethal effects of TNF-α.

It is clear, however, that futher work is needed in order to fully understand how cIAPs contribute to different types of cancers.

With regard to blocking IAPs as an anticancer therapy, currently there are five phase I clinical trials assessing the use of a range of compounds, many orally active, that are collectively referred to as "Smac mimetics" due to a shared similarity of action to Smac. They are currently being tested in the treatment of advanced solid tumors and lymphomas. Smac mimetics are small pharmacological drugs that have been developed as IAP antagonists that can selectively bind and block the interaction between type II BIR domains and caspases.

In vitro studies show that Smac mimetics are most efficient at blocking cIAP1 and cIAP2 rather than XIAP, such that within minutes of exposure they trigger auto-ubiquitylation and proteasomal degradation of cIAP1 and cIAP2. The outcome is cell death by necrosis or apoptosis. The molecular events that lead to such cell death following IAP inhibition include the spontaneous activation of NF-κB, NF-κB-mediated enhancement of TNF-α production, and autocrine stimulation of the TNF receptor (TNFR1). It appears to be the production of TNF-α that is necessary to induce cell death by binding to its receptor, as resistant cancer cell lines fail to produce TNF-α. However, when the culture medium is supplied with TNF-α, these cancer cells rapidly succumb to TNFR1-mediated apoptosis. Why certain cancer cells behave differently by not producing TNF-α in response to Smac mimetics is not known, but it is believed that *in vivo* this difference may not matter given that cancer cells are flooded with TNF-α produced by the tumor microenvironment. In this scenario, it is expected that a wide variety of tumor cell types would be sensitive to Smac mimetics *in vivo*. Moreover, Smac mimetics are well tolerated *in vivo* and, importantly, do not appear to sensitize normal cells to TNF-α-induced death and also potentially to the effect of immunomodulatory treatments.

At present, there are grounds for optimism about the development and use of IAP inhibitors in cancer treatments. The coming years will be very important in gaining information from clinical trials.

The central role of MOMP and its regulators in apoptosis – the BCL-2 family

As discussed already, MOMP is a key activator of caspases and cell death because it allows the escape of a collection of villainous proteins, such as cytochrome c, from their incarceration within the mitochondrial intermembrane space. The extremely dangerous nature of these proteins requires the deployment of a tight network of guardians that maintain the integrity of the outer membrane barrier. In this context, the antiapoptotic members of the B-cell lymphoma 2 (BCL-2) protein family are particularly important as it falls upon them to counteract the proapoptotic proteins, such as BCL-2-associated X protein (BAX) or BCL-2 antagonist or killer (BAK), that are seeking to engineer a breakout.

It is also worth noting that MOMP in mammalian cells results in mitochondrial dysfunction and fission which may prove fatal even in the absence of caspase activation. For example, when APAF-1 or caspase-9 activity was blocked by genetic ablation, cell death was delayed but not prevented. The caspase-independent variant of cell death produces a more lingering death, more akin to necrosis, as mitochondrial dysfunction produces a more gradual loss of viability. However, it is likely that under most usual conditions mitochondrial function is shut down in parallel with the release of caspase-activating proteins, giving a two-pronged assault on cell survival.

It has widely been believed that all stress-induced apoptosis (regulated by the BCL-2 family) proceeds through mitochondrial disruption and activation of procaspase-9 via APAF-1. However, further studies argue that the apoptosome and cytochrome c release is not always necessary. For example, the absence of APAF-1 or caspase-9 in the hematopoietic system caused no effect on cell numbers, but these cells underwent apoptosis in response to diverse stimuli with activation of various initiator caspases.

It was surprising that the predominant phenotype shared by the knockout of *Apaf-1*, *caspase-9*, and *caspase-3* genes in mice was the severe developmental defect in the CNS that resulted in the protrusion of brain tissue from the forehead and perinatal lethality. These findings indicate an essential role of the apoptosome in brain development but surprisingly not in other tissues. Given that apoptosis is known to occur in many tissues during development, such as the immune system, this suggests that other tissue-specific pathways are involved in the development of these organs, such as the "death receptor"-mediated pathway.

Given that the conventional knockout mice mentioned above resulted in lethality following birth, it is still possible that cytochrome c-mediated apoptosis is involved in the homeostasis of many adult tissues. We await the generation of more tissue- and time-specific genetically modified mouse models.

The BCL-2 family

In mammals, the BCL-2 family consists of at least 20 members, all of which share at least one conserved BCL-2 homology (BH) domain (Fig. 8.8). It is comprised of both proapoptotic and antiapoptotic members, the balance of which determines whether or not a cell commits suicide – apoptosis. These proteins are pivotal regulators of MOMP, necessary for cytochrome c release from the mitochondria, and therefore of apoptosis. Arguably they are among the most critical determinants of life–death decisions for cells and are emerging as new drug targets in cancer. The proapoptotic members are regulated by, among others, key tumor suppressors and DNA damage responses and in this context are discussed in Chapter 10.

Upregulation of antiapoptotic BCL-2 family proteins is common to many human tumors (see also Chapter 6), as is loss of function of proapoptotic proteins such as BAX and BAK (see below). Dysregulation of these proteins can confer a survival advantage,

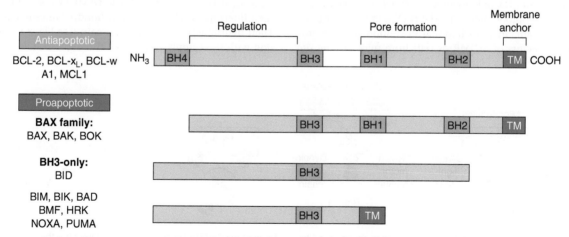

Figure 8.8 The BCL-2 family. The BCL-2 family of proteins can be subdivided into three categories according to their structure and function: the antiapoptotic members promote cell survival, whereas members of the BAX family and "BH3-only" categories are proapoptotic and thus promote apoptosis. The BCL-2 family members possess at least one of four conserved motifs known as BCL-2 homology domains (BH1–BH4). Most members have a C-terminal hydrophobic domain (TM) that aids association with intracellular membranes. Pore formation is enabled by residues BH1, BH2, and BH3. During apoptosis, the proapoptotic members are activated, and presumably undergo a conformational change leading to exposure of the BH3 domain – an interaction domain that is necessary for their killing action. In this way, BH3-only proteins promote apoptosis by directly binding to their antiapoptotic relatives.

preventing cell death in response to a variety of apoptotic stimuli that usually function to prevent tumorigenesis (such as DNA damage, hypoxia, detachment from extracellular matrix). As discussed in the last section of this chapter, p53-inactivating mutations, as well as blockade of apoptosis by BCL-2 family proteins, reduces sensitivity to antineoplastic agents that kill cells via apoptosis, thereby limiting their clinical efficacy.

Prosurvival (antiapoptotic) members: BCL-2, BCL-X_L, BCL-w, MCL-1, and A1

These proteins have four BH domains (BH1–BH4) (Fig. 8.8). A hydrophobic groove is formed by residues from BH1, BH2, and BH3, which is vulnerable to binding by the BH3 domain of an interacting BH3-only (proapoptotic) relative, discussed below. In this way, the activity of antiapoptotic BCL-2 family proteins can be neutralized by the BH3-only proteins (with the possible exception of BID). Importantly, apoptosis requires the additional activity of proapoptotic proteins, BAX and BAK, to drive MOMP. Recent studies show that antiapoptotic BCL-2 proteins, which patrol the outer mitochondrial membrane, can block MOMP using two modes: by sequestering BH3-only proteins as mentioned above, and binding active BAX and BAK.

Prosurvival proteins can be divided into two subclasses (BCL-2, BCL-X_L, BCL-w, and another comprising MCL-1 and A1). Interestingly, BAK activation may be restrained by at least two of these, one member from each subclass, namely BCL-X_L and MCL-1. Many members of the BCL-2 family (except A1 and many of the BH3-only proteins) have a C-terminal hydrophobic domain that aids association with intracellular membranes. In fact, the appropriate targeting to and location of BCL-2 proteins on the mitochondria (and incidentally also the endoplasmic reticulum) is a crucial regulator of apoptosis.

Proapoptotic members

Proapoptotic members are divided into two distinct groups: the BAX family and the BH3-only family:

- **BAX family** (BAX, BAK, BOK): These have three BH domains (BH1–BH3) that are similar in sequence to those in BCL-2, and are thus sometimes referred to as BH123 or multidomain proteins. In contrast to BH3-only members that act as direct antagonists of the antiapoptotic members, the BAX family proteins act further downstream, in mitochondrial disruption. BH123 proteins are key effectors of MOMP, as cells from mice lacking the two major forms – BAX and BAK – are resistant to multiple apoptotic stimuli and fail to undergo MOMP. Thus, BAK- or BAX-mediated MOMP is recognized as the point of no return in mammalian apoptosis.

- **BH3-only family** (BID, BIM, BIK, BAD, BMF, HRK, NOXA, PUMA): As the name implies, these proteins have only the short BH3 motif – an interaction domain that is both necessary and sufficient for their killing action. These proteins are direct antagonists of the antiapoptotic members, and may also serve at least in part to activate BAX and BAK. Perhaps not surprisingly, numerous BH3-mimetic drugs are currently being evaluated as anticancer agents (see Chapter 16). To prevent inappropriate cell death, BH3-only proteins are subject to stringent controls at both the transcriptional and posttranslational level (Fig. 8.9). It is worth noting that BH3-only proteins also control the initiation of autophagy (discussed later). Broadly, BIM, PUMA, and truncated BID (tBID) bind to all prosurvival BCL-2 family proteins; NOXA binds only to MCL-1 and A1; BAD and BMF bind only to BCL-2, BCL-X_L, and BCL-w. BIM, BID, and PUMA are believed to be direct activators of BAX and BAK, with weaker potential direct activating effects recently reported for BMF and NOXA (if BIM and BID were absent). At present, there are two competing models of activation of BAX by BH3-only proteins: (i) direct and (ii) indirect activation models. As mentioned above, in the direct activation model, the BH3-only activator proteins such as BIM and BID directly bind to BAX and BAK to trigger their conformational change and oligomerization. In the indirect mode of activation, the BH3-only proteins bind to prosurvival proteins and cause them to release activated BAX or BAK. It is this interaction of BCL-2 to BAX and BAK that opened an avenue for designing therapeutics that target the apoptotic machinery.

How does the BCL-2 family regulate apoptosis?

Antiapoptotic members BCL-2, and its closest homologs BCL-X_L and BCL-w, are potent inhibitors of apoptosis in response to many stress signals (the less well-studied A1 and MCL-1 seem to have weaker antiapoptotic activity). The widespread view is that these members block apoptosis by guarding mitochondrial integrity and preventing the "early release" of various noxious proteins, such as cytochrome *c*, Smac/DIABLO, and Omi/Htra2, from their internment within the mitochondrial intermembrane space. Their hydrophobic C-terminal domain (see Fig. 8.8) aids association with the cytoplasmic face of three intracellular membranes:

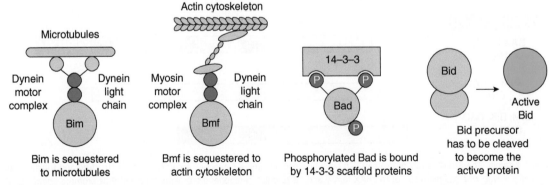

Figure 8.9 BH3-only proteins kept inactive in healthy cells. BIM and BMF are sequestered by binding to dynein light chains that are associated with the microtubules and actin cytoskeleton. BAD is kept inactive when phosphorylated by kinases such as AKT and protein kinase A, through being bound by 14-3-3 scaffold proteins (proteins that provide a platform for the assembly of other proteins). BID is inactive until proteolytically cleaved, for example, by caspase-8 or granzyme B.

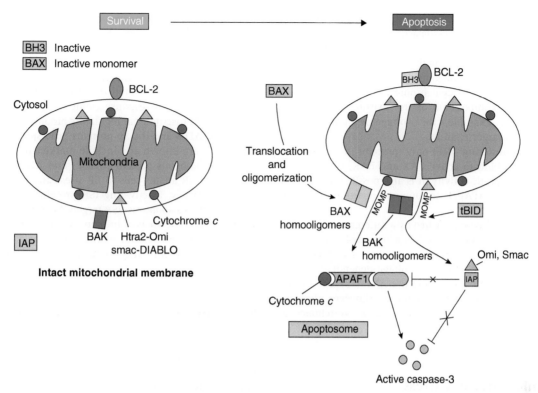

Figure 8.10 Regulation of mitochrondrial outer membrane permeabilization (MOMP) by the BCL-2 family. In healthy cells, mitochondrial membranes remain intact. Antiapoptotic member, BCL-2, and its closest homologs BCL-X_L and BCL-w, are potent inhibitors of apoptosis in response to many stress signals. They protect mitochondrial integrity and thus prevent the release of cytochrome c, as well as other proapoptotic molecules (DIABLO/Smac and Omi/Htra2). BCL-2 is an integral membrane protein in all cells, whereas BCL-X_L and BCL-w only become tightly associated with the membrane after a cytotoxic signal, possibly through conformational change. During apoptosis, proapoptotic "BH3-only" proteins neutralize the activity of BCL-2. Then, proapoptotic members BAX and BAK form homo-oligomers within the mitochondrial membrane and induce MOMP, allowing efflux of cytochrome c and the formation of the apoptosome. BAX is normally a cytosolic monomer but translocates to the mitochondrial membrane during apoptosis and changes conformation, forming homo-oligomers within the outer mitochondrial membranes. BAK, however, is an oligomeric integral mitochondrial membrane protein but also changes conformation during apoptosis and might form larger aggregates. The release of DIABLO/Smac and Omi/Htra2 from the mitochondria bind to and inhibit IAPs – proteins that prevent caspase-9 and caspase-3 activation.

the outer mitochondrial membrane, the endoplasmic reticulum (ER), and the nuclear envelope.

BCL-2 is an integral membrane protein in all cells, whereas BCL-X_L and BCL-w only become tightly associated with the membrane after a cytotoxic signal, possibly through conformational change. These antiapoptotic (or prosurvival) proteins prevent the release of cytochrome c from the mitochondria and hence prevent formation of the apoptosome. In addition, they also prevent the release of the protein Smac/DIABLO from the mitochondria, thereby forestalling inhibition of prosurvival IAPs, and helping maintain caspase-9 and caspase-3 in inactive states (discussed in the IAPs section) (see Figs 8.3 and 8.10).

Apoptosis is also prevented by antagonizing the ability of proapoptotic proteins BAX and BAK to induce MOMP and efflux of cytochrome c (Figs 8.3 and 8.10). The mechanism for MOMP is however, controversial (see below). In healthy cells, BAX exists as a cytosolic monomer but changes conformation during apoptosis, presumably flipping out its C-terminus and then translocating to form homo-oligomers within the outer mitochondrial membranes. BAK, however, is already an oligomeric integral mitochondrial membrane protein in healthy cells but also changes conformation during apoptosis and might form larger aggregates. Although it is not completely understood how the oligomers form, BAK and BAX activation involves exposure of the BH3 domain, which can then bind to the hydrophobic surface groove of another activated BAK molecule. Such BH3 groove interactions may be generally important in oligomerization of these proteins.

In healthy cells, individual BH3-only proteins are held in check (see Fig. 8.9), whereas NOXA, PUMA, and HRK, are controlled primarily at the transcriptional level (see p53 section). The BH3-only protein BID promotes death by activating BAX and BAK. Once cleaved by caspase-8 or granzyme-B, activated BID is lipid modified and migrates to the mitochondrion, where in the presence of BAX and BAK, BID rapidly triggers cytochrome c release and apoptosis (Fig. 8.10). However, despite more than a decade of intense investigation, controversy still surrounds the activation, regulation, and function of BAK and BAX during apoptosis. Further work is needed to define the stepwise activation and oligomerization of BAK and BAX because each step represents a potential therapeutic target for blocking apoptosis.

Mitochondrial outer membrane permeabilization (MOMP)

As mentioned, proapoptotic BCL-2 members BAX and BAK are instrumental in puncturing the mitochondrion (MOMP), releasing a powerful cocktail of proapoptotic proteins from the intermembrane space into the cytoplasm. Although the mechanism is not fully understood, one model is that BAX and BAK form channels in the mitochondrial membrane (Fig. 8.10). In support of this model, BAX oligomers can form pores in liposomes that allow passage of cytochrome *c*. Alternatively, BAX might interact with components of the existing permeability transition pore, for example, the voltage-dependent anion channel, to create a larger channel.

Apart from sequestering BH3-only proteins, antiapoptotic BCL-2 proteins also bind to active BAX and BAK, thereby preventing their proapoptotic activity. It is also possible that BCL-2 and BCL-X_L bind to APAF-1 to prevent it from activating caspase-9.

Lastly, MOMP may also be activated by a different mechanism, not involving BAX/BAK, but rather the formation of a permeability transition pore complex (PTCP) consisting of two putative pore-forming proteins – VDAC (voltage-dependent anion channel) in the outer membrane and the ANT (adenine nucleotide translocator) in the inner membrane. However, the role of this in physiological apoptosis induction in vertebrates is controversial.

Endoplasmic reticulum stress

The ER serves many specialized functions in the cell, including calcium storage and gated release, biosynthesis of membrane and secretory proteins, and production of lipids and sterols. When misfolded or unfolded proteins accumulate in the ER (**ER stress**), an intracellular signaling pathway termed the unfolded protein response (UPR) is activated, with the presumed purpose of alleviating the stress. If the stress inducer is not corrected then the UPR activates apoptosis. ER stress may result from defects in chaperone-mediated folding and numerous diseases may in part involve such defects, including diabetes. Moreover, cancer cells exposed to hypoxia or treated with some chemotherapeutic agents also trigger the UPR. The proteins involved in triggering the ER stress-responsive cell death pathway are not well understood. UPR gene expression is induced following exposure to ER stress agents in the presence or absence of caspase-9 and caspase-12 and even in cells that overexpress BCL-X_L. However, ER stress-induced apoptosis is caspase-dependent and is prevented at least in some cells by overexpression of BCL-X_L or a dominant negative caspase-9.

There are clearly other parallels between ER stress-induced apoptosis and that induced by other triggering stimuli, including an inhibitory effect of the PI3K–AKT signaling pathway. BIM is translocated to the ER in response to ER stress and plays a key role in activation of caspase-12 and initiation of the ER stress-specific caspase cascade. The proapoptotic BH3-only protein BIK, discussed earlier, targets the membrane of the ER but appears not to play a role in UPR. However, BIK is induced by genotoxic stress and overexpression of E1A or p53 and may link the ER with induction of MOMP. BAX and BAK also operate at both mitochondria and ER to regulate MOMP and the intrinsic apoptotic pathway. They play a key role in activation of the intrinsic pathway of apoptosis. In studies from the laboratory of Stanley Korsmeyer, employing *Bax/Bak* double knockout cells, no activation of CARD-containing initiator caspases was detected following either induction of ER stress or DNA damage. Nor were effector caspases activated, effectively ruling out a major role of some alternative activating pathway.

In response to ER stress, cells activate a translational control program known as the integrated stress response (ISR). Stress-induced genes are activated through assembly of transcription factors on ER stress response elements (ERSEs) in target gene promoters. Transcriptional regulators binding this region include TFII-I and ATF6 and the Sp proteins Sp1, Sp3, and Sp4. Various ISR targets are known and include markers of ER stress such as C/EBP homologous protein (CHOP) and ATF4 and their target TRB3. The ISR adapts cells to ER stress and may prevent apoptosis. Conversely, inactivating ISR proteins, such as PERK and eIF2α, increases apoptosis. ISR targets ATF4 and CHOP have been shown in hypoxic areas of human tumors, suggesting that ISR is involved in tumor cell adaptation to hypoxia, and raising the possibility of targeting this therapeutically in cancer.

Stress-inducible heat shock proteins

As will be referred to in subsequent chapters, diverse stresses such as DNA damage and nutrient starvation can trigger apoptosis, but must also activate survival pathways at least under conditions where the cell can cope with or repair the damage. Increasing attention is turning to the role of the heat shock proteins (HSPs) as regulators of apoptosis following damaging stimuli, and thereby key contributors to the recovery of cells from these insults. One way in which HSPs can protect the cell is as chaperones for other proteins; HSPs may be able to repair misfolded proteins which may contribute to avoidance of cell death that might otherwise result. The HSP27, 70, and 90 subfamilies are strongly implicated in avoidance of apoptosis following diverse triggers including DNA damage and ER stress. HSP can prevent release of cytochrome *c* and activation of caspases. HSP27 and HSP70 can inhibit activation of BID by caspase-8, and the extrinsic pathway of apoptosis and both HSP70 and HSP90 can prevent assembly/activity of the apoptosome, thereby inhibiting the intrinsic pathway. HSPs may also prevent MOMP by interacting with BAX.

Not surprisingly, HSP levels are increased in many cancers and are associated with resistance to chemotherapy. It is tempting to speculate that this may relate directly to the ability of such stress-inducible HSPs to block cancer cell apoptosis and facilitate their survival in the face of DNA damage and hypoxia. Geldanamycin is a therapeutic agent targeted to HSPs that, by binding to HSP90, prevents interactions with target proteins. Some proteins regulated by binding to HSP90 include AKT, RAS, and p53, suggesting that this may prove a potentially useful theraputic strategy in cancer.

Tumor suppressor p53

The *p53* tumor suppressor gene is one of the most frequent targets for mutation in human tumors. The majority of human tumors have either mutated *p53* itself, or have incurred mutations that disable function of the p53 pathway. Numerous mouse models have shown a critical role for p53 (and regulators of p53) in

Cell Death

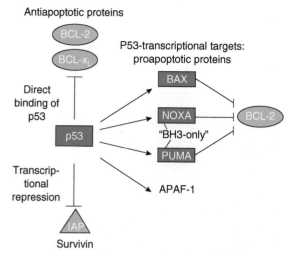

Figure 8.11 p53 and apoptosis. The tumor suppressor protein p53 can promote apoptosis in a number of ways. There are several proapoptotic transcriptional targets of p53, such as BAX and BH3 proteins, NOXA, and PUMA, that promote cytochrome c release from the mitochondria. p53 can also downregulate the transcription of genes that inhibit apoptosis, including the IAP family member survivin. p53 protein has also been shown to antagonize the antiapoptotic proteins BCL-2 and BCL-X_L at the mitochondrial outer membrane, by directly binding to them.

tumor suppression, such that inactivating p53 greatly facilitates tumor progression (see Chapter 7). In addition, p53 also has a profound impact on sensitivity to anticancer cytotoxic agents, such that disabling p53 function during tumorigenesis also reduces the sensitivity of these tumors to such antineoplastic agents, thereby limiting their clinical efficacy.

As described in Chapter 7, p53 guards against DNA damage, stresses, or a surfeit of proliferative signals, preventing cells from becoming malignant by inducing growth arrest or inducing apoptosis. How does p53 promote apoptosis? The precise mechanism of p53-mediated death is likely to differ depending on cell type and on the apoptotic trigger. There are several proapoptotic transcriptional targets of p53, such as BAX and BH3 proteins, and NOXA and PUMA that promote cytochrome c release from the mitochondria (Fig. 8.11), previously described in the BCL-2 family section. During apoptosis, BAX (normally an inactive monomer) can be induced to oligomerize and migrate from the cytoplasm to the mitochondria by various BH3-only proteins (e.g. BID). Once inserted into the outer mitochondrial membrane, BAX induces cytochrome c release by the creation or alteration of membrane pores (Fig. 8.10).

Aside from transcriptional activation, p53 can downregulate the transcription of genes that inhibit apoptosis, including the IAP family member survivin (see IAP section). p53 protein has also been shown to antagonize the antiapoptotic proteins BCL-2 and BCL-X_L at the mitochondrial outer membrane by directly binding to them.

Interestingly, APAF-1 is a transcriptional target of both p53 and the oncogene E2F. Given that loss of p53 and caspase-9 promoted fibroblast transformation, it may be that APAF-1 and caspase-9 mediate the apoptotic actions of wildtype p53, at least in certain cell types. Importantly, the *APAF1* gene is lost or silenced in some melanomas, indicating that loss of apoptosome function can be an important step in tumor development.

Oncogenic stress: MYC-induced apoptosis

Deregulated expression of c-*MYC* is present in most, if not all, human cancers and is associated with a poor prognosis. The c-*MYC* proto-oncogene is essential for both cellular growth and proliferation, but paradoxically may also promote cell death (see Chapter 6). The study of this "dual potential" of c-MYC over the last two decades has provided a paradigm for exploring the role of other proteins with mitogenic activity, such as E2F and RAS, many of which have now also been shown to have such intrinsic "tumor suppressor" properties. In fact, it may be a general feature of proteins that promote cell cycle that the oncogenic potential of deregulated expression is restrained by concurrent activation of processes, such as apoptosis or senescence, which effectively prevent propagation of the "damaged" cell. By implication, these in-built "failsafe" mechanisms must be overcome during tumorigenesis. However, once prevented, for instance by inactivation of the p53 or RB pathways (see Chapter 7) or upregulation of antiapoptotic proteins, the potentially devastating oncogenic potential of proteins such as c-MYC is unmasked (see Chapter 6).

Some 20 years ago, several laboratories made an exciting, although initially confusing, discovery: oncoproteins such as c-MYC and the adenovirus E1A – both potent inducers of cell proliferation – were shown to possess apoptotic activity (see Chapter 6, Fig. 6.4). The most widely held view of oncoprotein-induced apoptosis is that the induction of cell-cycle entry sensitizes the cell to apoptosis: cell proliferative and apoptotic pathways are coupled. However, the apoptotic pathway is suppressed so long as appropriate survival factors deliver antiapoptotic signals. Given this, the predominant outcome of these contradictory processes will depend on the availability of survival factors.

Since these early experiments, other promoters of cell proliferation (e.g. E2F) have been found to possess proapoptotic activity. The notion that cells acquiring growth-deregulating mutations *in vivo* possess an "in-built" tumor suppressor function, which hinders expansion of potentially malignant cells, is a fascinating one. Such a cell population would be unable to outgrow its environment unless apoptosis was inhibited. Indirect evidence supported this idea, as shown by the dramatic synergy between oncoproteins such as c-MYC and mechanisms that suppress apoptosis, for example overexpression of antiapoptotic proteins such as BCL-2 or BCL-X_L, or loss of ARF or p53 tumor suppressors (described in Chapter 6 under "Oncogene collaboration – from cell culture to animal models").

Interestingly, stimulation of apoptosis by c-MYC may not invariably be linked directly to cell cycling but might also be an indirect response to DNA damage. Thus, *in vitro* data links c-MYC to the accumulation of reactive oxygen species (ROS) that can damage DNA and generate double strand breaks (DSBs) (see also Chapter 10). The mechanisms are not fully understood, but may include inhibition of NF-κB. Interestingly, E2F1 may also be able to promote DSBs and apoptosis but independently from c-MYC and without ROS. The consequences of DNA damage – either apoptosis or growth arrest – may be critically dependent on cell type.

As already mentioned, at least two separate pathways are involved in the induction of apoptosis: an "intrinsic" pathway regulated by various factors such as DNA damage, stress and imbalances between growth promoting and survival promoting factors that act through mitochondrial permeability, and an

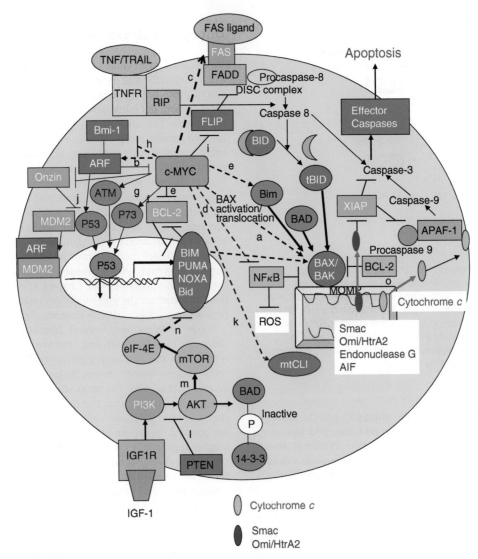

Figure 8.12 Legend on opposite page.

"extrinsic" pathway utilized for example by cytotoxic T cells, involving ligation of cell surface death receptors (e.g. FAS and TNFR) and activation of the "death-inducing signaling complex" (DISC). This distinction may, however, become "blurred," at least in some cells, where activation of caspase-8 in the extrinsic pathway may also co-opt the intrinsic pathway through activation of the BH3-only protein BID. Both pathways have been implicated in mediating the action of c-MYC on apoptosis in various studies and are in turn subject to various regulatory influences (summarized in Fig. 8.12). However, with the exception of the p19^{Arf} pathway, the relative contributions of these various potential mechanisms for any given tissue or cell type or critically within an individual developing tumor *in vivo* are generally not well known.

How does the c-MYC oncoprotein induce or "sensitize" cells to apoptosis? Most of the key experiments used to answer this question have come from cell culture. Although much of the apoptotic machinery has been identified, we are only beginning to define the mechanisms by which this is engaged by c-MYC. As is often the case for cell signaling events, there is not just one mechanism by which c-MYC may trigger apoptosis. The mechanism chosen, or even whether apoptosis will occur at all, most likely depends on factors, such as cell type and developmental stage, signals received from the cells environment, whether the cell has acquired DNA damage or not, as well as levels of MYC within the cell (highlighted in Chapter 6).

In different experimental models, expression of c-MYC may make cells vulnerable to a wide range of proapoptotic stimuli – such as hypoxia, DNA damage, and depleted survival factors – as well as enhancing sensitivity to signaling through the extrinsic pathway triggered by CD95, TNF, or TRAIL death receptors (Fig. 8.12). In fact, c-MYC can lead to downregulation of the inhibitor of caspase activation, FLICE inhibitory protein (FLIP). FLIP is an inhibitor of the extrinsic pathway that normally competes with caspase-8 for binding to the DISC. Releasing FLIP inhibition may at least in part explain the ability of c-MYC to sensitize cells to death receptor stimuli (Fig. 8.12).

MYC signaling through the BCL-2 family

First insight into c-MYC's role in apoptosis came from *in vitro* studies showing that c-MYC could induce the release of cytochrome *c* from the mitochondria during apoptosis, and that ectopic addition of cytochrome *c* sensitized cells to undergo apoptosis. As already discussed, once released into the cytoplasm,

Figure 8.12 c-Myc and apoptosis. c-MYC protein sensitizes cells to a wide range of proapoptotic stimuli (e.g. hypoxia, DNA damage, depleted survival factors, signaling through CD95, TNF, and TRAIL receptors). Moreover, as c-MYC can also transcriptionally activate other "dual function" mitogenic proteins, such as some E2F family members, the potential repertoire of mediators of MYC's role in apoptosis may be even more complex than suggested in this schematic. Broadly, three pathways are particularly implicated in driving MYC apoptosis, two involving activation of p53 either through DNA damage or through the protein ARF (p19 in mouse, p14 in humans) and another via activation of BID (which can be via either a p53-dependent or -independent route). During apoptosis, c-MYC induces release of cytochrome c from the mitochondria into the cytosol, possibly through activation of proapoptotic molecule BAX (**a**). Activated BAX within the mitochondrial membrane leads to creation or alteration of membrane pores, resulting in mitochondrial outer membrane permeabilization (MOMP). Once released into the cytosol, cytochrome c associates with APAF-1 protein and procaspase-9 to form the apoptosome ("wheel of death"). In the presence of ATP, caspase-9 is activated, leading to activation of downstream effector caspases including caspase-3, which ultimately leads to degradation of cell components and demise of the cell. Also released by MOMP, Smac and Omi can inhibit the action of IAPs, such as XIAP, that otherwise normally prevent activation of the apoptosome and also of effector caspases. Other pathways involving c-MYC-induced cytochrome c release and apoptosis include indirect activation of p53 tumor suppressor via p19^{Arf}, leading to transcription of BAX. ARF can also inhibit c-MYC by blocking transcriptional activation of genes involved in growth while not affecting c-MYC-mediated gene repression, resulting in cell cycle arrest. This feedback mechanism operates independently of p53 (**b**). In some tumor models, deregulated expression of the inhibitor of ARF, Bmi-1 can cooperate with MYC in tumorigenesis. Ligation of death receptor CD95/FAS triggers association of the intracellular adaptor protein FADD with the CD95 receptor and forms the DISC (**c**). FADD then recruits procaspase-8, resulting in autoactivation of the procaspase, which cleaves and activates executioner caspases. Activation of caspase-8 may be negatively regulated by FLIP, which competes for binding to the DISC. Caspase-8 may also activate the proapoptotic protein BID, which may promote MOMP, thereby linking the extrinsic and intrinsic pathways of apoptosis. c-MYC has been reported to mediate apoptosis in some cells by a mechanism involving generation of reactive oxygen species (ROS) and suppression of the survival-promoting activities of NF-κB (**d**). In some cell types c-MYC can induce expression of the BH3-only proapoptotic protein BIM and suppress expression of the antiapoptotic proteins BCL-2 and BCL-X$_L$ (**e**). c-MYC may increase expression of the p53 family member p73, which might act to direct p53 itself towards proapoptotic genes (**f**). It is likely that p53 is activated by either a DNA damage response mechanism involving the ATM kinase and downstream kinases such as CHK2 (**g**), or via ARF (see below); ATM is activated by DNA damage, such as DNA DSBs, known to accompany deregulated c-MYC activity (and can be both dependent and independent of DNA-damaging ROS production). p53 is also activated by the protein ARF (displaces HDM2/MDM2), which is not usually activated during normal cell replication, but is by deregulated oncogenes such as MYC, cyclin D and E2F (discussed in Chapter 10; see Fig. 10.11). ARF may also have p53-independent effects on MYC-transcriptional activity to induce cell-cycle arrest. A potential molecular link is through MYC inhibition of the protein Bmi-1 (**h**), though this is speculative. MYC can also directly activate the extrinsic pathway of apoptosis. The inhibitor of caspase activation FLICE inhibitory protein (FLIP) is transcriptionally repressed by c-MYC, providing a mechanism whereby c-MYC could sensitize cells to a wide range of death receptor stimuli (**i**). Recent studies suggest that downregulation of gene expression is important for MYC-induced apoptosis. The gene product Onzin, at least in myeloid cell types, normally inhibits p53 via activation of MDM2 and of AKT and thus confers apoptosis resistance when overexpressed. As *Onzin* is negatively regulated by c-MYC, this provides another means by which c-MYC may stabilize p53 and promote apoptosis via repression of *Onzin* expression (**j**), in a manner analogous to upregulation of ARF (**b**). Other recent studies suggest that a transporter mtCLIC may also be important for c-MYC-induced apoptosis (**k**). The links between cell-cycle transit and apoptosis is evident but at a molecular level remains confusing. Some studies show that at least in some cell types CDK2 is needed for apoptosis, though CDK2 knockdown or suppression by p27 did not prevent activation of p53 or BIM, so the mechanism is unclear. Survival signals that serve to block c-MYC-induced apoptosis (**m**), include signaling via the IGF-1 receptor and PI3K or activated RAS, which can lead to activation of AKT serine/threonine kinase and subsequent phosphorylation of the proapoptotic protein BAD. Phosphorylated BAD is sequestered and inactivated by cytosolic 14-3-3 proteins. This pathway may also block apoptosis in other ways. AKT also activates mTOR and its downstream target, the elongation factor eIF4E. In at least one model expression of eIF4E could block c-MYC-induced apoptosis and cooperate with c-MYC in tumorigenesis (**n**). The PTEN tumor suppressor acts as a negative regulator of AKT activation and thus would be expected to prevent survival signals via this route (**l**). Antiapoptotic proteins, such as BCL-2 and BCL-X$_L$, reside in the outer mitochondrial membrane and block cytochrome c release possibly through sequestration of BAX (**o**).

---

cytochrome *c* associates with another protein called apoptotic protease-activating factor 1 (APAF-1) to create the apoptosome, which acts as a scaffold for activating procaspase-9 (see Fig. 8.10). There are various ways in which c-MYC may provoke MOMP and the release of cytochrome *c*. One of the best known involves the expression of members of the BCL-2 family, discussed earlier. For example, c-MYC represses expression of the antiapoptotic proteins BCL-2 and BCL-X$_L$, and may induce expression/activation of proapoptotic members, such as BIM and BAX. This has the net effect of permeabilizing the mitochondria to release cytochrome *c* and other proapoptotic factors, which ultimately lead to activation of caspases. Moreover, c-MYC may actually increase expression of potentially dangerous mitochondrial genes, including cytochrome *c*, via the transcription factor nuclear respiratory factor-1 (NRF-1), a potential c-MYC target gene.

As previously discussed (see Fig. 8.10), interfering with BCL-2 family members can provoke MOMP. During apoptosis, BAX (normally an inactive monomer) can be induced to oligomerize and migrate from the cytoplasm to the mitochondria by various BH3-only proteins (those BCL-2 family members containing only the single BCL-2 homology-3 domain), such as BID. BAX then inserts into the outer mitochondrial membrane to create or alter the behavior of membrane pores (i.e. MOMP), through which cytochrome *c* can escape into the cytoplasm. Conversely, antiapoptotic proteins such as BCL-2 and BCL-X$_L$ reside in the outer mitochondrial membrane and suppress apoptosis by blocking MOMP, possibly through sequestration of activated BAX and also by preventing activation of the apoptosome. In this scenario, the balance of anti- and proapoptotic molecules present within a c-MYC-activated cell would determine whether it lives or dies.

The impact of such an imbalance has been highlighted in several mouse tumor models, and we urge the reader to refer to Chapter 6 (e.g. "Mouse models of Myc-induced tumorigenesis"). One example includes blocking Myc-induced apoptosis in pancreatic islet β-cells by overexpression of BCL-X$_L$. Rapid and uniform islet tumors develop, displaying many of the "hallmark" features of cancer (see Fig. 6.10). With this in mind, survival factors, such as IGF-1, have been shown to inhibit c-MYC-induced apoptosis *in vitro* by blocking cytochrome *c* release from the mitochondria (Fig. 8.12). In this case survival signals, for instance those mediated via the IGF-1 receptor or activated RAS, can activate the PI3K pathway and the AKT–PKB serine/

threonine kinase (see Chapters 5 and 6). Activated AKT can phosphorylate the proapoptotic BH3-only protein BAD, resulting in its sequestration and inactivation by the cytosolic 14-3-3 proteins. Referring back to the initial experiments *in vitro*, in which cells with deregulated expression of c-MYC die by apoptosis when grown in low serum, it becomes clear how growth factor signaling is able to regulate apoptosis. Given the importance of such growth factors in determining the survival or death of a cell, it is not surprising that elevated signaling through the IGF-1 pathway occurs in many tumors. Similarly, genetic mutations that activate the PI3K pathway dramatically collaborate with c-MYC during tumor progression.

MYC activates apoptosis via ARF/p53 tumor suppressor pathways

An important mechanism that links c-MYC and apoptosis is through the $p19^{Arf}$ (ARF) tumor suppressor protein, which acts in a checkpoint that guards against unscheduled cellular proliferation in response to oncogenic signaling (see Chapter 7; see also Chapter 10). ARF may contribute to c-MYC-induced apoptosis indirectly via activation of p53, or may engage apoptosis directly and independently of p53 (Fig. 8.12). The importance of ARF in c-MYC-induced apoptosis *in vivo* was underscored in genetically altered mice in which expression of c-*Myc* was targeted to B lymphocytes. When these mice were crossbred with another strain in which ARF was disrupted, MYC-induced lymphoma development was dramatically accelerated. This outcome, similar to that seen when p53 is disrupted, shows that c-Myc strongly collaborates with loss of p53 or ARF in murine lymphomagenesis, presumably by inhibiting c-MYC-induced apoptosis (extrapolating from *in vitro* experiments). In a similar fashion, deregulated expression of *Bmi1* oncogene (which encodes a polycomb group protein) accelerates c-MYC-induced lymphomas by inhibiting expression from the *Ink4a* locus, which encodes both ARF and another important tumor suppressor, $p16^{Ink4a}$ (see Chapter 7).

One key question that has intrigued many cancer researchers, is how do cells discriminate between normal and oncogenic MYC? We know that deregulated (oncogenic) MYC can induce apoptosis by triggering a variety of tumor suppressor pathways (Fig. 8.12) that serves to restrain tumor growth, but how is MYC-induced apoptosis confined only to tumor, and not normal, cells? Is it a matter of MYC levels or deregulation of MYC activity within the cell? This key question is discussed in more detail in Chapter 6 ("Levels of MYC matter"), but we can say briefly here that thanks to elegant work by Gerard Evan and colleagues, using a variant $MYCER^{TAM}$ mouse model in which $MYCER^{TAM}$ is driven by the very weak but constitutively active *Rosa26* promoter, we now have some answers. Only high levels of oncogenic *Myc* were able to induce apoptosis in pancreatic islet β-cells *in vivo* via ARF/p53 tumor suppressor pathways. In contrast, low levels of oncogenic MYC, although competent to drive β-cell proliferation, did not activate apoptosis or ARF/p53. These findings confirm that for MYC to activate apoptosis and ARF/p53, high levels of this oncoprotein are required.

One final but crucial point to make is that even at such low levels, MYC is still possessed of potent tumorigenic activity. The evidence arose from further studies using a conditional mouse model expressing the *RAS* oncogene in lung epithelium. Low deregulated levels of *MYC* cooperated with *RAS* to profoundly accelerate lung tumor progression. So, although high levels of MYC promote apoptosis in order to stall tumor progression, we are now beginning to realize that low-level MYC deregulation – that does not engage tumor suppression – is likely to serve as an important mechanism for the early stages of tumor growth.

MYC triggers apoptosis via the DNA damage response pathway

Although much remains to be learned, we can now attempt at a synthesis of how deregulated expression of c-MYC triggers MOMP (described in Fig. 8.12). An apical event in this seems to be activation of a DNA damage response, particularly involving the PI3-like kinases ATM and ATR (see Chapter 10). It also seems likely that this response might be triggered because c-MYC actually promotes DNA damage such as DNA DSBs, in part by excess production of ROS. However, it is also plausible that ATM is activated by a more ill-defined induction of "cellular stress," not necessarily requiring active DNA damage. ATM/ATR in turn activate various checkpoint pathways which culminate in apoptosis (with other oncogenes, e.g. *RAS*, growth arrest may be a more prominent response). This is discussed in further detail in Chapter 10, but in summary, phosphorylated ATM activates various downstream targets including another kinase, CHK2. Together, these events phosphorylate and activate p53 and p53 targets, such as PUMA and NOXA, which contribute to MOMP as previously outlined. ATM can also activate another BH3-only protein, BID, which can mediate both growth arrest (as non-truncated BID) and apoptosis, in its truncated form (tBID). As c-MYC can reduce expression of the prosurvival proteins BCL-2 and BCL-X_L, this, together with the inhibitory action of the various BH3-only proteins, allows activation of BAK and/or BAX and thereby MOMP (see Figs 8.10 and 8.12).

Linking in with the previous section on ARF/p53, although it is easy to see how ARF activation can contribute to apoptosis by activating p53, what still remains unclear is how ARF itself is activated by deregulated c-MYC. It is tempting to speculate that this might in some way depend on ATM or members of the MRN complex, such as NBS1 protein known to be upregulated by c-MYC, but this is conjectural.

Specific regions of MYC regulate its apoptotic activity

It seems likely that specific regions of c-MYC are responsible for its proapoptotic activity, at least in some systems. Thus, factors such as TIAM1 (T-cell lymphoma invasion and metastasis 1), which bind to the c-MYC box II (MBII) region of c-MYC (see Fig. 6.5), can inhibit the proapoptotic activity of MYC. Conversely, in other studies disruption of the MBIII region, important for transcriptional repression by c-MYC, increases the apoptotic activity of c-MYC, and may prevent transformation.

As has been discussed in Chapter 6, gene repression by MYC depends in many cases on its ability to bind and inactivate the transcription factor MIZ-1. MIZ-1, in association with its coactivator (P300 protein), can promote growth arrest/senescence by inducing expression of cell cycle inhibitors: $p15^{INK4b}$, $p21^{CIP1}$, and $p57^{KIP2}$. However, MYC–MAX heterodimers can prevent this action by binding to MIZ-1, thus enabling cell cycle progression. It was recently shown in transgenic mouse models that such interactions are critical for the induction and maintenance of T-cell lymphomas (Chapter 6). In contrast, in another study, albeit using cultured human fibroblasts, the ability of MYC to induce apoptosis in these cells in response to growth factor withdrawal was entirely dependent on its ability to inactivate MIZ-1.

Importantly, later work identified the prosurvival gene *BCL-2* as a downstream target gene of MIZ-1 and that inhibition of the MIZ-1/BCL2 signal is an essential event during the apoptotic response; targeting BCL-2 with short hairpin RNA or small molecule inhibitors restored the apoptotic potential of a c-MYC mutant defective for MIZ-1 inhibition. Such data may provide an explanation as to why we often see cooperation between c-MYC and BCL-2 overexpression in human cancer.

Important studies from Scott Lowe's laboratory have begun to unravel how different forms of mutant MYC may enable the disabling of the apoptotic "failsafe" mechanism that normally protects from deregulated MYC expression. This has also thrown further light on the regulation of apoptosis by p53-independent signaling. Two common mutant MYC alleles derived from human Burkitt lymphoma were found to uncouple proliferation from apoptosis and were unsurprisingly more able to drive B-cell lymphomagenesis in mice than wildtype MYC. Interestingly, mutant MYC proteins could still promote proliferation and activate p53, but were poor promoters of apoptosis, at least in part because they failed to induce BIM (as already discussed, BIM is an inhibitor of BCL-2).

MYC, apoptosis, and diabetes

As mentioned earlier, c-MYC-induced apoptosis may not always be a positive action as it undoubtedly is in restraining cancer. An example is the pancreas, which can adapt islet β-cell mass to meet changes in demand for insulin, as described in late pregnancy and obesity (a process known as "pancreas plasticity"). Individuals that fail to adapt become diabetic with time. In humans, during the progression to diabetes both functional defects and β-cell apoptosis contribute to defective insulin secretion generally described as β-cell failure. Several potential contributors to this have been identified, including c-MYC.

The notion that c-MYC may be involved in β-cell failure is supported by various studies. First, replicating β-cells seem to be most sensitive to apoptosis, which indirectly implicates proteins, such as c-MYC, known to be involved in β-cell replication. Moreover, c-MYC is upregulated in β-cells exposed to rising blood glucose levels. Several transgenic mouse models have demonstrated that the immediate consequences of c-MYC activation in β-cells *in vivo* include replication, but also β-cell apoptosis and loss of β-cell differentiation (see Chapter 6, Fig. 6.10). In fact, even when apoptosis is prevented by simultaneous overexpression of the antiapoptotic protein BCL-X_L, c-MYC activation still results in loss of β-cell differentiation and hyperglycemia, which is only reversed after considerable expansion in β-cell mass has taken place. Thus, a possible model is that β-cells activate c-MYC in response to a need for adaptive growth or later in the process because of rising blood glucose. Elevated c-MYC may then contribute to loss of function and apoptosis of β-cells.

Various other processes have been described as underlying glucose toxicity. These include activation of the extrinsic pathway of apoptosis by cytokines, such as IL-1β or FAS/FAS-ligand interactions and ROS. Intriguingly, c-MYC has been shown to trigger apoptosis through all of these pathways in other systems.

It is clear that the favored mechanism for c-MYC-induced apoptosis may be dictated by cell type as well as by tissue location and moreover modified by the presence or absence of additional mutations in other pro- and antiapoptotic genes. In order to further illuminate the complex interactions surrounding c-MYC and apoptosis, a role so sensitive to tissue location *in vivo*, we eagerly await the outcome of further studies in which apoptotic pathways are selectively manipulated in mouse mutants or by employing new pharmacological tools and phenotype determined in the all important context of the intact organism.

Autophagy – a different kind of cell death and survival

Until recently, apoptosis was considered the major form of programmed cell death occurring during normal development, tissue homeostasis, and in disease processes such as cancer. Extensive research over the last two decades has contributed to our vast knowledge of the apoptotic machinery and program, and the strategies by which cancer cells evade this form of cell death. However, recent studies have now brought to light another form of programmed cell death known as "autophagy," which, like apoptosis, plays an important role in development and tissue homeostasis and is thought to operate as a barrier to cancer. Paradoxically though, autophagy can mediate tumor cell survival as well, which will be discussed later.

The term "autophagy" (from the Greek, meaning eating of self) was first coined by Christian de Duve over 40 years ago based on the observed degradation of mitochondria and other intracellular structures within lysosomes of rat liver perfused with the pancreatic hormone glucagon. Like apoptosis, autophagy is a physiological process occurring at low levels in most cell types, where it is thought to play a housekeeping role by maintaining the integrity of intracellular organelles and proteins. In other words, autophagy mediates damage control by removing damaged or nonfunctional proteins and organelles. However, autophagy is strongly induced in certain states of cellular stress (e.g. nutrient deficiency or starvation), and once induced, the affected cell is able to break down its own organelles, such as mitochondria and ribosomes (this process is described below). Importantly though, byproducts of this degradation process, such as amino acids, are transported back out to the cytoplasm where they can be reused for building macromolecules and for metabolism. In this way, autophagy may be thought of as a cellular "recycling factory" that serves to maintain metabolism when the availability of external nutrients is limited. It is akin to burning the furniture if you run out of logs during a particularly cold winter – you can always replace them if you are alive in the spring.

The fact that many cancer cells are present in nutrient-limited environments *in vivo* implies that their survival may be supported by these metabolites generated via autophagy.

The importance of autophagy has been highlighted in mice that possess defects in this process such as loss of an autophagy gene, *Atg7*, in which there is an accumulation of cells containing protein aggregates and damaged mitochondria, elevated oxidative stress, and cell death. Moreover, without autophagy the persistence of such damaged cellular contents can ultimately lead to a proinflammatory response. Stress imposed on cells, such as oxidative damage from aging or hypoxia, damages proteins and organelles that require autophagy for elimination. So if such cells have defective autophagy the accumulation of damaged mitochondria and protein aggregates can lead to oxidative stress and DNA damage. It is of great relevance, then, that defects in autophagy are associated with many human diseases, including neurodegeneration, steatosis, Crohn's disease, infection, aging, and cancer.

The autophagic process – the basic machinery

Although autophagy is now recognized as an important process in both normal and tumor cells of mammals, it was, in fact, work on yeast (*Saccharomyces cerevisiae*) that provided the breakthrough in identifying many of the signaling molecules involved in autophagy. At present, 32 autophagy-related genes (*Atg*) have been identified in yeast and, importantly, many of these are conserved across phylogeny (plants, worms, flies, and mammals).

Once autophagy is triggered, the process begins with formation of a phagophore, an isolation membrane of uncertain origin, but is probably derived from lipid bilayer of the ER and/or the trans-Golgi and endosomes.

As the phagophore expands, it engulfs intracellular organelles, ribosomes, protein aggregates, etc., to become a double-membraned autophagosome which will eventually fuse with the lysosome. The autophagosomal contents can now be degraded by lysosomal acid proteases. As mentioned above, amino acids and other byproducts of degradation are exported back out of the cytoplasm (by lysosomal permeases and transporters), where they can be reused for building macromolecules and for metabolism.

The autophagic process is complex and is described in Fig. 8.13 in a simplified form to include: (1) formation of the phagophore by Beclin-1/VPS34 at the ER and other membranes; (2) Atg complex formation at the phagophore; (3) LC3 processing and insertion into the extending phagophore membrane; (4) capture of targets for degradation; and (5) fusion of the autophagosome with the lysosome, and degradation by lysosomal proteases.

Selective autophagy and cancer

Both random and selective capture of targets, such as a protein aggregate (e.g. p53) or an organelle, can occur at the autophagosome for degradation. Selective capture of targets is achieved via specific receptor molecules, such as p62 and NBR1, which can bind both to the target and to the autophagosome-associated protein LC3. By binding ubiquitin via their C-terminal domains, p62 and NBR1 mediate degradation of ubiquitinated targets by selective autophagy (e.g. p53). There is increasing interest in selective autophagy, as defects in this process have been linked to tumor progression. For example, in tumor cells that are defective in selective autophagy, there may be accumulation and aggregation of p62/NBR1 receptors and, as a result, accumulation rather than degradation of cancer-relevant proteins such as p53.

Similar signaling pathways connect autophagy and apoptosis

Both apoptotic and autophagic processes appear to use similar signaling pathways. For example, the survival signaling pathway involving PI3K, AKT, and mTOR kinases, which blocks apoptosis (see Fig. 5.18, Chapter 5) similarly inhibits autophagy; when survival signals are insufficient, this leads to downregulation of the PI3K signaling pathway and apoptosis and/or autophagy may be induced.

The protein VPS34 (vesicular protein sorting 34) and its binding partner Beclin-1 (Atg6) play a crucial role in phagophore formation and autophagy (Fig. 8.13). Although VPS34 is involved in various membrane-sorting processes in the cell, when it is complexed to Beclin-1 and other regulatory proteins it becomes selectively involved in autophagy. VPS34 is a unique class III PI3K using only phosphatidylinositol (PI) as substrate to generate phosphatidylinositol trisphosphate (PIP3), which is essential for phagophore elongation and recruitment of other Atg proteins to the phagophore. Importantly, it is the interaction of Beclin-1 with VPS34 that promotes its catalytic activity and increases levels of PIP3.

Another signaling protein linked to both apoptosis and autophagy is, in fact, Beclin-1 (Atg6) mentioned above. Beclin-1 is a member of the BH3-only subfamily of BCL-2 apoptotic regulatory proteins (discussed earlier) and its BH3 domain allows it to bind to antiapoptotic BCL-2/BCL-X_L proteins at the ER. By doing so, Beclin-1 activity in autophagy is inhibited as it is prevented from binding with VPS34. However, Beclin-1/BCL-2 interactions can be disrupted by the stress-signaling kinase JNK1-mediated phosphorylation of BCL-2 in response to starvation-induced signaling, thus enabling the liberated Beclin-1 to trigger autophagy (Fig. 8.13). In a similar way, proapoptotic BH3 proteins such as BAX and BAK, once released from their associations with BCL2/ BCL-X_L at the mitochondria can trigger apoptosis.

It is intriguing that, depending on its subcellular localization, BCL-2 plays a key role in determining cell survival by impacting on two different modes of cell death: (a) suppressing apoptosis by preventing cytochrome *c* release from mitochondria; and (b) suppressing autophagy at the ER by interacting with Beclin-1.

Autophagy – a barrier to cancer or a promoter?

Tumor cells experience elevated metabolic stress from nutrient, factor, and oxygen deprivation caused by an inadequate blood supply – the result of inadequate angiogenesis (see Chapter 14). To add to this, the high levels of cell proliferation experienced by many tumor cells make their metabolic demands higher than normal cells. As in normal cells, autophagy is activated in tumor cells by stress, such as starvation, hypoxia, and factor deprivation. It is then not surprising that in tumors, autophagy localizes to hypoxic regions most distal to blood vessels. The tumor suppressor effect of autophagy within a tumor would be either cell death or at least the inability to proliferate or move due to the greatly reduced tumor cell size.

The discovery that the *BECLIN1* gene is mono-allelically deleted in human breast, ovarian, and prostate cancer, and that inactivation of one copy of the gene in mice leads to increased susceptibility to cancer, has led biologists to postulate that autophagy has tumor suppressor properties. If so, potential cancer cells would need to develop strategies in order to block autophagy and ultimately allow tumor development. So autophagy, alongside apoptosis and cell senescence, appears to represent yet another barrier to cancer. Or does it? Somewhat paradoxically, tumor cells residing in hypoxic tumor regions and undergoing autophagy are, in fact, the tumor cells that resist radiation and chemotherapy. In other words, stress such as nutrient deficiency, radiotherapy, and certain cytotoxic drugs, although inducing high levels of autophagy, actually protect the cancer cell from autophagic cell death. Moreover, severely stressed cancer cells (e.g. following treatment with potent anticancer drugs) have been shown to shrink via autophagy to a state of reversible dormancy, making it possible for them to survive and regrow at a later stage. It is remarkable that tumor cells can progressively eat themselves under prolonged stress, becoming less than one third of their normal size but still retain the capacity to return to their normal size and resume cell proliferation within 24 hours of having normal growth conditions restored.

The reasons for such opposing outcomes are at present unclear and so future research is needed in order to clarify why autophagy

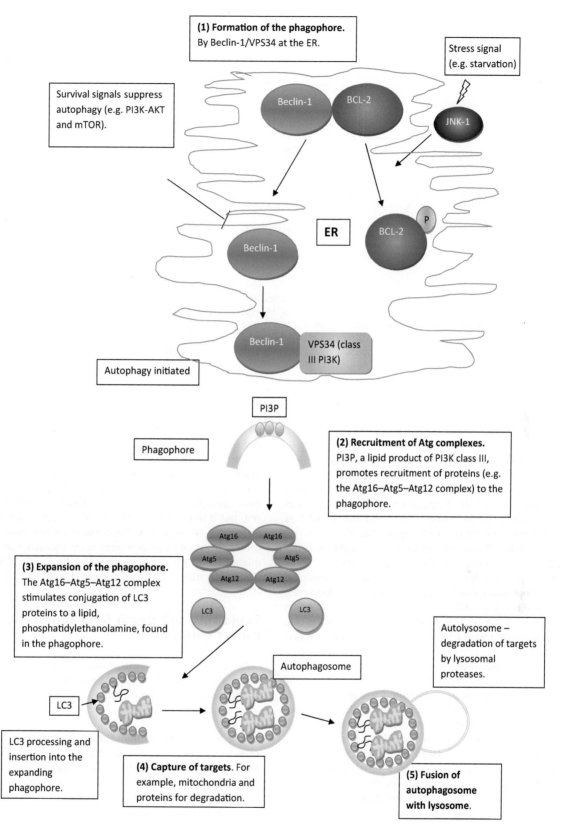

Figure 8.13 The autophagic process. Autophagy is a complex cellular process of self-degradation. When a cell is under stress such as starvation or hypoxia, important stress-signaling pathways can be activated, including the kinase JNK1, which can activate the formation of the isolation membrane (IM) or phagophore (**1**). Beclin-1 is a BH3-only protein which binds to BCL-2 at the endoplasmic reticulum (ER) when the cell is not under stress. However, the stress-signaling kinase JNK1 phosphorylates BCL-2, thereby liberating Beclin-1 and allowing it to bind to VPS34. VPS34, a class III PI3K, uses phosphatidylinositol (PI) as substrate to generate phosphatidylinositol trisphosphate (PIP3) which is essential for phagophore elongation, and recruits other Atg proteins (e.g. Atg16–Atg5–Atg12 complex) to the phagophore (**2**). This complex stimulates conjugation of LC3/GABARAP proteins (LC3) to phosphatidylethanolamine, a lipid found in the phagophore membrane. LC3 inserts into the phagophore membrane and may stimulate its expansion (**3**). The extending phagophore membrane captures targets such as mitochondria and proteins for degradation, and the autophagosome is formed (**4**). Autophagosomes must eventually fuse with lysosomes for their contents to be degraded (**5**).

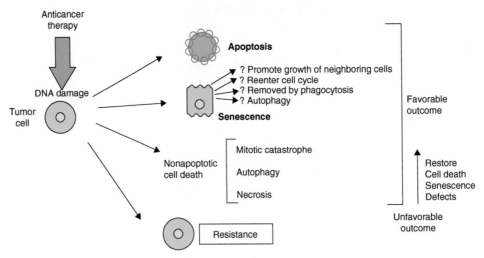

Figure 8.14 Potential responses to anticancer therapy. Anticancer therapy causes DNA damage to tumor cells. Depending on cell type and the genetic mutations acquired, a given tumor cell may respond to such damage in one of the following ways: die by apoptosis, mitotic catastrophe, autophagy, or necrosis. Elimination is of course the most favorable outcome. Alternatively, the cell may enter a state of "permanent" cell-cycle arrest (cellular senescence). However, it remains debatable as to whether the senescent cell promotes growth of neighboring tumor cells by secreting growth factors, or whether it will eventually re-enter the cell cycle. The most unfavorable outcome is obviously resistance to therapy as the cell is unable to undergo apoptosis or cellular senescence.

is tumor suppressive in some cancer cells and tumor promoting in others.

Cell death in response to cancer therapy

Apart from apoptosis and senescence, there are indeed alternative outcomes in response to drug-induced DNA damage: cells may still be induced to die by nonapoptotic mechanisms, such as necrosis, autophagy, and mitotic catastrophe (Fig. 8.14).

As discussed earlier, tumor cells treated with anticancer drugs also undergo death by autophagy. For example, breast carcinoma cells treated with tamoxifen (the estrogen receptor antagonist) accumulate autophagic vacuoles shortly before dying. However, after reading the previous section, the reader will appreciate that autophagy does not always lead to the demise of a cancer cell, but often a state of "dormancy," after which its oncogenic properties are unleashed.

Tumor cells may also undergo mitotic catastrophe, a type of cell death that is caused by aberrant mitosis. Mitotic catastrophe is associated with the formation of multinucleate giant cells that contain uncondensed chromosomes. Cells that have acquired a defective G_2 checkpoint of the cell cycle (this checkpoint is responsible for blocking mitosis when a cell has sustained DNA damage – described in Chapters 4 and 10) can enter mitosis prematurely, before DNA replication is complete or DNA damage has been repaired. This aberrant mitosis causes the cell to undergo death by mitotic catastrophe. A key molecule that regulates mitotic progression is survivin. Although a member of the IAP family discussed earlier, the primary function of survivin is not in the control of apoptosis, but rather of mitotic progression. At present, there are few anticancer therapies aimed to induce tumor cell death by mitotic catastrophe. However, targeting this pathway might represent a novel approach to eliminating tumors, perhaps through targeting molecules such as survivin.

One key observation is that the fraction of tumor cells undergoing nonapoptotic death increases if apoptosis is inhibited in response to anticancer agents or radiation. With this in mind, a strategy that deliberately induces nonapoptotic cell death might be considered a useful adjunct to standard cancer therapies and make a significant contribution to a favorable treatment outcome. However, there are also concerns with such an approach; cell death by necrosis can do further harm by recruiting inflammatory cells that promote growth of the surviving tumor cells. Furthermore, autophagy does not always lead to the demise of a cancer cell, but often a state of "dormancy," after which its oncogenic properties are exposed. With these considerations on board, cancer treatments that predominantly engage apoptotic cell death might present a superior strategy to achieve a reduction in tumor size and to minimize the chance of relapse (see Chapter 16).

Exploiting cell death (and senescence) in cancer control

Play not with paradoxes. That caustic which you handle in order to scorch others may happen to sear your own fingers and make them dead to the quality of things.

George Eliot

As previously pointed out in this chapter, the resistance of tumor cells to cell death is a crucial aspect of tumor development. Unfortunately, it is this same aspect that can cause some cancer cells to become resistant to anticancer therapies which will ultimately lead to the death of the individual. In other words, some of the mutations that allowed tumor progression can also confer chemoresistance, as we will see below. Cancer cells respond to drug-induced DNA damage in a variety of ways: apart from apoptosis, cells may still be induced to die by nonapoptotic mechanisms, such as necrosis, autophagy, and mitotic catastrophe (described previously), or enter a permanent state of cell cycle arrest – "cellular senscence." All these potential cellular responses to anticancer therapy are outlined in Fig. 8.14 and will be discussed next.

Many tumor cells are killed in response to cytotoxic therapies, such as chemotherapy, gamma irradiation or immunotherapy, predominantly mediated by triggering apoptosis (see Chapter 16). Such anticancer agents do not simply destroy cells directly, but rather trigger the cell's own apoptotic response programs. Although the underlying mechanisms for initiation of this apoptotic response is not always known and may differ depending on the cytotoxic agent, damage to DNA or to other critical molecules is considered to be a common initial event, which is then propagated by the cellular stress response (mitochondrial pathway – described earlier). Therefore, if a cell acquires defects in stress response programs during tumor formation, it may now be more resistant to drug-induced apoptosis. For example, preclinical data show that tumor cells that have acquired mutations in *p53* are also chemoresistant. Furthermore, *p53* defects have been shown to mediate multidrug resistance to numerous anticancer drugs. Similar observations have been made in clinical investigations. Figure 8.15 outlines potential pathways to such resistance.

T-cell or natural killer (NK) cells may contribute to tumor cell killing by releasing cytotoxic compounds such as granzyme B or by using "death ligands" such as FASL, which bind to FAS (CD95) receptors on tumor cells and subsequently activate downstream apoptotic pathways (see Chapter 13).

Since defects in apoptosis are likely to be universal lesions in tumor progression, restoring or activating apoptosis in tumors is an active area of cancer research. Preclinical trials have validated the antitumor efficacy of this approach, and clinical trials are currently ongoing for various apoptosis-activating strategies.

However, as we have seen, apoptosis involves many regulators and effectors which vary from tumor to tumor. This diversity can hamper determination of effective targets for individual tumors, and can contribute to drug resistance during treatment. The way forward for effective cancer treatment strategies is therefore likely to require combinatorial approaches that, together with individual tumor typing and matching with type-specific treatments, will allow effective tumor-specific therapeutics or personalized medicine (see Chapter 16).

In addition to inducing apoptosis, anticancer therapies can also induce cellular senescence – a permanent state of growth arrest (cellular senescence is described more fully in Chapters 6 and 9), or differentiation. What is the fate of drug-induced senescent tumor cells? Do they eventually re-enter the cell cycle or die? The answer to this remains unclear at present. Although senescent tumor cells are locked into a nondividing state, it is conceivable that such cells could still promote growth of other tumor cells in their vicinity, as they remain metabolically active cells and can secrete growth-regulating cytokines into the environment. It is also possible that senescent cells will ultimately be cleared by phagocytosis (although unlike apoptotic cells, no "eat-me" signals have been identified as yet), or indeed be able to eliminate themselves by autophagy (described above).

What of differentiation of tumor cells? The emerging possibility that many "successfully" treated cancers, in which tumors have regressed via cell death leading to apparent clinical elimination of the malignancy by histological criteria, actually retain a population of dormant tumor cells is perhaps not too surprising. Decades of clinical experience support this notion. However, the recent use of conditional mouse models of cancer (e.g. Myc-induced hepatocellular carcinoma – see Chapter 6) now provide possible explanations for what has been seen in the clinic for many years:

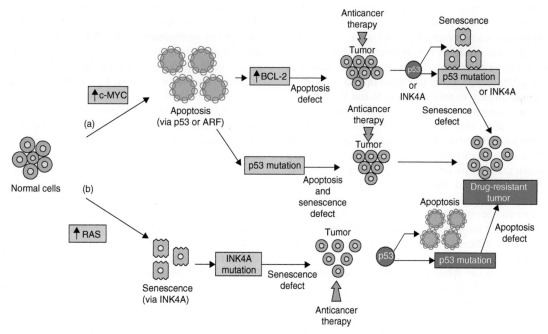

Figure 8.15 Pathways to resistance. Whether a tumor cell responds favorably (cell death or cellular senescence) or unfavorably (resistant) to anticancer therapy, will depend on the genetic mutations acquired during tumor development. For example, (a) in the early stages of tumorigenesis, if the initiating activating oncogene is c-*MYC* (where apoptosis is the "in-built" fail-safe mechanism), an apoptosis defect such as BCL-2 overexpression will now prevent cells from dying by apoptosis. Despite this, such tumor cells may respond favorably by entering a state of cellular senescence. Further mutations, however, may also prevent cellular senescence (e.g. inactivating *p53* or *INK4A* mutations), leading to resistance. Another example (b) shows tumor cells that have acquired an initiating oncogenic *RAS*, where cellular senescence is the fail-safe mechanism. Inactivating mutations in *INK4A* tumor suppressor gene prevents senescence in response to anticancer therapy. However, cells can still undergo apoptosis due to the presence of intact p53. Not surprisingly, the loss of p53 prevents tumor cells dying by apoptosis and the tumor is now resistant to anticancer therapy.

a population of cancer cells respond to therapies by differentiating into normal-looking cells (histologically) but in fact retain neoplastic properties that can be unleashed at a later time.

In a regulatable transgenic mouse model, in which mice develop Myc-induced hepatocellular carcinoma, switching off Myc initially induces differentiation which is followed later by gradual apoptosis of the majority of the tumor cells. Remarkably, some surviving cells behave like stem cells as they are able to differentiate into many cell lineages that form structures characteristic of normal hepatocytes and biliary cells. However, if Myc is reactivated in these "normal-looking" differentiated liver cells they rapidly become tumorigenic. A possible explanation for this outcome is that some of the differentiated tumor cells are, in fact, cancer stem cells and thus are capable of re-exposing their neoplastic properties. In this scenario, Myc inactivation induces a state of tumor dormancy. This could be likened to "successfully treated" cancers in patients, in which tumors that have completely regressed clinically and appear histologically normal often reappear at a much later time. What in fact could be happening is that a population of cancer cells respond to therapies by differentiating into normal-looking cells (histologically), but retain neoplastic properties that can be unleashed at a later time. Decades of clinical experience support this notion.

While treatment of many types of cancers, including breast cancer, lymphoma, leukemia, and sarcomas, can lead to long-term regression, this is usually followed by relapse at a much later time. Such cancers are likely to remain in an occult, dormant state that has the potential to re-emerge The specific type of cancer can often predict the outcome of treatment. For example, with regard to hematopoietic cancers, high-grade lymphomas such as Burkitt lymphoma and diffuse large-cell lymphoma frequently respond to chemotherapy with rapid apoptosis. Such cancers are readily cured and remain in remission. In contrast, low-grade lymphomas, such as follicular lymphoma, respond less well to therapy, and are rarely seen to regress completely.

Drug-induced cellular senescence has been elegantly shown *in vivo* using a mouse lymphoma model (Eμ-Myc) in which the oncogene c-*Myc* is constitutively expressed in the B-cell lineage. This mouse model has allowed the study of treatment responses at naturally occurring tumor sites. For example, if the antiapoptotic gene *Bcl2* is introduced into the B-cell lineage of Eμ-Myc mice, animals fail to undergo remission following therapy – that is, a reduction in tumor mass below detectable levels – due to resistance to undergo apoptosis. Instead, they preserve a constant tumor burden as a consequence of tumor cells entering senescence. Importantly, those tumors that started to grow again (progressive disease) were frequently found to have acquired defects in either *p53* or *INK4A* genes, indicating that these two gene products control drug-induced senescence *in vivo*. Thus, tumors that have acquired defects in both the apoptosis and senescence programs, such as overexpression of BCL-2 plus loss of p53, are chemoresistant and so severely impair the long-term survival of the host.

Importantly, sequential disruption of apoptosis and senescence-controlling genes during tumor formation and subsequent therapy have also been reported in human cancers. For example, INK4A defects correlate with relapses in non-Hodgkin lymphomas, while patients diagnosed with follicular lymphomas that overexpress BCL-2, due to the characteristic t(14;18) translocation, typically achieve long-term disease control following chemotherapy. Poor prognosis is associated with acquisition of *p53* or *INK4A* mutations that were not detectable at diagnosis. This pathway to resistance is outlined in Fig. 8.15(a).

So, on a final note, for cellular senescence and differentiation to be an effective anticancer response, its therapeutic potential strongly relies on the irreversibility of this process. Unlike apoptosis, which eliminates a potentially harmful cell, cellular senescence or differentiation only represents an altered state of a tumor cell, which if not eliminated (e.g. by phagocytosis or autophagy) may at some point revert into a dividing cell and cause tumor relapse. Therefore, at present the potential use of cellular senescence or differentiation in treating cancers is not clearly known, and is likely to depend on the various genetic lesions within the tumor cells.

Conclusions and future directions

Cell death by apoptosis (or programmed cell death) is now known to be at least as important as growth and replication in achieving appropriate cell numbers and tissue size during development and in maintaining these during adult life. It seems remarkable now that modern biology has only been focused on this process for the last 20 years or so, and yet despite this, progress has been nothing short of spectacular, as evidenced by the astounding numbers of key publications in this area. Much is now known about some of the key intracellular proteins that can regulate the delicate balance between life and death of the cell (e.g. BCL-2 family, IAPs, caspases, MYC, p53) and some of the environmental influences, which in turn regulate these (e.g. FAS, IGF-1, hypoxia). Not surprisingly, many of these regulatory processes are the subject of mutations or epigenetic changes in cancer and are increasingly being targeted in cancer treatments. Thus, we now have the opportunity to use various high-throughput techniques (see Chapter 18) on human cancer samples to detect abnormalities in many of the key signaling pathways in apoptosis and even begin to predict how these will influence disease progression and therapeutic response.

It may prove relatively straightforward in those extreme cases, in which the cancer cell is found to clearly overexpress antiapoptotic BCL-X_L or BCL-2, or proapoptotic BAX, to equate these straightforward findings with resistance or sensitivity to cell death, respectively. However, the same cellular outcomes might also arise by more subtle alterations in the balance of multiple factors and thereby be much harder to detect or predict. Moreover, the abnormalities in any given cancer cells will also have to be placed in the environmental context of that individual tumor as this will determine the balance of survival cues acting on the cell and also strongly influence the outcome.

The recent discovery that tumor cell death by autophagy (alongside apoptosis and cell senescence) might represent yet another barrier to cancer is still far from clear. This stems from the further findings that tumor cells residing in hypoxic tumor regions can undergo autophagy, in which such cells "eat" themselves to become less than one third their original size, but are able to survive in a state of reversible dormancy, making it possible for them to regrow at a later time. Furthermore, these cells are, in fact, tumor cells that are resistant to chemoradiotherapy. Clearly, this fascinating area of research requires further work in order to provide answers as to why autophagy is tumor suppressive in some cancer cells and tumor promoting in others.

Importantly, most chemoradiotherapy triggers apoptosis that will ultimately induce caspase activity (e.g. DNA-damaging drugs activate MOMP via p53). Unfortunately, as is the case with many cancers, some tumor cells are resistant to such therapies, and so will survive and continue to replicate or migrate. There are ongoing efforts to develop drugs that can directly activate apical or effector caspases in order to trigger apoptosis in cancer cells directly, or to target any number of regulator processes affecting caspases.

Of increasing interest in cancer research are the IAPs (inhibitors of apoptotic proteins), originally identified by their ability to suppress apoptosis by inactivating caspases. We now know that these proteins are involved in many cellular processes important in cancer cell behavior (e.g. survival, proliferation, metastases) and are associated with resistance to current cancer therapies, disease progression, and poor prognosis. Not surprisingly, this has led to much interest in developing small therapeutic compounds termed "Smac mimetics." It is clear, however, that further work is needed in order to fully understand how cIAPs contribute to different types of cancers.

A final note on cell death is directed to our recent understanding on how oncogenes, such as MYC, are able to induce apoptosis in tumor cells and not normal cells. For a long time, the question of whether this was due to unregulated oncogene expression or particular oncoprotein levels within the cell, or indeed other intracellular factors specific to the tumor cell, remained elusive. Now we have some answers thanks to the use of sophisticated mouse tumor models – high levels of MYC are required to activate apoptosis via ARF/p53. Perhaps more surprising was the finding that even at very low levels, oncogenic MYC still possesses potent tumorigenic activity. Thus, at high levels, MYC may promote apoptosis in order to stall tumor progression, whereas low-level MYC deregulation, which does not engage tumor suppression, is nevertheless able to promote early growth of tumors. Future research in this fascinating area of oncogene-induced cell death (or survival) within different tumor types and at various stages will no doubt be of great value to the understanding of tumor growth and of cancer therapeutic design.

I am of the opinion that my life belongs to the whole community and as long as I live, it is my privilege to do for it whatever I can. I want to be thoroughly used up when I die, for the harder I work the more I live.

George Bernard Shaw

Outstanding questions

1. Appropriate cell number and organ size in a multicellular organism are determined by coordinated cellular processes such as growth, proliferation, senescence, and death. Given that all of these processes are in themselves subject to regulation by multiple external factors, including nutrient availability, growth factors, and cell–cell and cell–matrix interactions as well as intrinsic signaling pathways, one of the key challenges is to more fully understand how tissue growth is constrained during development and maintained in the adult.

2. In the context of the balancing act between growth/replication and cell death, we need to progress in our understanding of how key proteins such as c-MYC, E2F, E1A, LATS, and others participate in signaling pathways regulating both opposing cellular outcomes.

3. As disturbing the balance of these processes at any of a myriad of potential target sites can result in cancer on the one hand or degenerative diseases on the other, we also need to further our understanding of the underlying apoptotic machinery and how individual components may be successfully manipulated in the treatment of cancer and other diseases.

4. Autophagy, the process whereby a cell "eats" itself, is a growing area of research in the cancer field. Further work is needed in order to provide answers as to why autophagy is tumor suppressive in some cancer cells and tumor promoting in others.

Bibliography

Historical and general

Ellis, H.M., and Horvitz, H.R. (1986). Genetic control of programmed cell death in the nematode *C. elegans*. *Cell*, **44**: 817–29.

Kerr, J.F.R., Wyllie, A.H., and Currie, A.R. (1972). Apoptosis: a basic biological phenomenon with wide-ranging implication in tissue kinetics. *British Journal of Cancer*, **26**: 239–57.

Kroemer, G. and Martin, S.J. (2005). Caspase-independent cell death. *Nature Medicine*, **11**(7): 725–30.

Levi-Montalcini, R. (1987). The nerve growth factor 35 years later. *Science*, **237**: 1154–62.

Raff, M. (1998). Cell suicide for beginners. *Nature*, **396**(6707): 119–22.

Shamas-Din, A., Brahmbhatt, H., Leber, B., and Andrews, D.W. (2011). BH3-only proteins: orchestrators of apoptosis. *Biochimica Biophysica Acta*, **1813**(4): 508–20.

Sulston, J. and Horvitz, H.R. (1977). Post-embryonic cell lineages of the nematode, Caenorhabditis elegans. *Developmental Biology*, **56**: 110–56.

Apoptosis and cancer

Brown, J.M. and Attardi, L.D. (2005). The role of apoptosis in cancer development and treatment response. *Nature Reviews Cancer*, **5**(3): 231–7.

Cotter, T.G. (2009). Apoptosis and cancer: the genesis of a research field. *Nature Reviews Cancer*, **9**(7): 501–7.

Lowe, S.W., Cepero, E., and Evan, G. (2004). Intrinsic tumor suppression. *Nature*, **432** (7015): 307–15.

Zhivotovsky, B. and Kroemer, G. (2004). Apoptosis and genomic instability. *Nature Reviews Molecular Cell Biology*, **5**(9): 752–62.

Linking cell growth and apoptosis, c-Myc

Christofori, G., Naik, P., and Hanahan, D. (1994). A second signal supplied by insulin-like growth factor II in oncogene-induced tumorigenesis. *Nature*, **369**(6479): 414–18.

Colombani, J., Polesello, C., Josue, F., and Tapon, N. (2006). Dmp53 activates the Hippo pathway to promote cell death in response to DNA damage. *Current Biology*, **16**(14): 1453–8.

Evan, G.I., Wyllie, A.H., Gilbert, C.S., *et al*. (1992). Induction of apoptosis in fibroblasts by c-myc protein. *Cell*, **69**(1): 119–28.

Hemann, M.T., Bric, A., Teruya-Feldstein, J., *et al*. (2005). Evasion of the p53 tumor surveillance network by tumor-derived MYC mutants. *Nature*, **436**(7052): 807–11.

Lai, Z.C., Wei, X., Shimizu, T., *et al*. (2005). Control of cell proliferation and apoptosis by Mob as tumor suppressor, Mats. *Cell*, **120**(5): 675–85.

Murphy, D.J., Junttila, M.R., Pouyet, L., *et al*. (2008). Distinct thresholds govern Myc's biological output in vivo. *Cancer Cell*, **14**: 447–57.

Pelengaris, S., Khan, M., and Evan, G. (2002). c-MYC: more than just a matter of life and death. *Nature Reviews Cancer*, **2**: 764–76.

Pelengaris, S., Khan, M., and Evan, G.I. (2002). Suppression of Myc-induced apoptosis in beta cells exposes multiple oncogenic properties of Myc and triggers carcinogenic progression. *Cell*, **109**(3): 321–34.

Vaux, D.L., Weissman, I.L., and Kim, S.K. (1992). Prevention of programmed cell death in *Caenorhabditis elegans* by human bcl-2. *Science*, **258**: 1955–7.

Zindy, F., Eischen, C.M., Randle, D.H., *et al.* (1998). Myc signaling via the ARF tumor suppressor regulates p53-dependent apoptosis and immortalization. *Genes and Development*, **12**(15): 2424–33.

The apoptotic machinery and other forms of cell death

Certo, M., Del Gaizo Moore, V., Nishino, M., *et al.* (2006). Mitochondria primed by death signals determine cellular addiction to antiapoptotic BCL-2 family members. *Cancer Cell*, **9**(5): 351–65.

Chipuk, J.E., Bouchier-Hayes, L., Kuwana, T., Newmeyer, D.D., and Green, D.R. (2005). PUMA couples the nuclear and cytoplasmic proapoptotic function of p53. *Science*, **309**(5741): 1732–5.

Cory, S. and Adams, J.M. (2002). The Bcl2 family: Regulators of the cellular life-or-death switch. *Nature Reviews Cancer*, **2**: 647–56.

Crawford, E.D. and Wells, J.A. (2011). Caspase substrates and cellular remodelling. *Annual Review of Biochemistry*, **80**: 1055–87.

Crighton, D., Wilkinson, S., O'Prey, J., *et al.* (2006). DRAM, a p53-induced modulator of autophagy, is critical for apoptosis. *Cell*, **126**(1): 121–34.

Garrido, C. and Kroemer, G. (2004). Life's smile, death's grin: vital functions of apoptosis-executing proteins. *Current Opinion in Cell Biology*, **16**(6): 639–46.

Glick, D., Barth, S., and Macleod, K.F. (2010). Autophagy: cellular and molecular mechanisms. *Journal of Pathology*, **221**(1): 3–12.

Green, D.R. and Kroemer, G. (2004). The pathophysiology of mitochondrial cell death. *Science*, **305**(5684): 626–9.

Gyrd-Hansen, M. and Meier, P. (2010). IAPs: from caspase inhibitors to modulators of NF-kappaB, inflammation and cancer. *Nature Reviews Cancer*, **10**(8): 561–74.

Hengartner, M.O. and Horvitz, H.R. (1994). *C. elegans* cell survival gene *ced-9* encodes a functional homolog of the mammalian proto-oncogene *bcl-2*. *Cell*, **76**: 665–76.

Hockenbery, D., Nunez, G., Milliman, C., Schreiber, R.D., and Korsmeyer, S.J. (1990). Bcl-2 is an inner mitochondrial membrane protein that blocks programmed cell death. *Nature*, **348**(6299): 334–6.

Llambi, F., Moldoveanu, T., Tait, S.W., *et al.* (2011). A unified model of mammalian BCL-2 protein family interactions at the mitochondria. *Molecular Cell*, **44**(4): 517–31.

Lowe, S.W., Schmitt, E.M., Smith, S.W., Osborne, B.A., and Jacks, T. (1993). p53 is required for radiation-induced apoptosis in mouse thymocytes. *Nature*, **362**: 847–9.

Martin, S.J. and Green, D.R. (1995). Protease activation during apoptosis: death by a thousand cuts? *Cell*, **82**: 349–52.

Nachmias, B., Ashhab, Y. and Ben-Yehuda, D. (2004). The inhibitor of apoptosis protein family (IAPs): an emerging therapeutic target in cancer. *Seminars in Cancer Biology*, **14**(4): 231–43.

Salvesen, G.S. and Dixit, V.M. (1999). Caspase activation: the induced-proximity model. *Proceedings of the National Academy of Sciences of the U.S.A.*, **96**(20): 10964–7.

Spierings, D., McStay, G., Saleh, M., *et al.* (2005). Connected to death: the (unexpurgated) mitochondrial pathway of apoptosis. *Science*, **310**(5745): 66–7.

Westermann, B. (2010). Mitochondrial fusion and fission in cell life and death. *Nature Reviews Molecular Cell Biology*, **11**(12): 872–84.

White, E. and DiPaola, R.S. (2009). The double-edged sword of autophagy modulation in cancer. *Clinical Cancer Research*, **15**:5308–16.

Zou, H., Henzel, W.J., Liu, X., Lutschg, A., and Wang, X. (1997). Apaf-1, a human protein homologous to C. elegans CED-4, participates in cytochrome c-dependent activation of caspase-3. *Cell*, **90**(3): 405–13.

Questions for student review

1) Apoptosis:
   a. Usually provokes an inflammatory response.
   b. Involves nuclear condensation.
   c. Involves membrane blebbing.
   d. Is usually indicative of disease.
   e. Is suppressed in cancer.

2) Apoptosis:
   a. Is suppressed by inactivation of the p53 pathway.
   b. Is suppressed by inactivation of telomerase.
   c. Is executed by activation of caspase enzymes.
   d. Is inhibited by cytoplasmic cytochrome c.
   e. Is the most ubiquitous form of cell death in most organisms.

3) Apoptosis may be triggered by:
   a. Fas–Fas ligand binding.
   b. Mitochondrial insertion of Bcl-X_L.
   c. Overexpression of c-Myc.
   d. Loss of cell–cell contacts.
   e. Active AKT.

4) c-Myc may promote apoptosis in some cell types by:
   a. Increased MOMP.
   b. Induction of Fas–Fas ligand.
   c. Oxidative stress.
   d. Upregulation of WRN.
   e. Upregulation of Bcl-X_L.

5) Cancer cells may avoid apoptosis by:
   a. Inactivating checkpoint proteins.
   b. Producing survival factors.
   c. Activating the PI3K pathway.
   d. Upregulating BAX.
   e. Inactivating surviving.

6) IAPs (inhibitors of apoptosis proteins):
   a. Function only to inhibit apoptosis.
   b. Directly inhibit caspases.
   c. Possess a BIR domain that is crucial for mediating their antiapoptotic function.
   d. Can activate NF-κB transcription factors.
   e. Are not overexpressed in human cancers.

7) Autophagy:
   a. Is a specific form of cell death.
   b. Occurs through signaling pathways that are distinct from apoptosis.
   c. Is strongly induced during cellular stress, such as nutrient deficiency.
   d. Of tumor cells can lead to tumor cell survival and regrowth at a later stage.
   e. Does not occur in tumor cells that are resistant to chemo/radiotherapy.

9 Senescence, Telomeres, and Cancer Stem Cells

Maria A. Blasco[a] and Michael Khan[b]

[a]Spanish National Cancer Research Centre (CNIO), Spain and [b]University of Warwick, UK

Middle age ends and senescence begins the day your descendents outnumber your friends.

Ogden Nash

So much has religion done for me; turning the original materials to the best account; pruning and training nature. But she could not eradicate nature: nor will it be eradicated 'till this mortal shall put on immortality.

Charlotte Brönte, Jane Eyre

Key points

- Normal cells in culture divide a finite number of times before they experience growth arrest and stop dividing – a largely stable and irreversible state termed "replicative senescence."
- Senescence results from the cumulative negative effect of repeated cycles of replication on important structures at the ends of chromosomes (telomeres).
- Progressive shortening and loss of function of the telomeres result in chromosome ends being treated as damaged DNA with resultant activation of DNA damage responses (DDRs).
- The activity of a key enzyme, telomerase, can prevent telomere attrition.
- Cancer cells need telomerase to maintain functional telomeres and to divide indefinitely, and this has opened the possibility of novel therapeutic approaches based on using telomerase as a target.
- Telomerase is reactivated in over 90% of all types of human tumor.
- Senescence may result from mechanisms independent of telomerase activity and telomere attrition, particularly in response to environmental factors and DNA damage that share common downstream signaling involving the p53–p21^{CIP1} and RB–p16^{INK4a} pathways, which collaborate to stop cellular proliferation.
- The RB and p53 tumor suppressor pathways are required to maintain stable growth arrest in senescent cells *in vitro*. In fact, at least in cultured cells, senescence may be reversible by inactivating these pathways. Therefore, the possibility should be considered that senescent cells could still contribute to cancer at a future date by re-entering the cell cycle.
- Oncogene-induced senescence has been shown to restrain transformation in human cancer cells and cancer progression in animal models *in vivo*, and together with apoptosis is a key innate restraint on the potential of oncogene deregulation to give rise to cancer.
- Stem cells are a likely cell of origin for many cancers, as they already exhibit many of the properties required of cancer cells, including longevity, lesser degrees of differentiation, and advanced replicative potential. Cancer originating by acquired mutations in stem cells would appear to have a head start on those originating in more differentiated cells. One caveat is that deregulation of some oncogenes can confer stem cell–like properties on previously differentiated cells.
- Irrespective of their origin, cancers appear to contain a particularly malign side population of cells potentially responsible for metastases, recurrence, and resistance to treatments. These cells are often referred to as "cancer stem cells" due to their similar phenotype to normal stem cells, to which their relationship is still hotly debated.

The Molecular Biology of Cancer: A Bridge From Bench to Bedside, Second Edition. Edited by Stella Pelengaris and Michael Khan.
© 2013 John Wiley & Sons, Inc. Published 2013 by John Wiley & Sons, Inc.

Introduction

Billions of bilious barbecued blue blistering barnacles!

Hergé

This chapter will cover some of the most important and intriguing concepts in cancer biology. Prominent among these are the effect of senescence as a barrier to tumorigenesis and how this may be overcome and the role of stem cells and cancer stem cells in the initiation and progression of cancers. We will discuss several processes such as telomere maintenance, RB and p53 signaling, and various factors involved in regulating stemness and epithelial–mesenchymal transition (EMT). Inevitably there will be overlap with other chapters, in particular Chapters 7 and 12.

Cells are regarded as senescent when they lose replicative potential but remain living and metabolically active. Senescence is a tumor suppressor mechanism which may serve as a related but alternative option to apoptosis in preventing the propagation of a cancer-causing mutation. Many stimuli that can promote apoptosis can also result in senescence, including deregulated oncogene activity, DNA damage, and telomere attrition. Moreover, there is considerable overlap between the signaling pathways regulating apoptosis and senescence, particularly those involving p53. However, Rb and Skp2 appear to be more specifically involved in growth arrest and senescence as compared to apoptosis. Intriguingly, some senescence-specific signaling pathways appear to suppress apoptosis, suggesting that at least in some cases apoptosis may be a last-resort option when growth arrest or senescence fails to allow damage repair or appears to the cell to be a futile effort in the face of serious DNA damage or other stress. Recent studies, as has been the case with many other areas of cell biology, have revealed important roles for microRNAs (miRNAs) in regulating senescence. Senescence is associated with alterations in chromatin as well as specific secretory patterns, knowledge which has been employed to identify senescent cells in tissues and cell cultures. The importance of epigenetic mechanisms is underscored by the central role of the senescence-related histone methyltransferase Suv39h1 in mediating senescence induced by TGF-β and by deregulated oncogene activation.

Most human cells in culture can reproduce only a limited number of times before they lose the ability to divide and become what is called "senescent." This process has been associated with various aspects of longevity determination and aging, and it also may play an important role in preventing cancer.

Seminal work by Hermann Muller and Barbara McClintock identified that the ends of chromosomes are capped by structures, which serve to prevent chromosome fusions, the telomeres. As the mechanics of DNA replication were described, it was identified that DNA polymerase, the enzyme responsible for DNA replication, could not fully synthesize the 3′ end of linear DNA. This was described by James Watson as the **end-replication problem**. At about the same time in Russia, Alexey Olovnikov proposed that the end-replication problem would result in telomere shortening with each round of replication and that this mechanism could account for replicative senescence, which was compatible with studies by Leonard Hayflick and colleagues that control of replicative senescence was situated in the nucleus. Olovnikov's model is supported by numerous studies which show that telomere shortening is a major contributor to replicative senescence and that telomere length may act as a "mitotic clock" to determine the number of cell divisions allowed for a given cell (see Box 9.1, "Hayflick limit").

Telomerase is a reverse-transcriptase enzyme that elongates the telomeres, thus counteracting the normal telomere attrition occurring every time DNA is replicated. Telomerase has two components, an RNA component and a catalytic subunit, and is strongly implicated in the avoidance of senescence by "immortal" tumor cell lines. Studies have also confirmed that expression of the catalytic subunit of human telomerase (hTERT) in various cell types avoids replicative senescence by maintaining telomere length. Telomerase is not the only mechanism capable of elongating the telomeres, although the mechanisms underlying this alternative lengthening of telomeres (ALT) remain unknown. Whatever the mechanisms, stabilizing telomeres is a prerequisite for immortality and likely for tumor development, normally by telomerase. In contrast, telomerase inhibition can induce senescence in cancer cells, but if cells with shortened telomeres manage to divide, this may also under these circumstances be a strong supporter of cancer progression due to the increased rate of subsequent chromosomal aberrations (Fig. 9.1). Most, if not all, human somatic tissues have no detectable telomerase activity; in the bone marrow, hematopoietic cells express telomerase. Telomerase activity is higher in primitive progenitor cells and then downregulated during proliferation and differentiation. Telomerase activity has been detected in several normal human somatic proliferating cells, for instance skin and colorectal tissues. Expression of hTERT appears to be regulated by the proto-oncoprotein c-MYC, thus providing a potential link between cell-cycle progression and telomere maintenance. As is so often the case, increased study does not always result in increased simplicity. Telomerase reverse transcriptase (TERT) has been implicated as a regulator of cell proliferation independently of its actions at the telomere, and may influence key proliferative signaling pathways, such as WNT, directly by interaction with the protein BRG1.

Although telomere length regulates replicative senescence, it is not the sole regulator of cell senescence. Telomeres are not

---

Box 9.1 Hayflick limit

Hayflick and Moorhead (1961) defined the stages of cell culture as three phases. Phase I is the primary culture, when cells from the explant multiply to cover the surface of the culture flask. Phase II represents the time when cells divide in culture; once cells cover the culture surface, they stop multiplying, with growth continuing only if cells are subcultivated. Subcultivation requires removing culture medium and detaching cells from the culture substrate with a digestive enzyme called trypsin that dissolves the substances keeping cells together. Cells can then be replated at lower density, and further culture medium added. Cells reattach to the new culture substrate and start dividing again until a new subcultivation is required. These procedures are termed "passages." Lastly, phase III begins when cells start dividing slower. Eventually they completely stop dividing and die. Hayflick and Moorhead noticed that cultures stopped dividing after an average of 50 cumulative population doublings. This phenomenon is known as Hayflick's limit, phase III phenomenon, or replicative senescence.

Senescence, Telomeres, and Cancer Stem Cells

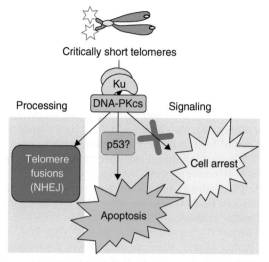

Figure 9.1 Critically short telomeres lead to end-to-end telomere fusions and loss of cell viability (apoptosis and/or cell-cycle arrest). The nonhomologous end-joining (NHEJ) complex DNA-PK has been involved both in mediating telomere fusions and in apoptosis triggered by critically short telomeres. In particular, these consequences of telomere dusfunction are rescued in DNA-PKcs-deficient and Ku86-deficient backgrounds. However, cell-cycle arrest due to short telomeres is not rescued by the absence of DNA-PKcs and Ku86 proteins. We have suggested that p53 may be downstream of DNA-PK in signaling apoptosis due to short telomeres.

linear, but form duplex loops called "t-loops" (Fig. 9.2), the formation of which is dependent on telomeric repeat-binding factors TRF1 and TRF2, and may be crucial for stabilizing telomere caps. Capping may protect the telomeres from being recognized as DNA damage and may be as important as telomere length in avoiding telomere dysfunction and preventing apoptosis and senescence.

In a recent publication, an intriguing link between telomere crisis and the progression of a form of breast cancer has been identified. It was found using an *in vitro* model that ductal hyperplasia may be associated with a critical shortening of telomeres in rapidly dividing cells resulting in crisis. Importantly, although the majority of such cells undergo apoptosis or presumably senescence, thus avoiding further risk of progression to cancer, a small number of cells may escape these usual consequences of chromosome instability. Instead rare cells may reactivate telomerase, thus preventing further telomere attrition and allowing such cells to progress to ductal carcinoma in situ and presumably thence to invasive cancer. Not surprisingly, the rate of genetic aberrations and chromosomal re-arrangements was highest during the crisis period. It is speculated that such a transition through a telomere crisis may be a crucial event in the development of most breast carcinomas.

Senescence can occur in fibroblasts in the absence of cell division and short telomeres. Although one hypothesis is that telomere dysfunction occurs in confluent cells despite a lack of telomere shortening, it is also apparent that viral genes such as SV40T-antigen, adenoviral E1A and E1B, and the human papillomavirus *E6* and *E7* genes can also immortalize cells. The E1B and E6 proteins bind and inactivate the tumor suppressor protein

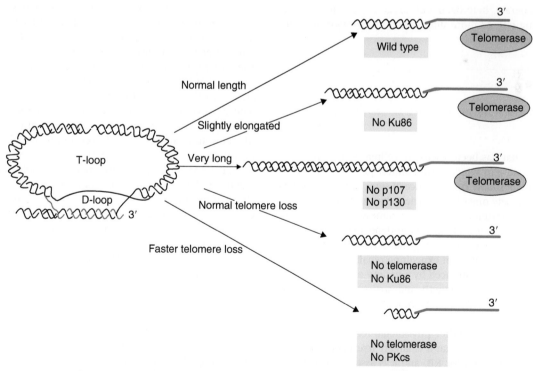

Figure 9.2 Different activities modulate telomere length, such as the telomere-binding proteins TRF1 and TRF2. More recently, activities involved in DNA repair (Ku86 and DNA-PKcs) and in cell-cycle regulation (p107 and p130) have also been shown to have a direct impact in regulating telomere length. Ku86 is a negative regulator of telomerase, while DNA-PKcs cooperates with telomerase in maintaining telomere length. Simultaneous absence of p107 and p130 results in a fast elongation of telomeres in the absence of changes in telomerase activity.

p53, while E1A and E7 bind and inactivate the retinoblastoma protein (RB, also spelled pRb). Immortalization depends on inactivation of p53, RB, or their pathways.

At present, it is accepted that the p53 and RB pathways are largely responsible for inducing senescence. Immortalization with viral proteins or oncogenes results in an extended lifespan after which cells enter a stage called "crisis," where cells proliferate but the rate of apoptotic cells gradually increases and thus cell numbers eventually diminish. Since both the p53 and RB–p16^{INK4a} pathways are inactive and chromosomal instability and fusions are abundant, crisis is thought to emerge due to extremely short telomeres. Occasionally, immortal cells emerge from crisis with stabilized telomeres, normally involving telomerase activation. In a sense, crisis can be seen as the ultimate consequence of telomere dysfunction. Whatever changes occur during telomere dysfunction, the mechanisms triggering growth arrest appear to involve the recognition of dysfunctional telomeres as DNA damage and that the p53 and RB–p16^{INK4a} pathways collaborate to stop cellular proliferation.

Importantly, even in some normal tissues a cell subpopulation exists, the stem cells, which already avoids many of the restraints posed by these processes. Not surprisingly, therefore, it is now widely assumed that many cancers may originate from these cells rather than more differentiated ones, because stem cells have a head start in the evolutionary race to malignancy. Closely allied to the role of stem cells as the origin of cancers is the role of cells within most cancers which have some stem cell properties (wherever they may have come from originally). These so-called cancer stem cells (CSCs) are now one of the most intensively studied areas in all of cancer biology. In brief, the CSC hypothesis states that cancers are maintained by a subpopulation of cells which manifest self-renewal (the motor for cancer) and differentiation (these cells generate heterogeneous progeny that proceed in some way along one or more differentiation pathways forming the bulk of the tumor).

Senescence

Getting older is no problem. You just have to live long enough.
Groucho Marx

To summarize what has been discussed, cellular senescence represents a stable state in which cells manifest proliferative arrest; it is a potentially desirable consequence of some anticancer treatments and conversely one of the barriers that must be overcome for malignant transformation to proceed. Senescence depends on alterations in chromatin and chromatin-dependent tumor suppression. However, it is worth noting that senescence may actually have evolved as a means of expediting repair. In senescence, terminal differentiation is induced by a variety of stimuli including alterations of telomere length and structure, some forms of DNA damage (e.g. oxidative stress), and activation of certain oncogenes.

Senescence differs from other physiologic forms of cell-cycle arrest such as quiescence in that it is largely irreversible, barring the inactivation of p53 and/or RB, and it is associated with distinctive phenotypic alterations such as cellular flattening, the expression of senescence-associated β-galactosidase activity, and the establishment of an unusual form of heterochromatin known as senescence-associated heterochromatic foci (SAHFs). These SAHFs may provide a chromatin buffer to silence proliferation-associated genes.

Senescence requires activation of the p16^{INK4a}–RB and the p53–p21^{CIP1} pathways. The CKIs, p16^{INK4a} and p21^{CIP1} (inactive in many cancers), inhibit the activities of CDK4 and CDK6 and of CDK2, CDK4, and CDK6, respectively, thus preventing cell-cycle progression (Chapters 4 and 7). Exogenous expression of p16^{INK4a} induces senescence in immortal cells lacking functional p53, and conversely cells may be immortalized by disruption of both p16^{INK4a} and p53. RB is needed for growth suppression mediated by p16^{INK4a} suggesting that it acts upstream of RB in regulating senescence. Telomere dysfunction can activate p16^{INK4a} and also p53. The p53 pathway is an important mediator of senescence that is induced by dysfunctional telomeres, in part by triggering DNA damage responses (Chapter 10). The cell-cycle can be blocked by p21^{CIP1} independently of p16^{INK4a} (Fig. 9.3).

Stabilization of p53 occurs in response to DNA damage and also to inappropriate activation of oncogenes such as c-MYC via induction of ARF (p19^{ARF} in the mouse), which can bind MDM2, thereby inhibiting the destruction of p53 (Chapter 7). Either p53–p21^{CIP1} or p16^{INK4a} can promote RB hypophosphorylation and initiate senescence. Some senescence-inducing stimuli (e.g. activation of the *RAS* oncogene) appear to induce both pathways, while others (e.g. DNA damage) appear to preferentially activate one or the other. Recent studies confirm that diverse potentially cancer-promoting genes can trigger senescence *in vivo*, including oncogenic *RAS* and *BRAF*, which appear to induce p16^{INK4a} via the ERK pathway and activation of AKT or inactivation of PTEN, which activate the ARF–p53–p21^{CIP1} pathway.

Numerous pathways can affect p16^{INK4a} and influence senescence. Thus, Hedgehog signaling and the GLI2 transcription factor can prevent senescence. Conversely, signaling via TGF-β can promote senescence. A recent study has shown that cells derived from mice with knockout of both p16^{INK4a} and p21^{CIP1} do not senesce in serial culture and can override Ras-induced senescence, which is not the case if only one or the other is inactivated. The authors also found these mice to be prone to skin cancer and suggest that loss of p16^{INK4a} promoted benign tumors which progressed to cancer under the influence of loss of p21^{CIP1}. The actual mechanisms involved will likely vary between different tumors but share a common final denominator, namely, prevention of RB phosphorylation and therefore inhibition of the G_0–G_1 to S transition. Intriguingly, RB may also promote senescence by interacting with various factors that can result in histone methylation and the formation of transcriptionally inactive heterochromatin. This may stably silence S-phase genes in senescent cells. This raises some questions over the use of drugs that can reverse such epigenetic changes (Chapter 11). Recently it has been shown in a mouse model of lymphoma that inactivation of a Suv39h1, a key histone methyltransferase often associated with RB, prevented senescence and fostered cancer progression. Many drugs targeting epigenetic modification (inhibitors of histone deacetylases and DNA methyltransferases) with the aim of reversing epigenetic silencing of tumor suppressors (Chapter 11) could conceivable also reverse or prevent senescence. I think that by now the reader will be well aware of the pervasive presence of functionally important miRNAs in essentially all areas of biology. Recent studies show that the RB pathway can upregulate miR-29 and miR-30; miRNA families are upregulated during induced and replicative senescence, and by inhibiting B-Myb expression contribute to cell-cycle arrest.

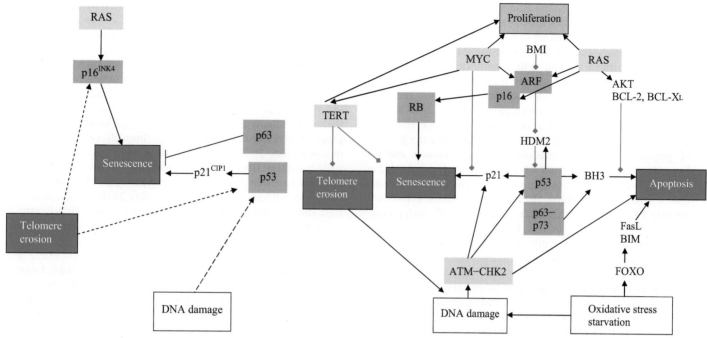

Figure 9.3 (a) Growth arrest can be induced by both p53–p21 and p16^{INK4a}–Rb-dependent mechanisms. (b) Linkage between signaling regulating replication, apoptosis, and senescence. Activation of RAS and MYC results in potential engagement of both replication and growth as well as of apoptosis and possibly senescence. If either MYC or RAS levels are excessive (as might occur during oncogenesis) or other pro-apoptotic signals are received, then the balance may be tipped away from replication. RAS can promote senescence through either p16^{INK4a} or ARF, which activate the RB or p53 pathways respectively. Senescence may also result from telomere erosion or dysfunction, which may activate DNA damage responses that are partly the same as those also activated by DNA DSBs. Functional telomeres must normally prevent DNA ends from being recognized and processed as DSBs. MYC can inhibit growth arrest or senescence by inhibiting p21^{CIP1} and inducing telomerase reverse transcriptase (TERT), which prevents telomere erosion and may also activate replication by other means. Although MYC may activate the apoptotic pathway (e.g. via ARF), RAS is able to suppress apoptosis by activating the PI3K pathway and, subsequently, AKT. It can readily be appreciated how oncogenic MYC and RAS may conspire in oncogenesis. The combination of RAS and MYC acting together provides a potential means of avoiding apoptotic and senescence mechanisms activated by either acting individually. Moreover, it can also be appreciated how inactivating mutations in RB or p53 (or their pathways involving p19^{Arf}, p16^{INK4a}, p21^{CIP1}, etc.) may contribute to tumorigenesis by enabling the cancer cell to avoid senescence, apoptosis, or both.

During advancing age p16^{INK4a} accumulates, providing a molecular link between aging and senescence; p16^{INK4a} inhibits CDK4 (Chapter 4), which is required for proliferation in adult cells. Recent studies show that defective replication and regeneration in mouse pancreatic β cells of adult transgenic mice with increasing age are mediated by increasing p16^{INK4a}. Interestingly, this effect is tissue specific and is not observed in the exocrine pancreatic cells of the same animals; moreover, only p16^{INK4a} of the CKIs tested was increased in an age-dependent manner. Finally, mice lacking p16^{INK4a} have improved regeneration and proliferation of β cells, raising the possibility that the effects of aging on cell growth could potentially be reversed therapeutically, though the effect this might have on cancer incidence would be a concern.

Much progress has been made on unraveling the complex crosstalk and other signaling pathways that affect cellular senescence. Thus, recent studies have implicated the activation of the NF-κB pathway and a variety of downstream gene targets as positive regulators of senescence. The tumor suppressor promyelocytic leukemia protein (PML), discussed in Chapter 11, is also an important regulator of senescence and terminal differentiation, in part by a well-described activation of p53 and through a more recently described PML–RB–E2F pathway that acts by localizing E2F transcription factors bound to their promoters and RB to PML nuclear bodies.

Senescent cells with irreparable DNA damage contain nuclear foci that contain DNA damage response (DDR) proteins, which persist and involve PML bodies in contrast to the transient complexes seen in cells with repairable DNA lesions that do not. Campisi and coworkers have recently described what they refer to as "DNA segments with chromatin alterations reinforcing senescence" (DNA-SCARS), in which DNA synthesis is absent. These contain histone H2AX, activated CHK2, and p53 and do not contain DNA repair proteins such as RPA and RAD51. Although not required for DNA-SCARS, p53 and RB are needed to trigger the senescence growth arrest.

A recent study has employed genome-wide expression profiling with genetic complementation in order to assess genes that were differentially expressed when conditionally immortalized human fibroblasts were induced to senesce by activating the p16–RB and p53–p21 tumor suppressor pathways. The expression of more than 1700 genes, of which 50% have been previously shown to be upregulated in cancers, was reversed when senescence was bypassed, leading the authors to note that cancers appear to devote considerable efforts to overcoming senescence. The study also suggested a central role of the NF-κB signaling pathway in maintaining senescence.

Cellular senescence, immortality, and cancer

I don't believe one grows older. I think that what happens early on in life is that at a certain age one stands still and stagnates.

T.S. Eliot

In 1961, Leonard Hayflick and Paul Moorhead made the unexpected discovery that human cells derived from embryonic tissues can divide only a finite number of times in culture (Hayflick and Moorhead, 1961 – and see Box 9.1 on the Hayflick limit). Hayflick and Moorhead worked with a connective tissue cell, the fibroblast, but similar findings have been observed for other cell types including skin keratinocytes, endothelial cells, and lymphocytes. Importantly, senescence also occurs in cells derived from embryonic tissues and in cells taken from mice and other animals. Early results suggested a relation between the number of "population doublings" that cells undergo in culture, and the longevity of the species from which the cells were derived. For example, cells from the Galapagos tortoise, which can live longer than a century, divide about 110 times, while mouse cells divide roughly 15 times. If cells are taken from individuals with premature aging syndromes, such as Werner's syndrome (the gene responsible for Werner's syndrome, *WRN*, is a member of the RecQ subfamily of DNA helicases), they exhibit far fewer doublings than normal cells.

Some cells never reach replicative senescence and are said to be "immortal." These cells include embryonic germ cells and most cell lines derived from tumors. Replicative senescence in human fibroblasts is characterized by growth arrest (i.e. the cells stop dividing, are larger and morphologically heterogeneous, and are more sensitive to cell–cell contact inhibition). Senescent cells growth arrest at the G_1–S transition of the cell cycle, which is in general irreversible (inactivation of p53 and RB pathways notwithstanding), and cells remain refractory to growth factors even though senescent cells can remain metabolically active for long periods of time.

Replicative senescence may limit the growth potential, but not necessarily survival, of a dividing cell as a consequence of telomere attrition. Senescent cells exhibit a particular phenotype when arrested in the G_1 phase of the cell cycle. They appear flattened and enlarged with increased cytoplasmic granularity, enhanced activity of senescence-associated lysosomal hydrolase β-galactosidase when assessed at an acidic pH. In 1995, Judith Campisi and her team discovered that the enzyme β-galactosidase is abnormal in senescent cells; senescence-associated β-galactosidase activity becomes active at a higher pH (pH 6 rather than 4). Early reports showed that lysosomes increase in the number and size in senescent cells, and the increased autophagy with aging in culture may be associated with an increase of lysosomal mass. While refractory to mitogenic stimuli, senescent cells remain viable and metabolically active and possess a typical transcriptional profile that distinguishes them from quiescent cells. At the protein level, numerous regulators of cell-cycle progression, checkpoint control, and cellular integrity such as p53 or $p16^{INK4a}$ have been found to be induced in response to various pro-senescent stimuli. Although the molecular mechanisms underlying the senescent phenotype remain largely unknown, there is increasing evidence that the formation of heterochromatin in the vicinity of promoters that control gene expression related to cell-cycle progression might be implicated in the maintenance of an irreversible growth arrest. This accumulation of SAHFs in senescent cells can act as a chromatin buffer able to prevent the transcriptional activation of proliferation-associated genes. Recent work from the laboratory of Scott Lowe has shown that SAHFs in senescent fibroblasts contain High-Mobility Group A (HMGA) proteins, which can cooperate with the $p16^{INK4a}$ both in the formation of SAHF and also in proliferative arrest. Interestingly, HMGA-mediated growth arrest is negated by coexpression of HDM2 and CDK4 oncogenes, which are often coamplified with HMGA2 in human cancers, and presumably limit activity of the Rb and p53 tumor suppressor pathways.

Normal human cells are diploid, yet with each passage in culture, the proportion of polyploid cells increases. Mutations in mitochondrial DNA also increase with age. Senescent cells are less able to express heat shock proteins, which may also increase their vulnerability to death triggers. The expression of multiple genes is altered during cellular aging. Normally, cell culture conditions include 20% oxygen, and these were the conditions initially used by Hayflick and Moorhead. However, if human fibroblasts are cultured at reduced oxygen (3%) closer to physiological conditions, they achieve a further 20 doublings. In contrast, different types of human cells cultured above 20% oxygen survive fewer doublings. Importantly, the same effect is not observed in tumor cell lines.

Cell senescence can be found without telomere shortening *in vivo*. One view is that replicative senescence is not a major factor *in vivo* and that cellular senescence instead depends on stress factors that trigger the RB and p53 tumor suppressor pathways. From a simple mathematical perspective, the existence of replicative senescence under normal physiological conditions *in vivo* has been questioned. The reasoning is along these lines: assuming human fibroblasts endure 50 population doublings, 2^{50} would seem more than enough cells to last several lifetimes under normal conditions. However, in cancer the issue has recently been somewhat clarified. First, deregulated expression of oncogenes seems to trigger cell cycles which in some way are detected by the cell as aberrant (cancer cell cycles). These cancer cell cycles trigger DNA damage responses, likely independently of telomere attrition, which culminate in growth arrest or apoptosis. These responses variously involve activation of the ARF–p53 or $p16^{INK4a}$–RB pathway. Whether the detection of DNA damage or the activation of DNA damage responses is essential for this remains to be shown.

Stem cells participate in tissue homeostasis by replacing differentiated cells that have undergone cell death or been lost (e.g. shed from the skin surface). Human stem cells express the enzyme telomerase suggesting that these actively dividing cells could avoid telomere attrition by replacing telomeric repeats. However, somatic stem cells can show senescence, and telomere-independent mechanisms of cellular senescence are increasingly being described.

The purpose of cellular senescence is unclear, but it may act as an anticancer barrier. Loss-of-function mutations in *WRN* lead to genomic instability, an elevated cancer risk, and premature cellular senescence. Interestingly, WRN is normally upregulated by c-MYC, suggesting that WRN may act downstream of c-MYC to prevent cellular senescence during tumorigenesis (Chapter 6). Recent studies suggest that in addition to WRN, two other MYC-regulated proteins, CDK2 and telomerase, may also mitigate MYC-induced senescence and contribute to the oncogene cooperation seen between MYC and other senescence-inducing oncogenes such as RAS. Interestingly, unlike WRN, loss of CDK2 does not contribute to oncogene stress and apoptosis.

It is also worth noting, as was discussed in Chapter 6, that the binding of MYC with MIZ1 is essential for the repression of various CKIs, required for efficient promotion of the G_1–S transition. It also appears that in some tumors related to deregulated MYC, such as lymphomas, TGF-β expression is greatly increased and in the absence of MYC activation (or if this is inactivated) will drive senescence associated with accumulation of trimethylated histone H3 at Lys 9 (H3K9triMe). Whilst MYC is active, however, H3K9triMe is suppressed. The authors propose that TGF-β expression may contribute to addiction to deregulated MYC in these cancers, as the high levels of expression would induce senescence once MYC was targeted or inactivated.

Epidemiological studies have indicated for a long time that cancer rates accelerate dramatically after a person reaches age 50, an observation usually attributed to an accumulation of harmful genetic mutations. However, Judith Campisi and others have suggested that the accumulation of cells with a senescent phenotype (persisting in certain tissues after undergoing changes in morphology, behavior, and function) might also contribute to cancer. Certainly, even in animal models it has been shown that telomere attrition may not invariably be an anticancer mechanism. Thus, if cells are able to avoid replicative senescence (e.g. by losing p53) despite telomere shortening, then they are at increased risk of undergoing major chromosomal aberrations during cell division, which may actually support cancer progression.

Senescent cells may promote cancer behaviors in other cells: the dangers of old stroma

The young are always ready to give to those who are older than themselves the full benefits of their inexperience.

Oscar Wilde

Senescent cells may secrete many different molecules, some of which have been shown by Campisi and colleagues to have a "field effect" that could encourage malignant behavior in nearby cells in a paracrine fashion. This has resulted in speculation that cellular senescence evolved as a cancer suppression mechanism at a time when the life expectancy for humans was far shorter than it is today. Given that most people now can expect a much greater life expectancy (at least in developed countries), senescence may be an example of "antagonistic pleiotropy." Namely, a trait selected to optimize fitness early in life turns out to have unselected deleterious effects later on. In fact, such mechanisms, including fetal programming, have been proposed as contributors to many of the more prevalent diseases of modern life including not only cancer but also obesity, diabetes, and atherosclerotic diseases. One version of this is the so-called Barker hypothesis, whereby exposure of a fetus to malnutrition during ontogeny may in some way foster the birth of a child "programmed" to successfully cope with malnutrition through adult life. The subsequent exposure of such an "adapted" individual to an unexpectedly energy-rich diet of burgers, pizzas, and sugar-containing drinks, for which that individual is now maladapted, may then result in disease, albeit usually at an older age.

These digressions aside, cellular senescence is an effective barrier to neoplastic transformation as it removes risky cells from the cell division cycle. If, however, senescent cells can secrete factors that stimulate the proliferation of nonsenescent cells – potentially including premalignant cells – then we have an intriguing situation in which a tumor suppressor response becomes a pro-tumorigenic one, termed "antagonistic pleiotropy." In a study from the laboratory of Judith Campisi, premalignant mammary epithelial cells exposed to senescent human fibroblasts in mice irreversibly lose differentiation and undergo malignant transformation. One of the secreted factors from senescent cells that may contribute to this behavior is the matrix-degrading enzyme MMP3. Thus, senescent cells in stromal tissues may contribute to age-related tumorigenesis by secreting factors that can alter the differentiation state of other cell types, such as epithelial cells in the evolving tumor. Intriguingly, MMP3 produced by stromal cells has recently been shown to influence various processes in a cancer (or would-be cancer) cell. In addition to fostering invasion and metastases at later stages in tumor development, MMPs may exert an early influence on cell transformation and cancer development. MMP3 can activate a member of the RAC family (RAC1b) promoting the formation of reactive oxygen species (ROS) with resultant DNA damage and genomic instability. Moreover, MMP3 can also induce EMT, a major factor in cancer formation and spread, possibly by cleaving E-cadherin and triggering signaling via the WNT–β-catenin pathway.

A further recent study has suggested that stromal-derived TGF-β (in this study, released from macrophages engulfing apoptotic lymphoma cells) can promote senescence in MYC-induced lymphomas.

Premature senescence, reversibility, and cancer

Whatever poet, orator, or sage may say of it, old age is still old age.

Henry Wadsworth Longfellow

Premature senescence recapitulates the cellular and molecular features of replicative senescence (see also Chapter 6 on oncogene-induced senescence). Extrinsic factors such as anticancer agents, γ-irradiation, and UV light have been shown to induce premature senescence as a DNA damage–mediated cellular stress response. DNA lesions are sensed and transduced via protein complexes involved in DNA maintenance and repair including the ataxia telangiectasia mutated kinase (ATM), among others (Chapter 10). Such kinases directly or indirectly phosphorylate the p53 protein at certain residues, which can then participate in apoptosis or in the induction of DNA damage–mediated senescence. The CKI p16^{INK4a} is implicated in both response to DNA damage and control of stress-induced senescence, although how p16^{INK4a} induces permanent G$_1$ arrest is not known. In fact, p16^{INK4a} is essential for the maintenance of cellular senescence. The p21^{CIP1} protein acts as a molecular switch that triggers telomere-initiated senescence. Interestingly, the mechanisms by which dysfunctional telomeres lead to p21^{CIP1} activation are similar, but not identical, to cellular responses to DNA damage. The p16^{INK4a} protein acts independently from telomeres (discussed further in this chapter).

Damage to DNA, the prime target of anticancer therapy, can trigger drug-induced senescence. Chemotherapeutic agents such as the topoisomerase inhibitors doxorubicin and cisplatin induce DNA damage and particularly DNA double-strand breaks (DSBs). DNA damage is followed by checkpoint-mediated cell-cycle arrest culminating in either repair or, if irreparable, apoptosis or senescence (Chapters 4 and 10). Chemotherapeutic drugs may trigger both senescence and apoptosis, but the former may be achieved at lower administered doses. Cancer cells must escape both senescence and apoptosis in order to become drug resistant.

The possibility that senescence may be one aspect of the inherent "tumor suppressor activity" of some oncogenes such as RAS was discussed in Chapter 6. In fact, in an important issue of the journal *Nature* (*Nature* 2005 Aug 4;436, 7051), four key papers confirmed the critical role played by senescence in tumor suppression and in particular suggest that this must be overcome in at least some RAS-dependent tumors for a transition from benignness to malignancy. Cellular senescence is an attractive therapeutic goal, but the value of this strongly relies on senescence being irreversible. Unlike apoptosis, senescence still leaves a viable cell, with reversion of a senescent cell into a dividing cell posing the threat of a tumor relapse, unless senescent cells were in some way cleared by other processes, such as shedding from the skin surface or phagocytosis. In an *in vivo* model of drug-induced senescence in mouse lymphomas, repeated anticancer therapy eventually selected against senescence-controlling genes such as *p16^{INK4a}* or *p53* resulting in tumor relapse and progression to a more aggressive tumor.

Several studies suggest that senescence can be reversible if key proteins involved in the maintenance of senescence are lost. Thus, replicative senescence in human lung fibroblasts was reversed by dual functional inactivation of p53 and RB, by the expression of simian virus 40 large T antigen (SV40T), or by a combination of p53 inactivation and knockdown of p16^{INK4a} expression using small interfering RNAs (siRNAs). Inactivation of p53 in senescent mouse embryonic fibroblasts enabled cells to resume proliferation. Overall, this supports the notion that p53 and p16^{INK4a}–RB variously cooperate not only in the induction but also in the maintenance of premature senescence *in vivo*. However, we still require experiments demonstrating a successful reversal of cellular senescence *in vivo*. While the acquisition of spontaneous mutations that disable p53 or RB in a resting cell without DNA replication seems rather unlikely, epigenetic changes such as promoter methylation to silence p16^{INK4a} expression might occur in senescent cells. Moreover, in any experiments conducted to address this important question, the adverse and potentially pro-tumorigenic effects that such senescent cells might have on their nonsenescent neighbors would also need to be examined.

SnoN which promotes cell proliferation by inhibiting TGF-β can, along with other oncogenes, induce senescence, in this case by stabilizing p53 in a PML-dependent manner. Interestingly, overexpression of SnoN inhibited oncogenic transformation induced by Ras and Myc *in vitro*.

Stem cells and cancer

We must, however, acknowledge, as it seems to me, that man with all his noble qualities ... still bears in his bodily frame the indelible stamp of his lowly origin.

Charles Darwin

Mammalian aging occurs in part because of a decline in the restorative capacity of tissue stem cells. These self-renewing cells are rendered malignant by a small number of oncogenic mutations, and overlapping tumor suppressor mechanisms (e.g. p16^{INK4a}–RB, ARF–p53, and the telomere) have evolved to ward against this possibility. These beneficial anti-tumor pathways, however, appear also to limit the stem cell lifespan, thereby contributing to aging.

Tissue homeostasis in the adult requires that cells which have died or been shed (blood cells, skin cells, liver cells, etc.) are replaced by newly generated cells. Even in organs once thought to be postmitotic such as the pancreatic islet and the brain, renewal has been demonstrated to occur and intriguingly in the former may usually actually take place from the existing mature β cells rather than from a stem cell precursor.

Adult mammals exhibit extensive proliferation and renewal even in the absence of disease. Thus, the intestinal lining replaces itself entirely on a weekly basis, and the bone marrow produces trillions of new blood cells daily. This obviously creates a potential risk of cancer, given the likely rate of somatic mutation in the stem cells giving rise to these new cells in some tissues. It may be unsurprising, therefore, that some 1% of neonatal cord blood collections contain significant numbers of myeloid clones harboring oncogenic fusions such as the *AML–ETO* fusion associated with acute leukemia and up to one in three adults possess detectable *IgH–BCL2* translocations, which are commonly associated with follicular lymphoma. Given that cancers are far less frequent, this gives testament to the efficiency of cancer-restraining mechanisms such as the tumor suppressors. As discussed in this book, various overlapping tumor suppressor barriers appear to be most important and include the p16^{INK4a}–RB pathway, the ARF–p53 pathway, and telomeres. A common endpoint for these major tumor suppressors is senescence or apoptosis.

Aging of stem cells may contribute to age-dependent defects in tissue maintenance and repair; examples are well described and include bone marrow failure. Certainly, in mice hematopoietic stem cells are less able to promote regeneration and self-renew, and they may be more prone to cell stresses. Recent studies suggest that, at least in part, such defects coincide with increased expression of p16^{INK4a} with aging (this occurs despite continued expression of the Bmi-1 protein, which normally inhibits p16^{INK4a} and ARF), and that some of these defects could be reversed in the absence of p16^{INK4a}. Telomere shortening also contributes to stem cell dysfunction, and conversely longer telomeres in the stem cell compartments are associated with an extended lifespan in mouse models.

Thus, networks of proto-oncogenes and tumor suppressors control stem cell self-renewal and stem cell aging as well as tumor cell growth. Proto-oncogenes support regeneration by promoting stem cell replication, but their activities must be restrained by tumor suppressors and also by the inherent tumor suppressor activity of these oncogenes themselves (C-MYC and apoptosis, and RAS and growth arrest) in order to avoid the risk of cancer. Conversely, these tumor-suppressing mechanisms may limit regenerative capacity by sensitizing stem cells to cell death or senescence. Many examples of this have been discussed in previous chapters, and many of the associated tumor suppressive mechanisms must be overcome in order for cancers to develop – for example, inactivation of the INK4A–ARF locus (by mutation, overexpression of cdc6 or the polycomb family proto-oncogene, Bmi-1, needed for stem cell self-renewal, and also proliferation of cancer cells) and loss of p53–RB signaling. As discussed in Chapter 11 in some detail, gene expression is also modified by epigenetic alterations among which DNA methylation, chromatin remodeling, histone variants, noncoding RNAs, and the activity of the previously noted polycomb group proteins are the best known to date. Such changes are inherited by cells resulting from mitosis (and meiosis) without any changes in the DNA's coding sequence and therefore are not detected by mutation screens. Epigenetic factors are believed to be very important in regulating behaviors of stem cells and cancer stem cells and can affect the progression of cancer. For embryonic stem cells (ESCs) to differentiate, they must swap an ESC self-renewal program for new tissue-specific programs. This process requires the activity of let-7 miRNAs, and in this regard they act in opposing fashion to ESC cell-cycle-regulating (ESCC) miRNAs.

Telomeres are re-elongated or rejuvenated during the induction of pluripotency, a phenomenon that coincides with dramatic changes at the telomeric chromatin. Telomere dysfunction severely impairs reprogramming coincidental with the elimination of DNA-damaged cells during reprogramming.

One area which has developed rapidly in recent years explores how organisms balance the need for stem cell maintenance and regeneration with energy metabolism, including during periods of stress and nutritional deprivation. A series of three studies in the journal *Nature* have examined the role of Lkb1 on stem cells. Hematopoietic stem cells (HSCs) can switch between growth states ranging from quiescence to proliferation depending on bioenergetic needs. Lkb1 is an evolutionarily conserved regulator of cellular energy metabolism acting in part through AMP-activated protein kinase (AMPK) and related kinases. Lkb1 regulates stem cell maintenance in hematopoiesis. In fact, Lkb1 deletion causes loss of HSC quiescence followed by rapid depletion of all hematopoietic subpopulations independently of AMPK and of mTOR complex 1 but instead through various routes converging on mitochondrial apoptosis, and by downregulation of PGC-1 transcriptional networks.

The intestinal epithelium which exhibits rapid self-renewal and regeneration is a model for a stem cell–supported tissue. Intestinal stem cells are characterized by high Lgr5 expression, are located between Paneth cells at the small intestinal crypt base, and divide every day. Recent studies suggest that intestinal stem cells follow a so-called neutral drift dynamic, with loss of a stem cell compensated for by a neighboring cell's replication. Of course, this model suggests that cell clones expand and contract randomly until they either take over the crypt or are lost. Using a lineage-tracing system, a recent study shows that the Lgr5(hi) stem cell population persists throughout life in mice, and crypts drift toward clonality over 1–6 months. Most Lgr5(hi) cell divisions occur symmetrically and not assymetrically (one Lgr5(hi) cell and one transit-amplifying cell per division). Resident stem cells double their numbers each day and stochastically adopt stem cell or differentiating fates.

Cancer stem cells

Cancer development has long been assumed to depend on the sequential acquisition of mutations, maybe due to genetic instability and carcinogens, affecting normal cells. One problem with this view is that prolonged periods would be predicted necessary for the first mutations to accumulate, but normal somatic cells within epithelia from which most cancer arises often have short lifespans. So how do they survive long enough to become cancer cells? Of course, one way around this is to postulate that cancers originate from cells that are long-lived or that must become so early in the tumorigenesis progress. The CSC model proposes that stem cells that are very long-lived or essentially immortal form cancers by accumulating mutations that deregulate normal self-renewal pathways.

As has been discussed, cancers are generally heterogeneous, consisting of noncancer and cancer cells of differing degrees of aberrant behavior. Differences in proliferation or in invasive or metastatic behavior of the cancer cells were traditionally attributed to a combination of genomic instability of cancer cells and resultant clonality, and also due to micro-environmental factors related to positional cues including proximity to blood vessels, inflammatory cells, or cancer-associated fibroblasts. Importantly, all cancer cells were believed to have an equal capacity to propagate the cancer – now termed the "stochastic model." More recently, the CSC model has become increasingly pervasive. The CSC hypothesis was first introduced to explain findings in hematologic malignancies in in the 1990s but is now accepted as important for most solid tumors in one way or another. CSCs are also variously referred to as cancer-initiating cells, tumor propagation cells, or a side population. In this model, the heterogeneity of tumors results from aberrant differentiation of tumor cells back toward cells of the original lineage from a subpopulation of CSCs. By implication, then, CSCs are uniquely capable of propagating a tumor as normal stem cells drive proliferation and differentiation in normal tissues. In both models, tumorigenesis is driven by Darwinian cycles of mutation and natural selection; the main difference is in the potential for specific therapeutics aimed at a minority population that may have unique characteristics compared to the bulk of cancer cells. The same thinking could be applied to a dominant clone in a cancer under the stochastic model, with one caveat: CSCs have common features across very different cancers. Given the CSC model's potential importance both conceptually and therapeutically, this will be discussed in some detail in this chapter.

Where CSCs originate is unclear but likely includes an origin from normal stem or progenitor cells, or possibly from other cancer cells. Due to their extended lifespan, stem cells represent the most likely target for the accumulation of epgenetic events required for tumorigenesis, but this has not been proven for most solid cancers. A recent study in prostate cancer is illuminating. Luminal cells are believed to be the cells of origin for human prostate cancer, because luminal cells are expanded and basal cells absent. However, it has now been shown that basal cells from primary benign human prostate tissue can initiate prostate cancer in immunodeficient mice, but such cells subsequently under the influence of AKT, ERG, and androgen receptor are gradually replaced by cells resembling luminal cells expressing prostate-specific antigen (PSA) and disappearance of the basal cells – the histology of an established cancer does not necessarily reveal its history and origin.

CSCs have common features with nomal stem cells and may in part be due to activation of similar signaling pathways. Thus, histologically poorly differentiated tumors, likely to be enriched for CSCs, manifest gene expression patterns similar to those found in embryonic stem cells including loss of Polycomb-regulated gene expression and activation of Nanog, Oct4, Sox2, and c-Myc. Moreover, the presence of these features conferred adverse outcome clinically. CSCs use many of the same signaling pathways found in normal stem cells, such as Wnt, Notch, and Hedgehog (Hh).

The CSC hypothesis of tumorigenesis is an attractive one, as it implies that even solid tumors are maintained by a minority side population of cancer cells which exhibit stem cell properties such as self-renewal and the ability to give rise to different lineages of more differentiated cells. This model does not presuppose that CSCs arise by epimutations in normal stem cells as they could also develop by EMT or other processes from differentiated cells or somewhere in between. CSCs are also often held responsible for metastases and for treatment resistance and recurrence. Markers of stemness and EMT correlate well with radio-chemoresistance and may actually be enriched in residual disease after conventional treatments, which eliminate those cells which do not have these characteristics.

CSCs have been defined following various isolation procedures by their enhanced ability to generate new tumors when xenotransplanted into immunocompromised mice. Though this view has been challenged by studies from the laboratory of Andreas Strasser, the existence of similar subpopulations of "über"-cancer cells in mouse cancers suggests that xenografting human cells does not merely select human cells able to grow in mice. Many, if not all, cancers contain a side population of CSCs, which may originate from stem cells in the relevant tissues or potentially by EMT in other precursor cells. Thus, induction of EMT can allow cancer cells to disseminate from a primary tumor and also promotes acquisition of a self-renewal capacity and other stem cell markers.

CSCs are unusual in their ability to manifest symmetric as well as asymmetric cell division, and can thus expand in number rather than just self-renew as is the case for normal stem cells. Of therapeutic relevance here is that restoring p53 can switch CSCs back from symmetric to asymmetric cell division. CSCs undergo clonal evolution, as has been discussed this chapter, and it is possible for different clones to be expanding, giving rise to more differentiated cells at the same time in a given cancer. It is also worth noting that these differences in behavior can change over time for individual cells and may be regulated not just by mutations but also by the action of tumor macrophages, cancer-associated fibroblasts, and EMT and WNT signaling. The existence of CSCs has profound implications for new cancer therapies, which might be concentrated on eliminating these cells specifically on the basis of their unique properties such as self-renewal pathways (Wnt, Notch, and Hedgehog, discussed in Chapter 5) or EMT. Various cell markers have been used to try to isolate CSCs for study, but as yet none have proved completely specific even when applied in combinations of up to four or five markers using fluorescence-activated cell sorting (FACS). Potential CSC markers are shown in Table 9.1.

As has been referred to in this book, a considerable body of evidence implicates stem cells or closely related progeny as the cell or origin for most cancers. Moreover, it is impossible not to notice that stem cells already normally exhibit certain behaviors essential for a would-be cancer cell, namely, self-renewal and longevity. Moreover, particularly in the skin and gut, where cells exiting the stem cell niche will be discarded by the body after a short lifespan in the outer reaches of the tissue, where is the time to acquire all those somatic mutations needed to become cancer cells? Mouse models of small bowel cancers directly support a stem cell origin as only a subset of lgr5 positive cells at the crypt base formed adenomas when *Apc* was deleted, whereas transit amplifying cells did not. Hematopoietic stem cells (HSCs) may give rise to many hematological cancers. For example, in chronic myeloid leukemia (CML) the Philadelphia chromosome (resulting in the BCR–ABL oncogenic fusion) is found in otherwise normal mature blood cells that have differentiated from an HSC carrying this translocation. However, leukemias also arise from more differentiated cells that have re-acquired stem cell properties under the influence of oncogenic fusion genes such as MOZ–TIF2 or elevated Wnt signaling. Melanomas are highly heterogeneous tumors that have been recently shown to contain a slow-cycling subpopulation of JARID1B-positive cells that are essential for tumor growth an from which highly proliferative progeny arise. Interestingly, this

Table 9.1 Markers for cancer stem cells

Cancer stem cell markers	Function and human cancer
CD24	Ligand for P-selectin, absent in some breast and prostate cancer
CD34	Myeloid leukemia
CD44	Hyalauronin binding, colon, breast, prostate, head, and neck squamous cell carcinoma (SCC)
CD90	Thy1, liver, and T-cell leukemia
CD105	Membrane glycoprotein and kidney
CD117	Also known as c-Kit, a receptor for SCFl; ovary
CD133	Prominin family of membrane proteins; colorectal, liver, lung, and ovary
CD166	Activated leucocyte cell adhesion molecule; colorectal
Bmi	Polycomb protein that inactivates the *INK4a–ARF* gene locus resulting in reduced p16 and ARF proteins
ALDH	Aldehyde dehydrogenase; leukemia, breast, and colon
ASCL1	Achaete-scute complex homologue 1, in non-small-cell lung cancer
EpCAM	Epithelial cell adhesion molecule; colon and pancreas
JARID1B	In melanoma
Podoplanin	Skin cancer

appears to be quite a plastic process as JARID1B-negative cells can become positive.

However, considerable direct evidence in mouse models demonstrates that cancers can be generated in previously terminally differentiated cells, particularly in those tissues where such cells are not automatically discarded from the body as the end point of differentiation.

EMT involves the loss of E-cadherin-dependent cell contacts between neighboring cells allowing epithelial cells to become fibroblast-like and motile. Transcription factors like Twist, Snail, and Slug can downregulate E-cadherin gene expression and can suffice to induce EMT in breast epithelium and intriguingly also the expression of CSC markers. In other models, Snail and Slug can de-repress genes directing stem cell behaviors such as Nanog and KLF4, and increase numbers of CSCs. Other proteins such as Six1, YB-1, and ZEB1 may also promote stem cell–like behavior, the latter possibly in part by blocking transcription of stemness-inhibiting microRNA-200 (miR-200). Adult stem cells are controlled by neighboring stromal cells, which together constitute the stem cell niche. CSCs appear to respond to similar cues and may co-localize with normal stem cells and at the tumor–stroma interface. In colon cancer, Wnt signaling in CSCs depends on hepatocyte growth factor (HGF) produced by stromal fibroblasts. The chemokine receptor CXCR4 is often expressed by CSCs, allowing cell migration down a SDF-1 gradient, originating from hematopoietic niches and elsewhere facilitating metastasis. Thus, by implication, antagonists of chemokine receptors could be used to treat cancer, which is supported by the inhibition of metastases in mice by the use of anti-CXCR4 antibodies. The more CSCs that are present in general, the worse the prognosis is likely to be, and it appears that the presence of CSC markers does at least in some cases predict a worse outcome; CD133 expression in colorectal and pancreatic cancers is associated with a notably worse prognosis and metastases, and raised ALDH predicts a worse outcome in breast cancer.

Resistance to radiotherapy may in part be explained by the presence of proliferatively quiescent CSCs located within protected hypoxic locations such as niches or molecular alterations in DNA damage responses. Therapies might even cause CSC enrichment, as has been noted in gemcitabine treatment of pancreatic cancer, cyclophosphamide treatment of colorectal cancer, and cisplatin, doxorubicin, and methotrexate treatment of lung cancer. Overcoming such resistance might require inactivating drug-detoxifying enzymes such as ALDH and/or inhibiting the multidrug-efflux pumps. As discussed, another approach is to target the signaling pathways involved in CSC self-renewal, in particular Wnt–β-catenin signaling. For instance, Dickkopf1 (Dkk1) binds to low-density lipoprotein receptor-related protein-6 (LRP6) preventing formation of the Frizzled–Wnt–LRP6 complex needed for Wnt signaling. More recently, it was shown that a small-molecule inhibitor of tankyrase prevents the enzyme from breaking down the scaffold protein axin needed for β-catenin destruction – with the net effect being an increased loss of β-catenin even in the absence of APC function.

The Notch pathway is frequently overactivated in cancer and can be inhibited by blocking the protease γ-secretase or by using Notch receptor–specific antibodies that block ADAM, thus inhibiting the release of icNotch. Hopefully, even fewer toxic agents may be developed based upon siRNA to Notch. Anti-Delta-like 4 ligand (DLL4) antibodies have shown early promise. The hedgehog (Hh) pathway is also involved in stem cell maintenance in colorectal cancer. Available drugs include cyclopamine target smoothened (SMO) downstream of the Hh receptor Patched 1 (PTCH1). Some are now in phase I clinical trials in basal cell carcinoma and metastatic medulloblastoma and CML. EMT can lead to the emergence of CSCs and the activation of downstream kinases such as Axl that are important for invasion and metastasis. Preventing EMT is thus an attractive option and has led to studies targeting TGF-β and relevant miRNAs such as miR-103/107 in combination with chemotherapeutic agents. Reduced expression of miR-199b-5p, which downregulates HES1, predicts worse outcome in medulloblastoma and its introduction reduces the proliferation of medulloblastoma CSCs in athymic mice; miR-128 is markedly downregulated in glioblastomas, and introducing miR-128 into glioma cells reduces their proliferation by downregulating BMI1. The miR-34a targets Notch1 and Notch2 promoting differentiation of glioma CSCs, and siRNA to Gli1 can reduce CSCs in ovarian cancer models. The miR-34a, which is a p53 target, can also inhibit prostate cancer spread in mice and this might be partly due to downregulation of CD44 and thereby of CSCs.

In most CRCs, mutations in APC or β-catenin drive constitutive activation of Wnt signaling, yet still cells differ in terms of β-catenin localization or activation. This heterogeneity is in part due to increased Wnt activity in colon CSCs, located close to stromal myofibroblasts, supporting the notion that CSCs are affected by environmental cues such as hepatocyte growth factor. Intriguingly, in one study myofibroblast-secreted factors could re-invoke a CSC phenotype in more differentiated tumor cells *in vivo*.

Many human cancer cell lines contain a subpopulation of highly drug-resistant cells dependent on signaling through IGF-1R and the activity of the histone demethylase KDM5a (JARID1A) that removes the active H3K4Me mark. Drug sensitivity can be restored by inhibiting IGF-1R or HDAC.

DNA damage stabilizes and activates p53 (Chapters 7 and 10), which in turn can trigger cell-cycle arrest, senescence, or apoptosis. Loss of p53 can allow cells to acquire aberrant self-renewal, and recent studies suggest that myeloid progenitor cells expressing oncogenic Kras and lacking p53 can become leukemia-initiating cells, resembling CSCs capable of maintaining acute myeloid leukemia (AML) *in vivo*. Recently, p53 and downstream mediators of apoptosis such as PUMA have been shown to promote depletion of adult stem cells in response to DNA damage resulting in defective tissue regeneration, and could be a significant contributor to aging.

Hypoxia promotes tumor progression via factors such as HIF1alpha and HIF2alpha, which promote angiogenesis, Warburg effect, and invasion. Hypoxia can help promote survival of normal and also cancer stem cells by enhancing the activity of stem cell factors like Oct4, c-Myc, and Nanog via HIF2alpha in particular. Targeting hypoxia niches may therefore improve therapy efficacy by eliminating CSCs.

In some pancreatic cancer cell lines, CSCs can be sensitized to TRAIL-induced apoptosis by sulforaphane, which targets NF-κB signaling.

One area related to CSCs which may become important in future clinical practice relates to stem cell therapy. A key aim in regenerative medicine, to treat disease such as diabetes where cells of a key type are deficient, is to replace the missing cells by transplantation of new ones. In some cases this may be done by deriving whole organs or mature cells from donors, but for many putative cell-based treatments and for common diseases the availability of donors is far too small. Thus, there has been a huge surge of interest in the idea of growing new cells in or without the body by using embryonic or adult stem cells as a potential source. Given the major ethical issues that have been raised about using embryonic stem cells, derived from aborted human fetuses or from "spares" arising in *in vitro* fertility treatments, great efforts have been made in the last 10 years to find and purify adult stem cells or to make stem cells by transforming adult somatic cells into a pluripotent state and then subsequently to generate the cells desired by controlling their differentiation into specific cell fates. One concern that has been supported by some recent studies is that such induced stem cells appear to share features in common with tumor cells and with CSCs in particular, suggesting that close monitoring of the fate of such cells or the mature cells derived from them will be required to avoid a potential cancer risk if these methods reach clinical use.

The epithelial–mesenchymal transition (EMT)

The epithelial–mesenchymal transition is an embryonic program which can be reactivated in adult cells and is critical for wound healing and repair. It has been alluded to already in the progression of cancer and formation of CSCs.

TGF-β is a major factor in inducing EMT, which is associated with several key alterations in signaling including isoform switching of fibroblast growth factor (FGF) receptors culminating in increased responsiveness to FGF-2. TGF-β and FGF-2 then cooperate in induction of EMT in various cells in the cancer micro-environment. Several transcription factors, including Twist1, Snail1, Snail2, ZEB1, and ZEB2, have been shown to induce EMT. Many of these interact with one another. Thus, Snail2 is essential for Twist1 to suppress E-cadherin transcription and induce EMT, cell invasion, and distant metastasis in mice. Zinc finger E-box binding homeobox 1 (ZEB1) is a recently identified activator of EMT and is important in tumor progression and metastasis. ZEB1 and miR-200 family members, whose members are strong inducers of epithelial differentiation, repress each other's expression in a reciprocal feedback loop. ZEB1 is essential for the tumor-initiating capacity of pancreatic and colorectal cancers by repressing expression of stemness-inhibiting miR-203, and therefore it promotes stemness. The miR-200 family members target stem cell factors, such as Sox2 and Klf4, and Notch pathway components, such as Jagged1 and the mastermind-like coactivators Maml2 and Maml3. Reduced miR-200 expression has been found in pancreatic adenocarcinoma and the basal type of breast cancer, both typified by an aggressive course. TGF-β1 stimulates transcription of metastasis-associated protein 1 (MTA1), which is also needed for EMT and for repression of E-cadherin via induction of a FosB–histone deacetylase 2 complex.

Telomeres

Telomerase activity is necessary to maintain the integrity of telomeres, which in turn prevent chromosome ends from being detected and processed as damaged DNA. As primary cells give rise to successive generations of progeny, so telomeres progressively shorten with each new generation, a process referred to as "telomere attrition." This is associated with the reproducible loss of proliferative potential, something that has been termed "replicative senescence." Similarly, increasing generations of a telomerase-deficient mouse model also demonstrate progressive telomere loss, which eventually leads to premature aging phenotypes in these mice. In contrast, more than 90% of all types of human tumors reactivate telomerase. Cancer cells need telomerase to maintain functional telomeres and to divide indefinitely, and this has opened the possibility of novel therapeutic approaches based on using telomerase as a target.

Telomeres and replicative senescence

Telomeres cap the ends of chromosomes by forming a higher-order chromatin structure that protects the 3' end from degradation and DNA repair activities (see Figs 9.4 and 9.5 on meiotic telomeres and metaphase telomeres, respectively, for a FISH visualization of telomeres; mouse chromosomes are shown in Fig. 9.6). Mammalian telomeres are composed of TTAGGG repeats bound to specialized proteins (see Fig. 9.7 on telomere proteins for the current model for telomere structure and telomere-binding proteins). Loss of telomere capping, due to either TTAGGG exhaustion or disruption of the telomere structure, results in end-to-end chromosomal fusions and loss of cell viability.

As cells proliferate, TTAGGG repeats are lost from telomeres unless they have activated telomerase, a reverse transcriptase that adds TTAGGG repeats onto preexistent telomeres. Most normal somatic cells do not have sufficient telomerase activity and thus suffer telomere attrition. Global telomere shortening eventually results in individual chromosomes with critically short telomeres that have lost their functionality. The most dramatic

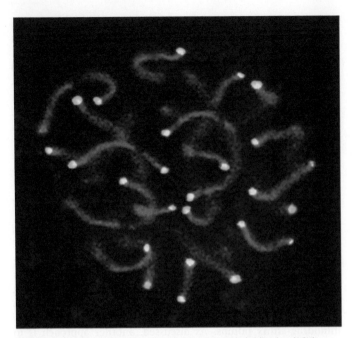

Figure 9.4 Meiotic telomeres. Fluorescence in situ hybridization (FISH) showing a meiocyte nucleous where chromosomes are stained in red with a synaptonemal complex protein, and telomeres are stained in yellow with a PNA-telomeric probe.

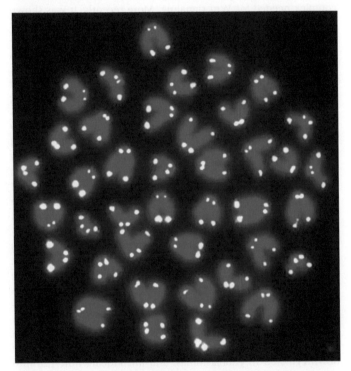

Figure 9.5 Metaphase chromosomes. Fluorescence in situ hybridization (FISH) of a mouse metaphase nucleus showing the chromosomes in blue (DAPI staining) and the telomeres in yellow (staining with a PNA-telomeric probe).

consequences of telomere dysfunction are the appearance of chromosomal fusions and the loss of cell viability, which occur with increasing passages of cultured primary cells and are thought to trigger what is known as "replicative senescence." Loss of viability due to critically short telomeres has been demonstrated in telomerase-deficient mice that show degenerative pathologies in various organ systems. These pathologies are coincidental with a decreased proliferative index and/or increased apoptosis of the affected cell types. Telomerase re-introduction results in elongation of the population of short telomeres and prevents telomere fusions, as well as pathologies, in the context of these mice. Similarly, telomerase reconstitution in most human normal primary cells is sufficient to immortalize them.

Telomeres and cancer

The observation that most human tumors reactivate telomerase has led to a working hypothesis that telomerase may be essential for sustaining the viability of cancer cells. Moreover, as normal cells lack telomerase activity whereas tumor cells depend on telomerase-mediated telomere maintenance to keep proliferating, telomerase might be another Achilles heel to be exploited in new drug developments against cancer. To date, telomerase is a potentially useful tumor marker, and selective expression by cancer cells is currently being exploited in newly emerging cancer immunotherapies directed against this protein. However, in terms of functional targeting of telomerase, caution is needed. Thus, lack of telomerase activity or failure of telomere maintenance may conversely contribute to cancer progression; in the absence of telomerase activity, short telomeres can trigger chromosomal instability, which could accelerate the rate of mutation as long as apoptosis or senescence is avoided.

Telomerase-deficient mice have been widely employed to explore the consequences of telomere attrition, as discussed in this chapter. Recently, studies from the laboratory of Ronald DePinho have assessed to what extent the progressive tissue atrophy, stem cell depletion, and poor regenerative capacity resulting from telomere attrition could be arrested or reversed by reactivation of endogenous telomerase activity. By using mice expressing a conditional TERT-ER chimeric protein, they showed that telomerase reactivation could extend telomeres, reduce DNA damage signaling, and remarkably also allow the re-entry of growth-arrested cells into the cell cycle with recovery of degenerative phenotypes in most organs examined and supporting the notion that at least some of the ravages of aging could be reversible on restoring TERT activity.

The consequences of critical telomere loss

Cancer cells generally have short telomeres and high levels of telomerase activity. Telomerase is able to rescue short telomeres, thus preventing activation of the DNA damage responses and allowing viability of cancer cells, which otherwise would arrest division or undergo apoptosis due to telomere damage and

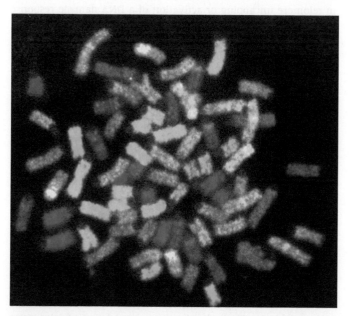

Figure 9.6 Mouse chromosomes. Spectral karyotying (SKY) to visualize mouse chromosomes.

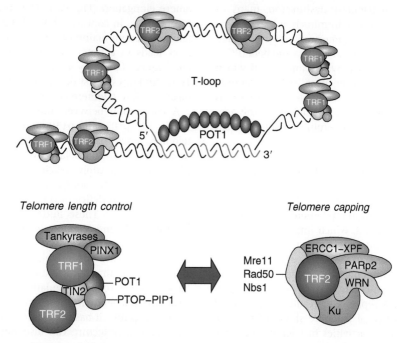

Figure 9.7 Telomere-interacting proteins. A number of proteins are bound to telomeres and have been involved in regulating telomere length and telomere capping.

catastrophic chromosomal rearrangements. In agreement with this, mice deficient for telomerase activity and with short telomeres are resistant to developing carcinogen-induced skin tumors, coincidental with p53 upregulation. If p53 is defective, however, the DNA damage response triggered by short telomeres is abrogated, and telomerase-deficient mice show a higher frequency of chromosomal aberrations and a higher incidence of epithelial tumors. Therefore, simultaneous absence of telomerase and p53 may favor survival of cancer cells harboring chromosomal aberrations. In contrast, if either RB–p16^{INK4a} or APC tumor suppressor pathways are abrogated, short telomeres have a negative affect tumorigenesis, suggesting that in the mouse, p53 is the main tumor suppressor that mediates cell arrest or apoptosis due to telomere dysfunction. Similarly, p53 but not RB–p16^{INK4a} is important in signaling telomere dysfunction produced by TRF2 mutation in mouse cells. In human cells, however, both pathways are important. Both ATM and the nonhomologous end joining (NHEJ) have been proposed to signal dysfunctional telomeres. Abrogation of the NHEJ activities of Ku86 or DNA-PKcs rescues both end-to-end fusions and apoptosis, but not cell-cycle arrest, triggered by critically short telomeres. These findings suggest that NHEJ could be involved in signaling apoptosis due to critical telomere shortening. Although it has been described that DNA-PKcs does not seem to phosphorylate p53 *in vivo*, it is possible, however, that DNA-PKcs may signal through p53 in the particular case of telomere dysfunction, since DNA-PKcs and Ku86 are located at mammalian telomeres. All together, these findings suggest that a dysfunctional telomere is probably detected as a DSB and signaled as such, and predict that mutations that abrogate NHEJ may affect the outcome of telomere dysfunction (Fig. 9.1). Recent data suggest that short and/or dysfunctional telomeres are indeed recognized as DSBs, as suggested by increased γ-H2AX at telomeres.

Activation of p53 by dysfunctional telomeres leads to atrophy and functional impairment of tissues with high turnover. Various "omics" studies have identified a variety of genes and proteins downregulated in tissues with telomere dysfunction, many of which are involved in metabolism and in mitochondrial biogenesis and function. These include peroxisome proliferator-activated receptor gamma (PPAR-γ) and coactivator 1 alpha and beta (PGC-1α and PGC-1β), which are repressed by p53 and loss of which is associated with reduction of gluconeogenesis, reduced production of ATP, increases in ROS, and heart muscle defects such as cardiomyopathy. This may provide an explanation for the age-related decline in metabolism which underpins the development of, for example, type 2 diabetes.

Telomere dysfunction in the absence of critical telomere loss

Mutation of telomere-binding proteins can disrupt telomeric capping in the absence of significant telomere shortening. A dominant negative mutant of TRF2, which impairs the binding of TRF2 to TTAGGG repeats, results in telomere fusions with long telomeres at the fusion point and in the absence of a significant loss of telomeric sequences. Similarly, mice deficient for either Ku86 or DNA-PKcs, both of which are essential components of the NHEJ machinery for DSB repair, also result in end-to-end telomere fusions in the absence of telomere shortening. These findings suggest a role for NHEJ activities in telomere capping, similar to that proposed for TRF2. Indeed, TRF2 and DNA-PKcs mutation share the outcome that the resulting end-to-end fusions involve telomeres produced via leading-strand synthesis, suggesting that these two activities could be required for the postreplicative processing of the G-rich strand and for the formation of a proper telomere cap.

The role of telomere-binding proteins in maintaining telomere function predicts that they may also be important in regulating cellular senescence, as well as cancer and aging. Recently, viable mouse models for Trf1, Tpp1, and Rap1 have been generated which have confirmed the importance of Trf1 and Tpp1 in cancelling telomere damage, telomere fragility, and the generation of end-to-end fusions, which lead to aging phenotypes as well as increased cancer. In contrast, deletion of Rap1 did not result in telomere uncapping or activation of a DNA damage response at telomeres. Instead, extra-telomeric roles for Rap1 have been recently proposed. In the case of primary mouse embryonic fibroblasts (MEFs), either telomere shortening to a critical length in the absence of telomerase (late-generation Terc$^{-/-}$ cells) or long and dysfunctional telomeres due abrogation of Ku86 (Ku86$^{-/-}$ cells) leads to similar frequencies of telomeric fusions. In both cases, dysfunctional telomeres trigger premature, senescence-like arrest and result in decreased immortalization frequencies of MEF. In addition, both Terc- and Ku86-deficient mice show premature aging phenotypes and are tumor resistant when in a p53 wild-type background, suggesting that dysfunctional telomeres impact both aging and cancer in the mouse. In turn, a dominant negative mutation of TRF2, which disrupts telomere capping, can influence the average telomere length at which senescence is triggered in human primary cells. Therefore, both telomere length and telomere state influence telomere function, which in turn affects senescence and immortalization in both human and mouse cells.

Multiple factors regulate telomere length and function

The telomere-binding proteins TRF1 and TRF2 can also influence telomere length. Similarly, simultaneous deletion of telomerase and Ku86 in doubly deficient Terc–Ku86 mice demonstrated that Ku86 acts as a negative regulator or telomerase-mediated telomere elongation (Fig. 9.2). The study of double Terc$^{-/-}$–DNA-PKcs$^{-/-}$ mice, in contrast, showed that DNA-PKcs is required for telomere length maintenance. In particular, an absence of DNA-PKcs leads to a faster rate of telomere loss and to an earlier appearance of phenotypes in telomerase-deficient mice (Fig. 9.2). These findings suggest that telomere-binding proteins simultaneously regulate telomere capping and telomere length.

A connection between the members of the retinoblastoma family, RB, p107, and p130, and the mechanisms that regulate telomere length has been recently made (Fig. 9.2). In particular, mouse embryonic fibroblasts (MEF) doubly deficient in p107 and p130 (DKO), or triply deficient in p107, p130, and RB (TKO), have dramatically elongated telomeres compared to those of wild-type or RB-deficient cells in the absence of changes in telomerase activity. These findings reveal a connection between the RB family and telomere length control in mammalian cells, which could be at the basis of the lifespan extension exerted by a number of viral oncoproteins that inactivate the RB family. It is also possible that the RB family proteins p107 and p130 could have a direct role in regulating telomere length or telomere structure independent of their role in cell-cycle regulation.

More recently, it has been described that activities that modify chromatin architecture can also influence telomere length. In particular, mammalian telomeres contain histone modifications characteristic of constitutive repressive chromatin domains,

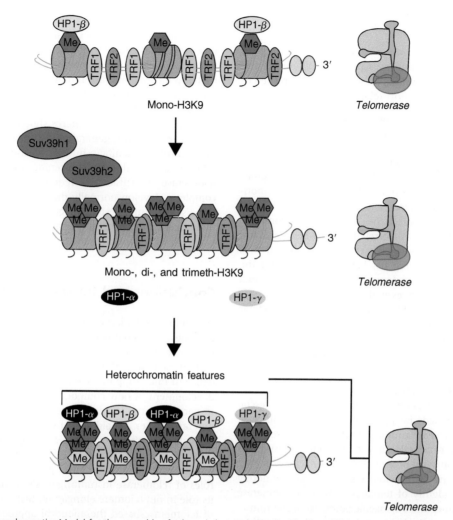

Figure 9.8 Telomere heterochromatin. Model for the assembly of telomeric heterochromatin. The Suv39h histone methyltransferases tri-methylate lysine 9 at histone 3 (H3K9), which in turn creates a binding site for the heterochromatin protein 1 (HP1). These modifications confer heterochromatic features to telomeric chromatin, which in turn may impede the access of telomerase to the telomere, thus regulating telomere length.

such as those of pericentric heterochromatin, and in the absence of these modifications telomeres suffer abnormal elongation (see Fig. 9.8 on telomere heterochromatin for a model on the regulation of telomere length by a heterochromatic higher-order structure).

Roles of telomerase beyond net telomere elongation

Telomerase activity is upregulated during mouse tumorigenesis despite mice having very long telomeres. A selection for tumor cells that are telomerase positive could be indicative of a novel role for telomerase in promoting survival independently of telomere length. The study of mice that either lack or have constitutive telomerase activity has provided evidence for this. In particular, telomerase-deficient mice with long telomeres are more tumor resistant than wild-type mice. Further evidence for a role of telomerase in promoting tumor growth independently of telomere length was first obtained from studying mice that express constitutive levels of the catalytic component of mouse telomerase, Tert, in basal keratinocytes. The epidermis of these mice is more responsive to mitogenic stimuli and shows a higher incidence of carcinogen-induced tumors than wild-type controls. In addition, these mice are more susceptible to develop tumors as they age, and this is further aggravated when in a $p53^{+/-}$ genetic background. Interestingly, many human cancers also show high telomerase activity and p53 mutations, suggesting cooperation between high telomerase expression and p53 mutation in both humans and mice. Mice with transgenic telomerase expression under a β-actin promoter also have an increased incidence of spontaneous mammary epithelial tumors as they age. Additional data obtained in epithelial cells and cells of neural origin, in which telomerase is reactivated by forced Tert expression, also indicate that telomerase activity has an active role in promoting growth and survival.

The mechanisms by which constitutive telomerase activity may be promoting cell growth and survival are still unclear, although they seem to require a catalytically active enzyme. Yeast lacking telomerase show wide-genome changes of gene expression, supporting the belief that telomerase may have other roles besides maintaining telomere length. A more recent genome-wide analysis showed that telomerase modulates the expression of growth-controlling genes and enhances cell proliferation. Telomerase could also have a role in suppressing or processing DNA damage in the genome, thus favoring cell survival and proliferation. In this regard, telomerase is modified by activities involved in DNA damage signaling, such as AKT, PI3K, c-ABL, and p53, among others.

In summary, telomerase activation could favor tumorigenesis by at least two different mechanisms: signaling proliferation and promoting growth independently of telomere length, and rescuing tumor cells with critically short telomeres.

Short telomeres and DNA damage repair

Telomeres are important biological determinants of sensitivity to DNA-damaging agents. In particular, late-generation Terc$^{-/-}$ mice show an enhanced mortality when irradiated with γ-irradiation. Intriguingly, the main DSB DNA repair pathways are not significantly altered in these mice. More recently, it has been demonstrated that shortened telomeres join with DNA breaks and interfere with their correct repair in the context of the telomerase knockout mouse (Fig. 9.1). A correlation between telomere length and radiosensitivity has also been found for cultured cells. In addition, telomere shortening has been shown to alter the chemotherapeutic profile of transformed cells and p53 is required for this. Telomeric dysfunction, even if telomeres are sufficiently long, also results in increased radiosensitivity. These findings have important implications for the therapy of cancer, as tumors treated with telomerase inhibitors could lead to telomere loss, thus increasing the sensitivity of these tumors to radiotherapy or to other genotoxic agents.

Telomerase inhibition in cancer therapy

The fact that cancer cells depend on telomerase-mediated rescue of short telomeres opened the possibility of using telomerase inhibitors to selectively cease tumor growth without interfering with the viability of normal cells. In turn, a telomerase inhibitor could also impair the survival or growth of tumor cells with long telomeres by interfering with telomerase's role in promoting tumorigenesis independently of telomere length. A telomerase inhibitor may also reduce the angiogenic potential of the tumor.

Telomerase inhibitors include antisense oligonucleotides against the telomerase RNA component, dominant negative mutants of the Tert subunit, molecules directed against G-quadruplex formation, as well as small pharmaceutical compounds highly specific for telomerase. Although the efficiency of these inhibitors has yet to be tested in clinical trials of human cancer, there is evidence from both cell culture models and mouse models to support that they would be efficient in ceasing tumor growth. In addition, as just discussed here, these inhibitors may be combined with genotoxic agents to potentiate their effects. The efficiency of a telomerase-directed drug may depend on the status of p53 in the tumor. However, when telomerase was inhibited in various human cancer cell lines, the status of p53 was irrelevant for the final outcome, suggesting p53-independent pathways to signal telomere damage.

In mouse tumors, inhibition of telomerase using a dominant negative mutant of Tert leads to a fast selection of tumor cells that upregulate the endogenous Tert gene, thus compensating for the presence of the inhibitor and allowing growth. Whether this also happens in human tumor cells has yet to be determined. A telomerase inhibitor could also result in the selection of tumor cells that maintain telomeres in the absence of telomerase, a phenomenon known as the alternative lengthening of telomeres. Although only 5% of human tumors seem to be sustained by ALT, inhibition of telomerase could generate a selection pressure to activate ALT within the tumor, which in turn would no longer be responsive to the inhibitor.

Finally, the efficiency of a telomerase inhibitor will depend on its toxicity on cells that express telomerase, such as germ cells and stem cells. Since these cells have longer telomeres than tumor cells, it has been reasoned that they will not be affected by temporary telomerase inhibition, although there is no direct evaluation of this. Data from the telomerase knockout model indicate that these cell types are not affected in the absence of telomerase until their telomeres shorten below a critical length after several mouse generations. Similarly, patients who suffer dyskeratosis congenita, a human disease that is characterized by a faster rate of telomere shortening due to mutations in either telomerase or Dyskeryn (required for an active telomerase complex), develop normally during the first years of life before the disease is manifested, suggesting that temporary telomerase inhibition during a chemotherapy treatment will not have deleterious effects for the organism.

Conclusions and future directions

Telomerase and telomere biology are intense fields of research in oncology. Much has been learned in the last 10 years about their roles in cell immortalization and tumorigenesis. On one hand, work done both with cultured human cells and with telomerase-deficient mice supports telomerase as an attractive target for the development of new compounds with anti-tumorigenic activity. In particular, the fact that short telomeres in the absence of telomerase trigger a rapid loss of cell viability, as well as aggravate the cytotoxic effects of DNA-damaging agents, suggests that telomerase inhibitors would be effective when combined with genotoxic agents in ceasing cancer growth. On the other hand, new roles of telomerase in tumorigenesis, which are independent of its role in net telomere elongation, further support the efficiency of telomerase-based therapeutical approaches in cancer. Therefore, we should expect that strategies that effectively inhibit telomerase are likely to be tested in cancer clinical trials in the near future. However, much is still unknown about the functional interactions between the different activities at the telomere and about the pathways by which telomerase activity is regulated and by which it regulates other processes.

Outstanding questions

1. Targeting telomerase is an effective treatment for which cancers?

2. What is the role of various telomere-interacting proteins in telomere maintenance?

3. Name alternative mechanisms (other than telomerase) for telomere maintenance.

4. Define the mechanisms underlying replicative senescence.

5. What is the potential "field effect" of senescent cells on other cell types during tumorigenesis?

6. What is the potential for inducing irreversible growth arrest in cancer cells as an effective therapeutic option?

Bibliography

Senescence

Beausejour, C.M., Krtolica, A., Galimi, F., et al. (2003). Reversal of human cellular senescence: roles of the p53 and p16 pathways. *EMBO Journal*, **22**: 4212–22.

Campisi, J. (2005). Senescent cells, tumor suppression, and organismal aging: good citizens, bad neighbors. *Cell*, **120**(4): 513–22.

Campisi, J. (2005). Suppressing cancer: the importance of being senescent. *Science*, **309** (5736): 886–7.

Campisi, J. (2010). Cellular senescence: putting the paradoxes in perspective. *Current Opinion in Genetics & Development*, November 17. [Epub ahead of print]

Cruickshanks, H.A., and Adams, P.D. (2010). Chromatin: a molecular interface between cancer and aging. *Current Opinion in Genetics & Development*, November 17. [Epub ahead of print]

Fridman, A.L., and Tainsky, M.A. (2008). Critical pathways in cellular senescence and immortalization revealed by gene expression profiling. *Oncogene*, **27**: 5975–87.

Fujita, K., Horikawa, I., Mondal, A.M., et al. (2010). Positive feedback between p53 and TRF2 during telomere-damage signalling and cellular senescence. *Nature Cell Biology*, **12**(12): 1205–12.

Fujita, K., Mondal, A.M., Horikawa, I., et al. (2009). p53 isoforms Delta133p53 and p53beta are endogenous regulators of replicative cellular senescence. *Nature Cell Biology*, **11**: 1135–42.

Hayflick, L., and Moorhead, P.S. (1961). The serial cultivation of human diploid cell strains. *Experimental Cell Research*, **25**: 585–621.

Janzen, V., Forkert, R., Fleming, H.E., et al. (2006). Stem-cell ageing modified by the cyclin-dependent kinase inhibitor p16(INK4a). *Nature*, September 6. [Epub ahead of print]

Krishnamurthy, J., Ramsey, M.R., and Ligon, K.L. (2006). p16(INK4a) induces an age-dependent decline in islet regenerative potential. *Nature*, September 6. [Epub ahead of print]

Lin, H.K., Chen, Z., Wang, G., et al. (2010). Skp2 targeting suppresses tumorigenesis by Arf–p53-independent cellular senescence. *Nature*, **464**: 374–9.

Martinez, I., Cazalla, D., Almstead, L.L., Steitz, J.A., and Dimaio, D. (2011). miR-29 and miR-30 regulate B-Myb expression during cellular senescence. *Proceedings of the National Academy of Sciences of the USA*, **108**(2): 522–7.

Narita, M., Narita, M., Krizhanovsky, V., et al. (2006). A novel role for high-mobility group A proteins in cellular senescence and heterochromatin formation. *Cell*, **126**(3): 503–14.

Nature. (2005). [Issue of August 4]. *Nature*, **436**: 7051.

Sage, J., Miller, A.L., Perez-Mancera, P.A., Wysocki, J.M., and Jacks, T. (2003). Acute mutation of retinoblastoma gene function is sufficient for cell cycle re-entry. *Nature*, **424**: 223–8.

Serrano, M., Lin, A.W., McCurrach, M.E., Beach, D., and Lowe, S.W. (1997). Oncogenic Ras provokes premature cell senescence associated with accumulation of p53 and p16INK4a. *Cell*, **88**: 593–602.

Takeuchi, S., Takahashi, A., Motoi, N., et al. (2010). Intrinsic cooperation between p16INK4a and p21Waf1/Cip1 in the onset of cellular senescence and tumor suppression in vivo. *Cancer Research*, **70**(22): 9381–90.

Vernier, M., Bourdeau, V., Gaumont-Leclerc, M.F., et al. (2011). Regulation of E2Fs and senescence by PML nuclear bodies. *Genes and Development*, **25**(1): 41–50.

Xue, W., Zender, L., Miething, C., et al. (2007). Senescence and tumour clearance is triggered by p53 restoration in murine liver carcinomas. *Nature*, **445**: 656–60.

Oncogene-induced senescence, and stem cells and stem cell renewal

Alison, M.R., Lim, S.M., and Nicholson, L.J. (2011). Cancer stem cells: problems for therapy? *Journal of Pathology*, **223**(2): 147–61.

Barker, N., Ridgway, R.A., van Es, J.H., et al. (2009). Crypt stem cells as the cells-of-origin of intestinal cancer. *Nature*, **457**: 608–11.

Baumann, M., Krause, M., and Hill, R. (2008). Exploring the role of cancer stem cells in radioresistance. *Nature Reviews Cancer*, **8**: 545–54.

Ben-Porath, I., Thomson, M.W., Carey, V.J., Ge, R., et al. (2008). An embryonic stem cell-like gene expression signature in poorly differentiated aggressive human tumors. *Nature Genetics*, **40**(5): 499–507.

Brabletz, S., Bajdak, K., Meidhof, S., et al. (2011). The ZEB1/miR-200 feedback loop controls Notch signalling in cancer cells. *EMBO Journal*, January 11.

Braig, M., Lee, S., Loddenkemper, C., et al. (2005). Oncogene-induced senescence as an initial barrier in lymphoma development. *Nature*, **436**(7051): 660–5.

Casas, E., Kim, J., Bendesky, A., Ohno-Machado, L., Wolfe, C.J., and Yang, J. (2011). Snail2 is an essential mediator of Twist1-induced epithelial mesenchymal transition and metastasis. *Cancer Research*, **71**(1): 245–54.

Chen, Z., Trotman, L.C., Shaffer, D., et al. (2005). Crucial role of p53-dependent cellular senescence in suppression of Pten-deficient tumorigenesis. *Nature*, **436**(7051): 725–30.

Christoffersen, N.R., Shalgi, R., Frankel, L.B., et al. (2010). p53-independent upregulation of miR-34a during oncogene-induced senescence represses MYC. *Cell Death and Differentiation*, **17**: 236–45.

Collado, M., Gil, J., Efeyan, A., et al. (2005). Tumour biology: senescence in premalignant tumours. *Nature*, **436**(7051): 642.

Gan, B., Hu, J., Jiang, S., et al. (2010). Lkb1 regulates quiescence and metabolic homeostasis of haematopoietic stem cells. *Nature*, **468**(7324): 701–4.

Gjerdrum, C., Tiron, C., Hoiby, T., et al. (2010). Axl is an essential epithelial-to-mesenchymal transition-induced regulator of breast cancer metastasis and patient survival. *Proceedings of the National Academy of Sciences of the USA*, **107**: 1124–9.

Graham, T.A., Jawad, N., and Wright, N.A. (2010). Spindles losing their bearings: does disruption of orientation in stem cells predict the onset of cancer? *Bioessays*, **32**(6): 468–72. [Review]

Gupta, P.B., Onder, T.T., Jiang, G., et al. (2009). Identification of selective inhibitors of cancer stem cells by high-throughput screening. *Cell*, **138**: 645–59.

Gurumurthy, S., Xie, S.Z., Alagesan, B., et al. (2010). The Lkb1 metabolic sensor maintains haematopoietic stem cell survival. *Nature*, **468**(7324): 659–63.

Hsu, Y.C., Pasolli, H.A., and Fuchs, E. (2011). Dynamics between stem cells, niche, and progeny in the hair follicle. *Cell*, **144**(1): 92–105.

Kelly, P.N., Dakic, A., Adams, J.M., Nutt, S.L., and Strasser, A. (2007). Tumor growth need not be driven by rare cancer stem cells. *Science*, **317**(5836): 337.

Krishnamurthy, J., Ramsey, M.R., Ligon, K.L., et al. (2006). p16INK4a induces an age-dependent decline in islet regenerative potential. *Nature*, **443**(7110): 453–7.

Liu, C., Kelnar, K., Liu, B., et al. (2011). The microRNA miR-34a inhibits prostate cancer stem cells and metastasis by directly repressing CD44. *Nature Medicine*, January 16.

Lopez-Garcia, C., Klein, A.M., Simons, B.D., and Winton, D.J. (2010). Intestinal stem cell replacement follows a pattern of neutral drift. *Science*, **330**(6005): 822–5.

Mani, S.A., Guo, W., Liao, M.J., et al. (2008). The epithelial–mesenchymal transition generates cells with properties of stem cells. *Cell*, **133**: 704–15.

Martello, G., Rosato, A., Ferrari, F., et al. (2010). A microRNA targeting dicer for metastasis control. *Cell*, **141**: 1195–207.

Michaloglou, C., Vredeveld, L.C., Soengas, M.S., et al. (2005). BRAFE600-associated senescence-like cell cycle arrest of human naevi. *Nature*, **436**(7051): 720–4.

Nakada, D., Saunders, T.L., and Morrison, S.J. (2010). Lkb1 regulates cell cycle and energy metabolism in haematopoietic stem cells. *Nature*, **468**(7324): 653–8.

Pece, S., Tosoni, D., Confalonieri, S., et al. (2010). Biological and molecular heterogeneity of breast cancers correlates with their cancer stem cell content. *Cell*, **140**: 62–73.

Quintana, E., Shackleton, M., Sabel, M.S., et al. (2008). Efficient tumour formation by single human melanoma cells. *Nature*, **456**: 593–8.

Raaijmakers, M.H., Mukherjee, S., Guo, S., et al. (2010). Bone progenitor dysfunction induces myelodysplasia and secondary leukaemia. *Nature*, **464**: 852–7.

Roesch, A., Fukunaga-Kalabis, M., Schmidt, E.C., et al. (2010). A temporarily distinct subpopulation of slow-cycling melanoma cells is required for continuous tumor growth. *Cell*, **141**: 583–94.

Rossi, D.J., and Weissman, I.L. (2006). Pten, tumorigenesis, and stem cell self-renewal. *Cell*, **125**(2): 229–31.

Sacco, A., Mourkioti, F., Tran, R., et al. (2010). Short telomeres and stem cell exhaustion model Duchenne muscular dystrophy in mdx/mTR mice. *Cell*, **143**(7): 1059–71.

Sanders, M.A., and Majumdar, A.P. (2011). Colon cancer stem cells: implications in carcinogenesis. *Frontiers in Bioscience*, **16**: 1651–62.

Shaker, A., Swietlicki, E.A., Wang, L., et al. (2010). Epimorphin deletion protects mice from inflammation-induced colon carcinogenesis and alters stem cell niche myofibroblast secretion. *Journal of Clinical Investigation*, **120**(6): 2081–93.

Sharma, S.V., Lee, D.Y., Li, B., et al. (2010). A chromatin-mediated reversible drug-tolerant state in cancer cell subpopulations. *Cell*, **141**: 69–80.

Shirakihara, T., Horiguchi, K., Miyazawa, K., et al. (2011). TGF-β regulates isoform switching of FGF receptors and epithelial-mesenchymal transition. *EMBO Journal*, January 11. [Epub ahead of print]

Snipper, H.J., van der Flier, L.G., Sato, T., et al. (2010). Intestinal crypt homeostasis results from neutral competition between symmetrically dividing Lgr5 stem cells. *Cell*, **143**(1): 134–44.

Takebe, N., Harris, P.J., Warren, R.Q., and Ivy, S.P. (2010). Targeting cancer stem cells by inhibiting Wnt, Notch, and Hedgehog pathways. *Nature Reviews Clinical Oncology*, December 14.

Vermeulen, L., De Sousa e Melo, F., van der Heijden, M., et al. (2010). Wnt activity defines colon cancer stem cells and is regulated by the microenvironment. *Nature Cell Biology*, **12**(5): 468–76.

Von Hoff, D.D., LoRusso, P.M., Rudin, C.M., et al. (2009). Inhibition of the hedgehog pathway in advanced basal-cell carcinoma. *New England Journal of Medicine*, **361**: 1164–72.

Vries, R.G., Huch, M., and Clevers, H. (2010). Stem cells and cancer of the stomach and intestine. *Molecular Oncology*, **4**(5): 373–84. [Epub 8 June 2010]

Wellner, U. (2009). The EMT-activator ZEB1 promotes tumorigenicity by repressing stemness-inhibiting microRNAs. *Nature Cell Biology*, **11**(12):1487–95. [Epub 22 November 2009]

Wend, P., Holland, J.D., Ziebold, U., and Birchmeier, W. (2010). Wnt signaling in stem and cancer stem cells. *Seminars in Cell & Developmental Biology*, **21**(8): 855–63.

Wu, Y., Cain-Hom, C., Choy, L., et al. (2010). Therapeutic antibody targeting of individual Notch receptors. *Nature*, **464**: 1052–7.

Yilmaz, O.H., Valdez, R., Theisen, B.K., et al. (2006). Pten dependence distinguishes haematopoietic stem cells from leukaemia-initiating cells. *Nature*, **441**(7092): 475–82.

Zhang, J., Grindley, J.C., Yin, T., et al. (2006). PTEN maintains haematopoietic stem cells and acts in lineage choice and leukaemia prevention. *Nature*, **441**(7092): 518–22.

Zhao, Z., Zuber, J., Diaz-Flores, E., et al. (2010). p53 loss promotes acute myeloid leukemia by enabling aberrant self-renewal. *Genes & Development*, **24**(13): 1389–402.

Telomeres

Blasco, M.A. (2002). Telomerase beyond telomeres. *Nature Reviews Cancer*, **2**: 627–33.

Blasco, M.A. (2005). Telomeres and human disease: ageing, cancer and beyond. *Nature Reviews Genetics*, **6**(8): 611–22.

Blasco, M.A., Lee, H-W., Hande, M.P., et al. (1997). Telomere shortening and tumour formation by mouse cells lacking telomerase RNA. *Cell*, **91**: 25–34.

Chin, L., Artandi, S.E., Shen, Q., et al. (1999). p53 deficiency rescues the adverse effects of telomere loss and cooperates with telomere dysfunction to accelerate carcinogenesis. *Cell*, **97**: 527–38.

Chin, K., Ortiz de Solorzano, C., Knowles, D., et al. (2004). In situ analyses of genome instability in breast cancer. *Nature Genetics*, **36**(9): 984–8.

d'Adda di Fagagna, F., Reaper, P.M., Clay-Farrace, L., et al. (2003). DNA damage checkpoint response in telomere-initiated senescence. *Nature*, **426**: 194–8.

Davoli, T., Denchi, E.L., and de Lange, T. (2010). Persistent telomere damage induces bypass of mitosis and tetraploidy. *Cell*, **141**(1): 81–93.

García-Cao, M., Gonzalo, S., Dean, D., and Blasco, M.A. (2002). Role of the Rb family members in controlling telomere length. *Nature Genetics*, **32**: 415–19.

González-Suárez, E., Samper, E., Flores, J.M., and Blasco, M.A. (2006). Telomerase-deficient mice with short telomeres are resistant to skin tumourigenesis. *Nature Genetics*, **26**: 114–17.

Goytisolo, F.A., and Blasco, M.A. (2001). Many ways to telomere dysfunction: in vivo studies using mouse models. *Oncogene*, **21**: 584–91.

Hemann, M.T., Strong, M.A., Hao, L.Y., and Greider, C.W. (2001). The shortest telomere, not average telomere length, is critical for cell viability and chromosome stability. *Cell*, **107**: 67–77.

Henson, J.D., Neumann, A.A., Yeager, T.R., and Reddel, R.R. (2002). Alternative lengthening of telomeres in mammalian cells. *Oncogene*, **21**: 598–610.

Karlseder, J., Kachatrian, L., Takai, H., et al. (2003). Targeted deletion reveals an essential function for the telomere length regulator Trf1. *Molecular and Cellular Biology*, **23**(18): 6533–41.

Lee, H.W., Blasco, M.A., Gottlieb, G.J., Horner, J.W., Greider, C.W., and DePinho, R.A. (1998). Essential role of mouse telomerase in highly proliferative organs. *Nature*, **392**: 569–74.

Munoz, P., Blanco, R., Flores, J.M., and Blasco, M.A. (2005). XPF nuclease-dependent telomere loss and increased DNA damage in mice overexpressing TRF2 result in premature aging and cancer. *Nature Genetics*, September 4. [Epub ahead of print]

Park, J.I., Venteicher, A.S., Hong, J.Y., et al. (2009). Telomerase modulates Wnt signalling by association with target gene chromatin. *Nature*, **460**: 66–72.

Sahin, E., Colla, S., Liesa, M., et al. (2011). Telomere dysfunction induces metabolic and mitochondrial compromise. *Nature*, **470**(7334): 359–65.

Schoeftner, S., and Blasco, M.A. (2010). Chromatin regulation and non-coding RNAs at mammalian telomeres. *Seminars in Cell & Developmental Biology*, **21**(2): 186–93.

Smith, L.L., Coller, H.A., and Roberts, J.M. (2003). Telomerase modulates expression of growth-controlling genes and enhances cell proliferation. *Nature Cell Biology*, **5**: 474–9.

Takai, H., Smogorzewska, A., and de Lange, T. (2003). DNA damage foci at dysfunctional telomeres. *Current Biology*, **13**: 1549–56.

Questions for student review

Mark all answers that apply.

1) Replicative senescence may be induced by:
   a. Telomere attrition.
   b. Inactivation of Rb signaling.
   c. Oncogenic Ras.
   d. DNA damage.
   e. Telomerase activity.

2) Telomeres:
   a. Protect the ends of chromosomes from DNA repair processes.
   b. Always shorten with every cell division in all cells.
   c. Are maintained entirely by action of telomerase.
   d. Can be maintained in the absence of telomerase.
   e. Are linear structures at the ends of chromosomes.

3) Telomerase:
   a. Is a reverse transcriptase.
   b. Comprises a catalytic and an RNA component.

c. In the adult is active only in stem cells and cancer cells.
d. Activity promotes senescence.
e. May be activated by c-Myc.

4) Cellular senescence:
a. Is always irreversible.
b. Might be reversed by mutations in the Rb pathway alone.
c. Might be prevented by c-Myc.
d. Can occur in the absence of telomere attrition.
e. Occurs only after 50 or more cell divisions.

5) Telomeres
a. Of adequate length are always effective at preventing genetic instability.
b. Require several interacting proteins in order to function effectively.
c. When critically short may cause premature aging.
d. When critically short always act to restrain cancer progression.
e. Are much longer in mouse cells compared to human cells.

10 Genetic Instability, Chromosomes, and Repair

Michael Khan
University of Warwick, UK

Without mutation, there can be no evolution. We may suffer for its faults, but if DNA were perfectly stable, we wouldn't even be here.
Phil Hanawalt

We share half our genes with bananas, something which is more apparent to me in some of my colleagues as opposed to others!
David Horobin

Key points

- Although DNA replication is usually carried out with fastidious precision, copying errors will creep in all the same simply because of the enormity of the task.
- Substantial DNA damage is also engendered under the combined auspices of extrinsic factors, such as ultraviolet (UV) light, ionizing radiation, and other genotoxic agents, that connive with radical elements from within, such as reactive oxygen species (ROS), to damage our DNA.
- Without efficient processes to monitor for and then repair the damaged or inaccurately copied DNA, our cells would not survive. From this, we can conclude that DNA damage responses (DDRs) are able to dramatically reduce the accumulation of damaged DNA even if they cannot prevent it entirely.
- Evolution and cancer may follow because, under some circumstances, DNA replication is sanctioned even though the DNA is improperly repaired. As a consequence, mutations are passed on to successive generations of:
  1. organisms, if the germ cells are involved; or
  2. successive generations of cellular clones, if mutations are confined to a somatic cell.
- An unresolved conundrum is why these cellular pariahs, with their potentially threatening unrepaired DNA damage, do not opt for one of the quixotic failsafe responses at their disposal and either commit suicide or permanently renounce replication. Moreover, the means by which a cell negotiates this final dilemma, apoptosis or senescence, are not fully understood.
- Another question that has long vexed cancer biologists is whether mutations arise by chance (stochastically) and are then selected for, or whether cancer development is consequent upon first acquiring a propensity to develop mutations. The answer for different cancers is likely a combination of both at different stages of progression.
- Aberrant DDRs arising through epigenetic or genetic changes may be present in premalignant colon cancer lesions, suggesting that, in this example of a multistage cancer pathway, a propensity to "epi-mutations" may come first.
- Sundry "caretaker genes" have evolved to tackle the surveillance and maintenance of DNA integrity; loss of these genes will give rise to cell death or to cells that survive with genomic instability and defective mismatch repair – a major cause or effect of carcinogenesis. In fact, around 200 proteins are now known to be involved in DDRs, and have to deal with substantive DNA damage and also an average point mutation rate of around 1.5×10^{-10} per base pair per cell generation.
- The caretaker genes are major barriers to the initiation and progression of cancer and, alongside the oncogenes and tumor suppressor genes, represent the third major class of genes subject to mutations and epigenetic silencing in cancer.
- Several, and arguably the most clinically important, hereditary cancer syndromes result from inactivating genes involved in DDR; these include, to name a few, *BRCA1* and *BRCA2* or *PALB2* in hereditary breast cancers, *ATM* in ataxia telangiectasia, and various mismatch repair (MMR) genes in Lynch syndrome–related colorectal cancers.

The Molecular Biology of Cancer: A Bridge From Bench to Bedside, Second Edition. Edited by Stella Pelengaris and Michael Khan.
© 2013 John Wiley & Sons, Inc. Published 2013 by John Wiley & Sons, Inc.

- DNA damage is either cytotoxic or mutagenic. Thus, severely damaged DNA, as exemplified by interstrand crosslinks and double-strand breaks (DSBs) that are not successfully repaired, will culminate in cell death because normal transcription, replication, and chromosome segregation are prevented, whereas lesser endurable damage that is not corrected before replication could cause miscoding resulting in mutations and subsequent carcinogenesis.
- By implication, the more extreme the degree of DNA damage that a cell can survive, the greater the risk of propagating a cancer-causing mutation.
- Conversely, apoptosis and senescence represent a *deus ex machina* if the situation cannot be resolved in any other way.
- DNA damage-signaling mechanisms seem to have evolved primarily to respond to cytotoxic lesions (and many of these overlap with those involved in telomere maintenance; see Chapter 9), whereas DNA repair mechanisms have evolved to deal with both classes of DNA lesions.
- Loss of telomeres results in tumors acquiring chromosomal re-arrangements with amplification and deletion of chromosomal regions (Chapter 9).
- DNA damage sensing likely takes place primarily during replication or transcription as the DNA strands may be more readily accessible to key "sensing" proteins at these times.
- DNA damage-signaling proteins, which are activated following detection of DNA damage, are the panjandrums of proliferation and are responsible for keeping damaged cells arrested at specific cell-cycle stages so that cells can attempt DNA repair before they are allowed to proceed to the S phase or mitosis.
- Once the DNA has been repaired and the crisis has blown over, replication can resume unabated. However, if effective DNA repair is impossible or damage is extensive, these signaling processes can also trigger irreversible growth arrest or apoptosis of the cell (see Chapters 8 and 9). The p53–p21^{Cip1} tumor suppressor pathway plays a central role in this.
- Recent work suggests that the DDR may also be a factor in regulating cell differentiation in stem cells.
- Four main types of DNA damage are recognized: base modification, often by methylation and de-amination; mismatches due to defective proofreading during replication of DNA; breaks in the DNA backbone (single-strand breaks (SSBs) or DSBs); and crosslinking between bases on the same DNA strand or adjacent strands.
- Assorted DNA damage repair mechanisms have been identified, but there is considerable overlap between them, and which is favored in a particular damage scenario is not always predictable: the simplest employs glycosylases and other enzymes to correct aberrant methylation and de-amidation; the remaining require excision of a damaged region and then repair, and these include base excision repair (BER), nucleotide excision repair (NER), and mismatch repair (MMR). DSBs are difficult to repair and require additional mechanisms because there is no intact template strand from which to restore a correct DNA sequence.
- DSBs are thus repaired by two major processes:
  1. In the S and G$_2$ phases, homologous recombination (HR), as suggested by the name, employs an homologous DNA molecule as a template and is therefore error-free;
  2. At any time of the cell cycle, nonhomologous end joining (NHEJ) joins up the two ends of DNA, giving an intact strand, but is self-evidently error prone, with the DNA sequence mostly altered. Maybe surprisingly, this more rapid method may be the default pathway, with overall choice determined by DNA-PKcs and BLM helicase among others.
- DNA mismatch repair (MMR) helps maintain genome stability by correcting DNA replication errors. MMR is initiated by mutSalpha (mutSα) or MutSbeta (mutSβ) dimers binding to mismatched DNA. Loss of MMR can drive a mutator phenotype and genetic instability. MMR increases the fidelity of DNA replication about 1000-fold.
- Although discussed in more detail in Chapter 11, ubiquitination (ubiquitylation) appears to be a key regulatory step in the recruitment of a variety of DNA repair proteins, including the core histone H2A and its variant H2AX in DSB repair.
- In response to diverse stresses, exemplified by DNA damage, the well-known tumor suppressor p53 prevents propagation of a potential cancer cell by inducing genes which arrest the cell cycle or promote apoptosis. Interestingly, p53 may also regulate genes required for successful DNA repair and moreover can reduce levels of ROS, thus protecting the genome from damage. This duality of p53 roles, as both protector and executioner, may in part be explained by the differences in genes regulated by p53 at low levels and at persistent or high levels, respectively. At low levels and early after activation, p53 mediates growth arrest, survival, and DNA repair, whereas high or persistent p53 activation mediates increased levels of p21^{Cip1}, ROS, and apoptosis.
- Failure of DNA damage responses, due to epi-mutations in "caretaker genes" or tumor suppressors, will allow the inappropriate survival and replication of cells with damaged DNA.
- Thus, once the usually impeccable fidelity of DNA replication is vitiated by, among other things, defective mismatch repair, further mutations can accumulate at a preternatural rate. This is a cardinal feature of cancer cells, culminating in one of the two major subtypes of genomic instability:
  1. Microsatellite instability (MSI), associated with a "mutator" phenotype in Lynch syndrome colorectal cancers;
  2. Chromosomal instability (CIN), recognized by gross chromosomal abnormalities.

Introduction

Human life ... is a sort of target – misfortune is always firing at it, and always hitting the mark.

Wilkie Collins

Given the intimate relationship between cell-cycle and DNA damage surveillance and repair (and, when this fails, DNA mutations), there is considerable overlap between this chapter and Chapters 3 and 4. Moreover, DNA damage represents the driving force behind tumorigenesis, and as such must contend with various cancer-restraining barriers such as apoptosis and senescence, which comprise the focus of Chapters 7, 8, and 9. With this comes a degree of repetition but with a different emphasis – we make no apology for this but ask you to indulge us as these are, after all, the central processes in tumorigenesis, and they are inextricably linked. This chapter will focus more on DNA damage and its repair and less on how the cell cycle is stalled in order for this to take place or what happens to cells if repair fails – this has been the focus of other chapters.

With around 2 m of DNA each, the 10^{14} cells that constitute the adult human contain a total length of DNA that if stitched end to end would be around 200 000 000 000 (that's 200 billion) km long and comprise a stretch of 10^{24} bases (6 billion base pairs per cell in the human genome), all subject to daily threat of damage and mutation. The Herculean task of patrolling this DNA, identifying damage, and then repairing it is delegated to each individual cell, which has to answer for its own 2 m and around 6×10^9 bases. Given that to all intents and purposes, all diploid cells contain the same DNA as each other, it might seem wasteful that this task could not in some way be coordinated between multiple cells. Well, in one circuitous way it is. DNA damage is most dangerous during cell replication (also, because DNA is unraveled, this is usually when it is detected) as the damage could be propagated, whereas even fatal damage to the DNA of most cells that cannot replicate is largely irrelevant to the organism – you have a myriad more cells, and the odd bad apple can usually be replaced. However, propagation of DNA damage can give rise to a new clone of cells that, if expanded, could compromise the survival of the whole organism. Coordination arises indirectly by means of one of the most startling examples of selfless cellular behavior observed in the context of a multicellular organism. Namely, a cell that determines that it has too much DNA damage to repair effectively either commits suicide or permanently declines to reproduce itself in order to serve the greater good – a cell will sacrifice itself rather than threaten the existence of the whole organism. In either case, the privilege of producing a new cell of the required type now falls to another cell without DNA damage, as long as it too can avoid and/or circumvent the various checkpoints already discussed in Chapter 4, a perfect outcome for the organism – *usually*. Not surprisingly, defects in any of these processes (detection, repair, apoptosis, or growth arrest) can result in cells with DNA damage replicating and propagating the defects – the essential platform for cancer.

It is always worth restating a key fact, and it is without apology that we repeat the mantra that **DNA damage is unavoidable**. Under the influence of extrinsic and intrinsic insults, mammalian cells are continuously acquiring DNA lesions. Extrinsic factors include exposure to UV light, anticancer drugs, and ionizing radiation, and intrinsic factors include reactive oxygen species (ROS), replication errors, and stalled replication forks. You can avoid some of the extrinsic causes, or mitigate them with sun cream, but nothing can prevent simple chance errors during DNA replication. On average we experience DNA damage at a rate between about 10 000 and 500 000 bases per cell per day, mainly due to chemical changes such as base loss by deamination or depurination and through oxidative damage. Fortunately, fewer than one in a thousand become permanent, because DNA repair mechanisms ensure that the genome remains intact, and normal cells can prevent the occurrence of mutations at the nucleotide sequence level and the chromosome level. These mechanisms include enzymes that repair damaged DNA, and signal transduction pathways (checkpoints) that induce cell-cycle arrest or apoptosis when individual stages in the cell cycle are not appropriately completed. In addition, cells with chromosomes not properly attached to the mitotic spindles are prevented from undergoing mitosis. In contrast to normal cells, most tumor cells acquire genomic instability resulting in multiple mutations – some of which may promote tumor development. The gene products that control genomic stability in normal cells have been defined to a certain extent, some of which will be discussed in this chapter. Not surprisingly, perhaps, defects in those genes that control genomic stability can contribute to tumor development.

DNA damage triggers a byzantine complex of signaling cascades which may be divided into (a) **DNA damage sensors**, (b) **checkpoint transducers**, (c) **checkpoint mediators**, and (d) **checkpoint effectors**. The checkpoints that are activated during the cell cycle when mutations are detected were discussed in depth in Chapter 4. Through the operation of these checkpoints, cells with damaged DNA are blocked from cell-cycle progression and DNA replication, DNA damage repair (DDR) processes can be recruited, and the damage is repaired. When DNA damage affects only one strand, then repair is far simpler as the normal strand can be used as a template, a nucleotide crib sheet, to make certain that the repair is accurate. If both strands are damaged, as in double-strand breaks (DSBs), then a more convoluted series of repair steps is required. In fact, the pervasiveness and variety of DNA damage on offer have resulted in the evolution of highly sophisticated systems to replicate and repair DNA accurately and efficiently or, if this fails, a variety of "damage limitation" mechanisms will kick in (Fig. 10.1). In other words, if the DNA cannot be fixed, then at least the potential cancer risk can be removed by the suicide or "celibacy" of the damaged cell. This does, however, leave the cell on the horns of a dilemma, the resolution of which may have profound implications to cancer biology and the development of targeted therapies. Dead cells cannot pose any future risk of cancer but must be replaced at some energy cost to the organism, whereas growth-arrested, or celibate, ones could conceivably recover their lost appetites over time.

A complete set of chromosomes in a human somatic cell contains around 6×10^9 base pairs of DNA. Although some of these comprise sequences representing genes and their regulatory elements, large stretches of DNA appear not to encode anything and are particularly prone to mutation (though, as discussed, some of this erstwhile "Junk" DNA is now known to encode, among others, microRNAs (miRNAs) that are subject to cancer-relevant mutations). The forces aligned against our genomes are legion, and DNA in every cell of our bodies is spontaneously damaged thousands of times every day. Specific sequences are particularly prone to this. For example, some repetitive DNA tracts such as microsatellites and mini-satellites are naturally unstable and, in

Cellular choices following DNA damage

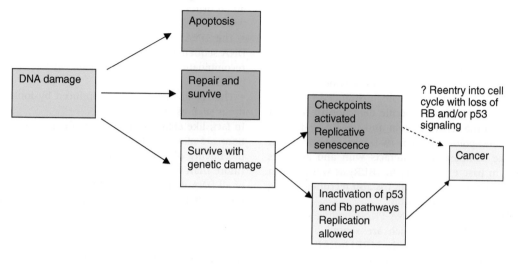

Cellular responses to DNA damage – molecular

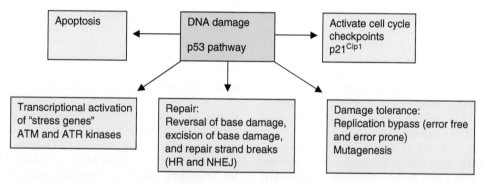

Figure 10.1 Cellular responses to DNA damage. Following DNA damage, the cell may undergo one of three fates: repair and survive, fail to repair, and apoptosis or replicative senescence. Rarely the cell with damaged DNA may still be able to replicate and thus pose a threat of cancer. The molecular processes activated by DNA damage are shown in the second part of the figure.

humans, alterations in these sequences are often associated with genetic diseases and cancer. There are many different types of DNA damage, including: conversion of bases by loss of an amino group (thus cytosine can be converted to uridine), mismatches during DNA replication (such as incorporation of uridine instead of thymine), breaks in the DNA backbone (either SSBs or DSBs), and crosslinking between bases on the same or opposite strands (may also be induced by chemotherapy).

To some extent, extrinsic DNA-damaging agents may be avoided or exposure to them minimized (sun screens and sun avoidance spring to mind). Importantly, it may also be possible to minimize the damage caused by some intrinsic factors; thus, in part the development of ROS may be reduced by healthy lifestyle choices, including diets low in saturated fat and cholesterol and high in anti-oxidants – dietary modifications are also convincingly linked to reducing risk of coronary heart disease. However, given that oxidative metabolism is a "nonnegotiable" and pivotal aspect of the life of a eukaryotic cell, any processes favoring release of superoxides from mitochondria may culminate in DNA damage, no matter how healthy the diet. During tumorigenesis, processes both intrinsic and extrinsic to the cancer cell can provoke oxidative stress. Thus, deregulated expression of oncogenes such as c-MYC can promote mitochondrial biogenesis and formation of ROS, which may be a contributory factor in the well-recognized propensity of oncogenic MYC to promote genomic instability. Recent studies suggest that neighboring stromal cells might indirectly provoke DNA damage in adjacent nascent cancer cells; matrix metalloproteinsases (MMPs) released by stromal cells may induce release of ROS from mitochondria in cancer cells.

Given that the DNA in a diploid human cell is very long (about 2 m in length), very convoluted, and very compacted (packaged into a compact chromatin structure by histones), monitoring and repairing damaged DNA are not trivial undertakings and involve a large and ever-expanding group of proteins encoded by what have been dubbed "caretaker" genes. But in an archetypal "snake eats tail" conundrum, these caretaker genes are themselves subject to mutations and epigenetic silencing. The resultant disruption of mechanisms that regulate cell-cycle checkpoints, DNA repair, and apoptosis result in genomic instability – a major contributor to tumorigenesis. Moreover, DNA damage responses take place within the context of the cell cycle and therefore affect and integrate with various checkpoints discussed in Chapters 4, 7, and 8. Not surprisingly, therefore, many tumor suppressor pathways, frequently subject to "epimutations" in cancer, such as the

ARF–p53–p21^{Cip1} and p16^{INK4a}–RB pathways, are also involved in DDR, alongside the caretaker gene products more specifically dedicated to this task.

Despite considerable redundancy and cross-talk, there are broadly two major pathways, plus a more recent addition, activated by different types of DNA damage:
- **The ATR–CHK1** pathway is triggered by a wide variety of insults, including but not restricted to single-stranded DNA and stalling of DNA replication forks. Specific detection of relevant DNA damage falls to the ATRIP protein and the RAD9–RAD1–HUS1 cell-cycle checkpoint complex (9-1-1 complex), which forms a clamp-like complex that interacts with and activates proteins involved in base excision repair (BER) as well as with various checkpoint proteins. ATM–CHK1 is also recruited during most other forms of DDR at some point.
- The **ATM–CHK2** (and parallel DNA-PKcs) pathway is triggered primarily by DNA DSBs, which are sensed by as-yet-unconfirmed means but likely include the MRE11–RAD50–NBS1 complex (MRN complex) and KU80 protein.
- A **MAPKAP kinase 2** (MK2) regulated pathway has been identified recently.

There are numerous examples of cross-talk, and it appears increasingly likely that the ATR–CHK1 pathway is in some way involved in most (if not all) DDRs. Thus MRN may activate some ATR responses, and it appears that, following initial ATM–CHK2 activation in DSB repair, ATR is also activated. It is worth noting that recognition of DSBs poses certain problems for the DDR; DSBs occur naturally during meiosis, so how do the ends of chromosomes avoid being recognized as a DSB, particularly in light of the presence of proteins involved in DDR, such as KU, at the telomere (discussed in Chapter 9).

The **sensors** (MRN and 9-1-1) detect DNA damage and recruit the **transducers**, ATM and ATR, which in turn activate further transducers, CHK2 and CHK1, respectively. ATM–ATR activation allows recruitment and activation of a plethora of mediators, including H2AX, 53BP1, Claspin, BRCA1, TopBP1, MDC1, SMC1, FANCD2, Timeless, and Tipin. These remain rooted to the site of damage, whereas CHK1 and CHK2 are liberated to unleash downstream signaling. The net result is the shackling of the cell cycle at several distinct phases by inactivation of CDC25s and/or activation of p53 effectors. Some of the complexity, the sequential rubs and bumps of partner proteins as they dance around the broken DNA, has been captured. Thus, RPA attaches to the damaged area and is joined by Timeless and Tipin, thereby enabling Claspin to associate with and phosphorylate CHK1, which in turn allows BRCA1 to join the melée. TopBP1 directly activates ATR–ATRIP and promotes ATR-mediated CHK1 phosphorylation.

ATM–ATR pathways appear to operate on two different time scales. Thus, growth-stalling inhibition of CDC25s proceeds rapidly, whereas the potentially more lethal p53-dependent pathway, which may be preferentially activated by nastier and difficult-to-fix DNA damage, is slower. This provides a period of respite during which repairs may be carried out before the suicidal urge becomes irresistible. Numerous other effectors have been identified, including a variety of kinases and repair proteins.

CHK1 can activate DNA repair kinases such as DNA-PKcs, which partners Ku70 and Ku80 in DSB repair. Phosphorylation of FANCE and Rad51 by CHK1 is needed for Fanconi anemia (FA)/BRCA-directed and homologous repair, respectively.

Genomic instability and cancer are held at bay by a series of different repair systems, but two are most frequently associated with cancer:
- The DNA mismatch repair (MMR) system, which corrects DNA sequence errors generated during DNA replication, during recombination, and by mutagenic agents, such as the drug *cis*-platinum;
- The repair of DNA DSBs, induced by ionizing radiation and drugs such as etoposide.

In fact, like ebbing tides and receding waves, small point mutations are eliminated by postsynthesis MMR, thereby improving the fidelity of DNA replication by several orders of magnitude (more than 100-fold!). DSBs are amongst the most indomitable of all DNA damage, and much space will be devoted to these in the chapter. Various proteins involved in mismatch and DSB repair also act to couple DNA repair to cell-cycle checkpoint regulation and apoptosis. It is worth noting here that promoting extreme genetic damage and instability with drugs or radiation may also paradoxically be a means of treating cancer; this is effective as long as growth arrest and/or apoptosis of cancer cells are appropriately triggered as a result. Not surprisingly, given everything that has been discussed regarding cancer cells so far, cancer cells are singularly accomplished at upregulating the efficiency of their DNA repair systems, thereby making them more capable of resisting DNA-damaging therapies. Both the ATM–CHK2 and ATR–CHK1 checkpoint pathways are activated after treatment with DNA-damaging agents; the relative contributions depend both on the agents used and on the nature of the damage ensuing. The result for the cell is either cell-cycle arrest (in the G_1, S, or G_2 and M phases) or death. These outcomes are all driven by various targets of the ATM–ATR kinases and include the tumor suppressor p53 (and in turn key p53 targets, in particular p21^{CIP1} for inducing growth arrest in G_1, and BAX, PUMA, and NOXA for promoting apoptosis) and the proapoptotic BCL-2 family member BID, which may mediate S-phase arrest (alongside various other ATM–ATR targets) as well as contribute to cell death and inhibition of CDC25, which causes arrest at the G_2 or M phase. Coordinated action between proteins involved in cell-cycle arrest, apoptosis, and DNA repair is evident at many levels, suggesting that a delicate balance between factors dictates if a cell with damaged DNA survives and repairs this or undergoes apoptosis, presumably with the survival option reviewed at various points in the process.

In the unlikely eventuality that you have somehow skipped over this, cancers are generally accepted as clonal in origin with the cancer cell arising through the stepwise accumulation of gene alterations in a process akin to Darwinian evolution and natural selection. At least for some cancers, it has been suggested that the iterative process of mutation and clonal evolution driving tumorigenesis may be compounded by various factors that together greatly increase the propensity of such nascent cancer cells to mutate. In fact, the prior acquisition of genomic instability is regarded by some as a *condicio sine qua non* in the progression of human cancer, because only then can mutations in other genes take place at a sufficient rate. It is worth noting that although cancers may be clonal, their chromosomes and genes may not be – in other words, genes and chromosomes may be quite heterogeneous, largely due to their inherent genomic instability.

However, the role of genomic instability in initiating cancers is still not fully elucidated. What is generally uncontested is that the majority of human cancers, when studied, exhibit genomic

Genetic Instability, Chromosomes, and Repair

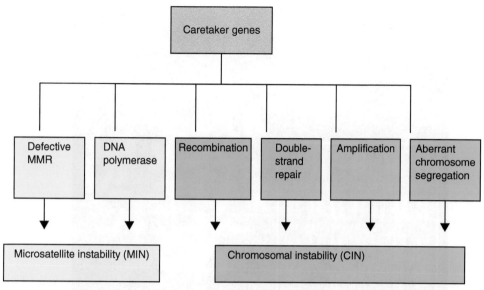

Figure 10.2 Potential causes of the "mutator" phenotype. Mutations in multiple pathways can result in the mutator phenotype in cancer cells.

instability. But when in the life of the cancer has this occurred, and is it a cause or effect of cancer? Possibly the best studied with regard to the role of genomic instability in patho-etiology are colorectal and breast cancers (discussed in this chapter). A number of recent studies using *in vitro* and *in vivo* models of breast cancer, among others, suggest that genomic instability (in some cases related to telomere shortening – see Chapter 9) might take place early or at the transition from benign to malignant cancer. Last year, two research projects employed sequencing to look at clonal populations in different cancers and concluded that genomic instability occurs early in the progression of cancer. They also demonstrated the remarkable genomic complexity of tumors, with the presence of multiple clones each of which continued to undergo parallel evolution in the primary and metastatic tumors. This is important as it suggests that natural selection does not eliminate all the older, less mutated clones; rather, cancers remain remarkably pluralist societies. In fact, all clones that endure continue to replicate and evolve. Whether these are general findings remains to be clarified, but we can now state that genomic instability plays a major role in the progression of most cancers at some stage in their life.

In addition to the acquired problems mentioned here, inherited factors may contribute to DNA damage and cancer. Thus, mutations or genetic variation at several alleles may either increase susceptibility to DNA-damaging agents such as UV light or compromise the effectiveness of DNA repair. The study of syndromes with inherited defects has greatly contributed to our understanding of how DNA repair is normally executed in mammalian cells. This will be further discussed in this chapter.

Types of genomic instability

Where you tend a rose, my lad, a thistle cannot grow.
Frances Hodgson Burnett, *The Secret Garden*

Most cancers are prone to mutations in DNA as well as more menacing gross chromosomal abnormalities. Genomic instability is not simply the presence of a number of defined mutations, as these are features of all cancers, but rather is the result of shortcomings in DDR that greatly increase the rate at which chromosomes and DNA are damaged and that allow cells with such damage to survive and to replicate. Genomic instability may result from abnormalities across a range of key cellular processes including cell-cycle regulation, DNA damage and repair, aging, and telomere function.

Genomic instability comprises several processes, of which chromosomal instability (CIN) recognized by gross chromosomal abnormalities (deletion and duplication of chromosomes or chromosome parts, re-arrangements, and aneuploidy) and microsatellite instability (MSI) (alterations in the length of short repetitive sequences – microsatellites) associated with a "mutator" phenotype (described in this chapter) have received the most attention (Fig. 10.2). These have already been discussed in the context of colorectal cancers (CRC) in Chapter 3. Aberrant chromosome number (**aneuploidy**) potentially affects cell populations that undergo multiple cell divisions, including yeast strains, cell lines, and tumor cells, and may result from genome instability. It has been appreciated for almost a century that advanced human cancers invariably contain cells with abnormal numbers of chromosomes, but the role of this in cancer pathogenesis and the likely causes are only recently being elucidated and in some cases remain controversial. The loss or gain of whole chromosomes results primarily from defects of segregation during mitosis, including chromosomal nondisjunction and failure of the mitotic (spindle) checkpoint pathways (Chapters 3 and 4). Normally, unattached kinetochores or those with inappropriate tension during mitosis activate this checkpoint and block mitosis. Mutations in key genes involved in this checkpoint, such as *NDC80* and *BUB1*, may disrupt the mitotic checkpoint and promote CIN. Loss of Rb or p53 may also cause CIN in model systems, possibly through upregulation of Mad2. RB and related pathways appear particularly important for avoiding CIN, as well as for cell-cycle regulation. Chromosome numbers can readily be visualized in skilled hands, by the application of a technique known as spectral karyotyping (SKY). An example of normal human chromosome

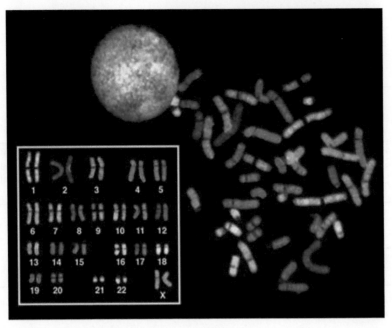

Figure 10.3 Normal Human Sky profile. Source: NIH Human Genome Resources.

numbers is shown in Fig. 10.3. Conversely, polyploidy (an increase in chromosome number) in mouse cells is shown in Fig. 10.4. Other types of genomic instability are characterized by an increased rate of small-scale mutations such as MSI. Amplification of individual loci on chromosomes is another form of genetic instability that causes overexpression of proteins, as often occurs for the oncogene c-MYC (Chapter 6).

A recent study from the laboratory of Titia de Lange has suggested a potential mechanism linking telomere attrition with tetraploidization, a putative precursor to aneuploidy in human cancers. This can be induced by persistent telomere dysfunction if the p53 pathway is defective. Despite undergoing a protracted G_2 phase due to activation of both ATR–CHK1 and ATM–CHK2 pathways and missing out on mitosis, cells still degrade the replication inhibitor geminin (by APC/C) and accumulate the origin licensing factor Cdt1. This scenario allows a second S phase and tetraploidy. Worryingly, if telomere protection is restored, such tetraploid cells proliferate.

As you will no doubt have become accustomed to reading by now, there is a lively debate in progress regarding whether genetic instability is a cause or effect of cancer. The CIN hypothesis contends that aneuploidy is the catalyst for transformation, whereas the gene mutation hypothesis asserts that cancer is driven by mutations to proto-oncogenes and tumor suppressor genes, with aneuploidy a side effect of tumorigenesis. The role of point mutations in human cancer is well established, and MSI is widely accepted as causal in tumorigenesis. However, the contribution of massive genomic changes resulting from CIN and aneuploidy is less certain. Aneuploidy is required for sporadic carcinogenesis in mice and may even collaborate with specific gene mutations during tumorigenesis. It has been argued that CIN contributes to cancer initiation because chromosome loss can provide the second hit in loss of a tumor suppressor gene. However, at the same time, CIN is costly for the cell because it destroys the genome and therefore compromises clonal expansion. Interestingly, either CIN or MSI alone may be a sufficient driver of tumorigenesis as, in general, individual tumors manifest one or the other but not both.

Mutator phenotype

The concept of a so-called mutator phenotype of cancers developed from the observation that mutations are rare in normal cells but typical in cancer cells (Fig. 10.2). The mutator phenotype model proposes that the mutation rate in the early stages of tumorigenesis must be greater than the normal spontaneous mutation rate of human somatic cells (1.5×10^{-10} per base pair per cell generation), because this mutation rate is deemed insufficient to produce the multiple mutations identified in many cancers. Note that the mutation rate is much less than the rate of DNA damage, due to the impact of DNA repair. The mutator phenotype implies that genetic instability provides the impetus for accumulation of multiple mutations by the cancer cell. However, only a small number of such mutations may actually confer a growth advantage (and will be selected for), whereas the rest (possibly the majority) have no bearing on the cancer and are merely "baggage" that is not selected for. This differs from the traditional view of cancer progression through clonal evolution driven by mutation and natural selection in so far as the traditional view implies that most mutations will confer some growth advantage and are selected for individually (or in small numbers) – there is little "baggage" in this model. Whatever the answer, most cancers probably develop through a combination of stochastic mutations (and epigenetic factors) as well as those arising through genomic instability (and maybe the epigenetic equivalent, the methylator phenotype, discussed in Chapter 11), which are then subject to natural selection and clonal expansion. However, as discussed in this chapter, particularly in colorectal and breast cancers there is increasing evidence that genetic instability may occur at the transition from premalignant to malignant tumors.

All this interesting debate notwithstanding, the most important issue (and probably one of the most important questions in the whole of cancer biology) is whether the plethora of mutations

Genetic Instability, Chromosomes, and Repair

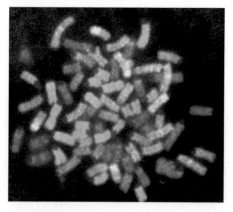

Normal mouse chromosomes

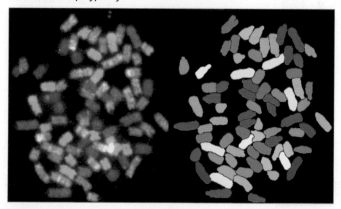

Chromosomal polyploidy

Figure 10.4 Polyploidy in mouse cells as shown by spectral karyotype (SKY). Images courtesy of Maria Blasco. SKY provides visualization of all an organism's chromosomes together, each labeled with a different color, and is readily employed to determine abnormalities in chromosome number. Left: The colors of the different chromosome pairs as seen under microscopy. Right: The polyploid cells have been artificially colored to more easily show the abnormal chromosome numbers – several chromosomes are present in excess numbers (polyploidy).

observed in tumors are all mission-critical or rate-limiting for tumor growth or whether many of the mutations are irrelevant to both the evolution and maintenance of the tumor. In other words, are most mutations in established cancers associated with tumorigenesis in a noncausal way? This has a direct bearing on the design of cancer therapies – new therapeutics should be designed to target cancer-relevant genes or proteins, which must first be distinguished from the rest.

The notion of a "mutator" phenotype is based on seminal studies with one form of a progressive multistage human cancer, a familial colon cancer known as hereditary nonpolyposis colon cancer (HNPCC). In HNPCC, mutation in one allele of the genes involved in DNA mismatch repair is inherited. Loss of the second wild-type allele occurs, frequently promoting the formation of colorectal carcinoma which exhibits MSI. Even sporadic colon cancers often exhibit MSI, and this is an almost ubiquitous finding in HNPCC. Firstly, described in the bacterium *Escherichia coli* (*E. coli*), the mismatch repair genes *mutS* or *mutL* were implicated in MSI, and subsequently mutations in two homologous human genes in the DNA mismatch repair pathway were found in HNPCC by linkage studies (Chapter 18): *MSH2*, encoded on chromosome 2, or *MLH1*, encoded on chromosome 3. These genes may also be inactivated or suppressed by epigenetic factors, though interestingly in human cancers this seems to involve the *MLH1* gene (Chapter 11). MSI occurs in several other nonhereditary (sporadic) cancers including, pancreas, colon, and ovary. It is likely that many other genes not necessarily involved in mismatch repair may also contribute to MSI. The human genome comprises vast numbers of microsatellites, but instability also manifests itself in the coding regions of various growth-regulating genes, including apoptosis regulators such as BAX. Thus, MSI may be a "readily measurable" marker of the existence of mutagenic mechanisms, but it must be noted that most sporadic tumors do not display high levels of MSI.

In the case of **entosis**, aneuploidy may arise by other mechanisms than the failure of cytokinesis due to aberrant mitotic pathways or checkpoints. Sometimes live cells may become internalized by another cell, entosis, resulting in the bizarre appearance of a cell within a cell. Such internalized live cells can persist throughout the cell cycle of the hapless host and cause severe disruption to cytokinesis, thereby resulting in binucleate cells that will father aneuploid cell lineages. Multinucleated cells in human breast cancers sometimes demonstrate such cell-in-cell structures.

Telomere attrition and genomic instability

The cowl makes not the monk.

Walter Scott

Increasing evidence indicates that dysfunctional telomeres likely play a causal role in the process of malignant transformation, in at least a fraction of human cancers, by initiating chromosomal instability. Critical telomeric shortening can lead to telomere "uncapping" and may occur at the earliest recognizable stages of malignant transformation in epithelial tissues. The widespread activation of the telomere-synthesizing enzyme telomerase in human cancers not only confers unlimited replicative potential but also prevents intolerable levels of chromosomal instability. See Chapter 9 for a detailed discussion of this area. There is a considerable overlap between DNA damage responses, particularly to DSBs, and proteins involved in telomere maintenance – see the "Double-strand breaks" section of this chapter.

The DNA damage response

The life of every man is a diary in which he means to write one story, and writes another; and his humblest hour is when he compares the volume as it is with what he vowed to make it.

J.M. Barrie

It is all too easy to assume that DNA synthesis is a slipshod affair, littered with errors in spelling and punctuation and successful only because of the imposition of intensive editorial scrutiny and amendment. At this point, I perceive a certain resonance with our own book editor! In fact, errors are remarkably few but are rendered inevitable only because of the enormity of the task.

The human genome, comprising 3 billion base pairs (remember that there are two copies in a somatic cell – so 6 billion bases per cell in total) coding for 30 000–40 000 genes, is under constant

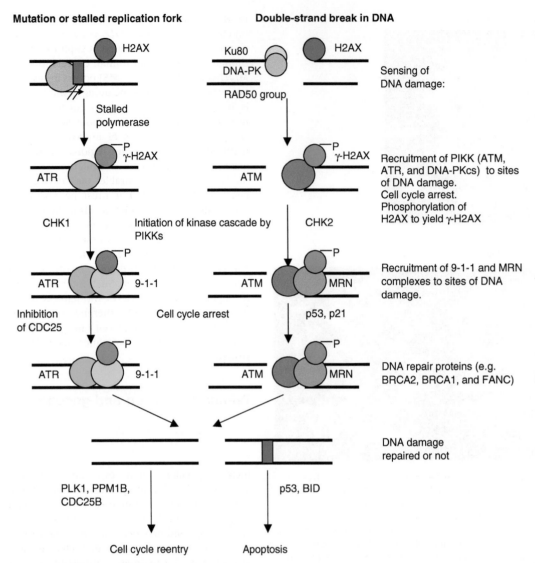

Figure 10.5 DNA damage and repair signaling. Signaling mechanisms are activated by DNA damage sensors, ultimately leading to a variety of potential cellular responses discussed earlier.

assault by endogenous reactive metabolites, therapeutic drugs, and a plethora of environmental mutagens that impact its integrity. Fortunately, the stability of the genome is under continuous surveillance by components of the DDR that will be described in more detail in this section. DNA repair mechanisms have evolved to remove or to tolerate pre-cytotoxic and pre-mutagenic DNA lesions in an error-free (or, in some cases, error-prone) way. Defects in DDR give rise to hypersensitivity to DNA-damaging agents, the accumulation of mutations in the genome, and finally the development of cancer. The importance of DNA repair is illustrated by DNA repair deficiency and genomic instability syndromes (described in this chapter), which are characterized by increased cancer incidence and multiple metabolic alterations. DNA damage, as outlined in Chapter 4, activates two major overlapping networks of proteins with two key apical PI3K-related proteins – the ataxia-telangiectasia mutated (ATM) and ATM- and Rad3-related (ATR) kinases. ATR and ATM, their respective substrates CHK1 and CHK2, as well as the molecules required for recruitment of ATR and ATM to sites of DNA damage, such as histone γH2AX and BRCA1 (see Figs 10.5 and 10.6), are all critical in activating checkpoints (and thereby key proteins involved in these checkpoints as described in Chapters 4 and 7).

The ATM–CHK2–p53 pathway responds to more substantive damage such as DNA DSBs, whereas the ATR–CHK1 pathway is activated by less extensive DNA damage, including mutations and lesions causing the stalling of polymerases. However, there is considerable overlap and redundancy. A simplified schematic of DNA damage signaling and repair is shown in Fig. 10.5. The effect of such DNA damage response signaling on the cell is to activate cell-cycle checkpoints leading to arrest in the G_1 or G_2 phase in order for DNA to be repaired before the cell cycle can continue. In certain cases, such as when DNA damage is extensive and/or irreparable, the outcome will be cell death (apoptosis) or irreversible growth arrest (senescence). The development of the concept of cell-cycle checkpoints is attributed to the pioneering work of Lee Hartwell and others starting in the 1960s and has been discussed in Chapter 4. However, given the key role played in avoidance of CIN, the mitotic (spindle) checkpoint will be discussed later in this chapter also. The DNA damage checkpoints coordinate a block in cell proliferation with the DNA repair

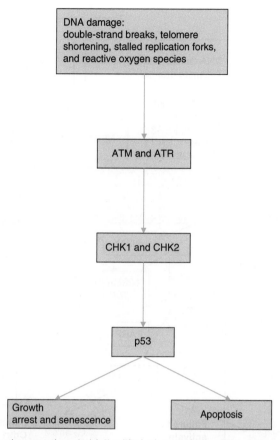

Figure 10.6 DNA damage-induced signaling pathways and repair. (**a**) Simplified schemata for key DNA damage-induced signaling pathways. Given that p53 functions in the checkpoint response to DSBs, it is not surprising that there is a selective pressure for premalignant tumor cells to mutate and thus inactivate the p53 gene – as is the case for most human cancers. The signaling pathways involved in DNA damage responses are complex and most certainly still incompletely described, but the major proteins involved are detailed here and in more complex detail in (**b**). Although it is known that various targets of ATM- and ATR-mediated phosphorylation participate in transmitting the DNA damage signal to CHK1/2 proteins, much is still not known.

process. However, once repair is accomplished (if possible), then the checkpoint is "released" and cell division can recommence; Polo-like kinase-1 (PLK1) and Cdc25B are essential components in this reactivation of cell division.

This is a complex and rapidly progressing area of cancer research, and one may be forgiven for approaching a Gordian knot with less trepidation. However, there are rewards awaiting the patient. The basic principles are straightforward and increasingly of very direct relevance to clinical practice. This is illustrated by recent studies which suggest that DDR may be invoked in very early human tumors (even before genomic instability and malignant conversion). Moreover, it is apparent that the DDR is intertwined with general stress responses and with what has been dubbed "oncogenic stress" in particular. Broadly, oncogenic stress may arise due to deregulated activity of a single oncogene, such as c-MYC or RAS, and trigger a failsafe response to eliminate or "hogtie" these potentially dangerous cells – see the "Oncogenic Stress" section of this chapter as well as Chapter 6.

Recent studies highlight the increasing complexity of the DDR, with distinct processes for differing types of damage and also common shared pathways. DNA DSBs are by far the most pernicious type of DNA lesion because, if inefficiently or inaccurately repaired, they can drive genomic instability and cancer. However, DSBs may also result if simpler single-strand lesions are not repaired or replication forks become stalled. Eukaryotic cells have two conserved mechanisms to detect and repair DSBs; homologous recombination (HR) repairs the break using genetic information recovered from the undamaged sister chromatid or chromosomal homologue, whereas nonhomologous end joining (NHEJ) involves the direct ligation of DNA ends. Various proteins are involved in ensuring that chromosomes have been accurately partitioned during mitosis, including those operating in the spindle checkpoint described in this chapter. Mutations in the "caretaker genes" encoding many of these proteins involved in DNA repair and in chromosome segregation and apoptosis have all been implicated in genomic instability. In fact, the caretaker genes represent the third major class of genes mutated in cancer, alongside the oncogenes and tumor suppressor genes.

Sensing DNA damage

Proofread carefully to see if you any words out.

William Safire

A major focus of recent attention has been the rather tricky issue of how the cell senses DNA damage. As mentioned already, the DNA is normally tightly packed into chromatin by histones – so how is the damaged DNA accessible to the damage-sensing machinery? An expanding body of evidence suggests that sensing

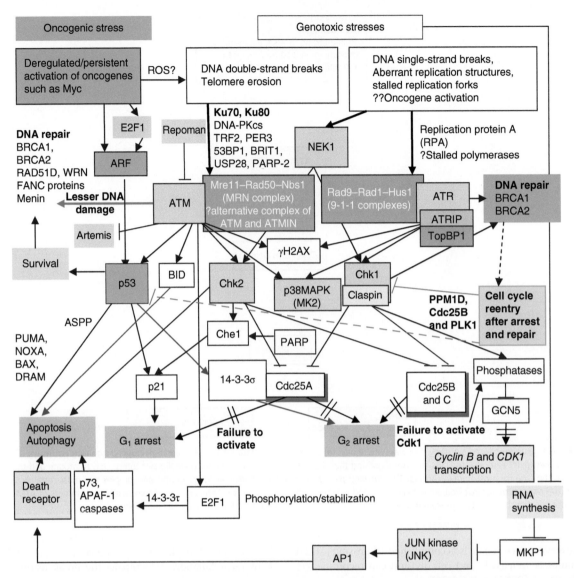

Figure 10.6 (Continued) (b) Complex regulation of DNA damage-induced signaling and repair. Signaling involved downstream of different types of DNA damage. To combat recurrent threats of genomic instability, numerous distinct enzyme systems sense DNA damage and coordinate its repair. Part of this coordination involves the activation of signal transduction cascades that target repair proteins, trigger DNA damage-dependent cell cycle checkpoints, and profoundly affect chromatin neighboring a DSB. Ultimately DNA damage results in cell-cycle arrest and attempted repair. If repair is successful, the cell will survive and may re-enter the cell cycle. There is obvious overlap between telomere maintenance and DNA damage response. Proteins directly involved in telomere maintenance and DDR include Ku, DNA-PKcs, RAD51D, PARP-2, WRN, and the MRE11–RAD50–NBS1 complex. The kinases ATM and ATR are key mediators of the DNA damage response and activate downstream effectors such as yH2AX, CHK2, and CHK1. Ultimately the activity of these kinases results in activation of p53 and BID and inhibition of CDC25, leading to growth arrest or apoptosis. Recent studies suggest that P53CSV and/or 14-3-3σ may prevent apoptosis if DNA damage is repairable; and activation of another protein, PPM1D, may allow cell-cycle reentry if DNA damage is repaired. TP53-induced glycolysis and apoptosis regulator (TIGAR) can inhibit glycolysis and production of reactive oxygen species (ROS) levels, both of which contribute to reduction of apoptotic response to p53, potentially also allowing survival in the face of mild or transient stress signals that may be reversed or repaired. Conversely, proteins of the ASPP (ankyrin-repeat-, SH3-domain-, and proline-rich-region-containing proteins) family may preferentially promote the apoptotic activity of p53. 14-3-3σ is a p53 target which may also help maintain G_2 arrest. Other 14-3-3 family members may sequester CDC25 and CDC2. Activation of p53 also occurs in response to oncogenic stress (deregulated oncogene activity of e.g. c-Myc), though uniquely in this case p53 is activated by ARF, which displaces the MDM2 negative regulator of p53. In addition to apoptosis, p53 is increasingly shown to also be involved in other forms of cell death. Thus, p53 can induce autophagy, and this is mediated in part at least by damage-regulated autophagy modulator (DRAM), a p53 target gene that encodes a lysosomal protein. DRAM is also essential for p53-mediated apoptosis in response to DNA damage.

may actually take place during the two occasions when DNA is normally stripped of its histone coat and the two strands are separated, namely, during either DNA replication or transcription. It might be argued that as DNA replication occurs during only the S phase of the cell cycle, whereas gene transcription is continuous (except during mitosis, when chromatin is condensed and gene expression is silenced), the latter would be a more suitable occasion to perform such damage sensing. As discussed in Chapter 3, eukaryotic gene transcription involves three different RNA polymerases: RNA polymerase I transcribes ribosomal DNA into ribosomal RNA for ribosome formation, RNA polymerase II transcribes genes into mRNA, and RNA polymerase III synthesizes transfer RNA and small nuclear RNA. Of these, RNA polymerase II reads a much greater proportion of the genome than the others and may moreover recruit DNA-repair proteins to undertake a form of nucleotide excision repair (NER) termed transcription-coupled repair (TCR). This is essential because if RNA polII progress remains stalled because DNA repair is not accomplished, then the cell may trigger other DNA damage sensors, activate the p53 pathway already described in Chapter 7, and even undergo apoptosis.

Potential candidates for the actual DNA damage sensors are discussed under the different types of DNA damage in this chapter, but they include, among other candidates, the so-called 9-1-1 complex (comprising RAD1–RAD9–Hus1) and ATRIP, the MRN complex (MRE11–RAD50–NSB1) and KU80, and the core MMR recognition protein complex, human mutS homolog 2 (hMSH2) and 6 (hMSH6). However, it is still unclear how stalling of DNA or potentially RNA polymerases may activate the DDR. Some proteins involved in transducing the DNA damage signal are described in detail in the "Signaling DNA damage" section.

Signaling DNA damage

Omnis traductor traditor.

Not surprisingly, mutations that compromise the DDR (e.g. in the ATR–CHK1 or ATM–CHK2–p53 pathway) could propel cell proliferation, survival, increased genomic instability, and thus tumor progression. Examples of such mutations occurring in human cancers are described in this chapter. The labyrinthine signaling pathways involved in the DDR are complex and still incompletely described, but the major proteins involved are detailed in Fig. 10.6. Although it is known that various targets of ATM- and ATR-mediated phosphorylation participate in propagating the DNA damage signal to CHK1 and CHK2 proteins, much is still not known. There is redundancy in the system, and both CHK1 and CHK 2 can inactivate CDC25, thus leading to cell-cycle arrest. Moreover, both CHK2 and ATM may act on the same substrates, providing a potential salvage mechanism if either were mutated. Phosphorylation of the histone variant H2AX by CHK proteins yields γH2AX, which is essential for retention (if not recruitment) of the MRN complex which rapidly accumulates at sites of DNA damage. The MRN complex is a central performer in human genome maintenance. It tethers the broken DNA ends across a strand break and then recruits various repair proteins needed to correctly process the DNA ends. Other important proteins, such as BRCA1, MDC1, and 53BPI (p53 binding protein 1), are also recruited following H2AX phosphorylation. Chromatin modifications are strongly associated with the repair of DSBs, and γH2AX helps prevent aberrant repair of damaged DNA. Mice deficient for both H2AX and p53 rapidly develop tumors. Moreover, even H2AX haploinsufficiency causes genomic instability in normal cells and, on a p53-deficient background, leads to the early onset of lymphoma. H2AX also maps to a cytogenetic region frequently altered in human cancers. Other histone modifications such as acetylation of histone H4 also have an important role in the response to DSBs, suggesting that DNA damage-induced histone modifications are not confined to phosphorylation of H2AX.

Although we know much about the target proteins phosphorylated by ATM, we know little about how ATM is itself phosphorylated and activated, though p53BP1 may be involved. The tumor suppressor protein p53 is a key target of the ATM–CHK2 pathway. Activated p53 induces cell-cycle arrest, which is described in detail in Chapter 7. CHK2 may also activate $p21^{CIP1}$ independently of p53, contributing to cell-cycle arrest in G_1. Given that the p53 pathway functions in the checkpoint response to DSBs, it is not surprising that there is a selective pressure for premalignant tumor cells to mutate and thus inactivate the p53 gene – as is the case for most human cancers.

14-3-3 proteins

These bind to and influence the activities of a diverse group of molecules involved in signal transduction, cell-cycle regulation, and apoptosis, including p53, RAF, PKC, and BAD. Interactions between 14-3-3 and target proteins are strongly influenced by the phosphorylation state of 14-3-3 and the target protein. The 14-3-3 proteins are remarkably promiscuous in their interactions with target proteins in part due to their ability to associate with both RSXpSXP and RXY–FXpSXP (pS=phosphoserine). In mammalian cells, seven highly conserved isoforms with distinct functions have been identified (14-3-3β, γ, ε, η, σ, τ/θ, and ξ). Several are involved in the G_2–M cell-cycle checkpoint; 14-3-3σ, a direct transcriptional target of p53, prevents initiation of mitosis by sequestering cyclin B1–CDK complexes in the cytoplasm, whereas the β and ε isoforms bind Cdc25C. The expression of 14-3-3σ may facilitate DNA damage repair by preventing apoptosis and by arresting cells in G_2. 14-3-3σ is silenced in a large number of invasive breast cancers and may operate as a tumor suppressor in humans. During a DDR, 14-3-3β and ζ actively promote the retention of activated CHK1 in the nucleus. 14-3-3γ is important for the ability of Chk1 to phosphorylate Cdc25A and target it for degradation.

Repairing DNA damage

We are the products of editing, rather than of authorship.

George Wald

In humans, at least 150 genes encode proteins involved in DNA repair. Various intracellular mechanisms are involved in repairing subtle mistakes made during normal DNA replication or following exposure to DNA-damaging stimuli. Damage due to alkylation can be reversed directly by a chemical process involving genes such as MGMT (no breaking of the DNA backbone is needed). But probably the most important mechanisms are the three excision repair processes – nucleotide excision repair (NER), base excision repair (BER), and mismatch repair (MMR) – all of which may contribute to avoiding MSI and a mutator phenotype. As an illustration of the likely importance of these mechanisms, it has been estimated that on average every somatic cell in humans may need to remove around 10 000–20 000 damaged bases or more every day. Fig. 10.7 summarizes the particular repair mechanisms and the type of damage to which they respond.

Fragile sites

Certain parts of the genome appear particularly vulnerable to damage. Chromosomal fragile sites, which are genomic regions prone to gaps or breaks in metaphase chromosomes during partial replication stress, are particularly vulnerable to DNA breaks and may contribute to chromosomal aberrations in cancer. This view is supported by the frequent identification of tumor suppressors or oncogenes, such as MYC and MET, at fragile sites;

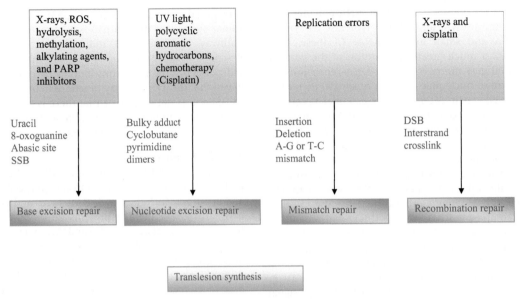

Figure 10.7 Various DDR pathways available for the repair of different types of DNA damage. If normal repair fails, tranlesion synthesis may prevent collapsed replication forks and resultant strand breaks.

the preferential integration of some oncogenic viruses, such as HPV, Hepatitis B, and EBV, at these sites; and the contribution to formation of chromosomal translocations, such as that found in human papillary thyroid carcinoma.

Fragile sites are a normal feature of chromosome architecture, but rare fragile sites can also be inherited. These sites differ in susceptibility to induction by different genotoxic agents and environmental factors. The ATR checkpoint pathway and downstream targets, such as CHK1 and BRCA1, are crucial for fragile site maintenance, suggesting that exposure of single-stranded DNA (ssDNA) during replication stress may activate this checkpoint.

Intriguingly, several putative carcinogens can increase common fragile site breakage in conjunction with agents such as APH and include caffeine, an inhibitor of ATR and ATM; ethanol; cigarette smoking; and pesticides and various other mutagens and chemotherapeutic agents (whose action is enhanced by the addition of caffeine). Recent studies have further supported a role for fragile sites and cancer-specific chromosomal aberrations. Thus, DNA breakage at fragile sites can lead to the formation of *RET–PTC* rearrangements and also deletions within the *FHIT* gene similar to those seen in the relevant human cancers.

The mechanisms of DNA repair

Various processes contribute to repairing different types of DNA damage and are described in this section.

Single-stranded DNA lesions

These are the commonest DNA lesions and may arise by a variety of insults ranging from alterations in a single base through more substantial structural alterations in DNA and chromatin. Unlike DSBs, which are discussed in this chapter, damage is confined to one strand. In general this enables repair by using the normal intact strand as a template, though an error-prone mechanism, known as translesional synthesis (TLS), allows DNA replication to proceed across a damaged template. Several distinct repair processes used to repair single-strand lesions are discussed here.

Base excision repair (BER)

Single DNA bases may undergo damage that compromises base pairing by several different chemical reactions, including deamination, oxidation, and alkylation. One noteworthy base substitution results from the incorrect incorporation of adenine opposite to an aberrantly oxidized base, 8-oxoguanine. This results in the conversion of a normal G–C base pair to a T–A one. Another important example is when uracil is incorrectly incorporated into DNA through deamination of cytosine. There are overlaps between BER and repair of SSBs, which can arise directly or as intermediates in BER. In fact, breaks in a single strand of the DNA molecule are repaired using the same enzyme systems that are used in BER.

Two main types of BER are described in humans: a DNA polymerase (Pol) β-dependent pathway and a proliferating cell nuclear antigen (PCNA)-dependent one (also referred to as short- and long-patch repair, respectively). The mechanism of choice may be influenced by various factors including the cell-cycle stage and how resistant the damage is to pol β lyase activity, which is required for short-patch repair. BER (Fig. 10.8) requires removal of the damaged base by one of several DNA glycosylases responsible for identifying and removing specific base damage, followed by activity of other enzymes that drive the removal of the deoxyribose phosphate in the backbone, producing a gap. The DNA glycosylase family of enzymes is important in the first step of BER: they remove damaged nitrogenous bases, often without altering the sugar–phosphate backbone. Glycosylases recognize specific DNA damage and then effectively flip out the damaged purine or pyramidine base and cleave the N-glycosidic bond to generate a nucleotide-free AP site. Different glycosylases recognize different lesions; thus some, such as members of the UNG (uracil–DNA–glycosylase, to give it its full name) superfamily, are specific for removal of uracil, and some for oxidized bases. A defective function of one 8-oxoguanine glycosylase, MUTYH, is associated with CRCs in humans and is recognized as a rare variant conferring inherited susceptibility to CRCs. One group of glycosylases, referred to as bifunctional, also acts as AP lyases and

Genetic Instability, Chromosomes, and Repair

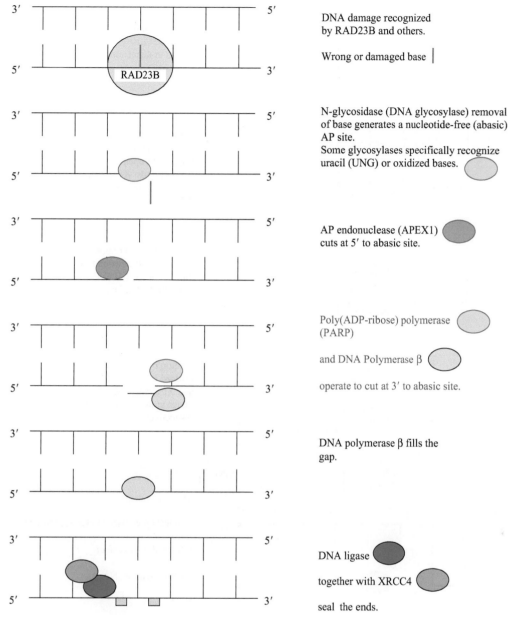

Figure 10.8 Base excision repair.

can cut the phosphodiester bond to generate a SSB. Glycosylases without this additional activity require the additional activity of an AP endonuclease, such as the human AP endonuclease 1 (APE1). Thymine DNA glycosylase (TDG) is in the UDG family and can remove thymine when mispaired with guanine, but may also interact with various enzymes involved in epigenetic alterations, including methylation, and in regulating gene expression during cell differentiation. Whether other glycosylases will turn out to have such wide-ranging roles is not known.

In humans, polynucleotide kinase–phosphatase (PNKP) phosphorylates 5′ hydroxyl ends and cleaves phosphates from 3′ ends in order to promote the formation of DNA strand breaks with a hydroxyl at the 3′ end and a phosphate at the 5′ end, which is necessary prior to ligation. The AP endonucleases, Endo1 in humans, open AP sites and cleave 3′ lesions to ensure the availability of a 3′ hydroxyl group for DNA polymerases. Pol β next inserts a single nucleotide at the gap site and repairs the 5′-deoxyribose phosphate end created by APE1.

This gap can then be filled by replacement with the correct nucleotide by DNA polymerases and the strand break ligated by DNA ligases (see Fig. 10.8). Pol β appears most important for short-patch repair, which is then subsequently ligated by DNA ligase III and XRCC1, whereas polymerases δ and ε and PCNA are required for long-patch repair, together with an endonuclease enzyme, FEN1. In the latter case, ligation is performed by ligase I. If SSB, in replicating DNA, are not properly repaired, they may result in DSBs.

Nucleotide excision repair (NER)

NER (Fig. 10.9) is a versatile and particularly important process for clearing substantive UV-induced DNA damage (such as thymine dimers and 6-4-photoproducts). NER differs from BER

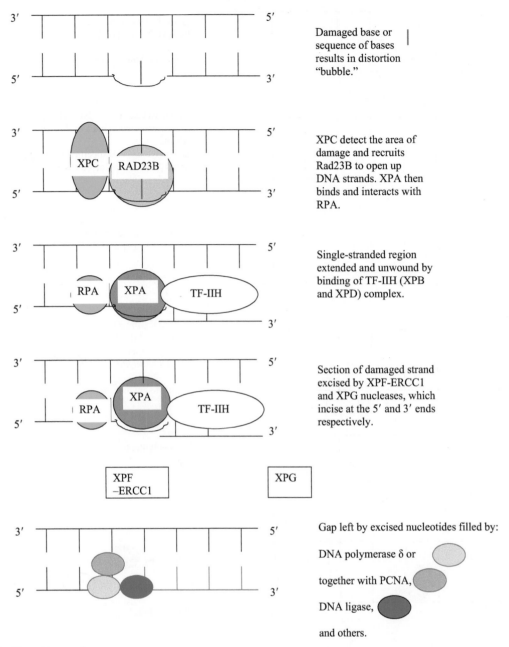

Figure 10.9 Nucleotide excision repair.

in that different enzymes are employed. These invariably remove not only the damaged base but also a number of adjacent bases even if these are undamaged. Moreover, whereas BER can work only if a relevant glycosylase exists to recognize the specific damage, NER is more flexible and can also respond to more widespread structural damage to DNA. There are 10 or more proteins required for NER. These include some named for the diseases in which they are defective, such as the Xeroderma pigmentosum proteins (XPA, XPB, XPC, XPD, XPE, XPF, and XPG) and the Cockayne syndrome proteins (CSA and CSB). Other important proteins needed for NER include RPA, RAD23A, RAD23B, DDB1, Cdk7, and ERCC1. NER is sometimes subdivided into two types, global genome repair (GGR) and transcription coupled repair (TCR), which are orchestrated by different groups of proteins. Global NER responds to damage that is not in the actively transcribed strand and is regulated by the XPC–Rad23B complex, whereas transcription-coupled NER is initiated by XPE and CSA/CSB.

TCR was identified in mammalian cells nearly 30 years ago by Mellon and Hanawalt. TCR operates during DNA transcription and clears damaged DNA from the transcribed sense strand of an active gene. TCR is thus intricately connected with transcription and may well be triggered by a stalled transcript, analogous to the stalled replication fork in DNA synthesis. Cockayne syndrome (CS) proteins, such as CSA and ERCC8 and CSB and ERCC6, act as sensors for stalled transcription because cells from patients with this condition cannot undertake TCR, whereas global repair is intact.

GGR can deal with a wide range of DNA lesions, including bulky ones and those which alter structure. Insults such as ultraviolet (UV) light induce bulky lesions such as cyclobutane pyrimidine dimers (CPD) and pyrimidine 6-4 pyrimidone photoproducts (6-4PP) in the minor groove of the DNA helix, which trigger GGR. Particularly lesions that result in local unwinding of DNA bases or intrastrand crosslinks are recognized by NER proteins. Chromatin structure is particularly important for NER in the nontranscribed strand of an active gene, because accessibility is more complicated. In GGR, DNA lesions are sensed by the XPC and human RAD23 homologue proteins (HR23). This allows the recruitment of XPA to sites of RPA expression on the exposed single strand. XPA can bind to and recruit the transcription Factor IIH (which contains the helicase proteins XPB and XPD), which unwinds the damaged DNA segment. Cuts are made on both the 5′ side and the 3′ side of the damaged area by the endonucleases XPF–ERCC1 and XPG, respectively. The section of ssDNA containing the altered base is thus removed. Next the resultant gap in DNA is filled in by DNA polymerases δ and ε, which make new DNA using the intact opposite strand as a template. This is again followed by the action of DNA ligase together with the protein PCNA.

The importance of NER is demonstrated by the severe phenotype of hereditary conditions in which NER genes are inactivated such as Cockayne syndrome and xeroderma pigmentosum (XP). The latter results from various mutations in any one of several NER genes, including *XPA*, encoding a damage recognition protein with an apical position in recruiting other NER proteins, *XPB* and *XPD*, which are part of TFIIH, and *XPF* and *XPG*, which cut the DNA backbone on the 5′ or 3′ side of damage, respectively (see Fig. 10.10). Defective TCR is characteristic of virtually all XP cells with the notable exception of the group C that cannot undertake global genome repair but can perform TCR.

Mismatch repair (MMR)

MMR is highly conserved in almost all organisms ranging from bacteria to humans. It is critical for preserving genomic integrity by correcting mismatches of the normal bases arising largely through a failure of polymerase proofreading during replication, that is, failure to maintain normal Watson–Crick base pairing (A–T and C–G). MMR can also detect insertion–deletion loops. MMR increases replication fidelity by up to 1000-fold and not surprisingly is often defective in cancers. In addition to a specialized and dedicated machinery of its own, MMR can trigger a DDR if specific DNA lesions are detected (such as those caused by insertion of false bases during certain types of chemotherapy). It can utilize enzymes involved in BER and NER.

MMR can involve the excision of a relatively lengthy oligonucleotide run or, in the case of some very specific mismatched pairs, simply excision of the immediate area long- and short-patch MMR.

To repair replication errors, the DNA ends must direct MMR to the newly synthesized strand containing the error. During DNA replication, strand discontinuity directs MMR to the discontinuous strand of a mismatched duplex, such as the 3′ ends of Okazaki fragments. The mismatch is recognized by numerous proteins including the core MMR recognition protein complex, human mutS homolog 2 (hMSH2) and 6 (hMSH6), and is followed by mismatch excision initiated by human MutL homologue MutLα (MLH1–PMS2 heterodimer), which nicks the

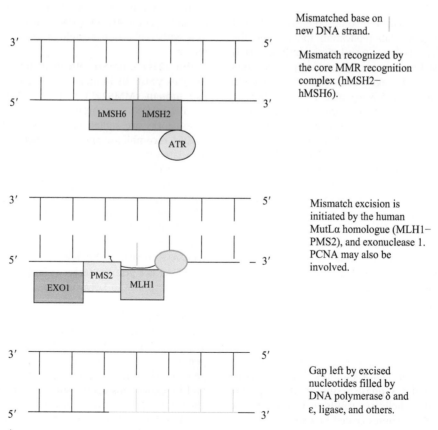

Figure 10.10 Mismatch repair.

discontinuous strand of the mismatched duplex. Excision does not appear to involve the preexisting strand break, which inevitably accompanies generation of Okazaki fragments during replication, but rather MMR is driven by nicks made by MutLα and exonuclease 1 (EXO1), which orchestrate 5' and 3' directed MMR, respectively. The excised area is again repaired by DNA polymerase δ and ε (see Fig. 10.10).

In man, MutSbeta complex (hMSH2 and hMSH3) may be important in certain types of MMR involving shorter repeats of mismatched pairs. A recent study has suggested that the MSH2 protein, which can bind ATR, may be a key factor in activating the ATR–CHK1 pathway after cisplatin-induced DNA crosslinking. MutSβ also recognizes and interacts with interstrand crosslinks induced by drugs such as cisplatin and also in the repair of DSBs – thus MSH6 can interact with Ku70 in repairing DSBs. Interestingly MMR proteins may also convert DNA nicks and point mutations into DSBs that then need repair by specific pathways discussed in this chapter. Inactivation of MMR genes causes microsatellite instability (MSI), a major contributor to tumorigenesis, which has been discussed.

Lynch syndrome, also known as hereditary nonpolyposis colon cancer (HNPCC), is an autosomal dominant cancer predisposition syndrome resulting from germline-inactivating mutations in MMR genes, including *MLH1*, *MSH2*, *MSH6*, or *PMS2*. Lynch syndrome is associated with greatly enhanced susceptibility to colorectal and also to endometrial, kidney, and stomach cancers. Somatic mutations in both *MSH2* and *MLH1* predispose to colon cancer and are found in a variety of other cancers. In some cases, *MSH2* can also be downregulated by overexpression of the miRNA miR-21.

Translesion synthesis (TLS)

Some DNA damage avoids repair and could potentially prevent any further replication (and stall replication forks – see the "Stalled replication forks" section). Translesion synthesis (TLS) allows the replication machinery to continue to copy DNA across and beyond an area of DNA damage, such as an AP site even in the absence of effective repair. TLS is regarded as a DNA damage tolerance process in which DNA polymerases are exchanged for those involved specifically in TLS (e.g. DNA polymerase η, ι, and κ from the Y polymerase family, as well as Pol ζ, θ, and ν). These are capable of inserting bases opposite damaged nucleotides; they share an unfortunate propensity for putting in the wrong bases where there is no damage, but are reasonably successful where there is damage. This may be accounted for by the lack of proofreading exonuclease activity (unlike Pol δ and ε). This switch is in part mediated by PCNA, which is ubiquitinated by RAD6/RAD18 to allow initiation of replication by the TLS polymerases. Why it is deemed preferable to risk creating point mutations during TLS rather than invoking other repair processes is a matter of conjecture. Deficiency in Pol η is found in one of the xeroderma pigmentosum types and is associated with an increased risk of UV-induced cancer.

Stalled replication forks

DNA replication represents the time of greatest vulnerability for the genome. The stalling of replication forks (Fig. 10.11), due to DNA damage, during the S phase results in the exposure of runs of ssDNA as MCM helicases continue to unwind the DNA template. This ssDNA becomes coated in a specific binding protein, replication protein A (RPA); a BRCA1 C-terminal (BRCT) domain-containing protein, TopBP1 (which interestingly appears important in the spindle checkpoint also); and the 9-1-1 complex (Rad9–Rad1–Hus1). RPA recruits the other major apical kinase ATR through interactions with its regulatory subunit ATRIP. ATR can also be activated by DSBs occurring in the late S phase or G_2 phase, but in this case via ATM because ssDNA is produced by the endo- and/or exonuclease activity of Mre11–CtIP complexes.

So what happens next?

The best-described substrates for ATM and ATR are the effector kinases CHK2 and CHK1, respectively. CHK2 ultimately activates the p53 pathway. Activation of CHK1 by ATR requires phosphorylation of the ssDNA binding protein Rad17 and recruitment of a mediator protein, Claspin. A huge number of other proteins are activated by ATM and ATR. The functional significance is often unclear but may be related to balancing cell-cycle arrest, repair, cell stress, and ultimately life–death choices for the damaged cell. The key functions of the ATM–CHK2 and the ATR–CHK1 pathways appear to result from the inactivation of Cdc25 phosphatases, needed for the activation of cyclin–CDK complexes and cell-cycle progression, in particular for dephosphorylation and activation of CDKs 1 and 2 at the G_1–S and G_2–M transitions. Cdc25A is the main CHK1 substrate for the intra-S-phase checkpoint and possibly also plays a minor part at the G_2–M transition, where CDC25B and CDC 5C are the key regulators of CDK1–cyclin B activation. Inactivation of CDC25B and C may result from activated Wee1 and 14-3-3 binding and exclusion from the nucleus. The E3 ubiquitin ligase Rad 18 recruits DNA polymerase eta (Polη) to sites of replication fork stalling where it ubiquitinates PCNA. This allows the binding of Y family trans-lesion synthesis (TLS) DNA polymerases. Rad18 is activated by DDK (also known as Cdc7). Hsk1, a yeast homologue of DDK, can phosphorylate Rad9 after DNA damage, which prevents interaction with replication protein A (RPA) and may facilitate DNA repair.

Mms22 is required for HR-mediated repair of stalled replication forks in yeast. In humans, a complex comprising the Mms22-like protein (MMS22L) and an MMS22L-interacting protein, NFκBIL2–TONSL, accumulates at regions of ssDNA where replication forks are stalled and is required for the loading of the RAD51 recombinase and HR-mediated repair of replication fork-associated DSBs.

The DDR signaling network spreads much wider than this and crosstalk with, amongst others, MAP kinases such as the p38MAPK–MK2 stress response pathway, PI3K–AKT, and IKK NFκB has been demonstrated. It appears that ATM, in addition to activating p53, inactivates CDC25B/C and may also contribute to inactivation of CDC25A. The kinase MK2 may work in parallel with ATR in DDR activation.

BRCA1 is a primary target of both ATR and ATM phosphorylation in response to DNA damage.

Double-strand breaks

DSBs are among the most serious and lethal types of DNA damage. They arise naturally through the action of oxygen radicals generated by normal metabolic processes. They can also be generated by oncogenic stress (see the "Oncogenic Stress" section) and by exogenous agents such as gamma irradiation and some chemotherapeutic drugs. DSBs represent dangerous chromosomal lesions that can lead to mutation, neoplastic transformation, or cell death. Mammalian cells possess potent and efficient

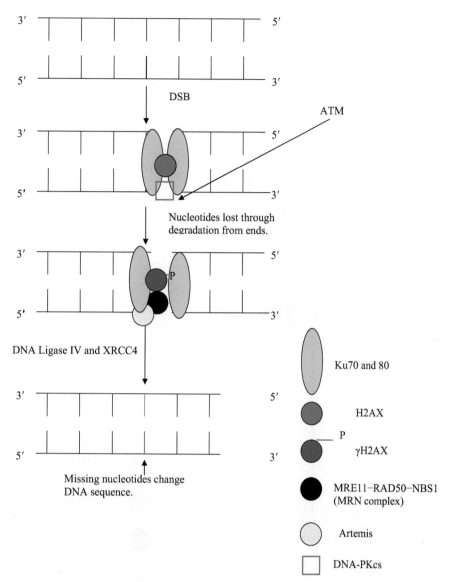

Figure 10.11 Repair of DSBs. During homologous recombination (HR), the damaged chromosome enters into synapsis with, and retrieves genetic information from, an undamaged DNA molecule with which it shares extensive sequence homology. In contrast, nonhomologous end joining (NHEJ) brings about the ligation of two DNA DSBs without the requirement for extensive sequence homology between the DNA ends and without synapsis of the broken DNA with an undamaged partner DNA molecule. In NHEJ (**a**), the broken ends of DNA are brought together and ligated. This produces a mutation at the site of joining as bases are often lost or a short segment inserted. This is the primary mechanism by which DSBs are repaired in humans, and likely produces problems only on rare occasions. Most human DNA is noncoding, and chance suggests that many repairs by NHEJ will be in such areas and have little if any phenotypic consequences, certainly if compared with the potential risks of fragmented chromosomes. The ends of a DSB are detected and bound by KU, a heterodimer consisting of Ku70 and Ku80 proteins. In mammals, KU forms a complex, known as DNA-PK, with DNA-PKcs. It is thought that KU holds the two ends together and facilitates end-to-end ligation by the complex of ligase 4 and XRCC4, and usually results in accurate repair of the DSB. Alternatively, binding of the ends by KU can be followed by resection of the free ends by the Mre11–Rad50–Nbs1 (MRN) complex. This pathway generally leads to error-prone repair of the DSB. HR (**b**) is the favored mechanism for DSB repair in yeasts and fruit flies. Diploid cells contain two copies of each double helix, and thus the intact strand contains the information required to effect an exact repair of the DSB – by effectively acting as a template. Although employed in human cells, this is probably much less frequently employed than NHEJ. When a DSB occurs in one of two sister chromatids, the ends of the DSB are recognized by the MRN complex, which processes the ends forming 3' single-stranded DNA (ssDNA) overhangs. The replication protein A (RPA) binds to the ssDNA overhangs, and recruits Rad51 and Rad52 to the DSB. Both RPA and Rad52 help to load Rad51 onto ssDNA to form ssDNA–Rad51 nucleoprotein filaments. This nucleoprotein filament searches for the homologous duplex DNA in the undamaged sister chromatid. A successful search results in strand invasion, strand exchange, and joint molecule formation involving BRCA1, BRCA2, and the Rad51-like proteins XRCC2, XRCC3, RAD51B, RAD51C, and RAD51D. DNA synthesis by DNA polymerases generates the genetic information that is required to seal the break. Ligation and the resolution of the two double helices joined by strand exchange complete this error-free repair event. (**a**) NHEJ results in a change in the original DNA sequence, either deletions (shown here) or insertions; (**b**) homologous recombination results in more accurate repair of DSBs but requires a more complex process.

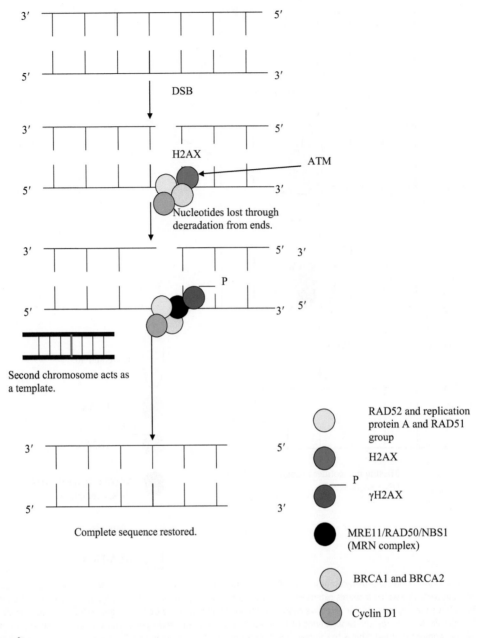

Figure 10.11 (Continued)

mechanisms to repair DSBs, and thus complete normal development as well as mitigate oncogenic potential and prevent cell death. Failure of DSB repair can lead to CIN and is often associated with tumor formation or progression. Most DSBs in G_1 are repaired by DNA NHEJ, whereas slower repair through HR mainly occurs during the S and G_2 phases. Choice is again determined by the cell-cycle stage. An interesting question is how many DSBs are needed to activate the checkpoint – the answer appears to be more than one and maybe as many as 10.

The Mre11–Rad50–Nbs1 (MRN) complex has numerous important roles, including telomere maintenance, processing of DSBs in meiosis, DDR checkpoints, and HR. Unprotected DSBs are recognized by two key protein complexes, MRN and KU70/80.

These are recruited to the site of damage, where MRN phosphorylates the histone H2AX across a surrounding area of up to several megabases. MRN and phosphorylated H2AX (γH2AX) then recruit a variety of other proteins such as Trrap and Tip60 (a tumor suppressor) that activate ATM. Broadly, MRN recruits ATM, whereas KU70/80 recruits DNA-PKcs. Members of the phosphatidylinositol-3 kinase-related kinases (PIKK) family such as ATM, ATR, and DNA-PKcs are critical regulators of DDR. Interestingly, activation of ATM may operate in parallel with initiation of DNA repair processes.

Thus, the Trrap–Tip60 proteins are also involved in positioning breast cancer early onset-1 (BRCA1) and p53-binding protein 1 (53BP1) to damaged chromatin. Other key ATM targets are checkpoint kinase 2 (CHK2), structural maintenance of chromosome-1 (SMC1), and mediator of DNA damage checkpoint protein-1 (MDC1).

ATM can also phosphorylate histone H2AX, allowing further interaction with another mediator protein, MDC1, which is also phosphorylated by ATM, and binds to γ-H2AX via its tandem BRCA1 C-terminal (BRCT) domains. Activated MDC helps retain active ATM and MRN complexes to γ-H2AX containing chromatin at the lesion site and also recruits an E3 ubiquitin ligase RNF8 which ubiquitinates chromatin-associated proteins, including H2AX. The ubiquitinated sites help recruit key repair proteins such as the Rap80, Abraxas, and BRCA1 complex and p53 binding protein 1 (53BP1) that mediate HR and NHEJ, respectively (see Chapter 10). The RAP80–Abraxas complex may help recruit BRCA1 to DNA damage sites in part through recognition of ubiquitinated proteins. The interactions between ATM, MDC1 and MRN further promote the phosphorylation of H2AX and recruit several other key factors to the lesion, including the RING finger proteins RNF8 and RNF168, 53BP1, and BRCA1.

Phosphorylation of H2AX is regulated by the opposing activities of the tyrosine kinase WSTF (Williams–Beuren syndrome transcription factor) and protein phosphatase EYA (eyes absent). WSTF is critical for keeping γ-H2AX at the DNA lesion. Before DNA repair can be initiated, γ-H2AX must be offloaded from DNA in part by dephosphorylation of serine 139 by the protein phosphatases PP2A and PP4.

The Nbs1 protein part of the MRN complex appears to be a critical factor in localizing proteins at the site of DNA damage, in which it is assisted by another protein CK2 and by the kinase PI3Kβ. MRN, along with its cofactor Ctp1 and ATM, is involved in 5′ resection of DSBs which results in the formation of an intermediate ssDNA that is required for activation of the ATR–CHK1 pathway and for HR. The ATR–CHK1 pathway is also vital to activate G_2–M arrest. The tumor suppressor Menin also appears to play a role in HR and interacts with CHK1.

Posttranslational modifications of DSB proteins by ubiquitination and Sumoylation appear to be critical regulatory steps in DSB repair. Local dimethylation of histone H4 lysine 20 (H4K20me2) by the histone methyltransferase MMSET is essential for recruitment of 53BP1 to the DSB. MMSET is recruited to the DSB by the γH2AX–MDC1 pathway. The BRCT domain of MDC1 phosphorylates MMSET on Ser 102. The serine–threonine protein phosphatase PP2A, and possibly WIP1, is also an important regulator of ATM activity. PP2A may help maintain ATM in an inactive state in the absence of DNA damage.

Different ubiquitin E3 ligases including RNF8, RNF168 (which interacts with UBC13), and BRCA1–BARD1 (which works with UBCH5c) regulate DSB repair and signaling. Once RNF8 is recruited by MDC1 to DSB sites, it collaborates with UBC13 to mediate K63-linked ubiquitination of the histones H2A and H2AX flanking the DNA lesion engendering structural changes which help recruit 53BP1 and receptor-associated protein 80 (RAP80). RAP80 facilitates assembly of the BRCA1-A complex containing BRCA1, BARD1, ABRAXAS–FAM175A, BRCC3–BRCC36, BRE–BRCC45, and MERIT40–NBA1. The SUMO E3 ligases PIAS1 and PIAS4 are also present at DSBs and may contribute to activity of RNF8 and RNF168.

A recent report suggests that ATM may also activate antioxidant defenses and DNA repair via activation of glucose-6-phosphate dehydrogenase (G6PD) and the pentose phosphate shunt pathway, which produces NADPH and nucleotides.

As we have said so many times before, few processes manage to escape regulation by miRNAs and the DDR is no exception, with almost 1 in 5 miRNAs upregulated by DNA damage. Thus, DSBs induce biogenesis of multiple miRNAs, a process activated by ATM-dependent phosphorylation of a key processing factor, KH-type splicing regulatory protein (KSRP).

Two primary mechanisms are employed to repair a complete break in a DNA molecule (Fig. 10.12a–b). The first of these, direct joining of the broken ends, requires proteins such as DNA-dependent protein kinase (DNA-PK), Ku70, and Ku80 that recognize and bind to the exposed DNA ends in order to bring them into apposition for joining. This does not necessarily require the presence of complementary nucleotides and is therefore also referred to as NHEJ (Fig. 10.12a). DNA-dependent protein kinase (DNA-PK) is a multi-subunit protein made up of DNA-PKcs, a very large catalytic subunit (cs) of 465 kDa, and Ku, a DNA-binding heterodimer which comprises the Ku70 and Ku80 subunits. DNA-PK is activated by DNA ends and plays a major role in DSB repair. As discussed in Chapter 10, many of the proteins involved in DSB repair are also involved in telomere maintenance, including Ku and the protein PARP (see Chapter 9). Errors in direct joining may cause or contribute to translocations in cancers such as Burkitt lymphoma and chronic myeloid leukemia (CML) (Chapter 6). DNA-PKcs, like ATM and ATR, has important kinase activity involved in activating downstream signaling after DNA damage.

The other means of repairing DSBs is by HR (Fig. 10.12b), whereby the broken ends are repaired using the information on the intact sister chromatid (in G_2 after chromosome duplication) or on the homologous chromosome (in G_1 before chromosome duplication). Two of the proteins used in HR are encoded by the genes *BRCA1* and *BRCA2*, mutations that contribute to breast and ovarian cancers (described in this chapter). In human cells a protein complex comprising MRE11, RAD50, and NBS1 (MRN) has exonuclease and endonuclease activities that may be vital for processing the DNA ends at the site of a DSB prior to relegation. NBS1 is defective in a human chromosomal instability disorder associated with an increased risk of cancer, called Nijmegen breakage syndrome (NBS). The MRE11 repair protein is involved in both HR and NHEJ, whereas the function of Ku70/80 is unique to NHEJ. Human cohesin subunits physically interact with RAD50 suggesting links with mitotic checkpoint control.

The *BRCA1* gene was first cloned about 10 years ago, and since then it has become clear that together with *BRCA2* these two genes are the major causes of hereditary predisposition to breast or ovarian cancers. Since that time, *BRCA1* has been linked to several key nuclear functions connected with the prevention of genomic instability. In particular, *BRCA1* functions in concert with *Rad51*, *BRCA2*, and other genes to control DSB repair (DSBR) and HR.

Protecting telomeres from repair processes

Chromosome end fusions (covalent connections between the C strand of one telomere and the G strand of another) occur when telomeres have become too short and when TRF2 is inhibited (Chapter 9). They predominantly involve the ends of two different chromosomes; resulting dicentric chromosomes can become

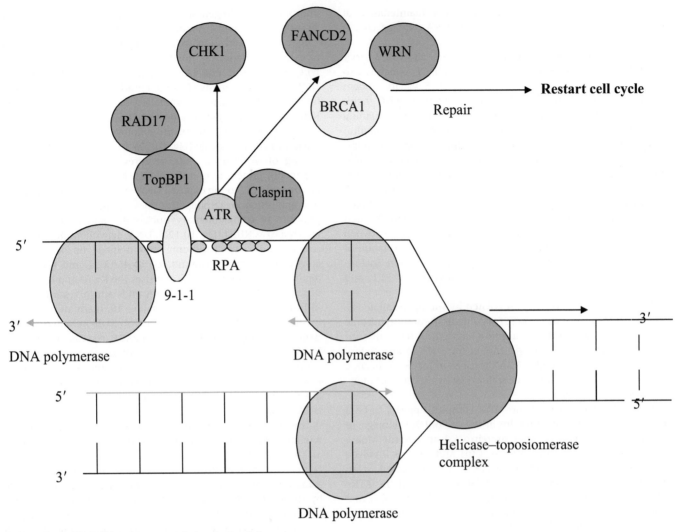

Figure 10.12 ATR–CHK1 pathway – a stalled replication fork.

attached to both spindle poles and lead to a problem for chromosome segregation in anaphase. The fusion of damaged telomeres requires the same factors as normal NHEJ, but with modifications due to the longer 3′ overhang. One way in which chromosome ends are usually protected from NHEJ is by the formation of t loops, as without an accessible end, the NHEJ proteins will not be able to form the synaptic complex needed for processing and ligating the ends. Telomeres also need to be protected from inappropriate HR. There are three types of HR that have detrimental outcomes at chromosome ends: telomere sister chromatid exchange (T-SCE), t-loop HR, and recombination between a telomere and interstitial telomeric DNA. One puzzling feature of telomere-interacting proteins has been emphasized in a recommended review by Titia De Lange. This relates to the role of telomere-associated DNA repair factors, DNA-PKcs and the Ku70/80 heterodimer, which promote NHEJ but are also associated with telomeres – in this case, how is the inappropriate triggering of DSB repair avoided? Moreover, how DNA-PKcs and Ku contribute to telomere function is not known; but what seems clear is that they would need careful control in order to avoid NHEJ of chromosome ends.

Breast cancer and the BRCA genes

Knudson's classic "two-hit" model describing the inactivation of a tumor suppressor gene (Chapter 7) appears to fit the behavior of both the *BRCA1* and *BRCA2* genes. In familial cancers, offspring inherit a germline mutation in one of the *BRCA1* alleles from one of the parents, thus this "first hit" is present in all cells of the body. A somatic mutation results in loss of heterozygosity (LOH) of the wild-type *BRCA1* allele and represents the "second hit" on a given cell, thus rendering both copies of the gene inactive. In sporadic cancers, inactivation of a tumor suppressor gene is accomplished by two somatic mutations that eliminate or inactivate the alleles on both chromosomes. Knudson's model accurately accounts for the early onset of familial cancers caused by a preexisting germline mutation, while the accumulation of two somatic mutations in a single cell may take several decades to produce a sporadic cancer. Women who inherit a defective *BRCA1* allele have about a 65–80% lifetime risk of breast cancer, which generally develops at a younger age and may affect both breasts, and a 20–50% lifetime risk of developing ovarian cancer, in contrast to the 2–3% risk in the general population. It is estimated that mutations in *BRCA1* alone account for 50% of all familial early-onset female breast cancers, while mutations in *BRCA2* may be responsible for up to 35% of the remaining hereditary breast cancers. LOH at the *BRCA2* locus has been observed in 30–40% of sporadic breast and ovarian cancers; however, very few somatic mutations have been found in the remaining allele. This suggests that either *BRCA2* is an infrequent target for somatic inactivation or that intron sequences or genomic deletions may be the target of somatic mutation.

BRCA2 carriers have similar risks, although the average age of onset for ovarian cancer appears to be similar to the age of sporadic cases. In addition, male *BRCA2* carriers are at increased risk for male breast cancer, with an approximately 6–8% lifetime risk, as compared with the general male population risk of 0.1%.

Women affected with breast cancer who are *BRCA* carriers also have increased risk of another primary breast cancer in the remaining or contralateral breast of 30–50% by age 70. Thus carriers who are also survivors of breast or other cancers warrant as much counseling for adherence to surveillance as non-affected carriers.

BRCA1 forms a stable heterodimeric E3 ubiquitin ligase with *BARD1*, and interfering with the interaction leads to *BRCA1* degradation. Once uncoupled from *BARD1*, *BRCA1* is degraded following ubiquitination by *HERC2*. *BRCA1* is involved in DSB repair and normal DNA recombination. Splice variants of *BRCA1* mRNA have been identified that exist normally in nonmalignant breast cells. These alternatively spliced mRNAs code for truncated proteins; however, it is yet undetermined whether these truncated proteins exhibit tumor suppressor activity. The *BRCA2* gene is similar to the *BRCA1* gene and also plays a role in DNA repair. Both genes have more than 20 exons, with a very large exon 11 and a code for highly charged proteins with a putative granin domain. *BRCA1* and *BRCA2* are key players in HR repair and, as will be discussed next, open up the possibility of a synthetic lethal interaction with PARP inhibitors as defective SSB repair converts these to DSBs that are themselves poorly repaired in the absence of *BRCA1* or *BRCA2*.

Recent studies reveal that many further examples of synthetic lethality may be awaiting discovery. In one, around 50 000 small molecules were tested in a synthetic lethal screen in rodent cells with a conditional oncogenic K-ras(G12D) allele. They identified lanperisone, which induces cytotoxic ROS that for unknown reasons are not effectively scavenged by K-ras mutant cells.

Poly (ADP-ribose) polymerase (PARP)

In the 1960s, a nuclear enzymatic activity that synthesizes an adenine-containing RNA-like polymer, now known to be poly(ADP-ribose), was identified in the laboratory of Paul Mandel. This poly(ADP-ribose) polymerase (PARP) was subsequently found to be activated after DNA damage, and PARP1 is involved in DNA repair. PARP1 hyperactivation after DNA damage can result in depletion of NAD+ and ATP, thereby potentially promoting a unique form of metabolic cell death called **parthanatos**.

PARP inhibition can synergize with DNA-damaging chemotherapy and can also kill cancer cells with inactive *BRCA1* or *BRCA2*, an example of synthetic lethality. Recent phase I and II clinical trials have confirmed potential benefits in *BRCA1*- and *BRCA2*-deficient breast and ovarian cancers. PARP1 has three functional domains, an amino-terminal zinc finger DNA-binding domain important for binding to DSBs and SSBs, a central domain that is auto-poly(ADP-ribosyl)ated and contains a BRCA homology region shared by many DDR proteins, and finally a C-terminal catalytic domain responsible for transferring ADP-ribose subunits from NAD+ to proteins to form pADPr. The PARP family is known to comprise 16 structurally similar proteins, though functionally they can be categorized into a related group of PARP1, PARP2, and PARP3 and another group of tankyrase 1 (TNKS) and TNKS2. PARP1 responds to DNA damage within seconds and becomes active with most of the generated pADPr remaining attached to it. This results in the recruitment of multiple proteins involved in DNA repair including XRCC1 used in DNA base excision repair (BER). The pADPr may also help remove PARP to allow access for repair proteins. PARP1 also regulates recruitment of MRE11 and ATM and the HR machinery, to DSBs. PARP1 can also promote NHEJ by activating a DNA-PK-dependent phosphorylation cascade and may also restart stalled replication forks. PARP2 appears to be important in mouse models, and TNKS and TNKS2 appear important in telomere maintenance by interacting with TRF1. PARP-3 is activated by DSBs and recruits poly(ADP-ribose)-binding protein APLF to assist DNA DSB repair. APLF in turn promotes retention of the XRCC4–DNA ligase IV complex to direct DNA ligation during NHEJ. We now known that PARP has other important functions including a key role in the interphase cytoplasm where it contributes to the formation of cytoplasmic "stress granules" containing RNA-binding proteins regulating mRNA translation and stability. PARP can affect the stress response by releasing miRNA-mediated gene silencing and by interfering with miRNA-binding Argonaute proteins. It may be concluded that PARP inhibition would prevent the increased translational of normally repressed genes through this means.

PARP inhibitors have been developed to target cancers with vulnerable genetic lesions and are particularly effective in killing cells with defective HR. For instance, BRCA-deficient cells or others with defective DSB repair would be highly dependent on PARP1 and BER to maintain genomic stability and would exhibit synthetic lethality with PARP inhibitors. Namely, cancer cells with defective HR would die, whereas normal cells would in theory be spared. On the other hand, as PARP contributes to DNA repair, PARP inhibition will sensitize cancer cells to the cytotoxic action of other DNA-damaging agents. In an interesting twist on the synthetic lethal story, a recent study has concluded that HR may be compromised in dividing cells within multiple solid tumors through a common and shared mechanism, namely, hypoxia. Thus, hypoxic tumors (most aggressive cancers) would also in theory show a synthetic lethal effect with PARP1 inhibitors. If this were so, then PARP inhibitors might be more widely beneficial than previously thought. In this study, hypoxic cells with PARP inhibition accumulated γH2AX and 53BP1 foci due to aberrant DNA replication firing during the S phase – a process described as microenvironment-mediated "contextual synthetic lethality" by the authors. PARP inhibitor treatment may also indirectly promote cancer cell death. Thus, PARP inhibitors activate error-prone NHEJ in HR-deficient cells, in part by phosphorylating DNA-PK substrates. In a recent study where NHEJ was disabled, in cell lines lacking BRCA2, BRCA1, or ATM, PARP inhibition was no longer toxic. They suggest that NHEJ may in itself contribute to genomic instability and the death of the cancer cells.

PARP inhibitors have been studied in a number of clinical trials and in particular in some breast and ovarian cancers lacking

BRCA1 and *BRCA2*. Most recently, a phase II trial has been completed in which the PARP inhibitor iniparib was added or not added to gemcitabine and carboplatin in patients with metastatic triple-negative breast cancer. Rate of clinical benefit improved from 34% to 56% (P=0.01) and prolonged the median progression-free survival from 3.6 months to 5.9 months without any increase in adverse events. A phase III trials is now underway. Other phase I and phase II trials of PARP inhibitors in combination with other agents are ongoing. PARP inhibitors can also enhance the action of methylating agents and topoisomerase I poisons. A clinical trial has examined the combined effect of a PARP inhibitor and the methylating agent temozolomide.

DNA polymerases

The mammalian genome encodes a bewildering array of at least 15 different DNA polymerases, the DNA-synthesizing enzymes, all of which appear to have particular roles in either normal genome replication (polymerases α, δ, and ε) or in repairing damaged DNA. Following BER and NER (discussed in this chapter), it has been noted that it falls to DNA polymerases to fill the gap previously filled by the excised base. Under certain circumstances the space left can be bypassed by TLS which employs a further DNA polymerase.

Polymerase α initiates DNA synthesis on both strands, leaving 20–30 base pair primers that are elongated by the polymerases δ and ε. These latter also contain proofreading exonuclease activity that prevents mutations.

Polymerase β is the main subtype responsible for gap filling during BER (Fig. 10.8) and has intrinsic dRP lyase activity needed to remove sugar–phosphate residues arising due to AP endonucleases. In contrast, in NER (Fig. 10.9) the around 27-nucleotide gap produced is filled primarily by the polymerases δ and ε (requiring both PCNA and replication protein A).

With respect to cancer, interest is increasing in the development of DNA polymerase inhibitory drugs and also in exploring the role of these enzymes in stem cells and potentially cancer stem cells. Drugs targeting polymerases, including folate inhibitors and pyrimidine analogues, are already in widespread use as chemotherapeutic agents but are rather toxic. Future development of polymerase β and TLS-specific inhibitors may be less toxic to normal cells.

Stem cells and the DDR

CSCs, as has been previously shown for ES cells and hematopoietic stem cells (HSCs), are more able to survive DNA damage than their more differentiated counterparts. In part, this reflects a greater capacity for repairing such damage. In practical terms, this translates into a resistance to DNA damaging chemo-radio therapy and comparative genomic stability under such conditions, which can kill most cancer cells. The oncogenic polycomb protein BMI-1 is an early DDR-expressed protein that may account for the increased resistance to radiation and DNA damage, observed in stem cells and cancer stem cells. BMI accumulates at DSBs due to PARP activity and not γH2AX. Together with RING2 and PRC1 histone H2A E3 ubiquitin ligase partners, BMI-1 ubiquitylates histones H2A and H2AX during the DDR resulting in the accumulation of RAP80, 53BP1, and BRCA1, which all act to increase resistance to DNA damage.

Checkpoints

The acknowledgment of our weakness is the first step in repairing our loss.
Thomas Kempis

(See also Chapter 4 for a discussion.)

Aneuploidy may not inevitably arise by selection during tumorigenesis but may instead arise as an inadvertent consequence of defects that uncouple mitotic control from normal cell-cycle progression. Data obtained from a variety of model organisms have revealed that disruption of the cell-cycle controls required for homeostasis results in the acquisition of genomic instability. Defects at the G_1–S transition have been discussed in other chapters, so in this chapter we will concentrate on the role of aberrant checkpoint signaling during mitotic progression in cancer. Cells contain numerous pathways designed to protect them from the genomic instability or toxicity that can result when their DNA is damaged. The p53 tumor suppressor is particularly important for regulating passage through the G_1 phase of the cell cycle (Chapter 7), while other checkpoint regulators are important for arrest in the S and G_2 phases. Tumor cells often exhibit defects in these checkpoint proteins, many of which have been discussed. This is illustrated by the predisposition to cancer seen in various inherited syndromes where checkpoint or caretaker genes are mutated: *p53* and *CHEK2* mutations in Li–Fraumeni syndrome, inactivating *ATM* mutations in ataxia telangiectasia, defective DNA damage repair with breast cancer susceptibility genes 1 and 2 (*BRCA1/2*), and *NBS1* mutations in familial breast cancer and NBS, respectively. Other checkpoint genes such as *53BP1*, Meiotic recombination 11 (*MRE11*), and *H2AX* have been identified in rodent cancer models.

Although some key gaps remain in our understanding, DNA damage is rapidly followed by a binding of protein complexes to areas of strand breaks in DNA. These complexes include MRN and DNA-PK, discussed in the "The mechanisms of DNA repair" section. The protein kinases ATM and ATR (and DNA-PKcs), as well as their downstream substrates CHK1 and CHK2, are important for checkpoint activation in response to DNA damage. Histone H2AX, ATRIP, as well as the BRCT-motif-containing molecules 53BP1, MDC1, and BRCA1 function as molecular adapters or mediators in the recruitment of ATM or ATR and their targets to sites of DNA damage (Fig. 10.6). The increased chromosomal instability and tumor susceptibility apparent in mutant mice deficient in both p53 and either histone H2AX or proteins that contribute to NHEJ indicate that DNA damage checkpoints play a pivotal role in tumor suppression. During tumorigenesis, tumor cells frequently lose checkpoint controls, and this facilitates the development of the tumor. However, these defects also represent an "Achilles heel" that can be targeted to improve current therapeutic strategies.

Mitotic entry depends on activity of the cyclin B–CDK1 complex. As described in Chapter 4, during G_2, CDK1 is kept inactive via inhibitory phosphorylation by WEE1 and MYT1 kinases; this phosphorylation is removed at the onset of mitosis by the CDC25 phosphatases, which are in turn themselves activated by PLK1. Cyclin B–CDK1 is the main target of the G_2 DNA damage checkpoint; following DNA damage, CDC25A, CDC25B, and CDC25C are inhibited by phosphorylation and/or ubiquitin-mediated degradation; PLK1 is inhibited; and cyclin B–CDK1 complexes are sequestered in the cytoplasm.

The spindle assembly checkpoint normally ensures accurate progression through mitosis in order that each new cell receives one copy of every chromosome. If even a single chromosome is improperly attached to the mitotic spindle (sensed by unattached kinetochores), then the mitotic checkpoint will arrest the cell-inhibiting activity of the APC/C ubiquitin E3 ligase (and its CDC20 cofactor – sometimes referred to as APC/C^{CDC20}) discussed in Chapter 4. Major proteins targeted for proteasomal degradation by the APC/C^{CDC20} include cyclin B and securin, which are both key regulators of mitosis. Unattached kinetochores generate a durable and powerful inhibitory signal (a single unattached kinetochore can delay anaphase by 3 hours). This signal is believed to comprise a series of proteins first identified in yeast but all with homologues in mammalian cells. These proteins include the BUB (BUB3 and BUB1) and MAD (MAD1 and MAD2) family and MPS1. Additional proteins involved in the mammalian checkpoint include the kinases BUBR1 and MAPK (see Chapter 5) and CENPE. Not surprisingly, this is complex and still not fully understood, and the identity of the "inhibitory signal" is still not known. Though MAD2, BUBR1, BUB3, and MPS1 are readily released by unattached kinetochores, MAD2 and BUBR1 can bind to CDC20, and it is possible (but unproven) that they could thereby prevent it from activating APC/C – thus, cyclin B and securin would remain undegraded. Time will tell whether these proteins or some variant conformation of them are the inhibitory signal for mitosis. The failure of cell-cycle regulatory checkpoints is a common event in human cancer. Human tumor cells have been shown to contain mutations to BUB1, BUBR1, MAD1, and MAD2 and also members of the so-called ZW10–ROD–Zwilch complex (involved in recruiting MAD1/2 to unattached kinetochores). Recently, haploinsufficiency of MAD2 has been associated with tumorigenesis. The actual effects of these mutations on the checkpoint are still untested, but the suggestion is that even a weakened (not necessarily absent) checkpoint response may be enough to facilitate tumorigenesis. In keeping with this view, various mouse models with inactivating mutations in single components of the mitotic checkpoint have a modest increase in risk of some cancers but often a long latency.

It is worth remembering that often the primary defect giving rise to CIN may not be in the checkpoint proteins. Although defects in mitotic checkpoint proteins are very commonly seen in cancers, MAD2 haploinsufficiency notwithstanding, this is not usually due to mutations in the encoding genes, but rather predominantly a secondary effect on protein levels due to aberrant activation of oncogenes or tumor suppressors. Loss of the tumor suppressors APC, p53, or BRCA1/2, or deregulated expression of oncogenic proteins such as c-MYC, E6, and E7, may all result in microtubule instability or centrosome amplification (both causes of CIN). Deregulation of the cell cycle may be a common feature in these cases, but whatever the underlying mechanism, overexpression of c-MYC or loss of RB may provoke CIN. In the latter case, this is associated with activation of E2F and paradoxically also of MAD2.

In order for such DNA damage to propagate, cells must overcome the checkpoints and tumor suppressor pathways discussed in this chapter; the most obvious explanation is that these aforementioned components are in some way inactivated. This is undoubtedly the explanation in many cases and in fact the identification of new lesions, which can cooperate with c-MYC and allow propagation, may help identify new genes and proteins involved in checkpoint control. The Werner Syndrome protein WRN may allow c-Myc-activated cells with genomic instability to propagate. Genomic instability may be caused by specific genetic alterations even before malignant conversion. Mutations in hCDC4 (also known as Fbw7 or Archipelago) have recently been identified in human colorectal cancers and their precursor lesions, and targeted inactivation of hCDC4 in karyotypically stable CRC cells results in micronuclei and chromosomal instability, largely through cyclin E–dependent metaphase defects and subsequent transmission of chromosomes. The checkpoint may also be overridden by mutations in Aurora-A (Chapter 4), which can prevent MAD2 from effectively preventing APC/C activation.

Most solid cancers are aneuploid (have lost or gained chromosomes or bits of them), indicating how common defects in this checkpoint must be in cancer. Yet, many traditional chemotherapeutic agents act by deliberately disrupting the mitotic spindle and preventing segregation of chromosomes. The apparent conflict is resolved if one imagines that in the context of therapeutic disruption, one is relying on the cell doing the decent thing when its mitotic checkpoint is triggered and thence undergoing apoptosis or growth arrest, whereas aneuploidy may be tolerated by some cancer cells that have acquired the means to evade checkpoint barriers. It is tempting to believe that such cancers may be more resistant to drugs which impair chromosome segregation, in which case it would be more likely that a proportion of cells, those where damage is not so severe as to directly compromise viability, would survive chemotherapy and the cancer would continue or recur.

As stated, most solid tumors are aneuploid, but most also contain mutations in oncogenes and tumor suppressors; thus, we must again revisit our constant companion – the chicken–egg conundrum. Namely, in this case. is aneuploidy the cause or consequence of cancer? Whatever the answer, there are many possible explanations for the near-ubiquitous presence of aneuploidy. CIN would certainly increase the chances of acquiring other mutations. Thus, directly, loss of a chromosome containing a haploinsufficient tumor suppressor might be enough to promote tumor progression. Alternatively, the chance of LOH discussed in Chapter 3, whereby a cell with a single mutant copy of a tumor suppressor gene loses the remaining normal copy, would be greatly increased by deletion of that chromosome or duplication of the one with the mutant gene. This is exemplified by the frequent loss of PTEN activity by LOH in cancers thus taking the brakes off PI3K–AKT signaling.

Life–death decisions after DNA damage

When God desires to destroy a thing, he entrusts its destruction to the thing itself. Every bad institution of this world ends by suicide.

Victor Hugo

Failure to repair DNA lesions properly after the induction of cell proliferation arrest can lead to mutations or large-scale genomic instability. Such changes may have tumorigenic potential if the cell cycle is reactivated inadvertently (possibly by mutations in PLK1 or CDC25B, or by other causes of premature checkpoint abrogation). Elevated levels of PLK1 correlate with metastatic potential and poor prognosis. Thus, in many ways it may be preferable for cells with damaged DNA to be eliminated via apoptosis. As described in Chapter 8, loss of this apoptotic response is actually one of the hallmarks of cancer. However, it

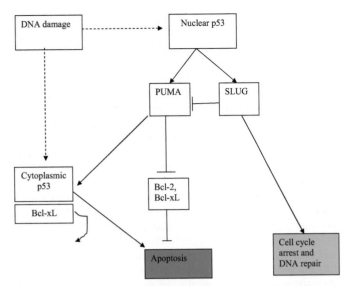

Figure 10.13 Switching between p53-mediated apoptosis and growth arrest and DNA repair; the role of SLUG and PUMA. p53 acts in the nucleus to regulate proapoptotic genes, whereas cytoplasmic p53 directly activates proapoptotic Bcl-2 proteins to permeabilize mitochondria and initiate apoptosis. After DNA damage, Bcl-xL sequesters cytoplasmic p53, whereas nuclear p53 regulates expression of PUMA, which then displaced p53 from Bcl-xL, allowing p53 to activate BAX and induce mitochondrial permeabilization. Mutant Bcl-xL that bound p53, but not PUMA, rendered cells resistant to p53-induced apoptosis. SLUG is also transactivated by p53 and inhibits PUMA; the balance of these may determine if a cell with DNA damage will arrest and repair or die.

should also be appreciated that avoidance of apoptosis may be coupled in a controlled way to growth arrest and DNA repair. There are now numerous examples of regulators of p53 and others that may allow "switching" between death and growth arrest depending on the extent of DNA damage (Fig. 10.13). Most recently, it has been shown that the antiapoptotic protein BclxL, in addition to blocking MOMP at the mitochondrion (see Chapter 8), may also stabilize the G_2–M checkpoint after DNA damage to promote growth arrest. This could be an inherent safeguard against blocking cell death without in some way limiting growth potential. Defects in DNA damage-induced apoptosis contribute to tumorigenesis and to the resistance of cancer cells to a variety of therapeutic agents. The intranuclear mechanisms that signal apoptosis after DNA damage overlap with those that initiate cell-cycle arrest and DNA repair, and the early events in these pathways are highly conserved. In addition, multiple independent routes have recently been traced by which nuclear DNA damage can be signaled to the mitochondria, tipping the balance in favor of cell death rather than repair and survival (see Chapter 8 on apoptosis). A conflict in cell-cycle progression or DNA damage can lead to mitotic catastrophe when the DNA structure checkpoints are inactivated, for instance when the checkpoint kinase CHK2 is inhibited. In these circumstances, cells undergo p53-independent apoptosis during metaphase. Suppression of apoptosis leads to the generation of aneuploid cells. DNA damage which fails to arrest the cell cycle together with inhibition of apoptosis can favor the occurrence of cytogenetic abnormalities and tumorigenesis.

The last few years have witnessed a large number of studies that have addressed the complex issue of how and at what point a cell attempting to fix damaged DNA is finally pushed into apoptosis. In response to DNA damage, p53 regulates cell-cycle checkpoints and induction of apoptosis; however, the molecular mechanisms responsible for committing to these distinct functions are only now being unraveled and appear to in part involve the transcriptional activity of p53 and gene targets such as PUMA and BID. But what switches a cell from cell-cycle stalling, maybe largely through inhibition of CDC25s, activation of p21, and death through new gene expression? How is the delicate balance between high sensitivity and tolerance to intrinsic damage achieved? Moreover, how is the DDR aborted once repair is deemed to be successfully completed? Much of this relates to the very strict regulation of p53 levels and activity in the cell. MDM2 and MDM4 (human double minutes (HDMs) in humans) are pivotal regulators of p53 activity and are themselves subject to a series of regulatory signaling pathways, including ARF (Fig. 10.14). It appears that under particular circumstances, E2F transcription factors can promote expression of ARF (Fig. 10.15). MDM proteins regulate p53 function after DNA damage and numerous causes of cell stress and after recovery, by targeting p53 for degradation. Phosphorylation of p53 at Ser46 appears important for expression of various apoptosis mediators such as p53AIP1.

A recent study implicates the HIPK2 protein, which phosphorylates p53 at serine 46 allowing activation of apoptotic genes. HIPK2 levels are normally low due to proteasomal degradation, but following serious DNA damage a LIM domain protein Zyxin helps stabilize HIPK2, protecting it from its ubiquitin ligase Siah-1 and allowing expression of p53 apoptotic genes. Dual-specificity tyrosine–phosphorylation-regulated kinase 2 (DYRK2) can also phosphorylate p53 at Ser46. Inhibitors of p53 apoptosis must also be considered. The KRAB-type zinc finger protein Apak (ATM and p53-associated KZNF protein) inhibits p53-mediated apoptosis and is inactivated by ATM kinase, thus activating p53 after DNA damage. In response to oncogenic stress, Apak is dissociated from p53 by ARF. Factors that may push a cell toward senescence have also been identified and include the polybromo-associated BRG1-associated factor components BRD7 and BAF180, which drive p53-mediated gene expression toward replicative and oncogenic stress senescence. Other p53-regulated genes are involved in the termination of a DDR necessary before the cell cycle can be restarted. The wild-type p53-induced phosphatase 1 (Wip1) overexpressed in various cancers can dephosphorylates gamma-H2AX, thereby preventing effective recruitment or retention of DNA repair factors. By implication, activation of WIP1 might be a means of limiting the DDR but, if inappropriately activated, it could compromise DNA repair in cancer. In an interesting recent study, time-lapse microscopy of individual human cells was employed to look at p53 in real time. It was found that proliferating cells exhibited spontaneous pulses of p53 in response to cell-cycle-related intrinsic DNA damage. Importantly, unless the DNA damage was sustained, p53 was kept inactive and did not induce p21 expression or arrest the cell cycle.

Oncogenic stress

Quod me nutrit me destruit. (What nourishes me destroys me.)
Christopher Marlowe

Inappropriate deregulated expression of oncoproteins driving cell-cycle progression can generate a signal that is detected by the

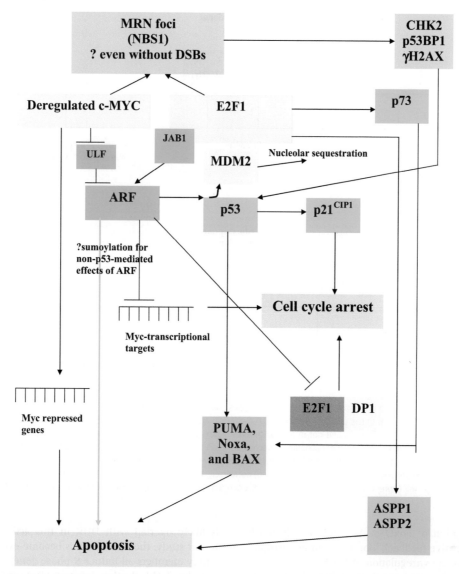

Figure 10.14 Tumor suppressor activity of the ARF protein in oncogenic stress. ARF expression is increased directly by E2F1, and in part by inhibition of the ULF ubiquitin ligase by oncogenic c-MYC. ARF activates or stabilizes p53, by displacing MDM2. ARF also inhibits the transcriptional activity of c-MYC, thus preventing MYC-regulated expression of genes required for cell-cycle entry or progression and causing reduced replication. However, ARF does not prevent MYC from inhibiting gene expression; as inhibition of gene expression (and not transactivation) is required for MYC to promote apoptosis, this may be indirectly favored by ARF. ARF may also directly promote apoptosis independently of p53. E2F1 may also activate the p73 protein and thereby apoptosis. The cofactor JAB1 appears to be required for E2F1 to activate the p53-dependent apoptotic response. Though speculative, it is tempting to speculate that suppressing the p21 growth arrest pathway or preventing growth arrest in some other way might favor apoptosis. In human cells, if ARF growth arrest is prevented by knockdown of p21, then apoptosis is enhanced (Hemann et al., 2005). That the situation is complex is highlighted by another recent study; loss of ARF did not prevent acute MYC-induced apoptosis in pancreatic beta cells of transgenic mice *in vivo*; in fact, both apoptosis and replication were enhanced – suggesting that at least in this case apoptosis might be primarily driven through the DNA damage–ATM–CHK2 route, but also that ARF activation was mediating some degree of inhibition of replication. Here one might speculate that inhibition of MYC-transcriptional activity was dominant, allowing relatively unfettered MYC effects on inhibition of gene expression and accelerating apoptosis.

cell triggering growth arrest or apoptosis (see also Chapter 6). This inherent tumor suppressor activity of oncogenes is a relatively recently appreciated phenomenon yet may be one of the major barriers to neoplastic progression since the organism would be protected from further mutations and the threat of cancer as the early tumor cell would either die or at least be prevented from propagating the oncogenic mutation. Intriguingly, to generate the inhibitory signal the would-be cancer cell must first distinguish a "cancer cell cycle" involving aberrant oncogene activation from a "normal" cell cycle involving more usual levels or persistence of the same protein (proto-oncoprotein counterpart). This has been referred to as oncogenic stress, and there are parallels to genotoxic stress induced by DNA damage as both involve activation of p53, though by different and overlapping routes. In response to stress signals, such as DNA damage, Myc is downregulated in a p53-independent manner, contributing to arrest of cell cycle and avoidance of apoptosis. The mechanism may involve upregulation of miR-34b and miR-34c via a route

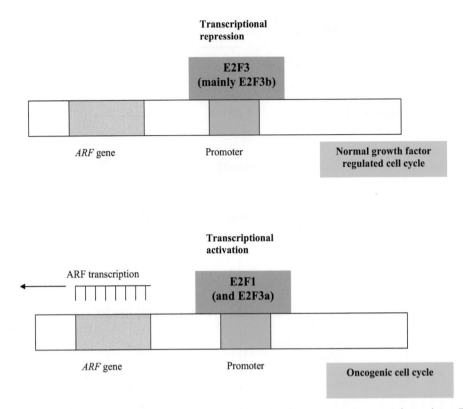

Figure 10.15 Regulation of ARF expression in normal and oncogenic cell cycles. Two pivotal tumor suppressor networks regulate cellular proliferation and form a crucial tumor surveillance mechanism: (i) the p16^{INK4A}–Cyclin D–CDK4–RB–E2F and (ii) the ARF–MDM2–p53. Recent studies suggest that activation of ARF, which occurs only in oncogene-driven cell cycles and not in normal growth factor–regulated cell cycles, is a pivotal suppressor of tumorigenesis, and in fact this means of p53 activation may be a more critical cancer-avoidance mechanism than the DNA damage response. How these pathways may interact and in particular result in expression of ARF remains unclear. E2F family members differentially regulate ARF transcription; in normal cells, the ARF promoter is occupied by E2F3b, not by other E2F family members, and transcription is repressed; conversely, oncogenic activation of ARF involves recruitment of E2F1 and E2F3a to the ARF promoter, thus increasing ARF expression. It is worth noting that in oncogenic stress driven by c-Myc, stability of the ARF protein is also important.

involving MK5 activation of FoxO3a. As MK5 is activated by Myc, this forms a negative feedback loop that can be disrupted, for example in CRC by downregulation of MK5.

Recent studies have thrown some light on how this key antitumor mechanism operates and have identified important links between DDR and activation of tumor suppressors (summarized in Fig. 10.16). Thus, it now appears that the tumor suppressor ARF–p53–p21 (see Chapters 6 and 7), BIM (Chapter 8), and ATM–ATR kinase pathways may all cooperate in various contexts in promoting apoptosis in response to elevated expression of oncogenes such as c-MYC. Recent studies of a variety of human tumors demonstrated that the early precursor lesions even prior to genomic instability express phosphorylated kinases ATM and CHK2, and phosphorylated histone H2AX and p53 – all markers of an activated DDR. Similar checkpoint responses could be evoked in cultured cells by deregulated expression of different oncogenes. This is particularly intriguing as it suggests yet another potential tumor suppressive mechanism triggered by inappropriate activity of an oncogene; apoptosis and senescence have already been described in this context in other chapters. Moreover, there may be connections. DDR to oncogene activation could act as an anticancer barrier in the same way as activation of ARF–p53 or BIM (another parallel apoptotic pathway activated by oncogenic Myc), namely, by in turn promoting either growth arrest or apoptosis of the cancer cell. ARF and BIM are activated by deregulated expression of oncogenes such as *c-Myc*, and there is likely overlap with DDR in this activity. Interestingly, in a recent study, the situation has become even more tortuous: E2F family members all induce S-phase genes and cell-cycle progression, but E2F1 also has pro-apoptotic activity. In a recent study, a putative mechanism has been unveiled that shows major parallels with c-Myc. The inactivation of RB and resultant deregulation of E2F1 in cultured cells led to an accumulation of DSBs, but not ROS. Moreover, E2F1 did not contribute to c-MYC-associated DSBs, suggesting that Myc and E2F1 (a c-MYC target) have independent effects on DDR. It is tempting to speculate that E2F1 induces apoptosis by activating DDR and oncogenic stress, but independently of upstream proteins such as c-Myc (which is important as a cell with mutated RB and deregulated E2F1 will not necessarily have elevated c-Myc from the outset). One mechanism that may link oncogenic stress and DDR is that the oncogenic proteins may promote recurrent initiation and collision of replication forks which might increase the risk of strand breaks.

As mentioned, ARF is a crucial mediator of oncogenic stress signals. ARF is a nucleolar protein that activates p53-dependent checkpoints by binding and displacing MDM2, which is sequestered in the nucleolus. ARF may ordinarily be sequestered and held inactive in the nucleolus by the nucleolar phosphoprotein, nucleophosmin (NPM). NPM may in turn compete with MDM2 and possibly other ARF targets. Thus, NPM overexpression antagonizes ARF function while increasing its nucleolar localization. In addition to activation of the p53 pathway, ARF can also

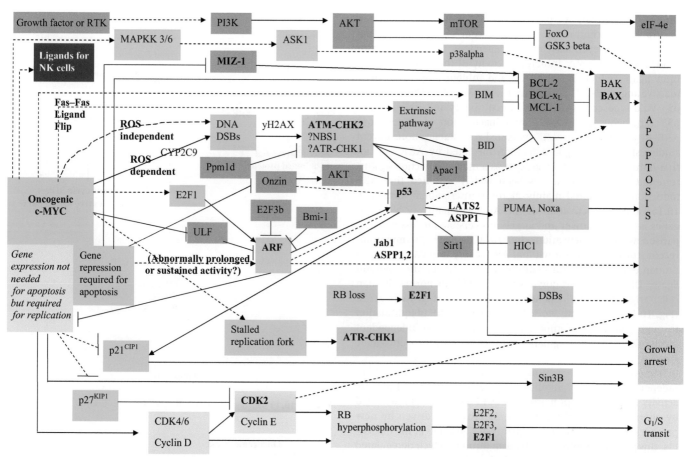

Figure 10.16 During tumorigenesis, the developing cancer cells experience "oncogenic stress," which evokes a counter-response to eliminate them. Although some of the key signaling proteins involved in oncogenic stress have been known for some time, notably ARF, the nature of all the key signals distinguishing between regular cell cycles and cancer cycles is still not fully characterized. This figure summarizes what is currently known. Deregulated expression of c-Myc can activate various BH3-only proteins such as BIM and BID, which by sequestering antiapoptotic BCL-2 proteins cooperate with the ARF–p53–PUMA–Noxa–BAX pathway in inducing apoptosis. At least in part, the activation of p53 and BID may be via DNA damage responses activated following c-Myc activation, suggesting that this may play a key role in activating these "failsafe" mechanisms not otherwise activated by c-Myc during normal cell cycles. BID may also contribute to arrest in the S phase. Oncogenic c-Myc can result in excess production of ROS and DNA DSBs (though the latter can occur independent of ROS formation), which may be key triggers for MYC-induced apoptosis in this context. Downregulation of p21^{CIP1} by c-Myc may also push cells from growth arrest toward apoptosis. DNA damage results in transcriptional induction of p53 target genes, including p21^{CIP1} as well as the proapoptotic BCL-2 family member p53 upregulated modulator of apoptosis (PUMA), NOXA, and BAX. p21^{CIP1} and PUMA mediate cell-cycle arrest and apoptosis, respectively. Importantly, p21^{CIP1} may trigger cell-cycle arrest at the expense of apoptosis; therefore, suppression of p21^{CIP1} by c-Myc may result in the predominance of apoptosis over cell-cycle arrest as an inherent tumor suppressor activity. Recent studies suggest that CDK2 may be important for Myc-induced apoptosis, but this action is independent of BID, p53, and likely DNA damage responses. Interestingly, Myc inhibits p27^{KIP1} (normally an inhibitor of CDK2 activity), thus potentially mitigating apoptosis through this route *in vivo*. It is possible that other CKIs such as p16^{INK4a} which can also growth-arrest cells might also prevent apoptosis by reducing MOMP; potentially p16^{INK4a} could push cells from apoptosis toward growth arrest or senescence. Loss of RB, which also deregulates cell-cycle control, can also culminate in DSBs and apoptosis, probably via aberrant activation of E2F1. E2F1 also induces ARF and apoptosis if deregulated. However, although E2F1 is regulated by c-Myc (via both transcriptional activation and then negatively via translational inhibition by two Myc-regulated microRNAs (miRNAs), miR-17-5p and miR-20a), E2F1 is dispensable for induction of ARF. It is suggested that E2F1 must be deregulated for activating apoptosis, and that this does not necessarily arise when c-Myc is deregulated. Recent studies suggest that, at least following growth factor withdrawal *in vitro*, Myc-induced apoptosis requires inhibition of gene expression; inhibition of Miz-1-regulated gene expression has been particularly identified as important, and loss of Miz-1-regulated expression of Bcl-2 may be a key factor in apoptosis with deregulated Myc. Onzin is the product of a gene negatively regulated by c-Myc, expressed at high levels in myeloid cells which can activate both Akt1 and Mdm2. Thus, onzin can downregulate p53, signaling an action lost when c-Myc is activated. Hatched lines indicate the presence of intermediate proteins (known or unknown). Sources: Bouchard, C., Marquardt, J., Bras, A., Medema, R.H., and Eilers, M. (2004). Myc-induced proliferation and transformation require Akt-mediated phosphorylation of FoxO proteins. *EMBO Journal*, **23**(14): 2830–40; Eischen, C.M., Roussel, M.F., Korsmeyer, S.J., and Cleveland, J.L. (2001). Bax loss impairs Myc-induced apoptosis and circumvents the selection of p53 mutations during Myc-mediated lymphomagenesis. *Molecular Cell Biology*, **21**(22): 7653–62; Eischen, C.M., Weber, J.D., Roussel, M.F., Sherr, C.J., and Cleveland, J.L. (1999). Disruption of the ARF-Mdm2-p53 tumor suppressor pathway in Myc-induced lymphomagenesis. *Genes & Development*, **13**(20): 2658–69; Fogal, V., Kartasheva, N.N., Trigiante, G., *et al.* (2005). ASPP1 and ASPP2 are new transcriptional targets of E2F. *Cell Death & Differentiation*, **12**(4): 369–76; Jacobs, J.J., Scheijen, B., Voncken, J.W., Kieboom, K., Berns, A., and van Lohuizen, M. (1999). Bmi-1 collaborates with c-Myc in tumorigenesis by inhibiting c-Myc-induced apoptosis via INK4a/ARF. *Genes & Development*, **13**(20): 2678–90; Patel, J.H., and McMahon, S.B. (2006). BCL2 is a downstream effector of MIZ-1 essential for blocking c-MYC induced apoptosis. *Journal of Biological Chemistry*, November 1; and Ruggero, D., Montanaro, L., Ma, L., *et al.* (2004). The translation factor eIF-4E promotes tumor formation and cooperates with c-Myc in lymphomagenesis. *Nature Medicine*, **10**(5): 484–6.

suppress cell growth in a p53- and MDM2-independent manner, illustrating the central role of ARF. Recent studies suggest that both the p53-dependent and -independent tumor suppressor actions of ARF involve inhibiting activity of the ARF–BP1 ubiquitin ligase, which is bound by and inhibited by ARF. Thus, inactivating ARF–BP1 suppresses the growth of p53 null cells; in p53 wild-type cells, ARF–BP1 directly binds and ubiquitinates p53 (analogous to MDM2); and inactivation of endogenous ARF–BP1 is crucial for ARF-mediated p53 stabilization. Recent studies showed that p53 activation mediated by ARF, but not that induced by DNA damage, acts as a major protection against tumorigenesis *in vivo*. It has been widely assumed that ARF induction is mediated mainly at the transcriptional level (Fig. 10.15) and that activation of the ARF–p53 pathway by oncogenes is a much slower and largely irreversible process by comparison with p53 activation after DNA damage. An important recent study has now shown that ARF is very unstable in normal human cells in part because of the activity of a specific ubiquitin ligase, ULF. ULF knockdown stabilizes ARF in normal human cells, triggering ARF-dependent, p53-mediated growth arrest (Fig. 10.14). c-Myc can block ULF-mediated ARF ubiquitination through showing the importance of transcription-independent mechanisms in ARF regulation during responses to oncogenic stress.

Deregulated c-Myc may also activate p53 through an ARF-independent pathway involving suppression of Mdm2 E3 ligase activity by ribosomal proteins such as RPL5 and RPL11 activated by aberrant ribosomal biogenesis. As growth is driven by new protein synthesis and by manufacture of ribosomes at the nucleolus, this provides another mechanism through which deregulated oncogene activity can be controlled and operates separate to DDR. Ribosome biogenesis is resource intensive and not surprisingly is therefore subjected to surveillance; if disturbed, it can give rise to yet another form of stress that the poor cell is subject to, "nucleolar stress," which leads to the activation of the RP–Mdm2–p53 stress response discussed in this chapter.

Many tumors harbor inactive mutant p53 or evade tumor suppression by mutations that impede the p53 pathway. ASPP1 (apoptosis-stimulating protein of p53-1) is a key mediator in the transcription of p53-regulated proapoptotic genes and is translocated to the nucleus in response to oncogenic stress. This shift from cytoplasmic to nuclear ASPP1 is driven by the tumor suppressor Lats2 (large tumor suppressor 2), which phosphorylates ASPP1. Therefore, the combined action of Lats2 and ASPP1 drives expression of p53-regulated proapoptotic genes during oncogenic stress. Conversely, this is antagonized by the Yes-associated protein 1 (Yap1) oncoprotein, which prevents Lats2–ASPP1 interaction.

In addition to the increasingly pervasive miRNAs, it is also clear that few processes in cell biology are unaffected by chromatin changes. A recent paper has intriguingly suggested that senescence-associated heterochromatic foci (SAHF), thought to be largely a mediator of senescence, may rather be activated by inappropriate oncogene activation. Oncogene-induced SAHF formation depends on DNA replication and ATR, linking this with oncogenic stress (Fig. 10.17).

Numerous apoptosis-related genes are regulated by p53, such as those encoding death receptors such as *FAS* and proapoptotic Bcl-2 proteins such as *BAX*, *BID*, *NOXA*, and *PUMA*. However, p53 also accumulates in the cytoplasm, where it directly activates *BAX* to promote *MOMP* (Chapter 8). *MOMP* causes release of

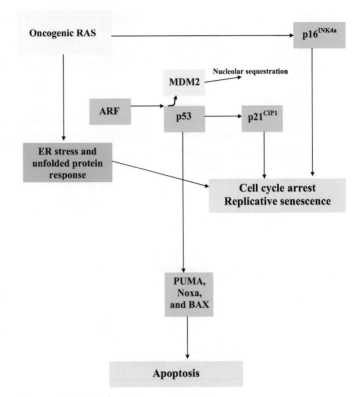

Figure 10.17 Oncogenic stress and senescence.

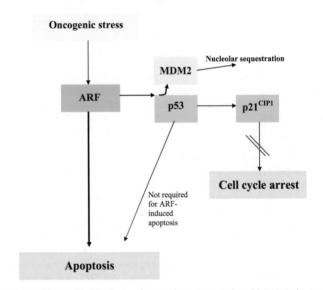

Figure 10.18 Mitochondrial pathway of apoptosis induced by DNA damage or oncogenic stress. In the presence of antiapoptotic BCL-2 family proteins, BAX and BAK in the outer mitochondrial membrane are inactive. However, if these are displaced by BH3-only proteins (such as PUMA, NOXA, and BIM) in response to DNA damage, oncogenic stress, and so on, then BAX and BAK allow MOMP and release of cytochrome C and others which trigger apoptosis via activation of caspase-9.

cytochrome c from mitochondria, triggering activation of caspase-9 and apoptosis (Fig. 10.18).

In a recent study from Doug Green's group, DNA damage-induced apoptosis was found to include both nuclear and cytoplasmic functions of p53. After p53 stabilization, p53 accumulated in the nucleus and promoted expression of proapoptotic genes, such as *BAX* and *PUMA,* and accumulated in the cytoplasm, where it bound to Bcl-xL. Intriguingly, following nuclear activity of p53, *PUMA* displaced p53 from BclxL, enabling direct activation of *BAX* and *MOMP*. This study provides an explanation as to why loss of BclxL might sensitize to DNA damage, whereas *PUMA* deficiency promotes resistance to numerous p53-dependent apoptotic stimuli. Aside from loss of p53, it now appears that activation of genes may also allow cells to survive oncogenic stress. The innate immune-signaling kinase TBK1, which is deregulated in some cancers, can overcome proapoptotic signals and allow cells to survive oncogenic stress, in part by directly activating AKT.

This is likely to be a fertile research area in the next few years as we learn more about how oncogenes such as *c-Myc* determine cell fate and how opposing outcomes such as apoptosis and cell-cycle progression are balanced.

It is important to note that oncogenic stress, by deregulated oncogenes such as RAS, may promote senescence rather than apoptosis (Fig. 10.19). This is discussed in more detail in Chapters 6 and 9.

Microsatellites and minisatellites

Microsatellites are short sequences of 1–5 bp repeated in tandem throughout the genome; because of their polymorphic nature, they have been widely used as genetic markers. Mutations can occur within the sequence of these microsatellites as a result of the slippage of DNA polymerase during DNA synthesis, resulting in small expansions or deletions within the repeat sequence manifesting as loss or gain of simple repetitive units. This is a common phenomenon, and in healthy cells, this is repaired by the cell DNA mismatch repair (MMR) system. In cells defective in MMR (e.g. as a result of mutation or hypermethylation of one or more of the MMR genes), these expansions and deletions are not repaired. Microsatellite analysis of DNA from such MMR-deficient cells may show the appearance of a new microsatellite PCR product on an electrophoretic gel, either higher or lower than the normal alleles, and the concomitant loss of intensity of one of the normal alleles. This is defined as an allelic shift, and if this occurs at a substantial proportion of microsatellite loci, it

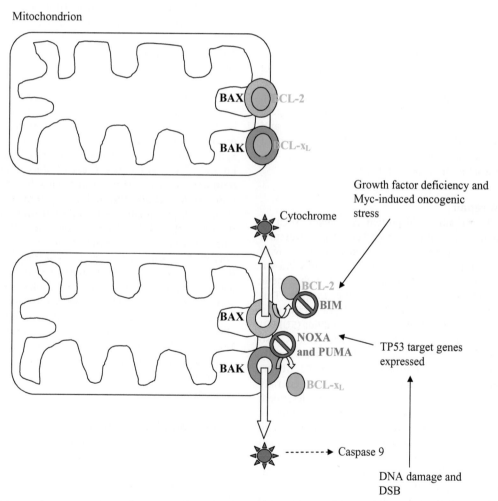

Figure 10.19 Blocking growth arrest enhances apoptosis induction by oncogenic stress and ARF. In some human cancer cell lines, loss of p53 or of p21 prevented ARF-induced growth arrest, but neither prevented ARF-induced apoptosis. In fact, lack of functional p21 resulted in the accumulation of cells in the G_2–M transition of the cell cycle and markedly enhanced p14[ARF]-induced apoptosis, suggesting that signals required for G_1 arrest and apoptosis induction by ARF diverge upstream of p53. In fact, it may be a general effect that, in cancer cells, oncogenic stress leads to either replicative arrest or apoptosis and, moreover, that inhibiting either will actively favor signaling down the alternative route of tumor suppression, namely, release growth arrest and apoptosis increases or *vice versa.*

is representative of microsatellite instability (MSI), often taken as diagnostic of loss of DNA MMR.

The human genome contains a unique class of domains, referred to as AT islands, which consist typically of 200–1000 bp long tracts of up to 100% A/T DNA. AT islands are inherently unstable (expandable) minisatellites that are found in various known loci of genomic instability, such as AT-rich fragile sites. The cellular function of AT islands may differ in cancer and normal cells, but appears to include DNA replication. AT islands are preferentially targeted by the potent DNA-alkylating antitumor drug bizelesin and may be a major factor in cytotoxicity of these agents.

Promyelocytic leukemia nuclear bodies (PML NBs)

PML NBs are generally present in all mammalian cells. Mice lacking PML NBs exhibit chromosome instability and are sensitive to carcinogens. PML NBs are implicated in the regulation of transcription, apoptosis, tumor suppression, and the antiviral response. They may be sites for the formation of multi-subunit complexes involved in posttranslational modification of regulatory factors, such as p53, and potentially also in DNA repair. Several repair factors transit through PML NBs, which may act as general sensors of cellular stress, and which rapidly disassemble following DNA damage into large supramolecular complexes, dispersing associated repair factors to sites of damage. The PML tumor suppressor protein potentiates p53 function by regulating posttranslational modifications, such as CBP-dependent acetylation and CHK2-dependent phosphorylation, in the PML NB. PML enhances p53 stability by sequestering the p53 ubiquitin–ligase MDM2 to the nucleolus. Following DNA damage, PML and MDM2 accumulate in the nucleolus in an ARF-independent manner. Nucleolar localization of PML is dependent on ATR activation and phosphorylation of PML by ATR.

SUMO and DNA repair

All eukaryotic cells contain multi-protein complexes, such as cohesin and condensin, that include members of the structural maintenance of chromosome (SMC) proteins, together with several non-SMC subunits. The third complex, currently known as the Smc5/6 complex, is poorly understood, though genetic analysis points to an essential role in genome integrity. This complex has been shown to have SUMO ligase activity in fission and budding yeasts. Small ubiquitin-like modifier (SUMO) is a small protein distantly related to ubiquitin and is also discussed in Chapter 11. Like ubiquitin, SUMO is covalently conjugated to lysines on target proteins using an E1 SUMO activator and an E2 SUMO-conjugating enzyme, generally referred to as Ubc9. E3 SUMO ligases have also been described, but there are far fewer such enzymes than the large family of E3 ubiquitin ligases. Modification by SUMO does not lead to protein degradation but may compete with ubiquitin. The main role of sumoylation appears to be in signaling pathways, transcriptional activities, and the regulation of subcellular localization. Mammalian cells have three distinct SUMO genes or polypeptides, providing further complexity. Many sumoylated proteins have been identified, including DNA repair proteins such as Rad51 and Rad52 involved in HR. Human homologues of both interact with SUMO in yeast two-hybrid assays. Cohesin and condensin are both implicated in DNA repair and contain SMC proteins.

Chaperones and genomic instability

Molecular chaperones, otherwise known as heat shock proteins, are key regulators of protein homeostasis within cells. Specifically, chaperones direct folding, localization, and proteolytic turnover of many key regulators of cell growth, differentiation, and survival. Moreover, they may contribute to inactivation of various nuclear receptors prior to ligand binding. Not surprisingly, therefore, chaperones might also be involved in facilitating or hindering malignant transformation at the molecular level. Thus, chaperones can serve as biochemical buffers at the phenotypic level for the genetic instability that is characteristic of many human cancers. Chaperone proteins could even enable progressing tumor cells to survive mutations that would otherwise be lethal. The recent discovery and study of several natural product anti-tumor agents that selectively inhibit the function of Hsp90 have confirmed the utility of these agents as relatively nontoxic inhibitors of multiple aberrant signaling molecules.

Viral oncogenes and instability

High-risk human papillomavirus (HPV)-associated carcinogenesis of the uterine cervix is a particularly useful model for basic mechanisms of genomic instability in cancer. Cervical carcinogenesis is associated with the expression of two high-risk HPV-encoded oncoproteins, E6 and E7. Aneuploidy, the most frequent form of genomic instability in human carcinomas, develops as early as in nonmalignant cervical precursor lesions. In addition, cervical neoplasia is frequently associated with abnormal multipolar mitotic figures, suggesting disturbances of the cell-division process as a mechanism for chromosome segregation defects. Spindle poles are formed by centrosomes, and the high-risk HPV E6 and E7 oncoproteins can each induce abnormal centrosome numbers. These two HPV oncoproteins, however, induce centrosome abnormalities through fundamentally different mechanisms and, presumably, with different functional consequences. High-risk HPV E7, which targets the RB tumor suppressor pathway, can provoke abnormal centrosome duplication in phenotypically normal cells. On the contrary, cells expressing the HPV E6 oncoprotein, which inactivates p53, accumulate abnormal numbers of centrosomes in parallel with multinucleation and nuclear atypia. These two pathways are not mutually exclusive, since coexpression of HPV E6 and E7 has synergistic effects on centrosome abnormalities and chromosomal instability. Taken together, these findings support the general model in which chromosomal instability arises as a direct consequence of oncogenic insults and can develop at early stages of tumor progression.

Radiation-induced genomic instability

Radiation-induced genomic instability encompasses a range of measurable end points such as chromosome destabilization, sister chromatid exchanges, gene mutation and amplification, late cell death, and aneuploidy, all of which may be causative factors in the development of clinical disease, including carcinoma. Clinical implications of genomic instability can be broadly grouped into two main areas: as a marker for increased cancer risk or early detection, and as a consequence of radiation therapy (IR) that may be causative of, or a strong marker for, the induction of a therapy-induced second malignancy.

Cancer susceptibility syndromes involving genetic instability

Integer vitae scelerisque purus. (Unimpaired by life and clean of wickedness.)

Horace

Given that many of these syndromes have given their names to the genes and proteins already discussed in this chapter, the reader should try to relate knowledge of these proteins' functions with the phenotype of the syndromes.

Ataxia-telangiectasia

Ataxia-telangiectasia (A-T) is a syndrome of cancer susceptibility, immune dysfunction, and neurodegeneration that is caused by mutations in the A-T-mutated (*ATM*) gene. ATM has been extensively discussed in Chapter 4 and this chapter, so you will need no reminder of the pivotal role of ATM in the DDR, including the activation of cell-cycle checkpoints and induction of apoptosis. Conversely, it is self-evident that defective cell-cycle checkpoint regulation, impaired activation of p53 by DNA damage, and associated genomic instability would contribute to cancer susceptibility in A-T, yet the mechanism of neurodegeneration in A-T is not well understood.

Nijmegen breakage syndrome

NBS is an autosomal, recessive, hereditary, chromosomal instability syndrome associated with a predisposition to tumor formation. Affected individuals also exhibit microcephaly, a "bird-like" facial appearance, growth retardation, immunodeficiency, and radiosensitivity. The gene defective in NBS has been cloned, and the gene product, NBS1 (nibrin), is a member of the MRN complex (Mre11, Rad50, and NBS1) important in DSB repair. There are obvious similarities between NBS and ATM-mutated fibroblasts with respect to radiation sensitivity, and this is explained by recent results suggesting that ATM kinase can phosphorylate NBS1. Although NBS1 is a putative tumor suppressor, it is also expressed in highly proliferating tissues developmentally and is located at sites of DNA synthesis through interaction with E2F. *NBS1* has been shown by some studies to be a c-MYC target gene, and constitutive expression of *NBS1* in cell lines paradoxically enhances transformation, at least in part by activating the PI3K–AKT pathway. Thus, *NBS1* can under some circumstances be oncogenic through the PI3K–AKT pathway. On the other hand, tumor suppressor activity lies in NBS1 function in the MRN complex required for DSB repair.

Bloom's syndrome

RecQ helicases are conserved from bacteria to humans and are required for the maintenance of genome stability in all organisms. RecQ helicases are caretakers of the genome and influence stability through their participation in DNA replication, recombination, and repair pathways. Mutations in three of the human RecQ family members give rise to genetic disorders characterized by genomic instability and a predisposition to cancer. Bloom's syndrome (BS) is associated with cancer predisposition and genomic instability. A defining feature of Bloom's syndrome is an elevated frequency of sister chromatid exchanges. These arise from crossing over of chromatid arms during HR, a ubiquitous process that exists to repair DSB and damaged replication forks. Whereas crossing over is required in meiosis, in mitotic cells it can be associated with detrimental loss of heterozygosity. BLM forms an evolutionarily conserved complex with human topoisomerase IIIalpha (hTOPO IIIalpha), which can break and rejoin DNA to alter its topology. Inactivation of homologues of either protein leads to hyper-recombination in unicellular organisms. The BS gene product, BLM, associates with the stress-activated ATR kinase discussed in this chapter, which functions in checkpoint signaling during S-phase arrest.

Xeroderma pigmentosum

Xeroderma pigmentosum (XP) is an autosomal recessive disease characterized by sun sensitivity, early onset of freckling, and subsequent neoplastic changes on sun-exposed skin. Skin abnormalities result from an inability to repair UV-damaged DNA because of defects in the nucleotide excision repair (NER) machinery. Xeroderma pigmentosum is genetically heterogeneous and is classified into seven complementation groups (XPA–XPG) that correspond to genetic alterations in one of seven genes involved in NER. The variant type of XP (Xeroderma pigmentosum V, or XPV), first described in 1970 by Ernst G. Jung as "pigmented xerodermoid," is caused by defects in the postreplication repair machinery, while NER is not impaired. The XPV protein, polymerase polη, represents a novel member of the Y family of bypass DNA polymerases that facilitate DNA translesion synthesis. The major function of polη is to allow TLS of UV-induced TT dimers in an error-free manner; it also possesses the capability to bypass other DNA lesions in an error-prone manner. XPV is caused by molecular alterations in the POLH gene, located on chromosome 6p21.1–6p12. Affected individuals are homozygous or compound heterozygous for a spectrum of genetic lesions, including nonsense mutations, deletions, or insertions, confirming the autosomal recessive nature of the condition.

Fanconi anemia

This is a hereditary chromosome instability disorder in which cells have a predisposition to chromosomal breaks, characterized by congenital abnormalities, retarded growth, early predisposition to cancer, and bone marrow failure. The pathogenesis is believed to be due to defective repair of strand breaks, but the genetic cause is complicated with defects in multiple distinct genes causing the condition in different families (*FANCA, FANCB, FANCC, BRCA2* (also called *FANCD1*), *FANCD2, FANCE, FANCF, FANCG, FANCI, FANCJ,* and *FANCL*). *BRCA2* was the first such gene with a clearly defined role in DNA repair (HR). More recently, *FANCJ*, which encodes a DNA-unwinding enzyme (helicase) called BRIP1, mutated in some cases of breast cancer, has been shown to interact with the BRCA1 tumor suppressor. BRIP1, like BRCA1, is involved in binding and unwinding forked duplexes in DSB repair. Ubiquitination of FANCD2 may be an important shared action of the Fanconi anemia core complex proteins (FANCA, FANCB, FANCC, FANCE, FANCF, FANCG, and FANCL), although speculative FANCD2 may then be able to stabilize structurally blocked and broken replication forks, thereby contributing to both trans-lesion synthesis and HR. Crosslinks can result in stalling of DNA replication forks, and must be unhooked by incision on both sides of an adducted base in order to break one parental strand at the replication fork. BRIP1 helicase activity might help expose the incision substrate by unwinding parental strands adjacent to the crosslink. The broken fork is subsequently restored by HR involving BRCA2 and potentially other proteins, such as BRIP1, that enable Rad51 to carry out strand transfer. Intriguingly, even though FA cells experience constitutive chromosomal breaks, activation of the G_2–M checkpoint, and increased apoptosis, some cells can acquire the ability to bypass the G_2–M checkpoint with reduced activation of the ATR–CHK1–p53 pathway, allowing the condition to stabilize clinically.

Werner's syndrome

This is an autosomal recessive disease characterized by premature aging, elevated genomic instability, and increased cancer incidence, resulting from inactivation of the *WRN* gene. Telomere attrition is implicated in the pathogenesis of Werner syndrome. c-MYC overexpression directly elevates transcription of the *WRN* gene, whose presence is required to avoid senescence during c-MYC proliferative stimuli. Recent work suggests that *WRN* may counteract the effects of genomic instability promoted by deregulated c-MYC activation and thereby enable propagation of mutant cells, which in the absence of *WRN* would be ablated.

Genomic instability and colon cancer

Colorectal cancer results from the progressive accumulation of genetic and epigenetic alterations that lead to the transformation of normal colonic epithelial cells to colon adenocarcinoma cells (see Fig. 3.3). The loss of genomic stability appears to be a key early pathogenetic event that may foster the subsequent occurrence of alterations in tumor suppressor genes and oncogenes; in particular, activation of Wnt target genes constitutes the primary transforming event in colorectal cancer. At least three forms of genomic instability have been identified in colon cancer: (1) microsatellite instability, (2) chromosome instability (i.e. aneuploidy), and (3) chromosomal translocations. Microsatellite instability occurs in approximately 15% of colon cancers and results from inactivation of the MMR system by gene mutations or hypermethylation of the MLH1 promoter. MSI promotes tumorigenesis through generating mutations in target genes that possess coding microsatellite repeats, such as TGFBR2 and BAX. CIN is found in the majority of colon cancers and may result primarily from defects in DNA replication checkpoints and mitotic-spindle checkpoints. Mutations of the mitotic checkpoint regulators BUB1 and BUBR1 and amplification of STK15 have been identified in a subset of CIN colon cancers.

Radiotherapy, drugs, and DNA damage

"To hold a wolf by his ears" or "To hold a tiger by its tail."

Proverb

Chemotherapy and radiation induce p53-dependent DNA damage in noncancer cells resulting in side effects that limit treatment and may in themselves be lethal. Most agents used to treat cancer (see Chapter 16) clinically target genomic DNA. The selectivity of these anticancer drugs for cancers is likely due to the presence of tumor-specific defects in cell-cycle checkpoints, compromising DNA repair or enhancing apoptotic responses in the tumor.

Drugs are now being developed to target components of these checkpoints. Activation of the ATM–CHK2 pathway is important in this context as it responds to more serious DNA damage such as DSBs. CHK2 can activate both apoptosis (via p53, E2F1, and PML) and cell-cycle checkpoints (via Cdc25A and Cdc25C, p53, and BRCA1). Thus, CHK2 inhibitors might be used to enhance the tumor selectivity of DNA targeted agents in p53-deficient tumors, and for the treatment of tumors whose growth depends on enhanced CHK2 activity.

In a recent study from the laboratory of Gerard Evan, using a model in which p53 status can be reversibly switched *in vivo* between functional and inactive states, it was unexpectedly shown that the p53-mediated pathological response to genetic damage (induced by whole-body irradiation) was irrelevant for suppression of radiation-induced lymphoma. Interestingly, delaying the reactivation of p53 until the acute radiation response had subsided dramatically curtailed undesirable radiation damage whilst only partially limiting protection from lymphoma. However, in this case intact ARF was essential in preventing lymphoma by protecting against oncogene deregulation.

Conclusions and future directions

Genomic stability is essential for normal cellular function, and highly conserved pathways have evolved to repair DNA damage and prevent genomic instability or, failing that, to trigger senescence or death of the cell. Should these mechanisms fail, this may promote genomic instability – a major driver of cancer.

Genomic instability is classed as microsatellite instability (MSI) associated with mutator phenotype or chromosome instability (CIN) manifesting as gross chromosomal abnormalities. Three DNA damage repair processes are implicated in the development of a mutator phenotype, NER, BER, and MMR.

The proteins ATR and RAD1 bind to chromatin and are needed to phosphorylate several other proteins that are critical to the DNA damage response, including BRCA1, which when mutated leads to hereditary breast cancer. Defects in DNA MMR and CIN pathways are involved in several hereditary cancer predisposition syndromes, including HNPCC, Bloom syndrome, and A-T. Epigenetic factors as well as genetic factors may compromise the operation of cell-cycle checkpoints and sensing of DNA damage. Telomere attrition can also produce genetic instability as discussed in Chapter 9.

At least three forms of genomic instability have been identified in colon cancer: (1) MSI, CIN, and chromosomal translocations. MSI occurs in approximately 15% of colon cancers due to defective MMR (mutations or hypermethylation of the MLH1 promoter) and generates mutations in genes with coding microsatellite repeats, such as *TGFBR2* and *BAX*. CIN is found in the majority of colon cancers as result of deregulation of the DNA replication checkpoints and mitotic-spindle checkpoints.

Much still needs to be learned about mechanisms that induce and influence genomic instability, in particular to understand more about how DNA damage is sensed and thence how mutations of the various mitotic checkpoint regulators (including BUB1, BUBR1, and others) may culminate in instability and cancer. It is hoped that eventually studies will lead to a deeper understanding of the pathogenesis of genetic instability and potentially new therapies exploiting this characteristic of cancer cells.

Tempora mutantur, nos et mutamur in illis. (Times change, and we change with them.)

Lothair

Key experiment

While Watson and Crick had suggested that DNA replication would be semiconservative, proof was provided by the experiments of M. S. Meselson and F. W. Stahl.

The bacterium *Escherichia coli* (*E. coli*) was grown in culture medium containing NH_4^+ as the source of nitrogen for DNA (as well as protein) synthesis. In these elegant studies, two different ammonium ions containing isotopes of either nitrogen ^{14}N (the common form) or ^{15}N were employed. Not surprisingly, after growing *E. coli* for several generations in a medium containing $^{15}NH_4^+$, they found that the DNA of the cells was heavier than normal because it contained ^{15}N. The difference in density could readily be measured by isolating the DNA and precipitating it in an ultracentrifuge – the denser DNA accumulates at a lower level in the centrifuge tube. If however, *E. coli* were then switched from growing in $^{15}NH_4^+$ to a medium containing $^{14}NH_4^+$ and were allowed to undergo a single cell division, the DNA was now found to be intermediate in density between that of the previous generation and the normal. In other words the newly synthesized DNA contains equal amounts of both ^{14}N and ^{15}N. Now, if *E. coli* were allowed to divide once more in normal $^{14}NH_4^+$-containing medium, two distinct densities of DNA would be formed in which half the DNA was normal and half was intermediate. This confirmed that DNA molecules are not degraded and then made anew from free nucleotides between cell divisions, but instead each original strand remains intact with a new complementary strand added by new synthesis. This is termed "semiconservative replication" as each new DNA molecule contains both an original strand and a new one.

Outstanding questions

1. Is the presence of MSI or CIN a general feature of cells destined to become cancer cells, and at what stage in carcinogenesis do these features of genetic instability arise?

2. Is the mutator phenotype a generally applicable model to cancers other than colon?

3. Further define the genetic and epigenetic changes in various genes that may contribute to defective DNA repair and chromosomal segregation.

4. How might knowledge about defective mismatch repair and other processes be translated into more effective anticancer therapies? For example, might further increasing genetic instability with agents such as taxols preferentially lead to death of cancer cells?

Bibliography

Note to reader: See also the bibliographies of Chapters 3 and 4.

General

Aggarwal, M., Sommers, J.A., Shoemaker, R.H., and Brosh, R.M., Jr. (2011). Inhibition of helicase activity by a small molecule impairs Werner syndrome helicase (WRN) function in the cellular response to DNA damage or replication stress. *Proceedings of the National Academy of Sciences USA*, **108**(4): 1525–30.

Albertson, D.G., Collins, C., McCormick, F., and Gray, J.W. (2003). Chromosome aberrations in solid tumours. *Nature Genetics*, **34**(4): 369–76.

Al-Ejeh, F., Kumar, R., Wiegmans, A., Lakhani, S.R., Brown, M.P., and Khanna, K.K. (2010). Harnessing the complexity of DNA-damage response pathways to improve cancer treatment outcomes. *Oncogene*, **29**(46): 6085–98.

Aylon, Y., Ofir-Rosenfeld, Y., Yabuta, N., et al. (2010). The Lats2 tumor suppressor augments p53-mediated apoptosis by promoting the nuclear proapoptotic function of ASPP1. *Genes & Development*, **24**(21): 2420–9.

Banerjee, S., Kaye, S.B., and Ashworth, A. (2010). Making the best of PARP inhibitors in ovarian cancer. *Nature Reviews Clinical Oncology*, **7**(9): 508–19.

Bassermann, F., Frescas, D., Guardavaccaro, D., Busino, L., Peschiaroli, A., and Pagano, M. (2008). The Cdc14B-Cdh1-Plk1 axis controls the G2 DNA-damage-response checkpoint. *Cell*, **134**(2): 256–67.

Bassing, C.H., Suh, H., Ferguson, D.O., et al. (2003). Histone H2AX: a dosage-dependent suppressor of oncogenic translocations and tumours. *Cell*, **114**(3): 359–70.

Bensaad, K., Tsuruta, A., Selak, M.A., et al. (2006). TIGAR, a p53-inducible regulator of glycolysis and apoptosis *Cell*, **126**(1): 107–20.

Boyd, S.D., Tsai, K.Y., and Jacks, T. (2000). An intact HDM2 RING-finger domain is required for nuclear exclusion of p53. *Nature Cell Biology*, **2**: 563–8.

Burrows, A.E., Smogorzewska, A., and Elledge, S.J. (2010). Polybromo-associated BRG1-associated factor components BRD7 and BAF180 are critical regulators of p53 required for induction of replicative senescence. *Proceedings of the National Academy of Sciences USA*, **107**(32): 14280–5.

Campbell, P.J., Yachida, S., Mudie, L.J., et al. (2010). The patterns and dynamics of genomic instability in metastatic pancreatic cancer. *Nature*, **467**(7319): 1109–13.

Castedo, M., Perfettini, J.L., Roumier, T., et al. (2004). Mitotic catastrophe constitutes a special case of apoptosis whose suppression entails aneuploidy. *Oncogene*, **23**(25): 4362–70.

Ceccaldi, R., Briot, D., Larghero, J., et al. (2011). Spontaneous abrogation of the G_2DNA damage checkpoint has clinical benefits but promotes leukemogenesis in Fanconi anemia patients. *Journal of Clinical Investigation*, **121**(1): 184–94.

Cha, H., Lowe, J.M., Li, H., et al. (2010). Wip1 directly dephosphorylates gamma-H2AX and attenuates the DNA damage response. *Cancer Research*, **70**(10): 4112–22.

Chen, D., Kon, N., Li, M., Zhang, W., Qin, J., and Gu, W. (2005). ARF-BP1/Mule is a critical mediator of the ARF tumor suppressor. *Cell*, **121**(7): 1071–83.

Chen, D., Shan, J., Zhu, W.G., Qin, J., and Gu, W. (2010). Transcription-independent ARF regulation in oncogenic stress-mediated p53 responses. *Nature*, **464**(7288): 624–7.

Chipuk, J.E., Bouchier-Hayes, L., Kuwana, T., Newmeyer, D.D., and Green, D.R. (2005). *Science*, **309**(5741): 1732–5.

Christophorou, M.A., Ringshausen, I., Finch, A.J., Swigart, L.B., and Evan, G.I. (2006). The pathological response to DNA damage does not contribute to p53-mediated tumour suppression. *Nature*, **443**(7108): 214–7.

Cosentino, C., Grieco, D., and Costanzo, V. (2010). ATM activates the pentose phosphate pathway promoting anti-oxidant defence and DNA repair. *EMBO Journal*, December 14.

Crighton, D., Wilkinson, S., O'Prey, J., et al. (2006). DRAM, a p53-induced modulator of autophagy, is critical for apoptosis *Cell*, **126**(1): 121–34.

Crone, J., Glas, C., Schultheiss, K., Moehlenbrink, J., Krieghoff-Henning, E., and Hofmann, T.G. (2011). Zyxin is a critical regulator of the apoptotic HIPK2–p53 signaling axis. *Cancer Research*, January 19.

d'Adda di Fagagna, F., Reaper, P.M., Clay-Farrace, L., et al. (2003). A DNA damage checkpoint-mediated response in telomere-initiated cellular senescence. *Nature*, **426**(6963): 194–8.

Davoli, T., Denchi, E.L., and de Lange, T. (2010). Persistent telomere damage induces bypass of mitosis and tetraploidy. *Cell*, **141**(1): 81–93.

Deisenroth, C., and Zhang, Y. (2010). Ribosome biogenesis surveillance: probing the ribosomal protein–Mdm2–p53 pathway. *Oncogene*, **29**(30): 4253–60.

Di Micco, R., Sulli, G., Dobreva, M., et al. (2011). Interplay between oncogene-induced DNA damage response and heterochromatin in senescence and cancer. *Nature Cell Biology*, February 20.

Duro, E., Lundin, C., Ask, K., et al. (2010). Identification of the MMS22L–TONSL complex that promotes homologous recombination. *Molecular Cell*, **40**(4): 632–44.

Frank, C.J., Hyde, M., and Greider, C.W. (2006). Regulation of telomere elongation by the cyclin-dependent kinase CDK1. *Molecular Cell*, **24**(3): 423–32.

Goldberg, M., Stucki, M., Falck, J., et al. (2003). MDC1 is required for the intra-S-phase DNA damage checkpoint. *Nature*, **421**(6926): 952–6.

Gery, S., Komatsu, N., Baldjyan, L., Yu, A., Koo, D., and Koeffler, H.P. (2006). The circadian gene per1 plays an important role in cell growth and DNA damage control in human cancer cells. *Molecular Cell*, **22**(3): 375–82.

Grandori, C., Wu, K.J., Fernandez, P., et al. (2003). Werner syndrome protein limits MYC-induced cellular senescence. *Genes & Development*, **17**(13): 1569–74.

Green, D.R. (2005). Apoptotic pathways: ten minutes to dead. *Cell*, **121**(5): 671–4.

Hasty, P., Campisi, J., Hoeijmakers, J., van Steeg, H., and Vijg, J. (2003). Aging and genome maintenance: lessons from the mouse? *Science*, **299**, 1355–9.

Hernando, E., Nahle, Z., Juan, G., et al. (2004). Rb inactivation promotes genomic instability by uncoupling cell cycle progression from mitotic control. *Nature*, **430**(7001): 797–802.

Hutchins, J.R., Toyoda, Y., Hegemann, B., et al. (2010). Systematic analysis of human protein complexes identifies chromosome segregation proteins. *Science*, **328**(5978): 593–9.

Jackson, A.L., and Loeb, L.A. (1998). The mutation rate and cancer. *Genetics*, **148**: 1483–90.

Jeyasekharan, A.D., Ayoub, N., Mahen, R., et al. (2010). DNA damage regulates the mobility of Brca2 within the nucleoplasm of living cells. *Proceedings of the National Academy of Sciences USA*. [Epub ahead of print]

Jirawatnotai, S., Hu, Y., Michowski, W., et al. (2011). A function for cyclin D1 in DNA repair uncovered by protein interactome analyses in human cancers. *Nature*, **474**(7350): 230–4.

Krajcovic, M., Johnson, N.B., Sun, Q., et al. (2011). A non-genetic route to aneuploidy in human cancers. *Nature Cell Biology*, Feb 20.

Lens, S.M., Voest, E.E., and Medema R.H. (2010). Shared and separate functions of polo-like kinases and aurora kinases in cancer. *Nature Reviews Cancer*, **10**(12): 825–41.

Lindahl, T. (1993). Instability and decay of the primary structure of DNA. *Nature*, **362**: 709–15.

Ljungman, M., Zhang, F.F., Chen, F., Rainbow, A.J., and McKay, B.C. (1999). Inhibition of RNA polymerase II as a trigger for the p53 response. *Oncogene*, **18**: 583–92.

Loewer, A., Batchelor, E., Gaglia, G., and Lahav, G. (2010). Basal dynamics of p53 reveal transcriptionally attenuated pulses in cycling cells. *Cell*, **142**(1): 89–100.

Macias, E., Jin, A., Deisenroth, C., et al. (2010). An ARF-independent c-MYC-activated tumor suppression pathway mediated by ribosomal protein–Mdm2 interaction. *Cancer Cell*, **18**(3): 231–43.

Mazumdar, M., Lee, J.H., Sengupta, K., Ried, T., Rane, S., and Misteli, T. (2006). Tumor formation via loss of a molecular motor protein. *Current Biology*, **16**(15): 1559–64.

McGlynn, P., and Lloyd, R.G. (2002). Recombinational repair and restart of damaged replication forks. *Nature Reviews Molecular Cell Biology*, **3**: 859–70.

Mellon, I., Spivak, G., and Hanawalt, P.C. (1987). Selective removal of transcription-blocking DNA damage from the transcribed strand of the mammalian *DHFR* gene. *Cell*, **51**: 241–9.

Mitchell, J.R., Hoeijmakers, J.H., and Niedernhofer, L.J. (2003). Divide and conquer: nucleotide excision repair battles cancer and ageing. *Current Opinion in Cell Biology*, **15**, 232–40.

O'Shaughnessy, J., Osborne, C., Pippen, J.E., et al. (2011). Iniparib plus chemotherapy in metastatic triple-negative breast cancer. *New England Journal of Medicine*, **364**(3): 205–14.

Ou, Y.H., Torres, M., Ram, R., et al. (2011). TBK1 directly engages Akt/PKB survival signaling to support oncogenic transformation. *Molecular Cell*, **41**(4): 458–70.

Pei, H., Zhang, L., Luo, K., et al. (2011). MMSET regulates histone H4K20 methylation and 53BP1 accumulation at DNA damage sites. *Nature*, **470**(7332): 124–8.

Rajagopalan, H., and Lengauer, C. (2004). Aneuploidy and cancer. *Nature*, **432**(7015): 338–41.

Reddy, S.K., Rape, M., Margansky, W.A., and Kirschner, M.W. (2007). Ubiquitination by the anaphase-promoting complex drives spindle checkpoint inactivation. *Nature*, **446**(7138): 921–5.

Reinhardt, H.C., Aslanian, A.S., Lees, J.A., and Yaffe, M.B. (2007). p53-deficient cells rely on ATM- and ATR-mediated checkpoint signaling through the p38MAPK/MK2 pathway for survival after DNA damage. *Cancer Cell*, **11**(2): 175–89.

Rouse, J., and Jackson, S.P. (2002). Interfaces between the detection, signaling, and repair of DNA damage, *Science*, **297**(2002): 547–51.

Rulten, S.L., Fisher, A.E., Robert, I., et al. (2011). PARP-3 and APLF function together to accelerate nonhomologous end-joining. *Molecular Cell*, **41**(1): 33–45.

Samuels-Lev, Y., O'Connor, D.J., Bergamaschi, D., et al. (2001). ASPP proteins specifically stimulate the apoptotic function of p53. *Molecular Cell*, **8**(4): 781–94.

Savitsky, K., Bar-Shira, A., Gilad, S., et al. (1995). A single ataxia telangiectasia gene with a product similar to PI3 kinase. *Science*, **268**: 1749–53.

Schramek, D., Kotsinas, A., Meixner, A., et al. (2011). The stress kinase MKK7 couples oncogenic stress to p53 stability and tumor suppression. *Nature Genetics*, February 13.

Scully, R., and Livingston, D.M. (2000). In search of the tumour-suppressor functions of BRCA1 and BRCA2. *Nature*, **408**: 429–32.

Slee, E.A., O'Connor, D.J., and Lu, X. (2004). To die or not to die: how does p53 decide? *Oncogene*, **23**(16): 2809–18.

Squatrito, M., Brennan, C.W., Helmy, K., Huse, J.T., Petrini, J.H., and Holland, E.C. (2010). Loss of ATM/Chk2/p53 pathway components accelerates tumor development and contributes to radiation resistance in gliomas. *Cancer Cell*, **18**(6): 619–29.

Stegmeier, F., Rape, M., Draviam, V.M., et al. (2007). Anaphase initiation is regulated by antagonistic ubiquitination and deubiquitination activities. *Nature*, **446**(7138): 876–81.

Ting, D.T., Lipson, D., Paul, S., et al. (2011). Aberrant overexpression of satellite repeats in pancreatic and other epithelial cancers. *Science*, **331**(6017): 593–6.

Trigiante, G., and Lu, X. (2006). ASPP and cancer. *Nature Reviews Cancer*, **6**(3): 217–26.

Wang, B., Matsuoka, S., Ballif, B.A., et al. (2007). Abraxas and RAP80 form a BRCA1 protein complex required for the DNA damage response. *Science*, **316**(5828): 1194–8.

Yachida, S., Jones, S., Bozic, I., et al. (2010). Distant metastasis occurs late during the genetic evolution of pancreatic cancer. *Nature*, **467**(7319): 1114–7.

Zhang, D., Zaugg, K., Mak, T.W., and Elledge, S.J. (2006). A role for the deubiquitinating enzyme USP28 in control of the DNA-damage response. *Cell*, **126**(3): 529–42.

Oncogenic stress

Aslanian, A., Iaquinta, P.J., Verona, R., and Lees, J.A. (2004). Repression of the Arf tumor suppressor by E2F3 is required for normal cell cycle kinetics. *Genes & Development*, **18**(12): 1413–22.

Bao-Lei, T., Zhu-Zhong, M., Yi, S., et al. (2006). Knocking down PML impairs p53 signaling transduction pathway and suppresses irradiation induced apoptosis in breast carcinoma cell MCF-7J. *Cell Biochemistry*, **97**(3): 561–71.

Bartkova, J., Horejsi, Z., Koed, K., et al. (2005). DNA damage response as a candidate anticancer barrier in early human tumorigenesis. *Nature*, **434**(7035): 864–70.

Baudino, T.A., Maclean, K.H., Brennan, J., et al. (2003). Myc-mediated proliferation and lymphomagenesis, but not apoptosis, are compromised by E2f1 loss. *Molecular Cell*, **11**(4): 905–14.

Chen, D., Shan, J., Zhu, W.G., Qin, J., and Gu, W. (2010). Transcription-independent ARF regulation in oncogenic stress-mediated p53 responses. *Nature,* **464**(7288): 624–7.

Denoyelle, C., Abou-Rjaily, G., Bezrookove, V., *et al.* (2006). Anti-oncogenic role of the endoplasmic reticulum differentially activated by mutations in the MAPK pathway. *Nature Cell Biology,* **8**(10): 1053–63.

Egle, A., Harris, A.W., Bouillet, P., and Cory, S. (2004). Bim is a suppressor of Myc-induced mouse B cell leukemia. *Proceedings of the National Academy of Sciences USA,* **101**(16): 6164–9.

Eischen, C.M., Roussel, M.F., Korsmeyer, S.J., and Cleveland, J.L. (2001). Bax loss impairs Myc-induced apoptosis and circumvents the selection of p53 mutations during Myc-mediated lymphomagenesis. *Molecular Cell Biology,* **21**(22): 7653–62.

Fajas, L., Annicotte, J.S., Miard, S., Sarruf, D., Watanabe, M., and Auwerx, J. (2004). Impaired pancreatic growth, beta cell mass, and beta cell function in E2F1 (−/−) mice. *Journal of Clinical Investigation,* **113**(9): 1288–95.

Frame, F.M., Rogoff, H.A., Pickering, M.T., Cress, W.D., and Kowalik, T.F. (2006). E2F1 induces MRN foci formation and a cell cycle checkpoint response in human fibroblasts *Oncogene,* **25**(23): 3258–66.

Hemann, M.T., Bric, A., Teruya-Feldstein, J., *et al.* (2005). Evasion of the p53 tumour surveillance network by tumour-derived MYC mutants. *Nature,* **436**(7052): 807–11.

Hemmati, P.G., Normand, G., Verdoodt, B., *et al.* (2005). Loss of p21 disrupts p14 ARF-induced G1 cell cycle arrest but augments p14 ARF-induced apoptosis in human carcinoma cells. *Oncogene,* **24**(25): 4114–28.

Lallemand-Breitenbach V., and de Thé, H. (2006). CK2 and PML: regulating the regulator. *Cell,* **126**(2): 244–5.

Lindstrom, M.S., and Zhang, Y. (2006). B23 and ARF: friends or foes. *Cell Biochemistry and Biophysics,* **46**(1): 79–90.

Milton, A.H., Khaire, N., Ingram, L., O'Donnell, A.J., and La Thangue, N.B. (2006). 14-3-3 proteins integrate E2F activity with the DNA damage response *EMBO Journal,* **25**(5): 1046–57.

O'Donnell, K.A., Wentzel, E.A., Zeller, K.I., Dang, C.V., and Mendell, J.T. (2005). c-Myc-regulated microRNAs modulate E2F1 expression *Nature,* **435**(7043): 839–43.

Pauklin, S., Kristjuhan, A., Maimets, T., and Jaks, V. (2005). ARF and ATM/ATR cooperate in p53-mediated apoptosis upon oncogenic stress. *Biochemical and Biophysical Research Communications,* **334**(2): 386–94.

Paulson, Q.X., McArthur, M.J., and Johnson, D.G. (2006). E2F3a stimulates proliferation, p53-independent apoptosis and carcinogenesis in a transgenic mouse model. *Cell Cycle,* **5**(2): 184–90.

Pusapati, R.V., Rounbehler, R.J., Hong, S., *et al.* (2006). ATM promotes apoptosis and suppresses tumorigenesis in response to Myc. *Proceedings of the National Academy of Sciences USA,* **103**(5): 1446–51.

Qi, Y., Gregory, M.A., Li, Z., Brousal, J.P., West, K., and Hann, S.R. (2004). p19ARF directly and differentially controls the functions of c-Myc independently of p53. *Nature,* **431**(7009): 712–7.

Rogulski, K., Li, Y., and Rothermund, K., *et al.* (2005). Onzin, a c-Myc-repressed target, promotes survival and transformation by modulating the Akt-Mdm2-p53 pathway *Oncogene,* **24**(51): 7524–41.

Russell, J.L., Powers, J.T., Rounbehler, R.J., Rogers, P.M., Conti, C.J., and Johnson, D.G. (2002). ARF differentially modulates apoptosis induced by E2F1 and Myc. *Molecular Cell Biology,* **22**(5): 1360–8.

Russell, J.L., Weaks, R.L., Berton, T.R., and Johnson, D.G. (2006). E2F1 suppresses skin carcinogenesis via the ARF–p53 pathway. *Oncogene,* **25**(6): 867–76.

Scaglioni, P.P., Yung, T.M., Cai, L.F., *et al.* (2006). A CK2-dependent mechanism for degradation of the PML tumor suppressor. *Cell,* **126**(2): 269–83.

Serrano, M., Lin, A.W., McCurrach, M.E., Beach, D., and Lowe, S.W. (1997). Oncogenic Ras provokes premature cell senescence associated with accumulation of p53 and p16INK4a *Cell,* **88**(5): 593–602.

Sharma, N., Timmers, C., Trikha, P., Saavedra, H.I., Obery, A., and Leone, G. (2006). Control of the p53–p21CIP1 axis by E2f1, E2f2 and E2f3 is essential for G1/S progression and cellular transformation. *Journal of Biological Chemistry,* **281**: 36124–31.

Stanelle, J., and Putzer, B.M. (2006). E2F1-induced apoptosis: turning killers into therapeutics. *Trends in Molecular Medicine,* **12**(4): 177–85.

Tago, K., Chiocca, S., and Sherr, C.J. (2005). Sumoylation induced by the Arf tumor suppressor: a p53-independent function *Proceedings of the National Academy of Sciences USA,* **102**(21): 7689–94.

Tolbert, D., Lu, X., Yin, C., Tantama, M., and Van Dyke, T. (2002). p19 (ARF) is dispensable for oncogenic stress-induced p53-mediated apoptosis and tumor suppression in vivo. *Molecular Cell Biology,* **22**(1): 370–7.

Wikonkal, N.M., Remenyik, E., Knezevic, D., *et al.* (2003). Inactivating E2f1 reverts apoptosis resistance and cancer sensitivity in Trp53-deficient mice. *Nature Cell Biology,* **5**(7): 655–60.

Questions for student review

Choose all that apply.

1) DNA damage:
   a. Results exclusively from extrinsic factors.
   b. Mostly results in apoptosis.
   c. May be detected by RNA polymerase.
   d. May be repaired.
   e. May be increased by certain gene mutations.

2) DNA repair:
   a. Is usually unimpaired in cancer cells.
   b. If ineffective, invariably results in apoptosis or senescence.
   c. Usually takes place while the cell cycle is arrested.
   d. Does not occur within microsatellites.
   e. Cannot correct double-strand breaks.

3) Chromosomal instability (CIN):
   a. Always implies abnormal numbers of chromosomes.
   b. May result from double-strand breaks.
   c. May result from premature mitosis.
   d. Can be prevented by a functioning spindle checkpoint.
   e. Is always a consequence rather than a cause of cancer.

4) Microsatellite instability (MSI):
   a. Is associated with a "mutator phenotype."
   b. Is characterized by aneuploidy.
   c. Is seen only in inherited forms of colon cancer.
   d. May arise through epigenetic silencing of mismatch repair genes.
   e. Is found in most sporadic tumors.

5) Genetic instability:
   a. Is present in most established human cancers.
   b. Always occurs once a cancer has developed.
   c. May result from mutations in single genes.
   d. May result from telomere attrition.
   e. Usually manifests as both CIN and MSI in the same tumors.

11 There Is More to Cancer than Genetics: Regulation of Gene and Protein Expression by Epigenetic Factors, Small Regulatory RNAs, and Protein Stability

Stella Pelengaris[a] and Michael Khan[b]
[a]Pharmalogos Ltd and [b]University of Warwick, UK

Experience is not what happens to a man; it is what a man does with what happens to him.

Aldous Huxley

Key points

- It was long assumed, wrongly as it happens, that gene expression was regulated exclusively by activators and repressors of transcription operating on noncoding regulatory elements within a given gene.
- More recently, it has been noted that gene expression is subject to multiple additional levels of control, all of which may become – in some way – defective in cancer. Thus, in addition to mutations in coding DNA and well-known regulatory elements, there are seemingly endless opportunities for an evolving cancer cell to skew gene and protein expression to serve its own ends, including:
  - Chromatin modifications that are transmitted from one **somatic** cell to all its descendants. Such a control is referred to as "epigenetic," as the DNA sequence is not altered.
  - Posttranscriptional modification of mRNA.
  - Stability and processing of mRNA.
  - Regulation of genes and mRNAs by small RNA molecules encoded by areas of the genome previously believed to be noncoding "junk."
- Patterns of DNA methylation and chromatin structure are often markedly disturbed in cancer cells potentially resulting in inappropriate silencing of tumor suppressor genes (hypermethylation) or conversely activation of oncogenes (hypomethylation). This can arise through the action of mutations in epigenetic regulators and environmental factors.
- Some cancers may be epigenetically unstable. Such cancers exhibit hypermethylation of usually unmethylated regions such as the CpG islands, found in up to half of all human gene promoters (CpG island methylator phenotype (CIMP)), resulting in aberrant silencing of hundreds of genes. CIMP is analogous to the mutator phenotype resulting from defects in DNA repair described in Chapter 10.
- It is now generally accepted that epigenetic modification can be inherited by the progeny of somatic cells, and therefore silencing of genes by this means will be passed on through successive generations of evolving cancer cells alongside any coincident alterations in the coding sequence of the DNA.
- It was also taken for granted that the epigenetic "slate" would be wiped clean in germ cells and at fertilization so that these experience-driven traits would not be passed on to successive generations of the organism.
- This view has, however, had to adapt in order to account for several observations which are hard to explain by genetic inheritance alone:
  - Firstly, the relationship between inadequate maternal nutrition and unhealthy lifestyle on the subsequent risk of heart disease and diabetes in the offspring. Although long ignored by the mainstream, one might say "the elephant in the womb," it is now generally accepted that fetal conditioning *in utero* can permanently alter metabolism and disease expectation for the child later in life.
  - Secondly, exciting recent studies suggesting that a father's experiences might exert a lasting influence on disease risk in the offspring are very hard to explain other than by the transfer of epigenetic information.
- Several miRNAs are aberrantly expressed in human cancers, where they can act as tumor suppressors or sometimes oncogenes and constitute an important new class of treatment targets and biomarkers. Endogenous small interfering RNAs (siRNAs) are broadly more homologous to specific mRNAs than their miRNA counterparts. As a result, siRNAs regulate individual genes whereas miRNAs may

---

The Molecular Biology of Cancer: A Bridge From Bench to Bedside, Second Edition. Edited by Stella Pelengaris and Michael Khan.
© 2013 John Wiley & Sons, Inc. Published 2013 by John Wiley & Sons, Inc.

operate as master switches controlling multiple genes in a coordinated fashion. The endogenous siRNAs are comparatively less well studied, but this mechanism is being exploited in the design of a range of novel therapies.
- Proteins are subject to a range of posttranslational modifications, including phosphorylation, ubiquitination, and sumoylation, which strongly influence their activity, stability, and location and may become aberrant in cancer cells.
- The ubiquitin–proteasome pathway is involved in degradation of numerous cancer-relevant intracellular proteins, including regulators of apoptosis and cell cycling. Mutations in ligases and other genes may compromise protein stability and degradation and contribute to tumorigenesis. Inhibitors of the proteasomal degradation of proteins are being examined as therapeutic agents against cancer.
- Aside from degradation, the biological activity of proteins may also be influenced by localization and partitioning within the cell.

Introduction

The discovery of oncogenes in the 1970s provided strong support for the contention that cancer was a genetic disease. Gradually this extended into a pervasive notion that within every cancer cell, and written into its genome, was the autobiography that would explain to anyone with the tools to read it exactly how a normal cell, against all the odds, achieved immortality. Every step in its evolution would be faithfully recorded from the precious first edition through all subsequent editions. Thus, even if the older editions become rare or out of print, their legacy remains visible within the most current. Thus, key events in the evolution of the cancer cell could be identified, and treatments and preventative strategies developed based on this knowledge. However, this requires the necessary tools – imagine having your ebook but no suitable reader. The human genome project and advances in systems biology have already allowed many of these cancer biographies to be read, and although much useful information has been acquired it has become very clear that the sequence of the DNA text is only part of the story. To avoid torturing this metaphor any further, we will conclude by noting that to fully comprehend what the biography means, we must also have the epigenetic "punctuation."

Every living being is also a fossil. Within it, all the way down to the microscopic structure of its proteins, it bears the traces if not the stigmata of its ancestry.

Jacques Monod

This chapter will explain how cancers may arise by changes in the expression of genes and proteins that are not primarily the result of alterations in the DNA sequence of protein-coding genes. Because mutations cause changes in expression of proteins, it does not follow that increased or reduced expression of a given gene or protein can result only from alteration to the DNA sequence. Rather, as we will see, chemical alteration of the DNA or of important associated proteins also has potent effects on gene expression, without any alteration of the DNA code. On the other hand, the genes that encode the regulators of methylation, acetylation, and other key processes that can influence gene expression are also subject to mutation. These mutations have the potential to influence expression of large numbers of genes by altering methylation patterns or expression of regulatory factors involved in protein translation and stability. Recent important examples of epigenetic changes resulting from mutations in regulatory genes include those in *ARID1A* and in numerous other genes encoding a variety of DNA methyltransferases, members of the SWI–SNF and miRNAs. However, in these cases the mutation is not in the gene encoding the protein concerned. This is not an abstract or semantic issue as exemplified by the long-overlooked role of MYC protein deregulation in the majority of cancers where, in complete contrast to Burkitt lymphoma, no chromosomal rearrangements or mutations in the *MYC* gene are present.

With the completion of genome-sequencing projects, one of the major challenges in modern biology is to understand what all the tens of thousands of identified genes actually do and moreover how their activity is regulated. To face this considerable challenge will require a much greater understanding of all the many processes that contribute to the regulation of gene expression and the ultimate formation of functional protein products. It has long been appreciated that gene expression is regulated by protein complexes that either promote or repress expression by binding to specific regulatory elements adjacent to but distinct from the coding region. We have also extensively discussed how gene expression is altered by mutations in these or the coding regions. One of the most exciting areas of cancer research now is describing and trying to interfere with other mechanisms, many alluded to in earlier chapters, which together conspire to influence whether a protein is made or not and how amounts in the cell are regulated. In fact, generation of functional proteins is determined by a remarkable variety of processes, many only recently appreciated. The term **epigenetics** (meaning "in addition to" genetics) was first used in the 1940s by Conrad Waddington to encompass nongenetic influences on phenotype. Epigenetics is now specifically used to describe stable mitotic or even meiotic inheritance of phenotype resulting from changes in a chromosome without alterations in the DNA sequence – in other words, non-Mendelian but heritable modifications such as chromatin remodeling that can alter gene expression in a cell and in its progeny.

Other important regulatory processes have come to light over the last decade. Once the gene has been transcribed, but before the mRNA is translated into protein synthesis, the gene transcripts are subject to alternative splicing, the inhibitory effect of noncoding RNAs (ncRNAs), and other factors affecting stability. Finally, translation of mRNA into protein is regulated at the ribosomes, and the proteins themselves are further modified in a bewildering number of ways that affect protein function, localization, and longevity. In fact, as the final functional arbiters, the proteins are subject to a variety of regulatory processes and when their usefulness has expired can be summarily destroyed or alternatively exiled, segregated, or excluded in order to prevent any further interactions with other proteins.

Without belaboring the point, as this has been covered in depth in this book, tumor cells are characterized by aberrant responses to cellular signals that normally regulate cell replication, differentiation, adhesion and motility, and apoptosis. As discussed in other chapters, these cellular processes are regulated by proteins and thereby by all of the aforementioned regulatory processes. In fact, essentially all links in the chain leading from an accurately maintained genome, through gene transcription and translation and ultimately protein modification, are in some way subject to gene mutations, and potentially cancer-causing mutations at that. But changes in gene and protein expression do not only arise as a result of gene mutations. In fact, protein levels can also be regulated by numerous other relatively more recently described processes operating at different stages of the gene–protein sequence, including:

- Changes in the methylation and structure of the DNA;
- Acetylation and numerous other posttranslational modifications of the associated histone proteins, which remodel chromatin;
- Alternative splicing of mRNA;
- Noncoding RNAs such as miRNAs; and
- Stability, transport, and partitioning of the key functional effectors – the proteins themselves.

Methylation of cytosine is a critical epigenetic alteration with profound regulatory effects on both the transcription and the replication of DNA and is also very well preserved through cell divisions. Methylation of a gene is effectively acting as a "mute button" that silences expression of that gene. Patterns of DNA methylation, which are passed on to cellular progeny, are responsible for achieving cellular differentiation and tissue-specific gene expression amongst cells which broadly all have identical genomes. Appreciation of the pivotal role played by methylation in cell biology and in particular how this can lead to imprinting of genes as well as the silencing of tumor suppressor genes during tumorigenesis have fueled a major research initiative into finding drugs that target this process. The main enzymes regulating methylation are the DNA methyltransferases (DNMTs), which can add methyl groups to the 5' position of cytosine rings within CpG dinucleotides. The other key epigenetic modification with powerful effects on control of gene expression is acetylation of histones. The dynamics of histone acetylation is dictated by balancing the opposing actions of histone acetyl transferases (HATs) and histone deacetylases (HDACs). In general, HATs activate and HDACs silence gene expression – "Stand still and silent with your HAT off and your HDAC on."

HDACs are important epigenetic regulators of gene expression through the remodeling of chromatin and have been discussed in other chapters in the context of c-MYC (Chapter 6) and RB (Chapter 4). It is important to be aware that histones are susceptible to the whole panoply of posttranslational modifications, and acetylation is just one of these. Methylation, phosphorylation, ubiquitination, and sumoylation are all present and may all contribute to chromatin structuring in one way or another. By implication, therefore, the enzymes that mediate the addition or removal of these modifications play important roles in normal cell processes and in cancer, and could all be potential drug targets. The combination of all of these various modifications in a given genomic region and the resultant effects on chromatin confirmation and gene expression have been dubbed the **histone code**. Another term that has become commonplace is **epigenome**, which refers to the sum of all epigenetic factors operating within a given cell at a particular time.

Histone modifications and methylation of cytosine are vulnerable to the effects of differing exogenous agents, including certain base analogues, radiation, tobacco smoke, hormones, and reactive oxygen species (see also Chapter 3), all of which can potentially influence gene expression and cellular phenotype epigenetically (methylation and/or acetylation, particularly of CpG islands in gene promoter regions) without changing their DNA sequence. It is worth noting that most of these factors could also cause mutations. Cancer cells typically manifest profound alterations in DNA methylation and histone modification patterns, including global hypomethylation and promoter-specific hypermethylation of DNA. Intriguingly, such epigenetic changes may already, in the main, be established in pre-malignant stages, powerfully arguing a case for causality in many human cancers. The mechanisms are becoming clear. Firstly, global hypomethylation can precipitate genomic instability by accelerating chromosomal rearrangements and translocations and could moreover activate large numbers of oncogenes and release imprinting of growth factors, such as IGF-2. Secondly, CpG island promoter hypermethylation could, as already discussed in Chapter 7, provide a "second hit" by silencing the remaining functioning allele of a tumor suppressor gene or miRNA. Don't worry – mutations are not off the hook here; the methylase-encoding gene *DNMT3A* is mutated in some cancers such as AML, and can contribute to the establishment of an aberrant cancer epigenome.

Over the last decade, both academic and industrial sectors have been driven to a frenzy of research activity, takeovers, and buyouts, prompted by the discovery that what was once embarrassingly referred to as "junk DNA" actually contains the blueprint for generating a new family of key regulatory RNA molecules. After a hasty rebranding exercise, what we now refer to as "noncoding DNA" (DNA that is not ultimately translated into protein) has been shown to encode an ever-expanding family of important regulatory factors, the imaginatively named noncoding RNAs (ncRNAs) that include the all-conquering microRNAs (miRNAs). So, far from being worthless junk, much of the genome is actually transcribed into thousands of ncRNAs, including not only miRNAs but also small interfering RNAs (siRNAs) and a variety of long ncRNAs that impose powerful transcriptional and posttranscriptional controls on protein synthesis. In fact, to show just how wide of the mark we were, miRNAs alone regulate at least 30% of all human genes by controlling translation and degradation of target mRNAs. Interestingly, the DNA encoding miRNAs is also subject to regulation by methylation – a form of "epi-epigenetics."

Aside from regulation of gene expression, the levels of oncoproteins and tumor suppressor proteins may also become abnormal through either increased or decreased degradation (Chapters 6 and 7). Lysosomes were historically regarded as the principal means of protein degradation within the cytoplasm. However, over the last 10–15 years, attention has increasingly focused on the role of ubiquitination of intracellular proteins; such ubiquitinated proteins are thereby targeted for degradation by a multiprotein complex termed the proteasome. In general it now seems that extracellular and transmembrane proteins are primarily degraded in the lysosomes, whereas intracellular proteins, including key regulators of the cell cycle and apoptosis, are normally degraded by the proteasome. Self-evidently, many proteasome substrates are involved in pathways that become deregulated in cancer, and proteasome inhibitors are now entering clinical practice. The partitioning and localization of proteins are also important determinants of biological activity, and these will also be discussed in this chapter.

As we have done in other chapters, we will start with a brief refresher on relevant cell biology which will make the mechanics of epigenetic regulation much easier to follow.

The language of epigenetics

Everything in life is speaking in spite of its apparent silence.

Hazrat Inayat Khan

Chromatin

DNA does not exist naked within the cell, but in association with histones, which constitute a protein scaffold that gives form to the complex tertiary structure referred to as chromatin (see Box 11.1). Chromatin was originally observed to exist in two different forms, during microscopic examination of cells during interphase. These observational differences are now known to correlate with gene expression activity – heterochromatin represents repressed segments of more tightly packed chromosomal DNA, while euchromatin represents a more open configuration with transcriptionally active segments. These two forms of DNA are inter-convertible. During cell differentiation and maturation, RNA synthesis declines and is accompanied by a corresponding conversion of euchromatin to heterochromatin – therefore, fewer genes are available for mRNA synthesis. A wide variety of cellular processes involve de-repression of previously repressed genes, and cells undergoing such gene de-repression often display a reversible transformation of heterochromatin to euchromatin within their nuclei. A similar alteration in DNA takes place during mitosis. At the onset of prophase, the nuclear membrane dissolves and the euchromatin seen during interphase condenses into large chromosomal masses in prelude to metaphase when the chromosomes will eventually become segregated and separated for completion of cell division. Strictly (although very similar at a molecular-level reduced mRNA synthesis), this condensation of interphase euchromatin into condensed chromosomal masses is not classed as heterochromatin because this terminology was restricted to describing appearances of chromosomes during interphase.

Nucleosomes form the basic units of chromatin, and each comprises 146 base pairs of DNA wrapped around an octamer of two molecules each of the histones H2A, H2B, H3, and H4. Neighboring nucleosomes are joined to each other by linker DNA, and progressive coiling of nucleomes leads to higher order structures. DNA contained within compacted chromatin, known as heterochromatin, is not available for transcription, unless appropriate activation of remodeling processes first takes place in order to enable access of the transcription factors and transcriptional machinery to individual genes. Chromatin remodeling requires the action of two classes of proteins: those that covalently modify DNA or histones (by methylation, acetylation, etc.) and those that mobilize nucleosomes such as the SW1–SNF complex.

SWI–SNF chromatin-remodeling complexes require energy generated by hydrolysis of ATP in order to influence gene expression through nucleosome remodeling. As will be discussed, these complexes function as tumor suppressors and are frequently inactivated in human cancers (see Table 11.1 for a list of SWI component-encoding genes altered in cancers). These include a variety of tumors suppressors such as BRG1, BRM, ARID1A, SMARCC1, and SMARCB1 that are often inactivated in human cancers.

CpG islands

Roll on, deep and dark blue ocean, roll. Ten thousand fleets sweep over thee in vain. Man marks the earth with ruin, but his control stops with the shore.

Lord Byron

CpG islands are areas of greatly increased density of a dinucleotide sequence, cytosine–phosphate diester–guanine, which can form regions of DNA several hundred to several thousand base pairs long. The human genome contains around 45 000 CpG islands (comprising a total of around 50 million CpG dinucleotides), mostly found at the 5′ ends of genes. They are widely accepted as unmethylated in normal somatic cells except for those on the inactive X chromosome and some associated with imprinted genes. This is in contrast to the majority of CpGs that lie outside of islands and are methylated in mammalian genomes. Around 60% of all human gene promoters contain CpG islands around the promoter regions, and these include housekeeping genes (essential for general cell functions) and many frequently expressed in a normal cell. Most CpG islands are unmethylated and can be either transcriptionally active or inactive depending on the balance of transcriptional regulators and histone modifications. As will be seen in this chapter, CpG islands are important sites for epigenetic regulation of gene expression, and are frequently aberrantly methylated in cancer cells. However, exactly how CpG-island promoters may become hypermethylated in cancer is not known. But recent papers have highlighted the potentially critical role of the enzyme TET1 in catalyzing hydroxylation of mC9, which together with hmC could regulate both DNA demethylation and gene expression. In this case, loss of TET1 would be predicted to result in promoter hypermethylation.

Epigenetics

Our deeds determine us, as much as we determine our deeds.

George Eliot

In many ways what is now referred to as epigenetics is not new science – it has long been recognized that traditional genetics theory, which implied a one-to-one relationship between genotype and phenotype, could not readily explain processes such as cell differentiation; here multiple, often very different cell phenotypes are produced, yet all bear ostensibly the identical genome. Thus, it was hypothesized that each undifferentiated cell underwent a crisis that determined its fate (a sort of cellular 11-plus), which was not inherent in its genes and was therefore, from the Greek, *epigenetic* – "in addition to" – the genetic information encoded in the DNA (see Box 11.2 for a historical overview). We now appreciate that the selective silencing of some genes and expression of others during development ultimately determine the phenotype of every cell, and that so-called epigenetic factors are key determinants of this. Epigenetics is one of the most exciting areas of modern biology, and particularly over the last few years we have developed a much clearer understanding of how this mechanism operates, and how epigenetic factors can control gene expression.

Changes to DNA and its associated proteins can alter gene expression without altering the DNA sequence. DNA is not found

Box 11.1 Chromatin

(The reader is also referred to an excellent overview of this subject in Alberts et al., *Molecular Biology of the Cell*, 2007.)

The length of the DNA molecule poses certain problems with respects to packing it all away in the cell nucleus, as anyone attempting the relatively trivial task of packing away a very long hosepipe will readily appreciate. Each human cell contains approximately 2 m of DNA if stretched end to end, yet the nucleus of a human cell, which contains the DNA, is only about 6 μm in diameter. Imagine trying to put an 80-mile long hosepipe into your garden shed. The packaging of DNA is accomplished by a truly remarkable array of specialized proteins that bind to and fold the DNA, generating a series of contorted coils and loops that provide increasingly higher levels of organization, preventing the DNA from becoming an unmanageable tangle. Yet, somehow, despite this unbelievable complexity, the DNA remains readily accessible to various enzymes needed for replication, repair, and gene expression.

In eukaryotes, the DNA in the nucleus (genome) is divided between chromosomes, of which there are 24 different pairs in humans. Each chromosome consists of a single, very long linear DNA molecule associated with the various proteins required to fold and pack the fine DNA thread into a more compact structure. The complex of DNA and proteins is called *chromatin* (from the Greek *chroma*, "color," because of its staining properties).

Two classes of DNA-binding proteins are recognized in eukaryotic chromosomes: histones and nonhistone proteins. Histones are very abundant and maintain the first level of DNA organization the nucleosome. Nucleosomes are arranged roughly like beads on a string – each bead is a "nucleosome core particle" that consists of DNA wound around a protein core formed from histones. Each individual nucleosome core particle consists of a complex of eight histone proteins – two molecules each of histones H2A, H2B, H3, and H4 – and double-stranded DNA that is 146 nucleotide pairs long. On average, nucleosomes repeat at intervals of about 200 nucleotide pairs interspersed by "linker" DNA. For example, a diploid human cell with 6.4×10^9 nucleotide pairs contains approximately 30 million nucleosomes. The formation of nucleosomes converts a DNA molecule into a chromatin thread about one third of its initial length, and this provides the first level of DNA packing.

Chromatin in a normal cell rarely adopts the extended "beads-on-a-string" form. Instead, the nucleosomes are piled on top of one another, generating regular arrays in which the DNA is even more highly condensed and forming what is referred to as the 30 nm fiber, which is wider than chromatin in the "beads-on-a-string" form.

As a 30 nm fiber, the typical human chromosome would still be around 100 times too big for the nucleus. Thus, a higher level of folding exists to fold the 30 nm fiber into a series of loops and coils. Each long DNA molecule in an interphase chromosome is divided into a large number of discrete domains organized as loops of chromatin, each loop comprising a folded 30 nm chromatin fiber. Interphase chromosomes are largely composed of euchromatin that is interrupted by stretches of heterochromatin, in which 30 nm fibers are subjected to additional levels of packing that usually render it resistant to gene expression.

Light-microscope studies in the 1920s distinguished between two types of chromatin in the interphase nuclei of many higher eukaryotic cells: a highly condensed form and all the rest, which is less condensed. Heitz (1929) originally described that portion of the nuclear chromatin remaining condensed throughout cell interphase as heterochromatin and the rest as euchromatin. Cooper (1959) suggested that heterochromatin and euchromatin differed in their biophysical conformations and in metabolic expression of their genes but not in their basic structure of DNA arranged within chromosomes. Since that time, increasingly detailed genetic studies have revealed that the genes within heterochromatin are repressed but can later be expressed when the heterochromatic region undergoes a transition to euchromatin. Similarly, heterochromatin displays little or no synthesis of RNA until it is converted to euchromatin.

In a typical mammalian cell, approximately 10% of the genome is packaged into heterochromatin. Although present in many locations along chromosomes, it is concentrated in specific regions, including the centromeres and telomeres. Most DNA folded into heterochromatin does not contain genes. However, those genes that are packaged into heterochromatin are not expressed, probably because heterochromatin is so compact. Some regions of heterochromatin are responsible for the proper functioning of telomeres and centromeres (which lack genes), and its formation may even help protect the genome from being overtaken by "parasitic" mobile elements of DNA. Moreover, a few genes require location in heterochromatin regions if they are to be expressed. Thus, heterochromatin should not be thought of as comprising only redundant DNA.

When a gene normally expressed in euchromatin is experimentally relocated into a region of heterochromatin, it is no longer expressed, and the gene is silenced. Such effects of location are referred to as "position effects," as gene activity depends on a position along a chromosome. The study of position effects has identified some intriguing properties of heterochromatin, namely, that it is dynamic and that the state of chromatin, heterochromatin or euchromatin, is inherited during cell division.

---

in isolation in the cell but is associated with proteins called histones to form a complex substance known as chromatin. Chemical modifications to the DNA or the histones alter the structure of the chromatin without changing the nucleotide sequence of the DNA. Such modifications are described as epigenetic. Changes to chromatin structure have a profound influence on gene expression: if the chromatin is condensed, the factors involved in gene expression cannot get to the DNA, and the genes will be switched off. Conversely, if the chromatin is in an "open" conformation, the genes can be expressed on demand for cellular activities. Unraveling these processes has been important as we now appreciate that in addition to the well-known role of DNA sequence changes (mutations), aberrant gene expression also results from more recently identified changes in gene silencing, due to epigenetic modifications.

Another useful metaphor, with which the reader will often be confronted, contends that genetic information provides a blueprint for manufacturing proteins necessary to create the organism, while the epigenetic information provides additional instructions on how, where, and when the genetic information

Table 11.1 List of genes encoding epigenetic regulators, known to be compromised by deregulated expression or mutation in human cancers

Functional class	Gene	Tumor	Role
DNA methyltransferases	DNMT1 DNA (cytosine-5-)-methyltransferase 1	Colon, bladder, breast, and head and neck cancer	Patterning of cytosine methylation and gene silencing
	DNMT3A	Acute myeloid leukemia (AML)	
	DNMT3B	Colorectal and breast	
Others involved in DNA methylation	MBD1 Methyl–CpG binding domain protein 1	Pancreatic cancer cell lines	Binding specifically to methylated DNA and inhibiting transcription
	MBD2	Cholangiocarcinoma and non-small-cell lung cancer (NSCLC)	Silencing of the p14–p16 locus and hTERT, among others
	MBD3	Gliomas and acute promyelocytic leukemia	Methyl CpG binding proteins
	MBD4	Colon and NSCLC	Linked to microsatellite instability
	TET1 Tet oncogene 1	AML	Methylcytosine dioxygenase that catalyzes the conversion of methylcytosine to 5-hydroxymethylcytosine
	TET2	Myeloid malignancies and some Wilms' tumors	Methylcytosine dioxygenase TET2
Histone methylation	EZH2 Enhancer of zeste homolog 2	Prostate, breast, lung, skin, lymphoma, liver, and colon	Promotes angiogenesis; part of polycomb repressive complex
	MLL myeloid–lymphoid or mixed-lineage leukemia	Infant leukemias, AML, and CLL	Histone H3 lysine 4 (H3K4) methyltransferase activity
	NSD1 Nuclear receptor-binding SET domain protein 1	Childhood AML, gliomas, and neuroblastoma	Translocation between nuclear receptor-binding Su-var, enhancer of zeste, and trithorax domain protein 1 and nucleoporinin; some AML
	PRDM1 PR domain containing 1, with ZNF domain provided	B-cell lymphomas and myeloma	Repressor of β-interferon gene expression
	PRDM2 PR domain containing 2, with ZNF domain	Breast	Tumor suppressor gene; member of nuclear histone–protein methyltransferase superfamily
	SMYD3 SET and MYND domain containing 3	Colon and hepatocellular	Histone methyltransferase
	WHSC1 (MMSET) Wolf–Hirschhorn syndrome candidate 1	Multiple myeloma and glioma	Histone methyltransferase, methylating Lys-9 of histone H3
	SUV39H1 Suppressor of variegation 3-9 homolog 1	Colorectal	Histone methyltransferase
	EHMT2 (G9a) Euchromatic histone–lysine N-methyltransferase 2 provided	Lung and colon	H3K9 histone methyltransferase
Histone demethylation	KDM1A Lysine (K)-specific demethylase 1A		Component of histone deacetylase complexes; silences genes by functioning as a histone demethylase; may help snail-mediated (EMT)
	KDM2B	Leukemias	
	KDMB4 (JMJD2B)	Breast	
	KDM4C (JMJD2C)	Squamous cell carcinoma, esophagus, breast, and lymphoma	Jumonji family protein
	KDM5A (JARID1A)	T-cell leukemias and melanoma	Interacts with Rb
	KDM5B (JARID1B)	Bladder, lung, prostate, and melanoma	Differentiation
	KDM5C (JARID1C)	Renal	
	KDM6A	Acute promyelocytic leukemia and renal carcinoma	
	ASXL1 Additional sex combs like 1	Chronic myelomonocytic leukemia	Chromatin-binding protein

(Continued)

Table 11.1 (Continued)

Functional class	Gene	Tumor	Role
Histone acetyl transferases	CREBBP CREB-binding protein	Various hematological malignancies such as AML	Coupling chromatin remodeling to transcription factor recognition
	EP300 E1A-binding protein p300	Colon, breast, and pancreas	Histone acetyltransferase
	KAT5 (TIP60) K(lysine) acetyltransferase 5	Stomach	MYST family of histone acetyl transferases (HATs)
	KAT2B (PCAF) K(lysine) acetyltransferase 2B	Intestinal-type gastric cancer	Transcriptional regulator; associates with ε-acetyl–lysine residues in nucleosomal histones
	MYST3, 4 MYST histone acetyltransferase (monocytic leukemia) 3	Various hematological malignancies	MYST family of histone acetyl transferases (HATs)
Histone deacetylases	HDAC1 Histone deacetylase 1	Stomach, liver, gliomas, and lymphoma	Component of the histone deacetylase complex that also interacts with RB to regulate cell proliferation
	HDAC2	Stomach, lymphoma, glioma, and colon	Forms transcriptional repressor complex
	HDAC3	Liver, lung adenocarcinoma, and lymphoma	Can downregulate TP53
Sirtuins (histone lysine deacetylases)	SIRT1 Silent mating type information regulation 2 homolog 1	Prostate and liver	Mono-ADP-ribosyltransferase activity
	SIRT2	Esophagus	
	SIRT3	Breast	Metabolic programming; regulation of hypoxia-inducible factor (HIF) and reactive oxygen species (ROS)
	SIRT7	Thyroid and breast	
ING family	ING1–5 Inhibitor of growth family, members 1–5	Glioblastoma, liver, NSCLC, and oropharyngeal cancer	Interact with TP53; histone modification reading
Chromatin remodeling factors	MTA1–3 Metastasis associated 1 family, members 1–3	NSCLC, larynx, breast, chorioncarcinoma, liver (MTA2), and endometrium (MTA3)	Component of NuRD, a nucleosome-remodeling deacetylase complex
	MEN1	Multiple endocrine neoplasia-associated cancers; sporadic neuroendocrine cancers	Component of histone methyltransferase complex
Parts of the switch mating type–sucrose nonfermenting (SWI–SNF) chromatin-remodeling complex	ARID1A AT-rich interactive domain 1A (SWI-like)	Ovarian and endometrial cancer	Part of the SNF–SWI chromatin-remodeling complex
	CHD family of chromodomain helicase DNA-binding protein, including CHD1–9	Gastric and colorectal cancers, neuroblastoma, oropharyngeal, prostate, and lung (CHD7)	Contains a SWI–SNF-like helicase–ATPase domain and a DNA-binding domain, and a chromodomain that directly modifies chromatin structure
	SMARCA2 (BRM) SWI–SNF-related, matrix-associated, actin-dependent regulator of chromatin, subfamily a, member 2	Melanoma, head and neck squamous cell carcinoma (SCC), and skin cancer (SCC and basal cell carcinoma (BCC))	Part of the SNF–SWI chromatin-remodeling complex
	SMARCA4 (BRG1) SWI–SNF-related, matrix-associated, actin-dependent regulator of chromatin, subfamily a, member 4	Melanoma, colorectal, NSCLC, and skin cancer	Part of the SNF–SWI chromatin-remodeling complex; can also bind BRCA1
	SMARCB1 (SNF5) SWI–SNF-related, matrix-associated, actin-dependent regulator of chromatin, subfamily b, member 1	Meningioma, rhabdomyomas, glioblastoma, and sarcoma	Part of the SNF–SWI chromatin-remodeling complex

Table 11.1 (Continued)

Functional class	Gene	Tumor	Role
	SMARCC1 (BAF155) SWI–SNF-related, matrix-associated, actin-dependent regulator of chromatin, subfamily c, member 1	Prostate	Part of the SNF–SWI chromatin-remodeling complex
	SMARCE1 (BAF57)	Breast	Part of the SNF–SWI chromatin-remodeling complex
	PBRM1 (BAF180) Polybromo 1	Renal carcinoma and breast	Part of the SNF–SWI chromatin-remodeling complex
	BARD7 Bromodomain containing 7	Nasopharyngeal	Part of the SNF–SWI chromatin-remodeling complex
ISWI chromatin regulating complex	RSF1 (HBXAP) Remodeling and spacing factor 1	Ovarian serous carcinoma	Part of the ISWI complex
	SNF2H (SMARCA5)	Stomach	Part of the chromatin-remodeling and spacing factor RSF together with RSF1
Transcriptional regulators	BRD2 Bromodomain containing 2; member of large family	Lymphoma	Member of the bromodomains and extra terminal domain (BET) family; binds the acetylated lysine-12 residue of histone H4 in association with transcription complexes
	BRD4 Bromodomain containing 4	Young-onset lung cancer	Contains 2 bromodomains (related to Brahma); chromatin targeting; chromosomal rearrangement results in BRD4–NUT fusion protein in NUT midline carcinoma
	TAF1 Member of the large TATA box binding protein (TBP)-associated factor family	Prostate	Part of the TFIID transcription factor with TBP; also has acetyl transferase activity; promotes p53 turnover
	Sin3A SIN3 homolog A	Possibly breast and lung	Transcriptional repressor in partnership with HDACs; mediates gene inhibition by MAD–MAX
	NCoR1 Nuclear receptor corepressor 1	Astrocytic gliomas	In PML, the repressive NCor–Sin3–HDAC complex may be replaced by a HAT and aberrant gene expression; involved in PML–RARa and PLZF–RARa fusion protein-related leukemias that bind NCoR less well and may result in disinhibition of gene expression

Modified from Esteller M. (2005).

is deployed. Epigenetic information is not contained within the DNA sequence itself, but can still determine mitotic inheritance of various characteristics as surely as modifications in the DNA sequence. Thus, epigenetic factors, which include DNA methylation and histone modifications, can dictate cell fate and gene expression patterns in the progeny after cell division and are important in the normal regulation of differentiation, aging, and senescence. Epigenetic factors can even turn environmental effects into heritable changes in cell phenotypes – which at face value challenges the central dogma of genetics. However, adult patterns of methylation are generally believed to be erased during early embryogenesis so that, in general terms, cells in a new organism are believed to start life with an epigenetically "clean slate." Subsequently, during development and adult life, cells progressively acquire epigenetic "chalk marks" that they can pass on to their progeny. Importantly, all epigenetic modifications are not invariably predetermined during ontogeny but are influenced throughout life by genetic and environmental forces. Intriguingly, such epigenetic factors may also be the explanation for differences in phenotype observed between genetically identical twins.

Over the last couple of years, important studies have convincingly demonstrated examples whereby epigenetic modification can under some circumstances result in the translation of life experiences in a parent into inherited alterations in gene expression in the offspring. Numerous examples, relating nutritional deprivation or excess in a parent to subsequent risk of obesity and diabetes in the offspring, have been demonstrated. Much of this was perceived to be *in utero* conditioning. However, intriguingly, a recent paper in *Nature* in 2010 suggested that a paternal high-fat diet (HFD) resulted in development of β-cell dysfunction in female rat offspring with altered expression of multiple pancreatic islet genes at least in some cases by altered methylation.

Box 11.2 Historical overview of epigenetics

It was first noted some time ago that traditional genetic theory, which implied a one-to-one correspondence between genotype and phenotype, struggled to explain processes such as cell differentiation (if all cells in an organism have the same genes, then how can so many cells be so very different from one another?). To accommodate this, it was suggested that each undifferentiated cell at some point reached a "crisis" that determined its fate and this was somehow separate from the genes and thus (borrowing from the Greek) **epigenetic**. The biologist C.H. Waddington may first have used the term in its modern context in the 1940s, when he defined it as "the branch of biology which studies the causal interactions between genes and their products which bring the phenotype into being." The notion that characteristics acquired during an organism's lifetime could be passed onto the offspring is, in honor of **Jean-Baptiste Lamarck**, known as Lamarckian. Not that long ago, this view was believed to be totally at odds with modern genetics and was often described in amusing terms in school biology curriculae. However, we all owe Lamarck an apology for, as we now appreciate, his theories are in many respects borne out by recent understanding of epigenetic inheritance.

Lamarck will be familiar to all biology students as the author of a widely discredited theory of heredity, the "inheritance of acquired traits." However, at the time his views almost certainly influenced other biologists wrestling with the emerging field of evolution, in particular Charles Darwin (see Box 3.1). In 1861, Darwin wrote, "Lamarck was the first man whose conclusions on the subject excited much attention. This justly celebrated naturalist first published his views in 1801. . . . He first did the eminent service of arousing attention to the probability of all changes in the organic, as well as in the inorganic world, being the result of law, and not of miraculous interposition."

Lamarck developed two laws:

1. "In every animal which has not passed the limit of its development, a more frequent and continuous use of any organ gradually strengthens, develops and enlarges that organ, and gives it a power proportional to the length of time it has been so used; while the permanent disuse of any organ imperceptibly weakens and deteriorates it, and progressively diminishes its functional capacity, until it finally disappears."

2. "All the acquisitions or losses wrought by nature on individuals, through the influence of the environment in which their race has long been placed, and hence through the influence of the predominant use or permanent disuse of any organ; all these are preserved by reproduction to the new individuals which arise, provided that the acquired modifications are common to both sexes, or at least to the individuals which produce the young."

Lamarck's own theory of evolution was based on the notion that an organism adapts to its environment during its own lifetimes and passes on traits that have been acquired to the offspring (in modern terms, the implication is that an organism will respond to events and environment by undergoing genetic alterations which can then be in some way passed on to the offspring). Offspring then adapt from where the parents left off, and evolution advances. Lamarck proposed that individuals increased specific capabilities by using them, while losing others through disuse. Lamarck believed in a teleological (goal-oriented) version of evolution, with organisms improving progressively as they evolved. Lamarck has become synonymous with pre-Darwinian ideas about evolution, now called "Lamarckism."

Modern evolutionary biology accepts that the environment plays a role during natural selection by dictating what characteristics are necessary for better reproduction opportunities. For natural selection to occur, individuals must differ somewhat genetically, in order that positive characteristics can amplify and negative ones can be deleted from the gene pool. These differences between individuals (or, for that matter, between cancer cells) arise from random mutations in genes – this is the mechanism underlying Darwinian evolution of individuals within a species or of cancer cells within a tumor. The environment can influence these variations (e.g. radioactivity and other mutagens will damage DNA), but probably only in a random manner. However, very recently multiple studies are indicating that we should revisit the notion that the environment may play a more direct and crucial role in evolution.

Epigenetic inheritance allows cells of differing phenotype but identical genotype to transmit their phenotype to their offspring, even when the original phenotype-inducing stimuli are no longer present. This is reasonably easy to understand with respect to somatic cells, and there is little debate any longer about the clonal evolution of cancer under the influence of somatic mutations or epigenetic changes, such as promoter methylation, and then natural selection of such changes which confer a growth advantage. However, the question still remains as to whether epigenetic inheritance plays a direct role in evolution of the organism; in this case, one must somehow postulate a means whereby information not encoded in the genome of the germ cells can be transmitted to the offspring. One possible explanation for such epigenetic inheritance might be the influence of uterine environment on the developing fetus – such fetal programming has been suggested as an explanation for the observed predisposition of malnourished fetuses to develop diabetes or heart diseases as adults (though this remains very contentious). Environmental factors are known to influence the emergence and reversion of epigenetic factors, allowing for the possibility that epigenetic variations at several loci and in several cells or organisms might play a role in evolution. Such an adaptive variation would be a Lamarckian form of evolution. A number of experimental studies seem to indicate that epigenetic inheritance can play a part in the evolution of complex organisms. Methylation differences between maternally and paternally inherited alleles of the mouse *H19* gene are preserved. There are also numerous reports of heritable epigenetic marks in plants.

Portrait of Jean-Baptiste de Monet Chevalier de Lamarck by Charles Thevenin (1764–1838).

At least in this case, offspring are inheriting the life experience of the father.

Methylation of DNA

Musicians paint their pictures on silence. We provide the music, and you provide the silence.

Leopold Stokowski

The prototypic epigenetic modification of DNA in mammalian cells is the covalent addition of a methyl (CH_3) group to the fifth position of cytosine within CpG dinucleotide islands, which can directly turn off gene expression ("silencing") (Fig. 11.1). The methylation of DNA is achieved by three DNMTs, termed DNMT1, DNMT3A, and DNMT3B, though exactly how these are targeted to specific DNA regions is unclear. The other major class of epigenetic modification involves posttranslational modification of histones and chromatin remodeling (see the "Epigenetics and cancer" section).

It is believed that DNA methylation may have evolved for silencing of repetitive elements, but has subsequently been adopted in order to effect transcriptional silencing in imprinting and X-chromosome inactivation. Imprinting, the phenomenon whereby expression of a gene may be silenced depending on whether it was inherited from the mother or the father, is thought to be due to differential methylation in maternal versus paternal genes (Fig. 11.2). Conversely, loss of imprinting refers to either the activation of normally silent imprinted genes or potentially the silencing of active imprinted genes, and is frequently observed in many different cancers. In particular, reactivation of the normally imprinted allele of the *IGF2* gene is often seen in human cancers and is associated with resistance to apoptosis and tumor progression in animal models.

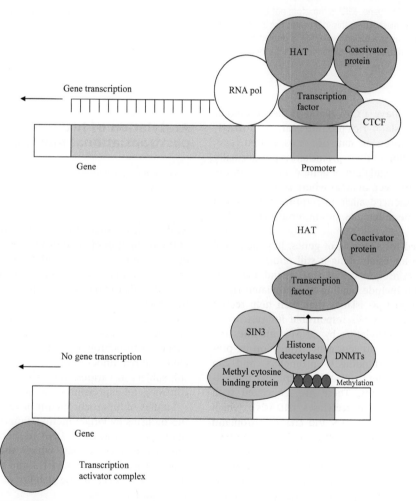

Figure 11.1 Methylation can inhibit gene transcription. In the presence of CTCF, the gene is "insulated" from methylation; in this case, the gene is transcriptionally active as it will remain in the unmethylated state. Conversely, in heterochromatin, CpG islands in the promoter region are methylated, and these regions are transcriptionally inactive. Gene silencing following methylation is reinforced by deacetylation and interactions with repressors of transcription. In fact, DNMTs, which further mediate methylation, may be actively recruited to help maintain silencing. Methylation in turn enables the binding of a complex comprising the methyl cytosine-binding protein (MBP) and histone deacetylase (HDAC); some MBPs (MECP2 and MBD1/2) can also associate with transcriptional co-repressors such as SIN3, which directly bind to HDAC and contribute to gene silencing. HDAC promotes deacetylation of histones, which contributes to organization of nucleosomes and also, more generally, repression of transcription. In order for gene expression to take place, the HDAC–SIN3 complex is displaced and a transcription activator complex (transcription factor, histone acetyl transferase (HAT), and coactivator protein) can associate with promoter elements; HAT acetylates the histone-reversing effects of HDAC. In cancer, many genes may be inappropriately inhibited by methylation in the promoter region, and this is a frequent cause of loss of tumor suppressor activity during tumorigenesis.

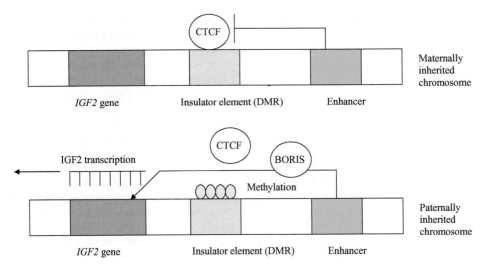

Figure 11.2 Imprinting of the gene for IGF2. On chromosomes inherited from the female, a protein called CTCF binds to an insulator preventing interaction between the enhancer and the *IGF2* gene. *IGF2* is therefore not expressed from the maternally inherited chromosome. Because of imprinting, the insulator on the male-derived chromosome is methylated; this inactivates the insulator by blocking the binding of the CTCF protein, and allows the enhancer to activate transcription of the *IGF2* gene. It is speculated that CTCF is displaced by another protein, BORIS. The methylation patterns (imprints) on the chromosome, inherited by the zygote after fertilization, are maintained in subsequent generations by maintenance methyl transferases. After Alberts *et al.* (2002).

Around half of all genes have a CpG island in their promoter region, but most such CpG island–rich promoters are not methylated, irrespective of the expression state of the associated gene. This suggests that methylation of these promoters is not normally involved in the day-to-day regulation of gene expression in the large majority of cases. However, in areas where gene expression is silenced, such as the silenced allele of imprinted genes and the inactive X chromosome in females, promoter-associated CpG islands are methylated.

Methylation is required for silencing of genes, but the actual mechanisms responsible for establishing it still remain unclear. However, some of the consequences of CpG island methylation are now known and include binding of methylated DNA-specific binding proteins to CpG islands that then help recruit various histone-modifying enzymes responsible for restructuring the chromatin. DNA methylation by DNMTs modifies the actual DNA itself, can directly prevent gene expression by preventing transcription factors binding to promoters, and can additionally exert a more general effect by recruiting methyl-binding domain (MBD) proteins. These are associated with further enzymes called histone deacetylases (HDACs), which function to chemically modify histones and change chromatin structure. Histones may also become methylated on lysine residues, and this may contribute to gene silencing and imprinting; in fact, it is increasingly likely that the machinery controlling DNA and histone methylation is linked and collaborates in gene silencing. Thus, methylation of a CpG island alters expression of a gene in two ways, directly by interfering with the binding of specific transcription factors to promoters and indirectly by recruiting proteins such as MBD that associate with HDACs, which function to deacetylate histones and change chromatin structure.

Thus, epigenetic regulation depends on two overlapping processes:
- methylation of the DNA; and
- posttranslational modification of histones.

Acetylation of histones and other posttranslational modifications

As we must account for every idle word, so must we account for every idle silence.

Benjamin Franklin

As discussed earlier, histone acetylation plays an important role in the regulation of gene expression and together with methylation can influence the binding and activity of transcriptional activator complexes (Fig. 11.1). In fact, it has recently become clear that deacetylation of histones by HDAC is a major means by which methylation results in formation of heterochromatin and in the suppression of gene expression. Most of the human genome is packaged up as transcriptionally inactive densely packed heterochromatin, and this chromatin is heavily methylated. The remainder of the genome is transcriptionally active but still subject to various stimulatory and inhibitory processes that control gene expression on a day-to-day basis.

Studies of heterochromatin have helped unravel the complex mechanisms by which methylation of DNA culminates in silencing of gene expression. DNA in methylated regions is packaged into dense nucleosomes, which also contain deacetylated histones such as deacetylated H3 and also H4. Histone acetylation has a direct effect on the stability of nucleosomal arrays and on chromatin structure.

Histone acetylation is a dynamic process that is regulated by two groups of opposing enzymes, the histone acetyltransferases (HATs) and the histone deacetylases (HDACs) – see Fig. 11.3. Through these effects on histone (and also nonhistone proteins – see later), these enzymes play key roles in regulation of gene expression (see also the section on c-Myc in Chapter 6), chromosome segregation, and development. Moreover, their deregulation has been linked to cancer. The HATs catalyze the covalent addition of an acetyl group from acetyl coenzyme A to the

Gene and Protein Regulation by Epigenetic Factors

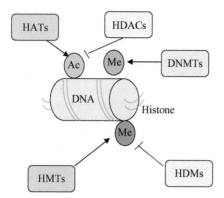

Figure 11.3 Methylation and acetylation in cancer.

N-terminal lysine residues of histones, whereas the HDACs remove such acetyl groups. Histone acetylation is a defining feature of transcriptionally active chromatin, and in the presence of appropriate transcriptional activators the gene will be expressed (note expression is still subject to regulatory factors). Conversely, the cardinal features of constitutive heterochromatin (transcriptionally inactive chromatin – genes not expressed) include deacetylation of histones as well as hypermethylation of DNA.

In keeping with this, inactive genes are associated with bound complexes containing HDACs that deacetylate histones, whereas active genes have strongly acetylated histones under the influence of HAT activity that reconfigure the chromatin to be open and accessible to the transcriptional activator complex. Histone deacetylation causes the condensation of chromatin, making it inaccessible to transcription factors, and the genes are therefore silenced (see Figs 11.1 and 11.3).

In fact, a number of other crucial posttranslational modifications of histones, in addition to acetylation, have been identified. These include methylation, phosphorylation, ubiquitination, sumoylation, glycosylation, and ADP ribosylation, all of which can help regulate the activity of many genes by modifying both core histones and nonhistone transcription factors. Also, we must not forget the importance of the variant histone yH2AX in the DNA damage response (see Chapter 10).

Recent studies have identified what has become known as a "methylation mark" that may help define and separate regions of transcriptionally active chromatin from transcriptionally inactive chromatin. These marks seem to involve methylation of lysine 9 in the tail of histone H3, which marks inactive and methylation of lysine 4 on histone H3 which marks transcriptionally active chromatin. The methylated lysine 9 appears to bind proteins required for maintaining a repressed state, but it remains to be shown how this leads to methylation of DNA. Possibilities include the facilitation of binding of DNMTs. What seems clear is that despite the complexity of epigenetic factors, methylation appears a dominant event over acetylation, as in cancer inhibiting HDAC alone does not reactivate aberrantly silenced genes and hypermethylated genes, whereas these same inhibitors can if cells are first treated with demethylating drugs.

The histone code

Since it became clear that the genome contains information in two forms, genetic and epigenetic, research efforts have been directed at trying to crack the "histone code" which is in many respects analogous to the DNA code which was unraveled many years ago.

As you will have gathered, histones not only are there to pack away the DNA, but also are pivotal regulators of chromatin structure and function, at least in part because they can integrate a variety of regulatory processes which operate through various posttranslational modifications of the histone tails. The "histone code" hypothesis, which postulates that these covalent histone modifications regulate gene transcription, was first proposed by Strahl and colleagues more than a decade ago. According to this hypothesis, histone modifications, such as lysine methylation, are "read" by specific binding proteins that enhance or suppress transcription depending on the site that has been "marked." In some ways, this histone code is complementary to the DNA code and determines how and when specific genes are transcribed. In addition to methylation and acetylation, other modifications used in the histone code include phosphorylation, ubiquitination, sumoylation, and ADP ribosylation within gene regulatory regions. Although the consequences of such modifications are not well known, they likely include chromatin restructuring and controlling the docking or function of transcription factors or other histone-modifying enzymes. This model offers an explanation for how modest single or small numbers of histone modifications might regulate chromatin functions, particularly gene transcription. Some progress has now been made in decoding histone modifications. For example, gene expression can be activated by monomethylation of lysines (K) at position 20 and 5 of histones H4 and H2B to form H4K20 and H2BK5 respectively or by trimethylation of lysines 4, 36, and 79 on H3 to form H3K4me3, H3K36me3, and H3K79me respectively. Genes are also activated by acetylation of H3K9 and H3K14. Conversely, gene expression is repressed by trimethylation of H3K9 and H3K27.

In many cancers, trimethylation of H4K20 and acetylation of H4K16 are reduced, and alterations in many of the enzymes responsible for histone posttranslational modifications and chromatin structuring have been demonstrated in a wide range of cancers and pre-malignant conditions. Thus, several HDACs, HMTs and HDMs, lysine acetyltransferases, sirtuins, and the JARID-1 family have been shown to be abnormally expressed in a variety of different cancers. See Table 11.1 for the lengthy list of genes encoding epigenetic regulatory proteins that are aberrantly expressed (up or down) or mutated in human cancers.

Epigenetic regulation of gene expression

Chromatin structure, nucleosome modeling, and promoter DNA methylation are amongst the most important determining factors in gene expression. We will now focus more on exactly how chromatin modification is recognized and acted upon by the transcriptional machinery, thereby turning histone coding into altered gene expression. The enzymes responsible for the posttranslational modification of nucleosomal histones, such as the DNMTs, have been described already (and are listed in Table 11.1, alongside the cancers that may result from their aberrant expression or function).

Histone modifications act by recruiting a number of modification-specific binding proteins, such as 14-3-3, that facilitate transcriptional activation. Gene expression is influenced by signaling pathways in several ways, including by histone modification, which is of relevance here, as exemplified by phosphorylation of histone H3. Expression of several key immediate-early (IE) genes is regulated in part by H3 phosphorylation at S10 (H3S10ph) at promoters and coding regions when MAPK signaling is activated. This might appear simple, but given the fact that a large number of kinases can directly phosphorylate H3, such as MSK1/2, PIM1, RSK2, and IKKα, it is distinctly possible that this might be a central regulatory node in the signaling-related activation of gene expression. H3 S10 phosphorylation is linked to acetylation of adjacent lysine residues (K9 or K14) suggesting rewriting of the histone code, which in turn can promote binding of various 14-3-3 proteins required for induction of gene transcription of several IE and *HDAC* genes.

How this actually works is exemplified by the *FOSL1* gene, where 14-3-3 binds to H3S10ph at the enhancer and then recruits a cohort of regulators including HAT, BRD4, males absent on the first (MOF), and the transcription elongation factor b (P-TEFb), whilst H3 S10 phosphorylation also restarts the preinitiated but paused RNA polymerase II. H3 S28 phosphorylation may also facilitate binding of 14-3-3 and has been observed at nucleosomes at IE gene promoters. A recent study by Lau and Cheung has shown that S28 phosphorylation, which induces a methyl–acetylation switch on an adjacent K27 residue, is important in the direct activation of the IE gene *c-fos* and the polycomb-silenced α-globin gene. The authors also suggest how histone coding differences might operate in different contexts; H3 S10 phosphorylation might facilitate transcriptional elongation of genes that are regulated by polymerase pausing, whereas H3 S28 phosphorylation might directly initiate transcription.

At least 90% of all human genes are subject to alternative splicing, which will be discussed in more detail later, but given the key role in determining the types of protein produced by a gene, it is important to consider how this is regulated. Chromatin as the template for nuclear transcription can influence splicing choices by altering structure and the way in which the histone code is interpreted. It is now known that nucleosomes are often located at exon–intron boundaries and are susceptible to specific histone modifications which can in turn regulate alternative splicing.

The mSin3A corepressor is a core component of a large multi-protein corepressor complex that links HDACs with chromatin targeting subunits such as Pf1 and MRG15. In man, an Rpd3S–Sin3S corepressor complex represses aberrant gene transcription from cryptic transcription initiation sites and blocks progression of RNA polymerase II in actively transcribed genes. A recent computational analysis by Ron DePinho and colleagues has confirmed the wide range of genes regulated by this means. Thus, several nodal points by which mSin3A influences gene expression have been identified, including the Myc–Mad, E2F, and p53 transcriptional networks.

The tumor suppressor RB is also a key player in the assembly of constitutive heterochromatin. As we have seen in Chapters 4 and 7, RB represses genes required for entry into the S phase and progression of the mitotic cell cycle, an action at least in part long appreciated to involve interaction with HDAC at the promoter regions of various genes activated by E2F family transcription factors. But recent studies, some from our coauthor Maria Blasco, suggest that RB may function as a global suppressor of gene expression and that moreover loss of RB could result in a generalized loss of repressive chromatin and reactivation of gene expression. Importantly, epigenetic factors may be a general way by which proteins and genes communicate with each other, and in particular histone acetylation is now known to contribute to the day-to-day regulation of gene expression. Thus, inhibition of gene expression by transcription factors such as c-MYC involves the recruitment of histone-modifying co-repressor complexes containing HDAC and mSIN3A (or, in other such complexes, N-CoR–SMRT), which can promote deacetylation of lysines in histone H4 tails. In many cases, it now appears that for transcriptional activation of multiple genes, epigenetic factors (possibly generally involving inhibitory complexes of RB and HDAC, among others) must first be overcome. In other words, the role of epigenetic factors in controlling gene expression is extending way beyond the previous notion of such factors predominantly mediating permanent or near-permanent inactivation of genes in development. This is graphically illustrated by the means by which some genes are inhibited and others activated during regulation of the G_1–S transition in the mitotic cell cycle, discussed in Chapter 4. To remove the brakes from the cell cycle, c-MYC, in partnership with MIZ1, inhibits expression of genes such as the cyclin-dependent kinase inhibitor (CKI) p21^{CIP1} by recruiting histone-modifying complexes containing mSIN3A and HDAC to the promoter regions and concurrently promotes expression of genes (cyclin–CDKs, etc.) in partnership with a different protein MAX. The net effect is hyperphosphorylation of the RB protein, which in turn displaces both RB and associated histone-modifying enzymes such as HDAC from the various promoters essential in order for E2F family transcription factors to drive expression of genes needed for the S phase.

Epigenetics and cancer

Epigenetic changes are key factors in several diseases, in particular cancer – and their importance is underscored by the devotion of a large part of this chapter to the topic. As discussed, the major forms of epigenetic modification that have been associated with cancer cells are aberrant DNA methylation of CpG islands located in gene promoter regions (including loss of imprinting of genes) and changes in chromatin conformation involving histone deacetylation.

Epigenetic regulation is often markedly abnormal in cancer cells, which typically manifest markedly disturbed DNA methylation patterns. In fact, cancer cells may display seemingly paradoxical region-specific hypermethylation of CpG islands, in particular genes, alongside a more generalized global decrease in DNA methylation (hypomethylation). DNA methylation is the most prominent epigenetic modification in humans, and most tumor types, including pre-malignant lesions, have abnormal DNA methylation patterns. Generalized loss of DNA methylation was the first epigenetic abnormality identified in cancer cells. Such loss of methylation affects nonpromoter regions of DNA and structural elements, but also numerous CpGs methylated in normal tissues are unmethylated in cancer cells. However, it should be borne in mind that the bulk of CpG islands are not methylated in normal cells. In fact, in concert with global hypomethylation, the most obvious aberration in cancer cells is the converse, namely, specific hypermethylation of some of the usually unmethylated CpG islands. Hypermethylation by DNMTs or other means may silence some tumor suppressor genes, because transcriptional activator complexes cannot bind to the methylated promoters, whereas conversely hypomethylation may activate selected proto-oncogenes.

Gene silencing by hypermethylation has been frequently alluded to in this book and is well known as a major factor in inactivation of tumor suppressor genes, such as that encoding p16^{INK4A}, in cancer. Epigenetic silencing of a tumor suppressor gene can provide the cell with a growth advantage as surely as inactivation by gene deletion or mutation (Fig. 11.1 and Chapter 7). The importance of gene silencing by hypermethylation is underscored by the observation of aberrant hypermethylation in the very earliest recognizable lesions in progression of colon cancer. Silencing of a tumor suppressor by hypermethylation of CpG in the gene promoters was first observed for the *RB* gene, but has now been shown for numerous other gene loci in cancer cells, including *CDKN2A* (encoding ARF and p16^{INK4a}), *VHL*, *E-cadherin*, *14-3-3σ*, *BRCA1*, and *MLH1*. In fact, in some cases genes are only or predominantly silenced by methylation rather than by mutations in the DNA, as for the DNA repair gene *MGMT*, the *CDKN2B* gene (encoding the p15^{INK4b}) CKI, *HIC1* (hypermethylated in cancer 1), and *RASSF1A*. Haploinsufficient tumor suppressor genes notwithstanding, Knudson's two hit-model, discussed in Chapter 8, predicts that abnormal phenotypes arise only if both gene copies are inactivated. In those cases where one allele is mutated in the germline, the second copy is frequently inactivated by methylation.

Some cancers appear to have multiple gene promoters methylated, which has led to the proposition of a CIMP by Issa and colleagues (discussed further in the "CIMP and MIN" section). The notion of CIMP is supported by observation of sporadic colon cancers with mismatch repair defects, though whether this model can be generally applied to other cancers remains controversial. In fact, it is still argued whether methylation is the primary cause of gene silencing in some cancers or merely contributes to the maintenance of gene silencing initiated by genetic mechanisms. Several lines of evidence confirm that maintenance of gene silencing in cancer cells requires DNA methylation, not least the findings that demethylating drugs or elimination of DNA methyltransferases can reactivate gene expression. However, it has recently been shown that histone modification and silencing of the *P16^{INK4a}* can precede promoter methylation, suggesting that at least in this case, methylation is not the primary cause of loss of tumor suppressor activity.

As mentioned in this chapter, most cancer cells manifest global hypomethylation alongside region-specific areas of hypermethylation in CpG islands. Although hypomethylation of genes was the first epigenetic change identified in cancer cells, this area has been comparatively neglected until recently. Cancers involve not just gene silencing but also, self-evidently, gene activation. In contrast to hypermethylation, hypomethylation of genes may provoke inappropriate gene activation. A notable example of this in human cancers is deregulated overexpression of *c-MYC* in Burkitt lymphoma, as described in Chapter 6. Normally, the *c-MYC* gene is located in a region of repressed chromatin and is expressed at a low level. However, in Burkitt lymphoma due to a chromosomal translocation in lymphocytes, *c-MYC* is "relocated" into a region of open and active chromatin with resultant overproduction of large amounts of c-MYC protein, deregulated proliferation, and ultimately lymphoma. It is now known that aberrant demethylation can activate the following genes in cancer cells: maspin (gastric cancer), *S100A4* (colon cancer), and the so-called cancer–testis genes (CT), such as *MAGE* and *CAGE*. Global demethylation might be an early mechanism favoring genomic instability (based on recent studies, exploring the role of loss of RB function might prove interesting in this context!), whereas region specific hypermethylation might be later events involved in progression and spread. The causes of hypomethylation in cancer remain unclear but may include altered function of the SW1–SNF chromatin-remodeling complexes. The mammalian SWI–SNF chromatin-remodeling complex is composed of more than 10 protein subunits, and plays important roles in epigenetic regulation. The DNA methyltransferase, DNMT1, accounts for most methylation in mouse cells, but both *DNMT1* and *DNMT3b* genes have to be silenced in order to eliminate methyltransferase activity in human colon cancer cells and reactivate expression of the tumor suppressor p16^{INK4a} (in fact, cells now lose around 95% of their total genome-wide methylation). Interestingly, all three DNMTs are moderately overexpressed in a number of cancers, and overexpression of DNMT1 can also promote cellular transformation under some circumstances. It may therefore not come as a surprise to hear that DNMTs are now regarded as more complex than previously appreciated. Depending on the context, either inactivation or overexpression of DNMTs might contribute to tumorigenesis. Progress has been made in unraveling how these enzymes interact. DNMT3A and B are responsible for initiation of methylation which is maintained by the action of DNMT1. However, recent studies suggest that these enzymes might silence some genes even without methylation changes either by an as-yet-unrecognized direct action or through interactions with other proteins such as HDACs.

So far more than 600 genes, including tumor suppressor genes and oncogenes, are known to be regulated by epigenetic mechanisms (see also Table 11.2), and in many cases this may be altered in cancers. Interestingly, it also appears that a large proportion of cancer mutations involve the regulation of methylation, acetylation, and epigenetic regulation of gene expression in some way, suggesting strong links between the classical genetic and epigenetic models of tumorigenesis. As you have no doubt noted by now, a recurring theme in much of cancer biology is the quintessential "chicken-egg conundrum," in this case whether epigenetic factors are a cause of cancer or whether instead cancer cells produce epigenetic changes. Strong support for a causal role of epigenetic factors in cancer originates from the study of patients with Beckwith–Wiedemann syndrome (BWS). BWS is caused by epigenetic defects, and these are specifically associated with increased cancer risk. Moreover, the presence of epigenetic alterations at early stages of CRC and pre-malignant conditions also argues for a causal role in some cases. In mouse models, DNA hypomethylation may contribute to tumorigenesis possibly by promoting chromosomal instability. Mice carrying a hypomorphic DNA methyltransferase 1 (*Dnmt1*) allele that reduces DNMT1 expression exhibit substantial genome-wide hypomethylation in all tissues. These mice also develop aggressive T-cell lymphomas with frequent chromosome 15 trisomy. Hypomethylation may contribute to cancer through several mechanisms including gene activation, defective DNA mismatch repair, and chromosomal instability; moreover, hypomethylation may result from dietary and carcinogen exposure linking such abnormalities to environmental factors.

Not surprisingly, there is great interest in the therapeutic targeting of aberrant or undesirable methylation patterns in cancer, and DNA methylation inhibitors are now being examined as potential anticancer agents (see Chapter 16) as these are reasonably predicted to restrain tumors by reactivating various, otherwise silenced tumor suppressor genes. However, given the recent recognition of the opposing role of hypomethylation in the inappropriate activation of various oncogenes involved in tumorigenesis, metastasis, and invasion (see Chapter 12), the application of broad-spectrum treatments that affect global methylation

Table 11.2 Genes known to be silenced by CpG island hypermethylation in human cancer

Gene	Normal role	Tumor	Consequences
hMLH1	DNA mismatch repair	Colon, endometrium, and stomach	Frameshift mutations
BRCA1	DNA repair and transcription	Breast and ovary	Double strand-breaks?
$p16^{INK4a}$	Cyclin-dependent kinase inhibitor (CKI)	Multiple types	Cell-cycle entry
$p14^{ARF}$	MDM2 inhibitor	Colon, stomach, and kidney	Degradation of p53
$p15^{INK4b}$	CKI	Leukemia	Cell-cycle entry
MGMT	DNA repair of O6-alkyl-guanine	Multiple types	Mutations and chemosensitivity
GSTP1	Conjugation to glutathione	Prostate, breast, and kidney	Adduct accumulation?
LKB1–STK11	Serine–threonine kinase	Colon, breast, and lung	Unknown
ER	Estrogen receptor	Breast	Hormone insensitivity
PR	Progesterone receptor	Breast	Hormone insensitivity
AR	Androgen receptor	Prostate	Hormone insensitivity
PRLR	Prolactin receptor	Breast	Hormone insensitivity
RARβ2	Retinoic acid receptor β2	Colon, lung, head, and neck	Loss of hypoxic response?
VHL	Ubiquitin ligase component	Kidney and hemagioblastoma	Entrance in cell cycle
RB	Cell-cycle inhibitor	Retinoblastoma	Neovascularization
THBS-1	Thrombospondin-1 and antiangiogenic	Glioma	Dissemination
CDH1	E-cadherin and cell adhesion	Breast, stomach, leukemia	Dissemination?
CDH13	H-cadherin and cell adhesion	Breast, lung	Dissemination?
FAT	Cadherin and tumor suppressor	Colon	Activation β-catenin route
APC	Inhibitor of β-catenin	Aerodigestive tract	Activation WNT signaling
SFRP1	Secreted Frizzled-related protein 1	Colon	Anti-inflammatory resistance?
COX-2	Cyclo-ox genase-2	Colon and stomach	JAK2 activation
SOCS-1	Inhibitor of JAK-STAT pathway	Liver and myeloma	JAK2 activation
SOCS-3	Inhibitor of JAK-STAT pathway	Lung	Silencing of target genes
GATA-4	Transcription factor	Colon and stomach	Silencing of target genes
GATA-5	Transcription factor	Colon and stomach	Resistance to apoptosis
DAPK	Proapoptotic	Lymphoma, lung, and colon	Resistance to apoptosis
TMS1	Proapoptotic	Breast	Unknown
TPEF–HPP1	Transmembrane protein	Colon and bladder	Unknown
HOXA9	Homeobox protein	Neuroblastoma	Resistance to apoptosis
IGFBP3	Growth factor-binding protein	Lung and skin	Cellular detachment
EXT1	Heparan sulphate synthesis	Leukemia and skin	Cellular detachment

Modified from Esteller (2005).

patterns should be approached with some caution. Theoretically, drugs able to reduce methylation might reactivate tumor suppressors (desirable) but could also activate oncogenes (undesirable) – the ideal scenario would be to understand how individual genes are epigenetically regulated and then selectively target therapeutically the desired gene repertoire without adversely affecting others, but this goal is some way in the future. Such therapeutic endeavors notwithstanding, there is now also a major interest in new diagnostic technologies able to rapidly screen the genome for DNA methylation and histone acetylation patterns (see Chapter 18).

Novel therapeutic strategies are also being developed based on the inhibition of histone deacetylation.

Chromatin remodeling contributes to probably all cancers in some way, as this process is critically involved in the inhibition of gene expression by a variety of transcription factors including c-MYC (see Chapter 6), the E2F family, and others. Aberrant function of histone-modifying complexes can have profound effects on the patterning of the "histone code" and on essential chromatin-related processes such as gene transcription, chromosome structure, and stability, among others, and it can set the stage for cancer. The commonest disruptions in the histone code in cancer are global reductions in the trimethylation of H4K20 (H4K20me3) and acetylation of H4K16 (H4K16Ac) together with DNA hypomethylation. Moreover, there are extensive examples of chromatin-modifying factors involved in cancer by aberrant expression or mutation (a fairly comprehensive survey is given in Table 11.1). The importance of chromatin structure in cancer is emphasized in a recent study of rare pancreatic neuroendocrine tumors; exome sequencing demonstrated that the major class of mutated genes encoded proteins implicated in chromatin remodeling, including MEN1, which encodes menin, a component of a histone methyltransferase complex, and death-domain-associated protein (DAXX) and ATRX (α thalassemia–mental retardation syndrome X-linked).

Aberrant histone acetylation is implicated in a large number of hematological cancers. In some cases this is due to a switching of the normal inhibitory nuclear co-repressor (NCoR)–Sin3–histone deacetylase (HDAC) complex for activating HAT activity, as exemplified by some cases of acute PM and by the functional alterations generated by the oncogenic AML1–ETO fusion protein in AML. Several prominent chromosomal translocations in hematological cancers can misdirect HATs in this way. The t(8;16) (p11;p13) results in translocation of CBP (an acetyltransferase), and t(11:16)(q23;p13.3) generates the MLL–CBP fusion protein, with aberrant HAT activity, in subtypes of AML. Missense mutations in *EP300* have been identified in GI and breast cancers, and many cancers exhibit increased activity of HDACs.

The fusion of MLL (an H3K4 histone methyl transferase) to others is found in a variety of myeloid and lymphocytic leukemias.

The proteins of the BRD family, which carry double bromodomains, are also important in cancers. The human *BRD4* gene, located at 19p13.1, affects breast cancer and is involved in the very rare NUT midline cancer in which translocation of *BRD4* with the nuclear protein in the testis gene (*NUT*) at 15q14 produces the BRD4–NUT oncoprotein.

A large number of so-called reader proteins carry distinct structural domains that recognize histone modifications and translate this code into changes in gene expression. Different readers, as exemplified by the tripartite motif-containing 24 (TRIM24) protein that recognizes both H3K4me0 and H3K23ac, can recognize different patterns of histone modification and allow a large degree of complexity to be integrated at the level of chromatin modification.

CIMP and MIN and the "mutator phenotype"

As discussed in Chapter 10, genetic instability may manifest primarily as either numerous gene mutations or major structural rearrangements of chromosomes (CIN). Broadly two classes of genome "caretaker" systems may be identified that normally either (a) prevent gene mutations, particularly during DNA replication or during gene expression, including the DNA mismatch repair pathway; or (2) detect aberrant chromosome segregation during mitosis. Any mechanisms that disrupt these pathways may cause genetic instability and thereby greatly accelerate the accumulation of further potentially cancer-causing mutations – the "mutator phenotype." The mutator phenotype hypothesis was discussed in Chapter 10 and was derived largely from studies of inherited colon cancer, hereditary nonpolyposis colorectal cancer (HNPCC, also known as Lynch syndrome). As mentioned earlier, in HNPCC mutation in one allele of a gene involved in DNA mismatch repair (e.g. *MLH1*) is inherited, but loss of the second, wild-type allele occurs with high frequency, promoting microsatellite instability (MIN). MIN is associated with mutations in numerous other genes, which contribute to progression of colon cancer.

Aberrant CpG island methylation for various individual genes has been repeatedly demonstrated in cancers, but only relatively recently has it been widely appreciated that in some tumors, groups of genes have consistently increased methylation, suggesting the existence of a more widespread defect in methylation. In sporadic colon cancer, where this has been particularly well studied, the methylation of two or more distinct genes is often strongly correlated, particularly in the subset of these cancers that also have MIN. Thus, in sporadic colon cancers promoter methylation of *CDKN2A* (encoding the p16^{INK4a} and p14^{ARF} proteins), *THBS1*, (thrombospondin 1), and *HPP1* has been shown, and most have silencing of the mismatch repair gene *MLH1* (mutL homologue 1).

Many genes become methylated with increasing age, and in fact many genes observed as methylated in cancers are not that

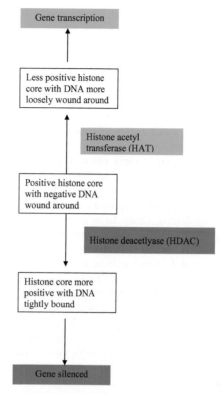

Figure 11.4 Regulation of gene expression by histone acetylation. Gene expression is also regulated by acetylation as well as by methylation. Acetylation of histones results in a more open configuration of chromatin, allowing access of transcriptional activators, and the like, to the gene to be transcribed. Conversely, deacetylation results in a more closed conformation of DNA denying access to genes. In fact, displacement of HDAC may be one way in which RB phosphorylation enables access of E2F for transcription of S-phase genes in the cell cycle.

different in, for example, normal aging colonic mucosa. However, a subset of genes, including several key tumor suppressors (p16^{INK4a}, MLH1, etc.), are now known to be more extensively methylated in cancers than in normal aged tissues, and this phenomenon has been referred to as CIMP. One obvious possibility is that epigenetic silencing of *MLH1* might actually be a causal factor for MIN in some sporadic colon cancers, which is compatible with studies showing that methylation-inhibiting agents can reverse the mismatch repair defect in cultured colon cancer cells. Thus, parallels can readily be seen between inherited colon cancers (HNPCC), where MIN results from germline mutations in mismatch repair genes, and sporadic colonic tumors, where MIN may result from aberrant methylation of *MLH1* (Fig. 11.4).

In fact, *MLH1* methylation could contribute to as many as 70% of sporadic colon cancers with MIN. Hypermethylation of the mismatch repair gene *MLH1* is commonly found in mismatch-repair-defective tumors. Such tumors are also more likely to manifest abnormal imprinting. Since these important observations were first made, CIMP has now been demonstrated in numerous other cancers, including glioblastoma, stomach, liver, pancreas, and ovary. Intriguingly, promoter hypermethylation of MLH1 is seen in pre-malignant polyps even before they develop mutations.

Tumor hypomethylation has recently been strongly linked to chromosomal instability; pericentromeric satellite sequences are hypomethylated, and many cancers contain unbalanced chromosomal translocations with breakpoints in pericentromeric DNA of chromosomes 1 and 16. Demethylation of satellite sequences could contribute to breakage and recombination.

These results are intriguing because they suggest that genetic instability and MIN in particular may not necessarily derive from mutations, but also from epigenetic factors that alter gene expression. That different cancers have significantly different rates of tumor suppressor gene silencing suggests that promoter methylation is not random and that aberrant methylation may be selected for during neoplastic transformation and might be the cause of cancer development in some cases. Although much aberrant DNA methylation may arise as a spin-off from tumorigenesis and aging, the presence of CIMP argues that at least in some circumstances DNA methylation may be a causal event.

Cancers manifesting CIMP have certain characteristics. Not surprisingly, CIMP-positive cancers are more frequent in older individuals but are also more common in women. CIMP-positive cancers also tend to have particular patterns of altered gene expression and mutations; *p53* mutations are rare, whereas mutations in *KRAS* or *BRAF* are ubiquitous. In part this may be due to links between CIMP and MIN, but such aberrant features may occur in tumors without apparent MIN also.

The actual causes of aberrant methylation and CIMP are not fully understood but include nonspecific factors that might accelerate methylation such as inflammation and exposure to various carcinogens or "epimutagens" as they may be referred to in this context. In fact, CIMP is more likely in colonic tumors arising on a background of inflammation such as ulcerative colitis, and particular lifestyles may also show correlations. One other possibility is that CIMP arises through the abnormal spreading of methylation from one gene to others; methylation spreading in *cis* is probably a normal feature of gene silencing, with abnormal spread contained by various proteins that operate in concert with CpG islands. This protective mechanism may become less effective with aging or potentially due to mutations in genes encoding "protective" proteins.

CIMP, as manifested by the simultaneous methylation of multiple genes, predicts a poorer prognosis in many cancers, including head and neck, lung, prostate, and acute leukemias. This observation fuels interest in employing such knowledge in the profiling of cancers (see the "Cancer profiling" section).

Imprinting and loss of imprinting

Imprinting is the conditioning of the maternal and paternal genomes during gametogenesis, so that a specific allele is more abundantly or exclusively expressed in the offspring. At present around 80 genes are known to be imprinted, and differential expression of the two alleles can occur in all or only some cell types or different stages of development. Imprinting is regulated by epigenetic changes such as DNA methylation, and methylation of specific localized areas, termed differential methylated regions (DMR), are characteristic of imprinted genes. It seems that the DMR contains a region known as the imprinting control region (ICR) essential for controlling expression of genes lying within an imprinted domain. CCCTC-binding factor (CTCF) is an important regulator of expression of imprinted genes. By binding to the unmethylated parental allele of genes containing an ICR, CTCF allows allele-specific gene expression. CTCF effectively "reads" DNA methylation marks and directs expression of the allele without DMR methylation. CTCF may also protect bound DMR from being methylated and, as stated in a recent review article by Keith Robertson, may even operate as a more or less general "insulator" from methylation for large regions of the genome. Loss of genomic imprinting (LOI) involves loss of the normal imprinting controls either by methylation or demethylation of the DMR or potentially by failure of CTCF function. Interestingly, with this latter in mind, cancer cells have been shown to express increased levels of a potential CTCF antagonist, BORIS. The result is either abnormal activation of the normally silent inherited allele or, conversely, inappropriate silencing of the normally expressed allele – examples of both scenarios have been described in human cancers.

Imprinting was first identified as disease relevant in humans through the demonstration of paternal uniparental disomy of 11p15 in Beckwith–Wiedemann syndrome (BWS) and by maternal uniparental disomy of 15q11–12 in Prader–Willi syndrome. BWS is characterized by multi-organ overgrowth and predisposition to embryonal tumors such as Wilms' tumor (Box 11.3), but loss of heterozygosity of 11p15 is also noted in numerous other tumors, including lung, bladder, ovarian, liver, and breast cancers. Two genes have been identified in this locus, *H19* and *IGF2*, which undergo reciprocal imprinting in the mouse and in humans, with maternal expression of H19 and paternal expression of IGF2. The maternal *IGF2* gene is imprinted in normal tissues (Fig. 11.2), whereas in some Wilms' tumors and other tumor types this imprinting is lost, leading to bi-allelic transcription of *IGF2*. Intriguingly, this mechanism contributes to Wilms' tumors primarily in Caucasian children as loss of imprinting is generally absent in Wilms' tumors in the East Asian population. It is, therefore, likely that this epigenetic mechanism may in part explain the difference in incidence of Wilms' tumor between populations. In a very recent study, it has been shown that loss of imprinting of IGF2 in a mouse model doubles the risk of developing intestinal tumors and that this is associated with a less differentiated phenotype of cells even in the normal colonic mucosa.

The converse scenario, whereby LOI could result in silencing of a normally expressed allele of a growth inhibitor, has also recently been implicated in cancer. Thus loss of the maternally expressed allele of the gene encoding the p57^{KIP2} CKI involved in G$_1$–S arrest can also provide a cancer cell with a growth advantage as shown in some cases of Wilms' tumor.

Insulin-like growth factors (IGFs) regulate growth and apoptosis through interaction with the IGF1 receptor, and overexpression of the human IGF1 receptor promotes ligand-dependent neoplastic transformation in a variety of cell types. Two main subtypes of IGF proteins are known, IGF1 and IGF2; the latter appears to play a predominant role during development and is believed to be largely redundant in the normal adult. IGF1 is an important mediator of survival signaling and is involved in growth regulation in the adult and also potentially during embryogenesis, where it may compensate to some extent for inactivating mutations in the IGF2 gene. Both proteins are ligands for the IGF1 receptor, and have been implicated in cancer.

The ING tumor suppressors

The inhibitor of growth 1 (*ING1*) gene is a member of the ING tumor suppressor family that includes at least five related genes, involved in diverse cellular processes including senescence, DNA repair, and apoptosis. ING proteins regulate gene expression by

Box 11.3 Wilms' tumor and neuroblastoma

Wilms' tumor (WT) is an embryonic tumor originating from the undifferentiated renal mesenchyme. Around 2% of WT are inherited, but a genetic component is believed to also contribute to sporadic tumors. Familial WT cases generally have an earlier age of onset and an increased frequency of bilateral disease. One WT gene, *WT1* at 11p13, has been cloned, but only a minority of tumors carry detectable mutations at that locus, and it has been excluded as the predisposition gene in most WT families. Two familial WT genes have been localized, *FWT1* at 17q12–q21 and *FWT2* at 19q13.4; lack of linkage in some WT families to either of these loci implies the existence of at least one additional familial WT gene.

The Wilms' tumor 1 gene (*WT1*) plays an essential role in urogenital development and malignancies, including breast cancer and leukemia. *WT1* acts to either enhance or repress transcription, either of which may predominate depending on the cell type and the context of the DNA-binding sites. This is exemplified by various reports in the literature suggesting that *WT1* either inhibits or stimulates the production of the protooncogene c-*Myc*, though at least in the case of breast cancer *WT1* may function as an oncogene in part by stimulating the expression of c-*Myc* (Han et al., 2004). *WT1* is overexpressed in a number of human cancers, and is associated with a poor prognosis.

The WT1 protein is normally expressed in the developing genitourinary tract, heart, spleen, and adrenal glands and is crucial for their development; however, its function at the molecular level is yet to be fully understood. Alternative splicing, RNA editing, and the use of alternative translation initiation sites generate a multitude of isoforms, which seem to have overlapping but also distinct functions during embryonic development and the maintenance of organ function. The protein is predominantly nuclear, and there is evidence that the two different isoforms of WT1 (−KTS and +KTS) are involved in two different steps of gene expression control: transcription and posttranscriptional processing of RNA.

Desmoplastic small round cell tumor (DSRCT) is defined by a chimeric transcription factor, resulting from fusion of the N-terminal domain of the Ewing's sarcoma gene *EWS* to the three C-terminal zinc fingers of WT1. Intriguingly, this chimeric protein activates a unique set of genes as compared to the normal WT1 protein.

---

regulation of chromatin remodeling probably by acting as cofactors for HAT and HDAC. One well-known target of ING2 is the lipid signaling molecules, phosphatidyl inositol phosphates (PIPs), also known to regulate apoptosis and motility. PIPs are also implicated in responses to DNA damage and in tumorigenesis. The PHD finger of the ING2 tumor suppressor interacts with nuclear PIPs, and may be involved in the p53-mediated apoptotic response to DNA damage. Thus, the ING family may link chromatin regulation with p53 function. However, further roles for ING independent of p53 are increasingly being appreciated.

Gliomas are a form of primary brain tumor in which levels of ING4 are known to be reduced. Interestingly, recent studies suggest that those tumors with the lowest levels of ING4 may be the most aggressive (highly malignant glioblastoma) and that this may in part be due to increased angiogenesis. One possible mechanism to explain this link has been suggested: ING4 normally acts to suppress NF-kB, and of ING4 leads to increased expression of NF-kB target genes, such as IL-8, involved in angiogenesis.

Polycomb group proteins

Polycomb group proteins (PcG) are involved in epigenetic regulation of gene expression, particularly in the determination of cell fate during development. The dysregulation of PcG genes, such as *Bmi1*, *Pc2*, *Cbx7*, and *EZH2*, has been linked with deregulated proliferation in cancer cells. PcG proteins are important for maintaining the silenced state of homeotic genes silenced during development and, together with Trithorax group proteins, regulate coordinated gene activity in self-renewal of some stem cells. Biochemical and genetic studies in *Drosophila* and mammalian cells indicate that PcG proteins function in at least two distinct protein complexes: the ESC–E (Z) or EED–EZH2 complex, and the PRC1 complex. Recent work suggests that the ESC–E (Z) complex can mediate gene silencing through methylation of histone H3 on lysine 27. In addition to being involved in *Hox* gene silencing, the complex and its associated histone methyltransferase activity are important in other biological processes including X inactivation, germline development, stem cell pluripotency, and cancer metastasis.

Overexpression of the PcG gene *Bmi1* promotes cell proliferation and induces leukemia through repression of *Ink4a–Arf* tumor suppressors. Conversely, loss of *Bmi1* leads to hematological defects and severe progressive neurological abnormalities. *Bmi1*-null mice manifest clonal expansion of granule cell precursors, which may result in development of medulloblastomas. The expression of BMI1 and Patched (PTCH) are linked in many human medulloblastomas, suggesting involvement of the Sonic Hedgehog (SHH) pathway.

Clinical use of epigenetics

It is not the organs, that are the character and form of the animal's bodily parts, that have given rise to its habits and particular structures. It is the habits and manner of life and the conditions in which its ancestors lived that have in the course of time fashioned its bodily form, its organs and qualities.

Jean-Baptiste Lamarck

Cancer profiling

As with genetic information, it is hoped that DNA methylation and histone modification patterns may provide new biomarkers. Much interest is developing in the concept of cancer profiling, whereby a particular observed "molecular signature" in an individual tumor might be employed to make more accurate predictions about prognosis and treatment responsiveness. The idea is that such information may enable more "individualized" or "tailored" approaches to treatment for each individual patient (the assumption made is that traditional disease classification is too broad and nonspecific, based on clinical observations and rudimentary laboratory testing, and patient heterogeneity is insufficiently recognized). At a simple level, further subclassification of a given tumor type, using "molecular profiling" of the tumor or potentially of body fluids, might allow individuals traditionally ascribed to a given diagnostic category (e.g. ductal carcinoma in situ) to be assigned to high- or low-risk groups and to specific treatments. Such molecular profiles might involve analysis of global gene or protein expression patterns in the tumor, using gene microarrays or proteomics; analyses of multiple-gene alleles in the patients genome; and potentially also epigenetic factors.

It is becoming increasingly apparent that tumors may have unique profiles of hypermethylated CpG islands, waiting to be exploited by the development of rapid assays for these in tumor samples and now that cancer-specific methylation has been detected in serum, from body fluids, and in stool samples also. Some epigenetic biomarkers are in clinical development and include hypermethylation of the glutatahione S-transferase gene (*GSTP1*) and the O6-methylguanine–DNA methyl transferase gene (*MGMT*) in diagnosis of prostate cancer and drug selection in glioma, respectively. Hypermethylation of particular genes may also have prognostic value as exemplified by *DAPK* and *CDKN2* in lung and colon cancers respectively. Again, as is now being realized with development of gene expression arrys in breast cancer, the global profile of CpG island hypermethylation in various malignancies might reveal an "epigenetic signature," which may be unique for each cancer "subtype."

Chromatin modification may also yield biomarkers; Pos translational modifications of histones, such as acetylation and methylation of lysine and arginine residues, also occur over large regions of chromatin including coding regions and nonpromoter sequences, which are referred to as global histone modifications. In a recent study, changes in such global levels of individual histone modifications were found to occur in prostate cancer and could help predict clinical outcome and recurrence.

Drug selection is already being determined by the presence of specific gene mutations, such as *HER*, *EGFR*, and *BCR–ABL* translocations for which specific drugs are available. It is now predicted that markers for epigenetic modifications of DNA repair genes, starting with *MGMT*, *MLH1*, and *BRCA1*, will herald a new age of pharmacoepigenetics. In fact, the presence of MGMT hypermethylation is already used to predict enhanced sensitivity to carmustine and temozolomide therapy in gliomas, and MLH1 status can guide the use of cisplatin in ovarian cancer and BRCA1 use of PARP inhibitors in breast cancer.

Drugs targeting epigenetic factors

A physician is obligated to consider more than a diseased organ, more than even the whole man – he must view the man in his world.

Harvey Cushing

The potential for hypomethylation to activate oncogenes in cancer cells obviously raises the possibility that drugs aiming to reactivate silenced genes could also inadvertently activate oncogenes. The risk of this remains uncertain, however, as to date administration of neither inhibitors of methylation nor those of deacetylation has activated oncogenes and affected growth of normal cells.

Given the key role played by gene silencing of tumor suppressor and caretaker genes by epigenetic factors in cancer, it is not surprising that drugs designed to reactivate gene expression by reversing DNA methylation and inhibiting histone deacetylation have been developed. Some of these agents are already in clinical trials used alone and in combination. Moreover, the recognition that particular patterns of DNA methylation and histone acetylation may predict disease onset and outcome has fueled initiatives aiming to exploit this information to screen patient blood, urine, or biopsies in the early diagnosis of cancer and precursor states and also to individualize treatment of those with established cancers. In fact, studies of tumor-derived free DNA in the circulation (presumably resulting from tumor cell apoptosis or lysis) or from epithelial cells shed into the lumen have revealed cancer-specific methylation patterns.

It has become apparent during the last decade that reversal of aberrant epigenetic silencing might be an effective therapeutic strategy for cancer. To date, a handful of drugs have been approved for clinical use. Agents such as 5-azacytidine (vidaza)) and 5'-deoxy-azacytidine (decitabine), which inhibit DNMTs, may reverse aberrant methylation of CpG islands and thereby reactivate tumor suppressor genes, which have been silenced. They are currently used in myelodysplasia and in acute leukemias. Inhibitors of HDAC, vorinostat and romidepsin, are used in hematological cancers including cutaneous T-cell lymphoma. HDACs and HATs catalyze the deacetylation and acetylation of lysine residues in histones, thus helping to regulate chromatin structure and gene expression, and nonhistone proteins, thus influencing cell-cycle progression and apoptosis. Their activity is also frequently disturbed in cancer. Several HDAC inhibitors are in clinical trials with significant activity against different hematologic and solid tumors. They have variously been shown to induce growth arrest, differentiation, apoptosis, and autophagocytic cell death of cancer cells. Early clinical results suggest a potentially useful role for HDAC inhibitors in the treatment of certain forms of lymphoma (e.g., cutaneous T-cell lymphoma) and acute leukemia. Interestingly, recent studies suggest that it is acetylation of nonhistone proteins that may be the most important means of action of these drugs and may explain lack of correlation between histone acetylation and induction of cell death. HDAC inhibitors have been shown to exert a wide range of effects on cancer cells, such as disruption of co-repressor complexes, induction of oxidative injury, upregulation of the expression of death receptors, generation of lipid second messengers such as ceramide, interference with the function of chaperone proteins, and modulation of the activity of NF-κB. Together, these actions may promote cell-cycle arrest in G_1 or G_2–M, differentiation, and even cell death, all potentially beneficial against cancer. Various other drugs are in development, including thopse targeting sirtuins, HATs, HMTs, and a number of the other proteins listed in Table 11.1. SIRT inhibitors might activate p53 signaling. Numerous patent applications describing new HDAC inhibitors have been filed. Recent studies indicate that DNA demethylating agents and HDAC inhibitors synergistically induce gene expression and apoptosis in cultured lung cancer cells, prevent lung cancer development in animals, and can induce immunogenicity and apoptosis of lung cancer cells in clinical trials. Various trials investigating combinations of these agents and other chemotherapies or even radiotherapy are being tested.

Regulation of translation

Life . . . is a relationship between molecules.

Linus Pauling

We do not intend to give an exhaustive description of how translation initiation and regulation of the eIF4F complex and its inhibitor 4E-BP1 can influence protein synthesis, as this is covered in Chapter 5, in connection with mTOR signaling and in more general terms, and along with regulation of mRNA stability later in this chapter and in many excellent general molecular biology texts, including *Molecular Biology of the Cell* (Alberts et al., 2007). Deregulated control of translation is a feature of many oncogenic signaling pathways and is implicated in cancer initiation and progression by increasing protein synthesis and via mRNA networks that promote transformation. Many cancer-relevant proteins may be regulated at this level as well as at transcription and protein stability.

Deregulation of RNA polymerases can increase transcription, whereas these are generally downregulated during cell-cycle G_2–M. Increased RNA pol III (Pol III) transcription can contribute to proliferation and transformation and may be the result of increased EGFR signaling, epignetic regulators such as H3S28ph and transcription factors such as Brf1. Accelerated translational activity depends on Pol III–dependent genes for transfer RNAs (tRNA) and the 5S ribosomal RNA (rRNA). Massive parallel sequencing of myeloma genomes has recently shown that around half of cases examined had mutations in genes involved in protein translation, and were the commonest class found.

In fact, in a very recent study Kastan and colleagues demonstrate how important this mechanism is for p53. As described in other chapters, p53 protein levels increase after DNA damage and it is now clear that this occurs not just through regulation of p53 degradation (MDM2 activity) but also through increased translation of p53 mRNA. In fact, translation of p53 depends on at least two proteins that bind to the 5′ untranslated region (UTR) of p53: ribosomal protein L26 (RPL26) and nucleolin increase or decrease the rate of p53 translation respectively. Thus, RPL26 can induce G_1 cell-cycle arrest and augment irradiation-induced apoptosis.

The GU-rich element (GRE) is a conserved sequence enriched in the 3′ UTR of those mRNAs subject to most rapid turnover. Regulatory proteins such as CELF1 that bind GREs can promote degradation and even splicing of whole networks of mRNAs in key growth-regulator processes and, if this breaks down, can contribute to cancer. Drugs inhibiting translation are already in preclinical and early clinical development. Inhibitors of eIF4E have been in clinical trials and are now being followed by trials of eIF4E directed RNAi.

Noncoding RNA and RNA interference

There is an electric fire in human nature tending to purify – so that among these human creatures there is continually some birth of new heroism. The pity is that we must wonder at it, as we should at finding a pearl in rubbish.

John Keats

Gene expression may also be regulated in other hitherto unexpected ways. In recent years, long-suffering biologists have once again been forced to take a quantum leap, experience a paradigm shift, and swallow a bitter pill in the face of new knowledge. It seems comparatively recently that we have accepted that transcriptional activator and repressor proteins were only part of the story and that processing of mRNAs and the aforementioned epigenetic factors were also of major importance. It now seems that gene expression is also determined by parts of the genome previously thought to be devoid of any meaningful information. It appears that more than 50% of the genome is transcribed. Thus, much of the genome is devoted to the production of a variety of RNAs that do not encode proteins and are thus grouped together as noncoding RNAs (ncRNAs). The ncRNAs are of varying lengths and functions ranging from long ncRNAs of over 200 nucleotides long, including ultraconserved genes (UCGs) and HOX antisense intergenic RNA (HOTAIR), to small ncRNAs derived from longer precursors, including siRNA and miRNA (discussed in this chapter). Included in this group of ncRNAs are many of long-appreciated importance, including the highly conserved ribosomal RNAs, transfer RNAs, and small nuclear RNAs. These will not be discussed further here; rather, we will focus on those previously unappreciated ncRNAs which are emerging as key regulators of gene expression, in particular the microRNAs (miRNAs).

Long noncoding RNAs (lncRNAs)

Before moving on to the now ubiquitous miRNAs, we will pause briefly to consider their less famous cousins, the long ncRNAs (lcnRNAs) and the related, even longer very long intergenic ncRNAs (vlincRNAs). Not much is known about the functional role played by lncRNAs, but many are being investigated as epigenetic regulators in cancer and neurodegeneration. Given that there are to date over 15 000 lncRNAs, encoded by both genic (overlapping with areas encoding mRNAs) and intergenic DNA, the task is daunting. A brief outline is given here, but this is an area likely to expand rapidly in the coming years. Among the first to be described was XIST, a 19 kb transcript involved in DNA methylation silencing of one of the two X chromosomes in placental females. Others located in imprinted regions have been described, including NESPAS, Airn, and H19, all of which silence gene expression in association with chromatin-modifying complexes. In cancer biology, a more recently identified lncRNA, HOTAIR, was shown to regulate methylation of the HOXD locus via the polycomb repressive complex 2 (PRC2). HOTAIR may play a role in cancer metastasis. A further lncRNA, ANRIL, located in the $p15^{CDKN2B}$–$p16^{CDKN2A}$–$p14^{ARF}$ locus, has been linked by GWAS to numerous disease states, including basal cell carcinomas and gliomas as well as diabetes.

MicroRNAs (miRNAs) and small interfering RNAs (siRNAs)

RNA interference (RNAi) is a highly conserved but comparatively recently recognized mechanism that operates through sequence-specific gene silencing by miRNAs encoded in the genome. This mechanism may also be harnessed by therapeutic delivery of oligomers designed to direct the RNAi pathway to silence a gene that may or may not be regulated endogenously in this way.

RNAi is a natural posttranscriptional process through which metazoan cells suppress the expression of genes when exposed to double-stranded RNA (dsRNA) molecules of the same sequence (sequence-specific gene silencing). RNAi occurs naturally in most organisms, including nematodes, plants, fungi and mammals, and probably evolved to combat viruses and rogue genetic elements that utilize dsRNA during their life cycle (Fig. 11.5). The discovery of RNAi has profoundly altered the way we think about RNA; we now appreciate that RNA not only acts as an intermediary between genes and protein synthesis but also plays an important role in regulating gene expression. Two main classes of endogenous regulators of RNAi are now known: microRNAs

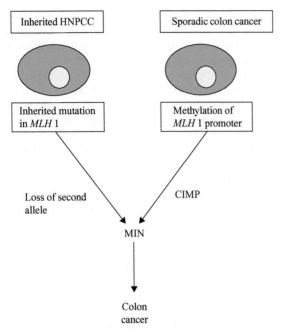

Figure 11.5 The genetic and epigenetic routes to colon cancer. Genetic instability (MIN in particular) contributes to many colonic cancers. MIN may arise through either genetic or epigenetic mechanisms (or possibly a combination).

(miRNAs) and endogenous small interfering RNAs (endo-siRNAs). The best studied are the miRNAs, originally discovered in *Caenorhabditis elegans*. The siRNAs were initially studied in the context of silencing of exogenous viral and other foreign and potentially pathogenic RNAs in plants, but are now being ascribed important roles in mammalian cell physiology also. Both are short single-stranded noncoding RNA molecules, averaging only 22 nucleotides in length, that are of complementary sequence to short segments of protein-coding genes. The main difference between them lies in their relative specificity (complementarity) for individual mRNAs. The siRNAs generally bind to a single-gene mRNA, whereas miRNAs may bind multiple mRNAs and can thus act as regulators of groups of genes. Both miRNA and siRNA can reduce gene expression by forming dsRNA duplex structures with mRNAs. However, due in part to the differences in complementarity between the guide and its mRNA target, miRNAs primarily inhibit translation, whereas siRNAs result in destruction of their target mRNA. Thus, the degree of base pairing with target mRNA determines the mechanism of silencing. If complementarity is strong, then the target mRNA is cleaved and then degraded, whereas less strong binding may only inhibit translation.

Endo-siRNAs are well known in plants, but the function of these regulators, which in general show much stronger complementarity to target mRNA sequences of single genes, in mammals is less well known. In a therapeutic context, manufactured siRNAs are also defined by the very strong homology to their designated target mRNA, with the result that siRNAs lead to destruction of their specific target mRNA and in general will affect only single genes (Chapter 16).

The miRNAs are now known to be involved in regulating a very wide range of cellular functions and play central roles in development, in physiological processes, and, of relevance to cancer, in the growth and differentiation of most cell types. Since each miRNA appears to regulate the expression of tens to hundreds of different genes, miRNAs are often considered as "master switches" that, in an analogous fashion to transcriptional regulators, can regulate multiple cellular pathways. Various miRNAs have been shown to be aberrantly regulated in cancers, and recent studies have shown that cancers can be regressed by replacement of downregulated, tumor-suppressing miRNAs.

Antisense transcription and production of dsRNA are increased in cancer cells, and many miRNAs are now known to be involved in various aspects of evolution and progression of cancers. Abnormal expression of miRNAs has been observed in most human cancers, and is expected to be a general phenomenon. These include lung cancer, leukemia, lymphomas, colorectal, prostate, and breast cancer and in some of these cases the expression signatures may correlate with disease behavior and prognosis.

Given these central regulatory roles, it will come as no surprise that RNAi is the subject of drug development for a number of diseases in humans, including cancer (Chapter 16).

miRNAs in animals differ from those of plants by being less specific to single mRNAs and capable of regulating expression of multiple genes, and by largely targeting the (3'UTR) of the mRNA. The miRNAs are highly conserved across eukaryotic species and may be one of the earliest evolved mechanisms for regulating gene expression. They represent a major novel class of regulatory molecules operating across a wide range of developmental processes in plants (where they are known to control flowering time and leaf polarity) and animals. Although many of the specific targets of miRNAs are not well known, we do know that they are remarkably pervasive. Around 5% of the human genome is devoted to encoding miRNAs. Together, the more than 1400 miRNAs described to date may regulate up to half of all genes and contribute to all biological processes (a running tally is available on http://microrna.sanger.ac.uk). A standard nomenclature catalogues miRNAs using the prefix "mir" followed by a number, which largely refers to the order of discovery. When the R is capitalized (miR), the mature form is being referred to. Lowercase letters are added to separate very closely related miRNAs.

Ongoing research on miRNAs is intense, and it is now known that these molecules may under some circumstances stimulate as well as inhibit transcription or translation. The DNA regions encoding miRNAs are largely located in intergenic regions or in antisense orientation to genes and have unique gene promoters and regulatory units. Some miRNA "genes" are situated within introns of regular genes, and as they can be in sense orientation may be co-regulated with the host gene. Transcription of miRNAs is mediated by RNA polymerases II and III that generate precursor molecules that must be subjected to a number of cleavage events, nuclear and cytoplasmic, in order to generate the mature miRNA (Fig. 11.5). A range of biogenesis pathways are available and may be selected dependent on the origin or sequence of the given miRNA. Diversity of miRNAs is further increased beyond that predicted by simply examining the genome, by editing of RNA sequences.

Circulating cell-free miRNAs are found in blood and other easily accessible body fluids, "naked" or in exosomes, and are promising new disease biomarkers, including for many cancer types. The presence of circulating miRNAs in exosomes raises the intriguing possibility that miRNAs may act as hormones and influence gene expression at sites distant from their production. That miRNAs may represent a primitive, conserved oligonucleotide hormone system could be exploitable in developing new treatments or delivery systems for miRNAs or siRNAs by incorporation into exosomes or synthetic analogs.

Maturation and function of miRNAs

As mentioned, miRNAs regulate gene expression by forming the RISC and activating it to target the specified mRNAs. Several distinct stages have been recognized in this process, involving RISC loading and activation, and then gene silencing ensues by a slicer-dependent or -independent translation inhibition pathway. Much of this process may take place in specialized compartments in which these enzymes are located, the P bodies.

Mature miRNA is generated by a combination of nuclear and subsequent cytoplasmic cleavage of the primary miRNA (pri-miRNA) under the influence of two ribonuclease III endonucleases, Drosha and Dicer. The actual sequence may differ depending on whether the miRNA is intergenic or coding intronic. Fot intergenic miRNAs, firstly, nuclear cleavage of pri-miRNA is conducted by the so-called microprocessor, incorporating Drosha and DGCR8 (dsRNA-binding protein DiGeorge syndrome critical region gene 8) as well as other proteins such as hnRNPA1. This process generates an intermediate precursor, pre-miRNA. In contrast, miRNA located within an intron of a protein-coding gene is transcribed by pol II as part of the pre-mRNA from which the miRNA (or mirtron) must then be excised. Further processing takes place after pre-miRNA has translocated to the cytoplasm, in complex with the transporter Exportin-5 and RanGTP. It is in the cytoplasm that a miRNA duplex is formed and from which mature miRNA is released. Much of these processes may take place in the cytoplasmic compartments known as P bodies, which comprise a number of specific proteins such as GW182, Dcp1, 4E-T, GE-1–hedls, p54–RCK, and Xrn1, and the RISC components described in this section. It is believed that pre-miRNA is incorporated into a RISC-loading complex complex, comprising Dicer and TRBP (human immunodeficiency virus transactivating response RNA-binding protein) with either Argonaute 2 (Ago2) or Protein Kinase R-activating protein (PACT) or maybe all of these together. In humans, a single Dicer isoform is responsible for cleaving dsRNA in both the miRNA and siRNA pathways. Dicer recognizes a two nucleotide 3′ overhang in pre-miRNA and initiates binding and then directs cleavage to the appropriate site. The resulting product is a double-stranded miRNA duplex with 3′ overhanging ends. One of these strands will become the mature miRNA and will incorporate into the RISC complex to act as a guide for silencing the target mRNA, whereas the other (passenger strand) will presumably be recycled. The RISC complex at this stage incorporates a series of further argonaute proteins, all of which are involved in gene silencing. However, it is notable that only Ago2 has "slicer" activity and can cleave mRNA. A further protein, RNA Helicase A (RHA), may also be involved in unwinding the dsRNA. RISC assembly and activation take place in P bodies, which explains the co-localization of miRNA duplexes and Ago 1–4 proteins with GW182. For gene silencing, the 3′UTR of the target mRNA hybridizes to the mature miRNA guide strand. One of two things now happens: if base pairing between miRNA and mRNA is near perfect, then the mRNA is cleaved by Ago2 endonuclease, whereas less extensive pairing may be the more usual and result in inhibition of translation by a slicer-independent process at both the initiation and elongation stages. Translation can also be inhibited by sequestering mRNA into cytoplasmic P bodies in which translational control factors such as Gemin5, p54, FMRP, and RAP55 are located. Once mRNA has been cleaved by slicer activity of Dicer, the mRNA is degraded by removal of the poly (A) tail and then completed 3′-to-5′ exonuclease within the exosome. Decapping of mRNA by Dcp1 and Dcp2 may also occur, thereby allowing degradation by the exoribonuclease Xrn1p. Degradation of mRNA is self-evidently irreversible, whereas translation inhibition can be released allowing translation to resume.

In a further interesting development, a potential link between RNAi and epigenetic regulation has been suggested. It has recently been shown in plants and yeast that the RNAi machinery may target genes for epigenetic silencing; dsRNA molecules targeted to gene promoter regions can induce transcriptional gene silencing in a DNA cytosine methylation-dependent manner (RNA-dependent DNA methylation). However, whether this also exists in mammalian cells is contentious still. It has been suggested that dsRNA arising from transcription of repetitive elements may be a mediator of gene silencing in mammalian cells by promoting methylation through as-yet-speculative interactions with DNMT or MDB. One possibility would be that in cancer cells, global low-level hypomethylation of repetitive elements in or near promoter regions could give rise to increased dsRNA homologous to the gene. The resultant activation of the RNAi machinery could then establish aberrant methylation at CpG islands. However, in one study from the laboratory of Stephen Baylin, small dsRNAs targeted exclusively to the one gene promoter (that for *CDH1* was chosen, as this is frequently inactivated by methylation in cancer) induced transcriptional repression with chromatin changes that was entirely independent of DNA methylation. Moreover, they confirmed these findings in a cancer cell line modified to lack almost all capacity to methylate DNA.

Therapeutic and research potential of RNAi

No doubt, man will continue to weigh and to measure, watch himself grow, and his Universe around him and with him, according to the ever growing powers of his tools.

Albert Claude

RNA interference (RNAi) is one of the major discoveries of recent years and is predicted to revolutionize the emerging areas of systems biology and functional genomics. RNAi represents a powerful new tool for the undertaking of functional genomics (analyzing gene functions) in eukaryotes, with some advantages over traditional genetically altered knockout models, particularly the potential for use in high-throughput functional screening. In fact, several national initiatives are currently underway to knock out multiple genes in cancer cells by use of RNAi.

Moreover, RNAi is now featuring strongly in the development of therapeutic gene silencing and several clinical trials (Phase I and II) are underway looking at siRNA in cancer and other diseases (Chapter 16). In fact, some predictions have been voiced that once delivery and stable levels of inhibition can be achieved with siRNAs, this will replace traditional chemistry approaches to drug design in favor of the far less complex production of siRNAs to potentially any gene which we would wish to silence (Chapter 16). Closely allied to siRNA is the development of replacement techniques for disease-downregulated miRNAs or for antagonizing action of upregulated miRNAs by use of antagomirs.

Even though dsRNA could induce gene-specific interference in mouse embryos and oocytes, early attempts employing siRNA in mammalian systems proved difficult. These experiments probably failed because they used long dsRNAs that instead of mediating RNAi actually resulted in global decreases in mRNA and apoptosis, a response mediated by the dsRNA-dependent protein kinase. However, this nonspecific response can be avoided by employing synthetic 19–23 nucleotide dsRNAs, which elicit strong and specific suppression of gene expression in different mammalian cell lines. However, this effect is transient only due to the short half-life of synthetic RNAs. These technical problems have limited the applications of siRNAs to more complex studies, in particular *in vivo*. One potential means of overcoming this problem is to generate siRNAs within the cell with a hairpin structure called short hairpin RNAs (shRNAs). The shRNAs are more stable than synthetic siRNAs, and since they are continuously expressed within the cells, this method permits long-lasting silencing of your gene of interest. Hairpin RNA is a key player in the RNA-silencing pathway: when cleaved by a Dicer enzyme, it generates siRNAs, which in turn guide RISC complexes to degrade target RNA. Coupled with improved delivery techniques, such as lipid nanoparticles that overcome some of the practical and theoretical concerns over the use of viral vectors, a new era of targeted therapeutics is being ushered in.

RNAi and cancer

In normal cells, multiple miRNAs interact in order to regulate important cellular functions, including proliferation, differentiation, and cell death. Given that individual miRNAs can bind to and regulate multiple mRNAs, it will come as no surprise that miRNA dysregulation is associated with many diseases, including cancer. In cancer, the loss of tumor-suppressive miRNAs, such as let-7, results in increased expression of numerous oncogenes. Conversely, increased expression of oncogenic miRNAs can repress target tumor suppressor genes such as p53 and PTEN. These findings are being exploited in developing new miRNA-directed therapies and in the design of novel siRNAs to target many different disease-relevant genes (Chapter 16). There is also great interest in measuring miRNAs in tumors or body fluids as potential biomarkers.

The list of cancer-related miRNAs is growing rapidly, and we will discuss only a few here. Among the first oncogenic miRNAs to be discovered was miR-21. miR-21, overexpressed in human glioblastomas and cancers of the breast, colon, liver, brain, pancreas, and prostate, is an oncogenic miRNA that can downregulate key tumor suppressor genes encoding PTEN, mapsin, PDCD4, and TPM1. The importance of PTEN in regulating survival pathways involving AKT has been discussed in Chapters 5 and 7. Repression of TMP1 and maspin by miR-21 may promote cell invasion and metastasis.

Intriguingly, recent studies have suggested that miR-21 may suppress expression of PTEN and might fuel a potentially novel cancer cell–stromal cell interaction, in which tumor cells could release miR-21 in exosomes that are then taken up by stromal fibroblasts where they inhibit PTEN, thereby subverting these stromal cells to support rather than inhibit invasion and spread. Loss of PTEN by mutation in stromal fibroblasts has already been discussed as a way in which mutations in noncancer cells can also be cancer causing. If, as now seems plausible, though as yet unproven, cancer cells can directly recruit stromal cells, then directing drugs at miR-21 or other such recruitment factors in the stroma could be of great value. In mice, inducible overexpression of miR-21 can drive pre-B-cell lymphoma, which could be reversed, restoring miR-21 to normal levels and suggesting that cancers can manifest "oncomir addiction."

Overexpression of miR-155 in the B-cell lineage in mouse models induces a pre-leukemic phenotype in the spleen and bone marrow, with B-cell cancers developing later once secondary mutations have been acquired. This may be encouraged by miR-155, which can silence DNA damage response genes.

Both miR-373 and miR-372 can prevent p53-dependent senescence by suppressing *LATS2*.

Let-7 miRNAs (let-7s) were among the first identified miRNAs, and 12 let-7 genes encoding nine miRNAs are now known. Importantly, let-7 genes are tumor suppressors; firstly, they frequently map to regions altered or deleted in various cancers; and, secondly, they are known to suppress expression of target oncogenes such as *RAS*s and *HMGA2*. They can also inhibit a variety of other mitogenic regulators such as c-MYC, CDC25A, CDK6, and cyclin D2. Therapeutics based on delivering let-7 miRNA are in development (Chapter 16).

LIN28B represses production of let-7 microRNAs and is both implicated in development and overexpressed in CRC. Moreover, in patients, LIN28B overexpression correlates with reduced survival and increased recurrence. This may in part be due to increased colon cancer stem cells as marked by increased expression of LGR5 and PROM1. Interestingly, polymorphisms in the LIN28B gene are associated with risk of ovarian cancer. BRCA1 protein is reduced by mutations in the gene, but interestingly also in some sporadic tumors by targeting of the BRCA1 mRNA by miR-146a and miR-146b-5p, which are also expressed at high levels in triple-negative breast cancer and basal-like mammary tumor epithelial cell lines.

The miR-34 family members, miR-34a, miR-34b, and miR-34c, are among the best characterized tumor suppressor miRNAs identified so far. They are induced by p53 and are responsible for some of the tumor suppressor activity by regulating apoptosis and senescence pathways downstream of p53. They are missing or expressed at reduced levels in multiple human cancers, and reintroduction of miR-34 inhibits cancer cell growth *in vitro* and *in vivo*. The miR-34 family promotes senescence by targeting E2F signaling. Clinical trials with a miR-34 analogue are in development (Chapter 16). The ZEB1 transcriptional repressor promotes metastasis through downregulation of miRNAs, such as miR-34a, that would otherwise promote epithelial differentiation and inhibit stem cells.

Recent studies show that miR-205 is required for the antimetastatic effect of p63 in part by preventing EMT, defined by expression of markers such as ZEB1 and vimentin. Interestingly, delivery of miR-250 can prevent metastases in animal models. Moreover, low levels of p63–miR-205 were found in advanced human prostate cancers and may prove a useful biomarker to predict metastatic spread.

Often miRNAs are found in large clusters that are expressed polycistronically; in these cases, it is often of value to evaluate the function of the cluster and of the individual miRNAs within the cluster. There are four known tumor suppressors in the miR-15/16 family organized in two distinct clusters, miR-15a–miR-16-1 (frequently deleted in CLL) and miR-15b or miR-16-2. All can repress the antiapoptotic BCL2 protein. Circulating miRNAs are expressed at higher levels in blood than in normal individuals and can be sensitive biomarkers for CLL, because certain extracellular miRNAs are present in CLL patient plasma at levels significantly different from those of healthy controls and circulating miR-20a correlates reliably with diagnosis-to-treatment time and ZAP-70 status (i.e. it identifies a specific subgroup of CLL).

The miR-17–92 miRNA cluster in humans, contained within a fragile site in the genome, produces six mature miRNAs (miR-17, miR-18a, miR-19a, miR-19b-1, miR-20a, and miR-92-1). The cluster is amplified in both solid tumors and hematopoietic malignancies. As discussed in Chapter 6, miR-17–92 is directly regulated by c-MYC and by E2F3, and can act as both an oncogene and a tumor suppressor. The 13q31.3 gene locus, where this cluster is located, is a frequent site for gene amplification and can lead to elevated levels of miR-17–92 in multiple cancers including lymphomas, lung cancer, colon, pancreas, and prostate cancers. Additionally, overexpression of c-MYC is common in cancers and can also increase expression. The oncogenic activity of miR-17–92 may result from selective repression of the apoptotic activity of E2F1. Conversely the miR-17-5p member of the miR17–92 cluster may act as a tumor suppressor. Normally, the miR-17~92 cluster can limit MYC activation by dampening the E2F positive feedback loop. However, when MYC is activated, the miR-17~92 cluster prevents apoptosis by inhibiting E2F. Levels of miR-17-5p levels were increased in CRC alongside reduced levels of E2F1. *E2F1* is inhibited by miR-17-5p; miR-20a, miR-106b, and miR-92 were found to inhibit *E2F1*; whereas miR-20a inhibits *E2F2* and *E2F3*.

Several miRNAs are now known to be specifically involved in metastatic spread of cancer cells, including miR-31 and miR-10b, which appear to have minimal roles in the primary tumor, and miR-200, which is involved in both. Expression of miR-31 is decreased in metastatic human breast cancer and, when overexpressed in mouse models, can specifically regress metastases without shrinking the primary tumor. Upregulation of miR-10b also appears specific to metastatic tumors.

Importantly, loss of miR-200, a p53 target, is often found in aggressive prostate, pancreas, and non-small-cell lung cancers (NSCLCs), where it is linked to EMT and in animal models to metastasis. In part, this effect may be due to loss of inhibition on the mesenchymal markers zinc finger E-box-binding homeobox 1 (ZEB1) and ZEB2, with resultant loss of E-cadherin expression (Chapter 12). Interestingly, miR-200 may also prevent stemness, through inhibiting expression of transcription factors ZEB1 and ZEB2 that promote EMT.

LIN28, a MYC target, is also involved in stem cell maintenance and is activated in high-grade cancers. Conversely, RAF kinase inhibitory protein (RKIP) inhibits RAF and downstream MYC activation and prevents LIN28-mediated stem cell support. Recently, miR-34a has been shown to suppress the stem cell marker, CD44, and when administered to mouse models reduced prostate cancer stem cells and antagonized multiple stem cell properties *in vitro*, including self-renewal. Moreover, miR-34a is greatly reduced in human prostate cancer cells, expressing high levels of CD44.

Telomerase is a direct target of miR-138, miR-143, and miR-146a. In particular, miR-22 can induce senescence by targeting CDK6, SIRT1, and Sp1 genes. A further recent study reveals that miR-122 can inhibit translation of cytoplasmic polyadenylation element-binding protein (CPEB) needed for p53-induced mRNA polyadenylation or translation. An antagomir to miR-122 can stabilize CPEB and might be exploited as a means of inducing senescence in tumor cells. Various techniques can be employed to accurately characterize miRNAs from body fluids, such as blood in which they are quite stable (possibly in exosomes shed by the tumor), and in tissue samples. Thus, by employing techniques ranging from quantitative RT-PCR and a range of array- and bead-based techniques to next-generation RNA sequencing, miRNAs can be identified and quantified and are already showing promise as novel biomarkers in the clinic. Studies have recently supported the notion that miRNA-based biomarkers might be useful in early diagnosis, disease subclassification, and treatment monitoring. For example, in lung cancer, miRNA profiles can help determine prognosis.

Treatments based on miRNA

As discussed, the last 5 years have witnessed an explosion of research in cancer biology using RNAi. RNA-based gene therapy operates through two overlapping mechanisms:
1. RNA or DNA molecules are directed against the mRNA of genes involved in cancer; and
2. Direct targeting of ncRNAs, particularly miRNAs involved in tumorigenesis.

Several miRNAs and designed synthetic siRNAs have shown benefits in animal models, and around 10 phase I to III clinical studies using RNAi are in progress. Some have already provided interim reports with real promise in a variety of late-stage solid cancers. These include phase I trials of ALN–VSP, targeting kinesin spindle protein, KSP, and VEGF in liver cancer, and Atu027, targeting protein kinase N3 (PKN3) in a range of advanced solid cancers.

A different approach to effectively mimicking this way of regulating gene expression is to actively antagonize the endogenous actions of miRNAs themselves by use of **antagomirs** that inhibit miRNA target binding. As is being learned from siRNA therapy in general, chemical modifications to enhance stability of RNA oligomers and improved delivery techniques, with reduced immunogenicity and improved delivery into cells, are required for clinical use. Promising results have been demonstrated in model systems inhibiting miR-121 in this way to lower cholesterol. It is anticipated that liposomes with reduced immunogenicity and nanoscale delivery devices may improve delivery. Various such systems are in clinical trials for siRNA delivery in a variety of cancers.

Regulating the proteins

When the curtain falls, the best thing an actor can do is to go away.
Harold MacMillan

The timely removal of proteins, having done their job, is critical to the appropriate regulation of cell signaling and function. Thus, protein degradation and transport are as important as protein synthesis and activation in communicating and executing cell function. Moreover, in addition to removing excess enzymes or transcription factors, degradation provides recycling opportunities by supplying amino acids for fresh protein synthesis or energy production. Transport of proteins will be discussed in this chapter. There are two key intracellular processes by which proteins are broken down: those involving lysosomes and proteasomes.

Lysosomes degrade extracellular proteins and cell-surface membrane proteins such as growth factor receptors or those employed for receptor-mediated endocytosis. Proteasomes primarily degrade endogenous proteins such as transcription factors, cyclins during cell-cycle progression, and abnormal or incorrectly folded proteins arising from mutations or errors in translation. Proteasomes also produce peptides, which are presented by the major histocompatibility complex I to the immune system in order to induce antibodies. Generally the proteasome produces peptides that are between six to nine amino acids in length.

The proteasome

I have always looked upon decay as being just as wonderful and rich an expression of life as growth.

Henry Miller

Much has already been discussed about this key process in Chapter 4, where the role played by the degradation of proteins in cell-cycle control has been exhaustively described. A more general overview will be provided here.

The proteasome is an abundant multi-enzyme complex that represents the major route for degradation of intracellular proteins in eukaryotic cells (Fig. 11.6). Turnover of several key cellular regulatory proteins results from targeted destruction via ubiquitination and subsequent degradation through the proteasome. The first step in proteasomal protein degradation is ubiquitination, which then targets proteins to the proteasome complex. The proteasome comprises the core particle (CP) and two regulatory particles, which contain recognition sites for ubiquitin. Evidence suggests that the proteasome serves both as a disposal system for damaged cellular proteins and as a mechanism for degrading short-lived regulatory proteins that govern cellular functions such as the cell cycle, cell growth, and differentiation. Because these processes or their dysregulation comprise crucial steps in tumor formation, the proteasome pathway has recently become a target for therapeutic intervention.

Ubiquitin–protein ligases

The key components that regulate substrate ubiquitination are the ubiquitin–protein ligases (see Chapter 4). Ligases are a heterogeneous group comprising single proteins as well as large multiprotein complexes. Specificity of targeting is influenced by various factors including the requirement for posttranslational modifications such as phosphorylation, hydroxylation, or oxidation of the substrate and, in some cases, the ligase itself.

Many inherited diseases are now known to involve defective ligase function. Since the initial recognition that Angelman syndrome is caused by maternal deficiency of the E6-AP ubiquitin E3 ligase, several other disease-relevant E3 ligases have been identified, including autosomal recessive juvenile Parkinson

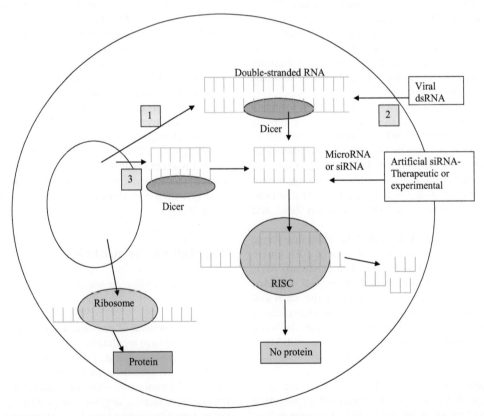

Figure 11.6 (a) miRNA synthesis and gene silencing. Cells can inhibit gene expression via small dsRNA. The enzyme Dicer cuts short interfering RNAs (siRNAs) from longer dsRNAs deriving from (1) self-copying gene sequences, (2) viruses, or (3) regulatory RNA sequences called microRNAs. All these RNAs are cleaved by Dicer into short siRNA segments that can suppress gene expression. The short siRNA pieces unwind to form single-stranded RNAs, which combine with proteins to form a complex called RNA-induced silencing complex (RISC). The RISC can capture native miRNA molecules with complementary sequences to the short siRNA sequence. If the pairing is perfect, the native mRNA is cleaved into untranslatable RNA fragments. If the pairing is not perfect, then the RISC complex halts translation by preventing ribosome movement along the native miRNA. (b) A detailed version of this process.

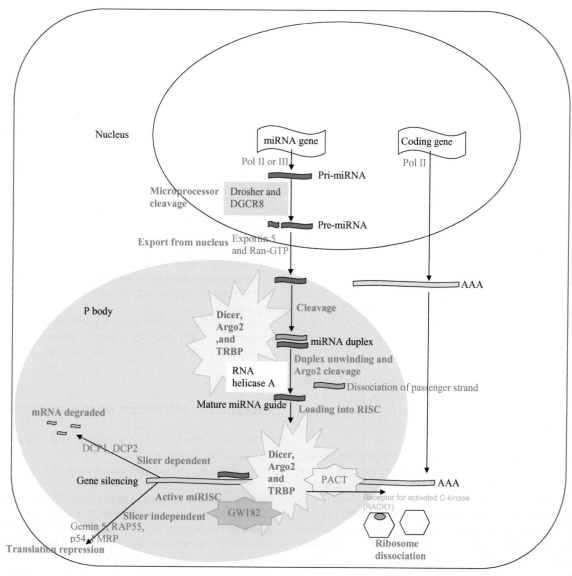

Figure 11.6 (Continued)

disease, von Hippel–Lindau syndrome, and congenital polycythemia. Furthermore, several disorders of abnormal ubiquitin regulatory signaling are also known including at least two subtypes of Fanconi anemia and the BRCA1 and BRCA2 forms of breast and ovarian cancer susceptibility. Many other disorders affect ubiquitin pathways secondarily.

Loss of von Hippel–Lindau (VHL) protein function results in an autosomal-dominant cancer syndrome known as VHL disease, which manifests as angiomas of the retina, hemangioblastomas of the central nervous system, renal clear-cell carcinomas, and pheochromocytomas. VHL tumor suppressor is a specific substrate-recognition component of the E3 ubiquitin complex, which regulates proteasomal degradation of the subunit of the hypoxia-inducible transcription factor (HIF). Impaired VHL complex function leads to accumulation of HIF, overexpression of various HIF-induced gene products, and formation of highly vascular neoplasia (Chapter 14). However, the ubiquitinating role of the VHL complex extends beyond its function in regulating HIF, as it appears to regulate the stability of other proteins that might be involved in various steps of oncogenic processes.

The HDM2 ligase, which negatively regulates p53 signaling, is also very important in cancer and has been discussed in Chapter 7. In unstressed cells, p53 is maintained at low levels by HDM2 (MDM2 in rodents) and the ubiquitin–proteasome pathway, and by blocking p53 transcriptional activity. Intriguingly, as *MDM2* is also a p53 target gene this creates a negative-feedback mechanism. In stressed cells, MDM2 activity is counteracted by ARF in response to strong proliferation signals, such as activated oncogenes. ARF then inhibits MDM2-mediated p53 degradation. This may explain why loss of ARF may exert similar effects in tumorigenesis to loss of p53. Recent data suggest that ubiquitination of p53 is also under control of enzymes induced by DNA damage that can directly deubiquitinate and stabilize the protein, such as the Herpes virus–associated ubiquitin-specific protease (HAUSP).

In fact, HAUSP stabilizes p53 even in the presence of excess MDM2, suggesting it could act as a tumor suppressor. Thus, a balance between ubiquitin ligases (Hdm2, COP1, and Pirh2) and ubiquitin proteases (HAUSP) determines the half-life of p53.

The F-box protein, Fbw7, a component of the SCF (Fbw7) ubiquitin ligase and a tumor suppressor, promotes proteasome-dependent degradation of numerous key proteins, including c-MYC (Chapters 4 and 6), thus c-MYC activation may be a key outcome of Fbw7 loss in cancer. Fbw7 also targets cyclin E, Notch, and c-JUN, all of which may contribute to tumorigenesis if Fbw7 is lost.

Ubiquitin inhibitors act at many levels to enhance apoptosis signaling. Proteasome inhibitors can decrease Fas-like inhibitor protein (FLIP) levels in tumors, activating the extrinsic pathway of apoptosis through caspase 8. The inhibitor of apoptosis proteins (IAP) E3 ligases (see Chapter 8) can directly bind to caspases, resulting in their degradation.

The proteasome and cancer

Great effort is required to arrest decay and restore vigor. One must exercise proper deliberation, plan carefully before making a move, and be alert in guarding against relapse following a renaissance.

Horace

Given the importance of the proteasome pathway in degrading various cancer-relevant proteins, including several cell-cycle regulators (see Chapter 4), such as cyclins, CDC25, and also some CKIs, there has been growing enthusiasm for manipulating this therapeutically.

Multiple other proteins of importance in regulating cell growth, survival, and cancer are also regulated by the proteasome, including the tumor suppressor p53, discussed in the previous section and in Chapter 7. Most human tumors have either mutations in p53 or alterations in positive (ARF) or negative (HDM2) regulators of p53. In the context of this chapter, it is important to note that HDM2 controls p53 stability via the ubiquitin–proteasome pathway. A few other pertinent examples will be described here. The NF-κB family are a group of transcription factors that are held inactive in the cytoplasm bound to the inhibitor protein IkB. Stimulation of cells by inflammation, chemotherapy, radiation, or oxidants initiates a signaling pathway culminating in the phosphorylation, ubiquitination, and degradation of IkB by the 26S proteasome; NF-κB can then translocate to the nucleus and activate the transcription of various genes, including growth factors, angiogenic factors, and antiapoptotic factors (such as inhibitor of apoptosis (IAP) and BCL-2).

β-Catenin, a member of the WNT signaling pathway, is downregulated by glycogen synthase kinase-3β (GSK3β)-dependent phosphorylation of serine and threonine residues in the N-terminus of the protein, followed by ubiquitination and proteasomal degradation. In human and rodent cancers, mutations that substitute one of the critical serine and threonine residues in the GSK3β region of β-catenin stabilize the protein and activate β-catenin–TCF–LEF target genes.

MEN1 is a tumor suppressor gene that is involved in inherited tumor susceptibility in multiple endocrine neoplasia type 1 (MEN1) and encodes a 610 amino acid protein called menin (see also Chapter 5). While the majority of germline mutations identified in MEN1 patients are frameshift and nonsense mutations resulting in truncation of the menin protein, various missense mutations have been identified whose effects on menin may largely be mediated via rapid degradation of mutant menin via the ubiquitin–proteasome pathway. Mutant, but not wild-type, menin may also interact with the molecular chaperone Hsp70 and with the Hsp70-associated ubiquitin ligase CHIP, which could promote ubiquitination.

A recent study has found that TRPC4AP (transient receptor potential cation channel, subfamily C, member 4-associated protein)–TRUSS (tumor necrosis factor receptor-associated ubiquitous scaffolding and signaling protein), a protein often downregulated in cancer cell lines, is important in degradation of MYC. TRPC4AP–TRUSS is a receptor for a DDB1 (damage-specific DNA-binding protein 1)–CUL4 (Cullin 4) E3 ligase complex that promotes ubiquitination and proteasomal degradation of MYC.

PTEN encodes a major lipid phosphatase, which acts as a negative regulator of the PI3K–AKT pathway to influence G_1 cell-cycle arrest and apoptosis (see Chapters 6 and 7). Several mechanisms of PTEN inactivation occur in primary malignancies derived from different tissues and may involve both inappropriate subcellular compartmentalization and increased or decreased proteasome degradation.

Smad4–DPC4 is a downstream mediator of TGF-β signaling, frequently inactivated in human cancer. The ubiquitin–proteasome pathway is important for inactivating Smad4 mutants in cancer.

Therapeutic inhibition of the proteasome

They will not force us, they will stop degrading us, they will not control us, we will be victorious.

Bellamy and Matthew, taken from the song "Uprising" by Muse

Inhibition of the ubiquitin–proteasome pathway or, more specifically, at the E3 ligases known to modulate apoptosis might be useful strategies in cancer therapy. Given the wide range and sometimes opposing actions of proteins degraded by the proteasome, it follows that any nonselective inhibition of proteasome-mediated degradation of proteins could either promote or inhibit replication depending on the balance of the various target proteins undergoing degradation at that time. However, in the context of cancer in which we assume that the balance is shifted in favor of proteins that drive replication and prevent apoptosis (many of the inhibitors having been inactivated at some point during tumorigenesis), even nonspecific blockade of the proteasome may be effective. In laboratory studies, proteasome inhibition can reduce cancer cell proliferation by preventing breakdown of proteins and transcription factors that normally restrict cell cycling. Moreover, the survival of cancer cells during chemotherapy can be reduced and cancer cell death enhanced by producing cellular stress.

The first proteasome inhibitor to enter clinical practice was Bortezomib, approved for the treatment of myeloma in 2003. The drug is a tripeptide boronic acid analogue, which can bind to and inhibit the catalytic site of the 26S proteasome. Theoretically, proteasome inhibitors could block the degradation of cancer-restraining proteins such as proapoptotic factors and tumor suppressors. Myeloma cancer cells appear to be uniquely sensitive to this approach, and Bortezomib is used alone or in combination with other drugs in refractory myeloma, where it has been shown to be effective in phase II and III studies, and in mantle cell lymphoma. It is being examined in other B-cell-related cancers, including some other lymphomas. Although the agent appears less effective in solid cancers, some early promise has been shown in combination with other agents in preclinical studies of colorectal and prostate cancer, among others. Such studies suggest the use alongside HDAC inhibitors and antiangiogenic agents such as thalidomide.

It is likely, based on some studies in cancer cell lines, that the benefits of proteasome inhibition will depend on the molecular underpinnings of the given cancer. Thus, in a neuroblastoma cell line, Bortezomib can help link the extrinsic and intrinsic pathways of apoptosis by amplifying TRAIL-dependent activation of Bid (increased levels of truncated Bid–tBid), and increasing levels of p53 and NOXA increases apoptosis (see Chapter 8). High levels of BCl2 conversely indicate resistance to these effects of Bortezomib.

Receptor degradation

Activation of receptor tyrosine kinase (RTK) signaling pathways is involved in tumorigenesis (Chapter 5), but it is increasingly apparent that aberrant RTK deactivation may be equally important. Reversible ubiquitination of activated ligand-receptor complexes and endocytosis can direct sorting between recycling pathways, on one hand, and lysosomal (and other sites such as the endoplasmic reticulum (ER)) degradation on the other. Protective mechanisms also exist, and ubiquitin-specific protease 8 (USP8) can prevent ligand-induced ubiquitination and lysosomal degradation of EGFR. There are also ligand-independent degradation pathways. Recent studies have shown that protein ubiquitination is pivotal in the vesicular trafficking of growth factor receptors. This degradative route is separate from the proteasome-mediated pathway, which involves polymeric chains of ubiquitin. Receptor ubiquitination takes place at the cell surface and directs internalized receptors to lumina of multivesicular bodies and the lysosome. The Cbl family of RING finger adaptors controls this late-sorting event, following ligand activation of receptors. The Cbl proteins appear important in regulating the stability of several cancer-relevant growth factor receptors, including MET in pancreatic cancer. Another group of E3 ubiquitin ligases, the Nedd4 family, regulates the initial sorting event, which targets receptors to clathrin-coated regions of the plasma membrane together with adaptor proteins such as epsins, which share ubiquitin-interacting motifs. Various genetic defects can prevent successful receptor downregulation through endocytosis as can certain viruses, culminating in growth factor-independent signaling as receptors are channelled to default recycling pathways. Recently, it has been shown that the E5 oncoprotein of the human papillomavirus type 16 (HPV16 E5) can influence the recycling of growth factor receptors, particularly promoting downregulation of the keratinocyte growth factor receptor or fibroblast growth factor receptor 2b (KGFR–FGFR2b) that acts as a tumor suppressor in epithelium.

The ErbB3 TKR may become overexpressed in some cancers, such as breast, through loss of the RING finger E3 ubiquitin ligase Nrdp1, required for ligand-independent receptor ubiquitination and degradation at the ER. The ubiquitin ligase activity of BRCA1 can promote ligand-independent and -dependent degradation of progesterone receptor. These processes are amenable to therapeutic manipulation through various means, not just by non-specifically blocking degradation pathways. Thus, the HDAC inhibitor, vorinostat, can potentiate actions of gefitinib even in resistant cells that have gone through EMT, by promoting ubiquitination and then lysosomal degradation of the TKRs, EGFR, ErbB2, and ErbB3.

Wrestling with protein transit – the role of SUMO and the promyelocytic leukemia (PML) body

The least movement is of importance to all nature. The entire ocean is affected by a pebble.

Blaise Pascal

The activity of a protein is also often constrained by processes that restrict its movements or banish it to cell compartments, where it is distanced from its natural partners, or conversely hold it in close proximity to those partners; the net effect is to avoid or incite functional interactions respectively.

The nucleus contains a number of distinct, non-membrane-bound compartments in which various different proteins are seen to concentrate. Among these, PML nuclear bodies have generated a lot of interest due to their involvement in cancer. PML bodies are present in all mammalian cells, but are disrupted in acute promyelocytic leukemia, in which the PML protein becomes fused to the retinoic acid receptor. This PML–RARα hybrid protein is therapeutically activated by the ligand retinoic acid, and the resultant disaggregation of the PMLNB might in part explain the benefits of such therapy. Mice that lack PML nuclear bodies (PMLNBs) have impaired immune function, exhibit chromosome instability, and are sensitive to carcinogens. Although their direct role in nuclear activity is unclear, PMLNBs are likely involved in the regulation of transcription, apoptosis, tumor suppression, and the response to viral infection. PMLNBs are sites at which multi-subunit complexes can form and mediate posttranslational modification of proteins such as p53 in response to DNA damage. Following DNA damage, several repair factors transit through PMLNBs, supporting the view that the PMLNB may function as a dynamic sensor of cellular stress. In one model, such stresses may favor disassembly of the PMLNB enabling dispersal of DNA repair proteins.

The PML tumor suppressor protein accumulates in the PML nuclear body, but cytoplasmic PML isoforms of unknown function have also been described. Recent studies suggest that cytoplasmic PML is an important modulator of TGF-β signaling. PML-null primary cells are resistant to TGF-β-dependent growth arrest, induction of cellular senescence, and apoptosis. These cells also have impaired phosphorylation and nuclear translocation of the TGFβ signaling proteins SMAD2 and SMAD3, as well as impaired induction of TGF-β target genes. The PML–RARα oncoprotein of APL can antagonize cytoplasmic PML function, and APL cells have defects in TGF-β signaling similar to those observed in PML-null cells. PML also potentiates p53 function by regulating posttranslational modifications, such as CBP-dependent acetylation and Chk2-dependent phosphorylation, in the PMLNB. PML interacts with the p53 ubiquitin–ligase MDM2 and can enhance p53 stability by sequestering MDM2 to the nucleolus. As nucleolar localization of PML may be regulated by ATR activation and phosphorylation of PML by ATR, this provides important links between DNA damage and apoptosis.

AKT is activated in many human cancers primarily due to inactivation of PTEN tumor suppressor action (see Chapters 5 and 7). Recent studies show that PML can inactivate AKT inside the nucleus. PML recruits the AKT phosphatase PP2a into PML nuclear bodies. Moreover, PML loss results in AKT-dependent loss of FOXO3A and of its transcriptional target, the proapoptotic protein BIM and the cell-cycle inhibitor $p27^{Kip1}$.

Study of the signals that may regulate PMLNB dynamics led Anne Dejean and colleagues to identify the small ubiquitin-related modifier (SUMO) pathway. SUMO family proteins are not only structurally but also mechanistically related to ubiquitin in that they are posttranslationally attached to other proteins. Like ubiquitin, SUMO is covalently linked to its substrates via amide (isopeptide) bonds formed between its C-terminal glycine residue and the varepsilon–amino group of internal lysine residues. The enzymes involved in the reversible conjugation of SUMO are similar to the ligases mediating the ubiquitin conjugation. The SUMO pathway uses an E1 activation enzyme (UBA2–AOS1 heterodimer), an E2 conjugation enzyme (UBC9), and three families of E3 ligases (RANBP2, PIAS, and PC2) that are believed to confer substrate specificity. Demodification is achieved by SUMO hydrolases (SENPs), of which seven isoforms exist in mammals (Fig. 11.7).

Since its discovery in 1996, interest in SUMO has increased rapidly, in part because its substrates include key proteins such as p53, c-JUN, PML, and Huntingtin. SUMO modification appears to play important roles in diverse processes such as chromosome segregation and cell division, DNA replication and repair, nuclear protein import, protein targeting to and formation of certain subnuclear structures, and the regulation of a variety of processes including the inflammatory response in mammals and the regulation of flowering time in plants. Unlike ubiquitination, sumoylation does not lead to the degradation of its target proteins but rather regulates subcellular localization.

One way in which SUMO acts is to target proteins to the PMLNB (analogous to ubiquitin targeting of proteins to the proteasome discussed here). Nuclear foci containing PML bodies occur in most cells and play a role in tumor suppression. Interestingly, proteins such as c-MYC may be regulated by several processes. Stability may be influenced by ubiquitination and acetylation, and interestingly now also by targeting to the PML body. PML bodies may also control the distribution, dynamics, and function of the checkpoint protein CHFR.

The SUMO modification pathway is now known to impact on various transcriptional regulators, predominantly through inhibiting signaling through recruitment of HDACs. Thus, for instance, sumoylation of Elk-1 promotes recruitment of HDAC-2. Recent studies suggest that activation of the ERK–MAP kinase pathway can activate Elk-1 signaling by promoting desumoylation and HDAC loss.

Arsenic, an ancient poison and constituent of many traditional medicines, has potent benefits in acute promyelocytic leukemia (APL) by promoting UBC9-mediated SUMOylation and degradation of the key causative PML–RAR alpha fusion protein.

There are numerous other ubiquitin-like proteins aside from SUMO which may have important roles in protein stability and localization within the cell. These include NEDD8, Atg8, Atg12, ISG15, FAT10, FUB1, Urm1, and UFM1. Of these, NEDD8 has been discussed.

Chromatin, senescence, and PML bodies

Senescent cells form so-called senescence-associated heterochromatic foci (SAHF), rich in the transcription-silencing histone H2A variant macroH2A. Heterochromatin is in general transcription-

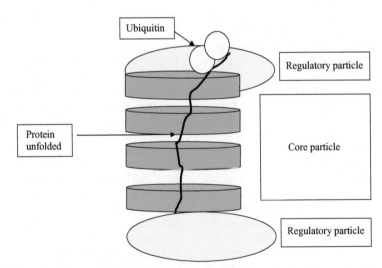

Figure 11.7 Structure of the proteasome. The core particle comprises two copies of each of 14 different proteins, assembled in groups of seven forming a ring. There are four rings each stacked on the other. There are two identical regulatory particles at each end of the complex, each consisting of 14 different proteins distinct from those in the core particle, six of which are ATPases, and some with sites that recognize ubiquitin. Proteins destined for destruction have been conjugated to a molecule of ubiquitin by ligases onto lysine residues. Additional molecules of ubiquitin bind to the first, forming a chain. The complex binds to ubiquitin-recognizing site(s) on the regulatory particle, and the protein unfolds, which utilizes energy from ATP. The unfolded protein translocates into the central cavity of the core particle. Several active sites on the inner surface break specific peptide bonds to form peptides averaging about eight amino acids in length. These peptides are extruded and may either be further degraded in the cytosol or incorporated in a class I histocompatibility molecule to be presented to the immune system as a potential antigen.

Gene and Protein Regulation by Epigenetic Factors

ally inactive, and SAHF contain genes associated with cell growth. Therefore, incorporation of cell-cycle-related genes into heterochromatin could be a contributor to senescence. Recent studies suggest that two proteins, HIRA and ASF1a, known to be involved in chromatin structure, also contribute to the formation of SAHF. As cells approach senescence, HIRA enters PML nuclear bodies, where it interacts with HP1 proteins, prior to incorporation of HP1 proteins into SAHF. Another chromatin regulator, ASF1a, is needed to complex with HIRA for the formation of SAHF and for senescence. HIRA and ASF1a may drive formation of macroH2A-containing SAHF and senescence-associated cell-cycle exit, via a pathway requiring the transit of heterochromatic proteins through PML bodies (Fig. 11.8).

Heat shock proteins and cancer

Heat shock protein 90 (Hsp90) interacts with a number of highly cancer-relevant oncoproteins and thereby contributes to regulation of many of the growth regulatory processes that we might wish to antagonize in the treatment of many tumors. Hsp90 is deregulated in several cancers, normally functions as a molecular chaperone facilitating protein folding into functional conformations, and may also act together with E3 ubiquitin ligases, particularly Cul5, to

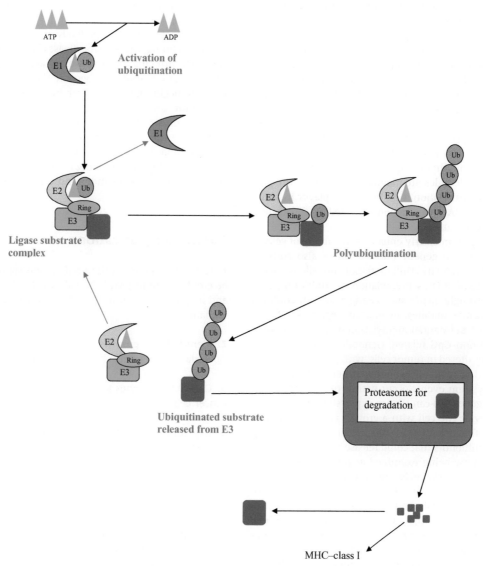

Figure 11.8 The ubiquitin–proteasome system. Degradation of protein substrates by the UPS is a sequential and highly regulated process. Step 1 involves ubiquitin activation by E1 (ubiquitin-activating enzyme) in which ATP is hydrolyzed and a ubiquitin molecule is adenylated. Step 2 involves loading of ubiquitin onto E2 (ubiquitin-conjugating enzyme). In step 3, the E2–Ub group is joined by the E3 ubiquitin ligase complex (comprising a ring domain and various other co-factors), and the substrate is recognized by the E3 ligase. Next, E3 catalyzes the transfer of ubiquitin from E2 to the target substrate, and then a polyubiquitin chain forms. Finally, after release from E3, the substrate is recognized by the proteasome due to the polyubiquitin chain and is degraded. Ubiquitin is recycled, whereas the peptides resulting from the degraded substrate generated are presented as antigens by the major histocompatibility class I system or are used for new protein synthesis.

allow E2 ubiquitin-conjugating enzymes and proteasome degradation of client proteins, such as ERBB2 and HIF1a.

Conclusions and future directions

In the first edition of this text, epigenetics was described as a new frontier in biology. Well, now it is well and truly part of the hinterland even if mapping remains incomplete. It is now clear that the genome contains information in two forms, genetic and epigenetic, both of major importance in cancer biology. The genetic information provides the blueprint for the manufacture of all the proteins necessary to create the organism, while the epigenetic information provides additional instructions on how, where, and when the genetic information should be used. Epigenetic information is not contained within the DNA sequence itself, but can specify mitotic inheritance of cell fate and gene expression patterns, can turn environmental effects into heritable changes in cell phenotypes, and intriguingly may even help in translating parental experiences into altered phenotype in the offspring – still probably not in a giraffe's neck, though. Epigenetic changes include modulation of chromatin structure, transcriptional repression, X-chromosome inactivation, and genomic imprinting, all of which may be altered in cancer cells and might be amenable to therapeutic targeting. The major form of epigenetic modification in mammalian cells is DNA methylation – the covalent addition of a methyl group to the fifth position of cytosine within CpG dinucleotides predominantly located in the promoter region. DNA methylation is an important player in many processes, including DNA repair, genome instability, and regulation of chromatin structure.

While DNA methylation clearly enhances the ability of cells to regulate and package the genetic information, it also adds an additional level of complexity. Differentiation of cells is clearly associated with extensive DNA methylation as irrelevant genes are silenced. Interestingly, many such changes are not confined to CpG islands, but may manifest in adjacent regions – referred to as island "shores." Substantial methylation in embryonic stem cells appears to be non-CpG related. Genomic methylation patterns are frequently altered in tumor cells, with global hypomethylation accompanying region-specific hypermethylation events. When hypermethylation events occur within the promoter of a tumor suppressor gene, this can silence expression of the associated gene and provide the cell with a growth advantage similar to that of gene deletion or mutation (Chapter 7). Not surprisingly, DNA methylation inhibitors that could release tumor suppressors from bondage are now being examined as potential anticancer agents (see Chapter 16). However, interest has again turned to the potential role of hypomethylation in the activation of oncogenes, involved in metastasis and invasion (see Chapter 12). Self-evidently this suggests that drugs which globally inhibit methylation may have both beneficial (reactivating tumor suppressors) and adverse (activating oncogenes) actions – these balancing issues still require much fundamental research. It is important now to make progress in understanding how methylation of specific genes is regulated and how this may lead us to develop newer, more selective therapeutics targeting specific genes only. Another area of interest is the role of methylation in non-CpG island promoters and the links between methylation of CpG in large repetitive sequences such as centromeres and maintainance of chromosome stability.

It is undoubted that RNAi technologies will continue to further research progress into the functional genomics of cancer and will ultimately yield new therapeutic agents that will be particularly important in cases where it has proved difficult or impossible to develop suitable small-molecule inhibitory agents. At least two siRNA-based therapies are already in clinical trials and are showing early promise against advanced cancers.

While functional understanding of the role played by miRNAs in regulating gene expression is rapidly progressing, application of miRNA profiling in developing novel biomarkers in body fluids is still in its infancy. Many issues still need to be resolved, not least the development of standards and a greater understanding of inter-individual variation and over time. Dynamic models need to be developed which can allow tracking of changes during treatments and also release of miRNAs into body fluids. It is already known that CRC with high miR-21 may respond less well to fluorouracil.

The next few years will likely witness startling progress in further defining the specific ligases and regulators that target individual proteins for proteasomal degradation (Fig. 11.9), and some of these will in time become the targets of new drugs. We are only beginning to discover how proteins are regulated by localization and partitioning, and various processes including sumoylation are increasingly being investigated in cancer biology. Finally, the role of chaperones such as HSP90 in regulating the function and levels of numerous key proteins implicated in cancer, including p53, is being recognized, and new drugs targeting HSPs are in development and some in trials.

Outstanding questions

1. Can information on the methylation status of CpG islands be employed to improve the subclassification of tumors and help predict clinical behavior and chemosensitivity? Explain.

2. Identify those enzymes directly responsible for mediating CpG island hypermethylation of tumor suppressor genes.

3. Can silenced genes be reactivated as a form of therapy in cancer cells?

4. Explain the role of hypomethylation in the activation of oncogenes.

5. Discuss non-CpG-related methylation events and the role of oxygenases in converting 5-methylcytosine into 5-methylhydroxycytosine in stem cells.

6. Explain the role of protein modification by ubiquitination and sumoylation in degradation and transit of cancer-contributing proteins.

7. What specific therapies have been developed targeting individual ligases or their components?

8. Describe the further develop techniques, such as enhanced stability and deliverability, for application of RNAi technology *in vivo*.

Gene and Protein Regulation by Epigenetic Factors

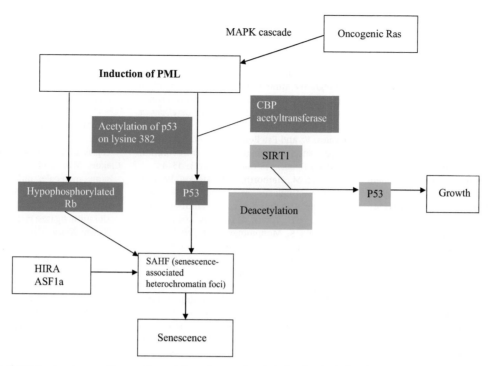

Figure 11.9 The role of PML in senescence pathways. Upregulation of promyelocytic leukemia protein (PML) results in relocation of CBP and p53 to the PML body. CBP acetylates p53, increasing the activity and expression of p53-target genes promoting senescence. An intact PML body may not be a prerequisite for this. Conversely, SIRT1 expression may antagonize PML by promoting p53 deacetylation. PML is also associated with hypophosphorylated RB and reduced E2F transcriptional activity. As cells reach senescence, a change in chromatin structure, called senescence-associated heterochromatin foci (SAHF), silences the genes that promote growth. In addition to PML, two other proteins, HIRA and ASF1a, are also involved in the formation of SAHF. After Langley et al. (2002).

Bibliography

General

Alberts, B., Johnson, A., Lewis, J., et al. (2002). *Molecular Biology of the Cell*. 4th edn. New York: Garland Science.

Alberts, B., Johnson, A., Lewis, J., et al. (2007). *Molecular Biology of the Cell*. 5th edn. New York: Garland Science.

Darwin, C. (1861). *On the Origin of Species*. 3rd edn. London: John Murray.

Epigenetics

Bachman, K.E., Park, B.H., Rhee, I., et al. (2003). Histone modifications and silencing prior to DNA methylation of a tumor suppressor gene. *Cancer Cell*, (1): 89–95.

Campos, E.I., Chin, M.Y., Kuo, W.H., and Li, G. (2004). Biological functions of the ING family tumor suppressors. *Cellular and Molecular Life Sciences*, 61(19–20): 2597–613.

Chan, T.L., Yuen, S.T., Kong, C.K., et al. (2006). Heritable germline epimutation of MSH2 in a family with hereditary nonpolyposis colorectal cancer. *Nature Genetics*, 38(10): 1178–83.

Egger, G., Liang, G., Aparicio, A., and Jones, P.A. (2004). Epigenetics in human disease and prospects for epigenetic therapy. *Nature*, 429: 457–63.

Esteller, M. (2005). Dormant hypermethylated tumor suppressor genes: questions and answers. *Journal of Pathology*, 205(2): 172–80.

Feinberg, A.P., and Tycko, B. (2004). The history of cancer epigenetics. *Nature Reviews Cancer*, 4(2): 143–53.

Feinberg, A.P., and Vogelstein, B. (1983). Hypomethylation distinguishes genes of some human cancers from their normal counterparts. *Nature*, 301(5895): 89–92.

Fukuzawa, R., Breslow, N.E., Morison, I.M., et al. (2004). Epigenetic differences between Wilms' tumours in white and east-Asian children. *Lancet*, 363(9407): 446–51.

Gaudet, F., Hodgson, J.G., Eden, A., et al. (2003). Induction of tumors in mice by genomic hypomethylation. *Science*, 300(5618): 489–92.

Han, Y., San-Marina, S., Liu, J., and Minden, M.D. (2004). Transcriptional activation of c-myc proto oncogene by WT1 protein. *Oncogene*, 23: 6933–41.

Hitchins, M.P., Wong, J.J., Suthers, G., et al. (2007). Inheritance of a cancer-associated MLH1 germ-line epimutation. *New England Journal of Medicine*, 356(7): 697–705.

Huang, J., Deng, Q., Wang, Q., et al. (2012). Exome sequencing of hepatitis B virus-associated hepatocellular carcinoma. *Nature Genetics*, August 26. doi:10.1038/ng.2391. [Epub ahead of print]

Issa, J.P. (2004). CpG island methylator phenotype in cancer. *Nature Reviews Cancer*, 4: 988–93.

Kinzler, K.W., and Vogelstein, B. (1996). Lessons form hereditary colorectal cancer. *Cell*, 87, 159–70.

Kirmizis, A., Bartley, S.M., Kuzmichev, A., et al. (2004). Silencing of human polycomb target genes is associated with methylation of histone H3 Lys 27. *Genes & Development*, 18(13): 1592–605.

Lau, P.N., and Cheung, P. (2011). Histone code pathway involving H3 S28 phosphorylation and K27 acetylation activates transcription and antagonizes polycomb silencing. *Proceedings of the National Academy of Sciences of the USA*, 108(7): 2801–6.

Loeb, L.A. (2001). A mutator phenotype in cancer. *Cancer Research*, 61, 3230–9.

Meissner, A., Mikkelsen, T.S., Gu, H., et al. (2008). Genome-scale DNA methylation maps of pluripotent and differentiated cells. *Nature*, 454: 766–70.

Ng, S.F., Lin, R.C., Laybutt, D.R., Barres, R., Owens, J.A., and Morris, M.J. (2010). Chronic high-fat diet in fathers programs β-cell dysfunction in female rat offspring. *Nature*, 467(7318): 963–6.

Rainier, S., Johnson, L.A., Dobry, C.J., Ping, A.J., Grundy, P.E., and Feinberg, A.P. (1993). Relaxation of imprinted genes in human cancer. *Nature*, 362(6422):747–9.

Rhee, I., Bachman, K.E., Park, B.H., et al. (2002). DNMT1 and DNMT3b cooperate to silence genes in human cancer cells. *Nature*, 416 (6880): 552–6.

Ropero, S., Fraga, M.F., Ballestar, E., et al. (2006). A truncating mutation of HDAC2 in human cancers confers resistance to histone deacetylase inhibition. *Nature Genetics*, **38**(5): 566–9.

Turner, B.M. (2003). Memorable transcription. *Nature Cell Biology*, **5**(5): 390–3.

Turner, B.M. (2005). Reading signals on the nucleosome with a new nomenclature for modified histones. *Nature Structural & Molecular Biology*, **12**(2): 110–2.

Welcker, M., Orian, A., Jin, J., et al. (2004). The Fbw7 tumor suppressor regulates glycogen synthase kinase 3 phosphorylation-dependent c-Myc protein degradation. *Proceedings of the National Academy of Sciences of the USA*, **101**(24): 9085–90.

Yamashita, K., Dai, T., Dai, Y., Yamamoto, F., and Perucho, M. (2003). Genetics supersedes epigenetics in colon cancer phenotype. *Cancer Cell*, **4**: 121–31.

Zhang, R., Poustovoitov, M.V., Ye, X., et al. (2005). Formation of MacroH2A-containing senescence-associated heterochromatin foci and senescence driven by ASF1a and HIRA. *Developmental Cell*, **8**(1): 19–30.

Translation of mRNA

Stumpf, C.R., and Ruggero, D. (2011). The cancerous translation apparatu. *Current Opinion in Genetics & Development*, **21**(4): 474–83.

Takagi, M., Absalon, M.J., McLure, K.G., and Kastan, M.B. (2005). Regulation of p53 translation and induction after DNA damage by ribosomal protein L26 and Nucleolin. *Cell*, **123**(1): 49–63.

Noncoding RNAs, microRNAs (miRNAs), and RNA interference (RNAi)

Amaral P.P., Michael B.C., Dennis K.G., Marcel E.D., and John S.M. (2011). lncRNAdb: a reference database for long noncoding RNAs. *Nucleic Acids Research*, **39**: D146–51.

Brummelkamp, T.R., Bernards R, Agami R., et al. (2002). A system for stable expression of short interfering RNAs in mammalian cells. *Science*, **296**(5567): 550–3.

Cabili, M.N., Trapnell, C., Goff, L., et al. (2011). Integrative annotation of human large intergenic noncoding RNAs reveals global properties and specific subclasses. *Genes & Development*, **25**: 1915–27.

Davalos, V., and Esteller, M. (2010). MicroRNAs and cancer epigenetics: a macrorevolution. *Current Opinions in Oncology*, **22**(1): 35–45.

Elbashir, S.M., Harborth, J., Lendeckel, W., et al. (2001). Duplexes of 21-nucleotide RNAs mediate RNA interference in cultured mammalian cells. *Nature*, **411**(6836): 494–8.

Fire, A., Xu, S., Montgomery, M.K., et al. (1998). Potent and specific genetic interference by double-stranded RNA in *Caenorhabditis elegans*. *Nature*, **391**(6669): 806–11.

He, L., He, X., Lim, L.P., et al. (2007). A microRNA component of the p53 tumour suppressor network. *Nature*, **447**: 1130–4.

Ma, L., Reinhardt, F., Pan, E., et al. (2010). Therapeutic silencing of mir-10b inhibits metastasis in a mouse mammary tumor model. *Nature Biotechnology*, **28**: 341–7.

MacFarlane, L.-A., and Murphy, P.R. (2010). MicroRNA: biogenesis, function and role in cancer. *Current Genomics*, **11**(7): 537–61.

Matzke, M.A., and Birchler, J.A. (2005). RNAi-mediated pathways in the nucleus. *Nature Reviews Genetics*, **6**(1): 24–35.

Pfeffer, S., Zavolan, M., Grässer, F., et al. (2004). Identification of virus-encoded microRNAs. *Science*, **304**: 734–6.

Tian, Y., Luo, A., Cai, Y., et al. (2010). MicroRNA-10b promotes migration and invasion through KLF4 in human esophageal cancer cell lines. *Journal of Biological Chemistry*, **285**: 7986–94.

Ting, A.H., Schuebel, K.E., Herman, J.G., and Baylin, S.B. (2005). Short double-stranded RNA induces transcriptional gene silencing in human cancer cells in the absence of DNA methylation. *Nature Genetics*, **37**(8): 906–10.

Wianny, F., and Zernicka-Goetz, M. (2000). Specific interference with gene function by double-stranded RNA in early mouse development. *Nature Cell Biology*, **2**(2): 70–5.

Ubiquitination and the proteasome

Adams, J. (2004). The proteasome: a suitable antineoplastic target. *Nature Reviews Cancer*, **4**: 349–60.

Clague, M.J., and Urbé, S. (2010). Ubiquitin: same molecule, different degradation pathways. *Cell*, **143**(5): 682–5.

Nakayama, K.I., and Nakayama, K. (2006). Ubiquitin ligases: cell-cycle control and cancer. *Nature Reviews Cancer*, **6**(5): 369–81.

Promyelocytic leukemia (PML)

Bischof, O., Kirsh, O., Pearson, M., Itahana, K., Pelicci, P.G., and Dejean, A. (2002). Deconstructing PML-induced premature senescence. *EMBO Journal*, **21**(13): 3358–69.

de Stanchina, E., Querido, E., Narita, M., et al. (2004). PML is a direct p53 target that modulates p53 effector functions. *Molecular Cell*, **13**(4): 523–35.

Langley, E., Pearson, M., Faretta, M., et al. (2002). Human SIR2 deacetylates p53 and antagonizes PML/p53-induced cellular senescence. *EMBO Journal*, **21**: 2383–96.

Trotman, L.C., Alimonti, A., Scaglioni, P.P., Koutcher, J.A., Cordon-Cardo, C., and Pandolfi, P.P. (2006). Identification of a tumour suppressor network opposing nuclear Akt function. *Nature*, **441**(7092): 523–7.

Chromatin

Cooper, K.W. (1959). Cytogenetic analysis of major heterochromatic elements. *Chromosoma*, **10**: 535–88.

Heitz, E. (1929). Heterochromatin, chromozentren, chromomeren. *Berichte der Deutschen Botanischen Gesellschaft*, **47**: 274–84.

Questions for student review

1) Gene expression can be regulated by:
   a. Modifications in chromatin.
   b. Processing of mRNA.
   c. Small RNA molecules.
   d. Ubiquitination of transcription factor proteins.
   e. DNA methylation.

2) DNA methylation:
   a. Is important only in regulating expression of tumor suppressor genes.
   b. Is frequently altered in cancer cells.
   c. Affects only CpG islands in gene promoters.
   d. Usually alters the DNA sequence.
   e. Does not influence chromatin structure.

3) Imprinting:
   a. Is regulated by DNA methylation.
   b. Of some genes may be lost in cancer cells.
   c. Always involves inactivation of the maternally inherited allele of a gene.
   d. Is a normal feature of the IGF2 gene.
   e. Of the IGF2 gene is lost in all Wilms' tumors.

4) RNA interference:
   a. Does not occur naturally.
   b. May be a means of combating certain viral infections.
   c. May regulate gene expression during development.
   d. May be employed by viruses against the host cell.
   e. May be employed to manipulate gene expression in the laboratory.

5) The proteasome:
   a. Degrades proteins that have been sumoylated.
   b. May degrade proteins involved in cell-cycle regulation.
   c. May be inhibited therapeutically to treat cancer.
   d. Is responsible for ubiquitinating proteins.
   e. Is a major site for degradation of misfolded proteins.

12 Cell Adhesion in Cancer

Charles H. Streuli
University of Manchester, UK

Key points

- Adhesive interactions between cells and their extracellular environment occur through extracellular matrix (ECM) receptors such as integrins, and intercellular adhesion receptors such as cadherins. The sites of physical interaction are called adhesion complexes. Integrins bind matrix proteins, while cadherins join cells together. Both organize intracellular cytostructure and transmit signals to the cell.
- Major aspects of cell physiology, including cell cycle progression, suppression of apoptosis, cellular differentiation, and migration, depend upon signals being received through adhesion complexes. The regulation of these cell fate choices is perturbed in cancer, and altered cell adhesion is central to cancer progression.
- The normal spatial organization of cells within tissues is determined by adhesive interactions. The transition from a primary tumor to a malignancy requires loss of both the proliferative *and* positional controls that would otherwise maintain correct cell number and architecture. The homeostasis of a whole affected organ is disrupted in cancer.
- Malignant cells move from their site of origin and disseminate to distant organs, where they form metastases. This progression requires the acquisition of a large number of characteristics, which involve changes in cell adhesion. These include migration, an ability to invade the stroma and endothelial linings, dissemination through the circulation, survival in inappropriate tissue contexts, and environmental decoding to allow a tumor to grow in defined secondary target sites. However, the formation of metastases is inefficient and rare.
- Reduced adhesiveness at adherens junctions is a key stage in malignant progression. Loss of E-cadherin occurs through mutations as well as phosphorylation, degradation, and altered transcription. It is also lost during the epithelial–mesenchymal transition.
- Invasive cells acquire the ability to migrate by altering their repertoire of integrins, activating the intracellular signaling machinery that controls the cytoskeleton and motile processes, and expressing ECM-remodeling enzymes. Cancer cells frequently migrate in collective groups rather than as single cells.
- Although most of the genetic changes in carcinoma occur within epithelial cells, the supporting tissue, or stroma, has a critical role in malignant progression. The stroma contributes growth factors and other signaling molecules to carcinoma cells, and altered ECM composition can change the biomechanical forces exerted on cancer cells, both of which contribute to disease progression.
- Anoikis is a specific form of apoptosis that is induced when cells no longer receive appropriate signals from the ECM. To survive in the stromal bed of the primary tissue that malignant cells migrate through and secondary metastatic sites, they undergo activating mutations in the ECM survival pathway.

Introduction

Cellular adhesion to the extracellular matrix and to neighboring cells orchestrates tissues into functional units, and is required for cell migration (Fig. 12.1). Adhesion also regulates cell shape, via the cytoskeleton and intracellular tension, and has a key role in determining the responses of signaling ligands to control cellular fate and differentiation, as well as survival and proliferation. Adhesion is mediated through specific classes of cell surface receptors, which join cells to each other or to the ECM. In cancer, altered adhesion influences each of the hallmarks of tumorigenesis, and is the defining characteristic of malignancy. Changes in cell–cell adhesion and cellular interactions with the ECM have severe consequences on the ability of extracellular factors to signal properly, so that cells lose their positional identity. Altered adhesion signaling results in the unscheduled migration of tumor cells, their dissemination through the circulation, and their embedment, survival, and proliferation at distant sites, thereby forming life-threatening metastases.

As discussed in other chapters, cancer is largely caused by genomic instability that arises within stem cells. It involves the

---

The Molecular Biology of Cancer: A Bridge From Bench to Bedside, Second Edition. Edited by Stella Pelengaris and Michael Khan.
© 2013 John Wiley & Sons, Inc. Published 2013 by John Wiley & Sons, Inc.

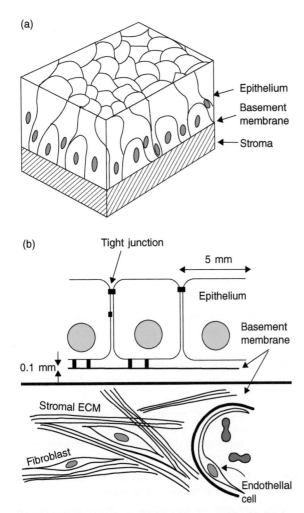

Figure 12.1 Tissue architecture. The key components of epithelial tissues shown as (a) a 3D representation of a pseudostratified epithelium, and (b) a schematic cross-section of a simple epithelium. They contain epithelial cells, separated from the stroma by a basement membrane; stromal cells such as fibroblasts and adipocytes; capillaries; neurons; and cells of the immune system. Although the main cell type to become altered genetically, and therefore neoplastic, is the epithelial cell itself, cancer progression depends on influences from other cell types. Thus, cancer is a disease where homeostasis of a whole organ is disrupted. It not only involves the cancer cells themselves, but also is characterized by disturbed stromal, endothelial, and immune cells, as well as altered ECM.

Box 12.1 Benign versus malignant tumors

- *Benign tumor*: a tumor that slowly grows at its site of origin and is well circumscribed. The cells within benign tumors frequently have normal morphology and are differentiated. These tumors do not normally cause the death of an organism, but sometimes can provide the seed cells for malignant tumors. Benign tumors can be large, but they are often operable.

- *Malignant tumor*: a rapidly spreading tumor that progresses through a number of stages. Initially they have diffusely infiltrative margins and can show differentiated characteristics. They evolve to poorly differentiated structures that sequentially invade basement membrane, stroma, and then vasculature. It is only once a tumor has spread that a person has contracted "cancer."

- *Metastasis*: a tumor growing within another organ of the body. Although individual metastases are usually clonal in origin, they are often present in large numbers within any given organ, making them inoperable. The excessive tumor burden resulting from multiple metastases leads to the failure of essential body functions and death.

---

escape of protective apoptosis and immune surveillance mechanisms, deregulated proliferative and aging controls, as well as inappropriate sampling of the environment so that tumor cells ignore normal positional cues and thereby migrate to and survive at distant metastatic sites (Box 12.1). This chapter is about how cells respond to their microenvironment, and how alterations in the way that cells perceive their neighbors and the ECM contribute to deregulated homeostasis, leading to the formation of benign tumors and subsequently malignant cancers that metastasize to distant organs.

The most common cancers arise in epithelial tissues, such as the epidermis, lungs, intestine, prostate, or breast, and are called carcinomas. This chapter considers the types of adhesive interactions that epithelial cells normally encounter, how such interactions break down in cancer, and the mechanisms that allow cells to migrate to distant sites, and then survive and grow there. Many of the references are to contemporary review articles, which cover each topic in much more detail.

Adhesive interactions with the extracellular matrix

There are two main types of adhesive interactions; those with the ECM and those between cells (Fig. 12.2). Normally, parenchymal cells (this includes epithelial cells) reside on a sheet-like ECM called the basement membrane (sometimes referred to as the basal lamina). This matrix is a continuous and flexible part of the basal surface of all epithelia, which is 40–120 nm thick and forms the interaction zone with stromal ECM. Basement membranes are also associated with endothelial cells lining capillaries and with muscle and Schwann cells. The stroma provides support for epithelia and is composed of the connective tissue matrix (synonymous with stromal ECM). Stromal cells, such as fibroblasts, myofibroblasts, adipocytes, endothelial cells and macrophages, are embedded within it, as are nerves. Blood and lymph vessels penetrate the stroma to provide nutrients and immune protection.

In cancer, epithelial cells come into contact with different types of ECM during their progression to malignancy. The basement membrane forms a boundary between epithelia and stroma in both normal tissue and premalignant lesions such as carcinoma in situ. In malignancy, the cells become invasive, breaking through the basement membrane and coming into contact with stromal or other ECM. Changes in epithelial cell-to-ECM interactions are required in order for the cells to do this.

Basement membrane
The basement membrane provides essential signals for the morphogenesis of epithelial structures during development as well as the survival, proliferation, differentiation, and cytostructural

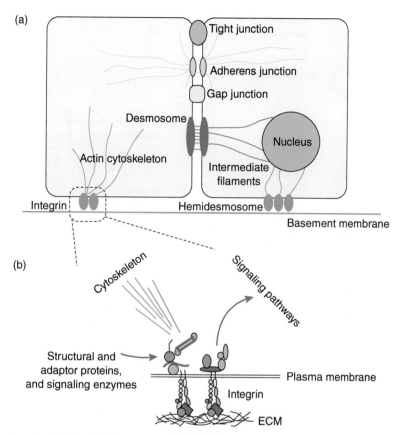

Figure 12.2 Key adhesive interactions of epithelial cells. (a) Several types of cell–cell adhesion systems bind adjacent cells together. Tight junctions form a ring around the apical surface of epithelia, and thereby ionically separate the extracellular spaces at the apical and basal surfaces. Adherens junctions connect cells via cadherins and organize the microfilaments, particularly toward the apical side of polarized cells. They also sequester the transcription factor β-catenin. Gap junctions provide ionic communication between adjacent cells. Desmosomes are large plaques that form strong intercellular bonds, again via cadherins. They are of key importance in maintaining tissue integrity under conditions of sheer stress. The basal surface of epithelial cells interacts with glycoproteins within the specialized ECM, the basement membrane, via both integrin receptors and transmembrane proteoglycan receptors (the latter are not shown). Integrins can be located either within multiprotein assemblies known as adhesion complexes, where they link to the actin cytoskeleton, or in very large adhesion plaques called hemidesmosomes, which link to the intermediate filament network, itself forming a bridge to the nuclear envelope. (b) In addition to binding cells to the ECM, integrin receptors regulate cell shape and polarity via the cytoskeleton, and they link to intracellular signaling pathways via structural adaptor proteins and signaling enzymes. Adherens junctions have a similar function in cytostructural regulation and signaling. *Polarity*: Most adhesive cells have an intracellular "direction," so that different components become spatially segregated to different sides of the cell. Epithelial cells have a basal surface, which contacts the basement membrane, and an apical surface at the opposite side of the cell. Plasma membrane lipids are different on apical and basal surfaces. Similarly, the cytoplasmic contents are different toward apical and basal poles, and nuclei are usually located within the basal half.

architecture of cells within them. ECM molecules, together with the growth factors (GFs) they harbor, deliver these signals. A key class of basement membrane glycoproteins is the laminins, which are cross-shaped molecules of an α–β–γ composition. 5α, 3β, and 2γ laminin genes exist, and their protein products assemble into at least 15 different heterotrimers. Some of the main laminin isoforms in epithelial basement membrane are laminin-1, laminin-5, and laminin-10. Laminin heterotrimers interact via their N-terminal head and arm domains to form a lattice-like network of interacting proteins. They bind to transmembrane cell surface integrin and proteoglycan receptors, resulting in organization of cell shape via the cytoskeleton, and triggering of intracellular signaling pathways.

An additional basement membrane protein is collagen type IV (collage-IV), which forms a self-interacting network that looks a bit like chicken wire. Although collagen-IV does bind surface receptors, its major function is to provide rigidity to the basement membrane. The laminin and collagen networks are linked together by nidogen, stabilizing the basement membrane. The fourth major component of basement membrane is perlecan, a large proteoglycan which binds a variety of growth and morphogenesis factors, and is involved with both sequestration of factors away from cells until they are released by proteases, as well as (in some cases) their presentation to appropriate surface receptors for signal transduction. Other minor components of the basement membrane include agrin, fibulin, and collagen types VII, XV, and XVIII.

Basement membranes are made through collaboration between epithelial and stromal cells, and are deposited to form a lamina at the interface between the two tissue compartments. Epithelia require direct signals from the basement membrane to function properly, indicating one way that the stroma influences how epithelia behave.

Basement membrane drives polarity

The interaction of epithelial cells with the ECM at their basal surface leads to adhesion-regulated organization of the cytoskeleton and, together with a contribution from cell–cell contacts, to

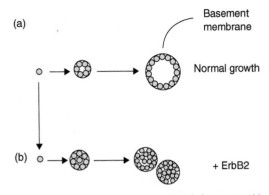

Figure 12.3 Three-dimensional (3D) culture models for cancer. Although many aspects of cell behavior, including survival and proliferation, are frequently studied in monolayer culture, these often do not reflect the responses of cells in 3D arrays *in vivo*, and extreme caution has to be taken to interpret experimental data. Fortunately, culture models are now available which mimic the 3D organization of tissues, and better models are therefore being developed to study carcinoma progression. ErbB2 is a member of the EGF receptor family and is frequently upregulated in breast cancer. However, an activated ErbB2 does not transform normal breast epithelial cells, as assessed by conventional methods such as growth in soft agar. In this example, (a) breast cells plated in 3D basement membrane gels grow to form hollow acini surrounded by a collagen IV–containing basement membrane, which resemble mammary alveoli *in vivo*. Under the same culture conditions, (b) the activated ErbB2 now causes the cells to proliferate so that they fill the luminal spaces and it alters their properties so that they form multi-acinar structures. These resemble some forms of early breast cancer, such as comedo ductal carcinoma in situ (Figure 12.15d) (Debnath and Brugge, 2005; Weigelt and Bissell, 2008).

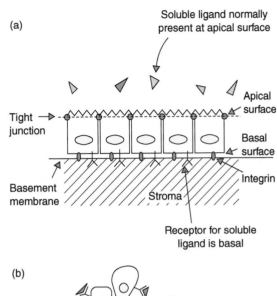

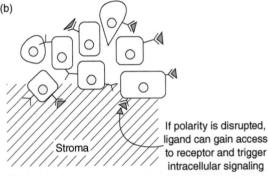

Figure 12.4 Polarity segregates signaling ligands from their receptors and thereby controls proliferation. GF receptor signaling in simple epithelia can be regulated by polarized segregation of the ligand to a different plasma membrane compartment than the receptor. Airway epithelia express the receptor ErbB1-4 as well as one of its ligands, heregulin-α. (a) The ligand is expressed on the apical cell surface and is secreted into apical medium, while the receptor is present only basally and cannot be activated unless heregulin-α is added ectopically to the basal cell surface or if tight junctions in intact monolayers are broken. Wounding the epithelium causes rapid heregulin-α-activated ErbB2 signaling. In this way, segregation of ligand and receptor provides an elegant mechanism for rapid GF receptor activation and tissue repair after epithelial injury. (b) Unfortunately, this has dire consequences if cellular polarity is disrupted in disease processes, because it can lead to unscheduled proliferation. For example, increased epithelial permeability in smoking-associated bronchitis can disrupt growth homeostasis and therefore contribute to tumor formation (Vermeer et al., 2003).

the establishment of cellular polarity (Fig. 12.2). ECM receptors control the orientation of epithelial polarity by causing the internalization of apical components and directing their trafficking to the opposite side of the cell.

The ability of ECM to regulate the cytostructure of individual cells also extends to the organization of multicellular structures. For example, the functional units of mammary glands are epithelial acini, which consist of a layer of polarized cells organized around a central lumen and with their apical surfaces facing inward (see, later in this chapter, Fig. 12.15b–c). Since epithelia in this tissue secrete milk, the cells have to be topologically orientated in this way for milk to be delivered into a discrete extracellular compartment. Modeling cancer accurately in experimental tissue culture depends upon appreciating the three-dimensional (3D) nature of tissues (Fig. 12.3).

The ECM control on multicellularity and topology has important consequences for the response of cells to their local environment. For example, in a polarized epithelium, extracellular regulatory factors or cytokines that are supplied to one surface of an epithelium work only if their receptors are present on the same cell surface. The expression of basement membrane proteins is often reduced in malignant cancer, contributing to a loss of cell polarity and altered proliferation (Fig. 12.4).

Basement membrane orchestrates growth and survival signals

Integrin-mediated adhesion is essential for most cell types to respond to growth factors. Thus, in addition to organizing the cytoskeleton and cell shape, cell–matrix interactions in epithelial cells are crucial for orchestrating accurate responses to growth and survival factors. Receptors for basement membrane proteins can activate signaling pathways and are required in *trans* for many GF receptors and their downstream signaling enzymes to work properly, and thereby for normal cells to proliferate (discussed further in this chapter and Chapter 5). Integrins are also required in most normal cell types to protect them from apoptosis; death following altered or lost adhesion is a process called anoikis (see the "Anoikis" section of this chapter). This effectively means that basement membrane proteins provide essential contextual information for instructing the way that epithelial cells behave. Appropriate crosstalk between cell–ECM interactions and oncogenes is essential for tumor formation.

Stromal–epithelial interactions

The stroma has an essential supporting function for epithelia, contributing signals for epithelial morphogenesis and function, and providing a vital substratum at times of tissue repair. Stromal ECM also has certain key properties that are important for cancer progression. First, it is composed of different proteins to the basement membrane and it harbors soluble factors and cytokines. This means that the stromal ECM can contribute directly to cancer progression by providing growth factors.

Second, the stromal ECM stimulates epithelial cells coming in contact with it to behave quite differently from those contacting basement membrane, for example by changing their shape, and altering gene expression as well as proliferative and survival responses. Tumor stroma is often more crosslinked and denser than normal stromal ECM, so its physical characteristics change during cancer progression. This means that during the course of malignancy, when epithelial cancer cells invade stromal ECM, the inappropriate environment may affect cells in ways that contribute to tumor development.

Third, stromal ECM networks serve to impede the passage of cells that are not specially designed to migrate through it. An important characteristic of malignant tumor cells is that they acquire mechanisms to help them migrate through the stroma, including both phenotypic changes that activate the motile machinery and inappropriate expression of proteinases that degrade collagens.

Fibrillar collagen

Stromal ECM is largely composed of fibrous proteins such as collagens and proteoglycans, which are proteins with large glycosaminoglycan (GAG) chains covalently bound to them. Collagen fibrils are made of helical bundles of collagen type I or III (in epithelial tissue stroma), to which are attached small collagens (e.g. collagen type IX) and proteoglycans (e.g. decorin). The triple-helical collagens assemble as bundles of fibrils that are organized into lattices, and through their high degree of mechanical strength they provide the tensile rigidity to maintain tissue form.

GAGs

GAGs are chains of repeating disaccharides, which can be up to several hundred sugar residues long, and are negatively charged due to sulphate groups. This allows GAGs to form gels, because of the mutual repulsion of negative charges and the entrapment of large numbers of water molecules, and thereby permits the stromal ECM to withstand compressive forces. Some GAG molecules are not covalently bound to a protein core and can occupy huge spaces; a single hyaluronan molecule, for example, may consist of up to 25 000 disaccharides and can form a 300 nm cube. Because of this property, hyaluronan can separate tissue components and thereby provide a space within dense stroma for cells to migrate. Hyaluronan also activates its cell surface receptor, CD44, which is a signal-transducing receptor for promoting migration. Although important for regulated tissue morphogenesis during development, altered hyaluronan expression and its interaction with CD44 occur in tumor cells and contribute to malignant progression.

Other stromal proteins

Other secreted glycoproteins, such as elastin and fibrillin, are also part of the stromal ECM, and contribute in their own way to the biology of the stroma; for example, the elastin fibers (made up of crosslinked networks of elastin molecules) provide resilience to tissues. Tissue fibronectin is also present throughout the stroma. This protein, which is a dimer, can provide adhesive links between collagen and cells, although cells also bind to some stromal ECM components directly. Fibronectin is one protein that cells rely on to migrate their way through stroma.

Tenascin is a further ECM protein that has a role in cancer progression. However, rather than being adhesive, its main function is anti-adhesive, because it interferes with integrin-dependent cell spreading. By antagonizing stable links between cells and the ECM, tenascin promotes cell migration. It is expressed during embryonic development when considerable cell movement occurs, and in the adult it is re-expressed in areas of tissue remodeling such as at wound repair sites. Tenascin is frequently upregulated in cancer.

Altered stroma contributes to malignancy

Although carcinomas are epithelial in origin, they usually show a strong stromal reaction, reflected in disorganized composition and cellular content of the neighboring ECM. The gene expression profile of stromal cells adjacent to primary carcinomas is widely different from that of normal stromal cells, and many of these changes contribute directly to tumor progression. Importantly, genomic alterations are confined to carcinoma cells and absent from stromal cells. This means that the epithelial tumor cells have a dominant influence over the stroma, driving it to "assist" the cancer cell's mission for dissemination. However, in some situations, the stroma can contribute directly toward the advancement of epithelial malignancy (Fig. 12.5).

Stromal ECM harbors growth factors

An additional feature of the stroma is that it harbors most of the locally acting paracrine factors that control epithelial cell function: transforming growth factor-β (TGF-β) binds to latency-associated peptide (LAP), fibroblast growth factors and Wnts are sequestered by proteoglycans, and insulin-like growth factors (IGFs) bind to IGF-binding proteins, which themselves are matrix bound (Chapter 5 covers growth factor signaling). GFs are released from the stroma by a variety of enzymes including ECM-degrading proteinases, which are inappropriately expressed in cancer cells. Chemokines such as CXCL12 are secreted by stromal cancer fibroblasts, and they also have a profound role in promoting tumor growth and the formation of secondary tumors.

Most of these proteins, and the ECM proteins themselves, are synthesized by stromal cells. Thus they have a strong controlling influence over the fate and function of epithelia. This has a key role in guiding the progression of epithelial malignancy, since cells within the stroma can be influenced by tumor cells to secrete factors that contribute to the invasive behavior of tumor cells themselves (Fig. 12.6).

Integrins and adhesomes

ECM proteins bind to cells through a variety of transmembrane cell surface receptors. The most prominent are integrins, which are α–β heterodimers (Fig. 12.7). Adhesion to ECM occurs at the distal ends of the heterodimeric receptors, where the α and β subunits interact. On the cytoplasmic side, integrins bind adaptor proteins that link them both to the actin-based cytoskeleton and to enzymes that trigger signal transduction cascades. Integrins therefore integrate the extracellular anchoring elements of the

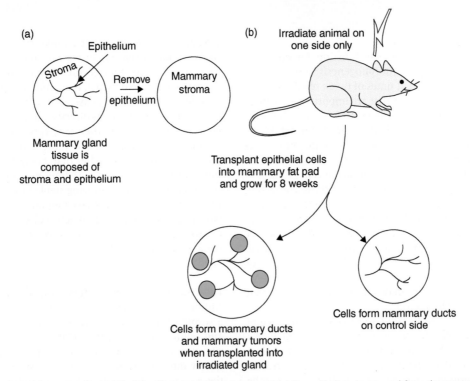

Figure 12.5 The stroma contributes directly to epithelial malignancy. In this experiment, (a) the epithelium is removed from the mammary gland of a 3-week-old mouse, leaving the mammary fat pads which consist of only the stromal components. (b) Eight weeks later, the mice are irradiated on one side only, and mammary epithelial cells are transplanted into both irradiated and control mammary fat pads. After a further 8 weeks, the cells in the non-irradiated glands proliferate and undergo developmental morphogenesis to form a normal-appearing ductal outgrowth. The cells in the irradiated side also form ductal outgrowths, but a large number of tumors also develop. The mammary cells injected are normal apart from harboring mutations in both p53 alleles, indicating that they are genomically unstable. Thus, ionizing radiation, which is a known carcinogen, causes changes in the stroma, which then facilitates the expression of neoplasia within the epithelial cell compartment. This type of communication between stroma and epithelium is often overlooked, but is of key importance in cancer progression (Barcellos-Hoff and Ravani, 2000).

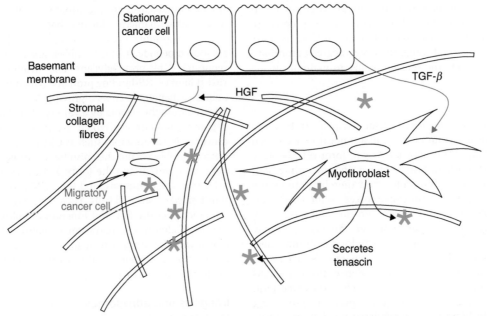

Figure 12.6 Stromal–epithelial interactions in invasive behavior of tumor cells. Tumor cells can subvert the normal function of stromal cells so that they contribute toward malignant progression, even though they don't become altered genetically. Here, tumor cells secrete TGF-β, which recruits myofibroblasts to differentiate from fibroblasts. These cells are also sometimes called cancer-associated fibroblasts (CAFs). CAFs have several effects on carcinoma cells that encourage them to migrate through the stroma. Firstly, they secrete HGF, which causes the carcinoma cells to undergo a phenotypic transition so that they take on features of migratory mesenchymal cells. Secondly, CAFs secrete tenascin into the stromal ECM: this is enhanced by TGF-β. As a consequence, the small GTPase Rac is activated within the tumor cells, while Rho is inhibited. These enzymes regulate the cytoskeleton so that the cells become more motile. Together, this epithelial–stromal–epithelial activation loop causes stationary carcinoma cells to become migratory (De Wever et al., 2004, 2008). CAFs also secrete SDF-1 (CXCL12), which acts on both endothelial cells and cancer cells through the CXCR4 receptor to promote tumor growth (Orimo et al., 2005).

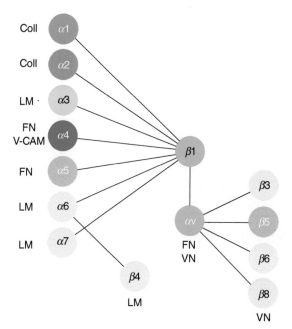

Figure 12.7 Integrin heterodimers. There are at least 25 distinct integrin heterodimer pairs in total, made up from 18 α and 8 β subunits. Each is specific for a unique set of ligands. This diagram shows only a subset of the various ECM proteins bound by integrins. Some integrin heterodimers are promiscuous and bind several ligands, while several ECM proteins interact with cells through different integrins, depending on the cell type and differentiation state. Coll: collagen I, III, or IV; LM: laminins; FN: fibronectin; VN: vitronectin.

ECM with intracellular proteins of the cytoskeleton. Because the interaction with ECM proteins occurs with low affinity, numerous ligand–receptor bonds are required to form stable adhesion sites for cells. The result is that macromolecular assemblies, visible in the light microscope, build up at the plasma membrane. This is rather like molecular "Velcro," where greater numbers of the same-sized bonds between two surfaces greatly increase the requirement for tensive forces to separate them. Large number of adaptor proteins, cytoskeletal elements, and enzymes are recruited to these sites, which are called adhesion complexes, or adhesomes (Fig. 12.8). Adhesomes are membrane-associated multifunctional machines, which are simultaneously sites of interaction between the cell and ECM, anchor points for the cytoskeleton, platforms for intracellular signaling, and the engines of cell migration.

Integrin subunit composition frequently changes during the progression to malignancy, providing carcinoma cells with the machinery to survive and migrate in the interstitial ECM, which they do not normally come into contact with. It is well known that signaling enzymes are upregulated in many cancers, but this actually extends to adhesion-activated enzymes such as FAK and ILK.

Integrin-containing adhesions exist in all cells, but some epithelia also contain a specialist type of assembly known as the hemi-desmosome. These are large structures that form strong bonds between epithelial cells and the underlying interstitial ECM through a chain of molecular interactions. Outside the cell, hemi-desmosomes link to the basement membrane protein laminin-5, via the signaling integrin, α6β4, together with a transmembrane collagen called BP180, or collagen type XVII. Inside the cell, rather than linking to the actin-based cytoskeleton, hemi-desmosomes contain adaptor proteins that bind to intermediate filaments. At their other ends, intermediate filaments bind to Nesprins, which directly link to Sun proteins and the structural laminins of nuclei. So hemi-desmosomes provide the ECM contact points that physically link the ECM with internal architectural proteins and nuclei. Hemi-desmosomal integrins are also altered in cancer, and for example are associated with advanced breast cancer.

Integrins integrate signaling networks

One important function of integrins is to control the activity of intracellular signaling pathways, including members of the Ras family discussed in Chapters 5 and 6, and other small GTPases such as Rho and Rac. These pathways in turn regulate proliferation, motility, polarity, differentiation, and survival. Integrins can trigger those pathways to some degree independently of GFs, while in the absence of adhesion GFs do not activate them efficiently (sometimes they fail to activate signaling at all). This means that in normal, nontumor cells, integrins are context-dependent checkpoints for signaling: cells monitor their local extracellular environment and only permit GF and cytokine responses if they are in the right place.

Perturbations in the GF arm of this signaling network lead to oncogenesis. Mutations or epigenetic changes in the level or activity of integrin-regulated proteins also upset the homeostatic signaling balance, and altered adhesion thereby has a profound effect on cell growth. Integrins are essential for GF responses, but their signaling pathways can be hijacked deleteriously in cancer (Fig. 12.9). Integrin mutations are actually even more damaging because they additionally affect migration, and consequently metastasis.

Transmembrane proteoglycan receptors for ECM

Some ECM proteins bind to other classes of transmembrane receptors that generate intracellular signals and are completely distinct from integrins. The important cell-regulatory ECM protein, laminin, binds to both a proteoglycan known as dystroglycan, as well as a lectin-like receptor β1,4-galactosyltransferase. Both of these receptors link laminin to the actin-based cytoskeleton as well as specific signaling enzymes. Dystroglycan has classically been studied in muscle, but has subsequently been found in epithelia. Both dystroglycan and galactosyltransferase have critical roles in regulating epithelial morphogenesis during development, and they are also involved with cellular polarity and differentiation. Dystroglycan is frequently lost in carcinoma progression, indicating a possible tumor suppressor role. Syndecan and glypican are additional proteoglycan adhesion receptors that have essential roles in epithelial cell function and neoplasia.

The elasticity of the cell niche

Different types of tissues have differing degrees of elasticity, which profoundly affects the behavior of cells within them. There are soft tissues such as brain and adipose, and stiff tissues such as cartilage. Most epithelial organs are in between. The degree of softness, or compliance, is largely determined by the ECM composition, and is measured in Pascal units. ECM compliance is a key contributor to developmental processes because it determines lineage selection of stem cells and therefore tissue identity. For example, identical mesenchymal stem cells embedded into

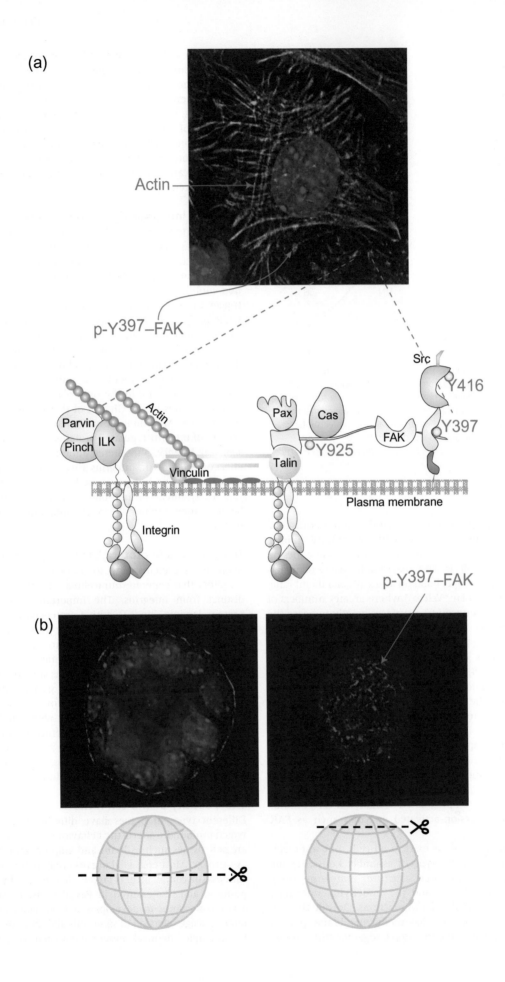

Figure 12.8 Adhesion complexes are multiprotein assemblies at the cell–ECM interface. (a) The schematic diagram of the adhesion complex includes transmembrane integrins, which link cells to the matrix on the outside and organize the cytoskeleton and cell shape on the inside. Integrin ligation activates proximal cytoskeletal adaptor proteins, such as talin and vinculin; signaling enzymes, for example FAK, Src, and integrin-linked kinase (ILK); and signaling adaptors such as Pinch and parvin, paxillin, and 130Cas that couple to further downstream pathways. These include the Ras–Erk and PI3K–PKB axes, PKC and PLCγ, and small GTPases including Rho, Rac, and Cdc42. A large number of other proteins (not shown) are also within adhesion complexes. See Humphries et al. (2009) for proteomic analysis of adhesion complex components. Note that adhesion-regulated signaling enzymes have key roles in controlling many aspects of cell behavior, including migration, polarity, survival, proliferation, and differentiation. The fluorescence image is of cells cultured on a tissue culture dish. It shows the integrin adhesion complexes where the cell interacts with ECM (red), the actin cytoskeleton (green), and nuclei (blue). In this case, the cell has been stained with an antibody specific for the phosphorylated, active form of focal adhesion kinase (otherwise known as pp125^{FAK} or FAK). Note that the skeletal components of the cell begin at the adhesion complexes (i.e. their "feet"). (b) Images of cells in 3D culture (Debnath and Brugge, 2005), also stained for p-FAK. Note the staining is located to the periphery of the "acinar" structures that form in 3D culture. In the fluorescence section, taken at the "top" of the ball of cells, are visible adhesion complexes similar to those that form in 2D culture.

(a)

Normal mammary ductal tree

(b)

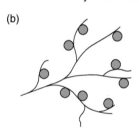

Tumors form if MT is expressed

(c)

Integrin ablation prevents tumor formation

Figure 12.9 Integrins are essential for tumor formation. (a) The normal mammary ductal tree consists of epithelial cells that form tubular arrays. In nonpregnant animals, these ductal trees remain stable. (b) If the polyoma middle-T antigen (MT) is activated *in vivo*, multifocal tumors form within a few weeks. This is because the middle-T antigen activates numerous signal transduction pathways that are normally associated with GF receptor signaling, such as those involving c-Src, phosphatidylinositol 3′-kinase (PI3K), and the Ras–Erk pathway. (c) If β1 integrin is deleted through conditional ablation of the gene, tumors do not form. This is most likely because the middle-T-driven signaling pathways need active integrins in order to function properly. Interestingly, the role of integrins varies among different cancer models. Thus, β1 integrin is dispensable for the growth of primary tumors in an ErbB2 animal model for breast cancer, but is required in order for those primary transformed cells to develop malignancies (Huck et al., 2010; White et al., 2004).

soft ECMs are specified to become adipocytes, while those in stiff ECMs take on chondrocyte cell fates.

The mechanisms for detecting ECM elasticity lie within adhesion systems (i.e. cell–cell and cell–ECM junctions) that sense forces within the cellular microenvironment, and convert them into intracellular chemical signaling pathways and reorganized cytoskeleton and actomyosin contractile machinery. For example, integrin adhesion complexes respond to sheer forces by triggering signaling enzymes (e.g. FAK), which in turn activate Rho GTPases and a downstream signaling pathway involving Rho kinase, myosin phosphatase, eventually leading to myosin contraction and the formation of f-actin bundles that create high levels of intracellular tension. One mechanism of force detection at the molecular level is through conformation changes of adhesion proteins. For example, the ECM protein fibronectin, integrins, and the intracellular adhesion complex proteins talin, vinculin, and p130Cas all expose cryptic binding sites for other proteins when they are stretched. It is not well understood how such tension determines cell fate. One possibility is that signaling through transcription factors, such as YAP, might be involved. Another is that the stretched cytoskeleton mechanically alters the shape of the nucleus, which might directly affect gene expression through epigenetic changes in chromatin architecture.

Extracellular forces applied to cancer cells have a significant effect on their behavior, and contribute to the progression of the disease. Breast cancer cells within stiff ECMs have more integrin clustering, activated adhesion complex proteins, and Rho-mediated cytoskeletal contraction which is reflected in greater potential for migration. This is now recognized to be an important factor in cancer progression. A high mammographic breast density, which equates to more fibrillar collagen, is one of the greatest risk factors for developing cancer. In animal models, collagen-dense stroma promotes the initial formation of mammary tumors, as well as their progression to form metastases. Collagen crosslinking increases the stiffness of the stroma and enhances tumor formation. Moreover, higher actomyosin contractility causes tissues to become stiffer and promotes tumors (Fig. 12.10).

Together, this discussion so far highlights the central role of the cell microenvironment in cancer formation and progression. Carcinomas are characterized by mutations within the epithelial component of the tissue; however, they are much more than just diseases of mutated "tumor" cells. In order to form tumors, changes in the cell niche occur which are required for cancers to progress. Thus, it is not appropriate to view cancer as a disease simply of "cancer cells"; rather, it is a dysfunction of a whole organ in which stromal cells and the ECM play profound and essential parts.

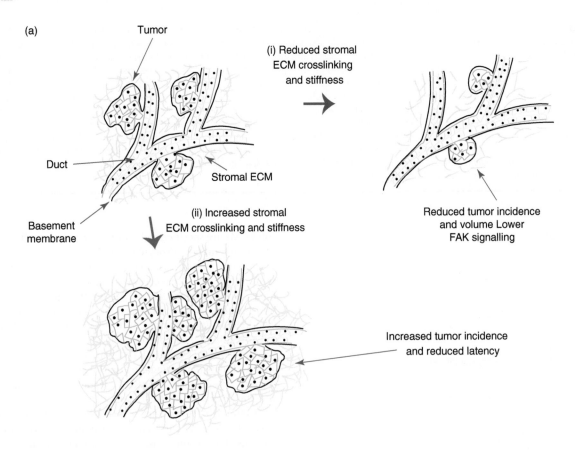
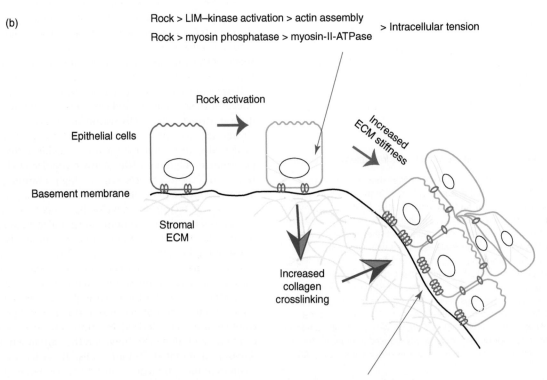

Figure 12.10 Tissue stiffness promotes cancer. (a) Two widely used animal models for breast cancer are the polyoma MT and the Neu–ErbB2 transgenic mice. In both cases, carcinomas arise sporadically from ductal epithelium. The ducts are surrounded by basement membrane and subtended by stromal ECM composed of collagen I and other ECM proteins, cells, and so on (see Fig. 12.1). Atomic force microscopy reveals that the stromal ECM is stiffer in the close vicinity of tumors than farther away, and there is increased collagen deposition and crosslinking, and higher levels of lysyl oxidase (Lox), a collagen crosslinking enzyme. Reducing crosslinking by using Lox inhibitors lowers the complexity of collagen crosslinking and tumor incidence. Conversely, increasing collagen crosslinking, either by using transgenic mice expressing collagenase-resistant collagen I, or by injecting glands with fibroblasts expressing excess Lox, promotes breast cancers. (b) Altering intracellular contractility also promotes cancer, and this occurs via changes in the stromal ECM. In a skin cancer model, genetically activating Rock alters the cellular microenvironment by increasing collagen crosslinking. The resulting extracellular forces impact back on the keratinocytes to increase proliferation in normal skin, and to enhance tumor formation in squamous skin cancer. The ECM forces most likely cause increased cell cycle by inducing integrin clustering and FAK activation, which promotes β-catenin nuclear localization and transcription of its targets. It is notable that the key players in both these models, Lox, collagen crosslinking, high Rock levels, and FAK and β-catenin activation, are all features of human carcinomas (Leventhal et al., 2009; Provenzano et al., 2008; Samuel et al., 2011).

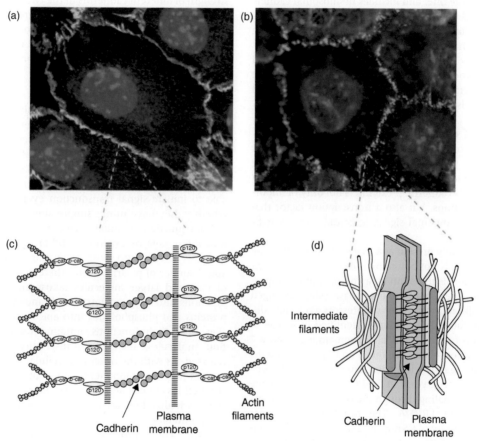

Figure 12.11 Adhesion complexes at the cell–cell interface. (a–b) These fluorescence images of epithelial cell sheets show the localization of adhesion molecules distributed to cell–cell contact sites. (a) Cells stained for E-cadherin (green), which is visible in a continuous ring at the lateral junction between cells. (b) Cells stained for plakoglobin (γ-catenin), which localizes only to desmosomes and therefore has a punctate appearance. (c–d) Schematic diagrams of (c) an adherens junction and (d) a desmosome. Adherens junctions connect to, and organize, the actin-based cytoskeleton via the adaptor proteins α- and β-catenin. p120 is an important regulatory molecule, while β-catenin has a dual function as it is also a transcription factor. Desmosomes are strong spot welds between cells, holding them together. They link to the intermediate filament network and thereby couple the scaffolding between hemi-desmosomes and adjacent cells.

Cell–cell interactions

Cell–cell interactions are critical for many of the cell processes elaborated in this book, including cell cycle and apoptosis. Cells interact and communicate directly with each other through a wide variety of mechanisms. Two different classes of cell–cell adhesive systems are involved. Multiprotein assemblies form the physical junctions between cells and provide contact points with the cytoskeleton. They are also sites of considerable signaling activity. These include adherens and tight junctions and desmosomes. Simple receptor–ligand pairs also have essential controlling roles at adhesive sites, and are involved with activating a variety of intracellular pathways with signaling nodes that include B-catenin, Erks, Smads, Nicd, Smoothened, and others.

Cadherins

Cell–cell adhesive junctions contribute to physically joining cells (Fig. 12.11). Two of these, adherens junctions (zonula adherens) and desmosomes, contain the most prominent cell–cell adhesion molecules, cadherins. Cadherins are transmembrane proteins that interact with each other both in *trans* and in *cis*. They bind to each other at their distal amino-terminal ends across the intercellular space, thereby connecting adjacent cells. They also form lateral interactions between neighboring molecules on the same cell. In this way, they form large multiprotein complexes that have sufficient numbers of adhesive interactions to hold cells together. These interactions are dependent on calcium cations. Although there are many cadherin species, their binding is usually homotypic.

The prototype, epithelial (E)-cadherin, is the founder member of a small collection of "classical" cadherins, present in most epithelia in junctions that are visible in the electron microscope. The family is a very large one, and includes desmosomal cadherins and protocadherins that mediate, respectively, strong connections between epithelial cells and interneuron junctions in the central nervous system (CNS).

Cadherins play a key role in both developmental processes and tissue homeostasis, and their altered expression in malignant cancer plays an important part in the dissemination of cells and the resulting metastasis.

The β-catenin–WNT connection

The cytoplasmic face of an adherens junction contains adaptor proteins that connect to actin filaments, and also to signaling molecules. One of these is β-catenin, a transcription factor that can be sequestered by E-cadherin. β-catenin is not only of importance at adherens junctions, but also a transcription factor that is regulated by a developmental signaling cascade, the Wnt–β-catenin pathway.

β-catenin is not normally present as a free molecule within the cytosol because any that is not bound by E-cadherin at adherens junctions is targeted for degradation by a complex of three proteins which include a serine–threonine kinase called glycogen synthase-3β (GSK-3β) and the tumor suppressors axin and adenomatous polyposis coli (APC, originally identified in a type of hereditary colon cancer called "familial adenomatous polyposis," in which its activity is lost – see Chapters 3 and 5). Wnt proteins are secreted signaling ligands whose distribution is spatially controlled by binding to proteoglycans in the ECM (Chapter 5). They stabilize β-catenin by signaling through the cell surface receptors comprising Frizzled and LRP5/6. The resulting intracellular pathway inhibits the activity of GSK3β, and so prevents β-catenin degradation. This leads to its nuclear translocation and interaction with a family of transcriptional repressors called Tcf–Lef. This interaction transiently switches Tcf–Lef from a transcriptional repressor to an activator, and thereby induces the transcription of numerous target genes including some involved in cell cycle regulation, apoptosis, and migration, such as c-Myc (Chapter 6).

Thus, there is a delicate relationship between extracellular signals that trigger the Wnt pathway and inter-epithelial interactions mediated by cadherins. Disruption of this equilibrium has a profound effect on tissue homeostasis, and leads to cancer progression.

Desmosomes

Cadherins also form the core of desmosomes, which provide structural integrity to cell–cell interactions in epithelia and cardiac muscle. Desmosomes are spot welds between adjacent cells, while adherens junctions are more diffuse in nature. As with hemi-desmosomes, the cytoplasmic faces of desmosomes contain adaptor proteins such as plakoglobin and desmoplakin that connect them to the intermediate filament system. Normal epithelia are networked together and to the basement membrane through contiguous intermediate filament-to-desmosome (or hemi-desmosome) links. Desmosomal cadherins are called desmocollins and desmogleins, and their composition differs between epithelial cell types. In complex epithelia, where more than one layer of cells lies on top of the basement membrane, the differential adhesiveness of specific desmosomal cadherins determines the cell's spatial positioning. For example, in epithelial tissues that have more than one layer of epithelial cells (e.g. the mammary gland), the relative positioning of the basal layer of cells to the ones above is partly controlled by the adhesion strength of different cadherins in the two cell layers. Desmosomes can sometimes be lost during the progression of epithelial cells to malignant carcinoma, and the introduction of desmosomal cadherins can suppress invasion. These cell–cell adhesion structures therefore have a tumor suppressor function.

Other adhesive junctions

In addition to cadherins, cells are physically cemented together by two other types of multiprotein assembly, gap junctions and tight junctions. Gap junctions form channels between cells, allowing the passage of ions and small molecules, and are a further type of junction that networks epithelial cells. These structures contribute to cellular differentiation and they may be able to initiate signal transduction events, but it is not certain whether they have tumor-suppressing or -promoting roles.

Tight junctions (zonula occludens) are present at the apical junction of polarized epithelia and bring the plasma membranes of adjacent cells into close apposition. They fully encircle the apical surface of polarized cells, thereby preventing the diffusion of ions and larger molecules across an epithelial layer. Tight junctions also form an intramembrane barrier, separating both proteins and phospholipids into apical and basal compartments (Fig. 12.2). Tight junctions contain a number of signal transduction molecules, including transcription factors that can be released. As with the other intercellular junctions, misregulation of tight junction components contributes to carcinogenesis, both through loss of cell adhesiveness and polarity, and also through the unscheduled release of transcription factors, which influence cell proliferation.

A final contributor to cell–cell adhesion is the cell adhesion molecules (CAMs). This is a large family of transmembrane glycoproteins that contain immunoglobulin repeats within their external domains, and are calcium independent for adhesion. Several types of CAM are important for cell signaling because they bind to GF receptors in the plane of the plasma membrane, and are therefore involved in cell signaling. The expression of many CAMs, including melanoma (M)-CAM and neural (N)-CAM, is altered in malignancy. In some cases, they are required for the formation of metastases.

Cell–cell signaling junctions

A separate set of intercellular junctions includes those that deliver direct signals through membrane-bound ligand–receptor pairs. Although these junctions are not adhesive, they can be activated when adhesive interactions between adjacent cells bring the ligand and receptors together. The Notch–Delta (see Chapter 5), Eph–ephrin, and TGF-α–ErbB cell–cell signaling systems are essential for patterning epithelia during development, and maintain tissue homeostasis in mature organs. The components of these systems can be mutated or disrupted in neoplasia. Moreover, their normal function is compromised when cell–cell contacts become altered in cancer. Together this can lead to abnormal signaling responses, which have knock-on effects contributing to the altered spatial disorganization and growth control of cancer cells.

Notch, Delta, and Stem cells

Notch receptors are used widely in development and regulate cell fate decisions (see also Chapter 5). They are also involved, together with many other factors, in maintaining stem cells within their appropriate niches, and are therefore very closely regulated in normal tissues. These receptors, and their various ligands including Delta and Jagged, are all present on most epithelia, and their discrete expression patterns are perturbed in cancer, leading to increased signaling via the Notch intracellular domain, NICD.

Since stem cells are the progenitors for many types of cancer, disrupting the Notch–Delta system has profound effects in the initial stages of tumorigenesis. This leads both to reduced sensitivity to apoptosis and to inappropriate expression of target genes, including the Hes and Hey classes of transcription factors. Altered Notch signaling can also lead to reduced sensitivity to radiotherapy, making tumors more resistant to treatments. This observation might end up being an advantage: One idea that is currently being tested for therapy is to inhibit Notch signaling at the same time as using conventional treatments. Maybe a combined approach would effectively eliminate some types of cancer.

Eph and ephrins

The Eph receptor family of receptor tyrosine kinases, the largest family of receptor tyrosine kinases, signals through Erks. This is an unusual type of system as both receptor and ligand act as signaling devices, and therefore responses can occur bidirectionally. Eph receptors provide repulsive stimuli to cells, rather than attractive ones, thereby preventing the mixing of cells between one environment and another. This leads to the formation of boundaries so that blocks of cell types become, and are maintained, separate from one another.

Ephs and eprhrins are critical in the progression of certain epithelial cancers, because if they become disrupted they alter the positional identity of cells, which leads to altered fate decisions and inappropriate proliferation. For example, in colonic epithelium, multipotent stem cells normally inhabit the lower part of the crypts of Leiberkuhn, and after they are specified to become absorptive epithelial cells, they migrate slowly along the crypt–villus axis before being sloughed into the lumen of the gut when they reach the villus tip. Localized expression of Wnt causes cells at the bottom of the crypt to express Eph. As the cells migrate into the villus, they go out of range of the Wnt signal and now express an ephrin, which defines the crypt–villus boundary. However, in cancer, epithelial cells can continuously express the crypt marker and therefore cannot progress into the villus because of Eph–ephrin repulsion; they continue to proliferate in the crypt, leading to hyperplasia.

Ephs and ephrins also regulate cell movement by affecting "contact inhibition." When normal cells move toward each other, they cease migrating once they "contact." However, cancer cells ignore other cells and keep going. This is one of the features of poor cell behavior in cancer, and it arises by altered expression of Ephs in cancer cells.

Critical steps in the dissemination of metastases

Benign lesions result from cellular disorganization

The earliest stages of carcinoma are the formation of hyperplasias, small lesions of disorganized cells within well-arranged epithelial tissues. These grow to form various adenomas and benign in situ tumors. However, benign tumors do not metastasize (Box 12.2). They form during the early stages of neoplasia, are not unusual, and often remain undetected. In some cases, benign tumors acquire additional characteristics that lead to tissue invasion. A key goal of modern cancer medicine is to detect benign tumors so that they can be removed prior to the formation of invasive, malignant tumors.

The two key characteristics of a benign tumor are that the mechanisms controlling normal cellular organization and cell number are lost. One of the earliest features of a carcinoma is that the spatial positioning of epithelial cells relative to each other is disrupted. This can occur through altered apical–basal polarity in epithelial cells, via changes in the expression of polarity genes, and lead to cells losing their ability to remain within the epithelial cell layer. The mechanisms controlling the mitotic spindle position can go awry, resulting in cell divisions occurring parallel to the plane of the epithelium rather than perpendicular to it. Cancer cells are also extruded from the rest of the epithelial monolayer, and they form small tissue masses (Fig. 12.12). When this tissue disorganization occurs in conjunction with the activation of oncogenic pathways and the inactivation of tumor suppressor genes, larger benign tumors can form. For benign tumors to grow efficiently, they need oxygen and nutrients, which are supplied through new capillaries; thus successful tumors also promote angiogenesis, often in response to hypoxia (Chapter 14).

Box 12.2 Four characteristics of cancer cells are of key importance to enable their transition to become invasive:

1. Malignant cells alter their repertoire of integrins, thereby enabling their migration through a stromal ECM and survival there.

2. The composition of the stromal ECM itself changes, producing matrix proteins that promote migration; tenascin is one example of a pro-migratory protein that is often upregulated in the stroma adjacent to neoplasias.

3. The cancer cells activate their intracellular signaling machinery that controls cytoskeleton and motile processes.

4. They also express or activate several classes of ECM-degrading proteinases and glycosidases that both remodel the ECM and release ECM-bound GFs; the latter has additional knock-on effects on the epithelial cells.

Some of these changes occur as a response to EMT or because the cell encounters a hypoxic, wound-repair environment. Others are achieved through acquisition of mutations.

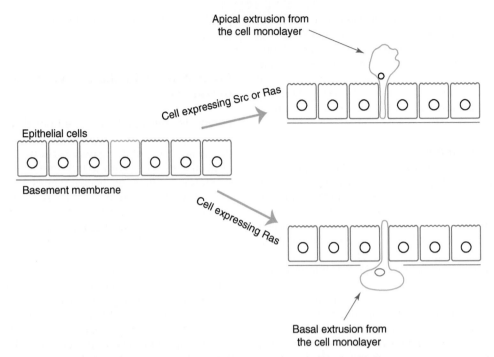

Figure 12.12 Normal epithelial cells extrude cancer cells from their epithelial sheets. One mechanism contributing to the early disorganization of benign tumors is the extrusion of cells expressing activated oncogenes. Cancer arises in single cells in which genetic changes occur. This has been modeled in polarized normal epithelial MDCK cells in which oncogenes are activated within individual cells. Normal cells monitor their neighbors and they cause any cells expressing an activated Src protein to become extruded from the epithelial monolayer on the apical side. This happens only to cancer cells surrounded by a sea of normal cells, because when all the cells are expressing Src, they don't extrude. This is probably a protective mechanism in which normal cells displace those expressing mutant genes. However, if the oncogene-expressing cells are also able to survive in the apical space, they could then form disorganized benign tumors. Different oncogenes have different effects because cells expressing an activated Ras are extruded both apically and basally (Hogan et al., 2009; Kajita et al., 2010).

Many epithelial tissues are normally able to undergo considerable amounts of proliferation, but the tissues remain organized. For example, skin wound healing involves extensive proliferation and tissue remodeling in a controlled manner. Colonic epithelial cells continuously proliferate in order to replenish cells that have detached from the villus tips. Perhaps the most spectacular example is the mammary gland, where pregnancy is accompanied by massive proliferation of the alveolar epithelial cells to fill the tissue with milk factories in preparation for lactation. All of these cells divide in a controlled way and remain perfectly organized as epithelial bilayers. This argues that the single most important defect in the early stages of cancer formation is the loss of normal tissue organization.

The hallmarks of malignancy

None of these features, however, is sufficient to cause cells within primary tumors to become malignant and to form secondary lesions in distant parts of the body. For this, further properties must be acquired. In particular, the spatial control mechanisms that keep cells within their proper "place" become disrupted and cancer cells acquire invasive and migratory properties. As discussed in this book, some of these properties may be conferred by environmental factors, not always resulting from mutations.

The hallmarks of malignant cells are that they can migrate away from the primary tumor, enter the circulation, extravasate at a secondary site, and grow there to form a metastatic lesion. Thus, the loss of both growth and spatial control is required for cancer. Even in *Drosophila*, the ability of oncogenic Ras^{V12} mutations to cause "metastases" depends on the additional alteration of genes involved in cell–cell and cell–ECM interactions, and invasion.

Fortunately, at the cellular level the metastatic process is extremely rare. In experimental models, fewer than 1 in 10 000 injected tumor cells are able to form metastases. The reason for this is that many changes within tumor cells need to occur for them to get into the circulation and survive in secondary sites. Indeed, the whole environmental awareness of a cancer cell is reprogrammed during the formation of metastases, and is a consequence of acquired genomic instability that permits the sequential accumulation of mutations to form successful seed cells.

Invasion and dissemination through capillaries

Malignancy is accompanied by altered epithelial intercellular adhesion and changes in cell–matrix interactions. Together with the inappropriate activation of matrix-degrading enzymes, this leads to the acquisition of invasive properties (Fig. 12.13).

In some cases, migratory cancer cells first break away from the primary tumor and enter the stroma. They continue to proliferate in that location as invasive lesions, before encountering suitable blood vessels or lymph nodes from which to escape into the circulation. In other cases, cells that have lost adhesivity within a primary tumor enter the bloodstream directly. The capillaries that are induced to grow into the tumor by angiogenic factors are poorly formed and have fenestrations, which provide an escape route. These cells, however, still require invasive proper-

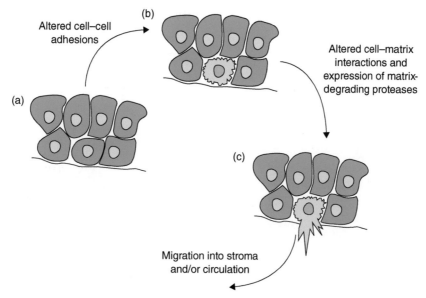

Figure 12.13 Steps in the transition between a benign and malignant tumor. (a) Cells become defective regarding DNA repair and apoptosis, and acquire oncogenic properties. Their proliferation is dysregulated, and normal tissue architecture is partially disrupted as a consequence. (b) Some cells may lose cell–cell adhesivity and have ragged borders. (c) Additional migratory properties lead to an exit of tumor cells from the primary neoplastic lesion. Disruption of the basement membrane is an important marker for the transition from a benign tumor to malignancy. These steps are essential to initiate tumor spread.

ties in order to cross the endothelial cell lining of vessels and its basement membrane at the secondary metastatic sites.

Providing that tumor cells are strong enough to withstand the considerable hydrostatic pressures of the blood system, transit through the vasculature allows them to settle in secondary sites (Fig. 12.14). Several factors determine the site at which metastatic cells are sown. Cancer cells are frequently much larger than the small capillaries (4–8 μm diameter) that form the vascular bed of tissues, and therefore do not travel far before they become trapped. This is particularly evident in cases where tumor cells travel as clumps of cells. Thus the vascular flow pattern is a primary factor in determining where metastases occur. For example, the first vascular bed that colon tumor cells encounter is that of the liver, and many people die of colon cancer through chronic liver metastases and the resulting lack of clotting factors, failure of glucose homeostasis, and blood toxicity.

Dissemination through the lymphatics

Some cancer cells that have exited the primary tumor follow the direction of the normal interstitial fluid flow to enter the lymphatic capillaries, and thereby drain into regional lymph nodes. One mechanism facilitating this is the expression of lymphangiogenic factors such as VEGF-C. The most prominent site for metastases is the primary lymph node that the tumor cells encounter. Here the cells proliferate extensively, and also can be carried slowly through the efferent lymphatic vessels toward the thoracic duct, where the lymph is emptied into the blood. The microenvironment of a lymph node is very different from that of the stroma or primary tumor, so cells must accumulate additional changes in order to survive. In addition, they must become resistant to the large population of cytotoxic CD8 T cells (Chapter 13). The lymph node is therefore a training ground for aggressiveness. Pathologists frequently determine the degree of tumor aggressiveness by examining sentinel lymph nodes (Fig. 12.15). Tumor cells travel through the lymph vessels to the heart, and then through the arteries to other organs of the body, which they colonize.

Seed and soil

Many metastases form at places where malignant cells become entrapped, but this alone does not explain why it is that certain organs are preferred as secondary sites to others. One possible explanation for this is that tumor cells (seed) are able to grow only if they are in a conducive environment (soil), which includes the presence of appropriate proliferation and survival factors, and adhesive interactions. Tumor cells therefore need to be biochemically compatible with a foreign environment in order to grow there. Such environments are tumor specific and multifactorial, and include the expression of diverse homing and invasion molecules.

For example, chemokine receptor transcription can be induced when the tumor environment becomes hypoxic. Chemokines are attractant cytokines that normally recruit lymphocytes to sites of infection. They encourage actin polymerization and the extension of lamellipodia, so that cells migrate along a concentration gradient that increases toward the chemokine source. Often, malignant breast cancer cells inappropriately express specific chemokine receptors (e.g. CXCR4 and CCR7), which can be activated only by the appropriate chemokines (CXCL12 and CCL21, respectively). These ligands are present in lymph nodes, thereby enticing the migration of breast cancer cells. Moreover, the CXCL12 ligand is also present at high levels in lung, liver, and bone marrow, which are all organs where breast cancers metastasize to, but not in other tissues such as skin and prostate where breast cancers do not form secondary lesions.

So, by subverting a homing mechanism normally used by other cell types, tumor cells of one tissue type can invade and colonize the stromal bed of another. This type of environmental reprogramming forms the molecular basis of the "seed and soil" hypothesis of metastasis, formulated in 1889.

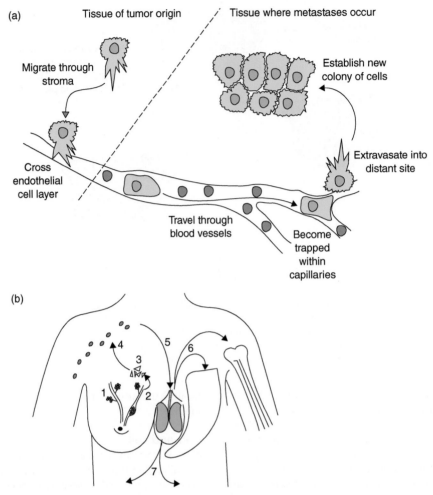

Figure 12.14 Mechanisms of tumor spread. (a) Passage of malignant cells from the primary tissue where a tumor forms to a secondary site, via the circulation. (b) Steps in the migration of malignant breast cancer to the organs they colonize and form metastases: (1) normal breast lobule, (2) formation of benign tumor (i.e. ductal carcinoma in situ), (4) early steps in malignancy (i.e. invasion into stroma), (4) high-grade malignant tumor (i.e. transfer via lymph vessels to lymph nodes), (5) passage of aggressive tumor cells to blood system, (6) proximal sites of distant spread to lung and bone, and (7) metastases to other organs such as liver. The appearance of breast cancer cells in the lymph nodes is a hallmark of malignancy. If the lymph nodes of breast cancer patients have enough tumor cells to be seen histologically by a pathologist, it is most likely that those individuals will have already acquired metastases, and their chances of still being alive within 5 years are not high.

Micrometastasis

Although the overt migration of mature tumor cells is a key mechanism in the formation of metastases, small numbers of cells may sometimes migrate to secondary sites early in the development of a cancer, and remain there in a dormant state for a long time (Fig. 12.16). These small lesions are micrometastases, which do not acquire all the hallmarks of aggressive tumor cells, often until years later.

Microarray studies have identified specific sets of genes that become overexpressed in specific cancers and which are all required for successful metastases (See also Chapter 20). One such metastatic gene–expression signature for breast cells that colonize bone includes not only CXCR4, but also metalloproteinases (MMPs), which facilitate invasion into the tissue, and osteopontin, an adhesive ECM protein. Importantly, this pattern of gene expression arises within a few cells of the primary breast tumor itself, arguing that the genetic changes necessary to progress to a malignant tumor preexist within the benign tumor. Moreover, different subfractions of the same primary tumors have expression signatures that promote metastases to other tissues such as the adrenal medulla. Thus, subpopulations of cells within a primary tumor have heterogeneous metastatic potential.

An implication of early acquisition of "metastases-enabling" mutations or inactivation of putative "metastasis suppressor" genes is that the seed cells of metastases may spread early in the progression of malignant disease, sometimes prior to the detection of the primary tumor itself. Further mutations and selection may occur within such "micrometastases," leading to the evolution of a growth-competent metastatic lesion within the local environment of the distant site itself. This can occur many years after the deposition of the original micrometastasis.

Metastasis suppressor genes

Understanding the molecular determinants that govern cancer invasiveness and metastasis will underpin the development of new therapeutic strategies aimed at diagnosing or treating metastatic cancers. It has long been appreciated that mutations in

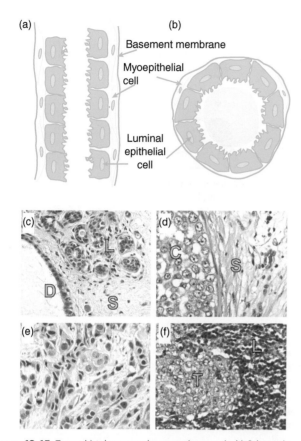

Figure 12.15 Tumor histology reveals aggressiveness. (a–b) Schematic diagram of a duct (a tube) and a lobule (a sphere) in mammary gland. Both are bilayered epithelia. The luminal epithelial cells (green) are subtended by myoepithelial cells (orange) and a basement membrane (red). (c–f) Some examples of the histology of breast tumors as they progress from benign to invasive carcinoma. (c) A normal breast has three main compartments: lobules, ducts, and stroma. Here, lobules of epithelia (L) with central lumina are shown, together with part of a large duct (D). Note the discrete organization of lobular epithelial cells, and that the duct is lined with a simple later of epithelium. Stroma (S) surrounding the epithelia. (d) Ductal carcinoma in situ. The simple bilayered epithelium of the duct has been replaced by large numbers of proliferating carcinoma cells (C). Note that the tumor has a discrete boundary with the stroma (S). (e) Advanced lobular carcinoma. This tumor arose from lobules. A large area of tissue is now taken up by proliferating cancer cells within the stroma (not visible). (f) Secondary tumor (T) metastasized to lymph node. The expansion of the tumor appears to be compressing the lymphocytes (Ly).

oncogenes and tumor suppressors are a prerequisite for tumorigenesis, but in addition distinct mutations are required, at least in some cases, for invasion and metastases. The metastasis suppressor genes are defined as those that can suppress metastasis without affecting tumorigenicity. There is growing evidence that loss of function of metastasis suppressor genes plays an important role in cancer metastasis. For example, the α2β1 integrin suppresses metastasis in breast cancer animal models, and its loss predicts metastasis and decreased survival of breast cancer patients. Interestingly, many metastasis suppressor genes have common roles in growth control, adhesion, and cytoskeletal reorganization, suggesting a common mechanism of metastasis suppression. Proposed candidate pathways include signaling through Src kinase and Rac GTPase.

E-cadherin downregulation in cancer leads to migration

One of the defining stages in the transition from a benign to a malignant tumor is modified cell–cell adhesion. This is primarily a consequence of reduced adhesiveness at adherens junctions, though in some cases it may be due to altered desmosomal function. The main culprit in this aspect of carcinoma progression is loss of cadherin function. Once epithelial cells lose the strong interactions that normally hold them together, they can invade the stroma, provided their migration machinery is also activated. In some carcinomas, E-cadherin loss is a prerequisite for invasion and metastasis. There are two ways in which adhesion can be lost. The first is through the direct genetic or epigenetic control of the E-cadherin gene or its protein product. The second is through gross changes in the phenotype of epithelia, which can be triggered when they are induced to undergo epithelial-to-mesenchymal transitions (EMTs).

Mechanisms of E-cadherin loss

There is an inverse relationship between the loss of E-cadherin in malignant tumors and patient mortality. If E-cadherin is reintroduced into cell lines derived from such tumors, they lose their ability to invade ECM in culture and to form metastases in animal models. This demonstrates the importance of E-cadherin in cancer progression, and points to its critical role as a tumor suppressor gene (Fig. 12.17). Understanding the mechanisms of E-cadherin loss might lead to novel therapeutic strategies for restoring its expression and thereby suppressing invasion and metastasis.

There are several examples of progressive cancer correlating with mutations in the E-cadherin gene, and this may happen at an early stage of the disease. However, this occurs in only a few specific tumor types, such as gastric cancer. The majority of mechanisms for the loss of function of the adherens junction are epigenetic (i.e. occur in ways that do not affect the genomic sequence) and can be caused in several ways.

First, cadherins (and other adhesion molecules) can be degraded from the outside by matrix-degrading MMPs. These enzymes are frequently activated in malignancy (see later), resulting the truncation of E-cadherin and loss of its adhesive function.

Second, adherens junctions are acutely regulated by phosphorylation. Normally, components within the junctions such as p120–catenin are not phosphorylated. However, when tyrosine is phosphorylated by, for example, c-Src, a protein tyrosine kinase whose activity is often upregulated in cancer, they disassemble. This is because tyrosine phosphorylation can induce the ubiquitination of components in the complex via an E3 ubiquitin ligase called Hakai.

Third, the normal controls on *E-cadherin* gene transcription are frequently altered in cancer to cause its downregulation, even though the gene is not mutated. One mechanism is through hypermethylation, which occurs in a wide variety of carcinomas. Hypermethylation prevents access of the transcription machinery, resulting in transcriptional inactivation.

Fourth, the levels of transcription factors that control E-cadherin expression can be altered. The *E-cadherin* gene is under strong transcriptional control, but many of the factors that have been identified to control its expression are repressors, rather than activators. The first cells to form during mammalian embryonic

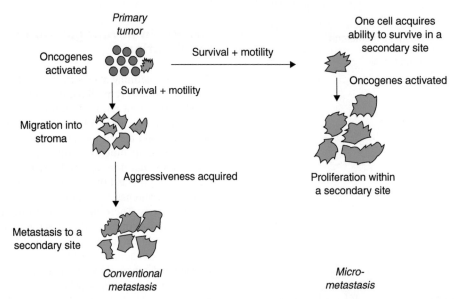

Figure 12.16 Tumor dormancy and micrometastases. Some cells break off from tumor prior to oncogene activation and settle in foreign sites. Although these cells might have altered adhesion mechanisms and are able to survive in an inappropriate microenvironment, they do not necessarily have a deregulated cell cycle. Thus, they have not acquired all the hallmarks of a malignant phenotype, and may remain dormant for many years. Subsequently, these cells evolve at secondary sites into metastatic lesions. An important implication of this is that an effective way to treat cancer may be to target altered proteins that occur within the metastatic lesion, rather than those that drive proliferation in the primary tumor.

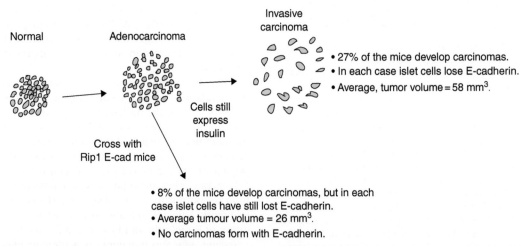

Figure 12.17 The loss of E-cadherin is rate limiting for tumor progression. In an experimental model of cancer progression, tumors form in the insulin-secreting islets of Langerhans in the endocrine pancreas. Transgenic mice that express the transforming SV40 T and t antigens under the control of the insulin promoter (Rip1Tag2 mice) develop cancer in a characteristic fashion that progresses through adenoma to carcinoma. The T antigens are expressed continuously from embryonic day 8.5, and by 7 weeks after birth, adenomas form. A few weeks after this, the tumors become angiogenic, and by the 11th week they become invasive. Tumorigenesis occurs because T antigen both promotes proliferation, by inactivating the retinoblastoma protein, and suppresses the proapoptotic activity of p53. E-cadherin is expressed by the normal pancreatic epithelial cells and in adenomas, but is completely lost once the tumor has progressed to the carcinoma stage. However, if the mice are crossed with those expressing E-cadherin under the control of the same promoter (Rip1E-cad mice), the incidence of carcinoma formation is dramatically reduced. Some carcinomas do form, but these have all lost E-cadherin expression. The experiments provide direct evidence that E-cadherin loss is required for the transition from benign adenoma to malignant carcinoma (Perl et al., 1998).

development are epithelial, and they express cell adhesion molecules, including E-cadherin, to keep them together and compact them to form the morula. Development is subsequently dependent on the conversion of some of these epithelial cells to mesenchymal cells (i.e. an EMT), so that different types of cells can form (e.g. trophectoderm and mesoderm from endoderm at the blastula stage of development) and morphogenetic cell movements can occur (e.g. during gastrulation). This requires continued suppression of the *E-cadherin* gene, which is carried out by several zinc–finger transcription factors that bind to E-boxes within its promoter called Twist, Snail, Slug, and SIP1. These are not expressed in normal epithelial cells, but are frequently upregulated in cancer cell lines, and in advanced cancers *in vivo*. Moreover, over expression of SIP1 leads normal epithelial cells to invade collagen gels more easily, most likely by preventing their ability to stick to each other.

The 3D structures of the interaction site between Twist, Slug, Snail, and SIP1 and the E-cadherin E-Box are currently being worked out. This knowledge will enable a hunt for small-molecule inhibitors that prevent this interaction, and may therefore be used therapeutically to activate the re-expression of E-cadherin.

In a mouse model of pancreatic islet β-cell tumorigenesis, deregulated expression of the oncoprotein c-Myc (Chapter 6) triggers loss of E-cadherin expression, alongside an invasion of β cells locally and into blood vessels. How c-Myc does this is the subject of ongoing studies. It is interesting to contrast this with β-cell tumors derived in mice overexpressing large T antigen, in which only a small percentage of islet tumors lose E-cadherin and become invasive.

In contrast to epithelial cadherins, some cadherins that are normally expressed in mesenchymal cells can be upregulated in cancer. For example, neuronal (N)-cadherin is induced in invasive breast cancers. This mesenchymal cadherin may encourage epithelial cells to migrate into a stromal environment. Moreover, it can bind in *cis* to the fibroblast GF receptor and thereby activate signaling pathways that lead to invasion. *De novo* expression of cadherins that are not normally expressed in epithelia, but that encourage invasion, is called a "cadherin switch" and extends to several other members of the family.

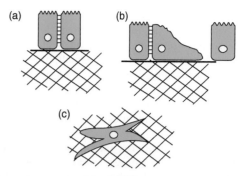

Figure 12.18 Epithelial–mesenchymal transitions. (a) Epithelia are normally stable, forming interactions with each other and with the ECM. These cells are polarized, and not particularly motile. (b) During EMT, signals provided by stromal cells, or oncogenes, cause epithelial cells to lose contact with one another and to alter their interactions with ECM. (c) The cells acquire a motile phenotype and take on many characteristics of fibroblasts (i.e. mesenchymal cells), such as their morphology and their expression of fibroblast markers including intermediate filaments that are characteristic of those cells. EMT is a highly conserved process that is required during embryonic development so that mesenchymal cells can be formed from the primitive epithelia of early blastocysts. This is necessary for the formation of a three-layered embryo during gastrulation. However, if the EMT program is reactivated in normal epithelial cells of adult tissues, it can contribute to malignant progression.

Epithelial–mesenchymal transitions

Despite the key role of altered cell–cell adhesion in carcinoma progression, the mere loss of E-cadherin does not lead to cancer. Studies with mice harboring conditional-null alleles of E-cadherin demonstrate that when the gene is deleted in normal epithelial cells *in vivo*, apoptosis arises and no tumors develop. When E-cadherin is lost in cancer, this is normally part of a larger program of alterations that occurs to change the phenotype of the cells. This program is similar to the EMT that occurs in development and converts cells from having a well-organized, stationary behavior to becoming separated from each other, acquiring motility, and forming metastases in organs distant from the primary tumor (Fig. 12.18). EMT is one of the programs that promote malignancy, but metastases may arise through other mechanisms that are not EMT dependent.

EMT is activated by factors that are secreted by stromal cells such as fibroblasts, providing another paradigm where the stroma has a huge influence on the progression of carcinoma. Several different types of GF are involved in EMT in cancer. These include TGF-β, Notch, Wnt, and RTKs such as hepatocyte growth factor (HGF, or scatter factor), all of which promote the expression of the Snail–Twist family of transcription factors. The EMT transcription factors are under multiple levels of control in addition to GR activation, including microRNAs (miRNAs) and the E3 ligase ubiquitination system. For example, the miR-200 family suppresses SIP1 and is lost in invasive cancer. In addition, EMT transcription factors are normally labile and are degraded by an E3 ligase called Partner-of-Paired.

EMT is primarily an embryonic mechanism to drive the formation of new cell types and tissues during development. Emerging evidence also suggests that aberrant EMT may enable epithelial cells to inappropriately acquire some properties of stem cells (Chapter 5). The activation of EMT in cancer therefore provides an ideal opportunity for therapeutic intervention, because inhibiting it is unlikely to affect normal tissues. Understanding the molecular basis for regulating EMT, particularly switching it off, will eventually lead to new cancer cures.

TGF-β can activate EMT

TGF-β normally inhibits the proliferation of epithelial cells. At higher doses, it induces apoptosis. However, if cells acquire resistance to the growth-inhibitory and apoptotic effects of TGF-β, for example if Ras becomes activated through either mutation or activation of receptor tyrosine kinases (Chapter 6), it can induce scattering and invasion. One of the ways it does this is through the expression of SIP-1, which reduces cell–cell interactions. Another is that TGF-β causes changes in the cytoskeleton through the small GTPase, RhoA, and thereby contributes to cell migrations.

TGF-β collaborates with oncogenes to induce invasiveness and cancer. In healing skin wounds, TGF-β becomes activated as part of the normal repair process, but if a viral oncogene such as v-Src (Chapter 6) is also present (e.g. after infection with the Rous sarcoma virus), tumors develop: in fact this virus, which was the first tumor virus to be discovered, induces tumors only at sites of wounding. Similarly, TGF-β cooperates with Ha-Ras to induce invasive spindle cell carcinomas of the skin, and moreover the TGF-β receptor signaling pathway is altered in several human cancers.

The apparent paradox that TGF-β is tumor suppressive in normal cells, but becomes oncogenic and potentiates EMT and metastasis in later-stage disease, means that therapeutic strategies to target this signaling pathway need to be designed with considerable care!

HGF drives EMT

HGF can also induce EMT by interacting with a receptor tyrosine kinase called c-Met (Chapter 5). In monolayer-cultured epithelial cells, HGF leads to cell migration and scattering of the cells away from each other. In cells placed in 3D ECM, an experimental paradigm where the cells can aggregate to form multicellular polarized structures resembling ducts and acini (Fig. 12.3), HGF induces branching morphogenesis. Normally this contributes to morphogenetic patterning, but if either HGF or c-Met is overexpressed or if the receptor is mutated, the signaling pathway can activate an invasion program, driving cancer. Alterations in either HGF or c-Met have been identified in a wide variety of human carcinomas.

Activation of c-Met induces several intracellular pathways, including a Ras-mediated pathway, and both phosphatidylinositol 3′-kinase (PI3K) and phospholipase Cγ. Together these pathways have several consequences on the cell. They promote proliferation, weaken cell–cell interactions by causing E-cadherin phosphorylation and reducing its expression, and activate invasion by altering integrins and MMPs.

Integrins, metalloproteinases, and cell invasion

Epithelial cells are normally restricted to their own compartment within a tissue and do not cross basement membranes or invade stroma. However, these rules of normal behavior are broken during a tumor's progression from benign to malignant. Cancer cells become motile and, together with the loss of cell–cell adhesion, this allows them to become invasive (Box 12.2).

Altered integrin profiles reflect a more migratory phenotype in cancer cells

Many cells migrate by extending plasma membrane protrusions called lamellipodia, in a polarized direction, into an ECM environment. The matrix provides mechanical support for migration. Lamellipodia engage with specific ECM molecules through integrins, which preferentially locate to the leading edge of migrating cells. This alters both the direct links with the cytoskeleton to generate intracellular tension and the formation of cell–matrix adhesions at the front of the cell, while at the same time it delivers signals to cause retraction at its rear coupled with myosin-driven movement.

Engagement is mediated by integrin activation and conformational changes within its extracellular domain so that the receptor binds with high affinity to its recognition epitopes on the ECM protein. Signals that initiate within the cytoplasm can cause the molecular reconfiguration of integrin extracellular domains so that their affinity for ligand is increased. This is called "inside-out signaling." Clustering of several integrin heterodimers within the plane of the membrane to form adhesion complexes leads to the triggering of signal cascades, involving numerous adaptor proteins and enzymes such as FAK, Src, and ILK (integrin-linked kinase), and thereby results in the activation of downstream effector proteins that alter the cytoskeleton (Fig. 12.8). These are "outside-in signals." The integrin requirement for invasion is both to mediate adhesion and to regulate signaling.

Of key importance are small GTPases of the Rho family, including Rho, Rac, and Cdc42, which promote actin reorganization. These enzymes are switched on by guanine nucleotide exchange factors (GEFs). For example, two different Rac GEFs are αPIX, which is indirectly bound to ILK in adhesion complexes, and DOCK180, which is bound to FAK via the adaptor protein, p120^{Cas}. Once activated, Rac and Cdc42 alter the function of cytoskeletal regulatory proteins, such as WASP and Arp2/3, to cause membrane ruffles and the extension of membrane processes. Rac and Rho, via their effector kinases p21-activated kinase and p160^{ROCK}, regulate the phosphorylation of proteins such as myosin light-chain kinase, which controls myosin II and the motile machinery. Rac and Cdc42 have a direct effect on invasion because when activated forms of these proteins are expressed in epithelial cells, they induce cellular depolarization and migration into the ECM.

Retraction at the tail of the cell occurs through a switch of integrins to a low-affinity state and the dissolution of adhesion complexes. This occurs in part through a reverse inside-out signal. In addition, Erk-mediated signaling through the FAK–Src complex disassembles adhesion complexes, and together with intracellular proteases such as calpain, which cleave proteins such as FAK, this results in detachment from the ECM.

Epithelial cells tend to be fairly stationary in their normal tissue environment, but they alter their expression of integrins in cancer to a profile that reflects a more migratory phenotype. The profile of integrins normally expressed is frequently reduced, while others, such as the αv integrins, which drive cell migration on stromal fibronectin and the ECM protein vitronectin, are often upregulated in malignancy, thereby contributing to invasion. α6β4 integrin is also upregulated in several carcinomas. This integrin is usually associated with hemidesmosomes, where it contributes to epithelial integrity. However, it can be released and recruited to lamellipodia during wound healing and in malignancy, thereby increasing the ability of cancer cells to become invasive.

Integrins and GFs can work in collaboration to promote cell movement, which results in aberrant migration if either integrins or oncogenes are expressed, or if tumor suppressors are mutated. For example, integrin heterodimers can interact directly with the receptor tyrosine kinases c-ErbB2 and c-ErbB3, the insulin signaling adaptor protein IRS, and Shc to promote PI3K and Ras signaling. GF signaling through the ErbB2 receptor promotes integrin-mediated migration through an adaptor protein called Memo. The tumor suppressor *Tp53* also has a direct role in cell migration. Mutant p53 elevates both the α5β1 integrin and GF receptor recycling through endosomal vesicles, which encourages cell migration.

Although changes in integrin expression can be caused by transcription factors that are altered in cancer, they occur as a consequence of the cell being in a different environment. Stromal ECM, for example, strongly induces integrin expression in mammary epithelial cells. The acquisition of motile integrins is therefore self-enhancing once an epithelial cell escapes its basement membrane constriction and enters a collagenous matrix. Many of the enzymes that are involved in signaling the motile machinery, such as FAK and Rac GEFs, as well as the Ras–Erk and PI3K pathways, are upregulated in malignant cells, sometimes through mutation, which also contributes to increased migration (Chapters 5 and 6).

The altered integrin profiles of invasive cancer cells inform therapeutic avenues to prevent migration in malignancy, and a number of inhibitors are currently in clinical trials (Fig. 12.19). Importantly, though, some integrins are tumor and metastasis suppressors, which underscores the requirement to understand precisely how different integrin–ECM interactions are involved with cancer cell behavior.

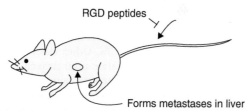

Figure 12.19 The key role of integrins in cell migration during metastasis. Melanoma cells are highly metastatic and can cause rapid patient mortality. In order for the cells to migrate and form metastases, they require integrin-mediated adhesion. ECM proteins contain specific amino acid recognition motifs to be bound by integrins, and one of these contains the sequence arginine–glycine–aspartic acid (RGD). This recognition sequence is contained within ECM molecules that promote migration such as fibronectin and vitronectin, which are bound by α5 and αv integrins. In this example, melanoma cells cause experimental metastasis when injected into the tail veins of mice, but injecting an inhibitory RGD peptide at the same time can inhibit metastasis. This not only shows that integrin-mediated adhesion is necessary for metastasis to occur but also indicates that strategies to block cell–matrix adhesion might be useful for cancer therapy. Anti-integrin drugs are also being used to prevent angiogenesis, thereby inhibiting the growth of tumors (Bhaskar et al., 2008; Robinson and Hodivala-Dilke, 2011).

Serine and metalloproteinases degrade ECM to allow invasion

Activating the cellular motile machinery by itself is not sufficient for migration, particularly through a 3D ECM. Cells need to re-landscape their local environment to forge their way through the woven network of basement membrane proteins and the dense network of fibrils such as occurs in the stroma *in vivo*. To do this cells require the help of ECM-degrading proteinases, of which there are several classes, including MMPs, serine proteases, and cathepsins. These enzymes are activated progressively during malignancy and promote migration by dissolving cell–cell cadherin contacts, breaking encapsulating ECM barriers, and clipping ECM proteins to expose pro-migratory epitopes and release migration-stimulating fragments.

Most MMPs and serine proteases are expressed by stromal cells, but perform their tumor-promoting functions in the neighborhood of epithelial cells. In addition, they have an essential role in new blood vessel growth (angiogenesis), which is required for tumors to grow. The mechanisms of tumor angiogenesis are dealt with in Chapter 14.

The ECM-degrading proteinases are normally present as inactive pro-enzymes, or zymogens, and become activated in the peri-cellular environment. This is important as it ensures that ECM remodeling does not occur throughout the stroma but instead is linked closely to aspects of cell phenotype. For example, urokinase plasminogen activator (uPA), a serine protease, is activated at the cell surface through a balance between its receptor (uPAR) and several regulators including plasminogen activator inhibitor-1, a protease nexin, and vitronectin. This system directs proteolysis at adhesion complexes and is therefore pivotal in altering the ECM at active sites of migration. Many of the components are upregulated in malignant tumors, and their levels correlate with tumor aggressiveness and poor patient prognosis.

The large family ($n=28$) of matrix-degrading MMPs is also regulated close to the cell surface. This occurs through the removal of pro-domain, which masks the active site of the enzyme. Pro-MMP-2, for example, is activated via a complex pathway involving a transmembrane MMP (MT1-MMP or MMP-14) and the tissue inhibitor of metalloproteinases-2 (TIMP-2). However, MT1-MMP can act in the absence of other MMPs, and MMPs are rarely altered by mutation in cancer; rather, their levels are frequently increased at the transcriptional level. This can occur through transactivation of MMP genes in stromal cells under the influence of cytokines secreted by the epithelial tumor cells. This is another important example of the cellular environment influencing tumor cell behavior. Occasionally MMPs are expressed within the cancer cells themselves, via polymorphisms within the promoter region so that they are recognized by inappropriate transcription factors, or the combined activity of multiple oncogenes.

During cell migration, both MMPs and serine proteases are recruited to lamellipodia and control local ECM-remodeling events that occur as the migrating cells make and break their contacts with ECM. Some ECM proteins contain cryptic domains that are recognized by integrins, and stretching the molecules or discrete MMP-catalyzed cleavage events exposes them.

Many of the MMPs are involved in cancer progression, but one of these, MMP-2, serves to illustrate the ways that they may work. MMP-2 contributes to both the aggressiveness of tumor cells as well as the angiogenic response. Mice lacking this enzyme do not support the ability of tumor cells to colonize lungs. MMP-2 is a broad-spectrum gelatinase that degrades many ECM proteins, including laminin and collagen type IV. A key function of this enzyme, together with MT1-MMP, is to unmask cryptic sites within specific ECM proteins. Both MMP-2 and MT1-MMP cleave the γ2 chain of laminin-5 in basement membrane to yield an armless molecule, together with some small fragments that have potent properties. One of these, DIII, is present in breast cancer, and its levels correlate with the stage of the disease. It contains an EGF-like domain and binds to the EGF receptor, triggering signal transduction and promoting motility. Other ECM proteins can be cleaved by MMPs to produce similarly active biofragments involved in cell migration and angiogenesis.

A further function of MMPs is to cleave the surrounding ECM molecules so that the tumor cells can stretch out and respond to GFs. A frequent stromal host reaction to an encroaching carcinoma is excessive fibrosis, which serves to encapsulate the tumor and prevent the malignant cells from migrating. Such cells are also squeezed into a rounded geometric conformation that inhibits proliferation. Cleaving the matrix proteins will allow the tumor cells to escape their constraining mesh (Fig. 12.20).

Collective cell migration

Histology of patient tumors reveals that the invasive front of carcinomas is rarely composed of individual cells that have breached the basement membrane and entered the stromal tissue. Rather, cells are usually present in aggregates within the stroma. Culture and animal models, and primary tumor explants, have revealed that many cancer cell lines actually migrate as groups of cells in "collective" cell migration. Cells migrating together interact with each other via cadherins, gap junctions, and desmosomes, and tight junctions can be present. In these cases, EMT is not required for cell migration in cancer.

Collective cell migration is normal in many situations of embryonic development and adult tissue repair; however, the

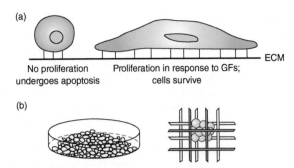

Figure 12.20 MMPs control the proliferative response of tumor cells constrained within a 3D ECM. (a) Cells that are attached to a substratum, but are forced into a rounded configuration, are unable to respond to GFs and don't proliferate. Rounded cells are, instead, prone to undergo apoptosis. This indicates that cellular geometry is critical in determining whether cells can proliferate or not. (b) When tumor cells are plated in two dimensions on a culture dish coated with collagen (left panel), the cells are stretched out and they can proliferate well; but inside a 3D collagen gel (right panel), the cells are rounded and their proliferation is severely compromised. This is because of low cyclin D3-kinase activity. However if the cells express a surface-bound matrix metalloproteinase, MT1-MMP, they can clip the collagen fibers that enclose them, and after doing so they reorganize their cytoskeleton, stretch out, and now respond to proliferative signals. Thus, MT1-MMP allows cells that would otherwise be entrapped within a stromal or dermal matrix to proliferate, thereby providing a growth advantage. Indeed, the tumor cells expressing MT1-MMP grow much more quickly *in vivo* than those without it. This highlights the dramatic differences in the growth potential of cells growing on a 2D surface compared to the same cells embedded with a 3D gel of the same matrix. As carcinoma cells exit a primary tumor and enter the stroma, they need to adapt to foreign signals and survive in the new ECM environment. They also need to proliferate, but can be prevented from doing so by the physical constraints of the stroma. The expression of MMPs provides a mechanism for carcinoma cells to proliferate at foreign sites that are frequently rich in dense networks of collagen or fibrin, and this is necessary for them to progress to full metastasis. This also supports the possibility that MMP inhibitors may have a dramatic therapeutic potential for cells growing in a 3D environment similar to that *in vivo*, even if they have little effect on monolayer-cultured cells (Chen et al., 1997; Hotary et al., 2003).

mechanisms are not fully understood. Some studies have indicated that clusters of cells migrate in similar ways to single cells, with actin-rich protrusions and the Rho-, Rock-, or myosinII-dependent generation of traction forces at integrin-containing adhesion complexes (Fig. 12.21). The combination of adhesive interactions, with localized MT1-MMP proteolysis, creates tracks within the 3D ECM that cells can migrate through. It is possible that leading individual cells form an initial track, while all the adjacent cells follow and form macro-tracks behind.

This model doesn't apply to all collectively migrating cells because groups of epithelial cells forming mammary gland ducts do so without lamellipodia or other plasma membrane protrusions. Currently the mode of carcinoma collective-cell migration is not well understood, but novel live-imaging technologies combined with genetic fluorescent marks promise to reveal the mechanisms.

MMPs release tumor-promoting growth factors

In addition to their direct effects on cell migration and indirect effects on proliferation, MMPs have other key functions in cancer progression because they release GFs that are sequestered by ECM or ECM-bound proteins. This influences angiogenesis, immune surveillance, EMT, and survival. For example, IGFs are important survival and proliferation factors for epithelia but are sequestered by IGF-binding proteins (IGF-BPs). MMPs cleave IGF-BPs to release the bioactive GFs. Other classes of growth factors are secreted in an inactive state until cleaved by MMPs; two examples are TGF-α and the heparin-bound form of EGF. Similarly, TGF-β is released from its sequestering molecule, LAP, by MMPs, thereby directing the transition to a mesenchymal, migratory phenotype.

Since ECM-degrading proteinases are upregulated strongly in cancer and have diverse effects on migration, proliferation, survival, and EMT, they represent ideal targets for therapy (Chapter 16). A number of potential treatments have been developed that range from agents that inhibit their synthesis to those that block enzyme activity. As with all treatments for cancer that will be really effective in the clinic, understanding the tumor-specific contribution of specific effectors is critical in therapeutic design. We therefore need to learn much more about the individual serine proteases and MMPs that contribute to the progression of specific neoplasms. It is likely that such inhibitors are most effective in treating early-stage disease, prior to the establishment of metastases, since this is the stage where active migration and stromal influence are most acute.

Survival in an inappropriate environment

Cancer cells are genetically unstable, and one might expect that the accumulation of mutations necessary to allow secondary tumor growth would be rapid and frequent. One of the reasons for metastasis being rare is that, as outlined, an unfavorably large number of steps are required for tumor cells to leave the primary tumor and settle in a distant organ, so the chances of all being acquired by the same cell are small. A further driving force to prevent cells from colonizing an inappropriate environment is anoikis, which is a specific form of apoptosis (Chapter 8) that is induced when cells no longer receive appropriate survival signals from the ECM. Although the trigger for this type of apoptosis may differ from that described in this book, the downstream effector pathways are similar (see the "Anoikis" section).

Anoikis

Anoikis was described initially in cells that had been experimentally deprived of all contact with ECM for a few hours. This situation does not occur for most cells during the transit of malignant cells through blood vessels, since cells that travel to other tissues via the circulation lose complete contact with ECM for only a few seconds, which is not long enough to trigger anoikis. However, anoikis can be induced when cells migrate into either the lymph nodes (where there is not much ECM to contact) or a matrix that they normally do not associate with (e.g. the stromal ECM), because any integrins that are activated there are not sufficient to maintain long-term survival. For example, normal epithelial cells from the mammary gland undergo delayed apoptosis when they come into contact with collagen I, which is the main type of collagen in the stroma. This process is not rapid; rather, it takes place over several days and is due to an altered sensitivity to ECM signals, resulting in stochastic apoptosis. Anoikis is one of the key factors to maintain normal epithelial cell positioning in adult tissue homeostasis, and if this mechanism becomes deregulated,

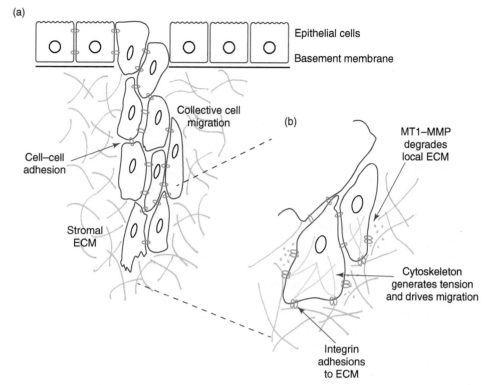

Figure 12.21 Collective cell migration. (a) The epithelial cell sheet is well organized, with polarized cells abutting a basement membrane on their basal sides. Malignant tumor cells penetrate this ECM and enter the stromal connective tissue matrix as groups of cells. The "leading" cells need to both form attachments with the ECM and degrade the entwining mesh of collagen fibers of the stromal ECM in order to migrate through it. Groups of migrating cells retain cell–cell interactions with their neighbors. It is not clear yet whether the cells in the vanguard are pulled along with the leaders, or if push them. (b) Detail showing integrin interactions and protease degradation of the matrix. See also Friedl and Gilmour (2009).

it provides cells with the opportunity to live in a spatially inappropriate environment.

Thus cancer cells (at least those of epithelial origin), which are derived from cells that normally contact a basement membrane, will have undergone further mutations in order to survive the stromal bed of either the primary tissue they migrate through or the secondary site to which they metastasize. The accumulation of mutations that allow cells to survive in inappropriate environments during malignant spread is as important as, if not more important than, the acquisition of oncogenic mutations in proliferation genes.

Circulating tumor cells

In some cases, epithelial stem cells are able to survive in the absence of cell–ECM adhesions, even though they paradoxically have increased levels of surface integrins. For example, mammary epithelial stem cells are enriched in suspension culture, where the more differentiated cells would normally undergo rapid apoptosis. The mechanism for this is unknown, but the implications are that cancer stem cells acquire properties that allow them to survive the circulation more readily than the normal cells from which they evolve. Such cells may provide the seed for secondary metastases that either remain dormant until appropriately activated by other triggers years later or proliferate to form potentially fatal lesions. Keratin-positive circulating tumor cells have been found in the serum of cancer patients, and such apoptosis-resistant cells have the potential to be used as a diagnostic biomarker for aggressive cancers.

Crosstalk between adhesion and growth factor receptors controls anoikis

Crosstalk of adhesion receptors with GF receptors and direct integrin-mediated signaling provide two separate mechanisms for the environmental sensing that determines whether cells live or die. In the mammary gland, IGF is a key determinant of epithelial survival that is synthesized by stromal cells in response to growth hormone. This is an example of a stromal–epithelial communication that keeps the latter alive. However, in epithelial cells the IGF type I receptor delivers efficient survival signals only when it resides on a basement membrane. Signaling through the IGF-triggered PI3K and protein kinase B (PKB) pathway is inefficient in cells embedded within collagen, and they die a slow death.

PKB has been implicated in the protection of other types of cancer cells from anoikis, though it is not always activated by an integrin switch. In some tumor types, it is triggered by the unscheduled expression of a receptor tyrosine kinase, the nerve growth factor (NGF) receptor, which can have dramatic effects on suppressing apoptosis and promoting metastasis.

A related crosstalk between adhesion receptors and receptor tyrosine kinases is critical for survival in other cell types too. For example, in capillaries, vascular endothelial (VE)-cadherin collaborates with the VEGF receptor to control survival and proliferation. This cell–cell adhesion molecule potentiates the PI3K–PKB signaling axis driven by VEGF, keeping the endothelial cells alive. Association with a third molecule, a phosphatase, causes specific tyrosine residues on the VEGF receptor to become dephosphorylated so that the Erk pathway is not activated, and this prevents excess proliferation in stable capillaries. When angiogenesis is occurring, the cells become motile as they invade new areas of stroma, and the cadherin junctions are dismantled, so that VEGF can also trigger a proliferative signal. Similarly, integrins cooperate with neuregulin receptors to regulate the switch between growth and survival when oligodendrocytes mature into differentiated myelinating cells.

Direct integrin signals control anoikis

Integrins can also deliver survival signals directly to block anoikis by inactivating the apoptosis machinery. For example, in fibroblasts, decreased adhesion results in Jnk signaling leading to p53 activation, while in detached endothelial cells the reduced activation of Erk leads to lower levels of the Fas antagonist, c-FLIP, and apoptosis triggered through the Fas pathway. In some cases, altered integrin expression within cancer cells contributes directly to the suppression of anoikis.

In mammary epithelial cells, a slightly different apoptotic pathway is activated when integrin signaling is abrogated or altered. Bax, a proapoptotic protein, is maintained in the cytosol by integrin signaling, but the loss of adhesion to ECM leads to rapid and synchronous translocation of Bax to mitochondria, thereby inducing apoptosis.

Many of the components linking adhesion to Bax in mammary epithelia (e.g. FAK) are altered in breast cancer, or their activities are enhanced. This provides cancer cells that have migrated into a stromal environment the ability to survive when they would otherwise have undergone apoptosis, and thereby gather further mutations that contribute to the final stages of malignancy.

ECM and the alteration of p53–ATM responsiveness

In many types of cancer cell, the accumulation of mutations is exacerbated by an inappropriate ECM environment because the apoptotic response to DNA damage is not fully functional. If DNA damage occurs through chemical attack or radiation, a normal cell responds either by activating a repair pathway or, in the case of heavily damaged DNA, by triggering a p53-dependent apoptosis response (Chapter 10). The levels of p53 are kept in check by a balance between Mdm2, which targets p53 to the proteasome, and p19Arf, which inhibits Mdm2. As outlined in Chapters 7 and 10, DNA damage normally results in phosphorylation of p53, so that Mdm2 cannot bind p53, and an increase in p19Arf that sequesters Mdm2. Together this leads to stabilized p53 and transcription of Bax, Noxa, and Puma – all proteins involved in apoptosis and discussed in Chapter 8. However, in some cancer cells (e.g. melanoma and sarcoma), integrins are necessary for the response. Lost or reduced adhesion to ECM leads to lower p19Arf levels, and the cells fail to stabilize p53 after treatment with DNA-damaging agents. As a consequence, less apoptosis occurs and the cells accumulate more chromosomal re-arrangements.

Moreover, a separate pathway for driving apoptosis in response to damage, mediated by c-Abl, is also compromised by loss of ECM adhesion. c-Abl is a protein tyrosine kinase that is present in adhesion complexes and is regulated by integrins. Its activity is increased by DNA damage, but only if cells interact with ECM and integrins are active. Certain cancer cells therefore escape the c-Abl-mediated damage response if they are not in an appropriate ECM environment.

Thus, conventional therapies for cancer, which kill cells by inducing apoptosis, are not likely to be very effective with some types of malignant cells that are displaced from their normal ECM and growing in the stroma, lymph nodes, or secondary sites, because the coupling between drug and apoptosis is less efficient in these micro-environments. In other words, cancer cells may be more resistant to apoptosis. Moreover, designer drugs such as those that target c-Abl (e.g. Glivec or STI571) reduce the apoptotic DNA damage response even further. However, treatments that activate integrins may be a positive benefit and enhance the usefulness of chemotherapy and radiation therapy; since integrins are also involved in cell motility, this strategy may need to be tempered with other approaches to block migration. This is discussed in greater detail in Chapter 16.

Conclusions

Cancer is the disease process whereby cells lose their normal controls not only on growth but also on their spatial location. Altered adhesion therefore has a critical role in cancer, and is the main characteristic of cells that become malignant and progress to form metastases. Genetic selection, or epigenetic changes affecting adhesion (e.g. EMT), reduce cell–cell adhesion, increase migration through the stroma, and activate ECM signaling pathways. Altered stroma contributes to this process, but the mechanisms of interaction between stromal cells and their local microenvironment are poorly understood. Epithelial cells are normally under strong homeostatic control, and mechanisms have evolved to prevent both the inappropriate accumulation of cells and their spatial positioning within tissues. However, a disruption of homeostasis causes malignant cells to survive in, and migrate through, diverse ECM environments, thereby forming life-threatening metastases.

Although the concepts underpinning mechanisms of tumor spread are becoming established, the details of how this occurs are still very uncertain. Much work has been done on a small set of cancer cell lines that have been acclimatized for efficient growth in tissue culture, but very little is known about the specific mechanisms of tumor invasion and survival *in vivo*. Still less is understood about the differences that diverse tumor types employ to metastasize.

Since metastasis is the major cause of cancer deaths, the key to keeping advanced disease under control (and therefore for keeping people alive longer) is to understand the detailed mechanisms of dissemination. Effective therapies need to be devised for targeting both migratory carcinoma cells (early-stage disease) and those that survive in distant sites (late-stage disease) but avoid normal cells. Given the large array of events that contribute to malignant progression, the future is bright for the introduction of designer drugs for adhesion-specific processes. Unwanted consequences of therapies may sometimes occur through surprising mechanisms, so it is important to guard against this.

Outstanding questions

1. What are the earliest events that cause epithelial cells to lose the controls on their positional identity? What are the detailed mechanisms?

2. Do integrins or other cell adhesion molecules contribute to the formation of cancer stem cells?

3. How do biomechanical tissue forces contribute to cancer onset and the spread of tumors?

4. What are the molecular changes that drive the transition between benign and malignant cancer?

5. Can targeting the stroma reduce tumor spread and suppress the growth of metastases?

6. Is EMT a universal mechanism of cancer metastasis, or is it limited to subsets of cancer types?

7. What reactivates dormant micro-metastases?

8. What are the adhesion characteristics of the metastatic niche? Can drugs target cell adhesion molecules at secondary tumor sites?

9. Can we develop synthetic lethal approaches to target altered adhesion signaling in malignant cancers?

Bibliography

Introduction
Excellent basic texts on the structure of cell types and tissues, the components of cell junctions and ECM, and the process of metastasis can be found in:

Alberts, B., Johnson, A., Lewis, J., et al. (2007). *Molecular Biology of the Cell*. 5th edn. New York: Garland Science.

Ruoslahti, E. (1996). How cancer spreads. *Scientific American*, **275**: 72–7.

Stevens, A., and Lowe, J.S. (2004). *Human Histology*. 3rd edn. Barcelona: Mosby.

Adhesive interactions with the ECM
Butcher, D.T., Alliston, T., and Weaver, V.M. (2009). A tense situation: forcing tumour progression. *Nature Reviews Cancer*, **9**: 108–22.

Cabodi, S., del Pilar Camacho-Leal, M., Di Stefano, P., and Defilippi, P. (2010). Integrin signalling adaptors: not only figurants in the cancer story. *Nature Reviews Cancer*, **10**: 858–70.

De Wever, O., Demetter, P., Mareel, M., and Bracke, M. (2008). Stromal myofibroblasts are drivers of invasive cancer growth. *International Journal of Cancer*, **123**: 2229–38.

DuFort, C.C., Paszek, M.J., and Weaver, V.M. (2011). Balancing forces: architectural control of mechanotransduction. *Nature Reviews Molecular Cell Biology*, **12**: 308–19.

Dupont, S., Morsut, L., Aragona, M., et al. (2011). Role of YAP/TAZ in mechanotransduction. *Nature*, **474**: 179–83.

Giancotti, F.G., and Tarone, G. (2003). Positional control of cell fate through joint integrin/receptor protein kinase signaling. *Annual Review of Cell and Developmental Biology*, **19**: 173–206.

Guilak, F., Cohen, D.M., Estes, B.T., Gimble, J.M., Liedtke, W., and Chen, C.S. (2009). Control of stem cell fate by physical interactions with the extracellular matrix. *Cell Stem Cell*, **5**: 17–26.

Hu, M., and Polyak, K. (2008). Molecular characterisation of the tumour microenvironment in breast cancer. *European Journal of Cancer*, **44**: 2760–5.

Katayama, M., and Sekiguchi, K. (2004). Laminin-5 in epithelial tumour invasion. *Journal of Molecular Histology*, **35**: 277–86.

Orend, G., and Chiquet-Ehrismann, R. (2006). Tenascin-C induced signaling in cancer. *Cancer Letters*, **244**: 143–63.

Sasaki, T., Fässler, R., and Hohenester, E. (2004). Laminin: the crux of basement membrane assembly. *Journal of Cell Biology*, **164**: 959–63.

Schedin, P., and Keely, P.J. (2011). Mammary gland ECM remodeling, stiffness, and mechanosignaling in normal development and tumor progression. *Cold Spring Harbor Perspectives in Biology*, **3**: a003228.

Streuli, C.H. (2009). Integrins and cell-fate determination. *Journal of Cell Science*, **122**: 171–7.

Streuli, C.H., and Akhtar, N. (2009). Signal co-operation between integrins and other receptor systems. *Biochemical Journal*, **418**: 491–506.

Toole, B.P. (2004). Hyaluronan: from extracellular glue to pericellular cue. *Nature Reviews Cancer*, **4**: 528–39.

Wang, N., Tytell, J.D., and Ingber, D.E. (2009). Mechanotransduction at a distance: mechanically coupling the extracellular matrix with the nucleus. *Nature Reviews Molecular Cell Biology*, **10**: 75–82.

Yu, H., Mouw, J.K., and Weaver, V.M. (2011). Forcing form and function: biomechanical regulation of tumor evolution. *Trends in Cell Biology*, **21**: 47–56.

Cell–cell interactions
Astin, J.W., Batson, J., Kadir, S., et al. (2010). Competition amongst Eph receptors regulates contact inhibition of locomotion and invasiveness in prostate cancer cells. *Nature Cell Biology*, **12**: 1194–204.

Berx, G., and van Roy, F. (2009). Involvement of members of the cadherin superfamily in cancer. *Cold Spring Harbor Perspectives in Biology*, **1**: a003129.

Bienz, M. (2005). Beta-catenin: a pivot between cell adhesion and Wnt signalling. *Current Biology*, **15**: R64–7.

Cavallaro, U., and Dejana, E. (2011). Adhesion molecule signalling: not always a sticky business. *Nature Reviews Molecular Cell Biology*, **12**: 189–97.

Guillemot, L., Paschoud, S., Pulimeno, P., Foglia, A., and Citi, S. (2008). The cytoplasmic plaque of tight junctions: a scaffolding and signalling center. *Biochimica et Biophysica Acta*, **1778**: 601–13.

Incassati, A., Chandramouli, A., Eelkema, R., and Cowin, P. (2010). Key signaling nodes in mammary gland development and cancer: β-catenin. *Breast Cancer Research*, **12**: 213.

Matter, K., and Balda, M.S. (2003). Signalling to and from tight junctions. *Nature Reviews Molecular Cell Biology*, **4**: 225–36.

Merlos-Suárez, A., and Batlle, E. (2008). Eph-ephrin signalling in adult tissues and cancer. *Current Opinion in Cell Biology*, **20**: 194–200.

Takebe, N., Harris, P.J., Warren, R.Q., and Ivy, S.P. (2011). Targeting cancer stem cells by inhibiting Wnt, Notch, and Hedgehog pathways. *Nature Reviews Clinical Oncology*, **8**: 97–106.

Wend, P., Holland, J.D., Ziebold, U., and Birchmeier, W. (2010). Wnt signaling in stem and cancer stem cells. *Seminars in Cell & Developmental Biology*, **21**: 855–63.

Steps in the dissemination of metastases
Balkwill, F. (2004). Cancer and the chemokine network. *Nature Reviews Cancer*, **4**: 540–50.

Barkan, D., Green, J.E., and Chambers, A.F. (2010). Extracellular matrix: a gatekeeper in the transition from dormancy to metastatic growth. *European Journal of Cancer*, **46**: 1181–8.

Chambers, A.F., Groom, A.C., and MacDonald, I.C. (2002). Dissemination and growth of cancer cells in metastatic sites. *Nature Reviews Cancer*, **2**: 563–72.

Hedley, B.D., and Chambers, A.F. (2009). *Tumor Dormancy and Metastasis*. New York: Academic Press.

Holopainen, T., Bry, M., Alitalo, K., and Saaristo, A. (2011). Perspectives on lymphangiogenesis and angiogenesis in cancer. *Journal of Surgical Oncology*, **103**: 484–8.

Huang, L., and Muthuswamy, S.K. (2010). Polarity protein alterations in carcinoma: a focus on emerging roles for polarity regulators. *Current Opinion in Genetics & Development*, **20**: 41–50.

Kang, Y., Siegel, P.M., Shu, W., et al. (2003). A multigenic program mediating breast cancer metastasis to bone. *Cancer Cell*, **3**: 537–49.

Langley, R.R., and Fidler, I.J. (2011). The seed and soil hypothesis revisited – the role of tumor-stroma interactions in metastasis to different organs. *International Journal of Cancer*, **128**: 2527–35.

Mantovani, A., Allavena, P., Sica, A., and Balkwill, F. (2008). Cancer-related inflammation. *Nature*, **454**: 436–44.

Müller, A., Homey, B., Soto, H., et al. (2001). Involvement of chemokine receptors in breast cancer metastasis. *Nature*, **410**: 50–6.

Talmadge, J.E., and Fidler, I.J. (2010). AACR centennial series: the biology of cancer metastasis: historical perspective. *Cancer Research*, **70**: 5649–69.

Weigelt, B., Hu, Z., He, X., et al. (2005). Molecular portraits and 70-gene prognosis signature are preserved throughout the metastatic process of breast cancer. *Cancer Research*, **65**: 9155–8.

E-cadherin downregulation in cancer leads to migration

Berx, G., and van Roy, F. (2009). Involvement of members of the cadherin superfamily in cancer. *Cold Spring Harbor Perspectives in Biology*, **1**: a003129.

Cavallaro, U., and Christofori, G. (2004). Cell adhesion and signalling by cadherins and Ig-CAMs in cancer. *Nature Reviews Cancer*, **4**: 118–32.

Epithelial–mesenchymal transitions

Gregory, P.A., Bert, A.G., Paterson, E.L., et al. (2008). The miR-200 family and miR-205 regulate epithelial to mesenchymal transition by targeting ZEB1 and SIP1. *Nature Cell Biology*, **10**: 593–601.

Kang, Y., Siegel, P.M., Shu, W., et al. (2003). A multigenic program mediating breast cancer metastasis to bone. *Cancer Cell*, **3**: 537–49.

Lander, R., Nordin, K., and Labonne, C. (2011). The F-box protein Ppa is a common regulator of core EMT factors Twist, Snail, Sluhg, and Sip1. *Journal of Cell Biology*, **194**: 17–25.

Perl, A.K., Wilgenbus, P., Dahl, U., Semb, H., and Christofori, G. (1998). A causal role for E-cadherin in the transition from adenoma to carcinoma. *Nature*, **392**: 190–3.

Roussos, E.T., Keckesova, Z., Haley, J.D., Epstein, D.M., Weinberg, R.A., and Condeelis, J.S. (2010). AACR special conference on epithelial-mesenchymal transition and cancer progression and treatment. *Cancer Research*, **70**: 7360–4.

Wendt, M.K., Allington, T.M., and Schiemann, W.P. (2009). Mechanisms of the epithelial-mesenchymal transition by TGF-beta. *Future Oncology*, **5**: 1145–68.

Yang, J., Mani, S.A., Donaher, J.L., et al. (2004). Twist, a master regulator of morphogenesis, plays an essential role in tumor metastasis. *Cell*, **117**: 927–39.

Integrins, MMPs, and cell invasion

Cox, D., Brennan, M., and Moran, N. (2010). Integrins as therapeutic targets: lessons and opportunities. *Nature Reviews Drug Discovery*, **9**: 804–20.

Friedl, P., and Gilmour, D. (2009). Collective cell migration in morphogenesis, regeneration and cancer. *Nature Reviews Molecular Cell Biology*, **10**: 445–57.

Gray, R.S., Cheung, K.J., and Ewald, A.J. (2010). Cellular mechanisms regulating epithelial morphogenesis and cancer invasion. *Current Opinion in Cell Biology*, **22**: 640–50.

Hood, J.D., and Cheresh, D.A. (2002). Role of integrins in cell invasion and migration. *Nature Reviews Cancer*, **2**: 91–100.

Kessenbrock, K., Plaks, V., and Werb, Z. (2010). Matrix metalloproteinases: regulators of the tumor microenvironment. *Cell*, **141**: 52–67.

López-Otín, C., and Matrisian, L.M. (2007). Emerging roles of proteases in tumour suppression. *Nature Reviews Cancer*, **7**: 800–8.

Madsen, C.D., and Sahai, E. (2010). Cancer dissemination – lessons from leukocytes. *Developmental Cell*, **19**: 13–26.

Marone, R., Hess, D., Dankort, D., Muller, W.J., Hynes, N.E., and Badache, A. (2004). Memo mediates ErbB2-driven cell motility. *Nature Cell Biology*, **6**: 515–22.

Muller, P.A.J., Caswell, P.T., Doyle, B., et al. (2009). Mutant p53 drives invasion by promoting integrin recycling. *Cell*, **139**: 1327–41.

Ramirez, N.E., Zhang, Z., Madamanchi, A., et al. (2011). The $\alpha_2\beta_1$ integrin is a metastasis suppressor in mouse models and human cancer. *Journal of Clinical Investigation*, **121**: 226–37.

Ridley, A.J., Schwartz, M.A., Burridge, K., et al. (2003). Cell migration: integrating signals from front to back. *Science*, **302**: 1704–9.

Rørth, P. (2009). Collective cell migration. *Annual Review of Cell and Developmental Biology*, **25**: 407–29.

Rowe, R.G., and Weiss, S.J. (2009). Navigating ECM barriers at the invasive front: the cancer cell–stroma interface. *Annual Review of Cell and Developmental Biology*, **25**: 567–95.

Selivanova, G., and Ivaska, J. (2009). Integrins and mutant p53 on the road to metastasis. *Cell*, **139**: 1220–2.

Yamaguchi, H., Wyckoff, J., and Condeelis, J. (2005). Cell migration in tumors. *Current Opinion in Cell Biology*, **17**: 559–64.

Survival in an inappropriate environment

Gilmore, A.P., Owens, T.W., Foster, F.M., and Lindsay, J. (2009). How adhesion signals reach a mitochondrial conclusion – ECM regulation of apoptosis. *Current Opinion in Cell Biology*, **21**: 654–61.

Katz, E., and Streuli, C.H. (2007). The extracellular matrix as an adhesion checkpoint for mammary epithelial function. *International Journal of Biochemistry & Cell Biology*, **39**: 715–26.

Krebs, M.G., Sloane, R., Priest, L., et al. (2011). Evaluation and prognostic significance of circulating tumor cells in patients with non-small-cell lung cancer. *Journal of Clinical Oncology*, **29**: 1556–63.

Lewis, J.M., Truong, T.N., and Schwartz, M.A. (2002). Integrins regulate the apoptotic response to DNA damage through modulation of p53. *Proceedings of the National Academy of Sciences USA*, **99**: 3627–32.

Nagaprashantha, L.D., Vatsyayan, R., Lelsani, P.C.R., Awasthi, S., and Singhal, S.S. (2011). The sensors and regulators of cell-matrix surveillance in anoikis resistance of tumors. *International Journal of Cancer*, **128**: 743–52.

Visvader, J.E. (2009). Keeping abreast of the mammary epithelial hierarchy and breast tumorigenesis. *Genes & Development*, **23**: 2563–77.

General

Barcellos-Hoff, M.H., and Ravani, S.A. (2000). Irradiated mammary gland stroma promotes the expression of tumorigenic potential by unirradiated epithelial cells. *Cancer Research*, **60**: 1254–60.

Bhaskar, V., Fox, M., Breinberg, D., et al. (2008). Volociximab, a chimeric integrin alpha5beta1 antibody, inhibits the growth of VX2 tumors in rabbits. *Investigational New Drugs*, **26**: 7–12.

Chen, C.S., Mrksich, M., Huang, S., Whitesides, G.M., and Ingber, D.E. (1997). Geometric control of cell life and death. *Science*, **276**: 1425–8.

Debnath, J., and Brugge, J.S. (2005). Modelling glandular epithelial cancers in three-dimensional cultures. *Nature Reviews Cancer*, **5**: 675–88.

De Wever, O., Nguyen, Q.-D., Van Hoorde, L., et al. (2004). Tenascin-C and SF/HGF produced by myofibroblasts in vitro provide convergent pro-invasive signals to human colon cancer cells through RhoA and Rac. *FASEB Journal*, **18**: 1016–18.

Hogan, C., Dupré-Crochet, S., Norman, M., et al. (2009). Characterization of the interface between normal and transformed epithelial cells. *Nature Cell Biology*, **11**: 460–7.

Hotary, K.B., Allen, E.D., Brooks, P.C., Datta, N.S., Long, M.W., and Weiss, S.J. (2003). Membrane type I matrix metalloproteinase usurps tumor growth control imposed by the three-dimensional extracellular matrix. *Cell*, **114**: 33–45.

Huck, L., Pontier, S.M., Zuo, D.M., and Muller, W.J. (2010). Beta1-integrin is dispensable for the induction of ErbB2 mammary tumors but plays a critical role in the metastatic phase of tumor progression. *Proceedings of the National Academy of Sciences USA*, **107**: 15559–64.

Humphries, J.D., Byron, A., Bass, M.D., et al. (2009). Proteomic analysis of integrin-associated complexes identifies RCC2 as a dual regulator of Rac1 and Arf6. *Science Signaling*, **2**: ra51.

Kajita, M., Hogan, C., Harris, A.R., et al. (2010). Interaction with surrounding normal epithelial cells influences signalling pathways and behaviour of Src-transformed cells. *Journal of Cell Science*, **123**: 171–80.

Levental, K.R., Yu, H., Kass, L., et al. (2009). Matrix crosslinking forces tumor progression by enhancing integrin signaling. *Cell*, **139**: 891–906.

Orimo, A., Gupta, P.B., Sgroi, D.C., et al. (2005). Stromal fibroblasts present in invasive human breast carcinomas promote tumor growth and angiogenesis through elevated SDF-1/CXCL12 secretion. *Cell*, **121**: 335–48.

Provenzano, P.P., Inman, D.R., Eliceiri, K.W., et al. (2008). Collagen density promotes mammary tumor initiation and progression. *BMC Medicine*, **6**: 11.

Robinson, S.D., and Hodivala-Dilke, K.M. (2011). The role of β3-integrins in tumor angiogenesis: context is everything. *Current Opinions in Cell Biology*, **23**(5): 630–7.

Samuel, M.S., Lopez, J.I., McGhee, E.J., et al. (2011). Actomyosin-mediated cellular tension drives increased tissue stiffness and β-catenin activation to induce epidermal hyperplasia and tumor growth. *Cancer Cell*, **19**: 776–91.

Vermeer, P.D., Einwalter, L.A., Moninger, T.O., et al. (2003). Segregation of receptor and ligand regulates activation of epithelial growth factor receptor. *Nature*, **422**: 322–6.

Weigelt, B., and Bissell, M.J. (2008). Unraveling the microenvironmental influences on the normal mammary gland and breast cancer. *Seminars in Cancer Biology*, **18**: 311–21.

White, D.E., Kurpios, N.A., Zuo, D., et al. (2004). Targeted disruption of beta1-integrin in a transgenic mouse model of human breast cancer reveals an essential role in mammary tumor induction. *Cancer Cell*, **6**: 159–70.

Questions for student review

1) Which of the following are true of the relationship between the cancer stroma and cancer cells?
   a. The stroma has no barrier effect to limit the spread of cancer cells.
   b. The stroma may actively support invasion and metastasis of cancer cells.
   c. Mutations in non-cancer cells in the stroma may promote tumorigenesis.
   d. Inhibiting cancer cell–stroma interactions could arrest cancer progression.
   e. Stromal cells are remarkably homogeneous.

2) Which of the following are true of local spread and invasion of cancer cells?
   a. Secretion of TIMPs enables cells to migrate through connective tissue.
   b. Drugs inhibiting MMP activity may be expected to reduce invasion.
   c. Loss of cell–cell contacts may can prevent anoikis.
   d. General expression of E-cadherin by cancer cells promotes invasiveness.
   e. Ras-family proteins are important regulators of cancer cell motility.

3) Which of the following are true of EMT?
   a. Increasing differentiation reduces the chance of formation of cancer stem cells.
   b. Increased expression of vimentin.
   c. Reduced expression of E-cadherin.
   d. Decreased chance of invasion and metastasis.
   e. May be present in normal wound healing.

4) True or false, the basement membrane:
   a. is also known as the extracellular matrix (ECM).
   b. forms a boundary between epithelia and stroma in normal tissue.
   c. is rigid due to the presence of integrins.
   d. drives the polarity of epithelial cells.
   e. forms unstable bonds with epithelial cells via simple protein–ligand binding.

5) True or false, increased migratory phenotype in epithelia cells is often associated with:
   a. loss of cell–cell adhesion.
   b. loss of cell–matrix interactions.
   c. integrin receptors binding with high affinity to epitopes on the ECM protein.
   d. inactivation of small GTPases of the Rho family, including Rho, Rac, and Cdc42.
   e. activation of stromal MMPs.

13 Tumor Immunity and Immunotherapy

Cassian Yee
University of Washington School of Medicine, USA

The Diabolical sometimes assumes the aspect of the Good, or even embodies itself completely in its form. If this remains concealed from me, I am of course defeated, for this Good is more tempting than the genuine Good.

Franz Kafka

Everything should be made as simple as possible, but not simpler.

Albert Einstein

Key points

- The immune system developed as a means for an organism to distinguish foreign invaders from normal host tissues.
- The tumor cell evolves from normal tissues that do not usually provoke an immune response. But through the expression of mutated products or the breaching of normal tissue barriers leading to an inflammatory milieu, the tumor cell may elicit an immune response.
- The immune response to a tumor cell initially activates early immune cells (the innate response) and then immune cells involved in antigen-specific recognition (the adaptive response).
- However, tumor cells can evolve and evade the immune response, leading to the clinical appearance of cancer.
- By dissecting the manner in which the immune response is activated, how immune cells recognize tumor cells, and the factors involved in augmenting or suppressing a tumor-specific immune response, effective immunotherapeutic strategies for cancer can be developed.

Introduction

The hypothesis that the immune system plays a role in the anti-tumor response and can be manipulated for the treatment of cancer was advanced as early as the 1890s, when William Coley used bacterial extracts to treat patients with sarcoma. It was thought that the ensuing inflammatory response successfully induced regression of large tumors by activating the immune system and inducing immune cells to attack the tumor. Since then, nonspecific forms of immunotherapy have been used with varying degrees of success for the treatment of a limited number of malignancies – BCG adjuvant for superficial bladder cancer, high-dose interleukin 2 (IL-2) for metastatic melanoma, and donor lymphocyte infusions for leukemic relapse after allogeneic stem cell transplant. However, these strategies were often accompanied by serious and potentially life-threatening toxicities. Advances in immunology, in the understanding of the requirements for T-cell activation and tolerance, and in the development of novel technologies to analyze and augment immune response now provide tumor immunologists with the opportunity to translate more broadly applicable principles in immunology to the practice of treating cancer patients in a more specific and effective manner.

In the last 5 years, several advances into the clinical arena have led to mainstream acceptance of immunotherapy as a rational treatment modality, led by reproducible clinical responses in the field of adoptive cellular therapy, with phase II and randomized phase III studies confirming the anti-tumor efficacy of vaccine therapy and the checkpoint inhibitor, anti-CTLA4.

This chapter on tumor immunology begins with a description of the endogenous immune response, followed by a discussion of the components involved, including T cells, dendritic cells, and tumor antigens, and finally a summary of immunotherapeutic strategies arising from an understanding of the anti-tumor immune response.

Endogenous immune response

The endogenous immune response is composed of two phases – the *innate* response, which provides the initial line of defense against tumors, and *adaptive* response, which provides longer lasting, antigen-specific immunity. These events are usually precipitated in response to a "danger signal" in the host that can be characterized by the introduction of foreign antigens or the disruption of the normal microenvironment by an invading tumor (see Box 13.1). As a first step, natural killer (NK) cells, representing the "rapid response" component of innate immunity, release proinflammatory cytokines such as interferon-γ leading to a cascade of soluble factors that include chemokines to recruit and activate macrophages and antigen-presenting cells (APC) such as dendritic cells (Fig. 13.1). Macrophages kill tumor directly through the release of lysosomal enzymes, reactive oxygen intermediates, and nitric oxide. Dendritic cells collect tumor fragments and tumor proteins and migrate to draining lymph nodes, where they process and present these proteins to antigen-specific CD4 and CD8 T cells, which lead to initiation of an adaptive immune response (Fig. 13.2). NK cells that can also mediate direct tumor cytotoxicity provide some degree of initial protection, but it is the adaptive response characterized by expanding populations of antigen-specific T cells and their differentiation into memory cells that provide the greatest potential for long-term immunoprotection against cancer. Naïve T cells, when they first encounter antigen (also known as priming), undergo a process of differentiation that leads to an effector population mediating tumor killing (see Box 13.2). A fraction of these cells persist as memory T cells. Memory T cells are T cells "at the ready," requiring a much shorter time to become fully activated and expand than naïve T cells. This population of memory antigen-specific T cells provides for a more robust and rapid response when antigens reappear.

Box 13.1 The "danger signal" in immune response

Activation of the innate and subsequently adaptive responses is believed to be precipitated by molecules that are released by cells undergoing stress or abnormal cell death, and have often been described as "danger signals." In the case of immunity to foreign pathogens, cell products from infectious organisms, for example lipopolysaccharides from gram-negative bacteria, double-stranded RNA from viruses, or zymosan from fungi, can all lead to a "danger signal" and a productive immune response. Empirical attempts by scientists and clinicians to activate an immune response by recapitulating endogenous immunity have been largely achieved through the use of adjuvants – which are an ill-defined mixture of bacterial components or irritants that are likely to induce inflammation and that, in the presence of a vaccine, for example, can lead to the desired antigen-specific immune response. Only recently has the molecular basis for this initiating response been defined. A family of receptors expressed by cells of the innate response – NK cells and antigen-presenting cells, in particular dendritic cells – allow these cells to perceive these "danger signals" and are known as "pattern recognition receptors."

These include the Toll-like receptors (TLR) of which there are 10 varieties in mammals, with each TLR recognizing specific ligands. What is more interesting is that these ligands are not limited to foreign pathogens, but may also be activated by host products, such as heat shock proteins. Other recognition receptors, not included within the family of TLR, include Fcγ receptors that recognize opsonized (antibody-bound) antigens, scavenger receptors, nucleotide–oligomerization domain family (NOD) receptors that bind bacterial peptidoglycans, and "stress"-activated molecules, such as MICA, which appear on virally infected or aberrant (i.e. transformed) cells and are believed to play an important role in anti-tumor immunity through the activation of stimulatory ligands on NK cells and CD8+ T cells. It is believed that transformed tumor cells by upregulation of surface MICA or the release of HSP following cell death provide a "danger signal" to dendritic cells, activating dendritic cells, inducing the release of cytokines and chemokines, and drawing effector cells to the site of attack, resulting in the cascade of events that lead to an adaptive T-cell response.

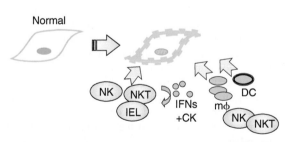

Figure 13.1 Innate anti-tumor immunity. See text for details. The development of tumor cells results in a proinflammatory milieu and activation of NK, NKT and IEL (intra-epithelial) cells. Release of cytokines and chemokines leads to recruitment of macrophages and additional innate effectors which provide an initial line of defense, and dendritic cells which initiate the adaptive immune response.

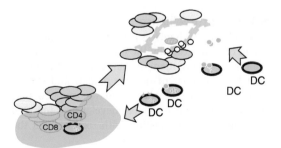

Figure 13.2 Adaptive anti-tumor immunity. Dendritic cells take up tumor antigen, and process and present antigen to CD4 and CD8 T cells in draining lymph nodes. Presentation of tumor antigen leads to CD4 and CD8 T-cell activation (cf. Fig. 13.3 for details). Activated CD4 and CD8 T cells traffic to tumor sites where they mediate an antigen-specific effector response through the release of cytotoxic granules, Fas–FasL interaction, and the recruitment of secondary effectors (cf. Fig. 13.4 for details).

Box 13.2 T-cell priming and differentiation, and T-cell memory response

During development, the pluripotent stem cell gives rise to lymphoid progenitors that mature into T cells in the thymus. Mature T cells leaving the thymus into the circulation are each endowed with a singular antigen specificity determined by the T-cell receptor they bear on the surface. Together, these T cells form the repertoire of all possible specificities the organism can respond to and are represented by "naïve" T cells on the basis of not having previously encountered antigen. The presentation of antigen, for example by dendritic cells, selects for and activates a preexisting naïve T cell ❶, leading to its differentiation into an effector T cell and clonal expansion. Effector cells possess the ability to release cytokines and toxic granules (e.g. perforin and granzyme B), and mediate tumor cell killing ❷. Following an effector response, the majority of antigen-specific T cells undergo apoptosis ❸, but a small fraction develop into memory T cells ❹ that can survive in a quiescent state for many years after the antigen has been eliminated; these are found to reside in lymphoid and mucosal tissues ❺. Upon antigen re-exposure ①, these memory cells are quickly called into action, having the capacity to be rapidly activated and to expand ② and kill tumor ③.

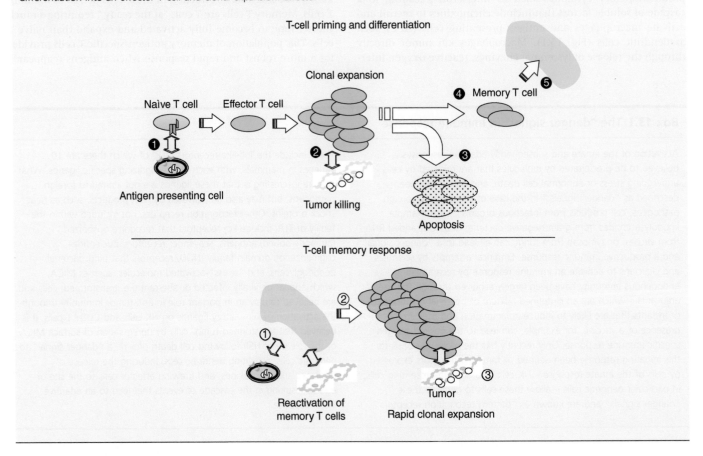

Cancer immunosurveillance

That these defense mechanisms provide an immunologic basis for suppression of tumor cells by the endogenous immune response was hypothesized more than 40 years ago and has recently been supported by a number of murine models and clinical observations. In 1957, Burnet and Thomas postulated that "small accumulations of tumor cells may develop . . . and provoke an effective immunological reaction with regression of the tumor and no clinical hint of its existence."

However, initial studies evaluating the incidence of chemically induced or spontaneous tumors in immunosuppressed mice treated by thymectomy or anti-lymphocyte sera led to conflicting results. Malignancies that did develop in immunodeficient mice were often virally induced or represented by lymphomas attributed to chronic antigenic stimulation from infectious agents. When athymic nude mice were evaluated for the development of spontaneous or chemically induced tumors, there was no increase in tumor incidence when compared to immunocompetent mice. Although not known at the time, athymic nude mice do retain a functional immune system composed of a small population of T cells and thymic-independent populations of NK cells. In the 1990s, mice were engineered with deficiencies in lymphocyte-specific recombinase, which was responsible for antigen receptor rearrangement (RAG-2 −/− mice). Such mice, which possessed no NK, T, and B cells, were found to develop sarcomas more frequently and with a shorter latency period than wild-type mice.

Interferon-γ was also shown to be protective against the growth of spontaneous and chemically induced tumors, and pivotal experiments using mice deficient for the interferon-γ receptor, or its signaling molecule, STAT1, demonstrated an increased incidence of chemically induced tumors that was over 10 times greater than that of wild-type controls. Mice deficient in both RAG-2 and STAT1 exhibited, in addition to inducible sarcomas, spontaneous intestinal and mammary cancers. Further, mice lacking perforin, which is released within cytolytic granules upon lymphocyte stimulation to effect tumor cell killing, were found to be more susceptible to sarcomas and spontaneous lymphomas. Interestingly, chemically induced tumors developing in immunodeficient (RAG-2 –/–) mice, when injected into wild-type, immunocompetent mice, were found to be more readily rejected than those tumors developing in immunocompetent mice. These results suggested that the innate immune response in immunologically intact hosts may be instrumental in sculpting the immunogenic phenotype of tumors and may, over time, result in immunoselection that leads to tumor immune escape.

Cancer and immunosuppression

The role of the immune system in suppressing the development and progression of human malignancies was suggested initially in observational studies evaluating infiltrating or intratumoral lymphocytes in melanoma biopsy specimens. A correlation between the extent of mononuclear infiltration and prognosis has been extended for breast, colon, and most recently ovarian cancer in which patients whose tumors contained T cells experienced a 6–8-fold greater 5-year overall survival rate than those whose tumors contained no T cells. Conversely, patients who are chronically and/or severely immunosuppressed, as a result of stem cell ablation, HIV, or induced immunosuppression following transplantation, experience a higher overall incidence of malignancy than the general population. Although melanomas and occasionally colon, lung, and bladder cancers are observed, most tumors that develop are virally related, for example Epstein–Barr virus (EBV) post-transplant lymphoproliferative disease progressing to lymphoma, Kaposi sarcoma, or human papilloma virus (HPV)–associated malignancies. This may be consequences of a shorter latency period for viral-associated malignancies and the shorter life span of immunosuppressed patients, which preclude the appearance of longer developing malignancies. In some cases, reversal of immunosuppression has resulted in dramatic responses.

Effector cells in tumor immunity

Effectors of adaptive immunity can be ascribed to "humoral" and "cellular" arms, represented respectively by B cells that mediate effects through the production of antibodies, and T cells that interact directly with target cells through the T-cell receptor. In humoral immunity, antibodies binding surface proteins on tumor cells can kill via complement activation or by bridging targets with cytotoxic cells through a process known as antibody-dependent cell-mediated cytotoxicity (ADCC). In this process, the Fc portion of antibody couples with receptors on macrophages or NK cells, which then effect cell killing. Although antibodies are highly effective *in vitro*, convincing evidence that antibody responses elicited *in vivo* play a critical role in anti-tumor immunity is weak. However, the significance of humoral responses with respect to tumor immunity has been supported by the identification of serum antibodies to potentially immunogenic tumor antigens (see the "SEREX" section of this chapter) and the successful therapy of patients using monoclonal antibodies (see the "Antibody therapy of cancer" section).

Dendritic cells

Dendritic cells are specialized or "professional" antigen-presenting cells that are activated during the innate immune response and are uniquely equipped to take up and present antigen to effector cells of the adaptive immune system – the antigen-specific CD4 and CD8 T cells. Dendritic cells are so-named because of pseudopods or "dendrites," which are processes that extend from the cell to facilitate antigen presentation. *In vivo*, the induction of an anti-tumor immune response may occur by tumor cells presenting antigen directly to T cells, but it is believed that a more common and robust pathway for tumor-specific T-cell activation *in vivo* is by cross-presentation. This is a process by which antigens released by necrotic, dying, or apoptotic tumors are taken up by dendritic cells and re-presented to T cells under more favorable stimulatory conditions in the tumor-draining lymph node.

Dendritic cells can be characterized in their immature or mature forms based on contrasting surface and functional phenotypes (Table 13.1). In their immature form, dendritic cells are well equipped to capture antigens through surface receptors such as the C-type lectins (e.g. DEC-205 and mannose receptors), $\alpha v \beta 5$ integrins, or CD36 for internal processing and presentation. Dendritic cell activation via "danger signals" mediated through some of these receptors and other surface receptors (e.g. the Toll-like receptors) leads to DC maturation in the presence of bacterial or viral products, TNF-α, or prostaglandins. It is also thought that in addition to DC activation by these receptors, cooperation of CD4 helper T cells is required to "license" dendritic cells with the capacity to stimulate CD8 T cells through interaction of the CD40 ligand on CD4 T cells with CD40 on dendritic cells. Upon maturation, further antigen uptake by dendritic cells is downregulated, and, in preparation for optimizing T-cell activation, dendritic cells upregulate surface expression of the T-cell costimulatory molecules (CD80, CD83, and CD86), and secrete cytokines such as IL-7 and IL-12 which facilitate full T-cell activation. In the case of tumor immunity, dendritic cells circulate through the blood and accumulate at tissue sites in response to chemokines arising from the site of tumor necrosis or inflammation. As immature dendritic cells, antigen is collected and processed for presentation on the surface in the context of major histocompatibility complex (MHC) molecules. Following maturation and upregulation of surface costimulatory molecules, lymphokines, and the chemokine receptor CCR7, dendritic cells traffic to lymph nodes where T-cell activation can occur.

T cells

Antigen presentation and T-cell stimulation

In contrast to B cells, which provide "humoral" immunity through the production of soluble antibodies, T cells mediate "cellular" immunity by interacting directly with their target cell. T cells achieve specificity for cells expressing the target antigen through the surface T-cell receptor which recognizes fragments of antigen (usually peptide fragments) presented by the MHC through either a Class I or Class II processing pathway (Fig. 13.3). Since proper antigen processing and presentation by the MHC are critical to antigen-specific immunity, a brief description of the MHC complex is presented.

The MHC is encoded by highly polymorphic genes that cluster on chromosome 6 in humans and are co-dominantly expressed. Human MHC molecules are called human leukocyte antigens

Table 13.1 Immature versus mature dendritic cells – surface and functional phenotypic differences

	Immature DC	Mature DC	Activation signals for DC maturation
Morphology		Increased "veil" and dendrite appearance	Bacterial products:
Phagocytosis	+++	–	LPS (lipopolysaccharide)
Costimulatory molecules			Teichoic acid
HLA-DR (*MHC CLASS II)			CpG DNA
CD40	+/–	+++	Viral products:
CD80 (B7-1)			dsRNA
CD83			
CD86 (B7-2)			CD40 ligand
Chemokine receptors			
CCR7	–	++	TNFα, PGE2
CCR2	++	+	
CCR6			

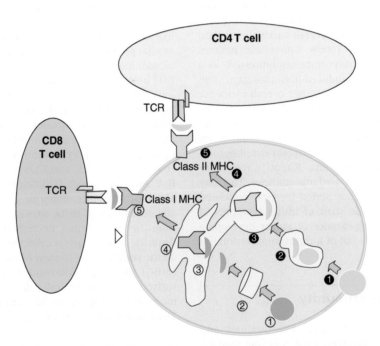

Figure 13.3 Antigen presentation. In the Class I pathway (①–⑤), cytosolic proteins are processed by proteasomes ② into peptide fragments, transported through the endoplasmic reticulum (ER, ③), complexed with Class I MHC ④ and β2-microglobulin, and brought to the surface ⑤, where they are presented to CD8 T cells.
In the Class II pathway, extracellular protein antigens are endocytosed ❶, degraded into peptide fragments ❷, combined with MHC ❸, and presented to the surface ❹ where they are presented to CD4 T cells ❺.

(HLA) and are divided into Class I and Class II HLA or MHC, which present peptides to CD8 and CD4 T cells, respectively. For the most part, Class I MHC molecules are represented in humans by the HLA-A, -B, and -C families of alleles, and Class II MHC by the HLA-DR, -DQ, and -DP families. The Class I MHC complex is composed of three parts: the MHC-encoded heavy α chain, a non-MHC-encoded β2-microglobulin chain, and an 8-11-mer peptide sitting in a groove formed by the polymorphic region of the α chain. The Class II MHC complex is composed of three parts, the polymorphic α and β chains and a 10-30+ mer peptide sitting in a groove formed by the two chains. These parts are assembled in cytosolic compartments together with peptides derived from tumor proteins that have undergone proteasome-mediated cleavage to peptides of the appropriate length. The peptide–MHC complex is then transported to the surface, where the immunogenic peptide sitting in the MHC groove is presented to the T-cell receptor. The T cell recognizes the peptide *only* in the context of the MHC complex; therefore, mutations affecting

Box 13.3 T cell–APC interaction

Tumor-derived peptides are processed within the APC and presented by the peptide MHC complex to their cognate T-cell receptor (TCR).

For CD4 T cells, optimal T-cell activation requires the ligation of costimulatory molecules such as CD28 with B7. Upregulation of CD40 ligand following TCR engagement delivers a "licensing" signal to APC that results in increased B7 expression and the production of T-cell modulatory cytokines such as IL-12. T-cell activation is downregulated by the inducible inhibitory receptor, CTLA-4, which blocks CD28-mediated signals and competes for B7 binding.

For CD8 T cells, activation following TCR engagement with the Class I MHC complex may be enhanced through costimulatory signals delivered by 4-1BB and other counterreceptors. IL-2 and other cytokines produced by activated CD4 T cells provide growth signals to cognate CD8 T cells.

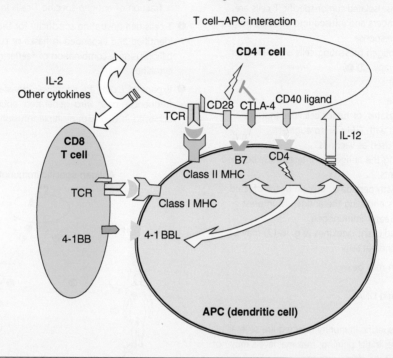

any component of the antigen0presenting machinery can abrogate specific T-cell recognition and killing of tumor cells.

Costimulatory and inhibitory T-cell signals

In addition to the interaction of the T-cell receptor (TCR) with the peptide–MHC complex, the activation of T cells can be modulated by the engagement of surface costimulatory or accessory molecules by their respective ligands on antigen-presenting cells (see Box 13.3). The most prominent of these is the signals provided by CD28 upon binding to B7-1 (CD80) or B7-2 (CD86) on APC. B7–CD28 interaction mediates signals that can fully activate an antigen-driven T-cell response, enhance T-cell survival by upregulation of antiapoptotic proteins such as BCL-xL, and drive proliferation. Absence of B7 has been associated with T-cell energy, while engineered expression of B7 in potentially immunogenic tumor cells can induce tumor rejection in murine models. While B7–CD28 interaction appears to be critical to the generation or priming of an effective anti-tumor response, it does not influence the effector or killing phase of T cells. Hence, T cells generated with a B7-transduced tumor vaccine can eradicate B7-negative tumor. Other costimulatory molecules that deliver a positive signal to T cells include ICOS (inducible costimulator), OX40, 4-1BB, and other B7 family members (e.g. B7–H3). Accessory or adhesion molecules such as ICAM-1 and LFA-1 are also critical to T-cell recognition. These molecules converge in and reinforce the TCR–peptide–MHC synapse by forming in aggregate with other molecules, a supramolecular activation complex to facilitate delivery of a longer lasting, more potent T–cell signal.

Cytotoxic T-lymphocyte associated antigen-4 (CTLA-4) delivers a negative regulatory signal to activated T cells and competes with CD28 for binding to B7 on target cells (see Box 13.4). Since the overall effect is to block cell-cycle progression, CTLA4 is considered a checkpoint inhibitor. CTLA-4 is present predominantly as an intracellular protein and appears on the surface as an inducible receptor with a greater affinity for B7 than CD28; however, in contrast to CD28, its surface expression is nonconstitutive and relatively short-lived. CTLA-4 is believed to provide an immunologic "brake" to prevent overly robust and potentially damaging overstimulation by suppressing T-cell proliferation through IL-2 inhibition and downregulation of cell-cycle activity. CTLA-4-deficient mice develop splenomegaly and a lymphoproliferative pathology. Since many tumor target antigens are also normal self-proteins, eliminating CTLA-4 inhibition may provide a means of breaking tolerance to self-antigens and augment an otherwise muted T-cell response to tumor. Administration of anti-CTLA-4 antibody in some murine models results in organ

Box 13.4 Antigen-specific immunotherapy

Augmentation of tumor antigen-specific cellular response can be achieved by vaccination or adoptive cellular therapy strategies. For vaccines, protein, peptides, or antigen-charged dendritic cells are manufactured *ex vivo* and used as immunogens to stimulate the immune response *in vivo*. An increased frequency of antigen-specific T cells is elicited *in vivo* after repeated immunizations. For adoptive therapy, T cells of defined specificity and phenotype are isolated *ex vivo* by iterative stimulations. Isolated tumor-specific T cells are expanded *ex vivo* to large numbers and infused into patients to augment the *in vivo* immune response.

A source of stimulator or antigen presenting cells is required for either vaccine or adoptive therapy. (① ❶)

For vaccine therapy (①–④):

② The vaccine reagent may be:
- Whole tumor cells, cell lysates, or tumor cell lines that are genetically modified (e.g. with GM-CSF to augment immunogenicity) can be used as vaccines.
- Peptides or DNA encoding the antigen of interest may be used directly as vaccine reagents.
- Dendritic cells charged with peptide or protein or transfected with recombinant vectors encoding the antigen of interest represent a potentially greater immunogen.
- Adjuvant (e.g. BCG) or adjuvant cytokines (e.g. IL-12) may be added to enhance immunoreactivity.

③ The route of administration may be:
- Intradermal.
- Intramuscular (e.g. plasmid DNA).
- Intralymphatic.

④ Augmentation of antigen-specific immunity may require several booster immunizations after initial priming; maximal levels may not be achieved until several weeks after vaccination.

For adoptive therapy (❶–❺):

❷ Leukapheresis:
- Is a procedure that collects white blood cells from patients and provides peripheral blood mononuclear cells as a source of dendritic cells and T cells for *in vitro* stimulation cultures.

❸ Isolation of antigen-specific T cells *in vitro*.
- Requires a source of stimulator and responder cells. Stimulator cells can be dendritic cells pulsed with peptide or engineered to express antigen following transfection or viral infection with recombinant vectors encoding the target gene of interest. Responder cells can be PBMC enriched for CD4 or CD8 T cells. Iterative restimulation cycles are performed to augment the fraction of antigen-specific T cells *in vitro*.

❹ T cells demonstrating specificity for the target antigen are identified and expanded in flasks or culture bags to several billion using a combination of feeder cells, TCR stimulation, and cytokines.

❺ Expanded T cells are adoptively transferred into patients; high frequencies of *in vitro*–generated, antigen-specific T cells may be achieved *in vivo* with multiple infusions.

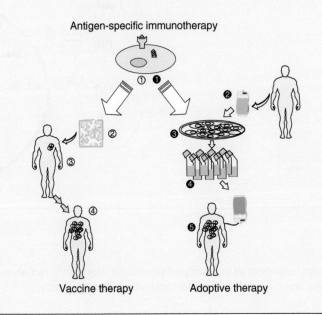

Antigen-specific immunotherapy

Vaccine therapy | Adoptive therapy

---

autoimmunity but can also lead to rejection of previously non-immunogenic tumors. In clinical trials, administration of anti-CTLA4 antibody has produced signs of autoimmune toxicity as well as tumor regression in individuals receiving a tumor-specific vaccine.

Programmed death-1 (PD-1) is another checkpoint inhibitor responsible for maintaining the balance between activation and suppression of T-cell responses. Its ligands PD-L1 (or B7-H1) and PD-L2 (B7-H2) are expressed on antigen-presenting cells, nonhematopoietic cells, and tumor cells. Bidirectional signaling between B7-1 and PD-L1 has also been observed, leading to an overall inhibitory effect. Blocking of PD-1–PD-L1 interactions has led to restoration of proliferative and cytokine capacity.

PD-L1 is aberrantly expressed on tumor cells and is a poor prognostic feature in patients with many solid tumor malignances. A PD-L1 blockade appears to enhance concomitant immunotherapies in animal models of vaccine and adoptive therapy. Anti-PD-1 therapy can also lead to reversal of effector cell exhaustion and increased cytotoxic T-lymphocyte (CTL) resistance to Treg-mediated inhibition.

T lymphocytes

T cells can generally be divided into helper CD4 T cells and cytotoxic or killer CD8 T cells. Helper CD4 T cells recognize antigen in the context of MHC Class II and can be further differentiated into Th1 and Th2 subsets on the basis of distinct cytokine and receptor profiles. Th1 CD4 T cells produce IL-2 and interferon-γ, and express IL-12 and IL-18 receptors and regulate T-cell immunity, while Th2 T cells produce IL-14, IL-15, and IL-13, and regulate B-cell immunity. It is believed that a Th1-type response would be beneficial in anti-tumor immunity since it mobilizes a T-cell-mediated response.

Cytotoxic CD8 T cells recognize antigen in the context of MHC Class I and, when activated, release perforin and toxic granules

Tumor Immunity and Immunotherapy

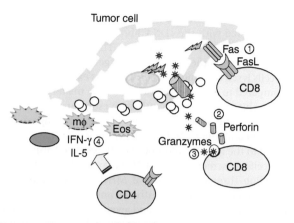

Figure 13.4 Effector mechanisms of T cells. Activated CD8 T cells deliver a "death" signal to tumor cells through Fas ligand–Fas interaction ①. CD8 T cells may also kill tumor cells directly through perforin and granzymes released upon engagement of the T-cell receptor ②. Perforin exocytosed in CTL granules forms pores in the tumor cell membrane. Granzymes enter tumor cells through pores and induce tumor cell death ③. CD4 T cells can mediate tumor death through Fas interaction. Activated CD4 T cells may also mediate cytotoxicity indirectly through the release interferon-gamma and IL-5 to recruit tumoricidal macrophages (mφ) and eosinophils (Eos) ④.

that mediate direct cell killing by punching holes in the cell membrane to facilitate entry of enzymatic packets (granzymes A and B). Although most studies have weighed in on a greater role for the cytotoxic CD8+ T lymphocyte (CTL) in tumor eradication, the helper CD4 T lymphocyte has also been shown to be a vital component in the induction and maintenance of a competent anti-tumor immune response. Not only have tumor antigen-specific responses been identified for CD4 T cells, but also the presence of CD4 T cells may be required for CD8+ CTL responsiveness. Acting in concert, both CD4 and CD8 T cells provide for synergistic mechanisms of tumor killing. CD8 T cells kill tumor cells through the release of perforin and granzymes A and B, or through engagement of the death receptor, Fas, through Fas ligand (FasL) expressed on activated T cells. FasL–Fas interaction leads to a form of cell death known as apoptosis. In contrast to necrosis or death due to cell injury, apoptosis or programmed cell death involves a stepwise cascade of events initiated by receptor engagement at the cell surface (in this case, Fas), leading to DNA fragmentation. CD4 T cells can kill tumor cells directly by FasL–Fas engagement, as well as through indirect mechanisms that involve the recruitment of nonspecific effectors such as macrophages and eosinophils, which can act even on MHC-negative tumors (Fig. 13.4).

Natural killer cells

NK cells are activated during the innate response by the inflammatory milieu that is established by invading tumor cells. These effector cells are not antigen-specific and do not express a TCR but do kill tumor through killer *activating* receptors (KARs) expressed on their surface. Engagement of KARs with tumor-derived ligands such as MICA and MICB, which are upregulated in infected or "stressed" cells such as tumor cells, leads to NK cell activation and tumor cell death. NK cells also engage self-MHC Class I molecules on target cells through *inhibitory* receptors (killer inhibitory receptors (KIRs)), perhaps as a means of preventing autoreactivity. The loss of MHC expression on tumor cells, a process which can develop during carcinogenesis and immunoselection, lends itself to preferential NK cell activation. The contribution of NK cells to the endogenous anti-tumor response *in vivo* may be best exemplified in athymic nude mice which have no T cells, but retain a population of functional NK cells which appears to be sufficient to mediate tumor resistance. In humans, NK-type cells can be expanded *in vitro* with high doses of IL-2 for adoptive transfer; however, in this setting, their efficacy is less well defined and treatment is often accompanied by serious toxicity. The *in vivo* augmentation of NK-type cells may also be one mechanism by which high-dose IL-2 therapy has shown some clinical effect in the treatment of patients with metastatic melanoma or renal cancer.

Regulatory T cells

A population of T cells with regulatory properties that control autoimmune and anti-tumor responses was postulated as early as 1975; however, convincing evidence for their existence has been elusive. Recently, a population of CD4+ CD25+ T cells that possess immunosuppressive function has been identified. This discovery has led to a renewed understanding of the role of regulatory cells. CD4+ regulatory T cells (Tregs) are represented by two subsets – naturally occurring Tregs representing 5–10% of peripheral T cells, and induced Tregs that develop from conventional CD4+ CD25– T cells. Naturally occurring Tregs mediate their suppressive properties through cell-to-cell contact by an unknown mechanism. Although activation is dependent on TCR engagement, their suppressive effects are nonspecific. They are known to express glucocorticoid-induced TNF receptor (GITR) and CTLA-4, a known T-cell inhibitor of T-cell costimulation. However, the role of CTLA4 and GITR in mediating the suppressive effects of naturally occurring Tregs is not well defined. Induced or adaptive Tregs can be generated from conventional CD4+ CD25– T cells following *in vitro* exposure to antigen and IL-10, and the induced Treg cells themselves appear to mediate their inhibitory properties through the production of IL-10 and TGF-β. Tregs have been found to be fundamental for the control of autoimmune responses in several murine models such as inflammatory bowel disease, and depletion of CD25+ T cells has been shown to mediate immune rejection of various murine tumors *in vivo*, presumably through the release of suppressive effects on T cells targeting shared tumor self-antigens. Elevated frequencies of CD4 Treg cells have also been described in cancer patients, leading to the design of clinical trials involving the administration of anti-CD25 antibodies to augment an endogenous anti-tumor immune response.

Tumor antigens

Methods for identifying immune targets

T-cell-defined antigens

Although it had been possible for some time to isolate tumor antigen-specific T cells from the peripheral blood or tumor-infiltrating population of lymphocytes by iterative *in vitro* restimulation with autologous tumor cells, it was not until 1991 that the first T-cell-defined human tumor antigen, MAGE-1 (for melanoma antigen-1), was discovered. In this strategy, a tumor-specific CD8+ CTL line was first generated *in vitro* following repeated stimulation of lymphocytes with autologous tumor

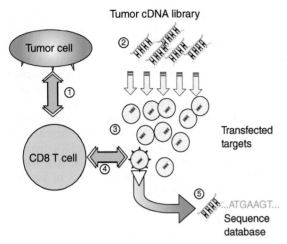

Figure 13.5 Identification of T-cell-defined tumor antigens. CD8+ T cells are stimulated *in vitro* with autologous tumor cells to generate tumor-specific CD8 T-cell clones ①. To identify the tumor antigen recognized by these T-cell clones, a cDNA library is first prepared from the tumor cell line ② and used to transfect HLA-matched nontumor target cells (in this case, COS cells) ③. These cells are screened using the tumor-specific T-cell clone ④ and those recognized by the T cell are isolated, their cDNA extracted, sequenced and compared to a genetic database ⑤ to identify the antigen targeted by the tumor-specific T cell.

cells. Target cells engineered to express the restricting allele were then transfected with the cDNA library of the target tumor, and the autologous tumor-specific CD8+ CTL line was used to screen pooled samples of the transfected target cells (Fig. 13.5). The cDNA cells lysed by tumor-specific CTL was extracted, sequenced, and compared to a gene database to identify the targeted antigen. Based on this strategy, many other investigators since have successfully isolated additional tumor-associated antigens in this T-cell-defined manner (Table 13.2). These studies have also been extended to the identification of Class II–restricted antigens using CD4 T cells to screen for a cDNA library modified to facilitate presentation through the Class II MHC alleles. Interestingly, a number of identified Class I–restricted antigens are shared among patients with similar tumors and even, in some cases, among different tumor types. Furthermore, several of these antigens are represented by normal nonmutated self-proteins such as tyrosinase, a melanosomal protein found in pigmented cells such as normal melanocytes.

SEREX

Serological recombinant expression cloning (SEREX) has been used to identify potentially immunogenic tumor antigens from a wide variety of human cancers. SEREX is a simple, robust immu-

Table 13.2 Tumor antigens

Antigen	Tissue expression	
	Tumors	**Normal tissues**
Self-antigens associated with normal differentiation		
Tyrosinase	Melanoma	Melanocytes
MART1/MelanA	Melanoma	Melanocytes
gp100	Melanoma	Melanocytes
Prostate-specific antigens	Prostate cancer	Prostate
CD20 and Idiotype	B-cell malignancies	B cells
Her-2-neu	Breast and ovarian cancer	
Self-antigens associated with tumorigenic phenotype		
Survivin	Most tumors	
Telomerase	Most tumors	
p53	Most tumors	
WT1	Leukemias, and lung and other solid tumors	
Cancer–testis and oncofetal antigens		
MAGE-1 and MAGE-3	Ovarian, breast, GI, and melanoma	Testis and placenta
GAGE and others	Head and neck (H&N) and melanoma	Testis and placenta
NY-ESO-1	Ovarian, GI, breast, melanoma, lung, and others	Testis and placenta
CEA	GI and breast	GI and breast
Mutated antigens		
CASP-8	Head and neck cancer	–
CDK4-kinase and MUM-3	Melanoma	–
Beta-catenin	Melanoma and lung	–
Viral antigens		
Human papilloma virus	Cervical cancer	–
Epstein-Barr virus	Hodgkin lymphoma, H&N, and post-transplant lymphoproliferative disorder (PTLD)	–

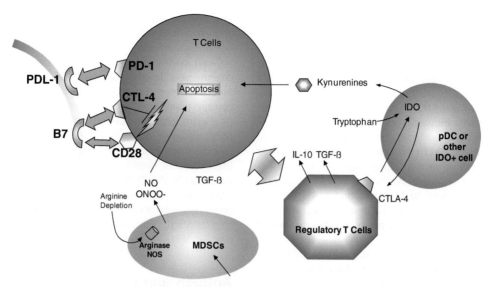

Figure 13.6 Mechanisms of immune evasion and sites of intervention. Immune checkpoint inhibitors, PD-1 and CTLA4 induced upon T-cell activation can be suppressed using anti-PD1 or anti-PD-L1 and anti-CTLA4 antibodies, respectively which individually have already demonstrated a robust clinical effect. Myeloid-derived suppressor cells (MDSC) which are expanded in tumor microenvironment, inhibit T-cell function through arginine depletion, production of peroxynitrites, ROS and TGF-beta; MDSCs can be targeted using ATRA to promote MDSC differentiation, COX-2 inhibitors and ROS inhibitors to decrease MDSC-mediated immune suppression. Regulatory T cells constitutively express high levels of the IL-2R (CD25) and Glucocorticoid-inducible TNF receptor (GITR) which may be targeted by anti-CD25 antibody or toxin-conjugated IL-2, and , anti-GITR antibody respectively. IDO (indoleamine dioxygenase)-expressing stromal or antigen presenting cells convert "T-cell-essential" tryptophan to kynurenines which mediate apoptosis through oxygen free radicals. The analog, 1-methyl tryptophan, can effectively inhibit IDO activity.

noscreening technique in which auto-antibodies present in patient serum samples are used to identify tumor proteins within tumor-derived cDNA expression libraries. Currently, over 2000 SEREX clones for several human cancers have been identified using this procedure. SEREX immunoscreening identifies tumor antigens on the basis of a T-cell-dependent IgG antibody response. The induction of an IgG response suggests that a cellular immune response was operative, and, indeed, both antibody and T-cell responses to several human tumor antigens such as NY-ESO-1, HER2/neu, PSA, and p53 have been detected within the same cancer patient. The majority of one class of tumor-associated antigens, the cancer–testis antigens, has been identified in this manner.

Gene expression profiling

The use of cDNA microarrays and more sophisticated methods including Serial Analysis of Gene Expression (SAGE) has helped to identify overexpressed genes among tumor cells compared to normal tissues. Combining this technology with immunomagnetic selection or laser capture microdissection to enrich for tumor cells represents a potentially powerful method for antigen mining. Once overexpressed or uniquely expressed genes have been identified, determining if such genes are in fact antigenically relevant targets is critical. This can be resolved by using algorithms to identify potential epitopes within the gene sequence that are predicted to bind given MHC alleles with high affinity or are more likely to be processed and presented by antigen-presenting machinery. These predicted epitope sequences would then be validated empirically by assessing their ability to generate antigen-specific T-cell responses *in vitro*. Alternatively, antigen-presenting cells can be engineered to express the target gene (by viral transduction, nucleotide transfection, or other means) and then validated through *in vitro* testing for T-cell reactivity (see the "Dendritic cell vaccines" section and Fig. 13.6).

Classification of tumor antigens

In the context of antibody therapy, tumor antigens may be represented by any overexpressed proteins on the surface of tumor cells, and antibody therapy has led by far immunotherapy advances into the clinical arena. However, such surface proteins may not necessarily be immunogenic (i.e. capable of inducing an immune response *in vivo*). For the purposes of vaccine studies or adoptive cellular therapy, immunization or *in vitro* generation of antigen-specific effectors can be readily achieved only against targets for which an immune repertoire exists and against which an effective immune response can be elicited. The list in Table 13.2 describes tumor antigens which are known to elicit a cellular or humoral response and represents potential targets for cellular immunotherapy.

Tumor antigens can be classified on the basis of tumor specificity and function.

Surprisingly, many tumor antigens initially described represented an increasingly important paradigm: the identification of normal self-proteins as tumor-associated antigens. Most of these differentiation type antigens were described in melanoma and play a role in melanin production. The reason for the preponderance of antigens that were initially identified in melanoma is perhaps due to the facts that melanoma is known to be amenable to immune manipulation (i.e. the use of nonspecific immunomodulators such as IL-2 and IFN can result in clinical responses), and that tumor lines and melanoma-specific T-cell lines for construction and screening of cDNA libraries were more commonly available. For melanoma, the appearance of vitiligo (depigmentation) in patients is predictive of a clinical response and suggests that such shared antigens are in fact therapeutically relevant. Targeting self-proteins is accompanied by the risk of autoimmune toxicity. In the case where the disease organ is surgically removed (e.g. prostate cancer) or toxicities are acceptable to the patient (e.g. vitiligo), targeting self-proteins may be a feasible strategy.

Those normal self-antigens which are overexpressed in tumors and are associated with a tumorigenic or proliferative phenotype provide the added benefit that immune escape through loss of antigen expression is less likely to occur. Among these include surviving, an antiapoptotic protein; telomerase, the ribonucleoprotein responsible for telomere synthesis; and WT-1, a transcription factor overexpressed in leukemia and other cancers. However, the safety and efficacy of targeting such self-antigens are presently unknown and are being evaluated in a number of current immunotherapy trials.

Truly tumor-specific antigens, such as the mutated cell-cycle regulator CDK4 kinase, or the virus-associated proteins HPV E6/E7 and EBV LMP proteins, represent favorable targets for immunotherapy but are often limited to a small number of cancer types or may be associated with inherent immune escape mechanisms, such as altered antigen presentation in the case of the E6 and E7 proteins associated with HPV+ cervical cancer.

Cancer–testis antigens represent a distinct family of tumor-associated antigens with normal expression limited to the adult testis and fetal placenta. These antigens are variably expressed in a wide variety of tumors, including melanoma, breast, lung, and ovarian cancers with generally increased expression in metastatic sites. Although their function is unknown, these genes are regulated by methylation, and agents that induce hypomethylation and/or histone acetylation increase cancer–testis antigen expression in tumors. Because of their relative immunogenicity and potential for inducible expression, cancer–testis antigens are being evaluated in a number of clinical trials as immunotherapy targets.

T-cell repertoire and affinity for tumor antigens

The repertoire of T cells in the adult immune system that can recognize the vast array of antigens and their peptide fragments is represented by the naïve T-cell population which is shaped during embryonic development by events in the thymus. During thymic development, potentially autoreactive T cells are deleted, while potentially beneficial T cells are retained. This process of thymic selection involves the presentation of auto-antigens to developing T cells and is dependent on the affinity of the TCR for its ligand such that high-affinity interactions lead to T-cell deletion, low-affinity interactions are ignored, and moderate-affinity interactions are retained. The resulting repertoire consists of a precursor or naïve T-cell population that is predominantly capable of responding to foreign antigens, and a lesser population of lower affinity, potentially self-reactive T cells.

Therefore, one overriding concern when targeting tumor antigens is whether T cells of sufficient affinity can be generated from a normal T-cell repertoire which not only harbor a very low precursor frequency of self-reactive T cells but also whose high-affinity responders may have been deleted as a consequence of thymic development. For differentiation antigens in melanoma, T cells responding to MART-1 or tyrosinase exhibit at least one log lower affinity for cognate targets than virus-specific T cells in the same individual. This is further compounded when peptides are used in vaccine therapy, such that supraphysiologic peptide concentrations are more likely to elicit low-affinity responses and delete high-affinity T cells in general.

T-cell avidity is largely dependent on the TCR affinity for its cognate peptide–MHC complex and is primarily a function of and inversely related to the dissociation constant (Kd). Altering epitope peptide sequences to enhance peptide binding to MHC or altering peptides that influence TCR contact with the peptide–MHC complex may elicit T cells with improved TCR affinity. Indeed, intentional mutations in the CDR3 (peptide–MHC binding) region of tumor antigen-specific TCR and amino acid substitutions in the TCR contact residues of the presenting epitope have been shown to lead to more robust, higher affinity T-cell responses. Engineering T cells to express such high-affinity receptors represents one rational strategy to the development of effective T-cell responses in adoptive cellular therapy.

Antigen-specific therapy of cancer

In general, there are two basic strategies to augment the antigen-specific immune response to tumor cells – vaccine therapy and adoptive therapy (see Box 13.4). Vaccine therapy involves the use of an immune stimulator – be it DNA, protein, or cell based – that is administered to induce a tumor-specific response in the host, often through repeated injections. It is anticipated that the immunogenic properties associated with the vaccine reagent will elicit an immune response that will be amplified *in vivo* with each boost. Adoptive therapy involves the *ex vivo* isolation and expansion of immune effector cells followed by their adoptive transfer into the host to augment the anti-tumor response. Generally, vaccination strategies are less labor intensive, especially in cases where the reagent is readily available (e.g. peptide-based vaccines). However, the magnitude and quality of the response may be limited by *in vivo* constraints. Adoptive therapy provides the opportunity to manipulate immune effectors *ex vivo* and thus more rigorous control over the intended immune response.

Vaccine therapy

Tumor vaccines represent an attractive modality for the treatment of patients with cancer because of their potential expediency and universal applicability. These vaccines can be tumor cell–based vaccines, derived from previously whole cells that have been digested, irradiated, or transfected and often combined with adjuvant to potentiate the immune response, or antigen-specific vaccine formulations that target known tumor-associated antigens.

Tumor cell–based vaccines

Tumor cell lines
Vaccines generated from established cell lines can provide multivalent anti-tumor responses against antigens that may share expression with patient tumor. This eliminates the requirement for "tailor-made" vaccines using autologous tumor cells, but may be less effective for any given individual. Allogeneic cell-based vaccines that have been most fully developed include a whole-cell vaccine and a tumor cell lysate vaccine admixed with adjuvant that are currently undergoing Phase III testing in clinical studies in patients with melanoma.

Autologous tumor

Autologous tumor cell–based vaccines are often more labor intensive to produce, requiring the isolation and short-term *in vitro* propagation of tumor cells from a biopsy sample. However, they offer the advantage that they are more likely to express antigens associated with tumor cells for that particular individual. These tumor vaccines may be represented by (a) whole tumor cells that are admixed with adjuvant or genetically modified to enhance immunogenicity; (b) cellular lysates produced by sonication, freeze-thaw, or other disruptive preparations; or (c) apoptotic bodies derived from tumor cells exposed to UV or gamma irradiation, antibody, or pharmacologic treatment (e.g. betulinic acid in the case of melanoma). Vaccines composed of cellular lysates or apoptotic products take advantage of the antigen uptake capacity of autologous dendritic cells *in vivo* followed by re-presentation of tumor antigens to stimulate an immune response. Whole tumor cells cultured for short-term *in vitro* use can be engineered to express adjuvant-type cytokines or costimulatory molecules to enhance the tumor-specific immune response. In a murine model of melanoma testing, the efficacy of syngeneic tumor transduced with various cytokines, GM-CSF, and IL-12 were found to be the most potent in eradicating tumor: IL-12 through the augmentation of TH1 responses, and GM-CSF through the activation of antigen-presenting cells such as dendritic cells. Phase I clinical studies utilizing surgically excised tumor cells engineered to express GM-CSF have been performed in patients with metastatic melanoma as well as renal, prostate, and advanced lung cancer. Clinical responses were observed in several patients accompanied by the presence of activated or mature DCs and granulocytes at vaccination sites, and a brisk infiltration of CD4 and CD8 T cells at tumor sites associated with tumor necrosis. Since efficient viral-mediated transfection of primary tumor lines can be problematic, methods to enhance local production of adjuvant cytokines have been addressed by peri-tumoral injection of recombinant virus or DNA encoding, for example GM-CSF, or by co-administration of irradiated autologous tumor cells with a cell line (e.g. K562) that stably expresses the cytokine of interest. Clinical studies utilizing these approaches are currently underway and may represent a more easily applicable vaccination strategy utilizing autologous tumor with cytokine adjuvants.

Peptide-based vaccines

Clinical-grade peptides, usually 9–10 amino acids long, corresponding to Class I–restricted epitopes of tumor-associated antigens, were one of the first reagents used in clinical trials to elicit antigen-specific anti-tumor immune responses. More recently, peptides of greater length (>14 up to 22 amino acids) selected on the basis of known Class II–restricted epitopes or representing consensus sequences predicted to bind Class II MHC, have been evaluated for their T-helper function. Clinical results from studies using peptide reagents, however, have often been inconsistent even when similar peptides were used and were confounded by a lack of correlation between measured CTL frequency and clinical responses. This may be due to imprecise tools for immunologic monitoring that were used in these earlier studies and a requirement for more robust adjuvants and means of antigen presentation – addressed in future studies with the use of dendritic cells, GM-CSF, CD40 ligand, and other reagents.

Dendritic cell vaccines

Since dendritic cells provide optimal antigen presentation and T-cell stimulation, they represent ideal vaccine reagents if a source of such cells can be identified and engineered to express target antigens of interest. Human DC can be procured by density-based elutriation, dedifferentiation of CD14+ monocytes, or treatment of CD34+ precursors. Density-based methods rely on the differential buoyancy of dendritic cells and are limited by the relatively low (<1%) frequency of DC among peripheral blood mononuclear cells. CD34+ cells enriched from the peripheral blood by immunomagnetic bead separation, for example, and treated with GM-CSF, IL-4, and TNF provide a purified population of human dendritic cells. The use of monocyte-derived dendritic cells perhaps represents the most easily achievable source of human DC. Adherent or CD14-selected peripheral blood mononuclear cell (PBMC), when cultivated in the presence of GM-CSF and IL-4, yields an enriched population of dendritic cells after 7 days. The subsequent addition of maturation factors (monocyte-conditioned media, TNF, LPS, CD40L, or a cocktail of IL-1, IL-6, TNF, and prostaglandin) results in the upregulation of adhesion and costimulatory molecules on the DC surface. With any of these approaches, the addition of FLT3-ligand, a hematopoietic cell growth factor, *in vivo* or *in vitro* can lead to further expansion of dendritic cells.

Engineering dendritic cells to present the target antigen of interest can be accomplished by a variety of methods. Dendritic cells can be charged with whole or partially processed protein, for example prostatic acid phosphatase (PAP) for the treatment of prostate cancer or immunoglobulin idiotype for the treatment of B-cell malignancies. Antigens for which the immunogenic epitope and restricting MHC allele are known can be targeted by pulsing autologous dendritic cells with peptides corresponding to these epitopes, as in the case of melanoma for which several HLA-A2-restricted epitopes have been characterized. Since clinical-grade peptides can be readily obtained, this represents the most widely used approach in current clinical trials. In cases where the epitope has not been identified, DC may be engineered to express the target antigen of interest following infection with recombinant viral (e.g. adenoviral) vectors; transfection with plasmid DNA introduced by lipid-based carriers or, when coated with gold pellets, impelled into cells by a "gene gun" and RNA in a "naked" form; or combined with specialized transfection reagents and/or electroporation. These strategies provide a highly efficient means of engineering antigen expression in dendritic cells with minimal cytotoxicity. Whole tumor–derived RNA has even been used to transfect dendritic cells and generate multivalent T-cell responses. Finally, custom vaccines can be generated using dendritic cells fused with autologous tumor cells to form tumor–DC hybrids. In this way, antigens known to be associated with a given patient's tumor can be optimally presented by autologous dendritic cells.

As with any vaccine reagent, the source (CD34-selected, monocyte-derived, or Flt3L-expanded), dose, schedule, route of administration (intralymphatic, intradermal, or intramuscular), and method of antigen delivery remain to be established. However, dendritic cell vaccines currently represent one of the most promising translational applications of tumor immunology.

Plasmid DNA-based vaccines

A surprising finding – that plasmid DNA encoding the target antigen when introduced intramuscularly can elicit an antigen-specific immune response – has been exploited in preclinical and clinical studies of tumor immunotherapy. DNA may be delivered by intramuscular injection or, when coated on inert beads, by a helium blast from a "gene gun." It is hypothesized that localized expression of injected DNA in myocytes or neighboring cells, together with the "danger signals" provided by local trauma, facilitates DC accumulation, antigen uptake, and cross-presentation of priming of T cells by DC in draining lymph nodes. To facilitate immunogenicity, adjuvants such as IL-12 or GM-CSF may be co-administered either in protein form or as DNA encoding these genes. Incorporation of immunostimulatory sequences, represented by methylated, palindromic CpG–oligonucleotide sequences, into the target DNA may also provide an adjuvant signal through Toll-like receptors expressed on dendritic cells. Methods to improve the efficacy of DNA vaccines include the use of cationic lipids to increase delivery as well as the use of this approach in conjunction with other antigen delivery methods in a prime-boost design that involves DNA immunization followed by peptide, viral, or protein-based immunization as a boost to initial priming.

Clinical trials in vaccine therapy

The first antigen-specific human vaccine was approved by the US Food and Drug Administration (FDA) in 2010 for the treatment of patients with castration-resistant prostate cancer. The vaccine, sipuleucel-T, consisted of autologous peripheral blood mononuclear cells activated *ex vivo* with a recombinant fusion protein composed of a prostate antigen, prostatic acid phosphatase (PAP) fused to GM-CSF, presumably acting in concert with antigen-presenting cells such as dendritic cells *in vivo*. In a double-blind placebo-controlled phase III trial, patients receiving sipuleucel-T experienced a 4-month improvement in median survival over the control group. It is anticipated that this vaccine will be widely used due to its minimal toxicities and evidence of clinical efficacy.

A more conventional antigen-specific cancer vaccine, composed of a recombinant tumor-associated protein and adjuvant, has finally reached large-scale phase III studies. MAGE-A3, a T-cell-defined cancer–testis antigen, was targeted in several phase I and II vaccine trials for the treatment of patients with advanced melanoma and non-small-cell lung cancer (NSCLC), where it is expressed in up to 65% and 50% of patients, respectively. One component critical to its success appears to be the adjuvant used; in a randomized trial of stage III and IV melanoma patients, an adjuvant containing a TLR9 ligand in vaccine given to patients was superior to an oil-in-emulsion adjuvant, resulting in a greater-than-expected median survival of more than 30 months for this patient population. Further, a gene signature defined from the original tumor appeared to predict for immunobiologic and clinical response. On the basis of these studies, large-scale phase III randomized clinical trials for the adjuvant treatment of patients with melanoma and NSCLC were implemented.

Adoptive therapy

Nonspecific cellular therapy

The *ex vivo* expansion and adoptive transfer of immune effector cells were pioneered in large part by work in the 1970s, first in murine models, and then in clinical trials. Exposure of peripheral blood mononuclear cells to supraphysiologic doses of IL-2 mediated the expansion of nonspecific effectors known as lymphokine-activated killer cells (LAKs), which included a population of NK and NK-like cells. LAK infusions provided only a modest response and significant toxicities related in part to the effector cells as well as to a vascular leak syndrome associated with the high doses of IL-2 that were needed to sustain LAK cells *in vivo*. Efforts to improve efficacy included the use of tumor-infiltrating lymphocytes (TIL) culled from tumor biopsy samples. It was hypothesized that these cells would be more effective than nonspecific LAK cells in eradicating tumor and, indeed, this was demonstrated in murine models. For clinical studies, TIL cells were expanded *in vitro* with high-dose IL-2 and infused in numbers as high as 10^{11} accompanied by high-dose IL-2 administration. Although initial results were promising, TIL therapy was also associated with significant toxicity and it was not clear if beneficial effects were any greater than with HD IL-2 alone in patients with melanoma.

For patients who develop lymphoproliferative malignancies due to severe immunosuppression or who relapse with certain leukemias following allogeneic stem cell transplant, peripheral blood lymphocytes from the donor (donor lymphocyte infusions (DLI)) have been successfully used to treat patients with a >75% response rate. However, such results were often accompanied by severe graft-versus-host disease (GVHD) and myelosuppression.

Antigen-specific T-cell therapy

Broader application of an adoptive T-cell therapy strategy with decreased toxicities would be desirable. With the advent of tumor antigen identification and newer methods of T-cell expansion, it became feasible to enrich for a population of antigen-specific effectors *in vitro* and expand these to a desired magnitude for infusion. Current methods for eliciting tumor antigen-specific T cells *in vitro* use, as stimulator cells, dendritic cells that are charged with protein or peptide, or transfected to express the target antigen of interest. The use of a TCR ligand such as anti-CD3 either in combination with irradiated feeder cells or cross-linked to reagents co-expressing T-cell costimulatory ligands such as anti-CD28 and 4-1BB provides an optimized T-cell signal for activation. When co-cultivated with T-cell lymphokines such as IL-2 or IL-15, a high rate of proliferation can be achieved *in vitro* to obtain several billion antigen-specific T cells from an enriched cell line or single T-cell clone. Initial studies using this strategy, although labor intensive, provided more rigorous control over the specificity, phenotype, and magnitude of the intended immune response. Transferred T cells can be enumerated from the peripheral blood or tumor site, and their duration of *in vivo* persistence determined using immunologic methods. Adoptively transferred T cells that had been isolated and expanded *in vitro* were found to persist *in vivo*, traffic to tumor sites, and mediate an antigen-specific immune response that in some cases led to tumor regression. Methods to enhance the efficacy of adoptive therapy include the co-administration of lymphokines known to promote T-cell proliferation and survival such as IL-15 or IL-7, the combined use of both T-helper CD4 T cells, as well as cytolytic CD8 T cells and their use in the post-lymphoablative setting where homeostatic response to lymphopenia may facilitate "engraftment" and *in vivo* expansion of transferred effector T cells. In phase II studies, more than half of patients with metastatic melanoma receiving adoptive T-cell therapy after nonablative or ablative conditioning experienced partial and occasionally complete responses. Unfortunately, most patients achieving a partial response subsequently relapsed. Long-term clinical responses appeared to correlate with the duration of *in vivo* persistence of transferred T cells; to this end, it was determined that adoptive transfer of early effector and/or helper-independent cytotoxic CD8 T cells would be conducive to an increased *in vivo* survival phenotype, and, in animal models, the eradication of large established tumors.

In an effort to bypass the requirement to isolate and enrich for the rare tumor-reactive T cell, strategies were developed using recombinant vectors expressing known tumor-associated antigen-specific TCR and antibody receptors to re-direct the specificity of a population of autologous effector (T and NK) cells to a desired target antigen. Methods were developed using viral and nonviral vectors engineered to ensure efficient transduction and expression of properly paired high-affinity TCR and antibody receptors, and robust downstream signaling, by incorporating the cytoplasmic regions of costimulatory molecules (e.g. CD28 or 4-1BB) or cytokine receptors (IL-2, IL-15, etc.) as a means to circumvent the requirement for an extant second or third signal. Redirecting specificity in *early* effector T cells using such vectors could lead to the rapid generation of a population of potentially long-lasting tumor antigen-specific T cells for adoptive therapy.

Antibody therapy of cancer

Antibody therapy represents one of the success stories of bench-to-bedside technology. The development of monoclonal technology led to the practical application of antibody-targeted therapy as an approved treatment for patients with cancer. Although the question of long-term immunoprotection that is not predicted with antibody therapy remains, this modality has proven to be highly effective in a number of malignancies, and greater use of this modality may be predicted on the heels of studies that use antibodies to target more fundamental mechanisms of tumorigenesis such as angiogenesis.

In contrast to T cells, antibodies can target only surface proteins. Upon binding, antibodies mediate tumor cell killing by various mechanisms (Fig. 13.7). Engagement of the Fc portion of the antibody by NK cells and macrophage can direct killing of antibody-coated tumor cell. Activation of complement by the Fc portion of the antibody leads to complement-mediated lysis. Binding of antibody to tumor antigen can initiate the apoptotic cascade, leading to tumor cell death, or induce cell cycle arrest. The conjugation of antibody to a radionuclide such as yttrium-90 and iodine-131 forms radioimmunoconjugates that selectively deliver radiation to tumor. The conjugation to biological toxins such as ricin or diphtheria toxin and cytotoxic drugs such as doxorubicin or calicheamicin mediate cell death following antibody internalization. Antibody binding to growth factor receptors leads to receptor blockade and inhibition of downstream-signaling events.

For B-cell lymphomas which are composed of a monoclonal tumor cell population, tumor-specific idiotype immunoglobulins or B-cell specific markers (e.g. CD20) expressed on the surface represent suitable antibody targets. The tumor-specific anti-idiotype strategy pioneered by Miller *et al.* provided a proof of principle for the safety and efficacy of such a strategy, but each reagent that was produced was, by definition, patient specific. The B-cell-specific antibody, rituximab, was the first anticancer monoclonal antibody licensed by the FDA in 1997 and was used for the treatment of patients with low-grade B-cell lymphoma. It

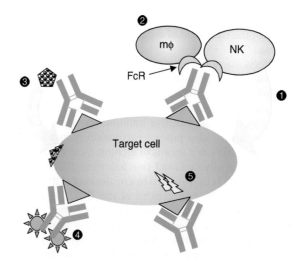

Figure 13.7 Mechanisms of antibody-mediated killing. NK cells ❶ and macrophages ❷ docking through their Fc receptors to the Fc segment of antibodies bound to tumor cell surface antigen mediate cell lysis, a process known as antibody-dependent cell-mediated cytotoxicity (ADCC). Alternatively, complement activation through the Fc portion of antibodies leads to a cascade of events resulting a cell lytic complex that inserts itself into the cell membrane leading to tumor cell death ❸. Antibodies bound to radionuclides or toxins mediate cell death following internalization ❹. Blockade of growth factor receptors and/or activation of downstream signals lead to apoptosis and cell death ❺.

is now an integral drug in the treatment of CD20-positive non-Hodgkin lymphoma. Antibodies targeting other surface markers on hematopoietic cell surfaces such as CD22 for B-cell lymphomas, CD52 for chronic lymphocytic leukemia, and CD33 and CD45 for acute leukemia have also been licensed for clinical use or are undergoing clinical studies.

Trastuzumab, an antibody targeting the HER2 tyrosine kinase receptor overexpressed in breast cancer, represents the first FDA-approved antibody for use against solid tumors. Mechanisms of action of trastuzumab include inhibition of cell proliferation by downregulation of HER-2 receptor expression, cell cycle arrest, and ADCC and CDC. Other antibodies targeting solid tumors include Cetuximab, an anti-epidermal growth factor receptor, for the treatment of colon cancer and head and neck cancer; and Bevacizumab, an antivascular endothelial growth factor receptor that has been FDA approved for metastatic colon cancer but, because of its broader mechanism of action (antiangiogenesis), can be considered for the treatment of other vascularized solid tumors, including renal cell cancer and lung cancer.

Cytokine therapy of cancer

Cytokine therapy represents a "nonspecific" approach to immunotherapy that does not target any particular tumor antigen. However, as a strategy to augment preexistent effector responses, when used alone, or as an adjuvant to antigen-specific approaches described in this chapter, the use of cytokines may play a greater role in optimizing the anti-tumor response.

IL-2

IL-2 is the prototypic anti-tumor cytokine. Its discovery in 1976 as a growth-promoting factor for T cells and subsequently

NK cells led to its eventual use in clinical trials and its FDA approval for the treatment of renal cell carcinoma and metastatic melanoma.

IL-2 affects many immune effectors; it can augment the cytolytic activity of NK cells, induce LAK activity, and enhance the direct cytotoxicity of monocytes against tumor targets. IL-2 plays a major role in expanding naïve and antigen-activated CD4 and CD8 T cells during the adaptive immune response; however, in later stages, it may also promote activation-induced cell death (AICD), presumably as a means of limiting T-cell expansion. IL-2 is produced primarily by CD4 T cells following TCR engagement, and its expression is upregulated by crosslinking of the costimulatory molecule CD28. Engagement of the CD28 co-receptor by antibody or its natural ligand, B7, results in tumor rejection in animal models, in large part due to its ability to enhance IL-2 production and upregulate anti-apoptotic signals (i.e. Bcl-X_L) leading to enhanced T-cell survival. The IL-2 receptor is composed of three subunits, IL-2Rα, IL-2Rβ, and IL-2Rγ. The γ–β heterodimeric complex alone can bind IL-2 with intermediate affinity and is capable of mediating an intracellular signal responsible for the downstream effects of IL-2 receptor engagement. Antigen activation upregulates expression of the α chain; the resulting heterotrimeric IL-2Rαβγ complex engages in higher affinity interactions during the amplification process of T-cell expansion. In tumor animal models, IL-2 is instrumental in tumor eradication and, in some cases, can supplant the use of CD4 T cells as a helper component when cytolytic CD8 T cells are used.

IL-7 and IL-15

The proliferative signals induced by IL-2 are believed to be mediated only when the IL-2Rβ and γ chains heterodimerize. It is not surprising, then, that the "common" γ chain is shared among the several cytokines that influence T-cell growth and function: IL-4, 7, 9, 15, and 21. Among these, IL-15 and IL-7 have a unique role in maintaining T-cell proliferation that is distinct from IL-2. Although IL-15 shares both the IL-2Rβ and γ subunits, its α chain is distinct. IL-15 receptor engagement promotes expansion and survival of memory T cells *in vitro* and *in vivo*, in contrast to the AICD effect induced in activated T cells with prolonged exposure to IL-2. The IL-15 receptor is expressed on activated T cells and T memory cells but not on naïve T cells, and appears to play an important role in the maintenance of T memory cell homeostasis. IL-7 also shares the common gamma chain receptor with IL-2 and IL-15. This cytokine plays a role in lymphopoiesis (i.e. during the development of precursor T cells) and is important in the survival of naïve T cells in the periphery. In concert with IL-15, IL-7 also supports the survival of the memory T-cell population. It appears that in the lymphopenic (lymphocyte-depleted host, e.g. following radiation), IL-15 and IL-7 are required for homeostasis of memory T cells, while in the normal (lymphocyte-replete) host, IL-15 has a predominant effect on this population. Preclinical studies have shown that IL-15 and IL-7 alone have a robust anti-tumor effect mediated by the *in vivo* expansion of T cells.

IL-21

In addition to a receptor that shares the common γ chain, IL-21 is also structurally related to IL-2, -4 and, -15. Its role in the immune response is related to both the innate and adaptive arms of immunity. IL-21 supports the expansion, effector function, and maturation of NK cells and augments the antigen-specific T-cell response *in vitro* and *in vivo*. Murine tumor models demonstrate a robust IL-21-mediated T-cell response that is tumor specific and superior in effectiveness compared to other γ-chain receptor cytokines, IL-2 or IL-15; its safety and efficacy in the clinical setting remain to be seen.

IL-12

IL-12, although not a member of the common γ-chain family, plays a vital role in NK cell activation, and in the differentiation of CD4 T-helper (Th) precursor cells into Th1 cells. IL-12 is produced by macrophages and dendritic cells. Its receptor is a heterodimeric molecule composed of an α chain and β chain expressed on activated T and NK cells. It is a proinflammatory cytokine known to activate NK cells resulting in secretion of IFN-γ. IFN γ in turn can activate macrophages (e.g. to produce nitric oxide), mediate adaptive immunity through Th1 differentiation, and induce CD8 T-cell cytotoxicity. A role for antiangiogenesis has also been ascribed to IL-12. Animal models have demonstrated a robust antitumor effect of IL-12 that is dependent on NK cells alone, CD8 T cells alone, or both CD4 and CD8 T cells. Clinical studies of IL-12 have been conducted alone or in combination with vaccines with modest success and occasional toxicities.

Interferon

Interferons, initially discovered as viral inhibitors, have proven to be important modulators of the immune response and therapeutic reagents for the treatment of specific viral and malignant diseases. Interferon-α (IFN-α), which is produced by macrophages *in vivo*, plays a role in the activation of NK cells during an innate immune response but has other effects including antiangiogenesis and upregulation of expression of MHC and adhesion molecules. In the clinical setting, interferon-α has demonstrated efficacy against chronic myeloid leukemia and had been a mainstay of treatment until the advent of newer agents such as the tyrosine kinase inhibitors (i.e. imitanib or STI-571). The use of IFN-α for the treatment of metastatic melanoma has been studied in well-designed randomized clinical trials, which demonstrate a significant increase in response rates and survival among patients with nonmetastatic high-risk disease when used alone. Due to its immunomodulatory properties, IFN-α has also been combined with chemotherapy (biochemotherapy) to induce responses in patients with metastatic melanoma. Other malignancies amenable to interferon therapy include hairy cell leukemia, Kaposi sarcoma, and non-Hodgkin lymphoma.

The initial enthusiasm in the use of cytokines such as IL-2 and interferon in patients with advanced cancer has been mitigated by dose-limiting toxicities, modest responses, and lack of a durable effect. More recently, cloned cytokines such as IL-12, IL-7, IL-15, and IL-21 have demonstrated potent anti-tumor efficacy and low toxicity in preclinical studies. Their use alone, or in combination with vaccine or adoptive cellular therapy to augment the frequency and *in vivo* persistence of antigen-specific T cells, represents a cause for renewed interest.

Tumor immune evasion

The process of immunosurveillance suggests that immunoselective pressure will eventually result in the outgrowth of tumors specially equipped to evade the immune response. This is known as tumor immune evasion or immune privilege. For example, cells among a heterogeneous tumor population that have lost expression of the target antigen, epitopes, or the machinery to present such antigen will escape immune detection. The outgrowth of such tumor cells under endogenous immunoselection or following specific immunotherapy was first reported in murine models, but has since been observed in patients with cancer. It has also been suggested that primary tumors may evade detection by a process of "immunologic ignorance" in which tumors are sequestered from immune recognition by a stromal barrier. Even if recognition is achieved by antigen-specific T cells, however, tumor cells may subvert the effector response.

In general, tumor immune-escape mechanisms can be divided into "intrinsic" and "extrinsic" factors. Intrinsic factors are (a) *alterations in tumor antigen presentation*, including loss of antigen expression, mutations affecting the immunogenic epitope, and defects in the antigen-presenting machinery that preclude effective presentation of the epitope to T cells; (b) *counterattack* or expression of molecules that inhibit T-cell effector function or viability, such as IL-10 and TGF-β (cytokines which inhibit T-cell function), galectin-1 (carbohydrate-binding proteins that induce apoptosis and inhibit effector function), kynurenines (tryptophan metabolites resulting from indoleamine dioxygenase upregulation that leads to T-cell apoptosis), and negative costimulatory molecules (B7-H1 and PD-L1, which induce IL-10 production and T-cell apoptosis); and (c) *inhibitors of apoptosis* which are immunoprotective mechanisms limiting T-cell-mediated lysis of tumor cells and includes the family of IAP (inhibitors of apoptosis proteins) such as survivin, c-FLIP (protects from Fas-induced apoptosis), and PI9 (a serpin protease inhibitor that inhibits granzyme B activity).

Extrinsic factors represent the influence of noncancer cells in the T-cell anti-tumor response. These include regulatory T cells as described in the "Tumor antigens" section of this chapter. By direct T-cell contact in the production of TGF-β, indirectly through CTLA-4-mediated upregulation of indoleamine dioxygenase or some other unknown mechanisms, these regulatory cells have been shown to have a profound T-cell suppressive effect and have been found at tumor sites in proportionately greater numbers. "Suppressor"-type dendritic cells have also been described in playing a modulatory role in the T-cell immune response and may act through upregulation of IDO activity leading to the production of tryptophan metabolites. Myeloid-derived suppressor cells are a heterogeneous population of "immature myeloid cells" characterized phenotypically in mice as Gr1+, CD11b+, and in humans as a lineage-negative CD14−, CD11b+. Normally comprising less than 0.5% of PBMCs, they are activated and expanded up to 10-fold under pathological conditions such as cancer. They exert their immunosuppressive effect through production of reactive oxygen species (ROS), Arginase (leading to depletion of arginine required for T-cell signaling), Peroxynitrite, TGF-β, and the induction of regulatory T cells.

This list of potential mechanisms is by no means comprehensive and remains an area of intense interest for tumor immunologists. As more mechanisms are discovered, several questions arise: what is the interplay among these varied actors in attenuating an otherwise effective antitumor response? What draws these suppressor or regulatory cells to the tumor site and activates their suppressive function? Do these mechanisms exist *in vivo*, and what is their relevance in subverting a clinical response?

Studies designed to address these mechanisms of immune escape include the use of multivalent vaccines or the transfer of effector cells targeting multiple antigens. This, together with the targeting of antigens representing proteins essential for tumorigenesis or tumor survival, may undermine the outgrowth of antigen-loss tumor variants as one mechanism of immune escape. The adjunctive use of chemotherapy or radiation therapy to reduce the tumor burden and pre-sensitize tumor cells to apoptosis may facilitate the effector response of immunotherapeutic strategies. Methods for selectively eliminating the impact of regulatory or suppressor cells are also now finding their way into clinical trials. As we gain a greater understanding of the molecular machinery responsible for tumor-induced immune escape, it may be possible to address these hurdles in a systematic and comprehensive fashion.

Clinical trials in immunomodulatory therapy

Anti-CTLA-4
Anti-CTLA4 antibodies undergoing clinical trials testing for FDA approval have been used to treat patients with advanced solid tumor malignancies; the majority of these patients have metastatic melanoma. Overall response rate (6%) and median survival (10 months) in phase II studies led to a randomized phase III study which recently demonstrated the superiority of an anti-CTLA4-containing regimen over the randomized control arm. A unique pattern of response has been observed in some patients characterized by delayed responses (up to 12 weeks or more from initiation) sometimes preceded by disease progression.

Anti-PD-1
In a phase I clinical trial of anti-PD-1 antibody therapy, treatment was well tolerated and immune-related adverse events (colitis, hypothyroidism, and arthritis) occurred later and were significantly milder and less frequent than those observed with anti-CTLA-4 therapy. Clinical outcomes yielded evidence of mixed, partial, and complete responses, which were more likely among patients whose tumor expressed the PD-1 ligand, B7-H1. Although early, this trial suggested that greater efficacy may be achievable with higher doses since the maximal tolerated dose was not reached and that, based on the non-overlapping mechanism of action and timing of adverse events, synergy with anti-CTLA-4 was a feasible strategy.

Conclusions

The immune response to tumor cells involves a complex interplay of antigen-presenting cells, effector cells, cytokines, and chemokines that evolves over time and space. An understanding of basic immunologic principles can lead to insights into the reasons for failure of the endogenous anti-tumor immune response and an opportunity to manipulate components of the immune system to augment an antigen-specific effect. The identification of tumor antigens capable of eliciting immunity was one of the first steps toward achieving this goal. Now, an understanding of the mechanisms of T-cell recognition and costimulation has led to the possibility of vaccinating patients using dendritic cells and the development of methods to isolate and expand antigen-specific CD4 and CD8 T cells *ex vivo* for adoptive transfer. The role of cytokines and chemokines in bringing together many of these effectors of innate and adaptive immunity yields yet another opportunity to augment the anti-tumor immune response. The use of monoclonal antibodies already in clinical use foreshadows the evolution of immunotherapy as a more broadly applied modality for the treatment of patients with cancer. The potential synergy of immunotherapy with chemotherapy, cytokines, and chemokines with a tumor-specific vaccine or antiangiogenic antibodies with adoptive T-cell therapy may provide additional weapons in the anticancer armamentarium. Furthermore, the recent resurgence in positive clinical trials for antigen-specific vaccines, immune checkpoint inhibitors, and adoptive cellular therapies heralds the development of more effective strategies through the combinational immune therapy. The application of immunotherapy at earlier stages of malignancy may provide the opportunity for more complete and durable response in patients for whom more conventional therapy would be ineffective. However, many questions remain unanswered. For example, what is the significance of regulatory T cells in tumor immunity? What is the best strategy for optimal vaccination? What phenotypic qualities are desired of effector cells for adoptive therapy? How can cytokines and chemokines be integrated into the use of vaccines in delivering immunogens to the site of activation and augmenting the ensuing response? What triggers cells down the path of immunologic memory to ensure long-term immunoprotection? How can we identify and address obstacles of immune escape? With more precise immunologic tools and preclinical models at our disposal, it is hoped that many of these questions can be answered.

Bibliography

General

Akira, S., and Takeda, K. (2004). Toll-like receptor signalling. *Nature Reviews Immunology*, **4**: 499–511.

Bromley, S.K., Burack, W.R., Johnson, K.G., *et al.* (2001). The immunological synapse. *Annual Review of Immunology*, **19**: 375–96.

Bluestone, J.A., and Abbas, A.K. (2003). Natural versus adaptive regulatory T cells. *Nature Reviews Immunology*, **3**: 253–7.

Gabrilovich, D.I., and Nagaraj, S. (2009). Myeloid-derived suppressor cells as regulators of the immune system. *Nature Reviews Immunology*, **9**: 162–74.

Grohmann, U., Fallarino, F., and Puccetti, P. (2003). Tolerance, DCs and tryptophan: much ado about IDO. *Trends in Immunology*, **24**: 242–8.

Kirkwood, J. (2002). Cancer immunotherapy: the interferon-alpha experience. *Seminars in Oncology*, **29**: 18–26.

Scanlan, M.J., Simpson, A.J., and Old, L.J. (2004). The cancer/testis genes: review, standardization, and commentary. *Cancer Immunity*, **4**: 1.

van der Bruggen, P., Traversari, C., Chomez, P., *et al.* (1991). A gene encoding an antigen recognized by cytolytic T lymphocytes on a human melanoma. *Science*, **254**: 1643–7.

Velculescu, V.E., Zhang, L., Vogelstein, B., and Kinzler, K.W. (1995). Serial analysis of gene expression. *Science*, **270**: 484–7.

Yee, C., Riddell, S.R., and Greenberg, P.D. (1997). Prospects for adoptive T cell therapy. *Current Opinion in Immunology*, **9**: 702–8.

Clinical immunotherapy

Atkins, M.B., Lotze, M.T., Dutcher, J.P., *et al.* (1999). High-dose recombinant interleukin 2 therapy for patients with metastatic melanoma: analysis of 270 patients treated between 1985 and 1993. *Journal of Clinical Oncology*, **17**: 2105–16.

Banchereau, J., Palucka, A.K., Dhodapkar, M., *et al.* (2001). Immune and clinical responses in patients with metastatic melanoma to CD34(+) progenitor-derived dendritic cell vaccine. *Cancer Research*, **61**: 6451–8.

Brahmer, J.R., Drake, C.G., Wollner, I., *et al.* (2010). Phase I study of single-agent anti-programmed death-1 (MDX-1106) in refractory solid tumors: safety, clinical activity, pharmacodynamics, and immunologic correlates. *Journal of Clinical Oncology*, **28**: 3167–75.

Brichard, V.G., and Lejeune, D. (2008). Cancer immunotherapy targeting tumour-specific antigens: towards a new therapy for minimal residual disease. *Expert Opinion on Biological Therapy*, **8**: 951–68.

Burnet, F.M. (1957). Cancer – a biological approach. *British Medical Journal*, **1**: 841–7.

Chapuis, A.G., Thompson, J.A., and Yee, C., (2012). Transferred melanoma-specific CD8+ T cells persist, mediate tumor regression and acquire central memory phenotype. *Proceedings of the National Academy of Sciences USA*, **109**(12): 4592–7.

Cobleigh, M.A., Vogel, C.L., Tripathy, D., *et al.* (1999). Multinational study of the efficacy and safety of humanized anti-HER2 monoclonal antibody in women who have HER2-overexpressing metastatic breast cancer that has progressed after chemotherapy for metastatic disease. *Journal of Clinical Oncology*, **17**: 2639–48.

Dudley, M.E., Wunderlich, J.R., Robbins, P.F., *et al.* (2002). Cancer regression and autoimmunity in patients after clonal repopulation with anti-tumor lymphocytes. *Science*, **298**: 850–4.

Dudley, M.E., Wunderlich, J.R., Yang, J.C., *et al.* (2005). Adoptive cell transfer therapy following non-myeloablative but lymphodepleting chemotherapy for the treatment of patients with refractory metastatic melanoma. *Journal of Clinical Oncology*, **23**: 2346–57.

Fong, L., Hou, Y., Rivas, A., *et al.* (2001). Altered peptide ligand vaccination with Flt3 ligand expanded dendritic cells for tumor immunotherapy. *Proceedings of the National Academy of Sciences of the USA*, **98**: 8809–14.

Hodi, F.S., Mihm, M.C., Soiffer, R.J., *et al.* (2003). Biologic activity of cytotoxic T lymphocyte-associated antigen 4 antibody blockade in previously vaccinated metastatic melanoma and ovarian carcinoma patients. *Proceedings of the National Academy of Sciences of the USA*, **100**: 4712–17.

Hodi, F.S., O'Day, S.J., McDermott, D.F., *et al.* (2010). Improved survival with ipilimumab in patients with metastatic melanoma. *New England Journal of Medicine*, **363**: 711–23.

Hsu, F.J., Benike, C., Fagnoni, F., *et al.* (1996). Vaccination of patients with B-cell lymphoma using autologous antigen-pulsed dendritic cells. *Nature Medicine*, **2**: 52–8.

Hunder, N., Wallen, H., Cao, J., *et al.* (2008). Treatment of metastatic melanoma with autologous CD4+ T cells against NY-ESO-1. *New England Journal of Medicine*, **358**(25): 2698–703.

Kantoff, P.W., Higano, C.S., Shore, N.D., *et al.* (2010). Sipuleucel-T immunotherapy for castration-resistant prostate cancer. *New England Journal of Medicine*, **363**: 411–22.

Kolb, H.J., and Holler, E. (1997). Adoptive immunotherapy with donor lymphocyte transfusions. *Current Opinion in Oncology*, **9**: 139–45.

Kolb, H.J., Mittermuller, J., Clemm, C., *et al.* (1990). Donor leukocyte transfusions for treatment of recurrent chronic myelogenous leukemia in marrow transplant patients. *Blood*, **76**: 2462–5.

Miller, R.A., Maloney, D.G., Warnke, R., and Levy, R. (1982). Treatment of B-cell lymphoma with monoclonal anti-idiotype antibody. *New England Journal of Medicine*, **306**: 517–22.

Mitchell, M.S., Kan-Mitchell, J., Morrow, P.R., *et al.* (2004). Phase I trial of large multivalent immunogen derived from melanoma lysates in patients with disseminated melanoma. *Clinical Cancer Research*, **10**: 76–83.

Robbins, P.F., Dudley, M.E., Wunderlich, J., *et al.* (2004). Cutting edge: persistence of transferred lymphocyte clonotypes correlates with cancer regression in patients receiving cell transfer therapy. *Journal of Immunology*, **173**: 7125–30.

Rosenberg, S.A., Lotze, M.T., Muul, L.M., *et al.* (1985). Observations on the systemic administration of autologous lymphokine-activated killer cells and recombinant interleukin-2 to patients with metastatic cancer. *New England Journal of Medicine*, **313**: 1485–92.

Rosenberg, S.A., Yang, J.C., Topalian, S.L., et al. (1994). Treatment of 283 consecutive patients with metastatic melanoma or renal cell cancer using high-dose bolus interleukin 2 [see comments]. *JAMA*, **271**: 907–13.

Rosenberg, S.A., Yannelli, J.R., Yang, J.C., et al. (1994). Treatment of patients with metastatic melanoma with autologous tumor-infiltrating lymphocytes and interleukin 2. *Journal of the National Cancer Institute*, **86**: 1159–66.

Schmitt, T.M., Ragnarsson, G.B., and Greenberg, P.D. (2009). T cell receptor gene therapy for cancer. *Human Gene Therapy*, **20**(11): 1240–8.

Sadelain, M. (2009). T-cell engineering for cancer immunotherapy. *Cancer Journal*, **15**: 451–5.

Soiffer, R., Hodi, F.S., Haluska, F., et al. (2003). Vaccination with irradiated, autologous melanoma cells engineered to secrete granulocyte-macrophage colony-stimulating factor by adenoviral-mediated gene transfer augments antitumor immunity in patients with metastatic melanoma. *Journal of Clinical Oncology*, **21**: 3343–50.

Talebi, T., and Weber, J.S. (2003). Peptide vaccine trials for melanoma: preclinical background and clinical results. *Seminars in Cancer Biology*, **13**: 431–8.

Yee, C., and Greenberg, P. (2002). Modulating T-cell immunity to tumours: new strategies for monitoring T-cell responses. *Nature Reviews Cancer*, **2**: 409–19.

Yee, C., Thompson, J.A., Byrd, D., et al. (2002). Adoptive T cell therapy using antigen-specific CD8+ T cell clones for the treatment of patients with metastatic melanoma: in vivo persistence, migration, and antitumor effect of transferred T cells. *Proceedings of the National Academy of Sciences of the USA*, **99**: 16168–73.

Yee, C. (2010). Adoptive therapy using antigen-specific T-cell clones. *Cancer Journal*, **16**: 367–73.

Scientific underpinnings

CD8 T cells

Berger, C., Jensen, M.C., Lansdorp, P.M., et al. (2008). Adoptive transfer of effector CD8+ T cells derived from central memory cells establishes persistent T cell memory in primates. *Journal of Clinical Investigation*, **118**: 294–305.

Hinrichs, C.S., Borman, Z.A., Cassard, L., et al. (2009). Adoptively transferred effector cells derived from naive rather than central memory CD8+ T cells mediate superior antitumor immunity. *Proceedings of the National Academy of Sciences of the USA*, **106**: 17469–74.

Townsend, S.E., and Allison, J.P. (1993). Tumor rejection after direct costimulation of CD8+ T cells by B7-transfected melanoma cells [see comments]. *Science*, **259**: 368–70.

Zhang, X., Sun, S., Hwang, I., Tough, D.F., and Sprent, J. (1998). Potent and selective stimulation of memory-phenotype CD8+ T cells *in vivo* by IL-15. *Immunity*, **8**: 591–9.

CD4 T cells

Hung, K., Hayashi, R., Lafond-Walker, A., et al. (1998). The central role of CD4(+) T cells in the antitumor immune response. *Journal of Experimental Medicine*, **188**: 2357–68.

Shimizu, J., Yamazaki, S., and Sakaguchi, S. (1999). Induction of tumor immunity by removing CD25+CD4+ T cells: a common basis between tumor immunity and autoimmunity. *Journal of Immunology*, **163**: 5211–18.

Wang, W., Lau, R., Yu, D., et al. (2009). PD1 blockade reverses the suppression of melanoma antigen-specific CTL by CD4+ CD25(Hi) regulatory T cells. *International Immunology*, **21**: 1065–77.

Other

Bauer, S., Groh, V., Wu, J., et al. (1999). Activation of NK cells and T cells by NKG2D, a receptor for stress-inducible MICA. *Science*, **285**: 727–9.

Butte, M.J., Keir, M.E., Phamduy, T.B., Sharpe, A.H., and Freeman, G.J. (2007). Programmed death-1 ligand 1 interacts specifically with the B7-1 costimulatory molecule to inhibit T cell responses. *Immunity*, **27**: 111–22.

Clay, T.M., Custer, M.C., Sachs, J., et al. (1999). Efficient transfer of a tumor antigen-reactive TCR to human peripheral blood lymphocytes confers anti-tumor reactivity. *Journal of Immunology*, **163**: 507–13.

De Bruijn, M.L., Schuurhuis, D.H., Vierboom, M.P., et al. (1998). Immunization with human papillomavirus type 16 (HPV16) oncoprotein-loaded dendritic cells as well as protein in adjuvant induces MHC class I-restricted protection to HPV16-induced tumor cells. *Cancer Research*, **58**: 724–31.

Freeman, G.J., Long, A.J., Iwai, Y., et al. (2000). Engagement of the PD-1 immunoinhibitory receptor by a novel B7 family member leads to negative regulation of lymphocyte activation. *Journal of Experimental Medicine*, **192**: 1027–34.

Leach, D.R., Krummel, M.F., and Allison, J.P. (1996). Enhancement of antitumor immunity by CTLA-4 blockade [see comments]. *Science*, **271**: 1734–6.

Maus, M.V., Thomas, A.K., Leonard, D.G.B., et al. (2002). Ex vivo expansion of polyclonal and antigen-specific cytotoxic T lymphocytes by artificial APCs expressing ligands for the T-cell receptor, CD28 and 4-1BB. *Nature Biotechnology*, **20**: 143–8.

Munn, D.H., Sharma, M.D., Lee, J.R., et al. (2002). Potential regulatory function of human dendritic cells expressing indoleamine 2,3-dioxygenase. *Science*, **297**: 1867–70.

Oelke, M., Maus, M.V., Didiano, D., et al. (2003). Ex vivo induction and expansion of antigen-specific cytotoxic T cells by HLA-Ig-coated artificial antigen-presenting cells. *Nature Medicine*, **9**: 619–24.

Parrish-Novak, J., Dillon, S.R., Nelson, A., et al. (2000). Interleukin 21 and its receptor are involved in NK cell expansion and regulation of lymphocyte function. *Nature*, **408**: 57–63.

Sharpe, A.H., and Freeman, G.J. (2002). The B7-CD28 superfamily. *Nature Reviews Immunology*, **2**: 116–26.

Walter, E.A., Greenberg, P.D., Gilbert, M.J., et al. (1995). Reconstitution of cellular immunity against cytomegalovirus in recipients of allogeneic bone marrow by transfer of T-cell clones from the donor [see comments]. *New England Journal of Medicine*, **333**: 1038–44.

Preclinical

Ikehara, S., Pahwa, R.N., Fernandes, G., Hansen, C.T., and Good, R.A. (1984). Functional T cells in athymic nude mice. *Proceedings of the National Academy of Sciences of the USA*, **81**: 886–8.

Kaplan, D.H., Shankaran, V., Dighe, A.S., et al. (1998). Demonstration of an interferon gamma-dependent tumor surveillance system in immunocompetent mice. *Proceedings of the National Academy of Sciences of the USA*, **95**: 7556–61.

Stutman, O. (1974). Tumor development after 3-methylcholanthrene in immunologically deficient athymic-nude mice. *Science*, **183**: 534–6.

Questions for student review

1) One of the following cell types is *not* commonly associated with an adaptive immune response:
   a. Dendritic cells.
   b. CD8 T cells.
   c. NK cells.
   d. CD4 T cells.

2) Proteins associated with the following viral pathogen are a potential immunotherapeutic target for cervical cancer:
   a. EBV.
   b. CMV.
   c. HSV.
   d. HPV.

3) Mature dendritic cells exhibit the following properties:
   a. Enhanced phagocytosis.

b. Increased expression of costimulatory molecules.
   c. a and b.
   d. Neither.

4) The following are all examples of a gamma (γ) chain receptor cytokine *except*:
   a. IL-2.
   b. IL-7.
   c. IL-12.
   d. IL-15.

5) Which of the following statements regarding CTLA4 is false?
   a. CTLA4 is an example of a positive costimulatory receptor on T cells.
   b. CTLA4 can compete for the same ligand as CD28.
   c. Treatment with anti-CTLA4 antibody can lead to autoimmune toxicity.
   d. Anti-CTL4 antibody is being used in clinical trials for the treatment of patients with cancer.

6) Which of the following statements regarding tumor immune evasion mechanisms are true? (Choose all that apply.)
   a. PD-1 can bind only to its ligand, PD-L1.
   b. Myeloid-derived suppressor cells mediate their effects through arginine depletion.
   c. IDO inhibition can be achieved using an IL-2 toxin conjugate.
   d. Regulatory T cells constitutively express high levels of CD25.

14 Tumor Angiogenesis

Christiana Ruhrberg
UCL Institute of Ophthalmology, University College London, UK

Key points

- New blood vessel growth (angiogenesis) plays a central role during the onset and progression of cancer, and the inhibition of vessel growth can therefore be exploited to control tumor growth and metastasis.
- The model of the angiogenic switch postulates that the transition from a normally quiescent blood vessel to a proliferative and invasive endothelial cell sprout during tumor growth is due to an imbalance between angiogenesis stimulators and inhibitors.
- The most potent known stimulators of tumor angiogenesis are the secreted growth factors VEGF and FGF; they cooperate with other signaling molecules such as angiopoietin, members of the EPH and ephrin families, and Notch ligands to control the patterning of vascular networks and vessel structure.
- Endogenous inhibitors of angiogenesis act either locally to modify vessel patterning or systemically to suppress new vessel growth.
- Endostatin and angiostatin are two examples of endogenous inhibitors that are secreted by the primary tumor to suppress tumor metastasis.
- Several endogenous and engineered angiogenesis inhibitors have been approved or are in clinical trials for cancer therapy; amongst these, an antibody-blocking VEGF function has proven beneficial for the treatment of several types of cancer in combination with chemo- and radiotherapy.

Introduction

New blood vessel growth plays a central role during the onset and progression of cancer. Accordingly, inhibiting vessel growth has become an important concept in the design of more efficient cancer therapies. In this chapter, we review our current understanding of the molecular mechanisms that govern vascular growth and differentiation and highlight their significance for tumor expansion. As it is increasingly evident that pathological blood vessel growth reactivates numerous signaling pathways that control tissue vascularization during embryonic and postnatal development, we will relate our current knowledge of vascular development to specific observations derived from studies of tumor vascularization. In addition, we will consider the notion that pathological vascularization depends on additional mechanisms that do not operate during developmental (i.e. physiological) vessel growth, and these mechanisms may affect the success of antiangiogenic therapies.

The vasculature is a flexible conduit that delivers and exchanges nutrients, wastes, hormones, and immune cells. During embryogenesis, blood vessels and the heart are the first recognizable organ system to develop. In the adult, the blood vessel network's continued maturation and maintenance are critical for tissue metabolism and homeostasis as well as repair processes such as inflammation and wound healing. Excessive vascular growth plays a central role during the onset and progress of a number of adult diseases including cancer and eye diseases.

Blood vessels are composed of two main cellular components, endothelial and mural cells (Fig. 14.1). The endothelium is a continuous, cylindrical epithelial sheet that creates the vessel lumen. The endothelial cells directly interface with blood and are responsible for maintaining a nonthrombogenic surface. A basal lamina separates the endothelium from the second component of vessels, the contractile mural cells. These cells surround the endothelial tube and are primarily responsible for the modulation of vascular tone. In large-caliber vessels such as arteries, veins, arterioles, and venules, the mural cells are called smooth muscle and are organized into multiple cell layers. At the microvasculature level, including capillaries and postcapillary venules, a single discontinuous layer of cells, called pericytes, surrounds the endothelium. Homotypic and heterotypic communication between the endothelium and mural cells contributes to the control of vascular physiology and growth.

The Molecular Biology of Cancer: A Bridge From Bench to Bedside, Second Edition. Edited by Stella Pelengaris and Michael Khan.
© 2013 John Wiley & Sons, Inc. Published 2013 by John Wiley & Sons, Inc.

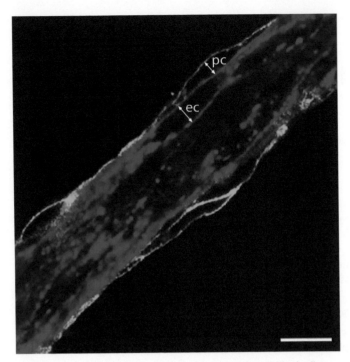

Figure 14.1 Blood vessels are composed of endothelial and mural cells. Double-label immunofluorescence reveals that blood vessels are formed from an inner layer of IB4-positive endothelial cells (ec, shown in blue) and an outer layer of NG2-positive pericytes (pc, shown in green). Scale bar: 10 μm. Image courtesy of Dr Alessandro Fantin.

Box 14.1 The main mechanisms of blood vessel growth

Vasculogenesis: Formation of new blood vessels from mesenchymal-derived endothelial cell precursors in the embryo. *Angiogenesis*: Formation of new blood vessels from preexisting vessels by sprouting growth. During angiogenesis, the insertion of endothelial pillars into existing vessels can also increase vascular complexity in a process termed "intussception." *Incorporation of circulating endothelial cell progenitors (EPCs) into preexisting vessels*: The bone marrow of adults contains progenitor cells, which have the capacity to differentiate into mature endothelial cells and have therefore been termed "endothelial progenitor cells." In response to growth factor signals, EPCs have been found to exit the bone marrow, travel within the vascular system to sites of neovascularization or tissue damage, and incorporate there into nascent vessels. The functional significance of EPCs is presently debated.

General principles of new vessel growth

In the adult, the vasculature normally represents an extreme example of a stable, quiescent population of cells (see Chapters 4 and 5), with an adult capillary endothelial cell having an estimated turnover time of approximately 1000 days. In contrast, the average turnover for a highly regenerative tissue, such as the gut epithelium, is 2–3 days. The growth of new blood vessels occurs in the adult normally only under the influence of specific physiological stimuli, such as hormonal fluctuations during the female reproductive cycle or chronic increases in tissue metabolism to stimulate new vessel growth.

In contrast to adult life, blood vessel growth occurs rapidly and in almost all tissues of the developing embryo. During embryogenesis, the two well-recognized mechanisms of blood vessel growth are vasculogenesis and angiogenesis (Box 14.1; Fig. 14.2). Vasculogenesis entails the *de novo* formation of blood vessels from endothelial cell precursors, termed angioblasts, which are derived from the primitive mesenchyme. Vasculogenesis first takes place extra-embryonically in the yolk sac. In the embryo, vasculogenesis is the mode of development for the major axial vessels such as the aorta and for the vascular plexus of some endoderm-derived organs, such as the spleen. Angiogenesis involves the formation of new vessels by sprouting growth from preexisting vessels (Fig. 14.3). This is the main mechanism for vascularization of neural and mesodermal tissues in the embryo, for example the brain and limb bud, and the main mechanism for the neovascularization of adult tissues. Extensive remodeling of blood vessels occurs subsequent or parallel to network formation by angiogenic sprouting, for example to increase vessel caliber or change network properties through vascular intussception, anastomosis, or selective regression of vessel segments. In addition, remodeling involves the recruitment of mural cells, the acquisition of venous or arterial properties, and an ill-understood increase in vessel stability and resilience that may be linked to changes in basement membrane composition and reciprocal signaling between mural and endothelial cells. Taken together, these mechanisms of vessel maturation appear to make the adult vasculature more refractory to angiogenic activation than is the case for embryonic vessels. In this context, it is interesting to note that some angiogenesis inhibitors regulate vascular growth in the adult, but are not essential for the development of the embryonic vasculature (see "Role of inhibitors in angiogenesis," this chapter). Finally, endothelial cells within the remodeling blood vessels undergo tissue-specific specializations, for example the acquisition of specialized cell–cell adhesions, termed tight junctions, or transcellular channels, termed fenestrae.

A third mode of vascular growth has been described, which postulates that circulating endothelial progenitors are incorporated into the tumor vasculature (Box 14.1). This mode of vascular growth may bear some mechanistic parallels to embryonic vasculogenesis, but its general significance and mechanisms are not clear as yet. It was originally thought that mobile endothelial progenitors derived from the bone marrow are recruited from the circulation into existing vasculature as a key step in tumor angiogenesis. More recently, bone marrow–derived precursor cells have been studied for their ability to contribute to tumor angiogenesis by adding to the mural complement of the vasculature or by providing transient accessory cells such as monocytes and macrophages, which secrete proangiogenic factors.

Pathological neovascularization: tumor vessels

In stark contrast to physiological neovascular responses, which consist of well-governed bursts of vessel growth limited in both time and space, blood vessel growth that accompanies diseases such as diabetes, psoriasis, and cancer is often characterized by chaotic growth, patterning, and dysfunction. During tumorigenesis, endothelial cell proliferation rapidly increases, with turnover

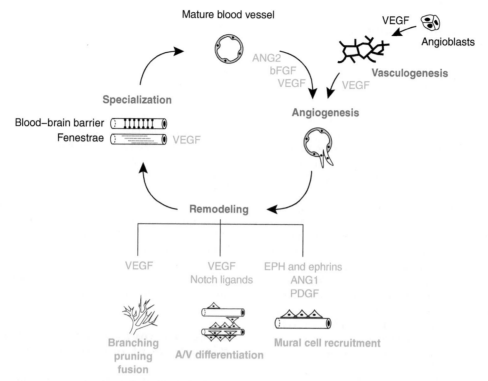

Figure 14.2 Several different classes of secreted molecules cooperate to stimulate vascular growth and differentiation. The life cycle of endothelium may begin with the assembly of angioblasts into a primitive vascular plexus (vasculogenesis) or with the reactivation of quiescent vessel endothelium. Through the process of sprouting growth (angiogenesis), the vessel network begins to expand. New vessels acquire mural cells and adopt an arterial or venous identity; concomitantly, the vascular network continues to branch and remodel into a hierarchical vascular tree. Lastly, vessels specialize according to local physiological needs, for example to form the blood–brain barrier or the fenestrated sieve plates found in the kidney glomerulus. With the notable exception of blood–brain barrier function, most of the processes illustrated are stimulated by VEGF; several other factors act upstream of VEGF or cooperate with VEGF to affect specific aspects of vessel growth and specialization.

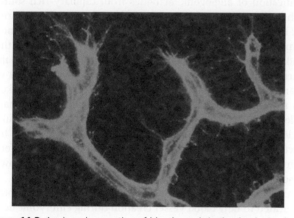

Figure 14.3 Angiogenic sprouting of blood vessels in the developing brain. Isolectin B4-positive blood vessels extend filopodia-studded sprouts to initiate new branch formation during brain vascularization in the midgestation mouse embryo.

time decreasing from approximately 1000 days in quiescent vessel beds to 50–60 hours. Nascent tumor vessels are also morphologically distinct, as they can be tortuous and of highly variable shape and size; moreover, they often lack a normal complement and juxtaposition of mural cells. In addition to assuming a chaotic structure, the vessels in tumors form disorganized networks bereft of the intricate pattern observed in normal tissues, and there is rarely a clear organization into arterial and venous compartments. Enhanced vascular permeability often accompanies tumor angiogenesis; it contributes to high interstitial pressures in the tumor microenvironment and chaotic blood flow. It has been suggested that both the tortuous nature of vessels and the high interstitial pressure inside the tumor impair delivery of conventional chemotherapeutics into the tumor tissue.

Several different factors likely contribute to the formation of the chaotic tumor vasculature. Firstly, host tissues influence the patterning of growing vessels by providing local cues. This is also observed during physiological angiogenesis, for example during development to control branching morphogenesis, the co-patterning of vessels and nerves, and the positioning and connectivity of the arterio-venous compartments. Such a well-orchestrated elaboration of vessel pattern is most likely not recapitulated during neoplasia, because a disorganized, heterogeneous, and chaotically expanding tumor mass could provide only similarly disorganized angiogenic cues. Secondly, the genetic alterations associated with tumor cell transformation and malignancy may trigger a gross imbalance in the levels of tumor-derived stimulators and inhibitors of angiogenesis (see "Basic concepts in tumor angiogenesis: the angiogenic switch," this chapter). Finally, an increase in activated immune cells and local tumor-induced changes in the stroma influence tumor vessel morphogenesis and permeability, in part simply by adding even more players to the local cytokine and growth factor "soup."

Basic concepts in tumor angiogenesis: the angiogenic switch

Pathological blood vessel growth occurs in tissues with elevated metabolic demands or reduced blood flow as a means to increase the capacity of the microvasculature. For example, the ischemic retina produces angiogenic factors to stimulate neovascular growth in diabetic retinopathy. In the case of cancer, an increased metabolic load and reduced oxygen provision to the tumor core are believed to be the primary trigger for angiogenesis. The original concept defining the importance of tumor angiogenesis postulates that the expansion of solid tumors beyond 1–2 mm^3 is dependent on the growth of new capillaries (Box 14.2). Importantly, the intimate relationship between tumor cells and blood vessels not only promotes the growth of the primary tumor, but also can contribute to metastatic spread. Moreover, neoangiogenesis appears to play a role in the establishment of "liquid tumors" (i.e. hematological malignancies); for example, leukemia is associated with neoangiogenesis in the bone marrow, and leukemic cells can be seen clustered on the new vessels.

A model explaining the transition from a quiescent capillary to a proliferative and invasive endothelial cell sprout during tumorigenesis is based upon the idea that the maintenance of vascular quiescence is due to a precise balance between angiogenesis stimulators and inhibitors. According to this model, a rise in the concentration of stimulators initiates an angiogenic response, whereas the return to quiescence is promoted by a decrease of stimulators. Vice versa, a reduction in the usual level of inhibitors would make endothelial cells more susceptible to stimulatory cues. Such changes in stimulator and inhibitor levels can be induced by a combination of genetic alterations within the tumor cells and physiological demands of the growing tumor mass, which together trigger the "angiogenic switch" of a slowly growing tumor.

This angiogenic switch requires that the tumor either produces its own stimulators of angiogenesis or induces its surrounding cells to secrete such factors. In parallel, tumor cells growing within an avascular environment, such as an epidermal tissue, need to induce the basement membrane's breakdown to allow invasion of blood vessels. In some instances, for example during metastasis, the switch to an angiogenic phenotype requires that tumor cells exit their host vessel and induce new vessel sprouts from neighboring vessels (co-option). Importantly, the term "switch" does not imply that a specific gene mutation is mandatory – the "switch" may be thrown early as an inevitable accompaniment of the growth-promoting activity of various oncogenes, whose main role may be in driving cellular replication (see the section on RAS and MYC collaboration in cancer in Chapter 6).

The idea that tumor growth and spread are intimately linked to an angiogenic switch, which therefore might be targeted to control cancer, catalyzed an intensive hunt for factors that are elaborated by tumor cells to stimulate angiogenesis. In the wake of this quest, the past three decades of research have shed much light on the complex molecular interplay between the endothelium and its environment during both tumor growth and embryonic development. This effort has given rise to new treatments for cancer that, in turn, validate the antiangiogenesis hypothesis first formalized by Folkman (Box 14.2).

Vascular growth and differentiation factors: stimulators of the angiogenic switch

Numerous stimulators of capillary endothelial cell growth, termed angiogenic factors, have been identified since the search began 30 years ago. Traditionally, assignment of angiogenic activity to biomolecules was dependent on their ability to increase endothelial cell proliferation and/or to stimulate new capillary growth using *in vivo* assays of neovascularization, such as the chick chorioallantoic membrane (CAM) or the rodent cornea pocket (Box 14.3). Using these assays, various laboratories have described the purification of angiogenic activities from capillary-rich sources, such as the brain and retina, and from various untransformed and tumor-derived cell lines. With the advent of reverse genetics in mouse and zebrafish models, we are now able to test these candidates' role in vascular development and determine their functional requirements for the elaboration of a functional vascular tree. In addition, the careful analysis of mouse knockouts has revealed unforeseen roles for several matrix and signaling molecules in vascular growth and differentiation. As a result, it is clear that no single signaling pathway alone controls both vascular growth and organization. Rather, a number of different pathways cooperate to build, branch, and mature the growing vessel network.

In this section, we review findings on several key proteins implicated in new blood vessel creation during embryonic development and highlight their relevance to tumor biology. The factors that act as negative regulators of vascular development and their relevance to tumor angiogenesis are less well understood. We will therefore discuss this topic only briefly. However, we will draw attention to the concept of endogenous angiogenesis inhibitors in the adult, because these factors, despite being non-essential for embryogenesis, can be specifically exploited in the clinic to block tumor angiogenesis.

Vascular endothelial growth factor and its receptors

The most potent and versatile angiogenic factor described to date is vascular endothelial growth factor (VEGF) (Fig. 14.2). VEGF is a secreted, homodimeric glycoprotein with endothelial cell-

---

Box 14.2 An experimental proof of principle – the importance of tumor angiogenesis for malignant growth

In 1972, Judah Folkman's group performed seminal experiments to prove the importance of neoangiogenesis for malignant tumor growth in the eye (Gimbrone et al., 1972): When small tumor fragments were implanted under the cornea, malignant tumors formed rapidly if the transplantation site was located adjacent to highly vascularized iris tissue. In contrast, when such tumor fragments were placed under the cornea, but within the avascular anterior chamber, tumor cells continued to proliferate briefly, but these tumors switched to a dormant state once they reached 1 mm in diameter. The modification of this original transplantation approach led to the development of the first widely used model system for the identification of tumor-derived angiogenic factors, the rabbit cornea pocket assay.

> **Box 14.3 Angiogenesis assays**
>
> **The CAM assay**
>
> The chick chorioallantoic membrane (CAM) assay is a relatively simple and inexpensive method to identify angiogenic factors and is suitable for large-scale screening. Grafts of tissue and polymers or sponges soaked in putative angiogenic factors are placed on the CAM of developing chick embryos through a window in the eggshell. If an angiogenic factor is released, vessel growth around the graft increases within 4 days after implantation. Blood vessels entering the graft can be visualized and counted under a stereomicroscope. Because young chick embryos lack a mature immune system, this assay is particularly useful to study angiogenesis elicited by tumor tissues. The major disadvantage of this assay is that the CAM contains a well-developed vascular network at the time of implantation, which can make it difficult to discriminate between new capillaries and existing ones.
>
> **The cornea pocket assay**
>
> The rabbit and mouse cornea are normally avascular. However, if sponges, polymers, or tissues containing angiogenic substances are implanted into pockets in the corneal stroma, vessels can grow out of the limbal region into the cornea. In albino animals, the vascular response can be readily quantified by stereomicroscopy and image analysis after perfusion of the cornea with Indian ink. This method is very reliable, but technically more demanding and more expensive than the CAM assay, and it is therefore not an ideal assay for screening. Moreover, it is not the assay of choice for ethical reasons where the CAM assay provides a suitable alternative. A more recent alternative to study factors in tumor angiogenesis involves the use of subcutaneous matrigel plugs in mice (not discussed here).

specific mitogenic activity and the ability to stimulate angiogenesis *in vivo*. VEGF also stimulates vascular permeability with an effect 50 000 times greater than that of the vasoactive substance histamine. In fact, the observation that tumor growth is associated with increased microvascular permeability provided the basis for VEGF's original discovery, and, accordingly, it was first named vascular permeability factor (VPF). In addition to the founding member, the VEGF gene family now includes five other members that have been implicated in various aspects of cardiovascular growth and function.

VEGF has two tyrosine kinase transmembrane receptors, the first one termed VEGFR2, KDR, or FLK1, and the second one termed VEGFR1 or FLT1. During development, VEGFR2 is first expressed in regions of the early mesoderm that are presumed to give rise to angioblasts, and it is the earliest known molecular marker for the endothelial cell lineage. Like VEGFR2, VEGFR1 is expressed by endothelial cells from the earliest stages of blood vessel formation. During later stages of embryogenesis, VEGFR1 and VEGFR2 messenger RNAs (mRNAs) are expressed in the hematopoietic lineages and blood vessel endothelium, with VEGF mRNA being expressed in adjacent embryonic tissues. The mRNA levels for VEGF and its receptors decrease significantly postnatally, but expression is upregulated in the endothelium of tissues with ongoing angiogenesis, such as tumors, and is also elevated at sites proximal to fenestrated endothelium. The observation that VEGF expression is regulated by hypoxia and glucose explains why VEGF levels are elevated in the ischemic regions of tumors and the retina, and it provides a molecular link between alterations in local tissue metabolism and growth factor control of angiogenesis.

Experimental data provide strong evidence for a necessary role of VEGF in developmental, adult physiological, and pathological blood vessel growth. In the mouse embryo, targeted disruption of VEGF or its receptors leads to early embryonic lethality due to severe defects in vasculogenesis. The postnatal blockade of VEGF function inhibits physiological neovascularization during bone growth and various aspects of the reproductive cycle. Consistent with a necessary role for VEGF in ischemia, neutralization of VEGF protein reduces the normal course of angiogenesis and disease severity in several experimental-model systems. VEGF's role as an angiogenic factor in diabetes, cancer, and chronic inflammation has made it a key target in the inhibition of angiogenesis and angiogenesis-dependent pathologies. Thus, anti-VEGF therapy with a humanized monoclonal antibody is now an approved method to treat colon cancer and some types of non-small-cell lung cancers and metastatic breast cancer. This type of antiangiogenesis therapy was found to work best in combination with traditional chemotherapy or radiotherapy.

Besides the VEGF antibody, two small-molecule receptor tyrosine kinase (RTK) inhibitors have been approved for cancer therapy; they both target VEGFR1 and VEGFR2, as well as the receptor for platelet-derived growth factor (PDGF) and other RTKs. Both anti-VEGF and anti-RTK drugs extend progression-free survival and overall survival in many patients with different types of metastatic carcinomas, for example colorectal, breast, or non-small-cell lung carcinoma. However, treated tumors eventually become nonresponsive, and in some cases do not respond at all, even though the drug targets are present and presumably contributed to tumor angiogenesis prior to treatment. The mechanisms that confer resistance to these antiangiogenic drugs are presently under intense investigation.

In contrast, the major clinical problem of rescuing the viability of tissues starved of a proper blood supply due to obstruction or circulatory disorders demands clinical methodology to stimulate ordered blood vessel growth, for example to alleviate coronary heart or diabetic disease in humans. VEGF is able to stimulate collateralization in experimental models of limb ischemia, and ongoing research efforts are focused on developing effective methods to deliver VEGF protein, cDNA, or virus to human patients suffering from peripheral and coronary ischemia.

A key feature of VEGF relevant to medical therapies is its expression in at least seven isoforms in humans; the predominant forms are secreted proteins of 121, 165, and 189 amino acids (VEGF121, VEGF165, and VEGF189; Fig. 14.4). These isoforms are generated by alternative mRNA splicing, which determines the presence or absence of heparin-binding domains; the splicing pattern seems to be subject to complex temporal and spatial regulation during development. The biological significance of the

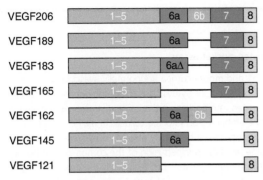

Figure 14.4 Schematic illustration of the human VEGF-A isoforms. The VEGF isoforms are generated by alternative splicing from a single copy gene, termed *VEGF-A*. The isoforms differ by the absence or presence of domains that confer binding to heparin sulfate proteoglycans and neuropilin receptors and are encoded by exons 6 (6a or 6b) and 7. The VEGF183 isoform contains a deletion within exon 6a. The VEGF121 isoform does not contain any exon 6 or exon 7 domain. The murine orthologs of VEGF189, VEGF165, VEGF145, and VEGF121 have been described; they are referred to as VEGF188, VEGF164, VEGF144, and VEGF120, respectively, because they are shorter by one amino acid.

VEGF isoforms is still under investigation. Two general ideas, by no means mutually exclusive, can be put forward to explain the function of the VEGF isoforms. The first model suggests that each isoform elicits a different signal within the responding endothelial cell through isoform-specific cell surface receptors. The interaction of VEGF165, but not VEGF121, with a nontyrosine kinase receptor termed neuropilin (NRP) 1 supports this model. The presence of NRP1 amplifies VEGFR2-mediated chemotaxis and mitogenesis in cultured endothelial cells, perhaps by directly increasing the affinity of VEGF165 for VEGFR2 or by enhancing receptor clustering or endocytosis. The second model suggests that differing affinities of the isoforms for heparan sulfate proteoglycans or other matrix proteins result in their distinct localization within the extracellular milieu, thus providing a mechanism to regulate VEGF availability to target cells.

While precise mechanistic details remain to be resolved, recent *in vivo* findings underscore the need to discern the relevance of VEGF isoform function to therapies aimed at human pro- and antiangiogenesis. Thus, the analysis of mice with targeted mutations that restrict VEGF expression to solely the 120, 164, or 188 amino acid isoform suggests functional specificity. For example, mice expressing only the 120 amino acid isoform, which lacks both NRP1- and heparin-binding capacity, are born with a grossly normal vascular tree, but have reduced capillary density and suffer from ischemic cardiomyopathy. On the other hand, mice expressing VEGF120 and VEGF188 (but lacking VEGF164) are perfectly viable, but show reduced inflammatory cell recruitment in eye pathology.

The expression of VEGF isoforms has been evaluated in various types of human cancer to determine the isoforms' significance as prognostic factors for tumor growth and metastasis and to correlate isoform expression profiles with the tumor vasculature's properties. However, a unifying theory of VEGF isoform function during tumor vascularization has not yet emerged. In contrast, in mouse models of tumor angiogenesis, differential VEGF isoform expression correlates with vascular network properties, as found in the embryo. In both situations, VEGF120 or VEGF188 alone do not support the formation of a robust vascular network.

In contrast, VEGF164 alone is sufficient for the formation of a functional vasculature, presumably because of its ability to bind heparin whilst remaining partially mobile. However, it is the combination of soluble and heparin-binding VEGF isoforms that maximizes tumor growth, presumably because full vessel network functionality is achieved only when all isoforms cooperate.

Whilst blockade of all VEGF isoforms may appear to be the best strategy to block tumor angiogenesis, this strategy may also maximize side effects, because some VEGF isoforms control homeostatic functions in adults. For example, specific VEGF isoforms may be important in preserving the fenestrated endothelia in the kidney glomerulus and in neuroprotection. In addition, the VEGF isoforms differ in their potency to promote inflammation, as VEGF164 is more proinflammatory than the other isoforms. It is not yet known if this VEGF164 property is related to its ability to bind VEGFR1 and NRP1, two VEGF receptors expressed by monocyte populations. The heightened inflammatory properties of VEGF164 may contribute to the onset and progression of some forms of pathological neovascularization. However, the relevance of this observation to tumor angiogenesis has not yet been fully addressed. This is an important issue, as both pro- and antitumorigenic subsets of immune cells have been described.

Fibroblast growth factors

The fibroblast growth factor (FGF) family of polypeptides was the first class of angiogenic factors described. FGF1 (acidic FGF, or aFGF) and FGF2 (basic FGF, or bFGF) are potent mitogens and trophic factors for cultured endothelial cells, and both have potent angiogenic properties in the CAM and cornea assays. The presence of FGFs, particularly bFGF, in tissues associated with new capillary growth made it an early favorite for a general angiogenic factor *in vivo* (Fig. 14.2).

Molecular cloning and biochemical analysis have led to the unexpected discovery that both FGF1 and FGF2 lack a signal sequence for secretion via exocytosis, therefore suggesting they normally reside inside the cell. Yet, in support of a role for these FGFs in extracellular signaling, a family of high-affinity tyrosine kinase cell surface receptors for FGFs has been described. This apparent contradiction might in part be explained by the finding that FGF2 is released from cultured cells and animal tissues *in vivo* upon mechanical injury. These data have led investigators to suggest that the FGFs may exert angiogenic activities in situations associated with cellular damage, thus acting as a "wound hormone." In situations where tissue injury does not play a role, the nonsecreted FGFs may act as intracrine signaling factors that translocate to the nucleus to exert their activity. Consistent with this idea, both FGF1 and FGF2 possess a nuclear localization sequence and FGF1 has been found in the nucleus of cultured vascular endothelial and smooth muscle cells.

Despite the fact that FGF1 and FGF2 are potent stimulators of angiogenesis in *in vitro* models, the targeted disruption of either FGF1 or FGF2 or indeed both molecules together does not appear to impair vascular development and results in only mild defects in hematopoietic and wound healing. It therefore appears that, in contrast to VEGF, neither one of these FGFs is required for vascular growth. However, FGFs may prime adult vessels to mount a VEGF response in certain pathological situations. In support of the latter idea, FGFs can induce VEGF expression in the cornea assay and in cultured vascular smooth muscle cells, VEGF and FGFs synergize to promote neoangiogenesis *in vitro* and *in vivo*, and blocking VEGF function inhibits FGF-induced angiogenesis. FGFs may also cooperate with PDGF to promote mural cell recruitment to blood vessels, perhaps by upregulating PDGF receptors. Thus, FGF2 and PDGF-BB were found to synergistically stabilize vessels.

Over the past decade, the FGF family has rapidly expanded in size and now includes 22 members, many of which are predicted to be secreted proteins. FGF3 and FGF4 can induce angiogenesis in the CAM assay, and FGF7 induces neovascularization in the rat cornea (Box 14.2). FGF5 and FGF7 have also been implicated in angiogenesis. However, like FGF1 and FGF2, none of these other FGF family members have essential roles in vascular growth or morphogenesis in mice.

Angiopoietin and the TIE2 receptor tyrosine kinase

The angiopoietins are ligands for TIE2, an endothelial-specific RTK, and are thought to act in a complementary and coordinated fashion with VEGF (Fig. 14.2). Four angiopoietin ligands termed ANG1 to ANG4 are known, and together they comprise a growth factor family that consists of both receptor activators and receptor blockers: whilst ANG1 and ANG4 act as TIE2 agonists, ANG2 and ANG3 behave as antagonists.

Angiopoietins are likely to play a later role than VEGF in vascular development, since both ANG1- and TIE2-deficient mice form a normal primitive vascular plexus and therefore develop further than VEGF and VEGFR2 mutants. However, failure to stabilize and remodel the primary vascular plexus subsequently causes embryonic lethality of ANG1-deficient mice around the time of midgestation. Ultrastructural studies suggest that these problems are due to a disruption of ANG1-mediated interactions between the endothelium and its supporting cells, such as smooth muscle cells. In support of the idea that ANG1 signaling is important for vessel remodeling, the mutation of its receptor TIE2 is responsible for a heritable human disease characterized by thin-walled blood vessels with markedly reduced smooth muscle layers. Overexpression of ANG1 in the embryo results in hypervascularization, presumably by decreasing the normal amount of vessel regression that accompanies development. In contrast, overexpression of ANG2 results in embryonic lethality with phenotypes similar to those of the ANG1 and TIE2 knockouts, presumably because ANG2 blocks ANG1 activity by competing for TIE2.

In support of the notion that angiopoietins also control the stability of mature vessels, ANG1 is widely expressed in adult tissues, whilst ANG2 expression is present at sites of active physiological angiogenesis, such as the female reproductive tract and the placenta. Detailed analysis of ANG2 expression in the ovary has revealed that vessel growth and sprouting occur at sites where ANG2 is coexpressed with VEGF. In contrast, expression of ANG2 in the absence of VEGF results in vessel regression. Taken together, these observations have led to the following model: Ang1 expression in the mesenchyme normally activates endothelial TIE2 to promote mural cell recruitment. Since ANG2 can compete for binding to TIE2, ANG1-dependent TIE2 activation is blocked in the presence of ANG2, and this leads to blood vessel destabilization. The exposed endothelium of destabilized vessels either degenerates, causing vessel regression, or, in the presence of VEGF, proliferates to yield net vessel growth. In support of the model of synergistic VEGF and angiopoietin signaling, their balance controls vascular permeability and tumor vascularization. It was shown that tumor vascularization is initially accompanied by high levels of ANG2, but these early vessels regress, causing necrosis in the tumor center. VEGF upregulation at the ischemic tumor margin then results in coexpression with ANG2, and a second wave of more stable vascularization is induced. Thus, sites of ANG2 expression correlate with vessel plasticity, and the outcome of vessel growth versus regression is decided by the presence of VEGF (Fig. 14.2). It remains to be elucidated how PDGF and angiopoietin signaling synergize to ensure mural cell investment of growing vessels, and what the role of TIE2-expressing monocytes may be in tumor vascularization.

Eph and ephrin signaling

The transmembrane Eph receptors and their membrane-bound ephrin ligands comprise a signaling pathway with well-characterized functions during the development of several different organ systems, including the vasculature. EPHs and ephrins are often reciprocally expressed at tissue compartment boundaries and are best known for their roles in axon guidance; however, in attempts to better understand their role in neural patterning, their role in vascular development was revealed. EphrinB2 is widely expressed in the embryo, but within the vasculature is restricted to arterial endothelium, whereas expression of its receptor EPHB4 predominates in the endothelium of veins. These findings provided the first molecular distinction between arteries and veins. Targeted inactivation of EPHB4 or ephrinB2 leads to failure in the remodeling of the primitive vascular plexus and subsequent lethality at midgestation.

Other EPH and ephrin family members are expressed in the vasculature, but expression is not always complementary between arteries and veins (e.g. ephrinA1 and ephrinB1 are expressed in both, and EPHB3 is present in veins and the aortic arches). Instead, ligand–receptor expression is often reciprocal between blood vessels and their surrounding tissues, suggesting paracrine signaling. Based on the findings to date, it seems reasonable to suggest that members of the EPH–ephrin family are acting in the vasculature much in the same way as they do in the nervous system, where the complementary ligand–receptor expression patterns provide guidance cues by defining spatial boundaries in the developing embryo. In addition, it was recently shown that ephrinB2 regulates VEGF signaling by promoting the internalization of VEGF receptors via endocytosis.

A number of different ephrins and EPH receptors are overexpressed in a wide variety of human cancers. For example, ephrinA1 and its receptor EphA2 have been found on tumor vessels and tumor cells, and may contribute to both tumor-induced angiogenesis and tumor growth or spread, possibly by stimulating an autocrine loop. EPH and ephrin expression by tumor cells might also attract or organize their vascular supply, in analogy to the role of this signaling system during embryogenesis. In support of this idea, it was recently shown that engrafted EPHB4-expressing tumor cells could attract ephrinB2-positive blood vessels to increase their blood supply. However, it is presently not known whether disregulated EPH and ephrin expression normally contributes to the vascularization of tumors or the chaotic organization of their vessel networks, and further research will be necessary before we can conclude whether EPH and ephrin signaling might provide a valid target for antiangiogenic therapy in cancer.

Notch signaling

Like the EPH–ephrin system, the Notch signaling pathway has been studied extensively in embryos, where it regulates the specification of cell fate through local cell interactions. The Notch proteins are transmembrane receptors that are activated by membrane-spanning ligands of the Delta, Serrate, and Jagged families. Both Notch receptors and ligands are expressed in specific vascular compartments (i.e. arterial versus venous endothelium or vascular smooth muscle cells and pericytes).

During vascular patterning, VEGF may act upstream of Notch to promote arterialization, but it also promotes other aspects of angiogenesis: loss of Notch4 and/or Notch1 impairs vascular morphogenesis in the embryo, and loss of function for Notch3 or Jagged 1 causes hereditary vascular diseases in humans. Recent mouse and human genetic studies provide compelling evidence for a role of Notch signaling in the adult vasculature, raising the possibility that this pathway may present a target for tumor angiogenesis. Most notably, Delta-like 4 (DLL4) is an endothelial-specific Notch ligand that is upregulated in tumor vasculature. See Chapter 5 for a more detailed discussion of Notch.

Role of inhibitors in angiogenesis

The paradigm of the "angiogenic switch" postulates that tumor angiogenesis may be stopped in its tracks if angiogenesis inhibitors are administered at levels that exceed those of angiogenesis promoters. Accordingly, the identification of angiogenesis inhibitors has become a subject of much research activity.

Embryonic vessel branching and physiological neovascularization both demand the existence of angiogenesis inhibitors that act at a short range to limit and refine vascular morphogenesis. Such factors would act in synergism with local angiogenic stimuli to control the balance between endothelial cell proliferation, apoptosis, and migration within growing tissues. The need for such factors is demonstrated in an analogous example of branching growth, that of lung-branching morphogenesis, where FGFs promote budding from the epithelium, whilst BMPs restrict budding to certain regions of the epithelium. The recent description of model systems to study vascular branching patterns should allow us to validate candidate antiangiogenic factors with local patterning capacity. Such candidate genes include the mammalian sprouty proteins, as some family members' overexpression inhibits endothelial cell proliferation *in vitro* and leads to defective vessel network formation in the mouse embryo.

In addition to short-range inhibitors of angiogenic growth, several naturally occurring compounds operate systemically in adults and have been termed "endogenous inhibitors of angiogenesis." Some of these endogenous inhibitors have been discovered because they are produced by primary tumors and inhibit the growth of secondary tumors (i.e. metastases). Because these inhibitors of angiogenesis are likely to be more stable than most angiogenic proteins, they are thought to remain in the circulation at relatively higher levels. Only a drop in endogenous angiogenesis inhibitors following removal of the primary tumor would increase relative levels of angiogenesis stimulators and thereby activate neoangiogenesis in previously dormant metastases. Accordingly, the systemic administration of such inhibitors should prevent metastatic growth after surgical removal of the primary tumor. Two such tumor-derived factors that have entered clinical trials for antiangiogenic cancer therapy are angiostatin, an internal fragment of plasminogen, and endostatin, an internal fragment of collagen XVIII. Interestingly, mice lacking plasminogen or collagen XVIII develop a normal embryonic circulation, suggesting that postnatal angiogenesis is in part controlled by mechanisms that do not operate in the embryo. Consistent with this idea, mice lacking the endogenous angiogenesis inhibitor thrombospondin 1 also develop normally; however, they are less sensitive to oxygen-induced vessel obliteration in the eye, and in the presence of the activated *neu/erbB2* oncogene, they are more susceptible to develop highly vascularized breast tumors.

The specific role of thrombospondin 1 in controlling postnatal but not developmental vascularization may be linked to its involvement in the maintenance of a quiescent and differentiated endothelial phenotype. At the cellular level, thrombospondin 1 inhibits endothelial cell migration and growth factor mobilization and promotes endothelial cell apoptosis. In contrast, VEGF is able to stimulate proliferation and is anti-apoptotic, and it has therefore been suggested that the relative balance of VEGF and thrombospondin 1 determines the outcome of oxygen-induced neovascularization in the eye. In an analogous mechanism, tumor angiogenesis may result when increased VEGF expression is accompanied by or is due to the de-repression of thrombospondin 1 following the oncogenic activation of genes such as *RAS* and *MYC*. Interestingly, mice lacking thrombospondin or its receptor are resistant to radiation damage, and targeting thrombospondin signaling in animal models of cancer appears to protect endothelial cells, soft tissue, bone marrow, and leukocytes during radiotherapy, but increases the sensitivity of tumors to radiotherapy. Like angiostatin and endostatin, thrombospondin 1 is in clinical trials.

Clinical outcomes and future directions

The growing list of experimentally validated angiogenesis stimulators and inhibitors suggests that there are many opportunities to develop new therapeutics to complement traditional cancer therapies. Accordingly, the number of antiangiogenic compounds in clinical trials has grown steadily in the past 20 years (www.cancer.gov/cancertopics/factsheet/Therapy/angiogenesis-inhibitors). These promising therapeutics fall into several different classes, including endogenous inhibitors of angiogenesis and engineered drugs such as those that diminish VEGF signaling, others that directly target and debilitate rapidly growing endothelium, or a third class that prevents extracellular matrix degradation and remodeling. The clinical success of an inhibitory antibody directed to VEGF in patients with colorectal cancer provided the first unequivocal evidence for the hypothesis that attacking their blood supply inhibits the growth of human tumors. However, none of the approved antiangiogenesis treatments turned out to be as effective as researchers had hoped, and further research is therefore necessary to understand the therapeutic limitations. Finally, these clinical experiences have, in turn, identified new areas for angiogenesis research in the basic science laboratory. For example, it will be important to elucidate the molecular mechanisms that mediate tumor-specific resistance to antiangiogenic molecules, define the role of immune cells in tumor angiogenesis and antiangiogenesis therapy, and identify novel targets present only on the tumor vasculature to increase safety. A better molecular understanding of tumor angiogenesis will in part be driven by the exchange of novel insight amongst scientists working on cancer, embryonic vascular development, eye disease, and other related disciplines. This interchange has recently suggested that, in some circumstances, vascular normalization may be superior over antiangiogenic therapy to aid the delivery of anti-tumor agents.

Acknowledgments

I am grateful to Dr. Alessandro Fantin for Fig. 14.1 and Professor David Shima for his contributions to the content of this chapter.

Bibliography

Mechanisms of angiogenesis

Carmeliet, P. (2005). Angiogenesis in life, disease and medicine. *Nature*, **438**: 932–6.

Engerman, R.L., Pfaffenbach, D., and Davis, M.D. (1967). Cell turnover of capillaries. *Laboratory Investigation*, **17**: 738–43.

Rhodin, J.A. (1967). The ultrastructure of mammalian arterioles and precapillary sphincters. *Journal of Ultrastructure Research*, **18**: 181–223.

Risau, W. (1997). Mechanisms of angiogenesis. *Nature*, **386**: 671–4.

Molecular control of angiogenesis

Andrae, J., Gallini, R., and Betsholtz, C. (2008). Role of platelet-derived growth factors in physiology and medicine. *Genes & Development*, **22**: 1276–312.

Pitulescu, M.E., and Adams, R.H. (2010). Eph/ephrin molecules – a hub for signaling and endocytosis. *Genes & Development*, **24**: 2480–92.

Powers, C.J., McLeskey, S.W., and Wellstein, A. (2000). Fibroblast growth factors, their receptors and signaling. *Endocrine-Related Cancer*, **7**: 165–97.

Ruhrberg, C. (2003). Growing and shaping the vascular tree: multiple roles for VEGF. *Bioessays*, **25**: 1052–60.

Thomas, M., and Augustin, H.G. (2009). The role of the angiopoietins in vascular morphogenesis. *Angiogenesis*, **12**: 125–37.

Thurston, G., and Kitajewski, J. (2008). VEGF and Delta-Notch: interacting signalling pathways in tumor angiogenesis. *British Journal of Cancer*, **99**: 1204–9.

Tumor angiogenesis

Folkman, J. (1992). The role of angiogenesis in tumor growth. *Seminars in Cancer Biology*, **3**: 65–71.

Hanahan, D., and Folkman, J. (1996). Patterns and emerging mechanisms of the angiogenic switch during tumorigenesis. *Cell*, **86**: 353–64.

Holash, J., Maisonpierre, P.C., Compton, D., et al. (1999). Vessel cooption, regression, and growth in tumors mediated by angiopoietins and VEGF. *Science*, **284**: 1994–8.

Knighton, D., Ausprunk, D., Tapper, D., and Folkman, J. (1977). Avascular and vascular phases of tumor growth in the chick embryo. *British Journal of Cancer*, **35**: 347–56.

Lyden, D., Hattori, K., Dias, S., et al. (2001). Impaired recruitment of bone-marrow-derived endothelial and hematopoietic precursor cells blocks tumor angiogenesis and growth. *Natural Medicine*, **7**: 1194–201.

Ruhrberg, C., and De Palma, M. (2010). A double agent in cancer: deciphering macrophage roles in human tumors. *Natural Medicine*, **16**: 861–2.

Senger, D.R., Galli, S.J., Dvorak, A.M., Perruzzi, C.A., Harvey, V.S., and Dvorak, H.F. (1983). Tumor cells secrete a vascular permeability factor that promotes accumulation of ascites fluid. *Science*, **219**: 983–5.

Anti-angiogenesis: concepts and target molecules

Carmeliet, P., De Smet, F., Loges, S., and Mazzone, M. (2009). Branching morphogenesis and antiangiogenesis candidates: tip cells lead the way. *Nature Reviews Clinical Oncology*, **6**: 315–26.

Folkman, J. (1972). Anti-angiogenesis: new concept for therapy of solid tumors. *Annals of Surgery*, **175**: 409–16.

Gimbrone, M.A., Jr., Leapman, S.B., Cotran, R.S., and Folkman, J. (1972). Tumor dormancy in vivo by prevention of neovascularization. *Journal of Experimental Medicine*, **136**: 261–76.

Kerbel, R., and Folkman, J. (2002). Clinical translation of angiogenesis inhibitors. *Nature Reviews Cancer*, **2**: 727–39.

Loges, S., Mazzone, M., Hohensinner, P., and Carmeliet, P. (2009). Silencing or fueling metastasis with VEGF inhibitors: antiangiogenesis revisited. *Cancer Cell*, **15**: 167–70.

Ruhrberg, C. (2001). Endogenous inhibitors of angiogenesis. *Journal of Cell Science*, **114**: 3215–6.

Angiogenesis assays

Gimbrone, M.A., Jr., Cotran, R.S., Leapman, S.B., and Folkman, J. (1974). Tumor growth and neovascularization: an experimental model using the rabbit cornea. *Journal of the National Cancer Institute*, **52**: 413–27.

Ribatti, D., Vacca, A., Roncali, L., and Dammacco, F. (1996). The chick embryo chorioallantoic membrane as a model for in vivo research on angiogenesis. *International Journal of Developmental Biology*, **40**: 1189–97.

Questions for student review

1) What phrase describes a shift in the balance of angiogenesis stimulators and inhibitors, leading to enhanced vascular and tumor growth?
   a. Neovascularization
   b. Angiogenic switch
   c. Arterialization
   d. Neoplasia

2) VEGF refers to a collection of protein isoforms that differ in their ability to bind to what?
   a. Angiopoietins
   b. Glucose
   c. Heparin
   d. Veins

3) Which of the following is thought to present the best target for antiangiogenesis therapy?
   a. VEGF
   b. Mural cells
   c. Thrombospondin-1
   d. Ephrins
   e. Endothelial progenitors

15 Cancer Chemistry: Designing New Drugs for Cancer Treatment

Ana M. Pizarro and Peter J. Sadler
University of Warwick, UK

The efforts of those, who are placed in a position fitted for the purpose, should be unceasing for the search after such a medicine; for nothing can be more unphilosophical than to conclude that it does not exist, because it has not yet been found.

Walter Hayle Walshe, *The Nature and Treatment of Cancer*

Key points

- The story of anticancer therapeutics begins during the First World War, when bone marrow suppression was observed in soldiers who had been killed by mustard gas.
- In 1942, a patient in the terminal stages of lymphosarcoma was successfully treated by the administration of a chemical, the nitrogen mustard tris(β-chloroethyl)amine.
- The discovery of methotrexate (an antimetabolite of folic acid) in the 1950s heralded the development of the antimetabolite class of anticancer agents.
- The more complete description of the molecular pathoetiology of cancers that has evolved in the last 20 years has identified multiple new potential targets for therapeutic intervention, as exemplified by the development of tyrosine kinase inhibitors such as imatinib.
- The drug discovery pipeline consists of the following steps:
  - Choosing a cancer-relevant "druggable" target;
  - Finding a series of chemical entities or biomolecules (**hits**) that are able to modulate that target by screening *in vitro*;
  - Finding the **lead** agent of the series;
  - Optimizing the lead candidate; and
  - Developing the candidate for clinical trials.
- Medicinal chemists have exploited a variety of sources to find leads, including: natural products (e.g. taxol), the application of pharmacological tools such as screening compound libraries *in vitro* (e.g. vatalanib) and *in silico* (e.g. HA14-1), established drugs (for "me betters," such as carboplatin), competitor patents, publications, and even serendipity (e.g. cisplatin).
- In addition to developing traditional compounds, new therapeutic approaches are needed in anticancer drug research, including wider exploration of metal- and metalloid-based compounds, vaccines, nucleic acid–based therapies (e.g. RNAi), and utilization of new techniques such as photodynamic therapy (PDT).
- The lead-development phase focuses on the metabolism and distribution of the drug in the body, on its efficacy against the target (pharmacokinetic (PK) and pharmacodynamics (PD) studies), and on the avoidance of toxicity. This may result in the development of a pro-drug that allows the active drug to form following metabolism in the body. The pro-drug approach has also been exploited so that the pro-drug is activated only near the tumor (staying innocuous in the rest of the individual to diminish off-target effects).
- The pre-clinical lead-development phase involves collating all necessary information required for a drug to enter clinical trials. Bioavailability, toxicity, safety data, manufacturing information, and other regulatory documentation are submitted to various regulatory agencies for ethical and scientific approval. Approval then allows the potential drug to enter clinical trials (see Chapter 16 on cancer clinical trials).

The Molecular Biology of Cancer: A Bridge From Bench to Bedside, Second Edition. Edited by Stella Pelengaris and Michael Khan.
© 2013 John Wiley & Sons, Inc. Published 2013 by John Wiley & Sons, Inc.

Introduction

Chemotherapy is defined as the use of chemicals to treat diseases, but it has become synonymous with cancer drug therapy. The key cellular processes involved in the pathoetiology and behavior of cancer cells have been outlined in other chapters of this book. Not surprisingly, many of these are or will become the target of anticancer drugs aiming to cure, arrest, or prevent cancer, on the one hand, or to prevent critical life-shortening behaviors such as invasion or metastases, on the other; Chapter 18 will provide an overview of those drugs already in clinical use. This chapter describes the processes by which these and future cancer drugs originate, and Chapter 16 concentrates on biologically targeted therapies and clinical trials. What will readily become apparent is that for a new compound to be developed and to successfully pass through the various stages described here is nothing short of miraculous. The route to the next blockbuster drug is strewn with the remains of tens of thousands of compounds which have failed on the way.

We describe here the various stages that constitute the general modern approach to drug discovery. Each developmental stage will be addressed with examples of specific therapeutics or potential future drugs. Where this chapter ends, the next on cancer clinical trials begins. The focus here is specifically on drugs to treat cancer. The development of vaccines to prevent infection with potentially tumor-causing viruses such as HPV, and drugs to assist smoking cessation, despite their potential value in preventing cancers, are not discussed.

A cautionary note – cancer cells have a nasty habit of treating cancer therapy as another opportunity for driving natural selection (or, indeed, even of accelerating their acquisition of potentially advantageous mutations). The net result is that cancers rapidly and often inevitably acquire resistance to the effects of drugs that target specific cancer pathways (see Chapter 5). A major task in the future of cancer drug development will be to overcome this using strategic combinations of agents and dosing schedules, or to target critical and nonredundant targets, such as c-Myc, where no "escape route" may be available.

Historical perspective

Cancer has been known for most of recorded history (Appendix 1.1), yet at the start of the 20th century there was not a single convincing report of a medical (nonsurgical) cure of cancer.

Ironically, the first steps toward medicines for cancer stem from an inadvertent consequence of the use of mustard gas as a chemical weapon during the First World War (1914–18). In 1919, Krumbhaar published results of hematological examinations of patients in a base hospital in France who had been poisoned by **mustard gas (sulfur mustard)**. Direct toxicity to the bone marrow, which peaked at 2 weeks after exposure, coincided with the highest mortality. It is interesting to note with hindsight how common this finding is as an unwanted side effect of so many chemotherapeutic agents.

In the United Kingdom during the late 1920s and early 1930s, Beremblum was investigating the effects of mustard gas on tar-triggered tumors in mice (tar was a known carcinogen), working on the then-plausible hypothesis that mustard gas would increase the carcinogenic effects of tar by inducing hyperemia. Somewhat to his surprise, tumors were inhibited in these mice, and by serendipity he had identified a potential anticancer compound.

James Ewing had been "impressed by the peculiar and specific nature of mustard-gas burns" while serving at the US Army Medical Museum, an observation that prompted Adair and Bagg to study effects of mustard gas on cancer-bearing animals and human cancer patients. In 1931 they published the regression effects of mustard gas (when topically administrated in alcohol solution) on several types of superficial tumors in 12 patients of the Memorial Hospital in New York. Although the nitrogen mustards were not a definite cure for cancers and the tumor regression was temporary, this was "proof-of-concept," and the search for agents that would be more selective toward specific types of malignancies began. Thus, the combination of "wartime" ingenuity, coupled with the application of suitable test systems such as animal models, opened up the era of cancer drug development.

Importantly, in the mid-1930s Alexander Haddow had published in *Nature* the paradoxical observation that three polycyclic hydrocarbons known as carcinogens, chemical substances that induced cancer in animal models, could also retard tumor growth in rats (Haddow's paradox), whilst noncarcinogenic but similar chemicals had no effect in delaying tumor growth (see Chapter 10 for a discussion about DNA damage as a cause and cure for cancer).

In the late 1930s Carl Voegtlin, the first director of the National Cancer Institute (NCI) of the National Institutes of Health in the United States, in a series of lectures on chemotherapy emphasized how little was known about the mode of action of chemicals in humans and, most importantly, the need to find chemical treatments for cancer, given the incomplete achievements of surgical and radiation treatments.

Nitrogen mustards

Regarding the discovery of DNA-damaging agents, it can be argued that the formal history of chemotherapy began in the 1940s, when **nitrogen mustards** were used for the treatment of lymphoma.

Goodman and Gilman, who will need no introduction to students of pharmacology, were recruited to test a series of chemical warfare agents developed earlier in the 20th century for their potential therapeutic value (a contract was signed by Yale University and the Office of Scientific Research and Development – an agency of the US government created in 1941 to manage scientific research for military purposes during World War II). They were joined on the project by Philips and Allen. Later, in the early 1960s, in narrating the events that led to the very first clinical trial in humans of nitrogen mustards, the chemical entities tris(β-chloroethyl)amine and methyl-bis(β-chloroethyl)amine, Gilman explained,

> Close contact was maintained between investigators by means of circulated research reports and frequent meetings. This accounted for the rapid elucidation of the unique and fascinating properties of the nitrogen mustards. Contrary to the present opinion of many, perhaps no compound had been more thoroughly studied prior to clinical trial than were the nitrogen mustards. The point to be emphasized is the collaborative nature of the basic investigations on the nitrogen mustards which led to their clinical trial.

The rapid progress of the chemists involved in the research shed light on the biotransformation of the nitrogen mustards. Scheme 15.1 shows the cyclization of β-chloroethylamines and the

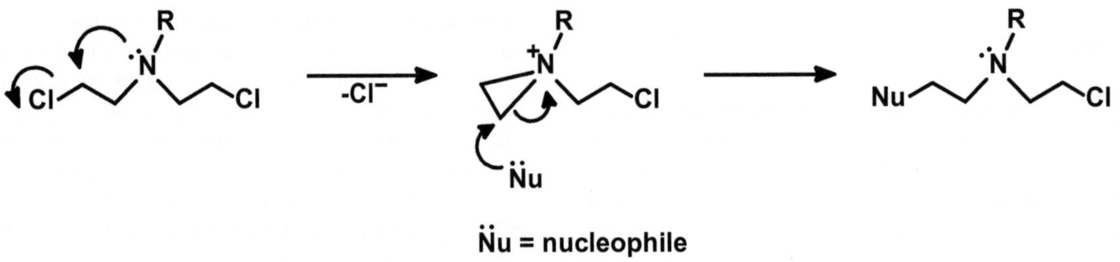

Scheme 15.1 Cyclization of β-chloroethylamines by formation of aziridinium rings.

Tris(β-chloroethyl)amine

Methyl-bis(β-chloroethyl)amine

Figure 15.1 Nitrogen mustards used in the first clinical trials in humans in the 1940s.

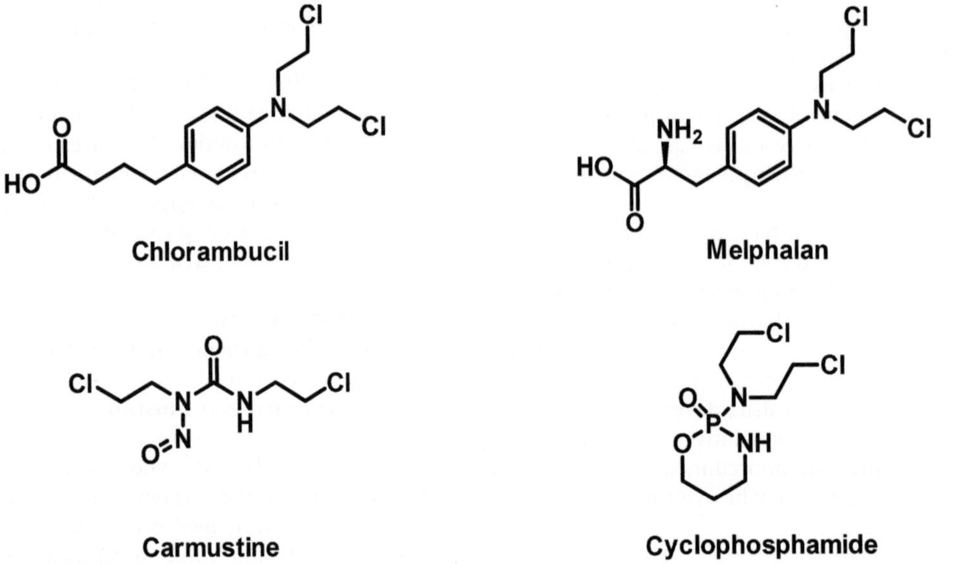

Chlorambucil

Melphalan

Carmustine

Cyclophosphamide

Figure 15.2 Some nitrogen mustards in the clinic today.

formation of the highly reactive aziridinium ring. In the case of tris and bis(β-chloroethyl)amines, this involves the formation and reaction of successive aziridinium rings. Much information was collected about the distribution, pharmacodynamics, and toxicity of the agent in animal models.

Finally, Gilman, Goodman, and Philips took the anatomist Dougherty on board, and efficacy experiments on lymphoma in mice were carried out. Strikingly positive results from this group convinced Gustav Lindskog (a thoracic surgeon) to treat a non-responding X-ray patient in the terminal stages of lymphosarcoma with tris(β-chloroethyl)amine (Fig. 15.1). This was in December 1942. The results, subject to a wartime secrets policy, were not published until 1946, when the restrictions were lifted. The same reason was behind the publication delay of independent but similar studies by Wilkinson and Fletcher in the United Kingdom. The authors tried tris(β-chloroethyl)amine hydrochloride and methyl-bis(β-chloroethyl)amine hydrochloride (Fig. 15.1) in 18 patients with one of the following malignancies: chronic myeloid and lymphatic leukemias, Hodgkin disease, or polycythemia. The level of response varied widely, with the best responses observed in chronic myeloid leukemia and Hodgkin disease. Today we know that the mechanism of action of nitrogen mustards involves alkylation of DNA (Scheme 15.1, where the nucleophile is DNA), which ultimately disrupts cell proliferation and leads to cell death.

These studies highlighted the importance of interdisciplinary collaboration and also suggested that finding a single "magic bullet" for the treatment of all cancers, even of the same type, might be unattainable.

As a result of these discoveries, other types of alkylating agents derived from the original nitrogen mustards were developed. Many of these are in clinical practice today (Fig. 15.2), most notably **chlorambucil**, **melphalan**, **carmustine**, and **cyclophosphamide** (discussed in this chapter).

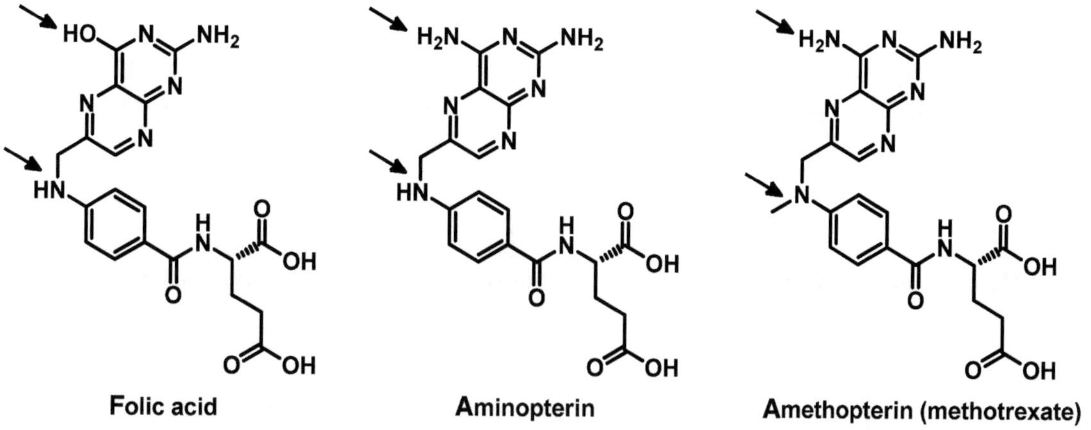

Figure 15.3 Structures of folic acid and its antimetabolites aminopterin and amethopterin (methotrexate). The arrows show the sites of the structural modifications.

Methotrexate

The antimetabolites: One of the earliest examples of hypothesis-led drug development resulted in the trial of an anti–folic acid metabolite by Sidney Farber and colleagues. Following the identification of folic acid as a requirement for leukemia development, they decided to administer an analogue of folate, which could compete with folate binding to the biological target. This became known as antimetabolite therapy, since the analogue of folate is an antimetabolite (a chemical entity similar enough to a natural metabolite to mimic it in a normally occurring biochemical reaction in the cell, but different enough to alter the cell's normal function). An antimetabolite drug inhibits a normal metabolic process involved in causing disease. They first tested the analogous compound, **aminopterin** (Fig. 15.3), which differs in structure from folate in that an OH group is replaced with an NH_2 group. Sixteen infants and children with acute leukemia were treated intramuscularly, and a marked improvement (manifested clinically, histologically, and pathologically) occurred in 10. The severe toxicity of the drug drove them toward the discovery of more efficient and less toxic analogues, such as **amethopterin**, now known as methotrexate (Fig. 15.3). Methotrexate differs from folate in having the same NH_2 group as aminopterin and, additionally, replacement of a hydrogen atom in the NH group by a methyl group (CH_3). Its clinical development was favored as it had a better therapeutic window. Today, methotrexate is still used in the clinic, and we know that it competitively inhibits dihydrofolate reductase (DHFR) at picomolar concentrations. DHFR catalyzes the reduction of dihydrofolate (DHF) to tetrahydrofolate (THF), an essential cofactor in the biosynthesis of thymidylate monophosphate (dTMP). Inhibition of DHFR leads to a deficiency of dTMP as DHF cannot be recycled, which ultimately causes cell death.

6-Mercaptopurine

Burchenal, an oncologist based in the Sloan-Kettering Institute in New York who had shown interest in the effects of methotrexate, extended the rational of antimetabolite therapy to nucleobases in the hope of disturbing nucleic acid synthesis. Burchenal established a collaboration with pharmaceutical chemist Hitchings and colleagues who had been investigating the relationship between the chemical structures for derivatives of purines and pyrimidines and their interference with the biosynthesis of nucleic acids and metabolism in bacteria. The discovery of the anti-leukemia activity of **6-mercaptopurine** (Fig. 15.4) is attributed to this collaboration. This extensive work was reported in a landmark publication in the journal *Blood* in 1953. Sixty years on, the concept is still alive, and a number of antimetabolites of nucleic acid building blocks are now used in the clinic against different types of cancer. Examples are **tioguanine, fludarabine, clofarabine, gemcitabine, capecitabine, cytarabine**, and **5-fluorouracil** (Fig. 15.4).

The toxicity of nitrogen mustards slowed their adoption into clinical trials. However, in the 1950s, thinking had begun to change, and both the public and researchers became interested in finding drugs that can affect cancer. The Developmental Therapeutics Program (DTP; originally called the Cancer Chemotherapy National Service Center (CCNSC)) was created in the United States in 1955 as a drug discovery and development arm of the National Cancer Institute (NCI). There was also a growing understanding that there are many types of cancers and there is no panacea for cancer therapeutics. In the 1960s, new strategies evolved using "drug cocktails." As seen in Chapter 18, drug combinations are now commonplace and responsible for much of the current clinical successes of chemotherapy for cancer.

The crucial early role of not-for-profit organizations, academia, and governments should not be overlooked – there was little incentive for the pharmaceutical industry (motivated primarily by the need to be profitable) at that time when the risk was highest, the need for investment was greatest, and the potential market was the most uncertain – a situation completely reversed now by the longevity of many cancer patients and the near-ubiquitous application of cancer drugs to those patients. However, a government-sponsored program allowed academic investigators to have access to resources only accessible to large pharmaceutical firms. The NCI and its Developmental Therapeutics Program played a key role in the development of anticancer drugs, in particular after the National Cancer Act in 1971, when President Richard Nixon declared "war on cancer" and the NCI was given a new mission: to support research in a way in which basic discoveries translate into actual applications to truly reduce cancer incidence, morbidity, and mortality.

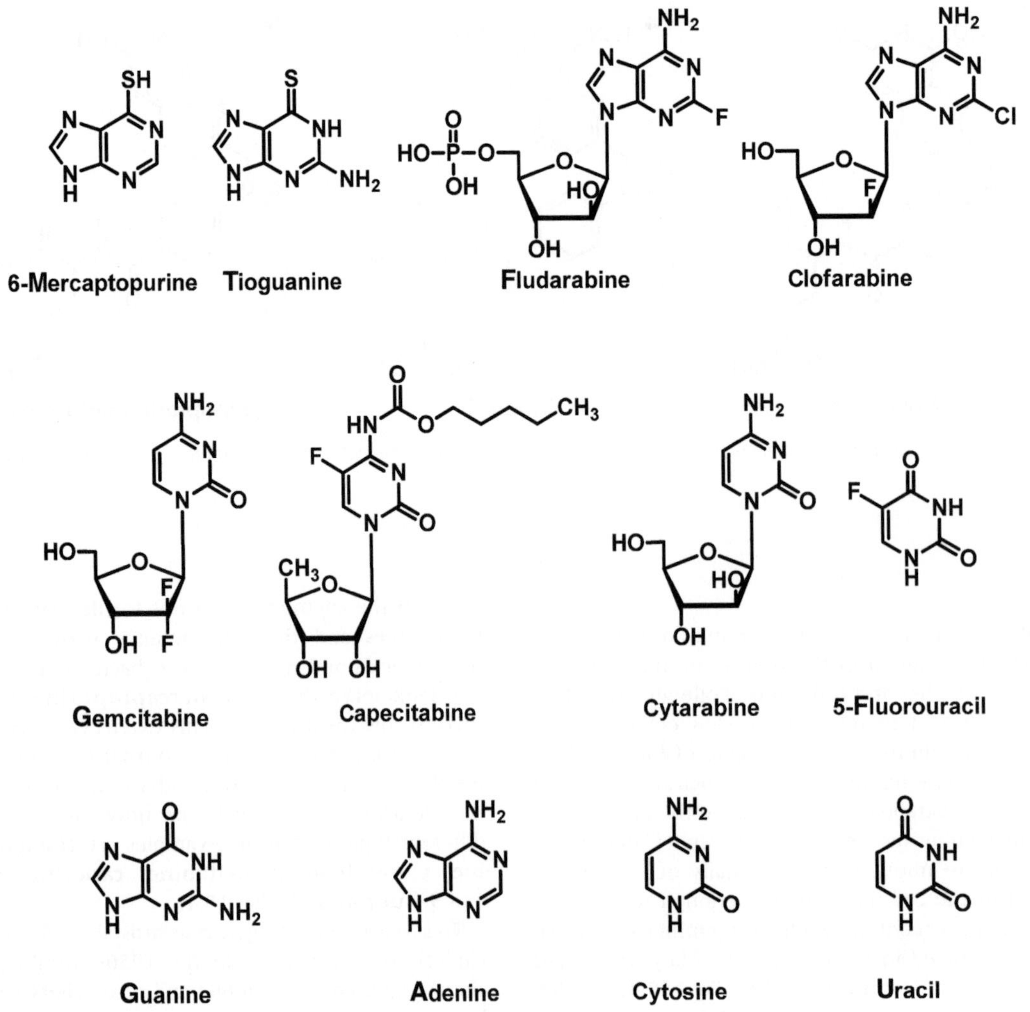

Figure 15.4 Structures of some DNA alkylating agents and the four nucleobases to which their structures relate.

The drug discovery process and preclinical development of a drug

It has been estimated that the discovery of a drug and its preclinical development take about 12–15 years and cost $0.8–1.7 billion. The result of this is that the patent life usually extends up to only 5 years before the generic competition starts. Ultimately, this results in pharmaceutical companies setting increasingly higher tariffs to make their new drugs sufficiently profitable to satisfy shareholders.

The second half of the 20th century changed views on cancer research for pharmaceutical companies. Discoveries with promising turnovers such as taxol or cisplatin did indeed spark a newly found interest in the pursuit of anticancer drugs. Cancer drug development has since transformed from a government- and charity-supported, low-budget research to a multibillion-dollar industry. Additionally, targeted therapies have driven a highly competitive race to bring ever-newer tyrosine kinase inhibitors and other targeted agents into clinical trials in unprecedented numbers.

According to the Wellcome Trust (a global charitable foundation based in the United Kingdom), the **drug development timeline** can be outlined as: discovery, preclinical, clinical trials, approval, and post-launch.

We will describe the first two, the drug discovery process and preclinical development of a drug, within the context of cancer research. These comprise a number of stages that can be summarized as follows:

• *Target identification and validation*;
• *Lead identification*: *in vitro* screening (toward the target and cancer cell lines) of large numbers of chemical entities (potential drugs) to find *hits* (modulators of the target and cell growth inhibitors) which result in *lead* identification. Filing a patent is likely carried out at this stage;
• *Lead optimization*: *in vitro* screening; PK, PD, and ADMET studies; *in vivo* efficacy; and acute toxicity – initial evaluation;
• *Development of candidate and progression to clinic*: bioavailability, duration of action, laboratory scale synthesis, toxicity, and safety. The differences in the time spent on the different steps in different medicinal chemistry projects are dictated by the relative resources available for each part of the process and the strategies employed by the researchers involved.

We describe these different steps and address them with several examples of anticancer drugs that have reached at least trial stages, and the history behind their discovery. Most of the names will be familiar, like cisplatin, taxol, and rituximab.

Selecting the target

There appears to be an established premise in drug research nowadays: no medicinal chemistry project readily attracts interest from either for- or not-for-profit agencies prior to target identification and – increasingly – validation (in cancer models). Put simply – first find and justify your target. Cancer targets in the present context are those identified biomolecules involved in the numerous cellular processes of carcinogenesis.

The medicinal chemist does not identify (i.e. discover and validate) molecular targets, although target discovery has become a conventional part of the drug discovery process. The choice of the target then governs the direction of the chemistry of the project. Knowledge of the existence and function of a target is generally provided by physicians, pharmacologists, geneticists, and molecular and chemical biologists. However, once a clear target has been identified in one or more of the multiple steps of carcinogenesis, it is compelling for the scientific community to try to find a drug or modulator for it, and indeed this can trigger a discovery program. In fact, despite the need for validation of the target, knowing that a gene is mutated and selected for in cancer cells might be persuasive evidence to initiate a drug discovery project.

The selection of the target is not trivial. From a medicinal chemist's point of view, the nature of the target is crucial for drug design, since this will affect the physical and chemical interactions of the pharmacophore (described by Paul Ehrlich in 1909 as the "molecular framework that carries the essential features responsible for a drug's biological activity") with the active site of the biomolecule. Additionally, the target is within a biological system, and reaching it with the "intact" drug or a derivatized drug (a pro-drug) is a major challenge for the medicinal chemist. For example, some targets will be accessible to the drug from the extracellular environment, but, on occasion, the target will be intracellular, and the drug must penetrate cellular membranes.

Since cancers are heterogeneous, involving different tissues of origin and differing molecular routes taken to achieve malignancy, the number of potential targets is considerable. Each chapter in this book contains a plethora of potential cancer drug targets. Because of the rapid progress made in identifying specific molecular targets in cancer and because of the central importance of this area to cancer biology, we have devoted a large part of Chapter 16 to a discussion of targeted agents and the new concepts that have arisen around their discovery and clinical use.

Rituximab (Rituxan), one of the most successful bench-to-bedside stories, was the first monoclonal antibody to gain FDA approval (1997) for the treatment of cancer, and the first single agent approved for lymphoma-targeted therapy (CD20-positive, β-cell, low-grade, or follicular non-Hodgkin lymphoma). Rituximab is a chimeric mouse–human antibody that recognizes and binds to cells expressing the CD20 antigen on the surface of malignant and normal B cells. Patients who do not respond well to rituximab treatment can be treated with antibodies labeled with radionuclides, additionally damaging cancer cells (90Y-ibritumomab, tiuxetan tositomomab, and 131I tositomomab).

Imatinib (Gleevec) is a small molecule and arguably the most successful drug in the young history of targeted therapies. A somatically mutated gene, BCR-ABL, is causally involved (through its product protein) in the carcinogenesis of a particular malignancy (chronic myeloid leukemia, CML). The inhibition of the aberrant protein BCR-ABL results directly in disease control, giving the first example of a success story of a targeted therapy. The discovery of imatinib is described in Box 16.1. Development of Imatinib is considered by some as the dawn of the molecular era of targeted therapy directed against oncogenic mutations. Second-generation BCR-ABL inhibitors include nilotinib, dasatinib, and bosutinib.

Other therapies based on this model include semaxanib (targeting VEGFR signaling), sorafenib, sunitinib, pazopanib, axitinib (targeting VEGFR), gefitinib, erlotinib (EGFR antagonists), flavopiridol (the first cyclin-dependent kinase inhibitor (CDKI) in trials), R-roscovitine (also a CDKI), bortzomib (proteasome inhibitor), cilengitide, Nutlin-3, and PRIMA-1. Some of these are discussed in more detail in Chapter 16.

Strategies to find lead compounds

Once the target has been chosen, the first phase of the drug discovery process is the so-called **hit-to-lead** phase. The issue was neatly summarized by the medicinal chemist Frank King, who wrote in 2002:

> It can be argued that the most important decision in drug discovery is that which is made every day by medicinal chemists; what compound to make next. Strategically, the two extremes to lead optimization and hence decision making are: rational design (careful design of single molecules using target structural information, pharmacophore identification and SAR [structure–activity relationships]) and random synthesis (make all possible analogues from readily available starting materials and trust luck). In practice, we do a combination of both, although the relative proportion of rational:random increases as the knowledge of SAR develops.

In this phase, hundreds of thousands of compounds are often screened against the target via at least one carefully selected test. A return of around 0.1% for hits is usually considered a success for molecular-based screening programs. The test is ideally carried out using high-throughput screening (HTS) where the appropriate choice of test is crucial, as interpretation of data may be difficult. Furthermore, poor choice of the screening assay may lead to overly positive evaluation that can be a carryover problem in the project.

In 1975, the NCI adopted the mouse P388 (leukemia) model as a primary screen. The model provided an indication of antitumor activity in a living system, yet because it represented only one tumor type, it was not always a good predictor of activity, did not guarantee reproducible success in humans, and, most importantly, could not be used in HTS. Human xenografts (human tumors growing in immunocompromised animals) presented similar problems. The need for an *in vitro* system that could provide researchers with solutions to these problems was addressed by the development of human tumor cell line screening. It was launched by the NCI in 1985, and used as primary routine screening since 1990. The screen includes 60 human tumor cell lines, representing leukemia, melanoma, and cancers of the lung, colon, brain, ovary, breast, prostate, and kidney. The panel provides the identification of compounds with activity against human cancers of a particular tissue. Most importantly, these systems allow for HTS, affording both time and economic value. Unsurprisingly, there are some drawbacks as laboratory-grown cells might undergo adaptive changes, raising concerns as to whether they are truly representative of the cancer phenotype. In addition, cell-culture systems are limited in their ability to mimic many aspects of the in situ tumor microenvironment, including hypoxia, stromal cell interaction, and vascularization (Chapters 12 and 14). Nevertheless, cell culture systems have been extremely valuable tools for the discovery and evaluation of potential hits and have fostered the development of enhanced analytical tools such as the COMPARE algorithm developed by Kenneth Paull and colleagues at the NCI. Data from *in vitro* cytotoxicity screening of large numbers of compounds provide easy-to-interpret dose–response curves. Sensitive cancer cell lines are readily identified, and more importantly it has been proven that two agents with similar selectivity patterns toward the panel are most likely to have a similar mechanism of action.

Screening compounds against selected molecular targets is a complementary technique in which interaction (inhibitory potency and selectivity) of the target with a potential hit is examined. The identification of compounds that hit the target is followed by retests and generation of dose–response curves which provide chemists with IC_{50} values for those hits. At this point quantitative assessment of binding affinities is very important, including stoichiometry of binding, conformational changes, and identification of promiscuous inhibitors. This can lead to an investigation of more hits and larger libraries. Usually, a different assay is carried out on confirmed hits in a cellular environment.

The next step following identification of a series of hits is the generation or determination of the lead or leads of the series. Analogues are synthesized (or purchased where possible), and a quantitative structure–activity relationship (QSAR) is determined. The lead must have improved potency, reduced off-target activities, and physiochemical properties suggestive of reasonable *in vivo* PK. The lead is determined through empirical modification of the hit structure.

But what is the origin of the hits, and where do the original compounds from which the hits are selected come from? In fact, many different strategies are used to identify leads. Our classification of these strategies is based on the class of compounds from which they were first screened in the search for hits.

The source of hits, and hence the **lead**, may be: **natural products** (less productive but still highly valuable in oncology), **established drugs** (for "me-betters," valid but commercially risky since this approach relies upon finding a novel, differentiated series of compounds, the benefits of which must translate into the clinic), **pharmacological tools** (HTS gives most of the leads), **competitor patents**, **publications**, or even **serendipity**. We have also included in this text a "Miscellaneous" section, where we describe light-activated therapeutics, RNA interference (RNAi), and vaccine therapies, and go on to explore the periodic table.

Serendipity

Although a serendipitous discovery of a hit series can hardly be classified as one of the intended strategies to find anticancer drugs, the importance of serendipity over the last 60 years in the discovery of drugs such as cisplatin and tamoxifen is undeniable.

Cisplatin

The discovery of **cisplatin** as an anticancer drug serendipitously arose from research on the effect of electric fields (generated with Pt electrodes) on the growth of bacteria in Barnett Rosenberg's laboratory at Michigan State University in the early 1960s. Dr Rosenberg was a biophysicist investigating whether the telophase stage of mitosis, which seemed to have some visual similarity with the lines of force between the poles of a magnet, might have a magnetic component. He speculated that cell division may be affected by the magnetic field created by an electric current.

There was indeed an effect on the growth of *Escherischia coli* (*E. coli*) bacteria. The application of an electric current to the culture made the bacteria grow in a spaghetti-like fashion; however, they did not divide. Replication was halted. The results were reproduced in a series of experiments in which various strengths of current were applied. What was inhibiting cell replication? Was it the current? The researchers discovered that the bacterial filamentous growth was not a result of the electric field but of a chemical formed in the cell culture medium due to electrolysis of the platinum electrodes in the presence of components of the medium (Cl^- and NH_3). This gave rise to *cis*-diamminetetrachloridoplatinum(IV), a pro-drug for what we now know as cisplatin (Fig. 15.5a). In 1968, the first tests of cisplatin in mice bearing Sarcoma 180 solid tumors were carried out. Instead of treating the mice on the day after the tumor was implanted (standard protocol at the time), Rosenberg and his assistant Van Camp waited for 7 days, until the tumor had grown to about 1 g in weight (a 20-fold weight increase). The results were exceptional, producing a high percentage of complete cures. Rosenberg presented the results to the NCI, who verified the potent anti-tumor activity of cisplatin. The drug progressed from activity in mouse models in 1968 to human trials in 1971 and FDA approval in 1978.

Although the discovery of the anticancer properties of cisplatin was serendipitous, we now have a wealth of information regarding the cellular processing and mechanism of action of the drug. DNA is accepted as its intracellular target, but it also binds to RNA and proteins. In its interaction with DNA, cisplatin targets guanine-rich sequences since it binds strongly to N7 in the guanine nucleobase (the most electron-dense and accessible site on DNA for electrophilic attack by platinum; Fig. 15.5b). Both chlorido ligands bound to platinum in cisplatin can be substituted by water molecules (hydrolyzed) in aqueous environments and these weakly bound water molecules can be further substituted by more strongly binding ligands such as guanines (Fig. 15.5c). Cisplatin readily forms bifunctional intrastrand GG crosslinks on DNA, believed to be lethal lesions. These intrastrand crosslinks cause DNA to bend. Such bent DNA is recognized by intracellular (HMG) proteins which can play a role in protecting the DNA from repair (Fig. 15.6). Additionally, this DNA perturbation may hijack proteins that recognize DNA kinks from their natural locus and contribute to the anti-tumor effects of cisplatin. Ultimately, the DNA damage results in apoptosis and cell death.

The story of John Cleland, *patient zero* for what would become the standard cisplatin treatment for men with advanced testicular cancer, was recorded in an interview for the magazine *Cure Today* in 2004 (winter issue), 30 years after he began the cisplatin treatment that cured him.

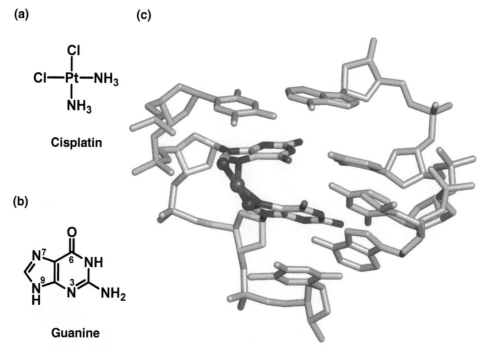

Figure 15.5 (a) Structure of clinically used drug cisplatin. (b) Guanine nucleobase. (c) Intrastrand cis-{Pt(NH$_3$)$_2$(GG)} crosslink in a head-to-head conformation as part of the adduct formed from the reaction of cisplatin with the duplex 5'-d(TCTCGGTCTC)·d(GAGACCGAGA). Platinum is light blue, nitrogens coordinating to platinum are deep blue, oxygens C6O in the crosslinked guanines are red, and other atoms are gray for clarity. (Coordinates taken from the Protein Data Bank, 1AIO.)

Tamoxifen

Tamoxifen (Nolvadex-D, now generic; Fig. 15.7) is an early exemplar of a targeted drug. It is a selective estrogen-receptor modulator (SERM), discovered during investigations of the role of estrogen in breast cancer and as a result of intense research in the 1950s to find a contraceptive. It evolved to become the first targeted therapy for the treatment of breast cancer and is considered by some as the most established case of tumor-tailored therapy. However, it had a rather tortuous journey to its pole position in cancer therapy.

Tamoxifen was first synthesized by Richardson in 1962 in the Alderley Park research laboratories of ICI Pharmaceuticals (now AstraZeneca). Walpole, who led the team in reproductive endocrinology where Richardson was working and who was trying to find a contraceptive, filed a UK patent primarily covering an invention for application on "the management of the sexual cycle." Although cancer treatments were not a corporate priority at the time, Walpole's personal interest in cancer therapies inspired him to include in the patent coverage an application for the "control of hormone-dependent tumours." After a rather quiet evolution, tamoxifen reached a critical point in its development in 1972, when ICI Pharmaceuticals almost terminated tamoxifen due to negative prospects for market exploitation. The crisis point was overcome by Walpole's tenacity, and ICI Pharmaceuticals marketed tamoxifen in the United Kingdom as a breast cancer treatment in 1973 and as an inducer of ovulation in 1975. The US patent for the treatment of advanced breast cancer in postmenopausal women was filed in 1977.

Certainly the 1970s were quite a dark era for tamoxifen. It enjoyed little interest from either medical advisors or the pharmaceutical industry right up to the mid-1980s. However, during those dark times, Jensen discovered the estrogen receptor and Jordan developed the application of the estrogen-receptor assay to predict endocrine responsiveness to endocrine ablation successfully. During those times, its mechanism of action was explained at the molecular level. It was discovered that tamoxifen, which has low affinity for the estrogen receptor, was in fact a pro-drug that accumulated and was then converted to 4-hydroxytamoxifen, the active metabolite with high affinity for the estrogen receptor and anti-estrogenic activity. A good example of how important patent protection becomes in drug development is the fact that the publication of these data in 1977 had been delayed for more than a year to secure patent protection for the metabolites, since tamoxifen did not have patent protection in the United States at the time.

Although tamoxifen was competing against other hormonal agents in a relatively slim market and was neither clinically nor financially remarkable at this stage, the 1980s observed an increasing acceptance of the drug as the endocrine treatment of choice. It was not until 1998 that the meta-analysis of the Oxford-based Early Breast Cancer Trialists' Collaborative Group showed definitively that tamoxifen saved lives in early breast cancer. It was most effective in preventing recurrence of breast cancer when administered for long periods (up to 5 years). These benefits are realized with minimal drug-related toxicity. Today (2011), there are over 100 active or recruiting tamoxifen clinical trials. This story owes its success to a few individuals – without their tenacity, over a period of more than 30 years, tamoxifen would not be in use.

Natural products

Natural products are those chemical compounds produced by living organisms and therefore found in nature. They emerge from a limited selection of a few building blocks, the combination

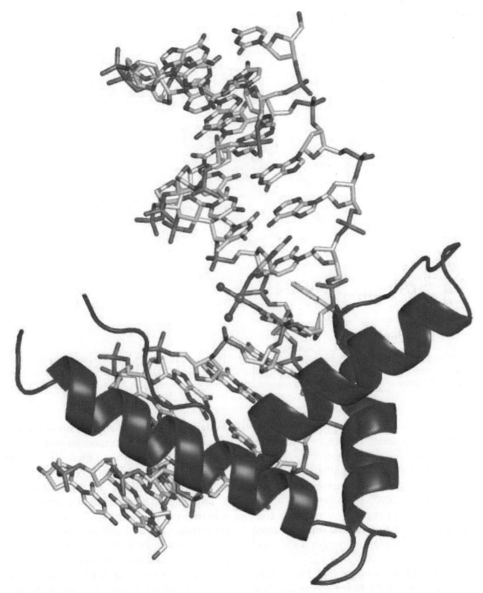

Figure 15.6 Adduct formed between the non-sequence-specific domain A of HMG1 and cisplatin-modified DNA. Platinum is light blue, nitrogen is deep blue, oxygen is red, phosphorus is orange, carbon is gray, and HMG is a purple ribbon, with a phenylalanine intercalating between two nucleobases highlighted in yellow. (Coordinates taken from the Protein Data Bank, 1CKT.)

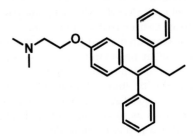

Tamoxifen

Figure 15.7 Tamoxifen, a selective estrogen-receptor modulator (SERM).

of which gives place to an excellent diversity in both structure and function, yet to be matched by synthetic compound libraries. This versatility is the reason why natural products have been a source of inspiration for chemistry, biology, and, naturally, medicine over the past decades.

Paclitaxel (taxol)

The history of **taxol** is a landmark in the story of the Developmental Therapeutics Programme (DTP). At the time of the discovery of taxol, pharmaceutical companies had a very limited interest in developing drugs to treat malignancies since no economic outcome was predicted for this class of drug. The US government-funded CCNSC (later integrated into the DTP), however, could provide the requirements for every step in a drug discovery project. From the early 1960s, the NCI invested money, time, and effort in the development of natural products through

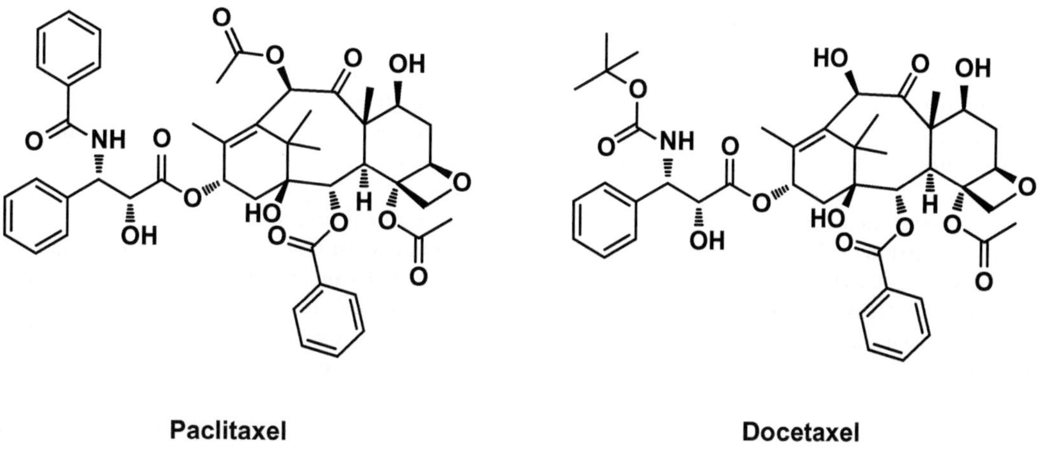

Paclitaxel

Docetaxel

Figure 15.8 Microtubule-targeting taxanes paclitaxel (Taxol) and docetaxel (Taxotere).

their Plant Screening Program, which supported the collection and testing of plant and marine sources all over the world. This program resulted in the discovery of taxanes, such as paclitaxel (taxol) and docetaxel (taxotere), and camptothecins in the 1960s.

In the summer of 1962, a natural product was extracted from the bark of the Pacific yew tree (*Taxus brevifolia*). In 1971, the molecular structure of taxol was disclosed (Fig. 15.8). In 1977 the NCI confirmed its anti-tumor activity in the mouse melanoma B16 model. The drug was selected as a candidate for clinical development. Activity was also observed in animal models against MX-1 mammary, LX-1 lung, and CX-1 colon tumors. Susan Horwitz of the Albert Einstein College of Medicine of Yeshiva University was awarded a grant by the NCI to study the mechanism of action of the compound. Horwitz and her student observed that the compound inhibited the de-polymerization of the microtubules formed during the mitosis of HeLa cells. The compound was blocking cell mitosis, by binding to and stabilizing the microtubule assembly so that it could not de-polymerize; shrinkage was prevented, and therefore segregation of the chromosomes was halted. This work was published in *Nature* in 1979, and paclitaxel (also known by its trade name, Taxol), a new anticancer drug, was born. We now know that taxanes stimulate tubulin polymerization and induce apoptosis via a novel G_2–M checkpoint that is independent of wild-type *p53* function. For 15 years, there were no other stabilizers of microtubules in the clinic.

However, the low availability of its natural source, the rare and slow-growing Pacific yew tree (18 months from bark to vial), together with undesired solubility, put the future of the drug in jeopardy. Scientists in the NCI were facing one of the most challenging moments in drug discovery: the finding of an effective drug with only a limited supply. Phase I clinical trials against a number of cancer types were initiated by the NCI in 1984, but environmentalists raised concerns about the destruction of ancient yew trees, further restricting supplies. This unfortunate situation led, however, to an excellent example of the synergy between a governmental research organization and the pharmaceutical industry, working together with the goal of producing the drug without destroying the tree. Large-scale synthesis became possible when a method to synthesize paclitaxel from a precursor, 10-deacetyl-baccatin III, extracted from the common yew *Taxus baccata*, was developed. The precursor was converted by chemical synthesis to taxol. This process also provided another taxane with anti-tumor activity, **docetaxel** (Taxotere; Fig. 15.8). Docetaxel was developed by Rhône-Poulenc Rorer following the discoveries of Pierre Potier at Centre National de la Recherche Scientifique (CNRS) during his work on taxol synthesis. Docetaxel is twice as active as paclitaxel. Currently, a cell culture method developed by Phyton Catalytic is used by Bristol-Myers Squibb (BMS) to produce paclitaxel. In December 1992, the FDA approved paclitaxel for the treatment for ovarian cancer. Researchers also tested the efficacy of paclitaxel as a treatment for advanced breast cancer. Subsequent clinical trials confirmed these results, and in 1994 the FDA approved taxol for use against breast cancer. Clinical trials to test paclitaxel against other types of cancer and in combination with other therapies are currently in progress.

By the early 2000s, taxol had become the best-selling anticancer drug ever manufactured. In fact, it has been suggested that it was taxol (the first billion-dollar drug for the NCI) that made drug companies realize there was financial gain in developing drugs for malignancies.

Other natural products with cytotoxic properties received attention in the 1960s.

Vinca alkaloids

In the 1960s, the natural products group at Eli Lilly and Company found that a series of *Vinca* alkaloids, originally discovered in a screen for antidiabetic agents, blocked proliferation of tumor cells. **Vinblastine** (Velban) and **vincristine** (Oncovin) belong to this group of drugs. They are natural products found in the sap of *Catharanthus roseus* (formerly known as *Vinca rosea*). They are, like paclitaxel, microtubule-targeted drugs, although, in contrast to paclitaxel, the *vinca* alkaloids suppress microtubule dynamics. The effect of the disruption of the microtubule intrinsic dynamics is the same: inhibition of assembly of the bipolar spindle, and consequent activation of the mitotic checkpoint, thereby inducing a prolonged mitosis, ending in cell death.

Bleomycins

The bleomycins are a family of natural products released by the bacterium *Streptomyces verticillus* into its environment as glycopeptide-derived antibiotics. They have strong antineoplastic properties and so are widely used as anti-tumor drugs. The clinically administered form of bleomycin (BLM), **Blenoxane**, comprises mainly bleomycins A2 and B2. It is generally accepted that the most important cellular response to bleomycin and the pharmacological properties of the drug are derived from its ability to mediate DNA degradation. In this process, it is believed that the antibiotic chelates iron (as Fe^{2+}, although copper, Cu^+, has also been proposed) and forms a complex that reacts with oxygen to produce a superoxide-activated complex that ultimately triggers DNA double-strand scission. A detectable intermediate prior to DNA cleavage is $BLM-Fe^{3+}-OOH$. The BLM O–O bond scission is the rate-limiting step in DNA attack. The primary site for bleomycin-induced DNA cleavage is the pyrimidine nucleoside 5′-GPyr-3′ sequence (Pyr is C or T).

Anthracyclines

Anthracyclines (or anthracycline antibiotics) are a class of natural products derived from *Streptomyces* bacteria used in cancer chemotherapy. They were discovered in the 1950s and developed in the 1960s as cytotoxic agents. The first anthracycline discovered was **daunorubicin** (daunomycin), produced naturally by *Streptomyces peucetius*, a species of actinobacteria. Isolated from a mutated strain of *Streptomyces peucetius*, **doxorubicin** (adriamycin) was discovered shortly after (Fig. 15.9), and showed better anticancer properties and less toxicity than daunorubicin. The mechanism of action of doxorubicin derives from inhibition of topoisomerase functions. Since it is a nonspecific DNA-intercalating agent, it was puzzling that it had enhanced selectivity for cancer cells, as deduced from its clinical efficacy. In 1984, it was reported that doxorubicin induces protein-associated strand breaks by trapping and stabilizing the DNA–topoisomerase II cleavable complex in a cell-free system. This provided some insight into the basis of doxorubicin selectivity for cancer cells.

Despite the general acceptance that the DNA–topoisomerase II adduct is an important molecular target for doxorubicin, it is likely that other mechanisms, such as direct oxidative damage to DNA, might be involved in the overall efficacy of doxorubicin, and the relative contribution from hitting different molecular targets may vary from one cancer to another.

Other natural products

Camptothecin (Fig. 15.10) is a cytotoxic quinoline alkaloid that was isolated from the bark and stem of *Camptotheca acuminata* (*Camptotheca*). Camptothecin and its analogues bind to a complex formed by DNA with topoisomerase I. Their ability to inhibit this enzyme correlates closely with activity in *in vivo* mouse leukemia assays. The two camptothecin analogues **topotecan** and **irinotecan** (Fig. 15.10) have been used in cancer chemotherapy since the 1990s.

Other natural products have also found their way into the clinic for the treatment of malignancies. **Geldanamycin**, originally discovered in the organism *Streptomyces hygroscopicus* in 1970 by DeBoer and colleagues, binds and alters the function of heat shock protein 90 (Hsp90). **Halichondrin B**, originally isolated from the marine sponge *Halichondria okadai* by Hirata and Uemura in 1986, is a tubulin-targeted mitotic inhibitor. **Trabectedin** (ecteinascidin 743 or Et-743; discovered by the NCI in the 1960s as a product from the sea squirt *Ecteinascidia turbinata*, now marketed as Yondelis by PharmaMar) appears to trap DNA-binding proteins at sites where structural distortion of the DNA is recognized by direct readout of H bonding. The distortion of DNA following covalent modification is likely to be responsible for its clinical efficacy. **Pyrrolobenzodiazepine** (SJG-136, NSC 694501) is a synthetic dimer based on the naturally occurring anthramycin family of anti-tumor antibiotics developed by the NCI. It is a sequence-selective DNA-targeting agent that crosslinks DNA and forms monoalkylated adducts, currently (2013) in phase II clinical trials. **Telomestatin** is a macrocyclic natural product first isolated from the bacterium *Streptomyces anulatus*. It has been shown to interact specifically with a number of G-quadruplex structures and is still in development.

New-generation platinum drugs

In the search for lead compounds, there is a strong tendency in cancer drug discovery projects to focus on already-established drugs. The aim is to find a second generation of drugs with higher and more controlled activity, better selectivity toward the target, and subsequently less toxicity and improved pharmacokinetics and formulation. This is also known in medicinal chemistry as the

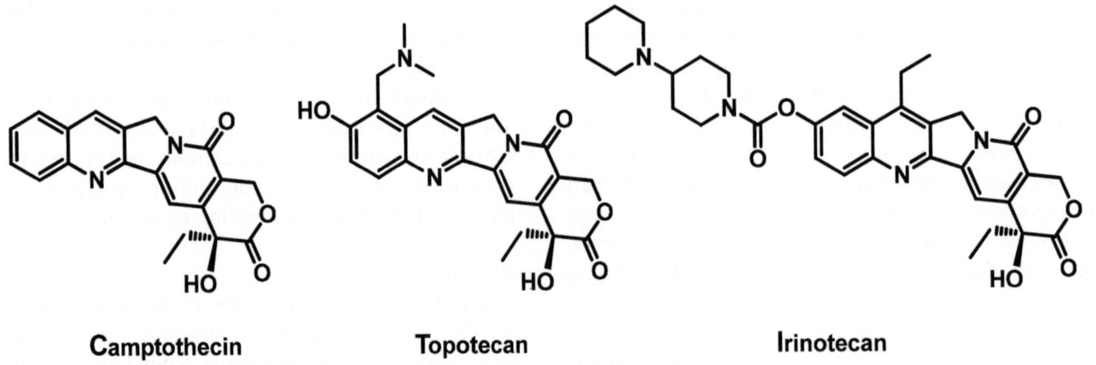

oxorubicin

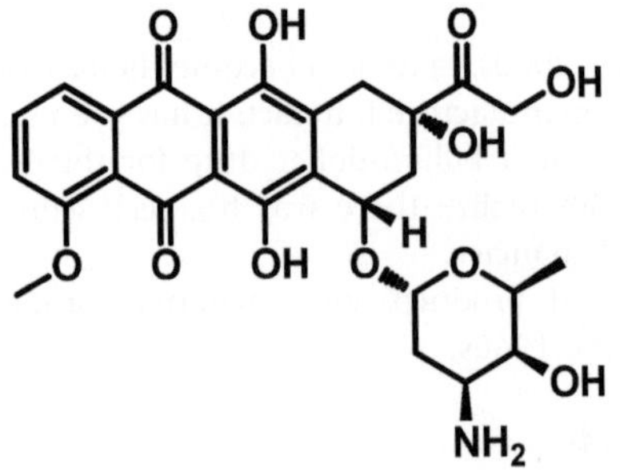

Figure 15.9 Anthracycline antibiotic doxorubicin.

Camptothecin   Topotecan   Irinotecan

Figure 15.10 Quinoline alkaloids, topoisomerase I inhibitors.

"me better" approach, and its high popularity in cancer research might well be due to the inherent limitations of most of today's anticancer therapies. This can be attributed to the fact that such a versatile illness tends to affect each individual differently.

The "me better" approach has advantages and disadvantages. Since the best validation of a target is clinical efficacy and safety data, this has a positive effect in second- and third-generation therapeutics. These are able to aim for better efficacy and side-effect profiles based on the clinical performance of first-generation drugs, due to an understanding of the target, the pharmacophore, and the mechanism of action. The main difficulty is to justify investment in a compound similar to one already in the clinic.

Since we have already seen a few examples (chlorambucil, melphalan, carmustine, cyclophosphamide, methotrexate, nilotinib, topotecan, and irinotecan), we will describe only how this applies to cisplatin derivatives.

Carboplatin and oxaliplatin

Carboplatin is a second-generation platinum anticancer drug approved for use in Europe in 1986 and by the FDA in the United States in 1989. The patent on carboplatin expired in Europe in 2000 and in the United States in 2004. Carboplatin (Fig. 15.11) is about eightfold more water-soluble than cisplatin and presents a similar qualitative spectrum of activity as cisplatin, but its toxicity is less severe. Carboplatin is much less reactive and more stable toward hydrolysis. Once the cyclobutane-1,1-dicarboxylate ring (the O,O-bidentate chelating ligand) opens and the chelated ligand is displaced, carboplatin can form the same lesions on DNA as cisplatin.

Oxaliplatin (Fig. 15.11), first made at Nagoya University, Japan, in 1976, was licensed to Debiopharm in 1989, and then to Sanofi-Aventis in 1994. It was approved for use in Europe in 1996 (the year the patent on cisplatin expired), and is sold by Sanofi-Aventis as Eloxatin. It was approved by the FDA in 2002. Patent protection on oxaliplatin expired in Europe in 2006 and will expire in the United States in 2013.

Oxaliplatin, like carboplatin, is also very stable in water, and DNA binding can occur by displacement of the chelated oxalate ligand. When this happens, as for cisplatin, 1,2-GG crosslinks can be formed on DNA. However, this crosslink now contains a more bulky chelated {Pt(diaminocyclohexane)}$^{2+}$ unit instead of a {Pt(NH$_3$)$_2$}$^{2+}$, which affects subsequent protein and repair-enzyme recognition. Oxaliplatin has performed better in combination with other agents for the treatment of platinum-resistant or -refractory cancers.

It is worth noting that the chemistry (and therefore the biochemistry and molecular pharmacology) of platinum is quite distinct from that of purely organic drugs. Platinum binds to ligands via coordination bonds. We note too that several metal ions are required by cells for natural biological functions, for example 10% of proteins coded for by the human genome are zinc proteins. Also, the mechanism of action of some organic drugs may involve modification of metal uptake, transport, and signaling pathways. This area merits further investigation.

Photoactivatable platinum compounds

In an attempt to overcome the high toxicity of platinum drugs while also benefiting from their potent anticancer activity, a series of photoactivatable platinum drugs is being developed that may provide promising candidates in future chemotherapeutic treatments. Still in the early stages of the drug discovery process, photoactivatable platinum compounds (Fig. 15.12) have been shown to be nontoxic in the dark and highly toxic when photoactivated. The photoactivity of these compounds is attributed to the presence, in the platinum first-coordination sphere, of the azido (N$_3^-$) ligands. The use of directed light to control where and when potent platinum cytotoxics are delivered to tissues may prove to be a valuable way of overcoming the drawbacks of current platinum-based drugs.

Pharmacological tools

In order to reduce the timescale of the drug discovery process, from target identification to clinical development, medicinal chemists have made a tremendous effort to develop tools to produce more compounds in a shorter time. Parallel synthesis and combinatorial methods allow large numbers of compounds (chemical compound libraries and microarrays) to be synthesized in an efficient manner regarding time and materials. Such methodologies are still evolving.

Additionally, in order to find leads, an increased implementation of high-throughput screening against the target of chemical libraries containing enormous compound collections (including known drugs, natural products, small molecules, and peptides) has resulted in enhanced research productivity. The overall aim of high-throughput screening is to identify a chemical lead amenable to further chemical and biological optimization. High-speed bioassays, robotics, and information technologies have changed the dynamics of drug discovery. Additionally, the medicinal chemist integrates the results from the *in vitro* screening with molecular modeling and computational chemistry data in order to determine "drug-likeness" and "lead-likeness" properties. *In silico* analyses of virtual libraries are often used prior to synthesis in order to design smarter libraries.

A few examples of the use of these pharmacological tools in success stories of clinical drugs are described here.

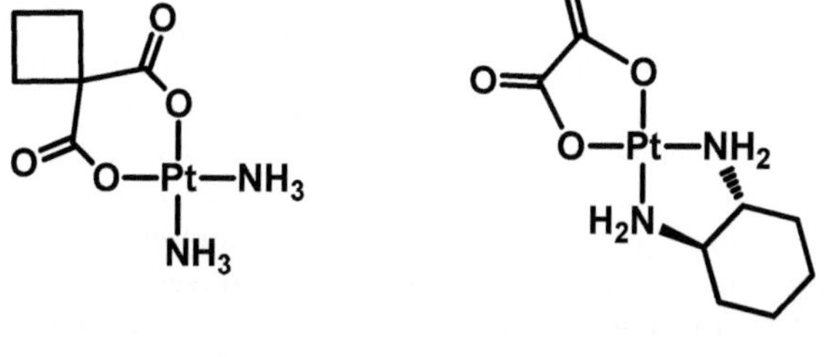

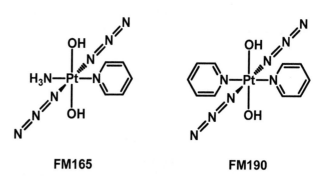

Carboplatin   **Oxaliplatin**

Figure 15.11 Cisplatin-derived anticancer drugs carboplatin and oxaliplatin.

FM165   **FM190**

Figure 15.12 Photoactivatable platinum compounds currently in development.

Vatalanib

Vatalanib (PTK 787/ZK 222584; Fig. 15.13) is a vascular endothelial growth factor receptor (VEGFR) tyrosine kinase–targeting small molecule that inhibits angiogenesis. The compound was identified from a high-throughput screening of about 120 000 compounds at the Novartis Institute for BioMedical Research, as presented by Traxler. Serendipity contributed to its discovery since the phthalazine compound had been on a shelf for 30 years! As noted by Traxler, a less stable compound would certainly have gone unnoticed. Vatalanib shows an outstanding chemical stability, and the results of the screening were promising. This compound is a potent tyrosine kinase inhibitor with good oral bioavailability and activity against the VEGFR family (by docking in the ATP-binding site of the VEGFR kinase), PDGFRβ, and c-KIT receptor kinases. It has been extensively investigated in phase I, II, and III clinical trials.

HA14-1

Wang and colleagues have used an *in silico* screening strategy that exploits new computational methods with large existing databases of compounds to identify new protein-binding molecules. Recently, the small organic molecule **HA14-1** (Fig. 15.14) has been reported as a result of a computer-aided design based on the high-resolution three-dimensional structures of specific molecules that interact with the target receptor protein. The interaction of the lead candidate with the surface pocket of Bcl-2 was further demonstrated by *in vitro* binding studies. This compound also induced apoptosis in HL-60 leukemia cells overexpressing Bcl-2 protein, associated with a decrease in mitochondrial membrane potential, and activation of caspase 9 and caspase 3. Since high expression of Bcl-2 is found in a wide variety of human cancers, Bcl-2 inhibitors will likely play an important role in the future of cancer-targeted therapeutics.

Shepherdin

The novel cell-permeable peptidomimetic **shepherdin** (H–Lys–His–Ser–Ser–Gly–Cys–Ala–Phe–Leu–OH) comes from computational and theoretical structure-based design, modeled on the binding interface between the molecular chaperone Hsp90 and the antiapoptotic and mitotic regulator survivin. It induces death of tumor cells through both apoptotic and nonapoptotic mechanisms.

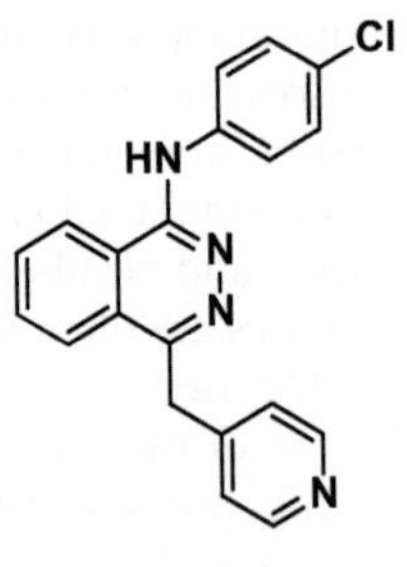

Miscellaneous

Some therapeutic strategies to find leads are driven by the need for the compound to have a particular chemical, biochemical, or physical characteristic. This is the case for photodynamic and photochemotherapeutic drugs. Additionally, even though gene-based therapies are beyond the scope of this chapter, we will describe briefly antisense oligonucleotide–related and RNA interference–related therapies. Finally, although Chapter 13 is fully dedicated to cancer immunotherapy, we mention here sipuleucel-T, an autologous cellular immunotherapeutic vaccine that has recently (April 2010) been approved by the FDA for the treatment of asymptomatic or minimally symptomatic metastatic castration-resistant prostate cancer (CRPC) – the first vaccine to gain FDA approval for cancer treatment.

Photodynamic therapeutics

In 1903, Niels Finsen was awarded the Nobel Prize in Physiology or Medicine "in recognition of his contribution to the treatment of diseases, especially lupus vulgaris, with concentrated light radiation, whereby he has opened a new avenue for medical science." This is considered the beginning of a new field of light therapy or phototherapy, in which a combination of light and specific chemicals can induce cell death when used together, but often have little intrinsic activity when used alone.

The first observation of photodynamic action is generally attributed to a German medical student, Oscar Raab, who in 1900 reported that certain light conditions were lethal to infusoria (*Paramecium caudatum*) in the presence of the chemical acridine. In 1900, neurologist J. Prime observed the development of dermatitis in sun-exposed areas of patients administered eosin orally for epilepsy treatment. Herman Von Tappeiner and Albert Jesionek treated skin tumors with topically applied eosin and white light in 1903. In 1907, Von Tappeiner and Jodlbauer coined the phrase "photodynamic action" for the phenomenon occurring in cells from the effect of not only a chemical reagent (or photosensitizer) and light but also oxygen.

After eosin, the first photosensitizer used in photodynamic therapy, a number of experiments to test combinations of reagents and light led to modern photodynamic therapy (PDT). PDT was first approved in 1993 in Canada, using the photosensitizer **Photofrin** for the prophylactic treatment of bladder cancer. But what is a photosensitizer, and how is PDT defined?

PDT involves three individually nontoxic components that induce cellular processes when combined. The first is the photosensitizer, a molecule sensitive to light, which localizes in target cells and/or tissues and is photo-activated by absorption of light. The second component involves the administration of light of a specific wavelength that activates the photosensitizer. Finally, PDT requires molecular oxygen. The photosensitizer, when exposed to specific wavelengths of light, converts ground-state triplet O_2 to excited-state singlet O_2. In turn, this generates a range of highly reactive oxygen species (ROS), such as hydroxyl radicals. These ROS have been shown to mediate cellular toxicity

Figure 15.13 Angiogenesis inhibitor valatanib (PTK 787/ZK 222584).

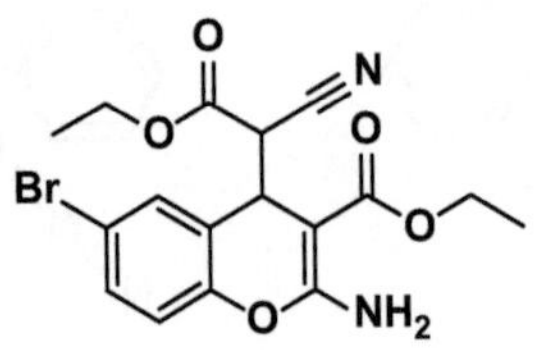

Figure 15.14 Bcl-2 inhibitor HA14-1.

by multifactorial mechanisms. In this way, PDT directly affects cancer cells in causing cell death by necrosis and/or apoptosis, but may also influence the tumor vasculature.

The most extensively studied photosensitizers are porphyrin derivatives. These compounds contain a porphyrin core – four pyrrole rings connected by methene bridges in a cyclic configuration – along with a side chain, and usually a metal such as iron (heme). After experiments performed on animals (Hausman in 1911) and humans (Meyer–Betz tested hematoporphyrin on his own skin in 1913) clearly indicated the potential application of these compounds as therapeutics, Lipson and colleagues at the Mayo Clinic in Minnesota in the 1960s initiated the modern era of PDT. Their studies involved a compound developed by Samuel Schwartz called hematoporphyrin derivative (HPD) in the 1950s. HPD is produced by acetylation and reduction of crude hematoporphyrin and was found to be twice as phototoxic as its precursor; to localize to tumors, emitting fluorescence; and therefore to hold promise as a diagnostic tool. The selective accumulation mechanisms are not fully understood.

The therapeutic application of PDT to cancer came from the realization by Diamond and colleagues that porphyrins combine advantageous properties of tumor localization and phototoxicity. In the 1970s, a number of *in vivo* studies revealed that PDT delayed glioma growth in rats, where HPD and red light eradicated mammary tumor growth and bladder carcinomas in mice. The first human trials with HPD, in patients with bladder cancer, began in 1976. In 1983, Dougherty proposed that the active component of HPD was composed of two porphyrin units linked by an ether bond, naming it dihematoporphyrin ether (DHE). Photofrin is partially purified HPD, a mixture of mono-, di-, and oligomers that all contain the porphyrin moiety.

Photofrin is the most commonly used photosensitizer in the clinic today. New photosensitizers with better characteristics than Photofrin (which is considered to be an "ill-defined drug") are under development. Ideally, new compounds should absorb light at longer wavelengths (to facilitate tissue penetration by light), have greater tumor specificity, be administered orally, and present less skin photosensitivity. Many other sensitizers have recently been developed (and approved in at least one country), such as **Foscan**, **Levulan**, and **Metvix**; or have entered clinical trials, including **Verteporfin**, **Benzvix**, **Hexvix**, **Purlytin**, **BOPP**, **Photochlor**, **Lutex**, **Pc 4**, and **Talaporfin**.

It will be interesting to see whether the development of agents such as the photoactive platinum(IV) complexes mentioned here, which do not require O_2 to cause cancer cell damage (a new form of "photochemotherapy"), can complement the O_2-requiring agents currently used in PDT.

Antisense therapy

Antisense oligo(deoxy)nucleotides (ASOs) are single strands of DNA or RNA (or a chemical analogue, e.g. peptide nucleic acid (PNA)) that specifically target (through complementarity), hybridize, and inhibit the messenger RNA (mRNA) sequence of a selected gene. In this manner, they prevent protein translation of the selected mRNA strand by binding to it. If the ASO is made of DNA, then the DNA:RNA duplex recruits RNase H endonuclease, which cleaves the RNA strand in the duplex and leaves the antisense DNA intact to hybridize other mRNAs of the target gene. The target mRNA is chosen on the basis of its direct causality to a cellular event that needs to be "turned off." The strategy of inhibiting gene expression by targeting messenger RNA (mRNA) with ASOs has been widely used to investigate the role of oncogenes in cancer development. ASOs targeted to inactivate oncogenes might have a therapeutic role in the treatment of human malignancies.

A major drawback to the use of ASOs as a feasible therapeutic alternative is their rapid degradation by nucleolytic enzymes present both inside and outside of cells due to their naturally occurring phosphodiester linkages. To avoid this, antisense technology has evolved to produce ASOs with increased functional stability and permeability, for example by replacing the phosphodiester backbone with a nuclease-resistant phosphorothioate linkage. Additional chemical modifications incorporated into the sugar, such as the electronegative substituents 2'-O-methyl or 2'-O-methoxy-ethyl, in the ribose ring at the 2'-position, have resulted in second-generation ASOs.

One of the first successful applications of antisense technology (1980s) was targeted to *Myc* expression in the HL-60 cell line. At present, the antisense molecule **LY2181308** (Eli Lilly and Company) is in phase II trials. This targets the protein survivin for the treatment of hepatocellular carcinoma. **Oblimersen** (Genasense; marketed by Genta/Aventis) is a Bcl-2 antisense drug currently in several trials (phases I–III), alone or in combination. An example of a second-generation ASO with significant promise for the future is **OGX-011** (OncoGeneX), which can potently suppress the target protein clusterin in humans and is also currently in trials (phases I–III).

Further series of nucleotide derivatives with the same therapeutic principle have been designed to target oncogene expression at the transcriptional level. For example, promising agents include triple-helix-forming oligonucleotides (TFOs) which bind to double-stranded purine-rich DNA within promoter regions and block the binding of transcription factors.

RNA interference

The phenomenon of RNAi is discussed in Chapter 11 and refers to an endogenous cellular pathway for gene silencing. At the beginning of the 21st century, Tuschl and colleagues showed that the phenomenon of RNAi in mammalian somatic cells silences gene expression with high specificity without activating a nonspecific interferon response. Translational researchers rapidly realized the implication of these siRNAs and their enormous therapeutic potential. Molecules that can specifically silence gene expression are powerful research tools. There are many human diseases where genes are involved that can be targeted and silenced by exogenous introduction of siRNA or by introduction of gene constructs expressing short hairpin RNAs (shRNAs) that are converted into siRNA by the RNA machinery. It is therefore not surprising that siRNA-based drug development has proceeded extremely rapidly. In 2003, Song and colleagues demonstrated for the first time the potential therapeutic use of siRNAs in a mouse model (for fulminant hepatitis). Human safety clinical studies began a year later, despite the lack of comprehensive understanding of the phenomenon of RNAi. This might be a reflection of the enormous support that RNAi has received in the scientific community. Some believe it will soon be the next major tool in targeted cancer therapy because of its impressive specificity and efficacy. For example, when compared to antisense approaches, siRNAs are 1000-fold more active without compromising activity in cell culture. Another major attraction of RNAi is that conceptually, while protein targets must be "druggable" in order to be modulated by small molecules or monoclonal antibodies, virtually any target is accessible by siRNA. However, there are important challenges to overcome: off-target effects (genes with imperfect complementarity might be unintentionally silenced), triggering innate immune responses, and, most importantly, obtaining effective and specific delivery into the cytoplasm of target cells. Some of them have been addressed, and delivery of siRNAs to tumors (accounting for transport into the target tissue and cellular uptake) remains a major obstacle for the development of RNAi-based therapeutics. The negative charge and the size of siRNAs hamper their penetration through the cell membrane. Additionally, RNA is quickly degraded in plasma, so chemical modification of the duplex for protection is a sought approach. Various delivery strategies include encapsulation in nanoparticles, cationic lipids, antibodies, cholesterol, aptamers, and viral vectors. Delivery strategies seem to fall into two categories – local delivery to the target tissue and systemic delivery – the latter of which can be divided into two classes: nonconjugated macromolecular assemblies (lipids, polymers, and biopolymers) and covalently bound siRNA conjugates (small-molecule conjugates, carbohydrate conjugates, peptide-mediated delivery, antibodies, and proteins). RNAi-based therapy with delivery into the cytoplasm of the cancerous cells in humans using targeted nanoparticles has been reported (Box 15.1).

A phase I study of a siRNA against PKNc, a novel putative Ser/Thr kinase, has been completed and may be the first in a new generation of liposome delivery systems and effective siRNA cargos.

Vaccine therapies

The use of vaccines in cancer therapeutic strategies has been explored for many years. Despite being thoroughly covered in Chapter 13 ("Tumor Immunity and Immunotherapy"), we will describe here three examples that deserve mention.

Historically, the Egyptians observed that the surgical opening of a tumor site could produce tumor regression, highlighting the possibility that the generation of infection activates the immune system. This gave rise to the concept of cancer immunotherapy. In fact, avoidance of immune destruction is one of the hallmarks of cancer. In the same context, over a century ago, W. Coley, a surgeon from New York, discovered that some infections could produce tumor regression. He created a "vaccine" based initially on erysipelas-causing bacteria to treat patients with sarcoma (Chapter 13). The bacillus Calmette–Guérin (BCG) vaccine, which is derived from a strain of the attenuated live bovine tuberculosis bacillus, *Mycobacterium bovis*, has been used to prevent tuberculosis since 1921, and has been applied for immune stimulation in tumors since the 1960s. It is most effective in superficial bladder cancer. The attractive concept of activating the host immune system to kill tumor cells and eradicate cancer has been exploited with varying degrees of success.

A recent example of a cancer vaccine is **sipuleucel-T** (Provenge; Dendreon), which exploits the approach of producing immunostimulatory dendritic cells specific for a particular tumor antigen, via maturing dendritic cells with the antigen *ex vivo* (Chapter 13). The vaccine consists of autologous peripheral blood mononuclear cells – including antigen-presenting cells (APCs) – that were activated during a defined period in cell culture with recombinant human PAP–GM–CSF, which is prostatic acid phosphatase (expressed in 95% of prostate cancers and largely limited to prostate tissue) and granulocyte-macrophage colony-stimulating factor (an immune cell activator). It has recently been approved by the FDA.

Exploring the periodic table

The use of platinum-based therapies with their unique mechanism of action is now well established. There is potential for the design of other metal complexes as anticancer agents. Indeed,

---

Box 15.1 RNAi therapies in humans

One of the major limitations to be overcome for a successful application of RNAi therapies in cancer is that of specific delivery into the cytoplasm of cells in the tumor tissue. This major hurdle has been successfully overcome using cyclodextrin and adamantine–PEG–Tf-based nanoparticles. A linear cyclodextrin-based polymer (CDP) is the core of the nanoparticle. Additionally, the particles contain molecules of polyethylene glycol (PEG) with the molecule adamantine (AD) at one end which can form inclusion adducts within the cyclodextrins. Transferrin (TF) added to the surface (via addition to the other end of PEG) plays the key role of targeting and binding to the receptor (believed to be overexpressed on the surface of "hungry" cancer cells) and therefore mediates the internalization of the nanoparticle into the target cell. These nanoparticles have proven successful in nonhuman primates, and most recently in humans (*Nature*, 2010, *464*, 1067). Under the name of **CALAA-01** in the clinical version, they entered Phase I trials in May 2008. There is evidence of RNAi pathways in humans that arise from siRNA systemically administered by these nanoparticles, and the successful inhibition of the target gene in patients from the trial. In CALAA-01, the RNA of choice was a siRNA sequence that is a potent inhibitor of the M2 subunit of ribonucleotide reductase (RRM2) and is active in mouse, monkey, and human. The reductase catalyzes the formation of deoxyribonucleotides from ribonucleotides, where inhibition of this pathway results in loss of cell proliferation.

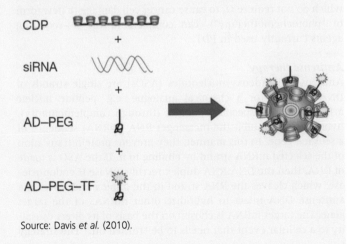

Source: Davis *et al.* (2010).

two complexes containing ruthenium are currently in clinical trials: NAMI-A, which is antimetastatic, and KP1019. Promising experimental data exist to suggest that organometallic compounds based on titanium, tin, and osmium, as well as gold porphyrins, for example, could also enter trials.

The arsenic-based drug **Trisenox** (As_2O_3; Cell Therapeutics, Inc.) is a first-line treatment for acute promyelocytic leukemia, and new arsenic compounds such as S-dimethylarseno-glutathione (**ZIO-101** or **darinaparsin**; Ziopharm Oncology, Inc.) are in phase I and II trials for myeloma, leukemia and lymphoma, hepatocellular carcinoma, and oral cancer.

Finally, it should be noted that the bonds from biomolecules to metals (coordination bonds) are special: they are weaker than covalent bonds but stronger than ionic or hydrogen bonds. The ability of metal compounds to exist with various metal oxidation states (the formal charge on the metal) and to bind to various types of ligand atoms (e.g. oxygen or sulfur) in various coordination geometries (the arrangement of ligand atoms around the metal) makes them highly versatile for tuning interactions with target sites. Metal–ligand bonds can make and break on a wide range of timescales (nanoseconds to years) and can be controlled by design features – a potentially powerful weapon in drug design.

Lead optimization

In general terms, in the lead optimization process a confirmed hit is structurally refined to improve its "drug-likeness" without compromising its efficacy. This stage frequently represents the bottleneck of any drug discovery program and is where struggles occur the most, particularly in academia, when progressing through the drug development timeline. The evaluation of potential for protection of the intellectual property associated with the leads is very important at this stage.

During studies of the mechanism of action (the interaction of the lead at the molecular, cellular, and animal levels), the lead can be structurally modified so as to cope with the different metabolic steps, reach the target, and modulate it. Lead optimization is a multistage process that implies iterative nonlinear experiments. These experiments and assays take into account the knowledge gained at every step, so the information is fed back into the lead to propose structural modifications to optimize the pharmacological properties. New improved analogues are analyzed, and the data fed back into the optimization cycle for the determination of potency, selectivity, and mechanism of action. Empirical information from the structure activity–function relationships of known drugs is combined with rational design in order to optimize the physicochemical properties of the lead.

Animal pharmacokinetic (PK) and pharmacodynamic (PD) assessments are carried out to judge the general pharmacology of the potential drug. The main objective in using animal models (usually rodents) is to understand the response of the whole organism to the new chemical entity (NCE), and to use this information to predict its effect in humans.

PK and PD studies rely greatly on analytical methods and instrumentation. A large number of parameters are assessed and taken into account as a whole, including ADME (absorption, distribution, metabolism, and excretion (also elimination)), bioavailability, protein binding, stability, half-life, maximum serum concentration, total exposure or area under the curve, clearance, and volume of distribution.

Additionally, the resulting drug should have minimal toxicity and side effects, as well as maximal efficacy. Depending upon the nature of the active sites, selectivity improvements can be achieved through increasing potency at the desired target or reducing potency at other sites. Assessment of toxicity, particularly in anticancer therapeutics, where drugs are expected to be particularly toxic, is a very important part of these studies. Early on, the toxicity investigation usually includes a series of standard assays such as inhibition of CYPs (using either recombinant cytochrome P450 enzymes or liver microsomes). Toxicity data from relatively simple *in vitro* assays go into the risk–benefit evaluation, being an important determinant for a lead to advance into preclinical studies. Animal models are used for escalating-dose studies aimed at determining a maximum tolerated dose (MTD). This step involves monitoring a series of parameters, such as body weight, food intake, blood chemistry, and liver activity. The kinetic data in the animal models, whether looking at data from an efficacy, pharmacology, or toxicology point of view, are used by the medicinal chemist to define the drug levels that are needed. For example, a high level of drug for a short period of time may be necessary, or a longer exposure at a lower concentration, with multiple doses at certain time intervals, may be required. ADME and toxicology studies are expensive and usually have limited throughput.

In addition, some PK and PD studies may require radioisotope labeling of the lead molecule. Some drugs might require specific formulations, or a pro-drug approach, all of which tend to weigh heavily on medicinal chemistry resources. The formulation and delivery of drugs are crucial parts of the drug discovery and lead development process.

Formulation problems and solutions also feed back into the iterative lead optimization cycle and in this way influence the design of the lead molecules. Formulation and delivery are closely linked. For example, intravenous delivery of a drug might require a different formulation compared to oral delivery. An example of the use of a special formulation is **Abraxane**. It was approved in 2006 as a paclitaxel nanoparticle formulation, and is essentially albumin containing bound paclitaxel. Liposomal formulations have been developed for a number of drugs. For example, a liposomal formulation of a highly lipophilic camptothecin derivative, **silatecan DB-67**, was found to be twice as cytotoxic as camptothecin, 25 times more lipophilic (and therefore readily incorporated into liposomal bilayers), and more stable in human blood than any of the camptothecins in clinical use. Increased cancer cell specificity can be achieved through conjugation with targeting molecules such as antibodies. For example, in **SGN-15**, doxorubicin is conjugated to a monoclonal antibody targeted to a tumor-specific antigen.

A pro-drug strategy is a common tool for the medicinal chemist to improve the ability of a compound to act as a drug, particularly with respect to stability, tolerability, and bioavailability. Pro-drugs are designed to undergo enzymatic and/or chemical transformation *in vivo* (e.g. in the bloodstream or in specific tissues such as the liver) into biologically active metabolites, which are the active drugs. The major problem with designing pro-drugs is the need for them to be sufficiently stable to provide an appropriate shelf life, but sufficiently unstable to allow rapid conversion into the active species *in vivo*. The pro-drug approach is becoming increasingly common: about 5–7% of approved drugs worldwide can be classified as pro-drugs, and approximately 15% of all new drugs approved in 2001 and 2002 were pro-drugs. Reasons for adopting the pro-drug approach have included improved physicochemical properties such as solubility, adsorption and distribution, drug targeting, improved stability, prolonged release, reduced toxicity, and improved patient compliance (e.g. better taste).

Next, we describe a few examples that illustrate **the pro-drug concept**.

Cyclophosphamide and ifosfamide

Cyclophosphamide and **ifosfamide** (Fig. 15.15) are nitrogen mustard alkylating agents from the oxazaphorine group. They were developed by Norbert Brock in ASTA-Werke, Germany (now Baxter Oncology). Brock and his team synthesized and screened over 1000 candidate oxazaphorine compounds. The idea behind their design was to transform the nitrogen mustard into nontoxic form (pro-drug) that could be transported through the body and activated enzymatically in cancer cells. The first clinical trials conducted by Rudolf Gross were published at the end of the 1950s.

Both pro-drugs are metabolized on passage through the liver by oxidase enzymes to generate active species (4-hydroxylation products). The main active metabolite of cyclophosphamide is 4-hydroxycyclophosphamide, which exists in chemical equilibrium with its tautomer, aldophosphamide (Fig. 15.16). Most of the aldophosphamide is oxidized by the enzyme aldehyde dehydrogenase (ALDH) to give carboxyphosphamide, which is believed to lack cytotoxic potential. However, some of the aldophosphamide can react further to produce phosphoramide

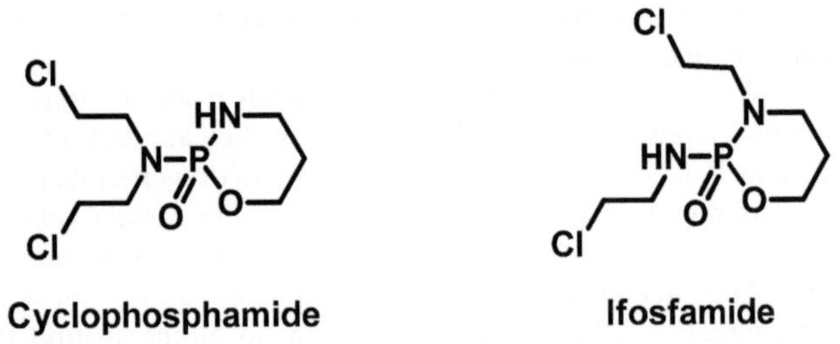

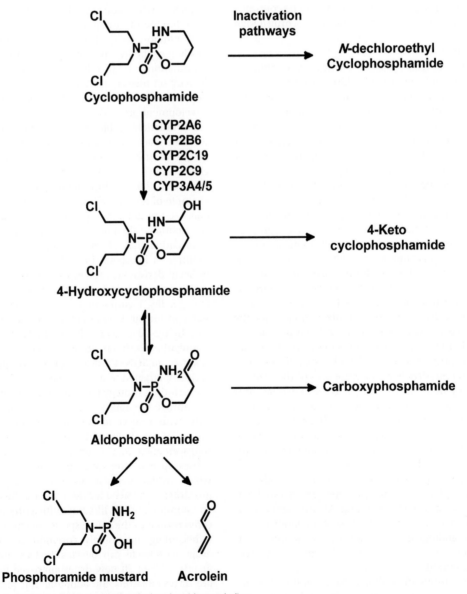

Figure 15.15 Structures of pro-drugs cyclophosphamide and ifosfamide.

Figure 15.16 Pathway of cytochrome P450-catalyzed cyclophosphamide metabolism.

mustard and acrolein. Phosphoramide mustard (Fig. 15.16) is believed to exert the cytotoxic effect, through the formation of inter- and intrastrand crosslinks in DNA.

Cyclophosphamide and ifosfamide (also called isophosphamide), despite sharing structural similarities, not only present a different therapeutic spectrum but also have important differences in their metabolism and toxicity. For example, only 10% of a therapeutic dose of cyclophosphamide is converted to renal and neurotoxic chloroacetaldehyde, whereas about 45% of ifosfamide is typically metabolized via N-dechloroethylation to chloroacetaldehyde. The involvement of different CYPs in their metabolism also varies. Although activation through 4-hydroxylation is accomplished mainly by CYP2B6, it is also mediated by CYP3A4, CYP2C19, and CYP2C9 for cyclophosphamide and by CYP3A4 for ifosfamide.

Temozolomide

Temozolomide (Fig. 15.17) is an example of an organic small molecule with anticancer properties, which is a pro-drug requiring activation through hydrolysis. It was first synthesized at Aston University (Birmingham, United Kingdom) over 20 years ago in the laboratories of Malcolm Stevens, who led a team of Cancer Research UK–funded researchers. In the late 1970s, it was known that some triazenes (molecules containing a chain of three nitrogen atoms) exhibited anti-tumor activity. Temozolomide was synthesized as a cyclic triazene, in particular an analogue of dacarbazine (DTIC), a drug already on the market against malignant melanoma. It was found to be cytotoxic. Temozolomide is converted at physiological pH to the short-lived active compound monomethyltriazenoimidazolecarboxamide (MTIC), whose cytotoxicity is due primarily to methylation of DNA at the O6 and N7 positions of guanine, mediated by a methyldiazonium ion (Fig. 15.17) and resulting in inhibition of DNA replication. Temozolomide benefits from chemical activation as opposed to DTIC-like metabolic activation. Additionally, temozolomide is administered orally and penetrates well into the central nervous system. It is used worldwide to treat the most common type of adult brain tumor, glioblastoma.

Phortress

Malcolm Stevens and the Cancer Research UK group, now at Nottingham University, investigated the cytotoxic activity of benzothiazoles in the 1990s. They solved a solubility problem by synthesis of the lysyl pro-drug **Phortress** (Fig. 15.18).

Phortress is a CYP-activated, DNA-damaging agent. Firstly, it transforms into its active component by losing the lysyl group that conferred the required solubility (Fig. 15.18). The active benzothiazole moiety binds to cytoplasmic arylhydrocarbon receptors (AhR) and is translocated into the nucleus, where it is further activated by the cytochrome P450 enzyme 1A1 (CYP1A1) to a metabolite that causes DNA damage. This DNA damage ultimately leads to cell death.

A number of other drugs are administered as pro-drug precursors. **Tamoxifen**, already described in this chapter (Fig. 15.7), is another example of a drug that requires activation to exert its therapeutic effect. The key metabolites of tamoxifen appear to be 4-hydroxytamoxifen and endoxifen (4-hydroxy-*N*-desmethyltamoxifen), formed primarily by CYP2D6; and *N*-desmethyltamoxifen, formed primarily by CYP3A4. **Capecitabine** (Xeloda, Roche; Fig. 15.4) is an orally administered chemotherapeutic agent used primarily in the treatment of metastatic breast and colorectal cancers. Capecitabine is also a pro-drug that is enzymatically converted into the DNA synthesis antimetabolite 5-fluorouracil (5-FU) *in vivo*. The activation of capecitabine to 5-FU occurs in several steps. Firstly, it is metabolized to 5-fluoro-5′-deoxycytidine (5′-DFCR) by carboxylesterase in the liver. 5′-DFCR is then converted to 5-fluoro-5′-deoxyuridine (5′-DFUR) by cytidine deaminase, and finally 5′-DFUR is converted to 5-fluorouracil in the tumor by thymidine phosphorylase.

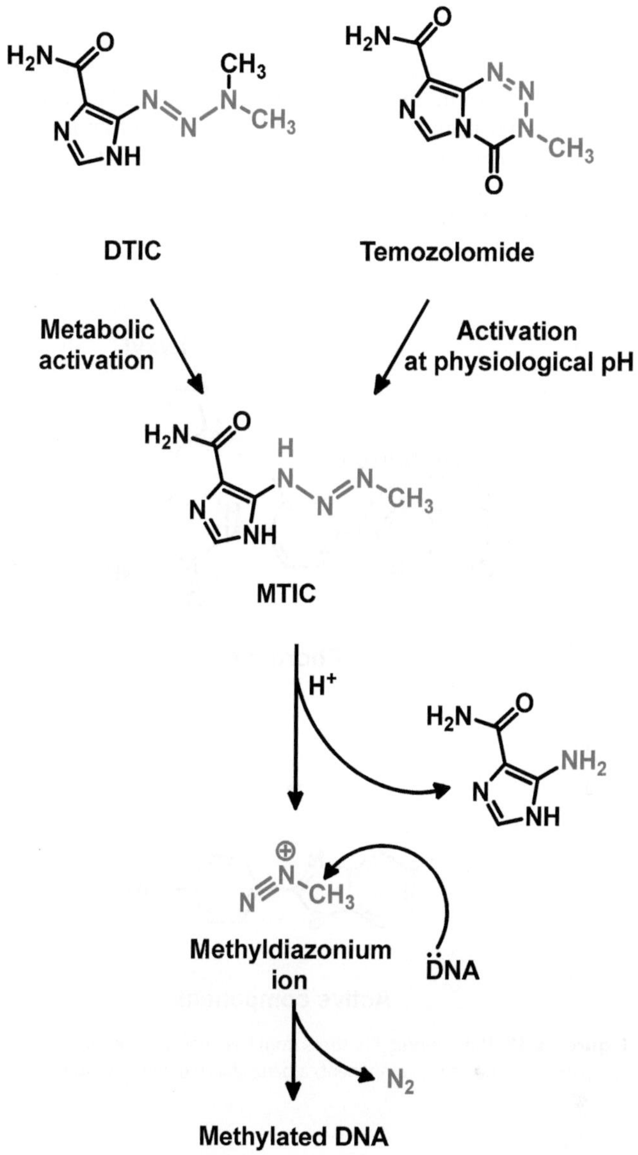

Finally, **SG2285** (Fig. 15.19) is a C2-aryl-substituted pyrrolobenzodiazepine dimer pro-drug that crosslinks DNA and exerts potent anti-tumor activity, when activated by hydrolysis. The pyrrolobenzodiazepines (PBDs) are naturally occurring anti-tumor antibiotics. SG2285 is a highly water-soluble pro-drug of the dimer SG2202, its active component. Two sulfonate groups

Figure 15.17 Dimethyl triazeno imidazole carboxamide (DTIC; top left) metabolizes to monomethyl triazeno imidazole carboxamide (MTIC). Top right: Temozolomide generates MTIC at physiological pH. MTIC alkylates DNA during all phases of the cell cycle, resulting in disruption of DNA function, cell-cycle arrest, and apoptosis.

in SG2285 inactivate the PBD N10-C11 imines (Fig. 15.19). Slow release of the bisulfite groups converts SG2285 to the active PBD dimer SG2202, in which the free imine groups can bind covalently in the DNA minor groove, forming an interstrand crosslink. The dimer has been found to be highly potent in tumor regression at nontoxic doses in a number of xenograft models, including ovarian, non-small-cell lung, prostate, and pancreatic cancers as well as melanoma.

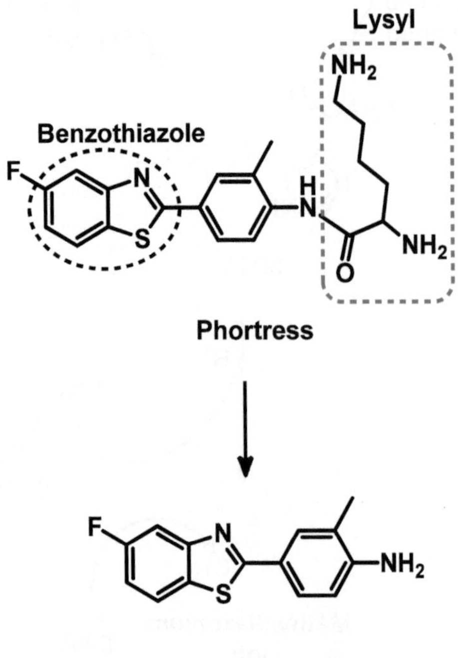

Figure 15.18 The pro-drug Phortress (top) has improved solubility in comparison to the active component, a benzothiazole derivative (below).

Development of candidate

The candidate development phase involves a significant commitment in terms of money, time, and resources. Drug development requires attention to Good Laboratory Practice (GLP), Good Manufacturing Practice (GMP), and standards and regulations on electronic data and electronic signatures.

This last stage involves candidate progression to clinical evaluation, and determines the ultimate safety profile of a drug. A Chemistry, Manufacturing, and Control (CMC) program is required for any drug to enter clinical trials. This program must contain the supply of the Active Pharmaceutical Ingredient (API), in which the synthesis is scaled up, and the pre-GMP (Good Manufacturing Practice) batch is analyzed and released. For preclinical and clinical development, it is of vital importance that the highest-quality compound is provided, repeatedly and consistently, at reasonable cost and in a timely manner. The process must be revised so the synthetic route, including accessibility of readily available and cost-effective starting materials, minimization of synthetic and purification steps, and feasibility of scale-up (preferably to the kilogram scale), is thoroughly optimized. The API physicochemical evaluation and pre-formulation activities are carried out at this stage.

There is often a realization early in this phase that the drug would benefit from special delivery, such as the **Gliadel Wafers** (NSC 714372), developed by the National Cooperative Drug Discovery Group (NCDDG) Program, in the NCI. Gliadel Wafer implants are a new approach to chemotherapy administration for brain tumors as an adjunct to surgery. The wafer is made of a gel that contains the anti-tumor drug **carmustine** (a mustard gas–related α-chloro-nitrosourea compound, an alkylating agent). During brain surgery, up to eight wafers are positioned in the place that had been occupied by the tumor. The wafers slowly release carmustine into the area for a few days, dissolving over a period of 2 to 3 weeks after implantation.

The regulatory preclinical studies include information on the ADME–PK studies, which include bio-analytical assays (to cover toxic species and refine and validate animal and human plasma

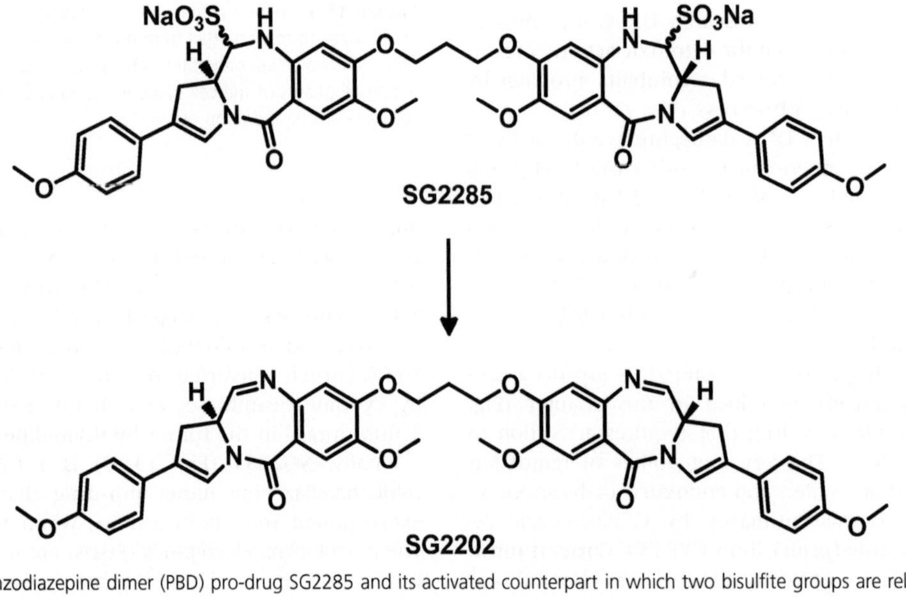

Figure 15.19 Pyrrolobenzodiazepine dimer (PBD) pro-drug SG2285 and its activated counterpart in which two bisulfite groups are released to generate PBD N10-C11 imines.

assays), PK evaluation in rodent and nonrodent animals, and ADME studies both *in vitro* and *in vivo*. The preparation of the Safety Pharmacology Package includes the toxicity data (including toxicity of the metabolites of the drug) and PD data. The Drug Safety Evaluation contains the mutagenicity test data together with local tolerance assessment, antigenicity potential, acute toxicity (both *in vitro* and acute oral studies), and repeat dose toxicology, including short-term toxicity studies ranging from 2 weeks to several months, depending on the proposed duration of use in the clinical study.

The regulatory preclinical compilation of data may be summarized as data collected in three areas. The first requires animal pharmacology and toxicology studies – preclinical data that permit an assessment of whether the product is reasonably safe for initial testing in humans. Secondly, manufacturing information is needed, which relates to the composition, manufacture, stability, and controls used for manufacturing the drug substance. Finally, clinical protocols and investigator information complete the package.

The compilation of regulatory documentation requires approval. Once the approval is given, the application for a clinical trial can be prepared and submitted to the relevant regulatory agency (such as the US Food and Drug Administration (FDA), European Medicines Agency (EMEA), or UK Medicines and Healthcare Products Regulatory Agency (MHRA)).

An Investigational New Drug (IND) application results from a successful preclinical development program and is the vehicle for advancing to the next stage of drug development – human clinical trials. Together with PK and PD studies, formulation and toxicology studies are essential for a successful IND application.

Here our description of preclinical development finishes, and clinical development of the drug begins – the progression through phase I–IV trials. This is discussed in Chapter 16.

Questions remaining

Cancer targets are continually being discovered, validated, and rediscovered. The genomics revolution has been the main engine driving target-based drug discovery over the last two decades. Technologies that allow us to analyze sequence data, copy number, and expression levels of individual genes within cancer cells in a colossal number of independent tumors and normal tissues have resulted in the belief that the differences between cancerous and noncancerous cells at the molecular level will allow clear and unequivocal identification of therapeutic targets. This approach has resulted in the expectation that the discovery and development of cancer drugs might become more predictable, effective, and efficient and that new drugs will cover the complete spectrum of cancers with minimal toxicity effects. However, although the sequences of approximately 23 000 protein-coding human genes are known, the linkage of gene sequence, and even gene expression, to human disease, in particular to cancers, is not so straightforward. Although there have been major advances in the molecular pathology of cancerous cells and tissues, and in cancer therapeutics over the last 60 years, there are still a number of unanswered questions.

The genomic revolution has resulted in the development of a large number of new biochemical tools and extremely advanced biotechnologies. It has therefore led to the anticipation that these developments will drive the discovery of new medicines. Has this anticipation created an exaggerated expectation for the outcome of the human genome sequencing project with regard to the future of cancer drugs?

Whilst the genome is static, the proteome is highly dynamic in its response to external and internal cellular events. Based on our understanding of the molecular biology of the cancer cell proteome (particularly dynamic in cancerous cells) from which we choose our targets, how can new therapies cope with the dynamic effects of complex protein networks?

Pharmacogenetics is the genetic study of drug metabolism, the impact of genetic variations on drug response. The growing knowledge of biological mechanisms in the cancer cell is paralleled by an expansion of our understanding of the complex interaction between the cancerous cell and its microenvironment and with the host. Is the combination of pharmacodynamics and pharmacogenetics the answer to the unsolved problem that all patients do not respond to the same treatment in the same way? New platforms, such as the Pharmacogenomics Knowledge Base (PharmGKB; NIH), are being created to produce tools that can automatically extract relevant information from the biomedical literature and databases, and intensify its focus on understanding the molecular basis for drug toxicity and multiple-drug interactions. Will this evolve into guidelines for doctors about the use of genetic tests to tailor dosage when prescribing medicines to cancer patients? It might be possible to profile patients so that specific metabolic pathways can be targeted for treating a cancer at different stages of the disease. Molecular testing might identify subsets of patients who will respond to selected agents, highlighting the importance of tumor genotyping to identify likely responders. To what extent can molecular diagnostics be used to provide real-time confirmation of the biological impact of the drug or drugs on the patient?

Few new molecular biology techniques have advanced to become practical applications as rapidly as RNAi. Advances in RNAi technology in mammalian cells have made it possible, for the first time, to interrogate systematically every cellular protein for potential addiction to cancer cells. Most importantly, RNAi is an emerging therapeutic approach with an enormous range of applications. RNAi was first described just over a decade ago, and its mechanism is still being unraveled. It is therefore notable that in such a short time, a number of clinical trials using RNAi-based therapies are already in progress. What is the real potential for using RNAi – not only as a tool to identify new drug targets but also, most importantly, as a therapeutic tool? The power of sequence-specific inhibition of gene expression is, of course, a goal worth achieving for the treatment of many diseases, but is it realistic to think that the RNAi therapeutic approach will provide access to an unlimited range of targets?

Conclusions and future directions

The pioneers of the development of anticancer drugs based their therapeutic strategies on the assumption that cancer cells divide more frequently than most normal cells. Since DNA is replicated in every cell cycle, cancer cells are therefore more sensitive to DNA damage. The hypothesis based on DNA damage and impaired cell division has driven the development of effective anticancer therapeutics over many years. However, molecular biology has now provided a new level of understanding of cellular processes.

We should strive to explain all cell events (in both healthy and cancerous cells) at the molecular level (i.e. at the atomic (chemistry) level). Attention must be given to understanding the dynamics of drug binding to target sites and to the potential differences in the behavior of targets (proteins) in vitro and in vivo. We must adapt and design drugs (chemical entities) that are targeted in a controlled manner, and allow the normal growth of tissue to recover.

The new sophisticated tools available for the discovery and validation of cancer targets and for the development of therapeutics have significantly improved the quantity and quality of information that can be collected. Furthermore, these tools have considerably shortened analysis times and resulted in a "faster" ability to fail or succeed in a drug discovery program. This suggests that progress will continue to accelerate as we learn how to incorporate these new technologies into the various steps of anticancer drug research.

The use of RNAi is an emerging therapeutic strategy with enormous potential. The ability to design siRNAs for any host gene or pathogen, once genetic sequences are known, and to test them rapidly highlights a key advantage of siRNA for drug development compared to more conventional drugs.

At a time when most drug discovery projects are focused on finding particular targets and designing agents to hit those targets, a reassessment of the drugs classified as cytotoxics is needed. Although many anticancer drugs in clinic were developed before the general adoption of the target-driven approach to drug design, they do have very distinct molecular targets. Although there are many successful targeted drugs in the clinic, effective therapeutic strategies based upon hypothesis-driven molecular targets are still in their infancy, and the development of resistance to this class of drug is an increasing problem. Additionally, consideration of only a single drug–target interaction in vivo has proven to be overly simplistic. Indeed, a multitargeted drug may be more effective over one with a single target (note the success of combination therapies mainly of "targeted therapies" with "cytotoxics"). The value of cytotoxic-defined drugs should not be underestimated, and their intracellular targets should be investigated where unknown. Their therapeutic value is clinically proven, and much can be revealed through elucidation of the targets they hit and the pathways they interfere with in cancerous cells and in the tumor microenvironment.

At present, we use both cytotoxics and targeted drugs together. Aside from rare exceptions, neither alone appears adequate for treating the major solid tumors. Clinical trials have demonstrated potent synergy between targeted molecules, particularly monoclonal antibodies such as rituximab, bevacizumab, and trastuzumab, and traditional chemotherapy. A major challenge ahead is the design of appropriate selective therapy.

We can envision a future in which a number of anticancer drugs target a much larger number of biomolecules involved in cellular pathways of each of the so-called hallmarks of cancer. These hallmarks, as proposed by Hanahan and Weinberg, include the acquisition of self-sufficient signals for growth, the capacity for extended proliferation, resistance to growth-inhibiting signals, the ability to evade cell death signals, the potential for tissue invasion and metastasis, and the power to induce blood-vessel formation (angiogenesis). Evasion of immune surveillance has also been proposed as a cancer hallmark by Kroemer and Pouyssegur. Finally, an additional set of hallmarks has been proposed by Luo, Solimini, and Elledge, which are based on recent analyses of cellular phenotypes. These depict the stress phenotypes of cancer cells and include metabolic, proteotoxic, mitotic, oxidative, and DNA damage stress.

"Translational research" is a model characterized by the transformation of scientific discoveries into new clinical modalities for oncology. It requires highly interdisciplinary groups, which include chemists, molecular biologists, pharmacologists, geneticists, physicians, oncologists, and statisticians. The interdisciplinarity of cancer research has benefited greatly from the revolution of the Internet, the massive expansion in the communication of findings. These are exciting times for clinicians and scientists working at the forefront of cancer research.

Summary

In this chapter, we have provided a summary of recent advances in the chemical design of anticancer drugs, revealing the stories (and sometimes the serendipity) behind current drugs, and have attempted to foresee what the future holds. We hope this has stimulated the reader's interest in this fascinating area, and a comprehensive reference list of selected reviews has been included in the bibliography section to aid further study.

We first described the drug discovery process, providing a historical perspective and describing the origins of anti-tumor drug research back to the middle of the 20th century. Next, we summarized the steps involved in a current drug discovery process, from the choice of the cancer target to the preclinical development of the drug, all benefiting from the recent revolution in molecular biology. Along the way, we have overviewed the emergence of over 50 different drugs, from conception and design through to compounds and up to mechanisms of action and clinical trials. We hope this summary provides a firm basis on which to establish further advances in anticancer drug design, an exciting and important area for further research.

Bibliography

Historical perspective

Adair, F.E., and Bagg, H.J. (1931). Experimental and clinical studies on the treatment of cancer by dichloroethyl sulfide (mustard gas). *Annals of Surgery*, **93**: 190–9.

Burchenal, J.H., Murphy, M.L., Ellison, R.R., et al. (1953). Clinical evaluation of a new antimetabolite, 6-mercaptopurine, in the treatment of leukemia and allied diseases. *Blood*, **8**: 965–99.

Chabner, B.A., and Roberts, T.G. (2005). Chemotherapy and the war on cancer. *Nature Reviews Cancer*, **5**: 65–72.

Djerassi, I., Farber, S., Abir, E., and Neikirk, W. (1967). Continuous infusion of methotrexate in children with acute leukemia. *Cancer*, **20**: 223–42.

Gilman, A. (1963). The initial clinical trial of nitrogen mustard. *American Journal of Surgery*, **105**: 574–8.

Gilman, A., and Philips, F.S. (1946). The biological actions and therapeutic applications of the β-chloroethyl amines and sulfides. *Science*, **103**: 409–36.

Krumbhaar, E.B., and Krumbhaar, H.D. (1919). The blood and bone marrow in yellow cross gas (mustard gas) poisoning: changes produced in the bone marrow of fatal cases. *Journal of Medical Research*, **40**: 497–507.

Wilkinson, J.F., and Fletcher, F. (1947). Effect of β-chloroethylamine hydrochlorides in leukaemia, Hodgkin's disease, and polycythaemia vera: report on 18 cases. *Lancet*, **250**: 540–5.

Medicinal chemistry and cancer drug discovery

Kamb, A., Wee, S., and Lengauer, C. (2007). Why is cancer drug discovery so difficult? *Nature Reviews Drug Discovery*, **6**: 115–20.

King, F.D., ed., for the Royal Society of Chemistry. (2002). *Medicinal Chemistry: Principles and Practice*. 2nd edn. Cambridge: Royal Society of Chemistry.

Varmus, H. (2006). The new era in cancer research. *Science*, **312**: 1162–5.

Yap, T.A., Sandhu, S.K., Workman, P., and de Bono, J.S. (2010). Envisioning the future of early anticancer drug development. *Nature Reviews Cancer*, **10**: 514–23.

Targeting drugs and their targets

Adams, J., Palombella, V.J., Sausville, E.A., et al. (1999). Proteasome inhibitors: a novel class of potent and effective antitumor agents. *Cancer Research*, **59**: 2615–22.

Altieri, D.C. (2008). Survivin, cancer networks and pathway-directed drug discovery. *Nature Reviews Cancer*, **8**: 61–70.

Baselga, J. (2006). Targeting tyrosine kinases in cancer: the second wave. *Science*, **312**: 1175–8.

Benson, J.D., Chen, Y.-N.P., Cornell-Kennon, S.A., et al. (2006). Validating cancer drug targets. *Nature*, **441**: 451–6.

Druker, B.J. (2008). Translation of the Philadelphia chromosome into therapy for CML. *Blood*, **112**: 4808–17.

Fedorov, O., Mueller, S., and Knapp, S. (2010). The (un)targeted cancer kinome. *Nature Chemical Biology*, **6**: 166–9.

Harley, C.B. (2008). Telomerase and cancer therapeutics. *Nature Reviews Cancer*, **8**: 167–79.

Herbst, R.S., Fukuoka, M., and Baselga, J. (2004). Timeline: gefitinib – a novel targeted approach to treating cancer. *Nature Reviews Cancer*, **4**: 956–65.

Hurley, L.H. (2002). DNA and its associated processes as targets for cancer therapy. *Nature Reviews Cancer*, **2**: 188–200.

Jordan, M.A., and Wilson, L. (2004). Microtubules as a target for anticancer drugs. *Nature Reviews Cancer*, **4**: 253–65.

Lessene, G., Czabotar, P.E., and Colman, P.M. (2008). BCL-2 family antagonists for cancer therapy. *Nature Reviews Drug Discovery*, **7**: 989–1000.

Meyer, N., and Penn, L.Z. (2008). Reflecting on 25 years with MYC. *Nature Reviews Cancer*, **8**: 976–90.

Overall, C.M., and Kleifeld, O. (2006). Towards third generation matrix metalloproteinase inhibitors for cancer therapy. *British Journal of Cancer*, **94**: 941–6.

Tamoxifen

Jordan, V.C. (2003). Tamoxifen: a most unlikely pioneering medicine. *Nature Reviews Drug Discovery*, **2**: 205–13.

Cisplatin and metal-based drugs

Hambley, T.W. (2007). Metal-based therapeutics. *Science*, **318**: 1392–3.

Mackay, F.S., Woods, J.A., Heringova, P., et al. (2007). A potent cytotoxic photoactivated platinum complex. *Proceedings of the National Academy of Sciences of USA*, **104**: 20743–48.

Wang, D., and Lippard, S.J. (2005). Cellular processing of platinum anticancer drugs. *Nature Reviews Drug Discovery*, **4**: 307–20.

Natural products

Johnson, I.S., Armstrong, J.G., Gorman, M., and Burnett, J.P., Jr. (1963). The Vinca alkaloids: a new class of oncolytic agents. *Cancer Research*, **23**: 1390–427.

Nicolaou, K.C., Chen, J.S., and Dalby, S.M. (2009). From nature to the laboratory and into the clinic. *Bioorganic & Medicinal Chemistry*, **17**: 2290–303.

Wall, M.E., and Wani, M.C. (1996). Camptothecin: discovery to clinic. *Annals of the New York Academy of Sciences*, **803**: 1–12.

Photodynamic therapy

Farrer, N.J., Woods, J.A., Salassa, L., et al. (2010). A potent trans-diimine platinum anticancer complex photoactivated by visible light. *Angewandte Chemie, International Edition*, **49**: 8905–8.

Juarranz, A., Jaen, P., Sanz-Rodriguez, F., Cuevas, J., and Gonzalez, S. (2008). Photodynamic therapy of cancer: basic principles and applications. *Clinical & Translational Oncology*, **10**: 148–54.

Antisense therapy and RNAi

Davis, M.E., Zuckerman, J.E., Choi, C-H.J., et al. (2010). Evidence of RNAi in humans from systemically administered siRNA via targeted nanoparticles. *Nature*, **464**: 1067–70.

Dykxhoorn, D.M., and Lieberman, J. (2006). Knocking down disease with siRNAs. *Cell*, **126**: 231–5.

Gleave, M.E., and Monia, B.P. (2005). Antisense therapy for cancer. *Nature Reviews Cancer*, **5**: 468–79.

Heidel, J.D., Yu, Z., Liu, J.Y.-C., et al. (2007). Administration in non-human primates of escalating intravenous doses of targeted nanoparticles containing ribonucleotide reductase subunit M2 siRNA. *Proceedings of the National Academy of Sciences of USA*, **104**: 5715–21.

Pro-drugs

Rautio, J., Kumpulainen, H., Heimbach, T., et al. (2008). Prodrugs: design and clinical applications. *Nature Reviews Drug Discovery*, **7**: 255–70.

Questions for student review

1) Which was the first group of chemotherapeutic drugs to be used?
   a. Targeted drugs.
   b. Monoclonal antibodies.
   c. Antimetabolites.
   d. Nitrogen mustards.
   e) Platinum drugs.

2) Define antimetabolite.

3) What is the main target for cisplatin?
   a. DNA, causing a kink in the double helix that attracts proteins, which protect the damaged DNA from repair.
   b. HMGB proteins, causing a conformational change in the protein that induces apoptosis.
   c. microtubules, stabilizing the microtubule assembly so that it cannot de-polymerize and induce apoptosis.

4) Please state whether the following statements are true or false.
   a. Taxol binds to and stabilizes the microtubule assembly so that it cannot de-polymerize; shrinkage is prevented, and therefore segregation of the chromosomes is halted.
   b. Vinblastine causes disruption of the microtubule intrinsic dynamics; it inhibits the assembly of the bipolar spindle, and consequently activates the mitotic checkpoint, thereby inducing a prolonged mitosis, which ends in cell death.
   c. Bleomycins do not require a metal ion to exert their cytotoxic effects.
   d. The mechanisms of action of doxorubicin and camptothecin involve inhibition of the functions of topoisomerases.

5) What is the meaning of the "me better" approach in drug development? Give examples.

6) Which of the following does PDT require? (Choose all that apply.)
   a. Light.
   b. A photosensitizer.
   c. A reducing agent.
   d. Oxygen.
   e. X-rays.

7) For what is RNAi used in drug discovery? (Choose all that apply.)
   a. Target identification and validation.
   b. As a therapeutic tool.
   c. For parallel synthesis of compounds.
   d. For modification of lipid bilayer structures.

8) Are the following true or false?
   a. Cyclophosphamide and ifosfamide are pro-drugs of nitrogen mustard alkylating agents.
   b. Some pro-drugs are metabolized by enzymes on passage through the liver.
   c. Temozolomide is a CYP-activated DNA-damaging agent that is transformed into its active component by loss of a lysyl group.
   d. Phortress is a pro-drug that is converted at physiological pH to the short-lived active compound, monomethyl triazeno imidazole carboxamide.

16 Biologically Targeted Agents from Bench to Bedside

Michael Khan[a], Peter Sadler[a], Ana M. Pizarro[a], and Stella Pelengaris[b]
[a]University of Warwick and [b]Pharmalogos Ltd, UK

No one is useless in this world who lightens the burden of it to anyone else.

Charles Dickens

To say the truth, every physician almost hath his favourite disease, to which he ascribes all the victories obtained over human nature.

Henry Fielding

Key points

- Cancers may be cured and life expectancy improved by surgery and/or local radiotherapy, but only if the entire tumor is accessible and as long as viable cells have not already relocated to other sites.
- Once cancers have spread, excepting rare cases where single metastases can be successfully removed, hope of a cure rests on the success of systemic treatments. However, even where no cure is possible, localized treatments remain a viable means of alleviating symptoms in order to improve quality if not quantity of life.
- Drugs may be used alone or as "adjuvants," given after surgery or before it (neoadjuvants) in order to improve long-term outcome by eradicating cancer cells that may already have escaped the primary site.
- With notable exceptions, such as testicular cancer and Hodgkin lymphoma, chemo- and radiotherapy treatments arrest but rarely eradicate the cancer, and the disease recurs or progresses.
- Treatment failure, through inherent or acquired resistance to drugs, is a major and common problem in cancer therapy. Thus, apparently successfully treated cancers can be repopulated by cancer cells that have endured the treatment either through their intrinsic resilience or by sheltering within oncogenic oases within the tumor, as exemplified by stem cell niches.
- Cancer stem cells (CSCs), a newly recognized subpopulation of "über"-cancer cells, may be very hard to kill from the outset, or new coteries of drug-resistant cancer cell clones may emerge during chemotherapy.
- More recently, cancer cells have been observed to enter a state of "dormancy," in which they may survive treatments and from which they may unpredictably awaken at some future time. Reminiscent of trench warfare, a protracted and seemingly unsurvivable barrage of chemo-radiotherapy lays waste to the tumor. Yet, still the whistle eventually sounds and the battle-hardened survivors emerge from their dugouts and, from no-man's land, sally forth and capture new positions.
- In fact, these post-apocalyptic cancer cells often prove remarkably resistant to further treatments. Thus, a pessimist might conclude that cytotoxic therapies directed against cancer cells, which only eliminate the camp followers, may simply clear the way for the expansion of more militant CSCs that have been left behind. The resultant revenant tumor, re-emerging phoenix-like from the ashes, will, compared to its forebear, be more aggressive, progress more rapidly, and be less sensitive to therapeutic attack.
- As was discussed in other chapters, the major limitation surrounding traditional chemotherapy is the extent of collateral damage suffered by normal cells, particularly those in bone marrow and epithelia. Thus the oncologist has to steer the patient between Scylla and Charybdis, between treatment failure and infection.
- It is often assumed that early diagnosis would improve the efficacy of most of our anticancer treatments. In particular, finding primary tumors that comprehend only *stay-at-home* cancer cells should enable treatment to be curative. Although likely to be true in most cases, this notion will be hard to prove until we achieve these aims and can actually test it.
- New analytical and imaging tools are under development that may allow us to pick up smaller and clinically silent tumors. However, we are a long way from being able to pick up the spoor of the first few cancer cells, at least until they have gone through a very large

(Continued)

The Molecular Biology of Cancer: A Bridge From Bench to Bedside, Second Edition. Edited by Stella Pelengaris and Michael Khan.
© 2013 John Wiley & Sons, Inc. Published 2013 by John Wiley & Sons, Inc.

- number of cell divisions. As each cancer cell division is a potential mutagenic event, such delays inevitably increase the chance that by diagnosis there will already be a vanguard of invading or metastasizing cancer cells, thereby precluding curative treatment.
- Progress in early detection and tracking down of cancer cells will offer little, if we lack the wherewithal to exploit this advantage.
- All too often, these ever resourceful foes manage to successively evade the finely honed edge of the surgeon's knife, find shelter from the withering blasts of radiotherapy, and finally ford a raging torrent of chemotherapy to reach and colonize a distant shore. It is the researchers who must ensure that clinicians are continually re-armed and able to contend with or preempt the evolving survival instincts and adaptive capabilities of the cancer cell, and they have responded magnificently to the challenge.
- Thus, it may once have been true to say that for cancer patients nothing was certain except *death and taxols*, but times are changing fast. In many cases, cancer has been relegated from the premier league of rapidly life-threatening conditions to the lower divisions along with other chronic diseases. This raises a series of new challenges including how to support patients living long term with cancer, prolonged periods of monitoring, chemotherapy, and symptom control (Chapter 19).
- Subtle refinements notwithstanding, modern approaches to cancer drug development are turning away from traditional chemotherapies, which paralyze cell division or damage DNA, in order to concentrate almost exclusively on targeted agents that are aimed at cancer-relevant molecules as exemplified by the oncogenic tyrosine kinases, BCR–ABL, EGFR, and HER2.
- Together, the molecular aberrations in a tumor constitute the *malignant manifesto* of the cancer cell in which the varied aspirations for the future are clearly set out and the key processes required to achieve them are underscored. By aiming for these essential processes we might, by the judicious placement of a few well-designed therapeutic obstacles, thwart the pretensions of the cancer cell and the realization of its manifesto promise.
- The ideal treatment will have an absolute predilection for cancer cells and would be entirely innocuous to normal cells. Given that traditional chemotherapy is often limited by toxic effects on noncancer cells, the closer we get to this ideal of "sectarian" cytotoxicity, the greater the chance of successfully completing a course of treatment that will eradicate the cancer.
- Cancer cells often become overreliant on certain aberrant pathways to support growth or even survival, a situation referred to as oncogene addiction. Oncogenically jaded cancer cells have little room left to accommodate other growth-supporting signals, which are often irrevocably suppressed.
- This situation offers unique opportunities to effectively sabotage the entire malignant manifesto by deploying targeted agents directed only against these central supports of oncogenic growth. Such maneuvers are also much less likely to be generally harmful to normal cells, in which these addictive pathways are supernumerary or nonexistent, because the alternative compensatory options have not been suppressed.
- Pioneering *fin-de-siècle* successes in the 1990s, exemplified by the introduction of imatinib to target the abnormal BCR–ABL fusion protein in chronic myeloid leukemia, have validated the concept of biological targeting. The ensuing flurry of research activity has now delivered a profusion of similar and even wildly dissimilar agents into current clinical use.
- However, despite early promise, long-term benefits have been limited by the near-universal development of resistance of cancer cells to these molecular targeted agents.
- It is anticipated that understanding the mechanisms through which cancer cells circumvent the growth-inhibiting actions of targeted agents will in future allow combinations of agents to be deployed that will avoid or overcome resistance.
- In many cases, resistance arises when the therapeutic blockade is bypassed by further mutations, which activate the same or another growth-supporting pathway independently of the targeted protein.
- Some cancer-causing proteins, such as RAS, have proved enormously difficult to target with small-molecule drugs. A variety of strategies are being explored to contend with this challenge. These include targeting of essential, but more accessible, oncogenic partners, using new agents such as siRNA, or adopting entirely new conceptual approaches to therapy. One such is synthetic lethality, whereby addiction to a particular oncogene (or loss of a tumor suppressor) also renders the cancer cell addicted to some other protein or signaling pathway, exemplified by PARP in BRCA1-deficient breast cancer cells, which may prove easier to target.
- Recent studies are revealing a new and intriguing feature of cancer cells: for many oncogenic mutations, either too much or too little of a good thing may not do. Rather, just the right amount of a given oncogene may be needed in order to produce cancer cells with exactly the right blend of endurance, fecundity, and freedom of movement. Too much and the cancer may fail as powerful anticancer processes are brought to bear.
- Another promising approach is to employ treatments to specifically prevent the spread of cancer cells, which is after all the cause of most deaths from cancer. As long as cancer cells remain hobbled to the primary tumor, the prognosis will generally remain good, with the most obvious exceptions being brain tumors where the primary alone may shorten life. Once cancer cells begin their peregrinations and distant organs become home to thriving colonies of émigré cancer cells, however, the outlook rapidly deteriorates.
- Antimetastatic therapies remain in their infancy, a state of affairs that appears surprising given that the vast majority of cancers would prove nonlethal if their spread were curtailed. A significant barrier has been the difficulty in testing such agents in clinical trials – long follow-up and large numbers of patients would be required to prove an antimetastatic effect, and the cost would prove prohibitive when compared to using tumor shrinkage as an endpoint. Surrogate endpoints such as numbers of circulating cancer cells or markers of vascular invasion may be needed to address this problem in future.
- There are also concerns over the long-term security of antimetastatic or cytostatic therapies, which leave primary tumors in place. After all, how durable is cancer cell rehabilitation? Maybe for cancer cells, only complete obliteration can offer reassurance.

- The emerging concept of tumor and cancer cell heterogeneity also poses new challenges to our therapeutic ingenuity. In fact, tumors have been rudely awakened from their Procrustean bed and can now be seen in their naked complexity. No two tumors are alike, and arguably at some level no two tumor cells are either.
- Given the extraordinary variegated complexity of cancer and the increasing number of therapeutic armaments at our disposal, efforts are increasingly directed toward finding new biomarkers to assist clinical decision making; predicting which patients will experience the worst outcomes, selecting the best combinations of treatments, and monitoring progress during and after therapy.
- Biomarker discovery is part of a broader ambition to reclassify cancers on the basis of a molecular taxonomy in preference to the current predominantly anatomical system, which places undue emphasis on tissue of origin, visible morphology, and extent of spread.
- Why do we need a new system? Because tumor molecular phenotype and mutation status are increasingly outperforming more traditional measures when it comes to subclassification of cancers in treatment selection and outcome prediction. Although not yet true for all cancers, clinical practice in breast cancer has already been transformed with treatment selection determined by the presence of specific drug targets such as hormone receptors, *HER2*, and others within the tumor. Conversely, the presence of downstream activating mutations in *BRAF* or *KRAS* in colorectal cancer (CRC) identifies patients who will not respond to TKIs targeting EGFR.
- To construct this new molecular taxonomy of cancer, we have an ever expanding array of analytical techniques at our disposal. In this way, the next decade will witness the molecular subclassification of cancers with ever greater precision. However, what is the purpose of unraveling the molecular *modus operandi* of cancer cells unless we have new therapeutic agents to exploit it? To paraphrase Thomas Paine, "We have discovered a world of windmills and our sorrows are that there are no Quixotes to attack them."
- Coupled with progress in non-invasive or minimally invasive techniques for sensitive detection of altered levels of proteins and various nucleic acids and in the isolation, quantification, and profiling of cancer cells from peripheral blood or other body fluids, it will become both practical and cost-effective to base future clinical decisions on the molecular biology of cancer. Taken together, one can sense that the realization of a new era of individualized medicine and tailored therapy is within our grasp.
- Ultimately, new treatments and biomarkers must be shown to be safe and clinically effective in cancer patients. Before a new treatment or diagnostic tool becomes generally available to clinicians, who wish to and can afford to prescribe it, it must first successfully negotiate a series of strictly regulated studies. Commencing with functional and toxicity studies in a number of species, safe and potentially effective agents then enter clinical trials in humans. These follow a well-defined pattern of phases, designated I–III, designed to confirm first safety and then efficacy in increasing numbers of patients. Agents progressing successfully through these trials will be afforded regulatory approval for marketing.
- Because of the expense and slow pace of clinical trials, a number of efforts are underway to find accurately predictive techniques that can suggest the optimal sequence, duration, dose scheduling, and combinations of treatments to be subsequently tested in trials. A variety of approaches are being developed, including *in silico* modeling, use of a range of *in vitro* reporter assays, and others that can determine both the PK and PD of drug activity rapidly and in such a way that meaningful predictions can be made for future activity in patients *in vivo*.

Introduction

"Men's courses will foreshadow certain ends, to which, if persevered in, they must lead," said Scrooge. "But if the courses be departed from, the ends will change. Say it is thus with what you show me!"

Charles Dickens

In one of the more surprising turnarounds in modern medical practice, cancer has for many sufferers become a chronic disease. Such a startling change in status brings new and unanticipated challenges for both carers and patients. Learning to live with cancer has, for many, taken the place of a coming to terms with the anticipated all-too-rapid death at its hands. For some cancers, exemplified by lung cancer, the outcome, though thankfully not the experience, has changed little in the last 50 years. The egregious epiphany, as one's worst fears are solemnly confirmed in the intimidating surroundings of a clinic room, is followed in quick-fire succession by a flurry of investigations, discussions, and finally a hope-giving full-on therapeutic onslaught or a brutally deflating "sorry." Once on treatment, matters progressed to one of three all-too-familiar outcomes, often with no time for reflection. Often in a matter of a few weeks, the patient was dead at the hands of the spreading cancer or the well-meaning clinician, or occasionally deemed to be cured. However, with all the relentless ill-fated irony of Thomas Hardy on a bad day, the celebrations would be cut short as the cancer recurred or a new one had arisen from the ashes of radiotherapy. Nowhere is this better demonstrated than in breast cancer. Fortunately, for numerous types of cancer, including breast, prostate, and colorectal, one can now anticipate a very different experience. Every day, thousands of women get up in the morning and take their breast cancers into work with them and after a normal day take them to bed with them at night – often year after year. How this *volte face* for cancer care came about is the subject of this chapter.

Ideally, all cancers would be cured by surgical resection and/or radiation ablation. However, such curative treatments are possible only where all of the tumor can be reached safely, and to be effective they must be deployed before cancer cells have spread. Regrettably, this is often not possible and localized or systemic radiation therapy and/or chemotherapy will be required. There are specific cases where surgery has been successfully employed to remove metastases, exemplified by local resection of solitary pulmonary metastases from colon cancer. These are however exceptional, because in most cases once cancer cells have colonized one distant tissue, they have colonized many. Moreover, even if it were possible to eradicate these established secondary deposits in some way, there will be numerous, surgically intractable cellular pioneers in transit or recent émigrés already arrived and ready to start a new secondary colony. This is not idle speculation; for many years, women underwent meretricious radical surgery for breast cancer in the vain hope that ever more drastic clearance of the breast, its surroundings, and its routes of spread might thwart the development of metastases. The blame for the failure of radical surgery can securely be attributed almost entirely to those small numbers of pioneering breast cancer cells that had slipped under the radiological and histological radar and evaded the surgeon's knife. Thus, in some cancers seemingly localized primary tumors may already be associated with distant spread and cannot be cured by local treatments alone. For this reason, progress was dependent on the development and application of systemic treatments that could seek and destroy even the most inaccessible of cancer cells – treatments that could be deployed alone or as adjuncts to surgery.

Treatments that aim to treat or even cure cancer should by definition be more toxic for cancer cells than they are for normal cells, and the wider this disparity the better the drug. By implication, the ideal therapy will exploit unique features that separate cancer cells from their normal pristine counterparts, such as the acquired hallmark features described in Chapter 1 or the aberrant biochemical pathways and mutated molecules that gave rise to them – *repetita juvant*.

Interest has also surrounded the development of drugs that might arrest or contain cancer growth or spread, rather than seek to eliminate them, as these may in general be less toxic to normal cells. However, as cancer cells might not be readily rehabilitated, "capital remedies" that will leave no chance of dormant cancer cells reoffending in the future might be preferable.

Traditionally we have set at cancer with "fire and iron," aiming to extirpate all cancer cells by surgery, radiation, and highly toxic chemicals. Most widely used chemotherapy regimens comprise drugs which exploit two basic properties of cancer cells that distinguish them (unfortunately not completely) from most normal cells: they proliferate and have aberrant DNA damage responses (DDRs) (Chapter 10). Giving drugs which interfere with the cell cycle, further interfere with DNA repair, or inflict further damage will kill cancer cells more readily than normal ones. This exquisite susceptibility to broad-spectrum agents is sometimes referred to as **genotype-dependent lethality**, because it is a manifestation of the entirety of the combined genotype and phenotype of the cancer cell. This is in large part the conceptual opposite of oncogene addiction and synthetic lethality, discussed in this chapter, that relate to specific individual mutations in oncogenes and tumor suppressors. Unfortunately, with traditional chemoradiotherapy, it is rarely possible to effectively kill all the cancer cells without causing unacceptable co-lateral damage, as normal replicating cells in the bone marrow and epithelia are withered by "friendly fire." Where sufficient normal cells can withstand the onslaught, cancers can be cured by traditional chemoradiotherapy, notable successful examples being *cis*-platinum in testicular cancer and combination chemotherapy in Hodgkin lymphoma. In almost all cases, it is the inadvertent destruction of normal cells and the resultant impaired immunity, bleeding, ulceration, and sickness that set the limits on the duration and intensity of traditional chemotherapy regimens. Treatment campaigns can be designed to mitigate this by scheduling periods of cease-fire, during which time is allowed for healthy tissues, such as the bone marrow, to recover from the drug onslaught before the bombardment resumes. However, these periods of respite can also be exploited by the cancer cells to resume growth and spread. At the present time, such cease-fires are arranged for a given tumor type and anatomical grade on a one-size-fits-all basis or are individually tailored in response to comparatively crude measurements such as a full-blood count and kidney function. With the recognition that cancer cells are not all Doppelgängers has come a more nuanced understanding of tumor heterogeneity at both the molecular and behavioral levels. Cancer cell heterogeneity between patients with ostensibly similar types and grades of tumor may well lead to a new subclassification of cancers on the basis of molecular differences rather than just anatomical ones. As yet, this knowledge has had little impact on the management of most cancers, but this situation is changing rapidly.

Cancer cell heterogeneity is now known to include important differences between cells within the same patient and tumor. With this in mind, traditional agents may have limited effects on so-called **dormant cancer cells** that are not dividing during the treatment period, but these cells might awaken at some point in the future and thereby rekindle an apparently successfully treated cancer. The recent recognition of a small cadre of elite and particularly malign cancer cells, referred to as cancer stem cells (CSCs), poses a range of additional challenges to the ingenuity of the pharmaceutical industry. CSCs are fitted with a more resistant molecular armor and are also more securely bunkered within protected niches, making them even harder to eradicate. Yet as the perceived *fons et origo* of all malign behavior, they represent the perfect treatment target. Finally, as we now think of tumors as a *Gesamtkunstwerk*, we can also consider how the whole installation could be dismantled by cutting off the cancer cells' support network. Thus, by subjecting the conniving stromal, vascular, and immune cells to a therapeutic onslaught, even where this has little direct effect on the cancer cells themselves, we might still bring down the entire ensemble.

So there we have the challenges, but how are we responding? It is easy to become despondent when facing the daunting complexity of tumor biology. In fact, the task of deciphering enough of it for cancer therapy to make any significant headway may seem on a par with resolving Jarndyce versus Jarndyce; no matter how many professionals earn their living in the attempt, the chance of any meaningful bequest at the end is slight. Progress has undoubtedly been made in the past, and if this owes as much to fortunate happenstance as to grand design, then we should be grateful all the same. However, the last decade has witnessed a remarkable and unequivocal shift in cancer drug development, so that with very few exceptions all new therapies are based on the fruits of unraveling the molecular complexity of cancer. The hunt is now well and truly on for drugs which, by targeting of specifically cancer-relevant proteins, have few adverse effects on normal cells. The advantages are self-evident and may extend to allowing cancer cells to be steeped in the toxic drug with minimal collateral damage. In fact, the ability to hit a cancer-critical target this hard might even be an essential prerequisite to eliminating generally treatment-resistant CSCs and dormant cancer cells.

Recent years have witnessed the addition of biologically targeted therapies to the lengthening catalogue of **paradigm shifts** in the natural sciences. It is uncertain that this hyperbole is truly justified as arguably we are at most simply restating Hippocrates, who suggested "**contraria contrariis curantur**" (the opposite is cured with the opposite) long before paradigms, **epistemological** or otherwise, first developed restless feet and began their now-inveterate shifting. What is, however, undeniable is that the last decade has witnessed a molecular definition of the "opposites" and, as a result, a radical reconfiguration of cancer drug development strategies. Fueled by advances in molecular biological techniques and the resultant unraveling of the molecular biology of cancer, we have largely left behind the search for nonspecific cell poisons in favor of drugs that specifically target aberrant cellular pathways – analogous to the replacement of carpet bombing everything even vaguely in the vicinity of a military target with computer and laser-guided missiles. Paul Ehrlich first introduced the term "magic bullet" in the 19th century in reference to chemicals specifically targeting microorganisms, but this idea has perfused many areas of medicine, not least oncology, in the last decade with the availability of a large and ever increasing number of new agents specifically targeting cancer cells. Of course, this strategy can succeed only if the selected targets prove **accessible**, **essential**, and **irreplaceable**. In the context of therapeutics, these properties far outweigh culpability, because the perfect drug target is anything that a cancer cell cannot do without and this is not always also its *raison d'être*. This is exemplified by the concept of synthetic lethality, whereby consequent upon the presence of various mutations, cancer cells are rendered co-dependent on noncausative genes or proteins. The appreciation of the vulnerability of cancer cells and in many cases the accurate prediction of their Achilles' heel has proved one of the most significant victories in the battle against cancer. Like the gloved fist of a heavyweight champion, targeted therapies have ended a reputation for invincibility; cancers are no longer the implacable foe, the more perfect versions of ourselves, but are revealed as having the same glass jaw as the rest of us. Lest we forget, cancer cells had come perilously close to a transfiguration in some eyes, so it has come as something of a relief that we can now once more see them in their true guise, not the "baddest" cells on the planet, just another idol waiting for a fall. So how did we get to this turning point in the story?

From chemotherapy to targeted drugs: Chapter 15 has covered the recent era of cancer chemotherapy, so we will skim over the details here. Since nitrogen mustards first broke a lance against cancer in the 1940s and 1950s, intensive research has forged an impressive armory of anticancer therapies. Firstly, a focus on metabolic enzymes resulted in the development of drugs targeting folic acid metabolism such as methotrexate, and subsequently unraveling of the intricacies of DNA structure and replication allowed the development of agents directed against DNA polymerases and topoisomerases (Chapter 15). The first genuinely targeted agents came from advancing research into hormone signaling and the discovery that some cancer cells may still exhibit the same dependencies on hormones that characterize the cell lineages from which they were derived – this is illustrated by the effectiveness of drugs targeting nuclear hormone receptors and their ligands in breast and prostate cancer. However, it must be appreciated that such treatments will have effects on normal cells and on the organism even if these are less pronounced than with traditional DNA-damaging treatments. Thus, tamoxifen treatment for breast cancer will also self-evidently prevent pregnancy. However, one has to think laterally and monitor therapies closely. Thus, androgen depletion in prostate cancer might increase risk of and progression of type 2 diabetes and cardiovascular diseases (CVDs), though this remains contentious. If this association proves genuine, then we must consider the indirect effects of treatment on obesity and insulin levels. In practical terms, until this issue is resolved it might be wise to intensify any concurrent risk-reducing therapies for CVD in those being treated with androgen depletion.

Most recently, we have begun to really exploit progress made in unraveling the molecular biology of cancer. As discussed, all cancers are caused by a finite number of cellular genes, oncogenes, tumor suppressors, and caretakers that have been corrupted by epimutations or altered expression. The stepwise accumulation of such epimutations allows the cancer cell to proliferate abnormally and to achieve near immortality (see the multistage model for CRC in Chapter 3). Given that cancer cells also appear remarkably resourceful and will repeatedly thwart our best therapeutic efforts by mutating around them, it is all too easy to imagine some intelligent (or should that be malign?) design behind this. Yet, in reality and to paraphrase Auden, cancer cells are just "irresponsible puppets of fate or chance," and what we are witnessing is a Darwinian process of evolution by natural selection, driven by the same forces but played out among somatic cells within the lifespan of a single organism. The essential randomness of mutations and natural selection has been underscored by rapid recent progress in molecular biology as a result of which we have now identified a truly daunting number of cancer genes, of which over 450 are thought to be causally related to the pathogenesis of different cancers. The obvious difficulty in separating next season's hot new therapeutic targets from all those paraded on the molecular catwalk is compounded by the fact that many of the protein products on display are themselves a small part of a complex signaling outfit, comprising numerous accessory drug targets. Furthermore, many cancer-relevant products are not even on show as they are not there by dint of mutation. Moreover, it is now generally accepted that many of the mutated genes found in tumors may never have been relevant or will have become outmoded or inessential by the time treatments are administered. To borrow from Winston Churchill, this leaves our ideal drug targets "a riddle wrapped in a mystery inside a tumor," but we also have a key at our disposal to unravel this conundrum. Thus, fortunately, we have become much better at selecting those genes and proteins that remain "mission critical" for any given cancer from this morass of potentially irrelevant passenger mutations. Step by step, we are narrowing down on a restricted and manageable range of the most tempting targets for new anticancer drugs. New concepts such as "oncogene addiction," first coined by Bernard Weinstein in 2000, have been introduced to encompass the startling discovery that cancer cells may actually become entirely dependent on an aberrant protein, even when that same protein or pathway may be dispensable in a normal cell. The protein, to which it has become so hopelessly addicted, thus represents an "Achilles' heel" just awaiting the inevitable sting of "Paris' arrow" to consign the cancer cell to Homeric oblivion. Such exquisitely specific weapons not only would prove devastating to cancer cells but also should be, by comparison, innocuous to normal cells.

Targeted therapies

Molecular differences between cancer cells and normal cells may also be exploited in other, less subtle ways. Thus, for instance, the uptake and concentration of iodine by some thyroid tumors, although shared by normal thyroid hormone-producing cells, are inconsequential in other tissues. Administering radioactive versions of iodine to patients results in the specific accumulation of radioactivity and hence restriction of damage to the thyroid, including thyroid tumor cells. In this scenario, the metabolic quirks render thyroid cancer cells vulnerable to a "targeted therapy." More modern variants of this approach include linking toxic cargoes, such as radioactive isotopes, to antibodies or proteins, such as receptor agonists, which specifically, or at least comparatively selectively, recognize proteins expressed preferentially by cancer cells. Linking somatostatin analogues, such as octreotide, to yttrium Y 90 (a radioactive form of the metal yttrium) in order to ablate somatostatin receptor expressing neuroendocrine tumor cells is an example drawn from current clinical practice. A variety of humanized monoclonal antibodies (mAbs) are being linked to toxic cargos, particularly for use in hematological malignancies. These can be radioactive (radio-immunotherapy), in which case often mouse, rather than humanized, mAbs are preferred as they are rapidly cleared from the body. One recent example is Tositumumab used for non-Hodgkin lymphoma. Therapies that manipulate or boost the immune system to combat cancer are often grouped together as **biological therapies** (Chapter 13). This can create some confusion as drugs such as herceptin may be regarded as both molecular targeted agents, as they interfere with EGFR family signaling, and biological therapies as they can direct an immune attack against cells expressing HER on the surface which is bound by herceptin. Biological therapies also include a variety of cytokines including interferons and colony-stimulating factors used to counteract the negative effects of drugs on the patient's white cell count.

Molecular targeted therapies

The last two decades have witnessed spectacular progress in unraveling the molecular roadmap for tumor development in many cancers. In particular, defining the central role played by kinases in growth-regulatory signaling pathways has been exploited in new drug developments. We have now entered the era of molecular targeted therapies. By exploiting the well-described differences between normal and malignant cells, new agents have been specifically developed to target gene expression and signaling pathways deregulated in the cancer cells.

So how far have we traveled along this new road? Naysayers may argue that the apparent victories of targeted therapies to date have often proved evanescent or even pyrrhic. Yet, once subjected to the spotlight of scientific scrutiny, the reasons for this, the resistance mechanisms, are being revealed. Taken together, clinical studies with targeted drugs have irrevocably altered our view of the cancer cell and have shown us how it might be vanquished. It is now hard to picture a cancer cell without superimposing a series of concentric circles and a bull's-eye. The cure for most cancers may still appear a distant possibility, but for cancer researchers it is too late to spit out the apple. Elegantly exemplified by targeting BCR–ABL, the knowledge that, against all previous convictions, targeted therapeutics can work has proved an epiphany for cancer researchers and a clarion call to the pharmaceutical industry.

Cancer cell heterogeneity

The application of extraordinarily powerful analytical tools for defining cell phenotype supports a view that you may feel is self-evident, namely, that no two cells are exactly alike. They may share a common ancestry and location, and even have several mutations in common, but invariably at the level of gene or protein expression there will be differences, and these differences will be reflected in different behavior and function. Many cancers, exemplified by breast cancer, are now known to represent a far more heterogeneous group than was previously appreciated (intertumoral heterogeneity). Thus, seemingly similar cancers may in different individuals be associated with widely differing clinical features, natural history, drug sensitivities, and prognosis. In part, such differences may reflect the different cellular origins of the cancer. Thus, stem cells or more differentiated cells will start at different points and follow a different roadmap to malignancy. In fact, it appears that even terminally differentiated cells can be encouraged back into the cell cycle. Stem cells may more readily be able to propagate newly acquired mutations, whereas differentiated cells would first have to forsake their vows. However, auspicious mutations can reawaken a latent reproductive appetite in even the most G_0-entrenched celibate. Subclassification of cancers, often described as grading or staging, has long been employed by oncologists to assist decision making. However, despite some improved codification, this process is still largely based on gross radiological and histological appearances, which may miss quite substantial and clinically vital differences at a molecular level. Biomarkers are discussed elsewhere, but it suffices to say that cancers can be further subdivided on the basis of molecular phenotype.

Once we examine cancer at a detailed molecular level, it rapidly becomes apparent that not only are no two tumors quite alike but also, probably, no two parts of the same tumor are (intratumoral heterogeneity). Moreover, the cells within a given single tumor show extensive variation. Not only do tumors comprise cancer cells, CSCs, stromal cells, endothelial cells, and immune cells, but also each of these different lineages encompasses cells of differing states of differentiation and clonal origin. Moreover, essentially all of these will show considerable dynamic variation in gene and protein expression over time. Recently, the theory, developed by Joan Massague and colleagues, that primary tumors may be self-seeded or reseeded by circulating cancer cells provides a further driver of cell heterogeneity. In fact, as our tools allow an ever more comprehensive survey of gene and protein expression, so the immense complexity as well as the unexpected heterogeneity of the cancer genome between and even within apparently histologically identical tumors are revealed. This does not automatically imply that earlier anatomical taxonomies of cancer are valueless. This is far from true, but we must acknowledge that they are imperfect and in some cancers unable to predict prognosis with even modest accuracy. In some cases, molecular phenotype and histology may be closely associated, implying that abnormalities in given signaling pathways may result in a reproducible observable phenotype.

Histological or molecular classification

In non-small-cell lung cancer (NSCLC), therapy decisions can be based on the gross histology (squamous vs. adenocarcinoma) and clinical parameters (smoking status, ethnic origin, and age). Particular mutations again show some relationship to these more traditional classifiers, which can be employed to direct targeted treatments more appropriately. Thus, squamous cell tumors respond poorly to first-line EGFR inhibitors due, for example, to the presence of downstream activating mutations in *RAS*. Thus, despite their limitations, currently employed analytical techniques may still serve practically until we can replace them with something better. However, histology will not identify the point at which signaling is aberrant and can be misleading where observable changes in phenotype are nonspecific. Therefore, if histology cannot be relied on to make sufficiently accurate treatment choices, then ultimately all cancers will have to be subjected to a molecular analysis in order to tailor treatments appropriately. This is exemplified by traditional histological categories of invasive breast cancer, such as luminal A, luminal B, and basal-like, which are often linked to specific mutations, but as these are not sufficiently reliable, it has already become commonplace to look for the presence of specific molecular changes, such as receptor status, before selecting treatments (discussed further in this chapter).

Molecular classification in breast cancer
This clinically heterogeneous cancer is currently classified into three broad molecular categories:
- Estrogen receptor (ER) and/or progesterone receptor (PR) positive tumors responsive to anti-hormone treatments.
- HER2 (ERBB2) amplified, responsive to mAb and TKIs directed against HER2.
- Triple-negative breast cancers (lacking expression of ER, PR, and HER2), with increased frequency of mutations in BRAC1. These tumors are also known as basal-like breast cancers and are currently treated with chemotherapy options, though PARP inhibitors may offer a targeted treatment.

Recent studies have increasingly focused on more complex molecular analyses, and expression of a variety of other pathways appears to be helpful in prognosis and may identify future targets. These include TP53, PIK3CA, and GATA3, common to multiple histological subtypes, and specific mutations in GATA3, PIK3CA, and MAP3K1 found in luminal A.

Future directions in molecular classification
This area is advancing rapidly, and it will soon be commonplace to look for key oncogenic drivers in NSCLC, such as amplification of TKRs, mutant KRAS, or EML–ALK fusions. Although still a very simplistic molecular phenotype, classification on the basis of presence of these mutations appears to identify mutually exclusive groups of tumors (though note recent studies in glioblastoma, discussed in this chapter, and the problem of intratumoral heterogeneity). Moreover, whatever the theoretical limitations, there is practical value where molecular classification is gainfully directed at targets for available and emerging treatments. But what of known mutations in other oncogenes, in tumor suppressors and differentiation markers, and in gene-protein expression and also in cellular heterogeneity?

Is there a clinical point to an ever more divergent subclassification of individual tumors based on an ever more complete description at a molecular level? Is it not sufficient to catalogue tumors on the basis of what treatments will eradicate them? In some cases this may suffice, and knowing that most cancer cells have mutations in a growth factor TKI or have lost a given tumor suppressor can be exploited in selection of a particular therapy. Put simply, subdividing cancers hierarchically on the basis of shared anatomical or molecular features may enable ever improving accuracy of clinical decision making, but in some cases the added gain in moving from broader, less comprehensive, but probably cheaper levels down to ever narrower subclasses based on detailed molecular analyses may be unnecessary. However, in most cases it is accepted that improved molecular subclassification of cancers is urgently needed. For example, at present assessment of prognosis in the intermediate stages of CRC is very imprecise, with widely varying 5-year survivals within apparently identical grades of tumor.

Revealing heterogeneity across single cells
Averaging expression or genome analyses across multiple cells may fail to reveal subtle differences between individual cells, including the existence of cancer stem cells, rare cells with additional mutations that might identify individuals likely to rapidly become resistant to particular targeted drugs, and so on. Moreover, where the population of cells is in the minority, their contribution may be masked.

What causes such heterogeneity? At a superficial level, the inherent diversity of biological systems is expanded during tumorigenesis by mutations, clonal evolution, and genomic instability. On top of these, we must also factor in epigenetic changes, ongoing differentiation along diverging lineages, and more transient effects of cell plasticity. Thus, interactions with the microenvironment, or niche (comprising other cancer cells; noncancer cells, including stroma, endothelium, and immunocytes; and those with cell-derived proteins and biophysical features such as low oxygen tensions), can all influence molecular phenotype and even encourage the expression of stem cell–like characteristics. Even cancer stem cells may display extensive plasticity that may allow them to alternate between cancer-expanding epithelial and metastatic or invasive epithelial–mesenchymal transition (EMT) phenotypes depending on their microenvironment, differentiation state, and metabolic parameters, including oxygenation and glucose metabolic pathways.

Thus, accurate analyses of cancers may be achieved only by analyzing single cells, if possible within their anatomical and spatial contexts (see Chapter 20). Improved understanding of the drivers behind tumor cell heterogeneity as well as of the resultant functional sequelae of diverse cell–cell interactions will prove an essential prerequisite to developing new treatment strategies for cancer. This follows from the desire to eliminate all cancer cells. Thus, individualized therapies may need to be tailored not just to the individual patient and tumor but also to its many varied and continually evolving cellular constituents.

Cancer heterogeneity and therapy
In fact, intratumoral heterogeneity may pose a substantial barrier to molecular targeted treatments, because their success will depend on the molecular and cellular diversity in individual tumors. This can be problematic for numerous reasons, and a few are suggested here. Not knowing what mutations are or may become present in other cancer cells in many cases could prevent us from thwarting the emergence of resistance by using a strategic combination of drugs. This view is receiving strong support from the improvements in outcome in breast cancer resulting directly from molecular phenotyping and the ability to select appropriate therapies based on that information. Furthermore, in glioblastoma several studies have now confirmed that driver mutations long known to coexist within tumors and contribute to progression are rarely present in the same tumor regions, let alone cells. Thus, EGFR, MET, and PDGFRA may all be amplified in a single tumor, but in a mutually exclusive pattern across differing single cells. Such cohabitation of multiple different clones, carrying activation of different pathways, within the same tumor, even where many may share a common ancestry, will affect the choice of targeted therapies and may predict the likelihood of resistance.

Finding the molecular targets

Let us now consider how we select biological drug targets from among the many potential candidates. Broadly, we must progress from a descriptive to a functional understanding of cancer cells by the application of a range of research techniques grouped together as "functional genomics," the executive arm of systems biology (Chapter 20). Why? Put simply, because identifying missing, overexpressed, or altered proteins in cancer cells does not necessarily mean they are mission critical for oncogenesis or would represent good treatment targets. Thus, by using combinations of cultured cell lines, genetically altered mouse cancer models, and implantation of human cancer cells into suitable recipient models, we can work out what all the myriad genes, proteins, and their mutant variants actually do, with whom, and for what results. This is no mean task, given the heroic numbers of potentially interacting protein partners. Of course, this pipeline can only suggest potentially druggable targets. Ultimately, these will need to be fully validated in clinical trials, but such functional studies can greatly reduce the numbers of targets that need to be explored in this way, with all the potential cost savings in financial and human terms. Other chapters discuss some of the methodologies employed to uncover the functional roles played by cancer-relevant molecules and to predict which may be the most optimum new drug targets (Chapters 6 and 20); next, we will specifically look at the all-important surrogates for testing the potential of targeting specific genes and proteins *in vivo* – conditional genetically altered organisms.

Tumor regression in mice by inactivating single oncogenes

What sense would it make or what would it benefit a physician if he discovered the origin of the diseases but could not cure or alleviate them?

Paracelsus

There is little doubt that functional studies offer the best hope of separating wood from trees and the propitious mutations from the capricious. New strategies targeting oncogenes (or other cancer-contributing mutations) can be investigated by employing model systems, even before a suitable drug has been developed. The use of mouse models in oncogene research follows on from two postulates: firstly, that mouse models provide valuable insights into how oncogenes work, particularly when combined with studies of human cancers; and, secondly, that they allow therapeutic strategies, including timing and dosing schedules, for oncogene-targeted agents to be evaluated.

Compelling results from rodent models and cell culture experiments suggest that several oncoproteins, including MYC, RAS, and RAS-signaling pathways, such as the RAF–MAPK and PI3K pathways, might prove effective treatment targets in cancer, with the important caveat that successful results in treating human diseases will have to be confirmed. Intriguing paradoxes have been revealed by several studies; thus, for example, the demonstration that transient inactivation of MYC may be sufficient to arrest and reverse certain tumor models, but not others, raises the importance of tissue microenvironment and epigenetic context in dictating the potential therapeutic responses of any given tumor.

With the emergence of conditional transgenic mouse models, in which the expression or activation of a given gene or protein could be regulated (Chapter 6), has come a new and more functional understanding of the role played by cancer mutations in cancer biology. The study of such models has resulted in the overturning of a number of articles of faith pertaining to the likely benefits of targeting single molecules in cancer. Thus, it was generally assumed that several oncogenic pathways would need to be targeted for tumor regression to occur. Surprisingly, this turned out not to be the case, and targeting only one specific gene or protein in a cancer often leads to tumor regression. The concept of "oncogene addiction" has gradually effloresced from such studies (discussed in this chapter); broadly, the consequences are that the survival of a cancer cell depends on the continued activation of the particular oncogene, such as *MYC* or *RAS* – inactivating the oncogene leads to death or senescence of cancer cells (Fig. 16.1). Of course, it is important to remember that although oncogene inactivation can cause tumor cell death at one particular stage of cancer, most tumors eventually recur, and we need to understand when and why this occurs in order to design therapeutic strategies to treat cancer (Fig. 16.1). Here, we will highlight some of the mouse models of cancer (Table 16.1) that have led to the discovery of "oncogene addiction" as well as cases in which tumor relapse occurs. It is also important to understand that animal models, along with cultured cancer cells, can provide unique opportunities to test the functional importance of a given cancer protein or gene without having to first develop a suitable drug. This is critical as it is often difficult to identify which of the many mutant genes or proteins in a cancer are actually "mission critical" – in other words, to separate the propitious from the capricious.

As was mentioned for MYC in Chapter 6, in general two major approaches have been used to generate conditional (regulatable) transgenic mouse models: the tetracycline ("tet") system and the modified estrogen receptor ER^{TAM} (see also Appendix 20.1). The tet system requires the drug doxycycline to regulate expression of the **gene** of interest, whilst the ER^{TAM} system relies on the administration of tamoxifen to regulate activity of the expressed **protein** (see Fig. 6.9).

The key advantage of these systems is that a given gene or gene product can be switched on in a specific tissue in a time-controlled manner, which allows the close monitoring of tumor development. Conversely, the gene–gene product can subsequently be switched off once tumors have developed, providing insight into mechanisms of tumor regression.

What we can derive from these studies is that inactivating the initiating oncogene (e.g. *MYC* or *RAS*), in tumors often results in tumor regression as a result of proliferative arrest, differentiation, senescence, and/or apoptosis. The outcome is dependent on the genetic alterations as well as the tumor or tissue type.

Conditional models systems to study c-MYC in cancer biology

Using the tet system, mice were generated that conditionally express MYC in their hematopoietic cells. In this mouse model, MYC was expressed in this cell lineage throughout development, resulting in the occurrence of highly invasive T-cell lymphomas and acute myeloid leukemias by 5 months of age. Remarkably, subsequent inactivation of the *MYC* transgene was sufficient to cause tumor regression in mice moribund with tumor burden; tumor burden was substantially reduced within 3 days, and after

Biologically Targeted Agents

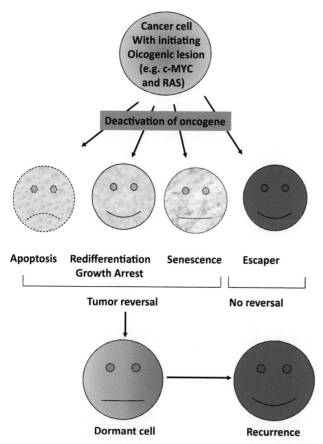

Figure 16.1 Tumor reversal following deactivation of initiating oncogenic lesion. Studies in numerous different transgenic mouse models of tumorigenesis have revealed that neoplastic lesions can be completely and often rapidly reversed when the initial causative oncogenic lesion is inactivated. Mechanisms of tumor regression include tumor cell apoptosis, growth arrest concomitant with differentiation and re-establishment of cell–cell contacts, and/or cellular senescence. The rapid onset of collapse of the vasculature, within tumor masses, may also contribute to tumor cell apoptosis. However, in some cases, a subset of tumors escapes reversal ("escapers") because they are no longer dependent on the initiating oncogenic lesion. In these cases, it is likely that additional oncogenic lesions have rendered the initial lesion redundant. In addition, cancer cells may avoid elimination by effectively lying dormant for a period of time before becoming reactivated in some way and resuming their oncogenic mission.

6 weeks the majority of mice exhibited a sustained remission for as long as 30 weeks. Downregulation of MYC expression resulted in both the elimination of tumor cells by apoptosis and the differentiation of others into mature lymphocytes with the restoration of normal host hematopoiesis. However, despite sustained tumor regression in the majority, 10% of mice develop recurrent tumors. Thus, some cancer cells in regressing tumors escape dependence upon the *MYC* transgene. Such "escapers" have probably acquired genetic lesions that substitute for the requirement for MYC and will repopulate the tumors (Chapter 6). In another cancer model, in which both MYC and the antiapoptotic protein BCL-2 are expressed in lymphocytes, switching off BCL-2 leads to complete reversal of lymphoblastic leukemia by apoptosis. Further examples are given in Table 16.1. The central importance of interactions between Myc and Miz1 in initiating and sustaining T-cell lymphomas has been demonstrated in a tet-regulatable mouse model, described in Chapter 6; reversing Myc expression restores the expression of growth-inhibiting genes, such as those for *cdkn2b* and *cdkn1c*, that were previously suppressed by Myc–Miz1 complexes leading to senescence. Tumor-derived TGF-β may contribute to this.

Regression of several different solid tumor types has also been shown, following MYC deactivation in regulatable mouse models, including skin papillomas, pancreatic islet β-cell, osteosarcoma, breast adenocarcinoma, and hepatocellular carcinoma. As described in Chapter 6 (see Fig. 6.9), continuous activation of MYC (using the c-MYCER^{TAM} system) in skin epidermis led to the growth of benign, angiogenic papillomas. Inactivating MYC led to the complete regression of papillomas over a 3-week period which coincided with skin keratinocytes exiting the cell cycle and resuming normal differentiation. Importantly, the complete regression of newly formed vasculature demonstrated that continuous MYC activation in keratinocytes is required to maintain neo-angiogenesis.

In a pancreatic islet β-cell tumor model, multiple angiogenic and locally invasive β-cell tumors, which develop following continuous activation of MYC and the antiapoptotic protein BCL-X$_L$ (described in Chapter 6 and Fig. 6.10), were also completely dependent upon active MYC. In this tumor model, regression appears to be mediated by a rapid collapse of vasculature that triggers the death of many β cells, which is presumably the result of hypoxia, lack of nutrients, and survival factors provided by the vasculature. Remarkably, despite the continued overexpression of BCL-X$_L$ in neoplastic β cells, deactivating MYC leads to general regression of tumors back to fully differentiated quiescent islets with restoration of normal cell–cell contacts concomitant with re-expression of E-cadherin (Fig. 16.2). Interestingly, tumor regression is also accompanied by substantial infiltration of inflammatory cells, which may play a role in phagocytosis and the clearance of apoptotic and necrotic debris. Recent studies suggest that interactions between activated oncogenes and immune cells may play a role in shielding the cancer from immune attack – a protection which may be lost when oncogenes are inactivated. The rapid disassembly of vasculature is consistent with a direct angiogenic role of MYC, described in this chapter, although it is also possible that β-cell growth arrest induced by MYC inactivation allows intrinsic anti-neoplastic mechanisms such as the immune response to initiate regression. With regard to ectopic β cells that had invaded pancreatic ducts, vessels, and local lymph nodes by 12 weeks of MYC activation, extensive histological analysis of pancreatic sections failed to reveal any ectopic β cells after several weeks of MYC inactivation. Moreover, other animals undergoing similar regression remained healthy after several months. It is not yet known, though, if longer term activation of MYC would lead to a progressively increasing proportion of tumors escaping dependence upon MYC, as observed in other tumor models.

An interesting study has shown that brief inactivation (10 days) of MYC is sufficient for the sustained regression of MYC-induced invasive osteogenic sarcomas in transgenic mice. Surprisingly, subsequent reactivation of MYC led to a "reversal of fortune" for the cancer cells as these now undergo extensive apoptosis rather than restoration of the neoplastic phenotype; for some reason, MYC had crossed the house and was now a committed anticancer representative. One possible explanation for this outcome is that a change in maturity has taken place. While MYC was inactivated, the epigenetic context has shifted and the

Table 16.1 Use of regulatable transgenic mouse models of cancer to test tumor regression

Initial oncogenic lesion	Target tissue or cell type	Phenotype	Deactivating initiating oncogene
SV40 T antigen	Embryonic submandibular gland	(a) Atypical cells (4 weeks old) (b) Transformed ductal cells (7 months old)	(a) Full phenotype reversal (b) Not reversed
BCR–ABL1	B cells	Acute B-cell leukemia	Most tumors regress by apoptosis, but some tumors recur.
HRAS (+INK4A$^{-/-}$)	Melanocytes	Melanoma	Rapid tumor regression by apoptosis Larger tumors (20%) not reversed
KRAS	Lung	Adeno-carcinoma	Tumor regression
KRAS (+p53$^{-/-}$ or INK4$^{-/-}$ or ARF$^{-/-}$)	Lung	Adeno-carcinoma	Rapid and complete tumor regression
c-MYC	Embryonic hematopoietic cells	T-cell lymphomas and acute myeloid leukemias (5 months old)	Rapid tumor regression by cell cycle arrest, differentiation, and apoptosis 10% tumor relapse
c-MYC	Adult suprabasal epidermis (keratinocytes)	Papillomatosis (carcinoma in situ)	Rapid and complete tumor regression by growth arrest and differentiation
c-MYC	Adult suprabasal epidermis (keratinocytes)	Papillomatosis (carcinoma in situ)	Tumor recurrence after transient c-Myc inactivation
c-MYC	Mammary epithelium	Invasive mammary adeno-carcinoma	Partial tumor regression Subset not reversed (Ras activated)
c-MYC (+BCL-X$_L$)	Adult pancreatic islet β cells	Invasive islet adeno-carcinoma	Rapid and complete tumor regression by apoptosis, growth arrest, differentiation, and vasculature collapse
c-MYC (+BCL-X$_L$)	Adult pancreatic islet β-cells	Invasive islet adeno-carcinoma	Tumor recurrence after transient c-MYC inactivation
c-MYC	Embryonic osteocytes	Malignant osteogenic sarcoma	Rapid and complete tumor regression by differentiation after transient c-MYC inactivation
c-MYC	Hepatocytes	Hepatocarcinoma	Tumor regression by apoptosis and differentiation Dormant tumor cells give rise to recurrent tumors when Myc is reactivated.
c-MYC	Lymphocytes	T- and B-cell lymphoma	Regression
(a) c-MYC (b) MYC(V394D) mutant – does not bind to Miz1	Lymphocytes	(a) T-cell lymphoma (b) Delayed onset of lymphoma	Regression by TGF-β-induced cellular senescence
MYC(V394D) mutant – does not bind to Miz1			
BCL-2 (+ c-MYC)	Lymphocytes	Lymphoblastic leukemia	Regression by apoptosis
NEU (and MMTV)	Mammary epithelium	Invasive mammary carcinoma	Essentially complete regression even of metastatic lesions. Eventually, however, some tumors spontaneously recur.
WNT (+ P53$^{-/-}$)	Mammary epithelium	Invasive mammary carcinoma	Essentially complete regression even of metastatic lesions. Eventually, however, some tumors spontaneously recur.
GLI2	Basal keratinocytes	Basal cell carcinoma	Reversal

Note: The potential for tumor regression following de-activation of the initial oncogenic lesion (c-MYC, SV40 T antigen, HRAS, KRAS, BCR-ABL, NEU, WNT, and GLI2) has been investigated using several regulatable transgenic mouse models of cancer. In general, the findings listed here have implications for therapy as they indicate that blocking oncogene function, even in advanced tumors, could lead to apoptosis or differentiation of tumor cells. Further to these findings, it has also been shown that transient, rather than sustained, inactivation of c-MYC is sufficient for full reversal of malignant osteogenic sarcoma in transgenic mice. Although transient inactivation of oncogenes could provide an effective cancer therapy limiting host cell toxicity, it has subsequently been shown in various other tissues that such transient inactivation is not sufficient for tumor reversal (see discussion in this chapter). In fact, in some tumors inactivation of the transgene is not followed by complete regression of all tumors or recurrence occurs spontaneously in regressed tumors, suggesting that mutations occur which can make the tumors independent of the initiating lesion.

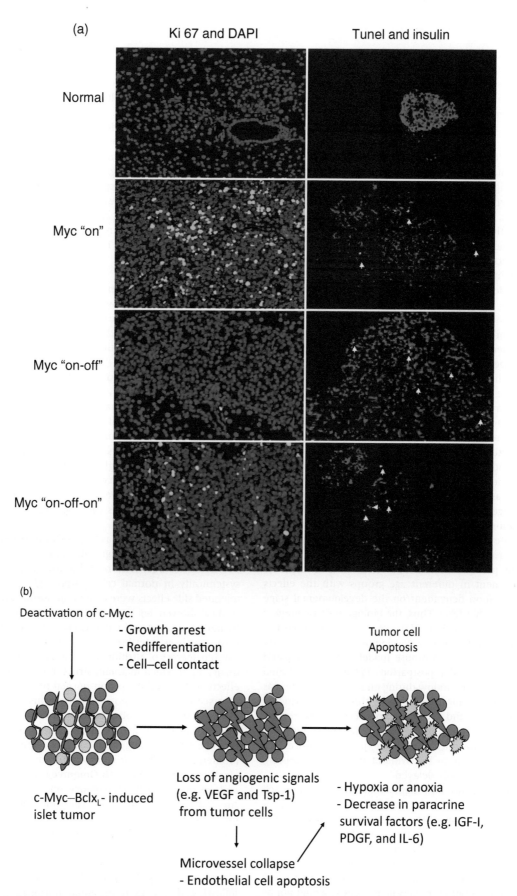

Figure 16.2 Reversal of murine pancreatic islet β-cell tumors. (a) Deactivation of c-MYC in pancreatic islet tumors results in complete reversal in the overwhelming majority of tumors by 4 weeks after deactivation. Here we show the effects of transient inactivation of c-MYC. After a period of 5 days of inactivation, β cells have stopped replicating (absence of Ki67 staining: green) and have acquired a more differentiated phenotype (enhanced expression of insulin: red). Moreover, some β cells are undergoing apoptosis during the period of c-MYC inactivation (TUNEL positive: green). Those cells undergoing apoptosis whilst c-MYC is on are not β cells as they do not express β-cell markers. (b) The processes of tumor reversal are shown schematically.

cell is no longer the immature one in which MYC was originally activated, but rather a more differentiated one with a resultant different response to MYC activation; the immature osteoblasts present during embryogenesis, as well as the osteosarcoma cells deriving from them, are very different from the mature osteocytes into which they differentiate during the period of MYC deactivation, and respond very differently to MYC. Therefore, reactivation of MYC now takes place in a different cellular context and induces apoptosis rather than neoplastic progression.

These intriguing findings suggested the novel possibility of employing transient inactivation of MYC as a therapeutic strategy in certain cancers, thus limiting potential toxic effects that result from prolonged therapeutic inactivation. But how general are these findings, and what happens in other tumor types? Unfortunately, brief inactivation of MYC failed to reverse tumor growth in several different mouse models. In contrast to the osteogenic sarcoma model described in this chapter, reactivating MYC in islet tumors does not lead to accelerated β-cell apoptosis, but rather restores the oncogenic properties of MYC, rapidly re-initiating β-cell proliferation, loss of differentiation, local invasion, and angiogenesis. In epithelial tumors, such as liver and breast cancer, reactivation of MYC in regressed tumors also led to tumor regrowth. Importantly, subsequent inactivation of MYC was less effective at regressing tumors, indicating that other oncogenic pathways are being activated in these cells.

In mice that had developed hepatocellular carcinoma, inactivating MYC promoted apoptosis as well as terminal differentiation of many tumor cells, giving rise to "normal" liver cells – hepatocytes and biliary cells. However, although such cells appeared and behaved as "normal," some indeed retained oncogenic properties, as was seen following MYC reactivation. So the important message here is that sustained inactivation of MYC is required for tumor regression, as some "normal" tumor cells are actually in a state of dormancy and upon reactivation of MYC can unleash their neoplastic behavior.

An important point to remember is that different types of cancers are prevalent in different age groups with the effects of oncogene activation dependent on the developmental stage of the target cell at that time. Thus, the biological consequences of activating oncogenes such as *MYC* are clearly influenced not only by environment but also by developmental stage. This has been elegantly shown in two mouse models, mammary gland and liver. MYC can inhibit postpartum lactation if activated within a specific 72-hour window during midpregnancy, whereas MYC activation either prior to or following this 72-hour window does not. In embryonic and neonatal mice, MYC overexpression in the liver immediately results in hyperproliferation and neoplasia, whereas in adult mice MYC overexpression induces cell growth and DNA replication but without mitotic cell division, and neoplasia is considerably delayed.

Taken together, these findings suggest that a cautious approach is required in considering cancer therapies aimed at brief, as distinct from sustained, oncogene inactivation. Firstly, a more comprehensive understanding of the genetic basis and environmental context of any individual tumor would be required in order to predict the likely success of such a treatment schedule. Secondly, at least under those circumstances where tumor cell differentiation and alteration of epigenetic context would not be predicted to reinstate apoptosis and no alternative mechanism exists for tumor cell removal, sustained inactivation of the offending oncogene would seem the desired therapeutic goal.

Conditional models systems to study RAS in cancer biology

Well, what of RAS? Much of the methodology employed has been discussed, so we will restrict ourselves to a brief overview of transgenic studies here. In a mouse model of melanoma, withdrawal of doxycycline-inducible oncogenic HRAS(G12V) expression in transgenic mice bearing melanomas causes apoptosis in both tumor cells and endothelial cells, followed by regression of the melanomas. Similarly, withdrawal of doxycycline-inducible activated KRAS(12D) expression from type II pneumocytes caused apoptosis and regression of both early proliferative lesions and lung cancers. Similar results were obtained in animals deficient in p53 or p19^{Arf}. Thus, activated HRAS or KRAS is required for tumor maintenance in mouse models of melanoma and NSCLC, respectively.

RAS and MYC together

Several studies have now looked at oncogene deactivation in mouse tumors related to combined deregulation of both MYC and RAS. Genetic context is clearly important in dictating the consequences of inactivating MYC – so what happens if RAS is also deregulated?

A well-established mouse model of NSCLC was used (*LSL–KrasG12D*) in which activation of the transgene *K–RasG12D* is initiated in mice by inhalation of adenovirus expressing Cre recombinase (for technical details on how this works, see Chapter 20). Multifocal lung tumorigenesis then develops, and by 18 weeks each lung harbors multiple independent tumors at all stages of evolution through to adenocarcinoma. The idea was to cross-breed Omomyc mice with *LSL–KrasG12D* mice in order to study both the therapeutic impact of blocking MYC activity in lung tumors as well as the side effects of inhibiting MYC in normal cells.

Remarkably, in as little as 3 days Omomyc expression triggered profound tumor shrinkage and, after 28 days, animals were overtly tumor free. With regard to the impact of inhibiting MYC systemically in normal cells, surprisingly, only mild and well-tolerated side effects were evident in normal proliferating tissues.

Mice showed no signs of distress; maintained their weight, hydration, and normal blood chemistry; and, while proliferating tissues such as intestine, bone marrow, skin, and testis exhibit varying degrees of attrition, cell death does not occur in any adult tissue, all of which maintain structural integrity. Importantly, all effects of MYC inhibition on normal tissue are fully and rapidly reversed upon restoration of endogenous MYC function.

These surprising observations strongly support rekindling interest in MYC as a therapeutic cancer target for two obvious reasons: firstly, endogenous MYC is required not only for the proliferation of RAS-driven lung tumors but also for their survival (inhibiting MYC with Omomyc leads to tumor regression); and, secondly, such dependency on endogenous Myc for cell survival is specific to tumor cells and absent from all normal proliferating cells.

In general, these findings have implications for therapy as they indicate that blocking oncogene function, even in advanced tumors, could lead to apoptosis, senescence, or re-differentiation of tumor cells (Fig. 16.1). The demonstration of tumor regression following deactivation of initial oncogenic lesions in mouse models (e.g. MYC, SV40 T antigen, RAS, BCR–ABL, and BCL-2) has provided valuable information and hope for future development of candidate drug molecules. To this end, recent *in vivo* studies show that re-engaging apoptosis pathways, which have become disrupted during tumor development, can indeed have a positive therapeutic effect. For instance, the inhibition of the antiapoptotic protein, BCL-2, or the restoration of p53 function has proven particularly lethal to particular tumor types. Even if targeting the oncogenes themselves may not be a viable strategy, various downstream targets mediating oncogenesis might. Over the last few years, a large number of studies have employed gene chip microarrays, deep sequencing, and other high-throughput techniques to identify oncogene-regulated genes in cultured cells *in vitro*. However, all these studies, despite their undoubted utility, do not give us any direct indication of which of these many oncogene-regulated genes are actually involved in tumorigenesis *in vivo*. Yet this is of great importance given the critical role of the tumor microenvironment as well as factors operating at the organism level. Taken together, these findings support the notion that a detailed understanding of the "road map to cancer" of a given individual tumor may be the essential prerequisite to selecting optimal therapeutic strategies in the future, thus providing some impetus for strategies aiming to achieve one of the great hopes of postgenome-era biology, namely, "individualized medicine" and "tailored therapy."

Resistance and recurrence in mouse models

As mentioned, there are some instances when tumors do not fully regress (Table 16.1). Although it is not clear as to why some tumors escape dependence upon the initiating oncogene, it has been reasoned that these tumors might have acquired further genetic lesions that in some way substitute for the requirement of the oncogene.

For example, in a mouse model of breast cancer, although reversal of MYC-induced invasive mammary adenocarcinomas occurred, a subset of tumors failed to reverse and was subsequently found to carry additional mutations in *RAS*. The observation that MYC may cause genomic instability in some model systems *in vivo* and *in vitro* might in the long term contribute to MYC-induced neoplastic progression. Further examples of mouse models in which tumors subsequently relapse are given in Table 16.1. Other specific examples are discussed in more detail later in the chapter in the sections on resistance to targeted treatments.

Targeted cancer therapies

A desire to take medicine is, perhaps, the great feature which distinguishes man from other animals.

Sir William Osler

The modern era of cancer therapeutics is increasingly based upon the exploitation of new knowledge deriving from the unraveling of cancer cell biology, supported by functional studies in surrogate models. Targeted cancer therapies that block the growth and spread of cancer by interfering with specific molecules involved in tumorigenesis and tumor growth effectively antagonize the hallmark features of cancer. Such agents are also known as "molecular targeted drugs." By targeting molecular and cellular changes that are specific to cancer, targeted cancer therapies may be more effective than current treatments and less harmful to normal cells. Although cancers are remarkably heterogeneous, they share certain key properties. Thus, all cancer cells exhibit uncontrolled and excessive replication; are generally very long lived; and, not surprisingly, as well-differentiated cells rarely if ever divide, are less differentiated than their normal counterparts. This is emphasized by the two broad functional categories into which new targeted agents in cancer may be separated: differentiation therapies (cytostatic) or destruction therapies (cytotoxic). Given that new therapies which simply arrest growth of the primary or act to prevent invasion or metastasis will not result in major radiographically apparent shrinkage or ablation of the tumor, therapeutic evolution will need to be matched by equally rapid progress in new measures of treatment success to run alongside more conventional measures of tumor volume, such as Response Evaluation Criteria in Solid Tumors (RECIST).

Targeting oncogenes to treat cancer?

Sane judgment abhors nothing so much as a picture perpetrated with no technical knowledge, although with plenty of care and diligence.

Albrecht Dürer

The recent appreciation that cancer cells may actually become critically dependent on an aberrant active oncogene (**oncogene addiction**; see Fig. 16.3) raises the possibility of developing agents specifically blocking the activity of these mutated or deregulated oncogenes with few harmful effects on normal cells. Moreover, as by definition oncogene-addicted cancer cells must

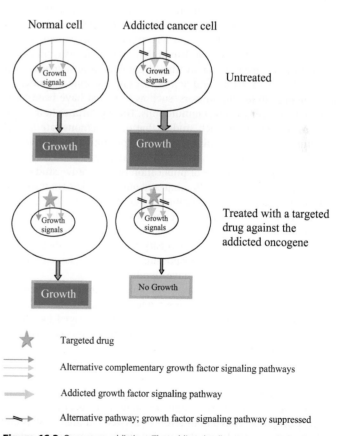

Figure 16.3 Oncogene addiction. The addicted cell suppresses alternative growth-stimulating pathways, making it more vulnerable than a normal cell to drugs blocking the addicted pathway.

have also acquired resistance to the intrinsic tumor suppressor action, apoptosis or senescence, invoked by deregulated oncoproteins, such cancer cells would also be vulnerable to the therapeutic reactivation of the interdependent inhibitory pathway. In fact, cancer cells may be very vulnerable to restoration of a missing or failing tumor suppressor, and this is sometimes referred to as **tumor suppressor hypersensitivity**, analogous to oncogene addiction. Finally, cancer cells may also become critically dependent on normal cellular processes that are redundant or supernumerary in normal cells; described as **non-oncogene addiction**, use is restricted to genes that are not themselves mutated in cancers. Such genes are disproportionately important to cancer cells, which have been dispossessed of otherwise interchangeable alternative pathways needed to complete some essential process. The *exemplar par excellence* of non-oncogene addiction is the complete reliance of *BRCA*-mutated cancer cells on an alternative DNA repair pathway involving PARP; such BRCA mutant cells are, therefore, ill equipped and lack the resources required to endure treatment with PARP inhibitors.

Identifying critical interdependent signaling pathways as therapeutic targets in mature cancer cells is a closely allied goal to the unraveling of the "cancer road map" that guided their development in the first place. However, with a critical difference that has considerable practical relevance, mutations that may have been essential in primordial cancer cells might no longer be at the point in time that treatments are mooted. This issue can be resolved only by functional studies in which the consequences of inactivating a cancer-causing target can be examined directly; simply describing whether a gene is mutated or not is insufficient, as in this particular cancer and at this particular time that gene may have ceased to be relevant. Progress has been made in this regard, and over the last few years numerous functional studies in animal models have confirmed that many cancers can be successfully reversed by inactivating the initiating oncogenes, such as *c-MYC*, *KRAS*, *HRAS*, and *BCR–ABL*, or by restoring the function of missing tumor suppressors, such as p53, that have been neutered by cancer-causing epimutations. The big surprise here is the dearth of failures in such studies that often demonstrate meaningful effects with small numbers of replicates, which seems in direct contrast to drug trials in humans. In part, this might reflect the pernicious influence of publication bias; negative studies are deemed boring and are often rejected out of hand or are relegated to seldom-consulted publications, but this applies even more strongly to drug trials in cancer patients. More likely, the remarkable genetic homogeneity of animal models encourages reproducibility, whereas the great diversity of outbred animals including humans largely ensures the opposite, necessitating large numbers of study participants to achieve significance.

Despite these shortcomings, the study of model systems has revealed several interdependent oncogenic signaling pathways (as well as the tumor suppressor pathways engaged by them), which must be overcome for tumorigenesis to proceed. Indeed, when oncogenes are deregulated without such other cooperating lesions already in place, the cells die or senesce rather than proliferate. Most cells are remarkably ambivalent about their own survival and will readily embrace apoptosis when this is mandated or scheduled. Although for some time believed to be a misconception related to differences between animal models and humans, we now know that under some circumstances cancers can develop through the conflation of as few as two oncogenic signaling pathways even if seven or more may be the norm. This is encouraging, as it supports the emerging view that cancer cells, despite their increasingly complex molecular divergence from their forbears, may at heart remain vulnerable to targeting of these primordial founding lesions. Proof-of-concept studies in humans for targeting the "dastardly duo" of *MYC* and *RAS* must await the arrival of suitable therapeutics. However, the notable successes in clinical trials of a number of TKIs directed at *BCR–ABL* and *EGFR* as well as antibodies directed at *HER2* and of a variety of others in various stages of clinical trials affirm that even well-established cancers are still in a state of *original sin*, and might be redeemed and restored to a prelapsarian purity by curing them of their originating molecular corruptions.

Tumor cells become addicted to mutations that activate growth-regulatory signaling pathways, possibly due to the resultant activation of compensatory inhibitory feedback loops that effectively repress all other growth-promoting pathways. Thus, cancer cells with mutations in growth-regulatory pathways are much more sensitive to specific drugs that inhibit those pathways than are cells in which these pathways are activated by physiological means (unless, of course, the drugs target the negative feedback loop as well; see the "Resistance to targeted therapies" section). However, each time the cancer cell loses a growth-promoting pathway to a new therapy, the mutational wheel of fate spins again and another deregulated growth pathway may take its place. These resistance pathways will be discussed in this chapter.

Targeted cancer therapies are being studied for use alone and in combination with each other and traditional chemotherapies or as adjuvants in combination with surgery or radiation (Chapter 18 discusses those that are in clinical use at the present time). Most of these new drugs target proteins that are involved in growth-signaling processes, in particular, though not exclusively, the tyrosine kinases. One great hope is that more detailed identification of the molecular phenotype of individual patients and their tumors will ultimately bring in the new age of **individualized medicine** with treatments ultimately "tailored" to fit the unique set of molecular targets present within any given individual patient's tumor. Targeted cancer therapies also hold the promise of being more selective, thus minimizing collateral damage to normal cells, improving tolerability, and improving quality of life.

These specific agents will be discussed in more detail in this chapter and in Chapters 15 and 18.

Oncogene addiction

You are invulnerable, you have no Achilles' heel.
You will go on, and when you have prevailed
You can say: at this point many a one has failed.

T.S. Eliot

The term "oncogene addiction" has been used to encompass a range of potential mechanisms whereby a cancer cell becomes critically dependent on a given activated oncogene for survival and/or proliferation (Fig. 16.3). This has evident implications for the potential success of targeted cancer treatments. Yet, if the number and complexity of targets are beyond our understanding, then will this hard-won knowledge be of any value? Well, there are good reasons for optimism. Firstly, despite the seemingly bewildering array of mutations, epigenetic alterations, and differential expression of genes and proteins between cancer cells and their normal forbears, cancer cell behaviors are much more limited. In fact, the essential banality of cancer is illustrated by how readily behaviors can be distilled into a short list of shared hallmark features. However, the limited repertoire belies the variety of molecular contributors; those directly involved approached 500 at last count.

Recent studies have begun to unravel the molecular basis behind oncogene addiction, and some common themes have emerged. Thus, the targeting of proteins to which cancer cells are addicted will have far greater impact on their growth than on that of normal cells, allowing the use of higher doses with fewer side effects. One can also target the ancillary factors that enable the addicted oncogene to direct aberrant growth. Thus, mutations in *MYC* and *RAS*, to name but a few, are actively oncogenic only if they are uncoupled from pathways regulating apoptosis and senescence; re-engaging these processes will allow cancer cells to kill themselves or to renounce replication. Addiction has also been shown for other growth-regulating pathways, such as mTOR, WNT–TCF, and Notch, and in mouse models to c-*MYC*, and even miRNAs.

Oncogene addiction implies that despite the multistage nature of tumorigenesis, targeting of certain single oncogenes can be remarkably effective in cancer treatment. The shaking off of growth factor dependence is a critical step in tumorigenesis, but often carries with it the addiction to those signaling pathways that have enabled it. Cancers with activating mutations in key signaling pathways, particularly those in *EGFR* or related *HER2*, such as non-small-cell lung and breast cancers, become "addicted" to these signaling pathways. In other words, therapeutic targeting of these pathways inevitably has major effects on the cancer or at least until resistance supervenes. Thus *HER2*-amplified breast cancers are highly responsive to trastuzumab, and lung cancers with activating mutations in *EGFR* are highly responsive to TKIs or mAb targeting *EGFR*. Cancer cells undergo apoptosis, because such treatments inactivate key downstream signaling pathways, PI3K–AKT–mTOR and MEK–ERK, which support growth and survival. At least in lung cancer, PI3K inhibition downregulates Mcl-1 and MEK inhibition upregulates BIM, showing key roles for both downstream signaling pathways in response to TKI.

Cancer cells are not necessarily addicted to the same extent even within a single tumor; cellular heterogeneity as exemplified by cancer stem cells, dormant cancer cells, and even the surviving descendants of earlier clones that have been supplanted is matched by differences at a molecular level. These subpopulations will behave differently and respond differently to treatments as a result. Moreover, in some cases these cells may endure and, after a short-lived renunciation of replication, serve to revivify the cancer sometime after apparently successful treatments have been concluded.

At the risk of belaboring the point, imatinib has not only irrevocably altered the treatment of CML but also unimpeachably changed the whole course of drug development for cancer. However, the imatinib story also illustrates the shortcomings of targeted therapies. Firstly, not all cells respond fully, which is explained by the presence of an inherently resistant subpopulation of CML stem cells, unlike their progeny not addicted to BCR–ABL, which can repopulate the disease if imatinib is stopped. Thus, even in the presence of imatinib, CML stems and progenitors persist and may even continue expansion, at least in part through the supportive presence of a network of locally available cytokines. Secondly, with continued use of imatinib, cancer cells eventually redraft their molecular roadmap; the acquisition of resistance-conferring mutations allows cancer cells to escape from growth inhibition.

Non-oncogene addiction

Adversity has the effect of eliciting talents, which in prosperous circumstances would have lain dormant.

Horace

Maybe not surprisingly, cancer cells are also dependent on a number of key genes and proteins that are not oncogenes. These include regulators of translational control such as 4EBP–eIF4E, a variety of DNA repair genes such as *PARP1*, heat shock proteins (HSPs), metabolic enzymes, proteases such as Taspase1, and others needed alongside various oncogenic mutations. Scaffold proteins that may be crucial for supporting the structure and interactivity of growth-regulating pathways are likely also important in many aspects of cell signaling. Scaffold proteins are becoming of great interest in cancer biology as they contribute to a variety of key processes such as signaling kinetics, crosstalk between signaling pathways, and conversely buffering against this in some cases. Scaffolding also decreases reaction times by co-localizing enzymes and substrates. Thus, for example, the three successive levels of the RAF–MEK–ERK pathway operate within a scaffold provided by a kinase suppressor of RAS (KSR) proteins, and transformation by mutant RAS can be blocked by knocking out KSR1 as this prevents recruitment of BRAF.

This reliance of oncogenes on subordinate non-oncogenes is the basis of the "non-oncogene" addiction hypothesis, whereby a variety of key non-oncogenes is essential to tumor maintenance and therefore comprises attractive treatment targets. Thus, for example, breast cancers with defective homologous recombination (HR) or glioblastoma cells overexpressing an oncogenic variant of EGFR are hyperdependent on base excision repair (BER) genes such as PARP1. Some mTOR-mutated lymphomas require 4EBP–eIF4E cap-dependent translation and respond to drugs such as PP42 which target this protein.

HR is an important process in the repair of DNA double-strand breaks (DSBs). DSBs can arise through aberrant cancer cell cycles or indirectly from many of the DNA-damaging treatments such as radiotherapy that are directed against cancer. HR is largely restricted to dividing cells during the S and G_2 phases of the cell cycle, so inhibiting this process will have comparatively less effect on normal cells. Moreover, as some cancers are bereft of alternative repair options, the resulting dependence on those remaining options generates the possibility for what is described as synthetic lethality. Combining the inhibition of two or more complementary pathways has a far greater effect than targeting either alone, as exemplified by PARP inhibition in BRCA-defective breast cancers (Chapter 10). This important area will be discussed in the "The concept of synthetic lethality" section.

The concept of synthetic lethality and collateral vulnerability

Do we not wile away moments of inanity or fatigued waiting by repeating some trivial movement or sound, until the repetition has bred a want, which is incipient habit.

George Eliot

Sometimes it isn't possible to target the actual mutant cancer-driving protein or to restore a missing anticancer protein. Mutant

RAS, for example, has proved notoriously difficult to directly target with small-molecule drugs. An alternative is to exploit the fact that given mutations may render the cancer cell addicted to some other protein which might represent an easier target. A simple analogy would be that removing a parachute would have little effect on a person's survival unless he or she had first been pushed out of an airplane.

Two genes are considered synthetic lethal if mutation of either alone does not compromise viability but mutation of both is fatal (Fig. 16.4). Synthetic lethality also occurs between genes and small molecules, and has been employed to study drug activity. In simple terms, targeting a gene that is synthetic lethal to a cancer-relevant mutation would be fatal only to cancer cells – another way of designing cancer-specific cytotoxic drugs. One scenario in which synthetic lethal interactions are likely to be critically important in cancer drug development is in the targeting of molecules that are synthetic lethal with loss of tumor suppressors.

One recent example of this is BRCA1, which regulates HR-mediated DSB repair. Breast or ovarian cancers with mutated *BRCA1* or *BRCA2* survive this defect because they utilize another DNA repair pathway regulated by PARP1 to repair DSBs and allow replication fork progression. PARP inhibition kills cancer cells by allowing accumulation of DSBs and is conditionally "synthetic lethal" with the repair defect. Obviously, it is simpler to give a PARP inhibitor than to try to put BRCA back into every deficient cell.

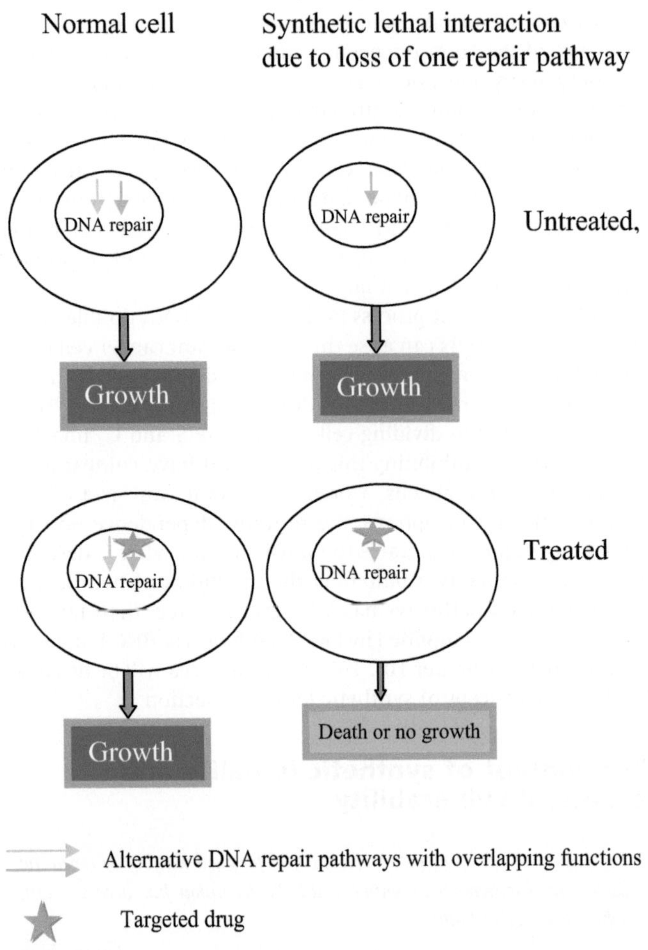

Other synthetic lethal interactions are being screened across a range of model organisms, so this is unlikely to be the end of the story. Thus, loss of RAD52 is also synthetic lethal with *BRCA2* and loss of Taspase-1 (threonine aspartase 1) with MYC and RAS in transformed cancer cells.

In addition to the DDR, cancer cells may depend on a range of "stress relievers" to prevent apoptosis or senescence when faced with potentially lethal oxidative stress, oncogenic stress, or altered metabolism. Drugs that incapacitate this "inner therapist" might be synthetic lethal for many cancer cells, with stress-inducing mutations such as deregulated MYC. Normal cells will not share this trauma, as exemplified by the synergism between chloroquine, an inhibitor of autophagy, and DNA damage in mice. Here, autophagy protects cancer cells from necrosis, but if autophagy is inhibited the cancer cell will die.

Another way to identify cancer cell vulnerabilities arises from the frequent presence of genomic instability. The premise is a simple one. The DDR for the loss of a known tumor suppressor will often also cause the "collateral" loss of other adjacent, or passenger, genes. If these other genes are critical and functionally redundant, their loss may render the cancer cell vulnerable in some previously unappreciated way. A number of such genes have been identified and given the acronym CYCLOPS (copy number alterations yielding cancer liabilities owing to partial loss) genes. This elite group include genes encoding spliceosome, proteasome, and ribosome components. One such gene, glycolytic gene enolase 1 (ENO1), is deleted in glioblastoma (GBM) but causes no obvious phenotype unless expression of ENO2 is also lost. This can be exploited therapeutically by administration of an enolase inhibitor, phosphonoacetohydroxamate, which is synthetic lethal in ENO1-deleted GBM cells.

Clinical progress in biological and molecular targeted therapies

Emancipation from the bondage of the soil is no freedom for the tree.
Rabindranath Tagore

Given the importance of oncogene activation in human cancers, specific targeting of oncogenic pathways can be a highly effective therapeutic strategy. The approval in 2001 by the FDA of the drug imatinib (Gleevec) for the treatment of CML was a crucial milestone, because it was the first agent aimed at a specific cancer target (the BCR-ABL tyrosine kinase). Since then a myriad of new targeted agents, small molecule drugs and mAb, have reached the market, including such household names as trastuzumab.

Fig. 16.5 shows an overview of some of the most prominent targets of growth signaling pathways in new cancer drug development. Given, that many oncogenes are tyrosine kinases it is not surprising that most efforts have been directed against them.

BCR-ABL is a typical example of an aberrantly activated tyrosine kinase, discussed in more detail in Chapter 6. Briefly, a reciprocal translocation creates a transcript encoding a novel fusion protein known as BCR-ABL, a deregulated Abl (Abelson

Figure 16.4 Synthetic lethality. A targeted drug against the synthetic lethal has little effect on the cell with an intact alternative pathway but is lethal for the cell without it.

Biologically Targeted Agents

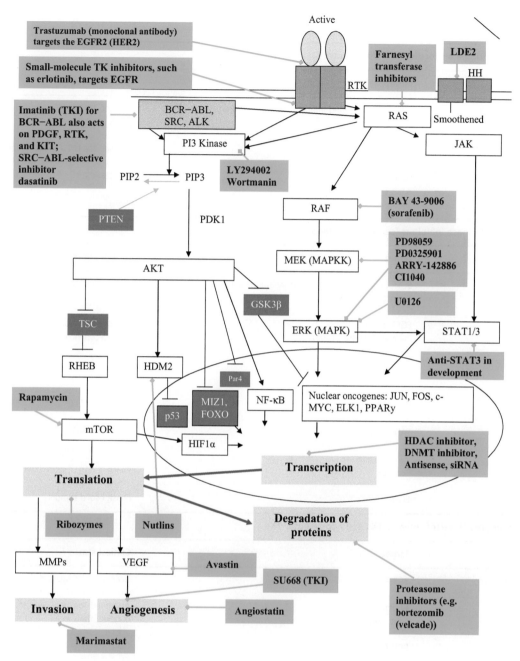

Figure 16.5 Major growth-regulating signaling pathways activated by RTK, showing site of inhibition by targeted therapeutics. It is readily apparent that many drugs that target tyrosine kinases and the RAS–RAF–MAPK at various points have been developed for cancer therapy. Of the two other major growth-signaling pathways, drugs targeting PI3K–AKT–mTOR have been developed, whereas those targeting JAK-STAT are awaited. These pathways eventually signal to the nucleus and regulate gene expression, and drugs are in development to target gene expression (HDAC inhibitors, DNMT inhibitors, antisense, and siRNA) and protein levels (ribozymes and proteasome inhibitors). Drugs able to reactivate tumor suppressors are also in development. Finally, several key processes that are to a degree at least specifically activated in cancers can be targeted, for instance several drugs preventing or treating angiogenesis in different ways are in clinical trials, inhibitors of telomerase or telomere maintenance may be used to impeded cancer cell "immortality," targeting HSP90 may be a means of reducing oncogene activity, and there is interest in drugs targeting invasion, for example by inhibiting the action of MMPs. The drug marimastat can target MMP activity but is still in early stages of evaluation.

murine leukemia viral oncogene) kinase where the endogenous autoregulatory domain is replaced by sequences from BCR (breakpoint cluster region). Attempts to target the BCR-ABL protein resulted directly in the groundbreaking discovery of imatinib mesylate (see Box 16.1). Imatinib is now known to also inhibit other TKs including KIT, constitutively activated by c-*KIT* mutations in the rare and otherwise treatment-resistant gastrointestinal stromal tumours (GISTs).

The success of imatinib has fuelled development of further kinase inhibitors as anticancer drugs (Table 16.2). As discussed

Table 16.2 Tyrosine kinase inhibitors

Anti-ERBB family monoclonal antibodies (see also Chapter 5)

Drug	Target	Clinical status
Trastuzumab and herceptin	ERBB2 (HER2) ligand interactions and downregulation of receptor	Licensed for ERBB2+ breast cancer and some gastric cancers
Cetuximab and Erbitux C225	EGFR (ERBB1)	Licensed for HNSCC and colorectal cancer (CRC)
Panitumumab (Vectibix)	EGFR (ERBB1)	Metastatic CRC
Pertuzumab	ERBB2	Breast

Small molecule tyrosine kinase inhibitors (see also Chapter 5)

Drug	Target	Clinical status
Gefitinib (Iressa)	EGFR (relatively monoselective with little action against other ERBB proteins)	Licensed for advanced NSCLC and phase III HNSCC
Erlotinib (Tarceva)	EGFR and ERBB2 (HER2) (relatively multiselective)	Approved for advanced NSCLC and for advanced pancreatic cancer
Lapatinib (Tykerb)	EGFR and ERBB2	Advanced breast cancer, renal cancer, and HNSCC
Dacomitinib	Pan-HER	Phase III NSCLC
Canertinib CI-1033	PAN-ERBB	Phase II breast
Dacomitinib	Pan-HER	Phase III NSCLC
EKB-569	EGFR, ERBB2	Phase I solid tumors

TKIs targeting other growth factor receptor TKs and non-receptor TKs

Drug	Target	Clinical status
Imatinib mesylate (Gleevec)	BCR-ABL, KIT, and PDGFR	Licensed for CML and GIST
Sunitinib (Sutent)	VEGFR-1,2,3, PDGFR, c-Kit, and FLT3	Approved for renal cell cancer and GIST; ongoing trials in NET and breast
Nilotinib (Tasigna)	BCR-ABL and others	CML
Sorafenib	B-RAF, VEGFR2, EGFR, and RET	Approved for advanced renal cell carcinoma; many trials in melanoma and others underway
Votrient	Multiple kinases	Renal cell carcinoma
Dasatinib (Sprycel)	Src and Abl	CML and some ALL
Vandetanib (zactima)	VEGFR-2, EGFR, and RET	Medullary thyroid cancer
Crizotinib (xalkori)	c-MET and EML4-ALK fusion protein	ALK-positive advanced NSCLC
Axitinib	VEGF	Advanced renal carcinoma

Box 16.1 Discovery of imatinib

The discovery of imatinib did not readily attract scientific confidence. It seemed unlikely that targeted inhibitors of protein kinases could be developed, given the ubiquitous nature of the protein. Tyrosine kinases mediate many of the signaling pathways by which cancer cells promote their own proliferation and survival. They are required for many essential cellular functions and were thought to be too similar in structure to be safely inhibited *in vivo*. On the other hand, it was hard to believe that targeting one single molecule would be enough to beat a disease that has such a heterogeneous pathophysiology.

However, by the early 1990s, again a multidisciplinary group of scientists tackled the concept of specifically inhibiting BCR–ABL kinase. The initial lead compound was identified by Nicholas Lydon's group, which performed high-throughput screenings of chemical libraries searching for compounds with kinase inhibitory activity *in vitro*. The activity of the 2-phenylaminopyrimidine series (Fig. 16.13), tested by Druker at the Oregon Health and Science University Cancer Institute, was subsequently optimized for inhibition of ABL and PDGFR (platelet-derived growth factor receptor) by synthesizing a series of chemically related compounds and analyzing the relationship between their structure and activity. A very important finding in this optimization was that methyl substitution of the anilino–phenyl ring at the 6-position (Fig. 16.13b) led to potent inhibition of ABL and PDGFR kinases, but had no such effect on the activity of the protein kinase C (PKC) family. The 2-phenylaminopyrimidine series was further optimized for absorption, distribution, metabolism, and excretion (ADME) properties by the introduction of the N-methyl piperazine group (Fig. 16.13c). The final compound, imatinib, emerged from these studies as the lead compound for preclinical development, based not only on its high selectivity against CML cells *in vitro* but also on its "drug-like" properties, including pharmacokinetics and drug formulation.

In 1999, Druker *et al.* designed a phase I dose-escalation trial of what was then known as STI571. Despite the initial concerns, they failed to identify a maximum tolerated dose due to the low toxicity of the drug. The trial was highly successful. The drug was approved for clinical use in 2001.

The success story of imatinib is the first example of a somatically mutated gene discovered to be causally involved (through its product protein) in the carcinogenesis of a particular malignancy and whose protein product could be safely and effectively targeted for cancer treatment. However, imatinib treatment must be maintained chronically, which is unfortunate for a patient who develops resistance to the drug. A second generation of BCR–ABL inhibitors has afforded compounds such as **nilotinib** (Novartis), **dasatinib** (Bristol-Myers-Squibb), and **bosutinib** (Wyeth).

---

in earlier chapters and below, cancer cells may become "addicted" to aberrant oncogenic signaling pathways and may as a result be especially sensitive to drugs inhibiting these pathways.

The human genome encodes at least 500 protein kinases (the "kinome"), each of which is a potential drug target in cancer therapy. Numerous kinase inhibitors have been approved for cancer therapy to date (Box 16.2) and over a 100 are being clinically evaluated at present. Most such drugs are targeted against receptor and cytosolic tyrosine kinases, but newer drugs are also aimed at different families of serine/threonine kinases. More than 50 oncogenes have now been robustly identified in cancer, many of which are already the subject of established or developing drugs (see Fig. 16.5 and Tables 16.2–16.4). Some of these will be discussed, under the same categories used for growth signaling and oncogene classification, in this chapter and in Chapter 6 and also in individual chapters relating to the specific biological effects and targets of the drug concerned.

Molecular targeted drugs – an inventory

Before we embark on an exhaustive list of molecular targeted therapies, it is worth pausing briefly to consider a few important principles. There is far more to tailoring therapies to individual patients than just picking the specific targeted drug that best matches a given detected mutation. Although such an approach has been well validated in CML and breast cancer, among others, it lacks context and may prove misleading. For a more complete battle plan, we should also consider the type of mutation affecting the potential target and, moreover, place the mutations within the context of the whole genetic makeup of both the individual tumor and the individual patient. Thus, considering the impact of synthetic lethality, non-oncogene addiction, and functional protein networks greatly increases the range of potential drug targets for any single given mutation. Furthermore, the choice among different agents against the same target will be refined in light of the particular mutation detected and also by factors that will specifically enhance or detract from sensitivity to the drug (see the sections on pharmacogenetics and pharmacogenomics).

Growth factors

There are comparatively few new molecular targeted drugs specifically affecting the levels or production of growth factors themselves. This is in stark contrast to drugs targeting the receptors and, to a lesser extent, the signaling pathways through which they act. That said, numerous established therapies, including some widely used drugs, can exert powerful effects on hormone and growth factor levels. The most obvious examples are androgen-depleting therapies in prostate and aromatase inhibitors in breast cancer. Antagonists of growth hormone receptors, such as pegvisomant, decrease IGF-1 levels and may have anticancer actions. A variety of antibodies targeting IGF-1 (KM3168) and IGF-2 (KM1468) can reduce polyp formation in APC mutant mice and have entered early-phase clinical trials. The HGF/c-Met pathway is often deregulated in human cancers. Highly selective small molecule c-Met inhibitors are in early stages of clinical development, although most are aimed at the receptors for these growth factors. Clinical trials of a humanized mAb against FGF2,

Box 16.2 Targeting tyrosine kinases for cancer therapy

He that will not apply new remedies must expect new evils; for time is the greatest innovator.

Francis Bacon

Two types of agents are used to block or inhibit the protein–tyrosine kinases: monoclonal antibodies, which target the extracellular part of the receptor, and small-molecule drugs (molecular weight <600 Daltons), which inhibit the intracellular portion.

Monoclonal antibodies

Some surface antigens are present predominantly or exclusively on malignant cells and not the surrounding normal cells, and represent ideal targets for specific antibodies. Such antibodies, when bound to tumor antigens, can trigger a host immune reaction that leads to cell death, may interfere with the functions of the protein targeted, or may serve as transport vehicles to deliver attached radioisotopes, immunotoxins, or cytotoxins to tumor cells. Several monoclonal antibodies have been approved, most of them for treating hematologic malignancies and, most recently, breast cancer.

Herceptin (trastuzumab) is a humanized monoclonal antibody that inhibits cell growth by binding to the extracellular part of the HER2 protein–tyrosine kinase receptor, which is involved in the pathogenesis of breast and ovarian cancer. Trastuzumab induces antibody-dependent cytotoxicity by directing natural killer cells and monocytes against the tumor cells expressing the antibody target. Trastuzumab is used alone or in combination with paclitaxel for treating breast cancer that is refractory to standard chemotherapy.

Rituximab (Rituxan) is a chimeric monoclonal antibody that binds to CD20, a cell-surface protein found almost exclusively on mature B cells, including those in B-cell non-Hodgkin lymphoma.

Gemtuzumab (Myelotarg) is composed of a recombinant humanized monoclonal antibody linked to a molecule of calicheamicin, a cytotoxic anti-tumor antibiotic. It is the first such targeted antibody–chemotherapy complex available. The antibody binds specifically to CD33, an antigen found on normal and leukemic myeloid cells, including 90% of acute myeloid leukemia blasts. CD33 is not present on bone marrow stem cells, which therefore are not affected by the treatment. Gemtuzumab is approved for patients older than 60 years with CD33-positive acute myeloid leukemia (AML) who have relapsed after initial treatment and are not considered candidates for cytotoxic chemotherapy.

Alemtuzumab (Campath) is a humanized monoclonal antibody that binds to CD52, an antigen expressed on B and T lymphocytes. Alemtuzumab is approved for patients with B-cell chronic lymphocytic leukemia for whom other therapy has failed.

Bevacizumab (Avastin) is a humanized monoclonal antibody against vascular endothelial growth factor (VEGF), and may be used as a first-line treatment for metastatic colorectal cancer (CRC).

Erbitux is a chimeric monoclonal antibody against epidermal growth factor receptor (EGFR) that is indicated for use in EGFR-positive metastatic colorectal carcinoma, which is refractory to irinotecan.

Nucleotide based

Pegaptanib is a covalent conjugate of an oligonucleotide of 28 nucleotides in length that terminates in a pentylamino linker, to which two 20-kilodalton monomethoxy polyethylene glycol (PEG) units are covalently attached via the two amino groups on a lysine residue. Currently indicated for wet macular degeneration. This oligonucleotide targets VEGF.

Small-molecule drugs

Imatinib (gleevec), a small-molecule drug, inhibits the intracellular part of three protein–tyrosine kinases:

1. Bcr–Abl, an abnormal fusion protein involved in the pathogenesis of CML;
2. c-kit (CD117), a receptor overexpressed in gastrointestinal stromal tumors (GISTs); and
3. The platelet-derived growth factor receptor alpha (involved in chronic myeloproliferative syndromes characterized by eosinophilia).

Since the drug has a very low molecular weight, it can be given orally. Imatinib has become the standard of care in Philadelphia-positive CML, and it shows efficacy for treating advanced GIST. Caution is needed when gleevec is given together with strong CYP3A4 inhibitors (including ketoconazole, itraconazole, and clarithromycin). Grapefruit juice may also increase plasma concentrations of imatinib and should be avoided. Can markedly increase levels of simvastatin when used concurrently. Levels of warfarin and some calcium channel blockers may also be increased; the former is best avoided in gleevec-treated patients, and some alternative anticoagulant, such as heparin, is preferred.

Chemical name: 4-[(4-Methyl-1-piperazinyl)methyl]-*N*-[4-methyl-3-[[4-(3-pyridinyl)-2-pyrimidinyl]amino]-phenyl]benzamide methanesulfonate

Structural formula: $C_{29}H_{31}N_7O \cdot CH_4SO_3$

Molecular weight: 589.7

A variety of second-generation compounds have been developed to deal with the resistance – inherent or acquired – of some CML patients to imatinib.

Gefitinib (Iressa) is a small-molecule anilinoquinazoline that inhibits the intracellular phosphorylation of numerous tyrosine kinases associated with transmembrane cell surface receptors, including the tyrosine kinases associated with the epidermal growth factor receptor (EGFR–TK).

It was identified by Zeneca in 1994 as a lead candidate in a medicinal chemistry project for the discovery of EGFR tyrosine kinase inhibitors.

Box 16.2 (Continued)

Gefitinib was originally identified from structure-activity studies based around a 4-anilinoquinazoline lead series, as a competitive inhibitor of the ATP-binding function of the tyrosine kinase active site of EGFR. Gefitinib gained FDA approval in 2003 and is the first targeted agent to be approved for the treatment of patients with advanced non-small-cell lung cancer (NSCLC). The presence of specific mutations in the EGFR gene correlates with clinical responsiveness to gefitinib. Inducers of CYP3A4 activity, such as phenytoin, accelerate gefitinib metabolism and decrease plasma concentration; conversely, as discussed, co-administering CYP3A4 inhibitors may increase plasma levels.

Chemical name: 4-Quinazolinamine, N-(3-chloro-4-fluorophenyl)-7-methoxy-6-(3-(4-morpholinyl)propoxy)
 Molecular formula: $C_{22}H_{24}ClFN_4O_3$
 Relative molecular mass: 446.9

Erlotinib (Tarceva; Fig. 16.14) is also a small-molecule quinazolinamine and an EGFR antagonist. It gained FDA approval in 2005 for NSCLC and in 2005 for metastatic pancreatic carcinoma. It has been shown to reduce mortality in non-small-cell lung cancer (NSCLC) after failure of first-line therapy. It has therefore been approved for use as second-line therapy for lung cancer. Assays of cytochrome P450 metabolism showed that erlotinib is metabolized primarily by CYP3A4 and, to a lesser extent, by CYP1A2 and the extrahepatic isoform CYP1A1, so drug levels will be increased by some antifungals and grapefruit juice, and caution is needed. See the discussion of gleevec at the beginning of this section.

Nilotinib is an inhibitor of the Bcr–Abl kinase. Nilotinib binds to and stabilizes the inactive conformation of the kinase domain of Abl protein. *In vitro*, nilotinib inhibited Bcr–Abl-mediated proliferation of murine leukemic cell lines and human cell lines derived from Ph+ CML patients. Under the conditions of the assays, nilotinib was able to overcome imatinib resistance resulting from Bcr–Abl kinase mutations, in 32 out of 33 mutations tested. *In vivo*, nilotinib reduced the tumor size in a murine Bcr–Abl xenograft model. Nilotinib inhibited the autophosphorylation of the following kinases at IC50 values as indicated: Bcr–Abl (20–60 nM), PDGFR (69 nM), and c-Kit (210 nM).

Chemical name: N-(3-Ethynylphenyl)-6,7-bis(2-methoxyethoxy)-4-quinazolinamine
 Molecular formula: $C_{22}H_{23}N_3O_4 \cdot HCl$
 Molecular weight: 429.90

Sorafenib (Nexavar) is a multikinase inhibitor targeting several serine–threonine and TKRs, including intracellular (CRAF, BRAF, and mutant BRAF) and cell surface kinases (KIT, FLT-3, RET, VEGFR-1, VEGFR-2, VEGFR-3, and PDGFR-β), involved in angiogenesis and growth. It is currently indicated for advanced renal cell carcinoma or unresectable hepatocellular cancer. As with most other TKIs, it is metabolized by CYP3A4. Caution is also needed when coadministering sorafenib with others that are metabolized or eliminated predominantly by the UGT1A1 glucouronidation pathway (such as irinotecan).

Chemical name: 4-(4-{3-[4-Chloro-3-(trifluoromethyl)phenyl]ureido} phenoxy)-N^2-methylpyridine-2-carboxamide 4-methylbenzenesulfonate
 Molecular formula: $C_{21}H_{16}ClF_3N_4O_3 \times C_7H_8O_3S$
 Molecular weight: 637.0

Sunitinib (Sutent) inhibits multiple TKRs, including PDGFRα and PDGFRβ, VEGFR1, VEGFR2 and VEGFR3, KIT, Fms-like tyrosine kinase-3 (FLT3), CSF-1R, and RET, many implicated in tumor growth, pathologic angiogenesis, and metastatic progression of cancer. It is indicated for GIST in patients with disease progression or intolerance to gleevec and those with advanced renal cell carcinoma. As with other TKIs, metabolism is primarily through CYP3A4 – see the discussion of gleevec at the beginning of this section.

Chemical names: Butanedioic acid, hydroxy-, (2S)-, compound with N-[2-(diethylamino)ethyl]-5-[(Z)-(5-fluoro-1,2-dihydro-2-oxo-3H-indol-3-ylidine)methyl]-2,4-dimethyl-1H-pyrrole-3-carboxamide (1:1)
 Molecular formula: $C_{22}H_{27}FN_4O_2 \cdot C_4H_6O_5$
 Molecular weight: 532.6 Daltons

Lapatinib (Tykerb) is a 4-anilinoquinazoline inhibitor of the intracellular tyrosine kinase domains of both EGFR (ErbB1) and of HER2 (ErbB2) receptors. It is indicated in combination with capecitabine (5FU) for the treatment of patients with advanced or metastatic breast cancer whose tumors overexpress HER2 and who have received prior therapy including an anthracycline, a taxane, and trastuzumab. As with other

(Continued)

Box 16.2 (Continued)

TKIs, metabolism is primarily through CYP3A4 – see the discussion of gleevec at the beginning of this section.

Chemical name: N-(3 Chloro-4-{[(3-fluorophenyl)methyl]oxy}phenyl)-6-[5-({[2 (methylsulfonyl)ethyl]amino}methyl)-2-furanyl]-4-quinazolinamine bis(4 methylbenzenesulfonate) monohydrate
   Molecular formula: $C_{29}H_{26}ClFN_4O_4S$ $(C_7H_8O_3S)_2$ H_2O
   Molecular weight: 943.5

Dasatinib (Sprycel) inhibits BCR–ABL, the SRC family (SRC, LCK, YES, and FYN), c-KIT, EPHA2, and PDGFRβ and is indicated for the treatment of adults with chronic myeloid leukemia (CML) with resistance or intolerance to prior therapy including imatinib. It is also indicated for the treatment of adults with Philadelphia chromosome–positive acute lymphoblastic leukemia (Ph+ ALL) with resistance or intolerance to prior therapy.

Chemical name: N-(2-Chloro-6-methylphenyl)-2-[[6-[4-(2-hydroxyethyl)-1-piperazinyl]-2-methyl-4-pyrimidinyl]amino]-5-thiazolecarboxamide, monohydrate
   Molecular formula: $C_{22}H_{26}ClN_7O_2S$ • H_2O
   Molecular weight: 506.02

Nilotinib (Tasigna) inhibits Bcr–Abl, PDGFR, and KIT and is indicated for the treatment of chronic-phase and accelerated-phase Philadelphia chromosome–positive chronic myelogenous leukemia (CML) in adult patients resistant or intolerant to prior therapy that included imatinib.

Chemical name: 4-Methyl-N-[3-(4-methyl-1H-imidazol-1-yl)-5-(trifluoromethyl)phenyl]-3-[[4-(3-pyridinyl)-2-pyrimidinyl]amino]-benzamide, monohydrochloride
   Molecular formula: $C_{28}H_{22}F_3N_7O$•HCl • H_2O
   Molecular weight: 565.98

Axitinib (also known as AG013736) inhibits VEGFR-1, VEGFR-2, VEGFR-3, PDGFR, and c-KIT (CD117). It has been successful in trials with renal cell carcinoma (RCC) and several other tumor types. A phase II clinical trial showed good response in combination chemotherapy with gemcitabine for advanced pancreatic cancer.

---

which has shown promise in models of hepatocellular cancer (HCC), are in development. However, this area is set to grow rapidly. First, growth factors are in many ways ideal targets for mAb (they are extracellular and can prevent receptor binding) and second many existing therapies may exert benefits in part by altering growth factor levels. Thus, recent years have witnessed a reappraisal of the range of potential benefits of exercise and weight loss. Long advocated for managing diabetes and risk of heart disease, it now appears that such lifestyle interventions may also reduce risk of many common cancers. In part, the benefits may be due to reducing levels of circulating insulin and likely also those of a gaggle of adipocyte-derived growth factors, such as leptin, visfatin, and others that might otherwise stimulate growth of some types of cancer cell. In some cases it is possible that increasing levels of growth-inhibiting adipokines, such as adiponectin, could reduce risk of some cancers. Along similar lines, drugs such as metformin, long in use for reducing blood glucose levels, also lower the increased levels of insulin and some adipokines that accompany Type 2 diabetes and its precursor states of obesity and insulin resistance. In fact, an increasing body of evidence suggests that metformin might help prevent a number of common cancers. Metformin can also interfere with downstream growth factor signaling through AMPK, thereby reducing mitogenic signaling via insulin receptors. However, some caution is needed in interpreting the results linking adipokine levels and cancer and it is not known whether they are simply markers for obesity or might actually be functionally relevant in human cancer. Functional studies using genetically modified mice are supportive of the notion that leptin and visfatin might promote and adiponectin inhibit growth of some tumour types, such as breast and CRC. These results are consistent with the alterations of these adipokines in T2DM and precursor states in which cancer risk is increased. It is likely that this area will expand rapidly in the coming years, particularly as treatments based on adipokines might have commercial value in both diabetes and obesity as well as cancer.

Receptor and nonreceptor tyrosine kinases

These are overwhelmingly the most prominent current targets for new anticancer drugs. Therapies targeting receptor tyrosine kinases have shown efficacy in subsets of cancers, but cancers invariably develop resistance, which must be overcome if new targeted agents are to realise their full therapeutic potential.

Table 16.3 Other molecular targets (approved drugs in bold text)

Growth factors, hormones, and receptors		
Tamoxifen	Selective estrogen receptor modulator (SERM)	**Breast cancer**
Raloxifene	SERM	**Breast cancer**
Fulvestrant (Faslodex)	Estrogen receptor antagonist	**Advanced breast cancer**
Goserelin (Zoladex)	LHRH analogue	**Prostate cancer**
Abiraterone	17 α-hydroxylase inhibitor reduces androgen production	**Metastatic castration-resistant prostate cancer**
Finasteride	5-alpha reductase	**Benign prostatic hypertrophy; use in chemoprevention of prostate cancer under debate**
Avodart	5-alpha reductase	**Chemoprevention of phase III prostate cancer**
Signal transduction and enzyme inhibitors (see Chapter 5)		
Everolimus (Afinitor)	Binds to immunophilin FK-binding protein-12, forming a complex that inhibits mTOR	**Kidney cancer, giant cell astrocytoma, and pancreatic NET**
Temsirolimus (Torisel)	FKBP–rapamycin-associated protein (inhibits mTOR)	**Renal cell carcinoma**
Vemurafenib (Zelboraf)	BRAF V600E	**Melanoma**
Anastrazole (Arimidex)	Aromatase inhibitor	**Hormone-sensitive breast cancer**
Letrozole	Aromatase inhibitor	**Hormone-sensitive breast cancer**
Exemestane	Aromatase inhibitor	**Potential chemoprevention of breast cancer in high-risk women**
SCH66336 Lonafarnib and sarasar	FTase (RAS)	Phase III NSCLC
R115777 Tipifarnib and zarnestra	FTase	Phase III NSCLC
BMS214662	FTase	Phase II NSCLC
Salirasib	RAS	Phase II NSCLC-adenocarcinoma
Vismodegib	Hedgehog pathway	Filed and phase II
Perifosine	PI3K	Phase III CRC and myeloma
LY294002	PI3K	Phase I/II
Enzastaurin	PKCβ, AKT, and GSK-3β	Phase I
RG7440	AKT	Phase I solid tumors
	GSK-3β	In development
PD98059	MEK (MAPKK)	Phase I
Selumetinib AZD6244	MEK1 and MEK2	Phase II solid tumors
U0126	ERK (MAPK) and MEK	Phase I
RG7414	EGFL7	Phase II solid tumors
AZD1840	JAK1 2	Phase I solid tumors and myeloproliferative diseases
Lestaurtinib	JAK2	Phase I
WP-1034	STAT3/5	Phase I
Zoledronic acid	Bisphosphonate drug affecting as yet poorly described targets	Phase III **bone metastases** and breast cancer
Iniparib	PARP	Phase III breast and phase II NSCLC
Oliparib	PARP	Phase II
Amentoflavone	NFkB	Phase I
Stimuvax (L-BLP25)	Mucin-1	Phase III NSCLC

(Continued)

Table 16.3 (Continued)

Other monoclonal antibodies		
Rituximab (Rituxan)	**CD20**	**CLL and B-cell lymphoma**
Mylotarg	**CD33**	**AML**
Alemtuzumab (Campath)	**CD52**	**CLL**
Ofatumumab (Arzerra)	**CD20**	**CLL**
Ipilumumab	CTLA-4	**Melanoma**
Denosumab	RANK ligand	**Bone metastases**
RG3638	c-MET	Phase II metastatic CRC and breast
MEDI-575	PDGFR-alpha	Phase II solid tumors
RG7444	FGFR3	Phase I
Ch14.18	GD2	Phase III neuroblastoma

Cell-cycle targeted agents (see Chapter 4)		
Roscovitine	CDK (E2F)	Phase II
Flavopiridol	Pan CDK and STAT	Phase I
AZ703, an imidazo[1,2-a] pyridine	CDK1/2	Phase I–III solid tumors
P276-00	CDK1-4	Phase I
PD332991	CDK4	Phase I
AZD1152	Aurora kinase	Phase II hematological malignancies
PF-03814735	Aurora kinase A and B	Phase I
UCNO1	CHK1 and CHK2	Phase III
AZD7762	CHK1 and CHK2	Phase I
LY2606368	CHK1	Phase I
NU7026	DNA-PKcs	Research use
IC486241	DNA-PKcs	Research use
GSK461364	Polo like kinase 1	Phase I solid tumors
BI 2536	Polo like kinase 1	Phase II NSCLC
KU-60019	ATM	In development

Apoptosis regulators and tumor suppressors (see Chapter 8)		
Human recombinant (TRAIL)	TRAIL receptor	Phase I
Nutlins	MDM2 (activate p53)	Phase I
RG7112	MDM2	Phase I solid tumors
CH401 mAb	ERBB2; can activate apoptosis via p38 MAPK and JNK	Phase II
Roscovitine	MCL-1 (antiapoptotic BCl-2 family member)	Phase II
ABT-737	BH3 mimetic; inactivates antiapoptotic BCL-2 family proteins	Phase I
Obatoclax	BCL-2	Phase I NSCLC
RG7459	IAP	Phase I

Chromatin targeted agents and proteasome inhibitors (see Chapter 11)		
Vorinostat (Zolinza)	HDAC inhibitors	**Cutaneous T-cell lymphoma** (CTCL)
Romidepsin	Selective HDAC inhibitor	**CTCL**
Bortezomib (Velcade)	Proteasome	**Approved for advanced multiple myeloma and mantle cell lymphoma**

Table 16.4 Angiogenesis inhibitors

Monoclonal antibody (see Chapters 14 and 13)		
Drug	Target	Status
Bevacizumab (Avastin)	VEGF-A	**Licensed in metastatic colorectal cancer**

Soluble receptor chimeric protein (see Chapter 14)		
VEGF-trap	VEGF-A	Phase I

Small-molecule multitargeted TKI (see Chapters 5 and 14)		
Sunitinib (Sutent)	VEGFR-1,2,3, PDGFR, c-Kit, and FLT3	**Approved for renal cell cancer and GIST** Pancreatic neuroendocrine tumors (NET)
Sorafenib (Nexavar)	VEGFR2, EGFR, and RET	**Approved in for advanced renal-cell carcinoma;** hepatocellular cancer
Pazopanib (Votrient)	VEGFR, PDGFR, and c-KIT	**Advanced renal-cell carcinoma**
Axitinib	VEGFR	Phase III; advanced renal-cell carcinoma

Vascular targeted agents (see Chapter 14)		
Combretastatin A	Endothelial tubulin	Phase II
CD13	Tumor endothelial cells	In development
Asn–Gly–Arg (NGR) tripeptide	Endothelial cells	In development

Inhibitors of endothelial cell proliferation (see chapter 14)		
Angiostatin	Endothelial cells	Phase I
Thalidomide	TNF-α	Approved in US
ABT-510	Thrombospondin1-mimetic	Phase 1

As TKIs are rewriting the rule book for cancer treatment, are dominating new drug development, and represent the ideal prototypes for targeted treatment of cancer, much space will be devoted to discussing these agents (see Box 16.1 and Chapters 15 and 18). Because they are already in widespread clinical use, inhibitors of BCR-ABL and HER2 are discussed separately and here we will focus on other TK targets.

RTKs, as exemplified by the EGFR, activate numerous signaling pathways involved in tumorigenesis (Fig. 5.5), including the archetypal growth-regulating RAF–MEK–ERK cascade and the PI3K–AKT–mTOR pathway (Chapters 5 and 6). These pathways offer a number of targets for therapeutic agents at different levels and will be discussed in this and the following sections. Intriguingly, much data suggest that these agents are particularly effective when the signaling proteins are mutated rather than just "upregulated". This is self-evident where the agent specifically targets an aberrant or mutant protein such as the mAb herceptin targeted to the HER2 protein. However, even TKIs such as gefitinib and imatinib appear most effective when target kinases are activated by mutation rather than purely by physiological upregulation via a relevant signaling pathway. The favoured explanation is that cancer cells have become "addicted" to the deregulated protein and are hypersensitive to inhibitory agents (see earlier in this chapter).

As has been discussed, a variety of techniques are employed to screen for new molecular targets in cancer, including studying cell lines, human tumors, and animal models. Once a promising molecule has been selected and has undergone some functional testing (often in genetically modified cell lines or model organisms) that support its suitability as a drug target, a drug must be developed. Small-molecule drugs are usually identified in high-throughput drug screens, in which thousands of test compounds, produced by the pharmaceutical company, are tested against the specific target, such as EGFR. Often the best candidates (lead compounds) are then further chemically manipulated to generate a series of related compounds and the very best in terms of activity, distribution, ease of administration, and so on will then, if all goes to plan, move into preclinical testing for efficacy and then safety. The other major class of molecular-targeted drugs, the mAb, are employed to target extracellular parts of RTKs and will be discussed later.

EGFR is frequently upregulated in non-small-cell lung cancers (NSCLC) and EGFR TKIs, such as gefitinib (Iressa) and erlotinib (Tarceva), have achieved significant tumor regression when used in advanced disease after chemotherapy and are now being investigated as maintenance therapies. However, barely 20% of patients achieved tumor regression in studies completed to date. This lack of success may result from mutations in downstream signaling pathways, such as RAS, or activation of other RTKs that dilute the effects. Interestingly, recent trials of NSCLC suggest that the nature of the mutations in EGFR itself may also influence the response to small-molecule TKIs; activating mutations in the ATP-binding cleft of the EGFR kinase domain confers sensitivity of NSCLC to gefitinib but not to the mAb cetuximab, whereas the latter can effectively treat colorectal cancers with overexpression of effectively any mutant forms of EGFR. At present EGFR mutations are only assessed in clinical trials, with TKIs given empirically to NSCLC patients with or after traditional chemotherapies.

Females, Asians, never-smokers, and those with adenocarcinoma appear most likely to have EGFR mutations.

Many of the early TKIs prevented phosphorylation of target proteins b8y blocking binding of ATP to the kinase. However, newer drugs such as imatinib and sorafenib appear to interact with other regions of their target kinases. These, agents may be used to overcome some types of site-specific resistance to the older agents.

Many TKIs are relatively nonspecific and can target multiple different kinases at clinically used doses. In some cases this may increase efficacy by simultaneously interfering with more than one kinase important for a given cancer, but could also increase side effects. More specific targeting of individual kinases can be achieved by employing mAb, but these are restricted to cell-surface RTKs.

ALK was identified in studies of anaplastic large cell lymphoma (ALCL), in which the *NPM* gene on chromosome 5 is fused to the *ALK* gene on chromosome 2; t(2;5)(p23;q35) chromosomal translocation. Several other ALK fusions are now known, including the *EML4–ALK* fusion gene, in around 2–7% of NSCLC. They are mainly observed in adenocarcinomas and in nonsmokers. As these fusion genes are not present in normal cells, specific inhibitors may have very few side effects. Various ALK inhibitors, including adenotriphosphate (ATP)-competitive inhibitors of ALK activity, are in clinical trials. Among the most studied to date, Crizotinib is an inhibitor of ALK, as well as other RTKs. Crizotinib is showing early promise in locally advanced or metastatic ALK-positive NSCLC. Thus, potentially meaningful responses were noted in phase I and phase II trials, though must be confirmed in phase III studies (see below). Treatment-related adverse reactions are infrequent but include visual disturbances, gastrointestinal upset, and edema.

Interest is also growing in targeting of the insulin–IGF signaling pathways in cancer by various means. The IGF-1 receptor has also been targeted by antibodies that bind to the IGF-1 receptor α subunit and inhibit action of both IGF-1 and IGF-2. The antibody robatumumab has shown promise in preclinical models. Another antibody, figitumumab, has been promising in a phase II study in NSCLC. A variety of TKIs have been shown to inhibit signaling of both IGF and insulin. These agents have also shown promise in preclinical studies and one such, OSI-906, is in phase III clinical trials for the treatment of adrenocortical cancer. Phase II clinical trials are being conducted in patients with metastatic breast cancer (NCT01013506). Phase I clinical trials are ongoing to establish the tolerability and safety of another agent XL228 (NCT00464113, NCT00526838). As expected, blocking of the insulin receptor resulted in hyperglycemia in the mouse models when treated with BMS-554417 and BMS-536924, an affect also reported in trials of XL228 in humans. As mentioned AMPK activators, such as metformin, may protect against cancer.

An interesting paper in *Cell* has suggested that the PTPN12 tyrosine phosphatase is a tumor suppressor in triple-negative breast cancer (TNBC). PTPN12 suppresses transformation by inhibiting multiple oncogenic tyrosine kinases, including HER2 and EGFR, and might be a useful drug target.

Targeting BCR–ABL

Until the development of imatinib, standard therapy for CML was largely ineffectual and involved the application of multiple cytotoxic agents, often with terrible toxic side effects. The pioneering research that led to the discovery of the causal Philadelphia chromosome and thence the identification of the resultant *BCR–ABL* oncogene has been discussed in Chapter 6. By the mid-late 1990s, high-throughput screens for tyrosine kinase inhibitors within the pharmaceutical company Novartis had resulted in the development of the new, small-molecule imatinib (gleevec), now the exemplar *par excellence* of targeted therapeutics (Box 16.2).

Despite initial reservations and what with hindsight may seem an inexplicable reluctance to progress the agent into clinical trials, eventually imatinib was shown to inhibit proliferation of BCR-ABL-expressing hematopoietic cells in culture. In clinical trials, even though CML cells were not completely eradicated, imatinib restricted the growth of tumor cells and decreased the risk of the feared "blast crisis," a rapidly fatal sudden increase in immature myeloid cells in the bone marrow and blood. A brief history of the development of imatinib is given in Box 16.1. Imatinib was marketed in 2001 and has not only transformed the quality of life of patients suffering from CML but also spearheaded the renaissance in cancer therapy discussed in this chapter.

Imatinib is also used in the therapy of KIT or platelet-derived growth factor receptor-α (PDGFRA) positive gastrointestinal stromal tumors (GISTs).

We have been able to draw salutary lessons from the failures of targeted therapies, not least the elucidation of drug-resistant signaling pathways (Fig. 16.6). In the case of imatinib, the mechanism of resistance is direct and usually involves acquired point mutations in the kinase domain of BCR–ABL that bypass the inhibitory effects of the specific agent. Thus, BCR–ABL remains vulnerable to newer TKIs, such as dasatinib and nilotinib, which might not have been the case if mutations activated downstream or alternative signaling pathways.

Therapeutic antibodies against receptor TK

Essentially, all mAb are developed by immunizing mice (or sometimes other species) with a purified form of the desired target molecule. The various antibodies are each produced by a specific clone of B lymphocytes, many in the spleen. Single B cells making a single type of antibody are isolated and fused with a myeloma cell to generate an essentially immortal and rapidly proliferating cell that after expansion can produce large amounts of the given mAb. Efficacious clones are selected for preclinical testing. mAb are only effective against extracellular proteins as they do not cross the plasma membrane and enter cells. Most current mAb are humanized, by replacing the mouse protein in the nonvariable regions with the human equivalent, to prevent immune reactions due to rejection of the foreign protein.

The first mAb kinase inhibitor to be approved was trastuzumab (Herceptin), which targets the ERBB2 (HER-2–neu) receptor. Cetuximab (Erbitux) and bevacizumab (Avastin), targeting EGFR and VEGF, respectively, have also been approved for clinical use. Not only can mAb be effective where small molecules are made useless by mutations in their target proteins, but they can uniquely cause direct recruitment of an immune response, including cytotoxic responses, directed against the cells carrying the protein targeted by the antibody. These agents cannot pass through the cell membrane so they can only be used to target cell-surface or secreted proteins, unlike small molecules which could theoretically target any cancer-relevant proteins. They have to be given intravenously or subcutaneously, whereas small-molecule TKIs are often orally active. On the plus side, a longer half-life usually allows once-weekly rather than daily dosing. mAb may also be less able to pass the blood–brain barrier

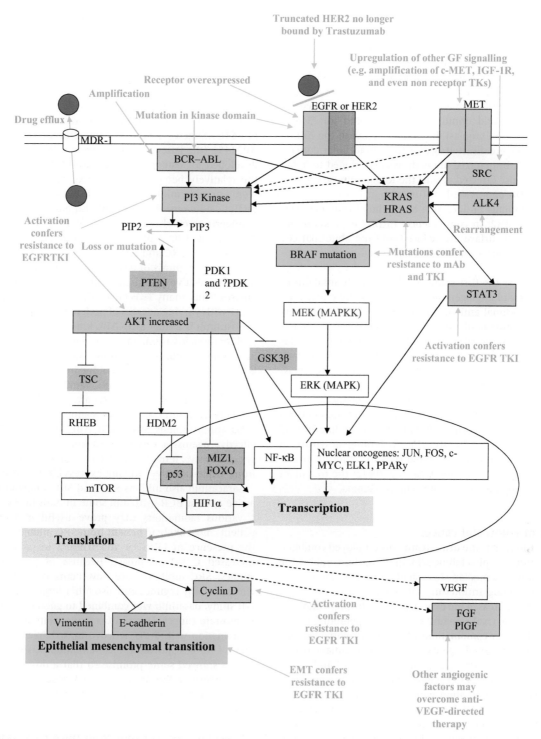

Figure 16.6 Resistance to targeted therapies. Several mechanisms contribute to the development of resistance, including secondary mutations in EGFR that prevent drug binding or render it resistant to inhibition. Mutations may also affect the activation of an immune response by monoclonal Ab, such as trastuzumab. The RAS–MAPK and PI3K–AKT pathways are major signaling networks linking EGFR activation to cell proliferation and survival. MET amplification or mutations and EML4–ALK fusions all constitutively activate RAS–RAF–MEK. RAS, RAF, PIK3, AKT, or cyclin D may also become constitutively activated or mutated, and even alternative pathways such as c-MET may become activated and help overcome EGFR inhibition. Signaling may also be activated downstream by loss of the tumor suppressor PTEN. Epithelial–mesenchymal transition may develop. Targets of resistance are highlighted in green boxes; means of resistance are shown in red text.

and might in general be most effective against circulating tumor cells and hematological malignancies, although many of these agents are licensed for use in solid tumors also.

The *ERBB2* (*HER2*) gene is amplified or aberrantly expressed in a quarter of all breast cancers. Trastuzumab was specifically developed to target this abnormality. Scientists at Genentech (San Francisco, CA, US) initially developed and evaluated 10 murine mAb that could immunoprecipitate (bind to) ERBB2 expressed in a mouse fibroblast line transfected with the *ERBB2* gene. These mAb were directed against the extracellular domain of the receptor. A humanized antibody was developed to avoid an immune response directed against the foreign mouse protein, with eventual FDA approval of trastuzumab in 1998.

Trastuzumab inhibits proliferation by binding to the extracellular region of ERBB2. In consequence, ERBB2 expression and function at the cell surface are reduced and the abnormal cell becomes the target for antibody-dependent cell-mediated cytotoxicity (ADCC).

Bevacizumab was the first commercially available inhibitor of VEGF and thereby of angiogenesis. It is a genetically engineered humanized monoclonal antibody approved by the FDA in 2004 as a first-line treatment of metastatic colorectal cancer. It has since gained FDA approval for second-line treatment of metastatic colorectal cancer, first-line treatment of NSCLC, metastatic HER2-negative breast cancer, second-line treatment of glioblastoma, and metastatic renal cell carcinoma.

Cetuximab is a recombinant, human–mouse chimeric monoclonal antibody that binds specifically to the extracellular domain of the human EGFR. The antibody inhibits the binding of EGF and other ligands to EGFR, thereby preventing downstream signal transduction, and targets bound cells for immune attack. Cetuximab was approved by the FDA in 2004 for the treatment of EGFR-expressing metastatic colorectal carcinoma.

Treatment of colorectal cancer

In metastatic CRC, first-line drug treatments employed combinations of 5-fluorouracil plus folinic acid. In trials, this combination was associated with a median survival of around 10–15 months. More recently, adding Irinotecan and Oxaliplatin to this traditional regime has increased life expectancy by a further 6 months. However, 5-year survival remains a rather disappointing 5–7%. To address this, newer combinations have been explored recently and include new targeted agents, in particular antiangiogenic antibodies, such as Bevacizumab, that target VEGF and others that inhibit EGFR, such as Cetuximab and Panitumumab. These have been used in combination with more traditional agents or even first line. Bevacizumab increases life expectancy when compared with standard therapies in some trials, but there are occasional undesirable side effects, including raised blood pressure, potentially life-threatening thromboembolic events noted in around 3–5% of treated patients, and more rarely GI perforation. Thus, further evaluation is needed. Refinements are rapidly introduced in this area, and it is now accepted that CRC patients who have activating mutations in *KRAS* or *BRAF* receive little benefit from therapies targeting EGFR. This is self-evident as EGFR is rendered essentially redundant when downstream signaling is activated directly by mutation. Why no benefit derives from the potential targeting of an immune response to EGFR expressing cancer cells, which should be independent of EGFR signaling, is an area that requires further investigation.

c-SRC as a target of drug therapy

SRC is an attractive target for new drug development in cancer. SRC is often overexpressed in a large number of human cancers, and there appear to be few adverse effects specifically related to its inhibition. Several SRC inhibitors are in preclinical and clinical trials, though most also have significant effects on other kinases. Several compounds targeting the SRC kinase domain have been described, including Bosutinib (previously SKI-606; Wyeth) and SU6656 (Sugen), and Ariad Pharmaceuticals – AP23464 and AP23451 (the latter a bone tissue–targeted SRC kinase inhibitor developed for osteoporosis therapy). One might anticipate that the relatively benign effects of SRC "ablation" in normal cells would provide minimal toxicity. However, the fact that the non-catalytic domains of SRC might also have biological effects, such as integrin assembly, raises the possibility that targeting the kinase domain might limit potential. Interestingly, recent data indicate that targeting SRC and FAK simultaneously might be effective in promoting apoptosis of colon cancer cells. It is also important to note the relative promiscuity of TKIs for a variety of different kinases. Thus, many TKIs target SRC and BCR–ABL and some also c-KIT with varying degrees of preference for particular kinases.

Bosutinib is a dual SRC–ABL kinase inhibitor that can inhibit proliferation, invasion, and migration of breast cancer cells *in vitro* and inhibit tumor growth in mouse models *in vivo*. This activity is believed to be at least in part due to inhibiting phosphorylation of AKT, FAK, and MAPK. SRC family kinases are activated in around a third of all NSCLC, and this is more prevalent in smokers. Bosutinib is effective against NSCLC cells *in vitro*.

Dasatinib inhibits SRC, BCR–ABL, c-KIT, and PDGFR and is effective against a variety of cancer cells *in vitro* and in model systems *in vivo*. It has been effective against breast cancer cell lines from "triple-negative" (do not express ER, PR, or HER2) tumors.

Saracatinib also inhibits SRC and SRC family kinases. It is also active against BCR–ABL and some mutant EGFR.

Various biomarkers may prove useful in identifying those patients most likely to benefit from SRC inhibition. These include the presence of LRRC19 and IGFBP2 in pancreatic cancer or increased CAV-1 in breast. Because of promising preclinical studies, SRC inhibitors are now being combined with various chemotherapy regimes and also other targeted agents. In a phase I–II study, dasatinib was combined to good effect with docetaxel in prostate cancer. Dasatinib is being examined in CRC used in combination with 5-FU, leucovorin, oxaliplatin, and cetuximab. Phase I studies showed the combination was well tolerated, and results showed some promise so that a dose escalation study is now underway. SRC inhibitors, such as saracatinib, are also being tested together with anti-VEGF therapies including the small-molecule VEGFR inhibitor, cediranib.

Molecular targets other than tyrosine kinases

These are listed in Table 16.3.

G protein-coupled receptors

Drugs that inhibit the GPCR-like (G protein-coupled receptor) Smoothened (Smo) are in phase I–III clinical trials. Smo activates Hedgehog (Hh) signaling, which you know from Chapter 5 is important in the growth control of numerous tissues. Smo is activated by members of the Hh ligand family, including Sonic Hedgehog (SHh), Desert Hedgehog (DHh), and Indian Hedgehog (IHh), which displace the inhibitory receptor Patched (Ptch). Hh signaling is deregulated in several cancers is can be blocked by

Smo antagonists such as LDE225, which are currently in clinical trials in basal cell carcinoma and for some other solid tumors in combination with other drugs.

Drugs targeting signaling molecules

The reader is directed to the extensive coverage of signal transduction in Chapter 5 and of the RAS–RAF–MAPK pathway in Chapter 6. Here we focus on the development of drugs to target these pathways in cancer therapies.

Signaling through RAS–RAF–ERK (MAPK) and PI3K–AKT

Contributors to most cancers, defects in the RAS–RAF–ERK signaling pathway appear particularly important in human melanoma and cancers of the colon, lung, pancreas, and thyroid. Mutant H-RAS, N-RAS, and K-RAS proteins are found with specific frequencies in different human cancers. RAS proteins have distinctive though overlapping functions and may have preferred oncogenic partners for promotion of tumor progression in different tissues. The RAS–RAF–ERK signaling cascade is subject to several activating mutations, operating at different levels in cancer cells; e.g. K-RAS is mutated in approximately 80% of pancreatic cancers; BRAF is mutated in several different cancer types, including approximately 65% of sporadic melanomas; and inactivating mutations are observed in the *PTEN* and *NF1* tumor suppressor genes, which normally restrain the pathway. RAS-regulated pathways are highly promiscuous and interact with multiple other pathways, including the PI3K–AKT–mTOR, c-MYC, and cell-cycle transit, among others. The PI3K–AKT–mTOR pathway is also activated in many cancers and is the focus of research. Given the central role of RAS and downstream signaling pathways in cancer (Chapter 6) there has been extensive activity to develop targeted drugs that inhibit them at various levels, including farnesyl transferase, RAF, and the MAPK pathway. Targeting of RAS per se has been ineffective, but inhibitors of BRAF and mitogen-activated ERK kinases 1 and 2 (MEK1 and MEK2) show early promise.

In 1995 *in vivo* studies in mouse models showed that farnesyl transferase inhibitors (FTIs) that prevent RAS activation (Chapters 5 and 6) induced regression of H-Ras-induced mammary carcinomas. However, despite this promising potential, and with many FTIs in phase II and III testing, the FTI strategy has failed for the following reasons. Firstly, at least for K-RAS and N-RAS proteins, prenylation may also be mediated by other enzymes such as geranylgeranyl transferase (GGT), which by adding geranylgeranyl groups to RAS can also help localize it to the membrane for activation. The concurrent inhibition of both FTase and GGTaseI proved too toxic. Secondly, it was discovered that FTIs can also block the farnesylation of several additional proteins, such as RHOB, prelamins A and B, and the centromere proteins (CENP-E, CENP-F). Thus, the earlier success of FTIs preventing RAS-driven cell transformation was attributed to these other proteins and not RAS. Alternative strategies to block RAS function were needed. A second class of inhibitors of RAS membrane association is currently under investigation, comprising two small molecules with farnesyl lipid groups (salirasib and TLN-4601) that are thought to compete with RAS-bound isoprenoid groups for binding to membrane-associated docking proteins. Statin drugs, widely prescribed to lower cholesterol, also reduce prenylation of various proteins, including RAS, because they reduce the availability of geranyl and farnesyl groups that are produced by the same sterol biosynthesis pathway, inhibited by the statin. Many statins are in clinical trials being given alongside other anticancer agents.

The inability to develop a successful, directly RAS-targeted therapy to date may reflect several challenges which have yet to be overcome. The mode of administration of agents, their therapeutic index, and not least their limited ability to cause sustained suppression of RAS activity *in vivo* have all been barriers. New approaches aimed at inhibiting RAS functions *in vivo* are being investigated and include the use of antisense RNA-based technology to inhibit RAS expression, antiRAS ribozymes (hammerhead models), and antiRAS retroviral therapy. Information on candidate antiRAS function inhibitors can be obtained from the ClinicalTrials.gov website.

A significant effort has, therefore, gone into developing low-molecular-weight protein kinase inhibitors of downstream effectors of RAS, such as those targeting the RAF–MEK–ERK (MAPK) signaling pathway. Thus, inhibitors of MEK and later RAF (e.g. Sorafenib) are now in clinical evaluation. In a recent phase III randomized clinical trial, comparing the BRAF inhibitor vemurafenib with dacarbazine in treatment-naive metastatic melanomas carrying the common BRAF V600E mutation has shown a very impressive 74% reduction in the risk of either death or disease progression compared with dacarbazine. The response to treatments is remarkably quick, within 1–2 weeks, though most patients develop resistance within a few months. Studies combining B-RAF inhibitors with cytokines, such as interferon; immunomodulators, such as CTLA-4 antagonists; or angiogenesis inhibitors are in development.

Other RAF inhibitors, many of which are orally active, are in various stages of development, many of which are being tested in melanoma. A variety of MEK inhibitors are in clinical development, and some are in phase I trials. Tumors with mutations in *KRAS* and *B-RAF*, which are resistant to EGFR-targeted drugs, are especially sensitive to MEK inhibitors.

Given the overlap between RAS and the PI3K–AKT pathway as well as the pivotal role of mTOR as a downstream mediator of cellular growth in both pathways, it is not surprising that inhibitors of PI3K activity (e.g. wortmannin) and of mTOR (rapamycin and CCI-779) are showing promise and may synergize with inhibitors of the RAS–RAF–ERK pathway. Mutations in PI3K are frequently observed in cancers, including CRC. More generally PKC activation, through various means, enhances the survival of cancer cells, making this pathway a tempting drug target. Rapamycin is the best known inhibitor of mTOR–Raptor signaling (see Chapter 5), but newer analogues have been developed as potential anticancer agents. In fact, two mTOR inhibitors, rapamycin and Temsirolimus, have been approved by the FDA for metastatic renal cancers. Interestingly, a recent study has shown that loss of the VHL tumor suppressor could be used as a biomarker for tumors most likely to respond to rapamycin. Other inhibitors of the PI3K–AKT–mTOR pathway are in clinical trials, including selective inhibitors of PI3Kα, pan-PI3K inhibitors, and a range of PI3K–mTOR dual inhibitors.

Studies in cell lines have shown that cancer cells might develop resistance to PI3K–mTOR inhibitors by acquired mutation or upregulation of either MYC or eIF4E. In both cases, the mechanisms may involve increased 5' cap-dependent protein translation.

RAS signaling is also important in cell-cycle transit, and drugs directly affecting this are in development – which is discussed further in this chapter.

Inhibitors of Rho and Rho kinases

The Rho family of Ras-related small GTP-binding proteins occupies a key position as regulators of cell motility and not surprisingly has been well documented as contributors to cancer cell invasion and metastases. Inhibitors of the RhoC kinases ROCK-1 and ROCK-2 have been shown to inhibit cancer cell invasion in matrigel chamber assays *in vitro* and are undergoing testing at present. Better understanding of their function is required before they become therapeutic targets. Despite this, one of their downstream effector molecules, endothelin-1 (ET-1), has been successfully targeted in early clinical studies in bladder cancer. ET-1 has two known receptors, endothelin receptor A (ETA) and endothelin receptor B (ETB). Two ETA inhibitors, atrasentan and zibotentan, are currently in phase II and III trials.

Inhibitors of JAK-STAT pathways

Janus kinase (JAK) inhibitors selective for JAK1 and JAK2 are in phase III studies. Deregulated JAK signaling is frequently observed in myeloproliferative tumors that are Philadelphia chromosome negative.

Cell-cycle inhibitors

Cell-cycle progression is controlled by multiple proteins (Chapter 4). In particular, the G_1–S transition is controlled by cyclin-dependent kinases 4 and 6 (CDK4/6) which phosphorylate and inactivate the tumor suppressor Rb to release E2F transcription factors to ready the cell for the S phase. Growth factor mitogens activate the important MAPK and PI3K pathways that converge on D cyclins and activation of CDK4/6. The Rb pathway is inactivated in essentially all cancers contributing greatly to aberrant replication, but in many cases this is secondary to various activated oncogenic signaling pathways or more through inactivation of $p16^{INK4A}$ or amplification of D cyclins or CDK4/6. Theoretically, drugs that inhibit activity of cyclin–CDK complexes, or enhance activity of the inhibitory CKIs or various checkpoint proteins, could be used to prevent aberrant proliferation of a very broad range of cancer types and carry almost any combination of epimutations.

Arguably, most DNA-damaging traditional anticancer treatments will activate checkpoints that arrest or kill the cancer cell, but do so indirectly. Recently, drugs have been developed that may arrest cell division in cancer cells by specifically targeting cell-cycle proteins. The discovery and cloning of the CDK proteins needed for cell-cycle progression (Chapter 4) has encouraged the design of novel drugs targeting CDK activity. **Flavopiridol** (alvocidib) was the first CDK inhibitor in clinical trials. Flavopiridol (Fig. 16.7) is a semi-synthetic *N*-methylpiperidinyl chlorophenyl flavonoid derived from rohitukine, an alkaloid isolated from the indigenous Indian plant *Dysoxylum binectariferum*. Edward Sausville brought this agent to the attention of the NCI in 1992 by showing activity against human cancer cell lines *in vitro* as part of an empirical screening program. It was found to bind to the ATP-binding pocket of multiple CDKs acting as a competitive antagonist of CDK activation. It may also downregulate expression of cyclin D1 and D3, thereby blocking cell-cycle progression at the G_1–S and G_2–M boundaries and triggering apoptosis. However, first-generation compounds, such as the pan-CDK inhibitors flavopiridol, and **R-roscovitine**, have been disappointing in early-stage clinical trials, despite early promise in preclinical studies.

Flavopiridol (Alvocidib) is a general inhibitor of CDKs, but novel drugs are in development targeting specific CDKs. For example, the Pfizer compound PD332991 inhibits CDK4 and 6, and the Astra Zeneca compound ZK304709 is unique as it is a multitarget tumor growth inhibitor of CDKs 1, 2, 4, 7, and 9, VEGFR 1, 2, and 3, and platelet-derived growth factor beta (PDGF-Rβ). Therefore, it can induce cell-cycle arrest and inhibit angiogenesis. Selective inhibitors of CDK4/6 are undergoing trials in solid tumors and in lymphoma.

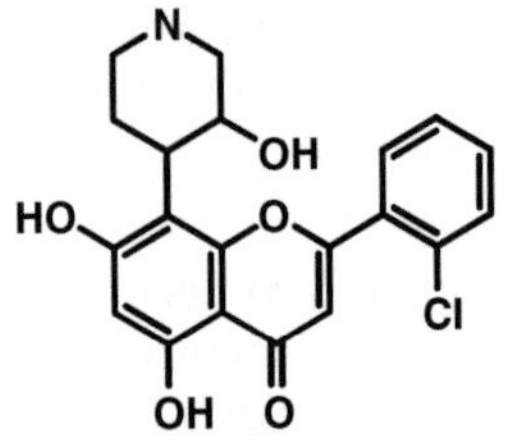

FLAVOPIRIDOL

Figure 16.7 Cyclin-dependent kinase (CDK) inhibitor.

DNA damage responses

Response to DNA damage involves a complex system of signaling pathways including checkpoints, DNA repair, and apoptosis. Given that most traditional chemotherapy induces DNA damage or interferes with DNA repair in some way, this important area is discussed in more detail in several other chapters (Chapters 4, 10, 15, and 18 in particular). Here we will confine ourselves to a brief outline of some drugs specifically targeting DDR proteins.

Cells can survive DNA damage only by the activation of relevant checkpoints that arrest the cell cycle until repairs can be completed. By definition, all cancer cells have circumvented the G_1 restriction point and may become very reliant on the activity of others checkpoints in S and G_2. As has been discussed in detail in Chapters 4 and 10, these checkpoints are p53 dependent and are activated by ATR–CHK1 and ATM–CHK2 signaling. In keeping with this, many cancer cells can be sensitized to DNA damage-induced apoptosis through inhibition of ATR or CHK1. Genetic inactivation of ATR and CHK1 abrogates cell-cycle arrest and enhances cytotoxicity following exposure to DNA-damaging agents. Small-molecule inhibitors of CHK1 are in preclinical studies and in early clinical trials data, where they prevent cell-cycle arrest in cancer cells exposed to DNA-damaging chemotherapy and promote cell death. Similar effects are seen with ATR inhibitors. ATR and CHK1 inhibitors are also synthetic lethal in some cancer cell lines with defective DNA repair pathways.

Inhibitors of various kinases, including aurora kinases and PLKs, have been discussed in other chapters, as has been the emergence of PARP inhibitors in breast cancer, so these will not be discussed again here.

Targeting of TP53

An obvious though challenging target for cancer therapy is to reactivate/restore the tumor suppressor p53. Meaningful restoration of missing tumor suppressors has so far proved beyond our capabilities, but we will discuss progress with p53 here. The protein product of the murine double minute (*MDM2*) oncogene binds with high affinity to the N-terminal transactivation domain of p53 to prevent p53-regulated gene expression and also reduces the stability of p53. Given that MDM2 is overexpressed in many cancers, inhibitors of MDM2–p53 interactions would be expected to increase p53 levels and activity and p53 stabilization is now the subject of a range of agents in development. A team at Roche (Nutley, NJ, US) identified a family of *cis*-imidazoline derivatives from a screen of a library of synthetic chemicals, termed Nutlins. Nutlins 1, 2, and 3 were all identified in the same screen. One of the enantiomers of Nutlin-3 (Fig. 16.8) was particularly active against cancer cells, inhibiting growth and triggering senescence. This targeting strategy has also identified a class of D-peptide antagonists of MDM2. These compounds, which rely on activating p53, are predicted to work best on tumors that harbor normal p53 (about 50% of all cancer patients).

Another approach, directed at the other 50% of cancers that have mutant or missing p53, is more challenging, as essentially p53 activity must be restored in most if not all tumor cells. If p53 is missing then either gene therapy will be required or drugs will need to target downstream or interacting proteins or miRNAs, such as Mir-34a. For mutant p53, small-molecule drugs that can interact with the altered protein and restore function by restoring a normal conformation have been identified. An example is the compound **PRIMA-1** (Fig. 16.9), found by screening a chemical library in cells carrying tetracycline-regulated mutant p53 (Saos-2-His-273).

There are numerous other tumor suppressors that could theoretically be targeted in cancer. However, the difficulty of restoring a missing gene/protein applies equally to these. The best hope lies with synthetic lethal interactions that exploit but do not directly target the missing or inactive protein. Other possibilities include targeting miRNAs involved in tumor suppressor pathways.

NSAIDS and COX-2 inhibitors

Prostaglandins and leukotrienes are important mediators of inflammation and may contribute to tumorigenesis and cancer progression in various ways, including in regulation of cancer cell growth and survival and angiogenesis. Evidence has long been accumulating for the role of nonsteroidal anti-inflammatory drugs (NSAIDs), in particular aspirin and the better-tolerated COX-2 inhibitors such as celecoxib, in the prevention of various cancers. These drugs can prevent polyp formation, a precursor to CRC, and have been reported to have potent antitumor activity in numerous studies. Recent trials have confirmed that prophylactic aspirin may reduce risk of future CRC. However, aspirin is known to increase risk of serious bleeding and COX-2 inhibitors, comparatively less prone to this complication, have been withdrawn from the market owing to an unexpected increased risk of cardiovascular disease (CVD). The search has been on for ways of using these drugs in ways that could minimize CVD risk. Currently, these agents are only used in patients with established cancer or with high-risk familial cancer syndromes, where the risk of CVD is deemed acceptable. Strategies to minimize CVD risk may be identified, for example, using COX-2 inhibitors topically in oral cancer prevention/treatments, or finding other targets in COX-2-controlled signaling that do not share the adverse risks.

Some encouragement is provided by the fact that aspirin is able to prevent both CRC and CVD. In this case use is limited by the increased risk of gastrointestinal upset and even serious gastrointestinal bleeding, although these can be minimised by assessing risk of this by clinical scoring and by use of drugs such as proton pump inhibitors (PPIs) that protect the stomach.

Epoxyeicosatrienoic acids (EETs) are also important lipid mediators, generated by cytochrome P450 epoxygenases. Like the related prostaglandins and leukotrienes, EETs also act as autocrine or paracrine factors regulating inflammation and the vascular system. EETs influence tumorigenesis and metastasis in several different rodent cancer models. Recent studies also suggest that endothelium-derived EET's may promote systemic metastasis in the relevant target organs, presumably in part by facilitating egress from the blood. In animal models, EET antagonists suppress tumor growth and metastasis, suggesting such drugs might have future promise in cancer treatment.

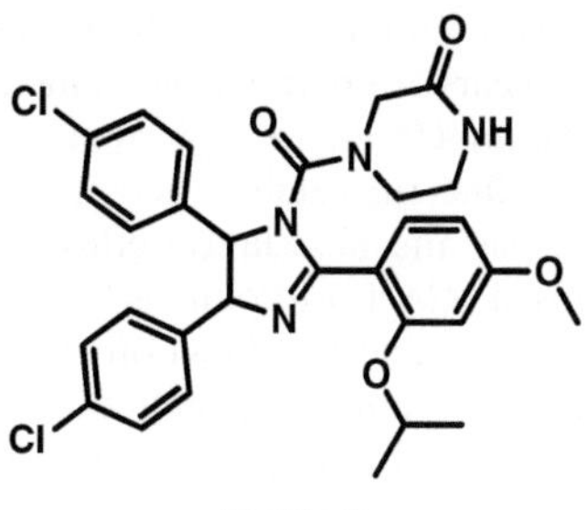

Transcription factors

Self-evidently, those transcription factors that promote cell cycle, growth, motility, and the other hallmark features of cancer should in theory prove ideal targets for anticancer drugs. However, this has proved very difficult to realize effectively in practice and almost all marketed targeted drugs in cancer are directed at signaling proteins. Several major transcription factor families are the subject of ongoing efforts and include NF-κB and AP-1, STAT and steroid receptors, as well as individual key factors, such as c-MYC. Drugs targeting peroxisome proliferator-activated receptors (PPARs), in particular agonists activating PPARγ such as rosiglitazone and pioglitazone, have been widely used clinically for metabolic diseases such as diabetes and have demonstrated benefits in cancer models. Rosiglitazone has been withdrawn from the market due to a greater than expected risk of CVD in those receiving the drug for diabetes, but some cancer studies using drugs of this class are still ongoing.

Another attractive approach is to target expression or stability of the oncogenic transcription factor. MYC proteins are involved in proliferative and apoptotic pathways vital for progression in cancer (Chapter 6). One of the strategies to target the expression of *Myc* turns out to be one of the first successful applications of antisense technology *in vitro* (see below and Chapter 15).

Figure 16.8 Nutlin-3: inhibitor of the MDM2–p53 interaction.

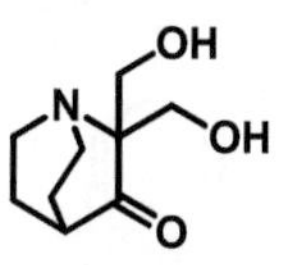

Figure 16.9 PRIMA-1: restores p53 activity in mutant p53-bearing tumor cells.

Another approach to regulating expression of multiple genes in a coordinated way is to target miRNAs and this is discussed later.

Survival pathways

It has been known for some time that tumors expressing high levels of anti-apoptotic BCL-2 family members (see Chapter 8) are resistant to cancer treatments. Conversely, downregulating anti-apoptotic genes can increase sensitivity of cancer cells to other drugs, such as chemotherapy and even TKIs such as lapatinib. Drugs targeting the anti-apoptotic proteins BCL-2, BCL-x_L, and MCL-1 are in clinical trials. A range of different techniques are being employed to target anti-apoptotic proteins, including inhibiting activity with small-molecule drugs and liposomal delivery of oligonucleotides, such as siRNAs or a mutant BIK gene, designated BikDD, to decrease expression.

Inhibitor apoptosis proteins (IAPs) prevent apoptosis by inhibiting caspases. During MOMP (Chapter 8), release of SMAC from the mitochondria releases bound IAPs from caspases allowing apoptosis to proceed. There are several ongoing phase I clinical trials assessing the use of a range of small-molecule drugs, many orally active, that are collectively referred to as "SMAC mimetics," due to a shared similarity of action to SMAC. SMAC mimetics are IAP antagonists that can selectively bind and block the interaction between type II BIR domains (Chapter 8) and caspases. They are currently being tested in the treatment of advanced solid tumors, including in triple-negative breast cancer, and in lymphomas.

SMAC mimetics have been shown to preferentially block cIAP1 and cIAP2 rather than XIAP *in vitro*. In fact, within minutes of administration cIAP1 and cIAP2 become auto-ubiquitylated and are degraded in the proteasome. In turn, IAP inhibition results in cell death through various processes, including spontaneous activation of NF-κB, and TNFα production and autocrine stimulation of the TNF receptor (TNFR1). In fact, in *in vitro* model systems TNFα may be essential for SMAC-mimetic-induced cell death. Thus, resistant cancer cell lines are not induced to produce TNFα and apoptosis can be restored by supplementation of TNFα. However, this resistance mechanism may have little bearing *in vivo*, where cancer cells are steeped in TNFα produced by the tumor microenvironment. SMAC mimetics are well tolerated *in vivo* and importantly do not appear to sensitize normal cells to TNFα-induced death. Other drugs targeting the IAPs (Chapter 8) such as survivin are also showing promise.

Targeting epigenetic regulation of gene expression

Many new drugs have been developed to alter the imprinting or silencing of tumor suppressor genes by altering chromatin modification or structure, and these are discussed in more detail in Chapter 11. Several different drugs targeting epigenetic factors, including those targeting HDAC and DNMTs, are in clinical trials. The class I zinc-dependent human histone deacetylases (HDACs) HDAC1 and HDAC2 are members of a large family that includes around 11 members. HDACs are frequently overexpressed in cancers, and currently more than 15 HDAC inhibitors have been tested. Inhibitors of these HDACs now in clinical trials show activity against several types of cancers. The HDAC inhibitor vorinostat (Zolinza) is the lead compound in class. It has been approved by the FDA for the treatment of refractory cutaneous T-cell lymphoma. The most frequent reported toxicities include fatigue, diarrhea, and nausea. There is some concern over cardiotoxicity which may prove to be a class effect of these compounds and limit their clinical application.

So-called pan-deacetylase inhibitors (pan-DACi) are in phase I–III clinical trials, having shown a number of potentially useful properties in cell culture models. Panobinostat is being tested against Hodgkin lymphoma, multiple myeloma (MM), and acute myeloid leukemia (AML).

Proteasome inhibitors

The proteasome is an essential regulator of cellular homeostasis (Chapter 11). It is responsible for the rapid and irreversible elimination of damaged and no-longer-required proteins that have been ear-marked for degradation. It plays a critical role in modulation (activation or repression) of multiple key signal transduction pathways, including the cell cycle (Chapters 4 and 11) and apoptosis. The most specific proteasome inhibitors developed to date share a pharmacophore that interacts with the active-site threonine in the proteasome and include peptide aldehydes, peptide boronates, peptide vinyl sulfones, peptide epoxyketones, and β-lactone inhibitors. A peptide moiety – found in all but the β-lactones – is required for binding to the substrate-recognition pocket in the proteasome.

Bortezomib, a peptide boronate derivative of tripeptide aldehydes, is a first-in-class proteasome inhibitor developed specifically for use as an antineoplastic agent in humans. Scientists at ProScript initially found that a tripeptidyl boronic acid was more than a 100 times as potent as the parent molecule tripeptidyl aldehyde. Specificity was improved by truncation of the tripeptide to a dipeptide. One of these dipeptides, bortezomib (Fig. 16.10), was shown not to compromise potency (K_i = 0.62 nM) and offered additional advantages in synthesis, stability, and specificity for the proteasome over thiol and serine proteases. In 1995 bortezomib and 18 analogs were submitted to the Development Therapeutic Program (DTP) for screening. Bortezomib showed high potency toward all cell lines in the DTP *in vitro* screen (average GI_{50} 18 nM), which correlated with proteasome inhibitory potency. Results from studies *in vivo* and further lead optimization studies resulted in bortezomib finally emerging as the best clinical candidate. The timeline of bortezomib development was less than 10 years from the development of proteasome inhibitors at ProScript (1994) to FDA approval (2003). Bortezomib reversibly inhibits the catalytic activity of 26S proteasome and induces apoptosis and has clinical efficacy in patients with refractory or relapsed MM. Bortezomib is believed to be effective in MM by interfering with normal function of the NF-kB signaling pathway, thus reducing tumor cell replication and angiogenesis and fostering apoptosis. The drug might interfere with proteins

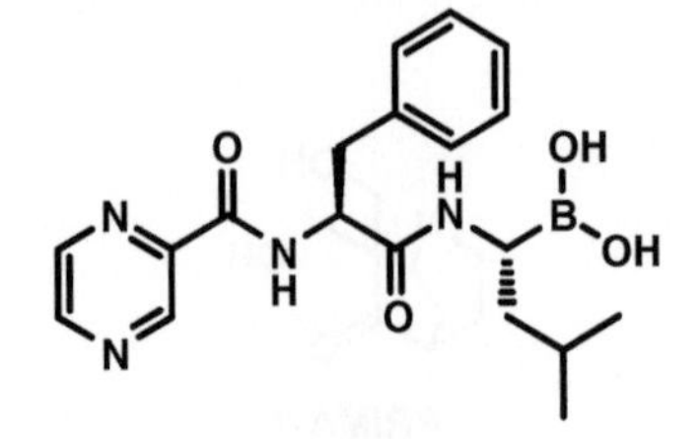

Figure 16.10 Bortezomib, a proteasome inhibitor.

that may be classed as non-oncogene-addicted. Recent trials have shown that bortezomib is also effective in patients with diffuse large B-cell lymphoma when added to CHOP chemotherapy. This effect is attributed to blocking activation of NF-kB.

Different molecules all prevent degradation of various proteins in the proteasome but show varying degrees of selectivity for those proteins, suggesting that further advances may enable the targeting of smaller numbers of desired predetermined proteins without impeding degradation of proteins where increased stability would be detrimental. Targeting the ubiquitination machinery also offers the potential of much more selective agents.

Targeting heat shock proteins

HSPs, such as HSP90, are molecular chaperones needed for optimal folding of numerous oncogenic client proteins, which they protect from proteasomal degradation, including estrogen receptor, HER2, NRAS, AKT, PDGFR, and B-RAF.

Cancers may be particularly reliant on HSPs because of oncogenic stresses. Inhibitors of HSP90 are active in many tumor models and are in clinical trials in hematologic malignancies, breast cancer, and multiple myeloma.

Hitting the extrinsic support network and preventing spread

Angiogenesis inhibitors

For cancers, the road to hell is paved with endothelial cells; cut off from this means of egress, this source and supplier of nourishment, and the cancer might be stopped in its tracks.

Discussed in detail in Chapters 14 and 18, angiogenesis is known to be pivotal in tumor progression and metastases and is now a major target for new drug development for cancer (Table 16.4). The VEGFs are critical and relatively specific regulators of angiogenesis and the growth factors and their RTKs have been the subject of several drugs, notably the mAb avastin, which is approved for treatment of CRC, though at the time this book was printed the indications for this drug were undergoing revision. Bevacizumab has also recently been shown to improve outcome in ovarian cancer.

It is worth noting that many of the nonspecific RTK inhibitors also inhibit receptors for VEGF or FLT3, and these multitarget drugs will have effects on angiogenesis as well as growth. Phase III trials have confirmed the efficacy of the RTK inhibitors, sunitinib and sorafenib, in the treatment of patients affected by GIST and renal cancer refractory to standard therapies, respectively. Other antiangiogenic drugs are being tested in clinical trials against a variety of cancers, including the notorious drug thalidomide, which is already approved for treatment of multiple myeloma. Relatively selective inhibitors of FGF receptors are in clinical trials and target angiogenesis as well as deregulated FGFR signaling in breast and bladder cancers. FGFR inhibitors may also be useful in targeting a known resistance pathway developing during anti-VEGFR therapies.

Despite progress, antiangiogenic therapies have not lived up to their original promise and in most cases have not improved survival for many cancer patients. In part, this may reflect the undesirable consequences of inducing hypoxia of the tumor. Antiangiogenic therapies may actually increase the risk of invasion and metastasis by increasing the CSC population within the primary tumor. In fact, studies suggest that intratumoral hypoxia activates hypoxia-inducible factor 1α, as well as a stem cell regulatory pathway involving Akt and β-catenin, which may in turn drive EMT and increase "stemness." Taken together, antiangiogenic agents might need to be combined with those targeting CSCs in order to fully realize the potential of these agents.

Targeting invasion and metastases

Given that over 90% of cancer deaths are due to metastases, it is often seen as surprising that so few treatments are specifically targeted against this hallmark of cancer cells. Given the numerous barriers that must be overcome for a cancer cell to colonize a distant site, it is maybe remarkable that metastases develop at all, and yet at least 90% of cancer deaths are caused by metastases. The challenges facing a would-be metastatic cancer cell are formidable and are illustrated by studies showing that only around 0.1 to 1.0% of cancer cells can give rise to metastases even if inoculated directly into a blood vessel.

Metastatic spread involves a number of recognisable steps that include ability to detach and invade through basement membranes; escape from the primary tumor by intravasation into blood vessels or lymphatic ducts; survival in the lymphovascular system by avoiding anoikis; evading of the immune system; soluble factors may prepare prior to the arrival of metastatic cells; attachment to microvessel endothelial cells; extravasation into the new tissue; establishing relationships with the new microenvironment, the metastatic niche; angiogenesis and expansion to form a new secondary tumor; and reseeding of the primary or new metastases.

Only a small subset of primary cancer cells may complete this adventure and may first have to undergo EMT and acquire stem cell characteristics. Many of these stages may be amenable to therapeutic targeting by metastasis-preventive agents, with few adverse effects on normal adult cells. Many metastasis-promoting genes have been identified, including several promoting EMT either by directly inhibiting expression of E-cadherin, such as Snail, ZEB, E47, and KLF8 factors, or by other means such as TWIST and Artemin or Goosecoid. Some agents interfering with metastasis-promoting proteins are in clinical trials. These include protease inhibitors, agents targeting integrins, inhibitors of cytokine signaling, and vascular stabilizing drugs. Small-molecule drugs interfering with metastasis signaling pathways are already extensively studied.

Metastasis suppressor genes are specifically involved in preventing metastases, distinct from preventing primary tumor expansion: Nm23-H1, P63, EZH2, metastasis suppressor 1 (MTSS1), BRMS-1 and LASS2/TMSG1, and SAM pointed domain containing ETS transcription factor (SPDEF), which downregulates MMPs, are the best known. Others include LKB1, RKIP, DAB2IP, SHARP1 (which degrades HIF1α and HIF2α), Nol7, and Elf5, but these may have other important roles. Restoring function of these may be a therapeutic possibility. Given that many of these are downregulated as a result of aberrant expression of miRNAs it may be possible to restore expression by using RNAi to block pathways involved in silencing of these genes (antagomirs are discussed below), though this is not the only approach. In the case of Nm23-H1, upregulation of expression by medroxyprogesterone acetate has been tested in a phase II trial. In preclinical studies gene therapy using nanoparticle delivery, cell-permeable NM23 protein, and inhibitors of lysophosphatidic acid receptor 1 (LPA1), all show some promise.

In animal models these agents share certain properties, in particular prevention of metastasis formation, but with little effect on shrinking primary or metastatic tumors already there. This raises a series of problems in conducting clinical trials, which are traditionally evaluated by RECIST criteria and unless a drug effects tumor shrinkage it is certain to fail during development. Also, many treatments directed at growth of the primary (or metastatic) tumor may actually increase the risk of spread. Thus radiotherapy may lead to emergence of small numbers of highly metastatic cells in the treated area: the so-called tumor bed effect. Similarly angiogenesis inhibitors may result in worsening tumor hypoxia, which may offset any benefits by increasing expression of metastasis-promoting factors such as HIF1α, fostering EMT and maybe further damaging already poor-quality blood vessels to allow intravasation by cancer cells. Other possibilities include the selective elimination of cancer cells, leaving the CTCs behind and in an environment now more likely to allow spread.

Migration across an endothelium is determined by adhesion molecules such as integrins and can be supported by a range of inflammatory mediators. Invasion is enabled by the secretion of matrix-degrading enzymes such as matrix metalloproteinases (MMPs). Several agents are being studied specifically as inhibitors of invasion and metastasis. Others may have effects on primary tumor cells as well as on spread, including antiangiogenic drugs, inhibitors of BRAF V600E, and EGFR inhibitors. Newer classes of vascular-stabilizing drugs may have effects mainly on spread.

The stroma is known to contribute substantially to tumor evolution. In part positive support from stromal cells can arise by mutations in stromal fibroblasts or more commonly in response to "recruitment" signals produced by cancer cells. These signals subvert their tissue microenvironment and may derive from soluble molecules or from extracellular vesicles, in particular exosomes, secreted by the tumor cells. Blocking exosome secretion can prevent invasion and spread of cancer cells. There is much interest therefore in what exosome constituents may be responsible for subverting stromal cells. Candidates include various enzymes, growth factors, and even miRNAs.

There are a number of enzymes involved in proteolysis of the extracellular matrix, which allows tumor cells to invade (Chapter 12). The family of MMPs, which include gelatinases, stromalysins, and collagenases, are the major group of enzymes involved. Their activity is ultimately terminated by tissue inhibitors of metalloproteinases (TIMPs). Synthetic MMP inhibitors have been developed. Of these, marimastat is most widely studied. Early studies showed these agents to have promising activity, but their potential has not been realised in randomized phase III controlled studies.

The urokinase receptor (u-PAR) is a critical molecule in migration, invasion, and metastasis. It is overexpressed in many solid tumors and is now a promising therapeutic target. NM23 is a histidine protein kinase that acts on various substrates including aldolase C and KSR, a scaffold protein in the RAF–MEK–ERK pathway. It may also play a role as a DNA exonuclease and is part of the set complex. Trials aiming to restore/increase NM23 with the steroid medroxyprogesterone acetate are underway. It is synthetic lethal with inhibitors of thymidine kinase, which may provide a more tractable target for drug design.

Bone disease is a major issue in advanced breast and prostate cancers, which often metastasize to bone, and in myeloma. Bone-targeting treatments are now being employed alongside more usual cancer therapies. Bisphosphonates which alter bone remodeling have been used in cancer for some time to control systemic complications of bone involvement, such as hypercalcaemia, and bone pain. The receptor activator of NF-κB ligand (RANKL), needed for osteoclast maturation, is a more recently evaluated target. RANKL is inhibited by the mAb danosumab, a strategy shown to have benefits beyond that of zoledronic acid (a bisphosphonate) in morbidity linked to bone metastases in prostate cancer. It is also of benefit in metastatic breast cancer, NSCLC, and myeloma.

In animal models ET-1 and its receptor have been shown to be required for lung metastasis in bladder cancer and possibly others. Moreover, ET-1 receptor inhibitors prevent development of new lung metastases, whereas they have no effect on established primary or metastatic tumors. These results support the notion that antimetastatic therapies will be particularly effective as adjuvant agents and not necessarily in treating established metastases. Clearly, this poses challenges for clinical trial design which largely tests new agents in advanced cancer.

Recently some exciting new concepts have begun to emerge and may in turn lead to new ways of either treating or most likely preventing metastases.

First, as discussed in previous chapters, the existence of genes known to be specifically involved in metastases, as exemplified by the metastasis suppressor genes, such as human nonmetastatic gene 23 (Nm23-H1). Traditionally, metastasis is regarded as a unidirectional process with cancer cells disseminating from the primary to seed metastasis in regional lymph nodes or distant sites. However, this view has been challenged by recent studies which suggest that a multidirectional process is more accurate and which includes the seeding of the primary tumor itself: "self-seeding." The possibility of interfering with the return of colonial cancer cells or those circulating cancer cells which have been in transit may offer unique targets.

Activation of the EMT allows cancer cells to invade and may support successful colonization of distant tissues. However, distant metastases more closely resemble the epithelial phenotype of the primary, suggesting the novel notion that disseminated cancer cells undergo a MET. Little is known about how this might be regulated but recent studies have identified versican, secreted by the microenvironment, as a potential regulator.

Integrin antagonists

A range of novel targets, not directly involved in TK signaling, are being studied. **Cilengitide** (Fig. 16.11) is a molecule designed and synthesized at the Technical University Munich in collaboration with Merck KGaA in Darmstadt. It is a cyclic arginine–glycine–aspartic acid (RGD) peptide that inhibits endothelial cell–cell interactions, endothelial cell–matrix interactions, and

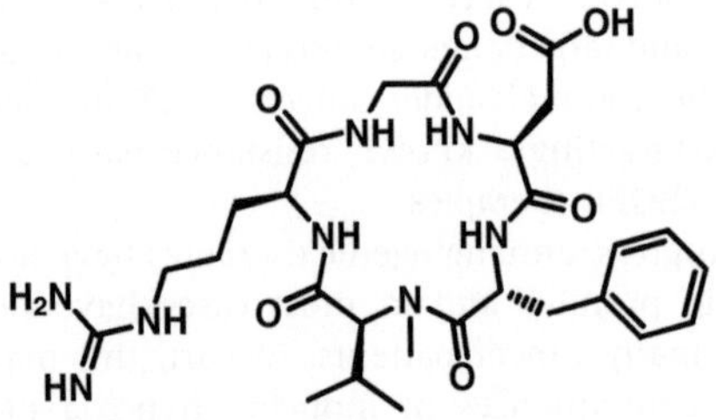

Figure 16.11 Cilengitide, an anti-tumor and antiangiogenic agent selective of both integrins αvβ3 and αvβ5.

angiogenesis. Cilengitide is both an anti-tumor and antiangiogenic agent, selective for both integrins αvβ3 and αvβ5. The involvement of integrins in angiogenesis is well described (Chapter 12), and recent studies have demonstrated that they also influence many other host cell responses to cancer. Therefore, integrin antagonists targeting the tumor microenvironment might significantly curtail tumor progression. Cilengitide is currently undergoing Phase III trials in patients with glioblastoma.

Stromal and immune targeting

The stroma plays a key role in supporting the tumor, and there are several recent studies in model systems confirming that mast cells, immune interactions with the cancer cell, and others are important in tumorigenesis and could therefore be therapeutic targets. Leukocytes and their secreted mediators contribute to most stages of malignant development and progression in some way. Various treatments, with a few side effects, designed to boost the immune response have been tested and include adjuvant vaccines, such as sipuleucel-T (Provenge), in prostate cancer and immunostimulatory chemicals in NSCLC. The survival benefits shown for sipuleucel-T when used with cabazitaxel in clinical trials have led to these being approved by the FDA. In a new approach, the cancer vaccine (PROSTVAC-VF) contains a virus genetically modified to contain prostate-specific antigen (PSA) that might stimulate the patient's immune system to recognize and destroy cancer cells containing PSA. Unfortunately, most immunotherapeutic strategies to date have failed to live up to their initial promise. At least in part, this may be due to the agents employed compromising or promoting the activities of several (including even antagonistic) subtypes of leukocytes, thereby diluting beneficial actions. It is hoped that advances and greater selectivity of new therapies will overcome these obstacles. Moreover, as immune-based mechanisms may influence the responses to conventional chemotherapy, further progress may follow testing of different combinations of cytotoxic alongside immune-based therapies. This area is described in greater detail in Chapters 13 and 18.

Gene therapy, antisense, and siRNA

Dysfunctional tumor suppressor genes are the most common genetic lesions identified in human cancers. Working versions of tumor suppressor genes can be restored by introducing them into cancer cells by gene transfer using adenoviral vectors. In fact, this approach has been well studied in cultured cells and in animal models, in which replication-defective adenoviruses expressing p53 have been directly inoculated. This treatment can induce growth arrest and apoptosis in cancer cells and has very low toxicity, but clinical studies are still awaited.

As has been discussed, miRNAs posttranscriptionally regulate expression of many genes involved in cancer, and various parts of the RNAi machinery and miRNAs themselves are altered in cancer. Almost all miRNAs act to silence gene expression by reducing the stability and translation of target mRNAs. Rare examples of miRNAs promoting gene expression have been identified in xenopus.

Much has already been mentioned about the role of miRNAs, such as miR-34 and Let7 in tumor suppressor pathways (see also Chapters 7 and 11), but some miRNAs can also act as oncogenes (oncomirs) to drive cellular transformation, tumor progression, and metastasis (see also Chapter 6). It is worth emphasizing that they are oncogenic by suppressing genes involved in restraining oncogene pathways and not, as far as we know, by activating oncogenes directly.

The small size of miRNAs limits the amount of sequence information available for specificity, which in part explains why miRNAs regulate large numbers of genes (via translation and stability of mRNA) and are often thought of as master regulators. Perfect pairing with target mRNAs is rare in mammals, whereas endogenous siRNAs are more common in plants. As discussed in Chapter 11, perfect pairing of siRNAs activates mRNA destruction via Argonaute-regulated endonucleolytic cleavage, whereas lesser degrees of complementarity may simply reduce translation of the mRNA at the ribosome. Homology is not the only factor of note, and it now seems that accessibility of AU-rich areas in the 3' UTR of mRNA, ribosomal shielding, and interactions within RISC may all contribute in defining the range of targets for any given miRNA.

A variety of treatments based on re-introducing missing tumor suppressor miRNAs are in development, including those based on miR-34, Let7, and miR-15–16, some of which are in clinical trials (Table 16.5). Moreover, as it is anticipated that human tumors will show addiction to oncomirs, a range of therapies aiming to silence them are also in development. Promising targets include miR-21, miR-155, and miR-17–92. In fact, oncomirs can themselves be silenced by RNAi, and miRNAs used to silence oncomirs are collectively referred to as antagomirs. If it needed re-iterating, it seems that no cancer-relevant process is unaffected by miRNAs. Recent studies have shown that miR-9 can decrease expression of MDR and transporter proteins which are involved in extruding chemotherapeutic agents from cancer cells. Using this miRNA alongside chemotherapy may thus overcome this resistance mechanism.

Another offshoot of RNAi research has been the development of novel therapeutics based on siRNA. As has been discussed in Chapter 11, RNAi activated by administration of small oligonucleotides that are very complementary to the mRNA to be silenced (siRNA) can confer great specificity. In fact, siRNAs can target individual mRNAs for destruction, as distinct from the effects of miRNAs which may inhibit translation of multiple mRNAs. Endogenous siRNAs may be comparatively rare, compared to miRNAs, in mammalian cells, but this does not preclude designing double-stranded siRNA guide molecules that can direct the cells' RNAi machinery to silence any given mRNA of interest. The challenge is delivering the siRNA to the desired tissues and cells.

Drugs based on siRNA and miRNA

A number of drugs based on siRNA and miRNA delivered within lipid vesicles (or other delivery systems) have entered clinical trials (Table 16.5). It is widely hoped that this approach will be successful, not least because effective and side-effect-free delivery systems for siRNAs essentially open up the possibility of targeting any overactive gene or protein without the difficult and time-consuming task of designing a small-molecule inhibitor. Rational and speedy drug development will be particularly useful in cancer and various orphan diseases for which drug design is traditionally complex. Once you know the gene sequence, you can have the "siRNA cargo" made within days and then it just needs to be delivered. At present the delivery issue has been reasonably well resolved *in vitro*, ushering in high-throughput screens in cell lines. However, *in vivo*, the challenges have proved substantial and have until very recently proved hard to resolve.

Table 16.5 Therapeutics based on siRNA

Target	Product	Indication	Company	Stage of development
Dual targets: kinesin spindle protein (KSP) and VEGF	ALN–VSP	Liver cancer	Alnylam	Phase I extension completed
PKN3 (a mediator of PKC signaling)	Atu027	Advanced solid tumors	Silence Therapeutics	Phase I completed
β-Catenin	CEQ508	Familial adenomatous polyposis (FAP)	Marina	Phases Ib–IIa initiated
M2 subunit of ribonucleotide reductase	CALAA-01	Solid tumors	Arrowhead	Phase Ib trial nearing completion
Replace dysfunctional miRNA34	Restore mir-34 tumor suppressor function	Various cancers	miRNA	In development
Replace dysfunctional let7	Restore let7 mir-34 function	Various cancers	miRNA	In development
ALN-RSV01	Nucleocapsid gene of RSV	Respiratory syncytial virus infection	Alnylam	Phase II completed
P53 (transient inhibition may prevent apoptosis due to reperfusion injury)	QPI-1002	Acute kidney injury at major cardiac surgery and delayed graft function after renal transplant	Quark	Phase II initiated
Transthyretin (TTR)	ALN-TTR02	TTR-mediated amyloidosis	Alnylam	Phases I and II initiated
Connective tissue growth factor	RXi-109	Prevention of scarring and keloid formation	RXI	Phase I initiated
PCSK9	ALN-PCS02	Hypercholesterolemia	Alnylam	Phase I trial initiated
Antithrombin III	ALN-APC	Hemophilia	Alnylam	Phase I
RTP801	PF-655	Diabetic macular edema and wet age-related macular edema	Quark	Phase IIb trial initiated
Caspase 2	QPI-1007	Non-arteritic ischemic optic neuropathy	Quark	Phase 1

However, this situation has changed rapidly over the last 4–5 years and there have been some notable recent successes by both Silence Therapeutics and Alnylam, which suggest that the field of siRNA in oncology may be about to deliver on promise.

Various delivery systems have been designed, many based on liposomes of different charge that can carry a variety of different siRNA (or other oligonucleotide) cargos. The present generations of these agents appear most suitable for delivering siRNAs to the blood vessel wall (they are injected into the bloodstream and do not readily pass into underlying tissues), but this area is advancing rapidly and soon it is anticipated that siRNAs can be delivered into liquid cancers (e.g. leukemias) and even ultimately into the cancer cells within solid tumors. Delivery to the liver and lung is showing promise, and several agents are in various stages of clinical trial testing (Table 16.5).

The challenges don't end with delivering the siRNA where it is needed. One must also ensure that the gene of interest is actually knocked down sufficiently, that this effect lasts, and finally that the cancer growth is inhibited. Phase I trial results with one such cargo, siRNA to PKN3, coupled to a very well-tolerated liposome delivery system shows real promise, as does a different approach combining two agents, against VEGF and a spindle protein.

An allied approach involving anti-sense RNA (larger constructs often comprising naked RNA are used to block translation) has recently shown very promising results in the inhibition of expression of the *Apo B* gene in liver. There are as yet no agents based on this mechanism in oncology.

Other novel approaches

Chemosaturation: Isolating the vascular supply to an organ such as the liver to allow greatly increased local concentrations of chemotherapy without spillover into the systemic circulation, thus reducing side effects, such as bone marrow suppression, elsewhere.

Biotherapy: Harnessing viruses or bacteria to attack cancers is another comparatively new way of treating cancers. Oncolytic viruses which specifically or preferentially reproduce and cause cell death in cancer cells have been used in early-stage clinical trials in metastatic CRC.

Bone-localizing alpha-emitting agents: radium-containing agents such as alpharadin have been shown to be effective in bone metastases.

Novel nonpharmacological treatments

New treatments are not simply based on advances in drug therapy. One exciting new treatment, known as high-intensity focused ultrasound (HIFU), employs highly focused ultrasonic beams to kill cancer cells by heating them up. Radiofrequency ablation (RFA) is already used as a palliative treatment for bone and liver metastases. Computed tomography (CT) or ultrasound is used to help direct a metal probe into the tumor where a high-frequency current can be passed to destroy part of the tumor. As will be discussed later in the book, we are now able to direct radiation more accurately by using new techniques such as conformal radiation therapy (CRT), intensity modulated radiation therapy (IMRT), and proton beam radiation, all of which reduce the exposure of normal tissues. Improved software and hardware are also supporting significant improvements in brachytherapy and extending use of radiation to intra-operative use.

Scheduling

Aside from picking the right drugs or combinations, it is also important to define the optimal duration of administration, interval between doses, and order of administration of given combinations. In many situations, this is still an area of some uncertainty and although advances in mathematical modeling and in *in silico* predictors of PK–PD interactions may help, ultimately all these possibilities need to be subjected to clinical trials. Sequential regimes, metronomic regimes, serial monotherapies, cavatina chemotherapy (simple brief regimes that are not repeated), or all together must all be considered.

Field cancerization

Field cancerization is a term that is increasingly employed, not as an affront to those with some lingering respect for the English language, but to describe the widespread genetic abnormalities found in a tissue that has been chronically exposed to a carcinogen. This provides an explanation for the appearance of multiple individual cancers or precancerous lesions within such affected tissues. A well-known example is the development of multiple nonmelanoma skin cancers and precancerous precursor conditions such as actinic keratoses in skin areas exposed to excessive sunlight. The often substantial areas involved require a different treatment approach, often referred to as field directed. Examples include photodynamic therapy (PDT).

Resistance to targeted therapies – intrinsic resistance and emergence of secondary pathways and tumor escape

Though you drive Nature out with a pitchfork, she will still find her way back.

Horace

As we have seen, molecular therapeutic approaches that target specific oncogenic signaling pathways are emerging as the new hope in cancer treatments and have already invigorated treatments of some cancers. However, the success of these approaches depends on the extent to which malignant cells are and subsequently remain dependent upon individual oncogenic mutations, and specific signaling pathways for maintenance of the transformed state, during targeted therapy. Some cancers are inherently unresponsive to certain treatments from the outset, and this may reflect diverse reasons including the nature of the mutations driving tumorigenesis in the given cancer, the presence of CSC, and various factors influencing drug or treatments responses including drug-resistant proteins, transporters, and DNA damage responses – the list is endless. In many cases, exemplified by the presence of mutant *KRAS* in CRC which predicts resistance to EGFR monoclonal antibodies, the reason for resistance becomes self-evident when detailed molecular analyses of a tumor are available. In this example, growth signaling is already cascading downstream of EGFR, rendering it irrelevant as a treatment target. However, it is increasingly apparent that in many cases the mechanisms remain occult – expect rapid progress in this area.

Another very important source of resistance is that acquired by cancer cells under the onslaught of chemoradiotherapy. Faced with **Stürm und Drang**, cancer cells evolve by mutation and natural selection to become resistant to those treatments being employed to treat them (Fig. 16.6). In this scenario, the very specific evolutionary pressure represented by a targeted therapy will bring forth equally specific mutations that restore aberrant growth via the same or a complementary pathway. Resistance is often observed when single agents are employed in cancer therapy. To overcome this, various strategies have been adopted, including switching to another different targeted agent or using combinations of targeted and traditional agents; it is evident that many parallels exist with antimicrobial therapy for diseases like TB.

Treatment resistance can arise when further mutations in DNA or epigenetic changes activate secondary tumorigenesis pathways that allow tumor cells to escape dependence on the initiating oncogene – this has been termed **molecular adaptation**. Molecular adaptation has been well described in treated cancer patients and following targeted oncogene inactivation in animal tumor models, and it can be very rapid – within days of initiating treatment. Put all these factors together and treating cancer appears a Sisyphian task; every time, some additional epimutations roll the boulder back down the hill again. A major goal in modern cancer research is to explore the nature of secondary pathways and identify new drug targets for use in rational combination therapies to avoid and overcome resistance (Fig. 16.12).

We will now describe some specific molecular adaptations that give rise to acquired emergence of drug resistance during treatment. Among this is the increasingly important and ubiquitous acquisition of resistance to TKIs. In fact, across the board, in order to realize the potential of targeted therapy in cancer, it is important to address the emergence of drug resistance in treated patients. We must contend with the emergence of not only drug-resistant mutations but also an increasingly malign cancer cell elite, as cancer treatments clear the way for the fittest cells by winnowing out the lesser cellular chaff.

This will most likely need to be addressed by combination therapy to reduce the risk of resistant cells emerging and by developing further agents which can still effectively inhibit growth of these resistant cells. Many of the epimutations that are engendered by targeted agents could also *a priori* confer primary resistance to the first use of a targeted drug.

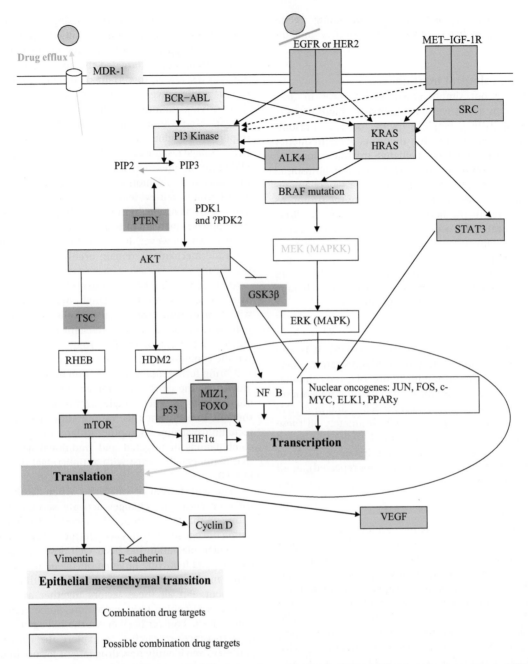

Figure 16.12 Combining targets to overcome resistance. The following can be combined with EGFR inhibition to overcome resistance: MET, ALK, IGF-1R, STAT3, SRC, or mTOR inhibitors (see light blue boxes), co-targeting more than one ERBB or HER receptor and inhibiting VEGF frequently overexpressed in carcinomas. Future potential targets are shown in light blue shading.

For simplicity, we will consider resistance conferring epimutations under four main headings:

1. Secondary resistance mutations or changes in expression in the drug target protein itself: These have been identified for BCR–ABL, KIT, and the EGF receptor (EGFR). In turn, this results in resistance to the drugs imatinib, gefitinib, and erlotinib, respectively. In some cases, resistance may arise due to amplification of the oncogenic protein kinase gene (increases in the target protein overcome the inhibitory effects of the agent) or by mutations in the catalytic active domain of the kinase that prevent binding or action of the drug. Certain drug-resistant mutants may still respond to alternative kinase inhibitors, thus the dual Src–Abl kinase inhibitor dasatinib (BMS-354825) and the imatinib derivative AMN107 are both effective against imatinib-resistant Abl kinase. It is not just kinases that are subject to these processes. A recent study has shown that truncating mutations in HDAC2 were present in some sporadic cancers with microsatellite instability (Chapter 10) and in some hereditary nonpolyposis CRCs, and could confer resistance to HDAC inhibitor therapies.

2. Activation of the targeted signaling pathway by mutations downstream of the treatment target: This is well documented in treatments directed at the EGFR. Activating mutations in BRAF or KRAS may be present from the outset or can be

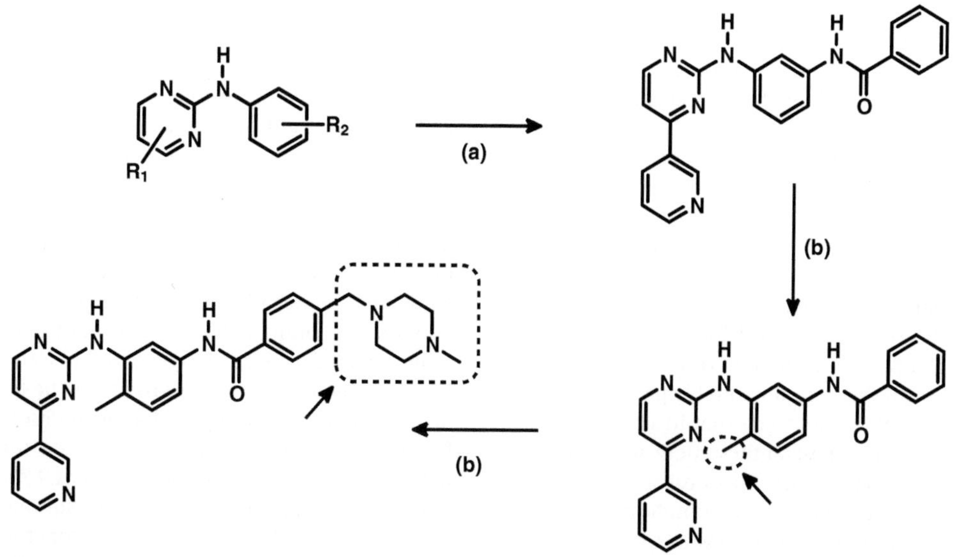

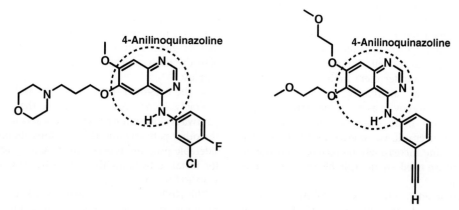

Figure 16.13 Optimization of the 2-phenylaminopyrimidine series. (a) Initial derivatization of the 2-phenylaminopyrimidine chemical class resulted in an ATP-competitive inhibitor of serine–threonine protein kinases. (b) Methyl substitution of the anilino-phenyl ring at the 6-position (arrow) led to potent inhibition of the ABL and PDGFR kinases, but did not inhibit the PKC family. (c) The 2-phenylaminopyrimidines class was finally optimized for ADME properties by the introduction of the N-methyl piperazine group.

Figure 16.14 EGFR tyrosine kinase inhibitors gefitinib (Iressa) and erlotinib (Tarceva).

acquired during treatments in CRC, rendering mAb such as cetuximab ineffective. Recent studies suggest that resistance to cetuximab, due to acquired KRAS mutations, might be tackled by employing simultaneous or sequential therapy with an inhibitor of MEK, which lies downstream of RAS signaling – an example of molecular chess!

3. Switching of oncogene dependence and evolution of new resistance pathways: This may develop and render the cancer cells independent of the oncogene being targeted and therefore also of treatments aimed at targeting that oncogene. Thus, following therapeutic targeting of an individual oncogene, epimutations affecting other genes may be acquired, and selected for, that allow the tumor to escape regression or recur. A well-known example is that of acquired activating mutations in *MET* that may bypass inhibition of EGFR in NSCLC and confer resistance to erlotinib. Other examples include the activation of ERBB2 signaling or amplification of *CRKL*, which activates MAPK signaling via both RAS and RAP1. The possibility of avoiding resistance by using multiple agents from the outset is being explored. One well-studied example is the use of combined inhibition of the RAS–RAF–ERK and PI3K–AKT–mTOR pathways; these show considerable overlap, and it has frequently been observed in animal models that targeting one pathway alone results in adaptation of the tumor and signaling increases through the uninhibited pathway, thus rendering treatment ineffectual. In these models, targeting both pathways is much more effective.

4. General resistance conferring epimutations: These can result in resistance to multiple agents. Thus, mutations in cell death pathways can allow cells to survive DNA damage; ironically, in these cases DNA-damaging treatments may actually foster evolution of more malignant clones of surviving cancer cells. Aside from cancer cells adapting and resisting the biological effects (pharmacodynamics) of the agent, changes that alter the metabolism, transport, and distribution (pharmacokinetics) of the agent are also described and exemplified by multidrug-resistant (MDR) transporter protein and cytochrome p450 family members. In these cases, cancer cells may resist agents by promoting efflux from the cell or altering important aspects of drug metabolism respectively. It must be emphasized that not all treatment resistance is acquired through mutations and natural selection – some patients are born that way. In fact, it is often predicted that future tests based on determining the genotype of an individual patient (in addition to genotyping his or her tumor to find somatic mutations) may help define the choice and dose of treatments – pharmacogenomics.

So we can conclude that under the exigencies of targeted drugs, the story of cancer therapy is often one of fluctuating success and failure as pharmacological ingenuity and acumen are pitted against the adaptive capabilities of the tumor and its allies in a game of molecular chess.

Lastly, it is now clear that targeted cancer drugs face all of the problems that have been well described for traditional chemotherapeutic agents. Thus, as noted, not only may they stop working or not work in all theoretically suitable patients, but also they may cause unacceptable side effects that in some cases may even prove fatal. Although in general these new agents are perceived to be safer (and in many cases, they probably are), recent studies have begun to identify very serious complications relating to the undesirable effects of drugs on noncancer cells. Imatinib, one of the longest established targeted agents, has recently been shown to cause toxic effects on heart muscle cells, enough to precipitate heart failure in some patients. This is important because it extends the potential problems of targeted agents from the expected low incidence of idiosyncratic allergy-type reactions to the unanticipated biological actions of the agent adversely effecting normal cell physiology – **off-target toxicity**.

Despite these formidable challenges, we have good reason for optimism. Adaptation to targeted therapies is not always a resounding success for the cancer cell. The addictive nature of cancer cells will continue to render them highly dependent on given oncogenic lesions, as alternative growth-promoting pathways are suppressed. Moreover, as we understand growth-signaling pathways in ever more detail, so the next moves available to the treated cancer cell, when faced by any particular targeted agent, will become increasingly predictable and therefore easily thwarted; all we need are the suitable drug combinations and its checkmate.

So, as noted, oncogene addiction offers up the Achilles' heel of cancer to our new targeted agents, but interestingly in some cases cancer cells have another, unexpected gambit up their collective sleeve. Thus, the negative feedback loop that appears to often accompany the growth-promoting activity of a deregulated oncogene may come to dominate – an unexpected ally in the midst of the tumor that we would wish to support. However, it appears that at least in some cases, our targeted agents are not adept at distinguishing friend from foe and abort both the growth-promoting activity and the beneficial negative feedback loop at the same time. In fact, depending on the balance between these oncogene activities at the time of treatment, as well as the relative proclivities of the targeted drugs themselves, could paradoxically make the tumor more aggressive as the negative feedback is lifted disproportionately. This will be discussed in more detail in the "Negative feedback loops" section.

Negative feedback loops and failure of targeted therapies

'Tis not always in a physician's power to cure the sick; at times the disease is stronger than trained art.

Ovid

Prolonged stimulation of a growth factor pathway can result in activation of time-delayed negative feedback loops that seek to limit the inappropriate or excessive activity. In fact, such feedback loops are likely a normal feature of all signaling networks and operate at different levels (intracellular signal transduction, receptor desensitization, and counterbalancing and antagonistic secreted factors). These processes can prevent deregulated growth but may also complicate therapies targeting the original causative mutant proteins, which gave rise to the excessive stimulation in the first place. That is, because by the time treatments are employed the negative feedback loop may be dominant over the growth-promoting activity and as both may be inhibited by the same targeted anticancer agent, the net effect of the agent may be to accelerate the tumor's growth as the feedback loop is inhibited. Such negative feedback has been observed for RAS and also for prolonged PI3K activation. In some tumors where PI3K is persistently activated, low activation of AKT has been described. In this case it appears that an mTOR–S6K mediated phosphorylation of IRS1 and IRS2 proteins (adaptors for some RTKs, such as the insulin receptor) at least in part reduces AKT activation. Importantly, it has been demonstrated that prolonged treatment with rapamycin can result in a "compensatory" increase in PI3K–AKT activation. This problem may be circumvented by using a combination of rapamycin with wortmannin (inhibits PI3K) as was shown by Scott Lowe and colleagues in rodent lymphomas expressing deregulated MYC and AKT.

Thus, a more detailed knowledge of the molecular "road map" for any given cancer can prove critically important in order to plan the most appropriate targeted agents and combinations to employ.

As the RAS–RAF–ERK and PI3K–AKT pathways overlap considerably, it may not be surprising that inhibiting one of these pathways alone often results in tumors increasing activity through the other pathway and thus resisting therapy. There is thus considerable promise in dual targeting both pathways concurrently, for instance by inhibiting the EGF, RTK, and PI3K (in cancers with PTEN inactivated).

The Nrf2 transcription factor is a key part of normal cellular defenses and is ubiquitously expressed in normal tissues, but at low levels due to Keap1-dependent ubiquitination and proteasomal degradation. Nrf2 is, however, expressed at high level in many cancers and is known to contribute to resistance to cancer drugs. A newly identified potential adjuvant drug, Brusatol, enhances proteasomal degradation of Nrf2 and can sensitize many cancer cell lines and xenografts to cisplatin and other chemotherapeutic drugs in part by reducing drug detoxification and drug removal. In the future, this may provide a novel adjuvant to traditional chemotherapy.

Defining the targets in resistance pathways

There will be little rubs and disappointments everywhere, and we are all apt to expect too much; but then, if one scheme of happiness fails, human nature turns to another; if the first calculation is wrong, we make a second better: we find comfort somewhere.

Jane Austen, Mansfield Park

So, as we have learned, many of those acquired mutations that enable TKI-treated cancers to escape from growth inhibition are being described. Such knowledge can then be exploited by combining agents or using them sequentially in order to overcome this problem (Fig. 16.12). However, if we are to play molecular chess with cancer cells and win, then we need to anticipate the ensuing mutational moves that may thwart our initial therapeutic gambit. We are progressing toward this goal and are aware of the options open to the therapeutically besieged cancer cell. Thus, in response to a targeted inhibitor, the cancer cell may evolve a drug-insensitive version of that protein, bypass the protein, switch to an alternative growth-promoting strategy, or find a way to inactivate or avoid the drug. Knowledge is power, and in theory, combined or sequential use of drugs that forestall these resistance mechanisms could be highly effective.

We will discuss some specific examples. CML patients treated with imatinib may become resistant by developing further mutations in the target gene (*BCR–ABL1* kinase domain). In this case, adding further different TKIs, such as nilotinib and dasatinib, may be effective and may also be tried in those resistant to the initial TKI from the outset. However, even this approach sometimes fails owing to the acquisition of highly resistant mutations. The epidermal growth factor receptor (EGFR) kinase inhibitors (EGFR–TKIs), such as erlotinib or gefitinib, have shown significant initial benefits in NSCLC patients, up to 40% of whom have somatic activating mutations in *EGFR* tyrosine kinase, but these tumors invariably develop drug resistance. Some patients are resistant from the outset, in part due to the presence of other somatic mutations, as exemplified in CRCs resistant to EGFR-targeted therapies that have downstream activating mutations in *KRAS*, and of those who do respond initially, most go on to develop resistance by 12 months of treatment.

To date, two main potential means of acquired TKI resistance have been identified: secondary resistance mutations and switching of oncogene dependence. The secondary T790M mutation is found in 50% of EGFR mutation positive patients who acquire TKI resistance. In fact, as some studies have found this mutation prior to TKI treatment, it has been suggested that a subclone of cancer cells carrying this mutation is selected for during subsequent TKI therapy. Mutations in genes encoding signaling proteins downstream of EGFR, such as RAS and RAF, may also be acquired during EGFR-targeted treatments. Focal amplification of the *MET* proto-oncogene occurs in 20% of cases but sometimes coexists with T790M. *MET* causes gefitinib resistance by driving ERBB3 (HER3)-dependent activation of PI3K. Inhibition of *MET* signaling in cells from these resistant cancers restores sensitivity to gefitinib. Increased *ERBB2* signaling has also been shown to cause resistance during *EGFR*-targeted therapies. Recent studies confirm that phenotype changes in cancer cells with TKI resistance have also been shown; thus, for example, resistant NSCLC are more likely to have features of EMT. A further mechanism of TKI resistance in NSCLC has now been identified. Environmental factors such as inflammation can also give rise to resistance. Thus, inflammation and TGF-β-dependent IL-6 secretion in particular can encourage emergence of a subpopulation of NSCLC cells which show features of EMT and are not addicted to EGFR signaling or resistant to erlotinib.

A recent study suggests that in some cases of breast cancer acquiring resistance to trastuzumab, this may be due to activation of SRC, suggesting that combined targeting of EGFR and SRC might be a way of overcoming this, though this waits to be tested.

Receptors for the scatter factors HGF and MSP that are encoded by the *MET* and *RON* oncogenes contribute to tumorigenesis and, as do many other receptors, exhibit crosstalk. Interestingly, *MET* amplification results not only in addiction to MET signaling (sustained MET activity is required for survival and proliferation) but also to constitutive activation of RON kinase. This crosstalk is not observed for other kinases such as EGFR or HER2. Resistance to MET inhibition can result from mutations in the MET activation loop (Y1230) or by activation of an alternative growth pathway involving increased expression of TGF-α. In both cases, activation of downstream PI3K–AKT and MEK–ERK pathways occurs despite the presence of the MET inhibitor. The latter could be overcome by combined EGFR and MET inhibition or potentially also inhibition of RON.

The RAF-selective inhibitor, PLX4032, demonstrated impressive early tumor response rates in patients with B-RAF (V600E)-positive melanomas. However, acquired drug resistance invariably supervenes, through either upregulation of PDGFRβ or reactivation of MAPK signaling by downstream mutations in *NRAS* or *MAP3K8*.

With notable exceptions, transcription factors have not in general proved good treatment targets. In cancer, the exemplar par excellence is the estrogen receptor-α (ER), which both drives breast cancer progression and is the archetypal molecular target in cancer therapy. In an interesting recent study, the application of CHIP and CHIP Seq technologies (Chapter 20) has revealed that breast cancer cells exhibit a degree of plasticity with respect to the repertoire of ER-regulated genes actually bound and expressed. Moreover, the pattern of expression correlated with outcome and also with drug resistance. Differential gene expression is hardly surprising given the plethora of potential factors acting to refine it (Chapter 11). What is interesting in this case is that binding appears to be strongly influenced by a single other factor, FOXA1. It is likely that many other important examples of heterogeneity in transcription factor networks and variant interactions with other regulatory elements in transactivating complexes may be important in progression and treatment responsiveness of cancer.

It is again important to note that specific molecular signaling does not necessarily imply that there will be simple solutions to resistance. As mentioned, various resistance mechanisms exist and may coexist with specific molecular changes as exemplified by activation of MET after treatments with EGFR TKI. It is worth noting that at least in mouse models, genetic instability and aneuploidy are known accelerators of tumor relapse after successful treatment.

Biomarkers to identify optimal treatments and tailored therapies

[B]ut it is not enough to wield a broadsword, one must also know against whom.

Nietzsche

The ability to inventory the molecular aberrations within tumors in almost impossible detail has led many to the conclusion that the writing is on the wall for the cancer cell; we simply need to get Belshazzar to turn around, and the means of the cancer cell's doom might be revealed.

To facilitate the clinical management of cancer, informed dialogue between clinicians, and consistent reproducibility across clinical trials, we have recourse to an ever-increasing vocabulary of malignant symbols of which molecular biomarkers and tumor node metastasis (TNM) staging are the most instantly recognizable. In fact, a number of molecular biomarkers are now in

Table 16.6 Biomarkers for screening and treatment selection

Biomarker	Cancer	Source	Clinical application
PSA	Prostate	Serum	Screening
			Disease monitoring
CEA	Colorectal	Serum	Disease monitoring
Pap smear	Cervix	Cervical smear	Screening
Fecal occult blood	Colorectal	Stool	Screening
			Disease monitoring
α-Fetoprotein	Testicular (nonseminoma)	Serum	Disease staging
Human chorionic gonadotrophin beta (β-HCG)	Testicular	Serum	Disease staging
Mammography	Breast	X-ray	Screening
Estrogen and progesterone receptors	Breast	Breast tumor	Treatment selection
HER2	Breast	Breast tumor	Treatment selection
HER2	Breast	Serum	Disease monitoring
CA125	Ovary	Serum	Disease monitoring
CA19-9	Pancreas	Serum	Disease monitoring
EGFR	Colorectal	Colon tumor	Treatment selection
KRAS	Colorectal	Colon tumor	Treatment selection
KIT	GIST	GI tumor	Treatment selection
NMp22	Bladder	Urine	Screening
			Disease monitoring
Cytokeratins	Breast	Breast tumor	Prognosis
Chromosomes (see Table 17.3)	Various leukemias	Blood cells	Diagnosis
			Treatment selection
CA15-3	Breast	serum	Disease monitoring
CA27-29	Breast	serum	Disease monitoring
Fibrin or FDP	Bladder	Urine	Disease monitoring
Bladder tumor-associated antigen	Bladder	Urine	Disease monitoring
Thyroglobulin	Thyroid	Serum	Disease monitoring
Calcitonin	Medullary carcinoma of thyroid	Serum	Screening
			Disease monitoring
Lactate dehydrogenase	Lymphomas	Serum	Disease monitoring
			Prognosis

widespread clinical use, and some are routinely measured (Table 16.6 and Fig. 16.15).

However, few current biomarkers have transcended their symbolic status, and most tell us little about the biology of the tumor even while they reveal much about the outcome for the patient. This is exemplified by the archetypal carcinoembryonic antigen (CEA) and PSA that are valuable contributors to outcome prediction and disease monitoring, but to date have little-known roles in the tumorigenic process. It is hoped that in future, biomarkers may be both disease relevant as well as predictive, but despite enormous progress in this area with tissue-based biomarkers, the greatest asset of CEA and PSA remains their ready visibility in peripheral blood.

Personalized medicine in the care of cancer patients is now an achievable goal. For some time, antibody-based biomarkers have assisted diagnosis and treatment monitoring as exemplified by PSA and CEA in prostate cancer and CRC, respectively. Although both techniques have significant drawbacks, they likely represent the best one could hope to achieve by the measurement of any single predictive biomarker. Treatment decisions in breast and colorectal cancer are already informed by detecting mutations in genes encoding *HER2* and ER or *KRAS* and *BRAF* within the tumor. These valuable biomarkers were identified and validated because of the publication of the human reference genome and continuing developments in high-throughput techniques for sequencing and transcriptional profiling. Similar successes in the application of mass-spectrometry-based proteomics are yet to be realized, but may follow new developments such as imaging mass spectrometry and peptidomics. However, these techniques share certain common features which limit potential: they are generally destructive, meaning that molecular phenotype cannot be mapped onto cellular or anatomical structure, or have relatively low resolution.

Traditional biomarkers are not related to individual cells or even to particular regions or structures within the tumor, but are derived from combined secreted output or by "mashing up" pieces of tumor. This has limited biological understanding of cancer behavior and has reduced discriminatory power because protein and gene expression and mutation status is averaged across many millions of often very different cells. The heterogeneity of most tumors has been elegantly demonstrated by recent studies of several cancer types that show not only the obvious expected differences between cancer cells and stroma, but also the often remarkable diversity of the cancer cells themselves. Thus, tumor residents comprise putative cancer stem cells as well as an unexpectedly large proportion of cells representing earlier "nondominant" clones that were there at successive points in the genealogy of that given cancer. In this "fossil record," the "fossils" are living cells that can potentially be identified by the use of appropriate biomarkers; unraveling the life history of the cancer cells in this way would be a powerful means of finding cancer-critical mutations and molecules for therapeutic targeting. More-

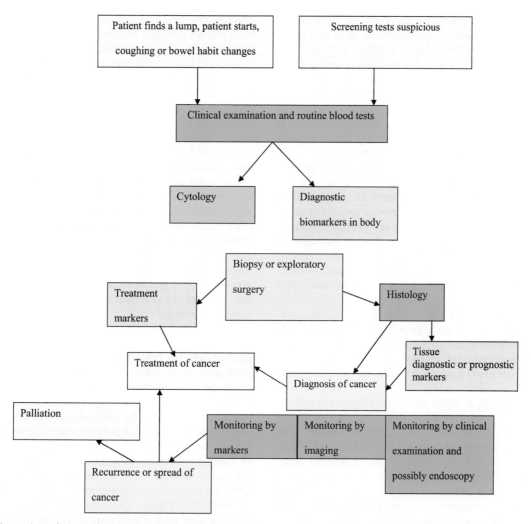

Figure 16.15 Diagnostic predictive pathway incorporating molecular biomarkers.

over, localizing molecular phenotype to single cells (or organelles) is a great step forward to describing functional molecular networks and identifying cell–cell interactions between cancer cells and their microenvironment.

Current clinical decision making is informed by clinical features supported by radiological assessment. If these reveal widespread metastatic disease, then subsequent treatment may be restricted entirely to chemotherapy. But which drugs would best suit which patient? This decision will at least in part be dictated by the presence or absence of specific mutations and the activation of dominant growth factor signaling pathways in the cancer (the so-called Achilles' heel of cancers, and subsequently the acquired resistance to Paris' arrow as cancer cells mutate around the therapeutic obstacle). In the presence of an apparently resectable primary tumor, it would be of enormous value to know what the likely behavior of that cancer is going to be, as it may range from clinically irrelevant to inevitably and rapidly metastatic or fatal, even with surgery. Moreover, should adjuvant chemotherapy be given and when – and if the answer is yes, which drugs, how often, and in what combinations? It is hoped that over the coming years, biomarkers will assist in decisions about most if not all of these parameters. However, what sort of analytical tools and biomarkers are most likely to cut the mustard?

There are a large number of potential biomarkers in various stages of development, mostly at the preclinical level. Many have shown promise in small numbers of patients or in the laboratory. The possibilities are limitless and, as will have been appreciated while reading the book, involve essentially all of the processes and molecular contributors to tumorigenesis. Measurements of any of these could potentially assist diagnosis, prognosis, and treatment selection, though ultimately very few will make it to the clinic. Thus, we can seek out germline mutations that correspond to potential drug targets or that predict how the body will respond to treatments; we can find somatic mutations and the aberrant signaling cascades that result from them. We can examine the immune status and vascularization of a tumor, and so on. The levels at which we measure these effects range from DNA, through RNAs of various types, to proteins. In most cases, we can measure these molecular changes in the patient, the tumor, or cells derived from the tumor.

Although to date relatively few biomarkers are used to guide therapies, some have already radically altered management of breast cancer and certain hematological malignancies, and are starting to affect CRC and NSCLC as well. Some of this has already been discussed in Chapter 3 and elsewhere, so it should now be apparent that detecting the presence of the Philadelphia chromosome, and thereby the *BCR–ABL* oncogene, leads inexorably to treatment with imatinib; mutations in *HER2* and *KRAS* are already used to dictate drug choices in breast and colorectal cancer, respectively. In breast cancer the presence of estrogen and progesterone receptors is also used to select specific treatments, including tamoxifen (discussed later and in Chapter 18). In testimony to the importance of these tissue-based biomarkers, a subtype of breast cancer (usually invasive ductal carcinoma) is now recognized and described by the absence of the aforementioned, routinely screened-for tissue-based biomarkers. Thus breast cancers which do not express estrogen receptors, progesterone receptors, or HER2 protein on the cell surface are referred to as triple-negative breast cancers, and these are (as one might expect) known to be unresponsive to hormone therapy or drugs targeting HER2. They may, however, respond to PARP inhibitors (Chapter 10). We will now discuss some further examples of biomarkers used to guide treatment selection (a more specific discussion of interactions between genotype and drug metabolism follows in the "Pharmacogenetics and pharmacogenomics" section).

Prospective studies confirm that women with germline mutations in *BRCA1* or *BRCA2* who undergo preventative risk-reducing mastectomy reduce their risk of future breast cancer and all-cause mortality. Risk-reducing salpingo-oophorectomy is associated with a lower risk of ovarian cancer and ovarian cancer–specific mortality.

A recent commercially available gene-based biomarker tool has been developed. The Oncotype DX Breast Cancer Assay screens for a number of genes simultaneously in tumor samples using RT-PCR. Trials investigating the potential of this test to predict recurrence and response to chemotherapy in ER-positive breast cancer have been promising. A similar approach is being tested for CRC also. The OncotypeDx approach for breast and colorectal cancer prognosis and treatment selection (kit-based commercial multiplex RT-PCR-based assays for gene expression of mutant and also cellular genes) will likely be reproduced for other cancers and for more genes. Briefly, such kit assays allow the simultaneous characterization of multiple mutations and expressed genes or proteins in cancer tissue samples. They will be particularly useful where several molecules need to be considered in choosing treatments, defining prognosis, and consider-ing the possibility of family screening.

Most of the current focus on cancer biomarkers has been on genes and, to a lesser extent, on RNA transcripts, but theoretically at least the proteins resulting from gene mutations would be ideal biomarkers, because they are ultimately the causes of tumorigenesis. This approach has, however, lagged behind because the sensitivity and specificity of equivalent techniques for proteins compared to those used to look at gene expression and mutations (PCR, microarrays, and sequencing) have not been good enough. Recent progress in immunoassays and in mass spectrometry is changing this, and recent studies are beginning to show promise in determining the presence of peptides resulting from normal and mutant alleles. Using this approach, it was recently shown that cancer cell lines contain around 1.3 million molecules of Ras protein per cell, with a ratio of mutant to normal Ras from 0.49 to 5.6.

The future will likely see an increase in the range of not just specific targets that may be screened for but also, one suspects, an appreciation of the combinatorial effects of maybe large numbers of low-frequency mutations, epigenetic changes, miRNAs, the presence and quantity of circulating cancer cells – well, you get the idea. We describe a selected number of illustrative possibilities in this section.

Tyrosine kinase inhibitors that target the *EGFR* are effective against NSCLC, but drug-resistant mutations emerge. One potential means of identifying such mutations non-invasively is to examine the genotype of CTCs isolated by using antibodies against epithelial cells. Such techniques, recently improved by using microfluidic devices, have been used successfully to find the T790M drug resistance mutation and to quantify CTCs, which may provide clinically useful prognostic information.

The miRNAs are showing promise as tissue-based markers for cancer classification, present in human plasma, and readily measured. Thus, serum levels of miR-141 (a miRNA expressed in prostate cancer) can distinguish patients with prostate cancer from healthy controls. As was discussed in this chapter for coding genes, single-nucleotide polymorphisms (SNPs) in pre-miRNAs could similarly alter the processing, expression, and/or binding of miRNAs to target mRNA. SNP in hsa-mir-196a2 was associated with survival in individuals with NSCLC and may prove a useful prognostic biomarker.

Biomarkers for colorectal cancer

Colorectal cancer (CRC) is responsible for an estimated 650 000 deaths per year worldwide and is among the major challenges facing overstretched healthcare resources. Locally confined CRC is potentially curable by early surgical resection, but is often diagnosed too late (25% have metastatic disease at presentation), in part because no suitable biomarkers for early diagnosis exist. Of those treated surgically, up to 50% will, unpredictably, still eventually develop recurrence or metastatic disease, a subgroup facing an aggressive disease course not identifiable by currently available biomarkers. TNM stage II disease poses particular unpredictable clinical outcomes as around 30% of patients will develop metastatic disease; although the cup-half-full crowd will claim that this means 70% will not develop it, we have no idea which ones.

All those with inoperable disease are potential candidates for systemic chemotherapy – but which patients require which drugs? We urgently need improved and more specific biomarkers for CRC subclassification on the basis of prognosis and treatment responsiveness, in particular those that can be acquired less invasively from blood or even stool. To date we have serum CEA, which can assist in screening symptomatic patients for early diagnosis and has proven helpful, when measured repeatedly, in screening for recurrent disease after surgery. Tissue-based biomarkers are also entering clinical practice. Screening for absence of mismatch repair proteins helps in identifying Lynch syndrome, but classification is still almost entirely anatomical. Currently the picks of the molecular marker crop are mutations in *KRAS* or *BRAF*, which signify that administration of *EGFR*-targeted therapies will be futile.

Recently an epigenetic signature has been identified that might help predict response to chemotherapy in advanced CRC. Thus, hypermethylation of the gene-encoding transcription factor AP-2 epsilon (TFAP2E) predicts a poor clinical response to chemotherapy in CRC. This effect has been attributed to activation of downstream signaling via Dickkopf homolog 4 (DKK4), suggesting a new target for the treatment of chemotherapy-resistant CRC in the future.

Pharmacogenetics and pharmacogenomics

What's in a name? That which we call a rose, by any other name would smell as sweet.

William Shakespeare

Pharmacogenetic testing is one specific area in which knowledge of the patient's genotype for a single gene can be used to tailor therapy. There is confusion of terminology in this area, and we will attempt to provide a guide that you can use to negotiate this area when approaching the literature. First, pharmacogenetics when used in cancer biology traditionally equates to unraveling the effects of inherited germline variants on how a specific drug will be metabolized, be distributed, and function in an individual patient, and what effects it might have on toxicity. We have included this separate section in this chapter despite the obvious parallels with using genotype to diagnose and screen for hereditary cancer syndromes, such as *BRCA*-related breast and ovarian cancers (see Chapter 3); this is because in this context, we are considering only how this information helps with selecting the right drug and dose for an individual patient, even from drugs with biologically identical actions.

Pharmacogenetics is already influencing clinical practice, and there are several examples of how SNPs and gene expression for a given gene can influence the availability (pharmacokinetics) or action (pharmacodynamics) of a particular drug(s). Variations in members of the cytochrome p450 (CYP) family of genes and their protein products are by far the most important, which is not surprising when you consider that this family is responsible for the metabolism of over 80% of all prescribed drugs.

So far, so good. However, pharmacogenomics means different things to different people, and this is reflected in the literature. Often this term is used in oncology to refer to the other important genome, namely that of the cancer cell, which as we have learned diverges from that of the germline to the tune of all the acquired mutations that have accumulated in somatic cells up to that point. In this case, *KRAS* status, in the tumor, is used to predict patient response to *EGFR*-targeted mAbs and would be an example of pharmacogenomics. We should note in passing that germline mutations will also be detected by analyzing tumor-derived nucleic acids; it is just far simpler to look for these in peripheral blood. So where is the confusion? Problems arise because pharmacogenomics is often also used interchangeably with the combinatorial effects of multiple-gene variants on drug metabolism and action; in this context, genomics is used to imply the whole genome. Our advice: use "pharmacogenetics" when discussing the effects of a single germline variant on a drug and "pharmacogenomics" when considering the effects of acquired somatic mutations or multiple variants.

Regardless of terminology, pharmacogenetics is likely to prove of substantial importance in cancer, because in general we use a greater number of drugs with a very narrow therapeutic window than in any other area of medicine. Even small differences in metabolism could have profound effects on the buildup of a drug and on the patient's well-being. Too much drug could prove highly toxic, whereas too little could impede efficacy. Across the board, those drugs most commonly implicated in adverse drug reactions are those metabolized by enzymes with known polymorphisms. In cancer, SNPs affecting the genes for Cyp2D6, TPMT, dihydropyrimidine dehydrogenase (DPD), CDA, and UGT1A1 are known to influence metabolism of tamoxifen, mercaptopurine–azathioprine, 5-FU–capecitabine, gemcitabine–capecitabine–AraC, and irinotecan, respectively. In most cases, detecting the presence of hypofunctioning alleles enables lower drug doses to be administered. Overall, high hopes are held out for pharmacogenetics; adverse drug reactions run to the millions every year and cost as much to put right as the entire drug budget, and anything which reduces this risk is worth investing in.

A selection of important and illuminating examples in cancer medicine will be discussed here. Historically, one of the earliest examples of genotype influencing treatment followed identification of the role played by thiopurine methyltransferase (TPMT) in the metabolism of the thiopurine drugs, 6-mercaptopurine and azathioprine used in leukemia, among others. When TPMT activity is compromised, thiopurines are instead metabolized by alternative pathways generating toxic metabolites. Although only 1 in 300 people have both variant alleles lacking in TPMT activity, this is highly important; such individuals require less than 10% of the usual amount of drug – the normal dose would be equivalent to taking 10 times the recommended amount and would cause severe bone marrow suppression. To confuse matters, 1 in 10 people are heterozygous for such variant alleles and are less efficient at metabolizing the drugs; they may experience more side effects but may actually have an enhanced response, as drug levels are on average higher than would be the case in those with two normally functioning alleles. In clinical practice, the cost and comparatively low incidence of TPMT deficiency mean that testing is not used routinely but is suggested for those who experience bone marrow suppression when treated with thiopurine drugs. In some cases, children with leukemia may be tested before commencing treatment.

One interesting recent example that neatly illustrates the concepts of pharmacogenetics relates to differences in the metabolism of the selective estrogen receptor modulator, tamoxifen. As has and will be discussed, tamoxifen has transformed treatment of breast cancer and improves mortality when used as an adjuvant therapy and when used in high-risk prevention. Tamoxifen is metabolized by various enzymes, including the rate-limiting cytochrome P450 2D6 enzyme (CYP2D6) to generate the more pharmacologically active metabolite 4-hydroxy tamoxifen (endoxifen). The CYP2D6 gene is highly polymorphic, and several alleles have been shown to reduce or even negate enzymatic activity. Clinical relevance, however, remains uncertain. A so-called poor **metabolizer phenotype** associated with inappropriately low levels of endoxifen in tamoxifen-treated women is well recognized and is, as expected, linked to the presence of alleles encoding proteins with reduced enzymatic activity. Yet recent data from clinical trials of tamoxifen and aromatase inhibitors, such as ATAC and BIG, have failed to show any impact of the CYP2D6 genotype on outcome. These trials were retrospective, so the jury is still out on this treatment marker at present. This situation should be resolved, however, because highly effective alternatives to tamoxifen, exemplified by aromatase inhibitors, are available and theoretically at least might be preferable in poor metabolizers of tamoxifen. Or one could consider the option of using the active agent, 4-OHT, as an alternative.

Irinotecan, used in the treatment of CRC, is metabolized by UGT1A1, and hypoactive variants result in higher than anticipated levels of drug and potentially lethal toxicity. This could be forestalled by UGT1A1 testing before treatment, but due to the cost and relative infrequency of these variants, testing is usually employed only if toxic problems arise.

There are plenty of other nongenetic variables that can affect how an individual might react to a drug, including unmodifiable ones such as age and gender and those that, in theory, could be manipulated such as smoking, alcohol intake, and diet. Probably most important clinically are interactions with other drugs that are prescribed or self-administered. Numerous online databases are available to help the wary physician avoid potentially toxic side effects, or sometimes loss of activity, when the wrong combinations of drugs are used.

Clinical trials in cancer

A gem cannot be polished without friction, nor a man perfected without trials.

Seneca

Once a new agent has been developed and has completed preclinical studies (Chapter 15), it is deemed ready for the first usage in human subjects – **first-in-man studies**.

The earliest clinical trials in patients are, not surprisingly, aimed at confirming the safety of the new agent in humans. Thus, although issues of toxicity will already have been addressed by extensive studies in different cultured cell lines and tissues and across a range of animal species, the ultimate proof has to come from use in human volunteers. These may be healthy recruits or may often be cancer patients for whom existing treatments are unavailable or no longer indicated. Usually these are small trials involving less than 30 patients and may recruit those with a variety of different cancers. Phase I clinical trials use a variety of standardized protocols, including progressive dose escalation and dosing intervals optimized for ensuring safety over any therapeutic considerations. They can be very time-consuming, with the first cohort receiving the lowest dose and then subsequent cohorts receiving higher doses if there are no problems at the lower one, up to a predetermined maximum. This is known as dose escalation. It is standard practice to measure a large number of parameters in blood tests to ensure that no damage is being done. Biomarkers may also be measured and finally pharmacokinetics, including drug elimination, are looked at. As these studies are often run in advanced cancers, one may gain useful clues as to the ability of the drug to shrink existing tumours (RECIST) and maybe on overall survival. However, phase I trials do not specifically address the issue of efficacy. They are not randomized controlled trials (RCTs) and the dose escalation and intervals that must be used are likely to be suboptimal, so potential beneficial responses can only be inferred if patients are noted to be responding clinically or surviving beyond that expected from the natural history of the stage and type of cancer being tested. Such findings, although speculative, can help when considering the types and stages of cancer to be included in the design of subsequent phase II and III trials, which increasingly include efficacy as the main endpoint, once safety has been successfully demonstrated.

Phase II trials are undertaken once phase I trials have confirmed that a drug appears safe. These are conducted in a larger cohort to assess the efficacy and to again look for potential side effects and toxicity in a larger number and under conditions more representative of the planned clinical usage. They often concentrate on a single type of cancer. The aim is to decide if the treatment looks good enough to take forward into much larger and more expensive phase III trials. The focus is on safety and side effects but it will use only a single or a small number of doses, but in an optimal schedule. Studies are usually larger, with up to a 100 recruited patients, and may compare the new treatment with an established one, or even with placebo. A trial may also compare two or more regimes of the same drugs, used alone or in combination with existing recommended treatments. If the results of phase II trials suggest the new treatment is equivalent or superior to an existing treatment then it can progress to phase III.

Phase III trials aim to demonstrate clinical benefits and are usually RCTs. They compare new treatments with the best currently available treatment and receive the new drug in addition to other treatments; in such studies it is considered unethical to withhold them. In other cases trials may be placebo-controlled if standard care is palliative and no longer involves active anticancer treatments. These studies may run across multiple centers to recruit sufficient numbers of participants (may be over 1000). The numbers needed are determined by power calculations to show the predicted benefit. The smaller the likely benefit over the control group, the more patients are needed and potentially the longer the follow-up. These studies can cost many millions of dollars.

Phase IV trials involve postmarketing surveillance and are an important route by which the adverse effects of new drugs can be picked up when these are of a low enough frequency to have been missed in earlier studies (large numbers of patients treated for longer are needed to reveal the problem). Obviously, it is hoped that no such problems will exist, but there are several examples of drugs which have been withdrawn only after such a potentially protracted period of monitoring. In general, this is less likely with diseases such as cancer where the natural history of the disease is often associated with a poor outcome.

The process of drug development is a long one. Even after a promising compound has been developed, a further 8 years will on average be needed to complete the clinical trials required for the drug to get regulatory approval and become licensed for general use by clinicians. This delay may be agonizing for cancer patients who may seek far and wide for clinical trials in which to enroll once existing treatments have failed or are discontinued. However, those with more responsive cancers may avoid participating in trials or are put off by the possibility that they will receive a placebo by chance. Overall, only some 5–10% of cancer patients are involved in clinical trials.

Conclusions and future directions

In addition to developing new agents with greater specificity for cancer-specific signaling, we must now unravel the most effective combinations and dosing schedules of targeted and nontargeted cancer treatments for any given cancer and patient. Recent studies suggest the importance of dosing; this is exemplified by the different phenotype induced by c-Myc at different levels in animal models and by the effects of inhibiting HIF-2a by different amounts in models of NSCLC, where "knockdown" is anticancer whereas "knockout" might actually promote tumorigenesis by downregulating tumor suppressors.

New techniques for determining the presence of mutations, as exemplified by *EGFR*, *HER2*, or *KRAS*, will increasingly be employed to identify suitable patients for drugs targeting the resultant aberrant protein TKs. After all, it serves little purpose to take an inhibitor of HER2 if this protein is not involved in your cancer and the presence of a mutation in *KRAS* in CRC identifies patients who will not benefit from targeting proteins upstream of this activating mutation. It is hoped that these examples will be rapidly followed by more biological targets and also by biomarkers that help in making the most appropriate treatment choices.

New targets are being identified, and in addition to those already discussed we are awaiting results with targeting of STAT (signal transducers and activators of transcription) proteins, telomeres and telomerase, thioredoxin (Trx) family members, and topoisomerases I and II.

New drug targets are emerging, and we are beginning to understand the ways in which mutations and natural selection allow cancer cells to evolve and, in doing so, to progress and to "escape" from or avoid the effects of various cancer treatments, in particular those very specifically targeting single signaling steps. Such knowledge will identify potential treatment combinations that can avoid or overcome the resistance to single targeted agents.

Finally, we all wait with interest the outcomes of trials and research studies using novel types of therapeutics, in particular new monoclonal antibodies and linked cargos, gene therapy, and, arguably the most exciting of all, miRNAs and siRNAs.

All things change, nothing is extinguished. There is nothing in the whole world which is permanent. Everything flows onward; all things are brought into being with a changing nature; the ages themselves glide by in constant movement.

Ovid

Bibliography

Biomarkers and tumor heterogeneity

Badve, S., and Nakshatri, H. (2012). Breast-cancer stem cells – beyond semantics. *Lancet Oncology*, **13**(1): e43–8.

Bertos, N.R., and Park, M. (2011). Breast cancer – one term, many entities? *Journal of Clinical Investigation*, **121**(10): 3789–96.

Carlsson, A., Wingren, C., Kristensson, M., *et al.* (2011). Molecular serum portraits in patients with primary breast cancer predict the development of distant metastases. *Proceedings of the National Academy of Sciences of the USA*, **108**(34): 14252–7.

Conley, S.J., Gheordunescu, E., Kakarala, P., *et al.* (2012). Antiangiogenic agents increase breast cancer stem cells via the generation of tumor hypoxia. *Proceedings of the National Academy of Sciences of the USA*, January 23.

Dalerba, P., Kalisky, T., Sahoo, D., *et al.* (2011). Single-cell dissection of transcriptional heterogeneity in human colon tumors. *Nature Biotechnology*, **29**(12): 1120–7.

Ferté, C., André, F., and Soria, J.C. (2010). Molecular circuits of solid tumors: prognostic and predictive tools for bedside use. *Nature Reviews Clinical Oncology*, **7**(7): 367–80.

Horst, D., Chen, J., Morikawa, T., Ogino, S., Kirchner, T., and Shivdasani, R.A. (2012). Differential WNT activity in colorectal cancer confers limited tumorigenic potential and is regulated by MAPK signaling. *Cancer Research*, February 8.

La Thangue, N.B., and Kerr, D.J. (2011). Predictive biomarkers: a paradigm shift towards personalized cancer medicine. *Nature Reviews Clinical Oncology*, **8**: 587–96.

Little, S.E., Edlow, A.G., Thomas, A.M., *et al.* (2012). Receptor tyrosine kinase genes amplified in glioblastoma exhibit a mutual exclusivity in variable proportions reflective of individual tumor heterogeneity. *Cancer Research*, February 6.

Peeper, D., and Berns, A. (2006). Cross-species oncogenomics in cancer gene identification. *Cell*, **125**(7): 1230–3.

Snuderl, M., Fazlollahi, L., Le, L.P., *et al.* (2011). Mosaic amplification of multiple receptor tyrosine kinase genes in glioblastoma. *Cancer Cell*, **20**(6): 810–7.

Vermeulen, L., de Sousa, E., Melo, F., Richel, D.J., and Medema, J.P. (2012). The developing cancer stem-cell model: clinical challenges and opportunities. *Lancet Oncology*, **13**(2): e83–9.

Preclinical development and drug screening

Barretina, J. (2012). The Cancer Cell Line Encyclopedia enables predictive modelling of anticancer drug sensitivity. *Nature*, **483**(7391): 603–7.

Benson, J.D., Chen, Y.N., Cornell-Kennon, S.A., *et al.* (2006). Validating cancer drug targets. *Nature*, **441**(7092): 451–6.

Garnett, M.J., Edelman, E.J., Heidorn, S.J., *et al.* (2012). Systematic identification of genomic markers of drug sensitivity in cancer cells. *Nature*, **483**(7391): 570–5.

Nguyen, L.V., Vanner, R., Dirks, P., and Eaves, C.J. (2012). Cancer stem cells: an evolving concept. *Nature Reviews Cancer*, **12**(2): 133–43.

Shi, Q., Qin, L., Wei, W., *et al.* (2012). Single-cell proteomic chip for profiling intracellular signaling pathways in single tumor cells. *Proceedings of the National Academy of Sciences of the USA*, **109**(2): 419–24.

Targeted therapies

Adams, G.P., and Weiner, L.M. (2005). Monoclonal antibody therapy of cancer. *Nature Biotechnology*, **23**(9): 1147–57.

Bild, A.H., Yao, G., Chang, J.T., *et al.* (2006). Oncogenic pathway signatures in human cancers as a guide to targeted therapies. *Nature*, **439**(7074): 353–7.

Boxer, R.B., Jang, J.W., Sintasath, L., and Chodosh, L.A. (2004). Lack of sustained regression of c-MYC-induced mammary adenocarcinomas following brief or prolonged MYC inactivation. *Cancer Cell*, **6**(6): 577–86.

Chabner, B.A., and Roberts, T.G., Jr. (2005). Timeline: chemotherapy and the war on cancer. *Nature Reviews Cancer*, **5**(1): 65–72.

Chapman, P.B., Hauschild, A., Robert, C., *et al.* (2011). Improved survival with vemurafenib in melanoma with BRAF V600E mutation. *New England Journal of Medicine*, **364**(26): 2507–16.

Demetri, G.D., von Mehren, M., Blanke, C.D., *et al.* (2002). Efficacy and safety of imatinib mesylate in advanced gastrointestinal stromal tumors. *New England Journal of Medicine*, **347**(7): 472–80.

Di Nicolantonio, F., Mercer, S.J., Knight, L.A., *et al.* (2005). Cancer cell adaptation to chemotherapy. *BMC Cancer*, **5**: 78.

Dornan, D., and Settleman, J. (2010). Cancer: miRNA addiction – depending on life's little things. *Current Biology*, **20**(18): R812–3.

Ebert, M.P., Tänzer, M., Balluff, B., *et al.* (2012). TFAP2E-DKK4 and chemoresistance in colorectal cancer. *New England Journal of Medicine*, **366**(1): 44–53.

Felsher, D.W. (2003). Cancer revoked: oncogenes as therapeutic targets. *Nature Reviews Cancer*, **3**: 375–80.

Felsher, D.W., and Shachaf, C.M. (2005). Rehabilitation of cancer through oncogene inactivation. *Trends in Molecular Medicine*, **11**(7): 316–21.

Hurwitz, H., Fehrenbacher, L., Novotny, W., *et al.* (2004). Bevacizumab plus irinotecan, fluorouracil and leucovorin for metastatic colorectal cancer. *New England Journal of Medicine*, **350**(23): 235–42.

Ilic, N., Utermark, T., Widlund, H.R., and Roberts, T.M. (2011). PI3K-targeted therapy can be evaded by gene amplification along the MYC-eukaryotic translation initiation factor 4E (eIF4E) axis. *Proceedings of the National Academy of Sciences of the USA*, **108**(37): E699–708.

Jonkers, J., and Berns, A. (2004). Oncogene addiction: sometimes a temporary slavery. *Cancer Cell*, **6**(6): 535–8.

Kantarjian, H., Sawyers, C., Hochhaus, A., et al. (2002). Hematologic and cytogenetic responses to imatinib mesylate in CML. *New England Journal of Medicine*, **346**(9): 645–52.

Khandekar, M.J., Cohen, P., and Spiegelman, B.M. (2011). Molecular mechanisms of cancer development in obesity. *Nature Reviews Cancer*, **11**(12): 886–95.

Kok, T.S., Wu, Y.L., Thongprasert, S., et al. (2009). Gefitinib or carboplatin-paclitaxel in pulmonary adenocarcinoma. *New England Journal of Medicine*, **361**: 947–57.

Maemondo, M., Inoue, A., Kobayashi, K., et al. (2010). Gefitinib or chemotherapy for non-small-cell lung cancer with mutated EGFR. *New England Journal of Medicine*, **362**(25): 2380–8.

Mehta, R.S., Barlow, W.E., Albain, K.S., et al. (2012). Combination anastrozole and fulvestrant in metastatic breast cancer. *New England Journal of Medicine*, **367**(5):

North East Japan Gefitinib Study Group. (2009). First-line gefitinib for patients with advanced non-small-cell lung cancer harboring epidermal growth factor receptor mutations without indication for chemotherapy. *Journal of Clinical Oncology*, **27**(9): 1394–400.

Pratilas, C.A., and Solit, D.B. (2010). Targeting the mitogen-activated protein kinase pathway: physiological feedback and drug response. *Clinical Cancer Research*, **16**(13): 3329–34.

Richardson, P.G., Barlogie, B., Berenson, J., et al. (2003). A phase 2 study of bortezomib in relapsed, refractory myeloma. *New England Journal of Medicine*, **348**(26): 2609–17.

Ross-Innes, C.S., Stark, R, Teschendorff, AE, et al. (2012). Differential oestrogen receptor binding is associated with clinical outcome in breast cancer. *Nature*, January 4, doi:10.1038/nature10730.

Sharma, S.V., and Settleman, J. (2010). Exploiting the balance between life and death: targeted cancer therapy and "oncogenic shock." *Biochemical Pharmacology*, **80**(5): 666–73.

Shepherd, F.A., et al. (2005). Erlotinib in previously treated non-small cell lung cancer. *New England Journal of Medicine*, **353**: 123–32.

Wagner, R.W. (1994). Gene inhibition using antisense oligodeoxynucleotides. *Nature*, **372**: 333–5.

siRNA

Liu, C., Kelnar, K., Liu, B., et al. (2011). The microRNA miR-34a inhibits prostate cancer stem cells and metastasis by directly repressing CD44. *Nature Medicine*, **17**(2): 211–5.

Mopert K., et al. (2012). Depletion of protein kinase N3 (PKN3) impairs actin and adherens junctions dynamics and attenuates endothelial cell activation. *European Journal of Cell Biology* **91**: 694–705.

Santel, A., et al. (2010). Atu027 prevents pulmonary metastasis in experimental and spontaneous mouse metastasis models. *Clinical Cancer Research* **16**(22): 5469–80.

Questions for student review

1) Which of the following are true of oncogene addiction?
   a. The addicted cancer cell is less sensitive than others to a therapy targeting the addicted protein.
   b. Resistance does not develop to agents targeting such oncogenes.
   c. NSCLC may be addicted to EGFR.
   d. Breast cancers may be addicted to HER2.
   e. Extreme sensitivity to another non-addicted protein may be present before therapy commences.

2) Which of the following may contribute to cancer cell heterogeneity in an individual tumor?
   a. Self-seeding
   b. Clonal evolution
   c. Genomic instability
   d. Cancer stem cells
   e. EMT

3) The presence of which of the following is currently used to guide treatment selection in breast cancer?
   a. Progesterone receptors
   b. BCR-ABL
   c. HER2
   d. Estrogen receptors
   e. Mutant RAS

4) The following are direct targets of tyrosine kinase inhibitors, true or false?
   a. BCR-ABL
   b. ERBB2
   c. P53
   d. HDAC
   e. Cyclin D

5) Which of the following are known to contribute to resistance to EGFR-targeted mAb therapies?
   a. Mutations in BRAF
   b. Inactivation of MET signaling
   c. Mutations in KRAS
   d. Upregulation of CYP450
   e. Activation of ERBB2 signaling

17 The Diagnosis of Cancer

Anne L. Thomas, Bruno Morgan, and William P. Steward
University of Leicester, UK

New ideas pass through three periods:
It can't be done.
It probably can be done, but it's not worth doing.
I knew it was a good idea all along!

Arthur C. Clarke

Key points

- Although there have been advances in the tests available to diagnosis cancer, the initial history remains the crucial first "investigation."
- Improvements in genetic tests and immunohistocytochemistry have ensured that the diagnosis can be made reliably and on increasingly small biopsy specimens. These tests also provide prognostic information about the diagnosis as well as predictive information, for example how the tumor will respond to specific biologically targeted therapies.
- Too many patients are still presenting with advanced disease, and effective screening programs exist for only a limited number of tumor types.
- Functional imaging in cancer with positron emission tomography (PET) and computed tomography (CT) is now standard of care for the main tumor types and provides superior information in discriminating malignant from benign disease. This ensures that patients are appropriately staged so that they are offered the best treatment available for them.

Introduction

In this chapter, presentations of some of the most common cancers are discussed. The range of diagnostic tests is reviewed, and the advantages and disadvantages of individual techniques are summarized. Future developments and potential new diagnostic techniques are reviewed (Box 17.1).

The diagnosis of cancer can be remarkably difficult to make. This is because patients may have nonspecific symptoms that mirror those found in common benign conditions. For example, a patient with colon cancer may have similar symptoms to a patient with the benign irritable bowel syndrome. In addition, some patients may hide their symptoms as a result of fear of the possible diagnosis of cancer and the likely treatment. Sadly, this means that many patients present late and already have advanced disease at the time of diagnosis.

It is essential to confirm that there is an underlying malignancy which is the cause of a patient's symptoms, and this ultimately relies on obtaining a representative sample of tissue in most instances. This will also provide information on the type of tumor so that the appropriate treatment can be chosen. Unfortunately it is frequently difficult to establish the histological diagnosis to confirm a suspected cancer. Tumors often contain areas which vary markedly in their concentration of malignant cells and frequently have large islands of necrotic and acellular tissue. As a result, the site from which a biopsy is taken may not actually contain tumor cells. A classic example of this arises when a clinician tries to make a diagnosis of pancreatic cancer. A lesion in the head of the pancreas can cause a desmoplastic reaction (an inflammatory local process), which, when biopsied, reveals reactive rather than neoplastic cells. Further attempts at biopsy may fail, and the clinician is left with the difficult decision of whether to make the diagnosis on clinical grounds alone, with all the implications this has for the prognosis that is given to the family and the potential for toxicity of the treatment that will be offered.

Many of the investigative procedures used in oncology are uncomfortable and inconvenient for the patient. They are often

---

The Molecular Biology of Cancer: A Bridge From Bench to Bedside, Second Edition. Edited by Stella Pelengaris and Michael Khan.
© 2013 John Wiley & Sons, Inc. Published 2013 by John Wiley & Sons, Inc.

Box 17.1 Concept figure for the chapter: how patients present

- Hematological malignancies
- Lung cancer
- Breast cancer
- Colorectal cancer
- CNS tumors

⇓

Investigations

Symptomatic patients

⇓

Blood tests

⇓

Non-invasive Imaging techniques
- X-ray
- Ultrasound
- CT scanning
- MRI

⇓

Tissue diagnosis
- Fine needle aspiration
- Needle biopsy
- Endoscopy techniques

⇓

Interventional radiology

⇓

New imaging techniques

⇓

Molecular techniques in cancer diagnosis

Asymptomatic patients
For example, screening tools
- Cervical cancer
- Breast cancer
- Colon cancer

⇒ Disease staging

---

ture of a tumor, it will be possible to individualize treatment using the increasing array of targeted therapies (see Chapter 16). This approach will increase the efficacy of anticancer therapy and also possibly reduce toxicity.

Clinical manifestations

Patients with cancer present in a variety of ways depending on the local effect of the primary tumor or the effects of any sites of metastatic spread, or from an indirect metabolic effect such as hypercalcemia. There may also be nonspecific effects which are frequently seen with many tumor types and are believed to be due to production of a variety of cytokines by the tumor cells. These symptoms include anorexia, weight loss, and lethargy. Finally, a variety of malignancies may produce symptoms and signs distant from the primary site or its metastases – termed "paraneoplastic syndromes" – and these are often related to the central nervous system (CNS). The clinical manifestations of the hematological malignancies (lymphoma and leukemia) and solid tumors involving lung, breast, bowel, and nervous system are discussed in more detail in this chapter. Further information on presenting symptoms in other tumors can be found in general texts (e.g. DeVita *et al.*, 2011).

Lymphoma and leukemia

The vast majority of patients with lymphoma will present with nodal disease. The patient may notice painless enlargement of a lymph node, most commonly in the neck. Sometimes the nodes will fluctuate in size, and therefore the patient may not consider this to be serious. For patients with Hodgkin disease, the initial site of presentation is most often in the cervical nodes (70% of all cases), axilla (25%), and inguinal area (5%). In non-Hodgkin lymphomas, the nodal disease tends to be more widespread. Occasionally the lymph nodes will grow very rapidly and patients may notice symptoms due to resultant compression of blood vessels. This may cause swelling of a leg when there is extensive inguinal lymphadenopathy or superior vena cava obstruction due to mediastinal nodes. In non-Hodgkin lymphomas, enlargement of the retroperitoneal lymph nodes can cause backache and even renal failure due to obstruction of the ureters. Patients may present with obstructive jaundice if the nodes are enlarged in the porta hepatis and, in the more extensive lymphomas, the liver and spleen may be involved causing their enlargement with associated abdominal discomfort. Lymphomas often cause constitutional symptoms which may be the presenting features. Fever is the most frequent of these and may occur with drenching night sweats. Weight loss is common and is usually more marked in those patients with bulky disease. The triad of constitutional symptoms (fever, sweats, and more than 10% loss in body weight) are known as B symptoms, and the presence of these is associated with more advanced disease and poorer prognosis. Some patients with Hodgkin disease develop pain in the enlarged lymph nodes on drinking alcohol.

In leukemias, the most common presenting symptoms are fever, infection, malaise, and bleeding. These occur due to neutropenia, anemia, and thrombocytopenia, which are secondary to bone marrow infiltration by leukemic blasts. Some patients will present with bone or joint pain, and others may notice lymphadenopathy or develop abdominal pain from hepatosplenomegaly.

---

difficult to interpret, with false positives and negatives. It is also important not to submit patients to unnecessary tests, which can cause morbidity (even mortality), huge costs to the healthcare system, and inappropriate distress. There should always be a logical sequence of investigations, and for many conditions there are now protocols to follow which should facilitate the ability to make an accurate diagnosis in the shortest time possible.

Thankfully, in oncology there has been a steady development of investigative procedures which are more reliable and less invasive. In the future, the hope is not only that diagnoses will be made swiftly with minimal discomfort to the patient, but also that more molecular information about the tumor can be obtained with the techniques available. By knowing the molecular signa-

Lung cancer

Most patients with lung cancer present with symptoms which are directly related to the tumor. These include hemoptysis, cough, dyspnea, and chest pain. These are common symptoms from cigarette smoking as it is often associated with background lung disease, and they may therefore be ignored by the patient and clinician, delaying the diagnosis of lung cancer. Direct tumor invasion into the mediastinum can cause severe pain which may be misinterpreted as being of cardiac origin. If the tumor invades the left recurrent laryngeal nerve, the patient will develop hoarseness of the voice, which may be the only presenting symptom. Classic pneumonias that fail to clear despite adequate antibiotics should always be considered as suspicious of an underlying malignant obstructing lesion.

A Pancoast tumor is situated at the apex of the lung and may cause severe pain in the shoulder, chest wall, and arm. This is due to the direct invasion of the brachial plexus. If the tumor is found in the right main or upper lobe bronchus, compression of the great veins of the neck may cause superior vena cava obstruction. Clinically a patient will have swelling of the face, neck, and upper arms with plethora or cyanosis and dilated veins over the upper part of the chest wall. Some patients may present with a lump in the neck where the tumor has metastasized to lymph nodes, or they may develop symptoms from other metastatic sites such as bone pain or headache from bone or brain metastases respectively. Finally, there may be weight loss, nausea, and jaundice from metastases to the liver. It is extremely common for patients with lung cancer to also have constitutional symptoms such as anorexia, weight loss, weakness, and fatigue.

Breast cancer

In breast cancer, the local effects of the disease may cause a lump in the breast, changes in the nipple, pain in the breast, or puckering of the skin overlying a lump. Some patients may notice a mass in the axilla, and if this is significant there may also be swelling of the affected arm. If the patient has an inflammatory breast cancer, the whole breast may be inflamed with firmness of the tissue and widespread erythema. Breast cancers metastasize most commonly to the bones, lungs, and liver. Therefore, the patient may present with symptoms from a secondary deposit, such as pain in the back due to a vertebral metastasis or upper abdominal discomfort from hepatomegally.

Colorectal cancer

The symptoms of colorectal cancer vary with the site of the tumor. When this arises in the rectum, the majority of patients will notice a change in bowel habit associated with rectal bleeding. Sometimes the patient may report the feeling of incomplete evacuation of the bowels, tenesmus, which is an extremely distressing symptom. Lesions affecting the left side of the colon often cause a change in bowel habit and also sometimes bleeding. If the tumor actually obstructs the bowel, patients will present with colicky abdominal pain, vomiting, and ultimately absolute constipation. Those lesions affecting the cecum and the right side of the colon most commonly present with ill-defined abdominal pain and anemia due to occult blood loss from the tumor. As with other solid tumors, patients may also present with constitutional symptoms or symptoms relating to sites of metastatic spread.

Tumors arising from the nervous system

Tumors of the nervous system can be divided into those which arise centrally and those in the periphery. Patients with brain tumors may present with symptoms of raised intracranial pressure such as headache, which is often worse in the morning; drowsiness; and nausea and vomiting. The actual site of the tumor may cause specific symptoms. For example, a tumor in the motor cortex can cause paralysis, tumors in the temporal region are often associated with epilepsy, and tumors in the cerebellum typically cause ataxia, nystagmus, and double vision. There may be endocrine effects of central nervous tumors. These usually result when they arise in the pituitary, which can cause elevated hormone levels with resultant manifestations such as Cushing's syndrome. These tumors may subsequently extend up and involve the cranial nerve pathways causing visual field defects. Sometimes the tumors may bleed and the patient will notice a sudden deterioration in vision associated with headache and signs of hypo-pituitarism.

Tumors involving the spinal cord cause symptoms depending on the site at which they arise. Patients may experience pain at that level and, if there is resultant compression of the spinal cord, weakness may occur below the lesion, with associated constipation, bladder dysfunction, and changes in sensation. This is a medical emergency requiring rapid intervention to reduce the risk of permanent paralysis.

Investigations in oncological practice

Interpretation of hematological and biochemical tests

In general, routine blood tests do not make the diagnosis of cancer but may indicate that there is a serious underlying illness. The most common changes which can accompany a malignancy are anemia, polycythemia, neutropenia, leucocytosis, thrombocytosis or thrombocytopenia, elevated liver enzymes, and reduced renal function. The erythrocyte sedimentation rate may rise in patients with cancer, but it is a nonspecific test and is also elevated in infections and several other conditions. In the leukemias, the peripheral blood film is essential as there may be a normocytic normochromic anemia, and occasionally leukemic blast cells are present. There may be an excess of white cells and a low platelet count. If the tumor has spread to the bone marrow in solid malignancies, a leukoerythroblastic picture will be seen with circulating primitive cells. In some solid tumors, the routine full blood count will show normocytic, normochromic anemia and, in those tumors where there has been blood loss, an iron-deficient anemia occurs.

Routine biochemistry may indicate varying degrees of renal failure if the tumor has obstructed the renal outlet. If the tumor has blocked the biliary tract, an obstructive picture in the liver enzymes will be seen with an elevated bilirubin and alkaline phosphatase. Bone metastases and some lung cancers will also cause a rise in alkaline phosphatase. The liver transaminases, aspartate and alanine amino transferase, will be increased in patients with liver injury which may be from alcohol, infection, or cancer. Lactic dehydrogenase (LDH) is an enzyme associated with tissue injury and is often increased in patients with a large tumor burden, lymphoma, or liver metastases. An elevated corrected calcium level indicates hypercalcemia, which may be due to bone metastases, or is a metabolic consequence of an underlying tumor such as lymphoma. Low sodium (hyponatremia) may

Table 17.1 Interpretation of blood results in the diagnosis of cancer

Abnormal blood test	Possible interpretation
Low hemoglobin	Anemia due to bleeding from the tumor
	Anemia due to bone marrow involvement by the tumor
Elevated white cell count	Infection
	Bone marrow involvement by tumor
Low white cell count	Neutropenia due to chemotherapy
	Bone marrow failure
Elevated urea and creatinine	Renal impairment; may be due to the kidneys being obstructed by the tumor, dehydration, or damage by chemotherapy
Elevated urea with anemia	Gastrointestinal hemorrhage
Hyponatremia and hyperkalemia	Addison's syndrome due to tumor involving the adrenal glands
Hyponatremia	Syndrome of inappropriate ADH secretion in small-cell lung cancer
Hypercalcemia	Indirect effect of metastatic cancer or bone metastases
Hypokalemia or hypomagnesemia	Renal tubular damage due to chemotherapy or diarrhea
Elevated C-reactive protein	Nonspecific indicator of infection or inflammation
Hypoalbuminemia	Liver impairment, or malnutrition or cachexia
Elevated alkaline phosphatase	Bone metastases or liver metastases
Elevated alkaline phosphatase with elevated bilirubin	Biliary obstruction
Elevated hepatic transaminases (AST or ALT)	Liver damage which could be due to chemotherapy or metastases

represent the syndrome of inappropriate ADH secretion, which is often associated with an underlying small-cell lung cancer. A low albumin may indicate liver involvement with tumor or be a general indicator of the poor nutritional state of the patient (Table 17.1). For further biochemical effects of tumors, see Pannall et al. (1997).

Tumor markers

Some tumors produce proteins which can be detected in the blood and can be used to indicate the presence of a malignancy. The value of different tumor markers varies; they can be useful in diagnosis, predicting prognosis or response to treatment and aiding follow-up of a patient. Tumor markers can be broadly divided into three groups: oncofetal proteins, cancer-related antigens, and hormones (Table 17.2).

An example of an oncofetal protein tumor marker is carcinoembryonic antigen (CEA). Although this marker may be produced in up to 80% of people with bowel cancer, it will also be elevated in patients with inflammatory bowel disease. Therefore CEA is usually not helpful in diagnosis but can be useful in measuring response to therapy or identifying the presence of recurrent disease at follow-up.

Cancer-related antigen tumor markers include prostate-specific antigen (PSA), CA125, and CA19-9. PSA is elevated in both inflammatory and malignant prostate disease. This means that it can support but not guarantee a diagnosis of cancer and can be very useful in monitoring the response to treatment. It has also been proposed as a marker to be used for population screening, but its utility as a screening tool remains controversial.

CA125 is elevated in approximately 80% of patients with ovarian cancer. Again, it can be extremely useful in monitoring response to treatment and detecting early relapse during follow-up. Indeed, in many patients the serum CA125 may become elevated before there is radiologically confirmed disease relapse, and as has been discussed at various points in other chapters in relation to prostate cancer, is this useful if the appropriate action is then uncertain? With this in mind, a randomized study has confirmed that early introduction of chemotherapy in ovarian cancer does not improve overall survival (Rustin et al., 2010).

CA19-9 is a tumor marker that is elevated in approximately 70% of patients with pancreatic malignancy. It is often a useful diagnostic tool, in combination with biopsy, to establish the etiology of pancreatic lesions. It can be particularly valuable to support a diagnosis in the significant number of patients for whom histological confirmation proves impossible.

Examples of hormones that are also tumor markers include human chorionic gonadotrophin (HCG) and calcitonin. An elevated HCG is pathognomonic of choriocarcinoma and in combination with alpha fetoprotein (AFP) is highly specific for testicular teratoma. These tumor markers are valuable as prognostic indica-

The Diagnosis of Cancer

Table 17.2 Tumor markers and their uses

Biomarker	Type	Cancer	Use
Carcinoembryonic antigen (CEA)	Oncofetal protein	Colorectal and breast	M
Alphafetoprotein (αFP)	Oncofetal protein	Hepatocellular	S, D, P, M
Alphafetoprotein (αFP)		Germ cell tumors	D, P, M
CA 19-9	Cancer-related antigen	Pancreas	D, M
CA15-3	Cancer-related antigen	Breast	D, M
CA 125	Cancer-related antigen	Ovary	D, M
Prostate-specific antigen (PSA)	Cancer-related antigen	Prostate	S, D, M
Beta choriogonadotrophin (βHCG)	Hormone	Germ cell tumors	D, P, M
		Choriocarcinoma	D, P, M
Calcitonin	Hormone	Medullary carcinoma of thyroid	D, M
5 Hydroxy indoleacetic acid (5-HIAA)	Hormone	Carcinoid	D, M
Lactate dehydrogenase	Enzyme	Germ cell tumors	P, M
		Lymphomas	

Note: D = diagnosis; P = prognosis; S = screening; M = monitoring of treatment.

tors in patients with teratoma, in addition to enabling the monitoring of response to treatment. Calcitonin is elevated in patients with medullary carcinoma of the thyroid and is used in detection, diagnosis, and follow-up after treatment.

There are now published recommendations for the use of tumor markers in clinical practice, for example through the American Society of Clinical Oncology. In the future, it is hoped that these markers will have greater specificity and sensitivity.

Genetic tests

Diagnostic DNA testing is now available for a number of hereditary cancers. These include von Hippel–Lindau disease, breast and ovarian cancer syndrome, and the Li–Fraumeni syndrome. DNA is usually extracted from blood cells and profiled for cancer predisposition genes and mutations. When a specific gene mutation has been identified in other family members with cancer, a relative can be screened to determine whether they carry that specific mutation and whether this places them at high risk of developing the malignancy themselves. Genetic testing can therefore provide some individuals with their personal lifetime risk of developing a cancer. These presymptomatic subjects can then enter intensive screening programs, or even elect to have prophylactic surgery performed. An example of this would be the decision to undergo bilateral mastectomy for a woman whose family carries the mutated *BRCA1* gene. (See Tables 17.3 and 17.4.)

Inherited genetic predisposition might contribute to the development of many cancers (see Chapter 3), but in the vast majority of cases may prove too complex to provide a readily applicable screening tool in the foreseeable future. In addition, mutations can occur during life in genetic sites which have not yet been identified. This therefore means that while DNA testing is currently available, it is actually helpful in only a small number of people, representing a minority of patients with cancer. The sequencing of the human genome has focused huge amounts of research in this area, and in the future the value of genetic testing is likely to increase for the whole population.

Screening tests

The objective of screening a population for cancer is to detect tumors at the earliest possible stage while they are potentially

Table 17.3 Chromosomal analysis in leukemia

Leukemia	Abnormality	Implication
CML	t(9;22)(q34.1;q11.2)	Diagnostic
CLL	Trisomy 12	Poor prognosis
	Deletion 13q14	Worst prognosis
ALL	t(9;22)(q34.1;q11.2)	Poor prognosis
	t(11;14)(p13;q11)	Poor prognosis
	t(8;14)(q24;q11)	Poor prognosis
AML M2	t(8;21)(q22;q21)	Good prognosis in young adults
M3	t(15;17)(q22;q11)	Best prognosis of all AMLs
M4	Inversion of 16	Good prognosis
M5	11q23 abnormalities	Poor prognosis

treatable and thus curable. This straightforward aim is actually a complex process. In an ideal screening program, the natural history of the disease should be well understood with a recognizable early stage that is treatable. Both the sensitivity (the chance that someone with the disease will test positive) and the specificity (the chance that someone without the condition will test negative) should be high. The test should be acceptable to the public, and the cost should be balanced against the benefit it provides. Unfortunately most proposed screening tests have failed to meet all of these criteria. Interpretation of the evidence is made difficult by two key biases in the data, firstly the so-called lead time bias. This occurs when a cancer is detected earlier than would be normally expected, but due to ineffective treatment the date of death remains the same. Screening in this case has produced no benefit, but the time from diagnosis to death has increased, giving an illusion of benefit. The second bias is related to a tendency to pick up more benign, slow-growing disease in screening tests. This has two consequences, firstly causing an average improvement in prognosis for the whole group and thus giving a false impression of the effect of screening, and secondly potentially overtreating the population with benign disease. This, along with the extra anxiety that a positive test causes (for every breast cancer detected at screening, 10 women are recalled for

Table 17.4 Familial cancer syndromes

	Defective gene	Chromosome location	Malignancy
Autosomal dominant			
Retinoblastoma	RB1	13q	Eye
Wilms' tumor	WT1	11p	Kidney
Li–Fraumeni	P53	17p	Sarcoma, CNS, and leukemia
Neurofibromatosis 1	NF1	17q	Neurofibromas
Von Hippel–Lindau	VHL	3p	Hemangioblastomas or renal cell
Familial adenomatous polyposis	APC	5q	Colon
Hereditary nonpolyposis colon cancer (Lynch syndrome)	MLH1 and MSH2	3p, 2p	Colon or endometrium
Breast ovary families	BRCA1 and BRCA2	17q	Breast or ovary
Multiple endocrine neoplasia 1	MEN1	11q	Pancreatic islet cell, adrenal cortex, or thyroid
Multiple endocrine neoplasia 2a	MEN2/RET	10q	Thyroid or pheochromocytoma

assessment), makes the analysis of cost-effectiveness for all screening programs a difficult one.

Cervical cytology

The screening test used in the early detection of cervical cancer is the Papanicolaou smear (Pap smear). This involves taking a scraping of cells from the transformation zone of the cervix (the area where columnar epithelium changes to squamous morphology). The precancerous lesion in the cervix is cervical intraepithelial neoplasia (CIN). Cytological studies can grade the CIN from I to III where CIN III is the most severe grade (approximately 30% of these precancerous lesions will develop into invasive carcinomas if left untreated). By treating patients with CIN changes, it is thought that the incidence of invasive cancers could be reduced by up to 90% if the screening test was offered to women every 3 years. There has also been interest in screening patients for the human papillomavirus (HPV), which is present in the majority of women with cervical malignancies. In the United Kingdom, a vaccination program has been in use since 2008 to vaccinate girls aged under 18 years against HPV 16 and 18 in an attempt to prevent them getting cervical cancer.

In practice there have been problems with the cervical cytology-screening program. Some of these difficulties have been related to the accurate reporting of CIN I to III. Quality assurance and audit have, to a large extent, rectified the interobserver variability that was a concern in early years. However, those most at risk of cervical cancer are women who have been sexually active with multiple partners at a young age, and who are generally of poor socioeconomic background. These are not typically the individuals who come forward for screening, and efforts are still ongoing to encourage these women to attend.

Mammography

Mammography involves an X-ray examination of the breast. There have been numerous randomized controlled studies investigating the effectiveness of mammography to detect breast cancers at an early stage. At the outset, the use of mammography as a screening test for breast cancer held much promise as this is one of the commonest malignancies, and mammography can detect tumors before patients are themselves aware of them. However, to date the overall results from screening programs worldwide have been the subject of much debate with doubt being raised over whether real reductions in mortality have been seen. There does seem to be an increased rate in mammographic detection of early cancer, but this does not necessarily correlate with a survival advantage. The false positive rate is considerable which makes the cost of screening high, both financially and in terms of the anxiety to women. A Danish study suggested no change in death rates in two populations with and without screening in Denmark, and the following correspondence rebutting this paper gives a feel for the controversy in this area. Whatever your individual conclusion as to the worth of breast-screening programs, it is clear that the benefit to the individual is not guaranteed, and there is a risk of overtreatment.

Currently in the United Kingdom, women are invited for screening on a 3-yearly basis, although the number of interval cancers (i.e. those cancers that develop in between the appointments) remains high. It is also contentious as to what age women should commence screening. In the United Kingdom, the national screening program invites women between the ages of 50 and 70 years. There is a take-up rate of 75%, and it is estimated that 10 000 women are diagnosed with screen detected breast cancer annually.

Prostate cancer screening and PSA

Prostate cancer is the most common cancer in men. It arises from the prostate gland, which lies just below the bladder. Unfortunately the symptoms of prostate cancer mimic those of a benign aging process called benign prostatic hypertrophy (BPH). These symptoms, such as a poor urine stream, having to pass urine at night (nocturia), and frequency and hesitance of urination, are very common. As a result, men may not seek medical advice. Prostate screening has been advocated to demonstrate prostate cancer at the early stage when local treatment options are available and so morbidity and survival from prostate cancer improve. It is possible to test for prostate cancer using the PSA tumor marker blood test, but this test is elevated in benign disease and increases with age as a natural process. Once an elevated PSA is found, men have to undergo a digital rectal examination to estimate the size and shape of the prostate gland and then undergo biopsies to identify adenocarcinoma cells.

Screening for prostate cancer is a matter of routine in the United States, but in the United Kingdom its role is contentious. A number of randomized studies have been carried out to see whether screening with PSA with or without rectal examination does improve survival. A recent meta-analysis suggests that routine screening for prostate cancer is not justified. There is a concern that screening merely increases the number of cancers diagnosed, but this does not cause an improvement in survival as these are low-grade tumors which were unlikely ever to have become life-threatening. There is no doubt that some men develop an aggressive form of prostate cancer, and therefore screening needs to identify these men specifically so that they can be offered appropriate early treatment.

Colonoscopy and fecal occult blood testing

Colorectal cancer should be an ideal tumor to screen. It is increasing in frequency in the Western world, and there is a clear precancerous lesion, the adenomatous polyp, which may form years before transformation to invasive malignancy. The genetic stages involved in the adenoma-to-carcinoma sequence have been well described. In the general population, there is a lifetime risk of about 20% of developing an adenoma, and approximately 2–5% of these will then become an invasive cancer. There have been large studies looking at the usefulness of fecal occult blood testing, flexible sigmoidoscopy, and colonoscopy as screening tools for colorectal cancer.

Colonoscopy involves passing a fiber-optic scope through the bowel, allowing both detection and treatment of abnormalities. Patients have to take strong laxatives to ensure that the bowel is clear, and the endoscope is not obscured by feces. It is then possible to examine the left side of the large bowel (flexible sigmoidoscopy) or large bowel from the rectum to the cecum and terminal ileum (colonoscopy). One of the major advantages of this test is that abnormal areas can be photographed and biopsied so that a definitive diagnosis may be made (Fig. 17.1). With the advance in technology in this area, it is also possible to perform endoscopic resections of small polyps such as adenomas in the bowel. This means that a potentially neoplastic lesion can be either biopsied or removed at the time of the investigation. The downside of the procedure is that it is uncomfortable and expensive, and relies on appropriately trained staff. As a result of these disadvantages, fecal occult blood testing is often used as a simple non-invasive test with only those who have positive results proceeding to colonoscopy.

A positive fecal occult blood test suggests that a tumor may be present in the colon. However, in the past there have been high false positive and negative results. With the advent of more sensitive fecal occult blood techniques, it does appear that this is a useful screening tool. Once there has been a positive fecal occult blood test, the patient needs to undergo flexible sigmoidoscopy or colonoscopy. A large randomized trial of patients aged 45 to 74 years was carried out offering people either 2-yearly fecal occult blood testing or routine medical management (Hardcastle et al., 1997). Those with positive results underwent colonoscopy, and a 15% reduction in mortality was seen in the screened population. This approach provides the basis for the UK screening program where people between the age of 50 and 79 years are invited to undergo fecal occult blood testing. In the United States, flexible sigmoidoscopy is used as the screening test for colorectal cancer in patients over the age of 50 years, with claims being made for 40% reductions in the incidence of malignancy in the population who attend. A recent large study in the United Kingdom has found that once-only flexible sigmoidoscopy in a patient group of 55–64 years reduced the incidence of colorectal cancer by 23% and mortality by 43% (Atkin et al., 2010).

Screening may be changed by virtual colonoscopy using CT. Although this is easier for the patient, it is not greatly so as bowel preparation is still required and air insufflations during the procedure. However, scans can uncover other causes of symptoms. We are awaiting the results of trials comparing virtual colonoscopy with colonoscopy and barium enema.

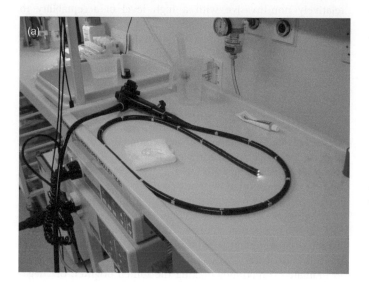
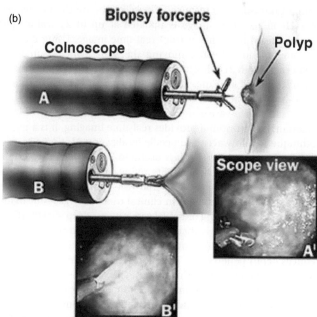

Figure 17.1 (a) Photograph of a colonoscope. (b) Principle behind polyp removal at colonoscopy, with view of polyp in situ down the colonoscope and biopsy forceps snaring the polyp off.

A.L. Thomas, B. Morgan, and W.P. Steward

Non-invasive imaging techniques

Plain film X-ray

Plain-film radiography (X-ray) uses the differential attenuation of X-rays passing through the body to obtain an image related to tissue density and atomic number. The film is a "negative" so dense tissues, or those of high atomic number, appear whiter. Technology has advanced over the last century since the first radiographs were created by Wilhelm Roentgen in 1895, and the process can now be completely digitized. By using different approaches, images can be optimized for soft-tissue contrast such as in mammography or spatial detail such as for subtle bony fractures. The information provided by plain radiographs in the detection of cancer is limited when compared with other available imaging techniques as it is a 2D representation of a 3D object, and areas of interest are often obscured by overlying structures. Plain radiographs of the skeleton can be used to detect lytic or sclerotic bone metastases, but up to half of cortical bone has to be destroyed to make some lesions visible. Therefore bone scintigrapy and/or magnetic resonance imaging (MRI) are preferred tests for early diagnosis of bone lesions. A skeletal survey, which includes plain views of the skull, thoracic spine, lumbar spine, and pelvis, can be useful to determine the extent of multiple myeloma, where bone scintigrapy is often negative. Although a normal chest X-ray (CXR) in a patient with cough and hemoptysis does not exclude lung cancer, the presence of a visible mass on CXR does allow follow-up studies, not only for tumor response but also to check for complications of therapy such as pneumonitis or infection. As bowel gas is well seen, an abdominal radiograph is useful in cancer patients who develop abdominal pain to rule out bowel obstruction.

Ultrasound

There have been major advances in ultrasonography over the past 10 years. The technique involves interpreting the different echo patterns returned when sound is passed through an organ from an ultrasound transducer. These echoes are then displayed as a 2D image. This image is updated at up to 25 frames per second, allowing a "flicker-free" real-time image. Using modern equipment and software, moving the scan plane through an organ can obtain 3D ultrasound images. As these images can also rapidly updated in time, this is referred to as 4D ultrasound. There are several advantages to this technique as it is cheap, quickl and non-invasive and does not subject patients to ionizing radiation. As ultrasound provides real-time imaging, it is a useful technique to allow accurate needle localization for tissue biopsy. The disadvantages are that it is useful to investigate only certain organs, it is operator dependent, and it is difficult to reproduce exact scanning parameters on serial visits, diminishing its use as a marker of tumor response in clinical trials. Any gas–soft tissue interface such as in lung or bowel gas prevents deep visualization. Another compromise is that the spatial resolution of ultrasound improves by increasing sound frequency, but this conversely reduces depth penetration. Therefore very high-resolution scans can be achieved for eye, skin, and testicular lesions and any body cavity where the ultrasound probe can be placed close to the tissue of interest. Ultrasound is very good at demonstrating blood flow (Doppler shift techniques) and fluid. This is useful for the detection of obstructive jaundice dilating bile ducts or obstruction to the kidney (hydronephrosis).

When a person presents with abdominal symptoms, an abdominal ultrasound is frequently the first screening test performed. While it is often effective at assessing the etiology of conditions such as obstructive jaundice, it may be less sensitive in imaging the pancreas due to overlying bowel gas. Ultrasound can reliably distinguish solid from cystic lesions and is therefore useful to differentiate between a benign and a malignant lesion in the kidney, liver, thyroid, and ovary.

Other areas where ultrasound is particularly useful include visualization and biopsy of breast lesions, characterization of testicular masses, detection of liver lesions, and evaluation of gynecological malignancy, particularly using a transvaginal ultrasound probe. Using endoscopic ultrasound techniques allows the probe to be much closer to the tissue of interest and avoids overlying lung or bowel gas. This has proved very useful in the local staging of esophageal and pancreatic malignancies.

Intravenous contrast agents are now available that enhance the ultrasound signal from highly perfused tissues. This provides both better visualization of tumors such as liver metastases and can also help monitor therapies that affect vasculature such as antiangiogenic therapy.

Computerized tomography (CT)

CT scanning (Fig. 17.2, Box 17.2) currently provides the cornerstone of imaging in cancer medicine. As for plain X-rays, the information obtained is related to tissue density but overcoming the problem of overlying structures. Information is presented a set of slices of predetermined thickness through the volume scanned. Advances in spiral (helical) and, more recently, multi-slice techniques have increased the scan speed and improved registration between images slices, allowing images to be reconstructed in any plane without overt loss in image quality (Fig. 17.3). In addition, new techniques in 3D imaging and virtual CT imaging have paved the way for virtual bronchoscopy and colonoscopy. The advantages of these techniques are that they are relatively non-invasive with a high level of acceptability to patients. They can also guide the radiologist to a site of interest when undertaking a biopsy. CT scanning is therefore useful in the diagnosis of cancer, in determining the anatomical extent of disease (stage), and also as the follow-up imaging technique when monitoring the anatomical tumor response to treatment. It is now common practice when performing CT to use intravenous contrast agents to give increased visualization of lesions and information on the vascularity of the region of interest. These agents are small-molecule iodinated compounds that are rapidly dispersed into the vasculature; leak freely into the extravascular, extracellular space; and are rapidly cleared by glomerular filtration. Further information about contrast enhancement is given in the "Magnetic resonance imaging" and "Future novel uses of imaging" sections of this chapter. Further contrast strategies include the use of oral contrast when examining the upper gastrointestinal tract and rectal contrast when imaging the lower gastrointestinal tract. This can be both "positive" contrast, such as iodinated compounds making the lumen brighter, or "negative" contrast, such as water or fatty drinks (e.g. cream). Generally a small amount of water before a scan is well tolerated by cancer patients, but a liter of iodinated oral preparation drunk over the hour before a scan, which is common in many CT protocols, is often the worst part of the scan experience for patients. As CT technology has improved and bowel is easier to identify

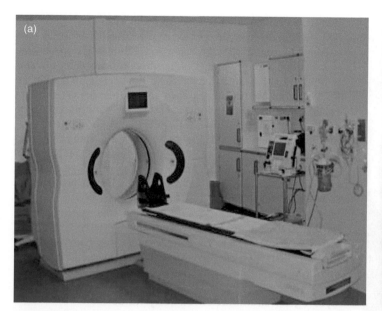
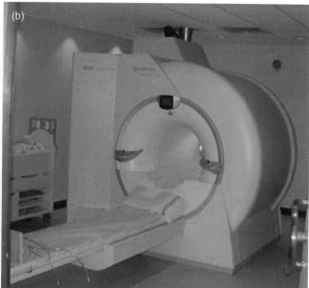

Figure 17.2 (a) CT scanner and (b) MRI scanner. Note the much longer imaging tube in which patient lies in the MRI scanner.

Box 17.2 CT and MRI acquisition

- *For computed tomography (CT)*: A collimated X-ray beam moves synchronously with its detectors across a slice of the body part of interest. A computer processes the transmitted X-irradiation for each element, or pixel within the slice. A value is generated (Hounsfield number, named after its inventor) depending on the density of the structure (bone is 1000 units, water is 0, and air is −1000). The difference in density (X-ray attenuation) between the different structures in the slice makes it possible to differentiate areas of normal and abnormal tissue on the computer-generated image.

- *For MRI*: The hydrogen nucleus is a proton whose electrical charge creates a local electrical field. Therefore it is possible to align protons with a strong magnetic field. When a radiofrequency wave is applied at 90 degrees to the alignment of the protons, they resonate and spin before returning to their previous alignment. During this process, images are made in the different phases of relaxation of the protons (e.g. T1 and T2 sequences).

By comparing these different sequences, it is possible to distinguish between normal and abnormal tissues.

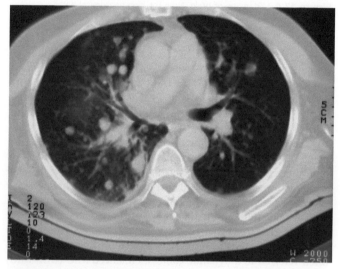

Figure 17.3 CT image of thorax showing heart in center with surrounding lung in which circular metastases are clearly seen.

as separate from other structures, many centers are abandoning the routine use of positive oral contrast.

Magnetic resonance imaging

MRI is an extremely valuable technique which provides excellent anatomic definition of structures and avoids the use of ionizing radiation. There are, however, some contraindications including the presence of cardiac pacemakers or other metallic foreign bodies. In addition, the procedure can be time consuming and claustrophobic for some people, and the refusal rate can approach 10%. MRI is of particular value in imaging the brain (Fig. 17.4), spinal cord, bones, and joints and can show areas of nerve compression and cerebral tumors with great accuracy. MRI works by imaging hydrogen protons, the majority of which are in water but they are also found in fat and other organic compounds. By altering sequence parameters, imaging can be made sensitive to different tissue physical or chemical properties, including T1, T2, T2* weighting, and fat and water suppression. Using different image weighting can both identify and characterize lesions. The multiple ways in which an MRI scan can be performed is clearly an advantage but adds complexity. Imaging protocols are always a compromise between information gained, time taken, and cost. Currently MRI is the investigation of choice for the local staging of rectal, prostate, cervical, endometrial, bone, and brain tumors.

Contrast agents are available that affect both the T_1 and T_2 image characteristics. The most commonly used are gadolinium chelate-based contrast agents. These have different physical

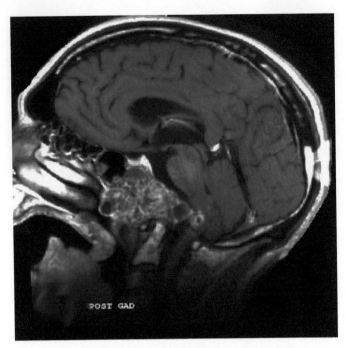

Figure 17.4 MRI image of brain. Note the high anatomical detail demonstrated. There is a large tumor in the pituitary region.

properties to CT iodinated contrast media, but the enhancement information obtained is similar. These are very good at identifying brain lesions where there is a breakdown of the blood–brain barrier causing increased contrast relating to the "non-enhancing" normal brain. Tumors in the body also have characteristic enhancement patterns, but this may be obscured by enhancement of surrounding normal tissues. Therefore, in abdominal indications, contrast is often given as a rapid bolus with rapid sequential imaging afterward. A typical imaging sequence may involve a precontrast image and then further acquisitions at 20, 70, 160, and 300 seconds after contrast administration. These are often referred to as "dynamic scans." When scans are obtained more rapidly, and the resulting information quantified, this is often referred to a dynamic contrast enhanced MRI (DCE-MRI) or, for CT, DCE-CT.

DCE-MRI can be useful for identification, characterization, and follow-up of tumors on therapy. These techniques are now being deployed in clinical practice, particularly in breast cancer, for both screening and lesion evaluation.

More specific MRI contrast agents are becoming available such as the supra-paramagnetic iron oxide–related media. These can also be useful in establishing the nature of liver lesions, as normal liver Kupffer cells will take up the media causing signal fallout and making abnormal liver cells "enhanced." This is useful for identifying lesions such as metastases with no Kupffer cells and also characterizing benign lesions such a focal nodular hyperplasia. May new contrast agents are in late stages of development and should further increase the value of MRI in the diagnosis, staging, and follow-up of malignancies.

Tissue diagnosis

The histology of cancer has been discussed in Chapter 3. Although imaging and tumor markers can provide strong evidence to support a diagnosis of malignancy, it is of paramount importance that tissue samples are obtained from abnormal areas within the body so that they can be sent to the pathology laboratory where a definitive diagnosis can be made. In addition, it is now possible to use an increasing number of immunohistochemistry markers to understand the biological nature of tumors (e.g. to provide an assessment of how aggressive they are likely to be). The ease of obtaining a biopsy depends on the site of the abnormality. There are some tumor deposits which are easy to biopsy by direct excision (e.g. subcutaneous deposits). However, the abnormal structure may be quite deep and as such poorly accessible. There are a number of techniques which can now be employed to obtain adequate tissue samples without repeated attempts being necessary. Concurrent imaging (usually with ultrasound) guides the biopsy needle with great accuracy to the appropriate site and minimizes the risk of damaging other organs. Methods of obtaining tissue or cells for analysis include cytology, fine needle aspiration, needle biopsy, incision and excision biopsy, and the use of endoscopy.

Cytological examination of body fluids

For some tumors, it may be appropriate for a small amount of body fluid (e.g. sputum, ascitic fluid, cerebral spinal fluid, pleural fluid, or urine) to be collected and spun down using a CytoSpin. Any cells present are then stained and examined under the microscope immediately after collection. It is often possible to establish whether cancer cells are present and give a rapid diagnosis. The main disadvantage is the frequency of false negative results, normally because the specimen taken is not large enough. An advantage of this technique is that it is a quick procedure that can be performed as an out-patient with minimal discomfort.

Fine needle aspiration (FNA)

A fine needle aspiration (FNA) involves the passage of a hypodermic needle into the abnormal area under investigation. Ultrasound or CT can be used to guide the needle to the suspected tumor if it is not possible to palpate it. Cells are then aspirated, put on a microscope slide, and sent to the cytology laboratory. This technique is particularly useful for assessing breast lumps in the outpatient setting. It can discriminate between reactive and malignant cells, and may sometimes provide further information about tumor type if specific staining techniques are employed.

Needle biopsy

This technique is more invasive than FNA as a larger biopsy needle (Trucut) is used under a local anesthetic. This means that an actual core of tissue can be sampled and sent to the pathology laboratory for fixing and mounting in wax. As this is a larger specimen, more information is potentially available. For example, an FNA of a breast lump can determine whether the lump is merely a benign fibroadenoma or malignant tumor. If a Trucut needle biopsy is used, it is possible to carry out immunohistochemical analysis and determine whether the breast cancer is expressing estrogen or progesterone receptors and what the tumor grade is.

Incision and excision biopsy

These techniques involve removing a sample of tissue under local anesthetic. For example, a punch biopsy using a small circular blade can be used to take a sample from a skin lesion. In practice, it is often preferable to excise an abnormal area totally. This is known as an excision biopsy, and in some cases the biopsy itself

The Diagnosis of Cancer

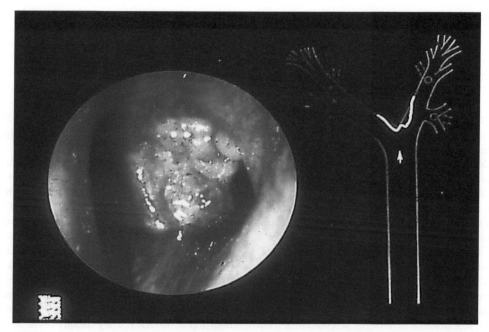

Figure 17.5 Photograph of carcinoma of bronchus as visualized down a bronchoscope. The site of the tumor is illustrated on the diagram to the right.

may be curative. When this form of biopsy has occurred, it is important to examine the periphery of the lesion to see whether the tumor has involved the microscopic margins. If this is the case, then further excision would usually be required.

With more advanced ultrasound and CT techniques, it is now often unnecessary for a patient to have to undergo a surgical procedure, such as laparotomy, to determine the nature of abnormal tissue. In most instances, a guided biopsy will provide sufficient tissue to make the diagnosis.

Bronchoscopy

When a patient presents with an abnormal area on a chest radiograph, the initial investigation is almost always a bronchoscopy. This involves having a fiber-optic instrument passed through the nose or mouth into the pulmonary tree. It gives extremely good access to lobar and proximal bronchi but is not usually helpful in assessing more peripheral lesions. When a bronchoscopy is performed, the abnormal area can be visualized (Fig. 17.5) and either washed out (and the washings sent for cytological examination) or biopsied when a sample of tissue can be sent to the pathology laboratory. In some cases, it is possible to actually establish, at the time of the bronchoscopy, whether a tumor is likely to be operable or not (e.g. tumors within 2 cm of the main carina would be deemed inoperable). Sometimes a rigid bronchoscope has to be used, and this is performed by a thoracic surgeon under general anesthetic. The main indication for the latter procedure is to visualize more difficult tumors and to have more control over blood loss at the time of biopsy. The rigid bronchoscope can also be used to study the mediastinum, and this again helps determine the operability of a tumor.

Endoscopy of the upper GI tract

Upper GI endoscopy is used to assess the esophagus, stomach, and first part of the duodenum. It involves the passage of fiber-optic instruments through the esophagus into the stomach. Abnormal areas can be visualized, and biopsies taken. When undertaken by a skilled operator, a positive biopsy is available in about 90% of patients. The areas where false negative biopsies occur are usually when submucosal gastric tumors are present. In this situation, a double biopsy may be necessary. Obviously this carries with it a higher risk of perforation, and this complication always has to be balanced against the potential value of the information to be gained.

Endoscopy is extremely useful to identify the source of hemorrhage. If this is from an ulcerated area where a bleeding vessel is noted, the vessel can be sclerosed using an injection (sclerotherapy). It is also possible to use laser techniques to ablate obstructing tumors. This is a simple method which can be carried out under light sedation and rapidly relieves symptoms of dysphagia.

For small lesions which are confined to the mucosa of the esophagus or stomach, it is now possible to remove them using a technique called endoscopic mucosal resection (EMR). This is a specialized technique where saline is injected under the lesion to raise it from the adjacent tissue, and then a snare is passed down the endoscope which is placed around the lesion to remove it. The area is then sent to the laboratory to ensure that the lesion has been removed in its entirety.

As mentioned in this chapter, ultrasound probes are now available that can be used endoscopically (EUS). These are very important in the staging of primary tumors of the esophagus and stomach. In some cases, it is even possible to use EUS to target a lymph node of interest and biopsy it.

Endoscopic retrograde cholangiopancreatography (ERCP) is a variation of standard endoscopy. The scope used has a side arm that allows cannulation of the pancreatic duct. Dye can then be injected to delineate the anatomy of the pancreatic and biliary system. Malignant strictures can be identified and brushings taken to produce cells for cytological examination. It may also be possible to alleviate an obstruction causing jaundice by placing a stent through the stricture.

Additional techniques used in diagnosis

There are some patients for whom a tissue or cytological diagnosis cannot be made. This may occur when the site of abnormality cannot be accessed or when biopsies fail to contain tumor. Other techniques may be employed to support a diagnosis of malignancy.

Interventional radiology

Imaging is used to support a wide variety of minimal access diagnostic and treatment procedures in cancer. Treatment procedures include image-guided insertion of stents and drains to deal with tubes blocked by cancer, including blood vessels, the GI tract, and the biliary and renal systems. Diagnostic CT and MRI scans have been improved in the past by insertion of cannulae into the vascular system to obtain blood samples or inject local contrast media directly into vessels such as hepatic artery or other splanchnic vessels to gain more specific tumor enhancement information. With the advent of superior CT and MRI techniques, these invasive diagnostic procedures have a diminishing role. Portal venous sampling is occasionally used to localize pancreatic endocrine tumors such as gastrinomas. These are "functioning" tumors and produce gastrin. It is possible to perform percutaneous transhepatic portal venous sampling and measure the hormone levels in the superior mesenteric and portal veins. The sensitivity of this test is over 95%, and the venous sampling gives precise anatomical mapping of the tumor so that radical resection can be planned.

Angiography

Angiography is used to accurately delineate the vasculature of the organ under study. It is most commonly used in the diagnostic evaluation of liver, renal, or pancreatic tumors. Again, with the advent of more sophisticated imaging techniques such as MRI, angiography is being used less frequently. It may sometimes be used to image the vascular blood supply around a pancreatic tumor and thereby allow surgeons to determine whether the tumor is resectable. Another situation is the use of angiography to determine the blood supply of liver lesions. It is now possible to administer intra-arterial chemotherapy or, more recently, radio-labeled microspheres for the treatment of liver metastases (Gray et al., 2001).

The technique of transarterial chemoembolization is being used more widely now to treat a number of tumor types including primary liver carcinomas. With this technique, high-dose chemotherapy is injected locally into the tumor bed and the blood supply to the tumor is also compromised. Usually this is a palliative procedure, but with technical advances this could be used with curative intent in the future. In these circumstances, embolic contrast material can also be injected in the tumor circulation. This lodges in the tumor and can remain visible for many weeks, allowing follow-up of tumor size response.

Radionucleotide imaging

Radio nucleotide studies use a radioactive tracer bound to a biochemically active molecule. The biochemical properties of the molecule dictate the information derived from the patient, which can be related either to a physiological process such as biphosphonates, linking to bone metabolism, or to molecular processes such as binding to specific cell surface receptors. The radioactive tracer and imaging equipment dictates the sensitivity to detection of the compound and the spatial resolution and accuracy of detection. The advantages of these studies is that the tracer can be injected at very low concentrations, reducing side effects and any unwanted physiological effects, but still provide a wealth of physiologic or molecular-specific information. The disadvantages relate mainly to low spatial resolution of detection. This has slowly improved over the last 30 years with improvement in gamma camera technology leading to single-photon emission computed tomography (SPECT). In recent years, the increasing use of PET has been considerably improved by concurrent registration with CT scans creating CT–PET images (Fig. 17.6). The basic principles of SPECT and PET are the same in that a biochemical compound is labeled by a radioactive tracer, a gamma ray emitter for SPECT, and a positron emitter for PET. PET has advantages in that the spatial resolution of detection is better for positron emitters and positron emitters tend to be of low atomic number, and can therefore be substituted into molecules without changing their biochemical properties. This has caused an explosion of interest in PET for oncological imaging.

The most common of the general scans images the bones. Radiolabeled bisphosphonates are administered intravenously to the patient, who is then scanned with a gamma camera. A positive bone scan involves the presence of "hot spots" or areas of high signal intensity. They usually occur due to increased levels of blood flow and osteoblastic activity (Fig. 17.7). The bone scan is therefore extremely useful in the detection of sclerotic bone metastases. It is now possible to use radioactive strontium to image and treat bone metastases as the radioactivity is localized to tumors and emits sufficient energy to kill neoplastic cells.

A highly tumor-specific radiolabeled scan targets thyroid tissue. ^{131}Iodine is taken up to a high level by the organ, whereas minimal activity is found in other sites. This can be used to stage a thyroid cancer or at a therapeutic level to treat thyroid malignancy. Another tumor-specific scan is the adrenal scan using ^{131}iodine metaiodobenzylguanidine (MiBG). This agent is valuable in assessing any metastatic deposits of pheochromocytomas. Octreotide is a synthetic somatostatin analogue that is used in the treatment of neuroendocrine tumors. ^{111}In pentetreotide is used in an "octrescan" and utilizes the characteristic of the agent to be taken up by tumor cells which express somatostatin receptors. The octreoscan is valuable to stage and evaluate whether the neuroendocrine tumor will respond to somatostatin as a therapeutic maneuver.

PET–CT scanning is currently mainly performed using the tracer ^{18}F fluoro deoxyglucose (18-FDG). This is a glucose analogue and is taken up by cells and then trapped in the cells after phosphoylation. Therefore 18-FDG concentration, after initial circulation, relates to glucose uptake. Although 18-FDG uptake is not tumor specific and is seen in heart, brain, and metabolically active fat, tumors often have high uptake due to the increased energy demands of high anaerobic metabolism. Currently 18-FDG PET–CT is expensive as the equipment is about twice the price of a standard CT scanner, appointment times are about three times as long, and, due to its short half-life (110 minutes), the tracer has to be prepared every day. Despite this, 18-FDG PET–CT is now in routine practice for the staging of malignancies such as lung, lymphoma, esophagus, head and neck, and melanoma. The benefit of PET is generally in identifying distant metastases not visible using CT or magnetic resonance techniques. Useful information is also obtained in the follow-up of tumors on treatment such as lymphoma where complete reduction of FDG uptake is associated with good outcome even in the presence of a residual mass.

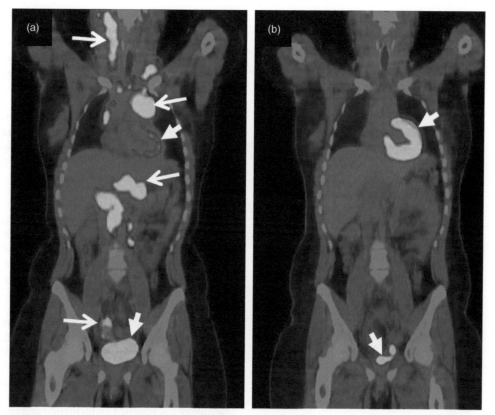

Figure 17.6 Radioisotope 18-FDG PET–CT scan. Co-registered image of CT and PET scan performed on the same table sequentially. (a) Extensive nodal disease in the neck, mediastinum, retroperitoneum, and pelvis (open arrows) and normal uptake in the heart and bladder (closed arrows). (b) Complete "metabolic" response post-chemotherapy.

One of the problems with FDG PET is the lack of specificity to tumors, leading to false positive uptake in metabolically active tissue, and also the disguising tumor response due to increase in FDG uptake secondary to inflammation. As fluorine substitutes for a hydroxyl group, it can be substituted into a wide range of molecules. Labeling of carbon and oxygen is also possible, albeit with a very short half-life. There are therefore numerous more tumor-specific strategies available, some of which are being introduced into clinical practice. This allows a wide range of biomarkers to be considered focused on hallmarks of cancer such as cell proliferation, angiogenesis, apoptosis, and hypoxia.

One such tracer is ^{18}fluoro thymidine or 18-FLT monitoring proliferation. This is potentially more specific than 18-FDG because it is not involved in inflammation and is not taken up by the normal brain. This has been used in humans to follow up therapy in breast cancer and is attractive as it may be useful for cytostatic drugs as well as cytotoxic ones as they inhibit proliferation. 18-FDG PET is often not useful in prostate cancer due to low uptake and high incidence of inflammation. ^{13}C or ^{18}F choline PET, however, can show both primary and recurrent prostate cancer. Molecular imaging is also possible using 18-F-based markers for integrin expression, such as 18-F-AH111585, which can show changes after anti-tumor therapy and is currently in early-phase trials.

Molecular imaging using PET markers can be very helpful in identifying the presence and location of therapeutic cell surface targets such as HER2 and VEGF receptor expression. PET studies also have the potential advantage of directly labeling drugs. This allows direct imaging of drug delivery by "micro-dosing," and chemotherapeutic agents, such as 18-F-fluorouracil, have been synthesized to assess their pharmacokinetics and metabolism.

Future novel uses of imaging

Generally, most imaging used in the diagnosis of cancer is focused on an anatomical approach, particularly to identify localized disease that is amenable to cure by surgery or radical radiotherapy. Even where functional imaging techniques are used, such as radionuclide bone scans or PET scans, the main aim is to identify previously occult metastatic disease. However, this fails to address three major issues of key interest to the patient. Firstly, is any **prognostic** information available beyond tumor volume and spread? Secondly, does any information **predict** which therapy may be optimal, especially with the recent rapid increases in therapeutic options? And, thirdly, can any tests show early whether therapy is failing (particularly important where second-line therapies are available)? Although non-imaging biomarkers exist for some tumors to answer these questions, imaging has the advantage of showing, at an identified time point, what is happening throughout the tumor volume and in all tumors in the body.

The recent explosion in new experimental cancer therapies over the last 15 years has been followed by an explosion in potential functional and molecular-imaging strategies to support these aims. Probably the main tests to attempt to answer

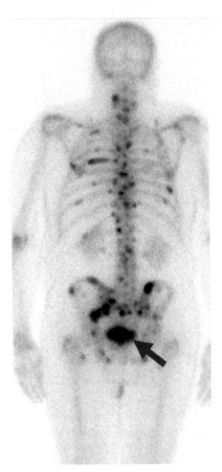

Figure 17.7 A nuclear medicine whole-body bone scan. Technetium-99m diphosphonate is preferentially taken up by active bone. Multiple black spots are metastases except the bladder (annotated).

these questions are FDG and non-FDG PET (described in this chapter), MRI strategies, and recently emerging contrast-enhanced ultrasound.

Functional MRI uses a variety of strategies to obtain valuable prognostic and predictive information. DCE–MRI, described in this chapter, and the similar DCE–CT provide information relating to tumor perfusion and vascular permeability. These parameters, as they relate to increased angiogenesis, may be a poor prognostic marker, as has been shown in cervical carcinoma, but they are also a good predictive marker for treatment, with better perfusion allowing better access to chemotherapy or better oxygenation to improve radiotherapy.

Diffusion weighted MRI (DW-MRI) was originally developed for brain imaging and can rapidly show the reduction in water molecular movement caused by cell death in stroke. In this case, diffusion is reduced when cell membrane pumps fail and cells swell; diffusion will then increase when necrosis develops. Tumors also tend to restrict diffusion, probably due to cellular edema, increased interstitial pressure, and tortuosity of the microenvironment. On treatment, especially when necrosis is induced, this diffusion may be expected to increase. Indeed, whole-body DW-MRI is being developed both as an aid to diagnose and identify tumors and to successfully predict response to therapy.

MR spectroscopy (MRS) was the original application of magnetic resonance *in vitro* but is now available *in vivo*. The technique relies on slightly different spin frequencies of metabolites containing protons (^{1}H spectroscopy), the main one of relevance to cancer being choline, which is increased in tumors with high proliferation. Other nuclei can also be imaged, including ^{31}P, giving metabolic information such as from adenosine triphosphate (ATP), and also injectable "nonradioactive" isotopes such as ^{13}C and ^{19}F. These nuclei can be linked to drugs such as 5-FU to show information about biodistribution.

MRI images can also be made sensitive to changes in blood oxygenation level (BOLD) imaging, which may be useful in drugs that cause hypoxia, and a variety of techniques are available to study the "elasticity" of tissues that may give information of normal tissue and cancer.

Ultrasound is a very useful technique in oncology, as described in this chapter, and its use in novel drug treatment has been improved considerably by contrast-enhanced imaging using micro-bubbles. This use has been further developed in animal models to target the bubbles to vascular endothelial markers of malignancy such as integrins. It is also possible *in vivo* to both target these bubbles and use them to deliver a chemotherapy payload directly to the tumor.

The role of molecular pathology in cancer diagnosis

In the past, the pathologist used light microscopy with hematoxylin and eosin staining to assess the morphological features of a tumor specimen (Chapter 3). The arrival of electron microscopy provided more detailed and higher resolution anatomical information relating to subcellular features. In recent years, molecular pathology has assumed an increasing and often fundamental role in establishing a definitive diagnosis. With more sensitive immunohistochemical and molecular biology techniques becoming available, the biological behavior of the tumor can be predicted. In addition, new molecular genetic assays can provide predictive information as to which treatment option would be superior. For example, the tumor can be stained to see if proteins coding for mismatch repair genes are present, as these would suggest resistance to platinum chemotherapy agents. Routinely the tumors of patients with breast cancer are tested for hormone receptor and *HER2* expression, and tumors of patients with colorectal cancer tested for *K-RAS* mutations. It is now also possible to use sensitive techniques that enable the detection of cancer using very small tumor specimens.

One of the hurdles in developing new molecular techniques has been that the majority of tumor specimens are fixed in formalin and then embedded in paraffin. It is only recently, with the development of techniques which allow the extraction of DNA from these specimens, that molecular biology can be more routinely utilized. When DNA is extracted, it can be directly sequenced by southern blotting and restriction fragment polymorphism. With the sequencing of the human genome, it is now possible to identify point mutations, translocations, amplifications, deletions, microsatellite length instability, and changes in methylation. If a tumor is known to harbor a specific mutation in a given gene, then the altered sequence can serve as a marker for clonality.

Flow cytometry

Flow cytometry is a useful tool to analyze the cellular DNA content in tumor specimens. The DNA index (abnormal cell DNA content relative to normal cells) may indicate a duplication (polyploidy) or loss (aneuploidy) of chromosomes. It is also possible to assess the DNA content of tumor cells and determine which phase of the cell cycle they are in. This can provide information regarding the prognosis of tumors. Numerous different fluorescent dyes are available and can be used to label cells and count the numbers with certain characteristics (e.g. the proportion dying from apoptosis), which can support a diagnosis and could help with the choice of therapy.

Fluorescent in situ hybridization (FISH)

FISH allows specific staining of any given region of the genome. It is possible to use a fluorescent technique to identify chromatin within tumor sections. The DNA probes used for this are a mixture of DNA fragments containing chromosome-specific repeat sequences. FISH is therefore able to identify changes in the genome, for example deletions in regions of the chromosomes, and provides quantitative data. A related technique called chromogenic in situ hybridization (CISH) again detects the number of gene copies in the nucleus; it has some advantages as it may be cheaper, and a permanent record of the result is available on a slide.

Powerful new tools for examining global gene expression changes have also greatly enhanced molecular pathological techniques. DNA microarrays consist of numerous cDNA probes which detect the expression of many thousands of genes in one experiment. This means that one sample can be screened to define which genes are up- or downregulated, or alternatively large numbers of tissues can be examined for the expression of selected genes. These techniques are themselves now being superseded by cheaper and more rapid sequencing techniques which can effectively identify and quantify transcripts for even weakly expressed genes and even noncoding RNAs in single assays. The availability of tissue arrays has also had an impact as it has facilitated the screening of sections from individual tumors for large numbers of potentially disease-relevant antigens and proteins. The results of these high-throughput molecular-phenotyping tools are already being translated into clinical practice and will no doubt increasingly contribute in the next few years.

Proteomics and microarrays

An area in cancer medicine that has expanded dramatically in the last few years is that of biomarker development. The definition of a biomarker is "a characteristic that is objectively measured and evaluated as an indicator of normal biological processes, pathological processes or response (pharmacological or otherwise) to a therapeutic intervention." There are predisposition biomarkers which identify an individual's risk of developing cancer, prognostic biomarkers which identify a patient's outcome from his or her cancer, and predictive biomarkers which identify how a patient is likely to respond to a specific anticancer therapy. The majority of these tests use both proteomics and microarray technology. For example, the MammaPrint test analyzes 70 genes in a breast cancer tissue sample to determine if the cancer has a low or high risk of recurring within 10 years after diagnosis. It was proteomic work studying the HER2 protein in breast cancer that identified it as a prognostic biomarker and also a predictive biomarker in breast cancer, identifying a subgroup of patients most likely to benefit from herceptin therapy. It is now very common for many clinical studies to have a translational substudy where both tumor and serum samples are collected for proteomic work.

Circulating tumor cells

There is great enthusiasm to develop biomarkers in oncology and hematology. These may be prognostic or predictive and in some cases may be a surrogate marker of efficacy of treatment effect (instead of a parameter such as overall survival, which may take years to reach in an adjuvant study). An approach which is currently subject to a huge amount of work is the development of circulating tumor cell (CTC) assays. CTCs can be measured in the blood of patients with cancer, and their presence may correlate with both prognosis and response to treatment. Currently a wealth of work has studied CTCs in breast, prostate, lung, and colorectal cancer. A number of commercially available platforms are available. In the Cell Search system, the whole blood sample is centrifuged and the tumor cells are first enriched immunomagnetically by means of ferrofluidic nanoparticles conjugated to epithelial cell adhesion molecules (EpCAMs). If cells are then stained CD45 negative (a blood cell marker) and nuclear and cytokeratin positive, they are deemed to be originating from the tumor. The number of CTCs depends on the tumor type and can be counted pre- and post-treatment to assess response. Given that cell surface markers can be readily used to isolate such cells, by FACS and allied techniques, they can also be isolated and functionally characterized.

Disease staging

This has been addressed in Chapter 3, so a brief overview is given here.

The accurate staging of a tumor provides information on its anatomical distribution and is essential to ascertain its most appropriate treatment. In general, tumors are staged using the tumor node metastasis (TNM) classification. This standardization of staging is recorded in a consensus publication by the American Joint Committee on Cancer (AJCC) and the Union Internationale Contre le Cancer (UICC). It is now in its seventh edition (2009). In general, the more advanced the stage of disease, the worse the prognosis and the greater the likelihood that systemic therapy will be necessary.

While anatomical staging gives important information regarding the extent of the cancer, it is also necessary to know the degree of differentiation (grade) of the tumor and other factors such as whether there is vascular or lymphatic invasion by malignant cells. Many of these factors are now being incorporated into more sophisticated staging systems to try to refine the appropriate treatment decisions so that only those most likely to benefit will be given specific treatments.

Now, a number of molecular markers can also be examined in tumors to help define prognosis. For example, in breast cancer the presence of the oncogene HER2 is associated with a worse outcome. In this situation, the use of the monoclonal antibody trastuzumab (herceptin), which recognizes this oncogene, can improve progression-free survival in the affected patient. In oncology, we are now entering the era of targeted therapies. In an ideal world, an oncologist would know the genetic makeup of an individual's cancer so that therapy could be individualized with drugs which were most likely to be effective. This would prevent unnecessary toxicity from an ineffective drug. As staging techniques become ever more sophisticated, they will incorporate

Conclusions and future directions

Making a diagnosis of cancer involves a process of first having a high index of suspicion based on the clinical history, which relies on knowledge of typical patterns of presentation, and then proceeding to a series of investigations aimed at confirming or refuting that suspicion. The investigations should be undertaken in a logical sequence, minimizing the number of tests, discomfort, and cost. Results should be available as rapidly as possible to allow early intervention when necessary and to reduce anxiety. Simple biochemistry and hematology measurements taken on blood specimens can provide an indication of malignancy, particularly if blood loss has occurred or if metastases have interfered with liver or renal function. Serum tumor markers can be valuable if elevated as they will support a diagnosis of malignancy and may provide information on prognosis, response, and relapse. Imaging is fundamental to the diagnosis, staging, and follow-up of most malignancies. Plain radiographs have a limited role but can be helpful to show bronchial neoplasms or bone metastases. Cross-sectional imaging with CT or MRI has revolutionized the anatomical resolution obtained and can highlight the likely site of a primary tumor and any metastases. The addition of contrast agents provides further information which can help distinguish abnormalities which indicate a greater likelihood of malignancy. Despite all these techniques, the final unequivocal diagnosis depends on pathological examination of a specimen (aspirated cells or biopsy) obtained from the area of suspicion. This is often facilitated by directing the biopsy needle with ultrasound, CT, or MRI.

Many new imaging and diagnostic techniques are being developed, often based on the increasing understanding of molecular changes which occur in malignancy. These are providing a more confident means of diagnosis on smaller specimens of tissue or numbers of cells and give more information on prognostic features and the potential to gain benefit from new targeted therapies. Increasingly treatment decisions are being driven by knowledge of the biology of each cancer, and information on this is obtained at the time of diagnosis, increasing information about the biology of individual cancers and play an even more pivotal role in treatment decisions.

Key significant event for this chapter

There is no doubt that the development of the TNM classification of malignant tumors has had a huge impact on the successful treatment of cancer. Until that time, there was no internationally accepted, uniform system to stage tumors accurately. This meant that it was difficult to compare the outcome of different treatments in a specific disease type as there was no clear idea of the heterogeneity of the patient population being treated. With the advent of new treatments and clinical studies, it became paramount to have one recognized staging system which would accurately record the anatomical extent of a tumor, thus ensuring physicians were comparing "like with like." In addition it means that patients can be given more accurate prognostic information about their disease. With the advances in radiological techniques, the TNM classification still is central to the interpretation of the radiological reports (Hermanek et al., 2002).

Ongoing questions are as follows:

1. There is a need for a reliable screening tool to detect non-small-cell lung cancer at such an early stage that curative treatments are possible.

2. Now that we are able to detect very small nodules (smaller than 5 millimeters) on multislice CT, we need to advance our technique so that we can characterize them into small metastases or benign disease.

3. We have used our knowledge in immunohistochemistry to identify genes that confer a worse prognosis for patients, for example *HER2* in breast cancer. We need to develop this understanding especially for the more common cancers, so that we can identify the patients "at risk" to direct our more aggressive treatments at them, thus reducing the treatment necessary for the "low-risk" patients.

4. More effort must be made to improve our education programs so that patients seek help early for symptoms and so that all cancers can be diagnosed at an early stage when curative treatments are available.

5. In this era of biological-targeted therapies, we are aware that conventional imaging techniques measuring tumor volume may not be helpful since the agents may be cytostatic rather than cytotoxic. Therefore the development of novel imaging techniques to determine the biological effectiveness of these agents is of paramount importance.

Bibliography

Atkin, W.S., Edwards, R., Kralj-Hans, I. et al. (2010). Once-only flexible sigmoidoscopy screening in prevention of colorectal cancer: a multicentre randomised controlled trial. *Lancet*, **375**: 1624–33.

Chan, J.K., and Berek, J.S. (2011). Impact of the human papilloma vaccine on cervical cancer. *Journal of Clinical Oncology*, **25**: 2975–82.

Crum, C.P., Abbott, D.A., and Quade, B.J. (2003). Cervical screening from the Papanicolaou smear to the vaccine era. *Journal of Clinical Oncology*, **21**: 224–30.

Devita, V.T., Lawrence, T.S., Rosenberg, S.A., Depinho, R.A., and Weinberg, R.A. (2011). *Cancer Principles and practice of oncology*. 9th edn. Philadelphia: Lippincott Williams & Wilkins.

Djulbegovic, M., Beyth, R.J., Neuberger, M.N. et al. (2010). Screening for prostate cancer: systematic review and meta-analysis of randomised controlled trials. *British Medical Journal*, **341**: c4543.

Forest, P. (1986). *Cancer Screening: Report to the Health Ministers of England, Wales, Scotland & Northern Ireland*. London: Department of Health & Social Science.

Friedberg, J.W. (2003). PET scans in the staging of lymphoma: current status. *Oncologist*, **8**(5): 438–47.

Gotzsche, P.C., and Nielsen, M. (2006). Screening for breast cancer with mammography. *Cochrane Database Systematic Reviews*, CD001877.

Hardcastle, J.D., Chamberlain, J.O., Robinson, M.H. et al. (1996). Randomised controlled trial of faecal-occult-blood screening for colorectal cancer. *Lancet*, **348**: 1472–7.

Hewitson, P., Glasziou, P., Watson, E., Towler, B., and Irwig, L. (2008). Cochrane systematic review of colorectal cancer screening using the fecal occult blood test (hemoccult): an update. *American Journal of Gastroenterology*, **103**: 1541–9.

Jørgensen, K.J., Zahl, P., and Gøtzsche, P.C. (2010). Breast cancer mortality in organised mammography screening in Denmark: comparative study *British Medical Journal*, **340**: c1241.

Olsen, A., Nijor, S.H., Vejborg, I. et al. (2005). Breast cancer mortality in Copenhagen after introduction of mammography screening: cohort study. *British Medical Journal*, **330**: 220–4.

Pannall, P., and Kotasek, D. (1997). *Cancer & Clinical Biochemistry*. London: ACB Venture.

Ransohoff, D.F., and Sandler, R.S. (2002). Screening for colorectal cancer *New England Journal of Medicine*, **346**: 40–4.

Ring, A., Smith, I.E., & Dowsett, M. (2004). Circulating tumour cells in breast cancer. *Lancet Oncology,* **5**: 79–88.

Rustin, G.J., van der Burg, M.E., Griffin, C.L. et al. (2010). Early versus delayed treatment of relapsed ovarian cancer (MRC OV05/EORTC 55955): a randomised trial. *Lancet,* **376**(9747): 1155–63.

Sobin, L.H., Gospodarowicz, M.K., and Wittekind, C. (2009). *UICC TNM Classification of Malignant Tumors,* 7th edn. Hoboken, NJ: John Wiley & Sons, Inc.

Sturgeon, C.M., Lai, L.C., and Duffy, M.J. (2009). Serum tumour markers: how to order and interpret them. *British Medical Journal,* **339**: b3527.

Questions for student review

Select as many correct answers as apply.

1) Biomarkers may be used to:
   a. Avoid CT scanning patients after treatment.
   b. Diagnose cancer reliably before symptoms arise.
   c. Predict a patient's response to some targeted therapies.
   d. Predict a patient's toxicity to chemotherapy.
   e. Predict prognosis for a specific cancer.

2) The tumor marker bHCG in patients with testicular tumors is useful in:
   a. Diagnosis.
   b. Monitoring the response to therapy.
   c. Prognosis.
   d. Screening.
   e. Therapeutic strategies.

3) A validated screening program exists for:
   a. Breast cancer.
   b. Cervical cancer.
   c. Colorectal cancer.
   d. Gastric cancer.
   e. Non-small-cell lung cancer.

4) PET–CT can be helpful in:
   a. Differentiating infection from malignancy.
   b. Differentiating malignant from benign lesions.
   c. Monitoring response to treatment in lymphoma.
   d. Staging patients with brain tumors.
   e. Staging patients with non-small-cell lung cancer.

18 Treatment of Cancer: Chemotherapy and Radiotherapy

Anne L. Thomas, J.P. Sage, and William P. Steward
University of Leicester, UK

Key points

- More patients than ever before are being cured of their cancer or are having ongoing treatment to allow them to live with their cancer with good quality of life.
- Major advances have been made in the delivery of radiotherapy to minimize the damage to normal tissues.
- The increased understanding of tumor biology has resulted in many biologically targeted agents being available in the clinic.
- For a number of the common tumor types, genetic testing of the tumors of patients is now standard practice to define which targeted therapies they should receive.
- Consideration regarding the future cost of cancer therapy needs to be made to ensure that the complex therapies developed are not prohibitively expensive and therefore not financially viable.

Introduction

Chemotherapeutic agents exert their effect by killing cells that are rapidly dividing. The agents are therefore not tumor cell specific and cause their toxic effect by killing dividing normal cells (e.g. hair follicle cells and gastrointestinal (GI) mucosa). In recent years, our understanding of the molecular pathways controlling the growth of both normal and tumor cells has improved significantly. By exploiting the differences between normal and malignant cells, we can target pathways and receptors unique to the cancer cells, thus avoiding the indiscriminate universal killing of dividing cells of conventional cytotoxics. We are therefore entering the era of biologically targeted therapies in cancer treatment (Chapter 16).

Radiotherapy is the application of *ionizing radiation* to treat disease. Ionizing radiation is electro-magnetic radiation and elementary particles, which deposit energy in materials through the processes of excitation and ionization events. The forms of ionizing radiation in common use are photon beams (X-rays and gamma rays) and electrons (β particles). Radiotherapy has been in use for over 100 years, and currently over 60% of patients diagnosed with cancer will receive it at some time during their illness. There have been major technical advances in delivery of radiotherapy in recent years.

Hormones are thought to be etiological agents in at least half of the cancers diagnosed in developed countries. Common cancers, such as breast and prostate, often express receptors for sex hormones (e.g. female hormones such as estradiol or male hormones such as dihydrotestosterone). The growth of such cancers is at least in part dependent on hormonal stimulation. The aim of hormonal therapy is to *withdraw the growth stimulus*, either by reducing hormone production or by interfering with the receptor–ligand interaction or signaling associated with agonist activity. Subclasses of estrogen and progesterone receptors are of particular interest with regard to breast cancer, and androgen receptors are relevant to the treatment of prostate cancer.

Radiotherapy physics

The biological effects of radiation are a result of absorption of energy from the radiation. Since absorbed dose is energy divided by mass, radiotherapy is prescribed in units of Grays, where one Gray = 1 joule of energy absorbed per kilogram of mass.

When photons pass through matter, they interact with that matter through a number of different physical processes. Low-energy photons (e.g. those used in diagnostic X-rays) interact with electrons which are bound within atoms, resulting in the *photo-electric effect*. This effect is highly dependent on atomic number, and therefore differentiates tissues such as bone very well from other tissues such as water. At higher photon energies, the photoelectric effect becomes less important, and photons tend to interact with unbound electrons within the matter. This physical process is called "Compton scatter." For photon energy

---

The Molecular Biology of Cancer: A Bridge From Bench to Bedside, Second Edition. Edited by Stella Pelengaris and Michael Khan.
© 2013 John Wiley & Sons, Inc. Published 2013 by John Wiley & Sons, Inc.

produced by linear accelerators (≥4 MeV), the effect of photoelectric absorption is negligible and other physical processes predominate.

External beam radiotherapy can be delivered using a number of different modalities:

X-ray tubes. These produce X-rays of energy in the range of 80 kV to 300 kV (superficial and ortho-voltage radiotherapy) and are used for the radical treatment of skin cancer and the palliative treatment of skin and bone metastases.

Cobalt gamma ray machines. These release gamma rays equivalent to 2 MV X-rays and are cheap to operate and service. However, due to safety concerns, and because of the low energy produced and the less sharp edges of the beam (called the penumbra), they have been largely replaced by linear accelerators.

Linear accelerators. These use electro-magnetic waves to accelerate electrons, which bombard a tungsten target resulting in the production of X-rays with energies of 4–25 MV. These machines enable the administration of high doses of radiation to deep-seated tumors and minimize the dose to normal tissues by careful treatment planning (discussed in this chapter). By removal of a tungsten target, they also permit treatment with electrons.

Electrons. These penetrate a short distance, dependent on their energy. They are useful where a superficial tumor overlies sensitive structures, for example skin and lip lesions. They may also be used to treat the whole body (e.g. mycosis fungoides).

In addition to external beam therapy, other modalities place the source of radiation internally:

Brachytherapy: This involves a radioactive source, typically cesium-137, iodine-125, or iridium-192, placed close to the tumor. The most widespread use of brachytherapy is the intracavity treatment of gynecological malignancies, although the implantation of iodine seeds to treat early prostate cancer is also common.

Systemic therapy: This is the use of radioactive material in pharmaceutical form, termed *unsealed sources*. The most common example is the use of iodine-131 in the treatment of thyrotoxicosis and thyroid cancers. This treatment has been used for over half a century, and hundreds of thousands of patients have received this therapy. Radioiodine therapy is remarkably safe and is not known to increase risk of other cancers, the only common side effect being hypothyroidism.

Radiobiology

Ionizing radiation causes extensive cellular damage, principally via the formation of free radicals. Cells have a considerable capacity for repair, and cell survival depends on a number of factors (see Chapters 8 and 10). In an average cell, one Gray (Gy) causes damage to over 1000 bases in DNA, approximately 1000 **single-strand breaks** (SSBs) in DNA, and approximately 40 **double-strand breaks** (DSBs). DSBs in DNA are considered the most significant types of molecular damage, and correlate with the probability of cell survival. Typically, survival curves are continuously bending, with a slope that steepens as the dose increases (see Fig. 18.1). Mathematically, a continuously bending curve is most simply described by a linear quadratic equation of the form: surviving fraction = exponential $(-\alpha D - \beta D^2)$, where D is the given single dose and α and β are parameters characteristic of the cells concerned. Since the *ratio of $\alpha:\beta$* gives the relative importance of the linear dose term and the quadratic dose term for

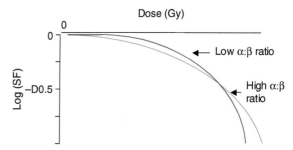

Figure 18.1 Survival curves for cells with different $\alpha:\beta$ ratios subjected to irradiation. The more steeply curving survival curve has a lower $\alpha:\beta$ ratio when fitted to the linear quadratic equation. Note log scale. SF = surviving fraction following irradiation.

Box 18.1 Radiobiological factors which influence response to fractionated radiotherapy (the five Rs)

1. *Radiosensitivity.* The intrinsic radiosensitivity of different tumor types determines the likelihood of survival following irradiation. Classically, lymphoma, myeloma, and small-cell lung cancer are considered more radiosensitive than most other tumor types.

2. *Repair.* Cells differ in their capacity to repair DNA damage, particularly SSBs and DSBs in DNA. Repair is considered more effective in nonproliferating cells, but it is now known that repair processes take at least 6 hours to be performed effectively.

3. *Repopulation.* Surviving tumor cells may proliferate more rapidly than those in normal tissues. The overall treatment time influences the repopulation of the tumor cells and the acute toxicity in acute responding normal tissues.

4. *Re-oxygenation.* Hypoxic cells are more resistant to radiation than oxygenated cells. Giving tumors cells time to re-oxygenate may improve their radiosensitivity.

5. *Redistribution.* Cells in certain phases of the proliferative cycle (e.g. the G_2 and M phases) are more radiosensitive than cells in other phases of the cell cycle. Given time, cells can redistribute themselves throughout the phases of the cells.

---

those cells, this ratio is used to characterize the radiosensitivity of a particular tissue type.

To improve the therapeutic ratio (the differential effect on malignant tissue compared to normal tissue), the total dose of radiotherapy can be divided into small daily amounts – "fractionation." Fractionated radiotherapy inflicts more damage on tumor cells than on normal cells for a given dose of radiation. Moreover, cell and tissue kinetics usually favor recovery and repopulation of damaged normal tissue over tumor tissue (see Box 18.1). Although tumors are very diverse, the radiobiological properties of most tumors are similar to those of acute responding tissues, that is, a high $\alpha:\beta$ ratio results in moderate sensitivity to changes in fraction size and some dependence on the total treatment times. It has been shown that extending total treatment time or allowing significant gaps during treatment can decrease the efficacy of the radiotherapy.

The timing and severity of injury to normal tissue depend on the rate of turnover of the mature cells in the tissue. Epithelial and hematopoietic tissues have rapid turnover and are therefore susceptible to *acute effects* over days or weeks (Box 18.2). Tissues with slower turnover are susceptible to late effects, with a timescale of months or years. In fact, risks of major late effects to spinal cord, lung, liver, and kidney are often the dose-limiting factors in radiotherapy. Since stem cells of late-responding tissues have more a curved survival curve (i.e. a lower $\alpha:\beta$ ratio) than cells of acute-responding tissues, the late-responding tissues are particularly sensitive to changes in fraction size. The probability of "long-term toxicity" can therefore be diminished significantly by using small fractions of radiotherapy rather than larger fractions, but this consideration must be balanced against the effect of the overall treatment time on the tumor tissue.

Treatment planning

Management plans are highly individualized and based on the nature of the cancer, the general health of the patient, and the goal of the therapy. Treatment planning for external-beam radiotherapy involves a number of steps, as shown in Fig. 18.2. The specification of a radiotherapy treatment will include the dose, the number of treatment fractions, and the volume that is to receive the dose. The gross tumor volume consists of all known macroscopic disease, demonstrated by either physical examination or imaging. The clinical target volume encompasses the gross tumor volume plus regions considered to harbor potential microscopic disease. The planning target volume provides a margin around the clinical target volume to allow for internal target motion (e.g. respiratory motion) and variations in treatment setup. Any nearby structures which may suffer from radiation damage are also defined as organs at risk. The challenge in defining the radiation beams which will be used for the treatment lies in maintaining dose coverage to the planning target volume, whilst ensuring doses to organs at risk are acceptable. These target volumes are shown in Fig. 18.3. Patient immobilization is used when highly sensitive structures are close to the target, for example spinal cord or brain. A thermoplastic shell is often used to immobilize the head and keep the neck position constant, with the advantages that the treatment fields can be drawn on the

---

Box 18.2 Acute side effects of radiotherapy

1. *Mucosa of the gastrointestinal tract.* Common side effects include mucositis (including loss of taste) and esophagitis (heartburn). Good oral hygiene and suitable analgesics are important during radiotherapy in order for adequate oral intake. A soft diet is advised, but if the patient is unlikely to tolerate this, a percutaneous gastrostomy tube should be inserted before the commencement of radical radiotherapy. This tube can then be used for the administration of feeds, liquids, and drugs.
2. *Brain.* Nausea, vomiting, alopecia, and transient worsening of neurological symptoms are common side effects of radiotherapy. Anti-emetics and corticosteroids are common treatments used to palliate these symptoms.
3. *Abdomen.* Nausea and vomiting are common side effects of radiotherapy, usually requiring strong anti-emetics such as serotonin receptor antagonists.
4. *Pelvis.* Diarrhea, dysuria, and urinary frequency are common side effects of radiotherapy to the pelvis. Anti-diarrheal agents are often prescribed, and midstream urine samples should be used to check for opportunistic bacterial or fungal infections.
5. *Skin.* Dry desquamation can advance to moist desquamation as a result of radical radiotherapy. Good skin clear and the avoidance of friction are important. Water-based moisturizers and corticosteroid creams can be applied topically.

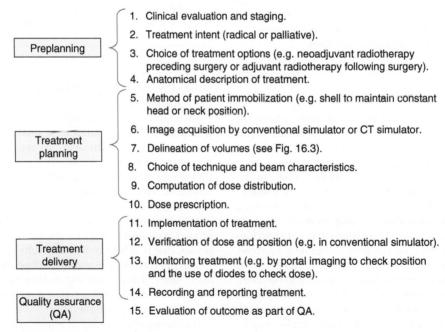

Figure 18.2 The radiotherapy planning process.

shell rather than the patient and the shell can be fixed to the radiation couch.

Recent advances

Intensity-modulated radiotherapy (IMRT)

Better methods for delivering radiation allow the area of high dose to more closely conform to the planning target volume. In IMRT, the pattern of radiation from each beam is manipulated to give greater control over the dose in the patient. These patterns would be too complex to design manually, and these treatments are created by a computer process called "inverse planning." With this technique, the defined volumes and specified acceptable dose levels are critical. IMRT has found widespread uptake in the treatment of prostate and head and neck cancers (Fig. 18.4).

Rotational IMRT has demonstrated an even greater improvement in control over the dose distribution, where the pattern of radiation is varied as the beam rotates around the patient in a continuous sweep. Alternatively, tomotherapy machines use a narrow slit of radiation, which rotates around the patient as they pass though the center of the machine, analogous to a spiral CT scanner.

Image-guided radiotherapy (IGRT)

Imaging systems integrated into the treatment device allow the position of the target to be monitored and corrected on a daily basis. As well as ensuring the treatment is accurately delivered as specified, this should also allow for a reduction in the planning target volume margin and hence reduce the volume of tissue irradiated. A whole range of techniques are available, but the challenge remains the visualization of the actual tumor rather than surrounding bony anatomy. Several techniques involve the implantation of markers into the tumor.

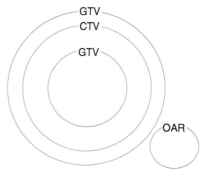

Figure 18.3 Schematic representation of "volumes" in radiotherapy. Gross tumor volume (GTV): demonstrable tumor; clinical target volume (CTV): GTV + suspected microscopic disease; planning target volume (PTV): CTV + margin for positional variation; OAR: organ at risk.

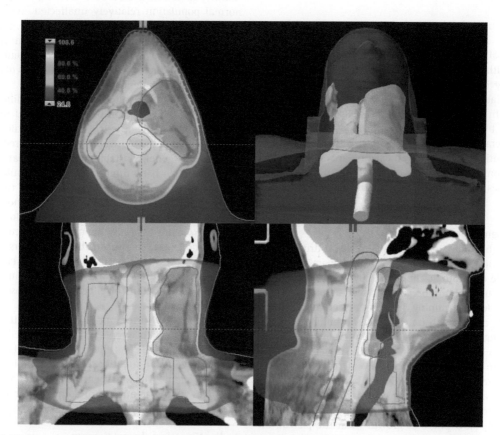

Figure 18.4 Dose distribution using intensity-modulated radiotherapy (IMRT). This dose distribution is for a modern IMRT treatment of a head and neck tumor showing the concurrent delivery of a high tumorcidal dose to the primary tumor and a lower prophylactic dose to the at-risk lymph nodes. The dose to the spinal cord and the contralateral parotid gland are kept below the tolerance dose for late effects.

Stereotactic radiosurgery

Highly accurate immobilization and positioning devices, coupled with the precise delivery of multiple small radiation fields, allow tightly conformed doses of radiotherapy to be delivered as a single dose or as fractionated doses. Stereotactic radiosurgery is a well-established technique for the treatment of small malignant lesions in the brain and benign arterio-venous malformations, although lesions must meet stringent size criteria. Advances in extracranial stereotactic radiosurgery now allow similar high-dose treatments for lung and hepatobiliary tumors as well as metastases in lung, liver, and spine.

Proton radiotherapy

The use of protons for radiotherapy can offer exquisite control over the dose distribution inside the patient. They are especially useful where the target is very close to critical structures and have an established role for neurological, pediatric, and ocular tumors. The immense devices that are used to produce proton beams, cyclotrons, are hugely expensive and can be funded at only a national level. In 2010 nearly 7000 patients received proton radiotherapy at one of around 30 centers worldwide. These low numbers often limit the application of protons to those cases with the greatest need. A number of research programs offer the promise of proton radiotherapy at greatly reduced cost, which would provide a huge advance for clinical radiotherapy.

Chemoradiation

The concomitant use of chemotherapy and radiotherapy is a curative modality in the treatment of anal, head and neck (H&N), and cervical cancer, and may be considered as an alternative to surgery. Although the concomitant use of chemotherapy with radiotherapy increases acute and late damage to normal tissues, it is an established standard of care for carcinoma of the cervix and anus. In squamous H&N cancer, meta-analyses have suggested a small but consistent survival benefit from the addition of chemotherapy to radical radiotherapy. The expanding use of chemoradiation is currently an exciting area of research, particularly the combination of newer chemotherapy drugs and biological therapies with radiotherapy in the treatment of GI cancers.

Rationale of chemotherapy

Much has already been discussed in this book about drug development in cancer, in Chapters 16 and 17 in particular. To understand the rationale of cytotoxic chemotherapy, it is important to recognize the features of tumor growth (see Chapter 5). The factors responsible for determining the growth of a tumor include the cell-cycle time and growth fraction (i.e. the number of cells that are undergoing cell division and the number of cells). Also, the more cancer cells there are, the more likely that some will become resistant to chemotherapy and prevent cure. The asymmetric sigmoidal growth curve ("Gompertzian" growth curve) describes the natural history of tumor growth. By the time they are clinically apparent (around 10^9 cells or more), most tumors are in the relatively slow phase of growth (near the plateau of the Gompertzian growth curve). Unfortunately, this is the time that chemotherapeutic agents are least likely to prove effective. However, if the tumor can first be resected, then subsequent chemotherapy should be more effective as remaining cancer cells are now likely to be in logarithmic growth.

Surgery

Historically, surgery was the only treatment option for cancer, and in the modern era it often still offers the best chance of a cure. There are six main areas where surgery plays a fundamental role in the treatment of cancer (Box 18.3).

More recently, laparoscopic or "keyhole" surgery has become an attractive proposition in cancer patients. There is good evidence that, with appropriate training, complete clearance of tumors can be achieved using this approach with the added benefit of reduced morbidity and hospital stay for patients.

Combination chemotherapy

In modern oncological practice, chemotherapeutic agents are used in combination rather than as sequential single therapies. For a particular tumor type, only drugs which are known to be effective as single agents should be incorporated, and ideally these drugs should be used at their optimal dose and schedule. It is imperative that the combination of drugs chosen should have a minimal overlap in toxicity to achieve this goal. The treatment should be delivered on an intermittent basis, with the shortest possible time between treatments that allows the recovery of the most sensitive normal tissue (e.g. the bone marrow or gut). This intermittent scheduling takes advantage of the observation that tumor cells recover slower from cytotoxic damage than do their normal counterparts. Thus, with each sequential cycle, the tumor population should be increasingly depleted, leaving the normal population relatively unaffected. Wherever possible, it is preferable to use drugs with known synergistic killing effects, for example the combination of oxaliplatin and 5-fluorouracil (5-FU). Another possibility is to use a combination of drugs which can kill cancer cells at different stages of the cell cycle (i.e. they are cell-cycle dependent).

Some regimens use alternating cycles of different drug combinations. One of the main principles behind this practice is the Goldie–Coldman hypothesis. The model predicted that the mutation of tumor cells to drug resistance occurred at a rate proportional to the genetic instability of a particular tumor. Therefore, the probability that a tumor at diagnosis would contain resistant clones would be proportional to both the tumor size and the inherent mutation rate. Theoretically, tumors should initially respond to treatment but recur as resistant clones expand. Therefore, the Goldie–Coldman hypothesis predicted that drug resistance could be present even with small tumors and the maximum chance of cure would occur if all available effective drugs were used simultaneously. This was the basis behind the development of using combinations of effective, non-cross-resistant drugs in *alternating* cycles.

In the 1980s, Norton and Day reviewed the Goldie–Coldman hypothesis. In their overview, they predicted that if two agents were used and if the less effective one was given first, the overall outcome would be superior to initial administration of the more effective agent. This has been the rationale for using sequential therapies in the treatment of some tumor types, for example breast cancer.

Indications for chemotherapy

Chemotherapy is a systemic treatment most commonly used to treat advanced cancer. In a limited number of situations, it may

Box 18.3 Role of surgery in oncology

1. *Surgery for diagnosis and staging*. Historically, surgery was important in both diagnosing patients with cancer and also staging them. Before the advent of accurate computed tomography (CT) scanning, surgery was necessary for staging, such as in the staging of lymphoma. This indication for surgery is now declining, although accurate surgical staging of patients with ovarian cancer remains of fundamental importance. With ultrasound and CT-guided techniques, it is now possible to biopsy areas of the body that previously were inaccessible. Again, the need for surgery still exists, for example in the excision biopsy of lymph nodes potentially involved by lymphoma.

2. *Surgery as treatment of the primary tumor*. When a tumor is confined to the anatomic site of origin, surgery usually is the main form of treatment. Obviously, the extent of that surgery is important to ensure that the tumor has been completely excised locally. The intention of curative surgery is to remove the entire solid tumor "en bloc." There are really only very limited indications now to perform de-bulking surgery. One of these indications would be in the treatment of extensive ovarian cancer.

3. *Surgery to resect metastases*. Nowadays, surgery is used not only used to resect primary tumors but also to perform metastasectomies. An example of this would be to remove isolated liver metastasis in the treatment of colorectal cancer or pulmonary metastasis in the treatment of soft-tissue sarcoma. Studies have shown extremely good survival rates for patients with liver involvement from colorectal cancer where chemotherapy has been used to downstage metastasis prior to surgical resection.

4. *Surgery as preventative treatment*. For a patient who is at high risk of developing a malignancy, surgery can be used as a preventative technique. The rationale here is to remove the organ at risk prior to the development of cancer. An example of this would be to perform a panproctocolectomy in patients in their teens that have polyposis coli. The issue of prophylactic surgery in women who carry the *BRCA1* and *BRCA2* genes is more complicated. To fully reduce their risk of malignancies, prophylactic bilateral mastectomies need to be performed in combination with salpingo-oopherectomy. Clearly, this is extremely major surgery and the patients need to have undergone genetic counseling before considering this.

5. *Reconstructive surgery*. Once a patient has undergone radical excision of his or her primary tumor, surgery may have a role in reconstructing the organ removed. Women are now offered a number of reconstruction options following a radical mastectomy. The cosmetic results are excellent, and part of this has been due to the development of better microsurgical techniques. In patients that undergo panproctocolectomy, pouches can be made to produce a false rectum so that patients do not have to have a permanent stoma bag. The success of this type of bowel surgery has been greatly aided by the development of highly accurate stapling devices that can reliably perform and can reduce the leak rate.

6. *Palliative surgery*. Finally, there still remains a palliative role for surgery. This could be in performing a toilet mastectomy for a patient with a large ulcerating breast cancer or a gastric bypass procedure for a patient with a distal gastric cancer causing gastric outlet obstruction. With the development of endoscopy techniques, sometimes it is possible to place a stent through the scope. It is likely that we will see more endoscopic treatments being performed in the next few years.

---

be used as the primary or sole modality of treatment (e.g. chemosensitive cancers such as leukemia or teratoma). For the majority of the common solid tumors, chemotherapy is frequently used to reduce the volume of disease and palliate symptoms caused by the cancer. A further indication for chemotherapy is to use it as an adjuvant after the primary tumor has been controlled by either surgery or radiotherapy. Here the rationale is to eradicate subclinical micrometastatic disease and reduce the risk of recurrence. An example where adjuvant chemotherapy is extremely useful is in patients with breast cancer who have nodal involvement.

Neoadjuvant chemotherapy is being used increasingly to de-bulk or downstage primary tumors prior to the definitive treatment, which could be either surgery or radiotherapy. An example of this would be to use neoadjuvant chemotherapy to downstage a large breast primary tumor, so that conservative breast surgery could be offered rather than radical mastectomy.

In the majority of cases, chemotherapy is used systemically either intravenously or orally. It may, however, be administered regionally, for example intrathecally in the treatment of hematological malignancies with a high likelihood of central nervous system involvement. It can also be infused directly into the blood supply of liver metastases from colorectal cancer, or into the peritoneum in the treatment of advanced ovarian cancer.

Classification of cytotoxic drugs

Historically, anticancer agents have been classified according to the phase of the cell cycle during which they work. With the increasing number and complexity of chemotherapeutic agents, classification according to their mode of action is far more relevant.

Alkylating agents

These were the first drugs to successfully treat cancer, and they comprise a diverse group of highly reactive molecules that react with DNA via the alkyl chemical group ($R\text{-}CH_2$) to promote single-strand breaks (SSBs) or DSBs and DNA cross-linking. These effects result in apoptosis or growth arrest (Chapter 10).

Some alkylating agents have two available alkyl groups which form cross-links and are known as bi-functional. These are more cytotoxic than alkylating agents with only one available alkyl group. The most commonly used alkylating agents bind in the major groove of DNA at the N_7 position of guanine. Minor groove binders are currently in development. In addition, the alkylating agents also bind to RNA and other cell proteins. Exactly how much these interactions contribute to the overall toxic effect remains the subject of further research.

Types of alkylating agents include:

- Nitrogen mustards (e.g. melphalan, chlorambucil, cyclophosphamide, and ifosfamide)
- Alkyl sulphonates (e.g. busulfan)
- Aziridines (e.g. thiotepa)
- Tetrazines (e.g. dacarbazine and temozolomide)
- Nitrosoureas (e.g. BCNU and CCNU)
- Metal salts (e.g. cisplatin, carboplatin, and oxaliplatin)

Melphalan is a nitrogen mustard derivative of the amino acid phenylalanine; it is mainly used intravenously, in the treatment of myeloma. Chlorambucil is a nitrogen mustard administered orally to patients with lymphoma. The nitrosoureas are a group of compounds that were developed in the 1960s by the US National Cancer Institute Screening Programme. They alkylate DNA through formation of chloroethyl diazonium hydroxide and isocyanate. BCNU and CCNU are particularly lipophilic and as such are commonly used to treat brain tumors. The tetrazines, dacarbazine and temozolomide, are small molecules that release a reactive diazonium ion that alkylates DNA. Dacarbazine is used to treat malignant melanomas, sarcomas, and Hodgkin disease, and temozolomide is licensed for use in malignant gliomas.

Platinum compounds

Cisplatin, carboplatin, and oxaliplatin (1,2-diaminocyclohexane platinum) are widely used in the treatment of malignancies. They all inhibit DNA synthesis through the formation of intrastrand cross-links in the DNA and formation of DNA adducts. Serendipity played a major part in the discovery of cisplatin in 1969. At that time, ongoing studies were looking at the effect of electric currents on the growth of bacteria. Bacterial growth was reduced, and it transpired that platinum from the electrodes was responsible. Cisplatin is very effective in the treatment of testicular tumors, but use can be limited by kidney damage (nephrotoxicity), though this can in part be reduced by adequate hydration protocols. Carboplatin is an analogue of cisplatin and has less nephrotoxicity and neurotoxicity. It has many applications, particularly in the management of lung and ovarian cancer.

Oxaliplatin is a third-generation platinum analogue, and it differs from the other platinum drugs by having a large 1,2-diaminocyclohexane (DACH) ligand in its structure. It is thought that the presence of the DACH group renders the molecule nonrecognizable by the mismatch repair process. This means that the DNA repair mechanisms responsible for platinum resistance are not so effective in repairing oxaliplatin-induced DNA damage. Oxaliplatin is therefore useful in tumors where cisplatin has failed, and also in colorectal cancer, a tumor type not classically treated by platinum drugs. Oxaliplatin does have neurotoxicity, although this differs from that typically seen with cisplatin. Patients develop a peripheral neuropathy with cumulative drug exposure, but, in addition, they also develop an acute neuropathy that is characterized by cold-induced tingling (paresthesiae) in the fingertips and toes after infusion. This can also affect the throat after cold drinks, and the patient develops a feeling of choking. Studies have demonstrated that oxaliplatin has a synergistic effect when administered concomitantly with 5FU chemotherapy, and therefore this is the most common oxaliplatin regimen used.

Antimetabolites

These structurally resemble the naturally occurring purines and pyrimidines involved in nucleic acid synthesis. Since they are mistaken by the cell for normal metabolites, they have two modes of action. The first is to inhibit key enzymes required for DNA synthesis, and the second is to become incorporated into the DNA and RNA and therefore produce incorrect codes and cause strand breaks or premature chain terminations. The antimetabolites are S-phase-specific drugs (Chapter 4), because their primary effect is on DNA synthesis. They can be divided into four groups.

Folate antagonists

These are exemplified by methotrexate, which inhibits dihydrofolate reductase (DHFR), an essential enzyme in folate metabolism. Dihydrofolate reductase maintains intracellular folate in the fully reduced tetrahydrofolate form required for synthesis of thymidine monophosphate and other purine nucleotides. Methotrexate is structurally related to folic but has a greater affinity for dihydrofolate reductase, which it binds and inactivates. It is possible to overcome the metabolic block and the activity of dihydrofolate reductase by administering folinic acid. This is a tetrahydrofolate and therefore provides an alternative source for the synthesis of nucleic acids. This has importance, particularly when high-dose methotrexate is used and folinic acid can be administered to the patient as a "rescue" from the toxic effect of the treatment. Clinically, the drug can be given orally, intravenously, or at a low dose intrathecally. It is excreted and changed in the urine and widely distributed in body fluids. The acute side effects of methotrexate are similar to those of the other S-phase agents, namely, bone marrow suppression, mucositis, and diarrhea. Methotrexate is most commonly used in the treatment of non-Hodgkin lymphoma, acute lymphoblastic leukemia, choriocarcinoma, and breast cancer.

Over the last 10 years, new antifolates have been developed including raltitrexed and pemetrexed. Raltitrexed is a specific inhibitor of thymidylate synthase (TS), thus reducing the nucleotides necessary for DNA synthesis. Due to toxicity, it is generally reserved for patients with colorectal cancer who have coexistent cardiac disease. Pemetrexed is a multitargeted pyrroloprimidine-based antifolate. When polyglutamated, it targets a number of folate-dependent enzymes, for example DHFR and TS. It has activity in a number of tumor types, including non-small-cell lung cancer (NSCLC) and mesothelioma.

Arabinosides

Cytosine arabinoside is an analogue of deoxycytidine. It is a competitive inhibitor of the enzyme DNA polymerase. Its principal action is as a false nucleotide competing for the enzymes, which are responsible for converting cytidine to deoxycytidine and for incorporating deoxycytidine into DNA. It is used mainly in the treatment of acute leukemia. Gemcitabine is a second-generation pyrimidine analogue. It is a difluoronated analogue of cytosine arabinoside and is metabolized by nucleotide kinase intracellularly to the active diphosphate and triphosphate. The triphosphate derivative of gemcitabine inhibits the enzyme deoxycytidine diaminase, which is responsible for deactivating gemcitabine. This increases the intracellular half-life of gemcitabine. It is administered intravenously, is the standard therapy for pancreatic tumors, and is also used in combination with platinum in the treatment of NSCLC. Fludarabine is another analogue of cytarabine. It is an adenosine analogue, which is incorporated into DNA and inhibits enzymes important in DNA synthesis such

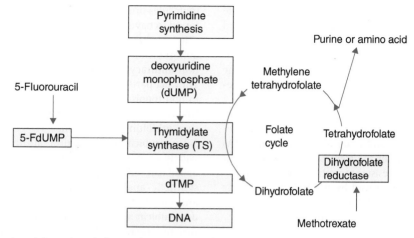

Figure 18.5 Mechanisms of action of the antimetabolites methotrexate and 5-fluorouracil.

as ribonucleotide reductase and DNA polymerase. It is a highly effective chain terminator. Fludarabine is also incorporated into RNA and its proteins, although the cytotoxic effect of this is not known. It is the most active single agent in the treatment of chronic lymphatic leukemia.

Antipyrimidines

5-Fluorouracil (5FU) is a fluoropyrimidine that inhibits DNA synthesis by inhibiting the main enzymes involved in the synthesis of cytosine and thymine. 5FU enters the cell via the uracil transport mechanism. It is converted to fluorouridine diphosphate (FudR) by the enzyme thymidine phosphorylase, and then to its active metabolites 5-fluoro-2 deoxyuridine monophosphate (FdUMP) by the enzyme thymidine kinase. In the presence of its co-factor 5,10-methylene tetrahydrofolate, FdUMP forms covalent bonds with TS, creating a complex that inhibits the formation of thymidine from deoxyuridine monophosphate (dUMP), thus inhibiting DNA synthesis. In clinical practice, it is usual to administer folinic acid in combination with 5FU therapy as this has been shown to increase the effectiveness of the drug, presumably by stabilizing the binding of FdUMP to TS. 5FU is also erroneously incorporated into RNA instead of uracil and thus inhibits RNA synthesis. One cause of chemoresistance to 5FU is that increase of dUMP may overcome the inhibition of TS by FdUMP. In an attempt to overcome this, specific TS inhibitors are being developed.

5FU is a prime example of a chemotherapy agent whose activity and toxicity depend upon its scheduling. In the past, it was possible to give 5FU only intravenously due to its unpredictable oral bioavailability. There are now oral 5FU analogues available: capecitabine is activated to 5FU in a three-step enzymatic reaction with tumor-selective generation by exploiting the higher level of thymidine phosphorylase in tumor compared with normal tissue. The pharmacokinetics of orally administered capecitabine essentially mimic a continuous-infusion, rather than bolus, 5FU. Clearly, it is a significant advantage to be able to administer tablets to patients, rather than use infusional chemotherapy with central venous catheters and all their potential complications. The second oral fluoropyrimidine is UFT, a combination of ftorafur and uracil. It needs to be given in combination with folinic acid. S1 is a combination of ftorafur and potassium oxonate. The ftorafur inhibits dihydropyrimidine deaminase, and the oxonate accumulates in the GI tract and inhibits the breakdown of 5FU, which theoretically should increase the availability of the active metabolites of 5FU in the tumor. Antimetabolite pathways are simplified in Fig. 18.5.

Antipurines

There are two purine antagonists; 6-mercaptopurine and 6-thioguanine. The latter is an analogue of guanine and is now rarely used in clinical practice. 6-mercaptopurine inhibits purine synthesis by being erroneously incorporated into DNA and thus interfering with DNA replication. It is normally given orally and used in the treatment of leukemia.

Mitotic inhibitors

These include the vinca alkaloids, vincristine, vinblastine, and vinorelbine, and the taxanes. The vinca alkaloids are either naturally occurring (from the periwinkle plant) or semisynthetic. They bind to tubulin and prevent assembly of the mitotic spindle (Chapter 4) required for chromosome segregation in mitosis. Vincristine is used in the treatment of lymphomas, leukemias, and breast cancer, and its major dose-limiting toxicity is peripheral neuropathy. Vinblastine is used in lymphoma and NSCLC, but may cause deficiency in platelets (thrombocytopenia). Vinorelbine is the most recently discovered vinca alkaloid and tends to cause less neuropathy than the older agents. The taxanes, in contrast to the vinca alkaloids, do not promote the assembly of the mitotic spindle but instead inhibit its disassembly. Paclitaxel is an extract from the bark of the pacific yew, and initially its development was restricted due to limited supplies and also its solubility properties. Docetaxel was synthesized from the needles of the tree and also a more commonly found yew species. Both are now made semisynthetically and are used in treating many solid tumors including prostate, breast, ovarian, and NSCLC. Bone marrow suppression is one of the major limiting side effects, though paclitaxel may cause neuropathy. Hypersensitivity reactions are well recognized, and it is now standard practice to pre-medicate patients with steroids and H_2 inhibitors to prevent this. For paclitaxel, the infusion reactions have been related to cremophor, the solvent used in its production. Using nanoparticle technologies, an albumen-bound formulation of paclitaxel called Abraxane has been developed and reduces these hypersensitivity reactions. Docetaxel is characterized by a fluid retention syndrome with the development of edema and weight gain. The etiology of this is unclear, but again pre-medication usually prevents the overall fluid retention effect.

Another group of drugs in this class are the epothilones. These are microtubule stabilizers and appear to be active in taxane-resistant tumors. The US Food and Drug Administration (FDA) approved ixabepilone in 2007 for the treatment of metastatic or locally advanced breast cancer resistant to other standard chemotherapies. The drug was not approved by the EMEA, and there are concerns regarding the tolerability of this group of drugs generally as they cause significant neuropathy and diarrhea.

Topoisomerase inhibitors
Topoisomerases are enzymes that control the coiling and uncoiling of DNA and are important during DNA replication. Topoisomerase I binds to double-stranded DNA and causes a SSB allowing super-coiled regions of DNA to unravel. Following re-ligation, the DNA is able to complete replication. Interestingly, levels of topoisomerase I are known to be higher in malignant tissues compared with normal ones, especially in the colon and ovaries. Inhibitors of topoisomerase I, such as irinotecan and topotecan, interfere with the topoisomerase–DNA bond and prevent re-ligation of the SSB. Thus, during replication, the replication fork re-ligation interacts with the SSB in the DNA forming an irreversible DSB in the DNA. The majority of cells will then die by apoptosis or growth arrest in the G_2 phase. The topoisomerase II enzyme binds to complementary double-strand DNA, causing both strands to cleave and move apart, allowing a second segment of double-strand DNA to pass through the cleavage site. This site then re-ligates. Topoisomerase II inhibitors prevent the re-ligation and therefore maintain DSBs which impede DNA replication.

Camptothecin was the first topoisomerase I inhibitor to be discovered. Its further development was hindered by toxicity. Irinotecan is now widely used in the treatment of colon cancer. Following intravenous administration, it is activated by carboxylesterase to 27-ethyl-10-hydroxycamptothecin (SN-38). The major route of excretion of SN38 is through the liver, and therefore care has to be taken when prescribing irinotecan for patients with abnormal liver function tests. Another side effect of irinotecan therapy is diarrhea, but this responds well to prophylactic administration of atropine and if treated promptly will respond to constipating agents such as loperamide and codeine. Failure to introduce these agents quickly can result in extreme diarrhea, which, particularly if it coincides with the myelosuppression, can be life threatening.

Etoposide is the most widely used topoisomerase II inhibitor for Hodgkin disease, non-Hodgkin lymphoma, small-cell lung cancer, and testicular teratoma. It is another example of a drug that is schedule dependent and is more effective given over 5 days than on 1 day. It can be given both orally and intravenously, and the major dose-limiting side effect is myelosuppression. In addition, alopecia and mucositis are common.

Miscellaneous
There are a group of chemotherapy agents that were initially developed as antibiotics. They are a group of related antimicrobial molecules produced by *Streptomyces* species in culture. They were found to have significant cytotoxicity and therefore are now classified as anti-tumor antibiotics. They include the anthracyclines, doxorubicin, daunorubicin, epirubicin, and idarubicin, and the non-anthracycline antibiotics, bleomycin and actinomycin D. They act partly as alkylating agents but also intercalate DNA and inhibit topoisomerase II function. In addition, they can generate free radicals, which cause oxidative damage to cellular proteins, thus inhibiting their action.

Doxorubicin is the oldest anthracycline and remains very widely used in clinical practice. Doxorubicin, daunorubicin, and epirubicin are all given intravenously. Idarubicin is an oral synthetic analogue of daunorubicin. The main toxicity after anthracycline treatment is myelosuppression. Daunorubicin and doxorubicin and, to a lesser extent, epirubicin cause cumulative cardiotoxicity, and there is a maximum recommended total dose. Consequently, anthracyclines should be avoided in patients with a significant history of cardiac disease. Dexrazoxane is an agent that reduces the incidence of cardiac toxicity in patients with breast cancer receiving doxorubicin by preventing the formation of free radicals.

Actinomycin D is a DNA intercalator and blocks the transcription of DNA by intercalating base pairs and preventing RNA from recognizing and binding to the DNA. It is a very toxic drug and is usually given intravenously as a single injection, mainly in the treatment of childhood cancers.

Bleomycin is another antibiotic and causes SSBs and DSBs through the production of free radicals. It is most active on cells during the G_2 phase and causes cell cycle arrest in G_2. In may be administered either intramuscularly or intravenously. It is renally excreted and appears to concentrate mainly in the skin and lungs. This is important since the main side effect is pulmonary fibrosis, particularly after high-flow oxygen therapy. Mucocutaneous side effects are also seen, and these may vary from hyperpigmentation to severe mucosal ulceration. Bleomycin plays a central role in the combination therapy of testicular tumors and lymphomas. It causes very little myelosuppression and is thus an ideal agent to be used in combination regimens.

Enzymes
Asparaginase is an example of this class of cytotoxic drug. It is an enzyme that degrades asparagine, an essential amino acid for protein and nucleic acid synthesis. Tumor cells, unlike their normal counterparts, have little or no asparaginase synthetase, the enzyme necessary for making asparagine. They therefore rely on exogenous sources of asparagine. Following treatment with asparaginase, the lack of this essential amino acid inhibits protein and nucleic acid synthesis. It is used most commonly in the treatment of acute leukemias, and the main side effect is hypersensitivity reaction.

Hydroxyurea blocks the action of ribonucleoside diphosphate reductase and interferes with the synthesis of DNA. It causes leucopenia and therefore is used in the treatment of myeloproliferative disorders such as polycythemia.

Trabectedin
During the 1950s and 1960s, the National Cancer Institute carried out a program of screening plant and marine organism material for anticancer activity. As part of that program, extract from the sea squirt *Ecteinascidia turbinata* was found to have promising activity. After many years, trabectedin was found to be the active

moiety. Clearly the production of the compound was challenging, but now a semisynthetic process is used and starts from Safracin B, an antibiotic obtained by fermentation of the bacterium *Pseudomonas fluorescens*. The actual mechanism of action is not known but is thought to be due to the reduction of molecular oxygen into superoxide via an unusual auto-redox reaction on a hydroxyquinone moiety of the compound, causing DNA strand breaks and apoptosis. The European Medicines Agency (EMEA) and the FDA have granted orphan drug status to trabectedin for soft-tissue sarcomas and ovarian cancer.

Biologically targeted agents

This area has already been extensively discussed in Chapter 16, so it will be only briefly addressed here. In recent years, there has been a paradigm shift in cancer treatment away from relatively nonspecific cell poisons toward drugs that target specific cellular pathways. In part, this has been due to the emergence of new molecular technologies which have allowed us to understand the differences between normal and malignant cells, specifically in terms of target gene expression and signaling pathway deregulation, but also the sequencing of the human genome has had a positive influence. Of course, this principle of targeted therapy is not entirely new. The first genuinely targeted agents came from advancing research into hormone signaling and the discovery that some cancer cells may still exhibit the same dependencies on hormones that characterize the cell lineages from which they were derived. This is illustrated by the effectiveness of drugs targeting nuclear hormone receptors and their ligands in breast and prostate cancer.

We are entering an era of "personalized medicine" in the clinic. Already in some tumor types such as breast and bowel cancer, we screen the tumors of patients for the expression of certain genes and their mutant counterparts. We then can recommend the use of selected targeted therapies with the confidence of optimized tumor response. There is a need to develop this practice further, and therefore there is a current wealth of research studying predictive biomarkers for response. These biomarkers may be measured in the serum (proteins, circulating tumor cells, or nucleic acids) or be some form of dynamic measure of the tumor itself using tissue biopsies or sophisticated imaging tools such as PET and dynamic contrast MRI.

More than 50 oncogenes have now been robustly identified in cancer, many of which are already the subject of established or developing drugs. Given that many oncogenes are tyrosine kinases, it is not surprising that in developing signal transduction inhibitors, most efforts have been directed against them. Protein phosphorylation is regulated by kinases which phosphorylate and phosphatases which dephosphorylate; phosphorylation, particularly of tyrosine or serine–threonine residues, plays a central role in signal transduction and regulation of protein activity. The enzymes that regulate phosphorylation are thus potential drug targets for the treatment of many diseases and cancer in particular. Kinases may become active in cancers by mutations or through activation of upstream signaling pathways. We will now describe the clinical application of some of these key signaling pathways and gene targets.

The tyrosine kinase inhibitor: imatinib

The development of imatinib to target the aberrant BCR-ABL tyrosine kinase in chronic myeloid leukemia (CML) has been discussed (see Box 16.1). This drug has revolutionized the treatment of CML and gastrointestinal stromal tumors (GISTs). Not surprisingly, the success of imatinib has fueled development of further kinase inhibitors as anticancer drugs. The human genome encodes at least 500 protein kinases (the "kinome"), each of which is a potential drug target in cancer therapy. Eight kinase inhibitors have been approved for cancer therapy to date, and over 100 are being clinically evaluated at present. Most such drugs are targeted against receptor and cytosolic tyrosine kinases, but newer drugs are aimed at different families of serine–threonine kinases also. These newer agents have been discussed in detail in Chapter 16, but a brief description of the processes targeted is provided here.

EGFR signaling pathway

This pathway plays a key role in tumorigenesis and therefore has been exploited in drug discovery (Table 18.1). Of the four known members of the family, EGFR (ERBB1) and ERBB2 (HER2) are overexpressed or amplified in many malignancies including non-small-cell lung, breast, H&N, pancreatic, and colorectal cancers. Licensed anti-EGFR therapies include monoclonal antibodies (cetuximab, panitumumab, and trastuzumab) and oral small-molecule inhibitors of the TK activity (gefitinib, erlotinib, and lapatinib) (Fig. 18.6).

Trastuzumab (herceptin) is a humanized anti-HER2 (ERBB2) monoclonal antibody that has been approved for monotherapy or combination therapy (with paclitaxel) for women with HER2 overexpressing breast cancer. Recently it has also been licensed for HER2 positive gastric cancer. Generally speaking the treatment is well tolerated, but infusion-related hypersensitivity reactions may occur and there is a risk of cardiac dysfunction. Cetuximab (Erbitux) is a chimeric IgG1 antibody and is licensed for use in combination with irinotecan-containing chemotherapy in patients with metastatic colorectal cancer. There is now good evidence that this treatment is most effective in those patients with wild-type *KRAS*. Those with mutant *KRAS* are resistant to EGFR-targeted treatments, because signaling is already activated downstream of the receptor. *KRAS* status should be established to ensure that the most appropriate patients are selected for this therapy. Approximately 55% of tumors will be *KRAS* wild type. This treatment is complicated by a unique rash which is acneiform – indeed, it is suggested that the rash may actually be a predictive biomarker of response to cetuximab. Other side effects include diarrhea and infusion reactions. Panitumumab (Vectibix) is a fully human IgG2 monoclonal antibody that binds to the ectodomain of EGFR. It is also licensed for use in *KRAS* wild-type colorectal cancer. It has a similar side effect profile to cetuximab, although as it is fully humanized there are fewer infusion reactions. Other monoclonal antibodies in development are shown in Table 18.1.

Gefitinib (Iressa) is a TKI which blocks the binding of adenosine triphosphate (ATP) to the intracellular domain of EGFR. Interestingly, gefitinib has monotherapy activity in NSCLC, but it is not beneficial when combined with chemotherapy. The trials of gefitinib were disappointing, with improved response rates not conferring a survival advantage. Further work has identified that only patients with activating EGFR mutations respond well to the drug, and therefore gefitinib is approved for use in only metastatic NSCLC with activating mutations of the EGFR-TK. The most common side effects seen with gefitinib are rash, diarrhea, and fatigue, although increasingly electrolyte disturbances such as hypomagnesemia are observed. Erlotinib (Tarceva)

Table 18.1 Selected EGFR inhibitors

Anti-EGFR monoclonal antibodies (see also Chapter 5)			
Drug	Target	Tumor	Clinical status
Trastuzumab Herceptin	HER2	Breast Gastric	Licensed for HER2-positive breast and gastric cancer
Cetuximab Erbitux C225	EGFR	Head and neck squamous cell carcinoma (HNSCC) Colorectal	Licensed in head and neck (H&N) and colorectal cancer
Panitumumab Vectibix ABX-EGF	EGFR	Colorectal	Licensed for colorectal cancer
Pertuzumab Omnitarg 2C4	HER2	Solid tumors	Phases II and III
Matuzumab EMD72000	EGFR	Solid tumors	Development stopped
Nimotuzumab H-R3	EGFR	Solid tumors	Phases II and III
Anti-EGFR tyrosine kinase inhibitors			
Gefitinib Iressa ZD1839	EGFR	Non-small-cell lung cancer (NSCLC)	Licensed in NSCLC
Erlotinib Tarceva OSI-774	EGFR and HER2	NSCLC Pancreas Breast HNSCC	Licensed in NSCLC and pancreas cancer
Lapatinib GW-572016	EGFR and HER2	Breast	Licensed in breast cancer
Afatinib BI2992	PAN–ERBB	Solid tumors	Phases II and III
Neratinib HKI-272	PAN–ERBB	Solid tumors	Phase II

is another EGFR TKI and is licensed for use in NSCLC and pancreas cancer. It should be said that the survival advantage in pancreas cancer is marginal, and therefore erlotinib has not been approved by National Institute for Clinical Evidence (NICE) for this indication. Lapatinib (Tykerb) is a dual-specific TKI targeting both EGFR and HER2. It is licensed for use in combination with capecitabine chemotherapy in patients with metastatic breast cancer who have progressed after trastuzumab and other chemotherapy therapy.

Signaling through RAS–RAF–ERK (MAPK) and PI3K–AKT

As has been discussed, little progress has been made in targeting RAS or the RAF-ERK pathway (Chapter 6). Some progress has been made in targeting the PI3K–Akt–mTOR pathway, however. Rapamycin is the best-known inhibitor of mTOR–Raptor signaling (see Chapter 5), but newer analogues have been developed as potential anticancer agents. Temsirolimus is the first mTOR inhibitor licensed for use in metastatic renal cancer.

VEGF pathway

This is also discussed in detail in Chapter 14, and see Table 18.2. Angiogenesis is known to be pivotal in tumor progression and metastasis and is now a major target for new drug development for cancer (Table 18.2). Since VEGF is the driving force behind angiogenesis, it is not surprising that it has been the target of many approaches. Agents have been developed that target the VEGF ligand, blocking its interaction with the VEGF receptors (Bevacizumab and Aflibercept) and agents that directly block the receptor (tyrosine kinase inhibitors such as sunitinib). Bevacizumab (Avastin) is a recombinant humanized monoclonal antibody to VEGF-A. The proof of concept of antiangiogenesis therapy was realized when the study of bevacizumab in colorectal cancer was presented. In this pivotal study, bevacizumab in combination with irinotecan-based chemotherapy improved the survival of patients with metastatic colorectal cancer. As a consequence, bevacizumab was licensed in this indication, although its application has been hampered by its huge cost, and therefore it has not been endorsed by NICE. The side effect profile of bevacizumab includes hypertension, a class effect of VEGF inhibitors; thromboembolic events, both arterial and venous; and gastrointestinal perforation. It is now also FDA and EMEA approved in breast, non-small-cell lung, and renal cancers. VEGF-trap (Aflibercept) is a soluble decoy receptor protein that has a high affinity for all isoforms of VEGF-A as well as PlGF. It therefore "traps" the VEGF ligand before it can interact with the receptor. It is currently being evaluated in multiple phase III studies, and it will be interesting to see how it compares to the market leader, bevacizumab.

Since VEGF mediates its angiogenic effect through several tyrosine kinase receptors, many VEGF TKIs have been tested. Sorafenib (Nexavar) is licensed for use in renal cell and hepatocellular cancer, and Sunitinib (Sutent) is licensed for use in renal cell cancer and GIST. Numerous other TKIs are in clinical studies (Table 18.2). Some are highly selective, targeting only specific VEGF receptors, whereas others are more "promiscuous" and target other kinases such as PDGF and EGFR. Of interest, sunitinib and sorafenib can be used as monotherapy, whereas bevacizumab needs to be combined with chemotherapy. As such, our understanding of this complex pathway is not complete. Despite being "targeted" therapies, these drugs do have significant side effects including neutropenia, fatigue, rashes, diarrhea, and stomatitis.

Alternative pathways of angiogenesis can be targeted. Currently, there is significant interest in the Notch-signaling pathway. Two Notch ligands have a role in tumor angiogenesis, and therefore drugs are being developed to inhibit this pathway.

Drugs targeting epigenetic regulators

As these are discussed in Chapters 11 and 16, we will just mention some of the agents that are approved for use in this area.

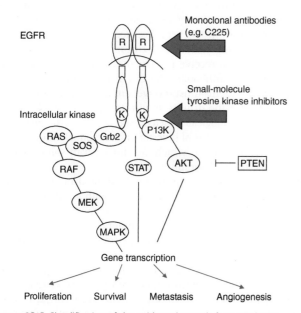

Figure 18.6 Simplification of the epidermal growth factor receptor-signaling pathway showing the site of action of the monoclonal antibodies and the small-molecule tyrosine kinases.

Table 18.2 Selected angiogenesis inhibitors

Anti-VEGF monoclonal antibodies (see Chapters 13 and 14)			
Drug	**Target**	**Tumor**	**Clinical status**
Bevacizumab Avastin	VEGF-A	Solid tumors	Licensed in colorectal, breast, renal, and NSCLC
Aflibercept VEGF-Trap	VEGF-A and PlGF	Solid tumors	Phase III
Ramucirumab IMC-1121b	VEGFR2	Breast Hepatocellular carcinoma (HCC)	Phase III
IMC-18F1	VEGFR1	Colorectal and breast	Phase II
Tyrosine kinase inhibitors			
Sunitinib (sutent) SU11248	VEGFR-1,2,3, PDGFR, c-Kit, and FLT3	Solid tumors	Licensed in renal cell cancer and GIST
Sorafenib Nevaxar BAY 43-9006	B-RAF, VEGFR2, EGFR, and RET	Solid tumors	Licensed for renal cell and hepatocellular carcinoma
Axitinib	VEGFR-1,2,3, PDGFR-α, PDGFR-β, and c-kit	Solid tumors	Phase III
Brivanib (BMS-582664)	VEGFR and FGFR	Solid tumors	Phases I and II
Cediranib AZD2171)	VEGFR1-3	Solid tumors	Phase III
Dovitinib lactate	VEGF-1,2,3, FGFR-1,2,3, PDGFR-β, c-KIT, and FLT3	Solid tumors	Phases I and II
Motesanib (AMG 706)	VEGFR, PDGFR, and c-KIT	Solid tumors	Phase III
Pazopanib (GW786034)	VEGFR PDGFR	Solid tumors	Phase III
Vandetanib	VEGFR-2,3, EGFR, and RET	Solid tumors	Phase III; licensed in medullary carcinoma of thyroid
Tivozanib AV-951	VEGFR-1,2,3	Solid tumors	Phases I and II

Bortezomib, the proteasome inhibitor, is approved for use in myeloma. The HDAC inhibitor, vorinastat, has been approved for use in refractory cutaneous T-cell lymphoma.

PARP inhibitors

In the DNA damage response mechanism, another excellent target for anticancer therapy is the nuclear enzyme poly(ADP-ribose) polymerase (PARP). It plays a critical role in DNA repair, and there has been some great success using PARP inhibitors in clinical studies both as monotherapy and in combination with chemotherapy. The PARP inhibitor Olaparib has been shown to have activity as a single agent in BRCA1 and BRCA2 mutated breast and ovarian cancer through synthetic lethality. In BRCA1- and BRCA2-deficient tumor cells, there is a defect in homologous recombination (HR) resulting in accumulation of DSBs. While this *per se* has little impact, when these tumors are then subjected to PARP inhibition, the SSBs become DSBs, and these persist due to the ineffective repair mechanisms. Ultimately this is lethal to the tumor cell, which dies through apoptosis. Olaparib generally is well tolerated, with mainly GI symptoms and fatigue seen. As such it is possible to combine it with chemotherapy agents, particularly the platinum agents and topoisomerase inhibitors.

Gene therapy

Dysfunctional tumor suppressor genes are the most common genetic lesions identified in human cancers. Working versions of tumor suppressor genes can be restored by introducing them into cancer cells by gene transfer using adenoviral vectors. This approach has been well studied with Onyx-015, a replication-defective adenovirus that expresses p53 and could be directly inoculated into tumors. This treatment caused growth arrest and apoptosis in cancer cells, but unfortunately the results from clinical studies have been disappointing and the Onyx-015 program has been discontinued. A more promising intra-tumor therapy is that using OncoVex GM-CSF. It is a second-generation oncolytic herpes simplex type 1 virus, encoding human GM-CSF. To date, promising results have been seen in melanoma patients and phase III studies are ongoing. Although OncoVex GM-CSF is administered locally by intratumoral injection, it provides systemic benefit by the induction of a potent anti-tumor immune response.

Immunotherapy

Essentially the main premise of immunotherapy approaches in cancer treatment is to stimulate the patient's immune system to attack the tumor. This can be achieved either by immunizing the patient with a cancer vaccine or by administrating therapeutic antibodies. To date there has been greater success with the latter approach, and approximately nine monoclonal antibodies have been approved by the FDA for the treatment of cancer. These include unconjugated (e.g. herceptin), immunoconjugated (e.g. gemtuzumab ozogamicin), and radioisotope-conjugated (e.g. iodine 131 tositumomab) antibodies. Historically the solid tumors treated by immunotherapy were renal cancer and melanoma, but now immunotherapy has more wide applications in other solid tumors and hematological malignancies.

Rituximab (Mabthera) is a chimeric monoclonal antibody targeting the CD20 antigen found on normal B cells and B-cell lymphomas. It is licensed for use by the FDA and EMEA for follicular lymphoma, diffuse large B-cell lymphoma, and chronic lymphocytic leukemia, and has remarkably improved the outlook for these patients. It can be administered either as monotherapy or in combination with chemotherapy such as the "CHOP" regimen. Iodine 131 tositumomab (Bexaar) and 90Y ibritumomab tiuxetan (Zevalin) are both radiolabeled CD20-targeted antibodies. They are licensed for use in some refractory lymphomas, but their costs are prohibitive and therefore neither is NICE approved.

Alemtuzumab (previously known as CAMPATH) is a humanized monoclonal antibody targeting the CD52 antigen found on B lymphocytes. It has been licensed for use by the FDA and EMEA for patients with B-cell chronic lymphocytic leukemia who are unable to receive the chemotherapy drug fludarabine. As with all these antibodies, there is a risk of infusion reactions and therefore patients must receive antihistamine and steroids as a premed before the drug is administered. It is associated with significant hematological toxicity, especially neutropenia. There are now a huge number of investigational monoclonal antibodies in clinical development, which are discussed in some excellent reviews (see the bibliography).

Lenolidamide is an immunomodulating drug that is both licensed and NICE approved for patients with advanced multiple myeloma (MM). It is a potent analogue of thalidomide and works in a number of ways: inhibiting proliferation of certain hematopoietic tumor cells (including MM plasma tumor cells and those with deletions of chromosome 5), enhancing T cell– and natural killer (NK) cell–mediated immunity, inhibiting angiogenesis, and inhibiting production of proinflammatory cytokines (e.g. TNF-α and IL-6) by monocytes. In studies, the most concerning side effects of treatment were venous thromboembolic events and myelosuppression. Not surprisingly, there is also a significant risk of teratogenicity, and therefore patients or their partners must not get pregnant while on treatment.

Cancer vaccines

DNA vaccines to treat cancer tend to fall in and out of fashion. Over the years, many of the strategies have failed to deliver for a variety of reasons. Often the tumor antigen targeted is only weakly immunogenic and not limited to tumor cells alone, thereby either having minimal effect or causing damage to normal cells in addition. For vaccination to be really effective, then, the patient needs to be in remission, which may limit the clinical application of this approach. A major hurdle has also been to actually deliver the vaccine to its target; however, improved technologies with liposomes and nanoparticles have gone some way to address this. The tumor types where cancer vaccines have been tried included melanoma (target gene Gp100, and HLA B7 NHC1), colon cancer (target protein CEA), prostate cancer (target protein PSA), and B-cell lymphoma.

Interest in cancer vaccination has been fueled with the success of the HPV vaccination program to prevent cervical cancer (see Chapter 13). HPV infection is the cause of cervical dysplasia which is the pre-malignant condition of cervical cancer. Worldwide, cervical cancer is a huge medical problem because 80% of deaths from cervical cancer occur in the poorest countries. Even in more developed nations, classically those women most at risk of the disease are not the ones who obtain cervical screening. Therefore, the HPV vaccines gardasil and cervirax have been developed against the HP16 and HP18 serotypes of the virus. The rationale is to vaccinate young women at an early age before they are sexually active and thereby prevent the chain of events starting from HPV infection and culminating in cervical cancer.

Hormonal therapy for breast cancer

In pre-menopausal women, biologically active estrogens are synthesized from androgens by the granulosa cells within the ovarian follicles. This conversion is regulated by the pituitary hormones, follicle-stimulating hormone (FSH) and luteinizing hormone (LH). A reduction of estrogen levels can be achieved in pre-menopausal women by oophorectomy (i.e. surgical removal of the ovaries), radiotherapy (artificial radiation menopause), or administration of drugs which suppress the secretion of LH and FSH by the pituitary (chemical castration).

Chemical castration can be achieved using gonadotrophin-releasing hormone (GnRH) agonists or antagonists. The secretion of LH and FSH depends on the pulsatile release of GnRH by the hypothalamus, and continuous stimulation results in downregulation of the pituitary response. GnRH analogues are modifications of the naturally occurring hypothalamic decapeptide and have greater potency and a longer half-life than the parent hormone. It is important to note that the GnRH agonists may cause an initial surge in pituitary levels of LH and FSH, which can be detrimental in the treatment of malignancy since it may result in a "tumor flare." Following this initial phase of stimulation, continued exposure to the analogues results in downregulation of GnRH receptors. The side effects of GnRH analogues include headache, hypersensitivity reactions, pruritus, skin rashes, changes in breast size, changes in scalp and body hair, edema of the face and extremities, weight changes, and mood changes including depression.

The estrogen receptor status of breast cancer predicts the usefulness of hormonal manipulation. In metastatic disease, more than 50% of patients with estrogen receptor positive (ER+) tumors experience regression of the cancer with hormonal therapy, compared to less than 10% of those with estrogen receptor negative (ER−) tumors.

Tamoxifen is a selective estrogen receptor modulator (SERM). It acts as a competitive antagonist for estrogen receptors, although it can have partial agonist activity for subclasses of estrogen receptors. Tamoxifen is capable of causing tumor regression in endocrine-sensitive tumors in both pre-menopausal and post-menopausal women. Both tamoxifen and its metabolites are highly protein bound, and it takes 8 weeks for serum levels of the principal metabolite to reach steady state. In addition to the treatment of metastatic breast cancer, tamoxifen is used as adjuvant therapy in women who have undergone surgery with or without radiotherapy as the primary treatment for their cancer. This is discussed in this chapter. The principal toxicities of tamoxifen are hot flashes, weight gain, nausea and vaginal discharge, cataracts, and a slightly increased risk of endometrial cancer following long-term usage.

In an attempt to improve efficacy and reduce the effect on bone and endometrium, more selective SERMs have been developed. These include those that are tamoxifen-like, including toremifene; fixed ring compounds, such as raloxifene; and selective estrogen receptor downregulators (SERDs), such as fulvestrant. A significant amount of work has been carried out to further understand the mode of action of these compounds, and this information will inform the use of these compounds both as chemoprevention agents, and as therapy for early and advanced breast cancer.

In postmenopausal women, androgens from the adrenal glands act as the principal source of estrogens by conversion in peripheral tissues, particularly subcutaneous fat. The aromatase enzyme system is the principle class of enzymes responsible for this conversion. Aromatase inhibitors are in common use in postmenopausal women with breast cancer. Aminoglutethimide was the first to be developed, but its use was associated with lethargy, hypotension, nausea, hypothyroidism, reversible agranulocytosis, and rash. More selective aromatase inhibitors have emerged over the past decade, which can be subdivided into nonsteroidal (anastrazole and letrozole) and steroidal compounds (exemestane). Common toxicities of these drugs include nausea, headache, hot flashes, and weight gain. Unlike tamoxifen, aromatase inhibitors may decrease bone density and predispose to osteoporosis. Although clear molecular differences exist between these two types of aromatase inhibitors, significant clinical differences have not been forthcoming from the studies to date. There may be some non-cross-resistance between the groups as switching from one to another can confer a modest improvement in outcome. A Cochrane analysis has found that in metastatic breast cancer, aromatase inhibitors show a survival benefit compared to other endocrine therapy, although limited data are available to identify which individual aromatase inhibitor has the greatest efficacy.

The overall response rate to hormonal manipulation in metastatic breast cancer is approximately 30%, although an additional number of patients will also have stabilization of their disease. Approximately 80% of ER+ and progesterone receptor positive (PR+) tumors will respond to hormonal therapy. Complete clinical remission is uncommon, and tumor regression may take 2 to 3 months. Partial responses and stable disease can be observed for many years in some patients. The median duration of response to first hormonal treatment is 18 months, and therapy is generally continued until disease progression. Second-, third-, and even fourth-line hormonal manipulation may be considered in these patients, either alone or in combination with palliative chemotherapy or radiotherapy. The progestins megestrol acetate and medroxyprogesterone acetate may also be used as late-stage therapies. Combinations of hormonal therapies are not generally used since no impact on survival superior in comparison to monotherapy has been shown.

A huge amount of work has explored the role of hormonal therapy in the adjuvant treatment of breast cancer. Studies have looked at the type of hormone used, the duration of treatment, and the sequencing of agents. Historically, tamoxifen for 5 years was the standard of care for patients with ER+ tumors requiring adjuvant treatment since it reduces the risks of relapse and death. More recent randomized controlled studies have suggested that for postmenopausal women, adjuvant therapy should include an aromatase inhibitor. This is because letrozole, anastrozole, and exemestane have showed statistically significant improvements in disease-free survival, with absolute risk reductions of between 2.5% and 8.8% of recurrence when compared with tamoxifen as either primary or sequential therapy and 4.6% when compared to placebo alone as extended therapy. Aromatase inhibitors in the adjuvant treatment of breast cancer have been endorsed by both the ASCO Guidelines and NICE guidance.

Hormonal treatment of prostate cancer

As discussed in this chapter, hormone treatments represent the first generation of biologically targeted therapies, though in most cases they substantially affect normal hormone-regulated physiology as well.

The growth of benign and malignant prostate cells is controlled by androgens. The most potent androgen is dihydrotestosterone, produced in the testis by the enzyme 5-alpha reductase. Androgen ablation is the primary treatment for men with metastatic prostate cancer. This can be achieved by orchidectomy or the use of GnRH analogues. Approximately 85% of metastatic prostate cancers are androgen sensitive at diagnosis, and response to androgen ablation is often dramatic. Androgen deprivation can also be used in localized prostate cancer to decrease the size of the prostate prior to definitive treatment with radiotherapy or surgery.

Since GnRH analogues are receptor agonists and can cause a transient stimulation of pituitary hormone production, it is important that androgen receptor antagonists are co-administered initially to prevent symptoms from tumor flare. Androgen receptor antagonists in common use include flutamide and cyproterone acetate. These drugs act at the androgen receptor by inhibiting the agonist activity of dihydrotestosterone. Common side effects include decreased libido, hot flushes, breast tenderness, and impotence. Treatment is continued until disease progression, and the median duration of response to first-line hormonal therapy is about 18 months. The prognosis is poor once the tumor develops androgen resistance – a situation now usually termed castrate-resistant or castrate-progressive disease. Diethylstilbestrol (stilboestrol) is sometimes used to treat metastatic prostate cancer, which is androgen resistant; however, more interest now lies in the novel hormonal agent abiraterone. CYP171A is a member of the cytochrome P450 family and is responsible for the conversion of pregnelonone and progesterone to their 17α metabolites. In addition, it can convert 17α-hydroxypregnelonone and 17α-hydroxyprogesterone to the androgens androstenedione and dihydroepiandrostenedione. Abiraterone binds CYP17A1 irreversibly and prevents the formation of androgens in the adrenal gland and peripheral tissues of castrate and noncastrate men. Early studies suggest that the drug has activity in advanced prostate cancer following chemotherapy and androgen deprivation failure, and randomized phase III studies are ongoing.

Androgen deprivation with GnRH analogues is also used in the adjuvant setting following radical radiotherapy for locally advanced disease in the hope of reducing the chances of tumor recurrence. The optimum duration of such therapy is unclear, and treatment is currently continued for 2 to 3 years in most countries. The side effects of GnRH analogue treatment include impotence, decreased libido, mild anemia, and decreased bone density. Following short-term therapy, the effects are reversible.

Conclusion

Options for the treatment of cancer are expanding at a great rate. Technological advances have facilitating the development of new radiotherapy techniques which enable the delivery of higher doses to the total tumor volume, sparing surrounding tissues. As a result, greater local tumor control appears possible and the limiting toxicity of radiotherapy – damage to adjacent normal cells – is minimized. With greater tolerability and efficacy, radiotherapy may take an increasing role in combination with surgery and chemotherapy to reduce the risk of local recurrence of a variety of tumors.

With the ever-increasing understanding of the biology of cancer, and the massive expansion of interest in oncology by the pharmaceutical industry, new systemic therapies are being developed at an unprecedented rate. Analogues of existing cytotoxics are in development and have the potential to reduce the toxicity of chemotherapy while retaining or increasing efficacy. There is also a paradigm shift with a focus on biologically targeted therapies, and it is encouraging to see more of these drugs available in the clinic. Over the next decade, research will focus on developing new targeted drugs and also on determining predictors of response using microarray assays, proteomics, and genomics. We may thus be able to tailor therapies to the individual depending on aspects of the biology of their cancer – so-called personalized medicine.

As many of the new therapies are cytostatic, we may also shift our paradigm of care from one that aims to eradicate all tumor cells into one that aims to stop progression and allow the patient to live with prolonged stable disease. The way in which we conduct clinical trials will need addressing to ensure that we use appropriate objectives. It is not adequate to just use a reduction in tumor volume as an efficacy outcome. For the targeted therapies, we need to develop surrogate endpoints to confirm that we are "hitting" the intended target. For some new agents, moving their indication from the metastatic to the adjuvant setting may take many years, with huge randomized studies required to establish improvement in overall survival. We need to accelerate this process so that patients are given the most active drugs as soon as possible; this may mean using progression-free survival as the primary endpoint of studies rather than overall survival.

It is hoped that with further development of biological agents, the side effect profile of treatment will improve. Greater understanding of chemoresistance is needed for both classical chemotherapy agents and the new targeted therapies. Within this chapter, some drugs described are only licensed by either the FDA or EMEA, and there are many reasons for this. More importantly, there is a transatlantic divide with most (if not all) drugs available in US clinics. This said, even in the United States, there is concern about the soaring costs of treating cancer patients. In the United Kingdom, there is a cost-effectiveness scheme in place through NICE. All high-cost therapies have to be approved by NICE before they can be routinely given to patients in the National Health Service (NHS). The decision of NICE has now also been recognized and adopted by 27 other countries. Indeed, there is an increasing number of countries either using the NICE recommendation or developing their own equivalent process. Essentially through quite complex health economic models, NICE defines a cutoff above which the cost of the drug is not offset by its potential benefit. If a drug falls below, then it is made available, but if the cost of the drug is deemed too high, then the drug is available to patients only through private insurance schemes or self-payment. There is obviously a concern that with the prohibitive costs of the new agents in clinical development, there is a serious risk that the current NHS will be unable to afford these novel treatments. As such, there is a real need for pharmaceutical companies to work with those who commission cancer services to ensure that the best drugs are available to treat patients with cancer. Table 18.3 shows a list of selected new drugs in development.

Table 18.3 Selected new agents in development (approved drugs in bold text)

Proteasome inhibitors (see Chapter 11)

Drug	Target	Tumor	Clinical Status
Bortezomib (velcade)	**Proteasome**		**Approved for advanced multiple myeloma**
Signal transduction and enzyme inhibitors (see Chapter 5)			
Everolimus (Rad001)	**mTOR**	**Novartis**	**Licensed in renal cell**
Temsirolimus (CCI-779)	**mTOR**	**Wyeth**	**Licensed in renal cancer and mantle cell lymphoma**
Dasatinib BMS-354825	**Src and Abl**	**BMS**	**Licensed in CML and ALL**
Nilotinib	**Abl**	**Novartis**	**Licensed in CML**
Crizotinib PF-02341066	Anaplastic lymphoma kinase (ALK)	Pfizer	Phase III
Brivanib BMS 540215	VEGFR and FGFR	BMS	Phases I and II
AZD6244 and ARRY-142886	MEK1 and MEK2	Astra-Zeneca	Phase I

P13K inhibitors

XL765	P13K and MTOR	Exelixis	Phase I
XL147	P13K	Exelixis	Phase I
BKM120	Pan P13K	Novartis	Phase I
MK-2206	AKT	Merck	Phase I

HGF and MET inhibitors

AMG102	HGF	Amgen	Phases I and II
ARQ197	cMET	Arqule	Phase II
Cabozantinib XL-184	MET, VEGF, and RET	Exelixis	Phase III

Cell-cycle targeted agents (see Chapter 4)

Seliciclib CYC202	CDK(E2F)	Cyclacel	Phase II
Flavopiridol	Pan CDK and STAT	Sanofi Aventis	Phases I and II
PD332991	CDK4/6	Pfizer	Phase I
AZD1152	Aurora kinase B/C	AstraZeneca	Phases I and II
AT9283	Aurora A/B	Astex	Phases I and II
GSK1070916	Aurora B/C	Glaxo Smith Kline	Phase I
Volasertib B16727	Polo-like kinase 1	Boehringer Ingelheim	Phase I
B12536	Polo-like kinase 1 and 2	Boehringer Ingelheim	Phase I
GSK 461364	All Polo-like kinases, Aurora A, and CDK2	Glaxo Smith Kline	Phase I
ON01910	Polo-like kinase 1	Onconova	Phases I–III
AT-9283	Aurora A+B, Jak2, Jak 3, and BCR–ABL	Astex	Phase I
LY2603618	CHK1	Eli Lilly	Phase I
AZD7762	CHK1 and CHK2	Astra Zeneca	Phase I
XL844	CHK1 and CHK2	Exelixis	Phase 1
MK1775	Wee-1	Merck	Phase 1

Apoptosis inducers and tumor suppressors (see Chapter 8)

Drozitumab	Apo2 ligand and death receptor 5 antibody	Genentech	Phase I
LY21813081	Survivin inhibitor	Eli Lilly	Phase I
HGS1029	IAP inhibitor	Aegera	Phase I
ABT-737	BH3 mimetic; inactivates anti-apoptotic BCL-2 family proteins	Abbott	Phase I

(Continued)

Table 18.3 (Continued)

HSP90 inhibitors			
Retaspimycin hydrochloride IPI-504	Ansamycin inhibitor	Infinity	Phase II
AUY922	Isoxazole-based	Novartis	Phases I and II
Gantetespib STA-9090	Non-ansamycin	Synta	Phases I and II
BIIB021	Non-ansamycin	Biogen	Phase II
HDAC inhibitors			
Vorinostat		Merck	Phase II
Panobinostat LBH589		Novartis	Phases I and II
Romidepsin		Celgene	Phases I and II
Parp inhibitors			
AGO14699		Pfizer	Phase I
Olaparib AZD2281		Astra Zeneca	Phase III
Veliparib ABT-888		Abbott	Phases I and II
Iniparib BSI-201		BiPar	Phase II
MK4827		Merck	Phase I

Bibliography

General

Devita, V.T., Lawrence, T.S., Rosenberg, S.A., Depinho, R.A., and Weinberg, R.A. (2011). *Cancer Principles and Practice of Oncology*. 9th edn. Philadelphia: Lippincott Williams & Wilkins.

Perry, M.C. (2007). *The Chemotherapy Source Book*. 4th edn. Philadelphia: Lippincott, Williams & Wilkins.

Surgery

Jayne, D.G., Thorpe, H.C., Copeland, J., Brown, J.M., and Guillou, P.J. (2010). Five-year follow-up of the Medical Research Council CLASICC trial of laparoscopically assisted *versus* open surgery for colorectal cancer. *British Journal of Surgery*, **97**: 1638–45.

Radiotherapy

Bhide, S.A., and Nutting, C.M. (2010). Recent advances in radiotherapy. *BMC Medicine*, **8**: 25.

Green, J.A., Kirwan, J.J., Tierney, J., et al. (2001). Systematic review and meta-analysis of randomised trials of concomitant chemotherapy and radiotherapy for cancer of the uterine cervix: better survival and improved distant recurrence rates. *Lancet*, **358**: 781–6.

Ma, C.M., and Maughan, R.L. (2006). Within the next decade conventional cyclotrons for proton radiotherapy will become obsolete and replaced by far less expensive machines using compact laser systems for the acceleration of the protons. *Medical Physics*, **33**: 571–3.

Martin, A., and Gaya, A. (2010). Stereotactic body radiotherapy: a review. *Clinical Oncology*, **22**: 157–72.

Steel, G.G., ed. (1993). *Basic Clinical Radiobiology*. London: Arnold.

Zelefsky, M.J., Fuks, Z., Hunt, M., et al. (2001). High dose radiation delivered by intensity modulated conformal radiotherapy improves the outcome of localized prostate cancer. *Journal of Urology*, **166**: 876–81.

Hormone agents

Danila, D.C., Morris, M.J., de Bono, J.S., et al. (2010). Phase II multicenter study of abiraterone acetate plus prednisone therapy in patients with docetaxel-treated castration-resistant prostate cancer. *Journal of Clinical Oncology*, **28**(9): 1496–501.

Dowsett, M., Cuzick, J., Ingle, J., et al. (2010). Meta-analysis of breast cancer outcomes in adjuvant trials of aromatase inhibitors versus tamoxifen. *Journal of Clinical Oncology*, **28**(3): 509–18.

Gibson, L., Lawrence, D., Dawson, C., and Bliss, J. (2009). Aromatase inhibitors for treatment of advanced breast cancer in postmenopausal women [review]. *Cochrane Library*, **4**.

Jordan, V.C., and O'Malley, B.W.O. (2007). Selective estrogen-receptor modulators and antihormonal resistance in breast cancer. *Journal of Clinical Oncology*, **25**(36): 5815–24.

Kesisis, G., Makris, A., and Miles, D. (2009). Update on the use of aromatase inhibitors in early-stage breast cancer. *Breast Cancer Research*, **11**: 211.

Stavridi, F., Karapanagiotou, E.M., and Syrigos, K.N. (2010). Targeted therapeutic approaches for hormone-refractory prostate cancer. *Cancer Treatment Review*, **36**: 122–30.

New cytotoxic agents

Cortes, J., and Saura, C. (2010). Nanoparticle albumin-bound (nab™)-paclitaxel: improving efficacy and tolerability by targeted drug delivery in metastatic breast cancer. *European Journal of Cancer Supplements*, **8**(1): 1–10.

Rivera, E., Lee, J., and Davis, A. (2008). Clinical development of Ixabepilone and other epothilones in patients with advanced solid tumours. *Oncologist*, **13**: 1207–23.

Targeted agents

Bergh, J. (2009). Quo vadis with targeted drugs in the 21st century. *Journal of Clinical Oncology*, **27**(1): 2–5.

Campoli, M., Ferris, R., Ferrone, S., and Wang, X. (2010). Immunotherapy of malignant disease with tumor antigen-specific monoclonal antibodies. *Clinical Cancer Research*, **16**(1): 11–20.

Chanan-Khan, A.A., and Cheson, B.D. (2008). Lenalidomide for the treatment of B-cell malignancies. *Journal of Clinical Oncology*, **26**(9): 1544–52.

Courtney, K.D. (2010). The PI3K pathway as drug target in human cancer. *Journal of Clinical Oncology*, **28**(6): 1075–83.

Dai, D., and Grant, S. (2010). New insights into checkpoint kinase 1 in the DNA damage response signalling network. *Clinical Cancer Research*, **16**(2): 376–83.

Grothey, A., and Galanis, E. (2009). Targeting angiogenesis: progress with anti-VEGF treatment with large molecules. *Nature Reviews Clinical Oncology*, **6**: 507–18.

Kitzen, J.J.E.M., de Jonge, M.J.A., and Verweij, J. (2010). Aurora kinase. *Critical Reviews in Oncology Hematology*, **73**: 99–110.

Lapenna, S., and Giordano, A. (2009). Cell cycle kinases as therapeutic targets for cancer. *Nature Reviews Drug Discovery*, **8**: 547–66.

Le Tourneau, C., Faivre, S., Serova, M., and Raymond, E. (2008). mTORI inhibitors: is temsirolimus in renal cancer telling us how they really work? *British Journal of Cancer*, **99**: 1197–203.

Lurje, G., and Lenz, H. (2009). EGFR signalling and drug discovery. *Oncology*, **77**: 400–10.

Ma, X., Ezzeldin, H.H., and Diasio, R.B. (2009). Histone deacetylase inhibitors current status and overview of recent clinical trials. *Drugs*, **69**(14): 1911–34.

Murukesh, N., Dive, C., and Jayson, G.C. (2010). Biomarkers of angiogenesis and their role in the development of VEGF inhibitors. *British Journal of Cancer*, **102**: 8–18.

Porter, R.P., Fritz, C., and Depew, K.M. (2010). Discovery and development of Hsp90 inhibitors: a promising pathway for cancer therapy. *Current Opinion in Chemical Biology*, **14**: 1–9.

Quintas-Cardama, A., Wierda, W., and O'Brien, S. (2010). Investigational immunotherapeutics for B-cell malignancies. *Journal of Clinical Oncology*, **28**(5): 884–92.

Rice, J., Ottensmeier, C.H., and Stevenson, F.K. (2008). DNA vaccines: precision tools for activating effective immunity against cancer. *Nature Reviews*, **8**: 108–20.

Roy, R., Yang, J., and Moses, M.A. (2009). Matrix metalloproteinases as novel biomarkers and potential therapeutic targets in human cancer. *Journal of Clinical Oncology*, **27**(31): 5287–97.

Sandhu, S.K., Yap, T.A., and de Bono, J.S. (2010). Poly (ADP-ribose) polymerase inhibitors in cancer treatment: a clinical perspective. *European Journal of Cancer*, **46**: 9–20.

Wood, J., Scott, E., and Thomas, A.L. (2009). Novel VEGF signalling inhibitors: how helpful are biomarkers in their early development? *Expert Opinion on Investigational Drugs*, **18**(11): 1701–14.

Workman, P., Clarke, P.A., Raynaud, F.I., and van Montfort, R.L.M. (2010). Drugging the PI3 kinome: from chemical tools to drugs in the clinic. *Cancer Research*, **70**(6): 2146–57.

Questions for student review

Choose all that apply.

1) Chemoradiation techniques have improved survival in the following tumor types:
   a. Anal cancer.
   b. Cervical cancer.
   c. Ovarian cancer.
   d. Sarcoma.
   e. Small-cell lung cancer.

2) The following cytotoxic drugs are alkylating agents:
   a. 5-Fluorouracil.
   b. Doxorubicin.
   c. Irinotecan.
   d. Oxaliplatin.
   e. Vinblastine.

3) Hormone therapy in breast cancer is commonly associated with the following side effect:
   a. Alopecia.
   b. Hot flushes.
   c. Myelosuppression.
   d. Vaginal dryness.
   e. Weight gain.

4) Which of the following are monoclonal antibodies targeting EGFR?
   a. Bevacizumab.
   b. Cetuximab.
   c. Gefitinib.
   d. Rituximab.
   e. Trastuzumab.

19 Caring for the Cancer Patient

Nicky Rudd[a] and Esther Waterhouse[b]
[a]Leicester Royal Infirmary, UK and [b]University Hospitals Leicester, UK

It is better to light a candle than to curse the darkness.

Chinese proverb

A man's dying is more the survivors' affair than his own.

Thomas Mann

Key points

- Not all patients with cancer will require specialist palliative care.
- Care can be shared between an oncologist and a palliative medicine physician.
- Patients may need intermittent palliative medicine input from diagnosis.
- Much can be done by those trained in the principles of general palliative care.

Introduction

The growth of palliative medicine and palliative care services (PCS) has been rapid in the last 10 years. The expansion of these services has led to increased specialization of palliative care. Not all patients with cancer will require specialist palliative care as their symptoms and problems can be dealt with by staff, such as oncology ward nurses, who are able to give general supportive care. Hospices and palliative care units are increasingly used as tertiary centers prioritizing those patients with complex physical or psychological problems. Oncology patients may need interventions from palliative medicine at any time during their illness, and in the United Kingdom they may be under the care of both the oncologist and palliative medicine physician simultaneously.

In the United States, there is an artificial boundary driven by the reimbursement system of Medicare and Medicare hospice benefit. This results in patients being transferred to the hospice system only when life-prolonging treatments are ineffective and death is imminent. This is beginning to change with the National Consensus Project Clinical Practice guidelines from the United States stating that "the effort to integrate palliative care into all health care for persons with debilitating and life-threatening illnesses should help to ensure that pain and symptom control, psychological distress, spiritual issues and practical needs are addressed with patient and family throughout the continuum of care."

Key concepts

The model shown in Fig. 19.1 illustrates the continuum of palliative care, which should be accessed in acute hospitals, hospices, or the community, and differentiates between palliative care and what is called "terminal care" in the United Kingdom and "hospice care" in the United States.

Communication with the cancer patient

Excellent care is difficult to achieve without good communication. By finding out what your patients are thinking, and tailoring information to what they want to know, communication can be markedly improved. Current training emphasizes key tasks in communication:

- Elicit the patient's main problems, the patient's perceptions of these, and the physical, emotional, and social impact of these problems on the patient and family;
- tailor information to what the patient wants to know, checking his or her understanding;
- determine how much the patient wants to participate in decision making (when treatment options are available);
- discuss treatment options so that the patient understands the implications;

The Molecular Biology of Cancer: A Bridge From Bench to Bedside, Second Edition. Edited by Stella Pelengaris and Michael Khan.
© 2013 John Wiley & Sons, Inc. Published 2013 by John Wiley & Sons, Inc.

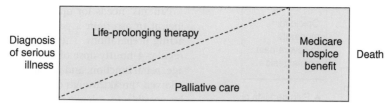

Figure 19.1 Palliative care's place in the course of illness (National Consensus guidelines).

- maximize the chance that the patient will follow agreed decisions about treatment and advice about changes in lifestyle.

In order to be able to do all these. Tasks, oncologists need to be able to communicate clearly and effectively.

It is now widely acknowledged that many doctors struggle with effective communication. They feel pressure of time and worry that if they explore distress during a consultation, they will be faced with a situation that they cannot handle. Consequently, many doctors use "blocking strategies" to prevent further disclosure.

Blocking strategies
- Offering "premature advice or reassurance"

Patient: "I am so worried... " (unsaid: *"about my wife"*)
Doctor: (assuming she knows what the patient is worried about) "The chemotherapy will work well and you'll be feeling better before you know it."
- Explaining distress as normal

Patient: "I'm so frightened."
Doctor: "Everybody I see feels frightened at first but that soon goes."
- Attending to physical aspects only

Patient: "I'm very worried.. ."
Doctor: "I see. How are your bowels?"
- Jollying patients along

Patient: "Oh I am upset about my cancer."
Doctor: "But we should be able to cure it."
- Switching topic

Patient: "My wife is so upset at the moment."
Doctor: "Now have you got your blood test forms for the next cycle of chemo?"

Doctors use strategies like these in the mistaken belief that they help, both the patient, by preventing them from getting upset, and themselves by minimizing the distress that they are exposed to. In fact the opposite is true: they prevent patients from disclosing their anxieties and problems and lead to increased distress for the patient and the doctor. Exploring these concerns can lead to better joint management of problems. It can also aid compliance with drugs and treatment programs.

When is palliative care appropriate for cancer patients?

Many oncologists will look after patients until their death. The increasing use of palliative chemotherapy has prolonged the palliative phase in many patients. However, this has also raised new problems, including when to stop palliative chemotherapy and how to ensure the patient's agreement with the decision.

Palliative care may be required at any time during the patient's treatment. Referral to the palliative care team (PCT) should also be considered as chemotherapy treatment is discontinued.

Palliative care assessment

The assessment of patients needs to be specific, be detailed, and encompass psychological, social, and existential issues as well as physical problems. Only by paying close attention to detail will the physician be able to identify the cause of problems and treat them effectively. For instance, a patient's pain may seem unresponsive to opioids not because the morphine is ineffective but because the patient is fearful of addiction and has not been taking the tablets.

Symptom control

A retrospective case study of 400 patients referred to PCTs showed that 64% had pain, 34% anorexia, 32% constipation, 32% weakness, and 31% dyspnea. The most common symptoms are covered in this chapter. For other problems, the reader may consult the reference list at the end of the chapter.

Pain
This may result from the tumor itself, indirectly from tissues related to the cancer, or from other unrelated causes. In a retrospective study in 1995, 2074 carers were asked about the patient's experiences in the last year of life. They reported that 88% of patients were in pain at some time, 47% of patients' pain was either only partially controlled or not controlled at all by general practitioners (GPs), and 35% had pain partially or not at all controlled by hospital doctors.

A study of 200 cancer patients referred to a pain clinic showed that 158 had pain caused by tumor growth; visceral involvement was the cause in 74 cases, bone secondaries in 68 cases, soft tissue invasion in 56 cases, and nerve compression or infiltration in 39 cases. Many patients had more than one type of pain.

Principles of pain control
The analgesic ladder (Fig. 19.2) is a simple scheme which emphasizes the stepwise approach to pain due to cancer and the need to take regular analgesics.

Step 1 – simple nonopioids, for example paracetamol/acetaminophen, and nonsteroidal anti-inflammatory drugs (NSAIDs)

Step 2 – opioids for mild to moderate pain such as co-codamol, codeine, and dihydrocodeine

Step 3 – opioids for moderate to severe pain, for example morphine

Move up the ladder if the current step is ineffective. All the medication has to be given regularly and orally unless unable to do so.

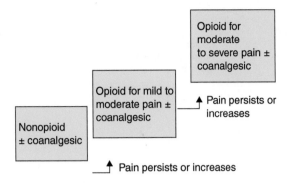

Figure 19.2 WHO analgesic ladder.

Table 19.1 Commonly used opioids

Morphine	4-, 12-, and 24-hour preparations (can be used parenterally)
Oxycodone	4- or 12-hour preparation (parenteral preparation available)
Hydromorphone	4- or 12-hour preparation
Methadone	Useful for patients with neuropathic pain and partially opiate-sensitive pain
Fentanyl	Transdermal patch. Change every 3 days. Reservoir of drugs for 12–24 hours after patch removed. Only use in stable pain. Oral transmucosal preparations are also available for buccal use. Intravenous analogs are becoming available.
Buprenorphine	Low-dose transdermal patch and sublingual preparation
Diamorphine	High solubility; can be given subcutaneously either as a stat dose or via a syringe driver (not available worldwide)
Levorphanol	Orally or parenterally
Oxymorphone	Suppository or parenteral preparation

Some patients will go directly from Step 1 to Step 3. Regular nonsteroidals or regular paracetamol can be used as opiate-sparing drugs.

Opioids

There are a variety of long- and short-acting opioids available that can be used for patients requiring step 3 analgesics; however, not all opioids are available in all countries (Table 19.1). Morphine remains the most frequently used opioid. Others can be used either for a different route of administration or in case of side effects.

Common side effects of opioids are drowsiness, nausea, constipation, xerostomia, vivid dreams, and itching. Opiate toxicity is characterized by pinpoint pupils, visual hallucinations, twitching, agitation and confusion, and respiratory depression. If patients are experiencing side effects from one formulation, it is worth trying another drug.

Patients' needs for opioids will vary markedly. It is therefore logical to ascertain a patient's needs before commencing long-acting preparations. The usual method is to start a patient on a regular 4-hourly dose of short-acting opioid judging the dose on age, renal function, and previous drug requirement. They are also allowed "breakthrough doses" for pain. After 48 hours, it should be possible to calculate their 24-hour opiate requirement and convert to a long-acting preparation. Whatever be the long-acting preparation, there should be a short-acting opiate given for breakthrough pain.

Cancer pain syndromes

Bone pain

This is a very common cause of cancer pain. Many patients will achieve good pain control with radiotherapy and bisphosphonates. NSAIDs may be effective or analgesics as per the analgesic ladder may be required.

Incident pain

This is most commonly seen in patients with spinal bone metastases and may be one of the most difficult pain syndromes to treat effectively. The patient will generally be pain-free at rest but experience severe and sometimes incapacitating pain when bearing weight. If the pain is treated with sufficient analgesia to cover the periods of activity, the patient will be sedated and overdosed at rest, and if the dose is only sufficient for rest pain, the patient will stay immobile.

Management includes optimum treatment for the cause, including chemotherapy, radiotherapy, and bisphosphonates. Opiates should be used as previously described; the use of short-acting oral fentanyl preparations may be helpful, with the addition of coanalgesics for bone pain and neuropathic pain. It may be necessary to change to methadone or add corticosteroids. Epidural or intrathecal analgesia may be required. Physiotherapists and occupational therapists may be useful in moderating either activity or the environment in order to reduce episodes of pain.

Liver capsule pain

This is due to the distention of the capsule and may be associated with diaphragmatic irritation. The pain may respond to NSAIDs or steroids; otherwise opiates may be required.

Bowel pain

Intra-abdominal cancer may present with poorly localized pain due to peritoneal inflammation, vascular or lymphatic blockage, or bowel spasm secondary severe to serosal or intraluminal disease. Constipation may be an additional problem due to drugs, bowel dysfunction, or a combination of both.

Nerve compression or plexopathy

Patients with pelvic tumors, pancreatic cancer, spinal secondaries, apical lung tumors, and head and neck tumors may have severe neuropathic pain (Table 19.2). Patients will typically describe their pain as "burning" or "stabbing" and may have associated loss of sensation or reflexes. The pain may be opiate-sensitive in approximately one third of patients; others will need the addition of drugs specifically for neuropathic pain such as gabapentin.

Patients with difficult pain should be referred to the palliative care or pain team.

Table 19.2 Descriptors in neuropathic pain

Allodynia	Increased sensitivity such that stimuli that are normally innocuous produce pain
Hyperalgesia	Increased sensitivity to a painful stimulus
Hyperpathia	Prolonged response to a stimulus even after its removal

Table 19.3 Commonly used coanalgesics

Nonsteroidal anti-inflammatory drugs (NSAIDs)	Bone pain or other pain, particularly if inflammatory component potentiates opiates
Tricyclic antidepressant (e.g. amitriptyline)	Neuropathic pain. Initial drowsiness may subside as analgesic effect becomes effective after 5 days.
Antiepileptics (e.g. gabapentin, pregabalin, and carbamazepine)	Neuropathic pain. Newer drugs have less interactions and side effects.
Steroids	Useful for short periods for pain due to nerve compression
Ketamine	Effective for neuropathic pain. Suppress hallucinations with haloperidol or diazepam.
Benzodiazepines	For muscle spasm and myofascial pain
Paracetamol/ acetaminophen	Used as a coanalgesic to potentiate the effects of opiates

Coanalgesics

These are drugs that are used, often in combination with opioids, for a variety of pains, often neuropathic in origin. They include antidepressants and antiepileptic agents such as gabapentin, though the mechanism of benefit is not fully understood (Table 19.3).

It may be necessary to give patients a combination of these drugs.

Anesthetic techniques in pain control

These techniques are used to either block the pain pathway or deliver analgesia through another route (e.g. spinally).

Nerve blocks

They are usually carried out in patients who have pain localized to a nerve distribution and do not respond to analgesia. They are usually carried out with local anesthetic in the first instance to assess response before the nerve is ablated with phenol, cryotherapy, or radio frequency.

Epidural and intrathecal catheters

They may be used to place opiates, local anesthetic, and steroids spinally for patients with back or pelvic pain.

Respiratory symptoms

Breathlessness is a common symptom and occurs in many cancers. It is thought that toward the end of life respiratory muscle weakness plays a large part, but earlier on in the illness there are many treatable causes. Thus, breathlessness may result from the cancer itself (pleural effusion, lung metastases, or massive ascites), but also from preexisting conditions or anticancer treatment (anemia).

There is evidence in lung cancer that nondrug treatments can be of help. Examples of such treatments include teaching breathing exercises and energy conservation techniques and explaining the cause of the breathlessness. Additional drug treatments, including β_2 agonists, should be used if required.

A recent systematic review has shown benefit of opiates orally or subcutaneously for the treatment of cancer-related breathlessness. Benzodiazepines may improve panic associated with breathlessness. Some patients who are hypoxic may benefit from supplemental oxygen.

Terminal breathlessness

Patients are often very frightened of suffocating to death. They and their family members need lots of explanation and reassurance. It is often easier to treat breathlessness at the terminal stage because sedation is less of an anxiety and patients and families often accept some drowsiness if symptom control is good. Paradoxically some patients become more alert once their breathlessness has been adequately treated.

Generally a syringe driver containing an opiate and an anxiolytic is used; many centers use diamorphine and midazolam. Extra doses should be given if the patient needs them.

Nausea and vomiting

Nausea and vomiting are common symptoms in cancer patients and, in the experience of the authors, are often poorly treated by nonspecialists. This seems to be because of a lack of understanding of the mechanisms causing the symptoms, and the different treatments available.

Assessing nausea and vomiting is essential. The history should include the time course of the problem, precipitating and relieving factors (including previously tried drug and nondrug treatments), the relationship between the nausea and the vomiting, and the amount of vomitus produced at each vomit. Knowing whether vomiting relieves the nausea can be very helpful. Examination and selected use of tests such as blood biochemistry are also helpful.

Some common syndromes and suggested treatments are outlined below.

Delayed gastric emptying

This encompasses a range of problems, from mechanical gastric outlet obstruction to opiateinduced reduced gastric motility. Patients often have little nausea, but may feel themselves "filling up" before having a large vomit. If they are managing to eat, the vomit will contain undigested food. The vomit may be effortless and often leaves patients feeling much better.

Dopamine antagonists, such as metoclopramide and domperidone, are the treatment of choice. An open-label study suggested the efficacy of metoclopramide, although it was not placebo controlled. Metoclopramide speeds up gastric emptying at the gastric level (acting on D_2 and $5HT_4$ receptors) as well as centrally, whereas domperidone acts mainly on D_2 receptors in the stomach. Therefore, most palliative medicine physicians will use metoclopramide. Patients with a mechanical outlet obstruction may benefit from steroids. Drugs should be given parenterally to start with, as oral absorption will be poor.

Chemically induced nausea

Common causes in patients with advanced cancer are uremia and hypercalcemia. If a patient who has been stable on opiates develops nausea, particularly with signs of opioid toxicity, it may be that renal function is impaired causing opioid accumulation and toxicity. Patients with chemically induced nausea complain of a constant nausea with intermittent retching, which usually fails to relieve the nausea. They rarely manage to eat much.

The most potent dopamine blocker is haloperidol, and this is usually the treatment of choice. Side effects are rare at low doses. If this is ineffective, when given parenterally, then levomepromazine should be used.

Emesis caused by chemotherapy and radiotherapy is usually treated with $5HT_3$ antagonists, according to written protocols. Despite revolutionizing chemotherapy-induced nausea and vomiting, $5HT_3$ antagonists have not found a clear place in the management of nausea and vomiting from other causes. Newer antiemetics which block the neurokinin (NK) receptors in the brainstem and gastrointestinal tract (GI) tract have been developed and are used for highly emetogenic chemotherapy protocols. They also have not as yet found a place for nausea and vomiting due to other causes.

Hypercalcemia is the commonest lifethreatening metabolic disorder related to malignancy (Table 19.4). It is seen in 10% of patients with malignancy, most frequently in patients with myeloma, breast, lung, and renal cancers. Up to 20% of cases occur in the absence of bone metastases. It is an indicator of a poor prognosis, with a median survival of 3–4 months. The underlying metabolic problems include increased osteoclastic bone resorption, decreased renal clearance of calcium, and enhanced absorption from the gut.

Some cancers produce parathyroid hormone–related protein, transforming growth factors, or osteoclast-activating factors.

Symptoms are often vague with most patients developing malaise and lethargy followed by thirst, nausea and constipation, and then confusion. It is important to look for hypercalcemia as treatment is often effective at resolving symptoms.

Treatment involves rehydration with intravenous (IV) fluids followed by IV bisphosphonates. Resolution of symptoms often takes several days. Some patients require repeated infusions at intervals to maintain normocalcemia; oral bisphosphonates, although effective in this context, have poor bioavailability and tolerability and are therefore not often used in the palliative setting. Other treatments such as calcitonin may be necessary in intractable cases.

Stretch and distortion of GI tract

Can be due to a host of causes, such as intraabdominal tumors, constipation, bowel obstruction, and retroperitoneal lymph nodes. It is usually necessary to treat the underlying cause; steroids can also be helpful in reducing peri-tumor edema; they may also have a direct antiemetic effect.

Cranial causes

Raised intracranial pressure can cause nausea. High-dose dexamethasone is usually indicated, and palliative radiotherapy may be used. An antiemetic acting at the vomiting center, such as cyclizine or levomepromazine, would be the symptomatic treatment of choice.

Anxiety

Can cause nausea and may cause vomiting.

Nondrug treatments

Should be remembered when assessing patients with nausea and vomiting. Sometimes simple changes to the environment, such as controlling odor or changing portion sizes, can be enough to significantly reduce symptoms. Remember that patients' tastes often change such that previously palatable food is no longer tolerated.

Intractable nausea

Not all nausea and vomiting are amenable to treatment despite best efforts. In these cases it is important to be realistic with the patients, most of whom will accept control of their nausea with intermittent vomits if that is the best that can be achieved.

Bowel obstruction

Malignant bowel obstruction (MBO) is a common problem in advanced abdominal and pelvic cancers. Estimates of frequency among patients with these cancers are difficult to ascertain. Some studies have quoted up to 51% of patients with gynecological malignancies and 28% of patients with GI malignancies developing bowel obstruction; the true may be higher.

Obstruction can be complete or incomplete, and the degree of obstruction, and therefore of symptoms, may change over time. Patients present with a mixture of distended abdomen, altered bowel habit (which may be constipation), colicky abdominal pain, nausea, and vomiting. They also often have a constant background abdominal pain.

The pattern of symptoms varies according to the level of obstruction:

Table 19.4 Humoral factors and hypercalcemia

Parathyroid hormone–related protein (PTH r P)
   Activates PTH receptors in tissue
   Stimulates osteoclastic bone resorption and renal tubular absorption
Transforming growth factors (TGF)
   TGF are functionally and structurally similar to epidermal growth factors
   Increases osteoblastic activity
   TGF-β increases production of PTH r P in breast cancer cells
Osteoclast-activating factors
   Tumor necrosis factor (TNF-α)
   Lymphotoxin (TNF-β) produced by activated Tlymphocytes and myeloma cells
   Interleukin 6 (IL-6)
   Interleukin 1 (IL-1)

High (stomach)	Large-volume vomits
	Persistent nausea; may be relieved by vomiting
	Little distension
	Constipation; develops late
Small bowel	Minimal distension
	Abdominal colic
Large bowel	Abdominal distension
	Colic
	Constipation

Managing bowel obstruction

Assess the cause of the obstruction. Is it amenable to surgery? Is chemotherapy appropriate? Would a stent be helpful? Stents can be useful for gastric outlet obstruction and, occasionally, for large bowel obstruction. Is decompression with a naso-gastric tube likely to help? In a few patients, it is the only way to maintain control of vomiting.

In advanced MBO, it is rarely possible to stop the patient vomiting and most palliative physicians will aim for control of nausea and reduction of vomiting to once every 1 or 2 days. It is important that all involved agree on the aims of the treatment.

Colicky abdominal pain may benefit from an antispasmodic such as hyoscine butylbromide (buscopan) in a syringe driver.

Most patients with MBO will be able to eat small amounts, although these may be vomited again later. Most palliative care units manage MBO without long-term parenteral hydration, although some patients find that symptoms of thirst and nausea ease with small amounts of fluid.

Constipation

63% of elderly people in hospital are constipated. Causes include debility, immobility, depression, inadequate fluid intake, and drugs, particularly opioids, tricyclic antidepressants, diuretics, and ferrous sulfate. Tumor effects include spinal cord compression and direct compression of bowel by tumor. Other concurrent problems may exist (e.g. diabetes and local anal pathology).

It is best to prevent constipation before it becomes established by ensuring adequate fluid intake and use of laxatives for all those on weak and strong opioids.

Fatigue

Fatigue is a commonly reported problem in advanced cancer but is rarely identified and treated. It can be caused by the cancer or by the treatment. Treatable causes of fatigue include:
- Anemia – it is not clear how much of the fatigue related to advanced disease responds to treatment of anemia;
- metabolic and endocrine disorders;
- psychological issues including depression;
- anticancer treatment.

Dexamethasone can help cancer-related fatigue, but the effect is often short-lived and side effects are common. Psychostimulants can be helpful and in the United States they are the treatment of choice for depression in the terminally ill, as they have a rapid onset of action. However, they can cause agitation and insomnia, which limit compliance. Trial evidence for the use of psychostimulants in cancer-related fatigue is minimal.

Table 19.5 Common problems causing anorexia or cachexia

Problem	Treatment
Mouth problems	
Dry mouth	• Consider side effects.
Intraoral infections	• Treat infections promptly.
Sore mouth	• Mouth washes
	• Analgesics (e.g. diamorphine mouthwash, NSAIDs, adcortyl in orabase, and gelclair)
Dysphagia	
Tumor	• Consider stent or palliative radiotherapy.
Infection	• Treat infections promptly.
Nausea	• See "Nausea and Vomiting" section (this chapter).
Malabsorption	
Enzyme deficiency	• Pancreatic supplements.
Chronic diarrhea	• Treat diarrhea.
Anxiety and depression	• See "Psychological Problems" section (this chapter).
Altered taste	• Increase spiciness of food.
	• Add lemon juice.

Cachexia and anorexia

Cachexia occurs in over 50% of patients with solid tumors. It is characterized by loss of both fat and muscle. The anorexia–cachexia syndrome (Table 19.5) is driven by production of cytokines, particularly tumor necrosis factor (TNF-α). There are also numerous changes in the humoral regulation of appetite, energy use, carbohydrate, and protein metabolism that cause loss of muscle as well as fat.

Drug treatment

- Metoclopramide may be helpful if gastric stasis is a contributing factor.
- Progesterone, for example megestrol acetate 160–800 mg (this may take 1–2 weeks to work). Long-term use of progesterones may induce adrenal suppression.
- Corticosteroids – use as short courses, which can be repeated if necessary.

Other drugs such as thalidomide (TNF inhibitor) or creatine may be useful in the future.

Psychological problems

The three main psychological problems causing morbidity in patients with cancer are depression, anxiety, and adjustment disorders.

Approximately 35–50% of patients with cancer will have a psychiatric disorder, the prevalence of which is approximately 22% in the general population. The skill for the oncologist and other health professionals involved in the cancer patient's care is to distinguish between sadness associated with a life-threatening illness and a psychiatric disorder.

Depression

Studies have found the prevalence of depression in cancer patients to vary from 3.2% to 42%. A more realistic figure is 5–15% fulfilling the criteria for major depression. Risk factors are previous psychiatric problems, young age, lack of social support, decreasing physical function, and the presence of pain.

Depression can be difficult to diagnose in cancer patients. A commonly used screening tool is the Hospital Anxiety and Depression Tool.

The UK National Institute for Clinical Excellence (NICE) guidance for palliative and supportive care states that "the psychological needs of all patients and carers are assessed on a continual basis throughout the patient pathway, with particular attention to points that are particularly challenging, such as the time prior to diagnosis and after bereavement."

This guidance mandates regular screening of all cancer patients, using some type of screening tool incorporated into routine cancer care.

Drug treatment of depression is as effective in this population as it is with those who are physically well. Selection of drugs may be determined by other symptoms, for instance a tricyclic antidepressant may be useful for those with neuropathic pain or insomnia, and selective serotonin reuptake inhibitors (SSRIs) should be used with care in those with preexisting nausea.

Anxiety

This is common in patients with cancer. However, some patients will present with debilitating symptoms. Much can be done with good communication and clear information. Some patients will require benzodiazepines, which are generally well tolerated. Patients should be reassured that these drugs are being used appropriately as even dying patients may require reassurance over addiction.

Adjustment disorders

These occur when patients are unable to adapt to a severe life event and cope inflexibly in response to their illness. The definition of an adjustment disorder is rather nebulous; it is a disorder characterized by a variety of clinically significant behavioral or emotional symptoms that occur as a result of some triggering event or stressor. It occurs in approximately 25–30% of patients with cancer but it is often overlooked. There may be overlap with depression but it can be diagnosed only if depression and anxiety disorders are excluded. Brief psychotherapy has been proven to be effective in this disorder.

The dying patient

Ongoing communication with the dying patient and their family remains essential. The physician needs to consider discontinuing ineffective treatments. In the United States, a high proportion of oncology patients die in intensive care. In the United Kingdom, the drive is to ascertain patients' choice in their place of death.

Relatives may have some difficulty in agreeing to discontinuing treatments such as IV fluids or antibiotics as they may perceive it as helping the patient to die. They need to be reassured that the patient is dying with or without active treatment and that most patients are more comfortable with good nursing care rather than invasive procedures.

All these decisions can be made much more easily if end-of-life issues have been discussed with patients and they have made their preferences known. Living wills can be very helpful to both the medical team and the family. Discussing death may seem a daunting task but many patients welcome a discussion on what is, after all, at the forefront of their minds. They will also be reassured that you are interested in not just their tumor response but also what happens to them when treatment is no longer effective. A care pathway for dying patients (an individualized care plan, or ICP) has been introduced into most hospitals in the United Kingdom in order to improve standards of care. It has a checklist format looking at physical, psychological, and social issues, prompting the regular checking of patients and their families.

Symptom control in the dying patient

Important principles in care of the dying patient are attention to individual symptoms and anticipation of problems such as agitation and inability to swallow. If the patient's symptoms have been previously under good control, they may not experience worsening of symptoms as they die. Problems occurring frequently are listed below.

Inability to swallow

This should be anticipated. Syringe drivers delivering drugs in a 24-hour subcutaneous dose are commonly used to replace necessary oral medications.

Agitation

This is common in the dying patient and has a variety of etiologies. It may be a manifestation of distress and inability to communicate but it is important to exclude and treat causes that are reversible (e.g. urinary retention, constipation, and pain). Discuss with the relatives what they think may be the cause and consider, with their agreement, sedation with benzodiazepines (e.g. midazolam) or levomepromazine, if appropriate. These can be given via the syringe driver.

Retained secretions

As patients become weaker, they are unable to clear oral secretions or cough effectively. This leads to accumulation of saliva at the back of the throat – "the death rattle." The patient may be relatively unaware of this, but it often causes distress to relatives sitting by the bedside. Hyoscine or glycopyrronium can be used subcutaneously, otherwise tipping the bed or, in the last resort, suctioning can be used. Other symptoms can be treated as previously described.

Palliative care services

These have expanded exponentially over the last 15 years in the United Kingdom particularly. Services prior to this were sporadic with much of the drive for development being from local groups, resulting in diverse services across the country due to a lack of regional planning. There have been recent efforts to remedy this and most regions will have services in both the acute sector and the community.

Supportive care

"Supportive care" is a relatively new phrase to describe an age-old concept, that of caring for patients and those close to them during difficult times. This concept has now been formalized in

NICE guidelines for supportive and palliative care. The definition from the National Council for Hospices and Specialist Palliative Care Services is that supportive care:

> helps the patient and their family to cope with cancer and treatment of it – from pre-diagnosis, through the process of diagnosis and treatment, to cure, continuing illness or death and into bereavement. It helps the patient to maximize the benefits of treatment and to live as well as possible with the effects of the disease. It is given equal priority alongside diagnosis and treatment.

Provision of supportive care is therefore not dependent on the stage or type of cancer, or indeed on the predicted treatment outcome. "Supportive care" can be seen as an umbrella term, covering all services, generalist and specialist, which may be needed to help patients and families.

An example of the care of a cancer patient

Mr. A has a chest X-ray suspicious of lung cancer. He attends outpatients, and the lung cancer nurse is present when the doctor is seeing him. After discussing the probable diagnosis, she gives him an information package. She then coordinates his diagnostic process, making sure that he is aware of what is going on and that he understands the need for further tests (information giving).

After the diagnosis is confirmed, she makes contact with him by phone – due to high levels of distress, she visits him and his family at home (psychological support). He agrees to attend the local lung cancer support group (user involvement, self-help, and support). As Mr. A is the main carer for his wife who has severe multiple sclerosis (MS), the nurse contacts other professionals to arrange for her to go into respite care while he is having treatment. She also refers him for specialist benefits advice (social support). Following his treatment he remains breathless and is referred to the local breathlessness program (rehabilitation).

He finds reflexology particularly helpful (complementary therapies). Later he develops significant psychological distress and she refers him to the community specialist PCTs for further assessment (palliative care). His distress seems to be due to a loss of faith in God and he is referred to the hospice chaplain (spiritual support). During the next few weeks he requires ongoing palliative care involvement to control his worsening pain.

His terminal illness is managed at home with support from the primary care and specialist palliative care teams, and this is where he dies (end-of-life-care). His wife requires a lot of support from the primary care team after his death (bereavement care).

Questions remaining

1. Increased understanding of the underlying mechanisms of the process of dying.
2. More large multicenter randomized controlled trials of analgesics for cancer pain; this is problematic for many reasons including:
   a. many patients in the early stages of their disease are participating in trials of anticancer treatment;
   b. there are numerous practical and ethical issues around research on patients in the last few weeks or months of life.

3. A better understanding of which services are most effective for this population.

Conclusions and future directions

This chapter has reviewed the common problems experienced by cancer patients and described a palliative approach.

Good cancer care, including palliative care, relies on the input from a diverse range of health professionals, access to whom should be on an as-needed basis rather than on the stage of disease. Effective communication between professionals as well as with the patient is paramount.

Palliative medicine works most effectively if integrated into oncological units and not limited to end-of-life care.

Work continues worldwide not only to increase recognition of palliative medicine and its achievements but also to educate patients and health professionals about available treatments. A first and important step would be worldwide availability of opiates for those in pain.

Underlying problems

i. Bone metastases in the left hip

ii. Renal failure

iii. Urinary tract infection

iv. Cauda equina compression

v. Constipation

Comment

1. Renal failure can occur secondary to urinary outflow obstruction. Beware of iatrogenic causes, (e.g. NSAIDs).

2. Constipation is very common in patients with cancer, especially if they are on morphine, and can cause urinary retention. "Diarrhea" may be due to fecal overflow.

3. Cauda equina compression can present gradually and with vague symptoms of weakness, numbness, urinary or fecal incontinence, a numb sacrum, and legs just feeling "odd."

2) A 65-year old patient with malignant pain is about to start morphine. Which of the following statements are correct?
   a. Patients should be observed for respiratory depression.
   b. All patients should start laxatives with opiates unless contraindicated.
   c. There is no ceiling dose of morphine.
   d. The majority of patients will experience nausea or vomiting.
   e. Addiction can be a problem for some patients.
   f. Doses can be limited by drowsiness.
   g. The dose will inevitably increase with time.
   h. The prescription of morphine will shorten the patient's life.

Comment These are all common worries expressed by patients and doctors. Approximately one third will experience nausea and vomiting. Tolerance can occur, but others will remain on a stable dose for some time. There is no evidence that morphine per se will shorten life.

Drowsiness can be a problem for some patients. Most patients require laxatives unless they are malabsorbing, ileostomy, or the like.

3) For each cause of nausea and vomiting, select the most useful drug:
   a. Initial opiate-induced nausea
   b. Gastric stasis
   c. Vestibular dysfunction
   d. Chemotherapy-induced nausea
   e. Associated with bowel obstruction

Underlying problems

i. Hyoscine hydrobromide

ii. Metoclopramide

iii. Ondansetron

iv. Haloperidol

v. Cyclizine

Bibliography

Service provision
Morrison, R.S., and Meier, D.E. (2004). Palliative care. *The New England Journal of Medicine*, **350**: 2582–90.

National Consensus Guidelines (2004). National Concerns Project for Quality Palliative Care. *Clinical Practice Guidelines*, April 2004; 57–60.

Communication
Maguire, P., and Pitceathly, C. (2002). Key communication skills and how to acquire them. *British Medical Journal*, **325**: 697–700.

General symptoms
Potter, J. (2003). Symptoms in 400 patients referred to palliative care services: prevalence and patterns. *Palliative Medicine*, **17**: 310–14.

Addington-Hall, J.M., and McCarthy, M. (1995). Dying from cancer: results of a national population-based investigation. *Palliative Medicine*, **9**: 295–305.

Pain
Mercadante, S., Radbruch, L., Caraceni, A. *et al.* (2002). Episodic (breakthrough) pain: consensus conference of an expert working group of the European Association for Palliative Care. *Cancer*, **94**(3): 832–9.

Sindrup, S.H., and Jensen, T.S. (1999). Efficacy of pharmacological treatments of neuropathic pain: an update and effect related to mechanism of drug action. *Pain*, **83**: 389–400.

World Health Organization (1986). *Cancer Pain Relief*. WHO, Geneva.

Foley, K.M. (1985). The treatment of cancer pain. *The New England Journal of Medicine*, **313**: 84.

Breathlessness
Bredin, M., Corner, J., Krishnasamy, M. *et al.* (1999). Multicentre randomized trial of nursing intervention for breathlessness in patients with lung cancer. *British Medical Journal*, **318**: 901.

Jennings, A.L., Davies, A.N., Higgins, J.P.T. *et al.* (2004). Opioids for the palliation of breathlessness in terminal illness (Cochrane Review). The Cochrane Library Issue 4.

Abernethy, A.P., Currow, D.C., Frith, P. *et al.* (2003). Randomised, doubleblind placebo-controlled crossover trial of sustained release morphine for the management of refractory dyspnoea. *British Medical Journal*, **327**: 523–8.

Booth, S., Wade, R., Johnson, M. *et al.* (2004). The use of oxygen in the palliation of breathlessness. A report of the expert working group of the Scientific Committee of the Association for *Palliative Medicine/Respiratory Medicine*, **92**(1): 66–77.

Bower, M., and Cox, S. (2004). Endocrine and metabolic complications of advanced cancer. In: Doyle, D. *et al.* (eds), *Oxford Textbook of Palliative Medicine*, 3rd edn. Oxford: Oxford University Press, pp. 687–702.

Bruera, E., Sweeney, C., Willey, J. *et al.* (2003). A randomized controlled trial of supplemental oxygen versus air in cancer patients with dyspnoea. *Palliative Medicine*, **17**(3): 659–63.

Gastrointestinal symptoms
Wilson, J., Plourde, J.Y., Marshall, D. *et al.* (2002). Long-term safety and clinical effectiveness of controlled-release metoclopramide in cancer associated dyspepsia syndrome – a multicentre evaluation. *Journal of Palliative Care*, **18**(2): 84–91.

Skinner, J. and Skinner, A. (1999). Levomepromazine for nausea and vomiting in advanced cancer. *Hospital Medicine*, **60**(8): 568–70.

Feuer, D.J., and Broadley, K.E. (2003). Corticosteroids for the resolution of malignant bowel obstruction in advanced gynaecological and gastrointestinal cancer (Cochrane Review). In: *The Cochrane Library*, Issue 4, 2003. Chichester, UK: John Wiley & Sons.

Complementary therapies
Pan, C.X., Morrison, R.S., Ness, J. *et al.* (2000). Complementary and alternative medicine in the management of pain, dyspnoea and nausea and vomiting near the end of life. A systematic review. *Journal of Pain and Symptom Management*, **20**(5): 374–87.

Cachexia and asthenia
Dein, S., and George, R. (2002). A place for psychostimulants in palliative care? *Journal of Palliative Care*, **18**(3): 196–9.

Knobel, H., Loge, J.H., Brenne, E. *et al.* (2003), The validity of EORTC QLQ-C30 fatigue scale in advanced cancer patients and cancer survivors. *Palliative Medicine*, **17**(8): 664–72.

Strasser, F. (2004) Pathophysiology of the anorexia/cachexia syndrome. In Doyle, D. *et al.* (eds), *Oxford Textbook of Palliative Medicine*, 3rd edn. Oxford: Oxford University Press, pp. 520–33.

Jatoi, A., Jr., and Loprinzi, C.L. (2001). Current management of cancer-associated anorexia and weight loss. *Oncology*; **15**(4): 497–502.

Loprinzi, C.L. *et al.* (1993). Phase, I.I.I evaluation of four doses of megestrol acetate as therapy for patients with cancer anorexia and/or cachexia. *Journal of Clinical Oncology*, **11**(4): 762–7.

Loprinzi, C.L. *et al.* (1999). Randomized comparison of megestrol acetate versus dexamethasone versus fluoxymesterone for the treatment of cancer anorexia/cachexia. *Journal of Clinical Oncology*, **17**(10): 3299–306.

Loprinzi, C.L., Fonseca, R., and Jensen, M.D. (1996). Induction of adrenal suppression by megestrol acetate. *Annals of Internal Medicine*, **124**(6): 613.

Depression and anxiety
Zabora, J., Brintzenhofenszoc, K., Curbow, B., Hooker, C., and Piantadosi, S. (2001). The prevalence of psychological distress by cancer site. *Psychooncology*, **10**: 19–28.

Alexander, P.J., Dinesh, N., and Vidyasagar, M.S. (1993). Psychiatric morbidity among cancer patients and its relationship with awareness of illness and expectations about treatment outcome. *Acta Oncologica*, **32**: 623–6.

Breitbart, W., Rosenfeld, B., Pessin, H. *et al.* (2000). Depression, hopelessness and desire for hastened death in terminally ill patients with cancer. *The Journal of American Medical Association*, **284**: 2907–11.

Minagawa, H., Uchitomi, Y., Yamawaki, S., and Ishitani, K. (1996). Psychiatric morbidity in terminally ill cancer patients – a prospective study. *Cancer*, **78**: 1131–7.

Endicott. J. (1984). Measurement of depression in patients with cancer. *Cancer*, **53**: 2243–8.

Zigmond, A.S., and Snaith, R.T. (1983). The Hospital Anxiety and Depression Scale. *Acta Psychiatrica Scandinavica*, **67**: 361–70.

Suggested reading

Simpson, K., and Budd, K. (eds). *Cancer Pain Management*. Oxford: Oxford University Press, 2000.

Faull, C., Carter, Y., and Woof, R. (eds). *Handbook of Palliative Care*. Oxford: Blackwell Science, 1998.

Lloyd-Williams. M. (ed). *Psychological Issues in Palliative Care*. Oxford: Oxford University Press, 2003.

Doyle, D., Hanks, G., Cherney, N.I., and Calman, K. (eds). *Oxford Text of Palliative Medicine*. Oxford: Oxford University Press, 2004.

Twycross, R., and Wilcock, A. *Symptom Management in Advanced Cancer*, 2nd edn. Abingdon: Radcliffe Publishing, 1997.

Twycross, R.G., Wilcock, A., Charlesworth, S. et al. (eds). *Palliative Care Formulary*, 2nd edn. Abingdon: Radcliffe Publishing.

Questions for student review

1) An 84-year-old man with metastatic carcinoma of the prostate presents with the following symptoms. Match them with the likely underlying problem:
   a. Acute confusion and vomiting
   b. Urinary retention and fecal incontinence
   c. Back pain, constipation, and urinary retention
   d. Pain in left knee and reduced mobility
   e. Confusion and fever

20 Systems Biology of Cancer

Walter Schubert[a], Norbert C.J. de Wit[b], and Peter Walden[c]
[a]Otto-von-Guericke-University Magdeburg, Germany, [b]Maastricht University Medical Center, The Netherlands, and [c]Universitätsmedizin Berlin, Germany

It is far more difficult to be simple than to be complicated; far more difficult to sacrifice skill and easy execution in the proper place, than to expand both indiscriminately.

John Ruskin

Key points

- The information flow in organisms moves from the genome to the transcriptome to the proteome. On various levels, there are control mechanisms. The system is highly dynamic, and the study of the total system is called systems biology.

- Cancers are systems diseases in which tumor cells are critically reliant on a strong supporting cast of stromal and immune cells and the whole performance is executed within the highly individualized setting of a specific organ in a particular individual and over a protracted time course during which the entire cast has lived through a variety of different environments. Each contributor is invaluable in realizing the whole, and if cancers were awarded Oscars then every tumor would get a myriad of nominations.

- New technologies colloquially summarized as "omics" (e.g. genomics, transcriptomics, proteomics, and metabolomics) aim at system-wide (viz., cell-, tissue-, or body-wide) recording of genetic, transcriptional, translational, posttranslational, and metabolic alterations relating to disease or any physiological process of interest. Omics data represent statistical averages of large populations of cells, or entire tissues or sections thereof. They lack spatiotemporal information on what happens when and where, and they lack information on the interactions of the recorded factors. Such information can in parts be obtained by cellular or subcellular fraction and time-lapse experiments which, however, are difficult to realize *in vivo*. Toponomics provides data with spatiotemporal resolution, including information on protein interactions directly in situ.

- Proteomics analyzes the protein content of cells, tissue, or plasma and derivatives thereof. It usually involves the separation of the proteins by SDS-PAGE to reduce the complexity of the protein mixtures or 2D electrophoresis (isoelectric focusing plus SDS-PAGE) to isolate individual proteins, followed by the identification of the proteins by mass spectrometry (MS). The identification is done by fragmentation of the proteins with enzymes such as trypsin, determination of the masses of the fragments by MS to obtain a peptide mass fingerprint (PMF) of every protein, and/or further fragmentation of the fragments by physical means such as collision with inert gases (collision-induced disintegration (CID)) or disintegration upon high-intensity laser excitation (postsource decay (PSD)) and mass spectrometry of the resulting subfragments with subsequent correlation of the spectra with theoretical spectra computed from a reference protein sequence database.

- Mass spectrometry is among the important tools for systems biology and is used in proteomics, peptidomics, glycomics, lipodomics, metabolomics, and so on to address the corresponding classes of biomolecules. The principles of this technology are ionization and vaporization of the analytes, separation of the resulting ions, determination of their mass-to-charge ratios, and detection of the ions. Accordingly, a mass spectrometer consists of three parts: an ion source, a mass analyzer, and a detector.

- Transcriptomics deals with the analysis of transcriptomes (i.e. the RNA content of cells). Transcriptomics provides information on the expression of genes and splice variations of the gene products. Most often, DNA arrays are used for genome-wide or focused transcriptome analyses. More recently, next-generation sequencing is used for the purpose. It has the advantages that genome-wide analyses are possible without prior definition of the target genes and that it provides sequence-level information on splice variants. In contrast, quantification of the expression levels is more indirect than with the microarray hybridization technologies. The main disadvantage, however, is the high costs compared to microarray hybridization.

The Molecular Biology of Cancer: A Bridge From Bench to Bedside, Second Edition. Edited by Stella Pelengaris and Michael Khan.
© 2013 John Wiley & Sons, Inc. Published 2013 by John Wiley & Sons, Inc.

- Next-generation sequencing is highly parallel sequencing of DNA which covers the human genome several-fold in a single run within a few days. The separation of amplified DNA by gel or capillary electrophoresis, as with conventional DNA sequencing, is replaced by stepwise sequencing reactions on templates fixed to plane surfaces or particles where light is emitted and recorded at every step. The reactions are done in successive rounds for the four nucleotides or with nucleotides carrying different color fluorochromes. The succession of light signals translates into the nucleotide sequences. Read lengths per run are, depending on the technology used, in the order of 35 to several hundred. Genome sequences are generated by projecting these short sequences onto a human genome reference sequence. Next-generation sequencing is employed for genome sequencing or resequencing for identification of mutations, for expression analysis, and for promoter and DNA methylation analysis. Because of the short reads, it is not well suited for the identification of chromosomal aberration, where array comparative genomic hybridization would be the method of choice.
- Systems biology seeks to integrate omics data into models for a disease or any other biological state or process of interest. This involves large-scale data collection and processing and the development of biomathematical models suitable for experimental testing.
- Research in systems biology is often undertaken on model organisms such as yeast, *Drosophila melanogaster* (fruit fly), and mouse or in cell cultures.
- Cancer genomes, as revealed by next-generation sequencing and comparative genomic hybridization, deviate to varying extents from the genomes of normal somatic cells.
- The number and nature of genetic or chromosomal aberrations differ greatly across different types of tumor. Thus, around 30–40 tumor-specific mutations have been found in childhood neuroblastoma compared to about 50 000 in breast cancer or colon carcinoma. Chromosomes may appear intact in many cancers, whereas in CML, the aberrant Philadelphia chromosome is both pathognomonic and a marker for the presence of the BCR–ABL fusion protein. On the other hand, in cutaneous T-cell lymphomas large numbers of chromosomal aberrations, sometimes involving all chromosomes, occur without known oncogenic products.
- Cancer cells express unique protein networks that allow them to organize themselves successfully in order to realize the hallmark features described in this book. Such networks are likely to be specific for each cancer cell type and to determine their specific behavior.
- Theoretically, different cancer cell types can form myriads of different protein networks with the identical proteins simply by differentially assembling them as different protein clusters to execute different cellular functionalities. For example, given 20 different proteins (#1 to #20), all expressed in two different cell types with the same abundance, cell type 1 may assemble proteins #1+#5+#20 to form a functional protein cluster at the cell surface membrane to initiate and control cell migration. Cell type 2 may assemble the same proteins plus proteins #15 and #16 (hence #1+#5+#20+#15+#16) to encode the opposite function, namely, stop migration and enter cell division, and an assembly of proteins #8+#9+#11 in this latter cell type may encode the opposite: initiation of the protein network that induces migration. Each of these protein clusters would have a unique subcellular topology that is essential for the execution of said functionalities.
- Protein arrangements and differential assembly rather than regulation of expression (up- or downregulation) may be the essential cell biological mechanism in (cancer) cells controlling their function.
- Protein networks cannot be predicted by current large-scale *ex vivo* expression-profiling technologies because these techniques average protein (proteomics) or messenger RNA (mRNA) abundances (transcriptomics) in millions of pooled and homogenized cells or in entire tissues. Subcellular topologies of the proteins and, in many cases, protein clusters of interest are destroyed in the process.
- Toponomics is a branch of systems biology addressing the toponome, defined as the entire protein network code of morphologically intact cells and tissues.
- Toponomics is based on the capability to map the subcellular combinatorial molecular structure of a large number of distinct proteins and other molecular species with subcellular resolution.
- Although it is hoped that in the future new developments in matrix-assisted laser desorption ionization (MALDI) imaging or other in situ proteomics techniques will help address these questions, at this time limited resolution limits their use.
- An exciting alternative possibility results from the high resolution of microscopy and the exquisite sensitivity of antibodies for their cognate protein epitope targets. Traditionally, these techniques have been invaluable in unraveling the spatiotemporal expression of one or a very small number of proteins but have been unable to study sufficient numbers of proteins concurrently to be genuine systems biology tools.
- Robotic workstations have now been developed to include microscopic technologies (Toponome Imaging System (TIS)) capable of running fluorescence cycles of protein or molecule tagging, imaging, and bleaching in situ on morphologically intact cells and tissues, thereby overcoming the spectral and wavelength limitations of traditional fluorescence microscopy.
- These techniques' data sets, which can be used for co-mapping 100 different molecular species or more in one experiment, contain information on protein abundance (regulation), protein arrangement or rearrangement, and subcellular or transcellular protein network structures at currently 120 to 200 nm subcellular resolution in up to 6000 cells simultaneously.
- A three-symbol code for rapid detection and quantification is one of several approaches to denominate multiplex protein-imaging data and detect functional hierarchies of the toponome code.

Introduction

The best way to get a good idea is to get a lot of ideas.

Linus Pauling

The sequencing of the human genome and the genomes of model organisms such as rat, mouse, the nematode worm *Caenorhabditis elegans*, and the fruit fly *D. melanogaster* as well as phylogenetically related organisms has revolutionized biology. The impact of this revolution will not only be felt in the biological sciences but also ultimately transform clinical medicine. Non-invasive diagnosis and treatment, disease prediction, and personalized medicine are expected developments resulting from the genomics revolution.

Interaction between chemistry, biology, bioinformatics, mathematics, statistics, and physics is necessity to deal with the complicated biological systems and the large quantities of data acquired with high-throughput techniques. To understand why a cancer cell differs functionally from a normal cell, among others, one must consider the complex interplay between molecular networks involved in regulating gene expression, signaling, protein synthesis, and metabolism across time and space, and how these processes are affected by mutations, and epigenetic and chromosomal alterations. The complexity of the networks, the existence of feedforward and feedback loops, and multiple interaction points provide ample opportunities for the application of the evolving discipline of systems biology comprising informatics, computational modeling, mathematics, and statistics in conjunction with large-scale omics-type data generation, and high-content molecule-imaging technologies. A quote by Paul Nurse, Nobel laureate, illustrates this best:

[P]erhaps a proper understanding of the complex regulatory networks making up cellular systems like the cell cycle will requires a ... shift from common sense thinking. We might need to move into a strange more abstract world, more readily analyzable in terms of mathematics than our present imaginings of cells operating as a microcosm of our everyday world.

Information flow in cells

Information is not knowledge.

Albert Einstein

It is a very sad thing that nowadays there is so little useless information.

Oscar Wilde

Human life starts as a single cell, a fertilized egg, which develops into a complex organism that contains trillions of cells and thousands of cell types. Cancer too starts as a single cell, a cell with Darwinian tendencies. The information needed for the creation and functioning of a human being is, as with other organisms, largely encoded in the genome. In essence, a genome is an encrypted encyclopedia of life, albeit a very difficult one. The mouse, human, and *Drosophila* genomes are remarkably similar, and most genes are conserved between species. It is not so much the genes in the genome but the regulatory networks that are responsible for the differences between species. Also, this information is encoded in the genome. The high homology between species explains why, for example, the mouse or the fruit fly can be a useful "model organism" to unravel human physiology and disease.

The genome is transcribed into mRNA, and the mRNA is then translated into proteins. The human genome codes for at least 30 000 genes (the number of functional genes is still a matter of much debate), which are transcribed into many more mRNAs that code for a suggested well over 100 000 proteins. At each level, many variations occur making the system very complex (Fig. 20.1). Individuals differ on the genome level by approximately 10 million DNA single-nucleotide polymorphisms (SNPs), but this amounts to only around 0.1% of the genome. However, small changes in the DNA can have fatal effects. Cystic fibrosis, for example, is an endocrine gland disorder that is caused by mutations in the cystic fibrosis transmembrane conductance regulator (CFTR) gene and affects several organs. The CFTR protein functions as an ATP-regulated chloride channel, and deletion of three base pairs in the CFTR gene results in deletion of the amino acid phenylalanine. The mutated chloride channel is dysfunctional, which causes the production of abnormally viscous mucus. In lungs, the viscous mucus is an ideal growth medium for bacteria, and the continuous bacterial growth results in chronic respiratory infections. Ultimately, the permanent tissue damage leads to a range of complications resulting in death.

Breast cancer, among other cancers, is also associated with specific mutations in the genome, but its relationship to a single-gene mutation is not as clear as in cystic fibrosis. Many mutations in *BRCA1* and *BRCA2* genes have been linked to increased chances of developing breast cancer, but these mutations do not predict all cases of breast cancer. This highlights our limited understanding of gene mutations and genetic variation between individuals (including SNPs and other polymorphisms). Many diseases are multifactorial, and different gene mutations conspire with environmental factors (e.g. lifestyle) to influence disease development and outcome. Attributing the role of SNPs and gene mutations to different diseases is a field of much research.

The next level of systems regulation is in the structure of DNA. Expression of genes is not just regulated by transcription factors. Chemical modifications of DNA can alter chromatin structure and therefore gene expression without changing the DNA sequence. The enzyme DNA methyltransferase methylates DNA, especially in CG nucleotide-rich areas called CpG islands, and methylation of DNA and in particular CpG islands alters chromatin structure. Methylated DNA forms condensed chromatin, which is not accessible to transcription factors and leads to gene silencing. Epigenetic regulation involves DNA methylation and demethylation, chromatin remodelling, methyl-binding domain proteins, and histone deacetylation and acetylation (see Chapter 11).

When chromatin is accessible to transcription factors, a gene can be transcribed into mRNA. Each gene can potentially give rise to a number of mRNAs which in turn can be posttranscriptionally regulated. Posttranscriptional regulation usually means splicing of the mRNA transcript, but lifetime and the translation of these transcripts into protein are also regulated. It is now increasingly apparent that much of what has been termed non-coding DNA may direct the production of small RNAs that have profound effects on gene expression and may silence the expression of many genes by a variety of mechanisms, including the formation of double-stranded RNA, which is targeted for destruction by enzymes such as Dicer as described in Chapter 11.

Transcripts are translated into proteins, but much debate still surrounds how this is related to protein levels. Synthesized proteins are often posttranslationally modified. The original protein product can be modified after synthesis by, for example, the attachment of a phosphate group (phosphorylation), methyl group (methylation), sugar moiety (glycosylation), or ubiquitin group (ubiquitination). There are hundreds of protein modifications possible which can be either permanent or temporary. Posttranslational modifications have a multitude of effects and control many cellular processes. Synthesized proteins have an N-terminal sequence tag or are tagged with a modification to ensure that they are transported to the correct cellular compartment. Such tag is akin to a postal code on a letter and guarantees punctual and correct delivery of the protein. As with postal services, this is not flawless. These modifications, ubiquitinylation and sumoylation in particular, can also determine when and how rapidly a protein is degraded, and thus control lifetime and expression levels in the cell. Proteins can also be modified to perform a specific action in a signal transduction pathway. Protein kinases play a crucial role in activating signal transduction pathways by phosphorylating tyrosine, threonine, or serine residues in proteins. Such posttranslational modifications are transient and activate the protein or induce its association with other proteins for a short period of time. Termination of this phosphorylation-induced cellular signal often occurs through de-phosphorylation by protein phosphatases.

Each posttranslationally modified protein is a different protein because it has a different function and may have different interaction partners. Hence, the roughly 30 000 genes probably encode for over a million functionally different proteins. However, proteins rarely function independently in a cell but usually form complexes that function as sometimes complex functional units. Proteins are the individual parts of these functional units, and protein complexes interact with each other forming protein networks. It is these networks that are essential for cellular communication. One of the largest functional units in cells is the well-known spliceosome. The spliceosome consists of approximately 145 proteins and five small RNA molecules. The spliceosome complex is dynamic and undergoes multiple assembly stages and conformational changes during the splicing process. Identifying and characterizing protein complexes, and unraveling pathways and cellular networks, are of critical importance to understand cell biology.

Fig. 20.1 illustrates the complexity of cells and organisms. The integrated study of a biological model is termed "systems biology." Although systems are often seen as cells, organs, organisms or even ecosystems, a systems approach can also focus on a cellular pathway or a tumor. Biological systems are highly dynamic, and organisms adapt and change constantly to respond to external influences. A sudden large amount of exercise leads to muscle ache due to muscle adaptation and the immune system is activated to deal with infections. This all leads to changes at various levels in, for example, an organism, and a biological system is therefore never static.

Studying dynamic systems such as cells, cancers, or organs is extremely challenging and requires much expertise. Fig. 20.1 summarizes the informational hierarchy and the flow of information in biological systems: the central dogma of molecular biology. This chapter roughly follows this flow of information and will focus on new techniques that play an important role in cancer research. Important discoveries and developments will be discussed and linked to previous chapters. Many technical advances in life science research are pioneered in cancer research. Moreover, cancer research plays a crucial role in the translation of laboratory techniques into clinical applications; key examples will be discussed. At the end of the chapter, a number of Internet resources are given as additional information.

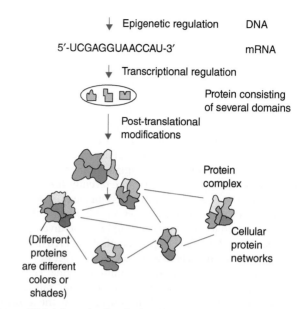

Figure 20.1 Information flow in the cell – from genome to proteome to cellular networks. On each level, variation has an effect on cell function, and abnormal variation can underlie cancer.

Model organisms and cancer models

Flies, worms, and flowers exceed me still.

Isaac Watts

Yeast, the fruit fly *D. melanogaster*, zebra fish, puffer fish, mouse, rat, and the nematode worm *C. elegans* are all established as model organisms. These organisms have been extensively used to study and understand biology. Much information is available on these organisms, and our understanding of human biology has increased by studying them. Each organism is used for different reasons, and an organism is often either a genetic, experimental, or genomic model. The mouse is a particularly interesting model because it is phylogenetically much closer to humans than other model species and its genome is organized similar to the human genome. Much has been learned about normal physiology and cancer using mouse models.

Two premises underlie the use of various animal models in cancer biology. Firstly, such models can contribute to our knowledge of cancer mechanisms, particularly when informed by knowledge derived from studies of human cancers, and thereby can serve for testing hypotheses on oncogenesis and other aspects of cancer. Secondly, such models can be crucial test beds for evaluating therapeutic targets and strategies, aimed at treating, curing, or preventing cancer. In fact, recent advances in regulatable genetic systems allow testing hypotheses of inactivating specific cancer genes or proteins even before a suitable drug has been developed (see Appendix 20.1).

Mice have been employed in cancer research since 1894. Initially, studies focused on same-species tumor transplantations and drug treatment studies. In the 1920s, inbred strains that had been shown to be prone to cancer were first disseminated among cancer researchers. Many more strains of mice were identified following the founding of the Jackson Laboratory in Bar Harbor, Maine, now a preeminent source of mouse cancer models for the research community. In 1962, the identification of the "nude" mutant mouse, which has lowered immunity, largely lacks T lymphocytes, and therefore has minimal tissue rejection capacity, enabled studies with transplanted human tumors which led to major advances in cancer research. More recently, breakthroughs in molecular biology resulted in the development of genetically altered mice in the 1980s. Oncogenes, or other genes that have been hypothesized to cause cancer, could now be investigated in much greater detail. Engineering of genetically altered mice made it possible to create models to address specific processes.

It has been suggested that transforming a normal cell into a cancer cell requires an estimated four to seven rate-limiting genetic events that confer specific properties to the evolving cancer cell. These have been elegantly construed as the hallmark features of cancer, described in previous chapters:
- Capacity to proliferate irrespective of exogenous mitogens;
- Refractoriness to growth-inhibitory signals;
- Resistance to apoptosis;
- Unrestricted proliferative potential (immortality);
- Capacity to recruit a vasculature (angiogenesis);
- Invasion of tumor cells into surrounding tissue;
- Ability to invade surrounding tissue, and metastatic dissemination of tumor cells to distant organs.

To these we may now add the Warburg effect, evasion of immune control, and tissue remodeling.

In humans, the molecular analysis of these multiple steps is rarely possible due to the limited availability of tumor specimens obtained prospectively from all tumor stages. Also, many of these properties involve the entire organism which cannot be directly accessed in patients. In contrast, mouse models of tumorigenesis allow the reproducible isolation of different tumor stages, in addition to normal tissue, which are then amenable to pathological, genetic, and biochemical analyses. The list of specific mouse models available is continuously expanding and includes those with various checkpoint and tumor suppressor proteins knocked out or mutated, with oncogenes overexpressed or with alterations in caretaker and apoptosis-regulating genes. Moreover, increasingly, modern systems enable the control of such genetic alterations in specific tissue- and time-dependent fashion (see Chapters 6 and 7 for discussions of cancer models in the study of oncogenes and tumor suppressor genes, respectively).

One of the most insidious and nefarious properties of scientific models is their tendency to take over, and sometimes supplant, reality.

Erwin Chargaff

Mice have proved invaluable as cancer models, in particular in the following areas:
- The many well-characterized inbred strains allow the mapping and identification of cancer susceptibility genes.
- They are a rich source of transforming viruses, carrying oncogenes.
- Transgenic mice carrying candidate oncogenes provided direct evidence for the oncogenic potential of these genes.
- Knockout (KO) mice of tumor suppressor genes have given insight into the normal (i.e. developmental) function of these genes and permitted assessment of their oncogenic potential when mutated.
- Cross-breeding of different genetically altered mice allows the identification of consecutive steps in the transformation process like oncogene collaboration.

During the last decade, much effort has been directed at improving mouse models to make them resemble human cancer more closely. In particular, we are increasingly paying attention to the introduction of similar mutations in a suitable genetic background, and also to achieving appropriate spatiotemporal control of gene expression and the extent to which a particular oncogene is overexpressed or a tumor suppressor gene is inactivated in a particular cell or tissue.

These newer systems may help to overcome some of the limitations of traditional transgenic and KO mice as models of sporadic cancers in the following ways:
- Most cancer-related genes are not mutated throughout the ontogeny of the organism but instead arise as somatic mutations in the adult.
- The widespread expression of the mutated gene creates a different microenvironment than can be expected during initiation of sporadic cancers where a few mutant cells are surrounded by many normal cells. In these situations, the effect of cell–cell contact or paracrine signaling may be very different.
- Many tumor suppressor gene knockouts show severe developmental defects or early embryonic lethality, precluding study of their role in cancer.
- Many knockouts show a very distinct tumor predisposition pattern not or not necessarily corresponding to the tumor spectrum seen in humans. This is exemplified by many earlier P53 knockouts and is being addressed by a number of researchers.
- One cannot test whether the genetic alteration is required for maintenance as well as initiation of the cancer. This requires the capacity to switch genes on and off.
- One needs to be able to follow the tumorigenic process *in vivo*. Therefore, imaging techniques permitting monitoring of tumor growth are useful – including use of luciferase or other light-emitting reporters that can be monitored in the living animal. This improves the utility of spontaneous mouse models for conducting therapeutic intervention trials.
- It is useful to be able to track cells during tumorigenesis *in vivo* in order to identify the originating cell for cancer formation and to address the requirement for stem cells in this process. Lineage pulse-chase tracking tools have been developed to achieve this. To identify multiple polymorphic regulators of cancer development, the effects of gene alterations can be examined in mouse strains of different backgrounds. In the ideal model, one should have a defined strain background (or several for comparison), and one should be able to switch genes on and off, ideally in a small subset of cells in a tissue in a time-dependent manner. Combining the Cre–Lox P techniques with hormone-inducible chimeric proteins (see Appendix 20.1) may allow this to be achieved, alongside new developments for gene knockdown with small interfering RNA (siRNA) that are already extensively used in cell culture systems and may soon be more readily employed *in vivo*.

All these important advancements notwithstanding, it should be remembered that a cancer model is a model and not the real human cancer. In the evolution of cancer cells, the kinetics of the genetic alterations and their manifestation, and not least the spatiotemporal manifestation of the disease, there are many differences between the species. These differences may relate to differences in the regulatory molecular networks. When working with animal models of human cancer, it is therefore important to establish systems-level marker sets for the cellular and molecular denominators of the human pathology and to monitor these markers in the model system and compare them with the information from the human disease.

Array-based technologies: genomics, epigenomics, and transcriptomics

Do not read, think!

Arthur Schopenhauer

The first array experiment was performed in 1995 and measured mRNA expression levels in the small flowering plant *Arabidopsis thaliana*. The array used contained 48 probes to measure differences in mRNA levels between wild-type plants and a transgenic line overexpressing the transcription factor HAT4. Current technologies allow measurement of mRNA levels covering complete genomes in a fraction of the time the first experiment took. The most established array types are DNA microarray chips for measuring mRNA expression levels and SNP microarray chips (SNP chips) to identify polymorphic variation in the population (see Box 20.1). However, other array types and applications have been developed like comparative genomic hybridization arrays (CGH arrays), CpG island microarrays (CGI arrays), and the ChIP-on-Chip approach (see Box 20.2). The latter array systems are mostly custom built, and their use is not as widespread as microarrays for mRNA expression analysis. Over the past few years, genome-wide sequencing technologies, called next-generation sequencing (NGS), have been developed that utilize array platforms for highly parallel sequence analysis. NGS is now in extensive use in cancer molecular biology to analyze mutation patterns, DNA methylation, and the specific patterns and levels of transcripts in tumor cells, including splice variants.

SNPs, the HapMap, and the identification of cancer genes

Quality is never an accident. It is always the result of intelligent effort.

John Ruskin

Linking a monogenic (single-gene) variation to a phenotype (i.e. disease) is relatively straightforward, and numerous examples have been described in Chapters 3 and 7. However, diseases are seldom the result of either single-gene defects or a single polymorphism. In fact, most common diseases are multigenic (i.e.

Box 20.1 Microarrays

DNA microarrays, gene chips, arrays, and single-nucleotide polymorphism (SNP) chips are all synonyms for the same technique. Microarray technology is based on the process of hybridization. Watson and Crick discovered that adenine (A) binds to thymine (T) and cytosine (C) binds to guanine (G). This biological phenomenon underlies hybridization, and as a result two complementary DNA (cDNA) sequences will hybridize to form double-stranded DNA (dsDNA). Equally, messenger RNA (mRNA) will hybridize to a DNA or mRNA strand as long as the strands are complementary. The hybridization process is so specific that a single strand of 25 nucleotides can pick out its matching partner from a mixture containing millions of different sequences. Microarrays take advantage of this process.

In essence, microarrays are a large number of parallel hybridizations on a chip. The chip surface (substrate) can be quartz wafer (silicon), glass, nylon, or nitrocellulose. Probes are usually cDNA or oligonucleotide strands which are attached to the chip surface. Microarrays can be produced by either spotting of the probes or by in situ synthesis of the probes on the substrate. The in situ synthesis of probes combines photolithography and combinatorial chemistry. The latest microarrays contain more than 45 000 probes on a single chip covering whole genomes and the probe density is still rising. Before hybridization to the chip, amplified mRNA or DNA (target) is labeled with a fluorochrome. After hybridization, each probe on the chip is scanned and the signal intensity is a semiquantitative measure for the amount of mRNA or DNA hybridized to each probe.

There are two different microarray systems available which are either one-color or two-color technology. If, for example, we are interested in breast cancer we could identify differences between breast cancer cells and normal cells. This would give us insight into cancer pathology. In a two-color experiment, mRNA extracts are labeled with either a green fluorescent Cy3 (healthy cells) or red fluorescent Cy5 (cancerous cells) label after mRNA amplification. Equally mixed Cy3–Cy5 labeled mRNA is hybridized to the microarray, and microarray chips are scanned. Color changes indicate differences in mRNA levels between healthy cells and cancerous cells. Yellow (equal intensity of red and green dye) means no change in expression, and green or red gradients indicate a change in expression.

One color experiment (Affymetrix Technology) uses a single fluorescent label. Labeled mRNA of healthy cells or cancerous cells is hybridized to separate microarray chips. Changes in mRNA levels are calculated by comparing the signal intensities of the separate microarrays.

Expression analysis is the most common application of microarrays. mRNA expression levels are measured in cells or tissue, and an array provides a snapshot of the cell's dynamic transcriptome. Changes between cell states can give information on pathway activation and other cellular processes occurring. Application of this technique ranges from fundamental cellular biology (e.g. cell cycle) to clinical applications (class prediction).

SNP chips are similar to the microarrays used for expression analysis, but are mainly used to screen for SNPs in the genome to link genotype to phenotype. The starting material is not mRNA but genomic DNA which is also amplified and labeled before hybridization. SNP chips have been used for different studies such as loss of heterozygosity in cancer.

Box 20.2 CGH arrays, CpG island microarrays, and ChIP-on-Chip

Nature, when she invented, manufactured, and patented her authors, contrived to make critics out of the chips that were left.

Oliver Wendell Holmes

Comparative genomic hybridization arrays (CGH arrays) are a chip-based version of the classical cytogenetic technique that allows for rapid detection and mapping of genome-wide changes in DNA sequence copy numbers between normal and abnormal genomes. CGH arrays detect relative DNA copy number changes which provide information on chromosomal deletion, duplication, and unbalanced translocation often seen in cancer. The advantage of CGH arrays over normal CGH is the much higher resolution and speed. CHG arrays have up to 1000 times higher resolution and generate information much faster.

Conventional cDNA microarrays have been used for comparative genomic hybridization, but only abnormalities in known genes (cDNA present on the chip) are detected and not those in other regions of the genome (Pollack et al.,1999). The use of normal microarrays is therefore limited. Current approaches use amplified genomic clones (pieces of human genome inserted into bacteria for replication) as probes on the array surface covering all nonredundant sequence of the genome instead of metaphase chromosomes used in conventional CGH.

CpG island microarrays (CGI arrays) are used to identify DNA methylation sites in a genome. Methylation occurs in CG-rich areas (CpG sites) of the genome, and these methylated regions are termed CpG islands. CpG islands are common in exons and promoter regions, but are rare in other parts of the genome. CGI arrays have probes that are selected on their potential methylation sites and are therefore CpG island enriched.

For genome methylation analysis, genomic DNA is purified and selectively amplified for methylated regions. The amplified DNA and control DNA are labeled with CyDyes and hybridized to the chip. CGI arrays are two-color technology, and labeled test (Cy5) and control samples (Cy3) are always mixed before hybridization.

The **ChIP-on-Chip** research tool combines chromatin immunoprecipitation (ChIP) and DNA arrays or CGI arrays. The ChIP-on-Chip approach is used to identify new promoter regions in genomes, but can equally be used to identify differences in transcriptional silencing of genes by methyl–CpG binding proteins in cancer or healthy cells.

Chromatin is immunoprecipitated using specific antibodies raised against transcription factors or methyl–CpG binding proteins. If, for example, binding sites of the E2Fα transcription factor are determined, chromatin fragments that bind an E2Fα transcription factor are precipitated using an E2Fα antibody. DNA is enriched for chromatin fragments that are bound to E2Fα, and these fragments are amplified, then hybridized on a chip against a control sample (not enriched DNA). This allows for the detection of E2Fα transcription factor DNA-binding sites.

This methodology for binding sites of methyl–CpG binding proteins is identical, and chromatin fragments are precipitated with a methyl–CpG binding protein–specific antibody.

---

having genetically complex traits), and the process of identifying these complex traits is difficult and involves several steps: (a) linkage or association analysis, (b) fine mapping of the genetic changes, (c) comparative sequence analysis, and (d) functional tests of the candidate genes. Identifying genes in the genome that underlie disease starts with either linkage or association analyses. Linkage or association studies identify regions, not a gene, on a chromosome – not necessarily on a single chromosome – that are linked or associated to disease or disease susceptibility.

Linkage studies are generally successful in familial monogenic disorders which are often high-risk but rare mutations or gene variants. The selection of families with a rare disease from the population serves as a filter and selects for the specific mutated allele. Genomic variation of the diseased family members is compared to members of the family without the specific disease. However, linkage studies are not very powerful for identifying alleles relating to common diseases such as cardiovascular disease, diabetes, and most cancers. Association studies are more useful in finding candidate disease genes in common diseases.

Genomic regions identified by linkage or association analyses, called "quantitative trait loci," are still large and can correspond to 200 to 300 genes in the DNA sequence. Fine mapping is required to reduce this large number of candidate genes that might be associated with the disease and to zoom into the region of interest. This reduction is needed to make functional studies feasible. Studying 300 genes extensively in relation to a disease with current technology would simply take too much time and resources. This fine mapping results in identification of what is called "minimal quantitative trait loci" covering several genes and numerous polymorphic variants. DNA sequence analysis is needed to identify these candidate variations. Each nucleotide variant and combinations of nucleotide variants in a single gene and between genes have to be identified, ranked, and ultimately tested in a population to confirm the association. All these steps can be undertaken by microarray analysis, and specific microarrays are commercially available. Because of the variations and the large number of polymorphisms in any genome, custom arrays are popular. The stepwise approach works as a filter reducing a full genome to a small number of genes and nucleotide variants. Identification of candidate genes and nucleotide variants is in itself not conclusive evidence, and functional tests are needed. This is perhaps the most difficult step because methods to demonstrate the link between phenotype and nucleotide variant are limited. The best conclusive experiment is experimentally changing phenotypes by genetic intervention. Phenotype "swaps" are successful only if the correct nucleotide variant or correct combination of variants has been identified and is changed, and if the pathological phenotype is caused only by the one altered candidate gene.

Association studies are an emerging field, and much effort has been directed at improving techniques and resources for researchers active in this area. The International HapMap consortium was set up to determine common patterns of DNA sequence variation in the human genome, the HapMap, to facilitate the identification of disease genes and mutations. Discovering genetic variation will contribute to advances in understanding pathogenesis, diagnosis, treatment, drug response, and metabolism. Identifying all possible SNPs in the genome at once by the before-mentioned approaches would be an impossible task and, given the ongoing evolution of the genome, never achievable. Taking advantage of our understanding of genetics, a genomic map with genetic markers can be constructed to facilitate the discovery of genetic factors underlying disease. These genetic markers are SNPs (tagSNPs), and a small number of tagSNPs identifies a haplotype.

Haplotypes are stretches of DNA sequence that are inherited as an uninterrupted block of sequence. These stretches of sequence contain many SNPs, and a combination of a small number of these SNPs (tagSNPs) can unequivocally identify a haplotype (Fig. 20.2). An understanding of meiosis is crucial to appreciate how haplotypes and therefore the HapMap can be used for discovery of genetic factors underlying disease. Meiosis is the generation of cells with a haploid (i.e. a single set of chromosomes and basis for the recombination of the paternal and maternal chromosomes during sex cell formation). This recombination event results in a new hybrid genome combining alleles of parental genomes, but the merging of the chromosomes does not happen from gene to gene. The recombination occurs in genomic blocks (i.e. large stretches of DNA coding for many genes), and SNPs in that block travel together. Therefore, a combination of a number of marker SNPs may suffice to identify a genomic block, a haplotype. Due to the relatively short existence of human species, the number of haplotypes per chromosomal region is limited. New haplotypes can be formed by recombination events and new mutations. Different haplotypes can be identified by determining which tagSNPs are present in the particular genomic region. By haplotype mapping people's genomes on the basis of tagSNPs, we can detect if certain haplotypes occur more frequently in a diseased population compared to a normal population. If one haplotype is predominantly present in the diseased population, the disease-associated polymorphism or mutation is on that specific haplotype. By identifying the haplotype and thereby the region of the genome that is associated with the disease of interest, more focused analyses of the genetic factors relating to a particular disease are possible.

As discussed in the introduction to this chapter, not all breast cancer cases can be explained by SNPs or mutations in the known hereditary breast cancer genes *BRCA1*, *BRCA2*, *TP53*, *CHK2*, and *ATM*. Yet, chance alone cannot account for the many cases of breast cancer that occur in some families. Although this suggests that genetic correlates should exist, the search for genes that account for the many cases of hereditary breast cancer not explained by known mutations has been largely unsuccessful. The HapMap will make association studies easier and will help to identify new genes that underlie common cancers such as breast cancer. Microarray chips will be used that contain all the tagSNPs selected by the HapMap project to identify the haplotypes. Subsequent microarray experiments will be undertaken to identify the SNPs or mutations in the haplotype region relating to the diseases followed by fine mapping, sequence analysis, and functional testing of candidate genes.

In the future, a patient's genome may be screened for all disease-associated SNPs by SNP arrays which will allow for personalized medicine. The very recently developed whole-genome NGS may soon complement or even replace SNP arrays for analysis of genetic variants and provide the means for even higher resolution diagnosis. Many diseases, in particular cancer, are thought to be combinations of genetic predisposition and environmental

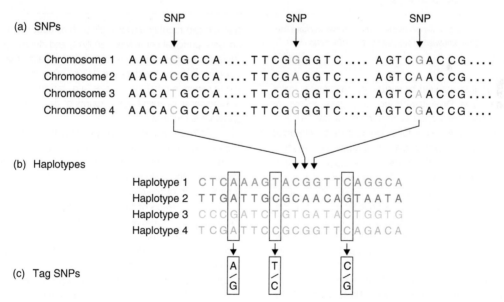

Figure 20.2 SNPs, haplotypes, and tag SNPs. (a) SNPs. Shown is a short stretch of DNA from four versions of the same chromosome region in different people. Most of the DNA sequence is identical in these chromosomes, but three bases are shown where variation occurs. Each SNP has two possible alleles; the first SNP in panel A has the alleles C and T. (b) Haplotypes. A haplotype is made up of a particular combination of alleles at nearby SNPs. Shown here are the observed genotypes of 20 SNPs that extend across 6000 bases of DNA. Only the variable bases are shown, including the three SNPs that are shown in panel (a). For this region, most of the chromosomes in a population survey turn out to have haplotypes 1–4. (c) Tag SNPs. Genotyping just the three tag SNPs out of the 20 SNPs is sufficient to identify these four haplotypes uniquely. For instance, if a particular chromosome has the pattern A–T–C at these three tag SNPs, this pattern matches the pattern determined for haplotype 1. Note that many chromosomes carry the common haplotypes in the population. From HapMap Consortium (2003). The International HapMap Project. *Nature*, **426**: 789–96. Reproduced by permission of *Nature*.

factors, including lifestyle. If patients are aware of disease susceptibility, they can act upon it by, for example, taking medication or changing their lifestyle. This already happens for breast cancer: women with a familial history of breast cancer are screened for *BRCA1* and *BRCA2* mutations. In case of increased risk as by disease-related gene variants, preventative mastectomy is often carried out and markedly reduces incidence of breast cancer.

Cancer mRNA expression analysis

Before the development of microarray technology, cellular mRNA levels were studied comparing expression levels between, for example, normal and diseased cells after amplifying the complementary DNA (cDNA) of mRNA by polymerase chain reaction (PCR). Identifying in this way gene transcription levels for many genes that might play a role in a cellular pathway or disease is very time consuming, especially if no prior knowledge is available of which gene should be chosen. The use of microarrays to study expression levels has revolutionized biology, but has come with many challenges. The advantage of microarrays is that all genes in a genome and, with some technologies, splice variants can be screened for expression. This gives a relatively complete overview of the cellular transcriptome. Most genes in the genome are known, but a number are bioinformatically predicted yet not confirmed genes. Less is known about splice variants that are often only predicted from DNA sequence. Coverage of the splice variants will be possible through more refined DNA arrays and NGS.

Expression analysis studies have confirmed many findings of and prediction from cancer research. Moreover, these studies have resulted in many new insights into cancer biology, and mRNA expression analysis is proving to be very useful for cancer classification, cancer diagnosis, and, in future, disease outcome prediction.

Challenges in mRNA expression analysis and other microarray-based techniques are extensive. The large data sets require new statistical methods to analyze the data (see Box 20.3), to extract useful information, to remove random variation or biases, and to confirm the findings. Highly important in arrays analyses is the proper selection and preparation of the cellular material to be analyzed. Tumors by and large are complex tissues comprising a number of different cell types where the tumor cells themselves can be present in low frequencies, sometimes below 1%. Often, the bulk of the tumors are tumor stroma cells, connective tissue cells, and immune cells. Bulk tumor mRNA preparations, therefore, may be very misleading, and approaches to specifically isolate the tumor cells are often stressful for the cells with the effect that the transcriptomes might be altered. Transcriptome analysis on single cells picked from tissue sections might be a way around this problem but is technically challenging. Cells in tissue sections are often cut open, and the fraction of cellular mRNA actually subjected to analysis is difficult to control, plus the cDNAs need to be amplified linearly to ensure sufficient detection levels without changing the ratio of the different cDNAs, which is difficult to achieve. To circumvent these difficulties, larger numbers of cells are commonly used for analysis. As cellular systems are highly complex, heterogeneous, and

---

Box 20.3 Microarray data analysis

Where is the Life we have lost in living? Where is the wisdom we have lost in knowledge? Where is the knowledge we have lost in information?

T.S. Eliot

The large amounts of data acquired by microarray experiments make data interpretation difficult. Dealing with such large data sets requires specialized expertise and a very good understanding of statistics and perhaps even mathematics. A successful experiment starts with good experimental design and ends with correct interpretation of results. When using high-throughput methods like microarrays, the principle of "Garbage in, garbage out" applies and statistical support is essential for getting useful data and results. The British geneticist and statistician Ronald Fisher already pointed out the need for good statistical support in research in 1938, writing that "to call in the statistician after the experiment is done may be no more than asking him to perform a post-mortem examination; he may be able to say what the experiment died of."

There are many software tools available, both commercial and in the public domain, for microarray data analysis. There is software for mRNA expression analysis and genetic variation analysis (e.g. SNP analysis), but there are numerous tools that focus on a single aspect of microarray experiments. There is much debate on how to correctly analyze microarray expression data. New algorithms and analysis techniques are being rapidly developed and there is still no consensus on the best methods. However, most biologists use software packages that contain roughly similar tools such as data normalization, statistical analysis, principal component analysis, and clustering. The main functions of these software tools are to visualize the data and reduce the complexity of the data sets. When more than 30 000 transcripts are screened in different samples (e.g. cancer and normal cells), making sense of such large amounts of data is difficult, so powerful software is required.

In mRNA expression analysis experiments, normalization of data sets is important as it corrects for variation in (a) experimental preparation of the mRNA, (b) fluorescent labeling, and (c) chip hybridization. Removal of this systematic bias is needed to ensure correct interpretation of the data.

Statistical analysis of the data is undertaken to filter out the interesting genes that show significant changes in expression as a result of the biological process studied. The statistical tools usually include parametric and nonparametric statistical tests often incorporating multiple testing corrections.

Clustering algorithms are applied to group genes that have a similar expression pattern. Genes that show similar expression patterns are thought to have a role in the same biological process or pathway. Clustered genes are therefore in theory biologically related. Clustering can be done only in large microarray experiments where, for example, there are several time points. The better known clustering methods are hierarchical clustering, k-means clustering, and QT clustering.

dynamic, transcriptome data acquired that way average many cells and cell states, and miss important information, especially information on the spatial and spatiotemporal organization of cells and tissues. On top of all this, the major challenge perhaps is to make biological sense of the large bodies of transcriptome data acquired. This can be partly addressed by undertaking time course experiments, but this is possible only with cultured cells, and the full dynamics of cells may never be captured. Research, therefore, increasingly incorporates *in silico* experimentation using the experimentally obtained data sets and prior knowledge of the biological system for mathematical modeling. Such models attempt to describe the biological system and derive biologically relevant predictions that can be tested using conventional laboratory techniques.

Comparing mRNA expression patterns of tumor cells from cancer tissue with appropriate control cells identifies gene expression changes associated with cancer. Changes in cellular growth pathways can be detected where growth signals increase and growth inhibitory signals decrease. As examples, changes in gene expression relating to evasion of programmed cell death (apoptosis) or to induction of angiogenesis and tissue invasion pathways, genes involved in the extracellular matrix remodeling and increased activity of proliferation genes can be detected. These changes can be identified by increased or decreased levels of single-gene transcripts, but often several genes are affected that contribute to the same biological process or pathway. Besides cellular pathways, cell types can be distinguished in microarray data when tissue biopsies are analyzed. Human tissues are heterogeneous, and cell types present can be identified by clustering cell types' specifically expressed genes.

Many microarray experiments have been undertaken to identify differences in mRNA expression between normal and cancerous tissue. In most cases, molecular signatures based on mRNA expression patterns can be deduced that are specific for cancer. Analyzing these datasets together in meta-analyses allows for identifying common transcriptional changes in cancers that relate to neoplastic transformation. Moreover, common differences between differentiated and undifferentiated cancers have been identified for some cancer types. Such differences in mRNA expression may translate into the clinical outcome. Undifferentiated cancers are more aggressive than differentiated cancers, and chances of survival are significantly lower for undifferentiated cancers. Microarray experiments have shed light on what the molecular origin of this phenomenon may be. Undifferentiated cancers maintain a more disordered state than differentiated cancers, and an increased level of cellular proliferation and invasion. The heterogeneous and unstable cell state results in a greater tumor aggressiveness ultimately leading to poor outcome. Three genes play a particularly important role in maintaining the undifferentiated phenotype: *EZH2*, *H2AFX*, and *H2AFZ* are overexpressed in undifferentiated cancers. The corresponding proteins are involved in chromatin remodeling and broad-spectrum transcriptional regulation. The protein EZH2 is involved in transcriptional memory, and H2AFX and H2AFZ are histone variant proteins known to relate to euchromatin–heterochromatin transition. These three genes modulate the expression of potentially hundreds of genes and likely play a role in maintaining the undifferentiated cancer phenotype.

Microarrays can be used for diagnostic purposes and are well suited for molecular classification of cancers. Breast cancer, for example, can have very different clinical outcomes, and although breast cancers cannot always be separated by morphology using histopathological techniques, these different cancers of the breast can be assumed to be different diseases. Using microarrays in combination with advanced statistical analysis techniques has improved cancer diagnosis in many cases. Microarray analysis not only allows for differentiation between cancer and normal tissue or cells, but also can subclassify cancers. There are ample examples that demonstrate the usefulness of microarrays in cancer diagnosis (Perou *et al.*, 2000; Shipp *et al.*, 2002; Sorlie *et al.*, 2003), and for breast cancer this subclassification has also been correlated to clinical outcome. Breast tumor gene expression profiles proved useful for predicting the overall survival and the probability of remaining metastasis free. The main advance of microarray approaches is their information content translating into better prediction of outcome compared to other criteria for classifying high- and low-risk cancer, and survival (see the "Structure, code, and semantics" section for a discussion of Zipf's law). Compared to the tumor gene expression signature, the standard National Institutes of Health (NIH) and St. Gallen diagnostic criteria misclassified a clinically significant number of patients leading to under- or overtreatment in many cases. Interestingly and importantly, microarray data suggest that the ability of a tumor to metastasize is probably an inherent genetic property of (breast) cancer and not a potential acquired in later stages of tumorigenesis.

The studies with breast cancers have highlighted the problem of cellular heterogeneity in many microarray studies. Studies that examine gene expression in tissue, even when areas of that tissue had been selected by techniques such as laser capture microdissection, will inevitably average gene expression across multiple cells. Particularly in the context of cancers which may contain multiple subclones of cells with genetic instability, the demonstration of a catalogue of mutant genes must not be assumed to mean that every cancer cell carries all these mutations. In other words, the presence of metastases-related gene mutations in the primary tumor prior to overt metastatic spread should be interpreted with caution until it can be confirmed that single cells in the original tumor carry all the same mutations as those in the early metastasis. This is important as we need to account for the well-documented presence of sometimes large numbers of cancer cells in the blood of patients (circulating tumor cells (CTCs)) with no secondary tumors and often considerable time before metastases develop. If mutations correlated with metastases were already present, then why are metastases not formed? These uncertainties notwithstanding, data from "whole-tumor" microarray studies will likely have a major impact on future therapies. Treatment of cancer will have to be tailored to the malignancy of the disease, and a more aggressive systemic treatment for the more malignant forms would be indicated compared to less aggressive localized tumors.

In conclusion, microarray analysis can facilitate cancer diagnosis and can classify cancers by their gene expression profiles. Tumor gene expression profiles thus allow for tumor subclassification and better prediction of tumor classes and outcome. Appropriate classification of cancer is essential for successful treatment. Microarray data have been used effectively for class prediction, but tumor class prediction by microarray is expensive and time consuming, and requires specialized expertise. Nonetheless, predictably, microarray-based diagnostic and prognostic testing will be more widespread in the future as costs decrease and supporting diagnostic software becomes available. In the meantime, translational research is ongoing to identify the genes that have the highest predictive potential. The objective is to develop simple multigene assays for routine clinical use. These can be quantitative reverse transcription polymerase chain reaction (RT-PCR) or specified diagnostic arrays to measure gene expression levels in tumors for sets of marker genes. Diagnostic and predictive tests will be designed to give identical answers to genome-wide microarrays but be more cost-effective and faster. Such tests have already been developed for diffuse large B-cell lymphoma (Lossos et al., 2004) and some other cancers. In the future, molecular data might supersede other nongenomic information. However, for all molecular tests, tumor biopsy is required which remains an invasive procedure.

CGH arrays, CpG island microarrays, and ChIP-on-Chip

Be careful about reading health books. You may die of a misprint.

Mark Twain

Comparative genomic hybridization (CGH) arrays, CpG arrays, and ChIP-on-Chip approaches are not as established as expression or DNA microarrays and certainly have not reached the stage of routine clinical application (see Box 20.2). However, they have enormous potential to advance our insights in cancer biology. CGH arrays are expected to be very useful in identifying new cancer-related genes and chromosomal abnormalities. Genomic mutations, unbalanced translocations, deletions, and duplications can be detected throughout the entire genome. All these genetic events can contribute to cancer development. No previous methodology was as advanced in detecting such genomic changes as CGH arrays. The main limitation has been insufficient genomic resolution to detect small changes. CGH arrays can detect DNA copy number aberrations which are indicative of loss or gain changes in the genome. Moreover, by the array setup, the location of an aberration on the chromosome can be defined within a relatively narrow stretch of DNA making the detection of breaking points and the identification of the affected genes much easier than ever before.

CGH arrays have already resulted in new insights in breast, ovarian, and gastric cancer indicating that they could become useful diagnostic and prognostic tools when the underlying chromosomal changes are correlated with clinical outcome. Unlike mRNA levels, genomic changes are less dynamic and are not influenced by stress, medication, or other factors that affect mRNA expression levels. Chromosomal changes may lead to changes in gene or protein expression necessary for cancer to arise. The further development of CGH arrays, in particular increased resolution, will contribute to understanding genomic aberrations in relation to tumorigenesis and to advances in diagnosis.

Not only quantitative changes but also functional protein changes can be the result of chromosomal abnormalities. The Bcr–Abl fusion protein is formed as a result of a chromosomal translocation. The first exon of the *abl1* gene is replaced by sequences of the *bcr* gene leading to the expression of the Bcr–Abl fusion protein which is linked to different forms of leukemia. Depending on the cell type, Bcr–Abl expression results in morphological transformation, enhanced proliferation, or abrogation of growth factor or adhesion dependence. These biological changes are the result of abnormal constitutive tyrosine kinase activity of the fusion protein Bcr–Abl. In ovarian and breast cancer, DNA copy numbers and mRNA levels of the *RAB25* gene, a small GTPase, are predictive of clinical outcome. Increased DNA copy number and mRNA levels of *RAB25* are prognostic for poor survival chances. Identification of chromosomal breakpoints, translocations, and loss or gain events are keys to further advances in cancer diagnosis and treatment. CGH arrays will be the best method currently available to identify cancer-related chromosomal changes.

CpG island arrays (CGI arrays) have been developed to map epigenetic changes in the genome. Epigenetics is a field that is evolving rapidly, and identification of methylated CpG sites in the human genome is essential to understanding chromatin structure and remodeling, and gene expression changes underlying disease. It is, for instance, known that Prader–Willi syndrome, Angelman's syndrome, and Beckwith–Wiedemann syndrome are caused by gene-methylation abnormalities. CpG islands are often found in gene promoter regions and are usually 0.5–4.0 kb in length. Hypermethylation of these promoter regions is the best categorized epigenetic change to occur in cancer, and hypermethylation is found in almost every type of human neoplasm. If epigenetic changes affect genes regulating cell proliferation, uncontrolled cell division may occur and result in cancer. Epigenetic abnormalities underlying cancer are described in Chapter 11.

ChIP-on-Chip arrays are a variation on CGI arrays. The array-based technology is identical to that of CGI arrays, but the preparation of DNA is different. Chromatin is precipitated with specific antibodies to chromatin-binding proteins such as methyl CpG-binding proteins. The antibody used can be changed according to chromatin-binding protein. Differences in chromatin precipitated provide information on chromatin remodeling and therefore potential differences in transcriptional activity. ChIP-on-Chip approaches have demonstrated that methylation-associated transcriptional silencing exists in cancer, and target genes have been identified. In breast cancer, the homeobox gene PAX6 and the prolactin hormone receptor gene show variation in transcriptional activity that have been correlated with changes in the methylation patterns in their promoter regions.

Both CGI array and ChIP-on-Chip methodology provide information on gene transcription and, in particular, on how DNA sequence and chromatin structure play a role in transcriptional regulation. There remains much unknown about transcriptional regulation in cancers, but both approaches will contribute to our understanding.

Next-generation sequencing

The Human Genome Project (HGP) was initiated in 1990. In 2000 the first draft of the human genome was reported, followed in 2003 by the complete genome. However, still today there are gaps in the assembly of the genome from numerous contigs that, as well as errors in the sequences, are being worked on and gradually filled or corrected. These painstaking efforts are crucial as the genome sequence is the reference for cancer genomics and other approaches discussed in this chapter. This notwithstanding, the publication of the human genome in 2000 has ushered in the systems biology revolution we are experiencing. The costs for the HGP were around US$3 billion (about US$1 per base pair) and involved a number of sequencing centers in the United States and Europe. Now with the highly parallel, array-based next-generation sequencing (NGS) technologies, the human genome can be sequenced in about 2 weeks in a single laboratory for 100 000 to 1 000 000 base pairs per dollar. This is still expensive compared to expression arrays, but the costs are bound to fall further as technologies advance and the numbers of sequences obtained from a single instrument run increase. NGS produces sequences of 30 to 600 base pairs depending on the technology employed. These short reads need to be assembled to the entire genome, a process which depends on a reference sequence. If this is available, as is the case for the human genome, NGS is the most powerful and most informative technology for cancer genome sequencing, mutation analysis, and transcriptome analysis, and, like no other technology, provides complete information on chromosome aberrations, mutations, splice variations, and the transcriptomes.

NGS is a very recent and highly dynamic field of technology development. Currently, there are three major technology platforms with already advanced instrumentation in increasingly widespread use (see Box 20.4). Each of these technology platforms has its strengths and its preferred areas of application. In addition, there are a number of new technologies in development, and it is likely that different NGS technology platforms will coexist, each with specific advantages for specific application fields. In broad outline, NGS technologies are integrated technologies that incorporate four basic steps: template preparation, sequencing, imaging, and data analysis.

NGS is done in array format, and the templates are prepared to ideally exclude any bias for specific sequence features or sequence representation. DNA is randomly broken into smaller pieces; the fragments are fixed to solid carriers and, in most technologies, amplified by PCR. The array format allows for parallel sequencing of billions of templates and, thereby, complete coverage of genomes in short times. For the sequencing step, current platforms incorporate four primary chemical or biochemical principles:

- Cyclic reversible termination (CRT)
- Pyrophosphate sequencing (PS)
- Real-time sequencing (RTS)
- Sequencing by ligation (SBL)

The first three of these approaches use DNA polymerases for sequence extension on the template, the fourth DNA ligase (see Box 20.4). The readout is by bioluminescence in case of pyrophosphate sequencing and by fluorescence imaging with different dyes for the different bases for the other three. The primary data are recorded by CCD imaging and translated into sequences. Pyrophosphate sequencing produces reads of 300 to 600 bp, CTR

Box 20.4 Next-generation sequencing

Next-generation sequencing (NGS) allows for parallel sequencing of billions of templates and can cover an entire human genome several-fold in a single run. This is achieved by fixing the template DNA onto a solid phase such as microbeads or glass slides arranged in array format, and running the sequencing reactions in parallel for all templates at the same time. The reactions are designed to produce light signals, bioluminescence, or fluorescence, which is recorded and translated into nucleic acid sequences. The templates are generated through random fragmentation of the DNA (e.g. by sonication). RNA has to be reverse transcribed into cDNA before this step. The resulting fragments are fixed to solid phase and, for most platforms, amplified by PCR. Great care is required to ensure that at these steps, no bias is introduced for or against certain sequences or against low-abundance copies in case of cDNA. Single-copy sequencing technologies that could omit such problems are in development and would, for instance, allow genome sequencing from single cells. Sequencing is done on the templates through enzymatic reaction with either a DNA polymerase (pyrosequencing, cyclic reversible termination, and real-time sequencing) or a DNA ligase (sequencing by ligation). The commercially available NGS platforms incorporate specific combinations of technologies for template preparation, sequencing, and data acquisition. The following are brief descriptions of the three currently leading technology platforms for NGS.

454 pyrosequencing is a technology developed by 454 Life Sciences, now acquired by Roche Diagnostics. DNA is attached to microparticles and amplified by PCR inside water droplets in oil solution so that each droplet contains a clonal colony of a single DNA template. The droplets are arranged in picoliter wells, each with a single particle and sequencing enzymes. Pyrophosphate released by the polymerase reaction is enzymatically converted to ATP, which drives the conversion of luciferin to oxyluciferin by luciferase, whereby light is emitted that is detected. The process is done successively for the four nucleotides to generate sequence read-outs. With this technology, reads of up to 600 or even more bps can be obtained.

Cyclic reversible termination is the basis for the Illumina Solexa sequencing technology. The templates for sequencing are attached in array format to slides and amplified by bridge amplification PCR to form clonal colonies. The sequencing reaction with the DNA polymerase is then done with reversible dye terminators with four different fluorescent dyes for the four nucleotides to determine the template-specific nucleotide coupled to the nascent DNA. After recording the fluorescent images, the terminal 3' blockers are removed chemically and the reaction is repeated for the next sequence position. The sequences are computed from the succession of fluorescence colors per bead. Read lengths with this technology are currently 75 to 100 bps.

Sequencing by ligation uses a DNA ligase for the sequencing reactions. This principle is the basis for the SOLiD technology of Applied Biosystems. The fragmented DNA is attached to beads and amplified by emulsion PCR. The beads, each with a clonal template colony, are then dispensed on glass slides and incubated with a universal primer. Sequencing is done with dinucleotide probes with an extension with degenerate bases and a fluorescence dye, one color for each of the four nucleotides. Properly annealed probes are ligated to the primer, a fluorescence image is recorded, the tag with the fluorescence dye is cleaved off chemically and removed by washing, and the process is repeated. With this technology platform, reads of about 50 bps can be obtained.

reads of 75 to 100 bp, and SBL reads of 25 to 50 bp. RTS has the potential for reads exceeding 1 kb but is not yet as widely used as the other technologies.

A major challenge in NGS is the processing of the huge quantities of data. This starts already with the extraction of sequence information from the video images and storage of the primary data. Next, sophisticated strategies are required to detect and correct random sequencing errors, which are done by multiple parallel sequencing of the genome (referred to as depth of sequencing). The short NGS reads then need to be aligned to a reference sequence to detect sequence variants of the genome. Finally, the most challenging task is to make biological sense of the data. The challenges discussed here for association studies, and detecting and validating genetic correlates of common disease, are even more acute with NGS and require dedicated bioinformaticians who are highly qualified for both the informatics and the biology involved. While technically NGS instruments can be run in small units, quality sequencing, and proper integration of the sequencing, bioinformatics, and biology, requires an elaborate infrastructure and sufficiently large and qualified groups of specialists.

Current fields of application of NGS are resequencing genomes to determine the genetic variations in populations, sequencing whole-genome expression libraries for expression analysis and genome-wide identification of splice variations, searching for disease correlation markers, and establishing cancer genomes to advance our understanding of cancer genetics. So far, a number of human genomes have been sequenced by NGS giving rise to personalized genomics as well as population genomics. Recently, the first cancer genomes established by NGS of tumor cells in breast cancer compared to normal breast tissue of the same individual were reported. The authors have identified some 50 000 mutations specifying the cancer genome. So far, breast cancer genomes have been most extensively studied and, besides point mutations, deletions, amplifications, tandem duplications, inter-chromosomal rearrangements, and inversions have been found. The potential role of these different types of aberrations in the oncogenic processes, and their implications for the course of disease and clinical outcomes, is subject to extensive clinical and experimental studies. As NGS is applied to an increasing range of human diseases, including cancers such as lung cancers, melanoma, small-cell lung cancer, and acute myeloid leukemia (AML), a new revolution in disease genetics is expected that will contribute to an accelerated discovery of new biomarkers and new targets, and the development of new diagnostics and targeted therapies, and will drive new research into the molecular basis of cancer.

Proteomics

Proteins are the final product of cellular synthesis starting with DNA transcription. Since Watson and Crick published their landmark paper on DNA, our understanding of cellular processes has improved, but much is still a mystery. Studying the protein end products of cellular synthesis is therefore central to understanding cellular processes. The proteome is usually defined as the protein complement of a cell, organ such as brain or heart, or genome. Tools mostly used in proteomics are one-dimensional (1D) and 2D gel electrophoresis, high-performance liquid chromatography (HPLC), and mass spectrometry (see Box 20.5).

Proteomics started off by identifying as many proteins of a cell or tissue as possible. Cataloguing proteins provided an insight into the complexity of the proteome. However, the proteome is highly dynamic and changes with respect to its environment, cell state, disease, and so on. Identifying proteins in itself does not provide a sufficient understanding of the cellular dynamics so that proteomics is extending toward quantitative proteomics. Relative increases and decreases of proteins in tissue as a result of, for example, cancer are being studied. Moreover, the presence and dynamics of posttranslational modifications and higher order structures such as protein interaction dynamics and inter- and intracellular communication networks have not been (fully) identified. This section deals with the various proteomic approaches that have increased our understanding of the proteome. Cancer is a biological system, and proteomic techniques have been applied to understand cancer cell dynamics at the protein level. Examples of proteome research will be discussed to illustrate the use of techniques. Cancer models vary from mouse models to yeast as well as human cells from cell cultures, and clinical specimens are used for cancer proteome analysis.

Quantitative proteomics

Studying relative changes in protein quantities in a cell or tissue can provide useful information about proteins that are involved in, for example, cancer, although it is difficult to determine if changes are a consequence of or responsible for the disease state. The main methods to identify protein changes are 2D gel electrophoresis (2D gels), chemical labeling of proteins like isotope-coded affinity tagging (ICAT), and metabolic labeling such as stable isotope labeling by amino acids in cell culture (SILAC) (see Box 20.6).

2D gel electrophoresis was the first method that allowed for quantitative proteomics and was developed independently by Klose and O'Farrell in 1975 (Fig. 20.3). Since then the technique has been improved, developed further, and standardized, including much-needed analysis software. Despite many limitations, 2D gels have been the mainstay for quantitative proteomics in the past 20 years. 2D gels resolve mainly the high abundant proteins in a cell which are usually of less interest. Moreover, 2D-gel electrophoresis has limited resolving power resulting in loss of proteins that are very small, very large, basic, or membrane proteins. However, a focused application of 2D gels can be very successful as for differently posttranslationally modified proteins or protein components of a purified protein complex. In essence, 2D gels have limitations in resolving power and sensitivity, but can still be a useful technique. Some interesting literature examples on 2D-gel based proteomics can be found in Lewis *et al.* (2000) and van Anken *et al.* (2003).

Proteomic research has shown that organ microenvironments are dynamic and change during tumor growth. Tumor cell membrane, endothelial cell membrane, and stromal proteomes, in particular stromal enzymatic activity, have been shown to change. The chaperone heat shock protein 90 alpha (HSP90α) plays an important role in cancer cell invasion in several implanted cancer cell lines and is likely to be of importance in other tumors as well. Aminopeptidase-P and annexin A1 have been found by a 2D gel proteomic approach to be differentially expressed at the blood–tissue interface of lung tumors which could be confirmed by gamma-scintigraphic imaging with labeled antibodies and related to differences in lung shape between normal and tumor-bearing rats (see Fig. 20.4).

Box 20.5 Tools in proteomics

Man must shape his tools lest they shape him.

Arthur Miller

The main tools in proteomics research are two-dimensional polyacrylamide gel electrophoresis (2D gels), liquid chromatography (LC), and mass spectrometry (MS). 2D gels and nanoflow high-performance liquid chromatography (HPLC) are protein separation techniques, and mass spectrometry is an analytical technique to identify and characterize proteins.

Cells contain thousands of proteins, and to identify these proteins powerful separation techniques are needed. 2D gels and liquid chromatography are mostly used to separate individual proteins in complex protein mixtures.

2D gel electrophoresis is a polyacrylamide gel–based technique that allows for the separation of proteins in two dimensions. Proteins are separated in the first dimension on isoelectric points using pH gradient strips. The pH gradient strips are then transferred to a polyacrylamide gel, and the proteins are separated on molecular weight in the second dimension. Proteins are visualized by staining the gel akin to developing a photo. The end result is a large gel with up to several thousand small spots that are all different proteins (Fig. 18.3). Before mass spectrometric analysis, proteins are digested by a protease into peptides. Proteases, such as trypsin, cut proteins into peptides at specific sites of a protein sequence. There are several reasons why proteins are digested into peptides, but in a few words it makes the identification process much easier. An alternative to 2D gels is liquid chromatography, which can also be used to reduce the complexity of a protein mixture by separating the (digested) proteins on certain characteristics like hydrophobicity or charge. The LC system is usually directly connected to a mass spectrometer (LC–MS).

Mass spectrometry is an essential technique to identify proteins and is used in any proteomics experiment. The three main parts of a mass spectrometer are the ionization source, the mass analyzer, and the detector. The two main ionization sources are electrospray and matrix-assisted laser desorption ionization (MALDI). The ionization source transfers nonvolatile compounds (peptides and proteins) into the gas phase. After peptide ionization, the peptides are guided into the mass spectrometer to the mass analyzer. The mass analyzer separates the peptides or proteins by mass before hitting the detector. The time a peptide or protein takes to travel through the mass analyzer before it hits the detector can be used to calculate the mass of the protein or peptide. The main mass analyzers in mass spectrometers are the time of flight (ToF) mass analyzer, the quadrupole mass analyzer, and the quadrupole ion trap. Mass spectrometers are used to obtain either (a) sequence information of a single peptide that allows for unambiguous identification of a protein, or (b) a mass fingerprint of the peptides from a digested protein that allows protein identification without sequencing the individual peptides. A peptide mass fingerprint is like a human fingerprint: it allows for unique identification of a protein.

An excellent basic introduction on peptide sequencing, protein identification, and mass spectrometers is written by Steen and Mann (2004).

To overcome some of the limitations of 2D gel–based proteomics, more robust techniques such as SDS-PAGE or gel-free protein separation combined with 1D and 2D HPLC and highly resolving mass spectrometry, or protein arrays which depart completely from the conventional proteomics technologies, have been developed over the past 5 or so years and are increasingly used for the various proteome-related problems addressed in cancer biology.

ICAT was one of the first chemical peptide labeling techniques that allowed for high-throughput proteomics by liquid chromatography–mass spectrometry (LC-MS). The main disadvantage of ICAT is that the labeling is dependent on a chemical reaction with cysteine, which has to be optimal for successful labeling. However, ICAT can be used on any biological system whether it is cell lines, (human) tissue, yeast, or bacteria. An example of the application of ICAT is the quantitative proteomic analysis of Myc oncoprotein function in mammalian cells (Shiio et al., 2002).

SILAC with heavy nitrogen (^{15}N) or carbon (^{13}C) labeled amino acids are the main types of metabolic labeling. Metabolic labeling is a very efficient method of proteome-wide labeling of proteins. Unlike chemical labeling like ICAT, the cell incorporates the heavy isotope form of the amino acids (see Box 20.6). This makes SILAC currently the most promising quantitative proteomics approach. However, SILAC can be used only for cell culture–based proteomic experiments. In contrast to SILAC, ICAT and 2D gel electrophoresis can be used on both tumor biopsies and cell culture–based experiments. The appropriate choice and application of proteomic techniques are important parts of the experimental design.

Metabolically labeled cells or organisms do not require a complicated 2D gel step to separate and semiquantify proteins prior to LC-MS analysis. Generally, SILAC protein extracts are mixed with a control in a 1:1 ratio, then separated by 1D gel electrophoresis, digested with an endopeptidase, and analyzed by LC-MS. The mass spectra are processed to identify and quantify proteins. SILAC has a large potential, and an increasing number of research using this technique is being published; an example will be discussed in the following section on posttranslational modifications.

Posttranslational modifications

Oxidation, sumoylation, methylation, acetylation, ubiquitination, phosphorylation, and glycosylation of proteins are all posttranslational modifications that determine cellular destination, protein function, interactions, turnover, and more. A change of function effectively corresponds to formation of a new protein. Knowledge of the modification site, kinetics, and regulation is important for understanding the cellular consequences of posttranslational modifications. All posttranslational modifications are of great importance in a cell even if they do not occur as frequently as, for example, glycosylation and phosphorylation. Glycosylation and phosphorylation are well-known modifications. Discussing

Box 20.6 Quantitative proteomics

Identifying all proteins in a cell is of limited importance as we know the majority of expressed genes through the genome sequence. Identifying changes in protein quantity or posttranslational modification – which also leads to a quantitative change of a functional protein – is far more interesting. These proteome changes result from altered normal cellular processes or disease, and identifying these changes helps us understand cellular processes.

Identifying quantitative changes in the proteome can be done by combining (a) 2D gel electrophoresis with mass spectrometry, and (b) HPLC techniques with mass spectrometry (LC–MS). Mass spectrometry is not a quantitative technique and is therefore combined with 2D gel electrophoresis or protein-labeling strategies such as ICAT or metabolic labeling.

After separating proteins on a 2D gel, the proteins are visualized with silver stain or fluorescent dyes. These staining techniques are semiquantitative and allow for identification of quantitative protein changes between proteins from, for example, cancerous and healthy breast tissue. Comparing 2D gels from diseased and normal tissue with specialized analysis software can lead to identification of proteins that play an important role in cancer. 2D gel electrophoresis technology is still improving. Difference in gel electrophoresis (DiGE) is one of these technological improvements and is the proteomics variation on the two-color microarray experiment also using Cy dyes. Proteins are labeled with different fluorochromes (Cy3 or Cy5) before 2D separation, and relative protein quantities can be measured.

The advantages of LC–MS methods over 2D gel electrophoresis are the increased sensitivity and the speed of acquisition, but mass spectrometry is not a quantitative technique. Using LC–MS methods for identifying proteome changes requires additional methodology for quantifying protein changes. This is done by differentially labeling proteins with a tag akin to a two-color microarray experiment. However, the proteins are not labeled with a different fluorochrome but with a different mass tag as mass spectrometers separate on differences in mass. Labeled proteins are digested into peptides by a protease before mass spectrometric analysis. A mass spectrometer can analyze only a small number of peptides simultaneously, and a powerful separation technique (HPLC) is needed before a complex mixture of peptides is analyzed.

ICAT is a chemical labeling method that uses heavy and light labels that react specifically with the thiol group of the amino acid cysteine in all proteins. The labels are identical except for the carbon (C) atoms, which in the heavy mass tag are replaced by ^{13}C atoms – the heavy isotopic form of ^{12}C. By replacing nine ^{12}C atoms by ^{13}C atoms, a mass gain of 9 Dalton is achieved which can be measured by mass spectrometry. Differentially labeled protein samples (e.g. from cancerous cells and normal cells) are mixed by equal amounts. After protease digestion, identical peptides with different mass tags can be separated by their mass in the mass spectrometer but will still co-elute from the LC column. The success of this approach depends on the existence of cysteine in a protein and the efficiency of the labeling. If a protein does not contain a cysteine or if the peptide is not labeled successfully, the protein cannot be quantified. Two labeled peptides for each protein are needed to identify and quantify protein changes by statistical means.

Because chemical labeling is never 100% efficient, new methods have been developed for *in vivo* labeling or metabolic labeling. Stable isotope labeling by amino acids in cell culture (SILAC) is an *in vivo* labeling method for cell culture. Essential amino acids needed for cell growth are substituted by heavy ^{13}C forms in the cell culture media. Essential amino acids used are ^{13}C arginine and ^{13}C leucine, which have a mass difference of 6 Dalton with the normal ^{12}C form. The cells are "forced" to consume these amino acids and incorporate them in the proteins that are synthesized. After a couple of days and several cell divisions, all ^{12}C arginines are replaced by ^{13}C-arginine in every single protein in the cell. By mixing normal ^{12}C-arginine cells with ^{13}C-arginine labeled cells (induced), proteome changes can be measured by LC–MS approaches.

Not only cells but also yeast, the nematode worm *C. elegans*, and *Drosphila melanogaster* have been labeled with heavy isotopes (^{15}N and ^{13}C). Completely heavy-isotope-labeled mice and rats are only a matter of time.

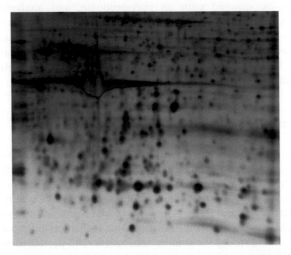

Figure 20.3 A section of a 2D gel from smooth muscle protein extract. The y-axis is separated on molecular weight. The x-axis is separated on isoelectric point.

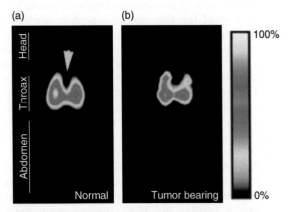

Figure 20.4 Gamma-scintigraphic images of (**a**) normal and (**b**) tumor-bearing rats. *In vivo* images were obtained after 125I-APP (125I-aminopeptidase-P) monoclonal antibodies were injected intravenously. A difference in lung shape indicates tumor growth. From Oh, P., Li, Y., Yu, J., *et al.* (2004). Subtractive proteomic mapping of the endothelial surface in lung and solid tumors for tissue-specific therapy. *Nature*, **429**, 629–35. Reproduced by permission of *Nature*.

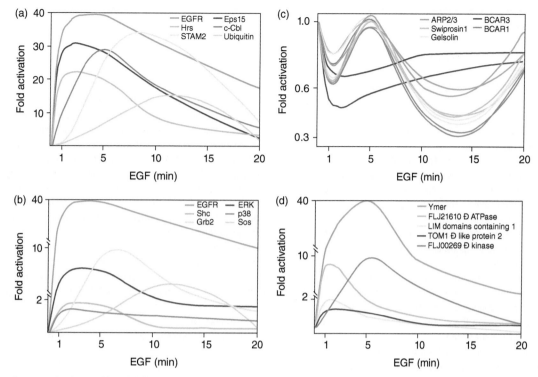

Figure 20.5 Continuous activation profiles of different categories of EGFR signaling effectors. (a) Key proteins involved in receptor internalization and endosomal trafficking. (b) Proteins from the Ras–MAP kinase pathways. (c) Proteins involved in actin remodeling. (d) Novel proteins found to be activated by the EGFR signaling pathway. From Blagoev, B., Ong, S.E., Kratchmarova, I., and Mann, M. (2004). Temporal analysis of phosphotyrosinedependent signaling networks by quantitative proteomics. *Nature Biotechnology*, **22**: 1139–45. Reproduced by permission of *Nature*.

all posttranslational modifications or even the modifications named in this book is beyond the scope of this chapter. Many posttranslational modifications can be found elsewhere in this textbook, in research papers, or on the Internet.

Glycosylation is the most complex posttranslational modification in eukaryotes. It plays a role in, for example, intracellular regulation, cell–cell communication, cell–substrate recognition, and secretory pathways. The sugars that form building blocks for glycosylation of proteins are fucose (Fuc), galactose (Gal), N-acetylgalactosamine (GalNac), glucose (Glc), mannose (Man), N-acetylneuroaminate (NeuNAc), sialic acid (Sia), and N-acetylglucosamine (GlcNAc). Although few sugars make up a glycan, the glycans that are attached to a protein as a posttranslational modification are highly complex due to the large number of sugar combinations formed and branching of the sugar chains. Depending on the site for protein glycosylation, N-glycosylation and O-glycosylation are distinguished. N-linkage occurs on the amide nitrogen atom in the side chain of asparagine, and O-linkage on the oxygen atoms in the side chain of serine or threonine. Potential N-glycosylation sites are detected in the amino acid sequence and are very specific sites. In contrast, O-glycosylation is not recognizable from its sequence context. Mass spectrometry is the most powerful technique for the investigation of protein glycosylation. Hereby, several points are of interest: the identification of the glycosylation sites, the detection and quantification of the glycosylation, and the structure of the glycan in relation to the cellular processes and states studied.

Proteins can be phosphorylated at serine, threonine, and tyrosine. The phosphorylation state of a protein is regulated through kinases and phosphatizes, and results in activation or de-activation of the protein which in turn often switches a cellular pathway on or off. These activities are highly regulated and lead to amplification of the signals. Overactivity of certain kinases is correlated with cancer. The Bcr–Abl fusion protein, as discussed in this chapter, is an example of kinase overactivity that underlies cancer. Temporal phosphotyrosine signaling of the EGF receptor pathway was studied by SILAC. Multiple SILAC experiments were performed to study the phosphorylation dynamics of the EGFR pathway over a time course of 20 minutes. Fig. 20.5 shows how the phosphorylation states change for the functionally clustered proteins. These temporal changes demonstrate how elaborate and complicated signal transduction pathways are, bearing in mind that this represents only phosphotyrosine-dependent signaling networks for one receptor in a specific cellular system.

Protein complexes and cellular networks

Proteins rarely act alone; they more often form complexes with other proteins to execute their cellular functions. Proteins can be activated by a posttranslational modification (see Box 20.7) and can travel between cellular compartments to interact with other proteins to perform a specific task. Also, ATP-driven conformational changes are important in complex formation. Hence the importance of 3D structure studies.

Identification of protein complexes is a key to elucidating and understanding intracellular protein networks. Protein complexes

W. Schubert, N.C.J. de Wit, and P. Walden

Box 20.7 Protein–protein interactions and networks

Protein–protein interactions are the basis of cellular communication. All proteins have interaction domains which are mostly conserved between species. This explains why findings in yeast can explain much about human intracellular signaling. Protein domains can be constitutionally active or can be activated by a posttranslational modification. Kinases, for example, can activate protein domains by phosphorylation which leads to binding of the activated domain with a recognition domain. Phosphotyrosine sites in proteins (phosphorylation-activated tyrosine domain) bind to the Src homology-2 domain (the SH2 recognition domain). Identifying posttranslational modifications and protein–protein interactions is needed to unravel interaction networks. Not only identification of protein complexes, which are essentially multiple protein–protein interactions, but also the temporal kinetics of these biological processes are of interest.

Identification of posttranslational modifications can be done by mass spectrometry. Modifications of a proteins or peptide lead to a mass change, and this mass change corresponds to the posttranslational modification. By sequencing the protein by mass spectrometry, the modification site can be identified.

To identify cellular protein complexes, molecular techniques need to be used to "fish" the protein complexes out of the cell. This is done by inserting a specific affinity tag at the 3' end of the gene in the genome. When the protein is expressed, it is expressed as a fusion protein with an affinity tag attached to the C-terminal end of the protein. This affinity tag allows for very selective purification of that particular protein out of the complete cellular extract, and interacting proteins are purified at the same time. This tandem affinity purification (TAP) method is a two-step purification technique and uses specific affinity columns to purify the protein complex (Fig. 20.6). The TAP tag consists of two binding domains of protein-A and a calmodulin-binding peptide with protease cleavage sites in between. Protein-A binds very strongly to immunoglobulin-G (IgG), and the tagged protein including interacting proteins binds to the IgG affinity column. The complex is eluted of the column by cleaving off a part of the TAP tag with a protease. The eluens is purified further by a second affinity column to which the calmodulin-binding peptide binds. The protein complex is eluted of the column after several wash steps, and purified proteins can be identified by mass spectrometry. The identified proteins are part of a functional protein complex.

There are some other tagging approaches, but the principle is identical.

---

are ordered but dynamic structures that assemble and de-assemble to store and pass on biological information. Tandem affinity purification strategies have been used to purify protein complexes from yeast, *Drosophila*, and human cell lines (Fig. 20.6). Identified protein complexes were used to develop models for cellular communication networks using mathematics and bioinformatics. A connectivity map of a yeast protein network is shown in Fig. 20.7. More than 80% of the yeast proteome is highly connected through sharing proteins in different complexes. Some proteins are highly connected and essential for cellular communication structures. Most research on protein complexes is done on yeast, but by screening human protein orthologues, our understanding of human protein networks is increased. (**Orthologues** are proteins from a common ancestor and have an identical function in different species.)

Different tag approaches have been used to label proteins and localize them in the various cellular compartments of yeast or human cells. By labeling proteins with a green fluorescent protein tag (GFP tag), the subcellular compartment in which the protein is located can be detected by fluorescence microscopy and may provide information related to the function of the protein. For yeast, this was done as a large screen that provided a wealth of information about thousands of proteins of the yeast proteome. Most research on protein networks has been done in yeast. This is mainly because yeast is a eukaryotic organism that can be very easily manipulated compared to mammalian cell lines. Nonetheless, due to evolutionary conservation, the acquired information is useful for research into the human proteome.

One of the few examples for protein interaction work in humans is the research on the TNF-α–NF-κB pathway in a TNF-α responsive HEK293 mammalian cell line. Thirty-two proteins known to be part of the TNF-α–NF-κB pathway were TAP tagged, and tagged proteins were co-purified with interacting proteins. Six hundred eighty proteins including 241 interactions were identified. Twenty-eight proteins were selected for validation, and 10 new functional modulators of the pathway were verified using a functional assay combined with RNA interference (RNAi). TRAF7 and TBKBP1 had no previous functional annotation and were among the 10 proteins verified by RNAi to be actively involved in the TNF-α–NF-κB pathway. TRAF7 formed a complex with heat shock protein 70 (HSP70), PFDN2, MARK2, MEK5b, and C20orf126. TBKBP1 interacts with TBKBP2 to form a complex with TBK1. TBKBP is an abbreviation for TBK-binding protein. More information on TAP tagging of the TNF-α–NF-κB pathway can be found in Bouwmeester *et al.* (2004) and Fig. 20.6.

Unraveling of a pathway allows for target selection to modulate pathway activation in disease with pharmaceutical compounds. Proteomics is therefore one of the key drives behind target selection for drug development. All this information about proteins from mammalian cells and yeast is being collated in databases and provides a valuable resource for researchers. Sequence information, structure, posttranslation modifications, interacting proteins, cellular function, and location can be found in databases. Some links to databases can be found at the end of this chapter.

Clinical applications of proteomics

Proteomic methodology can be used for identifying drug targets and biomarkers for development of diagnostic tests. In particular, unraveling pathways and the identification of diagnostic or prognostic markers have much potential. Plasma is particularly acces-

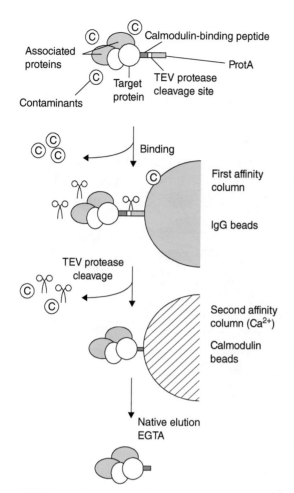

Figure 20.6 Schematic overview of TAP tag procedure. A protein of interest is expressed in the cell with a TAP tag. Cultured cells are lysed, and the complex is purified with the first affinity column (IgG beads). A specific protease, TEV protease, then cleaves the complex off the affinity column. The complex is eluted from the column onto the second affinity column and further cleaned up. Finally, the cleaned-up complex is eluted off the calmodulin beads by EGTA and can be analyzed by MS. All interacting proteins can be identified. From Puig, O., Caspary, F., Rigaut, G., et al. (2001). The tandem affinity purification (TAP) method: a general procedure of protein complex purification. Reprinted from *Methods*, **24**: 218–29. Copyright (2001), with permission from Elsevier.

sible for proteomic research as it is easily collected and contains many proteins or protein fragments that bear information on physiological and pathological states of the body. Therefore, mainly protein-based techniques and assays use plasma. For diagnostic purposes, there has been much interest in serum protein fingerprinting by proteomic techniques. The hypothesis is that cancer and other diseases (e.g. inflammation) release protein waste into the bloodstream. When plasma or serum is analyzed by mass spectrometric techniques, a fingerprint is obtained. This fingerprint can be specific for a disease state and would allow for diagnosis using specialized software. Such proteomic-based diagnosis does not require taking tumor biopsies and has a quick turnaround. However, serum is the product of a proteolytic process, blood clotting, which might generate artificial proteomic patterns that need to be sorted out. The development of proteomics-based diagnostic tools is at an early stage, and validation of proteomic-based tests is required before their introduction into the clinic.

Toponomics: investigating the protein network code of cells and tissues

A man's manners are a mirror in which he shows his portrait.
Johann Wolfgang von Goethe

Our present knowledge of the cell is the result of decades of cell biological research that has unraveled the biochemical reaction pathways and their compartmentalization in subcellular organelles. One of the biggest future challenges in cell biology and human medicine is to decipher the whole functional plan of a cell or a tissue (its biological code or protein network code). To detect and decipher the entirety of all protein networks in healthy and diseased organisms, we need technologies that address the protein network structure and function directly in the cell and tissue *in vivo* or in situ (in contrast to proteomics techniques, which rely on *ex vivo* protein analyses and lack spatiotemporal information). The number of corresponding problems to be solved in biology and medicine is quasi-unlimited. They all have in common that researchers rely on techniques allowing them to co-localize a very large number of molecular components in the same biological structure to detect the structure-bound architecture of protein networks in situ.

The robotic imaging technology MELC/TIS toponomics (Box 20.8) enables the investigator to locate and decipher functional protein networks consisting of hundreds of different proteins in a single cell or tissue section, and thus complements the omics approaches discussed here to provide direct systems biology information (Schubert 2003; Schubert *et al.* 2006). The technology reveals rules of hierarchical protein network organization in which state-specific lead proteins control protein network topology and function (the toponome). Detection of disease-specific protein nodes could be invaluable in developing specific drugs and diagnostics (Cottingham 2008; Schubert 2003, 2010; Schubert *et al.* 2006).

To provide a definition, the term "toponome" is derived from contraction of the ancient Greek nouns τόπος (place and position) and νόμος (law). Toponomics as field of research aims at the analysis of the functional architecture of large molecular networks directly in the morphologically intact cell or tissue section by using a robotically controlled microscopic imaging technology termed MELC/TIS. Corresponding high-dimensional TIS data sets are amenable for exact mathematical description of functionally interlocked protein clusters, each of which has a characteristic subcellular topology (for the definition of topology, see Box 20.9).

Processing the images from the cyclical imaging procedures

For data analysis, the protein location images (≥100) of one and the same visual field resulting from single incubation–imaging–bleaching cycles are aligned pixel-wise. All images are "fused," and thresholds are selected for each signal. This results in a toponome map in which protein combinations present in each

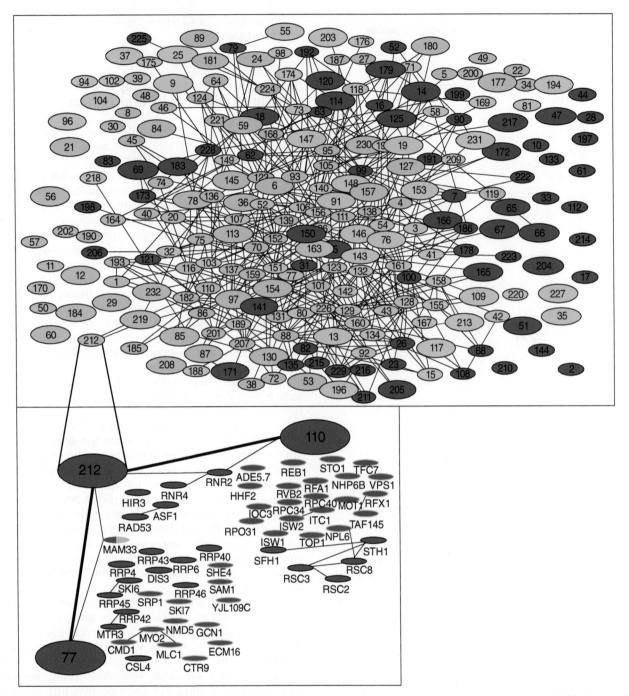

Figure 20.7 Connectivity map of the yeast proteome established through TAP-tagging approaches. More than 80% of the proteome is highly connected. Links were established between complexes sharing at least one protein. Individual complexes are color coded. (Please see our website for the color version of this figure: in the color version, red is cell cycle; dark green is signaling; dark blue is transcription, chromatin structure, and DNA maintenance; pink is protein and RNA transport; orange is RNA metabolism; light green is protein synthesis and turnover; brown is cell polarity and structure; violet is intermediate and energy metabolism; and light blue is membrane biogenesis and traffic.) From Gavin, A.C., Bosche, M., Krause, R., et al. (2002). Functional organization of the yeast proteome by systematic analysis of protein complexes. *Nature*, **415**, 141–7. Reproduced by permission of *Nature*.

pixel are mapped as "combinatorial molecular phenotypes" (CMPs) (Fig. 20.8). Each signal within a CMP is mapped as tag present (1) or tag absent (0), depending on whether the value for the fluorescence signal is above or below this threshold, respectively (1 bit information per protein at a pixel or voxel). CMPs assembled as a group will have unique features as defined by the assembly's lead proteins (L=1), absent proteins (A=anti-colocated, or 0), and wild-card proteins (W=variable occurrences of a protein 0 and 1 in a given CMP of a CMP group). We define such CMP groups as a CMP motif denoting a given functional region of a cell or tissue. By using a three-symbol code (LAW; Fig. 20.8), differences of cell types and cell states can readily be identified in studies comparing different experiments or diseases with normal conditions.

Box 20.8 Overcoming the spectral limitation of fluorescence microscopy

The recently described microscopic fluorescence robot technology MELC/TIS is capable of imaging at least 100 different molecular cell components (MCCs) in a cell or tissue section. MELC/TIS overcomes the spectral limitation of traditional fluorescence microscopy by using large dye-conjugated tag libraries; automatically bleaching a dye after imaging and re-labeling the same or another MCC in the identical sample with the same dye coupled to a tag (e.g. a specific antibody), with the same or another specificity; and repeating these cycles with other tags multiple times to assemble multidimensional colocation patterns. By this approach, one can address what is termed the toponome (the entirety of protein networks). After the first description of the technology in 1990 by W. Schubert (Schubert 1990), the relevance of this approach to cell function has been increasingly recognized and illustrated with several examples. Briefly, the technology has (a) uncovered a new cellular transdifferentiation mechanism of vascular cells giving rise to myogenic cells in situ and *in vivo*, a finding that has led to efficient cell therapy models of muscle disorders; (b) discovered a new target protein in sporadic amyotrophic lateral sclerosis (ALS) by hierarchical protein network analysis, a finding that has been confirmed in a mouse knockout model; (c) uncovered a lead target protein in tumor cells that controls cell polarization as a mechanism fundamental for migration and metastasis formation; and (d) found new functional territories in the central nervous system (CNS), as well in cells, defined by high-dimensional protein clusters.

Box 20.9 Definition of the term "topology" used for toponomics

The laws of connectivity, mutual position, and order of points, lines, surfaces, bodies, and their parts or aggregates in space, apart from the measures and scales (Listing, 1847).

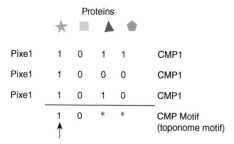

Figure 20.8 Illustration of the topological hierarchies of proteins within the toponome. L: lead protein (common to all CMPs of a CMP motif); A: absent protein signal (absent in all CMPs of a CMP motif); W: wild-card protein signals (proteins that are variably associated with the (L) and the (A) proteins of a CMP motif). From Schubert W. (2007) Cytometry A, **71**(6):352–60. Reproduced by permission of John Wiley & Sons, Inc

Extensive studies have focused on the reproducibility and robustness of the MELC/TIS procedures. For example, thresholds for each protein can be manually selected by an expert from an automatically generated range. It was shown that these thresholds were very similar among experts and are closely comparable to thresholds resulting from a fully automated system based on maximizing mutual information. Moreover, non-threshold-based methods can be also applied to MELC/TIS data sets so that independent methods of combinatorial molecular data mining can be used for data evaluation.

The position of a given protein in the labeling sequence might cause differences in the protein's apparent level, pushing its level in a pixel above or below the threshold and thereby affecting the CMPs. To address this possibility, the calibration of a new tag library involves (a) a given selected sequence of tags (cycles 1 to n), (b) the same sequence inverted, and (c) the same sequence randomly permuted resulting in a total of dozens to hundreds of informative labeling reactions. If sterical hindrance is observed, the cycle position of the corresponding tags can be changed until the optimal result is obtained.

Reproducibility of results can be demonstrated by performance tests, for example the R2 test which has proven specificity and robustness of MELC/TIS (e.g. >0.9 in R2).

Structure, code, and semantics of the toponome: a high-dimensional combinatorial problem

Every system has an inner structure, a code, and semantics. One of the most challenging problems in systems biology is to uncover the structure-bound functional organization, hence the coding structure of the toponome. This requires co-mapping a large number of proteins in morphologically intact cells and tissue sections. The following are some of the major questions of systems biology that can be addressed by toponomics: how many and which functionally relevant protein clusters (transient or stable) are expressed in a given cell type in its natural tissue environment or in culture, in health and disease? Is there a continuum of all theoretically possible protein combinations of the approximately 20 000 (core) proteins in humans rather than a large, albeit limited, number of combinations (a quasi-toponome dictionary) that clearly distinguish the different cellular functionalities? Are there rules behind a generative grammar? Is there a higher order of such protein clusters, with clusters interlocked as networks in which every cluster has a defined relative topology to other clusters? How predictive are such clusters? Can we assign to such clusters a decisive cellular function or dysfunction in disease?

Such questions have been addressed in previous studies. Although systematic studies are still in an early stage, recent evidence indicates that toponome structures uncover the features of highly organized systems with hierarchical properties that appear to be functionally predictive.

Firstly, co-mapping 100 proteins was performed in a skin tissue section across a psoriatic plaque and a non-involved, morphologically normal skin area of a patient suffering from chronic psoriasis (Schubert *et al.* 2006, 2012). When the mean abundance of each protein was measured and displayed in a protein profile, very little if any difference between the involved and the uninvolved skin was seen. However, when the 30 most frequent out of hundreds of thousands of expressed protein clusters were visualized *in situ*, clear-cut differences were readily identified: psoriasis-specific clusters were confined as a mosaic of protein clusters to the psoriatic epidermis, while these clusters were absent in the non-involved skin. Similarly, co-mapping of cell surface proteins in rhabdomyosarcoma tumor cells revealed that different functional states of these cells (spherical vs. exploratory state) were readily distinguished by protein cluster mapping (detailed in this chapter). These observations indicate that the proteins as components of the topologically organized protein network are less distinctive than their assemblies as clusters. This is consistent with observations in other systems such as written text in which words rather than letters distinguish sentences with different meanings.

Secondly, for example Zipf's law (Fig. 20.9) – a power law and measure of the hierarchical architecture of molecular systems – does not apply when the number of molecular components localized simultaneously is too low, for example <15, but applies when the number exceeds 45. When the number of protein clusters formed by co-mapped 49 different proteins was correlated with the individual frequency of these clusters, the relationship conformed to Zipf's law in which the probability $P(e_n)$ of the n most frequent individual event e_n is proportional to the negative power of n. Reducing the number of co-mapped proteins observed simultaneously weakened the relationship. Such relationship it was totally absent in randomly generated data (Fig. 20.9). The possibility given by MELC/TIS to co-localize many molecular components in cells has provided access to the systems character of cellular proteins and thus to the functional architecture of the cell by analyzing, on a large scale, which proteins belong together and which do not at a given site of the cell under given conditions.

Thirdly, it has been possible to solve the combinatorial 3D structure of 27 cell surface proteins in peripheral blood lymphocytes (Fig. 20.10; Friedenberger *et al.* 2007; Schubert *et al.* 2012). Surprisingly, these investigations have uncovered striking differences between CD4 and CD8T lymphocytes. CD4T lymphocytes showed a dense cell surface network of interlocked protein clusters, while CD8T lymphocytes displayed some non-interlocked clusters.

Together, these findings show that the features of highly organized protein systems in morphologically intact cells and tissue sections can be assessed only by co-mapping a large number of proteins. This level of complexity can be addressed only by overcoming the spectral limitation of traditional fluorescence microscopy with its limited number of dyes.

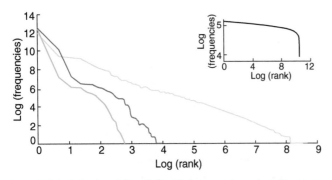

Figure 20.9 Zipf's plot of the relationship between the rank and frequency of CMP motifs in human skin. Data were obtained for 49 simultaneously recognized molecules (light green), or 10 subsets of these molecules (red) or five molecules (dark green). The (log–log) linearity seen with data for 49 molecules is progressively less apparent as fewer molecules are examined. Inset: Zipf's plot of the relationship between the rank and the frequency of motifs obtained with 49 randomly distributed molecules, showing that the linearity has disappeared. From Schubert *et al.* (2006).

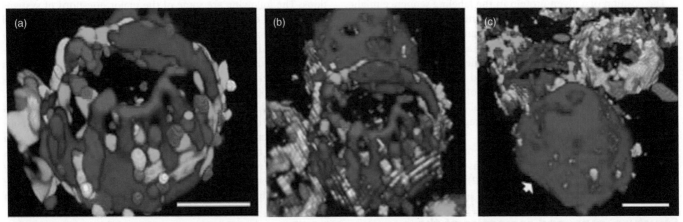

Figure 20.10 Three-dimensional toponome map of cell surface protein clusters in T lymphocytes. Note that the different colors represent differential local combinations of 27 different cell surface proteins co-localized by MELC/TIS. (a) CD4 T lymphocyte (high resolution); (b) the same CD4 T lymphocyte as in (a), but in the context with its environment. (c) Image from (b), rotated by 180°, shows a CD8 T lymphocyte (arrow). In the CD4 cell (a–b), numerous different protein clusters (circumscribed spherical structures) are embedded in the CD4 positive cell surface membrane (red). These clusters are interlocked as a network. By contrast, in the CD8 cell, the surface protein clusters, located in the CD8 positive membrane (blue), are largely unconnected. Bars: 5 μm. From Friedenberger *et al.* (2007).

Detecting a cell surface protein network code: lessons from a tumor cell

Recently, the cell surface toponome of a rhabdomyosarcoma tumor cell was partially mapped (Schubert *et al.* 2006, 2012). The data showed that specific protein clusters are arranged along the cell surface of cell extensions always showing the same topological sequence (array) of corresponding clusters (Fig. 20.11a, bottom). This sequence is observed only when the spherical cell elongates to form three cell extensions (the exploratory state) preceding the migratory state in which the cell withdraws one of the three cell extensions to form a long axis. The analysis of these clusters showed that one protein (aminopeptidase N, CD13), a proteolytic enzyme, is present in all clusters, while all other proteins are variably occurring in these clusters (absent or present). The protein that is present in all clusters is defined as the lead protein. The variably associated proteins are defined as wild-card proteins. Following the lead protein hypothesis, inhibition of the lead protein CD13 (by antibody SJ1D1 or small molecule RB3014) results in a complete disassembly of the cell surface clusters with the consequence that cells do not exert the transition from the spherical to the exploratory cell state (Fig. 20.11). Moreover, when the lead protein is inhibited in the exploratory state, cells

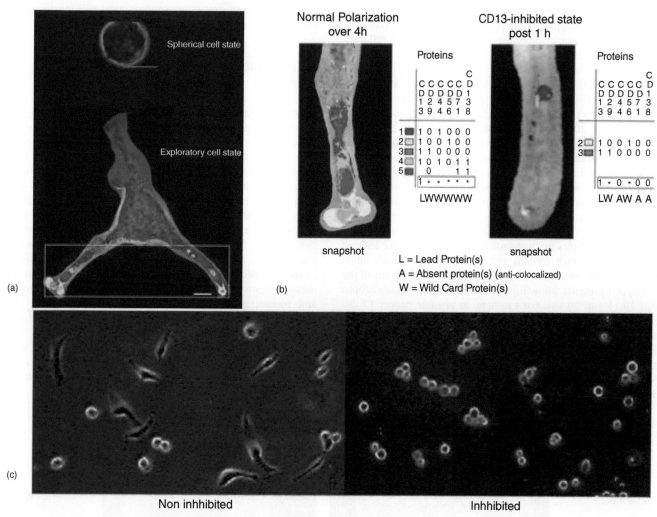

Figure 20.11 Identification of a molecular network in a single tumor cell. (a) Two cell states are seen, the spherical state (top) and exploratory state (bottom), which develop from the spherical cell state. On the cell surface of the lower cell (exploratory state), different colors are displayed, which are arranged as an array from the tip of the corresponding cell extensions toward the cell body: yellow, dark blue, light blue, red, and green. These colors indicate different protein clusters which were mapped by MELC/TIS (from Schubert et al., 2006). Note that the relative topology and sequential arrangement of these clusters are precisely corresponding in both lower cell extensions. (b) Left upper part: Detail of one of the cell extensions shown in (a, bottom left), together with the color-coding map for the corresponding protein clusters (CMPs). It is evident that these protein clusters have in common the proteolytic enzyme CD13 as their lead protein, while the other shown proteins are variably associated with the lead protein in the different clusters (present or absent). The common feature of all clusters can be expressed as a three-symbol code (Schubert, 2007): LWWWWW. When the lead protein is inhibited, a disintegration and total loss of most of the protein clusters are observed (b, upper right), and a new common motif occurs: LWAWAA. (c) Functionally, the latter inhibitory motif leads to complete loss of the ability of the cells to enter the exploratory state (compare the left part of c: non-inhibited cells rapidly undergoing cell polarization; and the right part of c: inhibited CD13: cells remain in the spherical cell state). Bar (a): 2 μm. From Schubert et al. (2006).

withdraw all cell extensions and fall back into the spherical state. This has shown, firstly, that wild-card proteins are dependent on the lead protein CD13; secondly, the detected protein clusters are functionally interrelated as a network; and, thirdly, the functional and topological integrity of the detected protein clusters are dependent on the control by the lead protein CD13. Hence, the detected network enciphers cell polarization, indicating that the array of cell surface protein clusters exerts control over intracellular mechanisms driving the cell into the exploratory state.

The molecular face of cells in diseases

As we have learned from toponome-imaging microscopy, not only a large number of different proteins must be co-mapped in individual cells or in tissue sections, but also the overall strategy must include co-mapping these proteins in a given cell compartment. If a sufficient number (meeting Zipf's law) of different proteins is expressed in or associated with a given cell compartment (e.g. cell surface), it is likely that one can detect a higher order of these proteins arranged as clusters (CMPs). Such clusters can exhibit characteristic geometric features and are often arranged as multicluster arrays along or within that compartment (Figs 20.11 and 20.12; Bhattacharya et al. 2010; Sage 2009; Schubert et al. 2009). These features make these geometries amenable to precise mathematical methods such as combinatorial statistics and geometry to explore, in conjunction with approaches to experimental alteration of the clusters (Figs 20.11 and 20.12), the rules that govern structure-bound protein networks. Analyses in blood, skin, and tumor cells in culture, and synapses in the CNS, are unequivocal and can readily be re-recognized in new samples for examples for diagnostics. Moreover, we have learned that the cell surface protein system expresses a large, albeit limited, number of the theoretically possible combinations of the co-mapped proteins, allowing us to draw functional conclusions (Figs 20.13 and 20.14). For example, in prostate cancer 17 different proteins belonging to different families were co-mapped. In this example, MELC/TIS worked at a power of combinatorial molecular discrimination (PCMD) of 2^{17} per data point (= 1 bit information per protein; 0 = absent, 1 = present). Only 2100 different protein clusters out of 2^{17} (= 131 000.72) CMPs were expressed. Out of these clusters, 40 were specifically expressed as a clusters motif only on the surface of cancerous cells inside prostate acini with features of intraductal metastasis. Similar observations were made in colon cancer (Bhattacharya et al. 2010; Schubert et al. 2009).

Deciphering the human toponome will be a huge future challenge, given the number of proteins to be co-mapped. However, the increasing number of tags binding human proteins and the availability of mature toponome-imaging technologies will make this vision achievable. Toponomics will progressively map not only the molecular face of cell types (Fig. 20.10) but also that of diseases as recently shown for prostate cancer, colon cancer, apoptosis, and rhabdomyosarcoma cell polarization (Fig. 20.11). This is the first step to completely decipher the protein network code of cancer cells. The ongoing establishment of toponome facilities launched by collaborative initiatives of academic and industry partners (www.toposnomos.com) will lay foundations for the human toponome project.

Individualized medicine and tailored therapies

Systematic classification of diseases including cancers, which mostly is based on clinically apparent criteria, allows the use of common frames of references for healthcare professionals, researchers, and patients, and is the mainstay of current clinical diagnosis, selection of treatments, and prognosis. However, disease is in all its aspects highly individual, and it is widely accepted that for many patients the current tools for classification do not accurately define prognosis, nor do they allow precise selection of treatments that will be both well tolerated and effective. In fact, disease classifications require updating as new knowledge is acquired and particularly when, as has happened in the last decade, we have developed new tools which allow measurements in patients at a previously unprecedented genomic and phenotypic (including proteomic and toponomic) level. This is exemplified by the development of large numbers of new molecular-targeted anticancer agents that are increasingly driving the need to subclassify cancers such as breast cancer in a more comprehensive way on the basis of variations in mutation status, gene expression, protein network patterns, and others – in other words, in a more individualized way.

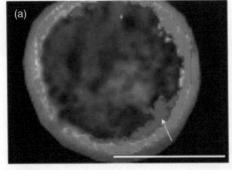
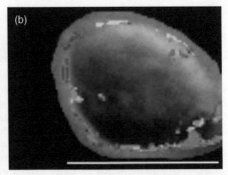

Figure 20.12 2D toponome maps of spherical rhabdomyosarcoma cells displaying cell surface protein clusters (different colors). (a) Spherical cell displaying the cluster (collocation of CD13, CD44, CD138, and CD71CD56) (a, arrow, dark blue color). Note that this node (cluster) corresponds to the "dark blue" protein cluster displayed by cells in Figure 20.11a (bottom). If CD13 in the cell of (a) is inhibited, the spherical cell in (a) disassembles the "dark blue" cluster (arrow in a) and remains in the spherical state (b). If CD13 in the exploratory cell state in Figure 20.11a (bottom) is inhibited, the cell falls back in the spherical state shown in (b). Bars: 10 μm. From Schubert et al. (2006).

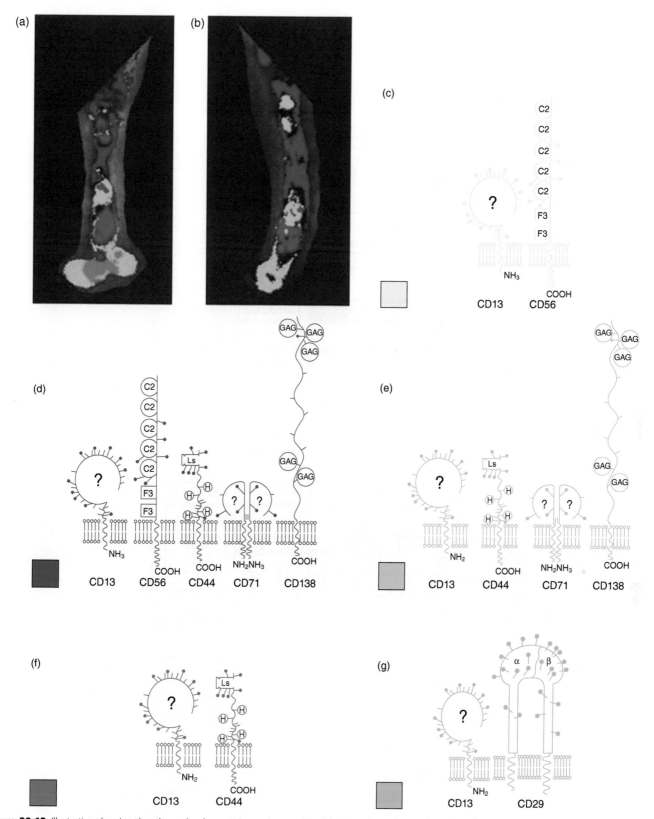

Figure 20.13 Illustration showing that the molecular protein species, combined in the various clusters along the cell surface of cell extensions (a–b), are highly dissimilar (dissimilarity rule) (c–g). By contrast, both the sequential arrangement and geometry of the corresponding clusters in the two cell extensions (a–b) of the same cell (whole cell; see Figure 20.11a, bottom) are highly similar (geometry rule). Note that the scheme of the molecular structure of the different protein species is depicted from Barclay *et al.* (1993). (c–g) Note the dissimilarity of the combined proteins within the clusters "yellow," "dark blue," "light blue," "red," and "green," corresponding to the colors in the toponome map in (a–b). From Schubert (2010).

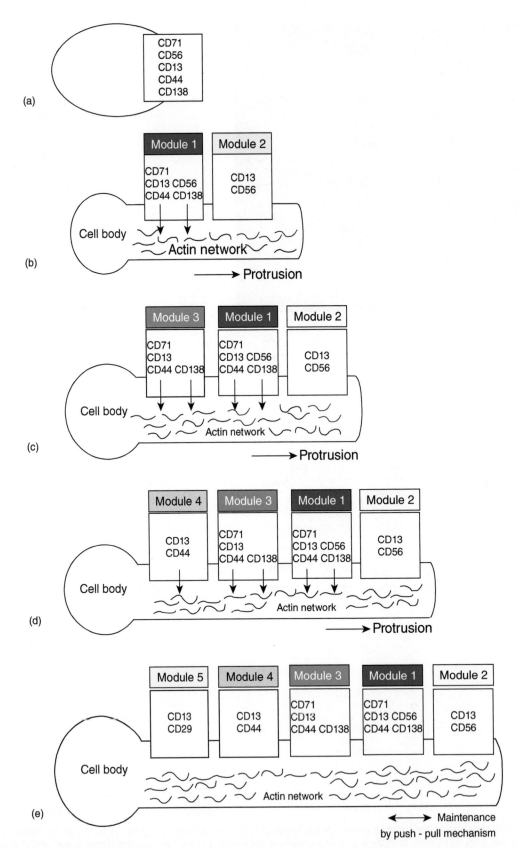

Figure 20.14 Dynamic model derived from experiments (Schubert et al., 2006) illustrating that the planning network (or node) in (a) allocates the actin network (executive network) by continuous proteolytic activity of CD13 (APN) activating interaction of cytoskeleton-binding molecules CD44 and CD138. (b–e) The node in (a) corresponds to Figure 20.12 (arrow), then continuously establishes the various cell surface modules (2–5) which initiate the growth of the actin network protruding the cell extension. From Schubert (2010).

Discussion and conclusion

With all the great and rapid advancements in the approaches to biological systems, technologies, IT and informatics tools, and biological concepts, systems biology in general and cancer systems biology in particular are just at the beginning of their developments. The prospects are tremendous, but there are also a number of tasks to be completed before these prospects can be fully realized. One very important issue in cancer research is the cellular heterogeneity within a tumor. Thus, in most tumors there are a large number of different cell types other than the cancer cells themselves, including inflammatory and immune cells recruited to the tumor, tissue stroma (now known to play a crucial role in tumorigenesis), vascular endothelial cells, and others. Moreover, the cancer cells are genetically and functionally not uniform and may vary significantly from one another due to genetic instability and clonal evolution. Even seemingly identical cells can vary in gene expression due to interactions between neighboring cells as, for instance, illustrated by lateral inhibition effects in Wnt signaling, recent exciting reports of Myc-activated cells affecting growth in neighboring identical cells, and others. All these effects can create major headaches for the investigator bent upon molecular and proteomic profiling. Profiling any combination of cells, no matter how seemingly homogeneous, will inevitably result in the averaging of gene expression across several cells. Results obtained must therefore be interpreted with caution. For example, mutations in *Ras*, *p53*, *TERT*, and *Myc* in a tumor would have very different ramifications if these mutations were present in a single cell as distinct from different combinations of these in different cells. One way of circumventing this issue is to profile single cells isolated from the tumor or from the circulation. Unfortunately, despite tools such as laser capture microdissection (LCMD), fluorescence-activated cell sorting (FACS), and immunopanning available to achieve this, the problems of analyzing gene and protein expression in single cells are still not resolved. Toponomics might point the direction for overcoming this problem with advanced image analysis of large protein networks and their hierarchical organization in histological sections.

Microarray and proteomic techniques will undoubtedly develop further in the coming years to become even more powerful research tools, and the new NGS technologies will produce genomic and transcriptomic information on as-yet-unseen scale and differentiation. Yet, the large amounts of data acquired with these high-throughput methods demand new approaches to processing and evaluating the data, confirming the results, and integrating the information in testable models for cancer as well as individualized diagnoses and treatment directives. Currently, only the genes or proteins of particular interest to the researcher involved are verified by standard laboratory techniques. Toponomics as a direct systems biology approach will become the major tool for direct visualization of complex protein networks in cells, tissues, and in particular clinical specimens, and thus usher in rapid new output of molecularly defined diagnoses and new targeted therapeutics of high specificity. Omics and toponomics are complementary and may operate in synergy. Omics can determine expression profiles on a genome-wide scale which is beyond the capacity of toponomics. Toponomics, however, resolves the molecular composition of cells and tissues with spatial and spatiotemporal resolution, and visualizes protein networks and the hierarchical organization of the proteins in these networks. Only together will these different sets of information produce what is truly systems biology.

Completion of the HapMap; sequencing of new genomes, in particular cancer genomes; and identification of the human transcriptomes and proteomes for different cells and organs will further increase our understanding of human biology and of the biology of diseases such as cancer. The ultimate goal here is to have mathematical models of cellular systems where one can navigate through the biological system from genome to protein network as depicted in Fig. 20.7. Such model mathematics, however, still bear quite some challenges that are being tackled but need to be and will certainly be addressed far more extensively in the near future.

As yet, validation of systems biology models and of targets or biomarkers produced by the approaches discussed in this chapter is the major bottleneck in systems biology. While the omics techniques are high throughput and produce vast amounts of data, validation is still done with conventional time- and resource-consuming technologies. Here major new developments are needed to transform the validation of targets, markers, and models to a qualitatively new advanced state that can operate on par with the omics technologies. Automated genome-wide RNAi screens are a first attempt on this and may direct future developments. Also, tight interlinks of clinical diagnosis and clinical research with basic and experimental science will expedite the validation process and the translation of new targets and markers into clinically applicable procedures and products.

All the discussed developments will contribute to the identification of new drug targets and diagnostic biomarkers. Personalized medicine will become reality with highly differentiated genotyping and phenotyping, disease outcome prediction, and predictive diagnostic tests. Cancer treatment that is currently very general will become more targeted as the effects of the developments discussed in this chapter and an understanding of cancer as a biological system advance.

At the end of this chapter and book, it is difficult not to be amazed by the complexity of cancer biology. It is, at times, worthwhile to step back and appreciate how controlled biological systems and organisms are. More important is how little actually goes wrong, considering the complexity of biology.

Bibliography

Information flow in the cell

Balmain, A., Gray, J., and Ponder, B. (2003). The genetics and genomics of cancer. *Nature Genetics*, **33**(Suppl.): 238–44.

Bell, J. (2004). Predicting disease using genomics. *Nature*, **429**: 453–6.

Brown, D., and Superti-Furga, G. (2003). Rediscovering the sweet spot in drug discovery. *Drug Discovery Today*, **8**: 1067–77.

Evans, W.E., and Relling, M.V. (2004). Moving towards individualized medicine with pharmacogenomics. *Nature*, **429**: 464–8.

Hood, L., and Galas, D. (2003). The digital code of DNA. *Nature*, **421**: 444–8.

Jones, P.A., and Baylin, S.B. (2002). The fundamental role of epigenetic events in cancer. *Nature Reviews Genetics*, **3**: 415–28.

Narod, S.A., and Foulkes, W.D. (2004). BRCA1 and BRCA2: 1994 and beyond. *Nature Reviews Cancer*, **4**: 665–76.

Ponder, B.A. (2001). Cancer genetics. *Nature*, **411**: 336–41.

Model organisms and cancer models

Chin, L., Tam, A., Pomerantz, J., et al. (1999). Essential role for oncogenic Ras in tumour maintenance. *Nature*, **400**: 468–72.

Jain, M., Arvanitis, C., Chu, K., et al. (2002). Sustained loss of a neoplastic phenotype by brief inactivation of MYC. *Science*, **297**: 102–4.

Moody, S.E., Sarkisian, C.J., Hahn, K.T., et al. (2002). Conditional activation of Neu in the mammary epithelium of transgenic mice results in reversible pulmonary metastasis. *Cancer Cell*, **2**: 451–61.

Pelengaris, S., Abouna, S., Cheung, L., Ifandi, V., Zervou, S., and Khan, M. (2004). Brief inactivation of c-Myc is not sufficient for sustained regression of c-Myc-induced tumours of pancreatic islets and skin epidermis. *BMC Biology*, **2**: 26.

Pelengaris, S., Khan, M., and Evan, G.I. (2002). Suppression of Myc-induced apoptosis in beta cells exposes multiple oncogenic properties of Myc and triggers carcinogenic progression. *Cell*, **109**: 321–34.

Pelengaris, S., Littlewood, T., Khan, M., Elia, G., and Evan, G. (1999). Reversible activation of c-Myc in skin: induction of a complex neoplastic phenotype by a single oncogenic lesion. *Molecular Cell*, **3**: 565–77.

Shachaf, C.M., Kopelman, A.M., Arvanitis, C., et al. (2004). MYC inactivation uncovers pluripotent differentiation and tumour dormancy in hepatocellular cancer. *Nature*, **431**: 1112–17.

SNPs, HapMap, and identifying cancer genes

Carlson, C.S., Eberle, M.A., Kruglyak, L., and Nickerson, D.A. (2004). Mapping complex disease loci in whole-genome association studies. *Nature*, **429**: 446–52.

Daly, M.J., Rioux, J.D., Schaffner, S.F., Hudson, T.J., and Lander, E.S. (2001). High-resolution haplotype structure in the human genome. *Nature Genetics*, **29**: 229–32.

Futreal, P.A., Coin, L., Marshall, M., et al. (2004). A census of human cancer genes. *Nature Reviews Cancer*, **4**: 177–83.

Glazier, A.M., Nadeau, J.H., and Aitman, T.J. (2002). Finding genes that underlie complex traits. *Science*, **298**: 2345–9.

Goldstein, D.B., and Weale, M.E. (2001). Population genomics: linkage disequilibrium holds the key. *Current Biology*, **11**: R576–9.

HapMap Consortium. (2003). The International HapMap Project. *Nature*, **426**: 789–96.

Zondervan, K.T., and Cardon, L.R. (2004). The complex interplay among factors that influence allelic association. *Nature Reviews Genetics*, **5**: 89–100.

Cancer mRNA expression analysis

Alizadeh, A.A., Eisen, M.B., Davis, R.E., et al. (2000). Distinct types of diffuse large B-cell lymphoma identified by gene expression profiling. *Nature*, **403**: 503–11.

Creighton, C., Kuick, R., Misek, D.E., et al. (2003). Profiling of pathway-specific changes in gene expression following growth of human cancer cell lines transplanted into mice. *Genome Biology*, **4**: R46.

Lossos, I.S., Czerwinski, D.K., Alizadeh, A.A., et al. (2004). Prediction of survival in diffuse large-B-cell lymphoma based on the expression of six genes. *New England Journal of Medicine*, **350**: 1828–37.

Ma, X.J., Salunga, R., Tuggle, J.T., et al. (2003). Gene expression profiles of human breast cancer progression. *Proceedings of the National Academy of Sciences USA*, **100**: 5974–9.

Perou, C.M., Sorlie, T., Eisen, M.B., et al. (2000). Molecular portraits of human breast tumours. *Nature*, **406**: 747–52.

Pittman, J., Huang, E., Dressman, H., et al. (2004). Integrated modeling of clinical and gene expression information for personalized prediction of disease outcomes. *Proceedings of the National Academy of Sciences USA*, **101**: 8431–6.

Rhodes, D.R., Yu, J., Shanker, K., et al. (2004). Large-scale meta-analysis of cancer microarray data identifies common transcriptional profiles of neoplastic transformation and progression. *Proceedings of the National Academy of Sciences USA*, **101**: 9309–14.

Ross, D.T., Scherf, U., Eisen, M.B., et al. (2000). Systematic variation in gene expression patterns in human cancer cell lines. *Nature Genetics*, **24**: 227–35.

Shipp, M.A., Ross, K.N., Tamayo, P., et al. (2002). Diffuse large B-cell lymphoma outcome prediction by gene-expression profiling and supervised machine learning. *Nature Medicine*, **8**: 68–74.

Sorlie, T., Tibshirani, R., Parker, J., et al. (2003). Repeated observation of breast tumour subtypes in independent gene expression data sets. *Proceedings of the National Academy of Sciences USA*, **100**: 8418–823.

van de Vijver, M.J., He, Y.D., van't Veer, L.J., et al. (2002). A gene-expression signature as a predictor of survival in breast cancer. *New England Journal of Medicine*, **347**: 1999–2009.

CGH arrays, CpG island microarrays, and ChIP-on-Chip

Albertson, D.G. (2003). Profiling breast cancer by array CGH. *Breast Cancer Research and Treatment*, **78**: 289–98.

Ballestar, E., Paz, M.F., Valle, L., et al. (2003). Methyl-CpG binding proteins identify novel sites of epigenetic inactivation in human cancer. *Embo Journal*, **22**: 6335–45.

Balog, R.P., de Souza, Y.E., Tang, H.M., et al. (2002). Parallel assessment of CpG methylation by two-color hybridization with oligonucleotide arrays. *Analytical Biochemistry*, **309**: 301–10.

Bird, A. (2002). DNA methylation patterns and epigenetic memory. *Genes & Development*, **16**: 6–21.

Chen, C.M., Chen, H.L., Hsiau, T.H., et al. (2003). Methylation target array for rapid analysis of CpG island hypermethylation in multiple tissue genomes. *American Journal of Pathology*, **163**: 37–45.

Egger, G., Liang, G., Aparicio, A., and Jones, P.A. (2004). Epigenetics in human disease and prospects for epigenetic therapy. *Nature*, **429**: 457–63.

Gitan, R.S., Shi, H., Chen, C.M., Yan, P.S., and Huang, T.H. (2002). Methylation-specific oligonucleotide microarray: a new potential for high-throughput methylation analysis. *Genome Research*, **12**: 158–64.

Mantripragada, K.K., Buckley, P.G., de Stahl, T.D., and Dumanski, J.P. (2004). Genomic microarrays in the spotlight. *Trends in Genetics*, **20**: 87–94.

Oostlander, A.E., Meijer, G.A., and Ylstra, B. (2004). Microarray-based comparative genomic hybridization and its applications in human genetics. *Clinical Genetics*, **66**: 488–95.

Pollack, J.R., Perou, C.M., Alizadeh, A.A., et al. (1999). Genome-wide analysis of DNA copy-number changes using cDNA microarrays. *Nature Genetics*, **23**(1): 41–6.

Weinmann, A.S., Yan, P.S., Oberley, M.J., Huang, T.H., and Farnham, P.J. (2002). Isolating human transcription factor targets by coupling chromatin immunoprecipitation and CpG island microarray analysis. *Genes & Development*, **16**: 235–44.

Yan, P.S., Perry, M.R., Laux, D.E., Asare, A.L., Caldwell, C.W., and Huang, T.H. (2000). CpG island arrays: an application toward deciphering epigenetic signatures of breast cancer. *Clinical Cancer Research*, **6**: 1432–8.

Yan, P.S., Chen, C.M., Shi, H., Rahmatpanah, F., Wei, S.H., and Huang, T.H. (2002). Applications of CpG island microarrays for high-throughput analysis of DNA methylation. *Journal of Nutrition*, **132**: 2430S–34S.

Next-generation sequencing

Metzker, M.L. (2010). Sequencing technologies – the next generation. *Nature Reviews Genetics*, **11**: 31–46.

Kircher, M., and Kelso, J. (2010). High-throughput DNA sequencing – concepts and limitations. *Bioessays*, **32**: 524–36.

Lee, W., Jiang, Z., Liu, J., et al. (2010). The mutation spectrum revealed by paired **genome** sequences from a lung cancer patient. *Nature*, **465**: 473–7.

Kan, Z., Jaiswal, B.S., Stinson, J., et al. (2010). Diverse somatic mutation patterns and pathway alterations in human cancers. *Nature*. [Epub ahead of print]

Meyerson, M., Gabriel, S., and Getz, G. (2010). Advances in understanding cancer genomes through second-generation sequencing. *Nature Reviews Genetics*, **11**: 685–96.

International Cancer Genome Consortium. (2010). International network of cancer genome projects. *Nature*, **464**: 993–8.

McPherson, J.D. (2009). Next-generation gap. *Nature Methods*, **6**: S2–5.

Kircher M., and Kelso J. (2010). High-throughput DNA sequencing-concepts and limitations. *Bioessays*, **32**: 524–36.

Proteomics and quantitative proteomics

Jessani, N., Humphrey, M., McDonald, W.H., et al. (2004). Carcinoma and stromal enzyme activity profiles associated with breast tumour growth in vivo. *Proceedings of the National Academy of Sciences USA*, **101**: 13756–61.

Krijgsveld, J., Ketting, R. F., Mahmoudi, T., et al. (2003). Metabolic labeling of C. elegans and D. melanogaster for quantitative proteomics. *Nature Biotechnology*, **21**: 927–31.

Lewis, T. S., Hunt, J. B., Aveline, L. D., et al. (2000). Identification of novel MAP kinase pathway signaling targets by functional proteomics and mass spectrometry. *Molecular Cell*, **6**: 1343–54.

Li, J., Steen, H., and Gygi, S. P. (2003). Protein profiling with cleavable isotope-coded affinity tag (cICAT) reagents: the yeast salinity stress response. *Molecular Cell Proteomics*, **2**: 1198–204.

Ong, S.E., Foster, L.J., and Mann, M. (2003a). Mass spectrometric-based approaches in quantitative proteomics. *Methods*, **29**: 124–30.

Ong, S.E., Kratchmarova, I., and Mann, M. (2003b). Properties of 13C-substituted arginine in stable isotope labeling by amino acids in cell culture (SILAC). *Journal of Proteome Research*, **2**: 173–81.

Shiio, Y., Donohoe, S., Yi, E.C., Goodlett, D.R., Aebersold, R., and Eisenman, R.N. (2002). Quantitative proteomic analysis of Myc oncoprotein function. *Embo Journal*, **21**: 5088–96.

Steen, H., and Mann, M. (2004). The ABC's (and XYZ's) of peptide sequencing. *Nature Reviews Molecular Cell Biology*, **5**: 699–711.

van Anken, E., Romijn, E. P., Maggioni, C., et al. (2003). Sequential waves of functionally related proteins are expressed when B cells prepare for antibody secretion. *Immunity*, **18**: 243–53.

Posttranslational modifications

Blagoev, B., Ong, S.E., Kratchmarova, I., and Mann, M. (2004). Temporal analysis of phosphotyrosine-dependent signaling networks by quantitative proteomics. *Nature Biotechnology*, **22**: 1139–45.

Larsen, M.R., and Roepstorff, P. (2000). Mass spectrometric identification of proteins and characterization of their post-translational modifications in proteome analysis. *Fresenius Journal of Analytical Chemistry*, **366**: 677–90.

Mann, M., and Jensen, O.N. (2003). Proteomic analysis of post-translational modifications. *Nature Biotechnology*, **21**: 255–61.

Protein complexes and cellular networks

Aloy, P., Bottcher, B., Ceulemans, H., et al. (2004). Structure-based assembly of protein complexes in yeast. *Science*, **303**: 2026–9.

Blagoev, B., Kratchmarova, I., Ong, S.E., Nielsen, M., Foster, L.J., and Mann, M. (2003). A proteomics strategy to elucidate functional protein-protein interactions applied to EGF signaling. *Nature Biotechnology*, **21**: 315–18.

Bouwmeester, T., Bauch, A., Ruffner, H., et al. (2004). A physical and functional map of the human TNF-alpha/NF-kappa B signal transduction pathway. *Nature Cell Biology*, **6**: 97–105.

Gavin, A.C., Bosche, M., Krause, R., et al. (2002). Functional organization of the yeast proteome by systematic analysis of protein complexes. *Nature*, **415**: 141–7.

Pawson, T., and Nash, P. (2003). Assembly of cell regulatory systems through protein interaction domains. *Science*, **300**: 445–52.

Puig, O., Caspary, F., Rigaut, G., et al. (2001). The tandem affinity purification (TAP) method: a general procedure of protein complex purification. *Methods*, **24**: 218–29.

Clinical applications

Petricoin, E. F., Ardekani, A. M., Hitt, B. A., et al. (2002). Use of proteomic patterns in serum to identify ovarian cancer. *Lancet*, **359**: 572–7.

Villanueva, J., Philip, J., Entenberg, D., Chaparro, C.A., Tanwar, M. K., Holland, E.C. and Tempst, P. (2004). Serum peptide profiling by magnetic particle-assisted, automated sample processing and MALDI-TOF mass spectrometry. *Analytical Chemistry*, **76**: 1560–70.

Toponomics

Bhattacharya, S., Mathew, G., Ruban, E., et al. (2010). Toponome imaging system: in situ protein network mapping in normal and cancerous colon from the same patient reveals more than five-thousand cancer specific protein clusters and their subcellular annotation by using a three symbol code. *Journal of Proteome Research*, **9**(12): 6112–25.

Cottingham, K. (2008). Human Toponome Project. *Journal of Proteome Research*, **7**: 1806.

Friedenberger, M., Bode, M., Krusche, A., and Schubert, W. (2007). Fluorescence detection of protein clusters in individual cells and tissue sections by using toponome imaging system: sample preparation and measuring procedures. *Nature Protocols*, **2**: 2285–94.

Listing, J.B. (1847). *Vorstudien zur Topologie*. Göttingen.

Sage, L. (2009). The molecular face of prostate cancer. *Journal of Proteome Research*, **8**(6): 2616.

Schubert, W. (1990). Multiple antigen-mapping microscopy of human tissue. In *Advances in Analytical Cellular Pathology*. Excerpta Medica (eds., G. Burger, G. Oberholzer, M. and Vooijs, G.P.). Elsevier, Amsterdam: 97–8.

Schubert, W. (2003). Topological proteomics, toponomics, MELK technology. *Advances in Biochemical Engineering/Biotechnology*, **83**: 189–209.

Schubert, W. (2010). On the origin of cell functions encoded in the toponome. *Journal of Biotechnology*, **149**(4): 252–9.

Schubert, W., Bonnekoh, B., Pommer, A.J., et al. (2006). Analyzing proteome topology and function by automated multidimensional fluorescence microscopy. *Nature Biotechnology*, **24**: 1270–8.

Schubert, W., Gieseler, A., Krusche, A., and Hillert, R. (2009). Toponome mapping in prostate cancer: detection of 2000 cell surface protein clusters in a single tissue section and cell type specific annotation by using a three symbol code. *Journal of Proteome Research*, **8**(6): 2696–707.

Schubert, W., Gieseler, A., Krusche, A., Serocka, P. and Hillert, R. (2012). Next-generation biomarkers based on 100-parameter functional super-resolution microscopy TIS. *New Biotechnology*, **29**(5): 599–610.

Internet resources

General background

www.wellcome.ac.uk/en/genome/index.html

This is an outstanding website with up-to-date news, interviews with scientists, and background information regarding the genome and proteome. The main sections are the genome, genes and body, genes and society, and tackling disease.

www.dnai.org

This is a beautiful website providing all the background on DNA both historically and biologically. All key scientists are discussed and placed in context in an interactive way. This DNA-interactive website is useful for both student and lecturer.

www.ygyh.org

The "Your Genes Your Health" website is a component of a DNA-interactive website. A number of genetic disorders, including cystic fibrosis, are discussed and illustrated.

www.ncbi.nlm.nih.gov/about/primer/

This up-to-date website gives an extensive primer on science and includes bioinformatics, genome mapping, microarrays, SNPs, genetics, and much more. This website gives a further background to this book chapter.

Research resources

www.geneontology.org
This database allows for searching the protein function of most known proteins. The gene ontology database archives protein information according to function. The database can be searched on protein name (e.g. RAB25, which is involved in ovarian and breast cancers). Useful for all genes discussed in this chapter and book.

www.expasy.ch
This website provides similar information on proteins as the gene ontology database, but focuses on the proteome. Searching for RAB25 gives three hits in different species. RAB25 human variant gives detailed information on this protein.

www.genome.jp/kegg/pathway.html
This is a wonderful site that contains most known pathways ranging from metabolism to cellular processes. By selecting one pathway (e.g. glycolysis), the complete pathway is shown, including all the proteins that are part of that pathway.

http://kinasedb.ontology.ims.u-Tokyo.ac.jp:8081/
This is a database that contains all information on human, mouse, and rat kinases. Interactions and pathway can be found for specific kinases.

www.abrf.org./index.cfm/dm.home
This website lists all known posttranslational modifications, including their weight. By typing in "All" in the box, all posttranslational modifications are listed.

Research institutes and organizations

www.hapmap.org
This is the official website of the HapMap Consortium. News and background information can be found on this website.

www.epigenome.org
This website provides some background on the epigenome project between the Sanger Institute, Epigenomics AG, and CNG.

http://cgap.nci.nih.gov
This website provides information on cancer research and diagnosis. The goal of this community is to determine gene expression profiles of normal, precancer, and cancer cells, leading eventually to improved detection, diagnosis, and treatment for the patient. The website is a repository for scientists interested in tools for cancer research.

www.sanger.ac.uk
This is the official website of the Sanger Institute and is one of the major contributors to genome research. A third of the human genome was sequenced at the Sanger Institute, and a lot of background information can be found on this website of previous and current research.

Questions for student review

1) What is the most common array type used in research?
   a. CGH arrays.
   b. mRNA expression array.
   c. Antibody arrays.
   d. ChIP-on-Chip.
   e. CGI arrays.

2) What is the correct flow of information in the cell?
   a. DNA, mRNA, protein, posttranslational modification, protein complexes.
   b. mRNA, DNA, protein complexes, posttranslational modification, protein.
   c. DNA, protein, mRNA, posttranslational modification, protein complexes.
   d. Protein complexes, posttranslational modification, protein, mRNA, DNA.

3) What is the most commonly used method to separate out a cellular protein extract?
   a. ICAT.
   b. Liquid chromatography.
   c. TAP tagging.
   d. Two-dimensional gel electrophoresis.

Appendix 20.1 Techniques for the generation of genetically altered mouse models of cancer

Gene knockouts (targeting) and transgenesis represent the two most direct and powerful approaches for analyzing gene function in higher organisms. These techniques allow the study of inactivation or activation of any given genes within the context of the whole organism during life. This is often essential as processes such as cellular growth, replication, and survival (central to diseases such as cancer) are critically dependent on networks of regulatory factors only accurately modeled *in vivo*. Thus, although wherever possible researchers strive to conduct studies in isolated cells or tissues, sometimes it is unavoidable to examine the consequences (phenotype) of a given gene in the animal. Both approaches have been successfully employed to derive strains of animals for the study of the molecular basis of cancer.

"Transgenesis" defines any process that involves the transfer of a gene from one species to another. However, the term is often used in order to describe the insertion in the mouse genome of modified mouse or human genes in order to dissect gene function.

Gene knockouts

The most widely known method for generating an abnormality in a mouse gene to establish its function is to make a "knockout" (i.e. using molecular biological techniques to make mice without a working copy of the gene). Despite the undoubted power of this technique, it is not without potential problems. Obviously, this approach can be used only when there is preliminary evidence that a particular known gene might be relevant to a particular known disease process. Moreover, functional redundancy (i.e. where the function of a gene can be replaced by the action of another gene) may prevent any obvious changes in phenotype when the gene under study is knocked out. This is particularly likely if the gene under study is knocked out throughout the ontogeny of the organism – where the knockout may even prove lethal if no "salvage" pathways can be invoked. Obviously, this may result in erroneous conclusions as to the role of a given gene in the adult. Similar problems may also pertain to transgenic mice, and for this reason considerable efforts are being devoted to refining mouse mutant technologies, such as developing regulatable systems which enable a gene (or its protein product) to be activated or knocked out in both tissue-specific and time-dependent manners.

Transgenesis

Transgenes can be directed to the germline by direct microinjection of the DNA encoding the gene under study in the pronucleus of a fertilized egg or by using embryonic stem (ES) cell technology.

A wide variety of vectors have been employed for transgenic studies including:

1. Regulatory sequences (e.g. promoter enhancer elements) of the gene under study are used to direct the expression of a reporter gene, for example the gene-encoding bacterial b-galactosidase for which simple and relatively sensitive histochemical detection exists. These studies can establish directly *in vivo* the stages and the tissue domains in which the gene under study is expressed.
2. By using several well-characterized promoters in order to drive the expression of the gene under study and alter the physiological pattern of gene expression. Typically these experiments lead to overexpression and/or mis-expression of the gene, and they enable the analysis of the ensuing phenotype. In a significant development of the latter approach, inducible or lineage-restricted promoters can be used in order to achieve further control of the pattern of expression of the gene.

The pronuclei of fertilized mouse eggs can be most easily injected at the one-cell stage when they are at their maximum size. For most strains of mice, this is between 24 and 28 hours after mating. Females used for the generation of fertilized eggs for pronuclear injection are typically induced to "superovulate," that is, to ovulate a larger number of eggs than normal (similar to techniques used for in vitro fertilization (IVF) in humans). Superovulation is achieved by sequential administration of follicle-stimulating hormone (FSH) and then luteinizing hormone (LH), or suitable analogues.

For microinjection, eggs are recovered from the oviduct and treated with hyaluronidase in order to free the egg from the cumulus cells, and then are transferred to a Petri dish where they are injected. Either of the two pronuclei can be injected. However, the male pronucleus is larger and is injected in preference. After injection, eggs are either transferred directly into the oviduct of a 0.5 dpc foster mother or cultured *in vitro* to the two-cell stage (overnight) and transferred the following morning.

The generation of targeted mutation in mammalian genomes that can be transmitted through the germline (and the effect of which can be assessed at the level of the whole organism) has been facilitated by the development of techniques for use of embryonic stem (ES) cells. ES cells were first established in culture in 1981 and within a decade had been employed for germline transmission of retroviral sequences and for germline transmission of ES cell mutations at various loci. Soon researchers had succeeded in achieving targeted mutations by homologous recombination in ES cells at various loci, including the loci for HPRT and β2-microglobulin. To date, several hundred mouse genes have been targeted in ES cells providing a unique tool for the dissection of gene function in the intact organism *in vivo*. Most of the ES cell lines currently in use are derived from the 129 mouse strain (129/Sv), which is homozygous wild type at the agouti locus. In order to generate chimeric mice, the 129 cells are injected in non-agouti blastocysts, for example from the C57BL/6 strain. In this way, the agouti and non-agouti offspring can be distinguished by eye at approximately 1 week.

ES cells are derived from blastocysts collected from the uterine horns and are cultured on "feeder" layers of immature fibroblasts. These feeder cells are prevented from growing by irradiation or treatment with mitomycin C (an inhibitor of DNA synthesis). After 1–2 days in culture the cells from the inner cell mass start to divide, and after another day or two the clump of cells derived from the inner cell mass is disaggregated with trypsin–EDTA and transferred to a new dish. Colonies are inspected, and those with stem cell–like morphology are recovered and propagated with regular subculturing in medium that minimizes the differentiation of the cells.

In most instances, gene targeting in ES cells by homologous recombination relies on some knowledge of the gene under study, and gene targeting is achieved by introducing a targeting vector in the form of naked DNA in the ES cells by electroporation.

The simplest targeting vectors consist of two gene fragments (of at least several hundred, and typically of several thousand, nucleotides) separated by an unrelated sequence that may replace the corresponding gene sequence through a double crossover event. It is essential that the unrelated sequence contains a genetic marker (e.g. the gene-encoding neomycin phosphotransferase) that, once expressed, confers resistance to neomycin or neomycin analogues in order to select for vector integration (whether random or homologous).

For blastocyst injection with targeted ES cells, normal 3.5 dpc blastocysts are collected and cultured briefly to promote expansion. For micro-injection, individual blastocysts are collected on a holding pipette and positioned such that the inner mass is located away from the injection needle. ES cells (10–15) are injected in the blastocoel cavity prior to re-implantation.

Regulatable systems

Controlled or inducible mis-expression of genes is a new and powerful tool for analyzing gene function. Various systems are now available for regulating gene expression in a temporally as well as spatially controlled manner. Genetically engineered mice are invaluable biological tools in biomedical research. The development of both transgenic and targeted mutant mice has allowed researchers to study gene function in the context of a whole mammalian organism. These studies, however, have limitations. For example, the deletion of a gene required during embryonic development often results in embryonic or perinatal lethality, making it impossible to study the effects of the gene ablation at later developmental ages. In addition, unregulated overexpression of transgenic gene products may have unwanted physiological or toxic effects. The development of inducible expression systems has allowed researchers to overcome some of the problems associated with transgenic and targeted mutagenesis studies. The two best-developed systems for this are the tetracycline-regulated systems developed by Hermann Bujard (University of Heidelberg) and the ER[TAM] system first described *in vitro* and more recently adapted to mouse transgenesis *in vivo*.

Tet-regulatable gene expression

The Tet-Off and Tet-On expression systems are binary transgenic systems in which expression from a target transgene is dependent on the activity of an inducible transcriptional activator. In both the Tet-Off and Tet-On systems, expression of the transcriptional activator can be regulated both reversibly and quantitatively by exposing the transgenic animals to varying concentrations of tetracycline or derivatives such as doxycycline (Dox).

The Tet-Off and Tet-On systems are complementary, and the decision to choose one over the other depends on the particular experimental strategy. In the Tet-Off system, transcription is inactive in the presence of doxycycline; conversely, in the Tet-On system, transcription is active in the presence of doxycycline.

In the Tet-Off expression system, a tetracycline-controlled transactivator protein (tTA), which is composed of the Tet repressor DNA-binding protein (TetR) from the Tc resistance operon of *E. coli* transposon Tn10 fused to the strong transactivating domain of VP16 from herpes simplex virus, regulates expression of a target gene that is under transcriptional control of a tetracycline-responsive promoter element (TRE). The TRE is made up of Tet operator (tetO) sequences fused to a minimal promoter (commonly the minimal promoter sequence derived from the human cytomegalovirus (hCMV) immediate-early promoter). In the absence of doxycycline, tTA binds to the TRE and activates transcription of the target gene, whereas in the presence of doxycycline, tTA cannot bind to the TRE and expression from the target gene remains inactive.

The Tet-On system is based on a reverse tetracycline-controlled transactivator (rtTA). Like tTA, rtTA is a fusion protein composed of the TetR repressor and the VP16 transactivation domain; however, a four amino acid change in the TetR DNA-binding moiety alters rtTA's binding characteristics such that it can recognize only the tetO sequences in the TRE of the target transgene in the presence of doxycycline. Thus, in the Tet-On system, transcription of the TRE-regulated target gene is stimulated by rtTA only in the presence of doxycycline.

The two-vector nature of the Tet-Off and Tet On systems allows tissue-specific promoters to drive tTA or rtTA expression, resulting in tissue-specific expression of the TRE-regulated target transgene. Further, the ability to strictly regulate the level of rtTA and tTA activity allows the investigator to regulate activation of the target gene both quantitatively and temporally. Successful applications of both the repressible and activatible systems have been achieved in mice and have provided important information on tumor formation and regression in particular. Problems have been reported including mosaic induction, background leakiness (particularly with the rtTA system), or no detectable expression of the transactivator. Many of the issues may be resolved by use of optimum promoters to drive expression of the tTA and characterization of suitable responder lines. An additional problem is the lengthy lag phase in the *in vivo* response of the system following drug administration or removal.

Transgenes encoding regulatable chimeric proteins

The general concept of utilizing the well-studied ligand-binding properties of the steroid hormone receptors in the design of inducible systems has proved highly successful. This approach offers several advantages over the "Tet" system, notably a much speedier regulation of "activity" of the transgenic protein.

Several groups have generated transgenic mice in which a transgene encodes a fusion (chimeric) protein between the ligand-binding domain of a modified estrogen receptor (now only responsive to synthetic estrogenic compounds such as tamoxifen and 4-hydroxy tamoxifen, but not to endogenous estrogens) and various proteins, which the researcher may wish to regulate. To date, the most prominent successes have been in regulating activity of c-Myc in various cancer models and in regulating the activity of a Cre recombinase. In the latter case, this potentially allows targeted gene knockouts in a time-dependent as well as tissue-dependent manner. In general, these fusion proteins are inactive except in the presence of the ligand, most likely because the ligand-binding domain is complexed with HSP90 proteins in the cytoplasm (or nucleus), preventing c-Myc or Cre recombinase from reaching their targets.

Regulatable mouse models of tumorigenesis using a "switchable" form of the c-Myc protein, MYCER[TAM]

The *c-MYCER*[TAM] construct encodes a chimeric protein composed of human c-MYC fused at its C terminus to the hormone-binding domain of a mutant murine eostrogen receptor. Directed expression of the chimeric protein to the suprabasal epidermis or pancreatic islet ß cells was achieved by placing *c-MYCER*[TAM] cDNA under the control of an involucrin promoter or an insulin promoter, respectively. Although c-MYCER[TAM] protein is expressed constitutively in these tissue compartments, in the absence of the specific ligand, 4-hydroxytamoxifen (4-OHT), c-MYCER[TAM] remains inactive through its association with HSP90. 4-OHT administration leads to HSP90 dissociation and activation of c-MYCER[TAM] as the protein is now able to bind its partner Max and become transcriptionally active.

Hit and run

Many mouse models of cancer have been accused of not accurately recapitulating the kinds of spontaneous mutations that characterize most human cancers – neither embryonic insertion nor tissue-specific transgenic mice can usually address this. What is required is a system whereby individual cells will acquire an oncogenic mutation, but are otherwise surrounded by normal cells.

One strategy for achieving this was used by Tyler Jacks and colleagues, who designed what they have termed a genetic "Trojan horse." Simply, they introduced latent *K-ras* genes into mice. The genes were inactivated because they had duplicated segments of DNA that prevented the genes from activating themselves. However, as the mice developed, individual cells undergo rare spontaneous, sporadic recombination events that deleted one copy of the duplicated sequence, such that the *K-ras* gene was activated and began to initiate tumor development. This technique is a variant of a "hit-and-run" gene-targeting technology developed by HHMI investigator Allan Bradley at Baylor College of Medicine. The two-part technique consists of "hitting" cells with an inserted mutated gene and then allowing the recombination event to "run," activating the inserted gene.

Glossary

5HIAA: 5-Hydroxyindoleacetic acid.

5HT: 5-Hydroxytryptamine.

ABL1: Protein tyrosine kinase; cellular homolog of the Abelson leukemia virus which can become oncogenic.

ACTH: Adrenocorticotropic hormone.

Adaptive response: Immune response which provides longer lasting, antigen-specific immunity to the host.

ADCC (antibody-dependent cell-mediated cytotoxicity): Part of an adaptive immune response in which antibodies binding surface proteins on foreign/tumor cells can kill by bridging targets with cytotoxic cells.

Adenocarcinomas: Malignant adenomas (tumors originating in glandular or secretory tissues).

Adenomas: Benign tumors originating in glandular or secretory tissues.

Adenoviruses: These viruses have never been associated with cancer in humans, but they transform cultured cells and cause cancer in animals. At least two gene products have been implicated in transformation studies (E1A and E1B).

ADH: Antidiuretic hormone.

Adhesomes: Membrane-associated adhesion complexes consisting of a large number of adaptor proteins, cytoskeletal elements, and enzymes. Adhesomes are multifunctional machines, which are simultaneously sites of interaction between the cell and ECM, anchor points for the cytoskeleton, platforms for intracellular signaling, and the engines of cell migration.

Adjuvant chemotherapy: Chemotherapy given after surgery in order to reduce risk of recurrence or spread by eradicating cancer cells that may already have escaped the primary site.

AFP: α-Fetoprotein.

AICD: Activation-induced cell death.

AKT: Protein kinase B, mediates survival signals downstream of RAS–PI3K.

Alemtuzumab (Campath): Monoclonal antibody that targets CD52 on the surface of B lymphocytes, which are then destroyed. It is used to treat chronic lymphocytic leukemia.

Alkylating agent: Chemotherapeutic that adds alkyl groups to the DNA of tumor and other replicating cells, triggering DNA damage and apoptosis.

ALL: Acute lymphoblastic leukemia.

Allele: One of a set of alternative forms of a gene in a diploid cell one will be on each of the two homologous chromosomes.

ALT: Alanine transaminase.

ALT: Alternative lengthening of telomeres, without telomerase.

Ames test: A widely used test for determining if a chemical is a mutagen. Named after its developer, Bruce Ames.

AML: Acute myeloid leukemia.

Anaphase: Stage in mitosis when duplicated pairs of chromosomes separate and move apart.

Aneuploidy: Deviation from the normal karyotype of a cell; loss or gain of chromosomes or segments of chromosomes often seen in cancer cells. Most solid cancers are aneuploid.

Angioblast: A primordial mesenchymal cell type from which embryonic blood cells and vascular endothelium differentiate.

Angiogenesis: Formation of new blood vessels from preexisting vessels by sprouting from existing blood vessels.

Angiogenic switch: A term derived from studies in mice postulating that the transition from a normally quiescent blood vessel to a proliferative and invasive endothelial cell sprout during tumor growth is due to an imbalance between stimulators and inhibitors of angiogenesis.

Angiography: Used to accurately delineate the vasculature of the organ under study. Most commonly used in the diagnostic evaluation of liver, renal, or pancreatic tumors.

Angiopoietins: Ligands for the endothelial-specific receptor tyrosine kinase TIE2 thought to act in a complementary and coordinated fashion with VEGF. Four angiopoietin ligands (termed ANG1 to ANG4) are known, and together they comprise a growth factor family that consists of both receptor activators and receptor blockers: while ANG1 and ANG4 act as TIE2 agonists, ANG2 and ANG3 behave as antagonists.

Angiostatin: An endogenous inhibitor of angiogenesis that acts either locally to modify vessel patterning or systemically to suppress new vessel growth. Similar to endostatin. It is also secreted by the primary tumor to suppress tumor metastasis.

Anoikis: Specific form of cell death engendered by depriving an epithelial cell of matrix attachment.

Antigen-induced cell death: Excess antigen density can lead to effector cell death following TCR engagement in the absence of an adequate costimulatory (i.e. CD28) signal.

APAF-1 (apoptotic protease-activating factor 1): An important cytosolic protein that binds to cytochrome *c* when apoptosis is triggered within a cell, leading to the formation of the apoptosome (see **Apoptosome**).

APC (adenomatous polyposis coli): Tumor suppressor often associated with colorectal cancer.

APC/C (anaphase-promoting complex/cyclosome): Ubiquitin ligase complex needed for degradation of cyclin B and securin for anaphase entry and exit from mitosis.

The Molecular Biology of Cancer: A Bridge From Bench to Bedside, Second Edition. Edited by Stella Pelengaris and Michael Khan.
© 2013 John Wiley & Sons, Inc. Published 2013 by John Wiley & Sons, Inc.

Glossary

APCs (antigen-presenting cells): Cells recruited and activated as part of an innate immune response. They collect and process foreign antigens (including some tumor fragments), presenting then to antigen-specific CD4 and CD8 T cells, thereby initiating an adaptive immune response.

Apoptosis: A specialized form of programmed cell death that serves as a natural barrier to cancer development. An active, energy-requiring process leading to a characteristic series of morphological changes (e.g. chromatin condensation, DNA fragmentation, cell shrinkage) that accompany the degradation of the cell brought about by caspases. Unlike necrotic cell death, apoptosis does not trigger an inflammatory response. There are two forms of apoptosis – "extrinsic" and "intrinsic."

Apoptosome: A complex that plays a crucial part in apoptosis. Once apoptosis is triggered within a cell, cytochrome c is released from mitochondria into the cytosol where it associates with APAF-1 to create the apoptosome. Activation of procaspase-9 on the apoptosome leads to activation via cleavage of downstream caspases (e.g. caspase-3) and the ultimate demise of the cell.

ARF (alternative reading frame) (also known as p19^{ARF} in rodents; p14^{ARF} in humans): Tumor suppressor protein encoded by a transcript that is read from the same exon as that encoding INK4a, but in an alternative reading frame. ARF acts in a checkpoint that guards against unscheduled cellular proliferation in response to oncogenic signaling. ARF may contribute to apoptosis (e.g. induced by oncogene *MYC*) indirectly via activation of p53 (by sequestering MDM2), or may engage apoptosis directly and independently of p53.

Arimidex (anastrazole): Aromatase inhibitor used to treat breast cancer.

AT (ataxia telangiectasia): A syndrome of cancer susceptibility, immune dysfunction, and neurodegeneration caused by mutations in the AT-mutated (*ATM*) gene.

Athymic nude mice: A mutant strain of mice with a genetic defect that affects the development of the thymus and hair follicles. These mice lack a thymus and therefore generally fail to develop T cells.

ATR: ATM- and Rad3-related.

Autocrine: Extracellular signals (e.g. those regulating replication or growth) released by the same cell on which those signals then elicit a response. For example the production of IGF by tumor cells then acting on those same tumor cells to support survival or growth.

Autophagy: A form of programmed cell death, meaning "eating of self," recently postulated as a barrier to cancer development. Paradoxically, however, autophagy can also enable cancer cells to survive a variety of stresses by temporarily degrading cell organelles in lysosomes and re-utilizing certain components until the stress is removed.

Avastin: see **Bevacizumab**.

B cells: Lymphocytes which provide "humoral" immunity through the production of soluble antibodies.

Barrett's esophagus: A precursor of esophageal cancer typified by dysplasia in which the squamous epithelium normally found in the lower esophagus changes gradually into a columnar type usually found in the lower intestine. This may be an adaptive response to prolonged exposure to refluxing stomach contents, but carries a high risk of progressing to cancer and for that reason is monitored by endoscopy.

Basement membrane: This matrix is a continuous and flexible part of the basal surface of all epithelia, which is 40–120 nm thick and forms the interaction zone with stromal extracellular matrix.

BAX family (BAX, BAK, BOK): These are proapoptotic members of the BCL-2 family possessing three BH (BCL-2 homology) domains (BH1 to BH3) that are similar in sequence to those in BCL-2, and are thus sometimes referred to as **BH123** or **multidomain proteins**. Members include BAX and BAK, key effectors of MOMP, which act in mitochondrial disruption, after which there is no point of return in mammalian apoptosis.

BCC: Basal cell carcinoma; also known as rodent ulcer.

BCG (Bacillus Calmette–Guerin): A proinflammatory mixture of killed mycobacteria (the organism responsible for tuberculosis) that acts as an adjuvant. When given to patients by injection, it promotes T-cell and macrophage activation.

BCL-2 (B-cell lymphoma 2): Human follicular lymphoma involves a chromosome translocation event that moves the *BCL2* gene from chromosome 18 to 14 (t14;18), linking the *BCL2* gene to an immunoglobulin locus.

BCL-2 family: Includes a large number of proteins with either anti- or proapoptotic activities, the balance of which determines whether or not a cell commits suicide – apoptosis. BCL-2 (B-cell lymphoma) protein is antiapoptotic and was the first member identified. It is localized to the inner mitochondrial membrane, endoplasmic reticulum, and nuclear membrane.

BCNU: Bis-colorethyl nitrosourea.

Benign: Tumor cells confined locally.

BER (base excision repair): Repair process for single DNA bases that have undergone damage (by several different chemical reactions, including deamination, oxidation, and alkylation) that compromises base-pairing.

Bevacizumab (Avastin; Genentech): First commercially available inhibitor of VEGF and thereby of angiogenesis. It is a genetically engineered humanized monoclonal antibody used as a first-line treatment of metastatic colorectal cancer. It has since gained FDA approval for second-line treatment of metastatic colorectal cancer, first-line treatment of NSCLC, metastatic HER2-negative breast cancer, second-line treatment of glioblastoma and metastatic renal cell carcinoma.

BH3-only family (BID, BIM, BIK, BAD, BMF, HRK, NOXA, PUMA): These are proapoptotic members of the BCL-2 family, that have only the short BH3 (BCL-2 homology) motif – an interaction domain that is both necessary and sufficient for their killing action. These proteins are direct antagonists of the anti-apoptotic members, and may also serve at least in part to activate BAX and BAK.

Biomarkers (or tumor markers): Some tumors produce proteins that can be detected in the blood and can be used to indicate the presence of a malignancy. The value of different tumor markers varies; they can be useful in diagnosis, predicting prognosis or response to treatment and aiding follow-up of a patient. Tumor markers can be broadly divided into three groups: oncofetal proteins, cancer-related antigens, and hormones. Examples are PSA and CEA.

Bleomycin: A glycopeptide antibiotic produced by *Streptomyces verticillus* that is used to treat lymphomas as well as testicular and head and neck cancers. It binds to DNA and generates free radicals, resulting in strand breaks.

BORIS: A CCCTC-binding factor (CTCF) antagonist; cancer cells have been shown to express increased levels of BORIS.

Bortezomib: First proteasome inhibitor to enter clinical practice, approved for the treatment of myeloma in 2003. The drug is a tripeptide boronic acid analog, which can bind to and inhibit the catalytic site of the 26S proteasome. Theoretically, proteasome inhibitors could block the degradation of cancer-restraining proteins such as proapoptotic factors and tumor suppressors.

Bosutinib: Several compounds targeting the SRC kinase domain have been described, including bosutinib (previously SKI-606; Wyeth) and SU6656 (Sugen), and Ariad Pharmaceuticals (AP23464) and AP23451 (a bone-tissue-targeted SRC kinase inhibitor developed for osteoporosis therapy).

BRCA: Tumor suppressor proteins that respond to DNA damage. They include the BRCA-1 and BRCA-2 proteins, which are recruited to DNA during homologous recombination and are found mutated in familial breast and ovarian cancers.

Bronchoscopy: Examination that involves passing a fiber-optic instrument through the nose or mouth into the pulmonary tree, lobar and proximal bronchi but is not usually helpful in assessing more peripheral lesions. The abnormal area can be visualized and either washed out (and the washings sent for cytological examination) or biopsied.

BS (Bloom's syndrome): Associated with cancer predisposition and genomic instability. A defining feature of Bloom's syndrome is an elevated frequency of sister chromatid exchanges. These arise from crossing over of chromatid arms during homologous recombination.

Burkitt lymphoma: An aggressive form of B-cell lymphoma commonly affecting children. The most common variant is the t(8;14) chromosomal translocation which involves translocation of the oncogene c-*Myc* to the IGH (immunoglobin heavy chain) locus, resulting in overexpression of c-*Myc* oncogene.

BWS (Beckwith–Wiedemann syndrome): Imprinting was first identified as disease-relevant in humans in BWS, which is characterized by multi-organ overgrowth and predisposition to embryonal tumors such as Wilms tumor.

CA-125: A protein biomarker that has been used in screening (produced by the mucin 16 gene) for ovarian cancer. However, CA-125 also has a number of problems that preclude its use in screening the general population because it is raised by a number of nonmalignant conditions and may be falsely positive. Despite these limitations, CA-125 may be used as an adjunct to imaging in women who have symptoms consistent with ovarian cancer.

Cachexia: Loss of body mass frequently seen in patients with chronic debilitating diseases.

Cadherins: Transmembrane proteins that play a major role in connecting adjacent cells. Cadherins not only bind to each other at their distal N-terminal ends across the intercellular space (thereby connecting adjacent cells) but also form lateral interactions between neighboring molecules on the same cell.

Caenorhabditis elegans: Nematode worm, established as a model organisms that has been extensively used to study and understand biology.

CAMs (cell adhesion molecules): Large family of transmembrane glycoproteins that contain immunoglobulin repeats within their external domains, and are calcium-independent for adhesion. Several types of CAM are important for cell signaling because they bind to GF receptors.

Capecitabine (Xeloda): Orally administered prodrug, converted enzymatically to 5-FU in the tumor. It is used as an adjuvant in colorectal cancer and in advanced breast and colorectal cancers.

Carboplatin: See also cisplatin. Variant on heavy metal-based DNA-crosslinking drugs, used in ovarian and lung cancers.

Caspases: Cysteine proteases that cleave hundreds of cellular proteins and ultimately lead to a series of morphological changes characteristic of apoptotic cell death. There are two classes of caspases: "initiator" caspases and "effector" (or executioner) caspases. Initiator caspases (e.g. caspase-2, -8, and -9) are activated as they bind to their appropriate adaptor molecules (APAF-1, FADD), after which they cleave and thereby activate downstream ("effector") procaspases and various proteins. The downstream "effector" (or executioner) caspases (e.g. caspase-3, -6, and -7) go on to degrade many cellular substrates that result in the characteristic features of apoptotic cell death.

CCNU (lomustine): cis-Chloroethyl nitrosourea; an alkylating agent used in brain tumors and lymphoma treatment.

CD4 helper T cells: Lymphocytes that recognize antigen in the context of MHC class II.

CD8 T cells: Cytotoxic or killer T lymphocytes that recognize antigen in the context of MHC class I and, when activated, release perforin and toxic granules that mediate direct cell killing by punching holes in the cell membrane to facilitate entry of enzymatic packets (granzymes A and B).

CDK (cyclin-dependent kinases): Family of protein kinases with an important role, with cyclins, in regulation of the cell cycle.

Cdk–cyclin-dependent kinases: Kinases that phosphorylate protein substrates on serine and threonine residues that form complexes with, and are activated by binding to, cyclins.

CEA: Carcinoembryonic antigen; a tumor biomarker elevated in serum of patients with a variety of cancers, including colon, where it was first identified.

Cell cycle: The mitotic cell cycle is composed of four phases: two gap phases: G_1 when the cells respond to the environment and make the decision to proceed through mitosis and G_2, a period between the DNA synthesis phase S and mitosis M, where cells check the integrity of their DNA before proceeding through division (M).

Cetuximab (Erbitux): Recombinant human/mouse chimeric monoclonal antibody used in the treatment of EGFR-expressing metastatic colorectal carcinoma. Cetuximab binds specifically to the extracellular domain of the human EGFR and inhibits the binding of EGF and other ligands to EGFR. This prevents downstream signal transduction, and targets bound cells for immune attack.

CGH (comparative genome hybridization): Cytological technique for detecting extreme genetic aberrations in cancer cells (loss or gain of chromosomal regions).

CGH arrays: Advanced technology compared to conventional CGH due to its much higher resolution and speed. Genomic mutations, unbalanced translocations, deletions, and duplications can be detected throughout the entire genome.

CGI arrays: CpG island microarrays are used to identify DNA methylation sites in a genome.

CHART: Continuous hyperfractionated accelerated radiotherapy.

Checkpoint: A point in the cell cycle during which cells pause and wait until the conditions are favorable for the cells to proceed through the next phase of the cell cycle.

Chemokine: Chemotactic factors that influence the migration of immune cells through interaction with specific receptors. They recruit immune cells to sites of inflammation and regulate traffic through lymphoid tissues. CCL-19 binds to the chemokine receptor CCR7 and is responsible for lymphocyte and dendritic cell migration to lymph nodes. CXCL8 (or IL-8) recruits neutrophils to sites of inflammation through interaction with the CXCR1 chemokine receptor.

Chemotherapy: The use of chemicals to treat diseases, but has become synonymous with cancer drug therapy.

ChIP (chromatin immunoprecipitation): Technique used to isolate/identify specific regions of DNA, such as promoters, that bind a protein of interest. Chromatin is immunoprecipitated using specific antibodies raised against transcription factors or methyl-CpG-binding proteins.

ChIP-on-chip: Research tool that combines chromatin immunoprecipitation (ChIP) and DNA arrays or CGI arrays. The ChIP-on-chip approach is used to identify new promoter regions in genomes, but can equally be used to identify differences in transcriptional silencing of genes by methyl-CpG-binding proteins in cancer or healthy cells.

CHOP: Combination regime of cyclophosphamide, doxorubicin, vincristine, and prednisolone (a corticosteroid) used to treat Non-Hodgkin lymphoma.

Chromatin: DNA does not exist naked within the cell, but in association with histones, which constitute a protein scaffold that gives form to the complex tertiary structure referred to as chromatin.

CIMP (CpG island methylator phenotype): Hypermethylation of usually unmethylated regions, such as the CpG islands, are found in up to half of all human gene promoters, resulting in aberrant silencing of hundreds of genes.

CIN: Cervical intraepithelial neoplasia.

CIN: Chromosomal instability recognized by gross chromosomal abnormalities.

CIP/KIP family: Important family of cyclin-dependent kinase inhibitors (CKI) that negatively regulate cyclin E– and cyclin A–CDK2, and cyclin B–CDK1 complexes. Members of the family include: p21^{CIP1}, also referred to as p21$^{CIP1/WAF1}$ (p21^{Cip1} non-human), which is encoded by the *CIP1(WAF1)* and *Cip1* genes, respectively. Importantly, p21^{CIP1} is one of the major transcriptional targets of p53. p27^{KIP1} (p27^{Kip1} non-human) is encoded by the *KIP1* (human) and *Kip1* (non-human) genes, respectively, and is a key inhibitor of the G_1/S transition and acts downstream of various inhibitory growth factors such as TGF-β. p57^{KIP2} (p57^{Kip2} non-human) is encoded by the *KIP2* (human) or *Kip2* (non-human) genes, respectively.

Circulating endothelial cell progenitor cells: The bone marrow of adults contains progenitor cells that have the capacity to differentiate into mature endothelial cells.

Glossary

Cisplatin: The prototype platinum-containing anticancer drug, a group including oxaliplatin and carboplatin. It is a DNA-binding and crosslinking agent that causes similar DNA damage to alkylating drugs, eliciting DNA damage and apoptosis. It is used in combinations to treat numerous different cancers, including lymphoma, small cell lung and ovarian cancers, and sarcoma. Resistance occurs largely by drug extrusion or enhanced DNA repair in tumor cells.

CKI (CDKI also used) (cyclin kinase inhibitor): Small polypeptide proteins that bind to and inhibit cyclin–cdk holoenzymes. They can inhibit cell cycle entry and progression in response to various checkpoints. They include two main families, the **INK4 family** and **CIP/KIP family**.

CLL: Chronic lymphocytic leukemia.

CML: Chronic myelogenous leukemia (often associated with the Philadephia chromosome).

COX2 (cyclooxygenase 2): Proinflammatory protein, involved in tumor angiogenesis. Inhibitors of this enzyme are being explored as chemopreventive agents in high-risk cancer syndromes such as familial adenomatous polyposis coli, where the potential risk of cardiovascular complications is deemed acceptable.

CpG islands: Dinucleotide sequence, cytosine–phosphate diester–guanine CG–nucleotide-rich areas of DNA often found in gene promoter regions. The enzyme DNA methyltransferase methylates DNA especially in CpG islands, giving condensed chromatin which is not accessible to transcription factors and leading to gene silencing.

CRT: Conformal radiation therapy.

CSCs (cancer stem cells): A minority side population of tumor cells, originally found in leukemias, that can reconstitute a new tumor comprising all the cell types present in the original cancer. CSCs are increasingly thought to be the main malignant cells among often heterogeneous cell populations in many cancers and attempts are underway to target them specifically in cancer treatments.

CT: Computerized tomography.

CTCF (CCCTC-binding factor): An important regulator of expression of imprinted genes.

CTCs: Circulating tumor cells are found in many patients with solid tumors and can be quantified, and in the future maybe also be functionally characterized as a potential biomarker.

CTLA-4 (cytotoxic T lymphocyte–associated antigen 4): Delivers a negative regulatory signal to activated T cells with the overall effect of blocking cell cycle progression. Provides an immunologic "brake" to prevent overly robust and potentially damaging overstimulation.

Cyclins: Proteins that exhibit expression cycles throughout the mitotic cell cycle, hence their name. Together with cyclin-dependent kinases, with which they form active holoenzyme complexes, they drive many of the key processes of the cell cycle. Cyclin D1 in human is encoded by the *CCND1* gene, cyclin D2 by *CCND2*.

CYP: Human cytochrome P450 enzymes play a key role in the metabolism of drugs and environmental chemicals.

Cytarabine (AraC): Antimetabolite with similarity to cytosine, but with altered sugar group, allowing it to incorporate into newly synthesized DNA, where it causes DNA damage and cell-cycle arrest/apoptosis. It is used in acute myeloid leukemia and Non-Hodgkin lymphoma.

Cytochrome c: Essential protein in energy production within cells usually located inside the mitochondria (in the space between the inner and outer mitochondrial membranes). Cytochrome c was found to play an important role in activating apoptosis when released from the mitochondria.

Cytokine: Soluble factors that generally influence the growth and function of immune cells through interaction with specific cytokine receptors on the cell surface. Cytokines such as interleukin -2 (IL-2), IL-7, or IL-15 can induce cellular proliferation. Other cytokines modulate T-cell development (e.g. IL-12 and interferon-γ promote the differentiation of T cells into becoming T-helper-1 (Th1) cells characterized by the ability to release IL-2 upon activation. Cytokines that act specifically on lymphocytes are called lymphokines.

DCC: "Deleted in colorectal carcinoma" tumor suppressor.

DCE-CT: Dynamic contrast-enhanced CT (computerized tomography).

DCE-MRI: Dynamic contrast-enhanced MRI (magnetic resonance imaging).

Dendritic cells: APCs that collect foreign fragments/proteins and process/present them to antigen-specific CD4 and CD8 T cells, leading to initiation of an adaptive immune response.

DIABLO/Smac (second mitochondria derived activator of caspase): These proteins are inhibitors of IAPs (see **IAPs**) released from the mitochondrial intermembrane space during apoptosis, alongside cytochrome c. They circumvent the action of the apoptosis-inhibiting IAPs and thus promote apoptotic cell death. Other IAP inhibitors include Omi/HtrA2.

Differentiation: Process by which cells become irreversibly committed to a specialized cell type.

DISC (death-inducing signaling complex): Complex associated with the "extrinsic" cell death pathway (see **FAS** and **FASL**).

DLC1 (deleted in liver cancer 1): RHO GAP that has a central role as a tumor suppressor in several different cancer types.

DMR (differential methylated region): Specific localized methylated areas of DNA that are characteristic of imprinted genes.

DNA (deoxyribonucleic acid): Formed by covalently linked deoxyribonucleotides that forms the genetic code; carries the genetic information.

DNA damage checkpoints: Enable cells to check the integrity of the chromosomes in the first gap phase (G_1) and the second gap phase (G_2) of the cell cycle before proceeding into DNA replication (S phase) or division (M phase). In the absence of these checkpoints, as a result of the loss of tumor suppressor function, the damage is not repaired and cells accumulate mutations that ultimately lead to cancer.

DNA damage repair (DDR) mechanisms: The simplest mechanism of DDR employs glycosylases and other enzymes to correct aberrant methylation/deamidation. The remaining require excision of a damaged region and then repair and include base excision repair (BER), nucleotide excision repair (NER), and mismatch repair (MMR). Double-strand breaks (DSBs) are difficult to repair and require additional mechanisms because there is no intact template strand from which to restore a correct DNA sequence.

DNA polymerases: DNA-synthesizing enzymes. The mammalian genome encodes at least 15 different DNA polymerases, all of which appear to have particular roles in either normal genome replication (polymerases α, δ, and ε) or in repair of damaged DNA.

DNMTs (DNA methyltransferases): The main enzymes regulating methylation. They can add methyl groups to the 5′ position of cytosine rings within CpG dinucleotides.

Docetaxel: One of the taxane drugs, first isolated from the yew tree, although this one is synthetic. Blocks mitosis and is used as treatment in many cancers including, breast, non-small-cell lung carcinoma, prostate, and stomach.

Donor lymphocyte infusions: To treat for leukemic relapse in the recipient of an allogeneic stem cell transplant, lymphocytes are collected from the original stem cell donor (by a process called leukapheresis) and infused into the recipient to induce a graft versus leukemia response. In a graft versus leukemia response, the donor lymphocytes kill host leukemic cells following recognition of minor antigens on the host leukemic cells.

Doxorubicin (Adriamycin): Inhibits the enzyme topo-isomerase 2 and is used to treat a wide range of cancers, including, breast, ovary, bladder, sarcoma, and lymphoma. Prevents correction of topological problems in DNA during replication, causing DNA double-strand breaks.

Drosophila melanogaster: Fruit fly, established as a model organism that has been extensively used to study and understand biology.

DSBs: DNA double-strand breaks.

DTIC: Dimethyltriazenoimidazole carboxamide (dacarbazine); an alkylating agent used to treat lymphoma, sarcoma, and melanoma.

E2F family: Important family of transcription factors; E2F1, 2, and 3a proteins and their DP partners activate the transcription of genes required for the cell to commit and proceed to the DNA synthetic (S) phase of the cell cycle. Some members can also promote apoptosis.

ECM (extracellular matrix): The connective tissue lying outside and in between cells in solid structures. It includes the basement membrane (sometimes referred to as the basal lamina).

EGF: Epidermal growth factor.

EGFR: Epidermal growth factor receptor.

EMT (epithelial–mesenchymal transition): Acquisition of features associated with tumor cell metastatic behavior.

Endocrine: Extracellular signals (e.g. those regulating replication or growth) which operate at a distance usually via the circulation. Such factors are termed hormones and include insulin produced by the pancreatic beta cell acting on distant muscle or liver cells.

Endoscopy: Upper GI endoscopy is used to assess the esophagus, stomach, and first part of the duodenum. It involves the passage of fiberoptic instruments through the esophagus into the stomach. Abnormal areas can be visualized and biopsies taken. Small lesions that are confined to the mucosa of the esophagus or stomach can be removed using endoscopic mucosal resection (EMR).

Endostatin: An endogenous inhibitor of angiogenesis that acts either locally to modify vessel patterning or systemically to suppress new vessel growth. Similar to angiostatin. It is also secreted by the primary tumor to suppress tumor metastasis.

Endothelial cell: Epithelial-type cells that line the cavities of blood vessels, lymph vessels and the heart.

EPH receptors: The transmembrane receptors and their membrane-bound ephrin ligands comprise a signaling pathway with well-characterized functions during the development of several different organ systems, including the vasculature. EPH receptors and ephrins are often reciprocally expressed at tissue compartment boundaries and are best known for their roles in axon guidance; although more recently a role in vascular development has been identified.

Epigenetic mutation: Alterations in chromatin structure that affect gene expression without altering the coding DNA sequences factors. Epigenetic regulation involves DNA methylation and demethylation, chromatin remodeling, methyl-binding domain proteins, and histone deacetylation and acetylation.

Epigenome: Sum of all epigenetic factors operating within a given cell at a particular time.

Epstein–Barr virus (EBV): The first human tumor virus identified in 1964. EBV, the etiological agent of infectious mononucleosis, has been associated with the genesis of Burkitt lymphoma and nasopharyngeal carcinoma (viral DNA, and sometimes virus, is present in these cancer cells). Furthermore, EBV has been detected in the Reed–Sternberg cells in high percentage of Hodgkin lymphomas.

ER: Endoplasmic reticulum.

ERBB: Family of proto-oncogenes comprising four closely related receptor tyrosine kinases, which include the epidermal growth factor receptor (EGFR), ERBB2 (also known as HER2), and ERBB3.

Erbitux: see **Cetuximab**.

ERCP (endoscopic retrograde cholangiopancreatography): A variation of standard endoscopy that allows canulation of the pancreatic duct for imaging/observation or removal of stones.

ER^{TAM}: Modified estrogen receptor used in the development of regulatable (or conditional) mouse models in which the administration of tamoxifen is required to regulate the activity of the expressed protein.

ES (embryonic stem) cells: ES cells are derived from blastocysts collected from the uterine horns of 129 mouse strain.

ESR: Erythrocyte sedimentation rate.

Etoposide: Inhibits topoisomerase and used in treatment of several cancers.

ETS (E-twenty six) family: One of the largest families of transcription factors, characterized by a conserved ETS DNA-binding domain. Under normal regulation, ETS family members are involved in a variety of cellular functions including cell cycle, differentiation, migration, apoptosis, and angiogenesis and are downstream nuclear targets of RAS–MAP kinase signaling.

EUS (endoscopic ultrasound): Ultrasound probes used endoscopically are very important in the staging of primary tumors of the esophagus and stomach.

Everolimus: Inhibitor of the growth-promoting mTOR signaling protein, used in renal carcinoma.

Extrinsic apoptotic pathway: Machinery of apoptosis involving distinct upstream regulators and downstream effectors and requiring activation of death receptors (e.g. FAS) independently of the mitochondria.

FADD (FAS-associated death domain): Adaptor molecule recruited to FAS/FASL to form a complex called death-inducing signaling complex (DISC).

FAK: Focal adhesion kinase.

Fanconi anemia: A hereditary chromosome instability disorder, in which cells have a predisposition to chromosomal breaks, characterized by congenital abnormalities, retarded growth, early predisposition to cancer, and bone marrow failure.

FAP: Familial adenomatous polyposis coli is an autosomal dominant inherited malignancy of the colon. Inherited loss of a single allele of the *APC* genes is inevitably followed by loss of the single remaining allele in some colon cells during adulthood, thereby removing the tumor suppressor function.

FAS: Death receptor on the cell surface of specific cells. Binding of the FAS/CD95 ligand (FASL) to FAS receptors activates downstream signaling cascades which in turn recruits the adaptor to form a complex called the death-inducing signaling complex (DISC), ultimately triggering caspases activation and the demise of the cell.

FASL (FAS ligand): Ligand that binds to the FAS receptor on the cell membrane of specific cells to induce a death response (see **FAS**).

FBC: Full blood count.

FdUMP: 5-Fluoro-2-deoxyuridine monophosphate.

FGF (fibroblast growth factor): Family of growth factors, which were the first class of angiogenic factors described. FGF1 (aFGF), FGF2 (bFGF) are potent mitogens and trophic factors.

Field cancerization: Term employed to describe the widespread genetic abnormalities found in a tissue that has been chronically exposed to a carcinogen.

FISH: Fluorescent *in situ* hybridization.

FLIP (Fas-like inhibitor protein): Protein associated with inhibition of the extrinsic pathway of apoptosis.

Flk1 (fetal liver kinase 1): Receptor for vascular endothelial growth factor; also known as KDR or VEGFR2.

Flow cytometry: Technique used to analyse the cellular DNA content in cells. Useful for tumor specimens. The DNA index (abnormal cell DNA content relative to normal cells) may indicate a duplication (polyploidy) or loss (aneuploidy) of chromosomes.

Flt1 (Fms-like tyrosine kinase 1): Receptor for vascular endothelial growth factor; also known as VEGFR1.

Fluorouracil (5FU): Chemotherapy drugs (given intravenously), commonly used after bowel cancer surgery. An antimetabolite inhibitor of thymidylate synthase, which can prevent formation of thymidine during DNA synthesis. Used primarily in pancreatic and colorectal cancer treatment.

FNA (fine needle aspiration): The passage of a hypodermic needle into the abnormal area under investigation. Ultrasound or CT can be used to guide the needle to the suspected tumor if it is not possible to palpate it. Cells are then aspirated and put on a microscope slide and sent to the cytology laboratory.

FOB: Fecal occult blood.

FOLFIRI: Combination of folinic acid, 5-FU, and irinotecan, used in colorectal cancer treatment.

FOLFIRINOX: Combination of folinic acid, 5-FU, irinotecan, and oxaliplatin, used in pancreatic cancer as an alternative to gemcitabine.

FOLFOX: A combination of folinic acid (leucovorin or calcium folinate), fluorouracil and oxaliplatin (given intravenously). Chemotherapy drugs commonly used after bowel cancer surgery.

FOS: The v-*fos* gene was identified as a transforming oncogene of the feline osteosarcoma virus. The product of the human homolog gene, FOS, interacts with a second proto-oncoprotein, JUN, to form a transcriptional regulatory complex.

FPC: Familial polyposis coli.

FSH: Follicle-stimulating hormone.

G_1/S transition: Once a cell passes this point in the mitotic cell cycle the cell will usually go on to replicate even in the absence of growth-promoting stimuli unless a problem is detected such as DNA damage or improper chromosome segregation in which case checkpoints can arrest the cycle or kill the cell. This key transition is mediated by hyperphosphorylation of the RB protein and is promoted by growth factors via cyclin D– and cyclin E–CDK complexes and inhibited by various CKIs including $p15^{Ink4}$, $p21^{Cip1}$, and $p57^{Kip2}$, some of which are triggered by inhibitory growth factors such as TGF-β.

G_1: Period in the cell cycle after mitosis during which cells sense their environment and decide to enter a second cell division cycle, exit cycle, or die.

GAPs: GTPase-activating proteins are key inactivators of RAS that facilitate nucleotide hydrolysis of RAS-bound GTP to GDP.

G-CSF: Granulocyte colony-stimulating factors.

GDNF (glial cell line-derived neurotropic factor): Ligand for the RET tyrosine kinase.

GEFs (guanine nucleotide exchange factors): Key activators of RAS signaling. They catalyze nucleotide exchange of RAS-bound GDP to GTP.

Gemcitabine (Gemzar): Antimetabolite used in pancreatic cancer and NSCLC, or in combination with platins or taxanes in several advanced cancers.

Gene expression signature: The analyses of changes/differences in expression of hundreds or thousands of genes/proteins between normal and cancer cells (e.g. a "poor prognosis" signature was identified for breast cancer that conferred a high risk of early development of metastases).

Genetic locus: In normal cells, genes usually occupy the same position on a given chromosome (genetic locus) in all individuals.

GFP (green fluorescent protein): Protein used in the labeling of cellular proteins (GFP tag). The subcellular compartment in which the protein is located can be detected by fluorescence microscopy.

GH: Growth hormone.

GIST: Gastrointestinal stromal tumor.

Glycosylation: Attachment of a sugar moiety to a protein. Such protein modification plays a role in, for example, intracellular regulation, cell–cell communication, cell–substrate recognition, and secretory pathways.

Gorlin syndrome: Inherited from a germline mutation of *PATCHED* (*PTC*), which predisposes patients to a childhood cerebellar tumor (medulloblastoma) and a high incidence of basal cell carcinoma, a skin cancer.

Growth factor (GF): Often used as a catch-all term for proteins encompassing all of the following functional categories: *Mitogens*: promote cell division largely by allowing G_1/S transition in the cell cycle. *Growth factors*: promote an increase in cell mass (cell growth), by enhancing protein synthesis. *Survival factors*: promote cell survival by suppressing apoptosis.

GSK-3β (glycogen synthase kinase 3 beta): Originally identified as a modulator of glycogen metabolism, GSK-3β is now known to play a key regulatory role in a variety of cellular processes, including initiation of protein synthesis, growth and proliferation, differentiation, apoptosis, and is essential for embryonic development as a component of the Wnt signaling cascade. GSK-3β interacts with ubiquitin E3 ligases to target various proteins for degradation through the ubiquitin–proteasome pathway. These include key growth and cell cycle regulators, such as c-MYC, c-JUN, cyclin D1. Conversely, GSK-3β activity is inhibited by signaling through the RAS–PI3K–AKT pathway.

GWAS (genome-wide association studies): Studies that look for the presence of variation in genes by detecting SNPs in large groups of patients. Many susceptibility alleles for most common cancers have been identified.

H2A and its variant H2AX: These histones are DNA repair proteins in DSB repair.

Haploinsufficiency: Defined by the appearance of a phenotype in cells or an organism when only one of the two gene copies (also called alleles) is inactivated. For some tumor suppressor genes, the loss of a single allele is sufficient to induce susceptibility to tumor formation. In other words, these are haploinsufficient tumor suppressor genes.

Haplotypes: Stretches of DNA sequence that are inherited as an uninterrupted block of sequence. These stretches of sequence contain many SNPs and a combination of a small number of these SNPs (tagSNPs) can unequivocally identify a haplotype.

HapMap: A haplotype map determines common patterns of DNA sequence variation in the human genome to facilitate the identification of disease genes and mutations.

HATs (histone acetyl transferases): In contrast to HDACs, these enzymes in general activate gene expression by enabling histone acetylation of specific target DNA. This leads to chromatin remodeling and increasing accessibility of DNA to transcription factors.

HCG: Human chorionic gonadotrophin.

HDACs (histone deacetylases): In general these enzymes silence gene expression. They are recruited to specific target DNA to result in local histone deacetylation within nucleosomes, thereby decreasing the accessibility of DNA to transactivation factors.

HDM2: Negative regulator of P53 tumor suppressor protein (MDM2 is the mouse homolog).

***Helicobacter pylori*:** Bacterium that is a causative agent in peptic ulceration and in gastric cancer.

Hemi-desmosome: Specialist type of adhesion assembly that exists in some epithelia. These are large structures that form strong bonds between epithelial cells and the underlying interstitial ECM through a chain of molecular interactions.

Herceptin (trastuzumab): The first monoclonal antibody kinase inhibitor to be approved in the treatment of breast cancers with HER2 mutations. Herceptin inhibits proliferation of such cancer cells by binding to the extracellular region of ERBB2 (HER2/neu) receptor. In consequence, ERBB2 expression at the cell surface is reduced and the abnormal cell becomes the target for antibody-dependent cell-mediated cytotoxicity (ADCC).

HIF: Hypoxia inducible transcription factor.

HIFU (high-intensity focused ultrasound): Employs highly focused ultrasonic beams to kill cancer cells by heating them up.

Histone code: All of the various modifications (acetylation, methylation, phosphorylation, ubiquitination, and sumoylation) in a given genomic region and the resultant effects on chromatin confirmation and gene expression.

HNPCC: Hereditary nonpolyposis colon cancer.

HNSCC: Head and neck squamous cell carcinoma.

Homeostatic regulation: Immune responses are tuned to return to a basal resting state in a process called homeostasis. Following antigen-specific proliferation, most effector cells undergo apoptosis or are inhibited by regulatory mechanisms as the antigen is eliminated and the immune response wanes. Conversely, in the lymphopenic state, induced by lymphodepletion or in the case of RAG knockout mice, homeostatic mechanisms involving the upregulation of lymphokines, such as IL-15 and IL-7, are induced to expand the pool of lymphocytes to a basal level.

HPLC (high-performance liquid chromatography): A tool used in the identification of proteins.

HPV (human papilloma virus): Virus believed to be strongly associated with cervical cancer. The most common is HPV subtype 16. HPV is also less commonly associated with head and neck cancer. It is believed that the HPV oncoproteins E6 and E7 may play a role in tumorigenesis.

HSPs (heat shock proteins): HSPs (e.g. HSP90) are deregulated in several cancers but normally function as a molecular chaperon facilitating protein folding of numerous oncogenic client proteins, which they protect from proteasomal degradation, including estrogen receptor, HER2, N-RAS, AKT, PDGFR, and B-RAF.

HtrA2/Omi: Reaper-related serine protease released during **MOMP**, encoded by the Prss25 gene.

Hydroxycarbamide (hydroxyurea): Antimetabolite used in polycythemia rubra vera. It was used in CML and as a neoadjuvant in head and neck cancers, though has been largely replaced in CML by imatinib. Inhibits ribonucleotide reductase and induces DNA double-strand breaks near replication forks.

IAPs (inhibitors of apoptosis proteins): A family of proteins that are able to inhibit apoptosis by directly binding and inhibiting specific caspases. IAPs are also involved in many important cellular processes that are often deregulated in cancer, such as proliferation, inflammation, immunity, cell migration, and metastases. Alterations in IAPs are found in many human cancers and are associated with resistance to current cancer therapies, disease progression, and poor prognosis, which have sparked great efforts into developing small therapeutic compounds termed "Smac mimetics."

ICR (imprinting control region): Region of differential methylated region (DMR) essential for controlling expression of genes lying within an imprinted domain.

IFN: Interferon.

Ifosfamide: A nitrogen mustard alkylating agent used in a wide variety of cancers.

IGF (insulin-like growth factor): Two types are known, IGF-1 and IGF-2.

IL-1β (interleukin 1β): A proinflammatory cytokine.

ILs (interleukins): Large family of cytokines involved in immune responses.

Imatinib mesylate (Gleevec): Tyrosine kinase inhibitor (TKI) used to treat CML and GIST. Although CMLs (which express BCR–ABL fusion protein) are not completely irradicated, imatinib restricts the growth of CML tumor cells and decreases the risk of "blast crisis."

Immune privilege: Term sometimes been applied to tumor immune evasion. Immune-privileged sites were initally used to describe normal tissues critical to the survival of the organism (e.g. reproductive organs) or tissues where an inflammatory response would be detrimental (e.g. CNS) and where the immune response is normally suppressed. In many instances, it has been found that mechanisms at such sites may also exist in tumors.

Immunologic ignorance: Primary tumors are sequestered from immune recognition by a stromal barrier.

Immunologic synapse: Region of interaction between the TCR and the peptide–MHC complex. In the mature synapse, which forms when cognate TCR–peptide–MHC interaction occurs, the immunologic synapse is reinforced by concentric outer ring of adhesion molecules and polarization of the microtubule organizing complex to the region of the synapse. This leads to downstream signals responsible for full T-cell activation and cell division.

Imprinting: Conditioning of the maternal and paternal genomes during gametogenesis, so that a specific allele is more abundantly or exclusively expressed in the offspring. Imprinting is regulated by epigenetic changes such as DNA methylation.

IMRT: Intensity-modulated radiation therapy.

ING1 (inhibitor of growth 1): Member of the ING tumor suppressor family that includes at least five related genes involved in diverse cellular processes including senescence, DNA repair, and apoptosis. ING proteins regulate gene expression by regulation of chromatin remodeling probably by acting as cofactors for HAT and HDAC.

INK4 (inhibitory protein of cyclin-dependent kinase 4): Small polypeptide proteins that bind to and inhibit cyclin–cdk holoenzymes (see **cyclin** and **cdk**).

INK4 family: Important family of **CKIs** (cyclin kinase inhibitors) that inhibit the CDK4/6 cyclin-dependent kinases. INK4 proteins have alternative names which can cause some confusion, and include the following: INK4a or $p16^{INK4a}$ ($p16^{Ink4a}$ non-human); $p15^{INK4b}$ ($p15^{Ink4b}$ non-human) encoded by the *CDKN2B* and *Ink4b* genes in humans and non-humans, respectively; $p18^{INK4c}$ ($p18^{Ink4c}$ non-human) encoded by the *CDKN2C* (human) and *Ink4c* (non-human) genes, respectively; and $p19^{INK4d}$, not to be confused with $p19^{ARF}$ (or **ARF**).

INK4a/ARF (CDKN2a): Gene locus encoding two proteins that regulate two of the most important tumor suppressor pathways represented by p53 and RB.

Innate response: Immune response which provides the initial line of defense against foreign antigens, including some tumor cells.

Integrins: Transmembrane cell surface receptors that exist as heterodimers and are important for adhesion to the ECM. On the cytoplasmic side, integrins bind adaptor proteins that link them both to the actin-based cytoskeleton and to enzymes that trigger signal transduction cascades. Integrins therefore integrate the extracellular anchoring elements of the ECM with intracellular proteins of the cytoskeleton.

Intrinsic (or mitochondrial) apoptotic pathway: Machinery of apoptosis involving distinct upstream regulators and downstream effectors and requiring the release of cytochrome *c* from the mitochondria.

Irinotecan: Inhibitor of topoisomerase 1, which causes DNA to become entangled and damaged during replication. It is used in combination to treat a wide variety of bowel cancers (see FOLFIRI, etc.).

JNK: Jun N-terminal kinase.

Karyotype: Number and configuration of chromosomes within a cell.

Lamellipodia: Plasma membrane protrusions that are extended during cell migration in a polarized direction, into an ECM environment.

LBC (liquid-based cytology): A sample of cells is removed by means of a special brush that is used to sweep around the cervix for cervical screening.

LC (liquid chromatography): An alternative to 2D gels which can be used to separate the (digested) proteins on certain characteristics like hydrophobicity or charge. The LC system is usually directly connected to a mass spectrometer (LC–MS).

LCM (laser capture microdissection): Tissue is viewed under a microscope and those areas containing the chosen cells, based on histologic appearance or specialized staining, are targeted by a near-infrared laser. A transparent transfer film immediately overlying the chosen cells is transiently heated by the laser, fuses with the cell(s) underneath, and is then lifted off the specimen so that the captured cells can be subjected to further analysis, In this way, homogenous cell populations can be selected away from other "contaminating" cells.

LH: Luteinizing hormone.

LHRH: Luteinizing hormone-releasing hormone.

Li–Fraumeni syndrome: Syndrome caused by loss of the *p53* tumor suppressor, in which children and young adults of the family develop an assortment of cancers, including sarcomas, brain tumors, acute leukemia, and breast cancer.

LMP: Latent membrane proteins of the Epstein–Barr virus are associated with Hodgkin disease and nasopharyngeal carcinoma. Some of these proteins (e.g. LMP-1) are believed to play a role in carcinogenesis due to its transforming properties in cell lines.

LOI (loss of imprint): Loss of the normal imprinting controls either by methylation or demethylation of the DMR or potentially by failure of CTCF function.

Loss of heterozygosity (LOH): Mechanism whereby a somatic cell, with an inherited mutated gene allele, can lose the normal gene copy and become vulnerable to cancer. LOH can occur by deletion of the normal allele, deletion of part of or the entire chromosome (the latter is referred to as aneuploidy).

Glossary

Lysosomes: Enzyme complexes that degrade extracellular proteins and cell surface membrane proteins such as growth factor receptors or those employed for receptor-mediated endocytosis.

MAD: see **MXD**.

MALDI (matrix-assisted laser desorption ionisation): Used in mass spectrometry and proteomics. The ionization source transfers nonvolatile compounds (peptides and proteins) into the gas phase.

Malignant: Cancer that has spread to distant sites (metastastic).

Mammography: X-ray examination of the breast for the presence/absence of tumors.

MAPK (mitogen-activated protein kinase): One of the most important signaling pathways activated by mitogens involves the RAS/MAP kinase cascade.

Mass spectrometry: Essential technique used to identify proteins and used in any proteomics experiment. The three main parts of a mass spectrometer are the ionization source, the mass analyzer, and the detector. After peptide ionization the peptides are guided into the mass spectrometer to the mass analyzer.

Matrigel: Matrix assay used *in vitro* to assess the ability of tumor cells to invade.

MDM2: Negative regulator of P53 tumor suppressor protein (HDM2 is the human homolog).

MDR: Multidrug resistant.

Mediastinal lymphadenopathy: Enlarged lymph nodes in the chest.

MEN: Multiple endocrine neoplasia.

***MEN1*:** Tumor suppressor gene involved in inherited tumor susceptibility in multiple endocrine neoplasia type 1 (MEN1). Encodes a 610-amino-acid protein called menin.

Mercaptopurine: Antimetabolite that inhibits purine nucleotide synthesis, thus resulting in DNA damage during replication and impaired repair. It is used in treatment of leukemias.

Mesothelioma: Cancer of the lung lining that is very rare in those not exposed to asbestos.

Metastasis suppressor genes: Genes that can suppress metastasis without affecting tumorigenicity.

Metastasis: Spread of cancer cells to distant sites.

Methylation: Attachment of methyl groups to DNA or protein which modifies behavior.

MHC (major histocompatibility complex): Allows for proper antigen processing and presentation and is critical to antigen-specific immunity. The T cell recognizes the foreign peptide *only* in the context of the MHC. Human MHC molecules are called human leukocyte antigens (HLA) and are divided into class I and class II HLA or MHC which present peptides to CD8 and CD4 T cells, respectively.

Microarrays: High-throughput technique based on the process of hybridization used mainly for gene expression analysis. mRNA expression levels are measured in cells or tissue and an array provides a snapshot of the cells dynamic transcriptome. **DNA microarrays**, **geneChips**, and **arrays**, are all synonyms for the same technique.

Micrometastases: Small numbers of tumor cells that migrate to secondary sites early in the development of a cancer, and remain there in a dormant state for a long time.

Microsatellites: Short sequences of 1–5 bp repeated in tandem throughout the genome. Because of their polymorphic nature, they have been widely used as genetic markers.

miRNAs (microRNAs): Small noncoding RNAs that can bind to specific complementary regions of target mRNAs through miRNA–mRNA base pairing, leading to accelerated destruction and translational repression of the target mRNA. They are often aberrantly expressed in human cancers where they can act as tumor suppressors or sometimes oncogenes.

Mitosis: Process of cell division during which one cell gives rise to two identical daughter cells.

MMPs (matrix metalloproteinases): Group of enzymes that can break down extracellular matrix proteins, and are involved in physiological processes such as wound healing, tumor cell invasion, and angiogenesis.

MMR (mismatch repair): Process critical for preserving genomic integrity by correcting mismatches of the normal bases arising largely through a failure of polymerase proofreading during replication, that is, failure to maintain normal Watson–Crick base pairing (A–T and C–G).

Molecular adaptation: Term used when further mutations in DNA or epigenetic changes arise, leading to activation of secondary tumorigenesis pathways that allow tumor cells to escape dependence on the initiating oncogene. Treatment resistance is evident as a consequence.

MOMP (mitochondrial outer membrane permeability): Key process involved in the mitochondrial pathway of apoptosis. Leads to release of key proteins such as cytochrome *c*, endonuclease G, AIF, Smac/DIABLO and Htra2/Omi. These can engage both caspase-dependent and -independent pathways of apoptosis.

Monogenic: Single-gene.

MRI: Magnetic resonance imaging.

MSI: Microsatellite instability associated with a "mutator" phenotype in Lynch syndrome colorectal cancers.

mTOR (mammalian target of rapamycin): One of the master regulators of protein synthesis and translation control which controls the translational apparatus through protein phosphorylation.

Multistep tumorigenesis: Concept in which cells have to acquire several oncogenic mutations in order to convert a normal cell into a cancerous one.

Mutation: Alterations in the coding sequence of the DNA.

MXD: Family of proteins (formerly known as the MAD family). Transcriptional repressors that can promote cell differentiation by dimerizing with MYC's partner, MAX. The MXD family includes MXD1 (MAD1), MXD3 (MAD3), MXD4 (MAD4), and MXI1 (MAX interactor-1; MAD2).

MYC: Originally identified as an oncogene in the avian myelocytomatosis virus. In its normal form as a proto-oncogene, it encodes a transcription factor that plays a key role in many normal cellular processes, in particular, cell growth and cell cycle. However, it is often found in its oncogenic form in human cancers. Paradoxically, at high levels, MYC protein can sometimes promote apoptotic cell death and thus can be said to possess tumor suppressor activity thought to prevent tumor growth.

Natural killer (NK) cells: Immune cells that represent the "rapid response" component of innate immunity, releasing proinflammatory cytokines ultimately leading to the recruitment/activation of macrophages and antigen-presenting cells (APCs).

NBS (Nijmegen breakage syndrome): Autosomal recessive hereditary chromosomal instability syndrome associated with a predisposition to tumor formation.

ncRNA (noncoding RNA): RNA that is not ultimately translated into protein, and includes ribosomal RNAs, transfer RNAs and small nuclear RNAs. In recent years, ncRNAs have been shown to encode an ever expanding family of important regulatory factors that includes microRNAs (miRNAs).

Necrosis: A form of cell death induced by external stimuli such as trauma, ischemia, and high-dose irradiation. This energy-independent form of cell death leads to loss of function of mitochondria and endoplasmic reticulum, resulting in a dramatic breakdown of energy supply culminating in cellular, nuclear, and organellar swelling, and ultimately rupture of the plasma membrane.

Neoadjuvant chemotherapy: Chemotherapy given before surgery to shrink the tumor and reduce risk of cancer cells having spread beyond the planned surgical resection.

Neoplasia (cancer = tumor): Result of inappropriate proliferation and loss of control.

Neovascularization: Vascularization of adult tissue in excess of developmentally predetermined vascularization.

NER (nucleotide excision repair): Versatile and particularly important process for clearing substantive UV-induced DNA damage, such as thymine dimers and (6–4)-photoproducts.

NF-1 (neurofibromatosis-1): GTPase that negatively regulates Ras signaling. It is mutated in patients with neurofibromatosis, hence its name, but also in sarcomas and gliomas.

NF-κB (nuclear factor-κB): The NF-κB family are a group of transcription factors that are held inactive in the cytoplasm bound to the inhibitor protein IκB. Stimulation of cells by inflammation, chemotherapy, radiation, or oxidants initiates degradation of IκB, allowing NF-κB to translocate to the nucleus and activate transcription of various genes including growth factors, angiogenic factors, and antiapoptotic factors.

NGS (next generation sequencing): Genome-wide sequencing technologies that utilize array platforms for highly parallel sequence analysis. NGS is used extensively in cancer molecular biology to analyze mutation patterns, DNA methylation, and the specific patterns and levels of transcripts in tumor cells, including splice variants.

NHEJ: Nonhomologous end joining occurs during the process of repairing DSBs.

NHL: Non-Hodgkin lymphoma.

NMSC: Non-melanoma skin cancer.

Notch signaling pathway: The Notch proteins are transmembrane receptors that are activated by membrane-spanning ligands of the Delta, Serrate, and Jagged families. Both Notch receptors and ligands are expressed in specific vascular compartments (i.e. arterial versus venous endothelium or vascular smooth muscle cells/pericytes).

NSCLC: Non-small-cell lung carcinoma.

Oncogene addiction: Aberrant activation of an oncogenic pathway leads to the suppression of alternative growth regulating pathways operating in normal cells, thus rendering cancer cells even more sensitive to treatments that block the oncogenic pathway.

Oncogene collaboration (or oncogene cooperation): Individual oncogenes do not cause cancer on their own but rather act in concert with other oncogenes or loss of tumor suppressors.

Oncogene: "Cancer-causing" gene which is an activated version of its normal counterpart (proto-oncogene) as a result of mutations or other means. Oncogenes either produce increased amounts of protein (oncoprotein) or an altered protein that is fixed in the "on" position.

Oncogenic stress: Concept whereby oncogenic mutations that drive uncontroled proliferation (e.g. MYC) are somehow recognized as aberrant by the cell and result in activation of intrinsic tumor-suppressing activity.

Oncomir: miRNAs acting either as oncogenes or tumor suppressors. Such oncogenic miRNAs would need to be silenced in cancer treatments. Agents directed against these are referred to as antagomirs.

Onco*type* DX Breast Cancer Assay: Commercially available gene-based biomarker tool used to screen for a number of genes simultaneously in tumor samples using RT-PCR. Trials investigating the potential of this test to predict recurrence and response to chemotherapy in ER-positive breast cancer have been promising.

Orthologs: Proteins from a common ancestor that have an identical function in different species.

p19^{ARF};p14^{ARF}: Important tumor suppressor (see **ARF**).

p21$^{Cip1/Waf1}$: Growth arrest gene encoding a cyclin-dependent kinase inhibitory protein, whose expression affects Rb function. In response to DNA damage, the tumor suppressor, p53, induces expression of p21 to promote growth arrest.

P53: The *p53* tumor suppressor gene, which encodes the transcription factor P53, is one of the most frequent targets for mutation in human tumors. The majority of human tumors have either mutated *p53* itself or have incurred mutations that disable function of the p53 pathway. P53 protein (53 represents the molecular weight of the protein) acts to guard against DNA damage, stresses, or a surfeit of proliferative signals, preventing cells from becoming malignant by inducing growth arrest or inducing apoptosis.

PAKs (p21-activated kinases): Effectors of RAC1 and CDC42 regulating actin cytoskeleton.

Palliation/palliative care: Cancer treatment/management aimed to target immediate symptoms of cancer or to delay the time until symptomatic progression, rather than in an attempt to cure.

PanIN: Pancreatic intraepithelial neoplasia, a pre-invasive neoplasm of pancreatic ductal adenocarcinoma (PDA).

Panprocto-colectomy: Removal of the entire colon and rectum.

Pap smear: The Papanicolaou smear is the screening test used in the early detection of cervical cancer.

Paracrine: Extracellular signals (e.g. those regulating replication or growth) which operate at a local level between adjacent cells, for example the production of angiogenic factors by tumor cells acting on adjacent endothelial cells.

PARP (poly(ADP-ribose) polymerase): Enzyme activated after DNA damage and involved in DNA repair.

PCD (programmed cell death): A normal physiological process in multicellular organisms that helps the efficient clearance of cells that are damaged, aged, or simply surplus to requirements. It is a prominent feature of normal development but also contributes to tissue mass homeostasis throughout the life of the animal (often termed **Apoptosis**).

PCR: Polymerase chain reaction.

PDA: Pancreatic ductal adenocarcinoma.

PDGF: Platelet-derived growth factor.

Pemetrexed (Alimta): A folate antimetabolite with a similar structure to folic acid. It inhibits both purine and pyrimidine synthesis and is used in mesothelioma and NSCLC.

Pericyte: Connective tissue-type cell containing smooth muscle actin, which is found in the outer layer of small blood vessels.

PET: Positron emission tomography.

Peutz–Jeghers syndrome: Characterized by intestinal hamartomas and increased epithelial cancers due to germline mutation in *LKB1* (serine/threonine kinase 11).

Pharmacodynamics: Biological effects of a therapeutic agent.

Pharmacogenetics: When used in cancer biology, the term traditionally equates to unraveling the effects of inherited germline variants on how a specific drug will be metabolized, distributed, and function in an individual patient and what effects it might have on toxicity.

Pharmacogenomics: In cancer biology, takes into consideration the effects of acquired (rather than inherited germline) somatic mutations or multiple variants on specific drug treatment.

Pharmacokinetics: Metabolism, transport, and distribution of a therapeutic agent.

Phase I clinical trials: Earliest clinical trials in patients aimed at confirming the safety of a new agent in humans. A variety of standardized protocols are used, including progressive dose escalation and dosing intervals optimized for ensuring safety over any therapeutic considerations.

Phase II clinical trials: Undertaken once phase I trials have confirmed a drug appears safe. Conducted in a larger cohort in order to assess the efficacy and to again look for potential side effects/toxicity in a larger number and under conditions more representative of the planned clinical usage.

Phase III clinical trials: Generally aimed at demonstrating clinical benefits. They are usually randomized controlled trials and may run across multiple centers in order to recruit sufficient numbers of participants (may be over a thousand).

Phase IV clinical trials: Involve postmarketing surveillance and are an important route by which adverse effects of new drugs can be picked up when these are of a low enough frequency to have been missed in earlier studies (large numbers of patients treated for longer are needed to reveal the problem).

Philadelphia chromosome: A truncated chromosome 22 (the Philadelphia chromosome) created following a translocation event between chromosome 9 and chromosome 22. This translocation event leads to juxtapositioning of the *ABL1* gene on chromosome 9 to a part of the *BCR* (breakpoint cluster region) gene on chromosome 22, to form the BCR–ABL fusion protein, leading to diseases such as CML.

Phosphatidylserine: Cells in the final throws of apoptosis display "eat me" signals, like phosphatidylserine, that are recognized by phagocytes such as macrophages that then dispose of the corpse, without provoking an inflammatory response.

Phosphorylation: Attachment of a phosphate group to a protein. Protein kinases activate signal transduction pathways by phosphorylating tyrosine, threonine, or serine residues in proteins. Such posttranslational modifications are transient and activate the protein or induce its association with other proteins for a short period of time.

PI3K (phosphoinositol-3 kinase): Enzyme that generates lipid messengers such as phosphatidylinositol 3,4,5-trisphosphate, which activate the kinase AKT/PKB, involved in survival.

PIPs (phosphatidylinositol phosphates): Lipid signaling molecules known to regulate cellular processes such as apoptosis, motility, as well as responses to DNA damage and in tumorigenesis.

PKC (protein kinase C): Family of serine/threonine kinases that regulates multiple cell functions including growth, proliferation, differentiation, cytoskeletal organization, motility, and apoptosis.

PKN3: Downstream target in the PI3K signaling pathway involved in metastasis, invasion, and vascular stabilization.

PLAP: Placental alkaline phosphates.

PLCγ (phospholipase Cγ): Enzyme that plays a key role in cellular growth and proliferation.

PML (promyelocytic leukemia): Tumor suppressor first identified in a mouse model for acute promyelocytic leukemia. Regulates responses of *p53* to oncogenic signals from RAS.

Polycomb group proteins (PcG): Proteins (e.g. Bmi1, Pc2, Cbx7, and EZH2) involved in epigenetic regulation of gene expression, particularly in the determination of cell fate during development.

Polyoma virus: Polyoma virus and SV40 (simian virus) are oncogenic and linked to a variety of animal tumors. Polyoma virus can transform human cells in culture. The latter is known to inactivate both the p53 and Rb tumor suppressors.

Pox virus: Viral disease of rabbits causing cancerous myxomas.

PR: Progesterone receptor.

Proteasome: Abundant multi-enzyme complex that represents the major route for degradation of intracellular proteins in eukaryotic cells. Proteins are targeted for destruction via ubiquitination and subsequent degradation through the proteasome.

Proteomics: Identification/study of the proteome, the protein complement of a cell, organ, or genome.

Proto-oncogene: The normal gene version of an oncogene (its product is known as a proto-oncoprotein), involved in a variety of cellular processes, such as replication, growth, differentiation, and motility.

PSA: Prostate-specific antigen.

PTC (Patched): Receptor protein for sonic hedgehog (Shh) ligand discovered in *Drosophila* and found to regulate patterning during development. Germline mutation of *PTC1* is responsible for a familial cancer, Gorlin syndrome, which predisposes patients to a childhood cerebellar tumor (medulloblastoma) and a high incidence of basal cell carcinoma, a skin cancer.

PTCH: Tumor suppressor gene, often inactivated in basal cell carcinoma.

PTEN (phosphatase and TENsin homolog on human chromosome TEN): Phosphatase that removes the phosphate groups from phosphatidylinositol 3,4,5-trisphosphate (PIP3) to generate a PI(3,4)P2 and counteract the action of the phosphatidylinositol-3 kinase (PI3K) that converts PIP2 to PIP3.

PTH: Parathyroid hormone.

PTLD (post-transplant lymphoproliferative disease): Often arises following stem or organ transplantation in heavily immunosuppressed patients as a B-cell neoplastic disease that may be causally related to EBV (Epstein–Barr virus) reactivation in the absence of a functional T-cell response.

Quiescence: A cellular state where cells are in a reversible growth arrest.

RAF: The *Raf* gene was originally identified as a transforming oncogene from the rat fibrosarcoma virus, from where its name derives. RAF is responsible for threonine phosphorylation of MAP kinase (MAPK) following receptor activation. Oncogenic RAF leads to constitutive activation of the downstream MAPK pathway.

RAG: Recombination-activating gene. The B-cell receptor is composed of immunoglobulin (Ig) chains responsible for binding antigen. The T-cell receptor is composed of a heterodimer of two polypeptide chains containing Ig-like domains. The *RAG1* and *RAG2* genes encode for lymphocyte-specific recombinases responsible for DNA recombination events required to form functional Ig and TCR genes. The absence of RAG1 or 2 function leads to the absence of mature B and T cells.

RAG-2–/– mice: Mouse strain engineered with deficiencies in lymphocyte-specific recombinase, an enzyme responsible for antigen receptor rearrangement. Such mice possess no NK, T, or B cells and develop sarcomas more frequently.

RAL (RAS-like GTPases): RAL proteins (RAL-A and RAL-B) can engage multiple effector proteins that direct various biological processes, such as trafficking of secretory vesicles to the plasma membrane, regulation of gene expression and protein translation.

RAS: Protein that binds GTP in the active state and GDP in the inactive state and relays signals from receptors at the surface of the cell to the nucleus. The RAS GTPases (H-RAS, K-RAS) are the founding members of the RAS superfamily and were among the first proteins identified that possessed the ability to regulate cell proliferation. First discovered as proteins encoded by retroviral oncogenes that had been hijacked from the host genome by the Kirsten (K-) and Harvey (H-) rat sarcoma (ras) viruses. N-RAS was the third member to be identified from neuroblastoma and leukemia cell lines.

RASSF: Family of RAS effector/tumor suppressors. Five members of this family have been identified (NORE1, RASSF1, RASSF2, RASSF3, and RASSF4) and are involved in cell cycle arrest and in apoptosis in response to RAS.

RECIST (response evaluation criteria in solid tumors): A set of agreed standards for defining when cancer patients respond, stay the same, or progress while being treated. This is now the gold-standard endpoint for clinical trials alongside mortality. It assesses tumor volume using imaging such as CT or MRI or, where easy and reliable, even plain X-ray. The important feature is that similar modalities are used for follow-up assessment to be certain of tumor progression.

Retinoblastoma: Tumor of the retina. Name of the tumor suppressor that when mutated or deleted induces this tumor type.

Retroviruses: RNA tumor viruses common in chickens, mice, and cats but rare in humans. The only currently known human retroviruses are the human T-cell leukemia viruses (HTLVs) and the related retrovirus, human immunodeficiency virus (HIV).

RFA (radiofrequency ablation): RFA of cancer cells is used as a palliative treatment for bone and liver metastases.

RHO: Members of the RAS superfamily of GTPases (e.g. RHO A/B, RAC1/2, and CDC42) that mediate a diverse range of cellular effects such as proliferation, motility and adhesiveness via cell–cell and cell–matrix interactions.

Rituximab: Monoclonal antibody against CD20 on mature B cells, which has little effect on newly forming B cells thus allowing the immune system to recover quickly after therapy. Used in lymphoma and leukemia.

RNAi: RNA interference is a natural posttranscriptional process through which many organisms, including nematodes, plants, fungi, and viruses, suppress the expression of genes when exposed to double-stranded RNA molecules of the same sequence (sequence-specific gene silencing). RNAi is an important tool for analyzing gene functions in eukaryotes and is now featuring strongly in the development of therapeutic gene silencing.

ROCK-I and II: Rho-associated coiled-coil domain kinases; downstream effectors of RHO A, involved in regulating actin cytoskeleton.

ROS: Reactive oxygen species.

RSV (Rous sarcoma virus): Virus identified by the eminent scientist Peyton Rous as the infectious agent responsible for causing sarcomas in chickens. The viral oncogene v-src was subsequently pinpointed as the key culprit. The cellular counterpart of this viral oncogene is c-src.

RTK: Receptor tyrosine kinase.

RT-PCR: Real-time reverse transcription-polymerase chain reaction.

S phase: Cell cycle phase during which cells synthesize DNA and double their chromosome content from 2N to 4N.

SAGE (serial analysis of gene expression): Provides quantitative expression profiling without a requirement for previous knowledge of genes that are to be detected. All genes, including novel genes, are tagged by a short 14–21 bp sequence (SAGE tags). These SAGE tags, which are uniquely associated with a given gene and can be used to identify it, are joined serially to allow efficient sequencing, amplified by polymerase chain reactions (PCR) and then digitally analyzed for levels of expression. This information forms a library that can then be used to analyze differential expression among cells that have been subjected to SAGE analysis.

SCC: Squamous cell carcinoma.

SCF: Ubiquitin ligase complex named after its three main protein subunits: SKP1/CUL1/F-box protein.

Self-seeding: Traditionally, metastasis is regarded as a unidirectional process with cancer cells disseminating from the primary to seed metastasis in regional lymph nodes or distant sites. However, this view has been challenged by recent studies which suggest that a multidirectional process is more accurate and which includes the seeding of the primary tumor itself – "self-seeding."

Senescence: An irreversible state reached by the cells after a certain number of cell divisions. Characterized by cell's flat morphology, irreversible growth arrest, and expression of the β-galactosidase protein.

SEREX (serological recombinant expression cloning): Process used to identify potentially immunogenic tumor antigens from a wide variety of human cancers. SEREX is a simple, robust immunoscreening technique in which autoantibodies present in patient serum samples are used to identify tumor proteins within tumor-derived cDNA expression libraries.

SFK (SRC family kinases): There are at least nine different known SRC family genes, including SRC, Blk, Fgr, Fyn, Hcy, Lck, Lyn, Yes, and Yrk, which through different mRNA processing can encode at least 14 different proteins, collectively referred to as SRC family kinases.

SH (SRC homology) domains: Domains in SRC required for the regulation and activation of SRC proteins.

SIADH (syndrome of inappropriate ADH secretion): A paraneoplastic syndrome associated with lung cancer, predominantly SCLC, associated with profound hyponatremia (due to water retention, secondary to the action of ADH on the kidney) causing lethargy and somnolence.

siRNAs (small interfering RNAs): Used in the development of novel therapeutics based on RNA silencing by the administration of small oligonucleotides complementary to the mRNA to be silenced (siRNA).

Smac mimetics: Small therapeutic compounds that are being developed to target IAPs that are often altered in human cancers and associated with resistance to cancer therapies.

Smac/DIABLO: Smac (second mitochondria-derived activator of caspase).

SNP chips: Similar to the microarrays used for expression analysis, but mainly used to screen for SNPs in the genome to link genotype to phenotype. The starting material is not mRNA but genomic DNA, which is also amplified and labeled before hybridization.

SNPs: DNA single nucleotide polymorphisms.

SPECT: Single-photon emission computed tomography.

SRC: First oncogene discovered. Originally identified as the transforming agent (v-src) of the Rous sarcoma virus (RSV), a retrovirus that infects chickens and other animals. It encodes a nonreceptor tyrosine kinase which activates downstream signaling through the addition of phosphate groups to tyrosine residues on target proteins.

STAT (signal transducers and activators of transcription): A family of signaling molecules associated with cytokine receptors. Cytokine receptor binding leads to phosphorylation of STAT proteins that migrate to the nucleus and bind to promoter regions of various genes.

Stroma: Provides support for epithelia and is composed of the connective tissue matrix (synonymous with stromal ECM). Stromal cells, such as fibroblasts, myofibroblasts, adipocytes, endothelial cells, and macrophages are embedded within it, as are nerves. Blood and lymph vessels penetrate the stroma to provide nutrients and immune protection.

SUMO (small ubiquitin-related modifier): Family of proteins that are both structurally and mechanistically related to ubiquitin in that they are posttranslationally attached to other proteins.

SVCO: Superior vena caval obstruction.

Synthetic lethality: Epimutations in the cancer cells, which may, for example, inactivate some DNA repair pathways, leave the cancer cell overreliant on any remaining alternatives. Blocking a remaining DNA repair pathway with drugs will affect cancer cells much more strongly than normal cells, as in the latter alternatives are functional and this drug target may be redundant.

T cells: Lymphocytes which mediate "cellular" immunity by interacting directly with a target cell (see also CD4 and CD8 T cells). T cells achieve specificity for cells expressing the target antigen through the surface T-cell receptor, which recognizes fragments of antigen presented by the major histocompatibility complex (MHC).

Tamoxifen: Selective estrogen receptor modulator. Tamoxifen has transformed treatment of breast cancer and improves mortality when used as an adjuvant therapy and when used in high-risk prevention.

Tegafur and uracil (Uftoral): Combination drug comprising a 5-FU prodrug and uracil, which is an inhibitor of 5-FU catabolism, producing higher levels of the active 5-FU in cancer cells.

Telomerase: A reverse transcriptase that adds TTAGGG repeats onto pre-existent telomeres.

Telomeres: These cap the ends of chromosomes by forming a higher order chromatin structure that protects the 3' end from degradation and DNA repair. Mammalian telomeres are composed of TTAGGG repeats bound to specialized proteins. Loss of telomere capping, either due to TTAGGG exhaustion or disruption of telomere structure, results in end-to-end chromosomal fusion and loss of cell viability.

TERT (hTERT for humans): Catalytic subunit of the enzyme telomerase, which in various cell types avoids replicative senescence by maintaining telomere length.

Tetracycline ("tet") system: Used in the development of regulatable (or conditional) mouse models in which the drug doxycycline is required to regulate expression of the gene of interest within the target tissue.

Tet-regulatable gene expression: The Tet-Off and Tet-On expression systems are binary transgenic systems in which expression from a target transgene is dependent on the activity of an inducible transcriptional activator. In both Tet-Off and Tet-On systems, expression of the transcriptional activator can be regulated both reversibly and quantitatively by exposing the transgenic animals to varying concentrations of tetracycline or derivatives such as doxycycline.

TGF-β (transforming growth factor beta): Normally inhibits proliferation of epithelial cells. At higher doses, it induces apoptosis (tumor suppressive) in normal cells, but becomes oncogenic and potentiates EMT and metastasis in later stage disease.

TIAM1 (T-cell lymphoma invasive and metastasis 1): RAC-specific guanine nucleotide exchange factor, which is also a downstream effector of RAS.

TICs (tumor initiating cells): Also known as cancer stem cells (CSCs).

TIMPs: Tissue inhibitors of metalloproteinases (MMPs).

TIS (toponome imaging system) or MELC (multi-epitope ligand cartography): New technology whereby thin tissue sections are examined for the colocalization of 30–100 proteins at a cellular and subcellular level.

Glossary

TK: Tyrosine kinase.

TNF-α: Tumor necrosis factor type alpha.

TNM: The tumor node metastasis staging system was originally devised in the 1940s and is a globally recognized classification system for staging cancer in patients.

Toponomics: Term derived from ancient Greek nouns *topos* ("place," "position") and *nomos* ("law"). Toponomics is the analysis of the protein network code directly in morphologically intact cells or tissue section using a robotically controlled microscopic imaging technology termed **TIS or MELC**.

TPMT (thiopurine methyltransferase): Important in the metabolism of the thiopurine drugs, 6-mercaptopurine and azathioprine, used in leukemia, among others.

Transarterial chemoembolization: Technique now widely used to treat a number of tumor types including primary liver carcinomas. High-dose chemotherapy is injected locally into the tumor bed and the blood supply to the tumor is also compromised.

Transformation: A cellular process whereby normal cells are transformed into tumorigenic ones. Various characteristic abnormal cell behaviours observed in the culture dish include uncontrolled cell proliferation and the "piling up" of cells on top of each other to form characteristic multilayered foci.

Transgenesis: Introduction of a transgene into the genome of an organism, most commonly mice. Transgenes can be directed to the germline by direct microinjection of the DNA encoding the gene under study in the pronucleus of a fertilized egg or by using embryonic stem cell technology.

Trastuzumab: see **Herceptin**.

TRRAP (transformation/transcription domain-associated protein): A MYC coactivator that is part of a complex containing histone acetyl transferase (HAT) activity. TRRAP acetylates nucleosomal histone H4 at promoter sites which alters the chromatin structure allowing accessibility of MYC–MAX complexes.

TS: Thymidylate synthase.

TSH: Thyroid-stimulating hormone.

TSP-1: Thrombospondin-1, an antiangiogenic factor.

Tumor immune evasion: see **Immune privilege**.

Tumor initiation: The result of exposure of a cell or cells to a carcinogen, which permanently alters its genetic material but does not immediately influence phenotype. Carcinogens in this category are described as mutagens or genotoxic.

Tumor markers: see **Biomarkers**.

Tumor promotion: Factors that can cause tumors, from cells that have already been initiated (see **Tumor initiation**). Promoters are non-genotoxic carcinogens. The best-known promoters are the phorbol esters, which activate the PKC signaling pathway and promote mitogenesis and survival.

Tumor suppressor: Protein present in all cells at low levels but induced in response to stress to counter inappropriate cell proliferation by inducing growth arrest or apoptosis. Its loss predisposes cells to cancer.

Two-dimensional gel electrophoresis (2D gels): A polyacrylamide gel-based technique that allows for separation of proteins in two dimensions.

Ubiquitination: Attachment of a ubiquitin group to a protein, which can determine when and how rapidly a protein is degraded.

UPR: Unfolded protein response.

Vasculogenesis: Formation of new blood vessels from angioblasts in the embryo.

VEGF (vascular endothelial growth factor): Also referred to as VEGF-A.

VENICE: Vaccine European New Integrated Collaboration Effort.

VHL (von Hippel–Lindau) syndrome: Hereditary form of renal carcinoma.

Vincristine: The prototype vinca alkaloid derived from the periwinkle plant. This chemotherapeutic is used in breast cancer, NSCLC, NHL, and some leukemias. It is an antimitotic drug that binds to tubulin dimers preventing microtubule formation and causing metaphase arrest and apoptosis.

VLS (vascular leak syndrome): *In vivo* administration of lymphokines, such as IL-2, and immunotoxins can lead to a condition involving damage to vascular endothelial cells, leakage of intravascular fluid into tissues, and organ failure, known as vascular leak syndrome or VLS. VLS represents a dose-limiting adverse effect in patients receiving high-dose IL-2. It is believed that a three-amino-acid consensus sequence in IL-2 is responsible for this initiating damage.

WBC: White blood count.

Werner syndrome: An autosomal recessive disease characterized by premature aging, elevated genomic instability, and increased cancer incidence, resulting from inactivation of the *WRN* gene. Telomere attrition is implicated in the pathogenesis of Werner syndrome.

WNT: The WNT family is composed of conserved secreted proteins that play a crucial role in patterning during development by regulating cell–cell contacts.

XELOX: A combination of oxaliplatin and capecitabine. Chemotherapy drugs commonly used after bowel cancer surgery.

XP (xeroderma pigmentosum): An autosomal recessive disease characterized by sun sensitivity, early onset of freckling and subsequent neoplastic changes on sun-exposed skin. Skin abnormalities result from an inability to repair UV-damaged DNA because of defects in the nucleotide excision repair (NER) machinery.

Zoladex (gosarelin): Inhibitor of luteinizing hormone release that thereby reduces estrogen production in women and testosterone production in men. It is used in breast and prostate cancers.

Zytiga (abiraterone): An inhibitor of Cyp17, needed for testosterone synthesis. It is used in prostate cancer.

Answers to Questions

Chapter 2

1. a. False
   b. True
   c. True
   d. False

2. a. False
   b. True
   c. False
   d. True

3. a. False
   b. False
   c. True
   d. True

Chapter 3

1. a. False
   b. True
   c. False
   d. True
   e. True

2. a. False
   b. True
   c. False
   d. True
   e. False

3. a. True
   b. True
   c. True
   d. True
   e. True

4. a. True
   b. True
   c. True
   d. False
   e. True

5. a. True
   b. False
   c. True
   d. True
   e. False

Chapter 4

1. a. False
   b. True
   c. True
   d. True
   e. False

2. a. False
   b. True
   c. True
   d. True
   e. False

3. a. True
   b. False
   c. True
   d. True
   e. False

4. a. False
   b. False
   c. False
   d. False
   e. False

5. a. False
   b. True
   c. True
   d. True
   e. False

The Molecular Biology of Cancer: A Bridge From Bench to Bedside, Second Edition. Edited by Stella Pelengaris and Michael Khan.
© 2013 John Wiley & Sons, Inc. Published 2013 by John Wiley & Sons, Inc.

Answers to Questions

Chapter 5

1. a. True
   b. False
   c. False
   d. True
   e. False

2. a. True
   b. False
   c. True
   d. False
   e. True

3. a. True
   b. True
   c. False
   d. True
   e. False

4. a. False
   b. True
   c. False
   d. True
   e. True

5. a. True
   b. False
   c. False
   d. True
   e. False

Chapter 6

1. a. True
   b. False
   c. False
   d. True
   e. False

2. a. False
   b. True
   c. False
   d. True
   e. True

3. a. False
   b. True
   c. False
   d. True
   e. True

4. a. False
   b. True
   c. True
   d. True
   e. False

5. a. True
   b. False
   c. False
   d. True
   e. True

Chapter 7

1. a. True
   b. True
   c. True
   d. True
   e. True

2. a. False
   b. True

3. a. True
   b. True
   c. True
   d. True
   e. True

4. a. False
   b. True

5. a. True
   b. False
   c. False

Chapter 8

1. a. False
   b. True
   c. True
   d. False
   e. True

2. a. True
   b. False
   c. True
   d. False
   e. True

3. a. True
   b. False
   c. True
   d. True
   e. False

4. a. True
   b. True
   c. True
   d. False
   e. False

5. a. True
   b. True
   c. True
   d. False
   e. False

Answers to Questions

6 a. False
   b. True
   c. True
   d. True
   e. False

7 a. True
   b. False
   c. True
   d. True
   e. False

Chapter 9

1 a. True
   b. False
   c. True
   d. True
   e. False

2 a. True
   b. False
   c. False
   d. True
   e. False

3 a. True
   b. True
   c. False
   d. False
   e. True

4 a. False
   b. False
   c. True
   d. True
   e. False

5 a. False
   b. True
   c. True
   d. False
   e. True

Chapter 10

1 a. False
   b. False
   c. True
   d. True
   e. True

2 a. False
   b. False
   c. True
   d. False
   e. False

3 a. False
   b. True
   c. True
   d. True
   e. False

4 a. True
   b. False
   c. False
   d. True
   e. False

5 a. True
   b. False
   c. True
   d. True
   e. False

Chapter 11

1 a. True
   b. True
   c. True
   d. True
   e. True

2 a. False
   b. True
   c. False
   d. False
   e. False

3 a. True
   b. True
   c. False
   d. True
   e. False

4 a. False
   b. True
   c. True
   d. True
   e. True

5 a. False
   b. True
   c. True
   d. False
   e. True

Chapter 12

1 a. False
   b. True
   c. True
   d. True
   e. False

Answers to Questions

2 a. False
   b. True
   c. False
   d. False
   e. True

3 a. False
   b. True
   c. True
   d. False
   e. True

4 a. True
   b. True
   c. False
   d. True
   e. False

5 a. True
   b. False
   c. True
   d. False
   e. True

Chapter 13

1 c

2 d

3 b

4 c

5 a

6 b and d

Chapter 14

1 b

2 c

3 a

Chapter 15

1 d

2 An antimetabolite is a chemical entity similar enough to a natural metabolite to be able to mimic it in a normally occurring biochemical reaction in the cell, but different enough to alter the cell's normal function. An antimetabolite drug inhibits a normal metabolic process involved in causing disease.

3 a

4 a. True
   b. True
   c. False
   d. True

5 It finds second-generation drugs (based on already established drugs) with higher and more controlled activity, better selectivity towards the target, and subsequently less toxicity, improved pharmacokinetics and formulation. Examples are: carboplatin, oxaliplatin, chlorambucil, melphalan, carmustine, cyclophosphamide, methotrexate, nilotinib, topotecan, and irinotecan.

6 a, b and d

7 a and b

8 a. True
   b. True
   c. False
   d. False

Chapter 16

1 a. False
   b. False
   c. True
   d. True
   e. True

2 a. True
   b. True
   c. True
   d. True
   e. True

3 a. True
   b. False
   c. True
   d. True
   e. False

4 a. True
   b. True
   c. False
   d. False
   e. False

5 a. True
   b. False
   c. True
   d. False
   e. True

Chapter 17

1 a. False
   b. False
   c. True
   d. False
   e. True

2 a. True
  b. True
  c. True
  d. False
  e. False

3 a. True
  b. True
  c. True
  d. False
  e. False

4 a. False
  b. True
  c. True
  d. False
  e. True

Chapter 18

1 a. True
  b. True
  c. False
  d. False
  e. False

2 a. False
  b. True
  c. False
  d. True
  e. False

3 a. False
  b. True
  c. False
  d. True
  e. True

4 a. False
  b. True
  c. False
  d. False
  e. True

Chapter 19

1 (a) ii; (b) v; (c) iv; (d) i; (e) iii

2 a. False
  b. True
  c. True
  d. False
  e. False
  f. True
  g. False
  h. False

3 (a) iv; (b) ii; (c) i; (d) iii; (e) v

Chapter 20

1 b

2 a

3 d

Index

Page numbers in *italics* denote figures, those in **bold** denote tables.

A1 cyclin 140, 280
A2 cyclin 140
Abelson mouse leukemia **193**
abiraterone **483**, 540
ABL **98**, **195**, **196**
abraxane 453, 533–4
Abraxas 333
ABT-510 **485**
ABT-737 **484**, 541
ABT-888 **542**
ACTH 44
  ectopic production 46
α-actinin 230
actinomycin D 534
activation-induced cell death (AICD) 424
active pharmaceutical ingredients (APIs) 456
acute lymphoblastic leukemia 62
  chromosomal analysis **513**
acute lymphoblastic lymphoma 263
acute myeloid leukemia (AML) 62, 170, 305
  chromosomal analysis **513**
acute promyelocytic leukemia (APL) 157, 378
adamantine 452
adaptive immunity 411
adenocarcinoma 11
adenoma 11
adenoma-carcinoma sequence *80*, 106, 107
adenoviruses 99
adherens junctions 230
adhesomes 387, 389, *389*
adjustment disorders 550
adjuvant chemotherapy 531
adoptive therapy 422–3
adriamycin 61
adult T-cell leukemia (ATL) 100
afatinib **536**
Afinitor *see* everolimus
aflibercept **537**
age, and cancer development *10*
agitation 550
AGO14699 **542**
agrin 385
AGTRL1 89
air pollution 96
Airn 369
AKT 24, 141, *153*, *154*, 155, 160, **161**, 169, 173, 226, 378
  see also PI3K–AKT pathway
AKT substrates 176
AKT2 174

alanine transaminase **512**
alcohol 96–7
aldehyde dehydrogenase *see* ALDH
ALDH **304**, 305, 454
ALDH2 87
alemtuzumab 63, 480, **484**, 538
ALK growth factor receptor 168, 169, 170, **197**
  molecular targeting 486
ALK oncogene **196**
alkaline phosphatase **512**
alkyl sulphonates 532
alkylating agents 52, 441–2, *442*, 531–2
  see also individual drugs
all-*trans*-retinoic acid (ATRA) 157, *419*
alleles 71
allodynia **547**
ALN-APC **496**
ALN-PCS02 **496**
ALN-RSV01 **496**
ALN-TTR02 **496**
ALN-VSP **496**
5-α reductase inhibitors 57, **483**
α-fetoprotein **502**, 512, **513**
alternative lengthening of telomeres (ALT) 296, 310
alternative reading frame *see* ARF
Alvocidib *see* flavopiridol
aml1 **197**
amentoflavone **483**
Ames test 98
amethopterin *see* methotrexate
AMG102 **541**
aminoglutethimide 539
aminopeptidase-P 566
aminopterin 441, *441*
AML1-ETO fusion protein 364
AMN107 499
AMP-activated protein kinase (AMPK) 177, 303
analgesic ladder 545, *546*
anaphase *116*, 133
anaphase-promoting complex/cyclosome *see* APC/C
anaplastic lymphoma kinase (ALK) 169
anastrazole **483**, 539
anchorage **148**
androgens 172
  deprivation 540
  receptor antagonists 540
anesthetic techniques in pain control 547
aneuploidy 22, 70, 82, 319, 320, 336
Angelman's syndrome 564
angioblasts 121, 430, *431*, 433

angiogenesis 158, *207*, 209, 403, 429–37
  clinical outcomes 436–7
  future directions 436–7
  general principles 430, *431*
  pathological 430–1, *432*
angiogenesis inhibitors 436, 450, *450*, **485**, 493, 536–7, **537**
angiogenic switch 158, 432
  stimulation of 432–6
angiography 520
angiopoietin 435
angiostatin 18, 429, 436, **485**
aniline dyes 94
animal models *see* cancer models; mouse models of tumorigenesis; and *specific animal models*
annexin A1 566
anoikis 25, 27, 222, 271, 404–5
  control of
    cell adhesion and growth factor receptors 405–6
    integrins 406
anorexia 549, **549**
ANRIL 369
antagomirs 373, 495
antagonistic pleiotropy 301
anthracyclines 52, 448, 534
anti-androgens 57
anti-CD20 antibodies 443
anti-CTLA-4 antibodies 425
anti-Delta-like 4 ligand (DLL4) antibodies 305
anti-MiRNAs 262
anti-PD-1 antibodies 425
anti-STAT3 drugs 477
antibody therapy 423, *423*
  see also specific antibodies
antibody-dependent cell-mediated cytotoxicity (ADCC) 413, 443, 488
anticancer drugs *see* chemotherapy; and *specific drugs*
antidiuretic hormone (ADH) 44
antiepileptics **547**
antigen presentation 413–15, *414*, *415*
antigen-presenting cells (APC) 337, 411, *415*
antigen-specific T-cell therapy 422–3
antigen-specific therapy 416, *416*, 420–2
antimetabolites 305, 441, *441*, 532–3, *533*
  see also individual drugs
antipurines 533
  see also individual drugs
antipyrimidines 533, *533*
  see also individual drugs

The Molecular Biology of Cancer: A Bridge From Bench to Bedside, Second Edition. Edited by Stella Pelengaris and Michael Khan.
© 2013 John Wiley & Sons, Inc. Published 2013 by John Wiley & Sons, Inc.

Index

antisense oligo(deoxy)nucleotides (ASOs) 451
antisense therapy 451
antithrombin III **496**
anxiety 550
AP endonuclease 1 (APE1) 327
AP23451 488
AP23464 488
APAF-1 130, 273, *274*, 275, 283, 285
Apak 338
APC gene 22, 53, *80*, 83, **84**, **87**, 88, 105, 106, 116, **124**, 228, **241**, **364**, **514**
APC protein 179, 224, 225, 227, 259, *260*, 305
APC specificity factors **124**
APC/C *123*, 126, 128, 129, 130, 132, 133, 135, 138, 320, 337
APC/C^CDC20 337
apoptosis 7, *24*, 26–7, 135, **148**, *154*, *201*, *207*, 239, 267
　　as barrier to cancer formation 271
　　BCL-2 protein family 82, **98**, **197**, 199, 209, 235, 279–81, *279*, *280*
　　caspases *see* caspases
　　context 267–71, *268–70*
　　in diabetes 287
　　IAPs 274, 276–9, *277*
　　mitochondrial pathway 342
　　MOMP 235, 272, 273, 274, 279–82, 285
　　MYC 209, 283–7, *284–5*
　　p53 protein 282–3, *283*
　　pathways 272–4, *273*, *274*
　　　　extrinsic pathway 26, 272–3, *273*
　　　　intrinsic pathway 26, 272, 273–4, *273*, *274*
　　RAS superfamily 222–3, *223*
　　type I cells 272
　　type II cells 272
　　see also cell death; necrosis
apoptosis inducers 425, **541**
apoptosis regulators **197**, 198–9, **484**
apoptosis-inducing factor 26
apoptosome 274, *275*
apoptotic bodies *270*, 272
apoptotic protease-activating factor 1 *see* APAF1
AR **364**
Arabidopsis thaliana 559
arabinosides 532–3
　　see also individual drugs
ARF gene 130, 245
ARF protein 254, 338
　　functions 254–6, *256*
　　in oncogenic stress *339*, 340–1, *340*
ARF-BP1 342
ARF-HDM2-p53 pathway 133
ARF-Mdm2-p53 pathway *256*
ARF-p53 tumor suppressor pathways 286
arginase 425
ARID1A 351, 353
Arimidex *see* anastrazole
aromatase inhibitors **483**, 539
aromatic amines 95
Arpp19 133
ARQ197 **541**
ARRY-142886 *477*, **541**
arsenic compounds 45
arylamines/amides 95
Arzerra *see* ofatumumab
asbestos 45, 95
ASCL1 **304**
ASF1a 379
Asn-Gly-Arg tripeptide **485**
asparaginase 534
aspartate transaminase **512**
aspirin 90
ASPP1 342
ASXL1 **355**

asymmetrical cell division 156
AT islands 344
AT9283 **541**
ataxia telangiectasia **85**, 141, 252, 336, 345
ATF4 282
Atg7 287
Atg8 378
Atg12 378
athymic nude mice 412, 417
ATM gene **85**, 86, 88, 130, 141, **241**, 561
ATM protein 252, 318, 322, 333
　　inhibitors 142
ATM-CHK2 pathway 318, 320, 322–3, 330, 346
ATM-CHK2-p53 pathway 322, 325
ATM/ATR *124*, 136
ATR 318, 322, 333, 346
ATR-CHK1 pathway 318, 320, 325, 330, *334*
ATRIP complex *118*, 136, 318, 325, 330, 336
Atu027 **496**
AURKA 134, 138, 140–2, 337
aurora mitotic kinases 134–5
　　and cancer 140–1
　　inhibitors 142
Aurora-A *see* AURKA
autoimmune lymphoproliferative syndrome (ALPS) 272
autologous tumor cell-based vaccines 421
autophagy 27, 287–90, *289*
　　and cancer 288–9
　　mechanism 288
　　signal pathways 288
AUY922 **542**
Avastin *see* bevacizumab
avian myelocytomatosis **193**
avian sarcoma 17 **193**
avodart **483**
axitinib **478**, *482*, **485**, **537**
AZ703 **484**
5-azacytidine 368
AZD1152 **484**, **541**
AZD1840 **483**
AZD2281 **542**
AZD6244 **483**, **541**
AZD7762 **484**, **541**
aziridines 532

B-cell leukemia 82, **470**
B-cell lymphoma 23, 181, 185, **196**, **197**, 206, 222, 235, 243, 279, 287, **355**, 372, 423, **470**, **484**, 493, 538, 564
B-RAF **198**
B7 family 415
B12536 **541**
B16727 **541**
bacteria 101
BAD 169, 173, *174*, *175*, 176, 222, *223*, 233, *234*, 280, *285*, 325
BAF180 338
BAI3 89
BAK 222, *223*, 235, 273–4, 279, 280, *281*, 282, 286, *288*, *342*
Bannayan-Zonana syndrome 259
BARD7 **357**
Barker hypothesis 301
Barrett's esophagus 12, 29
basal cell carcinoma 59
base excision repair (BER) 135, 325, 326–7, *327*, 335, 475
basement membrane
　　adhesive interactions 384–5
　　growth and survival signals 386, *386*
　　and polarity 385–6, *386*
BAT-25 106
BAT-26 106

BAX family 273–4, 280, 283, 285, 318, 342, 343, 406
BAY 43–9006 **537**
4-1BB 415
Bbc3 252
BCA2 128
BCG vaccine 452
BCL-1 **195**
BCL-2 inhibitors 450, *450*
BCL-2 oncogene 259, 271
BCL-2 protein family 82, **98**, **197**, 199, 209, 235, 279–81, *279*, *280*, **470**
　　MYC signaling through 284–6
　　proapoptotic members 280
　　prosurvival (antiapoptotic) members 280
　　regulation of apoptosis 280–1, *281*
BCL-w 280
BCL-xL 226, 280
BclxL 338
BCNU *see* carmustine
BCR ligase 129
BCR–ABL oncogene 368, 474, 504
BCR-ABL protein 32, 82, 171, **196**, **198**, 232–5, *233*, *234*, **470**, 475, 564
　　inhibition of 480
　　molecular targeting 486, 488
BCR-JAK2 178
Beckwith-Weidemann syndrome 23, **85**, 363, 366, 564
beclins 27, 288
benign prostatic hypertrophy (BPH) 514
benign tumors 384
benzodiazepines **547**
Benzvix 451
bevacizumab 49, 52–3, 57, 480, **485**, 488, 493, **537**
Bexaar *see* tositumomab
BH3-only family 280, *280*, 283
　　see also BAD; BID; BIK; BIM; BMF; HRK; NOXA; PUMA
BI 2536 142, **484**
bi-orientation 133
BID 272, 280, 342
BIIB021 **542**
BIK *279*, 280, 282, 492
bilirubin, raised **512**
BIM 130, 233, 280, 282, 285, 378, 475
biochemical tests 511–12, **512**
biological therapies 410–28, 466
biomarkers 2, 13, 15, 102–3, 368, 501–5, **502**, *503*, 512–13, **513**
　　cancer stem cells **304**
　　colorectal cancer 504–5
　　screening 104
　　and treatment selection 104–5
　　see also specific biomarkers
biopsy 518–19
　　fine needle aspiration 518
　　incision and excision 518–19
　　needle 518
biotherapy 497
BIR domains 277, *277*
BIRC2 278
BIRC3 278
bisphosphonates 57
bizelesin 344
BKM120 **541**
bladder cancer, susceptibility 88–9
bladder tumor-associated antigen **502**
blast crisis 486
blenoxane 448
bleomycin 61, 447–8, 534
BLM **85**
blood oxygen level (BOLD) 522

Index

blood vessels 429, *430*
  cancer spread through 31
  see also angiogenesis
Bloom syndrome **85**, 345
BMF 280
BMI **304**
BMI-1 gene 367
BMI-1 protein 21, *154*, 302, 336, *341*
BMP 178, 182, 184, **241**
BMS214662 **483**
BMS354825 *see* dasatinib
BMS536924 **483**
BMS540215 **541**
BMS554417 486
BNC2 89
Bnip3 130
body fluids, cytological examination 518
BOK 130, 280
bone metastases 494
bone pain 546
bone-localizing α-emitting agents 497
BOPP 451
borealin 134
BORIS *360*, 366
bortezomib 376–7, **484**, 493, *493*, 538, **541**
bosutinib 479, 488
bowel obstruction 548–9, **549**
bowel pain 546
53BP1 318, 336
brachytherapy 527
  prostate cancer 57
Bradford Hill, Austin 73–6, 78
BRAF 32, 58, 90, 106, 194, 198, 220, 366
  mutations 224
brain tumors 511
BRCA1 2, 14, 15, 22, 52, 71, 77, **84**, 91, 104, 105, **241**, 318, 333, 334–5, 363, **364**, 504, **514**, 556, 561
  synthetic lethality 476
BRCA1-A complex 333
BRCA1-BARD1 333
BRCA2 2, 14, 15, 22, 52, 71, 77, **84**, 91, 104, 105, **241**, 333, 334–5, 363, **364**, 504, **514**, 556, 561
BRD2 **357**
BRD4 **357**, 365
BRD4-NUT oncoprotein 365
BRD7 338
breast cancer 49–53, 334–5
  BRCA mutation *see BRCA1*; *BRCA2*
  clinical features 50, *50*, 511
  diagnosis and investigations 50–1
  familial **84**, 91, **514**
  genetic testing 513
  HER2 mutation 14, 467
  incidence 8, 49
  molecular classification 467
  mortality 43, 49
  pathology 50
  RB pathway **248**
  risk factors 49–50
  staging **51**
  susceptibility *83*, 86
  treatment 51–3
    chemotherapy 52–3
    hormone therapy 53, 539
    PARP inhibitors 2
    radiotherapy 51–2, *52*
    surgery 51, 463–4
    triple-negative 467, 486, 488
breathlessness 547
Breslow staging of malignant melanoma 58
BRG1 353
BRIP1 86, 345

brivanib **537**, **541**
BRM 353, **356**
bronchoscopy 519, *519*
BSI-201 **542**
Btg proteins 252
BUB proteins *123*, 138, 337
BUB1 gene 319
BUB1 protein *118*, *123*, *127*, 134, 137, 337, 346
BUB1-3 137
BUBR1 gene 130, 138
BUBR1 protein 346
buprenorphine **546**
burden of cancer 43–66
Burkitt's lymphoma *61*, 82, 255
  MYC overexpression in 363
busulfan 63, 532

c-ABL 82, 195, 232, 310, 406
c-ErbB2 402
c-ErbB3 402
c-FOS gene 362
c-FOS protein *173*, 184, *218*, *234*, 235
c-kit 168, 170, 232, **304**, **478**, 479, **502**
  inhibition of 480, 482, **485**, 537
c-MET 149, 168, 401, 402
  antibodies **478**, 484, 485
c-MYC *see* MYC
c-MYCER system 469
C-reactive protein **512**
c-sis 164
c-SRC *see* SRC
C/EBP homologous protein (CHOP) 282
C20orf126 570
CA-15-3 **502**, **513**
CA-19-9 **502**, 512, **513**
CA-27-9 **502**
CA-125 104, **502**, 512, **513**
CAAX motif 215
cabozantinib **541**
cachexia 44, 549, *549*
cadherins 393–4
  see also specific cadherins
Caenorhabditis elegans 267, 268–9, 370, 556, 557
CAGE 363
CALAA-01 452, **496**
calcitonin 44, **502**, 512, **513**
calpains 208
Campath *see* alemtuzumab
camptothecin 448, *448*, 534
cancer biology 3–42, *5*
cancer cells 1
  heterogeneity 466–7
  life history 9
cancer environment 28–9
cancer genes 559–62
  see also specific genes
cancer models 557–9
  mouse *see* mouse models of tumorigenesis
cancer pain syndromes 546–7, **547**
cancer pathways 18
cancer profiling 367–8
cancer road-map 189, 474
cancer stem cells (CSCs) 1, 6, 18, 19, 20, 106, 178–9, 298, 303–6, 464
  markers for **304**
  targeting of 20–1
cancer susceptibility syndromes 345–6
  see also specific syndromes
cancer therapy *see* chemotherapy; radiotherapy; surgery; *and specific cancer types*
cancer-initiating cells (CICs) 106
cancer-testis antigens 420
canertinib **478**
canine transmissible venereal tumor (CTVT) 31

capecitabine 52, 56, 441, **442**, 455
capillaries, cancer spread through 396–7, *397*
carboplatin 49, 449, 532
carcinoembryonic antigen (CEA) 55, **418**, **502**, 512, **513**
carcinogenesis, multistage 16–18, *17*, 79, *80*
carcinogens 91–2, *92*, 93
  activation 94–5
  chemical 94, 95
  lung cancer 45
  skin cancer 58
  testing for 98
  see also environmental causes of cancer
carcinoma 11
carcinoma *in situ* 11
caretaker genes 24–5, *24*, 317, 324
carmustine (BCNU) 368, **440**, 449, 532
  Gliadel Wafers 456
CAS 230
CASP-3 130
CASP-7 130
CASP-8 **418**
caspases 26, 252, 272, 274–6, *276*, **276**
  effector (executioner) 274–5, *276*
  inhibition 277–8, *277*
  initiator 274–5, *276*
  molecular targeting **496**
  procaspases 275, **276**
Castleman disease 99
catching cancer 31–2
β-catenin 376, 393, **418**
  molecular targeting **496**
β-catenin-WNT 394
cause and effect 70–1
causes of cancer 16–21
  cells of origin 18–20, *19*
  clonal evolution theory 16
  multistage carcinogenesis 16–18, *17*
CAV-1 489
Cbl 128
Cbx7 367
CCCTC-binding factor (CTCF) 366
CCL21 397
CCND1 **87**, 140
CCND2 204, *205*, 207
CCNE1 88
CCNU *see* lomustine
CCR2 414
CCR6 414
CCR7 397, 413, **414**
CD4 411, *411*, 413, 415, 416
CD8 411, *411*, 413, 415, 416–17
CD13 **485**, 576
CD14 421
CD20 **418**, 423
  antibodies 443, **484**
CD22 423
CD24 **304**
CD25 417
CD28 415, 424
CD33, antibodies **484**
CD34 **304**, 421
CD36 413
CD40 413, **414**, 421
CD44 21, **304**, 373
CD52, antibodies **484**
CD80 413, **414**, 415
CD83 413, **414**
CD86 413, **414**, 415
CD90 **304**
CD95 *see* FAS ligand
CD105 **304**
CD117 *see* c-kit
CD133 21, **304**

605

Index

CD166 21, **304**
CDC5C 330
CDC6 130
CDC7 *see* DDK
CDC7/Dbf4 kinase 132
CDC20 129, 133, 337
CDC25 133, 336
CDC25A 330, 336
CDC25B 330, 336, 337
CDC25C 336
CDC42 30, 226, 402
CDH1 **364**
CDH13 **364**
CDK1 133
CDK2 300
CDK4 **195**, 246
CDK4 kinase 420, **418**
CDK6 246
CDK7 328
CDKN1B 140, **241**
CDKN1C 469
CDKN2 368
CDKN2A **85**, **87**, 132, **241**, 363, 365
CDKN2B 363, 469
CDKN2C **241**
cDNA 562
CDR3 420
Cdt1 115, 129, 132, 320
cediranib 489, **537**
celecoxib 90
cell adhesion 139, 179–80, *179*, 383–409, *384*
    control of anoikis 405–6
    extracellular matrix 384–93
cell adhesion molecules (CAMs) 394
cell cycle 111–45, *113*, *127*, *207*, *244*
    and cancer 139–41
    cancer therapy effects 141–2
    checkpoints 117, *118*, 119–20
    DNA replication 6, 115, 131–5
    entry 138–9
    global gene expression 139
    inhibitors **484**, 490, *490*, **541**
    mitosis 22, *116*, 133
    phases of 120–3
    proteins **124**
    regulation 117, 150–1, *151*
    translational control 184–5
    see also specific elements
cell cycle engine 123–4, 150
    inhibitors of 125–6
cell death 266–94
    apoptosis *see* apoptosis
    autophagy 27, 287–90, *289*
    and cancer control 290–2, *291*
    endoplasmic reticulum stress 220, 222, 282
    historical perspective 267
    mitotic catastrophe 290
    necrosis 26–7, 267, 270, 271–2
    oncogenic stress 26, 126, 209, 283–7, *284–5*, 323–4
    programmed 267, 269
    response to cancer therapy 290, *290*
cell differentiation 208–9
cell division, asymmetrical 112, 156
cell plasticity 158
cell proliferation *see entries beginning with growth*
cell senescence 295–313
cell size regulation 155
cell stress 126
cell suicide *see* apoptosis
cell surface toponome 575–6, *575*
cell-cell interactions 393–5, *393*
    cadherins 393–4
    β-catenin-WNT 394
    desmosomes 394

cell-cell signaling junctions 394–5
cells of origin 18–20, *19*
cellular differentiation 157–8
    terminal differentiation 158
cellular information flow 35
cellular networks 569–70, *571*, *572*
cellular senescence *see* senescence
CENP-A 137
centromere proteins 337, 489
centromeres 137–8
centrosomal cycle 118
centrosomes 133–4
CEQ508 **496**
cervical cancer 60
    screening 103–4, 514
cetuximab 56, **478**, 480, 488, 535, **536**
Ch14.18 **484**
CH401 monoclonal antibody **484**
chaperones 344
checkpoints 117, *118*, 119–20, 126, 135, 136, 336–43
    in cancer 141
    G1/S 120, *121*, 131–3, 136–7
    G2/M 120, *122*, 136
    inhibitors 142
    organ-size 154–5
    spindle 120, *123*, 137–8
    see also specific checkpoint genes/proteins
checkpoint effectors 316
checkpoint mediators 316
checkpoint sensors **124**, 318
checkpoint transducers 316, 318
CHEK2 86, 105, 336
chemical carcinogens 94, 95
Chemistry, Manufacturing, and Control (CMC) program 456
chemokines 211, 387, 397, 411, 413, 494
chemokine receptors 414
chemoprevention 90
chemoradiation 530
chemoresistant niche 20
chemosaturation 497
chemotherapy 15–16, 438–60
    breast cancer 52–3
    and cell cycle 141–2
    and cell death 290, *290*
    cervical cancer 60
    colorectal cancer 56
    combination chemotherapy 530
    cytotoxic drugs 531–5
    drug delivery 452
    drug development 442–57
    indications 530–1
    individualization of 33–4
    leukemia 63
    lung cancer 49
    neoadjuvant 531
    new agents **541–2**
    non-Hodgkin lymphoma 62
    rationale 530
    resistance to *see* drug resistance
    targeted *see* targeted therapy
    see also specific drugs
chick chorioallantoic membrane (CAM) 432, 433
chicken erythroleukemia **193**
chimeric proteins 584
ChIP 203
ChIP on Chip 559, 560, 564
CHK1 330
CHK2 130, 330, 333, 338, 561
chlorambucil 63, *440*, 449, 532
cholangiocarcinoma (CCA) 101
CHOP *see* C/EBP homologous protein
CHOP regimen 62, 538

chromatin 131, 203, 324, 353, 354, 378–9, 556
    molecular targeting **484**
    remodeling 364–5
chromatin immunoprecipitation *see* ChIP
chromatin remodeling factors **356**
chromatolysis 267
chromium 45
chromosomal instability (CIN) 106, 319
chromosomal passenger complex (CPC) 137
chromosomal translocations 81, 82
chromosomes 114, 115
    segregation in mitosis 133
chronic lymphocytic leukemia (CLL) 23, 62
    chromosomal analysis **513**
chronic myeloid (myelogenous) leukemia (CML) 82, 170, 195
    BCR-ABL signaling *234*
    chromosomal analysis **513**
CI1033 **478**
CI1040 477
cilengitide 494–5, *495*
CIMP 106, 363, 365–6, *365*
CIP/KIP family 125, 126, *126*, 246
circadian rhythms 31
circulating tumor cells (CTCs) 52, 405, 488, 523–4
cisplatin 49, 60, 305, 444–5, *445*, *446*, 532
CKS 204
Claspin 318, 330
classification of cancer 11
clinical features of cancer 510–11
    breast cancer 50, *50*, 511
    colorectal cancer 55, 511
    lung cancer 511
    lymphoma and leukemia 510
    neurological tumors 511
    prostate cancer 56–7
clinical trials 506
    first-in-man studies 506
    immunotherapy 425
    vaccines 422–3, *423*
clofarabine 441, *442*
clonal evolution theory 7, 16, 20
clusterin 451
coanalgesia 547, **547**
cobalt gamma ray machines 527
Cockayne syndrome 329
Cockayne syndrome proteins 328
Coco 182
cohesin **124**, 134
Coley, William 410
collagen 385
    fibrillar 387
collective cell migration 403–4, *405*
colon cancer *see* colorectal cancer
colonoscopy 515, *515*
colorectal cancer 53–6
    biomarkers 504–5
    clinical features 55, 511
    diagnosis 54, *54*, *55*
    epigenetics *370*
    familial 91
    and genomic instability 346
    incidence 8, 53
    mortality 43
    multistage carcinogenesis *80*
    risk factors 53
    screening 54
    staging 54–5, **54**
    susceptibility 82, 84
    treatment 55–6
        chemotherapy 56
        radiotherapy 55
        targeted therapy 488
combination chemotherapy 530

Index

combinatorial molecular phenotypes (CMPs) 572, *573*, 576
combretastatin A **485**
communication with patients 544–5
   blocking strategies 545
comparative genome hybridization 22
comparative genomic hybridization arrays 559, 560, 564
COMPARE algorithm 444
complement-mediated cytotoxicity (CMC) 443
complementary DNA *see* cDNA
Compton scatter 526
computerized tomography (CT) 497, 516–17, *517*
conformal radiation therapy (CRT) 497
connective tissue growth factor **496**
constipation 549
contact inhibition **148**, 395
continuous hyperfractionated accelerated radiotherapy (CHART), lung cancer 49
COP9 128
cost-benefit ratio 36
Cowden syndrome 83, **85**, 174, 177, 259
COX-2 **364**
COX-2 inhibitors 89, 491–2
CpG islands 242, 352, 353, 360, 362–3, 368, 380, 556
CpG island methylator phenotype *see* CIMP
CpG island microarrays 559, 560, 564
creatinine, raised **512**
CREB-binding protein (CBP) 220
CREBBP **356**
Cripto-1 182
crizotinib 169, **478**, **541**
crk **196**
crossing over 22
cryosurgery
   prostate cancer 57
   squamous cell carcinoma 59
CTGF 30
CTLA-4 415
   antibodies **484**
CTNNB1 106
CUL1/*CUL1* 128, 204
cullin-RING E3 ubiquitin ligases (CRLs) 128
CXCL12 397
CXCR4 305, 397
CYC202 **541**
cyclic reversible termination 565
cyclin B-CDK1 complex 336
cyclin D1 179, **198**, 226
cyclin-dependent activating kinase (CAK) *205*, 246
cyclin-dependent kinase inhibitors (CKIs) 119, 123, **124**, **125**, *127*, 142, 246, *248*, 362
   and cancer **125**, 141
   CIP/KIP 125, 126, *126*, 246
   INK4 125, 126, *126*, 204
cyclin-dependent kinases (CDKs) 117, *120*, 123, **124**, 125, **125**, *127*, 132, 150, 203, 204
   and cancer **125**, 140
cyclins 117, *120*, 123, **125**, *127*
   and cancer **125**, 140
cyclizine 548
cyclodextrin 452
cyclophosphamide 62, 305, *440*, 449, 454–5, *454*
CYP1A1 **87**, 95
CYP1B1 **87**
CYP2A6 **87**
CYP2B6 455
CYP2C9 455
CYP2C19 455
CYP2E **87**
CYP3A4 455
CYP17 **87**
CYP19 **87**

CYP171A 540
cystic fibrosis transmembrane conductance regulator (CFTR) gene 556
cytarabine 441, *442*
cytochrome *c* 235, 266, 272, 273–4, *273*, 275, 279, 280, *281*, 282–3, *283*, *284*, 288, 342
cytochrome P450-dependent monooxygenases (CYPs) 94
   see also specific CYPs
cytokeratins **502**
cytokines 162
   immune response 411
   signal pathways 177–8
cytokine therapy 423–4
   IL-2 423–2
   IL-7 and IL-15 424
   IL-12 424
   IL-21 424
   interferons 424
cytokinesis *116*
cytomegalovirus 99
cytoplasmic polyadenylation element-binding protein (CPEB) 373
cytoplasmic regulators **196**
cytosine, methylation of 352
cytostatic signals 150
cytotoxic (CD8) T cells 416–17
cytotoxic drugs 531–5
   alkylating agents 52, 441–2, *442*, 531–2
   antimetabolites 305, 441, *441*, 532–3, *533*
   enzymes 534
   mitotic inhibitors 533–4
   platinum compounds 448–9, *449*, 532
   topoisomerase inhibitors 534
   see also chemotherapy; and *individual drugs*
cytotoxic T lymphocyte-associated antigen 4 *see* CTLA-4

dacarbazine 59, 61, 489, 532
dacomitinib **478**
DALYs 36
damage-regulated autophagy modulators (DRAM) 27
DAPK **364**, 368
darinaparsin 453
dasatinib 63, **478**, 479, 482, 488, 489, 499, **541**
   molecular target 477
   resistance 501
daunorubicin 448, 534
DAXX 364
Dbl **196**
DCC 84
DCGR8 261
DDB1 328
DDK **124**, 132
death ligands 291
death receptor pathway 26, 272–3, *273*
death-inducing signaling complex (DISC) 272, 284
dedifferentiation 20
definition of cancer 8, 10–1
degradation of proteins 126, 128–9
degrons 128
delayed gastric emptying 547–8
deletions 17, 80, 82
Deltex 180
dendritic cells 411, *411*, 413, **414**
dendritic cell vaccines 421
5′-deoxy-azacytidine (decitabine) 368
depression 550
Dermo 255
Desert Hedgehog pathway 489
desmocollins 394
desmogleins 394

desmoplastic small round cell tumor (DSRCT) 367
desmosomes 394
detection of cancer 13–4
   see also biomarkers; diagnosis; screening
Developmental Therapeutics Program (DTP) 441, 446, 493
dexamethasone 549
diabetes 287
DIABLO/Smac 274
diacylglycerol (DAG) 170
diagnosis 44, 509–25
   angiography 520
   clinical symptoms 510–11
   flow cytometry 522–3
   fluorescent in situ hybridization 523
   interventional radiology 520
   investigations 511–15
   molecular pathology 522
   radionuclide imaging 520–1, *521*, *522*
   see also staging; and *specific cancer types*
diamorphine **546**
dibenzanthracene (DBA) 94, 95
diet 75, 76, 96–7
diethylstilbestrol 540
differential methylated regions (DMR) 366
differentiation therapy 157
diffusion weighted MRI 522
difluoromethylornithine (DFMO) 90
dihydrofolate reductase (DHFR) 130, 441, 532
dihydrotestosterone 540
dioxins 95
disability-adjusted life years *see* DALYs
DISC *see* death-inducing signaling complex
diurnal rhythms 31
DKK4 505
DLC1 (deleted in liver cancer 1) 227–6
DLL4 436
DMP1 **241**
DNA 77, 556
   complementary (cDNA) 562
   endoreplication 115
   junk 316, 352
   methylation 243, 359–60, *359*, *360*, *361*, 362
   noncoding (ncDNA) 23, 352
   repair 2, 135
   replication 6, 115, 131–5, 244
   synthesis *126*
DNA damage 24, 141, 316
   checkpoints *see* checkpoints
   life-death decisions 337–8, *338*
   radiotherapy 346
   response to *see* DNA repair
   signaling of 325
DNA damage sensors 316, 324–5
   9-1-1 complex 325
   ATRIP *118*, 136, *318*, 325, 330, 336
   KU80 141, *318*, *322*, *324*, *331*, 325, 333
DNA methyltransferases (DNMTs) 23, 205–6, 352, **355**, 556
   inhibitors of 477
DNA microarrays *see* microarrays
DNA mismatch repair 22, 23, 106, 135, *318*, 325, 329–30, *329*, 343
DNA polymerases 130, 336
DNA repair 136, 260, 286, 299, *317*, 321–4, *322–4*, 325–36, *326*, 527
   double-strand breaks 330–3, *331–2*
   fragile sites 325–6
   molecular targeting 490–2
   PARP 35, 333, 335–6
   short telomeres 310

Index

single-strand breaks 326–30
  base excision repair 135, 325, 326–7, *327*, 335, 475
  mismatch repair 22, 23, 135, 325, 329–30, *329*, 343
  nucleotide excision repair 135, 325, 327–9, *328*, 345
  stalled replication forks 330, *331*
  translesional synthesis 326, 330
stem cells 336
transcription-coupled repair 325, 328
DNA segments with chromatin alterations reinforcing senescence (DNA-SCARS) 299
DNA synthesis 204
DNA viruses 25, 98–100
see also specific viruses
DNA-dependent protein kinase (DNA-PK) 333
DNMT1 gene 363
DNMT1 protein **355**, 359
DNMT3a protein 352, **355**, 359
DNMT3b gene 363
DNMT3b protein **355**, 359
docetaxel 49, 57, *447*, 533
DOCK180 402
dominant 71, 189–90
domperidone 547
donor lymphocyte infusions (DLI) 422
dormant cancer cells 464
double strand breaks (DSBs) 283, 316, 318, 527
  repair 330–3, *331–2*
double-minute chromosomes 81
dovitinib lactate **537**
doxorubicin 62, 305, 448, *448*, 534
doxycycline 468
DPC4 **84**, **241**
Drosha 261
Drosophila melanogaster 556, 557
  Hippo signaling pathway 183–4, *184*
  Notch signaling pathway 180
drozitumab **541**
drug delivery 452
drug design 438–42
  historical perspective 439–42, *440–2*
drug development 442–57
  lead compounds 443–53
  lead optimization 453–6
  target selection 443, 468
  timeline 442
  see also individual drugs
drug resistance 2, 34–5, 473
  chemoresistant niche 20
  downstream mutations 499
  epimutations 500
  mutations or changes in drug target protein 498–9
  switching of oncogene dependence 499
  target definition in 501
  targeted therapy 487, 497–501, *498*
drug targeting *see* targeted therapy
DSBs *see* double-strand breaks
DTIC *see* dacarbazine
Dukes' staging 101–2
  colorectal cancer **54**
duplications 82
dyads 115
dying patients 550
dynamic contrast enhanced CT (DCE-CT) 518, 522
dynamic contrast enhanced MRI (DCE-MRI) 518
dynein 134
DYRK2 338
Dyskeryn 310

E-cadherin 30, 179, 229, 363, 394, 399–401
E-twenty six family *see* ETS family

E1 140
E2 140
E2F oncogene 190
E2F transcription factors (E2F1–5) 129–8, 244
E2F-associated phospho-protein (EAPP) 130
E2F1 oncogene 255
E2F3a 244
E2F3b 244, 245
E2F6 245
E2F7 245
E6 oncoprotein 344
E7 oncoprotein 344
early-response genes 204
4EBP-eIF4E 475
effector cells 413–17
  dendritic cells 411, *411*, 413, **414**
  natural killer (NK) cells 291, 411, 417
  T cells *see* T cells
  Tregs 29, 417
effector kinases **124**
EGF 148, **149**, 160
EGFR gene 90, 168, **193**, 194, **195**, **196**, 214, 224, 368, 475, 502
EGFR 21, 32, 45, 101, 105, 106, 147, **149**, 167, 168–71, 184, **195**, **196**, **197**, 227, 229, 377, 477, 569
EGFR inhibitors 89, 535–6, **536**, 537
EGFR-RAS-MAPK system 181
EHMT2 **355**
Ehrlich, Paul 443, 465
eIF3E 175
EKB-569 **478**
elastin 387
electrons 527
Elk-1 378
Eloxatin *see* oxaliplatin
embryonal rest theory 19
embryonic stem cells (ESCs) 113, 302–3
Emi1 130
EML4-ALK fusion gene 486
EMT *see* epithelial-mesenchymal transition
end-replication problem 296
endogenous inhibitors of angiogenesis 436
endometrial cancer 172
endonuclease G 26
endoplasmic reticulum stress 220, 222, 282
endoplasmic reticulum stress elements (ERSEs) 282
endoreplication 115
endoscopic retrograde cholangiopancreatography (ERCP) 519
endoscopic ultrasound 516
endoscopy 519
endostatin 18, 429, 436
endothelial cell progenitors (EPCs) 430
endothelial cell proliferation inhibitors 485
entosis 321
environmental causes of cancer 93–101
  individual risk factors 93–4
  see also specific causes
enzastaurin **483**
enzyme inhibitors **483**, **541**
EP300 **356**
EpCAM 304
EPH receptor family 395, 435
EPH-ephrin signaling pathway 394, 435
EPHA3 88
ephrins 394, 395, 435
EPIC (European Prospective Investigation into Cancer and Nutrition) study 75, 95–6
epidemiology 8, 73–6
epidermal growth factor *see* EGF
epidermal growth factor receptor *see* EGFR
epidural catheters 547

epigenetic factors 6, 23, 105–6
  drugs targeting 368
epigenetics 157, 242, 353–9
  and cancer 362–5, **364**
  chromatin 131, 203, 324, 353, 354
  clinical uses 367–8
    cancer profiling 367–8
    drug therapy 368
  historical perspective 358
  regulator genes **355–7**
  molecular targeting 492–3, *493*, 537–9
epigenome 352
epigenomics 559
epimutagens 366
epimutations 1, 2, 6, 8, 21
  critical 24
  drug resistance 500
epirubicin 534
epithelial cell transforming sequence 2 (ECT2) 227
epithelial-mesenchymal transition (EMT) 1, 178–9, 296, 305–6, 399, 401–2, *401*, 467
epothilones 534
epoxyeicosatrienoic acids (EETs) 492
Epstein-Barr virus 99, 413, **418**
  Hodgkin disease 60
  latent membrane proteins (LMPs) 420
ER **364**
ErbA1 **197**
ErbA2 **197**
ErbB gene 98
ErbB family 32–3, 168, 535
  anti-ErbB monoclonal antibodies 477–8, **478**, 488
ErbB1 *see* EGFR
ErbB2 *see* HER2
ErbB4 88
Erbitux *see* cetuximab
ERCC1 328
Erks 393
erlotinib 32, 49, **478**, 481, *499*, 535–6, **536**
  molecular target 477
  resistance 498, 499
ERTAM 210, *211*, 468, 584
erythrocyte sedimentation rate (ESR) 511
erythropoietin 44, **149**
estrogen receptors 467, **502**
estrogens 171–2
  and cancer 97
etoposide 534
ETS family 170, 172, **197**, 198, *216*, 217, *218*, 234, 261
euchromatin 353, 354
European Medicines Agency (EMEA) 457
European Randomized Study of Screening for Prostate Cancer 103
everolimus **483**, **541**
Ewing, James 439
exemestane **483**
exercise 76
expression analysis 559
EXT1 **364**
extracellular matrix (ECM) 139, 162
  adhesive interactions 384–93, *385*
    basement membrane 384–6
    fibrillar collagen 387
    GAGs 387
    integrins and adhesomes 387, 389, *389*
    stromal-epithelial interactions 387
  degradation 403
  elasticity 389, 391, *391*, 392–3
  p53-ATM responsiveness 406
  transmembrane proteoglycan receptors 389
extrinsic pathway of apoptosis 26, 272–3, *273*
EZH2 **355**, 367, 563

Index

F-box proteins 128
FADD 272, 275
FAK 230, 278, 391, 406
false-negative diagnosis 103
false-positive diagnosis 103
familial adenomatous polyposis coli (FAP) 53, 83, **84**, 259–60, **514**
   screening 105
familial cancer syndromes **514**
   see also specific syndromes
familial hypercholesterolemia (FH) 72
FANCA 345
FANCB 345
FANCA-FANCH **85**
FANCD1 gene see BRCA2
FANCD2 gene 345
FANCD2 protein 118, 136, 137, 318
FANCE 318
FANCF 318
FANCG 318
FANCI 318
FANCJ 318
FANCL 318
Fanconi anemia 83, **85**, 318, 345–6
Fanconi anemia proteins 345
Farber, Sidney 441
farnesyl transferase (FTase) 215, 489
farnesyl transferase inhibitors (FTIs) 215, 477, 489
FAS ligand 29, 291, 342
FAS receptor 272
FAS-associated death domain see FADD
Fas-like inhibitor protein see FLIP
Faslodex see fulvestrant
FAT **364**
FAT10 378
fatigue 549
Fbw7 128–9, 376
Fbxw7 128
FdUMP 533, 533
fecal occult blood **502**, 515
fentanyl **546**
fes **196**
FGF **149**, 160, 306, 434–5
FGFR 170
   antibodies **484**
FGFR1 194
FGFR3 194
fgr **196**
fibrillar collagen 387
fibrillin 387
fibrin **502**
fibroblast growth factor see FGF
fibronectin 387, 391
fibulin 385
field cancerization 497
field effect 29–30
figitumumab 486
finasteride **483**
fine needle aspiration (FNA) 518
Finsen, Niels 450
first-in-man studies 506
flavopiridol **484**, 490, 490, **541**
Flemming, Walther 267
FLICE inhibitory protein (FLIP) 284
FLIP 376, 425
flow cytometry 522–3
Flt3 170
fludarabine 441, 442, 532–3
fluorescence microscopy 573
fluorescence-activated cell sorting (FACS) 579
fluorescent in situ hybridization (FISH) 523
5-fluoro-2 deoxyuridine monophosphate see FdUMP
5-fluorouracil 56, 441, 442, 533
   mechanism of action 533

flutamide 57
FM165 449
FM190 449
fms **196**
FMS-like tyrosine kinase 3 receptor (FLT3) 170
focal adhesion kinase see FAK
focal adhesions 230
folate antagonists 532
folic acid 441
Food and Drug Administration (FDA) 457
forkhead transcription factors 131, 176
FOS **98**, **197**, 198
Foscan 451
14-3-3 proteins 325
14-3-3δ 363
FOXC 178
FOXO 24, 153, 167, 173, 174, 176, 216, 221, 233, 234
FoxO3a 133, 340, 378
Fra 179
fragile sites 325–6
frameshift mutations 80
Frizzled 394
FUB1 378
fulvestrant **483**, 539
functional MRI 522
FWT1 367
FWT2 367

G-protein coupled receptors 89, **161**, 170–1, 194, **196**
   molecular targeting 489
G-proteins, membrane-associated **161**
G0/G1 transition 132
G1 cyclin **124**
G1/S checkpoint 120, 121, 131–3, 136–7
G2/M checkpoint 120, 122, 136
GADD45 136
GADD45A 130
GAGE **418**
GAGs 387
galactosyltransferase 389
galectin-1 29, 425
ganetespib **542**
gap junctions 394
GAPs see GTPase-activating proteins
Gardner's syndrome 53
gastric cancer, mortality 43
gastrin 171
gastrointestinal stromal tumor (GIST) 32, 170
GATA-4 **364**
GATA3 467
GATA5 **364**
GD2 antibodies **484**
gefitinib 32, 49, 90, **478**, 480–1, 485, 499, 535, **536**
   resistance 498, 501
GEFs see guanine nucleotide exchange factors
geldanamycin 448
gemcitabine 49, 52, 305, 441, 442, 532
gemtuzumab 480, **484**
gender and cancer incidence 8
genes 77
   cancer-causing see oncogenes
   cancer-relevant **161**
   caretaker 24–5, 24, 317, 324
   see also specific genes
gene amplification 81, 193
gene chips see microarrays
gene expression
   epigenetic regulation 361–2
   profiling 419
   regulatable systems 583–4
   signatures 33
   "tet" system 210, 211, 468, 584

gene expression signature 33, 398, 563
gene gun 421, 422
gene knockouts 558, 582–4
gene rearrangements 80
gene silencing 359, 359, 368
   hypermethylation 362–3, **364**
gene therapy 495–7, 538–9
gene-environment interactions 89, 91
genetic locus 70
genetic testing 14, 15, 513, **513**, **514**
genetics of cancer 21, 70, 78–89
   mutations see mutations
   see also oncogenesis
genome-wide association studies (GWAS) 72, 74, 88–9
genomic instability 260, 314–49
   cancer susceptibility syndromes 345–6
   chaperones 344
   colorectal cancer 346
   radiation-induced 229
   telomere attrition 27, 154, 306, 321–4
   types of 319–20, 320
   see also DNA repair; mutations
genomic mutations 105
genomics 87–9, 559
genotoxic stress 25, 126
genotype 14, 71
genotype-dependent lethality 464
geranylgeranyl transferase (GGT) 489
geranylgeranyl transferase type I (GGTase-I) 215
germline mutations 242
gip **196**
Gleason system 56
Gleevec see imatinib
GLI 198
GLI1 182, 259, 305
GLI2 182, 298, **470**
GLI3 182
Gliadel Wafers 456
glial cell line-derived neurotrophic factor (GDNF) 169
glial-derived neurotropic factor (GDNF) **85**, **149**, 169, **196**
glioblastoma 164
   RB pathway **248**
glioma 367
global genome repair (GGR) 328–9
glucuronosyltransferases 94
glutathione S-transferases (GST) 94
glycogen synthase kinase 3 (GSK-3) 176
glycogen synthase-3β (GSK-β) 394
glycolysis 159
glycosylation 557, 569
GM-CSF 421
Goldie–Coldman hypothesis 530
gonadotrophin-releasing hormone agonists/antagonists 539
Good Laboratory Practice (GLP) 456
Good Manufacturing Practice (GMP) 456
Gorlin syndrome **85**, 182, 259
goserelin 53, 57, **483**
gp100 **418**
graft-versus-host disease (GVHD) 422
granzymes 417
granzyme B 291
green fluorescent protein (GFP) 570
GRM8 89
growth
   arrest 299
   homeostasis 151–5, 152–4, 258–60, 258–60
   MYC 204–6, 205
   RAS superfamily 217
   regulated/deregulated 155–7, 156

Index

growth factors 139, 147, **148**, **149**, 160–4, *162*, *163*, 164, 194, **196**
  activation *175*
  and cancer 163–4
  cell cycle regulation 150–1, *151*
  mechanism of action *162*
  oncogenes 164
  signaling pathways 32, 160
    see also signal transduction pathways
  stromal 387
  targeted therapy 482–5, **483–5**
  vascular 432–6
  see also specific growth factors
growth factor receptor tyrosine kinases 165, 167, *167*
growth factor receptors 163, 194, **196**, **197**
  control of anoikis 405–6
GSK461364 142, **484**, 541
GSK1070916 541
gsp **196**
GSTM1 **87**
GSTP1 **87**, **364**, 368
GTPase exchange factors **196**
GTPase-activating proteins (GAPs) 165, 215
GTPases
  NF-1 260
  RAS superfamily 213, 214–5, 217
  RHO family 225–8, *226*
guanine 445
guanine nucleotide exchange factors (GEFs) 165, 215

H-RAS **98**, **198**
H2AFX 563
H2AFZ 563
H2AX 141, 318, 325, 332–3, 336
H3K9triMe 301
H19 369
HA14-1 450, *450*
Haddow's paradox 439
Hakai 399
halichondrin B 448
hallmarks of cancer 37, *37*
haloperidol 548
haploinsufficiency 22, 71, 190, 240, 242
haplotypes 561, *561*
HapMap 560–1, 579
Harvey rat sarcoma **193**
haspin 134
HATs *see* histone acetyl transferases
HAUSP 376
Hayflick limit 27, 296, 300
hCDC4 337
HDACs 208, 245, 252, 263, 352, **356**, 360
HDAC inhibitors 477, **542**
HDM2 128
HDM2 ligase 375
heat shock proteins (HSPs) 475, 570
  and cancer 379
  inhibitors of 493, **542**
  stress-inducible 282
HECT 128
Hedgehog signaling pathway 21, 59, 157, 178, 182, 298, 489
HeLa cells 27, 70
Helicobacter pylori 25, 93, 101
helper (CD4) T cells 416
hematological malignancies 60–3
  Hodgkin disease 60–1, *60*, *61*
  leukemia 8, 11, 23, 62–3
  non-Hodgkin lymphomas 61–2
hematological tests 511–2, **512**
hematopoietic stem cells (HSCs) 303, 304
hemi-desmosomes 389

hemoglobin, low **512**
hepadnaviruses 98–9
hepatocyte growth factor (HGF) 304
  EMT activation 401–2
HER 368
Her-2-neu **418**
HER1 see EGFR
HER2 14, 51, 168, 194, **195**, **196**, 467, 475, 488, **502**
HER2 tyrosine kinase receptor 423
HER2-Neu receptor tyrosine kinase 476
HER3 **196**
HERC2 335
Herceptin *see* trastuzumab
hereditary cancer syndromes 78, 82–6, *83*, **84–5**
hereditary nonpolyposis colorectal cancer *see* Lynch syndrome
hereditary papillary renal cancer (HPRC) **85**
herpesviruses 99
heterochromatin 353, 354, 360
heterogeneity of cancer cells 466–7
  effect on therapy 467
  single cells 467
heterozygosity 70
  loss of 22, *22*, 240
Hexvix 451
HGF **149**, 226
HGF inhibitors **541**
HGS1029 **541**
HIC1 363
HIF1 226
high-intensity focused ultrasound (HIFU) 497
high-penetrance mutations 71
high-throughput screening (HTS) 443
highly conserved genes 70
HIPK2 338
Hippo proteins 183–4, *183*, *184*, 271
HIRA 379
Hirschsprung's disease 169
histology 101–2, 403
  breast cancer 399
  Hodgkin disease *60*
  NSCLC 466–7
histone acetyl transferases (HATs) 352, **356**, 360–2, 559
histone code 352, 361, 364
histone deacetylases *see* HDACs
histone lysine deacetylases *see* sirtuins
histones 203, 324, 354
  acetylation 360–2, *361*, *365*
  demethylation 355
  methylation **355**, *361*
hit-and-run models 584
hit-to-lead phase 443
HIV-1 Rev-binding protein (Hrb) 100
HJURP 137
HLA-DR **414**
HMGA2 23, 372
hMLH1 53, **364**
HMN-214 142
hMSH2 53
HNPCC 91
Hockenbery, David 259
Hodgkin disease 60–1, *60*, *61*
homologous recombination 324, *331–2*, 333
homozygosity 70
hormone therapy
  breast cancer 53, 539
  prostate cancer 57, 539–40
Horvitz, H. Robert 267, *268–9*
HOTAIR 369
HOXA9 **364**
HOXD1 89

HPC1 **85**
hPMS1 53
hPMS2 53
HPP1 365
HRAS gene **87**, 95, 194, 201, 214, 474
HRAS protein 222, **470**, 472, 489
HRAS1 **193**, **195**
HRK 280
Hrk/DP5 130
HSPs *see* heat shock proteins
HST **196**
HSTF1 **195**
HtrA2/Omi 273, 278
human chorionic gonadotrophin β(β-HCG) **502**, 512, **513**
human double minutes (HDMs) 338
Human Genome Project 164, 564–5
human immunodeficiency virus (HIV) 100
human nonmetastatic gene 23 (Nm23-H1) 494
human papillomavirus (HPV) 8, 15, 99, 344, 413, **418**
  cervical cancer 60
human T-cell leukemia viruses (HTLVs) 100
hyaluronan 387
hydromorphone **546**
5-hydroxyindoleacetic acid (5-HIAA) **513**
hydroxyurea 63, 534
hyperalgesia **547**
hypercalcemia **512**, 548, **548**
hypermethylation 362–3, 368
  gene silencing 362–3, **364**
hyperparathyroidism, tertiary 155
hyperpathia **547**
hyperplasia 395–6, *396*
hypoalbuminemia **512**
hypokalemia **512**
hypomethylation 363
hyponatremia **512**
hypoxia 305
hypoxia-inducible transcription factor (HIF) 375

IAPs *see* inhibitor of apoptosis proteins
[90]Y-ibritumomab tiuxetan 443, 538
IC486241 **484**
ICAM-1 415
ICAT 566, 567, 568
ICOS 415
idarubicin 534
ifosfamide 49, 454–5, *454*
IGFs 44, **149**, 160, 366, 387
  and cancer 163
IGF-1 receptor 169
IGF-binding proteins (IGFBPs) 162, 164
IGF2 360
IGFBP2 489
IGFBP3 164, **364**
IKK-NFκB 330
IL-1β 210
IL-2 415, 423–4
IL-6 20, 178
IL-7 424
IL-8 29
IL-12 424
IL-15 424
IL-21 424
image-guided radiotherapy (IGRT) 529
imaging 516–8
  computerized tomography 497, 516–7, *517*
  future uses 521–2
  magnetic resonance imaging 517–8, *518*
  plain film X-ray 516
  radionuclide imaging 13, 520–1, *521*, *522*
  ultrasound 516, 522

Index

imatinib 32, 34, 63, 90, 170, 232, 475, **478**, 480, 485, 486, 535
   discovery of 479
   molecular target 477
   resistance 498, 501
IMC-18F1 **537**
IMC-1121b **537**
immortalization 258, 299–301
immune evasion 419, 424–5
immune privilege 29, 164
immune response 278, 411–3
   adaptive immunity 411, *411*
   cancer immunosurveillance 412–3
   danger signal 411
   immunosuppression 413
   innate immunity 411, *411*
   molecular targeting 495
immunocytes 29
immunodeficiency, and non-Hodgkin lymphoma 61
immunologic ignorance 425
immunosuppression 413
immunosurveillance 412–3
immunotherapy 410–28, 538
   antigen-specific 416, *416*
   clinical trials 425
imprinting 366–7
imprinting control region (ICR) 366
INCENP 134, 138
INCEP 134
incidence of cancer 5, 8, *10*
incident pain 546
incision/excision biopsy 518–9
Indian Hedgehog pathway 489
individualized medicine *see* personalized medicine
inflammation 27, 28–9
ING proteins **356**, 366–7
inheritance 21–2, 77
inhibitor of apoptosis proteins (IAPs) 274, 276–9, *277*, 376, 492
   caspase inhibition 277–8, *277*
   cell survival 278
   inhibition of 278–9
   and metastasis 278
inhibitory protein of cyclin-dependent kinase 4 *see* INK4
iniparib **483**, **542**
initiation 20, 94
INK4 proteins 125, 126, *126*, 246
INK4a gene **241**
INK4a protein 254
   functions 254–5
INK4a/ARF 29, 141, 254–7
   in cancer 256, *257*
   locus 254, *255*
   mouse models 256–7, **257**
INK4b 254
INK4c **241**
innate immunity 411, *411*
insertions 80
insulin receptor 169
insulin-like growth factors *see* IGFs
INT1 **195**
INT2 164, **195**, **196**
integrated stress response (ISR) 282
integrins **148**, 386, 387, 389, *389*, 391
   cell invasion 402–4, *403*, *404*
   control of anoikis 406
   in signal transduction 389, *390*
   in tumor formation *391*
integrin antagonists 494–5, *495*
intensity-modulated radiotherapy (IMRT) 497, 529, *529*
interferons 424
interferon-α 59, 63

interferon-γ 412–3
interleukins *see* IL
International Agency for Research on Cancer (IARC) 8
interphase *116*
interventional radiology 520
intra-S-phase checkpoint 120
intrathecal catheters 547
intrinsic pathway of apoptosis 26, 272, 273–4, *273*, *274*
intussception 430
invasion 395–6, *397–9*
   integrins in 402–4, *403*, *404*
   MMPs in 402–4, *403*, *404*
   molecular targeting 493–4
   see also metastasis
inversions 81–2
investigations 511–5
   angiography 520
   biomarkers 2, 13, 15, 102–3, 368, 501–5, **502**, *503*, 512–3, **513**
      cancer stem cells **304**
      colorectal cancer 504–5
      screening 104
      and treatment selection 105
   biopsy 518–9
   bronchoscopy 519, *519*
   colonoscopy 515, *515*
   endoscopy 519
   fecal occult blood **502**, 515
   genetic tests 513, **513**, **514**
   hematological and biochemical tests 511–2, **512**
   imaging 516–8
   interventional radiology 520
   mammography 50, *50*, 103, 514
   radionuclide imaging 520–1, *521*, *522*
   screening 12, 102–3, 513–4
      cervical cancer 103–4, 514
      colorectal cancer 54
      genomic mutations 105
      MEN 105
      opportunistic 12
      population 12
      prostate cancer 514–5
ionizing radiation 96
IPI-504 **542**
ipilumumab **484**
Iressa *see* gefitinib
irinotecan 56, 448, *448*, 449, 488, 506, 534
ISG15 378
isotopecoded affinity tagging *see* ICAT
ixabepilone 534

Jagged1 306
JAK-STAT pathway 160, 177–8
   inhibitors 490
JAK2 **194**
JAK–STAT–SOCS pathways 177
Janus kinases (JAKs) 177
JARID-1 361
JARID-1A 305
JARID-1B **304**
jelly belly ligand 169
Jesionek, Albert 450
JUN gene **98**, **197**, 198
JUN protein 128, 179, 184
JUN kinase 172
JUN N-terminal kinase interacting protein (JIP1) 1, 172
JUN N-terminal kinase (JNK) 230
junk DNA 316, 352

K-RAS **198**, 213–4, 216, 223, 224
KAI1 30

Kaposi's sarcoma herpesvirus (KSHV) 99
karyotype 70, 115
KAT2B **356**
KAT5 **356**
KDM2B **355**
KDM4 **355**
KDM4C **355**
KDM5A **355**
KDM5B **355**
KDM5C **355**
KDM6A **355**
KDR *see* kinase insert domain receptor
ketamine **547**
killer activating receptors (KARs) 417
killer inhibitory receptors (KIRs) 417
kinase insert domain receptor (KDR) 88, 170
kinesin 134
kinesin spindle protein **496**
kinetochores 115, *116*, 133, 134, 137–8, 337
King, Frank 443
kinome 537
KIP1 **241**
KIP2 **85**
KISS1 30
KIT 32, 170, **194**, **196**
KITLG **87**, 89
Klf4 **87**, 131, 209, **304**
knockout (KO) mice 558, 582–4
Knudson, Alfred G. 240
Korsmeyer, Stanley 259
KP1019 453
KRAS 14, 23, 45, 95, 106, 194, **196**, 366, **470**, 474, **502**
KRAS2 **195**
Kruppel-like factor 4 (KLF4) 182
KS3 **196**
KU-55933 142
KU-60019 142, **484**
KU70 136, 141, 318, 330, *331*, 332, 333, 334
KU80 141, 318, *322*, *324*, *331*, 325, 333
KU86 297, 308

L-BLP25 **483**
lactate dehydrogenase **502**, **513**
Lamarck, Jean-Baptiste 23, 358
lamellipodia 402
laminin 385, 389
lapatinib 52, **478**, 481–2, 535, **536**
laser capture microdissection (LCMD) 579
late S-phase cyclin **124**
latency-associated peptide (LAP) 387
latent niche 21
LATS/WARTS protein kinase 271
LATS1 271
LATS2 271
LBH589 **542**
LDE225 489
lead compounds 442, 443–53
   selection of
      antisense therapy 451
      natural products 445–8
      pharmacological tools 449–50
      photodynamic therapeutics 450–1
      RNA interference 451–2
      serendipity 444–5
      vaccines 452
   see also individual drugs
lead optimization 442, 453–6
lenolidamide 538
leptin 139, 482
lestaurtinib **483**
Let7 495

Index

letrozole **483**, 539
leukapheresis 416
leukemia 11, 62–3
   chromosomal analysis **513**
   clinical features 510
   incidence 8
   see also specific types
levomepromazine 548
levorphanol **546**
Levulan 451
LFA-1 415
Lhermitte-Duclos disease 259
Li-Fraumeni syndrome 22, 78, 83, **84**, 91, 253, 254, 336, **514**
   genetic testing 513
   screening 105
lifestyle factors 96–7
   see also diet; smoking
lifestyle interventions 14–5, 75
LIN28 23, 373
LIN28B 23, 372
linear accelerators 527
linkage (association) studies 560
lipid kinases **161**
liquid chromatography 567
liquid-based cytology (LBC) 103
lithocholic acid 172
liver cancer, mortality 43
liver capsule pain 546
liver kinase B1 (LKB1) 158
LKB1 177
LKB1–STK11 **364**
LMO1 **87**, 89
LMYC **197**
lncRNA 369
lomustine (CCNU) 532
lonafarnib **483**
long noncoding RNA *see* lncRNA
long terminal repeats (LTRs) 100
loss of heterozygosity (LOH) 22, *22*, 240
loss of imprinting (LOI) 23, 366–7
loss-of-function mutations 17, 300
low- to moderate-penetrance mutations 71
LPHN3 89
LRP1B 88
LRRC19 489
LTA **87**
lung cancer 45–9
   clinical features 46–7, *46*, 511
   diagnosis and staging 47, **48**
   incidence 8, 45
   mortality 43
   non-small-cell *see* non-small-cell lung cancer
   pathology 45–6
   small-cell *46*, 47, **248**
   treatment 47, 49
      chemotherapy 49
      radiotherapy 47, 49
      surgery 47
luteinizing hormone-releasing hormone (LHRH) antagonists 53
Lutex 451
LXCXE motif 245, 248
LY294002 477, **483**
LY2181308 451
LY2603618 **541**
LY2606368 **484**
LY21813081 **541**
lymphadenopathy 44
lymphatics, cancer spread through 31, 397
lymphokine-activated killer cells (LAKs) 422
lymphoma 11
   clinical features 510

Lynch syndrome 22, 53, 71, 83, **84**, 106, 321, 330, 365, **514**
   screening 105
lysosomes 352

M-CSF **149**
M-phase cyclin **124**
M-phase-promoting factor 133
MabThera *see* rituximab
McDonough feline sarcoma **193**
MAD *see* MXD family
mad1–3 137
MAD2 130, 137
MAGE 363
MAGE-1 417, **418**
MAGE-3 **418**
MAGE-A3 422
magnetic resonance imaging (MRI) 517–8, *518*
   diffusion weighted 522
   functional 522
major histocompatibility complex (MHC) 31, 413, 415
malignant tumors 384
   hallmarks of 396
Maml1 306
Maml2 306
mammalial target of rapamycin *see* mTOR
mammography 50, *50*, 103, 514
mantle cell lymphomas, RB pathway **248**
MAP2K4 89
MAPK pathway **161**, 172, 216
MAPKAP kinases 130, 318
mapsin 372
Marek's disease virus 99
MARK2 570
MART1 420
MART1/MelanA **418**
mas **196**
maspin 363
mass spectrometry 567
mast cell growth factor 170
Mastermind-like (MAML) family 181
Matrigel 228, 433, 490
matrix metalloproteinases *see* MMPs
matrix-assisted laser desorption ionization (MALDI) 567
matuzumab **536**
MAX 202, *202*, 362
maximum tolerated dose (MTD) 453
MBD1 **355**
MBD2 **355**
MBD3 **355**
MC1R **87**
MCL-1 226, 280
MCM2–7 helicase 132
MDC1 318, 333
MDM2 26, 251, *251*, 255, 338, 406, 491
MDM4 250, 251, 338
Mdmx 250
MDR 81
MEDI-577 **484**
mediator of DMA damage checkpoint protein-1 *see* MDC1
medical treatments and cancer 97
Medicines and Healthcare Products Regulatory Agency (MHRA) 457
medulloblastoma 259
Meier–Gorlin syndrome 132
meiosis 22, 112, 561
MEK-ERK pathway 475
MEK5b 570
melanoma, malignant 58–9, *58*, 257
   familial **85**
MELC/TIS toponomics 571, 573

melphalan *440*, 449, 532
membrane-associated G-proteins **161**
membrane-associated nonreceptor tyrosine kinases **161**
memory T cells 411, 412, *412*
MEN *see* multiple endocrine neoplasia
MEN **241**
MEN1 15, **85**, 105, 170, **356**, 364, 376, **514**
MEN2 **85**, 170, **514**
Mendel, Gregor 77
menin 333, 364
66-mercaptopurine 441–2, *442*, 533
Merkel cell carcinoma (MCC) 100
mesothelioma 45, 96
messenger RNA *see* mRNA
MET inhibitors **541**
MET oncogene **85**, **195**, **196**, 501
MET tyrosine kinase 168–9
metabolic switch 159–60, *159*
metabolism *207*
metabolizer phenotype 505
metaphase *116*
metaphase plate *116*, 134
metastasis 1–2, *9*, 10, 30–1, 44, 384, 395–9
   hyperplasia 395–6, *396*
   IAPs 278
   invasion and dissemination 395–6, 397–9
   micrometastasis 398, *400*
   molecular targeting 493–4
metastasis suppressor genes (MSGs) 30, 398–9
metastasis-associated protein 1 (MTA1) 306
metformin 90, 97, 162, 482
methadone **546**
methotrexate 305, 441, *441*, 449
   mechanism of action 533
methylation 557
methylation mark 361
metoclopramide 547, 549
Metvix 451
MGMT 363, **364**, 368
MICA 411, 417
MICB 417
microarrays 523, 559
   data analysis 562
microenvironment 28–9
micrometastasis 398, *400*
microorganisms and cancer 98–9, **98**, 99–101
microRNAs 23, 131, 203, 260–3, 296, 352, 369–71
   biology 260–2, *261*
   molecular targeting 495, 497
   and MYC 206, *207*
   and p53 262, *262*
   and RB *262*, 263
   treatments based on 373
microsatellite instability (MSI) 83, 106, 319, 344, 346, 365–6, *365*
microsatellites 343–4
midkine 169
minichromosome maintenance proteins (MCMs) 129
minimal quantitative trait loci 560
minisatellites 343–4
mir-2 372
mir-10b 373
mir-31 373
mir-34 372, 495
mir-121 373
mir-146a 372
mir-146b-5p 372
mir-155 372
mir-205 372
mir-372 372
mir-373 372
mir-449 263

miRNAs *see* microRNAs
mismatch repair (MMR) 22, 23, 135, 325, 329–30, *329*
missense mutations 80
mitochondrial outer membrane permeabilization *see* MOMP
mitogens 147, 246
mitogen-activated protein kinase *see* MAPK
mitosis 22, *116*, 133, 244
mitotic catastrophe 290
mitotic cell cycle *see* cell cycle
mitotic centromere-associated kinesin (MCAK) 134
mitotic inhibitors 533–4
   see also individual drugs
mitotic kinases 133–5
   and cancer 140–1
   inhibitors 142
mitotic spindle 133
MIZ-1 204, *205*, 286–7, 300, 362
MK1775 **541**
MK2206 **541**
MK4827 **542**
MLH1 23, **84**, 105, **241**, 260, 321, 363, 365, **514**
MLL 133, **355**
MLN8237 142
MMPs 29, 30, 162, 230, 301, 398
   cell invasion 402–4, *403*, *404*
   ECM degradation 403
   growth factor activation *175*
   molecular targeting 494
   release of tumor-promoting factors 404
Mms22 330
MMSET 333
MNT-MAX 208
model organisms 557–9
molecular adaptation 497
molecular classification 88, 367–8, 466–7
   breast cancer 467
   future directions 467
molecular pathology 522
molecular signatures 33
molecular targeted therapies *see* targeted therapy
MOMP 235, 272, 273, 274, 279–82, 285, 342
   see also BCL-2 protein family
MONO-27 106
monoclonal antibodies 466, **484**
   anti-EGFR **536**
   anti-ERBB **478**
   anti-VEGF **537**
   tyrosine kinase inhibitors 480, 485–6, 488
   see also specific antibodies
morphine **546**
mortality rates 6, *10*, 43
   see also specific cancer types
mos **196**
MOSAIC trial 56
motesanib **537**
mouse embryonic fibroblasts (MEF) 308–9
mouse mammary tumor virus (MMTV) 164
mouse models of tumorigenesis 557–9
   athymic nude mice 412, 417
   INK4a/ARF 256–7, **257**
   knockout (KO) mice 558, 582–4
   MYC 209–12, *211*, *212*, 468–73, **470**, *471*
   MYCERTAM 210, *211*, *212*, 286, 584
   osteosarcoma 193
   p53 protein family 253–4, **254**
   P388 (leukemia) model 443
   RAS superfamily 224–5, *225*
   RB protein family 248–50, **249**
   resistance and recurrence 473
MPS1 337
MR spectroscopy 522
MRE11 333, 336

MRE11–RAD50–NBS1 (MRN) complex 332
MRI *see* magnetic resonance imaging
MRN 332
mRNA 556
mRNA expression analysis 562–4
MSH2 **84**, 105, **241**, 260, 321, **514**
mSin3A corepressor 362
Msm2 **195**, **197**
MTA1–3 **356**
MTHFR **87**
mTOR 27, 154, 158, 167, 176–7, 217
mTOR inhibitors 57
multiple endocrine neoplasia (MEN) **85**, 169–70, **514**
   screening 105
   see also MEN; MEN1; MEN2
multiple susceptibility loci 88–9
multistage tumorigenesis 16–8, *17*, 79, 80, 105–6, 213
MUM-3 **418**
Münchausen Trilemma 6
mustard gas 439
mutagens 94
mutation rate 25
mutations 6, 7, 21, 78–82, *79*, 89–90, 92, *319*
   frameshift 80
   high-penetrance 71
   loss-of-function 17, 300
   low- to moderate-penetrance 71
   missense 80
   nonsense 80
   point 80, *81*
   silent 80
   somatic 72, 105–6
   splice-site 80–1
mutator phenotype 320–1, *321*, 365–6, *365*
MutSβ complex 330
MUTYH 86, 326
MXD family 130, 137, 208–9
MXD-MAX *205*, 208
MYB oncogene **195**, **197**, 198
MYC oncogene 18, 19, 20, 23, *24*, 25, 28, 81, 128, 158, 179, 190, **195**, **197**, 198, 199–213, 255, 474, 579
   expression in human tumours 200–2, *201*, 363
MYC protein 89, **98**, 202–13, *202*, 226–7, 340, 346
   apoptosis 209, 283–7, *284–5*
   cell differentiation 208–9
   cell growth/proliferation 204–6, *205*
   and diabetes 287
   and microRNAs 206, *207*, 263
   regulation of 206, 208
   signaling through BCL-2 family 284–6
   translation 203–4
   tumorigenesis 209–12, *211*, *212*
      levels for 212–3
      mouse models 209–12, *211*, *212*, 468–73, **470**, *471*
MYC-MAX *205*, 208
MYC-RAS cooperation *221*
MYCERTAM 210, *211*, *212*, 286, 584
MYCLK1 **195**
MYCN **195**
Mycobacterium bovis 452
Myelotarg *see* gemtuzumab
mypomagnesemia **512**
MYST3 **356**
MYT1 336

N-acetyltransferases (NAT) 94
N-CoR–SMRT 362
N-RAS **198**
NAMI-A 453
Nanog 209, 304, 305

NAT1 **87**
natural killer (NK) cells 291, 411, 417
natural products 444, 445–8
   see also individual drugs
natural selection 21, 70
nausea and vomiting 547–8
   chemically induced 548
   intractable 548
NBR1 288
NBS1 130, 333, 345
ncDNA 23, 352
NCoR1 **357**
ncRNA 2, 351, 369
NCT00464113 486
NCT00526838 486
NCT01013506 486
necrosis 26–7, 267, *270*, 271–2
Nedd4 proteins 180, 377
Nedd8 proteins 128, 378
needle biopsy 518
negative feedback loops 500–1
neoadjuvant chemotherapy 531
neovascularization 430–1, *431*
   see also angiogenesis
neratinib **536**
nerve blocks 547
NESPAS 369
NEU gene 81, **196**, 436
NEU protein 224, 229, **470**
neuroblastoma 367
neurofibromatosis 215
   type 1 **84**, **514**
   type 2 **84**
neurofibromin 215
neuropathic pain 546, **547**
neuropilin 1 434
neutrophils 29
nevaxar **537**
new chemical entities 453
Nexavar *see* sorafenib
nexin 403
next-generation sequencing (NGS) 559, 564–6
   cyclic reversible termination 565
   454 pyrosequencing 565
   sequencing by ligation 565
NF-κB 31, 178, 198, 226–7, 299
   activation 278
NF-κBIL2–TONSL 330
NF1 **84**, 88, 215, **241**, 260, *261*, **514**
NF2 **84**, **241**
NFkBIA 168
NFY 131
NGF **149**
nibrin 345
Nicd 393
nickel 45
nidogen 385
Nijmegen breakage syndrome (NBS) 260, 333, 345
nilotinib 63, 449, **478**, 479, 481, 482, **541**
   resistance 501
nimotuzumab **536**
9-1-1 complex 325, 330
nitrogen mustards 439–40, *440*, 532
   see also individual drugs
nitrosoureas 532
NM23 30, 31
Nm23-H1 494
NMp22 **502**
NMYC **197**
Nodal signaling pathway 182
Nolvadex-D *see* tamoxifen
nomenclature 11
non-Hodgkin lymphomas 61–2
non-oncogene addiction 474, 475

613

Index

non-small-cell lung cancer (NSCLC) 27, 29, 32, 45
  histological and molecular classification 466–7
  RB pathway **248**
  staging 47, **48**
  surgery 47
non-steroidal anti-inflammatory drugs *see* NSAIDs
noncoding DNA *see* ncDNA
noncoding RNA *see* ncRNA
nonhereditary mutations 72
nonhistone proteins 354
nonhomologous end joining (NHEJ) 308, 324, *331–2*, 333, 334, 335
nonmelanoma skin cancer (NMSC), incidence 8
nonpharmacological treatments 497
  see also radiotherapy
nonreceptor tyrosine kinases 172, 194–5, **196**
  molecular targeting 485–8
nonsense mutations 80
NORE1 223
Notch pathway 21, 100, 128, 178, 180–2, *181*, 304, 306, 395, 435–6
Notch-Delta signaling pathway 394, 395
NOXA 130, 280, 283, 286, 318, 342, 406
NR-21 106
NR-24 106
NRAS 194, **195**, **196**
Nrf2 transcription factor 500
NSAIDs 90, 491–2, **547**
NSB1 260
NSCLC *see* non-small-cell lung cancer
NSD1 **355**
NU7026 **484**
nuclear factor *see* NF
nuclear proto-oncogenes 184
nuclear receptors **161**, 171–2
nuclear respiratory factor-1 (NRF-1) 285
nucleolar stress 342
nucleophosmin (NPM) 340–1
nucleosomes 353, 354
nucleotide excision repair (NER) 135, 325, 327–9, *328*, 345
nucleotide–oligmerization domain family (NOD) receptors 411
NUMB 180, 181
Nurse, Paul 556
Nutlins 250, **484**, 491, *491*
nutrients 139, **148**
  and cancer 158
NY-ESO-1 **418**

obatoclax **484**
obesity 15, 27, 36, 57, 73, 75, 86, 90, 92, 93, 96, 97, 169, 287, 301, 358, 465, 482
oblimersen 451
occupational cancers 94
Oct4 209, 305
octreotide 466
ofatumumab **484**
off-target toxicity 500
OGX-011 451
Okazaki fragments 132, 330
olaparib **483**, 538, **542**
Omi/HtrA2 26, 274
Omnitarg *see* pertuzumab
ON01910 142, **541**
oncogenes 17, 23–4, 71, 188–238, 351
  cellular **193**
  classification 194–9, **195–8**
  collaboration 199, *200*
  discovery 191
  growth factors *see* growth factors
  inactivation of 468, *469*
  nomenclature 201
  products *190*

senenescence induced by 27–8, 219–20, *221*
signal transducers *see* signal transduction pathways
targeting of 473–5, *473*
as tumor suppressors 28
types of 193–4
viral **98**, **193**
see also specific oncogenes
oncogene addiction 2, 32, 169, 235, 468, 473, *473*, 474–5
oncogene-induced senescence (OIS) 27–8
oncogenesis 67–110
  cancer prevention 14–5, 76–8
  cause and effect 70–1
  environmental causes 93–101
  gene-environment interactions 89, *91*
  genetics 21, 70, 78–89
  genomics 87–9
  mutations *see* mutations
  polymorphisms 22–3, 70, 86–7, **87**
  risk factors 73–6, 90–93, *91*, *92*
oncogenic stress 26, 126, 209, 283–7, *284–5*, 323–4, 338–43, *339–43*
  double-strand breaks 330–3, *331–2*
oncomirs 23, 372, 495
Oncotype DX Breast Cancer Assay 14, 504
oncoviruses 100–101
Onocovin *see* vincristine
opioids 546, **546**
opportunistic screening 12
Orf63 protein 99
organ-level checkpoints 154–5
origin recognition complex (ORC) 115
orthologues 570
OSI-906 486
ovarian cancer
  biomarkers 105
  familial 514
  genetic testing 513
  susceptibility 89
overdiagnosis 103
overtreatment 103
OX40 415
oxaliplatin 56, 449, 488, 532
oxidative phosphorylation 159
oxidative stress 97–8, 476
oxycodone **546**
oxygen supply 139, **148**
oxymorphone **546**

p arm 115
$p14^{ARF}$ 245, **364**
$p15^{INK4b}$ 130, 133, 135, 246, **364**
p16 58
$p16^{INK4a}$ 25, 130, 132, 135, 246, 254, *255*, 286, 363, **364**
  cellular senescence 298–9
$p18^{INK4c}$ 135, 240, 246
$p19^{ARF}$ 29, 254, 255, *255*, 286
$p19^{INK4d}$ 135, 246
p21-activated kinases (PAKs) 225
$p21^{CIP1}$ 26, 130, 135, 136, 252, 298
$p21^{Cip1/Waf1}$ 240, 246
$p27^{KIP1}$ 89, 204, 240, 246, 255
p38MAPK-MK2 stress response pathway 330
p53 gene 17, 21, 22, 23, 24, 25, 30, **84**, 91, 105, 129, 133, 135, 174, 240, **241**, 282–3, *283*, 366, 467, 561, 579
  mutations 253, **514**
  see also p53 protein family
p53 protein family 250–4, *250*, **418**
  activation 251–2, *251*
  and apoptosis 282–3, *283*
  in cancer 252–3, *253*, 257–8

  functions of 252
  and microRNAs 262, *262*
  molecular targeting 491, *491*, **496**
  mouse models 253–4, **254**
  senescence and immortalization 258, 298
  see also TP53
p57 **85**
$p57^{KIP2}$ 246
p62 288
p63 *250*, 252
p73 gene 130
p73 protein *250*, 252
p107 130, 135, 243
p110 **198**
p130 130, 243
p130Cas 391
P276-00 **484**
paclitaxel 49, 52, 446, *447*, 533–4
pain control 545–7, **546**, **547**
  analgesic ladder 545, *546*
  anesthetic techniques 547
  coanalgesia 547, **547**
PAK1 123
palliative care 544–53
  assessment 545
  case history 551
  communication with patients 544–5
  indications 545
  key concepts 544, *545*
  supportive care 550–1
  symptom control
    bowel obstruction 548–9, **549**
    cachexia and anorexia 549, **549**
    constipation 549
    dying patients 550
    fatigue 549
    nausea and vomiting 547–8
    pain 545–7, **546**, **547**
    psychological problems 549–50
    respiratory symptoms 547
pan-deactylase inhibitors (pan-DACi) 492
Pancoast tumor 511
pancreas plasticity 287
pancreatic β-cells 158
pancreatic cancer
  RB pathway **248**
  susceptibility 89
pancreatic ductal adenocarcinoma (PDA) 224
panitumumab **478**, 488, 535, **536**
panobinostat 492, **542**
Pap smear 502, 514
papillomavirus *see* human papillomavirus
Paracelsus 93–4, *93*
paracetamol **547**
paraneoplastic syndromes 510
parathyroid hormone (PTH) 44
PARP 35, 333, 335–6
PARP inhibitors 2, 51, 142, 335–6, 476, 538, **542**
PARP1 475, 476
parthanatos 335
passenger genes 88
Patched pathway 179–80, *179*, 305, 367, 489
pattern recognition receptors 411
paxillin 230
pazopanib **478**, 485, **537**
PBRM1 **357**
Pc2 367
Pc4 451
PCSK9 **496**
PD-1 416
PD-L1 416
PD-L2 416
PD98059 477, **483**
PD0325901 477

PD332991 **484**, **541**
PDCD4 372
PDGFR 170
PDGFRA 32, 194
peau d'orange skin 50, *50*
PEBP1 30
pegaptanib 480
pegvisomant 482
peptide-based vaccines 421
perforin 416, *417*
pericytes 429, *430*, 436
perifosine **483**
peripheral blood mononuclear cells (PBMCs) 421
PERK 282
permeability transition pore complex (PTCP) 282
peroxynitrite 425
personalized medicine 32, 33–4, 36, 102, 107, 474, 535, 576
pertuzumab **478**, **536**
Peutz-Jeghers syndrome 53, 83, **84**, 177
PF-655 **496**
PF-02341066 **541**
PF-03814735 **484**
PFDN2 570
phagocytosis 291, **414**
pharmacogenetics 105, 505–6
pharmacogenomics 505–6
Pharmacogenomics Knowledge Base 457
pharmacological tools 449–50
phenotype 34, 70
  mutator 320–1, *321*
2-phenylaminopyrimidines 499
Philadelphia chromosome 82, 227, 232–5, *233*, *234*, 304
Phortress 455–6, *456*
phosphatase and TENsin homolog on human chromosome TEN *see* PTEN
phosphatases **161**
phosphatidylinositol phosphates *see* PIPs
phosphatidylinositol 3,4,5-trisphosphate *see* PIP3
phosphatidylinositol-3 kinase *see* PI3K
phosphatidylserine 266, 275
phosphoenolpyruvate (PEP) 160
phospholipase Cγ (PLCγ) 219
phosphoramide mustard 454–5, *454*
phosphorylation 557, 569
photo-electric effect 526
photoactivable platinum compounds 449, *449*
Photochlor 451
photodynamic therapy 450–1, 497
photofrin 450
PI3K 17, 172–4, *174*, 176, **198**, 216, 224, 233
  autophagy 288
PI3K inhibitors **541**
PI3K–AKT pathway 31, 150, 158, 160, 167, 215, 224, *258*, 282, 330, 345, 500
  molecular targeting 489–90, 536
  RAS signaling 218–9
PI3K–AKT–mTOR pathway 21, 184, 475, 485
PI9 425
PIAS1 333
PIAS4 333
PIK3C 194
PIK3CA 467
pim1 **196**
PIN1 208
PIPs 367
PIP3 173, 288
pituitary adenoma 171
PKN3 **496**
plasmid DNA-based vaccines 421–2
platelet-derived growth factor (PDGF) 147–8, **149**, 160, 170, 433

platelet-derived growth factor receptor (PDGFR) 170, 229, **478**, 479, 481, 482, **485**, 488, 493, **537**
  antibodies **484**
platinum compounds 448–9, 532
  photoactivable 449, *449*
  see also individual drugs
pleiotrophin 169
plexopathy 546, **547**
PLK1 336, 337
ploidy 115, 133
pocket proteins 130, 243–4, *244*
podoplanin **304**
point mutations 80, *81*
polo-like kinases (PLK1) 133, 135, 141, 323
  drug targeting 142
polyacrylamide gel electrophoresis, 2D 566, 567, *568*
poly(ADP-ribose) polymerase *see* PARP
polycomb group proteins (PcG) 367
polycyclic aromatic hydrocarbons (PAHs) 94, 95
  in tobacco smoke 95–6
polyethylene glycol 452
polygenic diseases 86
polymerase chain reaction (PCR) 562
polymorphisms 22–3, 70, 86–7, **87**
  single-nucleotide *see* single-nucleotide polymorphisms
polynucleotide kinase–phosphatase (PNKP) 327
polyomaviruses 100
polyploidy *321*
polyubiquitylation 128
population screening 12
positron emission tomography (PET) 520–1, *521*
post-transplant lymphoproliferative disease (PTLD) 413
posttranslational modification 215–6, *215*, 567, 569, *569*
poxviruses 100
PPARs 172, 308, 492
PR **364**
prad1 **196**
Prader-Willi syndrome 564
pRB 130, 135
pRB2/p130 135
PRCA1 **85**
PRDM1 **355**
PRDM2 **355**
precancerous lesions 12
prednisolone 62
preRC 124
prereplication complexes 132
prevention 14–5, 76–8
  chemoprevention 90
PRIMA-1 **491**, *491*
prion proteins 101
PRLR **364**
pro-drugs 453–4
proapoptotic proteins 280
procaspases 275, **276**
progastrin 171
progesterone 549
progesterone receptors 467, **502**
programmed cell death (PCD) 267, 269
  see also cell death
programmed death-1 *see* PD-1
prolactinoma 171
prometaphase *116*
promoters, methylated 243
promotion 20, 94
promyelocytic leukemia nuclear bodies (PML NBs) 344, 377–9
  and senescence 378–9, *381*

promyelocytic leukemia protein (PML) 82, 157, 220, 299
prophase *116*
prostate cancer 29, 56–7
  clinical features 56–7
  diagnosis and staging 57
  hereditary **85**
  hormone therapy 57, 539–40
  incidence 8
  mortality 56
  pathology 56
  screening 514–15
prostate-specific antigen (PSA) 14, 56, 104, 303, **418**, **502**, 512, **513**
prostatic acid phosphatase (PAP) 421
PROSTVAC-VF vaccine 495
prosurvival proteins 280
proteasome 352, 374, *374–5*
  and cancer 376
  structure *378*
  therapeutic inhibition 376–7
  ubiquitin-proteasome system *379*
proteasome inhibitors **484**, 492–3, *493*
proteins
  cell cycle **124**
  degradation of 126, 128–9
  regulation of 373–6, *374*, *375*
  see also specific proteins
protein arginine methyltransferase 252
protein complexes 569–70, *571*, *572*
protein degradation 126, 128–9
protein expression profiles 34
protein kinase C 194
protein phosphatase EYA 333
protein phosphatase PPA2 208
protein poly(ADP-ribose) polymerase-1 (PARP-1) 164
protein tyrosine phosphatases (PTPs) 171, 229
protein-protein interactions 570, *571*
proteomics 523, 566–7
  clinical applications 570–1
  quantitative 566–7, *568*
  tools 567
proto-cancer cells 8
proto-oncogenes 191, 193, **194**, 271, 302
proton radiotherapy 530
Provenge *see* sipuleucel-T
psychological problems 549–50
PTC *see* Patched pathway
PTC1 259
PTCH 59, **85**, 241
PTEN gene 21, 23, **85**, 174, 224, **241**, 258–9, *258*, *261*
  germline mutations 259
PTEN protein 332, 372
ptosis, lung cancer *46*
PTPN12 486
PTPRD 88
puffer fish 557
pulse-chase techniques 18
PUMA 130, 252, 280, 283, 286, 318, *338*, 342, 343, 406
  see also Bbc3
Purlytin 451
454 pyrosequencing 565
pyrrolobenzodiazepines 448, 455–6, *456*
pyruvate kinase (PKM2) 160

q arm 115
13q14 deletion **513**
11q23 abnormalities **513**
QALYs 36
QPI-1002 **496**
quality-adjusted life years *see* QALYs

Index

quantitative proteomics 566–7, 568
quantitative structure–activity relationship (QSAR) 444
quantitative trait loci 560
quiescence 158, 220, 244, 246, 258, 298, 303, 432

R-RAS 230
R115777 **483**
Raab, Oscar 450
RAC 30, 216, 402
RAC1 225, 226
race, and cancer incidence 8
RAD1 346
RAD18 330
RAD23A 328
RAD23B 328
RAD50 333
RAD51 22, 35, 71, 83, 141, 318, 344
RAD52 344
radiation-induced genomic instability 229
radiobiology 527–8, *527*
radiofrequency ablation 497
radionuclide imaging 520–1, *521*, *522*
   whole-body scans 13, *522*
radiosensitivity 527
radiotherapy 16, 464
   advances in 529–30, *529*
   breast cancer 51–2, *52*
   cervical cancer 60
   chemoradiation 530
   colorectal cancer 55
   DNA damage 346
   fractionation 527
   Hodgkin disease 61
   lung cancer 47, 49
   malignant melanoma 59
   non-Hodgkin lymphoma 62
   physics 526–7
   prostate cancer 57
   resistance to 305
   side effects 528
   treatment planning 528–9, *528*
radon gas 45, 96
RAF gene 198
RAF1 gene **195**
RAF kinase inhibitory protein (RKIP) 373
RAF proteins 172, 216
RAF-ERK pathway 222
RAF-MAP kinase pathway 217–18, *218*
RAF-MEK-ERK pathway 172, 215, 216, 475, 485
RAF-MEK–MAPK pathway 31, 172
RAG-2 412, 413
RALGDS 216, 219
RALGDS-like gene 216
raloxifene 172, **483**
raltitrexed 532
ramucirumab **537**
RANK, antibodies **484**
RANKL 494
Rap1 308
Rap80 333
rapamycin, molecular target 477
raptor 176
RARβ2 **364**
RAS oncogene 2, 17, 19, *24*, 25, 158, 190, 193, **196**, 198, 213
   mutations 224
RAS superfamily 208, 213–28
   activation *214*
      receptor tyrosine kinases 215
   angiogenesis 225
   and cancer 224
   cell growth/proliferation 217
   cell survival/death 222–3, *223*
   cellular senescence vs. transformation 219–20
   differentiation 222
   endoplasmic reticulum stress 220, 222
   GTPase 213, 214–15, 217
   PI3K–AKT pathway 218–19
   posttranslational modification 215–16, *215*
   signaling pathways 216–17, *216*, 261
   therapeutic targeting 228
   tumorigenesis 224–5, *225*, 472–3
RAS-like GTPases 219
RAS-like guanine nucleotide dissociation stimulator *see* RALGDS
RAS-MAPK pathway 167, 184
RAS-MYC cooperation *221*
RAS–PI3K–AKT pathway 208, 222
RAS-RAF-ERK (MAPK) pathway 90, 150, 158, 160, 169, 171, *173*, 176, 234–5, 500
   molecular targeting 489–90, 536
RASs 372
RASSF family 223
RASSF1A 363
RasV12 255
RB gene 17, 21–2, 25, 71, 91, 105, 118, *119*, 240, **364**
   genomic locus 242–3
RB protein family 242–50, 362
   A-B pocket 243–4, *244*
   in cancer 247–8, **248**, 249, 257–8
   cell cycle and differentiation 244–6, *245*, *246*
   hyperphosphorylation 131–2
   and microRNAs 262, 263
   mouse models 248–50, **249**
   senescence and immortalization 258, 298
   signal pathways 246–7
   see also specific proteins
RB-binding protein 1 (RBP1) 245
RB1 gene **84**, 87, **241**, **514**
re-oxygenation 527
reactive oxygen species (ROS) 29, 97–8, 283, 425, 451, 476
reader proteins 365
receptor activated solely by synthetic ligands (RASSLs) 171
receptor activity modifying proteins (RAMPs) 171
receptor degradation 377
receptor tyrosine kinase (RTK) inhibitors 433, 493
receptor tyrosine kinases (RTKs) **161**, 164–72, *175*, 194, 377, 477
   anaplastic lymphoma kinase 169
   and cancer 167–8
   ERBBs 32–3, 168
   growth factor 165, *165*–7, 167, **196**
   HER2-Neu 476
   IGF-1 receptor 169
   insulin receptor 169
   MEN syndromes 169–70
   MET 168–9
   molecular targeting 485–8
   RAS activation by 215
   RET 169
   TIE2 435
receptors 164–70
   G-protein coupled *see* G-protein coupled receptors
   nuclear **161**, 171–2
   tyrosine kinase *see* receptor tyrosine kinases
   see also specific receptor types
recessive 71, 190
RECIST 473
RECK 30
recombination-activating gene *see* RAG
RecQ helicases 345
redistribution 527
Reed-Sternberg cells 60, *60*, 99
regulatory T cells *see* Tregs

REL **195**, 197
renal carcinoma 57
   susceptibility 88
replication licensing 132
replication origins 125
replication protein A (RPA) 330
replicative senescence 306–7, *306*, *307*
repopulation 527
respiratory symptoms 547
response evaluation criteria in solid tumors *see* RECIST
restriction point 131
RET 15, **85**, 105, 170, 194, **196**, **514**
RET tyrosine kinase 169
retaspimycin hydrochloride **542**
retinoblastoma 21–2, 83, **84**, 91, 118, 240, **514**
   mouse model 250
   screening 105
   therapy 250
   see also RB gene; RB protein family
retinoic acid receptor α (RARα) 157
retinoid X receptor (RXR) 171
retroviral integration-induced transformation 100
retroviruses 25, 100
   cis-activating 100
   trans-activating 100
   transducing 100
retuximab 62, 63
RG3638 **484**
RG7112 **484**
RG7414 **483**
RG7440 **483**
RG7444 **484**
RG7459 **484**
rhabdomyosarcoma 259
RHEB 225
RHEBL1 225
Rho proteins 30, 226, 227, 228, 402
   inhibitors 490
Rho GAPs 227–8
Rho GEFs 227
Rho GTPases 225–8, *226*, 391
   and cancer 227
   transcription factor regulation 226–7
Rho kinase 391
   inhibitors 490
Rho-associated coiled-coil domain kinases (ROCK) 225
RHOGD12 30
ribonucleoside diphosphate reductase 534
rictor 176
RING finger proteins 128, 333, 377
risk factors 75–6
   combined 90–3, *91*, *92*
   environmental *see* environmental causes of cancer
   see also specific cancer types
Rituxan *see* rituximab
rituximab 423, 443, 480, **484**, 538
RNA
   double-stranded 556
   long noncoding (lncRNA) 369
   messenger (mRNA) 556
   microRNAs *see* microRNAs
   noncoding (ncRNA) 2, 351, 369
   small interfering (SiRNA) 302, 352, 369–71, 495–7, **496**, 556
   transfer (tRNA) 203
RNA interference (RNAi) 371–3, 451–2, 458, 570
   and cancer 372–3
RNA polymerases 203, 324, 369
RNF8 333
RNF169 333
robatumumab 486
rodent cornea pocket 432, *433*

romidepsin 368, **484**, **542**
RON oncogene 501
ros **196**
roscovitine **484**, 490
Rous sarcoma virus (RSV) 25, 98, 192
RPA 328
rs6983267 88
RSF1 **357**
RTP801 **496**
RXi-109 **496**

S-phase cyclin **124**
S6 kinases (S6Ks) 155
S100A4 363
Saccharomyces cerevisiae 117, **124**, 288
Safety Pharmacology Package 457
SAGE 252, 419
salirasib **483**, 489
Salvador protein 271
saracatinib 489
sarasar **483**
sarcoma 11
scaffold proteins 475
SCF **124**, 126, 128–9
SCH66336 **483**
Schizosaccharomyces pombe 117, **124**
Schleiden, Jacob 267
Schwann, Theodore 267
screening 12, 102–3, 513–14
    cervical cancer 103–4, 514
    colorectal cancer 54
    fecal occult blood **502**, 515
    genomic mutations 105
    mammography 50, *50*, 103
    MEN 105
    opportunistic 12
    population 12
    prostate cancer 514–15
secretions, retained 550
securin **124**, 130
selective estrogen receptor downregulators (SERDs) 539
selective estrogen receptor modulators (SERMs) 172, 445, **483**, 539
    see also tamoxifen
self-antigens, over-expression 420
self-renewal 18
seliciclib **541**
selumetinib **483**
senescence 7, 135, **148**, 298–310, 378–9
    avoidance of 27
    and cancer 299–301
    and cancer behavior in other cells 301
    and cancer control 290–2, *291*
    drug-induced 292
    oncogene-induced 27–8
        RAS 219–20, *221*
    premature 301–2
    RB and p53 in 258
    replicative 306–7, *306*
    see also cell death
senescence-associated heterochromatic foci (SAHFs) 298, 300, 342
separase **124**
sequencing by ligation 565
SEREX 418–19
serial analysis of gene expression *see* SAGE
serine, ECM degradation 403
serine/threonine kinases **161**, 177, 188, **196**, 198
    RAF-MAP kinase pathway 217–18, *218*
serological recombinant expression cloning *see* SEREX
Sezary's syndrome 177
SFK *see* SRC family kinases

SFRP1 **364**
SG2202 *456*
SG2285 455–6, *456*
SGN-15 453
shepherdin 450
short telomeres 310
SHP-1 178
shugoshin 134
signal transducers and activators of transcription *see* STAT proteins
signal transduction pathways 160–84, 194–5, **196**, **198**
    AKT 176
    cytokines 177–8
    growth factors *see* growth factors
    Hedgehog 21, 59, 157, 178, 182
    Hippo 183–4, *183*, *184*
    inhibitors of **483**
    integrins in 389, *390*
    MAPK **161**, 172
    mTOR 27, 154, 158, 167, 176–7
    nonreceptor tyrosine kinases 172
    Notch 21, 100, 128, 178, 180–2, *181*
    phosphatidylinositol kinase 172–4, 176
    RAS superfamily 216–17, *216*, *261*
    RB protein family 246–7
    receptor tyrosine kinases *see* receptor tyrosine kinases
    see also individual pathways
signalosome (CSN) 128F
SILAC 566, 567, 568, 569
silatecan DB-67 453
silent mutations 80
simian sarcoma **193**
single gene defects 21–2
single-nucleotide polymorphisms *see* SNPs
single-photon emission computed tomography (SPECT) 520
single-strand breaks (SSBs) 326–30, 527
    repair
        base excision 135, 325, 326–7, *327*, 335, 475
        mismatch repair 22, 23, 135, 325, 329–30, *329*, 343
        nucleotide excision 135, 325, 327–9, *328*, 345
        stalled replication forks 330, *331*
        translesional synthesis 326, 330
sipuleucel-T 452, 495
SiRNA 302, 352, 369–71, 556
    molecular targeting 495–7, **496**
sirtuins **356**, 361
sis 98
sister chromatids 115, 133
SKAP1 89
ski **197**
skin cancer 58–9, *58*
Skp1 128
Skp2 296
Slug 304, *338*, 401
Smac mimetics 277, 278–9, 293, 492
Smac/DIABLO 273, 278, 280, 281
Smads 393
Smad receptors 164
SMAD4 gene 89
Smad4-DPC4 376
small interfering RNA *see* SiRNA
small ubiquitin-like modifier *see* SUMO
SMARCA2 **356**
SMARCA4 **356**
SMARCB1 353, **356**
SMARCC1 353, **357**
SMARCE1 **357**
SMC proteins 344
SMC1 318, 333

smoking 75, 95–6
    and lung cancer 45, 92
Smoothened 59, 393, 489
Smurfs 128
SMYD3 **355**
Snail proteins 178, 304, 306
SNF2H **357**
SnoN 302
SNPs 22, 74, 86–7, 88, 556, 559–62, *561*
    tagSNPs 561, *561*
    see also specific SNPs
SNP chips 559
SOCS-1 gene **364**
SOCS-1 protein 178
SOCS-3 **364**
somatic mutations 72, 105–6, 192, *242*
Sonic Hedgehog pathway 21, 179–80, *179*, 367, 489
sorafenib 57, **478**, 481, **485**, 493, **537**
    molecular target 477
Sox2 209, 306
SP/KLF 131
SP1 204, *205*
spectral karyotyping (SKY) 319–20, *320*
spinal tumors 511
spindle checkpoint 120, *123*, 137–8
spindle fibers 134
splice-site mutations 80–1
spliceosome 557
sporadic cancers 22
SPRY4 89
Sprycel *see* dasatinib
squamous cell carcinoma
    lung 45
    skin 59
SRC gene **98**, **196**, **198**, 228–32
    and cancer 231
    mode of action 230, *231*
    molecular targeting 232, 488–9
    phenotype 229–30
    regulation and activation 228, *229*
SRC family kinases (SFK) 228, 488, 489
SRD5A2 **87**
STA-9090 **542**
stable isotope labeling by amino acids in cell culture *see* SILAC
staging 101–2, 523
    breast cancer **51**
    colorectal cancer 54–5, **54**
    Dukes' classification **54**, 101–2
    Hodgkin disease 61
    lung cancer **48**
    non-Hodgkin lymphoma 62
    prostate cancer 57
    TNM classification 101–2, 523
stalled replication forks 330, *331*, 334
STAT proteins 177, **198**, 226, 230, 235
STAT1 412–13
stem cell factor 170
stem cells 156, 178–9, 395
    cancer *see* cancer stem cells
    and cancer 302–3
    DNA repair 336
    embryonic 113, 302–3
    hematopoietic 303, 304
stereotactic radiosurgery 530
steroids 547
stimuvax **483**
stress-inducible heat shock proteins 282
stroma
    in cancer 387, *388*
    epithelial interactions 387, *388*
    growth factors in 387
    molecular targeting 495
structural maintenance of chromosome-1 *see* SMC1

Index

SU6656 488
SU11248 **537**
subpolymorphism 70
substrate recognition factors (SRFs) 128
SUFU 182
sulfotransferases (SULT) 94
sulindac 90
SUMO 344, 377–9
SUMO E3 ligases 333, 344
sumoylation 333
sunitinib 57, **478**, 481, **485**, **537**
superior vena cava obstruction (SVCO) 46, *46*
supportive care 550–1
surgery 16, 530, 531
    breast cancer 51
    cervical cancer 60
    lung cancer 47
    renal carcinoma 57
survival factors 147, **161**
survival pathways 232–3
    molecular targeting 492
survival time 5
survivin 134, 226, 278, **418**, 425
susceptibility to cancer 78, 82–6, *83*, **84–5**
Sutent *see* sunitinib
SUV39H1 **355**
SV40 T antigen **470**
SW1-SNF complex 353, 363
syndrome of inappropriate ADH secretion (SIADH) 46
synthetic lethality 2, 35, 475–6, *476*
systemic therapy 527
systems biology 554–83
    epigenomics 559
    genomics 87–9, 559
    information flow in cells 556–7, *557*
    microarrays 523, 559
    model organisms and cancer models 557–9
    mRNA expression analysis 562–4
    next-generation sequencing (NGS) 559, 564–6
    posttranslational modification 215–16, *215*, 567, 569, *569*
    protein complexes and cellular networks 569–70, *571*, *572*
    proteomics 523, 566–7, 570–1
    toponomics 571–6
    transcriptomics 559
    see also specific elements

T cells 413–17
    antigen presentation 413–15, *414*, *415*
    antigen-specific therapy 422–3
    costimulatory and inhibitory signals 415–16
    cytotoxic (CD8) 416–17
    differentiation 412, *412*
    helper (CD4) 416
    memory 411, 412, *412*
    priming 411, 412, *412*
    regulatory *see* Tregs
    repertoire 420
T-cell acute lymphoblastic lymphoma (T-ALL) 100, 128–9
    RB pathway **248**
T-cell leukemia 100
T-cell lymphoma 62, 99, **196**, 205, 216, 227, 260, 286, 363, 368, 468, 469, **470**, 484, 492, 538, 555
T-cell lymphoma invasion and metastasis 1 (TIAM1) 216, 227
T-cell-defined tumor antigens 417–18, **418**
t-loops 297, *297*
t(8;14)(q24;q11) **513**
t(8;21)(q22;q21) **513**
t(9;22)(q34.1;q11.2) **513**
t(11;14)(p13;q11) **513**
t(15;17)(q22;q11) **513**
TACE/ADAM17 171
TAF1 **357**
tagSNPs 561, *561*
Talaporfin 451
talin 230, 391
tamoxifen 53, 172, 290, 445, *446*, 468, **483**, 539
    metabolites 455
tandem affinity purification (TAP) 570, *571*, *572*
tankyrase 305, 335
Tarceva *see* erlotinib
target cyclins **124**, 130
targeted therapy 20–1, 146–87, 228, 368, 461–508, 535–7, **536**
    angiogenesis inhibitors 436, 450, *450*, **485**, 493, 536–7, **537**
    biomarkers *see* biomarkers
    clinical progress 476–9, *477*, **478**
    DNA repair 490–2
    EGFR 535–6, **536**, *537*
    epigenetic regulation 492–3, *493*, 537–9
    PARP inhibitors 2, 51, 142, 335–6, 476, 538
    field cancerization 497
    G-protein coupled receptors 489
    gene therapy 495–7
    growth factors 482–5, **483–5**
    integrin antagonists 494–5, *495*
    metastasis and invasion 493–4
    oncogenes 473–5, *473*
    PARP inhibitors 2, 51, 142, 335–6, 476, 538, **542**
    resistance to **487**, 497–501, *498*
    scheduling 497
    stromal and immune targeting 495
    synthetic lethality 2, 35, 475–6, *476*
    target selection 443, 468
    transcription factors 492
    tyrosine kinase inhibitors *see* tyrosine kinase inhibitors
    VEGF **496**, 536–7
    see also individual drugs
Tasigna *see* nilotinib
Taspase-1 475, 476
taxanes 52, 142, 533
taxol *see* docetaxel; paclitaxel
taxonomy 35–6
taxotere *see* docetaxel
TBKBP1 570
telomerase 27, 296, **418**
    inhibition 310
telomerase reverse transcriptase (TERT) 296, *299*
telomeres 297, 303, 306–10, *306*
    and cancer 307
    critical loss 307–8
    dysfunction 308
    heterochromatin *309*
    length *297*, 308–9
    protection from DNA repair process 333–4
    replicative senescence 306–7, *306*, *307*
    roles of 309–10
    short 310
telomere attrition 27, 154, 306, 321–4
telomere-interacting proteins 307
telomeric repeat-binding factors (TRFs) 297, *297*
telomestatin 448
telophase 116
TEL-JAK2 178
temozolomide 59, 368, 455, *455*, 532
temsirolimus 57, **483**, 490, **541**
tenascin 387, 395
terminal differentiation 158
TERT/hTERT *154*, 221, 296, *299*, **355**, 579
testicular cancer, susceptibility 89
testosterone 172
"tet" system 210, 211, 468, 584
tet-regulatable gene expression 584
TET1 353, **355**
TET2 **355**
2,3,7,8-tetrachlorodibenzo-p-dioxin (TCDD) 95
tetracycline *see* "tet" system
tetrazines 532
TFAP2E 505
TGF-α 149
TGF-α-ErbB signaling pathway 394
TGF-β 28, 29, 148, **149**, 164, 305–6, 387
    and cancer 163
    EMT activation 401
TGFBR1 89
thalidomide **485**
THBS-1 **364**, 365
6-thioguanine 533
thiopurine methyltransferase (TPMT) 505
thiotepa 532
thrombospondin-1 (TSP-1) 225, 436
thymidylate synthase 130, 532
thymine DNA glycosylase (TDG) 327
thyroglobulin **502**
TIE2 receptor tyrosine kinase 435
tight junctions 394
Timeless 318
timing of cancer development 28
Timp-1 20
TIMPs 230, 494
tioguanine 441, *442*
Tip60 332
TIPARP 89
tipifarnib **483**
Tipin 318
tissue diagnosis *see* biopsy; histology
tissue mass homeostasis *11*, 156
tissue remodeling 28–9
tissue repair/regeneration 151–5, *152–4*
tivozanib **537**
TLN-4601 489
TMS1 **364**
TNF 87
TNF-α-NF-κB pathway 570
TNM staging 101–2, 523
    breast cancer **51**
    lung cancer **48**
Tob1 gene **241**, 252
Tob1 protein 252, 256
tobacco smoke *see* smoking
Toll-like receptors (TLR) 411, 422
TopBP1 318, 330
topoisomerase inhibitors 534
    see also individual drugs
topoisomerases 130
topology, definition 573
toponomics 571–6
    cell surface toponome 575–6, *575*
    protein location images 571–3, *573*
    structure, code and semantics 573–4, *574*
    topological hierarchies 573
    tumor cell mapping 575, *576*, 576–8
topotecan 49, 448, *448*, 449, 534
TOR 155
Torisel *see* temsirolimus
tositumomab 443, 466, 538
totipotent cells 156
TP53 see p53 gene
TPEF–HPP1 **364**
TPM1 372
Tpp1 308
TPX2 134
trabectedin 448, 534–5
TRAF1 277
TRAF7 570

Index

TRAIL 90, 233, 305
    human recombinant **484**
transarterial chemoembolization 520
transcription 129–31
transcription elongation factor b (P-TEFb) 362
transcription factors **124**, 184, **197**, 198
    molecular targeting 492
transcription-coupled repair (TCR) 325, 328
transcriptional regulators **357**
transcriptomics 559
transfer RNA *see* tRNA
transferrin 452
transformation/transcription domain-associated protein *see* TRRAP
transforming growth factors *see* TGF-α; TGF-β
transgenesis 582, 583
transition 80
translation, regulation of 369
translational control 184–5
translesional synthesis (TLS) 326, 330
transmembrane proteoglycan receptors 389
transthyretin **496**
transversion 80
trastuzumab 51, 52, 168, 423, **478**, 488, 535, **536**
    molecular target 477
treatment 526–43
    chemotherapy 15–6, 438–60, 526, 530–42
        and cell cycle 141–2
        and cell death 290, *290*
        combination chemotherapy 530
        drug delivery 452
        drug development 442–57
        indications 530–1
        individualization of 33–4
        new agents **541–2**
        rationale 530
        resistance to *see* drug resistance
        targeted *see* targeted therapy
        see also specific drugs
    hormone therapy
        breast cancer 53, 539
        prostate cancer 57, 539–40
    radiotherapy 16, 464, 526–30
        advances in 529–30, *529*
        chemoradiation 530
        DNA damage 346
        fractionation 527
        physics 526–7
        resistance to 305
        side effects 528
        treatment planning 528–9, *528*
    response to 14
    surgery 16, 530, 531
    targeted therapy *see* targeted therapy
    see also specific cancer types
Tregs 29, 417
Trf1 134, 308
tricyclic antidepressants **547**
TRIM24 365
triple-helix-forming oligonucleotides (TFOs) 451
Trisenox 453
trisomy 12 **513**
Trithorax group proteins 367
TRK **196**
tRNA 203
TRPC4AP–TRUSS 376
TRRAP 203, 332
ts1 **197**
TSC1 **84**, **241**
TSC2 **84**
tuberin 177
tuberous sclerosis **84**, 176
tuberous sclerosis complexes (TSC) 158
tumor angiogenesis *see* angiogenesis
tumor antigens 417–20, *418*, **418**
    classification 419–20
    SEREX 418–9
    T-cell repertoire 420
    T-cell-defined 417–8, **418**
tumor cell line-based vaccines 420
tumor cell metabolism 159–60, *159*
tumor cell-based vaccines 420–1
tumor immunity 410–28
    effector cells 413–7
        dendritic cells 411, *411*, 413, **414**
        natural killer (NK) cells 291, 411, 417
        T cells *see* T cells
        Tregs 29, 417
    immune response *see* immmune response
tumor markers *see* biomarkers
tumor necrosis factor receptor-associated factor 1 *see* TRAF1
tumor necrosis factor-related apoptosis-inducing ligand *see* TRAIL
tumor profiling 34
tumor protein (TP) p53 *see* p53 protein family
tumor recurrence 473
tumor suppressor hypersensitivity 474
tumor suppressors 2, 17, 21–2, 24, 25–7, 71, 118, 239–65, **241**, **484**
    cell proliferation control 258–60, *258–60*
    definition 242
    DNA damage response 260
    genomic stability 260
    see also specific tumor suppressors
tumor-infiltrating lymphocytes (TIL) 422
tumor-initiating cells *see* cancer stem cells
tumor-promoting growth factors 404
tumorigenesis 29
    mouse models *see* mouse models of tumorigenesis
    multistage 16–8, *17*, 79, 80, 105–6, 213
    MYC protein 209–12, *211*, *212*
    RAS superfamily 224–5, *225*, 472–3
Turcot syndrome 83
Twist 31, 255, 256, 304, 306, 401
two-hit hypothesis 22, 71, 240, *242*, 334, 363
Tykerb *see* lapatinib
tyrosinase **418**, 420
tyrosine kinase inhibitors (TKIs) 32, 90, 477, **478**, 480–2, 485–8, 535, **537**
    anti-EGFR **536**
    monoclonal antibodies 480, 488
    nucleotide based 480
    small-molecule drugs 480–2, **485**
    see also individual drugs

U0126 477, **483**
UbcH10 E2 ligase 129
ubiquitin ligases 126, 128, 180, 333, 342
ubiquitin-proteasome system 379
ubiquitin-protein ligases 374–6
ubiquitin-specific protease 8 377
ubiquitination 333, 557
UCNO1 **484**
UFM1 378
UGT1A 89
ULF 342
ultrasound 516, 522
unfolded protein response (UPR) 222, 282
urea, raised **512**
Urm1 378
urokinase receptor 494

v-ABL **193**, 255
v-erb-b 193
v-fms **193**
v-fos **193**, 198
v-jun **193**
v-myc **193**
v-ras **193**
v-sis **193**, **196**
v-Src 192
Vaccine European New Integrated Collaboration Effort (VENICE) 99
vaccines 416, 420–2, 452, 538–9
    clinical trials 422–3, *423*
    dendritic cell 421
    peptide-based 421
    plasmid DNA-based 421–2
    tumor cell-based 420–1
vandetanib **478**, **537**
vascular endothelial growth factor *see* VEGF
vascular growth factors 432–6
    see also specific growth factors
vascular permeability factor (VPF) 433
vascular targeted agents 436, **485**
vasculogenesis 430, *431*
vatalanib 450, *450*
Vav **196**
Vectibix *see* panitumumab
VEGF 31, 164, 226, 405–6, 432–4, *434*
    molecular targeting **496**, 536–7
VEGF receptors 31, 432–4, *434*
VEGF-Trap **485**, **537**
Velban *see* vinblastine
Velcade *see* bortezomib
veliparib **542**
vemurafenib **483**
Verteporfin 451
VHL **85**, **241**, 363, **364**, **514**
vimentin 372
vinblastine 61, 447, 533
vinca alkaloids 52, 142, 447, 533
vincristine 62, 447, 533
vinculin 230, 391
vinorelbine 49, 533
viral oncogenes **98**
    and instability 344
viruses 25
    DNA viruses 25, 98–9
    retroviruses 25, 100
visfatin 482
vismodegib **483**
vitronectin 402, 403
Voegtlin, Carl 439
Vogt, Carl 267
volasertib **541**
von Hippel-Lindau syndrome 57, 83, **85**, 129, 375, **514**
    genetic testing 513
Von Tappeiner, Herman 450
vorinostat 368, **484**, 492, 538, **542**
Votrient *see* pazopanib
VPS34 288

WAF-1 *see* p21*CIP1*
Warburg effect 16, 26, 159–60, *159*, 558
WARTS 123
WEE1 336
Werner's syndrome 83, 300, 346
Werner's syndrome protein *see* WRN
wheel of death (apoptosome) 274, *275*
white cell count **512**
whole-body scans 13, *522*
WHSC1 **355**
Williams-Beuren syndrome 333
Wilms' tumor 23, 83, **84**, 91, 366, 367, **514**
Wip1 338
Wnt pathway 21, 148, 178, 259–60, *260*, 470
Wnt-β-catenin pathway 21, 179–80, *179*
Wnt1 **195**
Wolf-Hirschhorn syndrome **355**
Women's Health Initiative Dietary Modification Trial 97
Wortmanin 477

619

Index

WP-1034 **483**
WRN 300, 337
WSTF 333
WT1 gene **84**, 91, **241**, 367, **514**
WT1 protein 367, **418**
Wyllie, Andrew 267, 269

X-ray tubes 527
X-rays 516
Xalkon *see* crizotinib
xenobiotic responsive elements (XREs) 95
xeroderma pigmentosum 58, 83, **85**, 329, 345
xeroderma pigmentosum proteins 328
XIST 369

XL147 **541**
XL184 **541**
XL228 486
XL765 **541**
XL844 **541**
XPA-XPG **85**, 345
XRNF185 128

YAP 391
yes **196**
Yes-associated protein 1 (Yap1) 342

Zactima *see* vandetanib
zarnestra **483**
ZBRK1 130

ZEB1 178, 306, 372, 373
ZEB2 306, 373
zebra fish 557
Zelboraf *see* vemurafenib
Zevalin *see* ^{90}Y-ibritumomab tiuxetan
ZIO-101 453
Zipf's law 574, *574*, 576
Zoladex *see* goserelin
zoledronic acid **483**
Zolinza *see* vorinostat
ZONAB 219
zonula adherens 393
zonula occludens 394
ZW10–ROD–Zwilch complex 337
zymogens 403